Human Design

molecular, cellular, and systematic physiology

Human Design

molecular, cellular, and systematic physiology

WILLIAM S. BECK

Harvard University

ORIGINAL DRAWINGS BY *Edith Tagrin*

HARCOURT BRACE JOVANOVICH, INC.

New York   *Chicago*   *San Francisco*   *Atlanta*

127917

ISBN: 0-15-539815-6

Library of Congress Catalog Card Number: 76-115862

Printed in the United States of America

COVER: *Ernst Trova,* Falling Man (5′ Wheelman).
*Collection The Solomon R. Guggenheim Museum, New York.
Photo, Ferdinand Boesch.*

TO THE MEMORY OF

Walter Bauer

In order to know whether a human being is young or old, offer it food of different kinds at short intervals. If young, it will eat anything at any hour of the day or night. If old, it observes stated periods, and you might as well attempt to regulate the time of high-water to suit a fishing party as to change these periods.

Oliver Wendell Holmes, in *The Professor at the Breakfast Table*, 1882

Preface

"Why write another book on the human body?" one is asked. The answer must be, "To advance a fresh point of view." We are living in a remarkable period in the history of biological ideas. A new biology burst forth after World War II, and the consequences of this intellectual flowering are still to be reckoned. These years have witnessed the resolution in near-ultimate terms of the coding ciphers that govern specificity in protein synthesis, of the fundamental mechanisms of heredity, of the major pathways of metabolism and energy generation, of the ultrastructure of living cells and their intricately ordered subcomponents, of the laboratory synthesis of hormones, proteins, and nucleic acids, and of the development of major new theories of immunology, information transfer, nerve function, membrane transport, and cellular

regulatory mechanisms. It is a remarkable time with implications deep and wide.

I have written this book with the notion in mind that those wishing to understand the structure and function of the human body will profit best from discussions formulated in the context of contemporary biological thought. One does, of course, need to know certain facts about body structure and function. My own experience, however, convinces me that students no longer have time for the leisurely acquisition of unrelated facts. Whenever possible, facts must illuminate the principles from which they could conceivably have been predicted or inferred. They must stimulate creative thought and convey a sense of the intellectual ferment in the scientific community. There must be awareness of the sources of available information, the men and machines, and their curious habits, techniques, and methods of reasoning. The material should comprehend its own philosophical presuppositions, its historical past, and its implications for mind, matter, body, and spirit. These are ambitious goals for any teacher. But they seem worth pursuing if instruction is to be synonymous with education. Much beyond facts commends the study of the human body to serious thinkers. The human organism is a fine curtain on which to project old and new visions, and the wish to do so sensibly has guided this writing.

These are interesting and turbulent times for teachers of human physiology—indeed for teachers of biology in any of its guises. Not many years ago the traditional approach to the teaching of human physiology tended to detach it from biology at large, from medicine, and from the thoughtful student's concern for the survival of man and the many obstacles thereto. Then, in the wake of great advances in molecular biology, human physiology in many curricula gave way to cell physiology. For there were unifying concepts that deepened biology's intellectual content and added important new dimensions to its meaning for students. More recently, interest in human physiology has had a renaissance. It has come from many sources, among them the contemporary student's search for relevance in his studies and the growing trend toward the transfer of medical school science courses into the undergraduate years in order to avoid redundancy and shorten the time needed to obtain a medical education. It has come also from the burgeoning interest in human physiology on the part of physical science and engineering students, who in some centers are well on their way to developing a new discipline, human bioengineering. An awareness of these trends and a considerable sympathy with the views of today's students as I understand them have had a good deal to do with the genesis of *Human Design.*

The chapters in *Human Design* follow a conventional outline, but the text is divided into two parts. In explaining Part I, I must state yet another conviction. Today's student is an intelligent being and the time has come to speak to him early and often in the language of biochemistry and genetics—to expect him, for

example, to know something about enzymes, ATP, DNA, amino acids, chromatography, oxidations, and hydrogen bonds. It is difficult enough to avoid these topics in the pages of *The New York Times.* Surely we cannot sidestep them in a book about the human body. Part I reviews these principles, and the reader is urged to attack them with zest and good humor. Let there be no doubt that these are the tools and triumphs of today's physiology, and there can be no excitement or awareness without them. Physiology may have been simpler before these things were known, but it made much less sense. These are not sources of difficulty. Rather, they are the sources of much of modern physiology's clarity, uniformity, and predictability.

Principles and techniques explained in Part I are systematically referred to in the chapters of Part II. It is assumed that the chapters will be read in sequence, but enough cross references have been supplied to make another sequence a reasonable possibility. Anatomy is offered in Part II in a quantity deemed necessary for the conduct of meaningful physiological discourse. This will perhaps be too much for some and not enough for others. Were the body less complex, such decisions would stand beyond controversy.

Chapters include lists of references and suggestions for further reading. They are not comprehensive lists of source material, nor could they be. Rather, they include selected source materials, an occasional historical reference, general texts, monographs, and reviews in which more complete bibliographies can be found.

What I have tried most of all to emphasize in the text are the regularities of body function. Repeatedly, we shall see that individual mechanisms are of general significance. We shall observe, for example, significant identities in the behavior of cell membranes in nerve, muscle, kidney, and blood cells. We shall see that all regulatory mechanisms operate upon the principles of negative feedback, and we shall learn how such mechanisms integrate all body functions in maintaining homeostasis. We shall see that all body cells draw upon funds of metabolic energy that must be generated locally or elsewhere, that all body systems contain devices for transducing energy, that all body structures have a developmental history, the human body itself having emerged in organic evolution less than a minute before midnight if the history of life is placed on a twenty-four-hour scale. Human architecture and organization are at one with nature, and, despite Pope's dictum, "The proper study of mankind is man," much can be learned of man's body from the study of lower forms of life. Much, too, can be learned from the study of human disease.

It should be evident, then, that I am a confirmed interdisciplinarian. The rush of progress makes quite meaningless all attempts to separate physiology into categories and dominions. Physiology's strength today lies in its interdisciplinary character. The functions of the living body are inextricably associated with its gross, microscopic, and molecular structure. We shall see clearly, I believe, that anatomy, histology, biophysics, genetics, and biochemistry underlie and subsume the science of physiology —and this is as it should be.

Throughout the text I have tried to point out the ingenuities of today's investigators, for here is where enchantment lies. It is interesting that the body makes antibodies to foreign substances, but it is utterly fascinating that when investigators recently began asking how the body distinguishes between "self" and "nonself," a revolutionary new theory of antibody formation took form. It is interesting that the kidney precisely controls the level of sodium in body fluids, but the ingeniously simple scheme recently proposed to account for this function has added new luster to one of physiology's most honored disciplines. Such examples are easily multiplied. It is a game for the young in heart, those who will accept food for thought

of different kinds, at short intervals, and at any hour of the day or night.

A final prefatory comment concerns my many debts. The writer of a book such as this must depend upon the literature at large— scientific papers, reviews, monographs, the recorded judgments of bona fide experts. In a sense, he is more editor than writer and his virtues and faults are describable in terms of choice, taste, and accuracy. One learns early how many are the opportunities for imprecision or error in expounding on matters beyond one's own expertise. Fortunately, many such faults were purged by the criticism and counsel of Doctors Edward F. Bland, Werner D. Chasin, Leslie J. DeGroot, Thomas J. Fitzpatrick, Anne P. Forbes, John Gergely, Melvin J. Glimcher, Mehran Goulian, Kurt J. Isselbacher, John B. Josimovich, John H. Knowles, Stephen W. Kuffler, Carl Kupfer, Alexander Leaf, Sidney Leskowitz, Paul L. Munson, Lot B. Page, Harold F. Schuknecht, and Michael Young, all of Harvard University, and Horace W. Magoun and Charles G. Craddock of the University of California at Los Angeles. These kind colleagues reviewed portions of the manuscript in preparation. In expressing my gratitude, I wish emphatically to exempt them from responsibility for faults that may still be apparent to the critical reader. For these I am alone responsible.

It has been a special joy to work with Edith Tagrin, who prepared the original drawings. Her taste and keen understanding of physiology and the problems of exposition taught me a great deal. I wish also to thank my loyal secretaries, Carol Ditmore, Mary Young, and Carole Spencer who prepared the manuscript with care, and my colleagues, students, and family who acquiesced in many thefts of time.

WILLIAM S. BECK

Contents

Human Design

molecular, cellular, and systematic physiology

Man has multiplied so rapidly, that he has necessarily been exposed to struggle for existence, and consequently to natural selection . . . His body is constructed on the same homological plan as that of other mammals. He passes through the same phases of embryological development. He retains many rudimentary and useless structures, which no doubt were once serviceable. Characters occasionally make their reappearance in him, which we have reason to believe were possessed by his early progenitors. If the origin of man had been wholly different from that of other animals, these various appearances would be mere empty deceptions; but such an admission is incredible. These appearances, on the other hand, are intelligible, at least to a large extent, if man is the co-descendent with other mammals of some unknown and lower form.

Charles Darwin, in *The Descent of Man,* 1871.

1 *The Human Body*

Man's Place in Nature

Life on Earth began several billion years ago. In this unimaginable interval of time, evolution has given rise to countless species of living organisms. Somewhat more than a million have survived; many more quite probably have not. Man, the greatest achievement thus far of an evolutionary process that is surely not ended, is a latecomer, the earliest human remains dating back only a few hundred thousand years.

It was the great Charles Darwin (1809–82) who first theorized that man had developed, like all other living forms, in the evolutionary struggle. Part of the evidence for his conclusion, which is stated above, is man's many resemblances to lower forms. And yet, though "Man

still bears in his bodily frame the indelible stamp of his lowly origin," though he is heir to much that went before, it was clear to Darwin as it is to us that man has surpassed his predecessors in extraordinary ways. His creative mind, his use of tools, his predilection for learning and culture-building have set him so decisively apart that long centuries passed before men were able to recognize themselves as part and product of nature and organic evolution.

WHAT IS MAN?

In *Man's Place in Nature,* a volume of controversial essays that appeared in 1863 (4 years after the publication of Darwin's *On the Origin of Species*), Thomas Huxley humorously suggested that we should imagine ourselves zoologically trained scientists of the planet Saturn who have been called upon to examine and characterize an "erect and featherless biped, which some enterprising traveller, overcoming the difficulties of space and gravitation, has brought from the distant planet for our inspection, well preserved, may be, in a cask of rum." After examining our preserved human being, we would, perhaps, decide that it belongs to the animal kingdom. Observing that it consists not of one cell only but of many cells, we would classify it as a metazoan rather than as a protozoan. Observing its spinal column, we would classify it as a vertebrate rather than an invertebrate. And, if we were acquainted with the five types of vertebrates—fishes, amphibians, reptiles, birds, and mammals—we would see immediately that our specimen is not a fish because it lacks gills with which to breathe and fins with which to swim. For equally conclusive reasons we would rule it out of the amphibian, reptile, and bird classes. Thus we would be left with the class of mammals.

Mammals are warm-blooded, air-breathing animals with a covering skin that bears hair. Actually we should quickly establish that man is a mammal merely by noting the breasts on the ventral side of the body, since the most characteristic feature of this class, from which it derives its name, is its method of reproducing and nourishing the young. Birds and most reptiles lay eggs, and may leave nature to incubate them and hatch them out. The young of mammals develop not within eggs but in the body of the mother. For a considerable period of time, the embryo grows within the mother's uterus. It is then born in a relatively helpless condition and for a while is nourished by suckling at the mother's breasts, or *mammae*.

Mammals may be further subdivided into three subclasses, depending upon their method of reproduction: Prototheria, Metatheria, and Eutheria. Prototheria represent a transition between reptiles or birds and mammals, for they lay eggs, hatch them out, and then suckle the young at the breast. The duckbill is an example of this subclass. The other two subclasses nourish their young in the maternal uterus by means of a special structure called the *placenta*. This is a disclike organ, one side of which is embedded in the wall of the uterus while the other side gives origin to the umbilical cord through which nourishment is carried to the embryo. After the young have been delivered, the placenta usually becomes detached and is expelled. For this reason it is called the *afterbirth*. This reproductive arrangement, typical of higher mammals, is lacking in animals that lay eggs and hatch them out.

Metatheria are the pouched mammals, or marsupials, such as the kangaroo and opossum. In marsupials the placenta either is poorly developed or functions only briefly. The young are born while still immature embryos and after birth are transferred to a *marsupium*, or pouch, in the abdomen of the mother, where they remain until about a quarter grown, receiving their nourishment from nipples conveniently located inside the pouch.

Eutheria are the mammals in which the placenta is well developed. Man is an example of this subclass. He has certain outstanding characteristics. One that places him within the order of mammals called primates is the prehensile nature of his hands and feet. Among the primates he is quite unusual. He walks upon two legs and possesses an exceedingly large brain. He is highly gregarious and is the only animal that has developed the symbolisms of speech and writing and perhaps rational thought. He is therefore the only living being that can hand down patterns of acquired learning from one generation to another.

To these advantages man owes his dominant position in the world. They have enabled him to utilize his environment and increase his numbers and range of distribution far more quickly and extensively than any animal of comparable size. He has eliminated some of his competitors and exploited others for his own use, and different races even compete with each other within the closed arena of a limited environment.

In entering upon the study of the human body, we must remember that man is a member of the animal kingdom. Although his growth is not completed by reproduction (nor is it fulfilled by death, since his nature and intellectual endowment make him self-surpassing), he is an animal having much in common with other animals. Like them, he seeks food, water, air, shelter, and security; he eliminates wastes; he mates and reproduces; and he fights off the encroachments of hostile environments and advancing age until it is possible to fight no longer. Then, like other animals, he dies. His body is an animal body of incredible complexity, a repository of ancient inventions and evolutionary remnants, mechanisms, and devices. In the following chapters, we shall encounter many examples of man's ancestral bonds with other animals and other living organisms. Each should deepen our awareness of the unity of biology and add meaning and coherence to much of the subject matter of human physiology.

Study of the Human Body: Past and Present

For long centuries before the comparatively recent dawn of experimental science, knowledge of the human body was interpreted by the literary men, theologians, and philosophers of each age. In those early germinal years, speculations on the nature of body structure and function were more cultural sophism than science. Much of this speculation was downright fanciful. Moral qualities were ascribed to various viscera, and demons and superstitions permeated the thought of many. The condition of the organs and of the body as a whole was explained in terms of the fundamental qualities—heat, cold, wetness, and dryness—and of four corresponding body juices or humors—blood, phlegm, yellow bile, and black bile. Normal function and health resulted when these elements were present in proper proportion. When harmony was disturbed, sickness followed. Many of these ideas are with us still. We still speak of an "overheated brain" in referring to temperamental behavior. In *sanguine* people the predominating factor was thought to be blood, in *phlegmatic* people phlegm, in *choleric* people yellow bile, and in *melancholy* people black bile.

SCIENTIFIC ELEMENTS IN ANCIENT THOUGHT

The early speculations concerning the human body were, of course, similar to those of Empedocles and other ancient natural philosophers who held that the four fundamental elements of the world were air, water, fire, and earth. Neverthe-

less, it must be conceded that those who believed in these bizarre causal mechanisms were operating on the assumption that body abnormalities must have natural explanations. Normal function, according to this school of thought, could give way to pathological function. This concept represented a major step forward from primitive times when body function and the manifestations of disease were considered wholly in the control of demons.

The earliest systematic study of the human body was undertaken by the physicians of the early Christian centuries. The great physician of antiquity was Galen, who lived from 130 to 200. A courtier of rank, a fashionable practitioner of imperial Rome, Galen began his career as surgeon to the gladiators and ended it as physician to Emperor Marcus Aurelius. For thirteen centuries he stood as the final authority on human anatomy although his closest approaches to direct study of the human body were dissections of the ape.*

Galen concerned himself both with *anatomy*, the science of body structure, and with what today we call *medical physiology*. He recognized that the physician had to comprehend normal body structure and function from conception to death so that he could recognize the effects of disease and determine the objectives of treatment. Such thinking clearly foreshadowed later views of the reciprocal relationship between body function and structure. Modern thought was also portended in the preoccupation of ancient physiologists with explaining how the body functions, with analyzing its processes and mechanisms, however erroneously. The physicians of antiquity wished to do what we wish to do today. But the differences in the investigative methods and conclusions of the ancients and moderns overshadow the similarity in their intentions.

THE HUMAN BODY IN THE ENLIGHTENMENT

During the long Middle Ages, all science prospered little, and what was done as a formal pursuit consisted of foolish compilations of ancient

---

* Galen was not always careful to avoid drawing conclusions concerning human anatomy from his observations of ape anatomy, although he warned other anatomists of the dangers of such practice.

writings in the service of theology and dogma. Beyond this, the medieval mind was enmeshed in an almost unbelievable tangle of astrology, alchemy, magic, and sorcery. Although it has been held that such preoccupations were perfectly reasonable in their time and that alchemy represented an advance over primitive occultism, it is difficult to acknowledge that science grew out of these intellectual errors, that it owes its existence to them. Rather, it seems enlightenment came despite these years, not because of them.

The Renaissance marked not only the birth of modern science but a return to respectability of intellectual pursuits. Science had not by 1700 become the most honored of occupations, but scientific awareness had begun at last to enter the minds of intelligent men. The new enlightenment swept the Western world like a wave of fire, and imaginations were stirred to their deepest foundations. All that had once been believed was now suspect, and for the first time men challenged the notion that ancient culture must forever symbolize the ultimate in human aspirations. It was in these years that the scientific method was first employed.

Although modern science began in the brilliant physical researches of Isaac Newton, Robert Hooke, and others, the new age found men inquiring for the first time in centuries into the nature of living organisms and of the human body. That there was a new closeness to nature is clearly evident in the art of the period. For the first time nature was portrayed with painstaking and unaccustomed accuracy, the skill of the painter gaining new dimensions from the first use of perspective and realistic treatments of light and shadow. The painter's landscapes, his nudes, all reflected the new interest in nature and the human form.

The biologists of the Renaissance faced more difficult problems than did the physicists whose distinguished works illuminated those years. The complexity of the living organism is built upon the complexity of matter, and hence a longer time was needed to scrape away the thicker shell of myth to get at the core of valid inquiry. For this reason alone, science in the sixteenth and seventeenth centuries was dominated by physics, while the study of living organisms lagged behind. Biology offered grave difficulties to the

early experimentalists who desired to apply the scientific method to living organisms but who possessed no unifying theories or hypotheses to guide their efforts. For long years the only generalization that biology could claim was that life exists in many shapes and guises, and it was precisely this meager conception that inspired most of the early work. Many decades were devoted to the task of description, for, as in all sciences in their formative phases, one must look and examine before seeking explanations.

Description flourished in the first century of scientific biology, and the hallmark of the new awakening was anatomy, the epitome of systematic description. If there must be a "father of anatomy," it was Andreas Vesalius, who lived from 1514 to 1564. The Renaissance was heir to the ancient medical orthodoxy of Galen, which had held supreme for over 1300 years. We cannot blame Galen for the idolatrous acceptance of his erroneous views. His successors had every opportunity to learn the truth for themselves. Practically no one thought of doing so, however, until Vesalius. It is difficult to fathom why the world had to wait so long for a man who could write down what his eyes perceived.

Vesalius was not the first to dissect the human body. Dissection and the witnessing of surgery became part of the curriculum of the Italian medical schools as early as the eleventh century, and by the middle of the thirteenth century the practice was fairly well organized in the universities of Salerno, Bologna, and Padua. One might suppose that the misleading errors of Galen could have easily been corrected by several hundred years of firsthand dissecting experience.* But though the old editions of Galen's work be-

---

* One reason for the stalemate should amuse anyone familiar with the inmost secrets of modern medicine, for sometimes the situation seems not to have changed very much: the antagonism between the physicians and surgeons. In those days, the physicians, having received an essentially literary and philosophical education, looked with contempt upon the surgeons, who were mere technicians. In the dissecting theater, the surgeon would wield the knife while the professor lectured platitudes and "demonstrated" items of interest as they were dug up by the surgeon. Occasionally, to the delight of the students, great flatulent professorial debates would arise between visiting philosophical disciples of Aristotle and the medical followers of Galen. And while the discourse ebbed and flowed, the poor surgeon hacked away with his miserable implements, unnoticed except for an occasional epithet hurled from the cathedra. Is there any wonder that anatomy made no progress?

came encrusted with marginal notes, the medieval respect for classic authority knew no limit. Progress, it seems, was not to take place.

Even before Vesalius, early sixteenth-century anatomists had begun drifting toward a reliance upon their own observations. This new spirit was evident in the works of Jacopo Carpi, Lorenzo Massa, and the brilliantly versatile Leonardo da Vinci, who was born in 1452 and died in 1519 when Vesalius was 5 years old. It is a curiosity of scientific history that Leonardo's superb anatomical studies had so little influence. Francis Moore has suggested that the crucial difference between Leonardo and Vesalius was that only Vesalius worked with the sick and taught in a medical school. Leonardo, in contrast, was isolated from this environment. The rest of this book will give frequent testimony to the historic association of biological science in general and human physiology in particular with the teaching of medicine, the care of the sick, and the study of disease.

Despite their meager influence upon the growth of science, the predecessors of Vesalius must surely have influenced him. If so, Vesalius was, in fact, continuing an existing trend. Nevertheless, his work was prosecuted on a grand scale, and his individual contribution was gigantic. A professor at Padua at the age of 23, Vesalius published his great volume *De humani corporis fabrica* (often referred to as the *Fabrica*) in 1543, when he was 28 years old. In it, Vesalius was determined not only to give a systematic and accurate description of all parts of the human body, but to present his work in as elegant a setting as graphic art would allow. He fretted over the publication and insisted that the engravers must be the finest, that the paper should be strong and of uniform thickness, that every detail of every picture must be clearly visible. The work was based on the experience of 5 years' dissections, some performed on decaying corpses taken from the gallows. The resulting volume contained 278 magnificent woodcuts (see Fig. 1.1) and numerous decorative historiated initials, one of which depicted a "resurrection" scene in the dissecting room. Anatomy books to this day stand or fall on the quality of their drawings. From this standpoint, the *Fabrica* remains the most superb anatomical treatise ever to have been published.

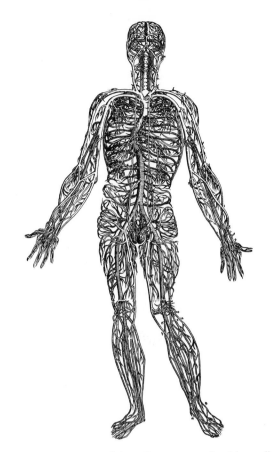

1.1 "A delineation of the entire vena cava freed from all parts" from *De humani corporis fabrica* by Vesalius (1543).

Another modern Galen was Jean Fernel,* the Parisian professor of medicine whose important treatise on body structure and functions, *De naturali parte medicinae*, was published in 1542, a year before the appearance of the *Fabrica*. Prior to this work, the word "physiology" was employed as a synonym of the more commonly used term "natural philosophy," and its subject matter was mainly what we now call physics. Fernel narrowed the province of physiology from the whole universe of natural phenomena to the structure and functions of the human body.

---

* Fernel's importance in the history of medicine and physiology are delightfully discussed in two volumes by the late Sir Charles Sherrington, *Man on His Nature* and *The Endeavour of Jean Fernel*.

If Vesalius was the father of modern anatomy, William Harvey (1578–1657) was undoubtedly the father of modern physiology. Harvey observed in the early 1600's that the beating heart expels the blood within it. Meditation upon the significance of this fact and of certain already known anatomical facts (e.g., the heart valves and the venous valves) led Harvey to postulate that blood expelled by the heart must circulate through the body, whence it returns to the heart. The experiments that he performed to test his hypothesis, an early utilization of the scientific method, resulted in the discovery of the circulation of the blood, a monumental and many-sided contribution to science.

Harvey's discovery discredited the belief of fourteen centuries that the blood passes through the septum of the heart between the right and left ventricles. But, more important, it was the first great step toward a *mechanistic* approach to biological problems. In visualizing the heart as a pump, Harvey heralded a new view of living organisms: life, like the rest of the universe, could be regarded as a material machine. Though there is evidence that Harvey did not fully appreciate the meaning of his work on this score, to his successors his work gave support to the concept that the body is kept alive not by vital life forces, spirits, or souls, as had once been believed, but by the interrelations of its mechanical parts. If nonmaterial factors did not need to be invoked, body function seemed, then, within range of human knowledge, its further understanding requiring only careful investigation of its mechanical processes. René Descartes was quick to praise Harvey for "having broken the ice in this matter," and in his own discourses he relied heavily on the work of Harvey to illustrate the mechanical nature of living objects. Their only difference from manmade machines, he insisted, was in the degree of complexity. With these assertions, sudden new excitement entered the realm of physiological thought. From this time on, we may properly speak of a science of physiology.

WHAT IS PHYSIOLOGY?

The subsequent history of physiology cannot be dealt with here, although historical notes relating to specific areas will appear in later chapters. However, brief comment is appropriate now on past and present views of the nature and scope of physiology. Soon after the work of Harvey, an important medical textbook was published in 1708 by Hermann Boerhaave of Leyden. It dealt almost exclusively with normal functions of the various parts of the healthy human body, and its great influence resulted in the establishment of physiology as a separate field of study in the academic curriculum. In 1726 the Edinburgh Medical Faculty established the first chair in physiology—not under that title, but as the "Institutes of Medicine," in conformity with *Institutiones medicae*, the title of Boerhaave's text. Gradually this term gave way to "physiology," the term used earlier by Fernel.

The early chairs of physiology also covered anatomy, with the emphasis on the latter subject. Even Boerhaave's most brilliant pupil, Albrecht von Haller, whose *Elementa physiologiae* (1757–66) was the great physiological text of the eighteenth century, had been professor of anatomy at Göttingen. Only when the mass of experimental physiological data became overwhelming in the late nineteenth century did the medical schools abandon the combination. Harvard established a separate chair of physiology in 1876; Oxford did the same in 1882, and Cambridge followed in 1883.

Today no one would seriously question the right of physiology to an independent place in the curriculum. Hence it is of some interest that contemporary physiologists have lately shown signs of uncertainty as to what physiology includes and what it ought to include. According to Ralph W. Gerard,* most physiologists consider physiology a "smokeless industry" that attempts to find out what makes organisms function. One group considers physiology the study of physical and chemical processes in living systems—in essence a paraphrase of the definition given by a founder of modern physiology, Claude Bernard: "[It is] the science whose object it is to study the phenomena of living things and to determine the material conditions in which they appear." Another group regards physiology

---

* In a 1958 publication of the American Physiological Society entitled *Mirror to Physiology: A Self-Survey of Physiological Science,* in which are reported the results of a comprehensive survey among physiologists designed to answer the question, What is physiology?

as a science primarily allied to medicine. For obvious reasons, much physiological study *has* been associated in some way with medicine, and it is a fact that human physiology has tended to overshadow the physiology of lower forms as a subject of investigation.

Gerard concluded that physiology is a great crossroads of the life sciences: its boundaries are poorly defined, and its scope is as broad as the living world. Though its central disciplines are animal, plant, and bacterial physiology, physiology merges with the totality of life science: biochemistry, general biology, biophysics, medicine, anatomy, and the rest. In this view, contemporary physiology is functional biology. To those seeking to learn how organisms function, how life goes on, physiology provides an approach that cuts across the lines of traditional biological disciplines.

THE TWO PHYSIOLOGIES AND THE SCIENTIFIC REVOLUTION

It has been held that the "human" or "mammalian" physiologist deals with systems so complexly organized that he is perhaps necessarily limited to the study of phenomena depending upon the integration or organization of living material—e.g., the circulation of the blood—whereas the province of the "general" physiologist is the study of basic endowments of living tissue, which must be pursued in simpler organisms. Often the mammalian physiologist comes from a medical or biological background, and the general physiologist springs from physics, chemistry, or mathematics—or, in the latter-day custom, biophysics or biochemistry. An example of this dichotomy will be found in Chapter 12, where the nervous system is discussed first in terms of the origin, nature, and propagation of the nerve impulse—most of these concepts having arisen from studies of isolated squid nerves—and second in terms of the integrative functions of the nervous system in the human body (reflexes, consciousness, etc.).

Although practitioners of the two physiologies have conceded that time is tending to erase the barrier that divides, textbooks have generally emphasized one approach or the other. This text is ordered along eclectic lines. We discuss the human body because we need to understand its workings, it is an important arena of contemporary research, and it is an excellent exemplar of certain of the current conceptions of "general" physiology, "comparative" physiology, and the other purer varieties of physiological and biological thought. But it is clear that we cannot expound on human physiology without ample recourse to the data and conclusions of general physiology.

One who chooses to write in this manner is forced to make choices, since an exceedingly large amount of material is thus admissible. In this book first choice will go to those aspects of human physiology that conceivably could have been predicted from a knowledge of first principles (be they physical, chemical, or behavioral) and that emphasize both the unity of physiology and man's enduring place in nature.

Goals of Physiological Study

EXPLANATION VERSUS DESCRIPTION

The complexity of man's body, his mind, his creativity, and his sense of values proclaim his supremacy and arouse in us a sense of poetry and wonder. We would be insensible were we to react otherwise. Therefore, in beginning our systematic examination of the human body, it will be well to state what we wish to achieve, for our goal is not the fostering of awe.

Our purpose is to describe the human body and to explain its functions. We shall try to answer the questions "what?" and, whenever possible, "why?" There is a considerable difference between these two levels of exposition, and it is important that the difference be understood.

A *description* simply formulates in some manner the results of many observations. Usually, in the history of science, description precedes explanation and, in fact, makes it possible. Suppose, for example, that we observe the motion of a particle and record its position at successive intervals. To answer the question "Where was the particle at such and such a time?" we have only to look at our descriptive table. Even if we find it possible to express the motion of the particle in a simple mathematical formula, we have still done no more than describe its movement.

Similarly, to invent a physiological example, suppose that we are concerned with the changes in color of a man's skin when exposed to sunlight. We might set up a table in which we list the color in one column and the degree of exposure in the other. From this we observe that, after exposure to sun, the skin becomes darker. Have we now achieved explanation? No, because the generalization is still purely descriptive. It is "what" and not "why."

Explanation consists in relating specific observations to separate laws of nature or valid generalizations that are outside the observations and from which they could potentially have been predicted. We did not explain our moving particle by describing its motion mathematically, but we would explain it if we showed that the motion is under the influence of a nearby magnet, for the law of magnetic fields would permit us in principle to predict the particle's movement. Similarly, we could explain the tanning of skin by sunlight by referring to the effect of sunlight on melanin synthesis in skin cells (Chapter 18). Perhaps our explanation would be strengthened if we also referred to the evolutionary concept that pigment formation is a protective mechanism with high survival value.

In general, then, we shall seek to explain complex physiological events in terms of the properties and behavior of the simpler components of each system—in short, we shall seek the elucidation of mechanisms. Even though much of human physiology has yet to be explained in mechanistic terms, we must avoid concluding that certain aspects are *beyond* such explanation. There is no evidence for this doctrine, and we reject it out-of-hand. The point remains, nevertheless, that huge areas of human biology are presently without acceptable explanations. It is likewise true that all the unexplained phenomena are as important as any we do understand and, in considering the human animal, we cannot gloss them over. For example, man has emotions. They affect the functioning of the body. But we do not know enough about the biochemistry and physiology of the nervous system to give an adequate explanation for the feeling of love or fear or anger—though we shall see in Chapter 12 that much scientific work is being done in this area. Although the answer is still lacking, we cannot ignore the emotions; nor can

we declare them eternally beyond understanding.

Even if we cannot explain a phenomenon, we may still be able to deal with it meaningfully. We drive automobiles, though many of us have no idea how engines operate. We recognize psychological forces and have devised valid means of studying them and manipulating them, though we do not know the mechanistic details of mental activity. Thus, while we shall stress the physicochemical basis of those physiological events that we understand (for herein is the ultimate in physiological explanation), we shall also acknowledge and describe those levels of human organization that have continued to elude explanation.

BARRIERS TO HUMAN EXPERIMENTATION

The contemporary study of the human body is many-faceted and complex. No longer can the student learn what there is to learn of the body in a setting such as that in Rembrandt's *Anatomy Lesson* (Fig. 1.2). However, a special obstacle complicates the study of human physiology. For obvious reasons, there are limitations upon the quantity and type of experimentation that can be performed upon the living human body. As a result, many physiological processes have been studied only in lower animals. In many instances, the relevance of these studies for the human species is not known.

Because of this difficulty, many ingenious techniques have been developed in recent years that do permit the study of physiological processes in living man without danger or great inconvenience. We shall see, for example, that the level of function of the thyroid gland can be assessed by the administration of radioactive iodine and the external measurement of the amount of radioactivity accumulating in the thyroid region of the neck (Chapter 15). Hydrostatic pressure within the heart can be determined by means of a tube inserted harmlessly into an arm vein and slid along until it can be seen (on the screen of a fluoroscope) entering the heart (Chapter 9). Many of the body's arteries can be readily studied by the technique of arteriography, in which a dense medium injected into the vessel is visualized by x-ray. Similarly, much can be learned from the chemical analysis of

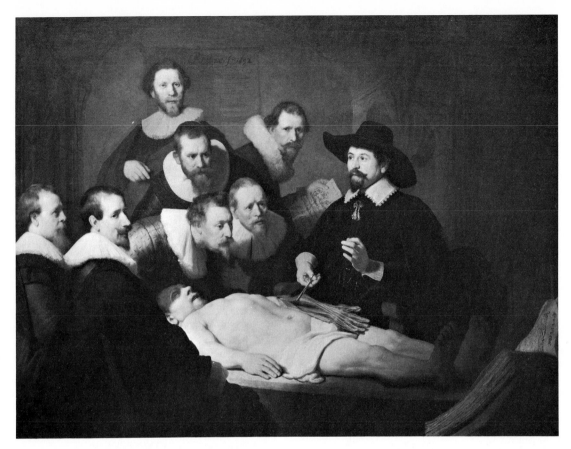

1.2 "Anatomy Lesson" by Rembrandt. The professor of anatomy graphically demonstrates on his own hand the concept of living function gained from dissection of the cadaver's dead hand. (Mauritshuis, The Hague.)

blood and organs removed at surgery, and more by the microscopic examination of small bits of tissue that are painlessly removed for this purpose.

Still, many problems remain because we cannot treat man as we would an experimental animal and because investigators have not yet devised a useful experimental attack. The remedy here is time and ingenuity.

UNIVERSALITY OF PHYSICAL LAWS

Despite the organizational complexity of the human animal, it has arisen like other living forms from more primitive organisms that themselves arose upon a lifeless planet. Body structure and function remain subject, therefore, to the same fundamental controls and limitations that the physical universe has placed upon matter in all its forms. This is a concept of some importance.

Since life arose in a world governed by certain physical laws, a world whose matter consisted of hydrogen, oxygen, iron, and the hundred-odd other elements now known to exist, we must conclude that the first living organisms were composed of some or all of these elements and, though these constellations of matter became possessed of a behavioral pattern called life, they retained their status as physical structures in and of the world with functions and properties according strictly with the laws of the physical universe. If heated, the water of a living creature will boil. Without apparatus capable of producing the forces needed to keep them aloft in accord with the laws of gravity and aerodynamics, birds would remain on the ground. In an organism so constructed that its survival depends upon the inward diffusion of nutrients through its surface, diffusion will occur more rapidly with small nutrient molecules or at high temperatures than with large nutrient molecules or at low temperatures, for the same physical laws govern the process of diffusion whether it takes place in or out of a living organism.

As we shall presently see, many phenomena in the living body superficially appear to violate the laws of the physical world. For example, our bodies remain warm, though the surrounding air may be cold. Our systems oxidize sugar to carbon dioxide and water, though only a hot flame accomplishes this at the laboratory bench. Actually, we can readily explain these phenomena in physical and chemical terms. The fact is that living systems must have some way of oxidizing sugar without recourse to flame temperatures, for at such temperatures body water will evaporate and other body ingredients will burn. In order to survive, living forms *had* to develop means of oxidizing substances at moderate temperatures. They have survived; therefore, they have developed such devices. And they are still physicochemical systems. This example bears out the conclusion that all phenomena are explainable in physicochemical terms and that any phenomenon that has not been so explained merely awaits an explanation in the scientific future.

REFERENCES AND SUGGESTIONS
FOR FURTHER READING

Beck, W. S. *Modern Science and the Nature of Life*, Harcourt Brace Jovanovich, New York, 1957 (pap. Natural History Library, Doubleday, New York, 1961).

Blum, H., *Time's Arrow and Evolution*, Princeton Univ. Press, Princeton, N.J., 1951.

Davson, H., *A Textbook of General Physiology*, 3rd ed., Little, Brown, Boston, 1964.

Huxley, T., *Man's Place in Nature*, Appleton, New York, 1896.

Moore, F. D., "Leonardo and Vesalius," in H. K. Beecher, ed., *Disease and the Advancement of Basic Science*, Harvard Univ. Press, Cambridge, Mass., 1960.

Nordenskiöld, E., *The History of Biology: A Survey*, Knopf, New York, 1928.

Romer, A. S., *The Vertebrate Body*, Saunders, Philadelphia, 1950.

Schaeffer, J. P., ed., *Morris' Human Anatomy*, 11th ed., McGraw-Hill, New York, 1957.

Sherrington, C. S., *Man on His Nature*, Cambridge Univ. Press, Cambridge, 1951 (pap. Doubleday, New York, 1953).

———, *The Endeavour of Jean Fernel*, Cambridge Univ. Press, Cambridge, 1946.

Simpson, G. G., *This View of Life: The World of an Evolutionist*, Harcourt Brace Jovanovich, New York, 1964.

———, *The Meaning of Evolution*, Yale Univ. Press, New Haven, Conn., 1949.

———, and W. S. Beck, *Life: An Introduction to Biology*, 2nd ed., Harcourt Brace Jovanovich, New York, 1965.

Smith, H. W., *From Fish to Philosopher*, Little, Brown, Boston, 1953.

Snow, C. P., *The Two Cultures and the Scientific Revolution*, Cambridge Univ. Press, Cambridge, 1959.

Wightman, W. P. D., *The Growth of Scientific Ideas*, Yale Univ. Press, New Haven, Conn., 1953.

PART ONE

Molecular and Cellular Physiology

In times past—and not so very long ago—biochemistry was a rather special field, to which some other groups of biologists paid relatively little heed. In the last decade, however, biochemistry has penetrated into the central core of biology and has transformed the outlook of all who are concerned with living beings. The boundary that might once have been drawn between biochemistry and genetics has vanished; these two disciplines have fused, and no one will ever disentangle them again.

The study of morphology, which some scientists half a century ago regarded as a dull and finished subject, is now deepening and revolutionizing our concepts of the fine structure of cells and tissues. Studies of form and function are advancing hand in hand through the associated enterprise of the electron microscopist and the biochemist; often the same man plays both roles.

Some have recently believed that taxonomy is a dead science, but we are already witnessing the first stages in the rise of a biochemical taxonomy . . . The biochemical study of evolution has only begun . . . but the great principles of biochemical embryology still await discovery. We have had brief glimpses of the promised land from afar, but we have not yet entered into it.

John T. Edsall, Opening Remarks, Sixth International Congress of Biochemistry, New York, 1964.

2 Materials and Principles

Materials of the Body

An organism is a functioning organization of physical matter. The outstanding feature of this living organization is its hierarchical character. The cell, the "unit of life" that will be discussed in the next chapter, is an organization existing at one level of complexity. It lives in a community of other cells, joining them in certain projects and competing with them for food and oxygen. But cells are components of more elaborate organizations: tissues, which in turn are parts of still higher organizations, organs. Organs interact in a community of other organs, not tissues or cells; and systems and the body at large supersede them on the organizational scale. Perhaps we could look upon the family and society as the two next higher levels of the hierarchy.

It is also possible to begin with the cell and descend the scale, for within the cell are substructures like the nucleus, its inner particles, and the particles within those particles. Descent may continue even below the molecules and atoms, although few modern biologists and physiologists have yet ventured into this realm of subatomic particles.* Since the nature and properties of a body structure at each organizational level depend entirely upon the shapes, positions, quantities, and behavior of its simpler components, understanding of a body structure ultimately rests upon knowledge of its components.

---

* An exception is Albert Szent-Györgyi, whose challenging ideas on energy transfer in living systems are mentioned in Chapter 17.

Hence we begin our study of the human body at the lowermost level.

CHEMICAL COMPOSITION OF THE BODY

Total body composition. Until 1945, the only available data on the constituents of the body came from the studies of several German workers more than 75 years earlier. A volume published in 1859 by Jacob Moleschott gave values for the amounts of protein, fat, salt, and water per 1000 parts of the human body but failed to describe the method of analysis. Four years later, Louis Bischoff carefully dissected the bodies of a man, woman, and boy who had died suddenly of injuries and published some measurements of the fat and water contents of the organs. And Richard von Volkmann, in 1874, reported that the body of a man weighing 62.5 kg (138 lb) contained 65.7% water and 4.7% total ash.

This paucity of information is doubtless due to difficulties in obtaining bodies for chemical analysis and to technical problems in analyzing such unwieldy specimens. Only recently have reasonably systematic studies been undertaken by E. M. Widdowson, R. A. McCance, C. M. Spray, and others. Using original methods for dissolving whole bodies under controlled conditions, these workers have reported the data on body composition summarized in Table 2.1. As might be expected, their subjects varied from 1 to 50% in percentage of fat (the values cited here range from 14.9 to 23.6%). For this reason some authors prefer to express body composition in terms of "fat-free" body weight. More than half the body weight is attributable to water, about one-sixth to protein, and 5 to 6.5% to minerals or ash—i.e., the noncombustible material remaining when the body is incinerated. The largest mineral components are calcium and phosphorus, the major ingredients of bone.

The body's four major elements. Of the body's many chemical elements, four have particular

TABLE 2.1 COMPOSITION OF THREE DEAD BODIES*

Component	4-yr-old male, 14.0 kg (31 lb) total body weight		25-yr-old male, 71.8 kg (158 lb) total body weight		42-yr-old female, 45.1 kg (100 lb) total body weight	
	% total body weight	% fat-free body weight	% total body weight	% fat-free body weight	% total body weight	% fat-free body weight
Water	53.8	69.7	61.8	72.8	56.0	73.3
Protein	18.5	23.8	16.6	19.5	14.4	19.2
Fat	22.7		14.9		23.6	
Minerals	5.0	6.5	6.5	7.6	5.8	7.6
Calcium	1.62	2.10	1.81	2.13	1.90	2.50
Phosphorus	0.81	1.05	1.19	1.40	0.99	1.30
Potassium	0.19	0.25	0.24	0.28	0.21	0.27
Sodium	0.18	0.23	0.17	0.20	0.17	0.22
Magnesium	0.028	0.036	0.041	0.048	0.033	0.043
Iron	0.0049	0.0063	0.0075	0.0088	0.0046	0.0060
Zinc	0.0017	0.0022	0.0028	0.0033	0.0017	0.0022
Copper	0.00025	0.00032	0.00014	0.00016	0.00014	0.00018
Other	0	0	0.2	0.24	0.2	0.26

* From data of Widdowson, McCance, and Spray. The child died of meningitis, the adult male of kidney failure, and the adult female of drowning.

significance because, in their diverse combinations, they form the protein, fat, and carbohydrate molecules making up the bulk of body substance. These are carbon, oxygen, hydrogen, and nitrogen.

Carbon (C) combines with oxygen (O_2) to form carbon dioxide (CO_2), a colorless, odorless gas comprising about 0.03% of the atmosphere and 50% heavier than air. In the body, gaseous CO_2 dissolves in water (H_2O) to form the weak acid carbonic acid (H_2CO_3), according to the following equation:

$$CO_2 + H_2O \rightleftharpoons H_2CO_3 \qquad (2.1)$$

We shall speak later of the processes that incorporate CO_2 into organic molecules such as sugar, fat, and protein.

Oxygen is a gas. Its great reservoir is the atmosphere, 20% of which (by volume) is free O_2. O_2 dissolves readily in water, but in doing so it remains uncombined. As we shall see in Chapter 11, O_2 is dissolved in water prior to entering the body tissues.* There O_2 combines with the carbon and hydrogen arising in the metabolic breakdown of organic compounds to produce CO_2 and water. Water thus formed may be further utilized by the organism, but the CO_2 is almost entirely eliminated.

Hydrogen, the lightest element, is unique in nature. Free H_2, a colorless, odorless gas, does not occur in living tissue. Instead, hydrogen atoms are extensively combined with carbon, oxygen, and the other elements. The ability of a compound to form hydrogen ions (H^+) is the property that identifies it as an acid.

The next most common body element is *nitrogen,* a characteristic constituent of proteins. Although the atmosphere is almost 80% elemental N_2 by volume, no animals and few plants are able to utilize it directly. Most plants accept N_2 from the environment only in the form of an inorganic compound: ammonia (NH_3), nitrates (salts containing NO_3^-), nitrites (salts containing NO_2^-). Certain bacteria and plants, however, can convert elemental N_2 into ammonia, nitrates, or nitrites by the process called *nitrogen fixation.* Animals cannot "fix" N_2 or convert inorganic nitrogen compounds to organic compounds. Hence they must obtain their nitrogen by ingesting plant or animal tissues containing nitrogen already in organic form. Thus withdrawal of N_2 from the atmosphere and its incorporation into animal tissues depend ultimately upon nitrogen-fixing plants and bacteria.

NATURE OF ORGANIC COMPOUNDS

The early belief that "vital forces" infuse living organisms, making them fundamentally different from nonliving objects, seemed to be bolstered when chemists discovered in living organisms countless substances that were previously unknown—e.g., uric acid in urine and citric acid in lemons. Such substances were apparently formed only by living creatures. Hence they were designated *organic* compounds in contrast to the *inorganic* materials of the lifeless rocks and minerals.

It is usually stated that the studies of the German chemist Friedrich Wöhler in 1826 revolutionized thought on this question. At first by accident and later by careful intent, Wöhler converted the familiar "inorganic" substance ammonium cyanate (NH_4OCN) to the "organic" substance urea (NH_2OCNH_2).† This achievement and the many test tube syntheses that followed gradually demolished the barrier between the two chemistries, and it became apparent that the same fundamental laws govern both.

The term "organic chemistry" remained in use, however, for it was found that almost all the important substances produced by animate nature contain carbon. Organic chemistry thus became the chemistry of the compounds of carbon. The remainder of chemistry is lumped together under the heading "inorganic chemistry." The division of chemistry that deals with the chemistry of living organisms is today called *biochemistry* or *biological chemistry.*

Much of the subject matter of physiology is related to the phenomena of biochemistry. In turn, biochemistry deals to a large extent with the transformations of organic compounds in the body. So that we may treat instructively those

---

* The fundamental laws of gas behavior are summarized in Chapter 11.

---

† Actually the inorganic substance ammonium cyanate was usually prepared commercially from animal horns and hoofs. Furthermore, George Wald has pointed out that in Wöhler's experiments a living organism was still essential for urea synthesis, the living organism being Wöhler.

principles of physiology that are based in biochemistry, we shall review briefly certain important properties of organic compounds.* Since there exist countless varieties of organic substances but only a hundred-odd kinds of atoms from which they can be constructed, it follows that the identity and uniqueness of an organic molecule depends upon the numbers, types, and spatial arrangement of its component atoms. If any one of these characteristics is altered, a different molecular species emerges. With ingenuity and skill, organic chemists have learned to determine the numbers, types, and spatial arrangement of atoms within the molecules of a given substance from studies of its chemical properties and, on the basis of such a *structural formula*, to predict the behavior of the compound in chemical reactions.

Significance of the structural formula. The ordinary chemical formula merely indicates the numbers and types of atoms present in each molecule; for example, ammonium cyanate is NH_4OCN. Much can be learned from such a simple formula: we know from H_2, Cl_2 (chlorine), HCl (hydrochloric acid), H_2O, CO_2, CCl_4 (carbon tetrachloride), CH_4 (methane), $CHCl_3$ (chloroform), and CH_2O (formaldehyde) that one hydrogen or chlorine atom combines with only one other atom, one oxygen combines with two, and one carbon with four. Thus carbon combines with four hydrogens but with only two oxygens, etc. We say that hydrogen or chlorine has a *valence* or combining power of one, oxygen of two, and carbon of four. Nitrogen's valence is ordinarily three, but in certain situations it can be five.

Molecular structure is determined by the valences of the component atoms. The two hydrogens of water, for example, could not be attached to one another, for, though the single valence of each would be satisfied, there would remain no means for attaching oxygen. The water molecule, therefore, must be H—O—H, with each of the two oxygen valences holding one of the hydrogens. Similarly, in methane each

of the hydrogens must attach to the carbon and not to each other, so that the atoms necessarily are combined and distributed in space as follows:

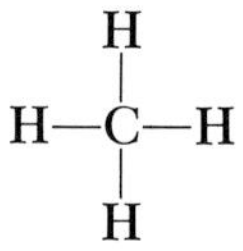

$$
\begin{array}{c}
\text{H} \\
| \\
\text{H—C—H} \\
| \\
\text{H}
\end{array}
$$

Chloroform is comparable,
$$
\begin{array}{c}
\text{H} \\
| \\
\text{Cl—C—Cl} \\
| \\
\text{Cl}
\end{array}
$$

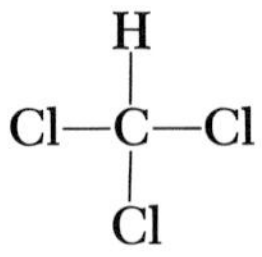

Likewise, the only possible arrangement of atoms in ethane (C_2H_6) is
$$
\begin{array}{c}
\text{H}\quad\text{H} \\
|\quad\ \ | \\
\text{H—C—C—H} \\
|\quad\ \ | \\
\text{H}\quad\text{H}
\end{array}
$$

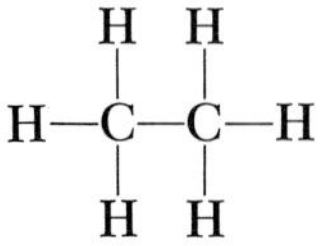

When molecular structures include more than one type of atom of valence higher than one, the possible arrangements become more elaborate. Carbon dioxide is straightforward enough, the four valences of carbon being attached in pairs to the two oxygens to give two double bonds, O=C=O. However, one of the two bonds in a double bond could open and still leave the oxygen and carbon firmly held together by the other. This happens when carbon dioxide is dissolved in water. The two molecules combine to form carbonic acid, H_2CO_3, or 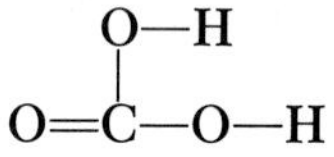
$$
\begin{array}{c}
\text{O—H} \\
| \\
\text{O=C—O—H}
\end{array}
$$
the H and OH of the water molecule being added to the O and C ends, respectively, of the opened bond.

In summary, the structure of an organic compound can be represented in several ways (Fig. 2.1). The ordinary *molecular* formula indicates the actual number of atoms of each element in the molecule.† *Structural* formulas of the two types shown in Fig. 2.1 suggest the spatial distribution of the atoms or atomic groupings. Even more precise representations are possible. In

---

* These concepts are fully developed in standard textbooks of organic chemistry. We note them here only to facilitate and introduce orderly discussion of the elementary principles of biochemistry.

---

† Sometimes a molecular formula is modified to give the ratios of the numbers of atoms of different elements to one another instead of the actual numbers. The result is an *empirical* formula. For example, the molecular formula of lactic acid is $C_3H_6O_3$, whereas its empirical formula is CH_2O. This kind of simplification is not feasible for alanine, the compound in Fig. 2.1.

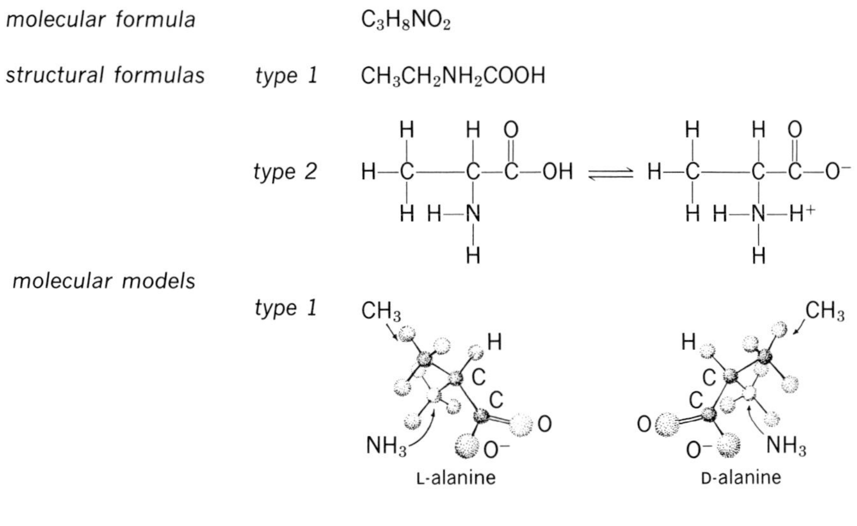

molecular formula $C_3H_8NO_2$

structural formulas   type 1 $CH_3CH_2NH_2COOH$

type 2

$$H-\underset{\underset{H}{|}}{\overset{\overset{H}{|}}{C}}-\underset{\underset{H-N}{|}}{\overset{\overset{H}{|}}{C}}-\overset{\overset{O}{\|}}{C}-OH \rightleftharpoons H-\underset{\underset{H}{|}}{\overset{\overset{H}{|}}{C}}-\underset{\underset{H-N-H^+}{|}}{\overset{\overset{H}{|}}{C}}-\overset{\overset{O}{\|}}{C}-O^-$$

molecular models

type 1

L-alanine   D-alanine

2.1 Various types of chemical formulas, all representing forms of the amino acid alanine. The significance of the D- and L-forms is explained on p. 18. (From *General Chemistry*, 2nd ed., by Linus Pauling, W. H. Freeman and Company, San Francisco, © 1953.)

type 2

reality, the atoms within a molecule do not lie within a single plane as in a printed structural formula. In three-dimensional space the four valences of carbon point to the vertices of an imaginary tetrahedron, a symmetrical pyramid whose three sides and base are identical equilateral triangles. To obtain a true picture of the structure of a molecule, then, chemists build models in which bond lengths and angles are portrayed in accurate proportion. Molecular models are also of two types. In one type (1 in Fig. 2.1), individual atoms are arbitrarily given small and equal radii so that bond arrangements are emphasized. In the other type (2 in the figure), the different atoms appear in proportion to their actual sizes. The importance of molecular models is simply stated: a structural formula for which a model cannot be built cannot be correct.

Isomerism. The organic chemist cannot be satisfied merely with molecular formulas because the laws of valence frequently permit more than one arrangement of a given group of atoms. From the formula C_2H_6O, for example, he readily concludes that all the hydrogens must

be attached to oxygen or carbon, which in turn must be bonded together, and that two structures are possible. If all six hydrogens were attached to carbon atoms, each carbon could hold three. Its fourth valence would then be left to combine with the oxygen, forming methyl ether,

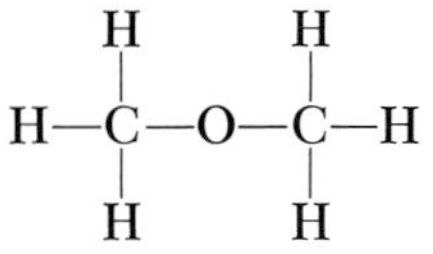

$$H-\underset{\underset{H}{|}}{\overset{\overset{H}{|}}{C}}-O-\underset{\underset{H}{|}}{\overset{\overset{H}{|}}{C}}-H$$

But, if one hydrogen were combined with oxygen, three hydrogens would be left for one carbon and two for the other. Only one oxygen valence would then be free to combine with carbon, so that the carbons would have to be attached directly together. The result would be ethanol,

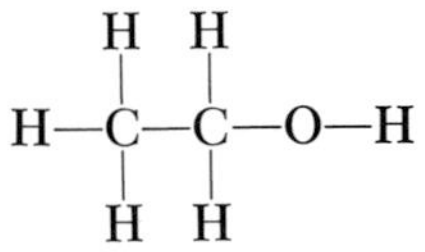

$$H-\underset{\underset{H}{|}}{\overset{\overset{H}{|}}{C}}-\underset{\underset{H}{|}}{\overset{\overset{H}{|}}{C}}-O-H$$

Two such compounds with differing structural formulas and identical molecular formulas are

isomers of one another. Given these two isomers, the only two known substances with the formula C_2H_6O, the chemist decides which is which by predicting the chemical behavior of each structure and then performing tests to confirm his hypotheses.*

Stereoisomerism. We shall now consider a special kind of isomerism that has had a significant place in the history of chemistry and biology, some of the classic work having been performed by Louis Pasteur. Molecules containing a carbon atom whose four valences are all attached to different atoms or groups of atoms may exist in two forms. These have similar structural formulas but are, in fact, mirror images of one another, or *stereoisomers* (sometimes called *epimers*). With the middle carbon of alanine (see Fig. 2.1), for example, two arrangements are possible:

$$
\begin{array}{cc}
\underset{\displaystyle \overset{\displaystyle \text{COOH} \quad \text{H} \quad \text{NH}_2}{}}{\overset{\displaystyle \text{CH}_3}{\text{C}}}
&
\underset{\displaystyle \overset{\displaystyle \text{NH}_2 \quad \text{H} \quad \text{COOH}}{}}{\overset{\displaystyle \text{CH}_3}{\text{C}}}
\end{array}
$$

Stereoisomers have identical chemical properties and hence cannot be distinguished chemically, as were methyl ether and ethanol in the preceding example. They are recognizable in the laboratory, however, chiefly on the basis of two properties: (1) each stereoisomer rotates a beam of polarized light in a particular direction;† and (2) many enzymes are active with one of the

pair and not the other—i.e., they are stereospecific. Stereospecificity is of great biological importance.‡

Functional groups. Certain groupings of atoms recur frequently in organic structural formulas; we have already encountered methyl ($-CH_3$), ethyl ($-C_2H_5$), hydroxyl ($-OH$), and carboxyl ($-COOH$) groups. This is a characteristic feature of organic chemistry. The chemical properties of these groups largely account for the chemical behavior of the molecules of which they are parts—hence the term *functional* groups. Ethanol, for example, reacts with sodium (Na) because of its $-OH$ group. Indeed, this reaction is typical of alcohols, each of which possesses the $-OH$ group.

Table 2.2 lists the more important organic functional groups, or *radicals*. We note that many occur in series. The open valences in the table are a reminder that functional groups do not ordinarily occur by themselves.

To illustrate the significance of functional groups in organic reactions—and as a background for later discussion—a summary of the major types of organic molecules and their basic reactions is given in Table 2.3. From a knowledge of such reactions, organic chemists have succeeded in synthesizing one after another the complex organic molecules found in living tissues, building ever more complex structures from simpler ones. Before a synthesis is possible, the structural formula of each compound must

---

* For example, in the structure
$$
\underset{\displaystyle \overset{\displaystyle \text{H} \quad\quad \text{H}}{}}{\text{H}-\overset{\displaystyle \text{H}}{\underset{\displaystyle \text{H}}{\text{C}}}-\text{O}-\overset{\displaystyle \text{H}}{\underset{\displaystyle \text{H}}{\text{C}}}-\text{H}}
$$
all six hydrogens are attached to carbons in like manner. Probably they would all behave similarly in chemical reactions. But, in the structure
$$
\text{H}-\overset{\displaystyle \text{H}}{\underset{\displaystyle \text{H}}{\text{C}}}-\overset{\displaystyle \text{H}}{\underset{\displaystyle \text{H}}{\text{C}}}-\text{O}-\text{H}
$$
one hydrogen might be expected to behave differently from the other five, since it alone is attached to oxygen. In fact, it is found in the laboratory that sodium metal (Na) does not react with methyl ether but does react with ethanol to yield sodium ethanolate, C_2H_5ONa. Obviously the sodium replaces the hydrogen linked to the oxygen.

† It was early discovered that the simplest sugar, the three-carbon molecule glyceraldehyde, has an asymmetric carbon and that its two stereoisomers rotate a light beam in opposite

directions. The isomer rotating the beam to the right was called *dextrorotatory*, and the isomer rotating it to the left, *levorotatory*. Thus it became customary to speak of *d*-glyceraldehyde and *l*-glyceraldehyde. Confusion arose with the later discovery that more complex molecules *structurally* related to *d*- or *l*-glyceraldehyde may be *optically* opposite to their glyceraldehyde analogues. Accordingly, the symbols D- and L- were adopted to refer solely to structural arrangements resembling *d*- and *l*-glyceraldehyde. To specify optical rotatory activity, the supplementary symbols (+)- for dextrorotatory and (−)- for levorotatory are used. For example, of two stereoisomers of a six-carbon sugar that are structurally related to *d*-glyceraldehyde, one could be a D(+)-sugar, and the other a D(−)-sugar.

‡ Compounds synthesized by organic chemists always contain equal quantities of D- and L-forms; hence they are optically inactive. Such mixtures are called *racemic* mixtures. Compounds synthesized by stereospecific enzymes are optically active D- or L-stereoisomers, not racemes. Note the stereoisomers D- and L-alanine in Fig. 2.1.

TABLE 2.2 SOME ORGANIC FUNCTIONAL GROUPS

Name	Molecular formula	Structural formula
Amino	$-NH_2$	
Alkyl	$-C_nH_{2n+1}$	
Methyl	$-CH_3$	
Ethyl	$-C_2H_5$	
Propyl	$-C_3H_7$	
Carboxyl	$-COOH$	
Hydroxyl	$-OH$	$-O-H$
Aldehyde	$-CHO$	
Keto (oxo)	$=O$	$=O$
Sulfhydryl (mercapto, thiol)	$-SH$	$-S-H$
	$-C_6H_5$	or

TABLE 2.3 TYPES AND PROPERTIES OF FUNDAMENTAL ORGANIC COMPOUNDS*

Class	Examples	Important reactions
Acid, R—COOH	Formic, HCOOH Acetic, CH_3COOH Propionic, C_2H_5COOH Fatty acids†	R—COOH + NaOH $\longrightarrow$ R—COONa + H_2O acid · sodium hydroxide · sodium salt of acid R—COOH + 2H $\longrightarrow$ R—CHO + H_2O acid · reducing agent · aldehyde R—COOH + R′—OH $\longrightarrow$ R—CO—OR′ + H_2O acid · alcohol · ester 2R—COOH $\longrightarrow$ R—CO—O—OC—R + H_2O acid · acid anhydride
Alcohol, R—OH	Methanol, CH_3OH Ethanol, C_2H_5OH Propanol, C_3H_7OH	R—OH + NaOH $\longrightarrow$ R—ONa + H_2O alcohol · sodium hydroxide · sodium alcoholate R—CH_2—OH + O $\longrightarrow$ R—CHO + H_2O alcohol · oxidizing agent · aldehyde R—OH + R′—COOH $\longrightarrow$ R′—CO—OR + H_2O alcohol · acid · ester
Aldehyde, R—CHO	Formaldehyde, HCHO Acetaldehyde, CH_2CHO Propionaldehyde, C_2H_5CHO	R—CHO + 2H $\longrightarrow$ R—CH_2OH aldehyde · reducing agent · alcohol R—CHO + O $\longrightarrow$ R—COOH aldehyde · oxidizing agent · acid R—CHO + H_2N—NH_2 $\longrightarrow$ R—CHN—NH_2 + H_2O aldehyde · hydrazine · hydrazone
Ketone, R—CO—R′	Acetone, CH_3COCH_3 Ethyl methyl ketone, $C_2H_5COCH_3$	R—CO—R′ + 2H $\longrightarrow$ R—CHOH—R′ ketone · reducing agent · alcohol R—CO—R′ + O $\longrightarrow$ poor reaction ketone · oxidizing agent R—CO—R′ + H_2N—NH_2 $\longrightarrow$ R—CN—NH_2—R′ + H_2O ketone · hydrazine · hydrazone
Ester, R—CO—OR′	Ethyl acetate, $CH_3COOC_2H_5$	R—CO—OR′ + NaOH $\longrightarrow$ R—COONa + R′—OH ester · sodium hydroxide · sodium salt of acid · alcohol
Acid anhydride, R—CO—O—OC—R	Acetic anhydride, $CH_3COOOCCH_3$	R—CO—O—OC—R + H_2O $\longrightarrow$ 2R—COOH acid anhydride · acid R—CO—O—OC—R + NH_4OH $\longrightarrow$ R—CO—NH_2 + acid anhydride · ammonium hydroxide · acid amide R—COOH + H_2O acid

* R and R′ represent alkyl groups. R′ represents the same or a different alkyl group.
† See Table 2.4.

be established from careful study of its empirical formula, the functional groups present in the molecule, and the character of the fragments produced upon decomposition. Although the details of this great labor are beyond our immediate interest, we should bear in mind that whenever a chemical formula is cited in the text, it has been demonstrated by just such intricate study of natural and synthetic substances.

Chemical bonds. The forces or bonds holding the atoms together in a molecule are of several distinct types. *Covalent bonds* are the ones usually represented by the short connecting lines in a structural formula. They occur when one or more electrons are transferred from the outer shell of one atom to that of another atom. Thus they are due to the sharing of electrons. They are so nearly universally present that G. N. Lewis, discoverer of the electronic basis of the covalent bond, called it *the* chemical bond. Functionally, covalent bonds are notable for the amount of energy needed to break them. This means that they hold atoms together very tightly.

Ionic bonds are the strong electrostatic forces existing between negatively and positively charged *ions.* Ions are atoms or groups of atoms that either have acquired one or more extra electrons and thus become negatively charged (*anions*) or have lost electrons and become positively charged (*cations*). We shall speak later of the properties of ions. Here we shall note only that ionic bonds hold anions and cations together, that such anion-cation combinations are called *salts,* and that salts are known for their solubility in water and the ability of their solutions to conduct an electric current—i.e., they are *electrolytes.*

Hydrogen bonds, though weaker than ionic or covalent bonds, hold electrically charged groups together under certain circumstances. They are interactions occuring between a hydrogen atom attached to a negatively charged atom in one molecule and another negatively charged atom in the same or a different molecule. A molecule of hydrogen fluoride (HF) is shown in Fig. 2.2. The positively charged hydrogen is a bare atomic nucleus resting on the surface of the larger and negatively charged fluoride ion. Because of its positive charge, it is able to attract another negatively charged fluoride ion. When one is found,

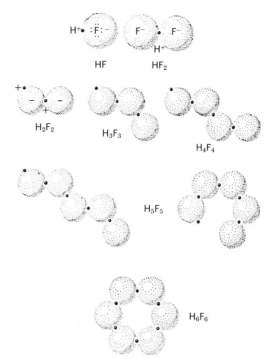

2.2 Some polymers of hydrogen fluoride, showing how hydrogen bonds function as connecting links. (From *General Chemistry,* 2nd ed., by Linus Pauling, W. H. Freeman and Company, San Francisco, © 1953.)

a new ionic grouping results with the structure $(F^- - H^+ - F^-)^-$ or HF_2^-. Such ions are stable, although the hydrogen bond holding them together can be broken with relative ease.* Hydrogen bonds are indicated in structural formulas by dashed lines: $F^- - H^+ - - - F^- - H^+$. Hydrogen bonding may also occur between a hydrogen-containing functional group and another functional group in the same or a different molecule (Fig. 2.3). Of particular physiological importance are the hydrogen bonds formed by carboxyl and amino groups:

$$
R{-}C\underset{O{-}H\cdots O}{\overset{O\cdots H{-}O}{\big\langle}}C{-}R' \qquad
R{-}C\underset{C{=}O\cdots H{-}N}{\overset{N{-}H\cdots O{=}C}{\big\langle}}C{-}R'
$$

---

* Linus Pauling has shown that hydrogen bonds may cause long chains to develop, so that the molecular species H_2F_2, H_3F_3, H_4F_4, H_5F_5, and H_6F_6 can arise (Fig. 2.2). The last of these is especially stable, probably because it forms a ring.

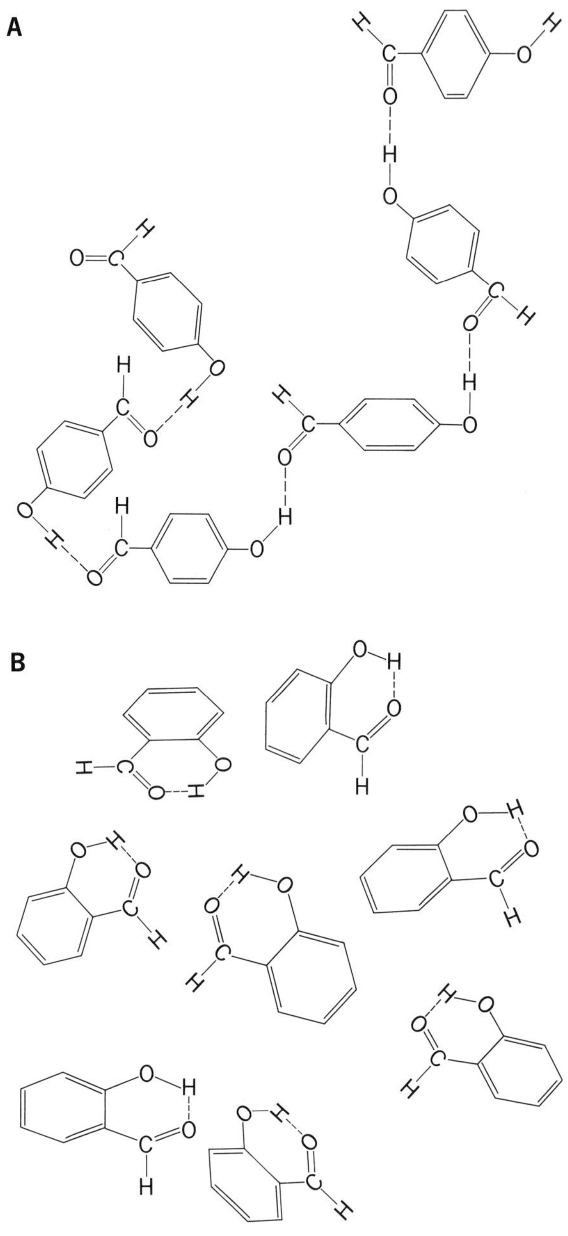

2.3 Hydrogen bonding. A. Intermolecular, between molecules of *p*-hydroxybenzaldehyde. B. Intramolecular, within individual molecules of salicylaldehyde. Intermolecular bonding can result in the formation of long chains.

When hydrogen bonds as well as other bonds join molecules or parts of a molecule, obviously the union is strengthened.

In the living organism, hydrogen bonds are as important as covalent bonds. Presently we shall see that they play a fundamental role in determining the properties of proteins and nucleic acids; and in Chapter 7 we shall learn of the tendency of water to form hydrogen bonds, which accounts for many of its unusual properties.

A fourth type of chemical bond comprises the very weak attractions known as *van der Waals' forces*, which exist among all molecules when they move sufficiently close together.

Why carbon? One might reasonably question why the bulk of living organisms is fabricated of the chemical compounds of carbon. Why were the earliest living organisms not made primarily of the compounds of silicon, another element of valence four and one present on earth in abundance far beyond that of carbon? Indeed, writers of science fiction have already reviewed the possibilities of siliceous beings in which silicon dioxide (SiO_2) is analogous to the carbon dioxide of nonfictitious life.

The answer is simply that for a number of reasons carbon atoms more readily join together in chains and rings than do silicon atoms. Hence larger and more varied molecular structures may develop from carbon. In addition, many (but not all) carbon compounds are readily soluble in water. The same is not true of silicon. The fact that most of the earth's rocks are silicates is testimony that silicon compounds formed insoluble precipitates when the newly formed earth cooled, whereas carbon, though present in many rocks as carbonates, remained in the earth's water and atmosphere as labile and reactive substances capable of conversion into the compounds we now call organic.

MAJOR CLASSES OF ORGANIC COMPOUNDS IN THE BODY

A survey of the organic molecules of the body suggests the existence of four major functional categories: (1) compounds essential to body structure; (2) those serving primarily as energy-rich fuels; (3) those conveying the hereditary information from generation to generation by which growth, differentiation, and biological specificity are controlled; and (4) those functioning primarily as catalytic agents in the body's chemical processes.

In general, *proteins* function as structural and catalytic elements, while *carbohydrates* and *lipids* (*fats*) serve as structural elements and fuels. The *nucleic acids* are the fundamental genetic materials, and the *vitamins* and *hormones* operate directly or indirectly in catalysis. We shall find, however, that these roles overlap extensively. Often the same materials act both as building blocks and energy sources, even simultaneously; proteins, for example, may at once be structural elements, energy sources, and chemical catalysts.

In examining these materials, we must bear in mind that in the body they participate in a constant, complex, interweaving flow of transformations, rather resembling the simultaneous building of a house, hauling in of fuel, and building of a hearth fire. First the body cells acquire the various essential organic substances,* and then they extract energy from them and use them in the synthesis and repair of tissues. In later chapters we shall consider the mechanisms by which the body plan persists while its chemical components undergo continuous modification and replacement.

Carbohydrates. The carbohydrates, comprising the sugars and their many derivatives, consist of carbon, hydrogen, and oxygen, the last two elements ordinarily present in the same proportions as in water. Most carbohydrates can therefore be represented by the formula $C_x(H_2O)_y$. This makes it seem that they are *hydrates* of carbon, whence the name "carbohydrates." In fact, the water molecule as such does not appear in the structural formula of a carbohydrate. Rather, a carbohydrate is fundamentally a carbon chain bearing many hydroxyl groups (hence they are alcohols) plus a single aldehyde or keto group.

Carbohydrates can be divided into two main categories: (1) the *monosaccharides*, or simple sugars; and (2) the *polysaccharides*, long chains of monosaccharides.

Monosaccharides, which cannot be split (hy-

---

* They may make these materials for themselves, receive them from other cells of the same organism, or ultimately derive them from outside sources.

drolyzed) to simpler forms, are usually named according to the number of carbon atoms in each molecule, with the *-ose* suffix characteristic of all carbohydrate nomenclature. Thus there are trioses ($C_3H_6O_3$), tetroses ($C_4H_8O_4$), pentoses ($C_5H_{10}O_5$), hexoses ($C_6H_{12}O_6$), heptoses ($C_7H_{14}O_7$), octoses ($C_8H_{16}O_8$), and nonoses ($C_9H_{18}O_9$). All the sugars containing an aldehyde group are known as *aldoses;* those containing a keto group are known as *ketoses*. Since the monosaccharide category includes aldoses and ketoses, we may speak of aldopentoses, ketopentoses, aldohexoses, and so on.

Nomenclature of the aldoses is summarized in Fig. 2.4, which gives the common names of the individual sugars. The series shown is built upon the dextrorotatory aldotriose glyceraldehyde; hence these are D-aldoses. As it happens, almost all the monosaccharides of the human body are dextrorotatory. It is noteworthy that each additional carbon atom increases the possibilities for stereoisomerism based upon the positions of the —OH groups relative to the C atoms.

The important aldoses of human physiology include D-glyceraldehyde, D-ribose, and D-glucose. The important ketoses are as follows:

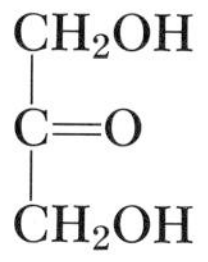

$$\begin{array}{c} CH_2OH \\ | \\ C{=}O \\ | \\ CH_2OH \end{array} \qquad \begin{array}{c} CH_2OH \\ | \\ C{=}O \\ | \\ H{-}C{-}OH \\ | \\ H{-}C{-}OH \\ | \\ CH_2OH \end{array}$$

dihydroxyacetone 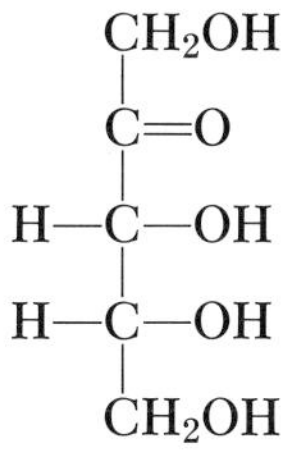   D-ribulose

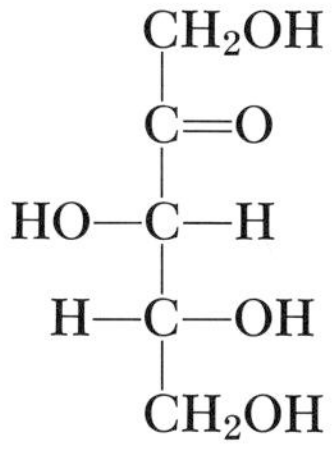

$$\begin{array}{c} CH_2OH \\ | \\ C{=}O \\ | \\ HO{-}C{-}H \\ | \\ H{-}C{-}OH \\ | \\ CH_2OH \end{array} \qquad \begin{array}{c} CH_2OH \\ | \\ C{=}O \\ | \\ HO{-}C{-}H \\ | \\ H{-}C{-}OH \\ | \\ H{-}C{-}OH \\ | \\ CH_2OH \end{array}$$

D-xylulose 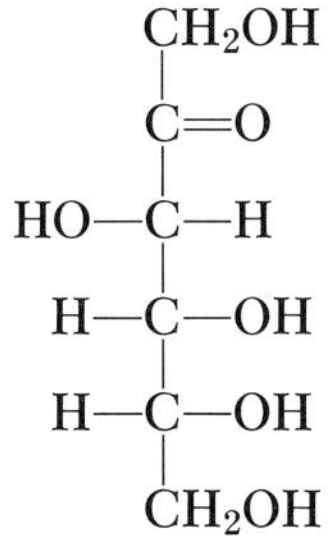   D-fructose

These formulas and those in Fig. 2.4 portray the monosaccharides as straight-chain compounds. As it happens, many chemical properties of the sugars can be explained only on the

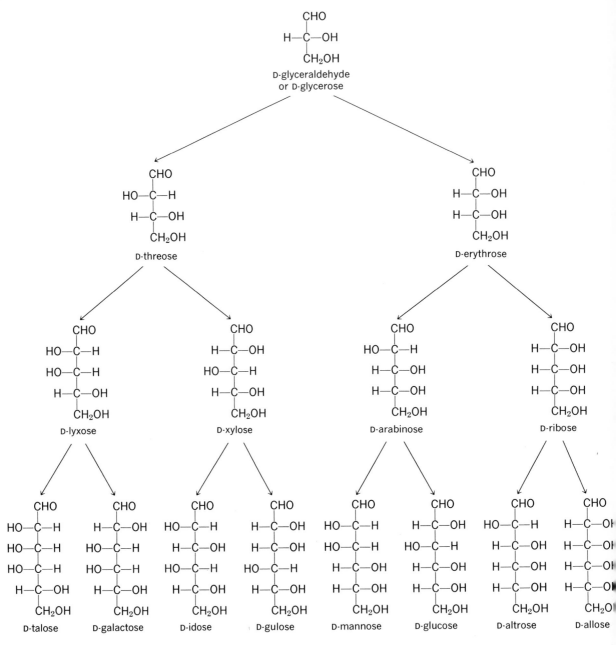

2.4 Structural relations of the D-aldoses.

basis of molecules arranged as rings. Fig. 2.5 shows the ring formulations of D-glucose and D-ribose. In the literature of biochemistry, straight-chain and ring structures are employed in different circumstances.

The simplest polysaccharides are the *disaccharides*, which consist of two linked molecules of the same or different monosaccharides. Examples are sucrose (glucose-fructose), lactose (glucose-galactose), and maltose (glucose-carbon numbers

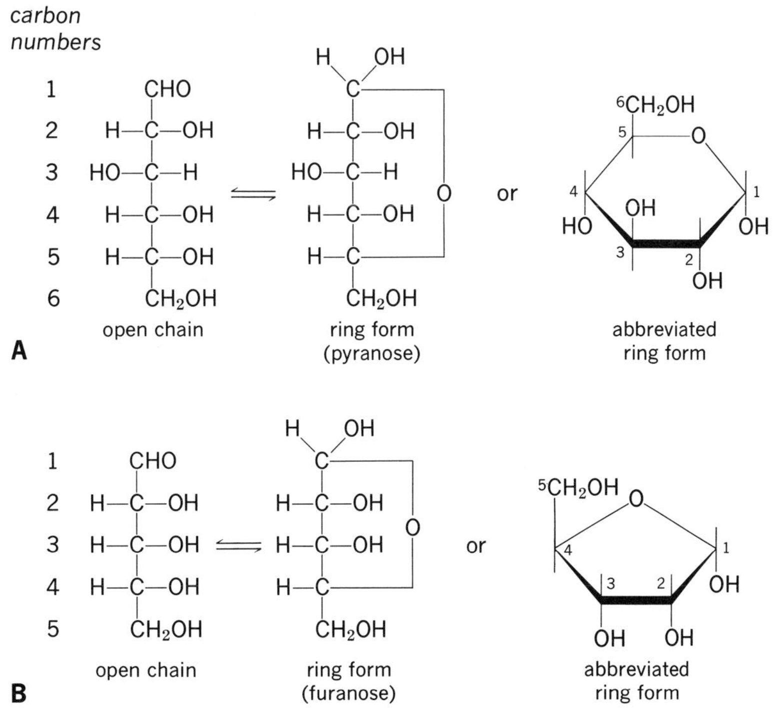

A

1	CHO
2	H—C—OH
3	HO—C—H
4	H—C—OH
5	H—C—OH
6	CH$_2$OH

open chain

ring form (pyranose)

or

^{6}CH$_2$OH

abbreviated ring form

2.5 Open-chain and ring formulations of D-glucose (A) and D-ribose (B). The open chain and ring form are actually in equilibrium with one another. The Haworth formulations, from which the C and H atoms of the ring are omitted, suggest the three-dimensional structure; the plane of the ring is intended to be perpendicular to that of the page.

B

1	CHO
2	H—C—OH
3	H—C—OH
4	H—C—OH
5	CH$_2$OH

open chain

ring form (furanose)

or

^{5}CH$_2$OH

abbreviated ring form

glucose). *Oligosaccharides* yield three to ten monosaccharide molecules upon hydrolysis, and larger polysaccharides yield many more. The long-chain polysaccharides, whose general formula is $(C_6H_{10}O_5)_x$, include glycogen (polyglucose), cellulose, and starches, and the dextrins. All of these differ in molecular weight, branching structure, and solubility. Cellulose is the tough, insoluble constituent of plant cell walls. Glycogen, soluble, is the main storage form of carbohydrates in the body. Starches, also soluble, are the chief source of carbohydrates in food. Dextrins, soluble, are breakdown products of starches. We shall presently find that some important tissue structural elements consist of polysaccharides combined with protein.

Lipids. The lipids comprise a heterogeneous collection (fats, oils, waxes, and related compounds) of organic substances that are insoluble or sparingly soluble in water and freely soluble in organic solvents such as ethanol or ether. The solubility properties are exploited in the extraction of lipids from tissues. Much of the biological importance of lipids is due to the fact that as a class they can bridge the gap from water-soluble to water-insoluble phases without sharp discontinuities.

The simplest lipids contain only carbon, hydrogen, and oxygen, though they have less oxygen relative to carbon and hydrogen than carbohydrates do. Upon hydrolysis, simple lipids yield glycerol and fatty acids. Glycerol contains three —OH groups and is thus an alcohol. When the acidic —COOH groups of the fatty acids react with the three alcoholic —OH groups of glycerol, a *triglyceride* results (Fig. 2.6).

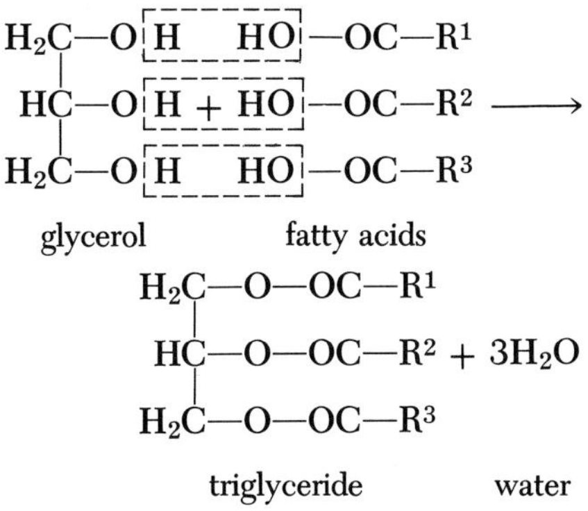

$$H_2C—O[H \quad HO]—OC—R^1$$
$$HC—O[H + HO]—OC—R^2 \longrightarrow$$
$$H_2C—O[H \quad HO]—OC—R^3$$

glycerol   fatty acids

$$H_2C—O—OC—R^1$$
$$HC—O—OC—R^2 + 3H_2O \qquad (2.2)$$
$$H_2C—O—OC—R^3$$

triglyceride   water

A

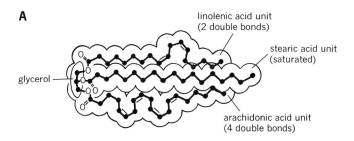

linolenic acid unit
(2 double bonds)

stearic acid unit
(saturated)

glycerol

arachidonic acid unit
(4 double bonds)

B

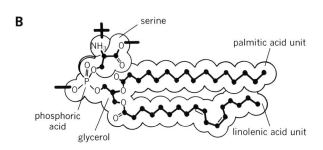

serine

NH₃

O

palmitic acid unit

phosphoric
acid

glycerol

linolenic acid unit

C

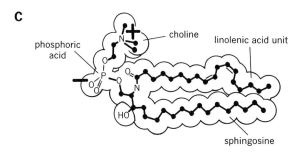

phosphoric
acid

N

choline

linolenic acid unit

P

HO

sphingosine

2.6 Structural formulas of several lipids. A. A triglyceride (1-linolenoyl 2-stearoyl 3-arachidonoyl glycerol). B. Phosphatidyl serine. C. Sphingomyelin. D. Cholesterol. Positively and negatively charged groups are indicated by + and −. (Adapted from J. L. Oncley, ed., *Biophysical Science,* Wiley, New York, 1959.)

D

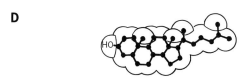

HO

R^1, R^2, and R^3 can be identical or different alkyl groups (see Table 2.2). A triglyceride is obviously an ester (see Table 2.3).

The term "fatty acid" is not easy to define accurately. Usually it refers to a straight-chain carboxylic acid having an even number of carbon atoms (Table 2.4). Some branched-chain fatty acids also occur as minor components of lipids. It is convenient to divide fatty acids into two groups depending on whether the carbon chain carries the largest possible number of attached hydrogens. *Saturated* fatty acids have structures like

$$-\underset{\underset{H}{|}}{\overset{\overset{H}{|}}{C}}-\underset{\underset{H}{|}}{\overset{\overset{H}{|}}{C}}-\underset{\underset{H}{|}}{\overset{\overset{H}{|}}{C}}-\underset{\underset{H}{|}}{\overset{\overset{H}{|}}{C}}-COOH$$

TABLE 2.4 FATTY ACIDS

Common name	Official name	Number of carbons	Molecular formula
SATURATED			
Butyric	n-Butanoic	4	C_3H_7COOH
Caproic	n-Hexanoic	6	$C_5H_{11}COOH$
Caprylic	n-Octanoic	8	$C_7H_{15}COOH$
Capric	n-Decanoic	10	$C_9H_{19}COOH$
Lauric	n-Dodecanoic	12	$C_{11}H_{23}COOH$
Myristic	n-Tetradecanoic	14	$C_{13}H_{27}COOH$
Palmitic	n-Hexadecanoic	16	$C_{15}H_{31}COOH$
Stearic	n-Octadecanoic	18	$C_{17}H_{35}COOH$
Arachidic	n-Eicosanoic	20	$C_{19}H_{39}COOH$
Behenic	n-Docosanoic	22	$C_{21}H_{43}COOH$
UNSATURATED*			
One double bond			
Myristoleic	Δ^9-Tetradecenoic	14	$CH_3(CH_2)_3CH{=}CH(CH_2)_7COOH$
Palmitoleic	Δ^9-Hexadecenoic	16	$CH_3(CH_2)_5CH{=}CH(CH_2)_7COOH$
Oleic	Δ^9-Octadecenoic	18	$CH_3(CH_2)_7CH{=}CH(CH_2)_7COOH$
Two double bonds			
Linoleic	$\Delta^{9,12}$-Octadecadienoic	18	$CH_3(CH_2)_4CH{=}CHCH_2CH{=}CH(CH_2)_7COOH$
Three double bonds			
Linolenic	$\Delta^{9,12,15}$-Octadecatrienoic	18	$CH_3CH_2CH{=}CHCH_2CH{=}CHCH_2CH{=}CH(CH_2)_7COOH$
Eleostearic	$\Delta^{9,11,13}$-Octadecatrienoic	18	$CH_3(CH_2)_3CH{=}CHCH{=}CHCH{=}CH(CH_2)_7COOH$
Four double bonds			
Arachidonic	$\Delta^{5,8,11,14}$-Eicosatetraenoic	20	$CH_3(CH_2)_3(CH_2CH{=}CH)_4(CH_2)_3COOH$

* In official chemical nomenclature, the symbol Δ is used with a superscript numeral to denote the lowest-numbered carbon atom to which double bonds attach, the carboxyl carbon always being number 1. Thus Δ^9 indicates a double bond between the ninth and tenth carbons from the right end of the molecular formula.

In *unsaturated* fatty acids, double bonds are attached to the carbon atoms that are not fully saturated with hydrogen, thus:

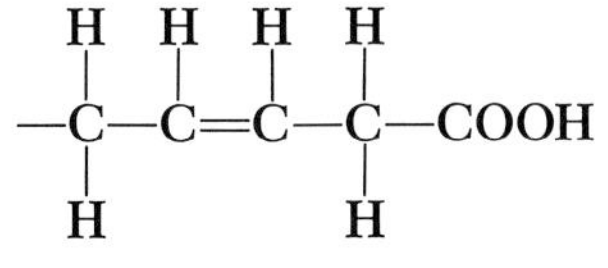

$$
\begin{array}{c}
\text{H} \quad\text{H} \quad\text{H} \quad\text{H} \\
| \qquad | \qquad | \qquad | \\
-\text{C}-\text{C}{=}\text{C}-\text{C}-\text{COOH} \\
| \qquad\qquad\qquad | \\
\text{H} \qquad\qquad \text{H}
\end{array}
$$

Table 2.4 also lists some *polyunsaturated* fatty acids, in which there is more than one double bond.

The higher members of the saturated series of fatty acids (lauric, myristic, palmitic, stearic, etc.) in combination with glycerol form the bulk of the body fat. The lower members (caproic, caprylic, and capric) are not widely distributed in nature. Like carbohydrates, lipids serve both as energy sources and structural elements. They are prominent components of cell membranes and membranous submicroscopic cellular particles, a role for which their peculiar structure ideally equips them.

A fatty acid molecule has two functional groups: the familiar carboxyl group, which contains carbon, hydrogen, and oxygen; and the chain, which contains only carbon and hydrogen and is thus a hydrocarbon chain. The presence of oxygen in a functional group usually signifies water-solubility; therefore, the carboxyl group

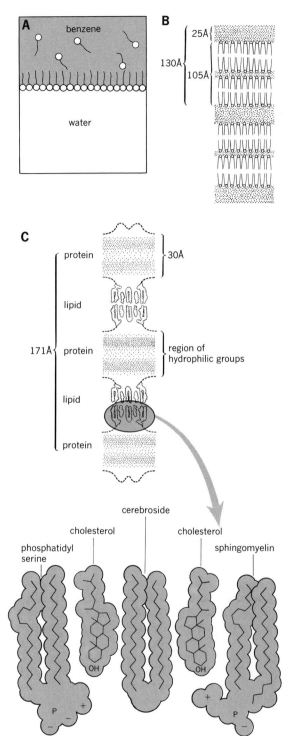

is water-soluble, or *polar*, whereas the hydrocarbon chain is water-insoluble, or *nonpolar*. The latter group accounts for the characteristic solubility properties of lipids. As a result of the difference in solubility behavior of the two ends of the molecule, fatty acids in an immiscible mixture of water and benzene orient themselves at the interface in a distinctive manner. The hydrocarbon chains project into the organic solvent while the carboxyl groups point toward the water layer (Fig. 2.7A). The affinity of the nonpolar hydrocarbon residues for one another tends to spread the fatty acids into a monomolecular film in which the hydrocarbon chains are parallel to one another. Such a configuration is frequently found in the functionally important surface layers of many tissue membranes (Fig. 2.7B,C).

In one group of simple lipids, one of the fatty acids of the triglyceride has been replaced in the ester linkage by compounds containing both phosphorus and nitrogen. The resulting substances are the *phospholipids,* or *glycerophosphatides.* Members of this category constitute a major fraction of the tissue lipids. As shown in Fig. 2.6, additional groups may be charged, increasing the solubility in both water and organic solvents. For this reason, phospholipids serve to bind water-soluble materials (such as proteins) to water-insoluble materials. Among the important phospholipids are phosphatidyl serine (see Fig. 2.6B), phosphatidyl choline, phosphatidyl ethanolamine, phosphatidyl inositol, phosphatidyl glycerol, and sphingomyelin (see Fig. 2.6C). *Cerebrosides* comprise another group of lipids. They contain the sugar galactose and are relatively uncharged (see Fig. 2.7C).

The *steroids* are fat-soluble derivatives of

2.7 Arrangement of lipids in molecular films and in living membranes. A. Monolayer arrangement of fatty acids at a water-benzene interface. B. Double-layer arrangement of phospholipids in one of the current models of the myelin nerve sheath. C. Another possible arrangement of phospholipids in the myelin nerve sheath. In A and B the polar (carboxyl) ends of fatty acid and phospholipid molecules are indicated by circles, and the nonpolar (hydrocarbon) ends by tails. In B and C the stippled areas represent protein molecules. Membrane thickness is given in Ångström (Å) units.

TABLE 2.5 MAJOR AMINO ACIDS:

$$\begin{array}{c} H \\ | \\ R-C-COOH \\ | \\ NH_2 \end{array}$$

Name	Abbreviation	Structure of R—	Category
Alanine	Ala	CH_3-	Neutral
Arginine	Arg	$\begin{array}{c}NH\\ \parallel\\ C-NH-CH_2-CH_2-CH_2-\\ \vert\\ NH_2\end{array}$	Basic
Aspartic acid	Asp	$COOH-CH_2-$	Acidic
Asparagine	Asp(NH$_2$)	$CONH_2-CH_2-$	Neutral
Cysteine	Cys	$SH-CH_2-$	Acidic
Glutamic acid	Glu	$COOH-CH_2-CH_2-$	Acidic
Glutamine	Glu(NH$_2$)	$CONH_2-CH_2-CH_2-$	
Glycine	Gly	$H-$	Neutral
Histidine	His	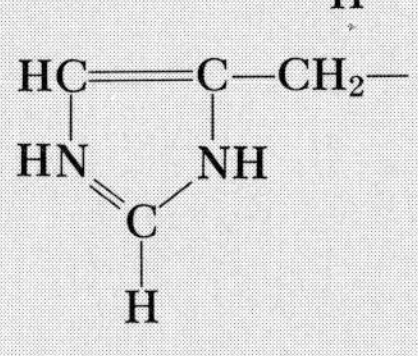 $\begin{array}{c}HC=C-CH_2-\\ \vert\quad\ \vert\\ HN\quad NH\\ \diagdown\ /\\ C\\ \vert\\ H\end{array}$	Basic
Isoleucine	Ileu	$\begin{array}{c}CH_3-CH_2-CH-\\ \vert\\ CH_3\end{array}$	Neutral
Leucine	Leu	$\begin{array}{c}CH_3-CH-CH_2-\\ \vert\\ CH_3\end{array}$	Neutral
Lysine	Lys	$NH_2-CH_2-CH_2-CH_2-CH_2-$	Basic
Methionine	Met	$CH_3-S-CH_2-CH_2-$	Neutral
Phenylalanine	Phe	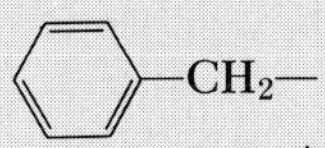 ⟨C$_6$H$_5$⟩$-CH_2-$	Neutral
Proline	Pro	*	Neutral
Serine	Ser	CH_2OH-	Neutral
Threonine	Thr	CH_3CHOH-	Neutral
Tryptophan	Try	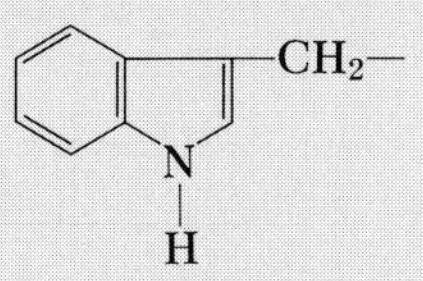 (indole ring)$-CH_2-$	Neutral
Tyrosine	Tyr	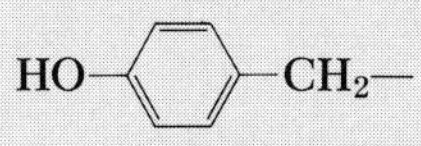 $HO-$⟨C$_6$H$_4$⟩$-CH_2-$	Acidic
Valine	Val	$\begin{array}{c}CH_3-CH-\\ \vert\\ CH_3\end{array}$	Neutral

* The structure of proline does not fit the prototype; it is

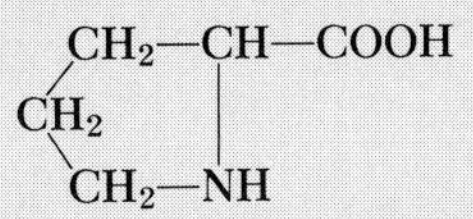

$$\begin{array}{c} CH_2-CH-COOH \\ | \qquad\quad | \\ CH_2 \qquad\quad | \\ \quad CH_2-NH \end{array}$$

cyclopentanoperhydrophenanthrene, a complex organic molecule of the following structure:

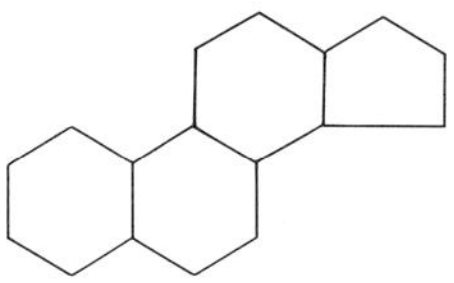

This large group includes cholesterol (see Fig. 2.6D), vitamin D, and many hormones, each of which will be discussed in later chapters.

Proteins and amino acids. Proteins are large, complex, and fragile chain molecules containing nitrogen as well as carbon, hydrogen, and oxygen. When they are broken down, it is found that their basic low–molecular weight components, the links in the chains, are simple *amino acids.*

Amino acids are organic acids with the general formula

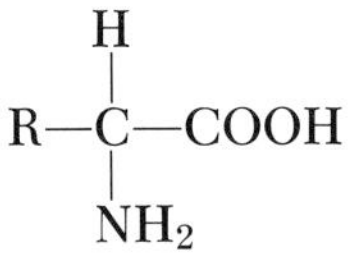

where the distinguishing cluster R may be any of the 20 different structures shown in Table 2.5. Amino acids possess an —NH_2 group on the carbon next to the carboxyl carbon; those designated *basic* possess an additional —NH_2, and those designated *acidic* an additional —COOH, elsewhere in the molecule.

When amino acids link together in a protein molecule, they join end to end, the carboxyl group of one amino acid combining with the amino group of the next to form a *peptide bond* (—CO—NH—) and a molecule of water, thus:

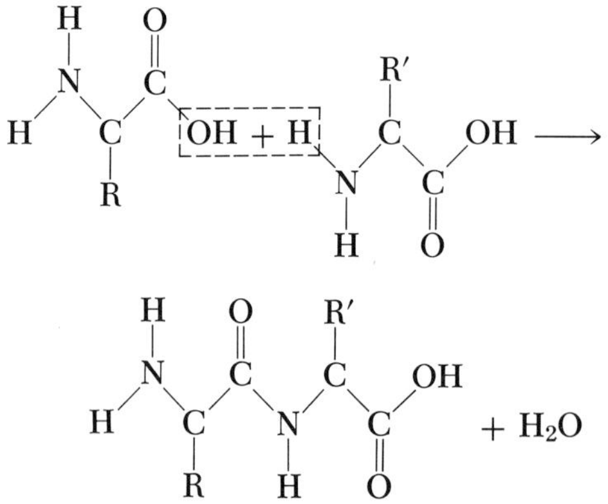

Since each component has lost part of a water molecule, it is now called an amino acid *residue.* Chains of amino acid residues are called *peptides,* two such residues linked together constituting a dipeptide, three a tripeptide, and a large number a polypeptide. Every peptide has a free —NH_2 group in one terminal amino acid residue, the *N-terminal* amino acid, and a free —COOH group in the other, the *C-terminal* amino acid. Most proteins contain several polypeptides; hence their molecular weights vary from 12,000 to 100,000. The long polypeptide chains of a protein are folded into complex and unique configurations, and, as we shall see, strategically located additional bonds or cross-linkages act to stabilize the folded structure. Strong acids, alkalies, and "proteolytic" enzymes can split (hydrolyze) a polypeptide into its component amino acids.

It would be difficult indeed to overstate the importance of the insights gained in recent years into protein structure. A protein molecule, with its long chains of simple amino acids arranged in diverse patterns, has a structure suited to extensive and subtle variation. It is now known that specific arrangements of amino acids are solely responsible for the enormous variety of living organisms, their uniqueness, and their biological specificity. Proteins are the structural elements of the body—skin, connective tissue, cell membranes, hair, and so forth—but they are very much more. Certain protein molecules are *enzymes,* the catalysts of the organism's chemical reactions. We shall speak of them later. The point to be emphasized here is the great specificity of enzyme action. In general, every one of the body's chemical reactions is controlled by a specific enzyme. This specificity is a consequence of precise amino acid sequences in the polypeptide chains and their three-dimensional configurations.

A most significant development of recent years was the discovery of methods for determining the entire amino acid sequence in a protein molecule. Because of the enormous amount of work involved, relatively few proteins have yet been examined completely. Insulin was the first to be fully analyzed (Fig. 2.8A), in a labor of 10 years for which Frederick Sanger received a Nobel Prize. It is seen that the insulin molecule consists of two unequal-sized polypeptide chains

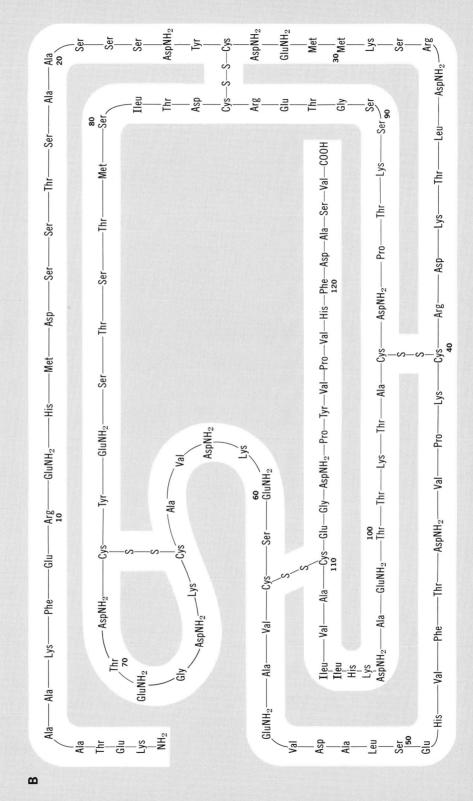

2.8 Amino acid sequences of two proteins. A. Insulin. B. Ribonuclease.

held together by —S—S— (disulfide) bonds formed between the —SH groups of opposed cysteine residues. Mild chemical treatment readily breaks —S—S— bonds into two —SH groups and, with insulin, releases two separate, intact chains. Obviously, the projecting —SH groups play a key role in establishing the higher structure of the protein, the disulfide linkages between —SH groups in different chains or in different regions of the same chain stiffening the higher folded structure.*

Since Sanger's study of the relatively small molecule insulin, the structures of a number of proteins, among them the enzyme ribonuclease (Fig. 2.8B), have been resolved. The sequence of the 124 covalently bonded amino acid residues in ribonuclease produces a more complex folding pattern than the sequence of amino acid residues in insulin.

The late Karl Linderstrøm-Lang introduced the concept of *primary, secondary,* and *tertiary* protein structure. The primary structure of a protein is its amino acid sequence, including the covalent bonds that form cross-linkages. Thus the primary structure is the backbone of the molecule. Physical measurements have shown that some portions of a protein are coiled, in helices, and that others, although folded, are uncoiled. The helices and nonhelical folds constitute the secondary structure. Stabilization of the helical coils is provided by hydrogen bonds between juxtaposed —CO and NH— groups in successive turns of the helix (Fig. 2.9). The tertiary structure comprises the foldings and turnings of the helical coil itself. This too depends upon various stabilizing bonds (Fig. 2.10), including covalent disulfide bonds, hydrogen bonds between basic —NH$_2$ and acidic —COOH groups, bonds between nonpolar side chains, resulting from mutual repulsion of a polar solvent, and van der Waals' forces. Protein molecules in which such stabilizing bonds have been irreversibly ruptured are said to be *denatured*.†

---

*Disulfide linkages may connect two or more polypeptides. When biochemists attempt to isolate and purify specific protein molecules from cells, large aggregates or complexes of proteins may form through such cross-linkages. In this event, the protein molecules appear larger than they actually are.

† The action of heat upon egg white is a classic example of denaturation.

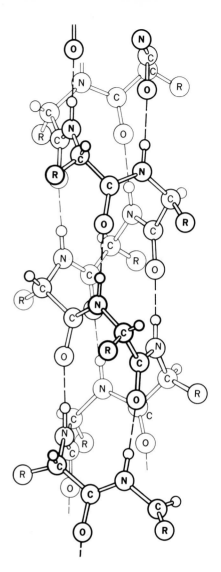

2.9 Alpha helix configuration of the polypeptide chain present in many proteins. R's represent side chains of the different amino acids; dashed lines represent hydrogen bonds. The polypeptide chain is coiled as a left-handed screw, with about 3.6 amino acid residues per turn of the helix. (Adapted from D. Green, *Currents of Biochemical Research,* Interscience, New York, 1956.)

One of the great achievements in the study of tertiary structure was the elucidation of the structures of myoglobin and hemoglobin. Using x-ray crystallographic techniques, J. C. Kendrew

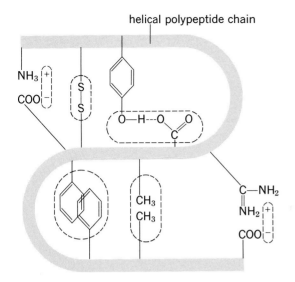

helical polypeptide chain

2.10 Bonds stabilizing the tertiary structure of a protein. Top, left to right: ionic bond between —NH_3^+ of an N-terminal amino acid and projecting free —COO^- of Glu; disulfide bond formed between —SH's of two Cys's by dehydrogenation; hydrogen bond between —OH of Tyr and —COO^- of Glu. Bottom, left to right: interaction of nonpolar side chains (of Phe) caused by mutual repulsion of a polar solvent; van der Waals' forces; ionic bond between —COO^- of a C-terminal amino acid and projecting free guanidyl group of Arg. Only the disulfide bond is covalent. (After Anfinsen.)

discovered that myoglobin has the three-dimensional structure shown in Fig. 2.11A, and M. F. Perutz found that hemoglobin has the structure shown in Fig. 2.11B,C.

The most recent addition to knowledge of protein structure is the concept of *quaternary structure*. As noted, a protein may contain two or more interacting polypeptide chains that differ in their primary, secondary, and tertiary structure. For example, insulin contains two, and hemoglobin contains four. This association of subunits is the quaternary structure. We shall later see that many enzymes owe certain of their physiological properties to their quaternary structure.

In addition to the *simple proteins*, which consist solely of amino acid chains in various structural configurations, there are many *conjugated proteins*. A conjugated protein contains protein united with a nonprotein substance by other than a salt linkage. The attached substance is termed a *prosthetic group*. Among the important conjugated proteins are (1) *nucleoproteins*, in which one or several molecules of protein are combined with nucleic acid; (2) *glycoproteins*, which have polysaccharide prosthetic groups; (3) *lipoproteins*, proteins conjugated to triglycerides or other lipids by poorly understood attachments; and (4) *chromoproteins*, such as hemoglobin and myoglobin, which have various chromophoric (i.e., colored) prosthetic groups.

Nucleotides, nucleic acids, and nucleoproteins. In addition to the inorganic-organic classification of body molecules, there is another basic dichotomy of great importance: small molecules and large molecules. Small molecules are the ones containing 10 to perhaps 100 atoms. Examples are the monosaccharides and amino acids. Large molecules, or *macromolecules*, contain hundreds, thousands, or tens of thousands of atoms. The leading members of this group are the polysaccharides, polypeptides, and polynucleotides. In discussion of the proteins, we have seen that only large molecules carry the rich configurational detail necessary for biological specificity. We shall learn in Chapter 3 that such specificity is achieved in protein synthesis under the guiding instructions of another group of large molecules, the nucleic acids. Here let us consider the molecular structure of these molecules.

The nucleic acids are the body's largest molecular species. Like proteins, they are *polymers* made up of repeating small-molecule *monomer* units. Unlike proteins, which may be viewed as long "words" composed from a 20-"letter" alphabet, nucleic acids are long words written from a 4-letter alphabet.* The four repeating monomer units are called *nucleotides*. Thus nucleic acids may alternatively be termed *polynucleotides*. In the organism, a nucleic acid, or

_____
* The terms "word" and "letter" are not idle metaphors, as we shall presently see. The information content of a word is determined by its letters and their sequence; the same is true of a protein and a nucleic acid. This explains why many polysaccharides containing only one repeating unit (e.g., glucose) carry no more information than does a word written with a one-letter alphabet. It is noteworthy that polylipids are not known.

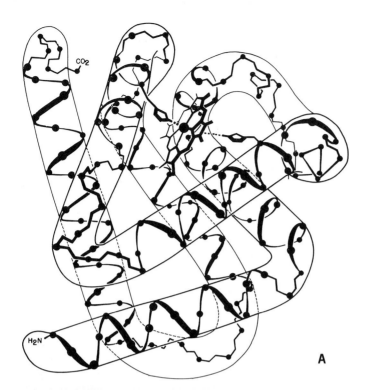

A

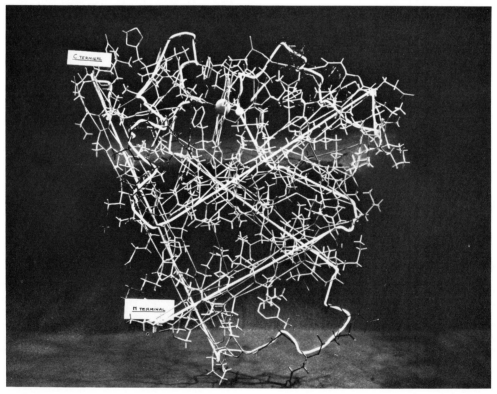

2.11 Myoglobin and hemoglobin molecules. A. Three-dimensional structure of myoglobin from X-ray crystallograph at a 2Å resolution. B. Model of a myoglobin molecule. (A from H. Neurath, *The Proteins*, 2nd ed., Vol. II. Academic Press, 1964. B, Edward Leigh from model by John Kendrew.)

B

2.11C Structures of a single molecule of human reduced hemoglobin and horse oxyhemoglobin at a 5.5 Å resolution showing β chains in black and α chains in white. (Muirhead, Cox, Mazzarella, and Perutz, *J. Mol. Biol.*, 1967, **28**, 117–156.)

polynucleotide, is attached to a protein, the resulting *nucleoprotein* comprising a conjugated protein whose prosthetic group is the nucleic acid polymer. It is important to understand the nature of nucleotides, for they have at least one major physiological function other than that of links in the nucleic acid chain: many *coenzymes*, essential in enzyme reactions, are nucleotides.

A nucleotide is more complex than most of the other small molecules, for it consists of three simpler molecules in direct linkage: a nitrogen-containing organic base, a sugar residue, and phosphoric acid (Fig. 2.12).* Since the identities of the base and sugar may vary, a variety of nucleotides exists. As shown in Table 2.6, the bases found in nucleic acid nucleotides are the *purines* adenine and guanine and the *pyrimidines* cytosine, uracil, and thymine. The two main nucleic acid sugars are ribose and deoxyribose, both five-carbon monosaccharides or pentoses, the latter differing from the former only in lacking one oxygen atom on carbon 2 (see Fig.

2.12). The deletion of an oxygen atom is implied by the *deoxy-* prefix. A nucleotide that has lost its phosphate group is called a *nucleoside*.

Nucleic acids fall into two major classes depending on whether their nucleotides contain ribose or deoxyribose; they are known respectively as *ribonucleotides*, or *ribotides*, and *deoxyribonucleotides*, or *deoxyribotides*. Thus there are ribonucleic acid, or RNA, and deoxyribonucleic acid, or DNA. Interestingly, RNA and DNA have one other fundamental chemical difference (in addition to the physiological differences to be described later). Both contain only

---

* Originally the term "nucleotide" was applied only to a nucleic acid constituent; through common usage it was extended to base-sugar-phosphate compounds whose nitrogenous bases do not occur in nucleic acids (e.g., the nucleotide coenzymes whose bases may be compounds such as nicotinamide).

TABLE 2.6   BASE AND SUGAR COMPOSITION OF RIBONUCLEIC ACID (RNA) AND DEOXYRIBONUCLEIC ACID (DNA)

	DNA	*RNA*
Purines	Adenine (A)	Adenine (A)
	Guanine (G)	Guanine (G)
Pyrimidines	Thymine (T)	Uracil (U)
	Cytosine (C)	Cytosine (C)
Sugar	Deoxyribose (dR)	Ribose (R)

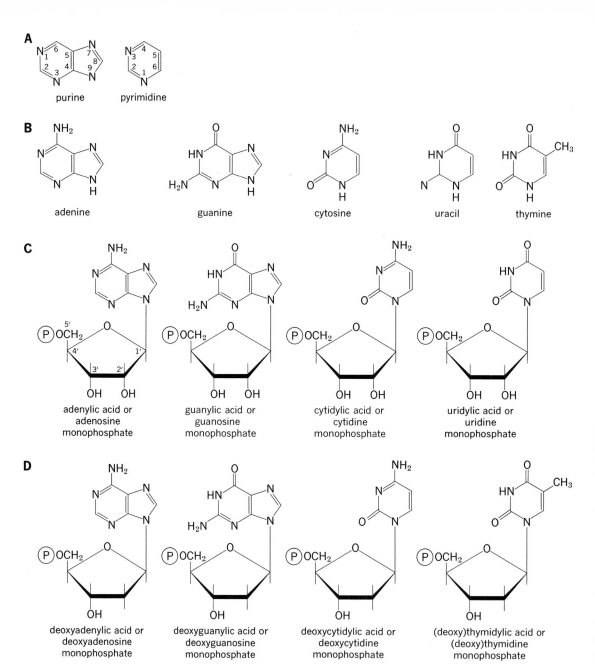

2.12 Chemistry of nucleic acids. A. Components and numbering systems (primes distinguish positions in sugars from those in bases). B. Bases. C. Ribonucleic acids (the sugar is ribose, which has an OH attached to carbon 2′). D. Deoxyribonucleic acids (the sugar is deoxyribose, which has an H attached to carbon 2′).

four nitrogenous bases, and although three are the same—adenine, guanine, and cytosine—the fourth is uracil in RNA and thymine (a methyl derivative of uracil) in DNA. RNA may be represented as a chain of the following general structure, the base sequence being variable:

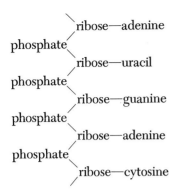

phosphate — ribose—adenine
phosphate — ribose—uracil
phosphate — ribose—guanine
phosphate — ribose—adenine
phosphate — ribose—cytosine

And DNA may be represented thus:

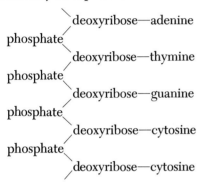

phosphate — deoxyribose—adenine
phosphate — deoxyribose—thymine
phosphate — deoxyribose—guanine
phosphate — deoxyribose—cytosine
phosphate — deoxyribose—cytosine

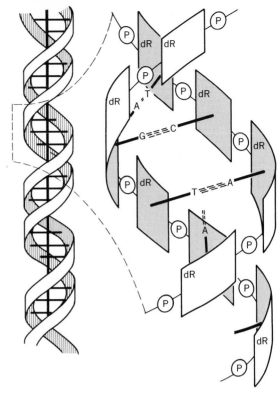

2.13 DNA molecule, a long double-stranded helical chain of nucleotides held together by hydrogen bonds between complementary purines and pyrimidines.

Before examining the remarkable features of these chains, one should carefully note the structural formulas of the purine and pyrimidine bases in Fig. 2.12. Purines have a double ring structure, which makes them slightly more complicated (and space-consuming) than pyrimidines. In addition to other differences in the attached functional groups, one purine and one pyrimidine have an amino group at the top of the ring (on carbon 6), whereas the others have a keto group in this position.

In 1953 J. D. Watson and F. H. C. Crick concluded from x-ray diffraction studies of pure DNA preparations that the DNA molecule is a *double-stranded* helix, resembling a long twisted ladder (Fig. 2.13), with two parallel nucleotide chains winding around an empty cylindrical space. The sides of the ladder consist of alternating deoxyribose and phosphate groups. The rungs between the sugars are *paired* nitrogenous bases. Base pairing is the key to the importance of the Watson-Crick discovery, for it was found that *bases can be paired only in a certain way:* adenine pairs only with thymine, and cytosine only with guanine (Fig. 2.14). Thus the structure of DNA must be represented as follows:

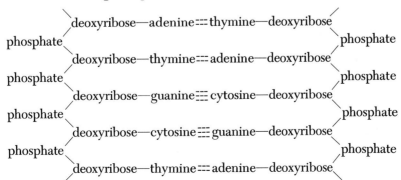

phosphate — deoxyribose—adenine ≡ thymine—deoxyribose — phosphate
phosphate — deoxyribose—thymine ≡ adenine—deoxyribose — phosphate
phosphate — deoxyribose—guanine ≡ cytosine—deoxyribose — phosphate
phosphate — deoxyribose—cytosine ≡ guanine—deoxyribose — phosphate
deoxyribose—thymine ≡ adenine—deoxyribose

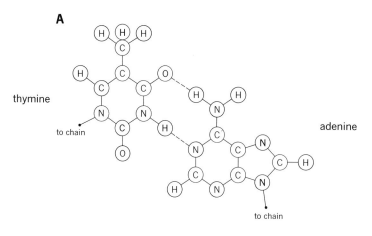

A

thymine

to chain

adenine

to chain

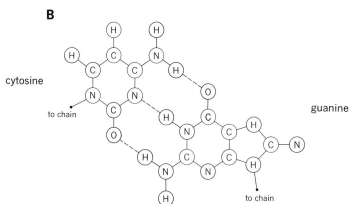

B

cytosine

to chain

guanine

to chain

2.14 Pairing of adenine and thymine (A) and cytosine and guanine (B) by means of hydrogen bonding.

The positions of the two sugar-phosphate chains are so fixed that only pairings consisting of a "large" purine and a "small" pyrimidine will fit in the confined space. Moreover, a base with a keto group on carbon 6 must stand opposite a base with an amino group on carbon 6 since only such a pair will form the hydrogen bonds required to hold the two strands together.* It is important to recognize that the two DNA strands are *complementary*, not identical. If one strand contains genetic information in its base sequence, then the other strand must contain the same information in a complementary sequence. In Chapter 3 we shall learn how

---

* These rules apply to pairing between a base in a DNA strand and a base in an RNA strand as well as to pairing between two bases in two strands of DNA. Thus cytosine of DNA pairs with guanine of RNA, and adenine of DNA pairs with uracil of RNA (the equivalent of DNA thymine).

"complementary double-strandedness" enables the DNA molecule to perform its most notable functions: (1) transmission of the genetic information handed down in the long processes of evolution and (2) accurate self-replication.

The facts just outlined imply that in any DNA molecule the ratio of 6-amino bases to 6-keto bases is always 1:1, as are the ratios of A to T and G to C. This has been verified by direct measurement. Unlike DNA, RNA may occur as a single-stranded nonhelix that lacks complementary base pairs. Hence the A:U and G:C ratios vary widely from one RNA to another, with an important exception to be noted in Chapter 3. The exception brings elegant order to the isolated facts.

The unusual properties of nucleic acids and proteins derive from their structures. They are long polymers constructed according to a basic plan from components that, with minor exceptions, are identical in all living species. Perhaps no other single fact offers better evidence of the unity of life and of the singularity of the evolutionary process.

Physicochemical Basis of Physiology

The study of molecular structure may seem a static discipline, but the study of the functional behavior of molecules is a dynamic one that deals with movement, process, and transformation, with the production, delivery, and utilization of energy, with the rates, mechanisms, and specificities that together constitute metabolism. If the fine structure of cells and tissues ultimately derives from the structure and arrangement of complex "biomolecules," so too the coordinated ordering of these molecules in time and space—the dynamic aspect of physical and biological chemistry—ultimately accounts for all physiological function. So that we may think in such terms in considering body function as it unfolds in later chapters, we shall now briefly review certain of the underlying principles and processes.

ACIDS AND BASES

Terminology. The degree of acidity in living tissues is extremely low and precisely controlled, small deviations from the norm being incompatible with life. According to the theory of J. N. Brønsted, an *acid* is any compound that can form a hydrogen ion, H^+ (also known as a *proton*). Thus HCl is an acid because it dissociates in water to form H^+ and Cl^-. Brønsted achieved a real simplification in emphasizing the central significance of the proton in acid-base reactions. An acid is characterized by its ability to furnish a proton in a chemical reaction, and a *base* is characterized by its ability to accept and combine with the proton furnished by an acid. In the following examples of acids and bases, H_2O can serve as an acid *or* a base. Any substance performing in such a dual capacity is said to be *amphoteric*. Acids and bases can be neutral molecules (e.g., HCl and NaOH) or charged ions

(e.g., NH_4^+ and HPO_4^{--}), since both can liberate or accept protons.

Acids

$$HCl \rightleftharpoons H^+ + Cl^- \qquad (2.4a)$$
$$H_2CO_3 \rightleftharpoons H^+ + HCO_3^- \qquad (2.4b)$$
$$NH_4^+ \rightleftharpoons H^+ + NH_3^- \qquad (2.4c)$$
$$H_2O \rightleftharpoons H^+ + OH^- \qquad (2.4d)$$

Bases

$$NH_3 + H^+ \rightleftharpoons NH_4^+ \qquad (2.5a)$$
$$NaOH + H^+ \rightleftharpoons Na^+ + H_2O \qquad (2.5b)$$
$$HPO_4^{--} + H^+ \rightleftharpoons H_2PO_4^- \qquad (2.5c)$$
$$H_2O + H^+ \rightleftharpoons H_3O^+ \qquad (2.5d)$$

Free H^+ is a highly reactive substance that combines readily with any available acceptor. Since H_2O can serve as a proton acceptor (Eq. 2.5d), it reacts immediately with free H^+ to form a *hydronium* ion, H_3O^+. If we wished to be rigorously correct, we would say that H^+ is almost completely converted by water to H_3O^+. If we did this, we would give up the conventional equation describing the dissociation of HCl to H^+ and Cl^- (Eq. 2.4a) in favor of

$$HCl + H_2O \rightleftharpoons H_3O^+ + Cl^- \qquad (2.6)$$

To avoid this needless complication, we shall continue to speak of hydrogen ions even though the ions exist in hydrated form.

The hydrogen ion concentration (abbreviated $[H^+]$) of pure water is 0.0000001 mole per liter.* This means that water dissociates to H^+ and OH^- (Eq. 2.4d) to only a small extent. For simplicity we represent H^+ concentration by the symbol pH, which, by definition, is the negative logarithm of H^+ concentration. In this way, we can cover the enormous range from $1N$ H^+ to $0.00000000000001N$ H^+ with a scale extending from 0 (the $-$ log of 1) to 14 (the $-$ log of

---

* A *mole*, or gram-molecular weight, is the amount of a substance equal to its molecular weight in grams. A *molar* (M) solution is one containing 1 mole of solute per liter of solution. An *equivalent*, or gram-equivalent weight, is the amount of a substance equal to its molecular weight in grams divided by its ionic charge. For example, an equivalent of Na^+ is 23 g, since the atomic weight of sodium is 23 and its ionic charge is $+1$; and an equivalent of Ca^{++} is 20 g, since the atomic weight of calcium is 40 and its ionic charge is $+2$. A *normal* (N) solution is one containing 1 equivalent (eq) of solute per liter of solution. Therefore, a solution containing 1 meq of a substance per liter of solution is $0.001N$, etc.

0.00000000000001). A change of 1 in pH therefore denotes a 10-fold change in $[H^+]$ and a solution of pH 4.7 has approximately twice the $[H^+]$ of a solution of pH 5.0.

Henderson-Hasselbach equation. An acid, symbolized as HA, dissociates reversibly in water to form H^+ and A^- (a hydrogen ion and an anion), but the degree of dissociation varies from acid to acid. *Strong* acids such as HCl dissociate completely. *Weak* acids dissociate to a small and varying degree. Thus when acetic acid (abbreviated HAc for convenience) is dissolved in water, only a small fraction dissociates to H^+ and Ac^-. A salt is symbolized as BA, B representing a cation and A an anion. All salts, both those of strong acids (e.g., NaCl) and those of weak acids (e.g., NaAc), dissociate completely in water to free B^+ and A^-.

The dissociation of acetic acid is a prototype for the ionization of any weak acid.

$$HAc \rightleftharpoons H^+ + Ac^- \qquad (2.7)$$

From the *law of mass action,* which states that the rate of a chemical reaction is proportional to the active masses of the reacting substances, we know that

$$K = \frac{[H^+][Ac^-]}{[HAc]} \qquad (2.8)$$

The term K is called the *ionization constant.* For purposes of calculation and instruction, it is useful to rearrange this equation. If we take the logarithms of both sides, we obtain

$$\log K = \log [H^+] + \log \frac{[Ac^-]}{[HAc]}$$

Rearranging gives us

$$- \log [H^+] = - \log K + \log \frac{[Ac^-]}{[HAc]}$$

If pH is substituted for $- \log [H^+]$ and pK for $- \log K$, this becomes

$$pH = pK + \log \frac{[Ac^-]}{[HAc]} \qquad (2.9)$$

which may be applied to any ionizing system, thus:

$$pH = pK + \log \frac{[\text{proton acceptor}]}{[\text{proton donor}]} \qquad (2.10)$$

This is the *Henderson-Hasselbach equation,** the fundamental relationship for dealing with the acid-base equilibria of weak acids and bases. Although it is merely a logarithmic restatement of the familiar law of mass action, it is a far more convenient form.

For example, we find that K for HAc is 1.86×10^{-5}. This means that pK for HAc is 4.73. Each weak acid or base has an unvarying pK value, which we can use in calculating the pH of a solution. What is the pH of a solution $1M$ in HAc and $1M$ in NaAc? The answer is 4.73. Since the bulk of the HA is undissociated (it is a weak acid) and *all* the BA is dissociated, both [HAc] and $[Ac^-]$ in Eq. 2.9 equal 1. The final term of the Henderson-Hasselbach equation thus equals 0, and the pH equals the pK.

Only when the final term of the Henderson-Hasselbach equation has a positive or negative value does the pH differ from the pK. What, for example, is the pH of a solution $2M$ in KAc (potassium acetate) and $3M$ in HAc?

$$pH = 4.73 + \log \tfrac{2}{3} = 4.73 + \log 0.66$$
$$= 4.73 + 0.82 = 5.55 \qquad (2.11)$$

What is the pH of a weak acid by itself—e.g., a $0.1M$ solution of HAc? In dealing with a solution of a weak acid *only,* we may employ a simplifying maneuver based on the fact that the only source of A^- is the poorly ionizing HA, there being no freely ionizing BA. Since HAc dissociates into equal portions of H^+ and Ac^- (Eq. 2.7), the ratio $[Ac^-]/[HAc]$ in Eq. 2.9 would be numerically equal to $[H^+]/[HAc]$. Eq. 2.9 can therefore be rewritten

$$pH = pK + \log \frac{[H^+]}{[HAc]}$$
$$= pK + \log [H^+] - \log [HAc] \qquad (2.12)$$

Since $pH = - \log [H^+]$, we can rearrange this to

$$2pH = pK - \log [HAc] \qquad (2.13)$$

For [HAc] we can substitute 0.1, the given concentration, ignoring the slight error introduced because some of the HAc is converted to H^+ and Ac^-. Thus the equation becomes

$$2pH = 4.73 - \log 0.1 = 4.73 + 1 = 5.73$$
$$pH = 2.87 \qquad (2.14)$$

---

* This formulation of the Henderson-Hasselbach equation demonstrates the value of the Brønsted theory of acids and bases. As we shall see, the critical term of the equation is the final one, the ratio between the concentrations of proton acceptor and proton donor. A base is always the numerator, and an acid is always the denominator.

Examining the Henderson-Hasselbach equation (Eq. 2.10), we see that, even when the pK of an acid is known, three unknown quantities remain. Hence the equation cannot be solved without precise knowledge of the composition of the solution. We can calculate the pH only if we know the concentrations of base and acid or their ratio. Conversely, fixing the pH fixes the ratio of the base and acid concentrations. This point deserves emphasis, for it is of profound physiological importance. For example, how does the pH affect a mixture of salts, acids, and bases? Obviously there can be only one value for $[H^+]$ at a given moment, no matter how complex the mixture. Once it has been determined, the dissociation behavior of all acids and bases present can be predicted from their individual pK values. Thus

$$
\begin{aligned}
\text{pH} &= \text{pK}_1 + \log \frac{[A_1^-]}{[HA_1]} \\
&= \text{pK}_2 + \log \frac{[A_2^-]}{[HA_2]} \\
&= \text{pK}_3 + \log \frac{[A_3^-]}{[HA_3]} = \ldots \quad (2.15)
\end{aligned}
$$

If the pH of the solution is higher than the pK of HA_3, then HA_3 will dissociate until the values of $[A_3^-]$ and $[HA_3]$ give the last term of the Henderson-Hasselbach equation a value equal to the difference between the pH and pK_3. The ratio in the last term is similarly readjusted for each of the ingredients.

Buffers. Mixtures of slightly ionized weak acids and their completely ionized salts are called *buffers* because they prevent sizeable pH changes upon addition of acid or base.

Their action can be readily shown with examples. The pH of 100 ml of a 0.1M solution of HAc (10 mmoles) is 2.87 (see Eq. 2.14). To what extent does the pH change if 1 mmole of NaOH, a strong base, is added (in a negligibly small volume)? NaOH ionizes completely to Na^+ and OH^-. Thus 1 of the 10 mmoles of HAc is converted to NaAc, leaving a solution containing 1 mmole of NaAc and 9 mmoles of HAc. The final pH is

$$
\text{pH} = 4.73 + \log \frac{1}{9} = 4.73 - 0.95 = 3.78 \quad (2.16)
$$

The pH change due to the NaOH addition, then, is $3.78 - 2.87$ or $+0.91$. How much does the pH change if 1 mmole of NaOH is added to 100 ml of a solution 0.1M in HAc *and* 0.1M in NaAc (a buffer mixture according to the definition)? As noted, the initial pH would be 4.73, since the last term of the Henderson-Hasselbach equation equals 0. The total initial $[Ac^-]$ arises almost entirely from the NaAc. Since it ionizes

completely, it contributes 10 mmoles of Ac^- per 100 ml. Since the HAc ionizes poorly, its negligible contribution of Ac^- can be overlooked. Addition of 1 mmole of NaOH converts 1 mmole of HAc to NaAc, adding 1 mmole of Ac^- to the 10 already present. The final solution contains 11 mmoles of NaAc and 9 mmoles of HAc per 100 ml, and the final pH is

$$
\text{pH} = 4.73 + \log \frac{11}{9} = 4.73 + 0.09 = 4.82 \quad (2.17)
$$

The pH change due to NaOH addition in this system is $4.82 - 4.73$ or $+0.09$. Comparison with the increase (ten times as great) when NaAc is missing from the initial solution makes the buffering function of the mixture apparent.

The pH of body fluids is maintained precisely at 7.4. In later chapters we shall describe the buffer mixtures responsible and the physiological systems that control them.

Ionization of amino acids. The presence in amino acids of one group that is acidic, the carboxyl group, and one that is basic, the amino group, results in some interesting properties that are reflected by proteins. We noted earlier that water is such an amphoteric compound, acting as an acid (Eq. 2.4d) and a base (Eq. 2.5d). Glycine, NH_2CH_2COOH, the simplest amino acid, dissociates as follows:

$$
{}^+NH_3CH_2COOH \underset{+H^+}{\overset{-H^+}{\rightleftharpoons}} {}^+NH_3CH_2COO^- \underset{+H^+}{\overset{-H^+}{\rightleftharpoons}}
$$
$$
\text{pH 1} \qquad\qquad\qquad \text{pH 6}
$$

$$
NH_2CH_2COO^- \quad (2.18)
$$
$$
\text{pH 11}
$$

When the pH is 6, the doubly charged ion ${}^+NH_3CH_2COO^-$ predominates. This is the *zwitterion* (hybrid ion) of classic biochemistry, a dipolar ion that can function as an acid or a base. When the pH is raised, the ion formed acts solely as an acid. When the pH is lowered, the ion formed acts exclusively as a base. It is evident, therefore, that the net charge on these molecules, and thus their solubility, is critically dependent upon the environmental pH.

OXIDATIONS AND REDUCTIONS

Among the most important chemical reactions occurring in the living organism are the *oxidation-reduction* reactions. Elementary chemistry tells us that when carbon burns in air, it forms carbon monoxide and carbon dioxide.

$$2C + O_2 \longrightarrow 2CO \qquad (2.19a)$$

$$2CO + O_2 \longrightarrow 2CO_2 \qquad (2.19b)$$

When hydrogen burns in air, it forms water.

$$2H_2 + O_2 \longrightarrow 2H_2O \qquad (2.20)$$

When iron (Fe) is heated in oxygen, it forms iron oxide, Fe_2O_3—the same product that forms slowly in the rusting of iron in air.

$$4Fe + 3O_2 \longrightarrow 2Fe_2O_3 \qquad (2.21)$$

Such combinations with oxygen were long ago termed *oxidations*.

Generalization of the term "oxidation." It was soon recognized by chemists that fundamentally similar combinations may occur with elements other than oxygen. For example, carbon, hydrogen, and iron may "burn" in fluorine.

$$C + 2Fe_2 \longrightarrow CF_4 \qquad (2.22a)$$

$$H_2 + F_2 \longrightarrow 2HF \qquad (2.22b)$$

$$2Fe + 3F_2 \longrightarrow 2FeF_3 \qquad (2.22c)$$

These reactions were early described as forms of oxidation. Later the definition of an oxidation was broadened to apply to any reaction in which an atom or group of atoms loses an electron. In the oxidation of metallic Fe to Fe_2O_3, the iron of iron oxide is in the form of a positively charged ion (cation), since $2Fe_2O_3$ consists of $4Fe^{+++}$ and $6O^{--}$. Hence the transformation undergone by iron may be written

$$4Fe \longrightarrow 4Fe^{+++} + 12e^- \qquad (2.23)$$

where e^- represents an electron. A *reduction* is defined as the acquisition of an electron by an atom or group of atoms.* While the reaction of Eq. 2.23 takes place, oxygen is being reduced.

$$3O_2 + 12e^- \longrightarrow 6O^{--} \qquad (2.24)$$

Thus the production of Fe_2O_3 requires both an oxidation and a reduction, with a net transfer of 12 electrons from Fe to O. It is apparent that oxidation and reduction must occur simultaneously, one at the expense of the other.

---

* This definition also grew from an older usage. Ore, such as Fe_2O_3, was said to be "reduced" to free Fe in refining.

Electromotive force and the emf series. If a strip of zinc (Zn) is placed in a solution of a copper (Cu) salt, a layer of metallic copper is deposited on the zinc, and zinc metal goes into solution, since zinc reduces copper and copper oxidizes zinc.

$$Zn \longrightarrow Zn^{++} + 2e^- \qquad (2.25a)$$
$$Cu^{++} + 2e^- \longrightarrow Cu \qquad (2.25b)$$

Placing a strip of copper in a solution of a zinc salt causes no visible deposition of metallic zinc. Such experiments show that metals vary in their ability to reduce the ions of other metals. Metals can therefore be arranged in a series, beginning with the metal with the strongest reducing power, with any metal being able to reduce the ions of any metal below it in the series. In part, the order of reducing power is as follows: potassium $>$ calcium $>$ sodium $>$ magnesium $>$ zinc $>$ iron $>$ cobalt $>$ nickel $>$ hydrogen $>$ copper.

This is called the *electromotive force*, or *emf, series* because the capacity of any metal to reduce the ions of another can be measured as the voltage or emf that passes between the two metals in an electric cell. This is a *potential difference*, not an absolute potential. In fact, absolute potentials of substances cannot be measured. Only potential differences between electrodes are measurable. For this reason the *normal hydrogen electrode* has been adopted as a reference standard of "zero potential" (by definition) to which potentials of other electrodes can be related. The hydrogen electrode consists of hydrogen gas at a pressure of 1 atm bubbling through a $1N$ solution of acid (pH = 0) over an electrode made of the chemically inert metal platinum (Pt). By means of a potentiometer, the potential difference (in volts) between the H_2–H^+ system and other systems can be determined. If the other system is Zn–Zn^{++}, the difference is found to be $+0.76$ v. This is known as the *standard potential* (symbolized E_0) of the Zn–Zn^{++} electrode. In the same way, the standard potential of a Cu–Cu^{++} electrode is found to be -0.34 v. Such experiments show that the emf series also follows the order of decreasing E_0 values. Knowledge of these values (all relative to the "zero potential" of the hydrogen electrode) makes possible calculation of the potential difference between any two items in the emf table. Thus the difference between Zn–Zn^{++} and Cu–Cu^{++} is 1.10 v.†

---

† A hydrogen electrode has another important function. Let us consider the potential difference between a hydrogen electrode and a system in which H^+ enters directly into the oxidation-reduction reaction, for example, the oxidation of hydroquinone to quinone.

$$(2.26)$$

In practice, it is unnecessary to use a hydrogen electrode. Any other oxidation-reduction system whose potential relative to the hydrogen system is accurately known can be used instead. Since hydrogen electrodes are indeed rather difficult to prepare and maintain, *calomel electrodes* are often substituted for them. These consist of metallic mercury (Hg) in contact with calomel (Hg_2Cl_2) in a saturated KCl solution. At 25°C the potential of the calomel electrode relative to a hydrogen electrode is +0.2458 v.

Determination of standard potentials. If a piece of platinum (or other inert metal) is immersed in a solution containing a mixture of Fe^{++} and Fe^{+++} ions, the potential developed upon the platinum surface depends upon the ratio between the concentrations of Fe^{+++} and Fe^{++} in the solution (i.e., the $[Fe^{+++}]/[Fe^{++}]$ ratio). If such an electrode were connected to a normal hydrogen electrode, it could be shown* that E_h, the actual potential difference between the electrodes at 30°C, is

$$E_h = E_0 + \frac{0.06}{n} \log \frac{[Fe^{+++}]}{[Fe^{++}]} \qquad (2.27)$$

where n is the valence change. The value of E_0, like that of pK in the Henderson-Hasselbach equation (Eq. 2.10) is constant. To measure it experimentally, we must set the last term of Eq. 2.27 equal to 0 so that E_h equals E_0. This setting holds when $[Fe^{+++}]/[Fe^{++}]$ equals 1, since log 1 equals 0. Thus the potential of a Fe^{+++}–Fe^{++} electrode relative to a normal hydrogen electrode equals E_0 when the concentrations of reductant, Fe^{++}, and oxidant, Fe^{+++}, are equal. The general equation covering such a determination is

$$E_h = E_0 + \frac{0.06}{n} \log \frac{[oxidant]}{[reductant]} \qquad (2.28)$$

The resemblance of Eq. 2.28 to Eq. 2.10, the expression determining the pH of a buffer system, is obvious. The latter shows that pH varies with the log of the [proton acceptor]/[proton donor] ratio. Similarly, E_h varies with the log of the [oxidant]/[reductant] ratio. The value of E_h for any oxidation-reduction system, its *redox potential* or *oxidation intensity*, is essentially dependent upon the values of E_0 and the [oxidant]/[reductant] ratio. *Oxidation capacity* depends upon the total amounts of oxidant and reductant in the mixture.

According to the convention employed by biochemists, E_h becomes more positive as the [oxidant]/[reductant] ratio increases.† If one oxidation-reduction system has a more negative E_h value than another oxidation-reduction system, the former tends to reduce the latter. A reaction takes place upon mixing of two solutions of different potentials (provided necessary activation takes place) and proceeds until equilibrium is reached—i.e., until both systems have arrived at the same potential as a result of alteration of the [oxidant]/[reductant] ratio of each.‡

Biological oxidations. The most important biochemical oxidation-reduction reactions involve transfers of hydrogen. Since the ultimate source of oxidizing power is atmospheric O_2, it could be said that the organism's fundamental task is to reduce O_2.

$$O_2 + 4H \longrightarrow 2H_2O \qquad (2.29)$$

Many hydrogen transfers occur in metabolism before hydrogen is passed to oxygen, the final oxidant. An example is the conversion of lactic acid to pyruvic acid.§

---

† The choice of sign is, of course, arbitrary, and there is much confusion between the two conflicting conventions. Most non-biochemists follow G. N. Lewis in writing reactions as oxidations (e.g., $Zn \longrightarrow Zn^{++} + 2e^-$), giving more easily oxidized compounds more positive E_0 values. Biochemists write reactions as reductions (e.g., $Zn^{++} + 2e^- \longrightarrow Zn$), giving more easily reduced compounds more positive E_0 values. In keeping with the 1953 recommendations of the International Union of Pure and Applied Chemistry, we shall hold with the biochemists.

‡ Since most biochemical oxidants and reductants are weak electrolytes, their redox potentials are usually pH-dependent, owing to pH effects on ionic equilibria. This factor adds complexity to the equation for a redox potential. It has been customary to define E_0 as the potential of the system at pH 0 when [oxidant] equals [reductant]. As noted, this condition makes E_h equal E_0. The redox potential at a given pH (usually 7) is symbolized by E_0'. Although this value is useless for the calculation of redox potentials at other pH's unless the manner in which the system changes with pH is known, it is useful in predicting what will happen at pH 7, a pH at which most biochemical reactions occur. Most lists of the redox potentials of biochemical systems give E_0' rather than E_0.

§ NAD and NADH₂ are now the recommended abbreviations for the coenzymes previously known as DPN (diphosphopyridine nucleotide) and DPNH₂ (reduced diphosphopyridine nucleotide). These compounds will be discussed later.

With such a system the extent of the reaction is obviously related to [H⁺]; that is, the number of electrons generated by the surface reaction (and responsible for the potential difference relative to the reference electrode) is proportional to [H⁺] or pH. Thus such a cell measures pH. Electrons flow from the electrode of higher potential—in this case the hydroquinone-quinone electrode (Eq. 2.26)—to the electrode of lower potential, the hydrogen electrode. As in the measurement of E_0 values, a calomel electrode may replace the hydrogen electrode if appropriate corrections are made.

* For this derivation, consult standard textbooks of physical chemistry.

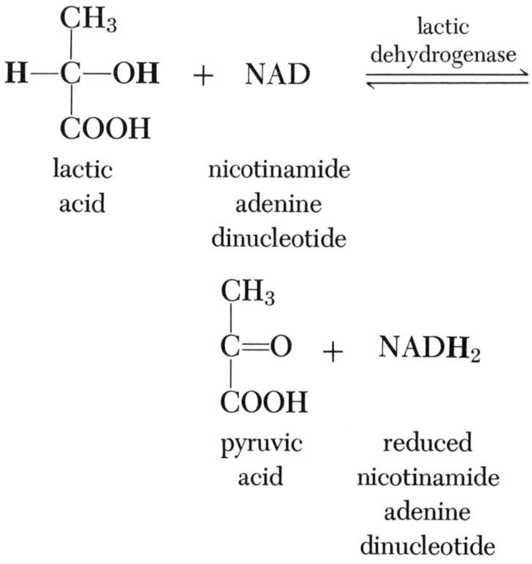

$$\begin{array}{c} CH_3 \\ | \\ H-C-OH \\ | \\ COOH \end{array} + NAD \quad \xrightleftharpoons{\text{lactic dehydrogenase}}$$

lactic   nicotinamide
acid   adenine
           dinucleotide

$$\begin{array}{c} CH_3 \\ | \\ C=O \\ | \\ COOH \end{array} + NADH_2 \qquad (2.30)$$

pyruvic   reduced
acid   nicotinamide
         adenine
         dinucleotide

Reactions of this type in no way negate the general definitions of an oxidation and a reduction as, respectively, a loss and a gain of an electron (we remember that the oxidation of any substance produces a simultaneous reduction of another substance, and vice versa). Such reactions can easily be broken into half-reactions or so-called *couples*, from which we can observe the path of the electrons as they transfer from reductant to oxidant. The two half-reactions of Eq. 2.30 are

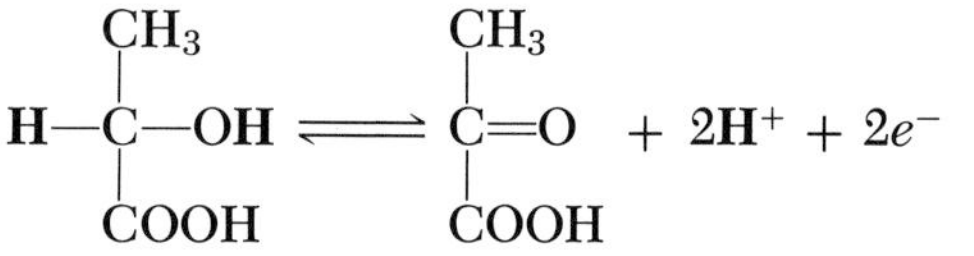

$$\begin{array}{c} CH_3 \\ | \\ H-C-OH \\ | \\ COOH \end{array} \rightleftharpoons \begin{array}{c} CH_3 \\ | \\ C=O \\ | \\ COOH \end{array} + 2H^+ + 2e^-$$
$$E_0' = -0.19 \text{ v} \qquad (2.30a)$$

$$NAD + 2H^+ + 2e^- \rightleftharpoons NADH_2$$
$$E_0' = -0.32 \text{ v} \qquad (2.30b)$$

It is clear that the oxidation-reduction is not just a dehydrogenation-hydrogenation, as suggested by Eq. 2.30, but an electron shift. The total redox potential is the sum of the two half-reaction potentials, or $(-0.19) + (-0.32) = -0.51$ v.

We shall discuss the physiological significance of biological oxidation later in more detail. We should note, however, that the body derives its energy from food by the process of oxidation. In cellular respiration, the carbon of sugar and fat is oxidized to CO_2, which is eliminated, and O_2 is reduced to H_2O.

PROPERTIES OF CHEMICAL REACTIONS

Enzymes are catalysts. Like nonbiological catalysts, (1) they accelerate chemical reactions, (2) they are not consumed in the reactions, and (3) they do not appear among the reaction products. Were it not for enzymes, many chemical reactions characteristic of living tissue would not occur. So that we may understand the nature of enzyme function, we must briefly consider certain properties of chemical reactions. What are chemical reactions? Why do they take place?

Elementary energetics. Energy and *work* are well enough defined in ordinary terms: work is something accomplished—ultimately, something moved—and energy is the capacity to accomplish work. The latter definition holds whether or not work is done. A boulder at the top of a hill has *potential energy* although it is at rest and not doing work. When it does work in rolling down the hill, part of the potential energy becomes *kinetic energy* (Fig. 2.15A). The potential energy inherent in the former position on the hilltop is therefore depleted in the performance of work.

Kinetic energy always involves matter in motion; work is movement completed. Kinetic energy may take many forms. That of the boulder is *mechanical* energy. *Light* is kinetic energy involving the movement of photons. *Electricity* is kinetic energy involving the movement of electrons from atom to atom in a conductor. *Chemical* energy involves the movement of atoms within or between molecules in chemical reactions. *Heat* involves the ceaseless random motion of atoms in all gases, liquids, and solids; as temperature rises, the rate of movement increases.

Generally speaking, all forms of kinetic energy are interconvertible. The *first law of thermodynamics* states that the total amount of energy in the universe is constant—energy cannot be created or destroyed. Hence energy is neither lost nor gained when it is converted from one form to another. The mechanical energy of a waterfall may be converted into electricity with a turbine, but never will more energy be extracted than was originally invested. Nor will any be lost, although some will be transformed

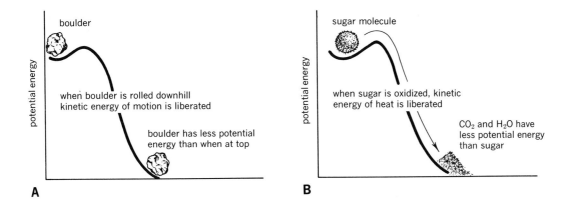

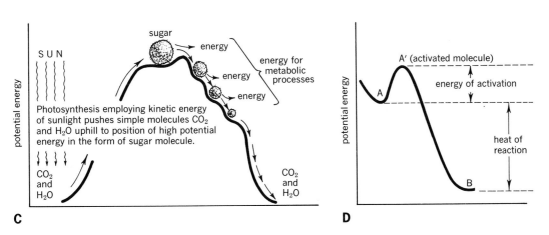

2.15 Potential energy and kinetic energy. A. Boulder on a hilltop. B, C. Sugar molecule. D. Activation energy for a chemical reaction in which a molecule A is converted to product P.

into forms that are practically useless in the performance of work (e.g., heat).

For atoms to be joined by specific bonds into a stable molecule, work must be done. This means that energy must be expended. In the chemical reaction

$$\text{A} + \text{B} \longrightarrow \text{C} + \text{D} \qquad (2.31)$$

$$\text{reactants} \qquad\qquad \text{products}$$

two molecules, A and B, interact in such a way that their atomic components recombine in new molecules C and D.* Affecting the reaction are

(1) factors relating to energy expenditures in the performance of the chemical work and (2) factors affecting the frequency of collision and contact between the molecules. We shall first consider the potential energy inherent in chemical structure.

When a boulder rolls downhill, part of its potential energy is converted to kinetic energy. When a chemical reaction changes a system from a higher to a lower energy level, the energy lost usually appears as heat.† We must conclude that before initiation of the reaction the poten-

---

* Other types of chemical reactions are possible: for example, one molecule may be broken into two (A $\longrightarrow$ B + C), two may combine into one (A + B $\longrightarrow$ C), and so forth. What is said about Eq. 2.31 is applicable to all types of reactions.

† Other forms of energy can also be liberated in chemical reactions. The chemical reactions in a storage battery generate electrical energy; the energy of exploding TNT is largely mechanical.

tial energy analogous to that of the hilltop boulder reposes in the molecular structure of compounds A and B. We are all aware that useful kinetic energy can be obtained from certain chemical compounds such as sugar, coal, gas, and wood. For example, in the combustion of sugar to CO_2 and H_2O,

$$C_6H_{12}O_6 + 6O_2 \longrightarrow$$
$$6CO_2 + 6H_2O + \text{energy} \qquad (2.32)$$

the molecular structure of sugar is destroyed, and energy is liberated in the form of heat and light. The potential energy must have been inherent in the structure of the large sugar molecule, and the smaller product molecules, CO_2 and H_2O, must represent a lower level of energy (Fig. 2.15B). Clearly the chemical reaction resulting in the formation of the simple products liberates energy held within the complex molecule. Conversely, the construction of a sugar molecule from CO_2 and H_2O by plant photosynthesis demands an investment of energy, the ultimate energy source being the sun (Fig. 2.15C).

It thus appears that there are two fundamental types of chemical reactions, one liberating energy and the other requiring a net input of energy. The former are termed *exergonic,* and the latter *endergonic.* We must now note an important feature of exergonic reactions. Once such a reaction is in progress, like a boulder rolling downhill, energy is freely produced. But a small expenditure of energy is necessary to initiate the reaction. As indicated in Fig. 2.15, some energy must be applied to push the boulder over a small hump of resistance onto its downward path. Similarly, in an exergonic reaction an investment of energy is necessary to start the process.

All molecules are in constant motion. Arrhenius was the first to point out that all the molecules in a given population do not possess the same quantity of kinetic energy. If we visualize in Fig. 2.15B a group of sugar molecules dancing about inside the "energy valley" that keeps them from uniting with O_2 and "rolling downhill" to CO_2 and H_2O, we can see that, if some, through collisions, acquire more energy than others, they may jump over the energy barrier and roll into the energy-yielding reaction. This occurs so rarely as to be negligible. But, if

the energy necessary for the molecules to overcome the energy barrier were supplied to the system, the reaction would be enormously accelerated—much as would be the combustion of dry timber exposed to the activating energy of an open flame.

Temperature effects, collision, and contact. In Eq. 2.31, molecules A and B are dispersed and in continuous random motion. Only when they collide can they react with each other. The dependence of chemical reactions on molecular collisions explains several of the basic laws governing chemical reactions. All factors that increase the frequency of collision increase the rate at which a reaction proceeds. For example, an increase in the concentrations of the reactants in solution increases the probability of collision and, consequently, the rate of reaction; similarly, an increase in temperature speeds molecular movement and raises the probability of collision and reaction.

The influence of temperature in accelerating chemical reactions is in part due to its effect in surmounting the energy barrier. The energy required to lift molecules over this barrier is called the *activation energy.* Fig. 2.15D depicts a reaction in which A is converted into B. First A must acquire the necessary energy to form the activated molecule A'. Then the rate of conversion of A' to B depends on the number of A' molecules existing at any one moment. By increasing the movement and the number of collisions of A molecules, a temperature rise increases the number of A' molecules. It is found that the rate of reaction increases logarithmically as the temperature increases. In general, for every $10°C$ rise in temperature, the rate of reaction doubles or triples. Systematic studies of this type permit experimental measurement of the activation energy.

Chemical equilibrium. It is important not to confuse the *rate* of a reaction with the *extent* of its progress toward completion. In those reactions that can proceed in both directions (i.e., *reversible* reactions), whether or not a reaction goes to completion depends upon the ratio between the rate constants of the forward and reverse reactions, k_1 and k_2. In a reversible reaction of the type

$$A + B \underset{k_2}{\overset{k_1}{\rightleftharpoons}} C + D \qquad (2.33)$$

the rate of the forward reaction is $k_1[A][B]$, and that of the backward reaction is $k_2[C][D]$. These rates obviously depend upon the concentrations of A, B, C, and D. We say that the reaction has reached equilibrium when the forward rate equals the backward rate—i.e.,

$$k_1[A][B] = k_2[C][D] \qquad (2.34)$$

At this point, [A], [B], [C], and [D] have reached fixed values, and the ratio $[C][D]/[A][B]$ is constant.

$$\frac{[C][D]}{[A][B]} = \frac{k_1}{k_2} = K \qquad (2.35)$$

The new constant, K, is the *equilibrium constant*. Clearly the final concentrations of the reactants would be the same regardless of whether we started with A and B or C and D. We now see why the reaction does not go to completion. If it did, the concentrations of A and B would equal zero, and so the value of K would be infinity. This is not the case since both the forward and backward reactions occur at finite rates. If the value of K is high, we can assume that a substantial reaction has taken place.

ENZYMES

The original names for enzymes (pepsin, trypsin, ptyalin, etc.) gave no indication of their function. A rigorous and complex systematic nomenclature was finally established, however.* Common names are still acceptable for everyday use, and these are usually derived by addition of the suffix *-ase* to the name of the *substrate* (i.e., the substance attacked by the enzyme) or to a term describing the enzyme's action. In the former category are names for the enzymes that split starch (Greek, *amylon*), the amylases; those that split fat (Greek, *lipos*), the lipases; the one that acts on uric acid, uricase; the one that acts on urea, urease; and so on. In the latter category are names such as oxidases, dehydrogenases, transferases, hydrolases, and isomerases.

The existence of so many special catalysts in living tissue accounts for the rapid execution and control of an exceedingly large number of biochemical reactions. Enzymes possess many of the same properties as inorganic catalysts, and, in addition, they have certain unique properties. Although enzymes are produced only by living organisms, they may be extracted from tissues,

purified, crystallized, and studied in isolation. Through such studies, knowledge of enzymes and their behavior has expanded dramatically in the last two decades.

Catalytic phenomenon. Catalysts, whether enzymatic or not, have three main properties: (1) they are not consumed or otherwise altered in the reactions that they promote; (2) they are potent in exceedingly small quantities; and (3) they affect only the rates of reactions. They have no effect on the equilibrium states of reactions—i.e., on the extent to which they proceed to completion. Therefore, in reversible reactions (and on theoretical grounds it is usual to assume that all reactions are reversible), they must increase the values of k_1 and k_2 in Eq. 2.35 to the same degree, so that K remains unchanged.

The basic nature of catalysis is still imperfectly understood. A catalyst appears to decrease the amount of activation energy required for a reaction. Presumably it acts by promoting instability of molecular structure in a single reactant or by bringing two or more reactants together on a surface and thereby promoting collision and contact. When substances are spread upon a catalyst's surface, the effect is similar to that of an increase in pressure or in the concentrations of reactants, although much greater. When the reaction is complete, the catalytic surface is free to act again.

Catalytic power is expressed numerically as the *turnover number*. This indicates the amount of reactant or reactants converted to product or products per unit time by a given quantity of catalyst. In the case of an enzymatic catalyst, the turnover number is specifically defined as the number of moles of substrate converted into product or products per minute by 1 mole of enzyme. Turnover numbers for known enzymes vary greatly, ranging from 100 to over 3,000,000.

Mechanisms of enzyme action. Enzymes constitute a special class of catalysts in that (1) they are all proteins; (2) they operate under mild conditions of temperature, pressure, and pH, such as those in the living body; (3) they demonstrate remarkable specificity as to the reactions they catalyze; and (4) they are outstandingly efficient

---

* See *Report of the Commission on Enzymes of the International Union of Biochemistry*, Pergamon Press, New York, 1965.

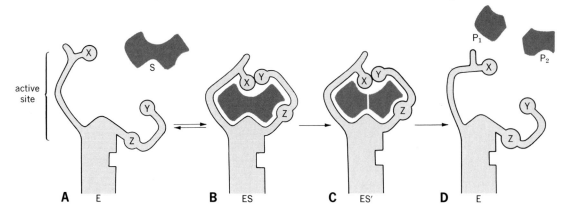

2.16 Theory of enzyme action according to which the substrate fits to the enzyme surface and thereby changes the enzyme shape or conformation. A. Enzyme (E) and substrate (S) molecules before coming together; X and Y are catalytic (and sometimes binding) groups in the active site of the enzyme that must be in proper alignment with the substrate molecule to alter it chemically, and Z is a group on the enzyme that binds the substrate in place. B. Formation of the enzyme-substrate complex (ES); the enzyme has changed its shape so that catalytic groups X and Y are in proper alignment. C. Formation of an activated enzyme-substrate complex (ES'); the catalytic groups X and Y have altered the substrate in a manner that facilitates its cleavage to final products. D. Liberation of the products (P_1 and P_2) and restoration of the enzyme to its original shape. (Adapted from D. E. Koshland, Jr., *Federation Proc.*, *23*, 719 (1964).) Compare with Fig. 2.18.

—molecule for molecule no inorganic catalysts approach them in this regard. Study of these four characteristics has provided some provocative clues to the mechanisms of enzyme action.

The ability to distinguish between similar chemical structures is one of the most noteworthy properties of biological systems. It is found not only in enzymes but in permeable membranes (Chapter 3) and antibodies (Chapter 19). The classic explanation for enzyme specificity is the famous *template* or *lock-and-key* theory of the nineteenth century pioneers Emil Fischer and Paul Ehrlich. A limited portion of the enzyme surface is known to be the *active site.* According to the template theory, this is a three-dimensional negative of at least a portion of the substrate; in other words, the substrate and the active site of the enzyme have complementary surface shapes that fit together like pieces of a jigsaw puzzle.

The main elements of this view are still accepted. We now know that an enzyme is a polypeptide and that the three-dimensional configuration of amino acid side groups at its active site is a precise and specific one capable of binding only the molecules of a particular substrate (and occasionally certain others of similar structure). The enzyme and substrate combine to form a short-lived enzyme-substrate complex.* What happens next is the subject of several current theories. The one summarized in Fig. 2.16 holds that the shape or conformation of the enzyme is changed when substrate is bound to it, so that catalytic groups in the enzyme structure are brought into the alignment essential for their action. Once in this alignment, they somehow make the substrate molecule more reactive, thus providing the necessary activation energy for its alteration. The substrate is then liberated from the complex in the form of reaction products, and the enzyme is restored to its original conformation. Here a glove and hand afford a better analogy than a lock and key because a glove is not a three-dimensional negative of a hand until the hand is introduced. An empty glove (enzyme) may have any of several shapes, and its fit is induced by the hand (substrate).

We shall presently see how the *conforma-*

---

* Although the existence of this complex was postulated long ago, final proof of it was lacking until recently, when a crystalline enzyme-substrate complex was actually isolated.

tional or *induced-fit* theory accounts for the action of some hormones, inhibitors, and drugs whose effects upon enzyme action are of great physiological importance; although it is not universally accepted, there is much evidence in its favor. Perhaps it will be found that different types of enzymes have different mechanisms of action. The fact that enzymes are proteins is undoubtedly fundamental to any interpretation of their mode of action and their specificity. For example, if the amino acid side groups of the active site were isolated as free amino acids, their catalytic power would be slight or absent. They acquire exceptional power only when they are parts of a macromolecule.*

Significance of the enzyme-substrate complex. According to Fig. 2.16, the first step in the enzyme-catalyzed conversion of a substrate to its reaction products is the combination of the enzyme, E, with the substrate, S, to form the unstable complex, ES. At this stage the substrate is activated by the enzyme to S′, and the intermediate ES′ gives rise to the final products, P, and free enzyme.

$$E + S \rightleftharpoons ES \rightleftharpoons ES' \rightleftharpoons E + P \qquad (2.36)$$

This equation has good experimental support. For example, when we observe the rates of conversion of different concentrations of substrate to products at a constant enzyme concentration, we see that reaction rates vary with substrate concentrations over a wide range of substrate concentrations (Fig. 2.17). The rate curve first rises steeply and then more gradually and finally levels off at a maximum value. When the reaction rate increases no further with an increase in substrate concentration, the enzyme surface is fully saturated with substrate. With the formation of ES′, sufficient activation energy is introduced to permit the conversion of S′ into its products. These dissociate from the enzyme sur-

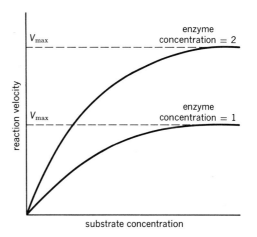

2.17 Effect of substrate concentration on the velocity of an enzyme reaction. Curves are plotted for two different enzyme concentrations (one twice the other). Note that the maximum velocity, V_{max}, at saturating substrate concentration is proportional to the enzyme concentration.

face, liberating free enzyme to combine with another substrate molecule. The conversions of ES to ES′ and then of ES′ to free enzyme and products occur rapidly and with great ease. The rate-limiting step of the reaction sequence is the initial one, the formation of ES from enzyme and substrate. At low concentrations of substrate, some enzyme remains free, and the maximum possible rate for the sequence is not achieved. But, when substrate is present in excess, all of the enzyme combines with it, and the sequence proceeds at maximum velocity (V_{max}). Under this condition V_{max} is determined by the amount of enzyme present.

Temperature affects enzyme-catalyzed reactions in the same way that it affects ordinary chemical reactions. At high temperatures, however, heat inactivates most enzymes by destroying (denaturing) their tertiary structure. Most biological systems thus have optimal temperatures above which their enzymes will not perform. For example, many cells lose the ability to carry on metabolic processes above 40°C. So that body metabolism and physiological processes may function independently of environmental temperature variations, man and other warm-blooded organisms have evolved remark-

---

* Studies of the amino acid sequence in the immediate vicinity of the active site in several types of enzymes have shown that serine, histidine, and certain other amino acids are frequently present. Moreover, they appear to participate directly in catalysis. It is also of interest that large portions (up to two-thirds) of the protein molecule can be removed from some enzymes (e.g., papain) without effect on catalytic activity. Of course, the removed portions cannot include the active site.

able systems for precise body temperature control. These will be described in later chapters.

Diversity of enzyme structure: polymers and isozymes. When biochemists attempted to obtain accurate values for the molecular weights of enzymes and other proteins, they found that many of these substances associate and dissociate reversibly, so that solutions of them in pure form are in fact mixtures of molecules in varying states of aggregation. As a result, determining the molecular weight of an individual molecule was extremely difficult. For an enzyme that behaves in this manner, the molecular weight is arbitrarily taken to represent the smallest unit characterized by catalytic activity. This may be regarded as a monomer, with higher values attributable to the dimers, trimers, and even higher aggregations formed. The aggregation into a polymer, or quaternary structure (see p. 33), is influenced by many substances, including some hormones, which may thereby affect catalytic power. We shall consider this subject in more detail later.

Another type of structural variation having implications for the control of enzyme activity within a living organism has been discovered. In addition to aggregating into polymers, some enzyme monomers dissociate into catalytically inactive subunits of lower molecular weights than the monomers. These subunits are individual polypeptides, and they can be obtained readily from proteins whose component polypeptides are linked together by noncovalent bonds rather than covalent bonds (see Fig. 2.10). Such is the case with the enzyme aldolase, which consists of three polypeptides. When they are rejoined, the original catalytic activity of aldolase is fully restored.

The enzyme lactic dehydrogenase (abbreviated LDH) provides a particularly instructive example of this phenomenon. It was observed that LDH isolated from ordinary (skeletal) muscle tissue differs strikingly from LDH isolated from heart tissue in its mobility in an electric field.* Extracts of other tissues showed LDH's of five separate mobilities, ranging from that of muscle LDH (the slowest) to that of heart LDH (the fastest). Because all five "species" are cata-

lytically similar, they were termed *isozymes.* It was then found that each LDH isozyme is made up of four subunits, which readily recombine into a whole active enzyme. Interestingly, (1) the four subunits of muscle LDH differ from those of heart LDH—hence the two LDH's can be represented as M_4 and H_4, respectively; and (2) the muscle LDH and heart LDH subunits recombine into hybrids. This accounts for the existence of five electrophoretically distinguishable LDH's: M_4, M_3H, M_2H_2, MH_3, and H_4. The discovery that LDH's consisting mainly of M subunits differ from those high in H subunits explains some of the important physiological differences between heart tissue and skeletal muscle tissue to be described in later chapters.

Substrate specificity and enzyme inhibition. Enzymes differ in the degree of their specificity. For example, lipase, which catalyzes the breakdown of a triglyceride to its constituents, glycerol and fatty acids, is specific only for the ester linkages present in the triglyceride, the identities of the fatty acids being of no moment. Other enzymes have far more specificity. Many catalyze the reactions of only a single variety of substrate molecule.

The phenomenon of *competitive inhibition* is an interesting consequence of the closely fitting complementary physical structures of enzyme and substrate. The enzyme succinic dehydrogenase catalyzes the oxidation of succinic acid as follows:

$$
\begin{array}{ccc}
\text{COOH} & & \text{HOOC} \\
| & \text{succinic} & | \\
\text{CH}_2 & \xrightarrow{\text{dehydrogenase}} & \text{CH} \\
| & \rightleftharpoons & \| \qquad\qquad + \ 2\text{H}^+ + 2e^- \\
\text{CH}_2 & & \text{HC} \\
| & & | \\
\text{COOH} & & \text{COOH} \\
\text{succinic} & & \text{fumaric} \\
\text{acid} & & \text{acid}
\end{array}
$$

(2.37)

If malonic acid,

$$
\begin{array}{c}
\text{COOH} \\
| \\
\text{CH}_2 \\
| \\
\text{COOH}
\end{array}
$$

a molecule whose structure resembles that of succinic acid, is added to the solution, the rate of the enzyme reaction is greatly diminished, al-

---

* This is determined by electrophoresis.

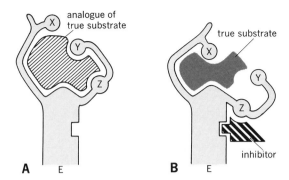

2.18 Forms of enzyme inhibition: A. Explanation of the competitive inhibition of succinic dehydrogenase by malonic acid according to the classic template theory; B. Explanation of competitive and noncompetitive inhibition according to the conformational theory. Inhibition is competitive if the Y group is involved in binding the substrate to the active site; it is noncompetitive if the Y group is exclusively catalytic and not binding. (Adapted from D. E. Koshland, Jr., *Federation Proc., 23,* 719 (1964).) Compare with Fig. 2.16.

though malonic acid itself undergoes no chemical transformation. According to the classic template theory of enzyme action, malonic acid attaches itself to the active site of succinic dehydrogenase and, by occluding the normal substrate, prohibits enzyme activity (Fig. 2.18A). It in fact *competes* with succinic acid for its position on the active site of the enzyme. Another possible mechanism of competitive inhibition, based upon the newer conformational theory of enzyme action, is shown in Fig. 2.18B. The inhibitor, I, attracts the C group and thereby prevents proper alignment of the B group. When the B group is involved in binding the substrate to the active site as well as in catalysis, inhibition is competitive (that is, the effect of the inhibitor decreases with increasing substrate concentrations). When the B group does not participate in binding, inhibition is *noncompetitive* (that is, enzyme activity is blocked irrespective of the availability of substrate).

Competitive inhibitors are known for many enzymes. Some are employed as drugs in the treatment of bacterial infection. One of these, sulfanilamide,

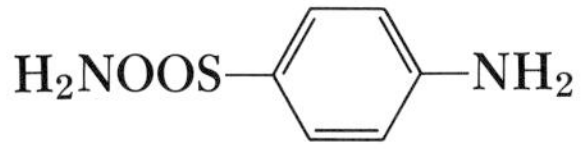

$$H_2NOOS-\langle\rangle-NH_2$$

acts by competitively preventing certain bacteria from utilizing in their nutrition the structurally similar compound *p*-aminobenzoic acid,

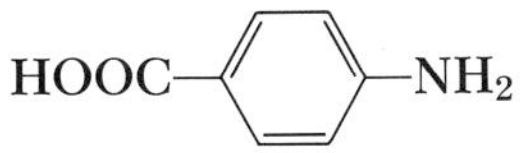

$$HOOC-\langle\rangle-NH_2$$

Noncompetitive inhibitors are also commonplace. Many are normally classed as *poisons.* For example, compounds such as cyanide and carbon monoxide, by combining with certain compounds essential in enzyme reactions, completely block electron transport in the final reduction of oxygen. This accounts for their highly toxic effect on the human body. Such agents have been extremely useful to biochemists studying the mechanisms of enzyme action.

Certain enzymes are inhibited or stimulated by low–molecular weight substances, called *effectors,* that are bound to *regulatory sites* on the enzyme molecules, remote from the active sites. The phenomenon has been termed an *allosteric effect* (from Greek roots meaning "other place") by its discoverers, Jacques Monod and François Jacob, to emphasize the fact that it involves a different locus on an enzyme from competitive inhibition by an analogue of the substrate. The interaction of effector and enzyme also causes a change in enzyme conformation—an *allosteric transition*—but not in the direct manner depicted in Fig. 2.18B because effector is bound so far from the active site. The conformational change produced by a *negative* effector decreases the affinity of enzyme for substrate; that produced by a *positive* effector increases the affinity of enzyme for substrate. Recent evidence indicates that the active site and the regulatory site may be on different subunits (i.e., polypeptides). How conformational changes are transmitted to the active site remains to be determined. Their importance in body function will be amply illustrated in later chapters.

Classification of enzymes. Enzymes are conveniently divided into five large groups* on the basis of the types of reaction catalyzed: (1) *hydrolases;* (2) *transferases;* (3) *adding enzymes;* (4) *isomerases;* and (5) *oxidoreductases.*

Hydrolases catalyze a hydrolytic splitting of their substrates.

---

* Other proposed groups will not be discussed here.

$$RX + H_2O \longrightarrow ROH + HX \qquad (2.38)$$

where X represents an anion. The bond between R and X is cleaved by water. The digestive enzymes of the alimentary tract are hydrolases, digestion consisting of an organized series of hydrolytic reactions in which the macromolecules of food, such as starches and proteins, are systematically broken down into small molecules. The hydrolases include many subgroups. For example, proteases hydrolyze proteins; peptidases, peptides; carbohydrases, sugars; lipases, lipids; nucleases, nucleic acids; nucleotidases, nucleotides; phosphatases, phosphate esters; etc.

Transferases catalyze a variation of the hydrolytic reaction in which a functional group is transferred from one molecule to another.

$$AB + C \rightleftharpoons B + AC \qquad (2.39)$$

Among the transferases are the phosphokinases, which catalyze the transfer of a phosphate group from one molecule to another.*

$$A-\text{℗} + B \rightleftharpoons A + B-\text{℗} \qquad (2.40)$$

We shall later learn that tissues contain nucleotides—e.g., the ribonucleotides of adenine, particularly the di- and triphosphates—that participate in enzyme reactions.

Structure	Name	Abbreviation
adenine—ribose—℗	adenosine monophosphate	AMP
adenine—ribose—℗—℗	adenosine diphosphate	ADP
adenine—ribose—℗—℗—℗	adenosine triphosphate	ATP

ADP frequently accepts —℗ in reactions catalyzed by phosphokinases, to form ATP. Conversely, ATP donates —℗, the result being ADP.

---

* The symbol —℗ designates the orthophosphate group, —OPO_3H_2 or

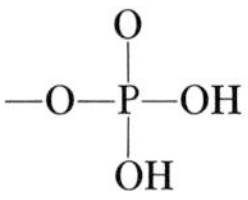

$$-O-\overset{\displaystyle O}{\underset{\displaystyle OH}{P}}-OH$$

—℗—℗ and —℗—℗—℗ denote pyrophosphate linkages,

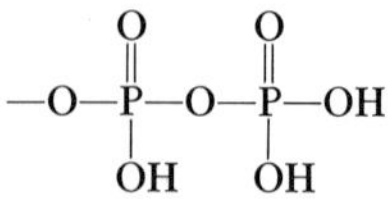

$$-O-\overset{O}{\underset{OH}{P}}-O-\overset{O}{\underset{OH}{P}}-OH \quad \text{and} \quad -O-\overset{O}{\underset{OH}{P}}-O-\overset{O}{\underset{OH}{P}}-O-\overset{O}{\underset{OH}{P}}-OH$$

$$A-\text{℗} + ADP \longrightarrow A + ATP \qquad (2.41a)$$
$$ATP + B \longrightarrow ADP + B-\text{℗} \qquad (2.41b)$$
$$\overline{A-\text{℗} + B \longrightarrow A + B-\text{℗} \qquad (2.41c)}$$

This acceptor-donor function of intermediary compounds is a common phenomenon in biochemistry. ADP and ATP, acceptor and donor of phosphate groups, are of great metabolic significance.

Adding enzymes catalyze the addition to a molecule of a group such as water or carbon dioxide. For example, an important reaction in the breakdown of a lipid involves the addition of water (catalyzed by hydrases) to an unsaturated portion of the carbon chain.

$$R-CH=CH- + H_2O \longrightarrow$$ 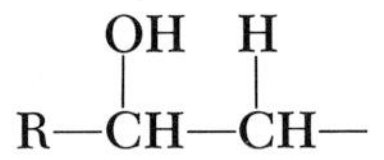
$$R-\overset{OH}{\underset{}{C}}H-\overset{H}{\underset{}{C}}H- \qquad (2.42)$$

Another important group of reactions involves the addition of carbon dioxide (catalyzed by carboxylases).

$$R-CH_2-COOH + CO_2 \longrightarrow$$ 
$$R-\overset{COOH}{\underset{}{C}}H-COOH \qquad (2.43)$$

One adding enzyme of special physiological interest catalyzes the addition of carbon dioxide to water.

$$CO_2 + H_2O \rightleftharpoons H_2CO_3 \qquad (2.44)$$

This is the ubiquitous carbonic anhydrase, about which we shall have much to say in later chapters.

Isomerases catalyze intramolecular rearrangements in which all atoms of the molecule are retained and no new ones are introduced. Such reactions are called *isomerizations,* substrate and product being isomers differing only in structural configuration (see p. 17).

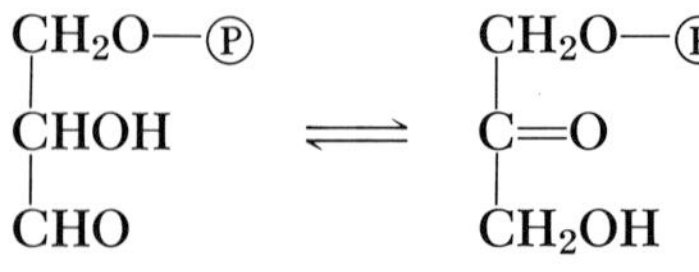

$$\begin{array}{ccc}
CH_2O-\text{℗} & & CH_2O-\text{℗} \\
| & & | \\
CHOH & \rightleftharpoons & C=O \\
| & & | \\
CHO & & CH_2OH
\end{array} \qquad (2.45)$$

glyceraldehyde   dihydroxyacetone
phosphate   phosphate

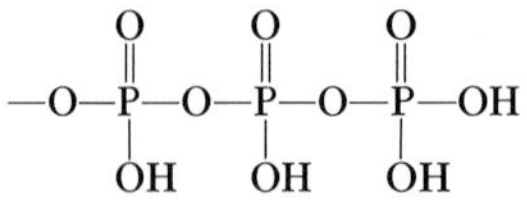

Oxidoreductases catalyze oxidative (or reductive) reactions. The two main types of oxidoreductases and the reactions in which each catalyzes the oxidation of a hypothetical substrate AH_2 are oxidases,

$$AH_2 + O \rightleftharpoons A + H_2O \quad (2.46a)$$

and dehydrogenases,

$$AH_2 + B \rightleftharpoons A + BH_2 \quad (2.46b)$$

It has long been known that each of these reactions occurs in two or more steps. Enzymes do not catalyze a direct transfer of hydrogen from donor to acceptor molecules. Rather, accessory substances called *coenzymes* intervene as intermediary acceptor-donors of hydrogen.

We noted earlier that oxidation of one substance must be attended by reduction of another. This fact points up an interesting property of oxidoreductases: they require the collaboration of coenzymes which are reduced (or oxidized) as the substrate is oxidized (or reduced). We shall now consider some of these compounds.

Coenzymes and activators. It was stated on p. 33 that body molecules can be classified as large and small. One of the oldest methods of separating the two groups is *dialysis* (Fig. 2.19). In this procedure, mixtures are placed in a bag made of cellophane, collodion, or some other porous membrane. When the bag is immersed in water, the small molecules pass out of the membrane into the surrounding water, leaving

behind the large molecules—proteins, nucleic acids, and polysaccharides.

When early biochemists exhaustively dialyzed tissue extracts containing active oxidoreductases, they observed a decrease in the catalytic power of the material in the bag. They found that activity was fully restored on the addition of a little boiled tissue juice or of the materials that had passed out of the bag—even though they had been boiled in the interim. Since boiling ordinarily denatures enzymes, they concluded that tissues contain other molecules that are (1) essential for the catalytic activity of oxidoreductases, (2) dialyzable and hence low in molecular weight, and (3) heat-stable and hence not proteins. These molecules are the coenzymes. The inactive protein moiety left in the bag is called the *apoenzyme,* and the active combination of apoenzyme and coenzyme is called the *holoenzyme.*

The coenzymes of a major group of dehydrogenases are these two dinucleotides:*

$$\text{P}—\text{ribose}—\text{nicotinamide}$$
$$|$$
$$\text{P}—\text{ribose}—\text{adenine}$$

nicotinamide adenine
dinucleotide (NAD)

and

$$\text{P}—\text{ribose}—\text{nicotinamide}$$
$$|$$
$$\text{P}—\text{ribose}—\text{adenine}$$
$$|$$
$$\text{P}$$

nicotinamide adenine
dinucleotide phosphate (NADP)

Some dehydrogenases operate with NAD, and others with NADP, but all employ one of these nucleotide coenzymes as an intermediary acceptor-donor of hydrogen. The catalytic function of the coenzyme is clearly seen in the simple notation of E. Baldwin:

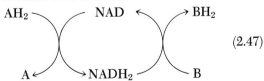

$$(2.47)$$

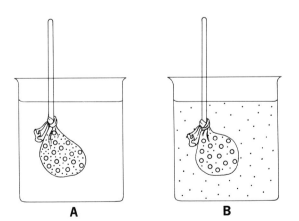

2.19 Separation of large (○) and small (·) molecules by dialysis. A. At start of dialysis. B. After equilibrium has been reached.

---

* These compounds were until recently referred to as diphosphopyridine nucleotide, or DPN, and triphosphopyridine nucleotide, or TPN, the base nicotinamide being a pyridine.

AH$_2$ transfers its hydrogen to NAD to form NADH$_2$, which regenerates NAD by passing its hydrogen to B to form BH$_2$. NAD is not consumed in the reaction but is continuously regenerated.

Other dehydrogenases utilize the flavin nucleotides and the cytochromes as hydrogen-carrying coenzymes. The flavin nucleotides include mononucleotides and dinucleotides, the latter containing riboflavin in the position analogous to that of nicotinamide in NAD (hence the abbreviation FAD). The cytochromes are conjugated proteins whose prosthetic groups are members of a large class of compounds called *porphyrins*. The cytochromes are more or less permanently attached to their apoenzymes, being extraordinarily tightly bound; they are therefore more difficult to remove by dialysis than are the loosely bound pyridine nucleotides. In our discussion of oxidative metabolism (Chapter 4), we shall see that hydrogen must be passed down a long series of acceptor-donor coenzymes—pyridine nucleotides, flavin nucleotides, cytochromes, and others in sequence—before it is transferred to the final acceptor, oxygen, to form H$_2$O.

Many enzymes other than the oxidoreductases also possess coenzymes, which accept and donate groups other than hydrogen. For example, ADP and ATP function as acceptor and donor of phosphate groups (see Eq. 2.41); coenzyme A accepts and donates acyl groups,

$$(R-\overset{\overset{\displaystyle O}{\|}}{C}-)$$

and so on.*

We have seen that enzyme reactions may involve more than the fitting of a substrate molecule to the active site of an enzyme. The participation of a positive effector or of a coenzyme may be necessary. Many enzyme reactions may also require the presence of certain inorganic substances. We call these *activators*. The best

known of them are metals. For example, copper is needed by certain oxidoreductases; magnesium is essential for the transfer of phosphate groups from ATP; and so on. The mechanism of action of metallic activators is currently under investigation in many laboratories. They too may have a role in maintaining the correct enzyme conformation during enzyme catalysis. Effectors, coenzymes, and activators are designated collectively as *cofactors*.

Methods of Biochemical and Biophysical Investigation

ANALYTICAL METHODS

Much of classic analytical chemistry was developed in industrial settings by chemists interested in such determinations as the metal contents of ores. Biochemists, however, are restricted by the delicate nature of their material. They cannot isolate proteins, for example, by boiling tissues in strong acids. Nevertheless, although their methods are limited to the mild conditions of pH and temperature found in living organisms, they have evolved techniques of unprecedented ingenuity and unparalleled precision.

Spectrophotometry. When light of different wavelengths strikes a solution, certain wavelengths are absorbed, and others pass through. Light wavelength is measured in millimicrons (mμ) or Ångstrom units (Å).† The identities of the absorbed wavelengths depend entirely upon the molecular structure of the compound in solution. In Fig. 2.20 a prism has dispersed ordinary white light into a spectrum of wavelengths, and measurements have been made of the ability of a solution to transmit each wavelength. The resulting graph is called an *absorption spectrum*.

---

* Although we shall discuss vitamins in a later chapter, it should be noted here that they are converted in the body to coenzymes. Thus NAD and NADP contain the vitamin nicotinamide; FAD contains riboflavin; coenzyme A contains pantothenic acid; etc. The symptoms of vitamin deficiencies stem from deficiencies of specific coenzymes needed by various enzymes.

† 1 mμ = 10^{-6} mm; 1 Å = 0.1 mμ. The eye can see light of wavelengths between 390 and 760 mμ (though at high light intensities the spectral range is wider). Light of longer (infrared) or shorter (ultraviolet) wavelengths is invisible. The longest visible wavelength appears as red, and the shortest visible wavelength appears as violet. Intermediate wavelengths appear as the other colors of the spectrum.

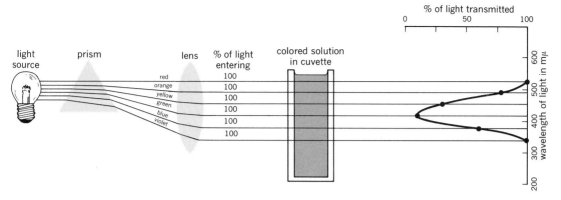

2.20 Absorption spectrum.

In the example, "peak" absorption occurs at a wavelength of 500 mμ.

Absorption spectra have two areas of usefulness. First, the shape and character of an absorption spectrum—particularly in the infrared and ultraviolet regions—are indicative of molecular structure. Second, the height of the peak is, in general, directly proportional to the amount of the substance in solution. Instruments called *spectrophotometers* permit an investigator to compare precisely the amount of light of given wavelength transmitted by a solution containing an unknown concentration of a substance with that transmitted by a "blank" containing no dissolved material and therefore assumed to transmit 100% of the entering light. He then matches the percentage of light transmitted by the sample solution to a table of data obtained with *known* concentrations of the same substance to find the concentration of the sample.

Many important biochemical substances absorb maximally in the ultraviolet region of the spectrum. Nucleic acids and nucleotides, for example, have absorption peaks near 260 mμ; proteins have peaks around 280 mμ. With these compounds ultraviolet absorption is attributable to organic ring structures: purines and pyrimidines in nucleotides and amino acids such as tyrosine, tryptophan, and phenylalanine in proteins.

Microbiological methods. Many substances are essential to bacteria as growth factors. When they are present in suboptimal concentrations, bacterial growth as measured by the turbidity of a culture is diminished. The proportionality between bacterial growth and the concentration of a limiting nutrient is an obvious basis for assay of the factor. Many quantitative procedures have been described for vitamins, amino acids, nucleic acid derivatives, and other compounds. The sensitivity and reproducibility of these methods are remarkable.

NEWER SEPARATION METHODS

Powerful new techniques of isolating substances from tissues in pure form are now available. Each takes advantage of a chemical or physical property of the material being isolated —e.g., solubility, electrical charge, volatility, molecular size, or density. Hard-won experience has shown that successful purification usually requires the sequential application of methods whose selectivities arise from widely different physical properties. For example, the isolation from a given tissue of all the molecules having a certain electrical charge is not likely to yield a homogeneous group of molecules, since several molecular species may have the same charge.

Unfortunately the isolation of natural products by selective methods is complicated by problems of stability, states of association, and conformational changes. Moreover, one can never be certain that he has achieved complete purification, since his criterion of purity remains his inability to demonstrate impurities with available techniques. The history of biochemistry is filled with instances in which new procedures have revealed "pure" substances to consist, in fact, of mixtures of substances. For obvious reasons such difficulties are more frequent in the purification of macromolecules than in that of smaller molecules.

Countercurrent distribution. In *countercurrent distribution,* similar substances are separated on the basis of their differing solubilities in two immiscible solvents. We consider it here because its simple principles are operative in some other methods to be discussed, notably chromatography.

If equal volumes of chloroform and water (two immiscible solvents) are placed in a separatory funnel, some compound A, which is fairly soluble in chloroform and only slightly soluble in water, is added, and the mixture is shaken and allowed to settle, the two final layers will contain more A and less A, respectively. The ratio of the two concentrations is the *partition coefficient* of compound A for a 1:1 water-chloroform mixture.

In this example, chloroform is the denser of the two solvents and consequently lies beneath the water. If we now place one volume of chloroform in each of a long series of consecutively numbered separatory funnels and transfer the water layer of funnel 1 (containing some compound A) to funnel 2 (containing one volume of chloroform), add one volume of fresh water to funnel 1 (which already contains one volume of chloroform and some dissolved A), and shake both funnels, the A in each distributes itself in the two solvents exactly as predicted by the partition coefficient. Subsequent transfers are diagramed in Fig. 2.21A. Each upper phase is transferred to the next funnel in the series, being added to fresh chloroform while fresh water is added to funnel 1.

Compound A at length will be distributed in a group of neighboring vessels whose numbers can be predicted with astonishing accuracy from the partition coefficient alone. If the first funnel contains a second compound, B, whose partition coefficient differs only slightly from that of A, it begins to separate from A after a number of transfers (Fig. 2.21B). The distribution curve of the component that is more soluble in the upper phase moves to the right more rapidly than that

of the other component. As the number of transfers increases, the degree of separation progressively improves.

If the partition coefficients of the substances to be separated are very close together or if the number of substances in the mixture is large, numerous transfers and equilibrations may be necessary. Accordingly, a labor-saving automatic apparatus has been designed. The 200-

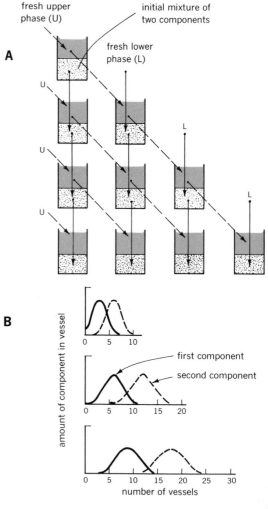

2.21 Countercurrent distribution. A. Transfer, with diagonal arrows representing fresh upper phase and vertical arrows representing fresh lower phase. B. Distribution of two components in extraction vessels after 10, 20, and 30 transfers.

2.21C A 30-tube hand-operated countercurrent distribution train. (Lyman C. Craig, Rockefeller University.)

vessel device shown in Fig. 2.21C can accomplish in a day as many transfers as would require many weeks of manual effort. Thus the purification of many hormones, nucleic acids, and other complex molecules has been greatly facilitated.

Chromatography. Among the most powerful biochemical methods, and oddly one of the simplest, is *chromatography*. In its infinite variations it permits the separation and identification of extremely small quantities of substances even in complex mixtures. Although chromatography takes many forms, it is always based on differing adsorption characteristics and partition coefficients of individual substances.

Adsorption is the adhesion of a thin layer of molecules to the surface of a solid body. Depending upon the natures of the solute, the solvent, and the solid, adsorptive forces may be weak or strong. *Partition,* as we have noted, is the distribution of a dissolved substance between two immiscible solvents. In *column chromatog-*

raphy a small amount of the mixture to be separated is placed at the top of a glass column that has been packed with an adsorbent substance such as aluminum oxide or silica gel (Fig. 2.22A). In *paper chromatography* the sample is dried in a small spot at one end of a strip of paper, which is then suspended in a sealed chamber, with the end of the paper near the "origin" dipped in a shallow trough containing solvent (Fig. 2.22B). The solvent is then allowed to percolate down through the column or travel down the paper by capillary action. The instant that it meets the adsorbed substances, the process of chromatography begins.

Adsorbed substances dissolve in the solvent to varying extents so that they migrate down the column or paper at different rates. The ratio between the rates of movement of a substance and the solvent front is designated the R_f. An R_f of 0.5 means that the substance moves half as fast as the solvent. The partition coefficient of the substance in a mixture of moving solvent and

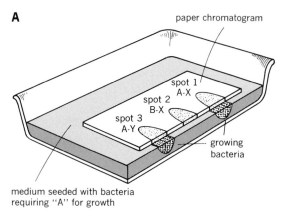

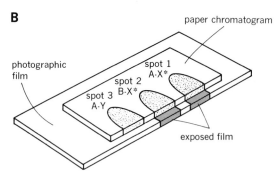

water determines the R_f value, just as it does in countercurrent distribution. This is due to the fact that water within the adsorbent molecules or the cellulose fibers of the paper acts as a stationary water phase like the lower phase in the countercurrent system. If the sample is much more soluble in water than in the moving solvent, it will tend to remain at the original location and have a low R_f. If the reverse is true, it will tend to migrate with the solvent front. With an appropriate choice of solvents, most mixtures can be separated. Fig. 2.22B shows the two-

2.23 Methods of localizing and identifying substances on a paper chromatogram. The diagrams show how a hypothetical substance A-X*, might be localized and identified (* indicates that only the X moiety is radioactive). A. Bioautography, which employs a culture of bacteria that grow only in the presence of A. Note that growth occurs under the A-Y and A-X* spots, but not under B-X. B. Radioautography, which distinguishes between radioactive and nonradioactive compounds. Note that the film is exposed by A-X* and B-X*, but not by A-Y. The combined results of the two techniques indicate that spot 1 is A-X*.

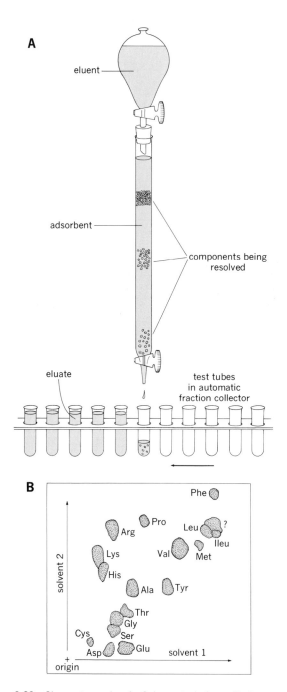

2.22 Chromatography. A. Column technique. B. Paper technique showing two-dimensional chromatogram of a mixture of amino acids.

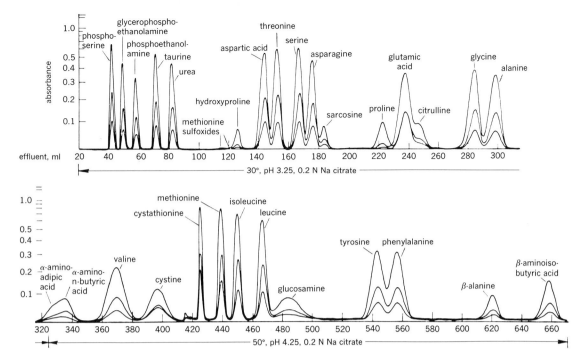

2.24 Pattern of elution of amino acids in the ion exchange chromatography of hydrolyzed ribonuclease. The name of each amino acid appears above its peak. (Spackman, Stein, and Moore, *Anal. Chem.*, 1958, *30*, 1190. © 1958 by the American Chemical Society.)

dimensional paper chromatographic separation of the major amino acids. Such strips are called *chromatograms.*

When the substances being separated are colored, they are easy to find on the chromatogram. The detection and location of colorless substances requires ingenuity. Many procedures have been used. Amino acids, for example, turn purple when sprayed with a reagent called *ninhydrin* (see Fig. 2.22C). Nucleotides absorb ultraviolet light (as in the spectrophotometer) and are thus rendered visible (see. Fig. 2.22D). If a growth factor for certain bacteria is thought to be present, the chromatogram can be placed on a culture dish containing the bacteria (Fig. 2.23A). Growth will occur only in the region of the factor. This technique is known as *bioautography.* If the material is radioactive, it shows up when the chromatogram is placed on a photographic plate and allowed to make a *radioautograph* (Fig. 2.23B).

One more important variation of the chromatographic method must be mentioned. Many ionizable compounds can be separated in columns containing charged materials that bind them, not by adsorption alone, but by ionic bonds. *Ion exchange resins,* often utilized for this purpose, consist of minute beads of polyvalent cations or anions. A column of cationic resin binds anions with varying degrees of strength. Bound anions are liberated when other anions of greater affinity for the resin are poured into the column. The reverse situation holds for cation-binding anionic resins. In ion exchange, separation depends upon the equilibration of ions between a mobile phase and a stationary phase—in analogy to other chromatographic techniques, in which separation depends upon partition or solubility behavior. The removal of bound ions is called *elution.* Many elaborate devices are available for systematically increasing the ionic strength of the eluting solvent, or *eluant*—the process being termed *gradient* elution—and for analyzing the fractions of *eluate* emerging from the column and recording their compositions. Amino acids and nucleotides are particularly susceptible to separation by ion exchange chromatography. Fig. 2.24 shows the

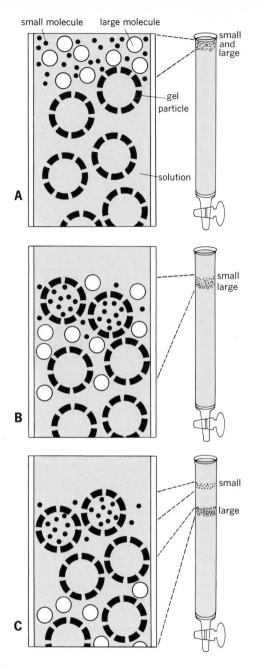

small molecule   large molecule

small
and
large

gel
particle

solution

A

small
large

B

small

large

C

2.25 Separation of large and small molecules by molecular sieving (i.e., gel filtration). A. Outside water initially contains mixture of large and small molecules at top of column. B. Most of the small molecules move into the inside of gel particles while large molecules remain outside. C. As outside water is replaced by eluent, large molecules move rapidly down column; small molecules move down column only after they have been washed out of the gel particles by eluent.

elution pattern of the amino acids obtained by hydrolysis of pure ribonuclease.

Gel filtration. Dialysis, the classic method of separating large and small molecules (see Fig. 2.19), has now been supplemented by *gel filtration.* A gel made of cross-linked strands of a high–molecular weight polysaccharide acts as a molecular sieve, admitting small molecules and excluding large ones (Fig. 2.25). Gels with accurately known pore sizes are available.

Differential sedimentation: the ultracentrifuge. Another technique that has facilitated separations of molecules on the basis of size and density is *analytical ultracentrifugation.* A centrifuge is nothing more than a mechanical means of rapidly rotating tubes so as to induce the heavier or denser particles to fall to the bottom. It is appropriate to speak of particles falling under the influence of centrifugal force, since their behavior is entirely analogous to that of particles falling to the ground under the influence of the earth's gravitational field. It is customary, in fact, to use a *gravity*, or g, as a quantitative unit of centrifugal force. If the force of the earth's gravity is 1 *g*, a centrifuge operating at 1000 *g* has a force 1000 times that of ordinary gravity. According to Stokes' law, the rate at which particles fall in such a field depends in large part upon their densities.*

In 1925 the Swedish investigator T. Svedberg developed a powerful centrifuge whose successors are capable of producing forces up to 500,000 *g*. In such an *ultracentrifuge* particles whose densities exceed that of the suspending medium, however slightly, fall or settle in a tube; those of lower densities rise or float, traveling against the field of force. The ultracentrifuge has been designed so that, as it whirls, photographs can be taken of the boundaries between the layers of particles at timed intervals. Hence sedimentation rates, or *sedimentation constants,* can be measured accurately (Fig. 2.26). Like molecular weights, sedimentation constants are peculiar to given substances or particle types under standard conditions. The subcellular particles

---

* Other determinants of the rate of fall are particle size, as we have noted, and the density and viscosity of the medium.

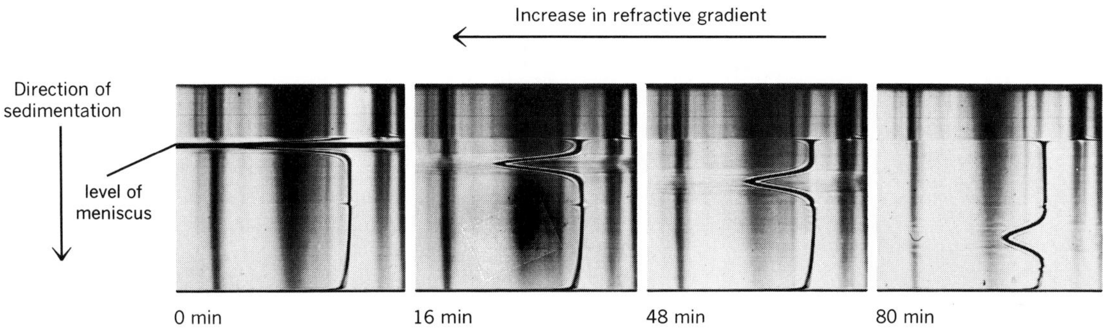

Increase in refractive gradient →

Direction of sedimentation

level of meniscus

0 min   16 min   48 min   80 min

2.26 Analytical ultracentrifugation. Appearance of protein peak in Schlieren photographs at various time intervals after centrifugation begins.

that we shall discuss in Chapter 3 have therefore been characterized extensively by ultracentrifugation.

Electrophoresis. The fact that many substances dissociate into ions provides the basis of a versatile separation procedure called *electrophoresis.* In an electric field, positively charged ions (cations) migrate toward a negatively charged electrode (cathode), and negatively charged ions (anions) migrate toward a positively charged electrode (anode). Differences in the migration rates permit separation.

These principles have been notably useful in the separation of proteins. The charge of a protein depends upon the numbers of basic and acidic amino acids in its chain (see p. 30), since these carry ionizable basic and acidic side groups. Thus proteins are amphoteric (see p. 41). The ionization of a protein is affected by pH in the same manner as the ionization of a simpler amphoteric compound such as glycine (see Eq. 2.18). The pH at which a protein's negative and positive charges neutralize one another, making the protein an effectively uncharged zwitterion, is the *isoelectric point.* A protein in an electric field moves toward a cathode or an anode at all pH's except the isoelectric point. Two proteins of different isoelectric points obviously have different mobilities. Indeed, when migration is observed at a fixed pH (usually 7.5 to 8.0, since the isoelectric points of most proteins are below 7), characteristic rates are found for each molecular species.

Fig. 2.27A shows the original apparatus used for electrophoretic separations. The mixture is placed in the U-tube, and electrodes are placed in its arm. Optical devices provide for the photographic recording of the migratory rate and the amount of each class of particles. Fig. 2.27B depicts a later modification, electrophoresis on a piece of filter paper. Fig. 2.27C illustrates a still later development, *disc* electrophoresis, a technique that can accurately separate the proteins present in only 0.003 ml of blood serum. We shall later encounter much evidence of the importance of electrophoresis.

ISOTOPIC METHODS

Every chemical element has atoms of different weights. These are the *isotopes* of the element. Though identical in chemical behavior, isotopes can be distinguished by their physical properties. They are either stable or unstable, the latter condition manifesting itself as *radioactivity.* Both types of isotopes have been used successfully in physiological investigation.

Radioisotopic techniques. A radioactive isotope, or *radioisotope,* is an unstable atomic species that, while seeking a more stable state, disintegrates and emits subatomic particles, or "radiation." These particles cause the molecules in their paths to ionize. This fact is exploited in the detection and measurement of radioisotopes. Since radioactivity is a process of decay, it is convenient to indicate the time needed for half

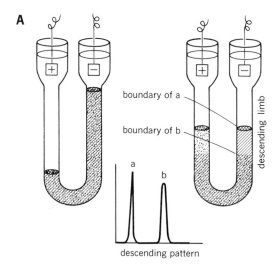

A

boundary of a

boundary of b

descending limb

a

b

descending pattern

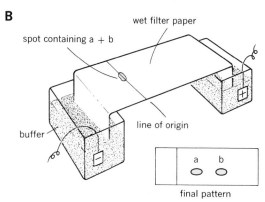

B

wet filter paper

spot containing a + b

line of origin

buffer

a   b

final pattern

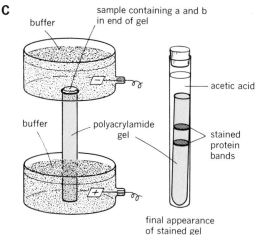

C

sample containing a and b
in end of gel

buffer

acetic acid

buffer

polyacrylamide
gel

stained
protein
bands

final appearance
of stained gel

of an isotope to decay as its *half-life.* Half-lives of isotopes vary widely. The isotopes of major biological interest and their half-lives are as follows: calcium-45, 164 days; carbon-14, 5568 years; cobalt-60, 5.3 years; hydrogen-3 (tritium), 12½ years; iodine-131, 8.1 days; iron-59, 45.1 days; phosphorus-32, 14.3 days; potassium-42, 12.5 hours; sodium-24, 15 hours; and sulfur-35, 87.1 days.*

It is possible now to incorporate radioisotopes into various molecules and follow their progress through the tissues of the body and their chemical transformations in metabolism. When a substance so "labeled" is eaten, tests show whether it is retained or excreted. When a compound containing carbon-14 is eaten and the CO_2 exhaled is found to be radioactive, it is obvious that the carbon of the compound has undergone oxidation. If tissue nucleic acid or protein is then isolated and also found to be radioactive, it must be inferred that the original compound has somehow been converted into this substance or its precursors. The rates of such transformations can be determined with precision.

Mass spectrometry. Nonradioactive isotopes do not decay. Since they are stable, they can be detected only on the basis of their differing masses. We can separate a small number of carbon-13 atoms from a large number of carbon-12 atoms, for example, with a *mass spectrometer,* which is little more than a giant magnet attached to coils that generate an electric field. The material to be treated is changed into the gaseous state, if it is not already a gas, and forced through a small opening, from which it emerges in a fine stream. An electron source converts the gaseous molecules into ions, in which form

---

* Symbols for these radioisotopes are ^{45}Ca, ^{14}C, ^{60}Co, ^{3}H (or T), etc.

2.27 Electrophoresis. A. Classic Tiselius apparatus, showing the separation of two proteins and resulting Schlieren photograph. B. Paper electrophoresis apparatus, showing the separation of two proteins on wet filter paper. C. Gel electrophoresis apparatus, showing the sharp separation of two proteins in a matrix of polyacrylamide gel.

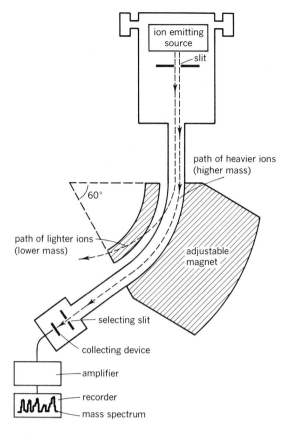

2.28 Principle of mass spectrometry.

they are susceptible to the accelerating influence of an electric field. As the stream gains velocity, it passes through the field in such a way that the paths of the heavier ions are deflected slightly more than those of the lighter ions (Fig. 2.28)—much as light is diffracted by a prism. Small slits and collecting cups at different locations along the stream sample the *mass spectrum*, and the isotopes are thus individually trapped and measured. The biologically important heavy isotopes are nitrogen-15, hydrogen-2 (deuterium), and oxygen-18. Though mass spectrometry is relatively difficult technically, it is the only method presently available for separating atoms existing only as stable isotopes.

REFERENCES AND SUGGESTIONS FOR FURTHER READING

Anfinsen, C. B., *The Molecular Basis of Evolution*, Wiley, New York, 1959.

Angyal, S. J., and D. Rutherford, "Carbohydrates—Mono- and Oligosaccharides," *Ann. Rev. Biochem.*, **34**, 77 (1965).

Atkinson, D. E., "Regulation of Enzyme Activity," *Ann. Rev. Biochem.*, **35**, 85 (1966).

Baldwin, E., *The Nature of Biochemistry*, Cambridge Univ. Press, Cambridge, 1962.

————, *Dynamic Aspects of Biochemistry*, 5th ed., Cambridge Univ. Press, Cambridge, 1966.

Benson, S. W., *The Foundations of Chemical Kinetics*, McGraw-Hill, New York, 1960.

Boyer, P. D., H. Lardy, and K. Myrbäck, eds., *The Enzymes*, 2nd ed., Vols. 1–8, Academic Press, New York, 1959–63.

Chargaff, E., and J. N. Davidson, eds., *The Nucleic Acids: Chemistry and Biology*, Vols. 1–3, Academic Press, New York, 1955–60.

Colowick, S. P., and N. O. Kaplan, eds., *Methods in Enzymology*, Vols. 1–12, Academic Press, New York, 1955–68.

Dixon, M., and E. C. Webb, *Enzymes*, 2nd ed., Academic Press, New York, 1964.

Edsall, J. T., and J. Wyman, *Biophysical Chemistry*, Academic Press, New York, 1958.

Fieser, L. F., and M. Fieser, *Organic Chemistry*, 3rd ed., Heath, Boston, 1956.

Fruton, J. S., and S. Simmonds, *General Biochemistry*, 2nd ed., Wiley, New York, 1958.

Jencks, W. P., "Mechanism of Enzyme Action," *Ann. Rev. Biochem.*, **32**, 639 (1963).

Kasha, M., and B. Pullman, eds., *Horizons in Biochemistry*, Academic Press, New York, 1962.

Kendrew, J. C., "Myoglobin and the Structure of Proteins," *Science*, **139**, 1259 (1963).

Koshland, D. E., Jr., "Conformational Changes at Active Site During Enzyme Action," *Federation Proc.*, **23**, 719 (1964).

————, and K. E. Neet, "The Catalytic and Regulatory Properties of Enzymes," *Ann. Rev. Biochem.*, **37**, 359 (1968).

Mahler, H. R., and E. H. Cordes, *Biological Chemistry*, Harper & Row, New York, 1966.

Meister, A., *Biochemistry of the Amino Acids*, 2nd ed., Vols. 1–2, Academic Press, New York, 1965.

Monod, J., J. Wyman, and J. P. Changeux, "On the Nature of Allosteric Transitions: A Plausible Model," *J. Mol. Biol.*, **12**, 88 (1965).

Neilands, J. B., and P. K. Stumpf, *Outlines of Enzyme Chemistry*, 2nd ed., Wiley, New York, 1958.

Neurath, H., and K. Bailey, eds., *The Proteins: Chemistry, Biological Activity and Methods,* 2nd ed., Academic Press, New York, 1964.

Pauling, L., *The Nature of the Chemical Bond,* 3rd ed., Cornell Univ. Press, Ithaca, N. Y., 1960.

———, *General Chemistry,* Freeman, San Francisco, 1958.

Perutz, M. F., "X-ray Analysis of Hemoglobin," *Science,* **140,** 863 (1963).

Pigman, W., ed., *The Carbohydrates: Chemistry, Biochemistry, Physiology,* Academic Press, New York, 1957.

Pimental, G. C., and A. L. McClellan, *The Hydrogen Bond,* Freeman, San Francisco, 1960.

Potter, V., *Nucleic Acid Outlines,* Vol. 1, Burgess, Minneapolis, 1960.

Rich, A., and N. Davidson, eds., *Structural Chemistry and Molecular Biology,* Freeman, San Francisco, 1968.

Rose, I. A., "Mechanisms of Enzyme Action," *Ann. Rev. Biochem.,* **35,** 23 (1966).

Sanger, F., and L. F. Smith, "The Structure of Insulin," *Endeavour,* **16,** 48 (1957).

Schachman, H. K., *Ultracentrifugation in Biochemistry,* Academic Press, New York, 1959.

Scheraga, H. A., *Protein Structure,* Academic Press, New York, 1961.

Stein, W. H., and S. Moore, "The Structure of Proteins," *Sci. Am.,* **204,** 81 (Feb., 1961).

Watson, J. D., The Double Helix, Atheneum, New York, 1968.

———, and F. H. C. Crick, "The Structure of DNA," *Cold Spring Harbor Symp. Quant. Biol.,* **18,** 123 (1953).

White, A., P. Handler, and E. L. Smith, *Principles of Biochemistry,* 3rd ed., McGraw-Hill, New York, 1964.

Widdowson, E. M., R. A. McCance, and C. M. Spray, "The Chemical Composition of the Human Body," *Clin. Sci.,* **10,** 113 (1951).

I took a good clear piece of Cork and with a Pen-knife sharpen'd as keen as a razor, I cut a piece of it off, and thereby left the surface of it exceeding smooth, then examining it very diligently with a Microscope, me thought I could perceive it to appear a little porous; but I could not so plainly distinguish them as to be sure that they were pores. . . . I with the same sharp Pen-knife cut off from the former smooth surface an exceeding thin piece of it, and placing it on a black object Plate . . . and casting the light on it with a deep planoconvex Glass, I could exceedingly plainly perceive it to be all perforated and porous, much like a Honey-comb, but that the pores of it were not regular; yet it was unlike a Honey-comb in these particulars: first, in that it had a very solid substance, in comparison of the empty cavity that was contained between; next in that these pores, or cells, were not very deep, but consisted of a great many little Boxes. . . .

Robert Hooke, in *The Micrographia* (1665)

3 *The Cell:*
Structure and Replication

Approach to the Cell

The hierarchical organization of living systems creates an interesting dilemma for the student and investigator. Somehow they must avoid narrowness of interests and remember that every aspect of the living organism is closely and critically related with phenomena occurring at both lower and higher levels of organization. Some biologists are proponents of a traditional or "organismic" viewpoint (dealing only with intact organisms and populations of organisms), and others subscribe to a "molecular" viewpoint (treating all problems in molecular or biochemical terms). It is clear that both of these approaches, if pursued without compromise, are self-defeating, for full explanation and understanding of a complex multilevel system can come only from a complex multilevel experimental attack. The power of contemporary biology derives entirely from the breadth of its interests. Only through a wide-ranging, "translevel" analysis has this power been realized.

These are useful considerations to have in mind as we approach the cell, the fundamental unit of which all organisms are made. Ever since the significance of the cell was recognized over a century ago, cell biology has traveled at least two different roads. One group of investigators equipped with increasingly powerful microscopes has developed our conceptions of cell structure. Once the cell appeared as little more than an amorphous jellylike blob of matter with a central nucleus; now we recognize it as a com-

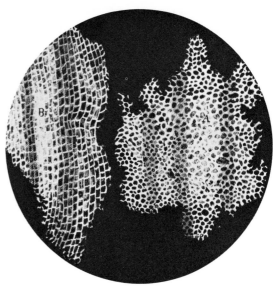

3.1 Hooke's microscope and his drawing of the microscopic structure of cork. (From Hooke, *Micrographia*, 1665, New York Public Library.)

plex and highly ordered organization of intricate substructures. A second group of cell biologists has been preoccupied with biochemical studies of the materials extracted from disrupted cells. Much of Chapter 2 dealt with the contributions of this school, which, in general, has been concerned with *function, mechanism,* and *process.*

Despite the apparent diversity of these two lines of cell study, we shall see in Chapters 3 and 4 how satisfactorily they have converged in recent years. The morphologist now seeks to explain in molecular terms what he sees in the microscope; the biochemist and biophysicist seek to explain how molecules are interlaced into functioning ultrastructures, and the physiologist profits from the insights of both.

GRADUAL EMERGENCE OF THE CELL THEORY

Hooke and Leeuwenhoek. Before the invention of the microscope, there was no reason to postulate the existence of a subworld of living forms too small to be seen by the naked eye, since no existing theory or data made it necessary. Even when the lens grinders and spectacle makers had produced reasonably powerful magnifying glasses of short focus, few, if any, set about deliberately to look for microscopic living forms. Rather, the early microscopes were more like toys for curiosity seekers.

It was the Englishman Robert Hooke (1635–1703) who first realized the importance of studying nature with these new instruments that so greatly extended the power of the human eye. In 1665 Hooke published a remarkable treatise, *The Micrographia; or Some Physiological Descriptions of Minute Bodies Made by Magnifying Glasses and Enquiries Thereupon,* which launched the study of microscopic anatomy. Using the primitive microscope shown in Fig. 3.1, he carefully examined a thin slice of cork and recorded his observations in a drawing and in the classic passage quoted at the beginning of this chapter. In fact, Hooke observed only the thickened walls of dead cells whose bodies had been lost, and, though he called them *cells*, he did not comprehend their nature. Accordingly, his use of the term "cells" has often been lamented. At that time cells were defined as "little rooms," and that is what Hooke thought the spaces were. In the light of later developments, the term "corpuscles," meaning "little bodies," would have been more appropriate.

The next major contribution came from the eccentric lens grinder of Delft, Anton van Leeuwenhoek (1632–1723). During his long life, Leeuwenhoek sent innumerable communications to the Royal Society of London describing the microscopic appearance of diverse objects. He observed many swimming creatures in rainwater and wrote the first description of bacteria. To doubters he offered affidavits from prominent citizens of Delft who had also seen the "wretched beasties." He did not, of course, relate bacteria to human diseases. He discovered red blood cells and human spermatozoa and declared that he had seen within a sperm a whole tiny man. Many of his discoveries—blood cells, bacteria, and protozoa—were single-celled living organisms, but Leeuwenhoek did not think in these terms. He described what he saw and made no attempt to fit his observations into a larger intellectual framework. The time for such a construction was not yet ripe.

Schleiden and Schwann. The death of Leeuwenhoek ended the first inquisitive period of microscopic biology, and little more progress was made until 1838, 173 years after Hooke's description of cork. In that year the German botanist Matthias Schleiden published a monograph in *Müller's Archiv* on the microscopic anatomy of plants. Theodor Schwann, a contemporary German physiologist, recognized the similarity of Schleiden's plant cell nuclei to structures that he had observed in animal nerve tissue. Thus both investigators are said to have been first to formulate the view that cells are individual organisms and that whole animals and plants alike are aggregates of these organisms "arranged according to definite laws."

Present-day historians of science tend to downgrade the contributions of Schleiden and Schwann because careful reading of earlier scientific literature discloses many statements directly or indirectly suggesting that cells are the units of life. By 1835 the cell was generally regarded as an entity with a life and structure of its own. Therefore, when the time arrived for a unifying statement of the cellular nature of all living things, similar insights came to several individuals at once, as happens frequently.

Furthermore, Schleiden and Schwann were in error. While recognizing the essential similarity of cells throughout the living world, they offered a peculiar and novel explanation of how cells arise in the first place, likening the process to crystal formation—in disagreement with certain contempories who believed (correctly, we now know) that new cells arise only by division of pre-existing cells. This feature of Schleiden's and Schwann's work is usually forgotten.

If for no reason other than that it suggested questions that could be put to nature with an expectation of meaningful answers, the cell theory began to prosper. In 1845 Karl von Siebold concluded that a protozoan is a simple animal consisting of one cell. By 1850 it was firmly established that cell division alone can originate new cells. In 1861, simultaneously with the beginning of the great period of microscopic staining research, Max Schultze proposed the essence of the modern protoplasm theory. Perceiving that a cell is more than a membrane surrounding fluid contents and a nucleus, he presumed a fundamental similarity of the jellylike substance composing plant and animal cells and, extrapolating to all living forms, concluded that the material called *protoplasm* is the "physical basis of all life," differing from species to species only in details of structure and composition. To Schultze we owe the oft-quoted definition of a cell as a "mass of protoplasm containing a nucleus"—a definition that, significantly, omits the surrounding membrane. Thus the concept of the universality of protoplasm came to be a part of the first general and unifying synthesis of biological science.

THE MODERN MICROSCOPE

Development of the microscope. In its simplest form, a microscope is little more than a metal tube with an *ocular lens* at one end and an *objective lens* at the other. Such was the instrument used by Hooke in 1665 (see Fig. 3.1). For

A

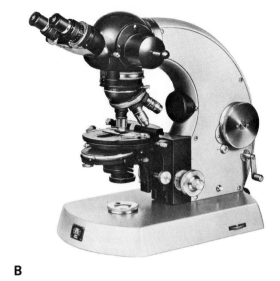

B

3.2 Microscopes. A. Tolles microscope (1873). B. Modern binocular microscope with built-in light source and automatic camera. C. Electron microscope. (A from Armed Forces Institute of Pathology. B from Carl Zeiss, Inc. New York. C from RCA)

many years the microscope changed little, but advances in the understanding of vision and optics finally led to rapid progress in microscopy.

As stronger lenses achieved higher magnification, images began to show distortion near the edges (spherical aberration), colored fringes (chromatic aberration), and blurring due to poor resolution. Spherical and chromatic aberrations were overcome by an objective lens combining two or more types of glass, so juxtaposed that defects in one offset those in the other. Superior resolution came with the use of improved *condensers* (combinations of lenses that effectively focus light rays). Fig. 3.2 compares a relatively advanced Tolles microscope of 1873 with a modern photomicroscope, which photographs fields automatically. The light microscope has now

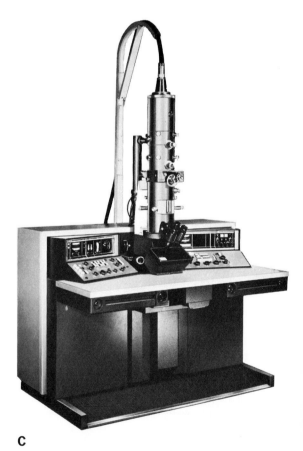

C

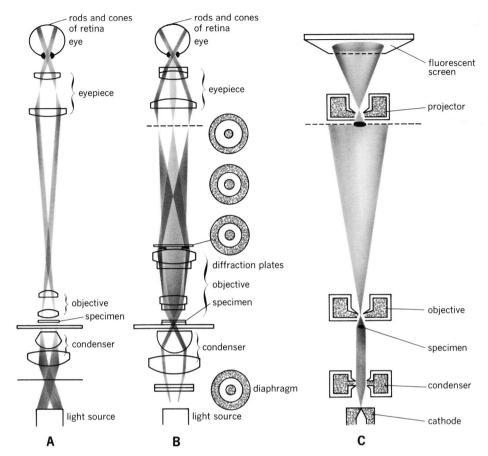

3.3 Optical systems of three types of modern microscopes. A. Conventional light microscope. B. Phase microscope. C. Electron microscope.

reached its highest development. Its optical system, along with those of other modern microscopes, is diagramed in Fig. 3.3.

Resolving power of the human eye. The eye is a remarkable optical instrument, thousands of times as sensitive to light as any man-made instrument. However, its resolving power is limited by certain structural features. The light receptor of the eye, the retina, is a mosaic of minute sensitive structures called *rods* and *cones,* each connecting to a nerve fiber. Together these form the optic nerve leading to the brain. Rods respond to gradations in light intensity; cones distinguish colors.

The diameter of a rod is the limiting factor in the resolving power of the human eye. Obviously

the eye cannot distinguish as separate objects points smaller than its smallest receiving unit. For example, when the retina receives the images of two quotation marks—dots about 1 mm in diameter with tails—it can distinguish them because the diameter of each image on the retinal surface exceeds the diameter of a single rod and the distance between the two images on the retinal surface also exceeds the diameter of a single rod. If the two marks were closer together, both images would fall upon the same rod, and the eye would be unable to distinguish them. This intrinsic limitation in the capacity of the eye to resolve the images of small objects explains why the discovery of cells and bacteria had to await the invention of the microscope.

But light microscopes also have a limit beyond

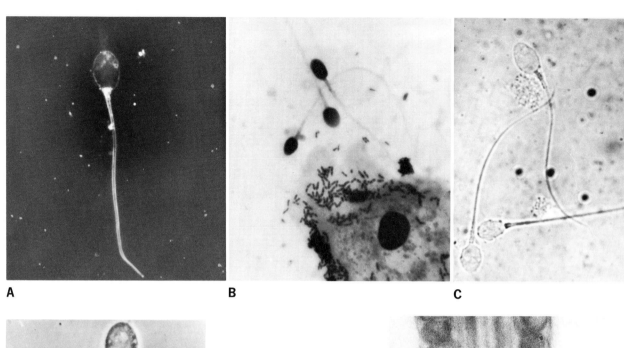

A

B

C

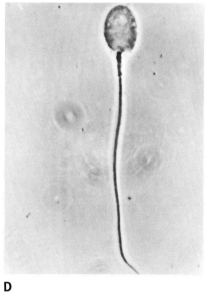

D

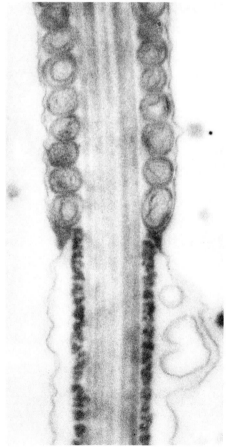

E

3.4 Human sperm cell as revealed with various types of microscopes: A, light, dark field (×3000); B, light, stained (×3000); C, light, unstained (×3000); D, phase (×3000). E, electron ultrastructure of the bovine sperm neck and tail (×44,500). (A–D from Upjohn Company, Kalamazoo, Michigan. E from R. G. Saake and J. O. Almquist, *Amer. J. Anat.* 115:163, 1964.)

which greater magnification is physically impossible. This limit is set by the wavelength of visible light. A light microscope's resolving power is limited to objects of diameter larger than one-half the wavelength of visible light. Since the wavelength range of visible light is from 4000 to 7000 Å, no object smaller than 0.0002 mm in diameter can be reflected or distinguished by light.

Staining techniques. A specimen in the field of a conventional light microscope can sometimes be made visible by illumination from the side. This is called *dark-field* illumination because the object appears against a dark background (Fig. 3.4A). More commonly, examination of a specimen with the ordinary microscope requires prior treatment of the specimen, first with a fixing or hardening agent and then with a dye or stain to render otherwise transparent structures visible (Fig. 3.4B,C).

The study of cell structure was greatly facilitated in the late nineteenth century by new progress in the chemistry of dyes. By skillfully applying to the chemically diverse substance of fixed cells principles learned in dyeing (and mordanting) wool, silk, and cotton fibers, microscopists were able to outline individual cell structures in vividly contrasting colors. More recently, investigators have used staining techniques to localize and identify specific chemicals and enzymes within the cell. Sulfhydryl groups and polysaccharides can be detected with dyes that react with them specifically. The locus of a certain enzyme within the cell can be determined by incubation of tissue sections in a substrate solution that is converted by the enzyme to a colored product. These procedures are the methods used in *cytochemistry* and *histochemistry*.

Newer types of microscopes. Contemporary knowledge of cell structure is largely the result of ingenious advances in microscopic technique that have circumvented many of the limitations inherent in the physical nature of visible light. With the shorter wavelengths of ultraviolet light, resolving power has been doubled. Fluorescent objects have been rendered visible by special optical techniques. The *polarizing* microscope reveals structures that transmit light in a single plane. The *phase* microscope utilizes an optical system (see Fig. 3.3B) that causes objects to be seen in bright or dark relief by virtue of slight variations in refractive properties (Fig. 3.4D). Since structures are apparent without prior staining, living cells can be observed in action.

In recent years the major tool of cytologic research has been the *electron* microscope (see Fig. 3.2C). In place of visible light, this instrument employs a beam of electrons whose wavelength is 0.05 Å—in contrast to 4000 to 7000 Å, the wavelengths of visible light. In principle the electron microscope is identical to the light microscope (see Fig. 3.3C), but instead of light-focusing optical lenses, it utilizes magnetic fields to focus electron beams and projects its image upon a fluorescent screen or photographic plate instead of the human eye. By this means, magnifications 100 times as great as those of the light microscope are easily attained (Fig. 3.4E). The main drawbacks of the electron microscope are the elaborate procedures involved in the preparation of specimens and the expensive bulky equipment.

The Generalized Cell

CELL STRUCTURE

The 1925 edition of E. B. Wilson's *The Cell in Development and Heredity*, an influential textbook in its day, contained a drawing that has since been reprinted many times. It was Wilson's diagram of a "generalized" cell as seen by light microscopy (Fig. 3.5), an idealized composite portrait of an object that probably does not exist in reality. Cytologists early learned that cells display an infinite variety of sizes and shapes. There are free-swimming ameboid white blood cells with curious polymorphous nuclei, nerve cells whose projecting fibers may be over a yard in length, contractile muscle cells, and many others. But for all this variety, each cell possesses features in common with the others, and it is these that are depicted in the Wilson drawing. We note, for example, that the "generalized" cell contains a central *nucleus* embedded in surrounding *cytoplasm*, the whole being encased in a *membrane*. Within the cytoplasm are various

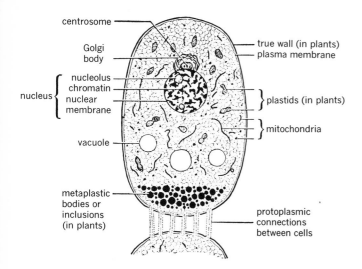

centrosome

Golgi body

nucleolus
chromatin
nuclear membrane

nucleus

vacuole

metaplastic bodies or inclusions (in plants)

true wall (in plants)
plasma membrane

plastids (in plants)

mitochondria

protoplasmic connections between cells

3.5 Wilson's "generalized" cell, based on conventional light microscopy. (From E. B. Wilson, *The Cell in Development and Heredity,* 3rd ed., Macmillan, New York, 1947.)

organelles—small particulate bodies of nondescript appearance in the field of the light microscope.

In recent years electron microscopy has disclosed exquisite ultrastructure in the subcellular particles, and it is now possible to distinguish clearly *endoplasmic reticulum, ribosomes,* a *Golgi complex, mitochondria, lysosomes, centrioles,* and other bodies. The contemporary "generalized" cell shown in Fig. 3.6 is the result. This diagram should be compared with actual electron micrographs accompanying the following brief descriptions of subcellular particles.*

Nucleus. Despite its prominence, the nucleus is one of the least-understood cell components. It is often easily seen by ordinary microscopy without preliminary staining; however, it has a strong affinity for numerous stains. The stained nucleus of a nondividing cell has a nonhomogeneous texture, the nuclear components obviously

differing in their staining ability. The tangled network of strongly staining material was long ago called *chromatin,* and this term is still in use, though it lacks specific biochemical or physiological implications. The well-known avidity of chromatin for *basic* dyes indicates that it must contain strongly acidic material.† In fact, this material is DNA. As we shall see, chromatin strands coalesce during cell division to form *chromosomes,* carriers of the *genes,* which we believe to consist of DNA. One or more small spherical bodies, the *nucleoli,* are easily distinguished within the nucleus. Specific staining reveals that a nucleolus contains substantial quantities of RNA.

Electron microscopy has shown that the nucleus is surrounded by an envelope consisting of an inner membrane and an outer membrane (Fig. 3.7). At certain sites the two membranes are joined together around small *pores* several hundred Ångstroms in diameter. Through these pores, the nucleus communicates directly with the cytoplasm. The interior of the nucleus, in sharp contrast with the cytoplasm, is free of membranes. Nucleoli appear in electron micrographs as dense accumulations of small granules similar to those scattered throughout the remainder of the nucleus.

Cytoplasm. The cell substance outside the nucleus is a complex, viscous material called cytoplasm. Of the several cytological discoveries creditable to electron microscopy, surely the most significant revealed the cytoplasm's rich content of membranes and membrane-limited elements—unsuspected from light microscopy. The only membranes shown in Wilson's diagram of the cell were the external cell membrane and

---

* Since experimental electron micrography is used most frequently with preparations of nonhuman origin, most of the sections to be pictured here come from species other than man. In light of the generalization stated by the cell theory, this should cause us no concern. The structures shown closely resemble those of human cells.

† The acidity of nucleic acids is due to their negatively charged phosphate groups (see Fig. 2.12). Within the cell, nucleic acid polymers exist as the anionic portions of saltlike nucleoproteins. The cationic halves of the salt pairs are strongly basic, positively charged proteins called *histones.* Thus nucleoproteins are "protein nucleates." Basic dyes are colored cations capable of forming "dye nucleates" by replacing the positively charged proteins. The capacity of anionic polymers such as nucleic acids to bind basic dyes is called *basophilia; acidophilia* is the capacity of basic cell components to bind acidic dyes. Most dye-binding methods do not distinguish between RNA and DNA. However, a number of cytochemical procedures take advantage of minor differences between them. For example, the familiar *Feulgen* stain reacts only with DNA.

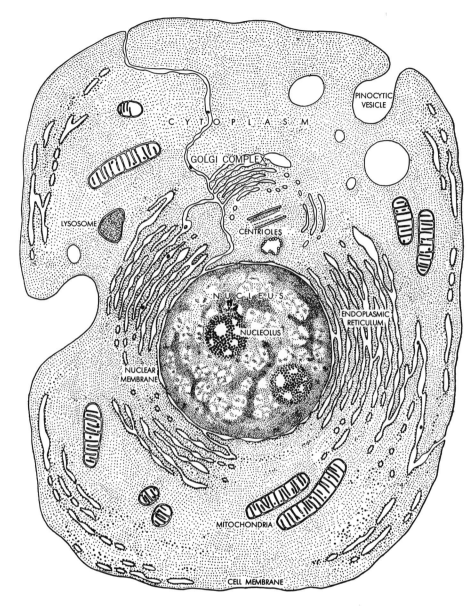

3.6 Contemporary "generalized" cell, based on electron microscopy. (From "The Living Cell" by Jean Brachet. Sept. 1961. Copyright © 1961 by Scientific American, Inc. All rights reserved.)

the nuclear envelope. The organelles were depicted as solid granules or filaments.

In disclosing the true membranous structure of each of the classic organelles, electron microscopy has made evident the existence within the cytoplasm of an *internal* phase separated from the continuous *external* matrix phase by mem-

branes. Since little is known of the matrix phase, we shall speak mainly of the membrane-sheathed elements.

Ribosomes and endoplasmic reticulum. Many cells, especially those engaged actively in protein synthesis, contain a labyrinthine system of

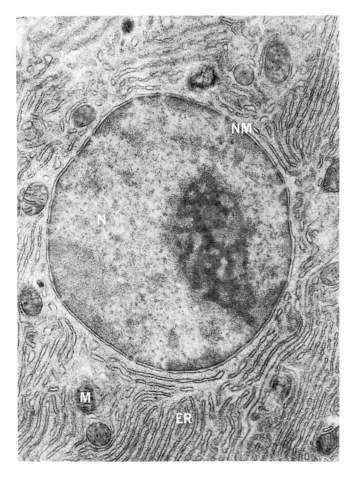

3.7 Cell nucleus (N) surrounded by nuclear membrane (NM), rough endoplasmic reticulum and ribosomes (ER), and mitochondria (M) (× 28,400). (Don W. Fawcett.)

tions seem to provide a means of communication between nucleus and cytoplasm and between cytoplasm and cell surface.

Two forms of endoplasmic reticulum can be distinguished: *granular* or *rough-surfaced* and *agranular* or *smooth-surfaced.* The membranes of typical granular reticulum are seen at high magnifications to be encrusted on the cytoplasmic side with dense particles about 150 Å in diameter. These are the ribosomes, the ribonucleoprotein particles whose role in protein synthesis has been elucidated in the past few years.

For some time, biochemistry-minded cytologists have fractionated fragmented cells containing granular endoplasmic reticulum by stepwise ultracentrifugation: as shown in Fig. 3.8, different quantities of gravitational force are required to deposit the various cell inclusions in the bottom of a centrifuge tube. The particles in the lightest fraction, which settles out only after prolonged centrifugation at extremely high rotor speeds, were first called *microsomes.* But examination by electron microscopy proved them to be fragments of endoplasmic reticulum and attached ribosomes. When microsomes were digested in detergents or other lipid solvents, the membranes of endoplasmic reticulum (whose high lipid content is reflected in a low density) dissolved away, leaving free ribosomes.†

Ribosomes are about 60% RNA and 40% protein by weight. They can be isolated, and their sedimentation constants can be measured by ultracentrifugation. An intact single ribosome of a bacterial cell has a sedimentation constant (S, for Svedberg unit) of 70 (Fig. 3.9). When exposed to altered conditions (notably a lowered concentration of magnesium ion, Mg^{++}, in the suspending medium), a 70 S ribosome dissociates into large and small 50 S and 30 S subunits. A ribosome from an animal body dissociates into comparable major and minor subunits.

membrane-limited channels throughout the cytoplasm. This system is the endoplasmic reticulum (see Figs. 3.6, 3.7). The membranes of the endoplasmic reticulum form concentric layers of tubelike passageways opening onto the surface of the cell membrane and into the space between the two membranes of the nuclear envelope (all these membranous elements being of notable structural similarity; see Figs. 3.6, 3.7). Conceivably the system is an outgrowth of the nuclear envelope.* In any event, the connec-

---

* Conversely, some authorities have suggested that the endoplasmic reticulum is an *ingrowth* of the outer cell membrane.

---

† Although ribosomes are located chiefly in the cytoplasm—indeed, most of the cell's RNA is present in cytoplasmic ribosomes—similar particles are found in the nucleus (see Fig. 3.7). The high RNA content of the nucleoli is attributable to ribosomes. This is of particular interest in view of the growing evidence that cytoplasmic ribosomes and ribosomal RNA are synthesized in the nucleoli.

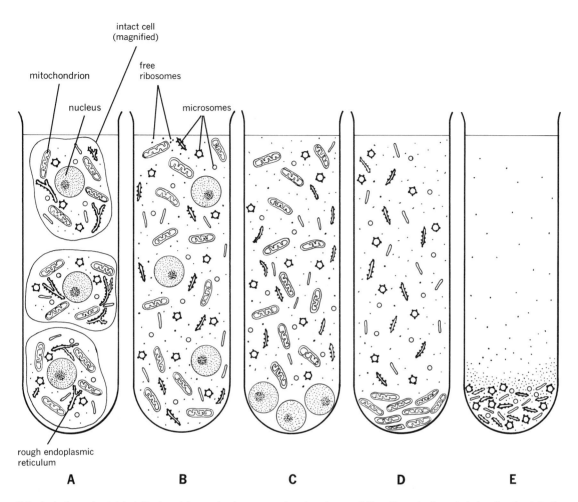

intact cell
(magnified)

mitochondrion

nucleus

free
ribosomes

microsomes

rough endoplasmic
reticulum

A   **B**   **C**   **D**   **E**

3.8 Isolation of nuclei, mitochondria, and microsomes by stepwise centrifugation. A. Suspended cells about to be broken up with a pestle. B. The fragments are suspended in sucrose solution prior to centrifugation. C. Heavier particles, such as nuclei, are thrown down. D. The remaining fluid is removed and recentrifuged at a higher speed, which throws down the mitochondria. E. The remaining fluid is centrifuged at a still higher speed, which throws down the microsomes and free ribosomes.

Intact 70 S ribosomes are often found free in clusters of two to six (Fig. 3.10), although in some animal cells the clusters contain up to 40 ribosomes. These are called *polysomes* (or *polyribosomes*).* We shall later see that the arrangement into polysomes is indispensable to the participation of ribosomes in protein synthesis.

Agranular endoplasmic reticulum is less well studied than the granular form but equally common. It displays smooth membranous surfaces devoid of ribosomal granules. Agranular reticulum may appear in the same cell with granular reticulum, and fragments of both may appear among the microsomes. Various theories implicating agranular reticulum in lipid and polysaccharide metabolism are as yet unsubstantiated.

---

* Polysomes should not be confused with microsomes. Polysomes are clusters of ribosomes held together by a single strand of RNA. Microsomes are heterogeneous fragments whose composition depends upon the proportions of granular and agranular endoplasmic reticulum in the original cell.

	sedimentation constant	molecular weight ($\times 10^6$)

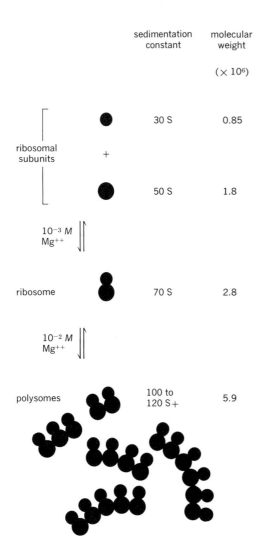

<table>
</table>

ribosomal subunits

+

30 S   0.85

50 S   1.8

$10^{-3} M$ Mg^{++}

ribosome   70 S   2.8

$10^{-2} M$ Mg^{++}

polysomes   100 to 120 S+   5.9

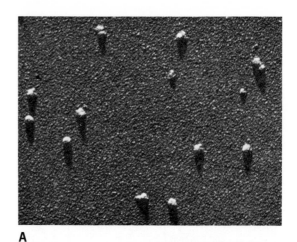

A

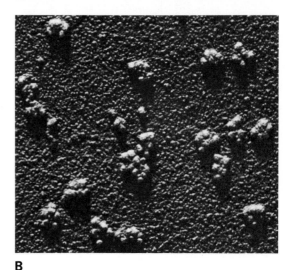

B

3.9 Ribosomes, ribosomal subunits, and polysomes. Ribosomes dissociate into 50 S and 30 S subunits on dilution of Mg^{++} concentration in the medium to 0.001M. The process is reversible, and in a 0.01M Mg^{++} solution, 70 S ribosomes reform. In the cell, ribosomes form into clusters called polysomes (i.e., polyribosomes) containing two to six or more ribosomes. The sedimentation constants correlate roughly with the molecular weights of the units, subunits, and clusters.

3.10 Ribosomes (A) isolated from *Escherichia coli* and polysomes (B) isolated from reticulocytes, red blood cell precursors ($\times 100,000$), showing that a 70 S ribosome is made up of a large (50 S) and a small (30 S) subunit and that most polysomes contain five ribosomes. (A from C. E. Hall; B from A. Rich.)

Golgi complex. The Golgi complex, or Golgi body, was first noted in the cytoplasm by the Italian cytologist Camille Golgi at the end of the nineteenth century. Light microscopists sus-

pected its involvement in secretory mechanisms* but were unable to delineate its ultrastructure. Electron microscopy revealed that the

---

* The process of secretion will be discussed in Chapter 4. We shall note here only that cells synthesize two classes of materials: (1) those that remain within the cell; and (2) those that are extruded through the cell surface, or secreted. Secretion is characteristic of certain cell types (e.g., gland cells).

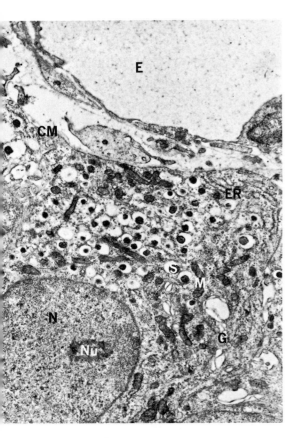

Mitochondria. The cytoplasm of nearly all cells contains small bodies called mitochondria (see Figs. 3.6, 3.7, 3.8, 3.11). Though barely visible with the light microscope, they show an intricate fine structure in electron micrographs (Fig. 3.12). Their numbers vary from cell to cell,

A

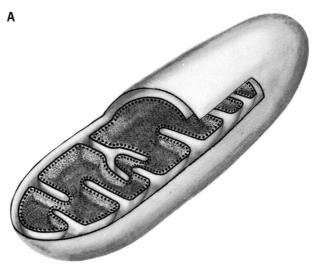

3.11 Insulin-secreting pancreatic cell, showing secretory granules and a Golgi complex: E, exterior of cell (capillary into which insulin is secreted); CM, cell membrane; ER, endoplasmic reticulum and ribosomes; S, encapsulated secretory granule containing insulin; M, mitochondrion; G, Golgi complex (note smooth membranes); N, nucleus; Nu, nucleolus. (Paul E. Lacy and The Upjohn Company, Kalamazoo, Michigan.)

multilayered complex is made of smooth membranes that are probably continuous with the endoplasmic reticulum (see Fig. 3.6).

The Golgi complex is now believed to participate in the secretion of proteins or polysaccharides—probably not as a site of synthesis but as a "collecting, wrapping, and packaging" station, which somehow encloses particles of matter to be "exported" (i.e., secreted) in membranous envelopes capable of fusing with the external cell membrane and ultimately passing through it. These *secretory granules* occur only in specialized secretory cells (Fig. 3.11).

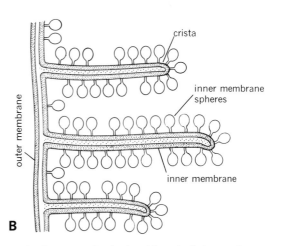

3.12 Structure of mitochondrion. A. Cutaway diagram of whole mitochondrion, showing the double-walled membrane with its intrastructure space and cristae. B. Enlarged diagram of portion of wall and crista, showing outer and inner membranes and inner membrane spheres.

a fact of interest since recent studies of ultracentrifugally isolated mitochondria (see Fig. 3.8) have identified them as centers of cellular respiration and energy-yielding metabolism. The occasional cell types containing no mitochondria (such as red blood cells) are biological curiosities, and they must rely upon notably less efficient means of energy production.

A mitochondrion is bounded by a double-walled surface membrane (Fig. 3.12A). Minute spherical bodies 85 Å in diameter stud the inside surface of the inner wall (Fig. 3.12B). Flattened infoldings of the inner wall form perpendicular platelike *cristae*, and the 70 Å space within the membrane contains fluid. The molecular arrangement of the proteins and lipids of the double-walled membrane was diagramed in Fig. 2.7B, and we shall presently encounter evidence indicating that the respiratory enzymes exist within the spherical bodies attached to the inside surfaces. We should note how the area of the mitochondrial membrane is expanded by its architectural plan. Recent experiments have shown that mitochondria may swell and contract in the course of physiological activity. We shall later consider the possible significance of these changes.

Lysosomes. The electron microscope also distinguishes between mitochondria and a heterogeneous group of bodies of similar size, the lysosomes. Lysosomes contain many of the hydrolytic enzymes that cleave macromolecules into smaller molecules capable of being oxidized in the mitochondria, such as phosphatase, ribonuclease, and various carbohydrases and proteases. The enzymes are sheathed within a lipoprotein envelope that isolates them from the rest of the cytoplasm. When the rupture of the envelope releases them, dissolution of the cell quickly follows. In white blood cells and other cells specially adapted to phagocytosis (i.e., the ingestion of foreign particles), lysosomes discharge their hydrolytic enzymes in such a way as to digest the ingested particles.

Centrioles and kinetosomes. The centrioles* are small paired cylindrical bodies lying near the

---

* Few cell structures have been given as many names as the centriole—e.g., centrum, centrosphere, and attraction sphere.

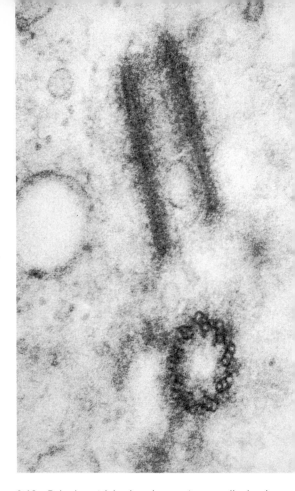

3.13 Paired centrioles in a human tumor cell, showing the positioning of the two bodies at right angles to one another and their fine structure (× 160,000). (From W. Bernhard, Institute for Cancer Research, Villejuif-sur-Seine.)

nucleus, usually in the middle of a zone cleared of other particles, the *centrosome*. Electron microscopy discloses that each centriole consists of a cluster of nine groups of delicate tubule-like fibrils, each group containing three fibrils (see Figs. 3.6, 3.13). Curiously, the two centrioles of a pair always lie at right angles to one another. The significance of this arrangement and of the paired nature of the centrioles in cell division will be discussed later.

A cell equipped with motile, hairlike processes on its surface, called *cilia* or *flagella*, has a basal body, or kinetosome, at the base of each process. This body is somewhat similar to a centriole but differs in having 11 groups of fibrils, nine containing two fibrils and two containing one fibril. The nine duplexes are arranged around the periphery of the filament, and the two singlets in the center. We shall encounter this so-called *9 + 2 configuration* later. It is of interest that both cilia and certain components arising in cell division are contractile. Conceivably kinetosomes and centrioles function similarly as "cell muscles."

Cell membrane. If it can be seen at all with the light microscope, the cell membrane, or plasma membrane, appears only as a thin de-

limiting boundary line (Fig. 3.14A). The electron microscope, however, reveals that it has a noteworthy fine structure. The membrane's complexity was long known simply because it could be shown that certain substances enter and leave the cell freely while others do not. Red blood cells and nerve cells, for example, distinguish between sodium and potassium ions despite their similarity in size and charge. Potassium ions can enter the cells freely, and somehow sodium ions are "pumped" outward. This implies the existence of both a selectively permeable barrier and an active "pumping" mechanism within the sur-

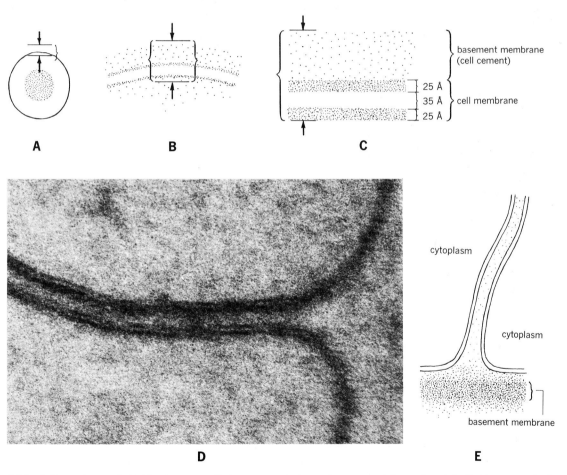

3.14 Cell membrane. A. Light microscopy. B. Low-power electron microscopy of the region delimited by arrows in A. C. High-power electron microscopy of the area delimited by arrows in B. D. Cell membranes of two adjacent cells, with cell cement between them (× 400,000). E. Fusion of the cell cement into the basement membrane, which forms the supporting surface for a sheet of cells. (D from J. D. Robertson.)

face membrane. Moreover, the membrane can be shown to transport actively certain large molecules and particles into the cell interior. Clearly it is more than a mere boundary.

By electron microscopy the cell membrane is seen to consist of two membranes each about 25 Å thick, separated by a 35 Å space (Fig. 3.14B,C). In many ways its structure resembles that of a typical membranous element of the cytoplasm.* The 150 Å gap between the surfaces

---

* The striking structural resemblance between the cell membrane and the cytoplasmic membranes has raised the interesting possibility that all of the cytoplasmic membrane systems (including the endoplasmic reticulum, Golgi complex, mitochondria, and nuclear envelope) arise by inward extension and folding of the cell membrane. If this is true, materials between the intracellular membranes are, topologically speaking, outside the cell.

of adjacent cells (Fig. 3.14D) contains extracellular material that presumably has the adhesive properties necessary to hold neighboring cells together. When cells are arranged in sheets of tissue, this substance appears to form the underlying *basement membrane* (Fig. 3.14E), a supporting layer that contains a delicate fibrillar network.

It should be emphasized that the cell membrane differs from the rigid plant cell wall, which lies outside the cell membrane. A plant cell wall is inert, serving only as support and playing no active role in controlling the passage of materials in and out of the cell. Because Hooke observed sheets of plant cell walls in his early microscopic studies of cork (see Fig. 3.1), plant cell walls were the first cell structures to be seen, and cell membranes the last.

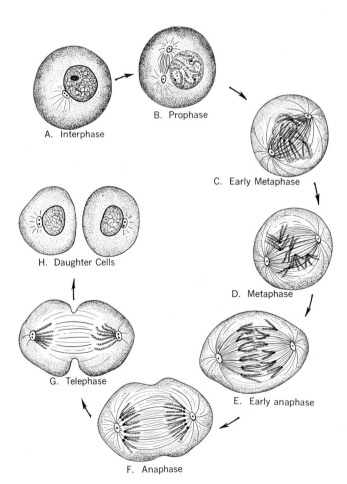

A. Interphase

B. Prophase

C. Early Metaphase

D. Metaphase

E. Early anaphase

F. Anaphase

G. Telephase

H. Daughter Cells

3.15 Mitosis.

CELL DIVISION

All cells come from cells, and each cell must begin its existence having the same features and potentialities as its parent. We here consider *mitosis,* the process by which cells reproduce. In mitosis, cells divide by a mechanism involving no union of parental cells with mixing and recombination of parental genes. Hence mitosis is an *asexual* process in which an individual cell is replaced by two daughter cells, each receiving some of the parental cell's components, each a faithful replica of the parent. In the course of division the parental cell ceases to exist.

Mitotic cycle. Cell division may be separated into *karyokinesis,* or nuclear division, and *cytokinesis,* or cytoplasmic division. The word mitosis is actually synonymous with karyokinesis, by far the most vivid aspect of cell division to the microscopist, but it is frequently applied to cell division as a whole.

The mitotic process has five stages or phases: *interphase, prophase, metaphase, anaphase,* and *telophase* (Fig. 3.15). Although the different stages are identified by certain landmark events, they actually merge smoothly into a continuous process.

In interphase (see Fig. 3.15A) the cell is in a resting state between divisions. In an interphase cell viewed with the light microscope, the nuclear chromatin appears to be randomly arranged in elongated strands within the nuclear envelope. Other than the nucleolus, there is little definable internal structure in the nucleus.

In prophase preparations for division begin. The nuclear membrane and nucleolus start to disappear, and the elongated chromatin strands condense into compact sausage-like chromosomes by coiling into tight helices (see Fig. 3.15B). Each chromosome is composed of two roughly parallel longitudinal helical threads called *chromatids*—yet another example of the helical design of information-containing biological structures. The chromatids are twisted about each other like the fibers in a cotton string, though they are held together by a small body called the *centromere* (Fig. 3.16). Thus the chromosome is a coiled coil. The coiling within an individual chromatid is called *major* coiling; the

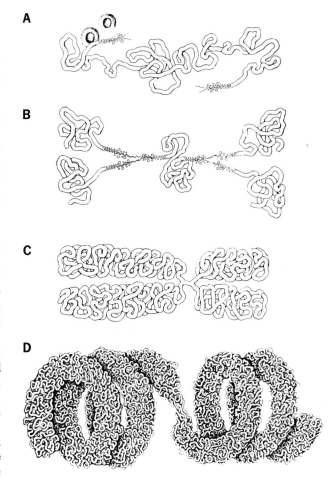

3.16 Diagram of the structure, organization, and replication of a single chromosome according to the "folded fiber" theory. A. Beginning replication of a single chromosome fiber, which consists of a single strand of DNA coated with protein. B. Replication proceeds from the ends to the middle. C. Daughter fibers fold up to form condensed metaphase chromosome. D. Detailed folding pattern of a single chromatid. (A–C from E. J. DuPraw, *Nature, 206,* 338 (1965); D from E. J. DuPraw, *Nature, 209,* 577 (1966).)

coiling of one chromatid about the other is called *relational* coiling. We shall presently see that the functional unit of a chromosome is the chromatid.

When the chromosomes have reached a maximal degree of condensation, systems of radiating fibers become prominent around each centriole. The resulting structures, which resemble stars, are named *asters*. The centrioles and their asters migrate to opposite poles of the cell. Then a remarkable event takes place.

Between the centriolar poles the *mitotic spindle* materializes. Its appearance marks the beginning of metaphase (see Fig. 3.15C,D). The spindle is a mass of delicate fibers (or tubules) forming an apparatus shaped like two cones placed together base to base. The plane formed by the bases of the cones is the *equatorial plate*. The chromosomes, heretofore arranged in seemingly random fashion, migrate onto the equatorial plate of the spindle. Somehow the centromere of each attaches to a spindle fiber. Though the dangling arms of the chromosomes may lie off the equatorial plate, the centromeres line up precisely on it. By this time the polar centrioles have almost duplicated.

Anaphase now begins. The two chromatids of each chromosome thicken and split lengthwise, each becoming a full chromosome containing two chromatids. The new chromosomes then move apart, presumably drawn by the spindle fibers toward opposite poles of the spindle. A chromosome is apparently pulled by its centromere so that its two arms characteristically form a V or J,* as though the entire structure were being towed against a current (see Fig. 3.15E). In this way two groups of chromosomes, each a sister set of the other, collect at opposite ends of the cell (see Fig. 3.15F).

The cell now begins to pinch in two, and telophase begins. This is characterized by the regrouping of the new chromosomes into new nuclear structures and the appearance of new nuclear membranes around each of the two new daughter nuclei (see Fig. 3.15G). The spindle vanishes, the chromosomes loosen their coils, their affinity for basic stains diminishes, and gradually there reappears the typical interphase nucleus containing a newly formed nucleolus. Telophase thus is prophase in reverse. When these events have transpired, the almost divided cell completes a new membrane and splits into two new cells.

---

* The structure of each chromosome is unique, and its centromere has a definite position nearer the middle or the end. Chromosomes with central centromeres resemble V's during anaphase; those with centromeres near the end resemble J's.

Recent studies of the mitotic apparatus. With few exceptions, most of the elements of the foregoing description are the contributions of classic light microscopy. Electron microscopy has afforded surprisingly little additional information.

One conclusion that was not derived entirely from morphological data is related to the replication of the centrioles. It had earlier been shown that the familiar laboratory reagent mercaptoethanol ($HSCH_2CH_2OH$) blocks mitosis if the cell is exposed to it before chromosomes begin to separate in metaphase. Mazia and his coworkers observed that, when mitosis was blocked for a long time and the block was then removed, division occurred, but four cells were produced instead of two. Careful examination of mercaptoethanol-inhibited cells revealed that during the block two new poles formed, making a total of four. Hence *tetrapolar* mitosis followed removal of the inhibitor. It became apparent that the two extra poles arose from a splitting of the normal centrioles, not from true duplications. Therefore, when a daughter cell "tried" to divide, it could not do so because its mitotic apparatus had only one pole. When the daughter cell completed its peculiar mitotic cycle involving one pole, half a spindle, and no division, its capacity for normal mitosis was restored.

From these results we conclude that each of the two normal centrioles has two components, an old portion and a new portion. Each duplex (combination of old and new units) forms a pole of the normal mitotic apparatus and reproduces early in mitosis, so that each daughter cell will contain two duplexes. However, in the presence of mercaptoethanol, the duplex splits into its component parts, so that the two daughter "centrioles" are not normal, though they are competent to serve as poles for an abnormal tetrapolar mitosis. When the inhibitor is removed, the single centriolar unit in each of the four developing daughter cells becomes a duplex, but still the cell has only one normal centriole, not the two needed for normal mitosis. Centriolar reproduction may occur without cell or chromosome reproduction. As a cell divides, it fabricates the centrioles for the divisions of its daughters. When centriolar reproduction was finally observed with the electron microscope, it was seen that the new portion grows out of the old portion at a right angle.

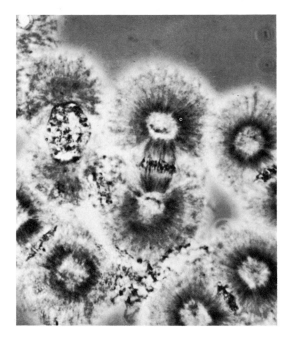

3.17 Isolated mitotic apparatus from sea urchin eggs (× 1000). (Photo from D. Mazia, University of California, Berkeley.)

Mazia and others have made other direct experimental approaches to the mitotic apparatus. They have, for example, isolated intact apparatuses, spindles and all (Fig. 3.17)—a notable achievement in view of their evanescence. An isolated apparatus contains about 12% of the total cell protein. This fact suggests that it is of cytoplasmic origin. The manner of mitotic apparatus construction, the nature of spindle fiber contraction, and the source of the energy underlying these elaborate activities are promising areas for future research.

Finally notice must be taken of the recent development of simple methods for studying the morphology of individual chromosomes. If we cultivate human cells briefly in tissue cultures and add an agent (i.e., colchicine) that blocks mitosis in metaphase, we can then literally squash the cells to disperse the chromosomes so that they can be photographed and studied (Fig. 3.18). In the few years since the "squash" technique was introduced, it has demonstrated (1) that human cells contain 46 chromosomes, not 48 as was earlier believed; (2) that

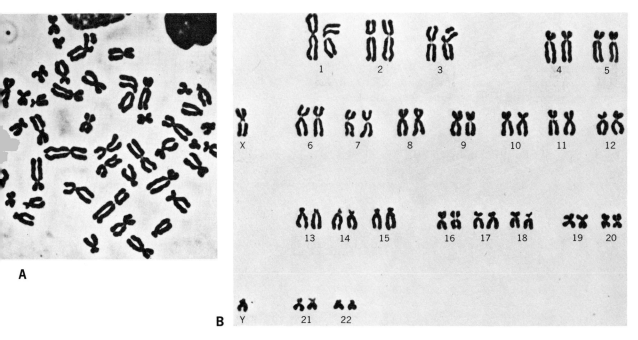

3.18 Chromosomes of a normal human male leukocyte in mitotic metaphase (× 3200). A. In a "squash" preparation. B. In a karyotype preparation. (J. L. Hamerton.)

the various chromosomes have unique, recognizable structures; (3) that a normal cell possesses a definite and reproducible constellation of chromosome structures, this pattern being termed the *karyotype;* and (4) that certain congenital diseases are regularly associated with chromosomal abnormalities attributable to defects of mitosis and of other mechanisms to be discussed later.

THE CELL AS AN ORGANISM

It should be kept in mind that fundamental differences exist between independent unicellular organisms and the individual cells of complex multicellular organisms. It is the whole multicellular organism that is alive in the usual sense of the word, and though it is a thriving federation of specialized single cells, each cell is dependent upon the presence and integrated functioning of the others—in short, upon the multicellular organism. Unlike a unicellular organism, a cell of a multicellular organism is not an independent life unto itself.

These considerations suggest that a living organism should be defined as an independent unit of integrated structures and functions. This definition is as applicable to a man as to an ameba. It does not apply to an individual human cell, which is a dependent unit of integrated structures and functions. If it becomes independent, we call it malignant, for its behavior is no longer subservient to the interests of the organism. Here independence becomes insubordination.*

Genetic continuity of life. The genetic continuity of life is one of the most important implications of the cell theory. Unicellular organisms come from unicellular organisms, and multicellular organisms come from a union of the germ cells of multicellular organisms. In both cases genetic mechanisms ensure the continuance of the species.

---

* A human cell growing in tissue culture presents no serious contradictions. Such a cell is neither a multicellular nor a unicellular organism. Since it belongs in neither category, we must conclude that it is not really a living organism. In fact, it is a dependent cell that under unusual circumstances has achieved independence, a structure behaving as though it were an organism.

The discovery that cells arise only from pre-existing cells raised a number of questions. One dealt with *spontaneous generation.* Between 1859 and 1861 Louis Pasteur had revealed the error of the long-accepted view that living organisms can arise spontaneously from nonliving materials such as soil or mud. In a historic experiment he showed that, until a flask of sterile broth is seeded with germs, none will arise in it. The unique exception to the dictum that spontaneous generation does not occur was the instance of life's origin on earth. It is interesting that until the works of Pasteur biologists were unaware that the origin of life on earth posed a major scientific problem.

The development of a second great contribution of the nineteenth century, the theory of evolution, was profoundly influenced by the implications of the cell theory concerning the continuity of life. Evolutionary thinkers were required to look upon an unbroken sequence of cell generations extending backward in time to life's beginning. When this was visualized, only a minor act of imagination and logic was needed to realize that all cells (and all species) must have common ancestry—in other words, evolution and change must have occurred. It also followed that all living cells, no matter what their differences, must have kinship, and we may consider their similarities as the evidences of common ancestry.

The concept that cells come from cells powerfully affected the development of another field, the science of pathology. Pathology had been an ill-defined adjunct to clinical medicine until Rudolf Virchow of Berlin (1821–1902) applied the cell theory to the study of diseased tissue. His 1858 volume, *Cellular Pathology,* presented a revolutionary view of the nature of disease by demonstrating the relation between the individual diseased cell and the manifestations of disease in the body at large. Virchow's contributions were largely responsible for the long dominance of microscopic anatomy in pathological thought.

Problems of differentiation and morphogenesis. Contemporary unicellular organisms—the protozoa, bacteria, etc.—are presumably descendants of the group from which multicellular organisms originally evolved. We may assume that at some point in the course of evolution individual cells found advantage in colonial life. It thus became desirable for evolution to "invent" a mechanism for transmitting the colonial habit to offspring of the individually dividing members. In the system that emerged, certain cells became keepers of the plan—specialized cells that somehow retained within their structures coded patterns both of themselves and of the colony at large. With a precedent set for specialization within the colony, newly formed cells soon became dependent upon one another.

Then means evolved whereby the community became the perpetuating entity, with the process of reproduction yielding new communities, not merely new groups of gregarious individual cells. Within the community, cells still divided individually, but reproduction of the whole community—actually a multicellular organism —required a higher level of organizational complexity. Body cells were now in the service of the organism, and reproductive cells had to give rise to diverse body cells while preserving the seed stock for future generations. Thus the phenomena of *differentiation* and *morphogenesis* were introduced.

The term "differentiation" has traditionally referred to all aspects of developmental biology, especially at the cellular level. In differentiating, a single fertilized egg cell, progenitor of the adult multicellular organism, must divide repeatedly to produce the new cells that growth requires, and these new cells must be transformed from the general to the specific. Though they remain true to species, they must acquire new forms and functions as they become liver cells, nerve cells, and blood cells in the course of embryonic development. The term "morphogenesis" has traditionally encompassed the interactions of these cells with each other and with their environment that lead to supracellular organizations such as tissues and organs.

Since the time of Aristotle (384–322 B.C.) the questions of differentiation and morphogenesis have been debated by the advocates of two theories—the theory of *orthogenesis* and the theory of *epigenesis*. The orthogenetic argument holds that, if cell or species A is transformed into B, then there must exist within A some miniature "infolded" form of B that is waiting to be "unfolded." This view has also been called *pre-formationism.* Probably the first "experimental" orthogeneticist was the seventeenth-century Dutch microscopist Niklaas Hartsoeker, who thought he saw a tiny but complete human figure, a *homunculus,* in the head of a human spermatozoon (the same little figure that Leeuwenhoek thought he saw; see p. 67). Had the homunculus been real, a great many difficult problems would have been simplified. If the adult human body came from a homunculus preformed in a germ cell, we should no longer have to worry about embryogenesis and differentiation. We should picture within each homunculus its own tiny germ cells, each containing a homunculus, each of which in turn contains more minute germ cells and homunculi, etc.

The alternative theory of epigenesis holds that a germ cell contains *potentialities* rather than a miniature adult. Development is thus viewed as an orderly sequence of actions and interactions among the parts of a developing organism, the whole organism, and its environment—for example, a limb bud arises under the influence of certain cells that themselves recently developed from less specialized cells, and limb bud cells subsequently develop into bone, muscle, and nerve cells at the proper times. Interestingly, this theory was expounded by Aristotle, who stated that organs and parts arise by gradual interaction of the simpler constituents of the system.

Current thought supports epigenetic viewpoints, and it has been the task of the investigator to explain how such complexity is packed into a single cell, a precursor of precursors, and in what form it exists. Little has been gained from morphological study of a developing embryo—until recently the only known approach to the problem. With this method an observer can only examine products, and, when he detects them, the process of differentiation has ended, and he has missed the critical causal events.

The chief difficulty for the student of differentiation has been to reconcile the paradoxical fact that a differentiating cell yields progeny differing from the parent with genetic laws (to be discussed later) ensuring that the offspring are *duplicates* of the parent. In short, how can a mitotic cell division produce daughter cells that differ from the parent cell?

Investigators of various disciplines have recently mounted vigorous new attacks upon the problem of differentiation—partly because of the availability of sensitive new biochemical and biophysical techniques but mostly because of extraordinary new advances in our understanding of genetics and their significance for those aspects of biochemistry concerned with macromolecule synthesis and its regulation. We shall review these developments later and shall see that their simplicity and generality encourage the belief that the mechanism of differentiation will soon be elucidated.

Asexual and sexual reproduction. Cell division is an *asexual* mode of reproduction in which the offspring inherits all the traits of the parent through genetic mechanisms to be discussed in the next section. *Sexual* reproduction involves a union of the germ cells, or *gametes*, of a male parent and a female parent. Since there is a *recombination* of the parental determinants of heredity, the degree of similarity between the offspring and one parent or the other may vary widely, and novel assemblages of traits can arise. Asexual reproduction does not produce such results.

Historically, the science of genetics developed from studies of sexually reproducing multicellular plants and animals. These studies showed how traits are transmitted from parents to progeny and accounted for the recombinations that are the hallmarks of sexual reproduction. They also provided an explanation for heredity in the simpler asexual systems, in which recombination does not occur.*

Principles of Heredity

The fundamentals of heredity may be considered in two categories: (1) the means by which hereditary materials are duplicated and transmitted from generation to generation; and (2) the means by which hereditary materials govern cell structure and function. The fact that heredity occurs was known before the details of mitosis were unraveled. When mitosis was finally understood, it was apparent that it afforded an ingenious method of transmitting to each daughter nucleus a set of chromosomes identical to that of the parent. Accordingly, this event and its variations belong in the first category. The second includes basic processes that have great significance for physiology and that, surprisingly, have all been discovered since 1961.

TRANSMISSION AND RECOMBINATION OF THE UNITS OF HEREDITY

Heredity and the gene. We begin to grasp the magnitude of the problem that must be solved in any attempt to explain heredity when we realize what is involved in the production of an offspring. It has been aptly pointed out that one cannot obtain two automobiles by cutting one in two, for this yields half-automobiles. To obtain a new automobile, we must build it from the ground up, and to do this, we need the same detailed blueprints (or copies of them) that were used in building the automobile we wish to duplicate. What sort of blueprint controls the duplication of cells?

The first real advance in this area came from the brilliant researches of the Moravian monk Gregor Mendel (1822–84).† In an attempt to determine how the cross-breeding of differing types produces variations in the traits of the progeny, Mendel conceived some remarkably simple experiments. He chose to investigate the ordinary garden pea plant because it has two outstanding traits that can vary widely, flower color and plant size. In his first experiment (Fig. 3.19A), he used varieties that differed only in

---

* It is noteworthy that sexuality has been demonstrated in bacteria and individual animal cells growing in tissue cultures. Large populations of the colon bacillus *Escherichia coli*, for example, contain a number of organisms that are capable of transmitting chromosomal material to other organisms, the act of conjugation having actually been observed by electron microscopy. Thus there are "male" and "female" bacteria, and the advantages of recombination are available even to these lowly organisms.

† Though Mendel's results were published in 1866, almost 50 years passed before any serious attention was given to his work. Historians of science still debate the reason for this neglect, but whatever it was, the curious fact remains that in 1950 the Genetics Society of America celebrated not the eighty-fourth anniversary of Mendel's discovery but the fiftieth anniversary of the discovery of his discovery!

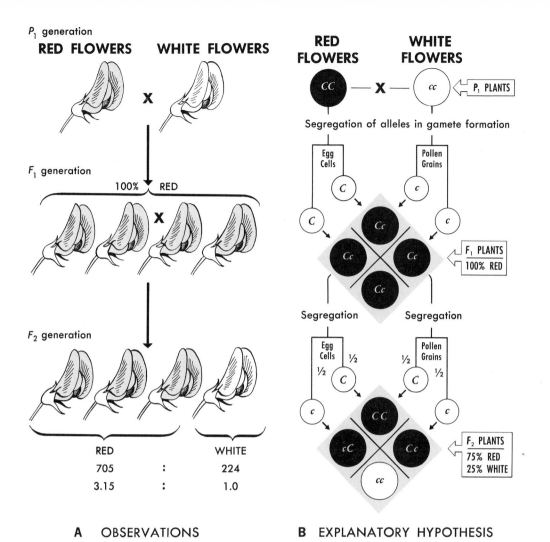

3.19 Mendel's first experiment (A) and his interpretation of its results (B). (From G. G. Simpson and W. S. Beck, *Life: An Introduction to Biology*, 2nd ed., © 1957, 1965 by Harcourt Brace Jovanovich, Inc. and reproduced by permission.)

color, crossing plants with red flowers with plants with white flowers. He found that the off-spring of this mating (which, collectively, are referred to as the *first filial* or F_1 *generation*) were always all red-flowered, regardless of whether the male or female parent had been red-flowered. However, when two red-flowered plants of the F_1 generation were crossed in a brother-sister mating, both red-flowered *and* white-flowered plants appeared in the next or F_2 *generation*. When Mendel counted them, he found almost exactly three reds for every white. In one experiment yielding 929 plants in the F_2 generation, there were 705 reds and 224 whites, or 75.9% reds and 24.1% whites. When seven different pairs of individual traits were so analyzed in different experiments, the average percentage frequency of one trait was 74.90, and that of the other 25.10.

This work demonstrated that, whereas an individual F_1 offspring showed a trait of one of its parents, it might nevertheless hand on to its own offspring the alternative trait of its other parent. To Mendel it suggested that a "hereditary factor" must exist for each color and, though one of these did not "express itself" in the F_1 offspring, it was still present, even though masked by the factor for the other color. Not only was the factor present though hidden in one generation, it was stable, so that when it could finally reassert itself in the traits of a member of the F_2 generation, it was unaltered from its condition in the parent generation.

To explain his results, Mendel postulated that each plant contains a *pair* of hereditary factors controlling flower color. One member of the pair is obtained from each parent. The two factors separate or *segregate* in the gametes (pollen grains or egg cells) so that an individual pollen grain or egg cell contains only one member of the pair. A pair of traits such as red and white color are two alternative forms of the same factor, and red, for some reason, dominates white.

This hypothesis is illustrated in Fig. 3.19B, which shows that heredity depends upon a mosaic of separate and discrete physical particles with a continuing existence in time. The Danish botanist Wilhelm Johannsen christened them *genes* in 1909.

In today's nomenclature, the two alternative forms of the same gene are called *alleles*. The two forms of the color gene, the form yielding red flowers and the form yielding white flowers, are alleles of each other, as are the two forms of the gene for size, the one for tallness and the one for shortness. A *dominant* allele manifests itself in the traits of an organism even when its partner *recessive* allele is present. In the following discussion, we shall indicate a dominant allele by a capital letter and its recessive partner by a small letter (using C or c for color and S or s for size), symbolizing a combination of alleles by a combination of letters. Accordingly, the C allele of the color gene would yield red flowers, and the c allele white. Since, as we have seen, all the offspring in the F_1 generation were red, all must have contained the C allele. Furthermore, they must also have contained the c allele. An organism containing a pair of identical alleles (such as CC or cc) is called a *homozygote;* one with both forms of the gene (such as Cc) is called a *heterozygote* or *hybrid*. Why must the F_1 offspring have been hybrids? Because one parent was white. Since plants containing c cannot be white unless C is absent, this parent must have been a cc homozygote. Since it had only the one

type of allele to give, its F_1 offspring had to contain c. And yet, from its external appearance, a red-flowered plant of the F_1 generation could not possibly be distinguished from its red-flowered parent, which was a CC homozygote genetically.

The important consequence of this concept is this: the detailed pattern of genes actually present in the organism, the so-called *genotype*, cannot necessarily be inferred from its external visible traits, or *phenotype*. Thus we see that a flower whose phenotype is red may arise in a plant whose genotype is either CC or Cc.

Mendel began his experiment with two homozygous plants of genotypes CC and cc and phenotypes red and white, respectively. The alleles were separated in the gametes of the plant so that the CC type produced only gametes containing C, while the cc type produced only gametes containing c. Union of pure populations of C egg cells and c pollen grains (or C pollen grains and c egg cells) can yield only heterozygotes of the Cc type in the F_1 generation; however, because C is dominant, the members of this generation will all be phenotypically red and therefore indistinguishable in external appearance from the red-flowered parent.

What type of gametes will arise from the F_1 heterozygotes when two of them are crossed? Unlike homozygotes, heterozygotes produce two types of gametes, one containing each allele. Thus half the gametes from a Cc heterozygote contain C, and the other half c. This situation permits more than one genotype in the F_2 generation: C cells can combine with both C and c cells, to yield CC and Cc offspring, respectively; and c cells can combine with both C and c cells, to yield cC and cc offspring, respectively. This is what the theory predicts and what was observed in a large number of experimental matings (Mendel used 1000).

The scheme of possible fertilizations is best depicted by the checkerboard shown in Fig. 3.19B. In a single mating the offspring may possess any of four genotypes, and in a large number of matings the four would be equally likely to appear. We would expect them in a ratio of 1:1:1:1 (or, since Cc and cC are the same thing, 1:2:1). If the F_2 generation is large enough, it will contain three kinds of genotypes and two kinds of phenotypes in precisely the following proportions:

		CC	:	Cc	:	cc
Genotypic ratio		1	:	2	:	1
Phenotypic ratio			3		:	1
			red			white

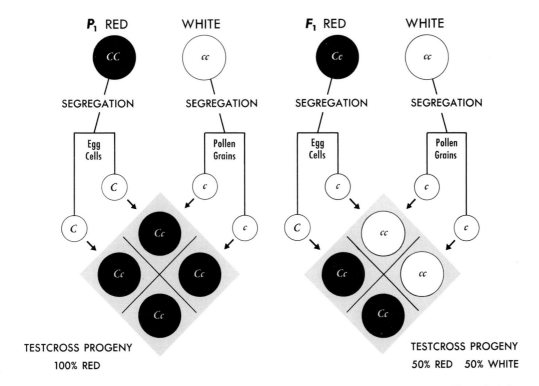

3.20 Test cross designed to determine whether the genotype of F₁ red-flowered plants is *Cc* or *CC*. (From G. G. Simpson and W. S. Beck, *Life: An Introduction to Biology*, 2nd ed., © 1957, 1965 by Harcourt Brace Jovanovich, Inc. and reproduced by permission.)

Mendel devised an ingeniously simple experiment to test this scheme. According to the theory, the red-flowered plant of the parental generation is a *CC* homozygote while that of the F₁ generation is a *Cc* heterozygote. The proof, Mendel realized, lies in the quantitative pattern of the offspring of a large-scale cross between an F₁ red and a white, which, if the theory is correct, is of genotype *cc*. If the theory is correct, the pattern of the offspring must be as diagramed in Fig. 3.20, with half the progeny red and the other half white. Otherwise, the theory must be discarded. When Mendel performed the experiment, 50.3% of the progeny was red, and 49.7% was white! It is from such evidence that we conclude that genes exist and that they are discrete and stable, and it is from such evidence that we have learned all we know of them.

Chromosome theory. From Mendel's experiments and those that followed came much of our information about genes. But the particles were real only in the abstract thinking of scientists. Despite the evidence of their existence derived

from breeding experiments, Mendel never actually saw them; nor did he have the remotest concept of the mechanisms or devices that living cells would need for his proposed schemes actually to operate. How indeed could these hypothetical particles decree or determine that a plant would have red rather than white blossoms?

The essential features of Mendel's theory are that genes occur in pairs, one gene from each parent, and that the two members of a pair segregate or separate in the process of gamete formation. How can these notions be reconciled with what we know of the cell's structure and its mechanism of division?

The insight that wedded these diversities came into the minds of three men almost simultaneously in 1902: W. S. Sutton, Theodor Boveri, and Hugo De Vries. Suddenly it seemed likely that Mendel's hereditary factors were carried in the chromosomes. This would explain why genes occur in pairs—because chromosomes also are paired structures—and why half of a gene pair derives from each parent—because half of a chromosome also derives from each parent.

It would also explain why gametes (egg cells and pollen grains in plants and egg cells and sperm cells in animals) all contain half the number of chromosomes found in the body cells of an organism. Biologists have known for a long time that every species has a characteristic number of chromosomes per cell. Mice have 20, certain lobsters 200, and fruit flies 8. It was generally believed that man has 48 until 1956, when more careful counting techniques showed the correct number to be 46 (see Fig. 3.18). It should be remembered that chromosomes also occur in pairs, which must be distinguished from the chromatid pairs making up the individual chromosomes. Thus man has 23 pairs of chromosomes in each of his body cells but only 23 chromosomes in his egg and sperm cells.

Each chromosome in a gamete is half of a pair of chromosomes in a body cell. If we have a hypothetical organism with two pairs of chromosomes in each of its adult body cells, each of its egg or sperm cells contains only two unpaired chromosomes, as in Fig. 3.21. When an egg cell is fertilized by a sperm cell, the resulting cell contains two pairs of chromosomes, and each of its offspring—i.e., each body cell of the new embryo—will contain the same two pairs of chromosomes. However, each of the new organism's gametes will contain only two unpaired chromosomes.

In referring to the number of chromosomes in a cell, biologists speak of the *ploidy*. A normal body cell contains the *diploid* number of chromosomes, and a gamete contains the *haploid* number. The occasional abnormal body cell with more than the diploid number is said to contain a *polyploid* number.

Mendel's conclusion that a gamete contains only half of a hypothetical gene pair received support from the discovery that a gamete contains only half the usual number of chromosomes in a body cell. And there is nothing abstract about this discovery. If specially treated cells are examined microscopically, the chromosomes can easily be counted (see Fig. 3.18). We can therefore see how the fertilization of a

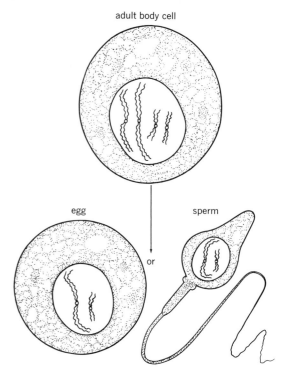

3.21 Segregation of chromosomes in the germ cells.

haploid egg cell by a haploid sperm cell gives rise to a diploid *zygote* (the sperm-fertilized egg cell that is the single progenitor of the entire animal).

We come now to the obverse question. How can haploid gametes arise in an animal or plant whose body cells are all diploid?

Importance of meiosis. In the preceding paragraphs, we have spoken of two general classes of cells: *body* cells (or *somatic* cells) and *germ* cells (or *gametes*). The former make up the bulk of the body mass, and the latter serve exclusively as progenitors of the next generation. This difference has a deeper significance. All the diverse body cells (muscle, nerve, bone, etc.) are descendants of a single sperm-fertilized egg cell, the zygote. Likewise, all the egg and sperm cells descend from the same single zygote. The nineteenth-century biologist August Weismann perceived that the gametes of each generation are direct descendants, through a long line of intermediate cells, of the gametes of the preceding generation; the body cells, though derived from gametes, cannot give rise to them.

A special form of mitosis occurs in the production of gametes that results in a reduction of the diploid number of chromosomes in the dividing cell to the haploid number in each daughter cell. This special form of mitosis, known as *meiosis*, consists of two successive divisions that resemble mitotic divisions except that they are accompanied by only one chromosomal division. The result is a reduction of chromosome number by one-half. Although all tissues of an adult human body, including those of the reproductive organs, are composed of diploid cells, haploid cells are produced in great numbers in the *testes* (male) and the *ovaries* (female). We must conclude that meiosis takes place in certain cells of these organs.

We shall defer a review of the anatomy of the testis until Chapter 16. Here we shall merely note that the testis contains cells called *spermatogonia*. These are the precursors of sperm cells, and they divide by mitosis. The number of spermatogonia remains constant because, when a spermatogonium divides, one of the daughter cells remains a spermatogonium while the other becomes something else (Fig. 3.22); otherwise, the number of spermatogonia would double with each mitosis. The changing daughter cell develops into a *spermatocyte*. It is the spermatocyte that undergoes the two divisions of meiosis.

The cell in which the first meiotic division occurs is diploid since it was derived by mitosis from a spermatogonium. It is the *primary spermatocyte*. The first meiotic division produces two *secondary spermatocytes*, each containing the haploid number of chromosomes. The secondary spermatocytes undergo the second meiotic division, yielding four haploid cells which are gametes, the *spermatozoa*, when they finish growing their tails.

Meiosis is compared to mitosis in Fig. 3.23. In the specialized process of meiosis, only one member of each pair of chromosomes, not both members as in mitosis, is transmitted to each new nucleus. In mitosis in an imaginary cell with two pairs (*bivalents*) of homologous chromosomes, pair A and pair B (each pair containing one chromosome from each parent, A^m and A^f and B^m and B^f), each member of a chromosome pair, behaving independently, moves onto the equatorial plate of the spindle at metaphase.

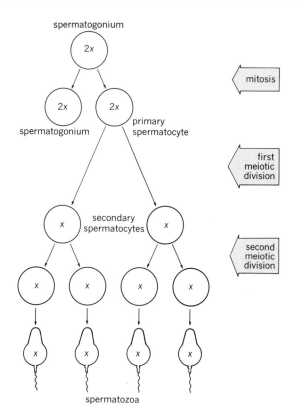

3.22 Meiosis. 2x represents the diploid number of chromosomes, and x the haploid number.

During anaphase each chromosome disjoins (splits in two), and an A^m, A^f, B^m, and B^f centromere and strand move to each pole. There are two notable differences between this sequence of events and that in the first meiotic division: (1) in meiosis the members of a chromosome pair do not behave independently of one another; and (2) in meiosis they do not disjoin but instead move intact toward the poles, one full member of each chromosome pair going to each pole.

It is clear early in the prophase of the first meiotic division (prophase I*) that the cells are

_____

* The highly specialized character of prophase I is evident from its five substages, distinguished on the basis of the appearance and behavior of the chromosomes: *leptotene* (the nucleus enlarges); *zygotene* (synapsis occurs); *pachytene* (the chromosomes thicken and shorten); *diplotene* (splitting of the bivalents begins); and *diakinesis* (the chromosomes migrate to the periphery of the nucleus, and the nuclear membrane disappears).

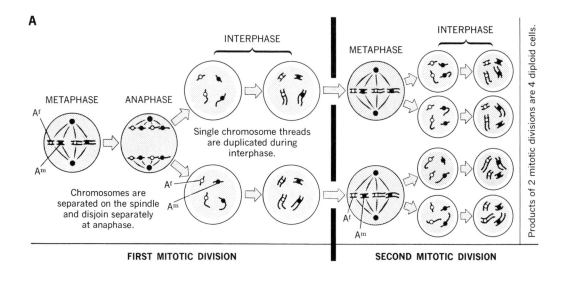

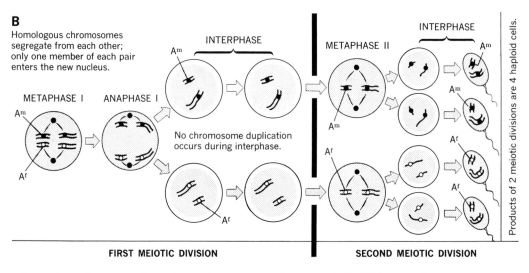

3.23 Comparison of mitosis and meiosis. A. Chromosome cycle in two mitotic divisions. B. Chromosome cycle in two meiotic divisions. (From G. G. Simpson and W. S. Beck, *Life: An Introduction to Biology,* 2nd ed., © 1957, 1965 by Harcourt Brace Jovanovich, Inc. and reproduced by permission.)

behaving in an unusual way, for their chromosomes remain single-stranded as they were at the telophase of the previous mitosis. In mitosis the single strands duplicate themselves during interphase. In meiosis the duplication is delayed, and the curious single-stranded chromosomes begin an unusual maneuver: they pair up, A^m joining with A^f and B^m with B^f. This pairing of

the two single-stranded chromosomes initially constituting a pair of otherwise independent homologous chromosomes is called *synapsis.*

The first meiotic division ensues. Each bivalent with its two centromeres and four chromatids moves onto the equatorial plate (metaphase I), though the arrangement is not as precise as in mitosis. The spindle then separates the bivalents, and anaphase I begins. The intact homologous centromeres, each with its two chromatids, move toward opposite poles. The new nucleus that forms around each end of the spindle therefore contains only one A centromere (Am or Af) and one B centromere (Bm or Bf), so that only one member of a chromosome pair is present.

In the interphase following this division, the cell has single A and B centromeres, each bearing two chromatids. The second meiotic division then begins. It is essentially like mitosis except that the centromeres divide for the first time. Thus the two meiotic divisions produce four cells, each with a haploid number of chromosomes—since each contains chromosome Am *or* Af and Bm *or* Bf (rather than both members of the A pair and both members of the B pair as in diploid cells).

In discussing the chromosome theory, we noted the striking parallel between the fact that genes occur in pairs and the fact that chromosomes occur in pairs. And we indicated that Mendel's hypothesis, which held that half of a gene pair derives from each parent, was substantiated by the discovery that half of a chromosome pair derives from each parent. In our hypothetical cell containing two chromosome pairs (Am with Af and Bm with Bf), the Am and Bm chromosomes arose from the male parent, and the Af and Bf chromosomes from the female parent. The point that should be emphasized is this: in meiotic division each chromosome acts independently, and so the gamete emerging from the second meiotic division that contains Am is as likely to contain Bf as Bm; the orientation of the B pair is quite independent of that of the A pair. Therefore, all possible combinations of chromosomes in gametes are equally probable and frequent: Am and Bm, Am and Bf, Af and Bm, Af and Bf. All that meiosis guarantees is that each gamete produced by an organism contains one member of each pair of homologous chromosomes. It does not guarantee that all the maternal or all the paternal representatives remain together.

Sex and the chromosome theory. We have now seen that cellular reproduction facilitates the transmission from one cell generation to the next of the information necessary for the devel-opment of the next cell generation's pattern of characteristics. An important question remains to be answered, however. What is the signifi-cance of the fact that the cell is usually diploid, carrying two equivalent sets of information, one of which seems to be redundant? A first answer would be that diploidy is a consequence of sexual reproduction, since it results from the fusion of nuclei of the gametes of two parents. Each parent contributes one complete copy of the necessary information. But this leads to a second question. Why would evolution so decisively indicate an almost universal preference for reproduction by sexual union when other perhaps less complicated methods might be possible?

Mendel showed that in crosses involving more than one pair of alleles, each pair behaved independently, sorting itself out in such a manner that a large number of progeny would contain different types in a ratio that would be expected if all possible types of gametes occurred in equal numbers and chance determined the frequency of their union. Mendel's discovery of *independent assortment* or *independent segregation* may be illustrated with an animal-breeding experiment.

We shall consider a cross between a black short-haired guinea pig and a white long-haired one. The actual chromosome number in the guinea pig is 64, but, to simplify, we shall assume that there are two pairs of chromosomes, that black color (*B*) is dominant over white color (*b*), and that short hair (*S*) is dominant over long hair (*s*). Fig. 3.24A shows the chromosome patterns in the cells of black short-haired (*BBSS*) and white long-haired (*bbss*) animals. The black-white alleles are in the long pair of chromosomes, and the short-long alleles in the short pair. The first row of Fig. 3.24B shows the gametes produced by the two parent strains, and below is the hybrid chromosome pattern emerging from a mating of the two strains. It is heterozygous for both pairs of alleles, containing one black and one white and one short and one long.

Fig. 3.24C depicts the event that is crucial to the independent assortment of alleles in the process of mating, the meiotic division of the gametes of the hybrid. Following the pairing or synapsis of homologous chromosomes and align-

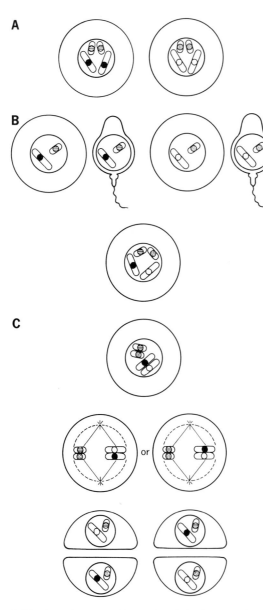

ment on the equatorial plate of the spindle, each chromosome separates from its partner. The orientation of the large pair of chromosomes on the equatorial plate is entirely independent of that of the small pair. Thus arrangements with white up and black down and with black up and white down are equally likely. This means that the allele B is just as likely to enter the same nucleus with s as it is with S. In a large population of cells, *four* types of gametes are produced in exactly equal numbers: black long-haired; black short-haired; white long-haired; and white short-haired. The random assortment of independently segregating alleles in the gametes of so-called "multiple hybrids" is a necessary consequence of the independence of the chromosomes in choosing their positions on the meiotic spindle, which determines which pole they approach.

If a male who produces these four types of gametes mates with a similarly endowed female, 16 types of zygotes are possible (Fig. 3.25). With such a Roman square, it is a simple matter to calculate the frequency of phenotypes in a large number of matings.

Genotypes	Phenotypes
1 *BBSS*	9 Black short-haired, or 56.2%
2 *BBSs*	
1 *BBss*	
2 *BbSS*	3 Black long-haired, or 18.8%
4 *BbSs*	
2 *Bbss*	3 White short-haired, or 18.8%
1 *bbSs*	
2 *bbSs*	
1 *bbss*	1 White long-haired, or 6.2%

When Mendel did this experiment with 561 pea plant matings, using round-wrinkled and yellow-green as the two paired alleles, he obtained the following results.

F_2 phenotype	% expected	% observed
Round yellow	56.2	56.2
Round green	18.8	19.2
Wrinkled yellow	18.8	18.9
Wrinkled green	6.2	5.7

3.24 Cross between a black, short-haired guinea pig and a white, long-haired guinea pig. A. Cells of the two strains, showing black-white alleles as black and white circles in long chromosomes and short-long alleles as shaded and stippled circles in short chromosomes. B. Gametes and hybrid offspring of the strains in A. C. Meiosis in the gamete of the hybrid.

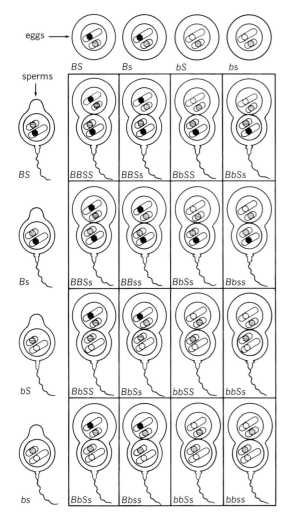

eggs

sperms

	BS	Bs	bS	bs
BS	BBSS	BBSs	BbSS	BbSs
Bs	BBSs	BBss	BbSs	Bbss
bS	BbSS	BbSs	bbSS	bbSs
bs	BbSs	Bbss	bbSs	bbss

3.25 Possible types of zygotes from the cross *BbSs* × *BbSs*. See Fig. 3.24.

Unfortunately the theory (which came to be known as Mendel's *law of independent assortment*) so well supported by these experiments was found to break down in individual cases because of numerous "exceptions." Later analysis of these exceptions led to great new advances, of which we shall speak momentarily.

Our example illustrates the importance of sexual reproduction. Black short-haired and white long-haired guinea pigs mate to produce offspring whose gametes include every possible combination of alleles. If the animals somehow

reproduced asexually, they could produce only black short-haired and white long-haired offspring. Thus sexual reproduction maximizes the possibility of variation among offspring, whereas asexual reproduction keeps variation at a minimum. Those species with the richest variation have been the ones best able to survive great changes in the environment and hence most likely to evolve new patterns when opportunities arose. Sex is almost universal because, like any other widespread biological device, it has favored the long-term survival of those species that possessed it.

There is one aspect of sex to which we have not yet alluded: members of a sexually reproducing species must be of one sex or the other. Since all traits of a living organism are determined by its gene-carrying chromosomes (and surely we agree that maleness and femaleness are traits), we may reasonably ask what chromosomes determine sex?

It was early found that there is a regular difference between male and female chromosome patterns in the fruit fly *Drosophila melanogaster*. In its relatively uncluttered field of four chromosome pairs, it was easily seen that despite the similarity in males and females of the two large V-shaped chromosome pairs and the single small dotlike pair, the fourth pair consists of paired rods in females and of a rod plus a curious J-shaped chromosome with its centromere near the bend in males (Fig. 3.26). A con-

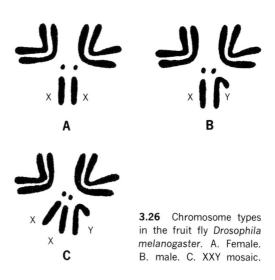

3.26 Chromosome types in the fruit fly *Drosophila melanogaster*. A. Female. B. male. C. XXY mosaic.

vention was adopted whereby the female fourth chromosome pair was termed the XX pair, while the corresponding male pair was labeled XY, the Y referring to the peculiar J-shaped member.

It was quickly recognized that these chromosome pairs have far more significance than as simple accompaniments of sex. They are, in fact, *determinants* of sex. Accordingly, all female cells contain the XX pair, and all male cells contain the XY pair. After meiosis, all egg cells contain X, but half the sperm cells contain X and the other half Y. If an X sperm fertilizes an egg, the zygote is XX, and the offspring will be female. If a Y sperm fertilizes the egg, the zygote is XY, and a male offspring will result. Thus sex determination, for 2000 years the subject of vague speculation and metaphysical dogma, is finally explained by the chromosome theory. Normally sex is determined at the moment of fertilization, and the decisive factor is the presence in the sperm of an X or a Y chromosome.

The XX-XY situation also applies to man, and in recent years it has had some interesting practical implications. *Sex chromatin bodies* (also called *Barr bodies* after their discoverer, M. J. Barr) are small (1 μ diameter) clumps of chromatin near the peripheries of the interphase nuclei of most cells in the female body (Fig. 3.27), though many cells show them poorly because their nuclei are small. Mary Lyon suggested in 1961 that one of the two X chromosomes in each cell of a female body is somehow inactivated. It becomes smaller and more aggregated in appearance, takes a darker stain, and synthesizes its DNA later than its partner. This late-replicating X chromosome is considered responsible for a sex chromatin body. "Lyonization" of one or the other X chromosome occurs early in embryonic life in each cell of the developing body. Thus the cells can be divided into two groups, in which one or the other X chromosome remains active.

A substantial number of human beings are born who, because of abnormalities either in the development of the gametes or in the fertilization process, are of indeterminate sex. These individuals have sex organs that are neither male nor female exclusively; for example, they may possess small testes as well as ovaries. Their disorders are due to abnormalities of the sex chromosomes. Fig. 3.26 shows a chromosome pattern sometimes seen in the fruit fly. This XXY type presumably arises in one of two ways: (1) the chromosomes of a female gamete undergoing meiosis fail to disjoin, and so one daughter gamete gets XX, the other gets nothing, and a Y sperm fertilizes the abnormal XX egg to produce XXY; (2) the chromosomes of a male gamete undergoing meiosis fail to disjoin, and so one sperm gets XY, the other gets nothing, and the XY sperm fertilizes a normal X egg to produce XXY. Cells bearing both male and female chromosomes are called *mosaics*. Interestingly, chromosome analysis of human subjects of doubtful sex has revealed that some of them are XXY mosaics (Chapter 16). Thus they exhibit aneuploidy.

In such cases, determination of the *genetic sex* by establishment of the presence or absence of sex chromatin bodies or by detailed study of metaphase chromosomes may be an important asset in the planning of hormonal, surgical, or psychiatric treatment. Work of this kind is increasingly significant in research on human genetics and genetically determined disease.

Linkage. The independent assortment that Mendel observed when he followed the inheritance of two pairs of genes turned out to be an exceptional case. It is true that two gene pairs segregate independently when they are on separate chromosomes. But we know that each chromosome carries many genes. Hence, where the chromosome goes, its genes go. They remain together through meiosis and in the gametes and in the fertilized egg. Each "chromosome-load" of genes segregates as a unit just as though it consisted of a single pair of alleles with many different effects. In fact, we observe a single 3:1 phenotypic ratio in the F_2 generation for all of these genes together. We say that such genes are *linked* because they are inherited as a single package.

In breeding experiments with fruit flies, four separate sets of genetic traits always appear together. These are called *linkage groups,* and the fact that there are four such groups in the fruit fly has obvious significance, since the species has four pairs of chromosomes.

The many known alleles of the fruit fly include the following pairs: (1) normal wings versus curved wings (normal is dominant over

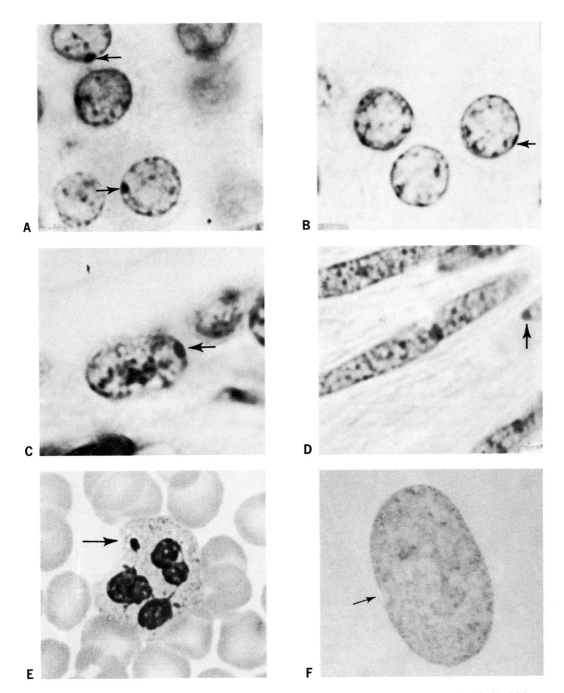

3.27 Appearance of sex chromatin in different cell types. Sex chromatin is indicated by arrows. A. Nuclei in zona fasciculata of male adrenal cortex. B. Nuclei in zona fasciculata of female adrenal cortex. C. Follicular cell of thyroid, female. D. Nuclei of smooth muscle cells, female. E. Nuclei of epithelial cells of the buccal mucosa. F. Fibroblasts from skin, female: (A–D from K. L. Moore, *Acta Anatomica, 21:* 197, p. 205 (1954). E from W. M. Davidson and D. R. Smith in Overzier, ed. *Intersexuality,* Academic Press, New York, 1963. Used by permission of Thieme Verlag, Stuttgart. F from Ursula Mittwoch, *Sex Chromosomes,* Academic Press, New York, 1967.)

curved, and the F_2 generation of a cross between normal and curved has 3 normal:1 curved); (2) normal legs versus bowlegs (normal is again dominant, and the F_2 generation has 3 normal:1 bowed); and (3) normal red eyes versus purple eyes (red is dominant, and the F_2 generation has 3 red:1 purple). These three genes can occur separately, each segregating independently in the usual way. Some flies have all three unusual traits: curved wings, bowlegs, and purple eyes. When we cross such a fly with one having normal wings, legs, and eyes—the common garden variety, or so-called *wild type*—the F_1 generation is all wild type, for all the normal traits are dominant. If the three pairs of traits are based upon three pairs of alleles located on different chromosomes, we would expect the F_2 generation to consist of eight phenotypes in a ratio of 27:9:9:9:3:3:3:1. But what is actually found upon mating one of the curved-winged, bowlegged, purple-eyed flies with a wild type is a 3:1 ratio of wild type to curved-winged, bowlegged, purple-eyed type. In other words, the three alleles remain together; they are linked. Despite this fact, their separate existence in other cases proves that they are from three different allele pairs and not merely different expressions of a single gene. We conclude, therefore, that these three pairs of alleles are located in the same pair of chromosomes.

It is of interest that a sex chromosome carries, in addition to the genes directly concerned with sex determination, a number of genes having no apparent connection with the process. Nevertheless, both types constitute one linkage group because both are carried by one chromosome. From this we can infer that certain genetic traits having nothing to do with sex will appear only in males or only in females.

Our inference is correct. An example is given in Fig. 3.28, which diagrams a cross between a white-eyed fruit fly and a red-eyed fruit fly. In one case the female parent is white-eyed, and the male red-eyed; and in the other the sexes are reversed. It is obvious that the traits displayed in the F_1 generation are entirely dependent on whether the white-eyed parent is the father or the mother. This is quite unlike the situations we have discussed previously. Furthermore, the distribution of eye color in the F_2 generation also depends on which parent had white eyes.

The figure shows how the results of such an experiment can be explained by the theory of *sex-linked heredity*. Every male derives its X chromosome from its mother; it receives from its father only the Y chromosome, which presumably is devoid of genes affecting eye color.* On the other hand, every female receives one X chromosome from its mother and one from its father. Since red eye color is dominant over white, all daughters of a cross between a white-eyed female and a red-eyed male must be red-eyed; all sons must be white-eyed because both the mother's X chromosomes carry the recessive allele.

Perhaps the most famous example of sex-linked heredity in humans is found in the family trees of patients with hemophilia, a disease characterized by a bleeding tendency due to defective blood coagulation. The genealogy of a hemophiliac family reveals that only males exhibit the disease although it is evidently transmitted from generation to generation by females. If the gene causing hemophilia travels on an X chromosome, and if the disease becomes clinically apparent only when a normal X chromosome is absent, then only males can show the disease. (If the X of the male XY pair is defective, the disease manifests itself. If one X of the female XX pair is defective, there is no sign of the disease, though it will appear in males of the next generation.) All the children of a normal mother and a hemophiliac father will be outwardly normal; the sons receive their only X chromosome from the mother, who has two normal X's. The male children of a carrier mother and a normal father will be afflicted. Only one combination, a carrier mother and a hemophiliac father, can yield a hemophiliac female. This is clearly an unlikely event, but it has occurred.

The presence of only one active X chromosome in each body cell of a female (see p. 96) probably explains a curious aspect of one sex-linked trait in man, the common type of color blindness: some females carrying the gene are seriously affected by it, whereas others are affected slightly or not at all. The gene is on the

---

* This presumption illustrates the generalization that Y chromosomes carry few or no genes. Some insects, such as grasshoppers, have dispensed entirely with Y chromosomes, with the male chromosomes being X and no X rather than X and Y.

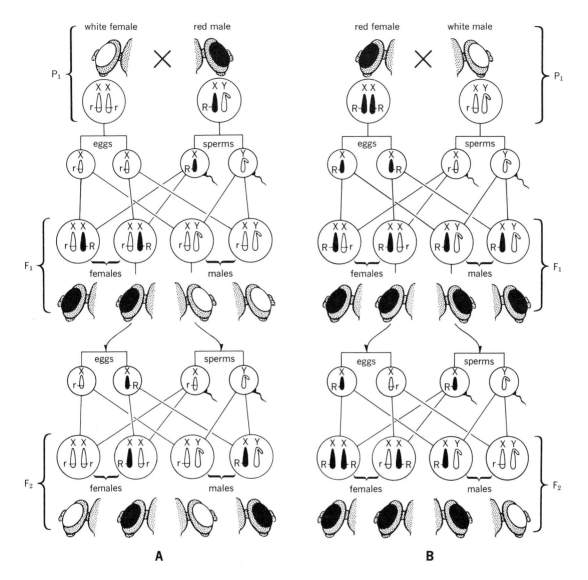

3.28 Sex-linked heredity. A. White-eyed female × red-eyed male. B. Red-eyed female × white-eyed male. (From G. G. Simpson and W. S. Beck. *Life: An Introduction to Biology*, 2nd ed., © 1957, 1965 by Harcourt Brace Jovanovich, Inc. and reproduced by permission.)

X chromosome. Since this may derive from either the mother or the father, the eye of a woman heterozygous for the gene may contain patches of color blind light receptors alternating with patches of normal ones.

Crossing over: mapping the chromosome. Experiments designed to determine which traits

are members of linkage groups and which are not occasionally had unexpected results. These were disturbing to theorists who had come to visualize the phenomenon of linkage in exactly the manner we have outlined. In experiments on corn, for example, repeated test crosses showed that the gene for kernel color and the gene for kernel texture were linked, *colored* generally

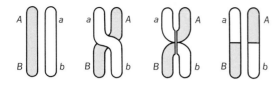

3.29 Exchange of parts by a pair of homologous chromosomes.

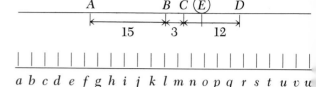

resents a chromosome carrying genes A, B, C, and D in the pattern shown.

appearing with *smooth* and *colorless* with *shrunken*. However, there were consistent exceptions to these pairings. In a typical experiment, 1.75% of the F_2 generation had colored shrunken kernels, and 1.75% had colorless smooth kernels. How could these exceptions arise if the concept of gene linkage is correct?

Discovery of an answer to this question was a brilliant achievement, partly for the elegance of its insight and partly for the massive support that it added to the basic theories of genetics. A close study of the behavior of chromosomes in the prophase of the first meiotic division revealed the reasons for the exceptions.

It was found that when the homologous chromosomes pair up in meiosis, the two homologous partners often exchange parts. This exchange of parts between homologous chromosomes in meiosis is called *crossing over*. It is represented in Fig. 3.29. What happens is literally the physical breakage of a chromosome and replacement of part of its structure with part of another chromosome.

The fact that the F_2 generation in the preceding corn-breeding experiment *regularly* contains 1.75% of each "irregular" type leads to another concept of theoretical significance. Why, we may ask, does colorless appear with smooth in only 1.75% of the F_2 generation? Why not 5% or 10%? Or why not 1.75% in one experiment and 10% in another? The answer to this question was discovered by Thomas Morgan between 1910 and 1915. Morgan recognized that each chromosome is a long thread on which the genes are located. Each gene, he reasoned, has a definite *fixed* position on the chromosome, and crossing over takes place constantly. Therefore, the frequency of crossing over between two given genes is proportional to the distance between them.

Suppose that the following straight line rep-

Let us assume that crossovers can occur by chance with equal probabilities at points a, b, c, and so forth. We see that the frequency of crossovers between A and B will be five times as great as that of crossovers between B and C. Exactly this is found experimentally. Armed with this concept (which gives us an experimental tool for comparing the distances between different pairs of genes on a chromosome), we can now set about determining where any gene is located on its chromosome.

Suppose that we wish to map the location of a gene E on this chromosome. We find that the frequency of crossing over between A and B is 10%, that between B and C is 2%, that between C and D is 8%, and that between D and E is 6%. Therefore, E could be six units to the right of D or six units to the left of D. How do we determine which? By measuring the frequency of crossing over between C and E, which proves to be 2%. We conclude, then, that E is located between C and D, and we know where.

Such experiments have shown the positions of hundreds of genes in the fruit fly and have established the concept that each gene has a specific unchanging locus on a given chromosome. We know little of the topography of human chromosomes, mainly because controlled breeding experiments with large litters are impossible. But we can assume that there is a characteristic human gene pattern even if investigators have so far failed to map it.

Mutation. Mendel's theory holds the genes of each species to be permanent and unvarying—genic stability is what guarantees that the offspring will be of the same species as its parents. But it fails to explain how different strains of the same species have arisen. The theory of evolution rests on the proposition that a species somehow develops variations that may improve its chances for survival and thereby alter its character. How can these two views be reconciled? The answer is by the concept of *mutation*, the occasional chance alteration of a gene.

When a gene mutates, it is abruptly and permanently changed in an all-or-none fashion. And, as we know of genes only by the results of their actions, by the so-called "test of progeny," so we know of mutant genes only from the appearance in the progeny of altered traits, which are transmitted to all subsequent generations.

Mutations are rare events, but in 1927 H. J. Müller found in experiments on fruit flies that the frequency of mutations can be artifically increased by exposure of egg or sperm cells to x-irradiation; the higher the dose, the greater the frequency of induced mutations. This discovery, in addition to providing a potent new tool for the experimental geneticist, led to the realization that the rare spontaneous mutation—caused primarily by the stray quanta of radiation always present in the environment—is the operating factor in evolution, producing the variation about which Darwin wrote. Thus is it possible to reconcile Darwin's concept of biological variation with Mendel's concept of genic stability. The gene *is* permanent unless it mutates. The mutant gene, once formed, is also permanent except for the small but definite probability that it will mutate back to its original state. For despite their rarity, mutations occur in cell populations with regular and predictable frequencies.

Perhaps the most important implication of the concept of mutation is the fact that mutations undoubtedly take place at a regular frequency in both the germ cells and body cells in man. However, for at least three good reasons, most mutant cells are hidden from notice so that the frequency of mutation appears to be lower than it actually is. First, many mutations are lethal in that they alter a cell so as ultimately to prevent its reproduction and therefore their own transmission to later generations. Second, many mutant genes are recessive and do not express themselves phenotypically. They are nevertheless handed down through the generations, and when a mating brings them together with identical recessive genes to form homozygous combinations, they manifest themselves in the phenotype. Third, many nonlethal muta-

tions cause changes so subtle that they are lost in the welter of nongenetic variations always produced in any organism by environmental influences. The latter, of course, are not transmitted to the offspring. In many cases a subtle mutation does not reveal itself because the environment does not force it to. If the environment changes, however, the mutation may become evident.

This last situation is best explained with an example. It was found during World War II that a new antimalarial drug, primaquine, produced severe and occasionally fatal blood disorders in a few individuals. The suspicion arose that there was a genetic basis for the adverse reactions since they occurred mainly in Negroes. Following the demonstration that approximately 10% of American Negroes and only a small number of Caucasians reacted unfavorably to primaquine, it was shown that these particular individuals had inherited a specific enzyme deficiency that rendered them vulnerable to the drug's harmful effects. Their red blood cells contained abnormally low levels of the enzyme glucose-6-phosphate dehydrogenase. This enzyme deficiency appeared to be due to a sex-linked gene.

As far as anyone knows, individuals with this enzyme deficiency were thriving until the environment was altered by the introduction of a new chemical agent. The mutation leading to the deficiency may have occurred many thousands of years ago, but it did not decrease the organisms' chances of survival in the congenial environment of that time. Only in the new environment did the mutation express itself phenotypically and prove to be injurious.

There is an important lesson in this example. We can never declare unequivocally that a given mutation is advantageous, disadvantageous, or neutral unless we can precisely predict what environmental changes may occur in the future. The mutation that made certain individuals sensitive to primaquine was not disadvantageous until primaquine had been synthesized and administered to them experimentally. A geneticist of 1000 B.C. could not have predicted the discovery of primaquine; hence, he could not have foreseen whether the mutation was harmful, beneficial, or neutral.

A growing list of mutations is known to produce in man either definite diseases or abnormalities under certain environmental conditions (obviously the abnormalities are not too severe to permit reproduction). It is likely that many unusual clinical phenomena, such as idiosyncratic drug reactions, are due to mutations. When these curiosities occur mainly in highly inbred ethnic groups, the chances of an underlying genetic mechanism increase. It is evident that gene-controlled enzyme deficiencies could explain why, despite identical exposures, only particular individuals are affected.

As for the role of mutations in evolution, we should bear in mind that there is no need for them to be *predominantly* advantageous in order for evolution to take place. Actually, far more mutations are harmful than helpful, since the probability is that any change in a system as highly ordered as the living organism will be for the worse instead of for the better. It is only necessary (1) for mutations to be highly diverse, so that occasionally a gene of a slightly improved type develops; and (2) for genes or groups of genes to be capable of accumulating mutations up to a practically unlimited degree of complexity. It must then follow that the extremely rare superior type, by virtue of its enhanced ability to survive and multiply, will tend to increase in number, thus providing a basis for further progress.* Through such a sequence of mutation supervening upon mutation, man arose on earth.

MECHANISM OF GENE ACTION

We have now seen how the units of heredity are transmitted to the offspring, how they are recombined and altered, and how certain regularities observed in breeding experiments can thereby be explained. We must now consider

---

* Bacterial cultures illustrate well the consequences of small mutational advantages. We may consider, for example, the effect of some unspecified genetic event that causes strain A to grow 1% faster than strain B in a given medium. If 10 cells each of A and B are inoculated together into a flask and allowed to undergo 30 successive divisions, a process that in many bacteria would take less than 15 hours, the differential of 1%, compounded 30 times, yields a ratio of A to B in the flask of 1.35:1. If a small amount of this culture is transferred to fresh medium and the process repeated every 15 hours, a differential growth advantage of 35% occurring with each transfer, not many transfers will be required before A outgrows B completely.

how the existence of a given trait is determined by gene action.

In recent years it has become apparent that the information-containing portion of a gene consists of DNA and that genes act primarily by determining the structures and rates of synthesis of specific proteins. In late 1961 discoveries were made of the form in which information is coded in the DNA molecule, the method of communicating the coded message to the sites of protein synthesis, and the mechanism by which the amino acid sequences upon which specificity depends are thereby established. Some of the evidence underlying these new and epoch-making concepts of gene action follows.

One gene, one enzyme hypothesis. Once a living organism has been conceived and its future traits have been determined, it becomes a self-regulating metabolic machine that is engaged primarily in extracting energy from nutrient materials and converting them to other compounds needed for growth and maintenance. We may illustrate the essential character of metabolism quite simply.

$$A \xrightarrow{e_1} B \xrightarrow{e_2} C \xrightarrow{e_3} D \xrightarrow{e_4}$$
$$E \xrightarrow{e_5} F \qquad (3.1)$$

A is a dietary ingredient or its derivative; F is a compound that the organism needs for some necessary purpose. The intermediate compounds are products and precursors in an orderly sequence of stepwise chemical reactions, each of which is catalyzed by a specific enzyme (designated e_1, e_2, etc.).

We may regard every trait of an organism as a visible consequence of one or another such sequence of enzyme reactions. For example, there are genes that cause brown skin; but skin becomes brown because a pigment is synthesized from simpler materials by a sequence of enzyme reactions in the skin cells. All other traits, such as height, hair color, and blood group, are similarly attributable to specific sequences of enzyme reactions.

The work of George Beadle and Edward Tatum, published in the 1940's, led to one of the first great simplifications in our concepts of how genes act: the *one gene, one enzyme hypothesis.*

These investigators studied a common red mold of bread, *Neurospora crassa*. In a typical experiment they examined organisms that were fully capable of synthesizing the amino acid arginine, the final steps of whose synthesis are known to be as follows:

$$\text{dietary carbon compounds} \longrightarrow \cdots \xrightarrow{e_1}$$
$$\text{ornithine} \xrightarrow{e_2}$$
$$\text{citrulline} \xrightarrow{e_3} \text{arginine} \qquad (3.2)$$

They found that if they exposed the organisms to x-irradiation, many of the offspring would not grow in a medium containing the usual dietary carbon compounds. These offspring could be divided into three groups on the basis of their growth response upon the addition of ornithine, citrulline, and arginine to the medium.

Group	Growth response in medium containing			Position of block in synthetic sequence
	ornithine	citrulline	arginine	
1	+	+	+	e_1
2	−	+	+	e_2
3	−	−	+	e_3

Evidently the radiated organisms still required arginine, but radiation had eliminated their ability to produce it from ordinary dietary carbon compounds such as glucose.

The absence (or nonfunction) of enzyme e_1 in group 1 was indicated by the normal growth in a medium containing ornithine (proving that the ornithine → citrulline → arginine sequence is intact); the absence of e_2 in group 2 was indicated by the normal growth in a medium containing citrulline (proving that the citrulline → arginine step is intact) and the lack of growth in a medium containing ornithine (proving that the ornithine → citrulline step is defective); and so on. Moreover, organisms of group 3 accumulated citrulline, and those of group 2 accumulated ornithine. When Beadle and Tatum directly analyzed the groups for e_1, e_2, and e_3, they found enzymes missing exactly as predicted from the nutritional patterns. If growth and reproduction were sustained by the artificial introduction of the product of a missing reaction, all subsequent generations displayed the same enzyme deficiency as their parents and hence were in perpetual need of a nutritional supply of the product.

The implications of this classic experiment are obvious. A mutation, induced by radiation, produces a strain lacking one enzyme. Such a strain is called an *auxotrophic mutant* because, through a single gene mutation, it has acquired one more nutritional requirement than the *prototrophic* parent strain. As long as the environment provides an ample supply of the compound that the organism can no longer make for itself, the heritable lack of an enzyme does not threaten its survival.* On the basis of many similar data, Beadle and Tatum postulated that a single gene governs the synthesis of a single specific enzyme—in other words, one gene, one enzyme.

Genetic significance of DNA. The experiments of Beadle and Tatum and of many later workers indicated that a gene acts as a kind of master pattern that somehow stamps specificity into the processes of protein synthesis. That the pattern is carried in the nucleotide sequence of the DNA molecule was established by a series of important scientific discoveries.

The very presence of DNA in chromosomes suggests such a function,† but the early data of viral and bacterial genetics greatly strengthened the presumption. Certain viruses were shown to

---

* For example, vitamin B_1 (thiamine) is just as necessary in *Neurospora* metabolism as in human metabolism, but *Neurospora* can make it from simpler compounds, whereas man cannot (thus human beings are vitamin B_1–requiring auxotrophs). Since vitamins are defined as essential substances that an organism cannot make for itself, thiamine is a vitamin for man but not for *Neurospora*. We may reasonably speculate that our remote evolutionary ancestors could make thiamine but that, through mutation, the enzymatic machinery for its synthesis was lost. Had there been no organisms in the environment to make it for the first mutants, the species might have perished then and there.

† Every living nucleus contains DNA, in proportion to the number of chromosomes, the DNA level of a haploid cell (e.g., egg or sperm) being half that of a diploid cell. The quantity of DNA doubles during mitotic interphase, just prior to cell division.

consist structurally of little more than minute quantities of DNA and an outer protein coating. It was found that when these viruses enter a host cell (as they must to reproduce), the protein remains outside and the DNA alone is injected into the cell, where it takes over the metabolic machinery, instructing it when and how to begin producing new viruses; hence viral DNA alone must carry the code that guarantees the similarity of new viruses to their parent.

Even stronger evidence that DNA is the actual genetic material comes from studies of the *transformation* of certain bacteria. Cultures were grown of two strains of the same bacterial species that differed in only one phenotypic trait: one was penicillin-sensitive, and the other penicillin-resistant. When purified DNA from the penicillin-resistant bacteria was added to a growing culture of penicillin-sensitive bacteria, the treated culture developed many colonies of penicillin-resistant bacteria. Moreover, their offspring were also penicillin-resistant; therefore, the newly acquired trait was hereditary. Since the *transforming principle* was DNA, the results suggest that donor-cell DNA entered a chromosome of a recipient cell, where it behaved like a new gene. Thus DNA must be the repository and conveyor of genetic information.

Replication of DNA. The discovery by Watson and Crick that DNA is a double-stranded molecule containing complementary base pairings held together by hydrogen bonds (see Chapter 2) disclosed a likely mechanism for the transfer of genetic information from generation to generation. In mitosis the two strands would separate, but each could serve as a template for the formation of a new partner strand, since each base in the chain would accommodate only the correct complementary base. Thus the daughter DNA would become double-stranded like the parent DNA, and the correct nucleotide sequences would be carried over into the new generation (Fig. 3.30).

Although this picture of the replication of DNA was first offered speculatively after the discovery of Watson and Crick, some critical experiments soon confirmed its accuracy. In one, J. H. Taylor, P. S. Woods, and W. L. Hughes made the DNA of bacteria radioactive by feeding the organisms radioactive thymidine.

They used tritium as the radioactive label because of its extremely low energy: when radioautographs were prepared (see p. 59), minute objects containing tritium exposed only the immediately adjacent particles of photographic emulsion, so that radioactive and nonradioactive fine structures could be distinguished with great sensitivity. When organisms containing fully labeled DNA were allowed to divide in a nonradioactive medium, half of the radioactive DNA present in each chromosome before replication appeared in the corresponding chromosome of each daughter cell.

M. Meselson and F. Stahl performed a similar experiment (Fig. 3.31), using heavy nitrogen, ^{15}N, a nonradioactive isotope of nitrogen, as a label for DNA and ascertaining its presence in the daughter DNA molecules by an ingenious technique of ultracentrifugation. Bacteria for the experiment were grown for many genera-

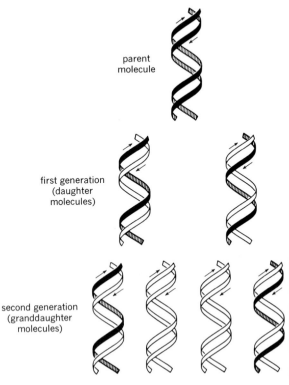

parent molecule

first generation (daughter molecules)

second generation (granddaughter molecules)

3.30 Proposed model of DNA replication. (From Meselson and Stahl, *C. S. H. Sym. Quant. Biol.,* 23, 10, 1958.)

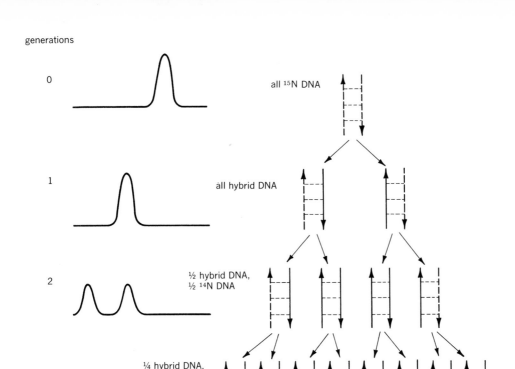

generations

0

1

2

3

all ^{15}N DNA

all hybrid DNA

½ hybrid DNA,
½ ^{14}N DNA

¼ hybrid DNA,
¾ ^{14}N DNA

^{14}N ^{14}N/^{15}N ^{15}N

Density scale

3.31 Chromosome duplication experiment of Meselson and Stahl.

tions in a medium containing ^{15}N as the only nitrogen, so that all of the ^{14}N normally present in the cells, including that of DNA, was replaced by ^{15}N. Then, the ^{15}N in the medium was replaced by ^{14}N, and the bacteria were permitted to multiply in the new medium for exactly two generations. At the beginning of the experiment and at various later times, the DNA molecules were isolated and characterized as to ^{15}N and ^{14}N content. Only three types of DNA molecules were found: those containing only ^{15}N, those containing only ^{14}N, and those containing both ^{15}N and ^{14}N (hybrid molecules).

Meselson's and Stahl's data supported the scheme shown in Fig. 3.30. Initially all the DNA molecules contained nitrogen in the form of ^{15}N. After doubling of the cell number, all the DNA molecules were hybrids, whose nitrogen was

half ^{14}N and half ^{15}N. When some of these were heated under conditions that break hydrogen bonds, pure single-stranded ^{14}N-DNA and ^{15}N-DNA molecules were obtained. After a second doubling of the cell number, equal amounts of normal ^{14}N-DNA molecules and hybrid DNA molecules were present. These results confirm that the DNA molecule contains two units (i.e., chains), one of which goes to each of the two daughter cells during cell division and serves as a template, or "primer," for the synthesis of a complementary new one (see Chapter 2).

Soon after the discovery of the double-strandedness of DNA, Arthur Kornberg discovered the enzyme that catalyzes the polymerization of DNA from its four component deoxyribonucleotides (in the form of their triphosphate derivatives), *DNA polymerase*. As might be anticipated, a strand of fully formed DNA is needed in this reaction to act as an information-containing, sequence-ordering template:*

x molecules of A·dR·PPP
x molecules of T·dR·PPP
y molecules of G·dR·PPP $+$ DNA (template) $\rightleftharpoons$
y molecules of C·dR·PPP

$$\begin{pmatrix} x\text{A·dR·P} \\ x\text{T·dR·P} \\ y\text{G·dR·P} \\ y\text{C·dR·P} \end{pmatrix} \cdot \text{DNA} + (2x + 2y)\text{PP} \quad (3.3)$$

It is significant that polymerization does not occur unless all four deoxyribonucleotides are present and that the DNA produced and the

---

* As noted in Table 2.6, the letters A, G, C, U, and T denote adenine, guanine, cytosine, uracil, and thymine, respectively. The letters R and dR denote ribose and deoxyribose, respectively. Hence A·dR·P is deoxyadenosine monophosphate (i.e., deoxyadenylic acid), and A·dR·PPP the corresponding triphosphate; and A·R·P is adenosine monophosphate (i.e., adenylic acid), and A·R·PPP adenosine triphosphate.

template have the same base ratio—and presumably the same base sequence, though this has not yet been proved in the laboratory.

The template DNA must be single-stranded. How double-stranded DNA is converted into single-stranded DNA in the cell (in contrast to the test tube) has not been established. According to one theory, duplication begins at one end of the DNA molecule and by itself separates the strands, rotating and unwinding and lengthening the daughter strand and shortening the parent strand (Fig. 3.32).

How DNA controls protein synthesis: messenger RNA. A few years after the discovery of DNA polymerase, Samuel Weiss and Jerard Hurwitz discovered a similar enzyme that catalyzes the synthesis of RNA, *RNA polymerase.* Interestingly, a DNA template is also necessary in this reaction. However, it must be double-stranded DNA. Although the DNA and the newly formed RNA combine into short-lived hybrids, the RNA is eventually released, and the DNA can be recovered.

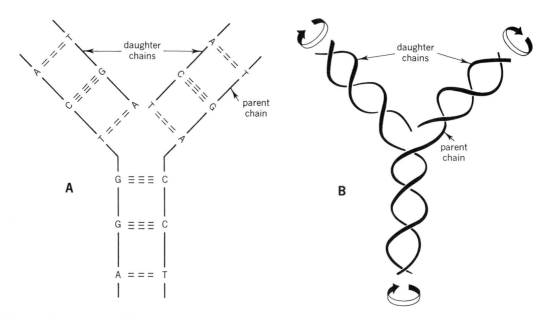

3.32 Replication of the DNA molecule, showing how parent double-stranded helix unwinds to provide separate single-stranded templates for the two daughter strands.

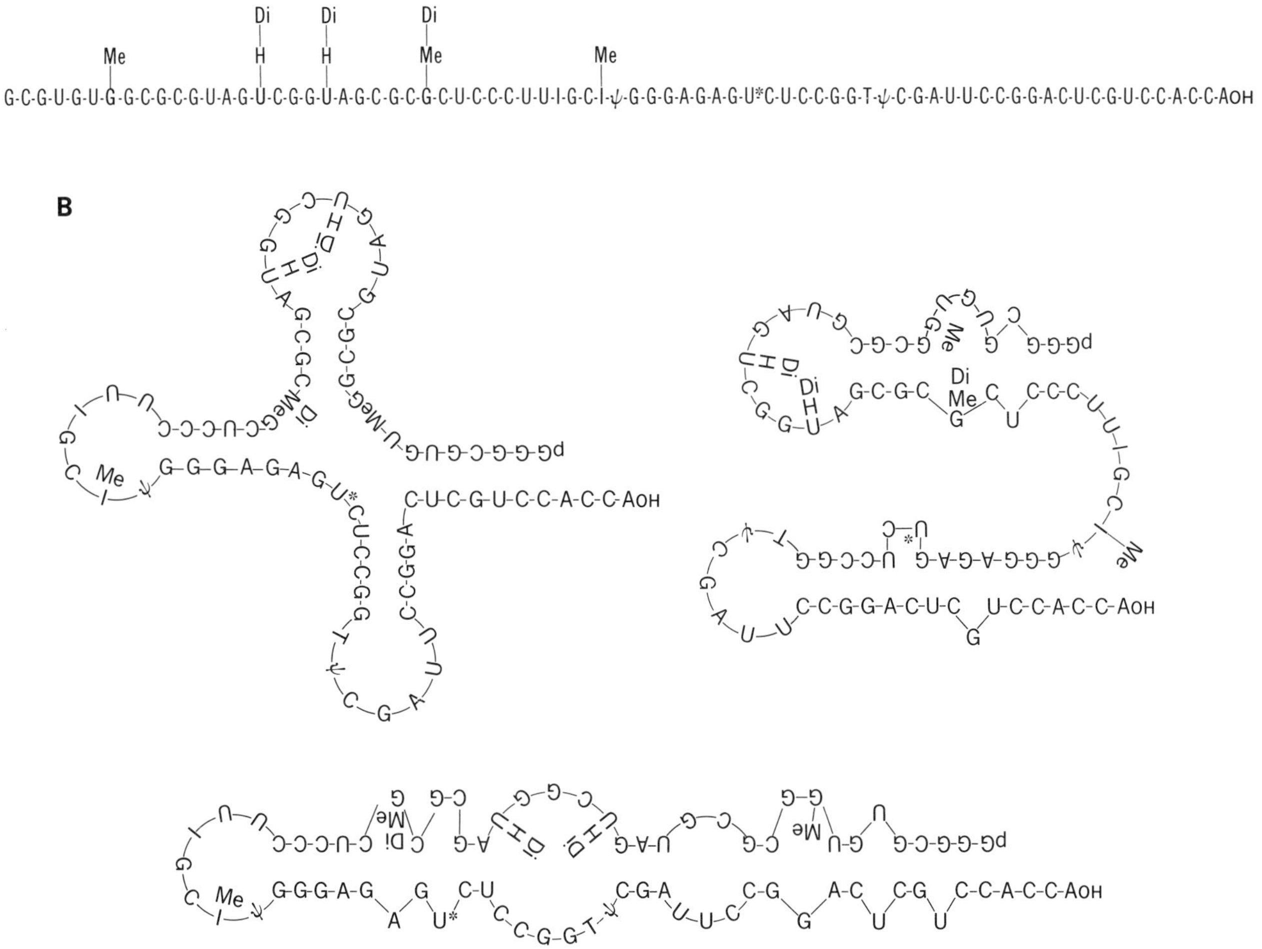

G-C-G-U-G-U-G-G-C-G-C-G-U-A-G-U-C-G-G-U-A-G-C-G-C-G-C-U-C-C-C-U-U-I-G-C-I-ψ-G-G-G-A-G-A-G-U*-C-U-C-C-G-G-T-ψ-C-G-A-U-U-C-C-G-G-A-C-U-C-G-U-C-C-A-C-C-AoH

3.33 Structure of alanine tRNA. A. Nucleotide sequence, with A, G, U, and C representing the four major nucleotides and other letters representing nucleotides with unusual bases (these so-called minor constituents occur mainly in tRNA, and their function is unknown). B. Three proposed conformations of alanine tRNA, based on possible complementary base pairings in short double-stranded regions. (From R. Holley, *et al., Science, 147,* 1462, 1965.)

a molecules of $A \cdot R \cdot PPP$
b molecules of $U \cdot R \cdot PPP$
c molecules of $G \cdot R \cdot PPP$ $+$ DNA (template) $\rightleftharpoons$
d molecules of $C \cdot R \cdot PPP$

$$\begin{pmatrix} aA \cdot R \cdot P \\ bU \cdot R \cdot P \\ cG \cdot R \cdot P \\ dC \cdot R \cdot P \end{pmatrix} + \text{DNA} + (a + b + c + d)\text{PP} \quad (3.4)$$

The requirement for DNA in RNA synthesis clearly suggests the following means of transmitting the information in the DNA nucleotide sequence from the nuclear genes to the cytoplasmic ribosomes, the sites of protein synthesis.

DNA directs the synthesis of RNA in much the same way that it directs its own replication. The RNA is complementary to only one of the two DNA strands and is itself single-stranded.* In this way the coded instructions for protein synthesis contained in one DNA nucleotide sequence are duplicated in the nucleotide se-

---

* If RNA and a single strand of DNA form a double-stranded RNA-DNA hybrid when mixed in vitro we conclude that long stretches of the two nucleotide sequences are complementary. Hybrid formation in vitro is the basis of a useful laboratory test for sequence complementarism—even when the sequence is unknown.

quence of the DNA-determined RNA strand. The RNA then carries these instructions from the nucleus to the ribosomes. The existence of this *messenger RNA* (abbreviated mRNA) and its template function were postulated in early 1961 by Jacques Monod and François Jacob, and its existence was confirmed experimentally a few months later by Hurwitz and Paul Berg.

Protein synthesis begins at the ribosomes only after another series of events has taken place. First the individual amino acids must be brought to the ribosomes. Crick and others speculated that an "adaptor" molecule of some sort accomplishes this task. The adaptor was found in 1957 to be another special type of RNA, much smaller than mRNA. Because of its function, it was

termed *transfer RNA* (abbreviated tRNA). Different tRNA's exist for each of the 20 amino acids. In 1965 Robert Holley and associates succeeded in elucidating the entire nucleotide sequence of alanine tRNA (Fig. 3.33). Although the molecule contains only 77 nucleotides—most other RNA molecules contain thousands—delineation of its structure required 6 years of intensive study. This was the first complete sequence analysis of any nucleic acid. Historically, then, alanine tRNA resembles insulin, the first protein whose amino acid sequence was resolved. The full nucleotide sequence of an 81 nucleotide serine tRNA was reported in 1966 by H. G. Zachau and co-workers.

Before an amino acid can be attached to its

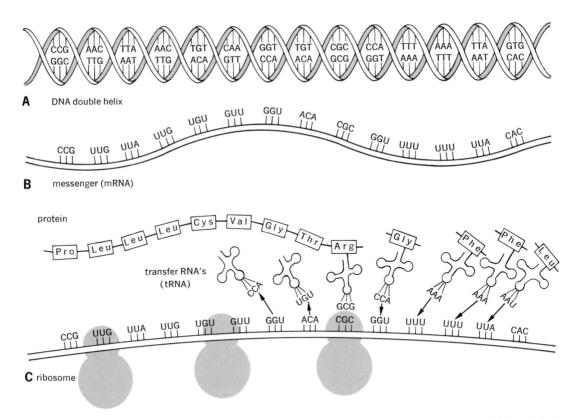

3.34 The mechanism of action of messenger RNA. A. The DNA molecule is a series of nucleotide triplets. Note the A-T and G-C pairings in complementary strands. B. Messenger RNA is a series of triplets complementary to one of the two DNA strands. C. Messenger RNA finds its way to a polyribosome, the site of protein synthesis. There amino acids are carried to the proper sites on messenger RNA by molecules of transfer RNA. The amino acids are then linked by peptide bonds, and the resulting protein "peel off." (Adapted from G. G. Simpson and W. S. Beck *Life: An Introduction to Biology*, 2nd ed., © 1957, 1965 by Harcourt Brace Jovanovich, Inc. and reproduced by permission.)

tRNA, it must be activated by an enzyme. In the presence of the proper activating enzyme, ATP reacts with an individual amino acid to form an amino acid–AMP complex whose acyl bond has a high energy content. The complex joins the end of the tRNA molecule, which terminates in the sequence -cytosine-cytosine-adenine (end) (see Fig. 3.33). The tRNA molecule removes the amino acid from the complex and carries it to the ribosomes. Transfer RNA thus provides the mechanism for properly aligning the amino acids on the messenger RNA.

Genetic code. How is specificity stamped into the amino acid sequence? How are the amino acids faultlessly arranged in a given order? Strands of mRNA issuing from the nucleus travel to the ribosomes and join them into polysomes (see p. 75). The amino acid sequence is determined by the nucleotide sequence of the mRNA on a polysome. The bases in each strand of the tRNA pair with complementary bases in the mRNA on the polysome (Fig. 3.34). For example, the base sequence uracil-guanine-uracil (UGU) in tRNA attaches to the complementary sequence adenine-cytosine-adenine (ACA) in mRNA. The amino acid is left dangling in a manner that allows it to form peptide bonds with the amino acids dangling on either side of it.

When a full chain of amino acids has formed in the sequence dictated by the mRNA, the resulting protein peels away from the polysome and passes into other regions of the cell, where it functions as an enzyme, structural element, or secretory product. Thus the amino acid sequence upon which protein specificity depends is strictly spelled out in the nucleotide sequence of the DNA.

The remarkable mechanism by which polysomes produce proteins is shown in Fig. 3.35. Single ribosomes attach themselves to a mRNA strand to make a polysome. They then move, one after another, along the strand, "reading" its coded information. Appropriate amino acids are brought by their tRNA's at each step of the way, and the polypeptide chain grows progressively longer. As a rolling ribosome drops off the end of the mRNA strand, a new one attaches itself at the other end. The synthesis of the single protein molecule takes only a few seconds.

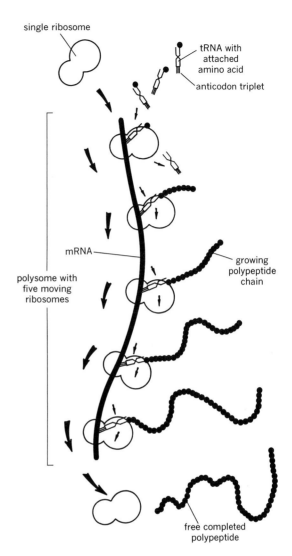

single ribosome

tRNA with attached amino acid

anticodon triplet

mRNA

polysome with five moving ribosomes

growing polypeptide chain

free completed polypeptide

3.35 Assembly of a polypeptide chain on a polysome. The mRNA molecule bearing a series of triplets codons attaches to and slowly moves across a group of ribosomes to form polysomes. Amino acids are transported to appropriate sites on the mRNA by their attached specific tRNA molecules. The amino acids are linked by peptide bonds, the polypeptide chain grows, and the resulting protein "peel off."

In late 1961 M. W. Nirenberg and J. H. Matthaei made a spectacular breakthrough by deciphering the code in mRNA. This achievement was made possible by the production of synthetic mRNA's containing arbitrarily chosen combinations of nucleotides. When a synthetic mRNA containing only one base, uracil, was added to a ribosomal protein-synthesizing system (complete with amino acids, ATP, and all necessary tRNA's), the "protein" synthesized consisted of only one repeating amino acid, phenylalanine. It was therefore clear that the base U (or some combination of U's, such as UU, UUU, or UUUU) was the "code word" (or *codon*) determining the insertion of phenylalanine in the chain.

There were already good theoretical reasons for believing that the correct code word for phenylalanine is the triplet UUU, and indeed that the code words for all amino acids are triplets. It had been evident for some time that the nucleic acid "alphabet" has only 4 letters, whereas code words are needed for 20 amino acids. A 2-letter word from a 4-letter alphabet yields only $4^2 = 16$ possible sequences. A 3-letter word offers $4^3 = 64$ possible sequences, more than enough. (Some amino acids have more than one code word. A code with more than one word for each term coded is "degenerate.")

Experiments of four main types were performed in deciphering the entire genetic code, and incidentally they proved that the code, in fact as well as theory, consists of nonoverlapping triplets. The first experimental approach was an extension of the original study using synthetic mRNA. When a synthetic mRNA containing one molecule of adenine for every two of uracil was added to the protein-synthesizing system,* the "protein" synthesized contained much phenylalanine together with small amounts of another amino acid, isoleucine. The 2:1 ratio of uracil to adenine in the synthetic mRNA indicated that one code word for isoleucine combines two U's with one A (i.e., UUA, UAU, or AUU). The incorporation of phenylalanine into the protein synthesized with 2U:1A-mRNA presumably reflects the UUU triplets that would frequently occur in such a polymer. When

enough adenine was added to the mRNA to give a 2A:1U ratio, the amino acid asparagine appeared in the protein synthesized—in addition to phenylalanine and isoleucine. Hence the code for asparagine was 2A:1U. The amounts of the three amino acids incorporated into the proteins were proportional to the statistical frequencies with which we would expect 3U, 2U:1A, and 2A:1U in the synthetic mRNA.

Studies of this type failed to reveal the sequence of bases within a code word. However, the second approach established that a synthetic short-chain mRNA containing only three nucleotides in a *known* sequence binds to ribosomes and attracts a specific tRNA charged with its own amino acid. Therefore, it has become possible to determine base sequences (Table 3.1).

The third approach to the nature of the code was by purely genetical experiments. It is known that certain chemicals (e.g., the acridines) cause mutations by adding or removing one or two bases from the DNA sequence and thereby destroying a cell's ability to synthesize a certain protein molecule. However, when *three* bases are added or subtracted, the ability to synthesize the protein molecule is retained, though the protein molecule formed has one erroneous amino acid. Presumably the insertion or removal of one or two bases throws the sequence of bases off by one or two places, or shifts the "reading frame" one or two places, causing a major distortion in the message,† whereas the insertion or removal of three bases shifts the reading frame three places—the length of a single code word—thereby preserving the original sequence except for a local jumbled segment too small to prevent the synthesis of a near-perfect specific protein. These results emphasize that the coding ratio of three bases to one amino acid is correct.

The fourth approach to delineation of the genetic code was by studies of the amino acid se-

---

* Though we can synthesize a long-chain mRNA containing any desired ratio of bases (e.g., 2U:1A or 2A:1U), we cannot yet order the entire base sequence. We can calculate, however, that 2A:1U-mRNA contains the sequences 3U, 2U:1A, 2A:1U, and 3U in a ratio of 8:4:2:1.

† The significance of the reading frame is best explained by an analogy. If the first letter were lost from the sentence, "The cat ate the rat," the reading frame would be shifted one place, and the sentence would read "Hec ata tet her at-."—nonsense. But, if the error were confined to one word, as in "The cat ate thr rat," the sense would be largely preserved. The existence of a reading frame in the genetic context indicates that sequences are always "read" from a fixed point.

TABLE 3.1 GENETIC CODE*

Code word	Amino acid	Code word	Amino acid
UUU	Phe	AUU	Ileu
UUC	Phe	AUC	Ileu
UUA	Leu	AUA	Met
UUG	Leu	AUG	Met
UCU	Ser	ACU	Thr
UCC	Ser	ACC	Thr
UCA	Ser	ACA	Thr
UCG	Ser	ACG	Thr
UAU	Tyr	AAU	Asp(NH₂)
UAC	Tyr	AAC	Asp(NH₂)
UAA	†	AAA	Lys
UAG	†	AAG	Lys
UGU	Cys	AGU	Ser
UGC	Cys	AGC	Ser
UGA	†	AGA	Arg
UGG	Try	AGG	†
CUU	Leu	GUU	Val
CUC	Leu	GUC	Val
CUA	†	GUA	Val
CUG	Leu	GUG	Val
CCU	Pro	GCU	Ala
CCC	Pro	GCC	Ala
CCA	Pro	GCA	Ala
CCG	Pro	GCG	Ala
CAU	His	GAU	Asp
CAC	His	GAC	Asp
CAA	Glu(NH₂)	GAA	Glu
CAG	Glu(NH₂)	GAG	Glu
CGU	Arg	GGU	Gly
CGC	Arg	GGC	Gly
CGA	Arg	GGA	Gly
CGG	Arg	GGG	Gly

* The code words for the amino acids indicated are nucleotide triplets. Since most amino acids are represented by more than one, the code is degenerate. The amino acid abbreviations are explained in Table 2.5.

† A code word for which an amino acid is not given is either a nonsense word or complementary to the terminal CCA sequence of tRNA (see Fig. 3.33).

quences of normal and mutant proteins. Vernon Ingram, for example, showed that sickle-cell anemia, a genetically determined human disease associated with an abnormal hemoglobin molecule that behaves distinctively upon electrophoresis (see Fig. 2.30B), is due entirely to the presence of one erroneous amino acid in one of the two types of polypeptide chains of 146 amino acids each that comprise the hemoglobin molecule (see Fig. 2.11). Instead of

-His-Leu-Thr-Pro-Glu-Glu-Lys-

the abnormal molecule contains

-His-Leu-Thr-Pro-*Val*-Glu-Lys-

Through mutation, valine has replaced glutamic acid in the amino acid sequence.* This discovery, made before the elucidation of the genetic code, was the first demonstration of a specific biochemical consequence of mutation, and it established the localized nature of the mutational event, for surely the localized error in a protein copies a localized change in the mRNA and its parent DNA. The genetic code accounts for Ingram's results, for each of the two code words for glutamic acid (GAG or GAA) differs by only one base from a code word for valine (GUG or GUA). Hence we conclude that the ancient mutation causing the synthesis of sickle-cell hemoglobin in fact altered no more than one base of the DNA sequence.

It is now believed that the genetic code is universal—i.e., that an amino acid is coded by the same triplet in all species—though this remains to be proved. "Nonsense" triplets (see Table 3.1) are not read, it is conjectured, because the cell lacks tRNA molecules with an *anticodon* (i.e., a triplet complementary to a mRNA codon) capable of reading them. Or perhaps the tRNA molecules are present, but the existing amino acid–activating enzymes are incapable of charging them with amino acids. Conceivably "nonsense" triplets act as spacers or "punctuation marks" along the DNA strand, separating the

* Glutamic acid possesses a negatively charged carboxyl group, whereas valine does not (see Table 2.5). This accounts for the slower movement of abnormal hemoglobin, as compared with normal hemoglobin, toward the anode of the electrophoresis apparatus, the property that led to its discovery by Harvey Itano and Linus Pauling.

sequences that comprise genes. The significance of code degeneracy is unknown. It has been pointed out that in a degenerate code replacement of one base within a triplet might occasionally lead to another triplet for the same amino acid. Thus, from an evolutionary point of view, extensive degeneracy of the genetic code would tend to stabilize species by decreasing the frequency of mutations.

Mutations and the genetic code. We have seen that the intracellular flow of genetic information takes place as follows:

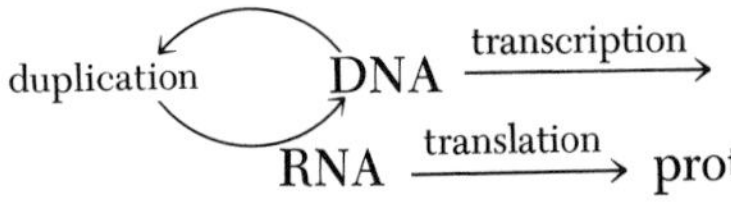

$$
\text{duplication} \; \overset{\curvearrowleft}{\underset{\curvearrowright}{\text{DNA}}} \xrightarrow{\text{transcription}}
$$
$$
\text{RNA} \xrightarrow{\text{translation}} \text{protein} \qquad (3.5)
$$

Three modes of information transfer occur: (1) *duplication,* which provides exact copies for hereditary transmission; (2) *transcription,* in which DNA generates complementary copies of RNA; and (3) *translation,* which begins with the 4-unit language of the nucleic acids and ends with the 20-unit language of the proteins.

Mutations fall into two categories: *chromosome* mutations, changes in chromosome number or structure (aneuploidy); and *gene* mutations, changes in DNA base sequences. The result of a gene mutation is an altered base sequence in the mRNA. The four possible consequences for the final protein molecule are summarized in the following scheme (which uses a glycine code word as an example).

A "missense" mutation yields a protein that contains amino acid Y instead of amino acid X and that may or may not be functionally active. If it is functionally active, it is difficult to detect. Testing for its activity as an enzyme is useless, and it is recognized only after laborious determination of its amino acid sequence and comparison of the sequence with that of a normal protein. For this reason many missense mutations may have been overlooked. When a mutation produces a "nonsense" triplet, protein synthesis stops. This event is also difficult to establish in the laboratory.

A change of a single nucleotide in the DNA chain can be brought about in several ways. Oddly, we do not yet know how x-rays, the most potent mutagenic agents known and the first to be discovered, cause base substitutions. Chemical mutagenic agents are better understood. These are of various kinds. For example, nitrous acid (HNO_2) is a powerful oxidizing agent that converts amino groups to hydroxyl groups. Nitrous acid therefore converts cytosine to uracil.

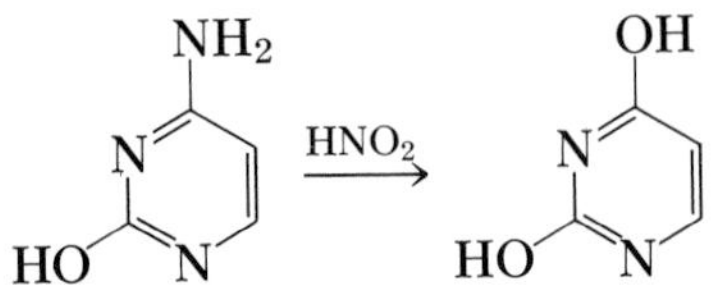

$$
\begin{array}{ccc}
\text{NH}_2 & & \text{OH} \\
\end{array}
$$

structure diagram with HNO_2 converting cytosine to uracil $\qquad (3.7)$

If a DNA molecule is exposed to HNO_2, the following events may occur. First, cytosine is converted to uracil.

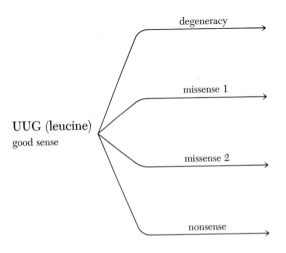

UUA (leucine)
no change in protein structure

UUC (phenylalanine)
amino acid substitution; protein is synthesized but enzymatically inactive

UCG (serine)
amino acid substitution; protein is synthesized and enzymatically active

UAG (no amino acid)
no recognizable protein is synthesized, though incomplete polypeptide chains may be formed $\qquad (3.6)$

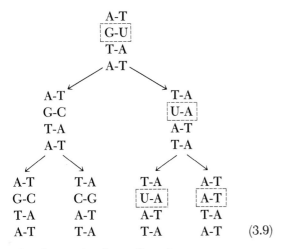

$$
\begin{array}{ccc}
\text{A-T} & & \text{A-T} \\
\text{G-C} & \xrightarrow{\text{HNO}_2} & \boxed{\text{G-U}} \\
\text{T-A} & & \text{T-A} \\
\text{A-T} & & \text{A-T}
\end{array} \qquad (3.8)
$$

Then the chains separate, complementary chains are synthesized, and the new chains separate.

$$(3.9)$$

The change (or "error") in base pairing is perpetuated, and the mutant DNA contains an A-T pair in place of a G-C pair.

Some of the many mutagenic agents now utilized are chemical analogues of natural purines and pyrimidines that "trick" DNA into making a pairing "error." Others, such as the acridines, cause the synthesis of DNA molecules containing one more or one less nucleotide. All, it appears, act by producing a localized alteration in the nucleotide sequence of DNA. In light of the ease with which experimental mutations can be induced, the rarity of natural mutations is surprising.

Structural and regulator genes; operator loci. In the classic formulation, genes are independent molecular blueprints that determine the structures of individual cellular constituents. Jacob and Monod have suggested that only *structural genes* function in this way. They produce cytoplasmic transcripts of themselves, the mRNA's that in turn govern protein structure.

But the functioning and survival of a cell require more than an outpouring of the structur-

ally correct cell ingredients. Coordination is also necessary, and it is now apparent that genes exist whose only function is to initiate or direct the activities of the structural genes. Instead of dispatching mRNA's to the ribosomes, these *operator loci* combine reversibly and specifically with products arising from *regulator genes.* The reactions control the formation of mRNA's by structural genes and so repress (or induce) protein synthesis. We shall learn later how repression (and induction) of enzyme formation serves as a metabolic regulatory device of prime importance in physiology. We must note here that the ability of a cell to perform such adaptive maneuvers is part of its genetic endowment. Hence it is represented in a genic constellation that must be viewed as a series of blueprints for protein structure *and* a coordinated program for protein synthesis.

Cytoplasmic heredity. We have spoken thus far of genes and the information-bearing DNA of the nucleus as though they were the only mechanisms of hereditary transmission. In fact, the self-duplicating cytoplasmic particles, which are transmitted independently from one cell generation to the next, provide genetic continuity and can be regarded as hereditary units. It was long suspected that certain cytoplasmic structures, notably the mitochondria, are capable of self-duplication. The supposition was based on the fact that mitochondria are unaffected by interruptions in the synthesis of mRNA on nuclear DNA by agents such as actinomycin D— hence are largely independent of mRNA-mediated control by nuclear genes. When it was found that mitochondria contain traces of a DNA that differs in base ratio and density from nuclear DNA, it was suggested that it participates in their duplication. F. S. Vogel searched carefully for evidence of mitochondria duplication in cultures of isolated mitochondria and in 1965 obtained the electron micrograph shown in Fig. 3.36. He also demonstrated replication of the mitochondrial DNA. A similar pattern doubtless holds for the reproduction of other cytoplasmic organelles during cell division. The cytoplasmic ribosomes, in contrast, arise in the nucleus—specifically, in the nucleoli.

A great deal of data points to the existence of nonnuclear genes in certain protozoans (e.g.,

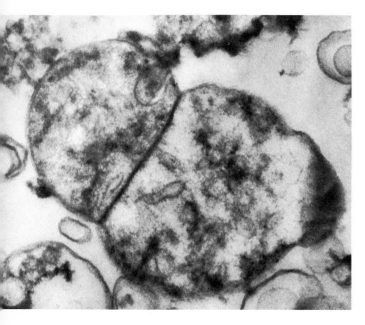

3.36 Budding of isolated mitochondria after 72 hours' incubation (×67,000). (From Stephan F. Vogel, *Laboratory Investigation, 14,* 1880, 1965.)

the kappa particles of *Paramecia aurelia*) and plant cells (e.g., the chloroplasts that govern leaf variegation). Like mitochondria, kappa particles and chloroplasts are self-duplicating, DNA-containing cytoplasmic particles that are transmitted from generation to generation.

There is much current debate on the prevalence and importance of nonnuclear inheritance. Some investigators speak of nonnuclear or nonchromosomal genes as if they constituted a general genetic system comparable to that of the chromosomes. Others believe that they appear sporadically, are under the ultimate control of the chromosomal system, and are of little or no special significance for long-range genetic studies. Perhaps the truth lies somewhere between these alternatives.

REFERENCES AND SUGGESTIONS FOR FURTHER READING

Allfrey, V. G., and A. E. Mirsky, "How Cells Make Molecules," *Sci. Am.,* **205,** 74 (Sept., 1961).

Anfinsen, C. B., *The Molecular Basis of Evolution,* Wiley, New York, 1963.

Beadle, G. W., "Genetics and Metabolism in *Neurospora,*" *Physiol. Rev.,* **25,** 643 (1945).

Benzer, S., "The Fine Structure of the Gene," *Sci. Am.,* **206,** 70 (Jan., 1962).

Bonner, D., and S. E. Mills, *Heredity,* 2nd ed., Prentice-Hall, Englewood Cliffs, N. J., 1964.

Bourne, G. H., *Division of Labor in Cells,* Academic Press, New York, 1962.

Brachet, J., "The Living Cell," *Sci. Am.,* **205,** 50 (Sept., 1961).

Bryson, V., and H. J. Vogel, eds., *Evolving Genes and Proteins,* Academic Press, New York, 1965.

"Cellular Regulatory Mechanisms," *Cold Spring Harbor Symp. Quant. Biol.,* **26** (1961).

Chargaff, E., and J. N. Davidson, eds., *The Nucleic Acids,* Vols. 1–3, Academic Press, New York, 1955–60.

Cohn, N. S., *Elements of Cytology,* Harcourt Brace Jovanovich, New York, 1964.

Crick, F. H. C., "The Genetic Code," *Sci. Am.,* **207,** 66 (Oct., 1962).

De Duve, C., "The Lysosome," *Sci. Am.,* **208,** 64 (May, 1963).

De Robertis, E. D. P., W. W. Nowinski, and F. A. Saez, *Cell Biology,* 4th ed., Saunders, Philadelphia, 1965.

Fawcett, D., *The Cell: Its Organelles and Inclusions,* Saunders, Philadelphia, 1966.

Gibor, A., and Granick, S., "Plastids and Mitochondria: Inheritable Systems," *Science,* **145,** 890 (1964).

Green, D. E., "The Mitochondrion," *Sci. Am.,* **210,** 63 (Jan., 1964).

Hartman, P. E., and S. R. Suskind, *Gene Action,* Prentice-Hall, Englewood Cliffs, N. J., 1965.

"Human Genetics," *Cold Spring Harbor Symp. Quant. Biol.,* **29** (1964).

Hurwitz, J., and J. J. Furth, "Messenger RNA," *Sci. Am.,* **206,** 41 (Feb., 1962).

Ingram, V. M., *The Hemoglobins in Genetics and Evolution,* Columbia Univ. Press, New York, 1963.

Jacob, F., and J. Monod, "Genetic Regulatory Mechanisms in the Synthesis of Proteins," *J. Mol. Biol.,* **3,** 318 (1961).

Jinks, J. L., *Extrachromosomal Inheritance,* Prentice-Hall, Englewood Cliffs, N. J., 1964.

Kornberg, A., "The Synthesis of DNA," *Sci. Am.,* **219,** 64 (Oct., 1968).

Levine, L., ed., *The Cell in Mitosis,* Academic Press, New York, 1963.

Levine, R. P., *Genetics,* Holt, Rinehart and Winston, New York, 1963.

Loewy, A. G., and P. Siekevitz, *Cell Structure and Function,* Holt, Rinehart and Winston, New York, 1963.

Mazia, D., "Biochemistry of the Dividing Cell," *Ann. Rev. Biochem.,* **30,** 669 (1961).

———, "How Cells Divide," *Sci. Am.*, **205**, 100 (Sept., 1961).

———, and A. Tyler, *The General Physiology of Cell Specialization*, McGraw-Hill, New York, 1963.

Meselson, M., and F. Stahl, "The Replication of DNA in *Escherichia coli*," *Proc. Natl. Acad. Sci. U. S.*, **44**, 671 (1958).

Moore, K. L., ed., *The Sex Chromatin*, Saunders, Philadelphia, 1966.

Neutra, M., and C. P. Leblond, "The Golgi Apparatus," *Sci. Am.*, **220**, 100 (Feb., 1969).

Nirenberg, M. W., "The Genetic Code," *Sci. Am.*, **208**, 80 (March, 1963).

———, and P. Leder, "RNA Codewords and Protein Synthesis," *Science*, **145**, 1399 (Sept., 1964).

———, and J. H. Matthaei, "The Dependence of Cell-Free Protein Synthesis in *E. coli* Upon Naturally Occurring or Synthetic Polyribonucleotides," *Proc. Natl. Acad. Sci. U. S.*, **47**, 1588 (1961).

Palade, G., "The Endoplasmic Reticulum," *J. Biophys. Biochem. Cytol.*, **2** Suppl., 85 (1956).

Petermann, M. L., *The Physical and Chemical Properties of Ribosomes*, American Elsevier, New York, 1965.

Porter, K. R., and M. A. Bonneville, *An Introduction to the Fine Structure of Cells and Tissues*, Lea & Febiger, Philadelphia, 1963.

Riley, M., and A. B. Pardee, "Gene Expression: Its Specificity and Regulation," *Ann. Rev. Microbiol.*, **16**, 1 (1962).

Robertson, J. D., "The Membrane of the Living Cell," *Sci. Am.*, **206**, 65 (April, 1962).

Sager, R., "Genes Outside the Chromosomes," *Sci. Am.*, **210**, 70 (Jan., 1965).

———, and F. J. Ryan, *Cell Heredity*, Wiley, New York, 1961.

Simpson, G. G., and W. S. Beck, *Life: An Introduction to Biology*, 2nd ed., Harcourt Brace Jovanovich, New York, 1965.

Singer, M. F., and P. Leder, "Messenger RNA: An Evaluation," *Ann. Rev. Biochem.*, **35**, 195 (1966).

Spiegelman, S., "Hybrid Nucleic Acids," *Sci. Am.*, **210**, 48 (May, 1964).

Stahl, F., *The Mechanics of Inheritance*, Prentice-Hall, Englewood Cliffs, N. J., 1964.

Swanson, C. P., *The Cell*, 2nd ed., Prentice-Hall, Englewood Cliffs, N. J., 1964.

———, T. Merz, and W. J. Young, *Cytogenetics*, Prentice-Hall, Englewood Cliffs, N. J., 1964.

"Synthesis and Structure of Macromolecules," *Cold Spring Harbor Symp. Quant. Biol.*, **28** (1963).

Taylor, J. H., "The Duplication of Chromosomes," *Sci. Am.*, **198**, 36 (June, 1958).

Vogel, H. J., V. Bryson, and J. O. Lampen, eds., *Informational Macromolecules*, Academic Press, New York, 1963.

Wagner, R. P., and H. K. Mitchell, *Genetics and Metabolism*, 2nd ed., Wiley, New York, 1964.

Watson, J. D., *Molecular Biology of the Gene*, Benjamin, New York, 1965.

———, "Involvement of RNA in the Synthesis of Proteins," *Science*, **140**, 17 (1963).

Wilson, E. B., *The Cell in Development and Heredity*, 3rd ed., Macmillan, New York, 1925.

Zamecnik, P. C., "Unsettled Questions in the Field of Protein Synthesis," *Biochem. J.*, **85**, 257 (1962).

———, "Historical and Current Aspects of the Problem of Protein Synthesis," *Harvey Lectures, Ser. 54 (1958–59)*, 256 (1960).

The most impressive feature of a cell is not the constant flux, reshuffling, and variability of its population of molecules and particles, but the fact that, in spite of this ever-present change, each cell remains so remarkably invariant in its total behavior; that indeed, as an entity each behaves so much like millions of other entities equally variable in inner detail, that one comes to recognize them as essentially alike. Such relative invariance of the whole presupposes the harmonious subordination of the behavior of the parts to the conditions of the collective group. It presupposes that the free interactions among the subunits are subject to restraints, the nature and direction of which vary adaptively with the state of the system as a whole.

Paul Weiss, in *Cellular Dynamics* (1959)

4 *Physiology of the Cell*

Introduction

The evolutionary advance from unicellular to multicellular construction permitted living organisms to become larger. But, as we have seen, individual cells of a multicellular organism paid a price for the new advantages: they became functional specialists and lost their independence. As organism size increased, there was a proportional increase in the need for methods of coordinating and controlling the specialized functions of the dependent cells and for communicating information from place to place within the organism. Thus tissues, organs, and body systems evolved. In later chapters we shall deal with these systems of specialized body cells, which serve the organism at large.

First, however, we must consider those basic functions that are displayed in some measure by *all* cells, be they the specialized cells of a multicellular organism or independent unicellular organisms. They are five in number: (1) the metabolic production of energy; (2) the biosynthesis of specific cell constituents; (3) the control of cell metabolism; (4) the selective behavior of the cell membrane (one of whose important consequences is the phenomenon of irritability); and (5) cell movement.

It is important to recognize that the successful execution of these five functions is essential to the survival of any organism. Energy must be generated, products and structures must be manufactured, competing enterprises must be controlled, borders must be guarded, and movement must occur. In this chapter we shall study the means by which a cell carries out these functions. In later chapters we shall learn that equivalent functions are performed by various body systems on behalf of the whole body. We shall also find that in the final analysis all of the complex functions of tissues, organs, and systems are attributable to the combined basic functions of individual cell components.

Production of Energy in Cell Metabolism

The molecular components of a cell are in a constant state of flux, in which they are continuously being transformed, renewed, and replaced. We speak of all this activity as *metabolism*. In its broadest sense, metabolism includes all of the chemical reactions taking place within a body or its individual cells, whether they are involved in growth, reproduction, maintenance, or energy production. To all cells come, in ever-varying amounts, nutrient materials. Some are rejected, and some are rapidly taken in and chemically changed. It is this seemingly purposeful pattern, with its building up and breaking down of molecules, that our study of metabolism aims to elucidate.

Metabolic processes can be divided into two categories. Those concerned with the *synthesis* of cell constituents and cell products from simpler substances are referred to collectively as *anabolism*. They do not supply the energy needed for cell function; indeed, they require energy. Those concerned with *energy production* are referred to collectively as *catabolism*.* In catabolism complex molecules are broken down to simpler ones. Most of these are then consumed in energy-liberating oxidative reactions. Hence measurement of the body's total oxygen consumption provides a rough index of the extent of catabolism.

Complex molecules yielding energy on oxidation are taken into the cell as food. Food also supplies the building blocks for anabolic reac-

---

* We shall later find that certain metabolic processes are both anabolic and catabolic. Bernard D. Davis has recently suggested that these be called *amphibolic*. He also proposed that processes resisting classification be called *diabolic*.

tions. Thus the substances needed for the nutrition of every cell are destined to support only two broad functions. The particular uses to which a food is put depend upon its chemical nature and upon the enzymes present in the cell. Many enzymes are common to all cells; some are found only in certain cells. In considering the processes of metabolism, we should keep in mind the extremely small volume that encloses them. Within the physical confines of a cell are the enzymes, the metabolic machinery, the various substances acted upon and their products, the apparatus for energy generation and utilization, and finally the mechanisms by which all of these processes are ordered, directed, and controlled.

NATURE OF METABOLIC ENERGY

In Chapter 2 we learned (1) that energy is defined as the capacity to do work; (2) that large quantities of potential energy are inherent in the chemical structures of sugars and other nutrient materials; (3) that potential energy is provided these molecules directly or indirectly by photosynthesis, the ultimate energy source being the sun; and (4) that exergonic oxidative reactions convert potential energy to free kinetic energy, although such reactions frequently require an initial investment of activation energy.

We must now consider in greater detail the metabolic processes that generate energy, the form in which this energy is transported, stored, and dissipated, and the physiological significance of this information.

Concept of free energy. When the boulder in Fig. 2.15A slid down the hill, energy was liberated and work was performed in amounts depending entirely upon the size of the boulder and the height of the hill. In any such system, however, only a portion of the initial potential energy is available for the performance of work. The sliding boulder wastes some of its potential energy in overcoming frictional forces. Likewise, in chemical reactions larger or smaller parts of the total energy are unavailable. Hence we must distinguish between the *free* or *available energy* and the *total energy* of the system.

In order to express this concept more precisely, let us consider the energy-liberating chemical conversion of substance A to substance B.

$$A \rightleftharpoons B \qquad (4.1)$$

Energy is liberated until the rate of conversion of B to A equals the rate of the forward reaction. At this, the equilibrium point, no more free energy is liberated. Thus, if we know the amount of A that is converted to B before the attainment of equilibrium, we can calculate the amount of free energy liberated from the relationship known to exist between the free energy liberated in a reaction and the extent of the reaction. This formula makes use of the equilibrium constant (see p. 47) in calculation of the change in free energy (symbolized ΔF, Δ meaning "change in").

$$\Delta F = -4.6T \log K \qquad (4.2)$$

where K is the equilibrium constant ([B]/[A]) and T is the absolute temperature. Obviously, if most of A has been converted to B when equilibrium is reached, K will be large, and ΔF will be large and negative, the negative sign denoting energy liberation. When $\Delta F < 0$, the reaction is *exergonic*. If little A has been converted to B, K will be less than 1, and ΔF will be positive (i.e., $\Delta F > 0$). This result indicates that the reaction cannot take place unless energy is supplied to the system; in other words, the reaction is *endergonic*. If $\Delta F = 0$, the system is at equilibrium, neither liberating nor consuming energy.

Thermodynamics has established the following general expression for the change of free energy in any system:

$$\Delta F = \Delta H - T \Delta S \qquad (4.3)$$

ΔH is the change of total heat when the system goes from one state to another. If heat is given off, ΔH is negative; if heat is absorbed, ΔH is positive. T is the absolute temperature. And ΔS is the change in a quantity called *entropy*. Entropy is a measure of the order in a system, high order always representing the most improbable condition. The more ordered or improbable the state of the system, the lower its entropy; the more randomly ordered or probable the system, the higher its entropy.* In simpler words, the

entropy of a system indicates how much the system is run down.

We have seen that free energy, the energy that is available for useful work, differs from the total energy of a reaction. Thus

change in utilizable energy = change in total energy − change in nonutilizable energy   (4.4)

Depending on the system, the change in nonutilizable energy may be small or large. In any case, its value in Eq. 4.4 is expressed quantitatively as the absolute temperature (T) times the change in entropy (ΔS) in going from one state to another. Therefore, the terms of Eq. 4.3 are interchangeable with those of Eq. 4.4.

These equations have important implications. A reaction whose ΔF is negative in a calculation based upon Eq. 4.2 is exergonic, but knowledge of the ΔF value (however indirectly arrived at) does not tell us the values of ΔH and ΔS. In fact, ΔH is the only term in Eq. 4.3 that is directly ascertainable in the laboratory. We could determine the ΔH of Eq. 4.1, for example, simply by separately placing some A and some B in a bomb calorimeter and determining the amount of heat given off by each upon complete combustion to carbon dioxide and water. If 5000 cal† were liberated from the combustion of A and only 2000 from the combustion of B, then ΔH would equal -3000 cal. If Eq. 4.1 occurs at 27°C (300°A) and if the entropy change from A to B is 10 (as determined by a formula not given here), ΔF can be calculated from Eq. 4.3.

$$\begin{aligned} \Delta F &= -3000 - (300 \times 10) \\ &= -6000 \text{ cal} \end{aligned} \qquad (4.5)$$

We note that ΔF is negative (i.e., energy is liberated) and that, in this instance, entropy change contributes significantly to the ΔF value. We note, too, that the heat evolved in the reaction (the ΔH value) is not an index of the change in

---

* A sugar crystal dissolving in water provides an example of a system progressing from order to disorder. Initially the sugar molecules are in a highly ordered array. When the sugar has

dissolved and diffused throughout the water, their arrangement is random and disordered. Statistically, the former situation is far more improbable than the latter. Hence solution and diffusion increase the entropy of the system.

† A *calorie* (cal) is the quantity of heat required to raise the temperature of 1 g of water 1°C, starting at 14.5°C. It is equivalent to 4.185 joules, the joule being the fundamental unit of energy. Biochemists commonly speak of a *kilocalorie* (Cal, or kcal), the amount of heat needed to raise the temperature of 1000 g of water from 14.5 to 15.5°C.

free energy (except in those reactions in which the change in entropy is extremely small).

The currency of energy exchange. Although energy liberated in exergonic reactions can take the form of heat, heat is no more useful to the cell as a prime energy source for the performance of physical, chemical, and electrical work than is a bonfire under its engine to an automobile. The whole body does make use of metabolically generated heat in the maintenance of a constant internal temperature. But, as energy, heat is entirely nonspecific, promoting all reactions at once instead of a specific reaction in a specific place at a specific time.

What is needed instead is a means of coupling energy-yielding reactions with energy-requiring reactions, of efficiently transporting energy from the site of its production to the site of its utilization, and of storing it for future use. In short, the cell needs a means of trapping the energy of exergonic reactions before it is dissipated as heat in a form that can serve the same function in metabolism that currency serves in the economy of a community.

Nature's currency of energy exchange is a class of compounds containing *energy-rich bonds.* Among the most important of these are adenosine triphosphate (ATP) and adenosine diphosphate (ADP). As shown on p. 52, the structures of these compounds are, respectively, A—R—(P)—(P)—(P) and A—R—(P)—(P), where A is adenine and R is ribose. Eq. 2.41 shows how ADP accepts phosphate to form ATP and how ATP in turn serves as a phosphate donor.

The ATP molecule has a remarkable property. When the terminal phosphate group is split off by simple hydrolysis, 8000 to 10,000 cal per mole is liberated. In contrast, hydrolysis of a variety of other phosphate esters yields only 2000 to 4000 cal per mole. In 1941 Fritz Lipmann suggested that phosphate ester bonds of the type in ATP be termed "energy-rich phosphate bonds" and that they be designated by ~(P), in contrast to ordinary energy-poor phosphate bonds, which are designated by —(P). It was soon demonstrated that ADP also contains one energy-rich phosphate bond. Thus the structure of ATP is properly represented as A—R—(P)~(P)~(P). Later it was found that some nonphosphate bonds are also rich in energy.

We cannot dwell here on the reason why certain phosphate ester bonds are energy-rich and others are not. Suffice it to say that this is the consequence of certain structural features of the entire molecule. When an enzymatic reaction produces these features, low-energy phosphate bonds become high-energy phosphate bonds. The two main types of reactions in which phosphate ester bonds are so transformed are *dehydrations* and *oxidations.* An instructive example of dehydration is the conversion of 2-phosphoglycerate to phosphoenolpyruvate.

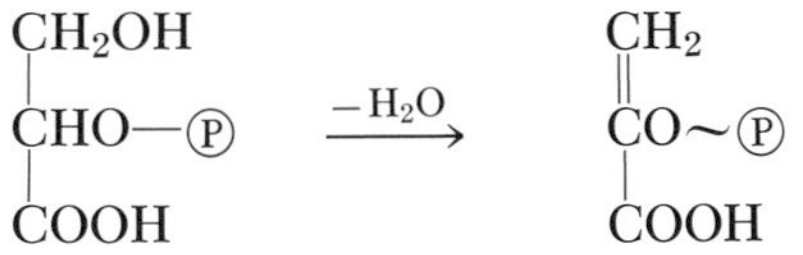

$$
\begin{array}{ccc}
\mathrm{CH_2OH} & & \mathrm{CH_2} \\
| & & \| \\
\mathrm{CHO}\text{—}\textcircled{P} & \xrightarrow{-\mathrm{H_2O}} & \mathrm{CO}\sim\textcircled{P} \quad (4.6)\\
| & & | \\
\mathrm{COOH} & & \mathrm{COOH}
\end{array}
$$

2-phosphoglycerate   phosphoenolpyruvate

The removal of water from the phosphoglycerate molecule alters the molecular structure and redistributes free energy, concentrating it in the region of the phosphate group.

We can now appreciate the great importance of group-transfer reactions of the type shown in Eq. 2.41, for it appears that when a group (such as the phosphate group) is transferred from one molecule to another, it frequently takes with it its full quota of potential energy. Thus we may properly rewrite such an equation in the following manner:

$$
\begin{aligned}
\mathrm{A}\text{—}\mathrm{R}\text{—}\textcircled{P}\sim\textcircled{P}\sim\textcircled{P} + \mathrm{B} &\longrightarrow \\
\mathrm{A}\text{—}\mathrm{R}\text{—}\textcircled{P}\sim\textcircled{P} &+ \mathrm{B}\sim\textcircled{P} \quad (4.7)
\end{aligned}
$$

where B is an unspecified phosphate acceptor. Similarly, when the ~(P) from a compound such as phosphoenolpyruvate (Eq. 4.6) is transferred to ADP, the high bond energy is retained, and a molecule of ATP is generated.

$$
\begin{aligned}
\mathrm{B}\sim\textcircled{P} + \mathrm{A}\text{—}\mathrm{R}\text{—}\textcircled{P}\sim\textcircled{P} &\longrightarrow \\
\mathrm{A}\text{—}\mathrm{R}\text{—}\textcircled{P}\sim\textcircled{P}\sim\textcircled{P} &+ \mathrm{B} \quad (4.8)
\end{aligned}
$$

where B~(P) represents phosphoenolpyruvate.

Thus ATP with its energy-rich phosphate bonds may be regarded as the currency of energy exchange. Indeed, it is the universal intracellular carrier of chemical energy (Fig. 4.1). Since we can equate energy production with ATP generation—and since glucose and other foodstuffs are the prime sources of metabolic energy—we must now determine precisely how

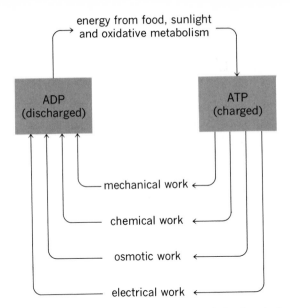

energy from food, sunlight and oxidative metabolism

ADP (discharged)

ATP (charged)

mechanical work

chemical work

osmotic work

electrical work

4.1 The ATP-ADP cycle. The energy of the terminal phosphate bond of ATP is used in the performance of diverse forms of physiological work. In turn, ADP is converted to ATP by oxidative phosphorylation.

the oxidative reactions of food combustion cause the efficient synthesis of energy-rich phosphate bonds in ATP.

METABOLIC GENERATION OF ENERGY-RICH BONDS

The breakdown and synthesis of an organic compound in a cell occur in orderly reaction sequences of the type shown schematically on p. 102. Each step is an individual enzyme reaction whose product is the substrate for the next reaction. We speak of such sequences as *metabolic pathways*.

It is beyond the scope of this discussion to describe the pathways of metabolism and their many intermediary compounds in detail. We shall point out those segments of the major pathways that are of notable physiological significance and those features of the pathways that determine whether they are concerned primarily with energy production, with synthetic processes, or with "amphibolism" (see footnote, p. 117).

Quantitative aspects of energy generation. The principal source of cellular energy is carbo-

hydrates. In seeking to understand how the oxidation of carbohydrates generates utilizable energy in the form of energy-rich phosphate bonds in ATP, we gain perspective from a few simple quantitative relationships. We learned (see Eq. 2.32) that glucose is oxidized according to the following overall reaction:

$$C_6H_{12}O_6 + 6O_2 \longrightarrow 6CO_2 + 6H_2O + energy \qquad (4.9)$$

Thermodynamics tells us that the amount of energy liberated in the combustion of a given substance is always the same, no matter what the combustion method. When a mole of glucose (180 g) is oxidized by actual burning in a bomb calorimeter, 690,000 cal is released. We may assume, therefore, that this much energy is liberated in the stepwise metabolic oxidation of glucose.

In the late 1930's Herman Kalckar incubated suspensions of ground muscle or kidney tissue with glucose in the presence of oxygen and inorganic phosphate and measured the rates of glucose disappearance and oxygen uptake (using a Warburg manometer). He observed that as glucose is oxidized, inorganic phosphate is converted to ATP phosphate. Moreover, it was found in these and later experiments that the uptake of oxygen and the disappearance of inorganic phosphate are coupled: for each atom of oxygen utilized, three atoms of phosphate are incorporated into ATP. It was this discovery of *oxidative phosphorylation* that led to the realization that ATP generation is dependent upon oxidative metabolism. The P:O ratio of 3 means that the utilization of 12 atoms of oxygen in the total oxidation of glucose (see Eq. 4.9) must be accompanied by the production of 36 molecules of ATP. Later experiments showed that, in fact, 38 molecules of ATP are formed, the extra two arising from the preliminary reaction sequence, called *glycolysis*. Accordingly, Eq. 4.9 should be expanded as follows:*

---

*In this and later equations, ℗ denotes phosphoric acid (H_3PO_4). The combination of H_3PO_4 with ADP actually yields a molecule of H_2O as well as a molecule of ATP. For simplicity, the $38H_2O$ arising from the reaction of 38℗ with 38ADP is omitted from Eq. 4.10. The H_2O arising from the reaction of ℗ with ADP is also omitted from Eqs. 4.11, 4.15, and 4.16.

$$C_6H_{12}O_6 + 6O_2 + 38ADP + 38\textcircled{P} \longrightarrow$$
$$6CO_2 + 6H_2O + 38ATP \quad (4.10)$$

These figures permit an interesting calculation. Since the terminal $\sim\textcircled{P}$ group of a mole of ATP contains about 10,000 cal, 38 moles of ATP represents an energy yield of 380,000 cal, or 55% of the 690,000 cal originally contained in the glucose. The remainder is dissipated as heat. Even the best modern steam-generating plants convert no more than 30% of the invested energy to useful work.

The four stages of glucose metabolism: a balance sheet. The many individual reactions summarized in Eq. 4.10 occur in four main stages, as diagramed in Fig. 4.2. We shall attempt to identify the stages at which each of the several components of Eq. 4.10 makes its entrance and exit, noting in particular what happens to the carbon skeleton of glucose and at what point electrons (or hydrogens) are transferred* and $\sim\textcircled{P}$ groups created.

In the first stage a six-carbon glucose molecule is split into two molecules of pyruvate containing three carbons each. This process, involving no less than 10 steps, is glycolysis. Although some ATP is utilized early in the sequence, enough new ATP is produced later to yield a net gain of two ATP molecules per molecule of glucose cleaved. Glycolysis also involves the oxidative transfer of four hydrogens. In this stage no CO_2 is produced, all the carbons being accounted for as pyruvate, and no O_2 is utilized.

Stage 1
$$C_6H_{12}O_6 + 2ADP + 2\textcircled{P} \longrightarrow$$
$$2 \text{ pyruvate} + 2ATP + 4[H] \quad (4.11)$$

In the second stage each of the three-carbon pyruvate molecules is converted to a two-carbon derivative of acetic acid, acetyl coenzyme A (or acetyl CoA). Coenzyme A (or CoA) is essential in a large number of group-transfer reactions. The group transferred is an acyl radical (i.e., the R—CO— portion of an organic acid, R—COOH). CoA, like most coenzymes, contains a vitamin. In this instance the vitamin is

pantothenic acid (Table 4.1). The active portion of the CoA molecule is a sulfhydryl group. In fulfilling its acceptor-donor function in a transfer reaction, CoA forms a thioester with the acyl group being transferred. The thioester bond is energy-rich, the energy being derived from the ATP needed to convert an acid to its CoA derivative.

$$\begin{array}{c} O \\ \parallel \\ R-C-OH \end{array} + HS-CoA + ATP \longrightarrow$$

$$\begin{array}{c} O \\ \parallel \\ R-C{\sim}S-CoA \end{array} + AMP + PP \quad (4.12)$$

Because the resulting thioester has an energy-rich bond, the formation of an acyl CoA compound is usually referred to as *activation*.†

We shall presently learn that the conversion of pyruvate to acetyl CoA in the second stage of glucose metabolism is somewhat more complex than as depicted in Eq. 4.12. In any event this stage involves (1) one oxidative reaction per pyruvate molecule (thus four hydrogens are removed per original glucose molecule) and (2) the liberation of one carbon atom per pyruvate molecule as CO_2. Again no O_2 is utilized, although two energy-rich thioester bonds are created. In sum, stage two involves the following overall transformations:

Stage 2
$$2 \text{ pyruvate} + 2CoA \longrightarrow$$
$$2 \text{ acetyl CoA} + 2CO_2 + 4[H] \quad (4.14)$$

In the third stage the two-carbon molecules of acetyl CoA enter a metabolic cycle in which their four carbons are converted to CO_2. First,

---

† The energy of the thioester bond drives a subsequent reaction, and the acyl group provides a package of carbon atoms that can serve as a biosynthetic building block.

$$\begin{array}{c} O \\ \parallel \\ R-C{\sim}S-CoA \end{array} + HR' \longrightarrow$$

$$\begin{array}{c} O \\ \parallel \\ R-C-R' \end{array} + HS-CoA \quad (4.13)$$

The energy-rich bond is consumed. In this way the energy of ATP is at last channeled into the endergonic synthetic reactions by which carbon chains are lengthened. As we shall see, acetyl CoA is the most important and versatile acyl CoA compound.

---

* We shall speak of the transfer of electrons and the transfer of hydrogens in oxidative reactions as if they were synonymous. Justification for this usage appears in Chapter 2.

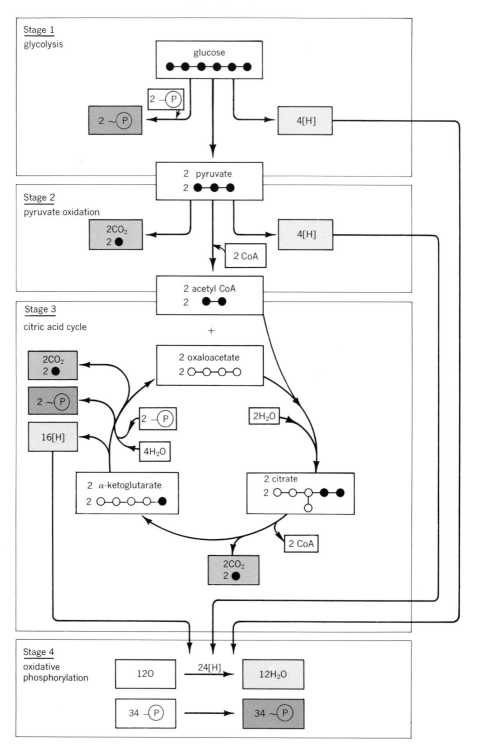

4.2 The four stages of glucose metabolism, showing at which stages energy is released as ~ⓅP, hydrogen is transferred, carbon is released as CO_2, and molecular oxygen is consumed.

each acetyl CoA molecule combines with a four-carbon molecule, oxaloacetic acid, to produce a new six-carbon molecule, citric acid. The citric acid is systematically oxidized, first to a five-carbon molecule and then to a four-carbon molecule. In the course of these reactions, two molecules of CO_2 are released, and four pairs of hydrogens are removed, for each acetyl CoA molecule entering the cycle. The four-carbon molecule emerging from this sequence is oxaloacetic acid, which is free to combine with another acetyl CoA molecule. Hence the cycle begins again.

Stage 3

$$2 \text{ acetyl CoA} + 6H_2O + 2ADP + 2\text{\textcircled{P}} \longrightarrow$$
$$4CO_2 + 2ATP + 2CoA + 16[H] \quad (4.15)$$

This stage has been variously termed the *citric acid cycle,* the *tricarboxylic acid cycle,* and the *Krebs cycle,* in recognition of Sir Hans Krebs, who demonstrated its existence in 1937.

Interestingly, the oxidations of the citric acid cycle convert all of the carbons of acetyl CoA to CO_2, but with the exceptions given in Eq. 4.15, they do not produce any ATP. Instead there is an accumulation, in the form of reduced coenzymes, of the hydrogens transferred in the oxidations (i.e., dehydrogenations), as in the first two stages. By the end of the third stage, 24 hydrogens have accumulated.

In the fourth stage all of these hydrogens are transferred through a complex chain of coenzymatic hydrogen donor-acceptors (including pyridine nucleotides, flavoproteins, and cytochromes; see Chapter 2) to oxygen, the final hydrogen acceptor. Water is formed, and the coenzymes are freed to function again as hydrogen acceptors. Since oxygen is utilized at last, this reaction sequence is termed the *respiratory chain.* During its course the bulk of the ATP is generated.

Stage 4

$$24[H] + 6O_2 + 34ADP + 34\text{\textcircled{P}} \longrightarrow$$
$$12H_2O + 34ATP \quad (4.16)$$

If we add up the equations describing the overall changes in the four stages of glucose metabolism (Eqs. 4.11, 4.14, 4.15, and 4.16), the net result is identical to Eq. 4.10.

We should now examine the four stages more carefully.

Aerobic and anaerobic glycolysis. The list of investigators who elucidated the individual reactions of glycolysis includes the names of the principal architects of modern biochemical thought: Büchner, Harden, Young, Robison, Embden, Meyerhof, Parnas, Needham, Carl and Gerty Cori, Warburg, and others. For, in the solving of this complex riddle, there emerged many of the techniques and concepts that are fundamental to our present understanding of cellular metabolism.

Chronologically, the explication of glycolysis came after studies of the fermentative conversion of glucose to alcohol by yeast. It appears that the desire for alcoholic beverages of quality was sufficiently strong in nineteenth-century society that fermentation technology was more intensively supported than many areas of physics, chemistry, biology, and medicine. Despite the practical goals, however, research on fermentation led also to Pasteur's microbiological discoveries and to modern biochemistry. As yeast was to fermentation, muscle tissue was to glycolysis, but not until the late 1920's was it discovered that, except for the final steps (involving transformations of pyruvate), the two metabolic pathways are identical.*

We shall briefly consider certain aspects and implications of glycolysis (Fig. 4.3), for it is a process whose fundamental importance for many body functions will be repeatedly apparent to us. The first step, a transfer of phosphate from ATP to glucose, is catalyzed by hexokinase, a typical phosphokinase of the type described in Chapter 2. The products of the reaction are glucose 6-phosphate and ADP. The aldohexose glucose 6-phosphate, after conversion to its ketohexose isomer, fructose 6-phosphate, reacts with a second ATP to form a hexose diphosphate, fructose 1,6-diphosphate. This six-carbon sugar splits into two triose phosphate molecules. Although the cleavage products are aldo-keto isomers of one another, only one of them, the aldotriose glyceraldehyde 3-phosphate, can proceed along the glycolytic pathway. Thus the other, the keto-triose dihydroxyacetone phosphate, is continuously converted to the aldo form so that all carbons can be conserved. Since the glucose skeleton has now been split in two, from this point on two molecules are acted

---

* For historical surveys see References and Suggestions for Further Reading. The historical chapters in Fruton and Simmonds, *General Biochemistry,* and Baldwin, *Dynamic Aspects of Biochemistry,* are especially recommended.

TABLE 4.1 THE MAJOR VITAMINS

The vitamins are classified in two groups. Those in Group 1 are water-soluble compounds. With the exception of vitamin C, they are converted in the body to coenzymes that participate in group transfer reactions. Those in Group 2 are families of water-insoluble (organic solvent-soluble) compounds whose biochemical functions are complex or unknown. None in this group appears to form a coenzyme that participates in group transfers.

GROUP 1

Structure and names	Conversion of vitamin to coenzyme	Mechanism of coenzyme action
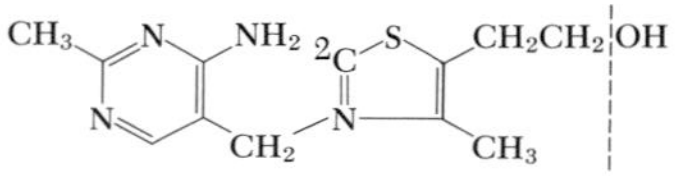 thiamine (T) vitamin B_1	$T + ATP \xrightarrow{Mg^{++}} AMP + T-\text{®}-\text{®}$ thiamine pyrophosphate (TPP)	TPP participates in (1) the transketolase reaction (see Fig. 4.14) and (2) in the removal of CO_2 (i.e., decarboxylation) from the α-keto acids, pyruvate and α-ketoglutarate (see Fig. 4.4). Thus TPP is also known as *cocarboxylase*. Carbon 2 of the 5-membered thiazole ring is the active site. 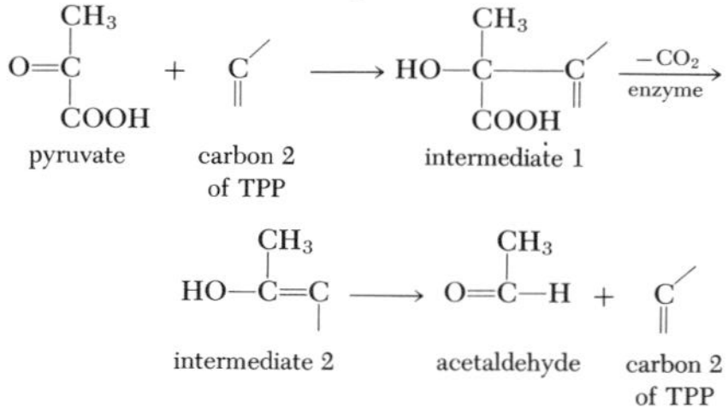
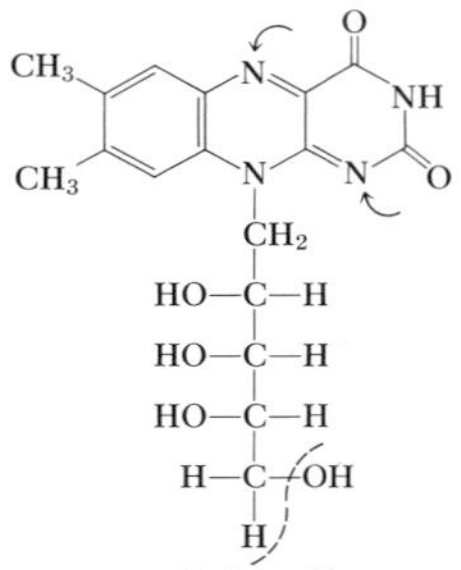 riboflavin (R) vitamin B_2	$R + ATP \longrightarrow ADP + \text{®}-R$ flavin mononucleotide (FMN) $FMN + ATP \longrightarrow \text{®}-\text{®} + \text{®}-\text{®}-\text{®}-\text{ribose-adenine}$ flavin adenine dinucleotide (FAD)	FAD is an acceptor-donor of hydrogens in oxidation-reduction reactions. $$AH_2 + FAD \rightleftharpoons A + FADH_2$$ $$FADH_2 + B \rightleftharpoons BH_2 + FAD$$ FMN functions similarly. The hydrogens attach to the nitrogens indicated by arrows (see left). EXAMPLES: hydrogen transfer in the respiratory chain (see Fig. 4.6); oxidative deamination (Eq. 4.25).

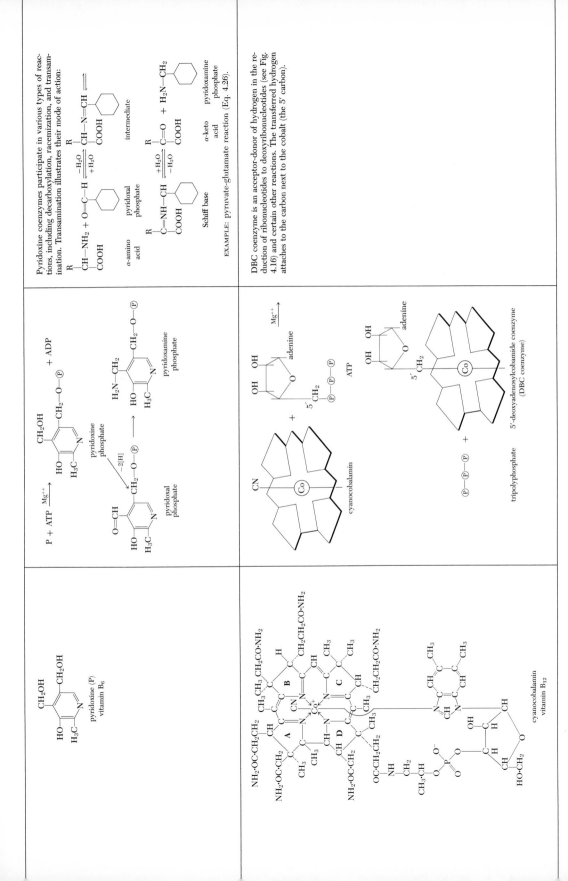

Pyridoxine coenzymes participate in various types of reactions, including decarboxylation, racemization, and transamination. Transamination illustrates their mode of action:

EXAMPLE: pyruvate–glutamate reaction (Eq. 4.26).

DBC coenzyme is an acceptor-donor of hydrogen in the reduction of ribonucleotides to deoxyribonucleotides (see Fig. 4.16) and certain other reactions. The transferred hydrogen attaches to the carbon next to the cobalt (the 5′ carbon).

pyridoxine (P) vitamin B$_6$

$$P + ATP \xrightarrow{Mg^{++}} \text{pyridoxal phosphate} + ADP$$

pyridoxal phosphate

pyridoxamine phosphate

cyanocobalamin

ATP

5′-deoxyadenosylcobamide coenzyme (DBC coenzyme)

tripolyphosphate

cyanocobalamin vitamin B$_{12}$

TABLE 4.1 THE MAJOR VITAMINS (cont.)

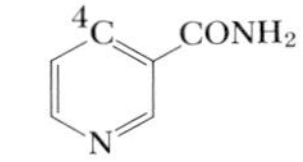

Nicotinamide

nicotinamide (N)

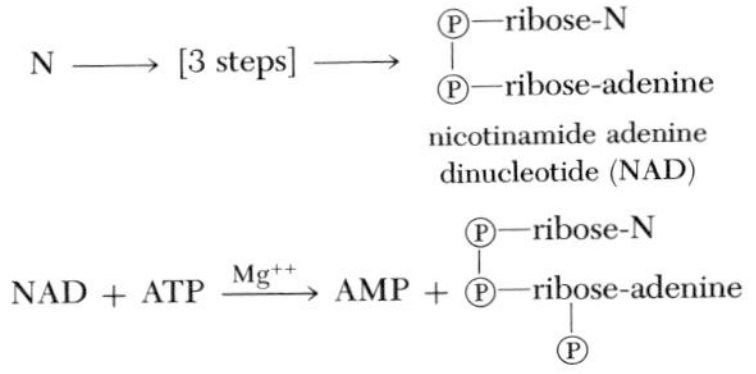

$$N \xrightarrow{\text{[3 steps]}} \quad \textcircled{P}-\text{ribose-N}$$

$$\textcircled{P}-\textcircled{P}-\text{ribose-adenine}$$

nicotinamide adenine dinucleotide (NAD)

$$\textcircled{P}-\text{ribose-N}$$
$$\text{NAD} + \text{ATP} \xrightarrow{Mg^{++}} \text{AMP} + \textcircled{P}-\textcircled{P}-\text{ribose-adenine}$$
$$\textcircled{P}$$

nicotinamide adenine dinucleotide phosphate (NADP)

NAD is an acceptor-donor of hydrogens in many oxidation-reduction reactions.

$$AH_2 + NAD \rightleftharpoons A + NADH_2$$
$$NADH_2 + B \rightleftharpoons BH_2 + NAD$$

NADP functions similarly. The hydrogens are attached to carbon 4 of the nicotinamide moiety (see left).
EXAMPLE: lactic dehydrogenase reaction (Eq. 2.30); glyceraldehyde-3-phosphate dehydrogenase reaction (Eq. 4.18).

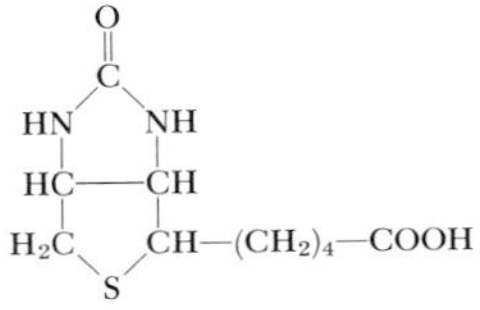

Biotin

biotin (B)

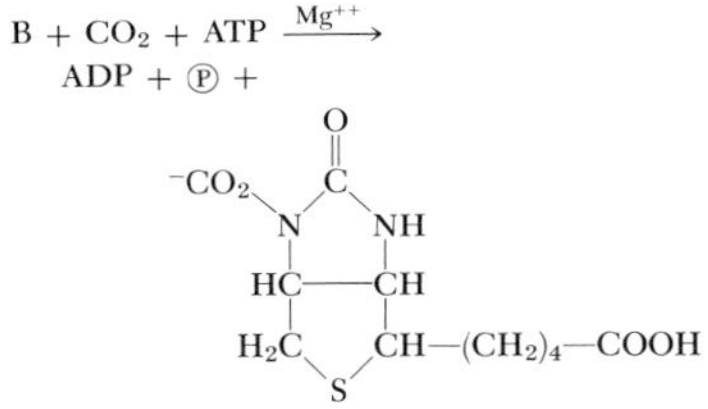

$$B + CO_2 + ATP \xrightarrow{Mg^{++}} ADP + \textcircled{P} +$$

1'-N-carbonylbiotin ($B-CO_2$)

B is an acceptor-donor of CO_2 in carboxylation reactions.

$$B-CO_2 + R-H \xrightarrow{Mg^{++}} R-COOH + B$$
$$\text{CO-S-CoA} \qquad\qquad \text{CO-S-CoA}$$

EXAMPLE: propionyl CoA carboxylase reaction (Eq. 4.24).

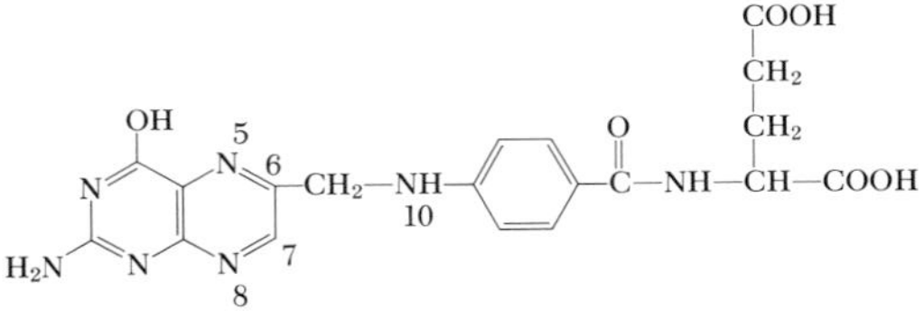

Folic acid (F) / pteroylglutamic acid

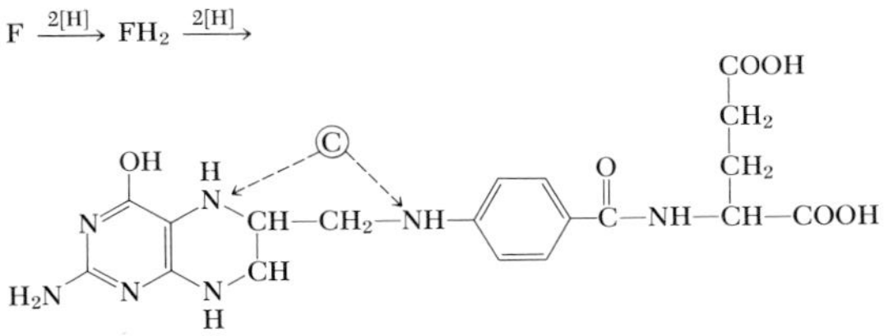

$$F \xrightarrow{2[H]} FH_2 \xrightarrow{2[H]}$$

tetrahydrofolic acid (FH$_4$)

FH_4 is an acceptor-donor of the following one-carbon units: formyl ($-CHO$), hydroxymethyl ($-CH_2OH$), methyl ($-CH_3$), and formimino ($-CHNH$). One-carbon units (symbolized by $\textcircled{C}$) attach to nitrogen 5 or 10. FH_4 participates in transfers of $\textcircled{C}$ from a donor compound, $X-\textcircled{C}$, to an acceptor, Y:

$$X-\textcircled{C} + FH_4 \rightarrow X + \textcircled{C}-FH_4$$
$$\textcircled{C}-FH_4 + Y \rightarrow Y-\textcircled{C} + FH_4$$
$$\text{Sum: } X-\textcircled{C} + Y \rightarrow Y-\textcircled{C} + X$$

EXAMPLE: thymidylate synthetase reaction (see Fig. 4.16)

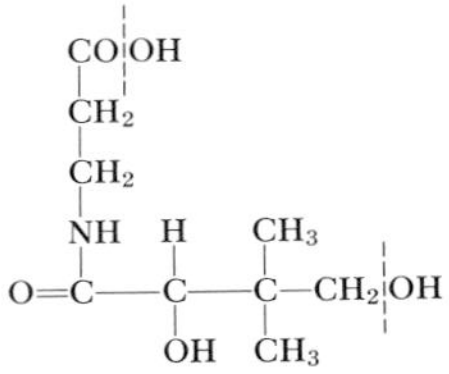

Pantothenic acid

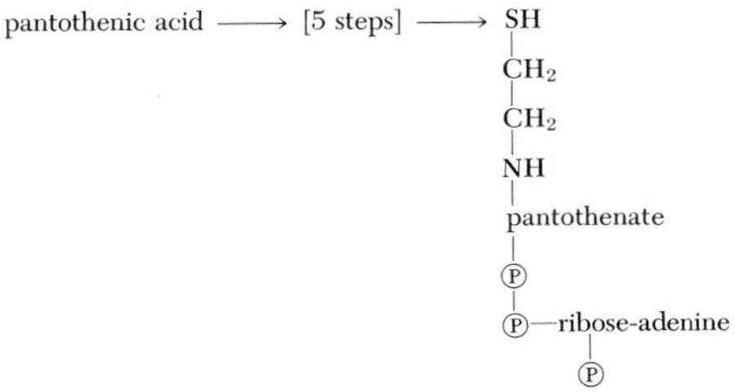

pantothenic acid $\xrightarrow{\text{[5 steps]}}$

coenzyme A (CoA)

CoA is an acceptor-donor of acyl ($R-CO-$) groups which condense with the $-SH$ group to form acyl CoA thioesters (see Eqs. 4.12 and 4.13).

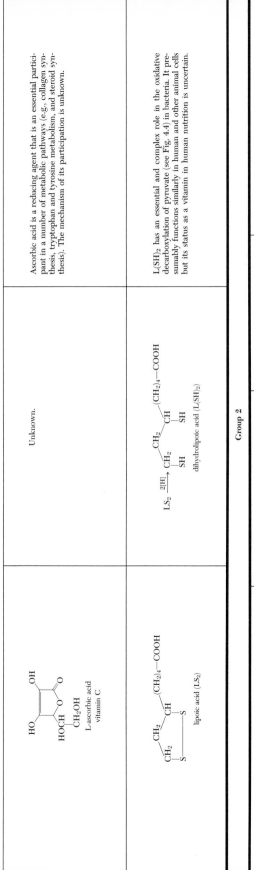

OH

HO

HOCH

CH$_2$OH

L-ascorbic acid
vitamin C

Ascorbic acid is a reducing agent that is an essential partici-
pant in a number of metabolic pathways (e.g., collagen synthesis, tryptophan and tyrosine metabolism, and steroid synthesis). The mechanism of its participation is unknown.

Unknown.

$$LS_2 \xrightarrow{2[H]}$$

CH$_2$
|
CH$_2$
|
CH —(CH$_2$)$_4$—COOH
|
SH   SH

dihydrolipoic acid (L(SH)$_2$)

CH$_2$
|
CH$_2$
|
CH —(CH$_2$)$_4$—COOH
|
S   S

lipoic acid (LS$_2$)

L(SH)$_2$ has an essential and complex role in the oxidative
decarboxylation of pyruvate (see Fig. 4.4) in bacteria. It presumably functions similarly in human and other animal cells
but its status as a vitamin in human nutrition is uncertain.

Group 2

CH$_3$ H CH$_3$ H
| | | |
C   C   C   CH$_2$OH
...
retinol
vitamin A$_1$

Retinol is one of several members of the vitamin A family. These compounds function mainly
in the photochemistry of vision (see Fig. 12.62).

O
CH$_3$—C$_{15}$H$_{33}$
CH$_3$
CH$_3$
HO   CH$_3$

α-tocopherol
vitamin E

α-tocopherol is one of 8 known vitamin E compounds. They are antioxidants, but their precise metabolic role is unknown.

CH$_3$
|
CH—CH$_2$—CH$_2$—CH$_2$—CH
| |
CH$_3$ CH$_3$

CH$_3$

CH$_2$

HO

cholecalciferol
vitamin D$_3$

Cholecalciferol is one of the many members of
the vitamin D family (see Fig. 15.22). They
function mainly in promoting normal calcium
metabolism, bone growth and development.

O
CH$_3$
C$_{20}$H$_{39}$
O

phylloquinone
vitamin K$_1$

Phylloquinone is one of many known vitamin K
compounds (see Fig. 8.35). They are believed to
participate in oxidative phosphorylation.

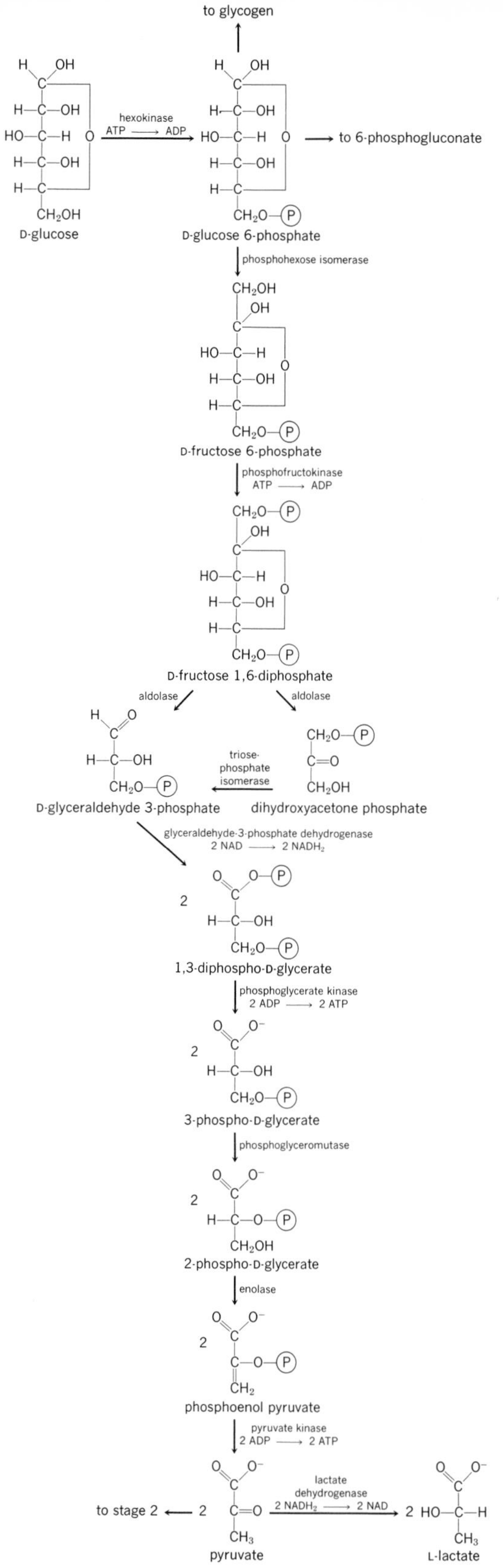

to glycogen

H OH
|__C
H—C—OH
HO—C—H   O
H—C—OH
H—C
CH₂OH
D-glucose

hexokinase
ATP ⟶ ADP

H OH
|__C
H—C—OH
HO—C—H   O
H—C—OH
H—C
CH₂O—Ⓟ
D-glucose 6-phosphate

⟶ to 6-phosphogluconate

phosphohexose isomerase

CH₂OH
OH
C
HO—C—H   O
H—C—OH
H—C
CH₂O—Ⓟ
D-fructose 6-phosphate

phosphofructokinase
ATP ⟶ ADP

CH₂O—Ⓟ
OH
C
HO—C—H   O
H—C—OH
H—C
CH₂O—Ⓟ
D-fructose 1,6-diphosphate

aldolase   aldolase

H   O
C
H—C—OH
CH₂O—Ⓟ
D-glyceraldehyde 3-phosphate

triose-phosphate isomerase

CH₂O—Ⓟ
C=O
CH₂OH
dihydroxyacetone phosphate

glyceraldehyde-3-phosphate dehydrogenase
2 NAD ⟶ 2 NADH₂

O   O—Ⓟ
2   C
H—C—OH
CH₂O—Ⓟ
1,3-diphospho-D-glycerate

phosphoglycerate kinase
2 ADP ⟶ 2 ATP

O   O⁻
2   C
H—C—OH
CH₂O—Ⓟ
3-phospho-D-glycerate

phosphoglyceromutase

O   O⁻
2   C
H—C—O—Ⓟ
CH₂OH
2-phospho-D-glycerate

enolase

O   O⁻
2   C
C—O—Ⓟ
CH₂
phosphoenol pyruvate

pyruvate kinase
2 ADP ⟶ 2 ATP

O   O⁻
2   C
C=O
CH₃
pyruvate

to stage 2 ⟵ 2

lactate dehydrogenase
2 NADH₂ ⟶ 2 NAD

O   O⁻
2   C
HO—C—H
CH₃
L-lactate

4.3 Pathway of glycolysis. Most of the individual enzyme reactions are reversible. For the few that are not, different enzymes exist that catalyze the backward reactions. Thus, as a whole, glycolysis is reversible.

upon in each reaction for every initial glucose molecule.

Glyceraldehyde 3-phosphate is oxidized to an acid by glyceraldehyde-3-phosphate dehydrogenase. This reaction, the only oxidation in glycolysis, illustrates the oxidative production of energy-rich phosphate bonds.* Glyceraldehyde-3-phosphate dehydrogenase contains a sulfhydryl group, which combines with the aldehyde group of the substrate to form an enzyme-substrate complex.

$$R-\overset{\overset{O}{\|}}{C}-H + HS-E \longrightarrow R-\overset{\overset{OH}{|}}{\underset{\underset{H}{|}}{C}}-S-E \qquad (4.17)$$

Oxidation takes place next, the withdrawn hydrogens being transferred to the coenzyme NAD (see p. 53); the energy of the molecule is redistributed; and an energy-rich carbon-sulfur bond is formed.

$$R-\overset{\overset{OH}{|}}{\underset{\underset{H}{|}}{C}}-S-E + NAD \longrightarrow$$

$$R-\overset{\overset{O}{\|}}{C}\sim S-E + NADH_2 \qquad (4.18)$$

If the compound with the new carbon-sulfur link were then merely hydrolyzed, the products would include an acid (note the carboxyl group), regenerated HS—E, and a large amount of heat.

$$R-\overset{\overset{O}{\|}}{C}\sim S-E + H_2O \longrightarrow$$

$$R-\overset{\overset{O}{\|}}{C}-OH + HS-E + heat \qquad (4.19)$$

However, energy is not wasted as heat in a useless hydrolytic reaction but is retained in an energy-rich phosphate bond. This means that the product of Eq. 4.18 reacts with phosphoric acid (H_3PO_4)—undergoing phosphorolysis, not hydrolysis.

$$R-\overset{\overset{O}{\|}}{C}\sim S-E + Ⓟ \longrightarrow$$

$$R-\overset{\overset{O}{\|}}{C}-O\sim Ⓟ + HS-E \qquad (4.20)$$

_____

* Dehydration, the other mechanism for producing energy-rich phosphate bonds, was described on p. 119.

The energy-rich phosphate bond is then transferred to ADP, generating ATP and 3-phosphoglycerate.

The latter compound is now converted to 2-phosphoglycerate, which, as we saw in Eq. 4.6, is dehydrated to phosphoenolpyruvate, the dehydration producing another energy-rich phosphate bond. In the presence of pyruvate kinase, it is transferred to ADP to form more ATP and pyruvate.

In summary, at the end of the first stage, (1) one molecule of glucose has been converted to two molecules of pyruvate; (2) two molecules of glyceraldehyde 3-phosphate have been oxidized, and simultaneously two molecules of NAD have been reduced to $NADH_2$; (3) two molecules of ATP have been employed in the early phosphorylations needed to launch the process; and (4) four molecules of inorganic phosphate have been converted to energy-rich phosphate bonds, which have been transferred to ADP to produce four molecules of ATP, thus repaying the ATP debt incurred earlier and yielding a net profit of two ATP molecules.

Pyruvic acid (or its ion, pyruvate) is one of several intermediates that stand at metabolic crossroads. Others in this category are glucose 6-phosphate and acetyl CoA. Under certain circumstances pyruvic acid is reduced to lactic acid,* the necessary hydrogens coming from the $NADH_2$ produced in the oxidation of glyceraldehyde 3-phosphate. This reaction regenerates NAD and completes glycolysis. The entire glycolytic process can occur in the absence of oxygen—in which case it is called *anaerobic glycolysis*—since its only oxidative step can be balanced by a later reduction. Thus oxygen is not needed as a terminal hydrogen acceptor. Though anaerobic glycolysis is a relatively inefficient if self-contained method of extracting energy from glucose, some cells (e.g., red blood cells) carry the metabolism of glucose no further. Most cells, however, complete the oxidation of pyruvate in the next three stages of metabolism, wherein oxidative dehydrogenation is offset ultimately by the reductive hydrogenation of oxygen.

Many cells of the latter type were early shown to be capable of surviving with anaerobic glycolysis if they were deprived of oxygen. Indeed, Pasteur was the first to observe that in the ab-

sence of oxygen the rate of lactic acid production increases. This phenomenon is termed the *Pasteur effect*. Its precise mechanism is not yet entirely clear.

Pyruvate oxidation and the citric acid cycle. In the presence of oxygen, the bulk of the pyruvate is converted to acetyl CoA. This complex multistep process requires four coenzymes: NAD, CoA, thiamine pyrophosphate (TPP), and lipoic acid (Fig. 4.4, Table 4.1). For each pyruvate molecule the reaction sequence of stage two removes one carbon as CO_2 and generates one molecule each of acetyl CoA and $NADH_2$. TPP and lipoic acid participate catalytically, emerging from the series of reactions in the form in which they entered.

Acetyl CoA is a versatile crossroads compound, as shown in Fig. 4.5. This explains why the citric acid cycle is often called the "final common pathway" of metabolism. Though most acetyl CoA normally comes from carbohydrate breakdown, the citric acid cycle cannot be considered a terminal pathway of carbohydrate metabolism alone. Furthermore, a substantial part of the acetyl CoA escapes oxidation to CO_2 and H_2O because acetyl CoA is a fundamental biosynthetic building block, a function for which its carbons and energy-rich thioester bond admirably suit it (see footnote, p. 121).

Acetyl CoA initiates the citric acid cycle by entering into a synthetic reaction (see Fig. 4.4). It condenses with oxaloacetate to form citrate, a six-carbon acid containing three carboxyl groups. The energy of the thioester bond is consumed in the reaction. Citrate is then converted, by the removal of water, to *cis*-aconitate; *cis*-aconitate is hydrated to isocitrate (an isomer of citrate); and isocitrate is oxidized to oxalosuccinate. Liberation of a CO_2 molecule produces α-ketoglutarate, which is subjected to a series of conversions almost identical to those occurring in the oxidation of pyruvate (also an α-keto acid), the products here being succinate and CO_2. Succinate is then oxidized to fumarate, which accepts water to form malate. The oxidation of malate regenerates oxaloacetate, which condenses with acetyl CoA to begin a new cycle. When all the necessary enzymes and coenzymes are present, only a continuing supply of acetyl CoA is needed to keep the cycle going.

---

* In yeast cells pyruvate is converted to ethanol.

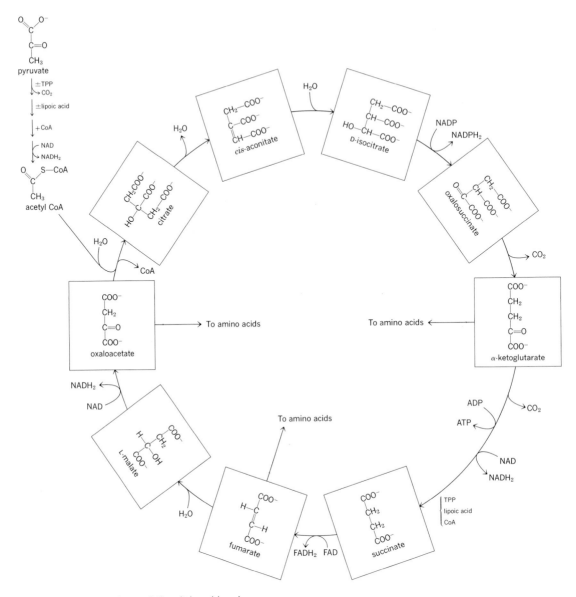

4.4 Pyruvate oxidation and the citric acid cycle.

Oxidative phosphorylation. At the completion of two turns of the citric acid cycle, the six original glucose carbons have been converted to CO_2, six molecules of H_2O have been utilized, and 12 pairs of hydrogens have been transferred to coenzymatic hydrogen acceptors. Only four molecules of ATP have been generated, however, and so stage four, oxidative phosphoryla-tion, must accomplish these objectives: (1) transfers of the 24 accumulated coenzyme-bound hydrogens to 12 oxygens so that coenzyme is regenerated for further hydrogen transport; (2) coupling of these hydrogen transfers to the conversion of inorganic phosphate groups to $\sim$℗ groups and their subsequent transfer to ADP to make ATP.

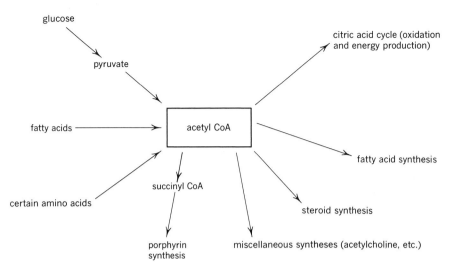

4.5 The central position of acetyl CoA in intermediary metabolism.

Though the mechanisms are not yet wholly understood, it is established that coenzyme-borne hydrogens are transferred from one carrier to another through a reaction sequence similar to that shown in Fig. 4.6. This *respiratory chain* is a series of interlocking oxidation-reduc-

tion cycles. Each carrier is reduced when it receives hydrogen and reoxidized when it transfers hydrogen to the next link in the chain. The transfer of hydrogen from one carrier to another takes place because each carrier has a lower redox potential, or E_0' (see p. 43), than the next

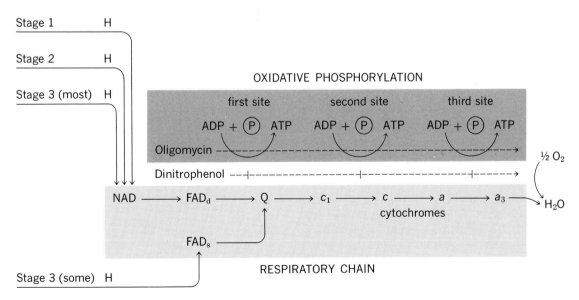

4.6 A current theory of oxidative phosphorylation, showing the coupling of oxidations in the respiratory chain to the several phosphorylations of ADP to ATP. X, Y, Z are postulated (still unidentified) enzymes.

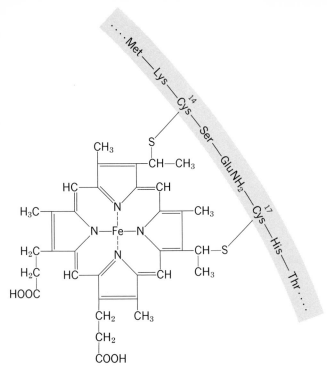

4.7 Human cytochrome c. The iron-porphyrin portion is joined to the protein portion by two thioether (—C—S—C—) bonds involving the sulfurs of two cysteine residues in the protein, a single polypeptide chain of 104 amino acid residues.

carrier in the chain and therefore tends to reduce it. The respiratory chain has been compared to a series of small waterfalls. The water (hydrogen) starts at a high energy level (at the substrate) and ends at a low energy level (at oxygen). The flow is one way, although, if energy is put into the system experimentally, its direction can be reversed.

The first step in the sequence is the transfer of hydrogen from the citric acid cycle intermediates of stage three. Most of it is transferred to NAD. The $NADH_2$ thus formed combines with that formed in stages one and two (glycolysis and pyruvate oxidation). The hydrogen carrier next in line is FAD_d (flavin adenine dinucleotide), which contains the vitamin riboflavin. As noted in Table 4.1, its structure is identical to that of NAD except that NAD contains the vitamin nicotinamide. FAD_d accepts hydrogen from $NADH_2$ to become $FADH_2$. Note in Fig. 4.6 that some of the hydrogens (actually one pair)

from the citric acid cycle (stage three) bypass NAD, entering the respiratory chain via another flavoprotein, FAD_s.

$FADH_2$ transfers hydrogen to coenzyme Q, a substance chemically related to vitamin K. Hydrogen is then transferred to a series of acceptors called *cytochromes*. A cytochrome is a chromoprotein (see p. 33) whose prosthetic group is a complex iron-porphyrin structure (Fig. 4.7) similar to that of the hemoglobin molecule. It consists of a large ring made up of four smaller pyrrole rings and a central iron atom. Though the various cytochromes (c_1, c, a, and a_3) differ in their protein structures, their iron-porphyrin groups all function similarly as hydrogen carriers, the iron atom being reversibly reduced.

$$Fe^{+++} + e^- \rightleftharpoons Fe^{++} \qquad (4.21)$$

The first cytochrome to receive hydrogen from $FADH_2$ is cytochrome c_1. In turn, cytochrome c, cytochrome a, and cytochrome a_3 are alternately reduced and oxidized. Finally hydrogen is transferred to oxygen in a reaction catalyzed by cytochrome oxidase.* Cyanide and carbon monoxide are powerful inhibitors of this reaction; this accounts for their action as respiratory poisons.

In the respiratory chain are three particular oxidation-reduction waterfalls in which oxidation is coupled with the synthesis of a $\sim\!\!\text{\textcircled{P}}$ group, which is then transferred to ADP to form ATP. The sites are (1) between $NADH_2$ and coenzyme Q, (2) between coenzyme Q and cytochrome c, and (3) between cytochrome a and a_3. Therefore, for every substrate transferring two hydrogens to NAD (and ultimately to one atom of oxygen), three ATP molecules are formed.†

---

* It is a striking fact that the process of photosynthesis, which involves a diametrically opposite series of events—uptake of carbon dioxide, reduction of carbon, and synthesis of sugar—also has a hydrogen-carrying chain containing NAD and FAD. But its iron-porphyrin compound is chlorophyll, and it dehydrogenates water to form oxygen, whereas the respiratory chain hydrogenates oxygen to form water.

† This picture has been modified by the discovery that one pair of hydrogens enters the respiratory chain at the middle and thus yields only two molecules of ATP. The deficit is made up, however, by the conversion of ADP to ATP during the oxidation of α-ketoglutarate in the citric acid cycle. The respiratory chain remains the primary site of $\sim\!\!\text{\textcircled{P}}$ production.

As shown in Fig. 4.6, two chemical substances profoundly affect oxidative phosphorylation. Accordingly, they have served as useful tools for the study of its mechanism. One is dinitrophenol, which uncouples the oxidations of the respiratory chain from the phosphorylations of ADP so that respiration proceeds normally, but it produces heat instead of ATP. The other is oligomycin, an antibiotic which acts by inhibiting the production of ATP. As long as phosphorylation and respiration are coupled, oligomycin also inhibits respiration. When dinitrophenol is added, respiration becomes normal but no ATP is produced.

How is the coupling of respiration and phosphorylation brought about? Two views are currently held on the nature of the mechanism. One, originally suggested by E. C. Slater in 1953, holds that during the transport of hydrogen, high-energy intermediate compounds are formed at the three coupling sites which are composed of hydrogen carriers of the respiratory chain and an unknown substance (X). For example, let us consider the first phosphorylation site in Fig. 4.6, though similar events are believed to occur in the second and third phosphorylations. The hydrogen carriers NAD and FAD_d are represented by A and B. The central idea is that the coupling between hydrogen transport and phosphorylation involves an intermediate compound, $A\sim X$.

$$AH_2 + X \rightleftharpoons AH_2{-}X$$
$$AH_2{-}X + B \rightleftharpoons A\sim X + BH_2$$
$$A\sim X + ⓅP \rightleftharpoons X\sim ⓅP + A$$
$$X\sim ⓅP + ADP \rightleftharpoons X + ATP \qquad (4.22)$$

AH_2 is eventually oxidized to A, its hydrogens being transferred to B. X enters as a free carrier and leaves in the same form. The high-energy bond created during the oxidation of the intermediate $AH_2{-}X$ is transmitted to inorganic phosphate, which emerges as the terminal $\sim$Ⓟ of ATP.

A second view has recently been proposed by Peter Mitchell. Instead of the high-energy intermediate compound postulated by Slater, Mitchell suggests that the energy for ATP production arises from an electric potential developing during respiration. Hydrogens in the form

of positively charged hydrogen ions (protons) move to one side of the mitochondrial membrane while negatively charged electrons move to the other side. The separation of charges results in the formation of a high-energy intermediate, $X\sim Y$, which powers the formation of ATP. Membrane integrity would be essential for this mechanism and Mitchell considers that uncouplers such as dinitrophenol act by making the membrane "leaky" to protons, thus preventing a separation of charges.

In summary, oxidative phosphorylation involves the transfer of the 24 hydrogens arising in the first three stages of metabolism to oxygen via the following sequence of carriers: NAD $\longrightarrow$ Q $\longrightarrow$ FAD $\longrightarrow$ cytochromes $\longrightarrow$ O_2. Simultaneously, 34 molecules of inorganic phosphate are converted to $\sim$Ⓟ groups of ATP (excluding the four formed in stages one and three). The exact nature of the mechanisms coupling these two processes is still uncertain.

Role of the mitochondria. The main reason for the great difficulty in elucidating the mechanism of oxidative phosphorylation lies in the physical arrangement of the respiratory and phosphorylative enzymes in the mitochondrial membrane. It has long been apparent that the enzymatic machinery of the citric acid cycle and oxidative phosphorylation is too complex to function efficiently if simply dissolved in the fluids of the cell. We now know that these enzymes exist in a highly ordered state within the mitochondria. That they are all there is indicated by the fact that isolated mitochondria can convert pyruvate to CO_2 and H_2O.

In the late 1950's, a number of laboratories made significant progress in breaking up mitochondria by means of intense sound waves or detergents. Internal fluids escape, and insoluble membrane fragments are isolated by centrifugation. The enzymes of the respiratory chain are found exclusively in the membrane fragments. Careful investigation of the catalytic properties of the fragments has strongly suggested that these enzymes are organized in precise spatial patterns.

These theoretical estimates received impressive support when the high-resolution electron micrographs of H. Fernández-Morán revealed ribosome-like round particles, each 85 Å in di-

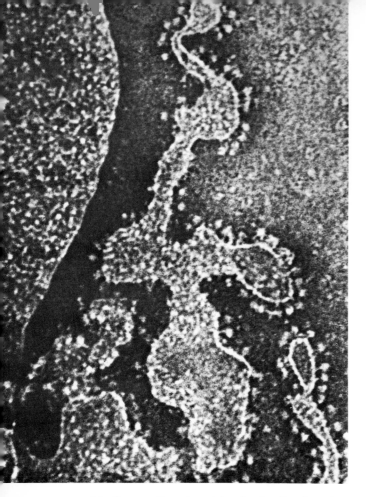

4.8 Cross section of a crista, showing the elementary particles of the mitochondrion. The base pieces of the particles form a continuous layer around the crista. (H. Fernández-Morán.)

ameter, on the inner surface of the inner mitochondrial wall (Fig. 4.8; see also Fig. 3.12). The particles, called *inner membrane spheres*, have stalks. It is now known from the brilliant studies of E. Racker, that these spheres contain an interesting ATPase (a specific phosphatase that cleaves ATP to ADP and $\textcircled{P}$). The inner membrane to which the spheres are attached contains the enzymes of the respiratory chain. The outer mitochondrial membrane carries the enzymes of the citric acid cycle and fatty acid oxidation. Released hydrogens shuttle across the fluid-filled intramembrane space to the inner wall and its stalks and particles in which are located the enzymes of the respiratory chain and its associated phosphorylations.

Racker successfully managed to remove the spheres from the inner membrane and put them back on again and in doing so he made the important observation that sphere ATPase is inhibited by oligomycin when it is attached to the inner membrane but resistant to oligomycin when separated from the membrane and solubilized. The term *allotopy* was coined to designate this phenomenon—the alteration of an enzyme's properties depending on whether it is in solution or bound to a membrane. Clearly, allotopic properties are importantly related to the functions of membranes. Racker suspects that ATPase of the spheres may be an important coupling factor that along with other coupling factors links respiration with phosphorylation. Conceivably, it participates in the formation of the $X \sim Y$ intermediate envisioned by the Mitchell hypothesis.

Like other membranes, that of a mitochondrion is semipermeable—that is, some substances (e.g., citrate) traverse it with difficulty whereas ions like K^+ are taken up against a concentration gradient (see p. 79). Moreover, a mitochondrion swells and shrinks, its membrane changing physical dimensions in the course of metabolic activity. Like a sheet of muscle tissue, it relaxes or contracts in apparent response to the local ATP concentration: it contracts when the ATP concentration is high and relaxes when it is low. Thus, like a muscle fibril, a mitochondrion contains a contractile protein that appears to respond to ATP addition. The significance of mitochondrial swelling and shrinking is still uncertain. Presumably it is part of a regulatory mechanism whereby local concentrations of ATP determine the total rate of new ATP production. Overproduction of ATP would automatically throttle down the system and gear its rate of power production to the cell's needs.

The efficient integration of function and structure at the molecular level are beautifully illustrated by the mitochondrial membrane. Certain of its features (e.g., two protein-phospholipid membranes separated by a space, a contractile protein, the capacity to transport ions selectively) are universal properties of all cell systems having to do with the transformation or utilization of energy, and we shall encounter them again later.

RELEASE OF ENERGY FROM
FATTY ACIDS AND AMINO ACIDS

The discovery in the early 1950's that acetyl CoA is the main product of fatty acid breakdown explained why lipid is an efficient energy source. This discovery, however, was foreshadowed by the classic experiment performed by F. Knoop in 1904. The Knoop experiment demonstrates, incidentally, how molecules were labeled in the days before the advent of radioisotopes in biochemistry.

The β-oxidation of fatty acids. Knoop prepared a series of phenyl derivatives of fatty acids, each having the phenyl group on the carbon farthest from the carboxyl group—e.g.,

$$\langle\rangle\text{---}(CH_2)_3COOH$$

was his phenyl derivative of the fatty acid $CH_3(CH_2)_2COOH$ (butyric acid)—and fed them to dogs. He observed that the dogs fed phenyl derivatives of fatty acids containing odd numbers of carbons excreted benzoylglycine,

$$\langle\rangle\text{---}CO\text{---}CH(NH_2)COOH$$

a compound formed from the conjugation of benzoic (or phenylformic) acid,

$$\langle\rangle\text{---}COOH$$

with glycine, $CH_2(NH_2)COOH$, and that dogs fed phenyl derivatives of fatty acids containing even numbers of carbons excreted phenylacetylglycine,

$$\langle\rangle\text{---}CH_2CO\text{---}CH(NH_2)COOH$$

a compound formed from the conjugation of phenylacetic acid,

$$\langle\rangle\text{---}CH_2COOH$$

with glycine. For our present purposes we can ignore the conjugation that occurred after fatty acid degradation was complete. The results were as follows:

Substituted acid fed	Carbons in original fatty acid	Compound excreted
⟨⟩—CH₂CH₂COOH	3	Benzoic acid
⟨⟩—CH₂CH₂CH₂COOH	4	Phenylacetic acid
⟨⟩—CH₂CH₂CH₂CH₂COOH	5	Benzoic acid
⟨⟩—CH₂CH₂CH₂CH₂CH₂COOH	6	Phenylacetic acid

. . .   .

Knoop and later workers interpreted these data as evidence that the organism is unable to remove carbon atoms from the chain one at a time. If it could, odd- and even-numbered acids would all yield benzoic acid. Presumably, then, carbon atoms are split off in even-numbered fragments, probably in pairs. Knoop suggested that the loss of each two-carbon unit was preceded by oxidation of the β carbon (the second carbon away from the carboxyl carbon)—i.e., that β-oxidation was followed by the loss of a two-carbon unit (the α and carboxyl carbons) and that the remainder of the chain underwent repeated β-oxidations and two-carbon removals.

$$CH_3\cdots\text{---}CH_2\text{---}CH_2\text{---}CH_2\text{---}CH_2\overset{\beta}{\text{---}CH_2}\overset{\alpha}{\text{---}CH_2}\text{---}COOH$$

$$\downarrow \text{β-oxidation}$$

$$CH_3\cdots\text{---}CH_2\text{---}CH_2\text{---}CH_2\text{---}CH_2\overset{O}{\overset{\|}{\text{---}C}}\text{---}CH_2\text{---}COOH$$

$$\downarrow +H_2O \text{ cleavage}$$

$$CH_3\cdots\text{---}CH_2\text{---}CH_2\overset{\beta}{\text{---}CH_2}\overset{\alpha}{\text{---}CH_2}\text{---}COOH \text{ plus a two-carbon unit}$$

$$\downarrow \text{β-oxidation}$$

$$CH_3\cdots\text{---}CH_2\text{---}CH_2\overset{O}{\overset{\|}{\text{---}C}}\text{---}CH_2\text{---}COOH$$

$$\downarrow +H_2O \text{ cleavage}$$

$$CH_3\cdots\overset{\beta}{\text{---}CH_2}\overset{\alpha}{\text{---}CH_2}\text{---}COOH \text{ plus a two-carbon unit}$$

$$\downarrow \text{β-oxidation}$$

. . .   (4.23)

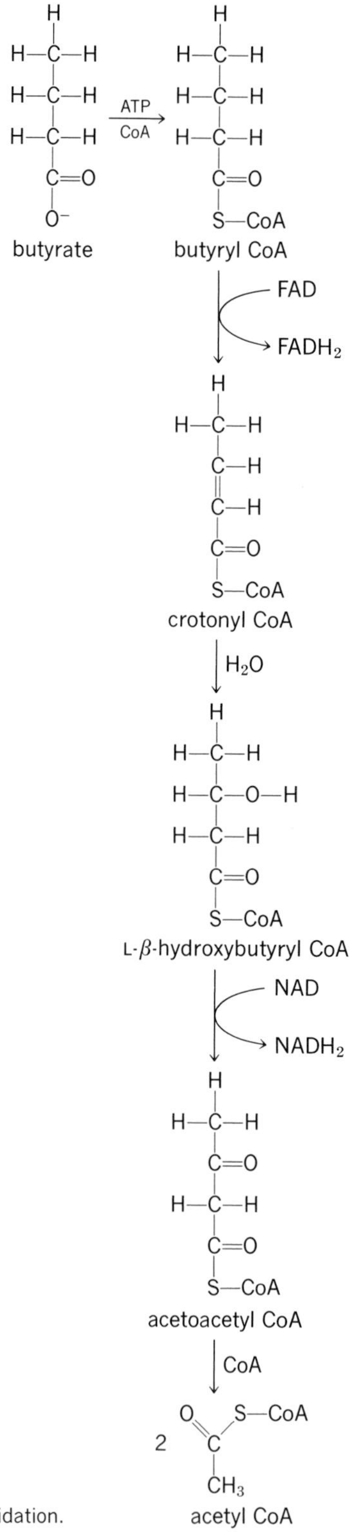

4.9 Fatty acid oxidation.

The acids used in the feeding experiment were artificially synthesized. Naturally occurring fatty acids almost always contain even numbers of carbons (see Table 2.4). Therefore, the scheme implies that most natural long-chain acids can be completely degraded into two-carbon units.

When the actual pathway of fatty acid breakdown was discovered over 50 years later, the essence of Knoop's hypothesis was verified. It is now known that a fatty acid must be "activated" —i.e., converted to its CoA derivative by the ATP-dependent reaction of Eq. 4.12—before it can be oxidized. For example, in the oxidation of butyrate (Fig. 4.9), butyrate is first activated to butyryl CoA. Then butyryl CoA is oxidized by FAD and a double bond appears between the α and β carbons. This is the β-oxidation. Water is added across the double bond, and an OH group attaches to the β carbon in a reaction analogous to the hydration of fumarate to malate (see Fig. 4.4). Oxidation follows, leaving a keto group attached to the β carbon. The compound formed reacts with free CoA to produce two molecules of acetyl CoA—the two-carbon unit predicted by Knoop. This sequence, as we have noted, takes place in the outer wall of the mitochondrial membrane.

The rare fatty acid with an odd number of carbons goes through a similar sequence. With an 11-carbon fatty acid, acetyl CoA units are liberated in regular steps ($C_{11} \longrightarrow C_9 \longrightarrow C_7 \longrightarrow C_5$) down to propionyl CoA (C_3CoA). It has recently been shown that propionyl CoA takes up a molecule of CO_2 to form a four-carbon compound, methylmalonyl CoA, that is an isomer of succinyl CoA.

$$\begin{array}{l} CH_3 \\ | \\ CH_2 \\ | \\ CO\!-\!SCoA \end{array} + CO_2 \xrightarrow[Mg^{++}]{ATP} \begin{array}{l} CH_3 \\ | \\ CH\!-\!COOH \\ | \\ CO\!-\!SCoA \end{array} \quad (4.24)$$

We shall later encounter a similar reaction in which acetyl CoA is carboxylated to malonyl CoA. This occurs early in fatty acid synthesis. Both reactions are mediated by the vitamin biotin (see Table 4.1). A specific isomerase converts methylmalonyl CoA to succinyl CoA, which enters the citric acid cycle as succinate (see Fig. 4.4).

The oxidation of fatty acids results in a rich harvest of energy. The oxidation of one molecule of butyrate, for example, produces two molecules of acetyl CoA. Their oxidation in the citric acid cycle yields eight pairs of hydrogens—an amount sufficient to reduce 8 atoms of oxygen and to generate 24 molecules of ATP in oxidative phosphorylation, since the P:O ratio is 3:1. The five reactions taking place before the entrance of acetyl CoA into the citric acid cycle yield two pairs of hydrogens. Since one is in $FADH_2$ and the other in $NADH_2$, only five molecules of ATP are formed (the $FADH_2$ bypassing the first phosphorylation). Thus the oxidation of a four-carbon fatty acid produces $24 + 5 - 1$ (the one used in the initial conversion of a butyrate to butyryl CoA) = 28 ATP molecules.

Oxidation of palmitic acid, a 16-carbon chain, yields a total of 131 ATP molecules.* Thus for butyric acid $^{28}\!/_4 = 7.0$ ATP's are generated per fatty acid carbon oxidized, and for palmitic acid $^{131}\!/_{16} = 8.2$ ATP's are generated per fatty acid carbon oxidized. When these yields are compared to the $^{38}\!/_6 = 6.3$ ATP's generated per glucose carbon oxidized, it is evident that more energy can be obtained from fatty acids than from carbohydrates of equivalent chain lengths.

Amino acid degradation. At least a portion of the total quantity of each amino acid needed in protein synthesis enters the cells ready-made. Indeed this is the only source of the 10 or 12 amino acids that the cells of the human body cannot make for themselves. We shall refer later to the mechanisms by which the other amino acids are synthesized. Here we shall take note of the contribution of amino acid catabolism to the production of energy.

Cells contain two major enzyme systems that remove the amino groups of amino acids and convert them to citric acid cycle intermediates.

---

* The β-oxidation of palmitate produces acetyl CoA seven times $(7 \times 5\text{ATP} = 35)$, the last two carbons remaining in the form of acetyl CoA. These acetyl CoA's produce eight turns of the citric acid cycle $(8 \times 12\text{ATP} = 96)$. Hence $35 + 96 = 131$.

The enzymes called *amino acid oxidases* carry out *oxidation deamination.*

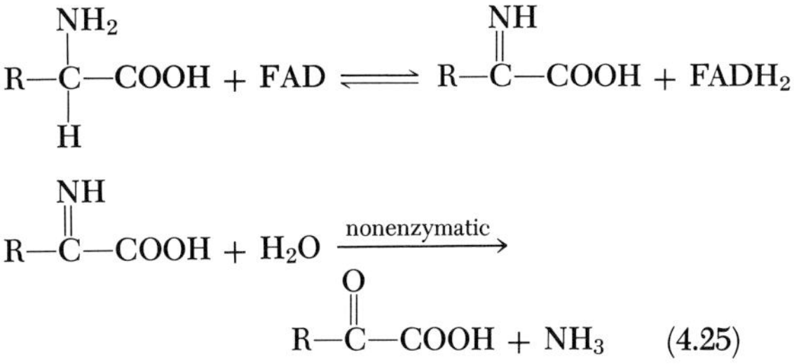

$$
\begin{array}{c}
NH_2 \\
\mid \\
R\!-\!C\!-\!COOH + FAD \\
\mid \\
H
\end{array}
\rightleftharpoons
\begin{array}{c}
NH \\
\parallel \\
R\!-\!C\!-\!COOH + FADH_2
\end{array}
$$

$$
\begin{array}{c}
NH \\
\parallel \\
R\!-\!C\!-\!COOH + H_2O
\end{array}
\xrightarrow{\text{nonenzymatic}}
\begin{array}{c}
O \\
\parallel \\
R\!-\!C\!-\!COOH + NH_3
\end{array} \quad (4.25)
$$

The keto acids formed are citric acid cycle intermediates that can be oxidized to CO_2 and H_2O.

The enzymes called *transaminases* carry out *transamination,* a reversible reaction between an amino acid ($RCHNH_2COOH$) and a keto acid ($R'COCOOH$) in which the amino and keto groups exchange places.

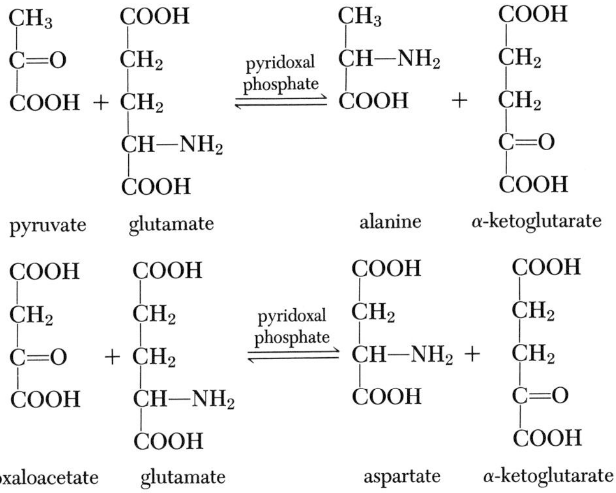

pyruvate   glutamate   →(pyridoxal phosphate)   alanine   α-ketoglutarate

oxaloacetate   glutamate   →(pyridoxal phosphate)   aspartate   α-ketoglutarate

(4.26)

The amino acid becomes a keto acid, and vice versa. Transaminations convert alanine to pyruvate, glutamate to α-ketoglutarate, and aspartate to oxaloacetate—the three resulting keto acids proceeding through the citric acid cycle. Since transaminations are reversible, it is evident that citric acid cycle intermediates may provide carbon skeletons for the syntheses of two amino acids. Thus the citric acid cycle is an "amphibolic" pathway. Vitamin B_6, pyridoxine, in the form of pyridoxal phosphate, is an essential coenzyme in all transaminations. The mechanism of its participation is shown in Table 4.1.

MECHANISMS OF ENERGY STORAGE

The preceding discussion has surveyed the energy-generating systems of the cell. We have seen that despite the diversity of the molecules degraded, the mechanisms for extracting their energy are simple and uniform. The fundamental process consists of oxidative hydrogen (or electron) transfers coupled to reactions producing energy-rich bonds, whose energy is available for biosynthesis and cell function. The basic reactions are oxidation, reduction, dehydration, hydration, decarboxylation, acetylation, phosphorylation, deamination, and transamination. All of the three main classes of cell constituents—carbohydrates, proteins, and lipids—are susceptible to energy-yielding catabolism, the three key oxidizable substances that they form being acetyl CoA, α-ketoglutarate, and oxaloacetate (Fig. 4.10). With their oxidation in the citric acid cycle, the bulk of the cellular energy is produced.

Glycogen synthesis. Although a large amount of potential energy can be stored in energy-rich phosphate and thioester bonds, cells require a reserve supply of oxidizable nutrient materials

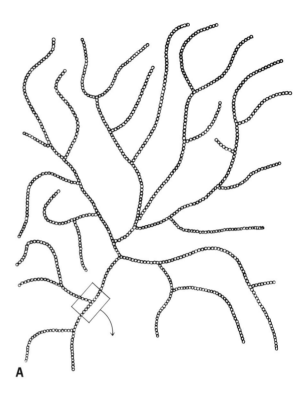

A

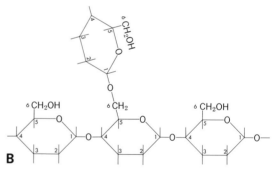

B

4.11 Structure of glycogen. A. General arrangement of the molecule, with each circle representing a glucose moiety; B. Enlargement of a segment of the molecule, showing bonds. An explanation of the glucose ring formula is given in Fig. 2.5. (A from E. Baldwin, *The Nature of Biochemistry*, Cambridge Univ. Press, 1962.)

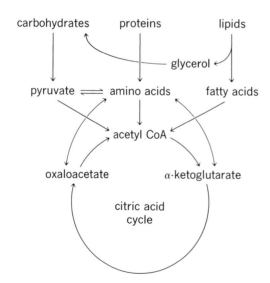

4.10 Interrelations of carbohydrate, protein, and lipid metabolism.

that can be drawn upon during periods of nutrient limitation or intense physiological activity. Lipid provides such a supply, and fat deposits in the body can grow to massive proportions.

Another important energy reservoir in the cell

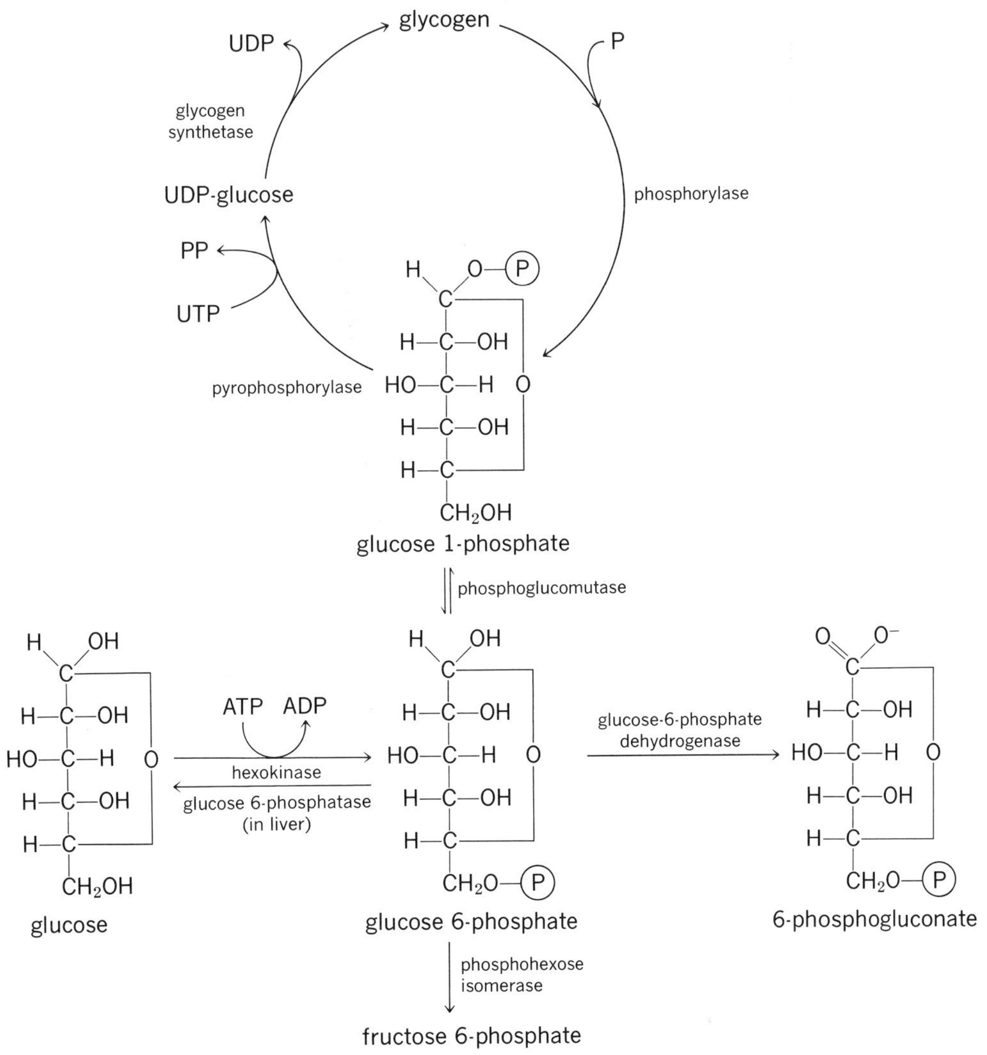

4.12 Glycogen synthesis and breakdown.

is glycogen. Glycogen is formed in every body cell but in largest quantities in liver and muscle cells. Structurally it is a branched polysaccharide whose monosaccharide units are glucose. An insoluble compound with a molecular weight of several million, it serves admirably as a store of glucose. As shown in Fig. 4.11, the glucose molecules in glycogen are linked together by (glycosidic) bonds between carbon 1 of one glucose and carbon 4 of the next glucose. Branching results from occasional bonds between carbon 1 and carbon 6.

Until 1960 it was thought that the pathway of glycogen synthesis is merely a reversal of the pathway of glycogen breakdown. The important discovery that this is not so resulted in part from studies of rare individuals in whom a mutation had eliminated the capacity for glycogen breakdown without affecting the capacity for its synthesis. The pathway of glycogen synthesis begins at the crossroads compound glucose 6-phosphate (Fig. 4.12). The enzyme phosphoglucomutase produces glucose 1-phosphate, which in turn is polymerized to polyglucose (glycogen) through the intervention of a uridine nucleotide coenzyme (UTP). The breakdown of glycogen is a phosphorylytic cleavage catalyzed by the enzyme phosphorylase.

$$\text{(glucose)}_n + n\text{P} \longrightarrow$$
$$n\text{(glucose 1-phosphate)} \quad (4.27)$$

where n is a large number. The phosphoglucomutase reaction is reversible, and, after the formation of glucose 6-phosphate, glycolysis occurs via the usual pathway to pyruvate. Since glycogen was split by inorganic phosphate, only one molecule of ATP is needed to produce a molecule of hexose diphosphate. Thus, when glycolysis utilizes glycogen instead of free glucose as the source of glucose 6-phosphate, there is a net gain of three molecules of ATP for each glucose unit of glycogen metabolized.

Biosynthesis of Cell Constituents

In the second of the major functions common to all cells, the small building blocks and energy-rich bonds generated in catabolic reactions are applied to the task of fabricating the countless compounds making up the cell—a process essential to cell function, cell repair, and cell division. In Chapter 3 we considered the means by which genetic specificity is "stamped" into the processes of macromolecular (i.e., protein) synthesis. Here we shall consider the methods by which the cell converts its small building blocks into the recognizable cell constituents—carbohydrates, lipids, proteins, amino acids, and nucleotides.

BIOSYNTHESIS OF RIBOSE, RIBONUCLEOTIDES, AND DEOXYRIBONUCLEOTIDES

As it happens, a principal function of carbohydrates is carried out in their breakdown: energy production. In addition, monosaccharides form the polysaccharide glycogen, and in certain cells distinctive polysaccharides such as heparin and hyaluronic acid are synthesized from various carbohydrate derivatives (e.g., acetyl-glucosamine and glucuronic acid).

Ribose synthesis and the pentose phosphate shunt. Undoubtedly the most important biosynthetic function of the sugars is the synthesis of ribose, the five-carbon sugar of RNA and all ribonucleotides. Interestingly, the pathway of ribose synthesis begins with glucose 6-phosphate, thus constituting a *third* alternative pathway for this compound.* Since this pathway includes several oxidative steps, it incidentally provides the cell with a physiologically advantageous means for the combustion of glucose to CO_2 that by-passes the citric acid cycle. For this reason it has been called the *oxidative* or *pentose phosphate shunt.*† The major glycolytic pathway is referred to as the *fermentative* or *Embden-Meyerhof pathway.*

In the first step of shunt metabolism, glucose 6-phosphate is oxidized to 6-phosphogluconic acid, with hydrogens being transferred to NADP by the enzyme glucose-6-phosphate dehydrogenase (Fig. 4.13) As we shall see later, the resulting $NADPH_2$ may be an essential hydrogen donor in the syntheses of fatty acids and other important molecules. If it is not used in such syntheses, it can be oxidized by the respiratory chain with the generation of ATP and the reduction of molecular oxygen.

A second oxidation follows the first: 6-phosphogluconic acid is oxidized by NADP to an intermediate compound. This is decarboxylated to yield ribulose 5-phosphate, CO_2, and another molecule of $NADPH_2$. The ketopentose ribulose 5-phosphate is readily converted to its isomer, the aldopentose ribose 5-phosphate, in a reaction paralleling the isomerizations of aldo- and keto-hexoses and aldo- and ketotrioses (see Fig. 4.3).

These reactions account for the synthesis of ribose, the compound with which we are here primarily concerned. As for the remaining reactions of the shunt, we shall note only that following the oxidative synthesis of ribulose 5-phosphate and ribose 5-phosphate, there occurs a

---

* The fourth and last alternative pathway for glucose 6-phosphate is cleavage to free glucose by a specific *glucose-6-phosphatase.* This occurs only in liver cells and is the reaction by which the liver produces the glucose present in blood and body fluids. It will be discussed in detail later.

† This pathway is also critically involved in the uptake of CO_2 in plant photosynthesis. In this situation the reactions proceed in the reverse direction, for the incorporation of CO_2 into carbohydrate is a reductive process.

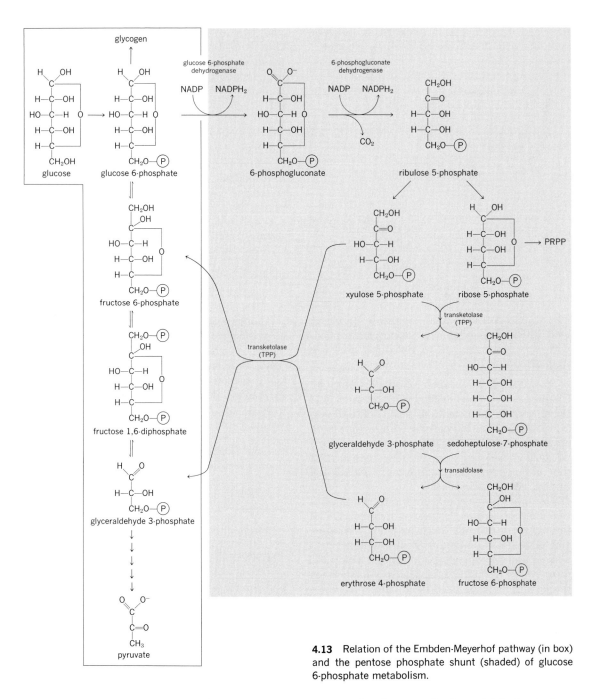

4.13 Relation of the Embden-Meyerhof pathway (in box) and the pentose phosphate shunt (shaded) of glucose 6-phosphate metabolism.

series of remarkable group-transfer reactions in which the carbons of three-, four-, five-, six-, and seven-carbon sugars are thoroughly "scrambled," the final products of the interconversions being the familiar glycolytic intermediates fructose 6-phosphate and glyceraldehyde 3-phosphate. These can be converted to pyruvate via the Embden-Meyerhof pathway. However, the reversibility of the early reactions of this pathway (see Fig. 4.3) means that neither of the intermediate compounds *must* go on to pyruvate. Instead, they can go back to glucose 6-phosphate and reenter the shunt. In this manner, glucose can be completely oxidized to CO_2 and H_2O independently of the citric acid cycle.

The ribose 5-phosphate produced in the shunt is not available for nucleotide synthesis until it is further activated by ATP in the following reaction:

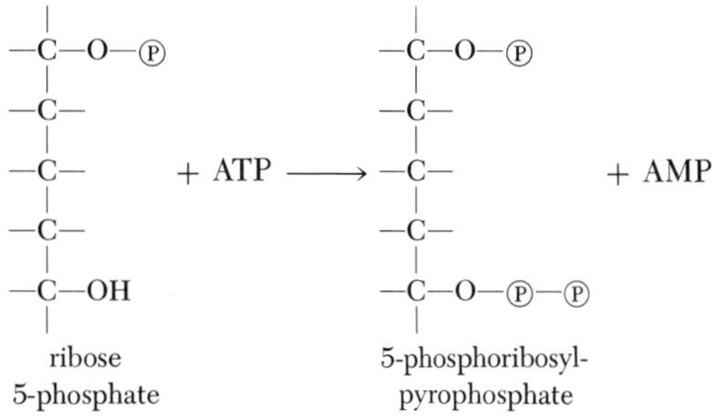

$$\begin{array}{c} | \\ -C-O-\text{\textcircled{P}} \\ | \\ -C- \\ | \\ -C- \\ | \\ -C- \\ | \\ -C-OH \\ | \\ \text{ribose} \\ \text{5-phosphate} \end{array} \quad + \text{ ATP} \longrightarrow \begin{array}{c} | \\ -C-O-\text{\textcircled{P}} \\ | \\ -C- \\ | \\ -C- \\ | \\ -C- \\ | \\ -C-O-\text{\textcircled{P}}-\text{\textcircled{P}} \\ | \\ \text{5-phosphoribosyl-} \\ \text{pyrophosphate} \end{array} + \text{ AMP} \qquad (4.28)$$

The 5-phosphoribosylpyrophosphate (abbreviated PRPP) is the form in which ribose reacts with free purines and pyrimidines to yield ribonucleotides.

Biosynthesis of purine ribonucleotides. In Chapter 3 we encountered the enzyme that polymerizes ribonucleotides (as triphosphates) into RNA, and we have just seen how the ribose destined for ribonucleotides is synthesized. Now we shall consider the biosynthetic pathways of the purine (and then pyrimidine) portions of ribonucleotide molecules.

The basic biosynthetic routes to the nucleotide molecule are (1) *de novo* pathways in which the cell requires only simple carbon, nitrogen, and phosphate compounds as biosynthetic precursors and (2) *salvage* pathways in which the purines and pyrimidines arise ready-made from the environment or from degradative processes within the cell. Purines are never synthesized *de novo* as free bases but as purine nucleotides, and except in unusual circumstances purine nucleotides are rarely synthesized by salvage pathways.

What is probably the universal pathway of *de novo* purine nucleotide synthesis is shown in schematic form in Fig. 4.14. It includes at least 9 separate enzymatic reactions. It is interesting to note how the successive steps liberate individual carbons and nitrogens from the various precursors (CO_2, formate, glutamine, aspartate,

and glycine) to build the complex double purine ring. It is also noteworthy that PRPP early contributes ribose phosphate, so that all of the intermediates are nucleotides, that coenzymes containing the vitamin folic acid (see Table 4.1) insert one-carbon formyl groups, and that many of the steps require the energy of ATP. The purine ribonucleotide emerging from this pathway, inosine 5'-phosphate, is not itself incorporated into RNA; however, it is the key intermediate from which the RNA precursors, adenosine 5'-phosphate and guanosine 5'-phosphate, are formed.

$$\text{inosine 5'-phosphate} \xrightarrow[\text{GTP}]{\text{aspartate}} \text{adenylosuccinate}$$
$$\longrightarrow \text{adenosine 5'-phosphate} \qquad (4.29)$$
$$\text{inosine 5'-phosphate} \xrightarrow[\text{NAD}]{-2\text{H}} \text{xanthine 5'-phosphate}$$
$$\xrightarrow{\text{glutamine}} \text{guanosine 5'-phosphate} \qquad (4.30)$$

Though the enzymes of the salvage pathways of purine nucleotide synthesis are present in most cells, these are not ordinarily major pathways. Free purines are utilized by many cells, being converted directly to nucleotides chiefly by reaction with PRPP.

$$\text{purine} + \text{PRPP} \rightleftharpoons \text{ribonucleotide} + \text{PP} \qquad (4.31)$$

Purines are converted to ribonucleotides via ribonucleosides as follows:

$$\left.\begin{array}{c} \text{guanine} \\ \text{adenine} \\ \text{uracil} \end{array}\right\} + \text{ribose 5-phosphate} \rightleftharpoons$$
$$\text{ribonucleoside} + \text{\textcircled{P}}$$

$$\text{ribonucleoside} + \text{ATP} \longrightarrow$$
$$\text{ribonucleotide} + \text{ADP} \qquad (4.32)$$

Specific kinases catalyze this reaction.

Biosynthesis of pyrimidine ribonucleotides. The general scheme of pyrimidine ribonucleotide synthesis is shown in Fig. 4.15. The principal *de novo* pathway of pyrimidine nucleotide synthesis, unlike that of purine nucleotide synthesis, involves the intermediate formation of a free base, orotic acid. Initially, aspartate combines with carbamyl phosphate to form carbamylaspartate. The removal of water yields dihydroorotic acid, which is oxidized by dihydroorotic dehydrogenase and NAD to free

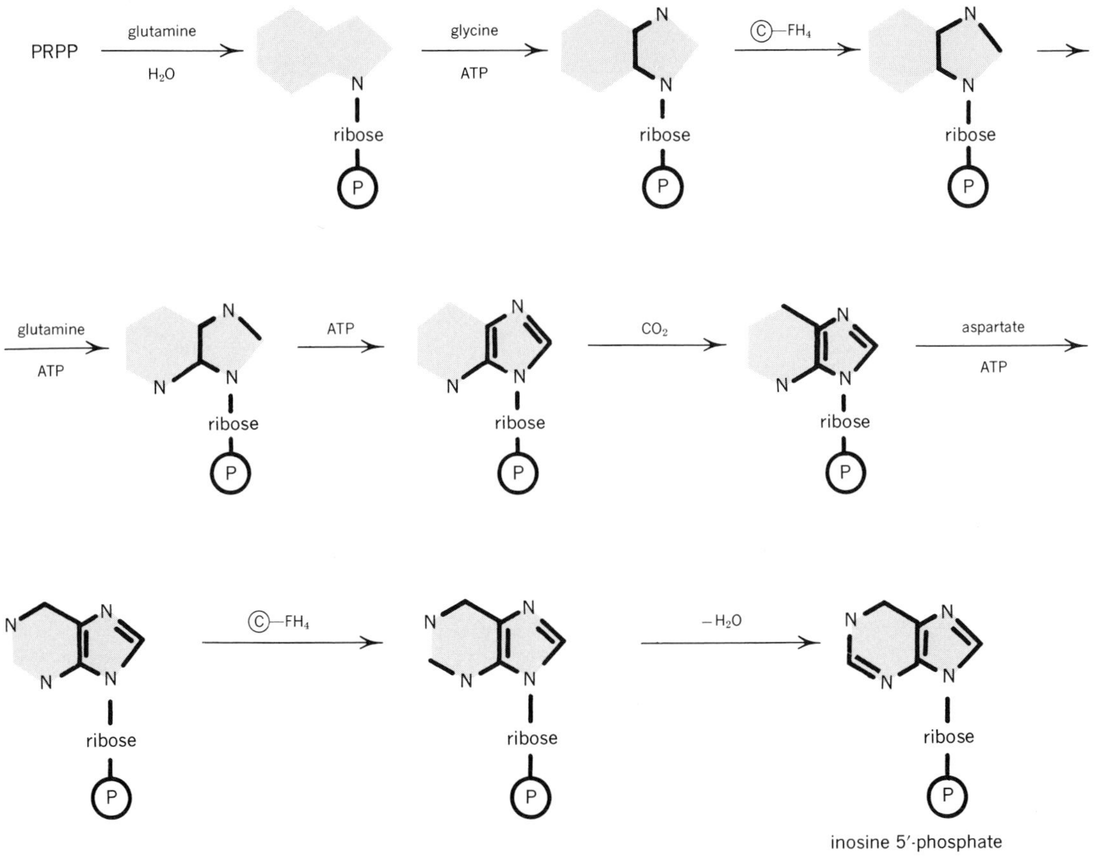

4.14 Biosynthetic pathway of the purine ribonucleotide, inosine 5′-phosphate. The folic acid derivatives, ⓒ—FH₄, are explained in Table 4.1.

orotic acid. This reacts with PRPP to produce the ribonucleotide orotidine 5′-phosphate. Subsequent decarboxylation yields uridine 5′-phosphate. Cytidine nucleotides apparently arise from uridine nucleotides by direct amination, with glutamine furnishing the necessary amino groups.

Several interesting human "auxotrophic mutants" were recently discovered whose kidneys excreted large amounts of free orotic acid. The subjects were shown to lack the enzymes catalyzing the reaction of orotic acid with PRPP and the conversion of orotidine 5′-phosphate to uridine 5′-phosphate. This observation provides additional evidence that free orotic acid lies on the biosynthetic pathway to uridine and cytidine ribonucleotides.

Biosynthesis of ribonucleoside di- and triphosphates. Ribonucleoside monophosphates cannot serve as nucleic acid precursors until they have been converted to the corresponding ribonucleoside triphosphates. A number of enzymes catalyze the phosphorylation of nucleoside diphosphates to the triphosphates, but until recently no mechanism was known (with one exception) for the phosphorylation of a nucleoside monophosphate to a diphosphate. The well-known exception was adenylate kinase or myokinase, which catalyzes the following reaction but is strictly specific for the adenine ribonucleotides.

$$\text{ATP} + \text{AMP} \rightleftharpoons 2\text{ADP} \qquad (4.33)$$

Enzymes are now known, however, which catalyze the following reactions:

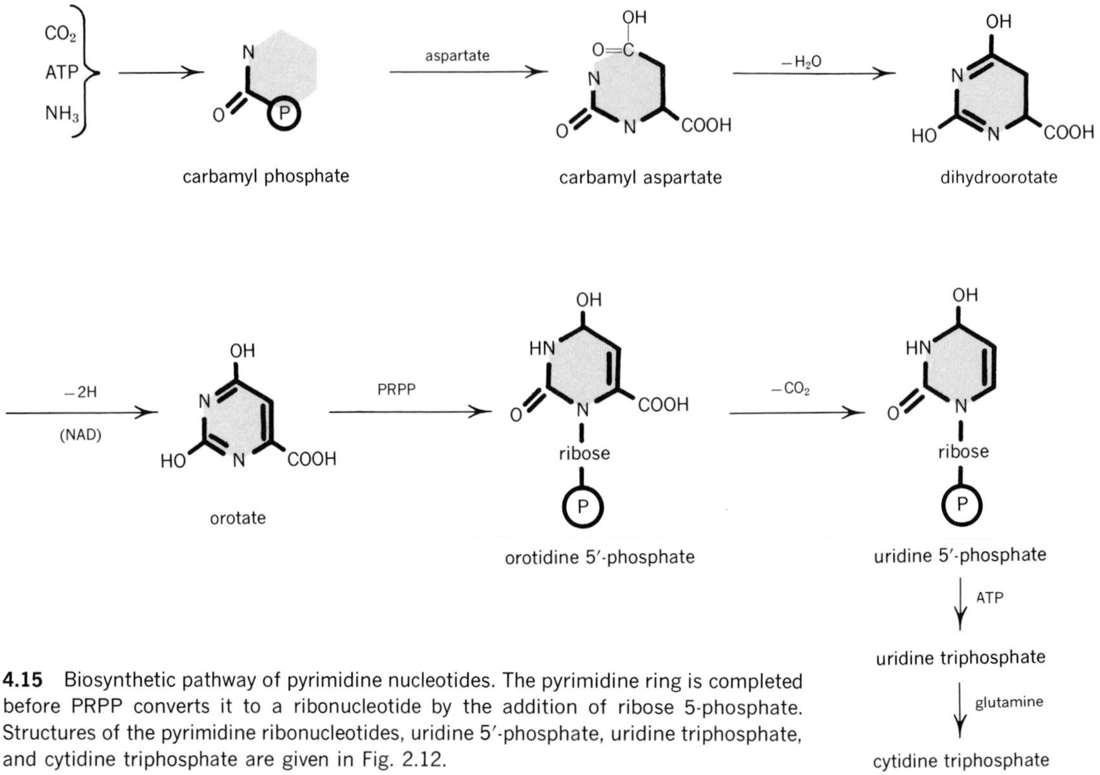

4.15 Biosynthetic pathway of pyrimidine nucleotides. The pyrimidine ring is completed before PRPP converts it to a ribonucleotide by the addition of ribose 5-phosphate. Structures of the pyrimidine ribonucleotides, uridine 5′-phosphate, uridine triphosphate, and cytidine triphosphate are given in Fig. 2.12.

$$\text{adenosine 5′-phosphate} + \text{UTP} \rightleftharpoons \text{UDP} + \text{ADP}$$
$$\text{uridine 5′-phosphate} + \text{ATP} \rightleftharpoons \text{UDP} + \text{ADP}$$
$$\text{uridine 5′-phosphate} + \text{UTP} \rightleftharpoons 2\text{UDP}$$
$$\text{guanosine 5′-phosphate} + \text{ATP} \rightleftharpoons \text{GDP} + \text{ADP}$$
$$(4.34)$$

Conversion of ribonucleotides to deoxyribonucleotides. It is surprising that the biosynthetic pathway for the distinctive sugar of DNA was not discovered until recently. Early isotopic studies indicated that ribonucleotides are directly reduced to deoxyribonucleotides—in a reaction involving the removal of one atom of oxygen from ribose (see Fig. 2.12). However, all attempts to demonstrate such a reaction in cell-free systems were unsuccessful, and not until 1959 were there convincing reports of the conversion of ribonucleotides to deoxyribonucleotides. We now know that the production of deoxyribonucleotides from ribonucleotides proceeds according to the scheme in Fig. 4.16. The figure also depicts the relations of each set of nucleotide precursors to the polynucleotides RNA and DNA.

The conversion of ribonucleotides to deoxyribonucleotides by the enzyme ribonucleotide reductase requires the presence of a coenzyme form of vitamin B_{12}. It is for this reason that vitamin B_{12}-dependent bacteria starved of the vitamin develop into nondividing filamentous forms (Fig. 4.17). In this situation RNA synthesis and protein synthesis continue normally while DNA synthesis is impaired. The result is cytoplasmic growth without cell division, a phenomenon called *unbalanced growth*. We shall see in Chapter 8 that the anemia of human vitamin B_{12} deficiency is caused by a similar defect in the red blood cell precursors.

The synthesis of DNA's unique sugar, deoxyribose (by the reduction of ribonucleotides to deoxyribonucleotides), could be a rate-limiting step in DNA synthesis. Since DNA, unlike RNA, is synthesized intermittently in tempo with the

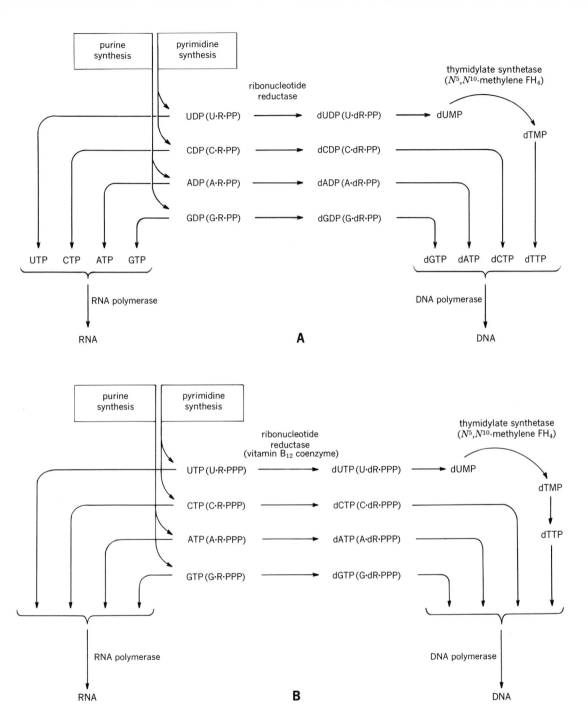

4.16 Biosynthetic pathways of RNA and DNA precursors. The abbreviations in parentheses identify the nucleotide sugar moiety (e.g., A·R·PP is adenine-ribose-diphosphate). For brevity, the pathways of *de novo* synthesis of purine ribonucleotides and pyrimidine ribonucleotides are shown to produce ribonucleoside di- or triphosphates. In fact, they yield monophosphates (see Figs. 4.14, 4.15), which are then phosphorylated to di- or triphosphates. A. Pathway in *Escherichia coli*. B. Pathway in *Lactobacillus leichmannii*, a vitamin B_{12}-requiring organism.

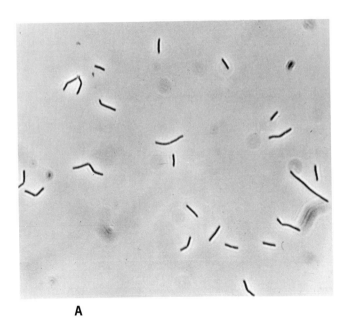
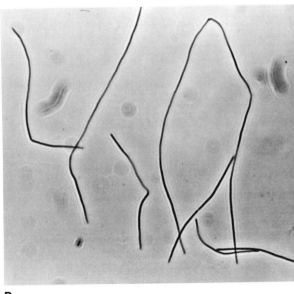

A

B

4.17 Effect of vitamin B$_{12}$ deprivation on the organism *Lactobacillus leichmannii* ($\times$660): A, optimal vitamin B$_{12}$ concentration in the medium: B, deficient vitamin B$_{12}$ concentration in the medium.

mitotic cycle, its synthesis is somehow turned on and off. How or why this occurs is not yet known.

SYNTHESIS OF PROTEINS, AMINO ACIDS, AND FATTY ACIDS

Amino acid activation in protein synthesis. Chapter 3 summarizes current views of the mechanism of protein synthesis with emphasis on the informational functions of transfer RNA (tRNA) and messenger RNA (mRNA). Fig. 3.44 indicates that amino acids must be activated before they can be transported to a ribosome and incorporated into a peptide chain. Now that we comprehend the significance of activation and its dependence upon ATP, we may look again at this essential preliminary step in the biosynthesis of proteins.

Amino acid activation consists in the conversion of a free amino acid to an *amino acid adenylate* under the influence of a specific activating enzyme.

$$\text{amino acid} + \text{ATP} \rightleftharpoons$$
$$\text{amino acid—AMP} + \text{PP} \quad (4.35)$$

The enzyme-bound amino acid adenylate then reacts with the terminal adenylic acid of a specific molecule of transfer RNA.*

$$\text{amino acid—AMP} + \text{tRNA} \rightleftharpoons$$
$$\text{amino acid—tRNA} + \text{AMP} \quad (4.36)$$

ATP contributes an energy-rich bond to amino acid—AMP that is carried on into the amino—tRNA complex (see Fig. 3.44). Presumably the energy is utilized finally in the synthesis of peptide bonds.

Amino acid synthesis. In studying amino acid breakdown, we encountered two processes that convert amino acids to keto acids: deamination and transamination. Since transamination is reversible, it provides one mechanism by which a cell can synthesize a needed amino acid (see Fig. 4.10).

Still other mechanisms for the biosynthesis of amino acids not provided by the environment are available. In some, nitrogen enters the reac-

---

* Transfer RNA always ends with a cytidylic-cytidylic-adenylic sequence, enzymes being present to prepare this sequence when necessary.

tion in the form of ammonia. For example, there is a *reductive amination* in which α-ketoglutarate is converted to glutamate.

$$
\begin{array}{ccc}
\text{COOH} & & \text{COOH} \\
| & & | \\
\text{CH}_2 & & \text{CH}_2 \\
| & & | \\
\text{CH}_2 + \text{NADH}_2 + \text{NH}_3 \rightleftharpoons & \text{CH}_2 & + \text{NAD} + \text{H}_2\text{O} \\
| & & | \\
\text{C}=\text{O} & & \text{CH}-\text{NH}_2 \\
| & & | \\
\text{COOH} & & \text{COOH} \\
\end{array}
$$

α-ketoglutarate   glutamate   (4.37)

It is apparent that NH_3 can be incorporated into many of the carbon skeletons arising in carbohydrate and fatty acid metabolism. Thus amino acid biosynthesis depends largely upon citric acid cycle metabolism. Different cell types synthesize amino acids by different means, but the scheme in Fig. 4.18 is a useful general summary of the pathways of synthesis.

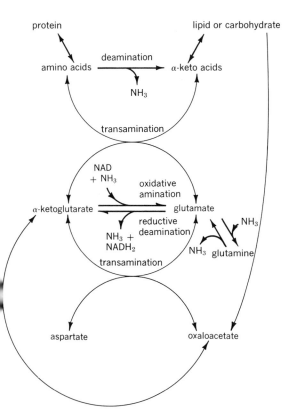

4.18 Some reactions in the synthesis and degradation of amino acids.

The ammonia removed in deamination may be reutilized in amination. In body metabolism, however, an excess of ammonia is highly toxic, and most of the ammonia is converted by liver cells to the innocuous compound urea, NH_2CONH_2, which is excreted by the kidneys. Urea is synthesized in an interesting metabolic cycle that resembles the citric acid cycle in its principle of operation and in the fact that it was also discovered by Krebs. The *urea cycle* is dependent on citric acid cycle activity and leads incidentally to the synthesis of the amino acid arginine.

As shown in Fig. 4.19, the citric acid cycle intermediate α-ketoglutarate can be converted to glutamate, and in turn glutamate can be converted to ornithine, by reduction and transamination. Ornithine reacts with carbamylaspartate —the same compound that serves as a precursor for pyrimidine synthesis (see Fig. 4.15)—to yield citrulline. Citrulline accepts an —NH_2 group from another molecule of aspartate to form arginine. If arginine is not used for protein synthesis, it is broken down by the hydrolytic enzyme arginase to urea and ornithine, which reenters the cycle. Thus the cycle results in the formation of an easily excreted compound containing two nitrogen atoms arising from the breakdown of an amino acid.

All pathways of amino acid synthesis and degradation utilize the reactions involved in the synthesis and breakdown of glutamate, aspartate, and arginine: amination, deamination, transamination, and the cyclic utilization of ammonia that has been fixed into the high-energy compound carbamyl phosphate.

Fatty acid synthesis. It was early established that fatty acids can be synthesized from acetate. After the discovery that acetyl CoA forms during the β-oxidation of fatty acids (see Fig. 4.9), it appeared likely that fatty acid synthesis involved nothing more than a reversal of the degradative pathway, and, indeed, evidence that some fatty acid synthesis occurs this way was soon forthcoming. Fatty acid oxidation, as we may have supposed, takes place in the mitochondria, and, to the extent that the oxidative enzymes act reversibly, fatty acid synthesis also takes place in the mitochondria.

However, in 1958 a major nonmitochondrial

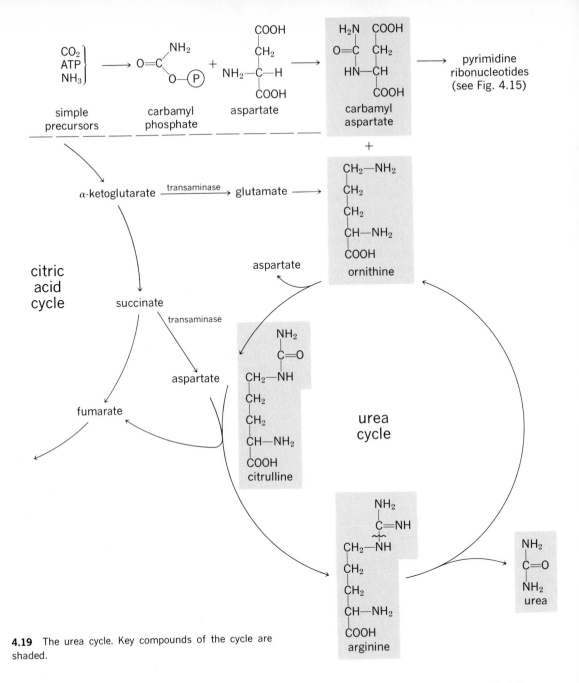

4.19 The urea cycle. Key compounds of the cycle are shaded.

pathway of fatty acid synthesis was discovered that begins with malonyl CoA, a three-carbon compound produced by the addition of CO_2 to acetyl CoA.*

$$\begin{array}{c} CH_3 \\ | \\ CO—SCoA \end{array} + CO_2 \xrightarrow[Mn^{++}]{ATP} \begin{array}{c} CH_2—COOH \\ | \\ CO—SCoA \end{array} \quad (4.38)$$

---

* This carboxylation of a two-carbon monocarboxylic acid to a three-carbon dicarboxylic acid resembles Eq. 4.24, in which a three-carbon monocarboxylic acid is carboxylated to a four-carbon dicarboxylic acid.

Malonyl CoA condenses with acetyl CoA or the CoA derivative of a longer acid so that the carbon chain is lengthened.

$$\begin{array}{cc} R & CH_2—COOH \\ | & | \\ CO—SCoA + & CO—SCoA \end{array} \rightleftharpoons$$

$$\begin{array}{c} R \\ | \\ CO—CH_2—CO—SCoA + HS—CoA + CO_2 \end{array}$$
$$(4.39)$$

The liberation of CO_2 in this reaction—the same CO_2 taken up in Eq. 4.38—helps to force the equilibrium in the direction of synthesis. This advantage is not available in systems utilizing acetyl CoA as chain-extender. The subsequent synthetic sequence is essentially the reverse of the degradative sequence except that (1) malonyl CoA instead of acetyl CoA adds two-carbon units to the chain, (2) NADP instead of NAD participates in the $—CO— \rightleftharpoons —CHOH—$ reaction, and (3) flavin mononucleotide (i.e., flavin—ribose—$\circledP$) instead of FAD (i.e., flavin—ribose—$\circledP$—$\circledP$—ribose—adenine) participates in the $—C=C— \rightleftharpoons —CH_2—CH_2 —$ reaction.

For the living cell there are obvious advantages in having separate synthetic and degradative processes instead of having synthetic and degradative processes that are alternate aspects of a reversible mechanism (see Fig. 4.13). In the case of fatty acid synthesis, however, the cell must pay a price for synthesizing fatty acids from malonyl CoA: it must utilize one extra ATP for each two-carbon unit added to the chain (the one required in Eq. 4.38).

Control of Cell Metabolism

Even a brief survey of the cell's diverse metabolic pathways must convince us of the necessity for control systems capable of coordinating them. Such control systems would need to be of a complexity appropriate to the complexity of the processes being regulated if the cell is to utilize energy efficiently in the performance of its various tasks, to function uninterruptedly during periods of nutritional deprivation, and to adapt to adverse or injurious conditions, resisting their encroachments and repairing their damaging effects.

It is an interesting fact that meaningful study of cellular control systems began relatively recently. Until the mid-1950's investigators were engaged chiefly in working out the reaction sequences of metabolic pathways. Now that most of these are known, analysis of the factors controlling the rates of flow through the pathways is possible. The exciting advances of late have shown that many regulatory mechanisms exist. In general, they can be divided into two groups: (1) those controlling the *levels of activity* of individual enzymes; and (2) those controlling the *rates of formation* of individual enzymes.

FACTORS AFFECTING THE LEVEL OF ENZYME ACTIVITY

Our earlier discussions have portrayed cell metabolism as a vast network of enzyme reactions whose complete description would yield a "road map" of formidable size. Although each reaction on the metabolic map depends upon the functioning of a specific enzyme, the activity level of that enzyme is a critical determinant of the rate behavior of the entire pathway only when the reaction in question is the rate-limiting step of the sequence. In fact, many enzymes are present in cells in amounts far larger than traffic through the pathway necessitates. Small fluctuations in the activities of these enzymes would have little significance in terms of control. The reaction with the lowest maximum velocity is the rate-limiting step of the pathway, and fluctuations in the activity (or quantity) of its enzyme are critically important. Hence, in studying the control of metabolism by factors affecting the functioning (or concentrations) of enzymes, we shall be mainly concerned with those strategically placed enzyme reactions that are the "bottlenecks" or "pacemakers" of metabolism.

Substrate and coenzyme availability. Since the rate of an enzyme reaction is affected by the concentration of the substrate (see Fig. 2.17), the rate behavior of a metabolic pathway varies with variations in the concentration of the initial substrate of the pathway. If the glucose supply is limited by starvation, glycolysis is depressed. Conversely, an excess of glucose accelerates glycolysis to a point at which the rate-limiting enzyme becomes saturated with substrate. Maximum velocity is attained at this point, and velocity can be increased further only if the amount of enzyme is increased.

Reactions requiring the participation of a coenzyme are similarly affected by the availability of the coenzyme. Since most coenzymes are derivatives of vitamins (the only coenzymes that do not contain vitamins are those coenzymatic nucleotides resembling the nucleotides of RNA —e.g., ATP, UTP), vitamin deficiency decreases coenzyme concentration and thereby depresses enzyme function.*

The local availability of coenzyme and substrate is of particular regulatory significance in the case of substances that stand at metabolic crossroads, because they are the substrates of more than one enzyme. As we have seen, the compound glucose 6-phosphate has at least three major pathways open to it: (1) isomerization to fructose 6-phosphate (see Fig. 4.3); (2) conversion to the glycogen precursor glucose 1-phosphate (see Fig. 4.13); and (3) oxidation to 6-phosphogluconate, a reaction requiring NADP as coenzyme (see Fig. 4.14). Ordinarily the first pathway is favored, in part because of the high affinity of glucose 6-phosphate for the isomerase. If, in addition, the supply of available NADP were low, the oxidation pathway would receive even less glucose 6-phosphate than usual.

Obligatory coupling mechanisms. Coupling mechanisms, inherent in the design of certain metabolic pathways, are perhaps best explained with examples. The transfer of hydrogens (or electrons) along the respiratory chain is possible only when ADP and P are simultaneously converted into ATP within the mitochondria.† In other words, O_2 cannot be utilized unless P and ADP are available. Thus demand controls supply in a simple interlocking manner. The supply of ADP and P in turn depends on the rate at which ATP is cleaved in the performance of work. Hence an accumulation of ADP and P is both a signal of the need for more ATP and a necessary preliminary to its synthesis. As a result, ATP is synthesized only when needed and in the correct amount.

Another of the many couplings between metabolic systems regulates the synthesis of fatty acids. This pathway requires $NADH_2$ and $NADPH_2$. Since the proportions of these coenzymes in the reduced form depend on the rate of oxidative metabolism and its associated dehydrogenations, a high rate of oxidative metabolism tends to promote fatty acid synthesis.

Actions of hormones in controlling enzyme activity levels. A *hormone* is a substance that is secreted in one region of the body and that acts in another as regulator of certain metabolic processes. Since hormones arise in special cells that are remote from the cells whose metabolism they control, it may seem inappropriate to mention them in a discussion of the means by which a cell regulates its own metabolism. However, hormonal control involves principles that have considerable significance for intracellular control systems.

Knowledge of how hormones control metabolism is only beginning to emerge. Since the rates of metabolic reactions are determined by their enzymes, it is generally assumed that hormones act by influencing enzyme behavior in some manner, direct or indirect. The following five mechanisms for hormonal control of an enzyme have been proposed: (1) alteration of the activity level of existing enzyme molecules; (2) alteration of the rate of synthesis of new enzyme molecules; (3) alteration of the permeability of the cell membranes or of certain intracellular membranes (e.g., the mitochondrial membranes); (4) participation by the hormone as a coenzyme in the reaction; and (5) competition by the hormone with a coenzyme for specific sites on the enzyme molecules.

The concept that hormones control enzyme activity levels is an old one. Indeed, hormones were the first rate-determinants recognized. It

---

* It is perhaps improper to speak of substrate and coenzyme "concentrations," as if the cell interior were a uniform solution containing fixed amounts of enzymes, substrates, and coenzymes. To a large extent, the availability of substrates and coenzymes depends upon both concentration and spatial distribution. We have seen that many enzymes are bound to organized intracellular particles such as ribosomes and mitochondria. Hence a substrate may be spatially separated from its enzyme. Also, many coenzymes (e.g., CoA and NAD) may be tightly bound to certain structures and therefore localized in the cell.

† Certain external agents (e.g., dinitrophenol) can artificially uncouple oxidation and phosphorylation, so that the two processes occur independently (see p. 133).

was early shown that epinephrine (adrenalin), the hormone secreted by the adrenal medulla, stimulates the breakdown of glycogen. We now know that epinephrine activates phosphorylase, the enzyme that cleaves glycogen. In resting muscle, phosphorylase is mainly in the form of a catalytically inert enzyme precursor (phosphorylase *b*, or dephosphorylase). Epinephrine functions by converting phosphorylase *b* to phosphorylase *a*, its active form. We shall deal with the mechanism of this conversion in a later chapter. Other hormones function by changing the physical structures of enzymes. The enzyme glutamate dehydrogenase, for example, is a large protein that reversibly dissociates into four identical, catalytically inert monomers (see p. 50). Certain steroid hormones promote its dissociation and thereby affect enzyme activity.

Although many hormones may control enzyme activity levels, an up-to-date survey reveals that only a few such interactions have been proved. Often there is difficulty in demonstrating a clear-cut effect of hormone on enzyme in vitro—or, conversely, in establishing an in vivo effect on the basis of an in vitro effect. Nevertheless, it is clear that hormonal control of enzyme activity levels is facilitated by the existence of separate pathways for the synthesis and breakdown of a single substance. We have seen such dual pathways for glycogen and fatty acid metabolism. They also exist for nucleotide, nucleic acid, carbohydrate, and protein metabolism. Often duality is ensured by the placement of the enzymes of synthesis and those of degradation in different parts of the cell or in different cells. Physiologically this means that the two pathways are distinct and thus subject to individual influences. Therefore, epinephrine can stimulate the breakdown of glycogen without stimulating its formation.

Feedback inhibition and activation. In feedback inhibition the catalytic activity of an early enzyme of a biosynthetic pathway is inhibited by the final product of the pathway. An example of this phenomenon is shown in Fig. 4.20. Certain bacteria contain metabolic pathways that convert the amino acid L-threonine to L-isoleucine. When excess L-isoleucine is added to the medium, however, the cells immediately cease to produce it and do not resume its production until its concentration is decreased. This control

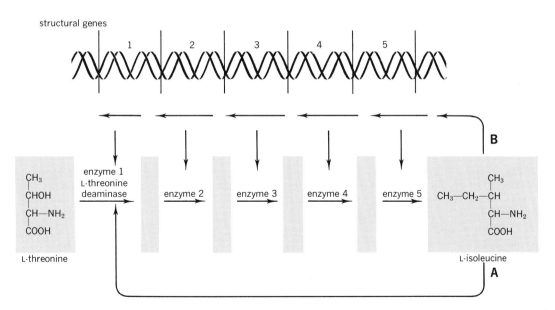

4.20 Two feedback systems controlling the five-enzyme pathway in which L-threonine is converted to L-isoleucine: A, feedback inhibition, with the end product of the pathway inhibiting the first enzyme, L-threonine deaminase; B, repression, with the end product repressing the gene-directed syntheses of all five enzymes.

is mediated by the powerful inhibitory effect of L-isoleucine upon L-threonine deaminase, the first enzyme in the pathway. The action is extremely specific: no other enzyme is inhibited (hence the inhibited enzyme is called a *regulatory* enzyme), and no other substance is an inhibitor—not even the stereoisomer D-isoleucine. Moreover, inhibition is instantaneous and easily demonstrated in vitro. Feedback inhibition takes place in animal as well as bacterial cells and controls the synthesis of certain other amino acids and the purines and pyrimidines of RNA and DNA.

Feedback activation also occurs, though fewer instances of it are known. It has been found, for example, that glucose 6-phosphate activates the enzyme that converts UDP-glucose into glycogen (see Fig. 4.12). Thus, when energy is plentiful and there is no shortage of glucose, a rising glucose 6-phosphate level stimulates the formation of glycogen and thereby the storage of energy.

An elegant example of the interaction of feedback inhibition and feedback activation is seen in the pathways of purine and pyrimidine synthesis. Since nucleic acids require these precursors in definite proportions, arrangements exist by which the two parallel "assembly lines" regulate one another. Aspartate transcarbamylase—an early enzyme of pyrimidine synthesis catalyzing the reaction of carbamyl phosphate with aspartate (see Fig. 4.15)—is *inhibited* by excess CTP, an end product of pyrimidine synthesis, and *activated* by excess ATP, an end product of purine synthesis. Therefore, the two pathways operate in tandem.

The implications of feedback control are evident. Its rapidity and precision mean that a cell capable of synthesizing a substance will stop making it at once when the supply becomes adequate. Control is automatic and involves no expenditure of energy. Thus the mechanism serves the interests of economy and avoids possibly toxic accumulations of synthetic intermediates.

The fact that certain biosynthetic enzymes are acted upon by certain inhibitory end products implies that the enzymes have two kinds of specificity—the usual specificity with respect to substrate and an additional specificity with respect to inhibitor. Furthermore, there need be little chemical similarity between the inhibitory end product and the substrate for which the enzyme has affinity.

We have already alluded to the significance of allosteric effects in the regulation of enzyme function (see p. 51). Regulatory enzymes are controlled by allosteric effects: positive or negative effectors (e.g., ATP and CTP in the case of aspartate transcarbamylase) interact with an enzyme at a site other than the active site, altering the conformation and thus the catalytic behavior of the enzyme. This concept evolved in the mid-1960's in part from a careful study of the curve describing the relationship between the rate at which an enzyme catalyzes the conversion of substrate to product and the concentration of substrate. For ordinary, or nonregulatory, enzymes, this substrate saturation curve is hyperbolic (see Fig. 2.17), revealing that the rate of conversion increases with the substrate concentration and finally levels off.

The curve reflects the fact that the first step in the catalytic transformation of a substrate is the binding of a substrate molecule to a specific site on an enzyme molecule. As the substrate concentration increases, substrate molecules occupy the binding sites of more and more enzyme molecules. The number of enzyme molecules is limited, and at high substrate concentrations nearly all the binding sites are taken. The rate of reaction then remains constant.

Interestingly, a regulatory enzyme (i.e., one capable of allosteric transitions) does not ordinarily yield a hyperbolic substrate saturation curve but a sigmoid, or S-shaped, one (Fig. 4.21).* This characteristic shape provides a clue to the operation of a regulatory enzyme. Apparently the enzyme molecule is made up of two kinds of subunits, one containing a single binding site for a substrate molecule and the other a single binding site for an allosteric effector (i.e., a feedback inhibitor or activator) molecule. In other words, substrate and effector fit into specific binding sites on separate parts of the enzyme molecule, one equipped for catalysis and the other for control (see Fig. 2.18).† Enzyme activity depends on the indirect interactions between those sites.

---

* A similar sigmoid curve describes the saturation of hemoglobin with oxygen and indicates that the binding of the first oxygen molecules picked up by hemoglobin somehow promotes the binding of others. This property, the *Bohr effect*, facilitates the transport of oxygen from the lungs to other tissues.

† Among the most convincing experimental evidence in support of this "two-site model" is the discovery by J. C. Gerhart that an aspartate transcarbamylase molecule has a binding site for its substrate on one subunit and a binding site of CTP, its inhibitor, on another subunit. When the subunits are split apart, one retains the ability to recognize and transform the substrate—indeed its substrate saturation curve is hyperbolic—and the other retains the ability to recognize CTP.

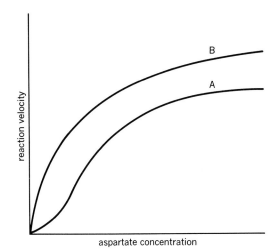

reaction velocity

B

A

aspartate concentration

4.21 Effect of the substrate concentration on the activity level of a purified regulatory enzyme, aspartate transcarbamylase. A. Typical sigmoid curve, obtained with intact enzyme. B. Hyperbolic curve, obtained after heat-denaturation of the inhibitor-binding subunit of the enzyme under conditions in which the substrate-binding subunit remains intact so that the cooperative effect is lost. The capacity for inhibition by CTP is also lost, as indicated by the fact that the curve is the same whether the enzyme is assayed with CTP or without CTP. A hyperbolic curve is typical of a nonregulatory enzyme. (Adapted from J. C. Gerhart, *Brookhaven Symposium in Biology. Subunit Structures of Proteins: Biochemical and Genetic Aspects.* No. 17. Brookhaven National Laboratory, Upton, N.Y.)

Regulatory changes are thought to arise from the ability of an enzyme molecule to shift back and forth between two conformational states differing in their affinities for substrate molecules (Fig. 4.22). In the more relaxed state, the enzyme has a high affinity for substrate. In turn, the binding of the substrate relaxes the enzyme, therefore promoting further binding. In the more constrained state, the enzyme has a low affinity for substrate and preferentially binds effector. The binding of a negative effector keeps the enzyme constrained. Hence the compound bound (substrate or effector) affects the equilibrium between the two states and tips it in favor of further binding of its type of molecule. This situation explains the sigmoid substrate saturation curve. As the substrate concentration increases, further binding of substrate is facilitated. A high concentration of inhibitor relative to substrate tips the balance the other way. Similar considerations apply to interactions between a substrate and a positive effector. Accordingly, the different kinds of binding

sites on the enzyme molecule may interact cooperatively or antagonistically, and the special capacities of a regulatory enzyme are contingent entirely on its quaternary structure.

REGULATION OF THE RATE OF ENZYME FORMATION

We noted in Chapter 3 that a structural gene determines the *structure* of a protein molecule whereas operator and regulator genes determine the *rate* of protein synthesis. We shall here review the two instances of variable rates of enzyme synthesis that have led to much of our knowledge of nonstructural genes: *induction* and *repression*.

Induction. Induced enzyme synthesis is defined as the increased rate of synthesis of a single enzyme relative to the rates of synthesis of other proteins resulting from exposure of a cell to substances called *inducers*, which are identical or closely related to the substrate of the enzyme. For example, a culture of the organism *Escherichia coli* in a medium with a carbon source such as succinic acid contains only trace

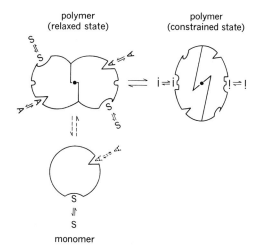

polymer (relaxed state)   polymer (constrained state)

monomer

4.22 Operation of a regulatory enzyme, according to Changeux. The relaxed polymer has a high affinity for substrate and activator and a low affinity for inhibitor, and substrate binding promotes relaxation. The constrained polymer has a high affinity for inhibitor and a low affinity for substrate and activator, and inhibitor binding promotes constraint. (After J.-P. Changeux.)

amounts of β-galactosidase, an enzyme that splits galactosides. The addition of a suitable galactoside (e.g., lactose, a disaccharide whose structure is glucose-galactose) is followed within minutes by a sharp increase of over 10,000-fold in the rate of synthesis of the enzyme. The high rate of synthesis is maintained as long as the bacteria grow in the presence of the inducing galactoside. When it is removed, the rate of synthesis drops to its original low level. No compounds other than galactosides induce the synthesis of β-galactosidase. Enzymes such as β-galactosidase have been called *inducible enzymes* in contrast to the *constitutive enzymes* normally present in the cell.

Repression. Repression has one feature in common with feedback inhibition. Both depend upon the concentration of the final product of a reaction sequence—i.e., both are feedback mechanisms. In repression, however, the end product does not inhibit the catalytic activity of an enzyme; rather, it represses its synthesis. Conversely, when the concentration of end product is low, synthesis of the enzyme is accelerated as a result of *derepression*. Derepression of the enzymes of L-isoleucine synthesis is illustrated in Fig. 4.20B.

Although the details of induction and repression are still under study, it is evident that the two processes are related aspects of the same phenomenon. A promising hypothesis explaining their relationship has been offered by Jacob and Monod (Fig. 4.23; see also Chapter 3), who contend that structural genes produce the mRNA that orders protein structure but that the *rate* of mRNA synthesis is controlled by regulator genes, which are in a neighboring region of the chromosome but quite distinct from the structural genes. Regulator genes function by producing a specific macromolecular substance (evidently a protein) known as a *repressor*.* It

* Repressor molecules were ingeniously detected by W. Gilbert in 1966 and later purified. This evidence suggests they are proteins that are products of a regulator gene. Repressor appears to be an allosteric protein capable of being activated by corepressor—or, with different results, by inducer.

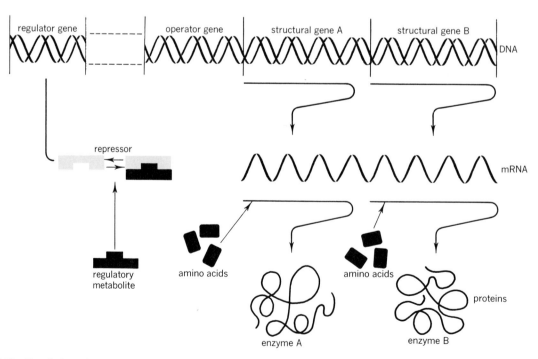

4.23 Regulation of enzyme synthesis, according to Jacob and Monod. (After J.-P. Changeux.)

can combine with either of two types of small cytoplasmic regulatory molecules, *corepressors* or inducers, which activate it. Activated repressor has an affinity for a special site on the chromosome, the operator locus. An operator locus may be regarded as a switch that turns on and off the synthesis of mRNA by adjacent structural genes. In most cases a single operator locus controls a cluster of structural genes,* locus and genes together being termed an *operon*. When activated repressor binds to the operator locus controlling one or more structural genes, it influences the rate at which they synthesize mRNA.

Corepressors are end products of biosynthetic pathways (e.g., L-isoleucine in the threonine pathway of Fig. 4.20). These arise only in certain pathways (those called *repressible*). When a repressor is activated by a corepressor it binds to an operator locus and *inhibits* the synthesis of all the enzymes normally produced by means of its structural genes. Under ordinary conditions many enzymes are synthesized at rates well below the maximum rates. These are repressible enzymes, and the operator loci controlling the structural genes responsible for their synthesis have been bound with corepressor-activated repressors.

In certain systems (those called *inducible*), the regulatory molecule is an inducer (e.g., lactose in the β-galactosidase system). When a repressor is activated by an inducer it binds to an operator locus and stimulates the synthesis of the enzymes produced by a neighboring structural gene. Enzyme synthesis then proceeds virtually without limit.

Physiological significance of induction and repression. Most of our knowledge of mechanisms regulating enzyme synthesis comes from studies of microorganisms. There are of course differences between bacteria and animal cells. An animal cell contains a thousand times as much DNA as a bacterium and takes 60 times as long to divide. Moreover, indications are that the mRNA synthesized in an animal cell is stable and long-lived compared to the rapidly destroyed mRNA of a bacterium. Nevertheless, the extrapolation of our concepts of genetic regulatory mechanisms from bacteria to animal cells is supported by increasing evidence that the two kinds of cells have identical mechanisms for translating genetic information into protein structure. The similarity even extends to the genetic code (see Chapter 3).

Repression has been demonstrated in most of the major biosynthetic pathways. Functionally it resembles feedback inhibition in preventing wasteful endogenous synthesis in the presence of an exogenous supply of a needed end product. Feedback inhibition, however, is by far more rapid and sensitive. It appears, therefore, that feedback inhibition is the chief regulator of small-molecule production, whereas repression mainly coordinates the processes of protein synthesis so as to provide an optimal combination of essential enzyme proteins with a maximal economy. Repression represents the coarse adjustment of a metabolic machine, and inhibition the fine adjustment.

By harmonizing competing protein syntheses in a cell, repression prevents what would otherwise surely be a lethal overactivity of the structural genes. It has been shown, for example, that when the structural gene for β-galactosidase, alkaline phosphatase, or one of several other enzymes is fully derepressed, the amount of the single enzyme protein synthesized may represent 8% of the total cell protein! Obviously even a few such derepressed syntheses would wreck the economy of the cell.

As for induced enzyme synthesis, it should be noted that only certain enzymes are inducible. Moreover, an enzyme may be inducible in one cell strain and constitutive in another. In cells where an enzyme is constitutive, its synthesis is uninfluenced by the presence or absence of external inducers. A few years ago it was believed that some unknown fundamental difference exists between the synthetic machinery of constitutive enzymes and that of inducible enzymes. According to the current theory, however, constitutive enzymes arise when an inducible system is altered through mutation. Mutation can alter the operator locus and prevent its response to a competent repressor (an "operator-constitutive" mutation), or it can alter the regulator gene and cause it to synthesize a totally or partially incompetent repressor (a "regulator-constitu-

---

* The cluster may include the pattern for the sequence of enzymes in a metabolic pathway.

tive" mutation). Either mutation results in constitutive enzyme synthesis. Such mutations comprise the evidence that regulator genes and operator loci exist.

The student of classic human and mammalian physiology has traditionally dealt with the many *inter*cellular feedback control mechanisms that will be described in later chapters (e.g., the actions of the nervous system and of the endocrine system). It is necessary for us to consider the role of *intra*cellular feedback control in the study of such processes as *differentiation*. Differentiation is usually defined as the complex of changes that occur as cells undergo progressive diversification of structure and function. It encompasses the development of the embryo, the appearance of new properties, the attainment of complexity. Each of these processes requires drastic changes at all subordinate organizational levels—cell groups, individual cells, cell organelles, molecular aggregates, molecular species. When we consider the differentiation of cells, we stress the appearance of new cell properties and cell types. When we consider the differentiation of subcellular components, on the other hand, we emphasize trends toward specialization in which diverse cell activities become focused on a new activity—for example, the synthesis of a new protein.

The studies of R. Briggs and T. J. King in 1952 established that functional changes take place in the genes in the course of differentiation, though it was not clear at that time whether the changes are in the structures of the genes or their rates of activity or both. It is now evident that differentiation involves coordinated sequences of selective repressions, derepressions, and inductions of operons. Existing genes are activated, and thus "new" functions emerge. These events probably follow a cascading pattern in which the products of "new" enzymes induce the synthesis of enzymes of an entirely different pathway, whose products turn other operators on, and so on. How this pattern begins is not yet known, but it is of interest that under certain experimental conditions animal cells respond to mRNA of foreign origin. Hence an exchange of mRNA molecules may initiate feedback circuits and cascades of gene activations.

The histones are possible agents of selective gene repression. These basic proteins form salts with the acidic DNA molecules (see p. 33) in the cell nucleus. The histones are of several chemical types, which vary in the ratio of arginine to lysine and in the extent of acetylation. These factors somehow determine the degree to which the histones influence the functioning of their partner DNA and thus offer another mechanism for programing gene action in the selective manner visualized in newer interpretations of differentiation.

Actions of hormones in regulating rates of enzyme synthesis. Although the actions of those hormones that directly affect enzymes (such as epinephrine) are independent of gene activity, the actions of numerous other hormones are not. As stated on p. 150, some hormones (perhaps many) control metabolism by regulating the rates of synthesis of specific enzymes. The behavior of these hormones has always had two puzzling features: the time lag between the administration of hormone and the initial appearance of effects; and the astonishing variety of effects (Chapters 15, 16). In fact, they exercise their control through the genes. Each is powerless to exert its effects when the genes of the cells on which it acts are prevented from functioning by the antibiotic *actinomycin D.** This clearly implies that hormonal stimulation normally causes the production of new molecules of specific enzymes.

A second type of observation also supports this concept. Study of individual chromosomes has recently revealed localized but shifting regions of puffing (Fig. 4.24A). These regions, called *Balbiani puffs* or *rings,* result from an unfolding of chromosomal material. They are rich in histone and are sites of active mRNA synthesis. Accordingly, they are sites of intense gene activity. The first effect of a hormone whose stimulation is mediated by gene activity is the formation of a puff (Fig. 4.24B) in which mRNA is synthesized from the usual nucleotide precursors. The mRNA travels to the ribosomes and there directs the synthesis of a specific enzyme.

---

* This agent penetrates the cell, forming a complex with DNA that keeps it from directing the synthesis of mRNA. Although actinomycin D thereby inhibits the synthesis of a specific protein, it has no direct effect on other metabolic processes.

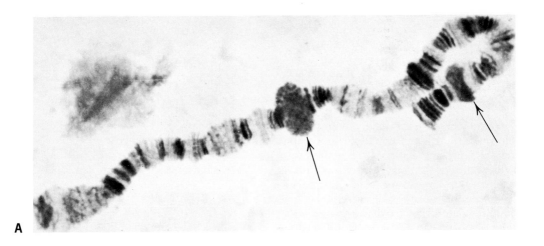

A

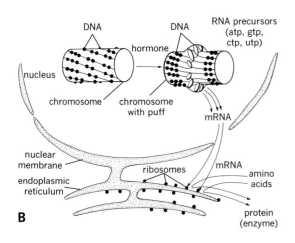

B

4.24 Balbiani puffs and their role in gene activation by hormones. A. Tip of a giant chromosome of a fruit fly, showing a small puff at the arrows. B. Scheme showing puff production and its consequences. (A from Grace Cannon, *Journal of Cellular and Comparative Physiology*, 65, 163, April, 1965. B adapted from Peter Karlson, *Perspectives in Biology and Medicine*, 6, 203, 1963.)

Exactly how hormones activate genes remains unknown. Do they act directly on DNA, exerting their effects, so to speak, in person? Or do they alter the character of that newly recognized regulator of DNA function, the chromosomal histone? Nor do we know why hormones affect mainly the genes of particular target cells, or why all the members of this large group of substances, ranging in chemical structure from small steroid to large protein, seem to act identically on genes.

Enzyme activation and molecular conversion. Before we leave the topic of metabolic control, we must refer again to the conversion of inactive enzyme precursors to active enzymes. We have already learned that epinephrine converts in-active phosphorylase *b* to active phosphorylase *a* (see p. 151). Other examples of this phenomenon are known, though they are relatively few. A familiar one is seen in the *zymogens*, precursors of the digestive enzymes that are secreted into the stomach and intestine before activation. Thus the inert stomach secretion pepsinogen is converted to the active protease pepsin (Chapter 13). This transformation involves conformational changes, with loss of most of the helical structure of pepsinogen.

In the case of powerful enzymes such as the hydrolytic digestive enzymes or the enzymes of blood clotting (Chapter 8), the existence of control mechanisms that delay "turning on" activity until the enzymes are in the proper physical locations is of obvious value to the organism.

Selective Behavior of the Cell Membrane

Chemical conditions in the environment of a cell ordinarily differ from those inside the cell and are much more variable. For these reasons maintenance of the correct internal composition requires that interchanges of substances between cytoplasm and cell exterior be meticulously regulated. The barrier responsible for this state of controlled isolation is the cell membrane, a structure less than 100 Å thick (see Fig. 3.14). Despite its thinness, it exhibits remarkable activity and selectivity.

Many substances traverse the cell membrane directly and easily and in both directions. If the forces driving substances across it arise outside the cell, the movement is called *passive transport*. If the driving energy originates from intracellular metabolic processes, the movement is called *active transport*. This interaction between cell and environment not only serves to control intracellular composition but also is concerned in the basic cell function known as *irritability*.

The cell membrane has one other important function. Many cells bring in materials from the outside by a process of ingestion that has been variously termed *phagocytosis* and *pinocytosis* (from the Greek roots for "eating" and "drinking," respectively, and for "cell"). In this process the membrane forms a pocket or invagination that draws material from the outside toward the cell interior, finally enveloping it in a vesicle or vacuole that pinches off and floats free in the cytoplasm.

Let us consider these phenomena in detail.

MEMBRANE TRANSPORT SYSTEMS

Passive transport. When a cell is immersed in a fluid containing more solute molecules than the intracellular fluid, a *concentration gradient* exists between the two fluids. This creates a powerful force tending to drive solute molecules into the cell (Fig. 4.25). In fact, any time two solutions of differing concentrations are brought into contact, solute diffuses from the region of

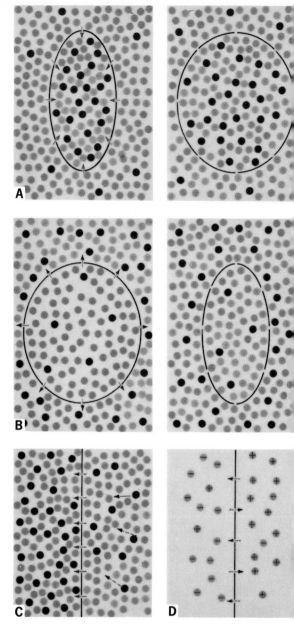

4.25 Passive transport; ●, molecule of solute; ●, molecule of water; ⊖, negative ion; ⊕, positive ion. A. When a cell contains more solute than the extracellular fluid does, water flows in, and the cell swells. B. When the extracellular fluid contains more solute than the cell does, water flows out, and the cell shrinks. C. When the water flowing across a membrane pulls along a few molecules of solute, despite the concentration gradient, solvent drag occurs. D. When the solution on one side of a membrane contains a high concentration of negative ions and that on the other side contains a high concentration of positive ions, the ions migrate across the membrane.

higher concentration to the region of lower concentration until its concentration is the same everywhere. This diffusion takes place even when the two solutions are separated by a membrane, provided the membrane is permeable to the solute. Because of the lipid nature of a membrane, only lipids permeate it freely. Water and small water-soluble particles permeate it through hypothetical pores or channels approximately 7 Å in diameter.*

When a membrane is permeable to solvent but not to solute, the concentration gradient creates an *osmotic* force. In *osmosis* solvent moves from the region of lower solute concentration to that of higher solute concentration. Concentration gradients and osmotic forces may be considerable across a cell membrane, since the intracellular concentrations of many substances can differ greatly from their extracellular concentrations.

In passive transport, then, the differential permeability of the cell membrane regulates the passage of materials across it. Permeability to a given molecule reflects the nature and properties of the molecule as well as its size. The membrane excludes certain substances completely and allows others to cross at different rates. A living fresh-water ameba, for example, has only 1% the permeability to water of a human red blood cell. This restriction in the ameba conserves energy, which would otherwise be expended in pumping water out. The red blood cell needs no such protection because its interior and its plasma environment are in osmotic balance. If the red blood cell is placed in water, however, it swells and bursts, owing to the inrush of water.

If, as is usually the case, the molecules of solute dissociate into ions, another force comes into play. In many and perhaps all cell membranes, a nonuniform distribution of positive and negative ions creates a difference of electric potential between the two sides of the membrane. Thus a *potential gradient* acts along with the concentration gradient in promoting passive transport across a cell membrane.

Still another possible force is *solvent drag*, which arises if the membrane is porous enough to permit the passage of large quantities of solution. Solute particles diffusing in the direction of flow are speeded up, whereas those diffusing against the flow are slowed down. This effect is normally a minor one.

Whether a concentration gradient, a potential gradient, or a solvent drag (or any combination thereof) is the driving force in passive transport, movement is always "downhill" (i.e., from a high solute concentration to a low one), and the membrane's role is only that of a passive barrier.

Active transport. Transfers of substances across a cell membrane that cannot be attributed to the forces just described are classed under active transport. Movement in these cases is "uphill," against the forces of passive transport, and so must involve an immediate expenditure of energy—though intensive research has failed thus far to reveal exactly how the necessary energy is transferred from ATP to the transport process. The membrane nevertheless functions in a dynamic manner.

Of the mechanisms postulated to explain active transport, the most popular one involves a "carrier" molecule. The theory (which has many variations) is that there exists within the membrane a carrier molecule that combines with the substance to be transported on one side of the membrane, forming a complex that passes through the membrane to the other side, where it dissociates, with free carrier then diffusing back to the first surface. The movement of the complex, as distinct from the transported substance, is assumed to be downhill, following a concentration gradient. It is also assumed that the reaction at one or both surfaces is coupled to the exergonic breakdown of ATP.

The substance transported can be an ion or an uncharged molecule, a fact that adds to the physiologist's difficulties in visualizing the actual chemistry of carrier complex formation. Moreover, if it is an ion, the question arises whether one ion can be transported without the movement of others. The active transport of ions is of great physiological interest because most cells contain much more potassium than sodium in

---

* Such pores could admit water molecules and certain ions, even some that, like sodium and potassium, are surrounded by a shell of water molecules (i.e., hydrated). A. K. Solomon believes that each pore is lined with positively charged ions. These would hinder the movements of positively charged ions but not those of negatively charged ones.

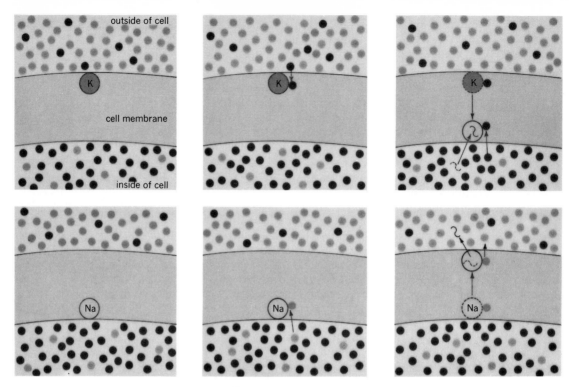

4.26 Active transport, according to T. I. Shaw: (K), potassium ion; ●, potassium carrier; (Na), sodium; ●, sodium carrier; ~, energy. The potassium carrier cannot cross the cell membrane until it combines with a potassium ion. The potassium complex migrates from the outer wall of the membrane to the inner wall, where it gives up the potassium ion to the interior of the cell and receives energy from the cell. The energy converts the carrier into a sodium carrier, which cannot cross the membrane until it in turn combines with a sodium ion. The sodium complex migrates to the outer wall of the membrane, where it releases the sodium ion and energy, to become once again a potassium carrier.

their cytoplasm. Yet they live in an environment in which the proportions of the two ions are the reverse. For example, plasma contains 35 times as much sodium as potassium, and a red blood cell contains 5 times as much potassium as sodium. To a small extent, the membrane of the red blood cell is passively permeable to both sodium and potassium ions. But, if its permeability were uncontrolled, sodium would leak into the cell, and potassium would leak out. To maintain its composition, therefore, the cell must constantly expel sodium and accumulate potassium against an aggregate 40-fold concentration gradient. Interestingly, it has been shown that, in a human red blood cell, three sodium ions pass outward per two potassium ions inward per one ATP high-energy phosphate bond hydrolyzed. The ratios indicate the events are related.

A number of ingenious models have been pro-posed for these events. The best known (Fig. 4.26) attempts to account for the movement of potassium and sodium ions across a membrane by suggesting a link between them. According to this hypothesis, complex formation and transport occur in alternating sequence. Potassium and sodium ions (K^+ and Na^+) are transported across the membrane by specific lipid-soluble carriers (X and Y, respectively). The complexes formed (KX and NaY) can diffuse through the membrane, whereas the free carriers and ions cannot. At the outside surface of the membrane, sodium carriers are converted to potassium carriers, losing energy in the process. At the inside surface potassium carriers are converted to sodium carriers, owing to ATP energy arising from cell metabolism. This theory is based on many assumptions difficult to prove experimentally and is by no means universally accepted.

Nevertheless, it illustrates the intricacy of active transport.*

Phagocytosis and pinocytosis. Long before biologists were troubled by the problem of passive and active transport across the cell membrane, they had observed cells in the act of "eating." In the late nineteenth century Elie Metchnikoff saw white blood cells engulfing bacteria and gave them the name of *phagocytes.* In 1920 Asa Schaeffer witnessed a similar phenomenon in an ameba. Then in 1931, while studying tissue culture cells by time-lapse photography, Warren Lewis saw the membranous fringes at the periphery of a cell occasionally close up like a clenched fist and trap a small portion of the surrounding fluid in a vesicle or vacuole. The event was so suggestive of drinking that Lewis coined the term "pinocytosis." Not until the 1950's was it discovered that pinocytosis occurs in all cells.

The essential features of phagocytosis and pinocytosis are the same: an area of the cell membrane detaches itself from the cell surface to form a vesicle around solid matter (phagocytosis) or fluid (pinocytosis), which then migrates toward the cell interior (Fig. 4.27). Vesicles vary widely in size. In tissue culture cells the diameter of a newly pinched-off pinocytotic vesicle is usually 1 to 2 μ, but the diameter of a vesicle discernible through an electron microscope may range from 0.1 to 0.01 μ. *Micropinocytosis* is pinocytosis in which the vesicles are extremely small.

Though materials can enter a cell by phagocytosis or pinocytosis, precisely how they leave their vesicles within the cell and distribute themselves throughout the cytoplasm is not known. Apparently secondary vesicles form, separate from the primary vesicles, and move into the cytoplasm. Physiologists have criticized the con-

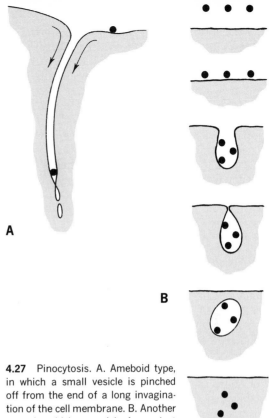

4.27 Pinocytosis. A. Ameboid type, in which a small vesicle is pinched off from the end of a long invagination of the cell membrane. B. Another type, in which a vesicle forms just inside the surface of the cell.

cept that phagocytosis and pinocytosis are basic cellular functions on the grounds that such bulk processes seem to offer no satisfactory explanation for the high degree of selectivity that characterizes the uptake of materials by cells. It has been suggested that the outer surface of a cell is coated with agents (e.g., polysaccharides) capable of selectively binding particular molecules to it for ingestion. Still the cell surface is often relatively indiscriminate, and on many occasions the cell must cope with useless or even obnoxious substances.

If selective mechanisms should be found to exist for them, phagocytosis and pinocytosis could be regarded not as alternatives to membrane transport but rather as supports for it. Their contribution would be the creation of large interior interfaces where the forces of passive and active transport could work with even greater efficiency than at the cell surface and with less danger of loss by leakage.

---

* Studies on bacteria have shown that the active transport systems in their cell membranes have certain properties typical of enzymes. Hence they have been termed *permeases.* Moreover, they appear to be subject to the same regulatory mechanisms as enzymes; for example, the capacity to transport specific substances can be lost through mutation, and transport can be induced and repressed. In 1965, C. F. Fox and E. P. Kennedy directly demonstrated the existence of a membrane protein, termed M protein, that is an integral part of the permease system.

Numerous physiological processes depend on phagocytosis and pinocytosis. For example, fertilization is believed to consist in the pinocytotic engulfment of a sperm-containing droplet by an egg, an act initiated by the interaction of specific substances at the surfaces of the two cells. The uptake of iron (in the form of ferritin) by the precursors of red blood cells and of the vitamin B_{12}–intrinsic factor complex by the intestine (Chapter 8) are also pinocytotic. It has been calculated that if all the intestinal secretions (7 liters per day) were reabsorbed by pinocytosis, each cell in the intestinal lining would form 1000 vesicles per second.

CELLULAR IRRITABILITY

Cells do not merely vegetate; nor do they squander their energy reserves capriciously. Rather their activities are directed—usually in a manner determined by changes in the environment, by external stimuli. In short, cells react to the environment, whether it be hostile or congenial. The reaction tends to preserve or restore the status quo. This is an expression of the protective function called *irritability*.

Three phases of irritability. When an ameba is touched with a needle, it moves away from it. When a white blood cell is exposed to a pathogenic bacterium, it engulfs it. An ameba has no organs for the perception of touch, no nerves, and no muscles. Yet it appears to "feel" the needle prick, and it reacts by extending itself on its opposite side, so that the "feeling" must somehow be conducted from one side to the other. The end result is movement away from the foreign, inedible, and possibly harmful object.

We see here a typical sequence of events. First there is a *stimulus*. We know from careful investigation that whatever its nature—light, pin prick, or electric shock—certain quantitative relationships can be observed. A certain minimal intensity of stimulus, the so-called *threshold intensity*, is required; weaker stimulation produces no reaction. Ordinarily, once the threshold intensity has been attained, stronger stimulation elicits no additional response; in other words, the threshold intensity provokes an *all-or-none reaction*.

Second, there is *conduction*. After the stimulus has been received, a signal of some kind is conducted or spread throughout the cell. The next section deals with the mechanism of conduction. Here we note only that the excitation wave initiated by the stimulus is transmitted to some *effector* region in the cell.

Finally there is a *response*. In light-sensitive cells of the retina of the eye, the response to light is the discharge of impulses into the fibers of the optic nerve. In muscle cells, the response to excitation is physical contraction. In the pricked ameba, the response is movement away from the offending object.

Irritability is a fundamental property of living cells. It might be said that a ball, too, responds to a stimulus when it moves through the air after being kicked, but there is an important difference between the two cases. The response of the ball depends entirely upon the stimulus and not upon reactions initiated inside the ball. In living cells the response comes from within and may have little relationship to the nature or strength of the stimulus. The pricked ameba does not rebound according to Newton's law of motion, and its response is not proportional to the strength of the needle's thrust.

The cell membrane and the mechanism of irritability. It is firmly established that local excitation by a stimulus causes negatively charged ions to accumulate outside a cell membrane and positively charged ions to accumulate inside the cell membrane, the opposite of the situation in an unstimulated cell. This results in local polarization (Fig. 4.28), a state that tends to spread throughout the cell membrane like a wave. As positively charged ions move in to neutralize or depolarize the highly negative area, they leave behind them a shortage of positively charged ions and an excess of negatively charged ions, the net effect being a movement of the negative area from its original location to a new one. By this mechanism excitation is conducted throughout the cell membrane. Evidence exists that membrane permeability also is altered at the site of local stimulation.

We shall consider irritability at length in our discussion of the human nervous system. The nervous system is essentially an elaborate irritability machine in which the three phases of ir-

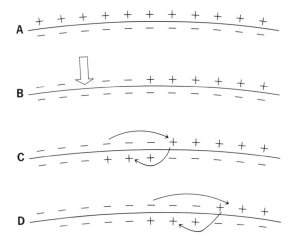

4.28 Conduction of an excitation wave along the cell membrane. A. Unstimulated cell membrane. B. Immediate effect of local stimulation. C, D. Redistribution of ions, causing movement of the polarized area.

ritability have been physically separated as a result of the evolution of multicellular organisms. Special cells, generally adapted to receive only particular kinds of stimuli, serve as receptors in the sense organs; the nerves function as conductors, carrying signals from the receptors; and the sense organs act as effectors, producing appropriate responses.

Cellular Movement

Cellular movement is perhaps the least well understood of the fundamental cell functions. It takes several forms. One of the most impressive is *protoplasmic streaming*, the constant churning within all living cells. *Ameboid movement* involves the actual locomotion of a cell from one place to another, resulting from an organized pattern of intense protoplasmic streaming. *Ciliary movement* is the whiplike action of cilia. A similar movement is displayed by the tails of free-swimming sperm cells. The beautifully precise turning, positioning, and separation of the chromosomes in cell division are other manifestations of cellular movement.

Finally there is the movement associated with muscular contraction, in which the expenditure of energy leads to physical shortening of a cell and the concomitant performance of work. This fascinating topic will be dealt with later.

PROTOPLASMIC STREAMING

Protoplasmic streaming should probably be distinguished from ameboid movement, but in many instances the distinction is difficult. In a nonmoving cell protoplasmic streaming accounts for the distribution of nutrients and building blocks throughout the cell. Hence it provides the cell with an internal transportation system.

Non-Newtonian character of protoplasm. The streaming of protoplasm raises a number of important biological and biophysical questions that are not frequently discussed. Although it may be explained by the physical laws that govern the flow of liquids, it involves certain phenomena that are unfamiliar to the physical chemist who deals only with nonliving systems. As a result, the science of *rheology*, which is concerned solely with the flow properties of fluids like protoplasm, has arisen.

Fluids whose behavior can be predicted by conventional laws of physics and chemistry are called *Newtonian fluids*. Protoplasm is a *non-Newtonian fluid*. This does not mean that living matter does not obey physical laws. Rather, it means that protoplasm has certain special properties that require for explanation unusually complex physical laws. Perhaps then we should say that protoplasm behaves according to rheological laws.

As an example, protoplasm is an elastic, contractile fluid with a substantial tensile strength, a measurable rigidity, and a tendency to develop what are called *states of torsion*. To a large extent these properties are due to the fact that protoplasm is a concentrated solution of macromolecular materials; thus there are complex interactions (hydrogen bondings, etc.) between molecule and molecule and between molecule and solvent.

When such a fluid flows, it moves differently from water. Compared to the flow of water through a pipe, the flow of protoplasm resembles the movement of a school of fish. Its motive force appears to lie within itself; it moves on its

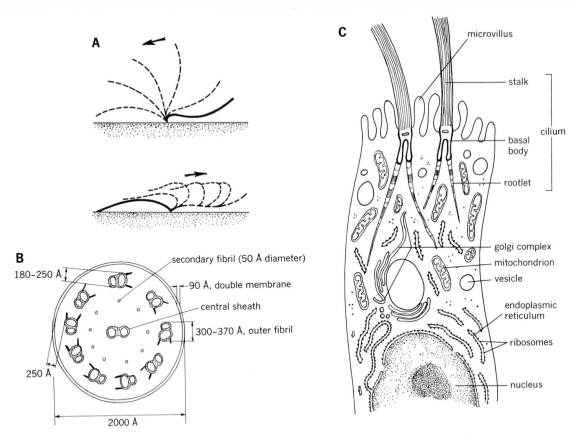

4.29 Cilia. A. Movement, showing that a stiff sweeping stroke in one direction is followed by a relaxed recovery stroke. B. 9 + 2 pattern of fibrils. C. Relations to the rest of the epithelial cell. (B after Gibbons and Grimstone.)

own. Although the source of energy for its movement is not yet known, it is obviously related to the structure of the component molecules and their stored elastic energy, which in turn are related to the metabolic processes within the protoplasm; for example, protoplasmic streaming depends upon the presence of oxygen and metabolizable sugar.

HOW CELLS MOVE

Muscle tissue is found only in multicellular animals, and so, as we have noted, we shall discuss its behavior later. Here we shall consider two outstanding forms of cellular movement that are nonmuscular in nature, ciliary and ameboid movement.

Ciliary movement. Many body cells are covered, at least in part, with small hairlike structures, the cilia. These constantly sweep back and forth, in one direction like a stiff rod and in the other like a whiplash (Fig. 4.29A), and in a rotary screwlike pattern. The cilia of adjacent cells operate in a coordinated manner, and their aggregate motion resembles breeze-blown waves in a field of grain. It is easy to see how such a "meadow" of cilia, beating up to 20 times per second, can move considerable quantities of material across the surfaces of ciliated cells. When cilia are less numerous, they are classified as flagella. When a flagellum is attached to a movable object such as a sperm cell, it propels it through its liquid medium.

The structure of a cilium has certain remarkable features. For one thing, it resembles that of a basal body (see p. 78). Both contain nine groups of outer fibrils plus two inner fibrils, in an arrangement that has come to be known as the 9 + 2 configuration (Fig. 4.29B,C). Interestingly, this pattern has been found in all cilia and flagella—whatever the cell, organism, or phylum. Hence its biological significance must be great.

Presumably the beating of a cilium in one plane can be explained by alternate contractions and relaxations of opposite fibrils. These would not account for screwlike motions, however, which are presumably due to coordinated contractions near the base of the cilium. The reason for the rhythmic movement and coordination among many cilia is unknown.

The structural similarity between a cilium and a centriole poses some interesting biological questions. Judging from the importance of the centriole in cell division, we can assume that it arose very early in evolution. In some cells a centriole serves as the basal body of a cilium. Accordingly, perhaps the centriole somehow developed a motile extension that was advantageous to the cell and thus was conserved. We shall find later that certain specialized cell structures, such as the rod cells of the retina, apparently evolved from cilia.

Ameboid movement. Although ameboid movement is named for the unicellular ameba, it is seen in body cells of multicellular organisms. The white blood cells, for example, propel themselves along the walls of capillaries and squeeze through tiny openings on their way to sites of injury or infection with motions identical to those of an itinerant ameba.

In ameboid movement part of the main body of a cell bulges out as a blunt irregular extension. This is called a *pseudopodium* ("false foot"). As the pseudopodium attaches itself to the surface at hand, the main body of the cell begins to stream in its direction, filling and enlarging the pseudopodium. When the pseudopodium contains most of the cell, it is no longer recognizable. Meanwhile, other pseudopodia form. In this way the entire cell moves, at a speed of, at most, a few centimeters per hour.

According to a recent theory of R. D. Allen, a relaxed stream of protoplasm traveling through the center of a cell abruptly reaches the end or front of what is to become a pseudopodium (Fig. 4.30). There it thickens, contracts, and everts, turning inside out like a cuff. As it travels backward like a tube toward the main body of the cell, it draws more protoplasm forward.

All of these instances of cell movement depend upon changes in protein structures. However, since their mechanisms are unknown, we cannot yet declare that they are not all different. We must therefore conclude that nature did not invent motility in one great leap but tried different methods. Its best method, as we shall learn, is muscular contraction. That it is best is apparent from the simple fact that it has been retained and given the widest distribution.

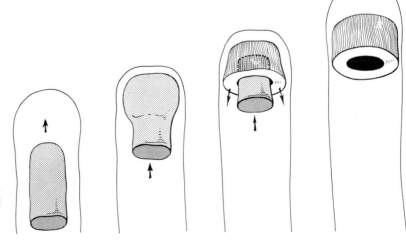

4.30 Ameboid movement, according to R. D. Allen. (Adapted from R. D. Allen, *Expt. Cell Res.,* Suppl. 8. (1961).)

REFERENCES AND SUGGESTIONS
FOR FURTHER READING

Allen, R. D., "Amoeboid Movement," in J. Brachet and A. E. Mirsky, eds., *The Cell*, Vol. 2, Academic Press, New York, 1961.

———, "Amoeboid Movement," *Sci. Am.*, **206**, 112 (1962).

Andersen, B., and H. H. Ussing, "Active Transport," in M. Florkin and H. S. Mason, eds., *Comparative Biochemistry*, Vol. 2, Academic Press, New York, 1960.

Atkinson, D. E., "Regulation of Enzyme Activity," *Ann. Rev. Biochem.*, **35**, 85 (1966).

Baldwin, E., *Dynamic Aspects of Biochemistry*, 5th ed., Cambridge Univ. Press, Cambridge, 1966.

Beadle, G. W., "Physiological Aspects of Genetics," *Ann. Rev. Physiol.*, **22**, 45 (1960).

Beck, W. S., "Deoxyribonucleotide Synthesis and the Role of Vitamin B_{12} in Erythropoiesis," *Vitamins & Hormones*, **26**, 413, 1968.

Beerman, W., and U. Clever, "Chromosome Puffs," *Sci. Am.*, **210**, 50 (April, 1964).

"Cellular Regulatory Mechanisms," *Cold Spring Harbor Symp. Quant. Biol.*, **26** (1961).

Chance, B., R. W. Estabrook, and J. R. Williamson, eds., *Control of Energy Metabolism*, Academic Press, New York, 1966.

Changeux, J.-P., "The Control of Biochemical Reactions," *Sci. Am.*, **210**, 36 (April, 1965).

Csáky, T. Z., "Transport Through Biological Membranes," *Ann. Rev. Physiol.*, **27**, 415 (1965).

Davidson, E. H., "Hormones and Genes," *Sci. Am.*, **212**, 36 (June, 1965).

Davis, B. D., "The Teleonomic Significance of Biosynthetic Control Mechanisms," *Cold Spring Harbor Symp. Quant. Biol.*, **26**, 1 (1961).

De Reuck, A., and M. O'Connor, "The Mechanism of Action of Water-Soluble Vitamins," *Ciba Found. Study Group*, No. 11 (1961).

Ebert, J. D., *Interacting Systems in Development*, Holt, Rinehart and Winston, New York, 1965.

Ernster, L., and Chaun-pu-Lee, "Biological Oxidoreductions," *Ann. Rev. Biochem.*, **33**, 729 (1964).

Fawcett, D., "Cilia and Flagella," in J. Brachet and A. E. Mirsky, eds., *The Cell*, Vol. 2, Academic Press, New York, 1961.

Fruton, J. S., and S. Simmonds, *General Biochemistry*, 2nd ed., Wiley, New York, 1958.

Hayashi, T., "How Cells Move," *Sci. Am.*, **205**, 184 (Sept., 1961).

Holter, H., "How Things Get into Cells," *Sci. Am.*, **205**, 167 (Sept., 1961).

Ingram, V. M., *The Biosynthesis of Macromolecules*, Benjamin, New York, 1965.

Jacob, F., and J. Monod, "Genetic Regulatory Mechanisms in the Synthesis of Proteins," *J. Mol. Biol.*, **3**, 318 (1961).

Karlson, P., "New Concepts on the Mode of Action of Hormones," *Perspectives Biol. Med.*, **6**, 203 (1963).

Lark, K. G., "Initiation and Control of DNA Synthesis," *Ann. Rev. Biochem.*, **38**, 569, 1969.

Lehninger, A. L., *Bioenergetics*, Benjamin, New York, 1965.

———, "How Cells Transform Energy," *Sci. Am.*, **205**, 62 (Sept., 1961).

———, "Energy Transformation in the Cell," *Sci. Am.*, **202**, 102 (May, 1960).

Litwack, G., and D. Kritchevsky, eds., *Actions of Hormones on Molecular Processes*, Wiley, New York, 1964.

Monod, J., J.-P. Changeux, and F. Jacob, "Allosteric Proteins and Cellular Control Systems," *J. Mol. Biol.*, **6**, 306 (1963).

Monod, J., and F. Jacob, "Teleonomic Mechanisms in Cellular Metabolism, Growth, and Differentiation," *Cold Spring Harbor Symp. Quant. Biol.*, **26**, 389 (1961).

Monod, J., J. Wyman, and J.-P. Changeux, "On the Nature of Allosteric Transitions: A Plausible Model," *J. Mol. Biol.*, **12**, 88 (1965).

Pinchot, G., "Mechanisms of Oxidative Phosphorylation—Observations and Speculation," *Perspectives Biol. Med.*, **8**, 180 (1965).

Racker, E., "The Membrane of the Mitochondrion," *Sci. Am.*, **218**, 32 (Feb., 1968).

Satir, P., "Cilia," *Sci. Am.*, **204**, 108 (Feb., 1961).

Schweet, R., and R. Heintz, "Protein Synthesis," *Ann. Rev. Biochem.*, **35**, 723 (1966).

Skou, J. C., "Enzymatic Basis for Active Transport of Na^+ and K^+ Across Cell Membrane," *Physiol. Rev.*, **45**, 596 (1965).

Solomon, A. K., "Pumps in the Living Cell," *Sci. Am.*, **207**, 100 (Aug., 1962).

Stumpf, P. K., "Metabolism of Fatty Acids," *Ann. Rev. Biochem.*, **38**, 159, 1969.

"Subunit Structure of Proteins: Biochemical and Genetic Aspects," *Brookhaven Symp. Biol.*, No. 17 (1964).

Wagner, A. F., and K. Folkers, *Vitamins and Coenzymes*, Interscience, New York, 1964.

Watson, J. D., *Molecular Biology of the Gene*, Benjamin, New York, 1965.

White, A., P. Handler, and E. L. Smith, *Principles of Biochemistry*, 3rd ed., McGraw-Hill, New York, 1964.

Wolstenholme, G. E. W., and C. M. O'Connor, eds., *Ciba Found. Symp., Regulation Cell Metab.*, 1959.

The advances of functional biology can be described by two vectors. The first evolves from the idea of a general physiology and is expressed in the advances of the fields loosely designated as cell biology, biochemistry, and molecular biology. Here the theme is Claude Bernard's *phénomènes de la vie communs aux animaux et aux végétaux* and much more: the intrinsic material design and operations of *all* entities having properties associated with life, past and present, terrestrial and possibly extraterrestrial. The other vector is directed by our experience of real organisms in a real world: the awful variety, subtlety, and adaptations of real organisms, cells, and communities of cells. However appealing unity may be, biological science has found no way to reject the perplexities of specialization and still to perform its mission.

Alternative modes of specialization can be considered: (1) that specialization represents extremes of modifications of the common design of cells or (2) that specialization involves the introduction of unique and uncommon attributes in some cells. The former appeals to our intellectual prejudices favoring unity; the latter implies that evolution in its more drastic sense was not suspended when the archetype of a modern cell came into being ... [In either case, most of us believe] the specialization of cells is not a delusive and incomprehensible play of shadows on the walls of a cave in which the Platonic idea of The Cell is concealed.

Daniel Mazia, in *The General Physiology of Cell Specialization*, 1963

5 Cell Types and Tissues

Introduction

We have dealt thus far only with the generalized cell—its structure, its basic functions, and its mode of replication. In this chapter we shall direct our attention to more realistic objects, the diverse cell types of the human body. They will remind us that the idealized and somewhat oversimplified model cell was formulated only as an essential aid in the investigation, analysis, and discussion of cell physiology. We must not allow the model cell to obscure the great differences in behavior and structure among individual cell types.

DIVERSITY OF CELL SHAPES

That the generalized cell of Fig. 3.6 displays an artlessly indifferent shape is evident when it is compared with certain nerve cells found in the cerebellum. Moreover, there are wide variations in shape among individual cells of a given type as well as among different cell types. In fact, no two cells are ever identical, nor is any one cell quite the same at different times in its life history. It is this diversity that textbook pictures fail to reveal. As Paul Weiss has pointed out, cell shape is determined by forces similar to those that determine the shape of a garden shrub. The shrub was not stamped out in the shape in which we find it. It grew into that shape from a seed. Thus what we see patterned in the shrub is the residual record of prior activities of a particular protoplasmic system. In other words, shape is simply an index of the antecedent differentiative and environmental processes by which it came about.

Most cells are immersed in a complex fluid medium and reside in a physical framework that integrates them into the structural continuum called *tissue*. They constantly interact among themselves and with their environment to produce various responses, some permanent, some transitory, whose visible expression is cell shape. The well-known size and shape variations observed when animal cells are grown in tissue culture attest to the modulating influences of environmental conditions. If the environment is reasonably stable for a number of cells of the same type, their behavioral histories will be sufficiently similar to yield the reasonably similar and classifiable shapes that make possible the science of histology.

Basic Types of Cells and Tissues

Tissues are organized collections of similar cells in recognizable groupings. There are not as many different types of tissues in the body as we might suppose. In fact, there are only four fundamental types, each with its own distinctive structural, functional, and developmental characteristics: *epithelial* tissue, *connective* tissue, *muscle* tissue, and *nervous* tissue. Blood and lymph may be considered as a type of connective tissue in which the intercellular substance is fluid. All of the organs to be studied in later chapters are constructed of one or more of these basic tissue types.

EPITHELIAL TISSUE

Epithelial tissue, or *epithelium*, consists of sheets of cells joined together in continuous membranes. The individual cells are closely attached to one another with little intercellular substance. In this respect epithelium differs

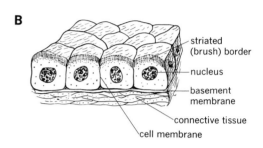

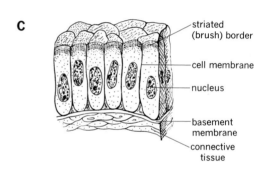

5.1 Simple epithelium: A, squamous (mesentery) transverse section; B, cuboidal (left, bronchus); C, columnar (intestine). (Adapted from W. Bloom and D. W. Fawcett, *A Textbook of Histology*, 9th Ed., W. B. Saunders Co., Philadelphia, 1968.)

from connective tissue, in which intercellular substance quantitatively predominates. Epithelium contains no blood vessels but is richly equipped with nerves and nerve endings. Its most prominent characteristic is that it forms the covering and lining membranes of the whole body and of its organs, cavities, and passageways. It also forms important parts of glands and sense organs.

Epithelium has different functions in different locations of the body, and its structure is adapted to these functions. In outer regions of the body, as in the skin, it provides protection against drying and mechanical injury and contains the various sensory nerve endings. It also has remarkable regenerative properties in the event of injury. In the body cavities, it frequently contains glands that secrete mucus, which lubricates surfaces, minimizing friction. The epithelium of the stomach and intestine also serves to absorb nutrient substances; that of the lungs, to absorb gases.

Types of epithelium. Epithelium may be classified according to the shape and structure of the individual cells as *squamous, cuboidal,* and *columnar.* If the cells are arranged in a single layer, it is called *simple.* If they are arranged in more than one layer, it is called *stratified.*

Simple squamous epithelium consists of extremely thin, scalelike, flat cells in a paving layer (Fig. 5.1A). The cellular outlines can easily be demonstrated by chemical treatments that blacken the intercellular substance. A squamous cell is only 1 to 3 μ thick although it may be 20 to 30 μ wide. In cross section it looks like a slender rod thickened in the region of the nucleus. This type of epithelium is found in the alveoli of the lungs. *Endothelium* is a form of simple squamous epithelium that lines the blood vessels and heart. Unlike the epithelium, derived from embryonic *ectoderm* or *endoderm,* endothelium comes from *mesoderm. Mesothelium,* also of mesodermal origin, is a form of simple squamous epithelium that lines the great body cavities, the pleura, pericardium, and peritoneum.

Simple cuboidal epithelium consists of cells in which height approximately equals width (Fig. 5.1B). From the free surface the cells appear as polygons. In sections perpendicular to the surface, they appear almost square. This type of epithelium is found in all but the largest bronchi and in certain glands and their ducts.

Simple columnar epithelium consists of cells in which height exceeds width (Fig. 5.1C). In sections parallel to the surface, the cells appear as small polygons, resembling cuboidal cells, but in sections perpendicular to the surface, they appear rectangular. This type of epithelium lines the stomach, intestines, and uterus.

In stratified epithelium, where several layers of cells make up the epithelial sheet, only the deepest layer touches the underlying tissue. In stratified squamous epithelium the outer layers

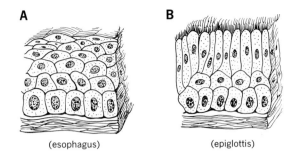

(esophagus) (epiglottis)

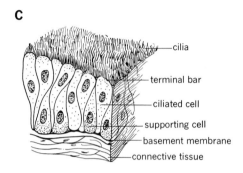

cilia
terminal bar
ciliated cell
supporting cell
basement membrane
connective tissue

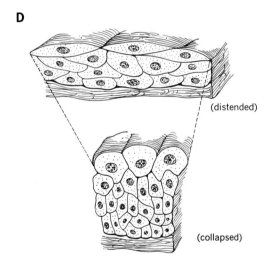

(distended)

(collapsed)

5.2 Types of epithelium. A, stratified squamous (esophagus); B, stratified columnar (epiglottis); C, pseudostratified (lung); D, transitional (bladder), distended and collapsed. (Adapted from W. Bloom and D. W. Fawcett, *A Textbook of Histology,* 9th Ed., W. B. Saunders Co., Philadelphia, 1968.)

are squamous while the deeper layers show a transition to cuboidal or columnar (Fig. 5.2A). Such an arrangement is typical of *epidermis,* the

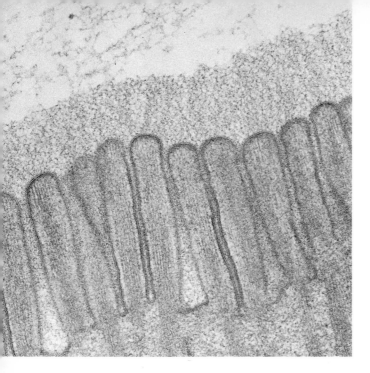

5.3 Brush micrograph of the striated border of a columnar epithelial cell (cat intestine), showing that the microvilli are covered with a thick glycoprotein surface coat (×70,000). This may have a role in determining the specificity of absorption (see p. 159). (Dr. Susumu Ito)

outer layer of skin. Stratified columnar epithelium is seen in Fig. 5.2B.

Two variants of stratified epithelium are *pseudostratified* epithelium (Fig. 5.2C) and *transitional* epithelium (Fig. 5.2D). In the former, found in the lungs, the nuclei of the individual cells are situated at different levels, giving the appearance of several cell layers although, in fact, each cell reaches from the outer side to the inner side of the sheet. The latter is intermediate in structure between stratified squamous epithelium and stratified columnar epithelium. It lines hollow organs (notably the urinary tract) in which there is frequent distention and stretching. Hence it may vary considerably in appearance.

Structure of epithelial cells. An epithelial cell may be equipped on its free surface with cilia or with a fringelike structure that light microscopists have traditionally called the *striated* or *brush border* (see Figs. 5.1, 5.2). Cilia, as noted in Chapter 4, flip rapidly in one direction and thereby move solid materials and fluids along the epithelial surface.

The striated border is especially well developed in the columnar cells of the absorptive epithelium lining the renal tubules (see Fig. 5.1B) and the intestine, although its specialized nature was not clear until these cells were examined under the electron microscope (Fig. 5.3). The border was then resolved into a series of fingerlike projections of membrane-covered cytoplasm called *microvilli*. In contrast to the ever-changing surface configurations of cells actively engaged in phagocytosis or pinocytosis, microvilli exist in stable, highly ordered arrangements. Their functional significance resides in the fact that they greatly increase the surface area exposed to substances to be absorbed into the cell.

Between the epithelium and the underlying connective tissue, the basement membrane (see Fig. 3.14E) is usually discernible. Between the adjoining lateral surfaces of two epithelial cells are discontinuous button-like junctional structures that hold the cell together. This structure, called a desmosome (or *macula adherens*), consists of two dense plaques on opposing cell surfaces, separated by an intercellular space about 250 Å wide.

One characteristic of all epithelial cells is *polarity*. Since they are arranged in sheets, and since the inside and outside surfaces of a sheet of cells are exposed to different environmental conditions, it is not surprising that the end of a cell toward one side differs functionally from the end toward the other side. This polarity is manifested structurally by the positions and shapes of the mitochondria at the two poles of the cell and by the positions of the centrioles and other particles in the cytoplasm. The proximal or deeper end receives the nutrient material. The distal or free end is exposed to a body cavity or to the air, depending upon the location of the epithelium.

Glands and glandular epithelium. Glands are special forms of epithelial tissue whose function is the production of certain chemical substances and the secretion of these substances into the immediate environment in the case of *exocrine glands* or into the blood in the case of *endocrine glands*. Some body organs include both exocrine

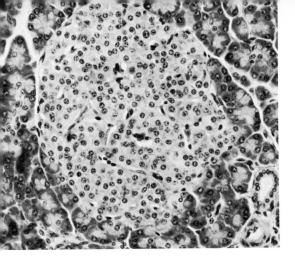

5.4 Section of pancreas, showing acinar glands surrounding an islet of Langerhans (× 300). (From W. Bloom and D. W. Fawcett, *A Textbook of Histology*, 9th Ed., W. B. Saunders Co., Philadelphia, 1968.)

and endocrine gland elements. The pancreas, for example, contains exocrine glands that secrete digestive enzymes into the intestine and many isolated endocrine glands that secrete insulin into the blood (Fig. 5.4). The latter are

called *islets of Langerhans*. Here we shall speak only of the exocrine glands. Endocrine glands will be discussed in Chapter 15. The three methods of classifying exocrine glands are given in Table 5.1.

The *mucous glands*, or *goblet cells*, are the best-known *unicellular* glands (Fig. 5.5). Since

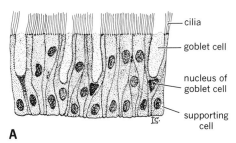

A

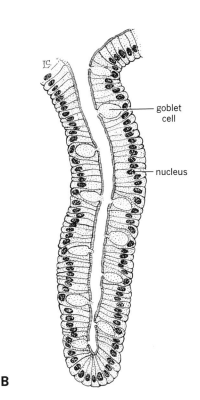

B

5.5 Mucous glands, or goblet cells: A, in pseudostratified ciliated columnar epithelium (trachea); B, in simple columnar epithelium (intestine). (From King and Showers, *Human Anatomy and Physiology*, 5th Ed., W. B. Saunders Co., Philadelphia, 1963.)

TABLE 5.1
TYPES OF EXOCRINE GLANDS

   I. Classification by number of cells
      A. Unicellular
      B. Multicellular
  II. Classification by structure
      A. Simple
         1. Tubular
            a. Straight
            b. Coiled
            c. Branched
         2. Alveolar
            a. Straight
            b. Branched
      B. Compound
         1. Tubular
         2. Alveolar
 III. Classification by mode of secretion
      A. Merocrine (eccrine)
      B. Holocrine
      C. Apocrine
      D. Cytocrine

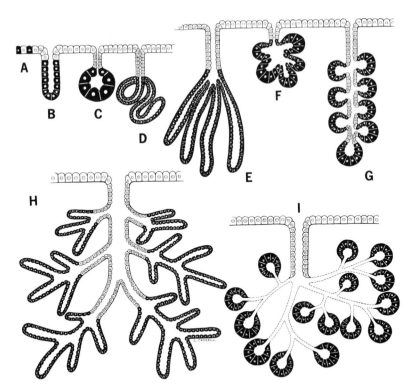

5.6 Various types of glands: A, unicellular; B, simple straight tubular; C, simple coiled tubular; D, simple branched tubular; E, simple straight alveolar; F, G, simple branched alveolar; H, compound tubular; I, compound alveolar. Shaded portions are secretory cells. (From King and Showers, *Human Anatomy and Physiology*, 5th Ed., W. B. Saunders Co., Philadelphia, 1963.)

they are scattered throughout epithelium, particularly that of the columnar type, its membranes are often called *mucous membranes*. A mucous gland resembles a goblet into which numerous secretory droplets are deposited by the surrounding cytoplasm. Its function is the formation and extrusion of mucus. The *melanocytes*, which synthesize melanin, are also unicellular glands. Most of the remaining glands are *multicellular*. The basis for their classification as tubular or alveolar and straight, coiled, or branched is evident from Fig. 5.6. As the figure shows, the common features are a duct opening to the surface and one or more blind cavities opening into the duct. In a straight tubular gland the duct is almost nonexistent.

In *merocrine* glands the secretory cells remain intact during the formation and discharge of secretory products. The salivary and exocrine pancreatic glands are of this type. By convention, merocrine sweat glands are called *eccrine* sweat glands. In *holocrine* glands secretory products accumulate within the cell bodies; the

cells then die and are themselves discharged, laden with the secretions; and new cells replace them. The *sebaceous* glands of the skin are holocrine glands. In *apocrine* glands secretory products accumulate near the free ends of the cells, which then pinch off, leaving the remainder of the cell bodies intact. The milk-secreting *mammary* glands and certain sweat glands are of this type. The only known *cytocrine* glands are the melanocytes. In the skin a granule-containing process of a melanocyte is "injected" into a neighboring target cell and pinched off.

CONNECTIVE TISSUE

Connective tissue binds together and supports all other types of cells and tissues. It is characterized by a large amount of intercellular substance and a relatively sparse cell population. The nature of the intercellular substance provides the basis for a classification in which connective tissue is divided into three main types: (1) *connective tissue proper*; (2) *cartilage*; and

A

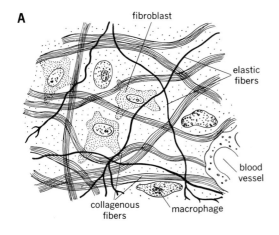

fibroblast

elastic
fibers

blood
vessel

collagenous
fibers

macrophage

B

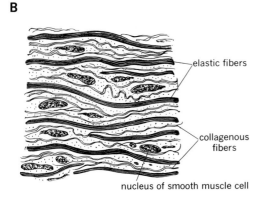

elastic fibers

collagenous
fibers

nucleus of smooth muscle cell

C

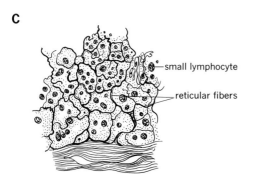

small lymphocyte

reticular fibers

5.7 Connective tissue: A, collagenous fibers in fibrous connective tissue; B, elastic connective tissue (middle coat of large artery); C, reticular fibers (lymph node). (Adapted from W. Bloom and D. W. Fawcett, *A Textbook of Histology,* 9th Ed., W. B. Saunders Co., Philadelphia, 1968.)

(3) *bone.* In most connective tissue proper the intercellular substance is soft, jellylike, and amorphous; in cartilage it is rubbery; and in bone it is dense, hard, and mineralized.

In all types of connective tissue, the intercellular substance includes fibers, which may be loosely arranged as in subcutaneous tissue or densely packed as in ligaments and tendons. Three types of fibers occur: *collagenous* or *white* fibers; *elastic* or *yellow* fibers; and *reticular* fibers. Collagenous fibers (Fig. 5.7A) are long, strong, nonelastic fibers that contain the protein called *collagen.* We shall discuss its properties later. Elastic fibers (Fig. 5.7B) predominate in tissue requiring elasticity, as in hollow distensible organs. When extended, they are straight; when relaxed, they are curled up in spirals and broad curves. Reticular fibers (Fig. 5.7C) make up the internal structures of the solid organs. As their name suggests, they form networks.

Connective tissue proper. The major types of connective tissue proper are *mucous, loose* or *areolar, white fibrous, elastic, adipose,* and *reticular.*

Mucous connective tissue is a loose connective tissue occurring chiefly in embryos, just beneath the skin and in the umbilical cord. It consists of large star-shaped *fibroblasts* (see Figs. 5.7A, 5.8A), whose processes interlace to form a network. The intercellular spaces are filled with a gelatinous matrix containing fine collagenous fibers.

Loose or areolar connective tissue consists of a semifluid matrix containing loosely organized collagenous and elastic fibers as well as a scattering of diverse cell types (Fig. 5.8B). Loose connective tissue is widely distributed: it forms the layer just beneath the skin known as the *superficial fascia;* it lies beneath most of the epithelial membranes; and it fills the spaces between organs, where its loose construction permits easy displacement and movement of the tissue to which it is connected. Besides its mechanical function, loose connective tissue takes part in the nutrition, defense, and repair of the body. It furnishes a bed for the capillary and lymphatic systems and mediates the exchange of fluids and dissolved materials between blood and tissue cells. In certain pathological states the amount of fluid permeating the loose connective tissue is

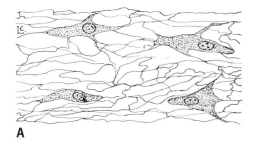

A

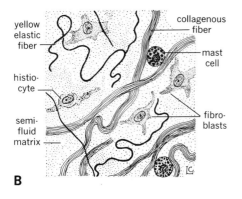

yellow elastic fiber

collagenous fiber

mast cell

histio- cyte

fibro- blasts

semi- fluid matrix

B

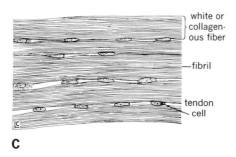

white or collagen- ous fiber

fibril

tendon cell

C

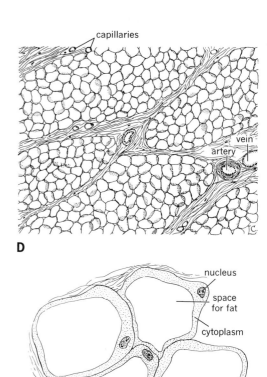

capillaries

vein

artery

D

nucleus

space for fat

cytoplasm

E

5.8 Connective tissue proper: A, mucous (Wharton's jelly from an umbilical cord); B, loose (from beneath the skin over the abdomen); C, tendon (left, cross section; right, longitudinal section); D, adipose, low magnification; E, adipose, high magnification. (From King and Showers, *Human Anatomy and Physiology*, 5th Ed., W. B. Saunders Co., Philadelphia, 1963.)

enormously increased. Also, at times many cells wander into it as part of the disease-combating process called *inflammation.*

White fibrous connective tissue consists of collagenous fibers densely packed together. In regions of the body where strong support is needed, the fibers form cords, bands, and membranes—tendons, ligaments, and fibrous membranes—and the intercellular substance has an extremely firm consistency. Tendons connect muscles to the skeleton (Fig. 5.8C), and ligaments connect bones to each other. In contrast to loose connective tissue, the glistening white

ligaments are notable for their strength and inelasticity.

Elastic connective tissue is composed chiefly of elastic fibers, with occasional collagenous fibers and fibroblasts. It is found in the walls of blood vessels (see Fig. 5.7B) and in the lungs and bronchi, where its stretching properties are of special value.

Adipose tissue is loose connective tissue in which fat cells have displaced most of the other elements (Fig. 5.8D,E). The narrow spaces between adjacent fat cells contain a few fibroblasts and lymphocytes, networks of collagenous and elastic fibers, and richly developed networks of blood capillaries.

Reticular connective tissue constitutes the structural framework of blood-forming lymphoid tissues such as the spleen and lymph nodes (see Fig. 5.7C) and bone marrow, as well as other solid organs.

Cells of connective tissue. We have spoken of the cells of normal and inflammatory connective tissue. These are of great physiological importance though relatively poorly understood. They include the following types: *fibroblasts, macrophages, lymphocytes, mast cells,* and *fat cells.*

Fibroblasts are the predominating cells of connective tissue and are believed responsible for the formation of collagen and intercellular fibers. Their appearance varies in different locations and conditions, but generally they are long, flat, stellate cells that resemble in profile slender spindles (see Figs. 5.7A, 5.8A). They arise from the early embryonic tissue called *mesenchyme.*

Macrophages, known also as *fixed macrophages* and *histiocytes,* are almost as numerous as fibroblasts. They too vary in shape, ranging from flat rounded forms to elongated spindles with branching extensions. They ingest foreign particles such as bacteria and cellular debris by phagocytosis and hence play an important role in body defenses.

The lymphocytes of connective tissue are similar to the lymphocytes of blood. We shall discuss them in Chapter 8.

Mast cells are curious structures scattered throughout the connective tissue. They contain large amounts of histamine, which actively contracts the small blood vessels, and heparin, which, when isolated and added to blood, powerfully prevents clotting. These effects of heparin and histamine do not occur while the substances remain within the cells, and it is not yet clear under what circumstances they are released. There is some evidence, however, that mast cells influence the functioning of the tiny capillaries coursing through the loose connective tissue.

Fat cells are large, brilliant, spherical cells typically so distended with lipid that the cytoplasm is only a thin ring encircling the displaced nucleus (see Fig. 5.8D). As fat is laid down in the body, fat cells increase in number, size, and

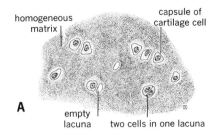

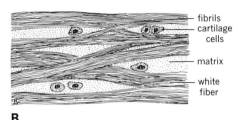

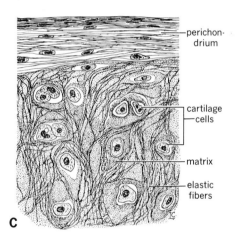

5.9 Cartilage: A, hyaline; B, white fibrous; C, yellow elastic. (From King and Showers, *Human Anatomy and Physiology,* 5th Ed., W. B. Saunders Co., Philadelphia, 1963.)

lipid content. They accumulate in massive numbers in adipose tissue. When adipose tissue is consumed, as in starvation, they are depleted of their lipid and come to resemble fibroblasts.

Cartilage. Cartilage is the tissue commonly known as "gristle" because of its firm, resilient texture. It is composed of round or oval cells, which tend to be arranged in groups of two or more, and a massive tough intercellular matrix (Fig. 5.9). The spaces in the matrix occupied by

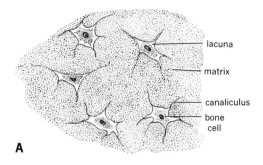

A

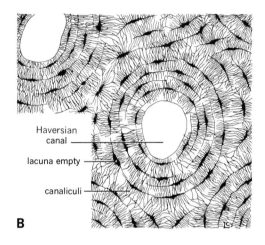

B

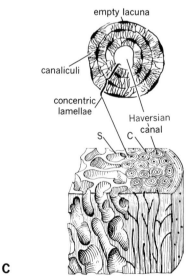

C

5.10 Bone: A, at high magnification; B, at lower magnification; C, spongy (S) and compact (C). (From King and Showers, *Human Anatomy and Physiology*, 5th Ed., W. B. Saunders Co., Philadelphia, 1963.)

the cells are called *lacunae*. Except on bare surfaces in joint cavities, cartilage is covered by a sheath of dense connective tissue called *perichondrium*. The fibroblasts of the perichondrium are transformed into cartilage cells and deposit the matrix. Cartilage contains no blood vessels. Three types of cartilage are distinguished on the basis of variations in the texture of the intercellular substance: *hyaline* cartilage, *white fibrous* cartilage, and *yellow elastic* cartilage.

Hyaline cartilage (see Fig. 5.9A) is the simplest and most widespread type. It has the blue-white color of skimmed milk. The matrix is homogeneous, but, when stained, the portion immediately surrounding a cartilage cell may be darker than the remainder of the matrix. Hence this area may be referred to as the *capsule* of the cell. Hyaline cartilage covers the surfaces of bones within joints and forms the rib cartilages, the cartilage of the nose, and the ring cartilages of the trachea and bronchi; in the embryo it constitutes most of the temporary skeleton.

White fibrous cartilage contains a large number of collagenous fibers arranged in parallel rows in the matrix (see Fig. 5.9B). It is found in the discs between the bodies of the vertebrae, between the two pubic bones, and in the articular discs in many joints.

Yellow elastic cartilage develops from hyaline cartilage. Characterized by a network of yellow fibers through the matrix (see Fig. 5.9C), it forms the cartilages of the external ear, Eustachian tube, and epiglottis and some small cartilages of the larynx.

Bone: a mineralized connective tissue. Certain tissues must perform a variety of mechanical functions, such as maintaining the form and shape of the body against the forces of gravity, protecting delicate organs by encasing them in vaults, and providing the body with a system of movable but structurally rigid levers.

Nature has devised several methods of meeting this need. In some insects the cross-linked and polymerized carbohydrates and proteins of the procuticle join to form a structurally rigid *exoskeleton*. In certain Elasmobranchii, such as the shark, a cartilaginous skeleton combines resiliency and rigidity. In other organisms large quantities of crystalline inorganic matter are deposited in an organic matrix to produce a shell, as in clams or oysters, or the bone of an *endoskeleton*. This process, widespread in nature, is known as *tissue mineralization*.

Like cartilage, bone (or *osseous tissue*) is a dense connective tissue containing cells sparsely distributed throughout a massive intercellular matrix. In bone, the matrix is made hard by deposits of crystalline calcium phosphate salts; on a dry-weight basis 65 to 70% of bone consists of inorganic crystals—chiefly hydroxyapatite, $Ca_{10}(PO_4)_6(OH)_2$—and 30 to 35% consists of organic matrix, of which 95 to 99% is collagen.

Bone has a highly complex structure that must be examined at several magnifications. At the highest magnification we see the bone cells, or *osteocytes*, lying in their lacunae (Fig. 5.10A). Fine projections of the cell bodies grow out into the matrix in *canaliculi*, channels that radiate from the lacunae in all directions to form networks. Thus there exists a continuous system of communicating cavities.

At lower magnification we see that bone is composed of layers, or *lamellae*. Bone may be divided into two types, depending upon the lamellar pattern: *spongy* or *porous* or *cancellated;* and *compact* or *hard*. In spongy bone the lamellae are arranged in a latticework with large spaces. In compact bone the lamallae are arranged in layers around central canals (Fig. 5.10B). These latter organizations are called *Haversian systems* or *osteones*. As shown in Fig. 5.10C, they are oriented longitudinally in the shafts of long bones. The canals of the Haversian systems contain blood and lymphatic vessels.

The collagen fibers in the lamellae of a Haversian system are collected in small bundles that encircle the canal repeatedly, crossing one another to produce a trellis-like arrangement. The hydroxyapatite crystals are embedded in this highly ordered collagen matrix. We shall discuss the fascinating problems of calcification and bone formation in Chapters 15 and 17.

The ends of long bones consist of spongy bone with a thin covering of compact bone (Fig. 5.11). The shaft is made up almost entirely of compact bone. Extending through the center of the shaft is a cavity containing *bone marrow*. A thin, delicate membrane, the *endosteum*, lines this cavity. The bone surface is covered by peri-

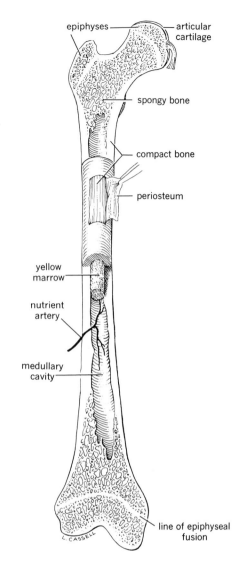

5.11 A long bone. (From King and Showers, *Human Anatomy and Physiology*, 5th Ed.', W. B. Saunders Co., Philadelphia, 1963.)

osteum, whose dense outer fibrous layer contains blood vessels and whose inner layer contains the fibroblasts that play a key role in bone formation. Periosteum is attached to bone by the *fibers of Sharpey*, which extend into the bone tissue. If periosteum is stripped from a piece of fresh bone, blood oozes from minute pores on the bone surface. These are called *Volkmann's canals* (see Fig. 5.10C).

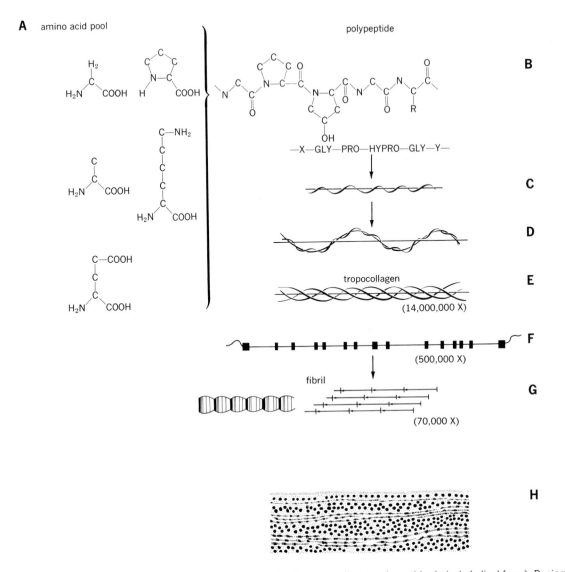

5.12 Collagen. A, free amino acids; B, polypeptide chain; C, single collagen polypeptide chain (a helical form); D, single polypeptide coiled in major helix; E, three such chains coiled about each other to form rope-like super helix (this is tropocollagen); F, tropocollagen molecule, showing asymmetrically placed discontinuities as determined by electron microscopy; G, staggered arrangement of tropocollagen molecules in a collagen fibril and microscopic appearance of a collagen fibril; H, grid-like arrangement of collagen fibrils in connective tissue. (From J. Gross. In *Comparative Biochemistry*, Vol. 5., Academic Press, N.Y., 1963.)

Physical and chemical properties of collagen. Collagen possesses remarkable properties that make it unique among the proteins and fit it admirably for its physiological role. In tendon it has the tensile strength of light steel wire; in cornea it has the transparency of water; and it alone accounts for the toughness of leather, the tenacity of glue, and the viscosity of gelatin. It is not surprising that collagen, whose main functions are protection and support, constitutes one-third of the body protein.

Collagen is synthesized mainly by fibroblasts.

The molecule contains about a thousand amino acid residues. In ordinary proteins the amino acids undergo no further alteration once they are linked together. In collagen, however, two unusual amino acids, hydroxyproline and hydroxylysine, are formed *after* the molecular chain has been assembled, the new amino acids being created by the addition of hydroxyl groups to some of the proline and lysine units in the chain. Proline and hydroxyproline, which together make up as much as 25% of the collagen molecule, have the curious capacity to prevent easy rotation of the regions in which they are located; thus they impart rigidity and stability to the molecule. The collagen molecule has yet another distinctive feature: every third amino acid is glycine. Finally, it contains a small amount of sugar (0.6% hexose by weight).

Collagen owes its properties both to its chemical composition and to its physical structure (Fig. 5.12). The molecule consists of three peptide chains. Each of these is a left-handed helix, and three such helices are coiled around each other and hydrogen-bonded (and occasionally cross-linked by covalent bonds) into a right-handed superhelix.

Examination of collagen fibrils by electron microscopy revealed a regular series of bands 700 Å apart (see Fig. 5.12E). It was assumed that the bands marked the places where individual molecules were joined end to end. Hence it was concluded that the collagen molecule is 700 Å long. That the length of the collagen molecule is $4 \times 700 = 2800$ Å was learned from observations of a most unusual phenomenon. If native collagen is dissolved in acid and the solution is then neutralized and warmed, fibrils of normal structure automatically reappear (Fig. 5.13A–D). It was found that variations in the treatment of dissolved collagen yield reconstituted fibrils with strikingly different band patterns in electron micrographs. One type has major bands spaced 2800 Å apart and a *symmetrical* series of fine bands (Fig. 5.13E). This is designated "fibrous long spacing," or FLS, to indicate that it is fibrous and has a long period. Another type lacks the characteristic beltlike appearance of the two other forms and occurs in short pieces instead of long fibrils. Superficially it resembles isolated segments of FLS, but closer study discloses an *asymmetric* series of fine bands (Fig.

5.13F). This is called "segment long spacing," or SLS, to indicate that it is nonfibrous and has a long period.

Each of the three different forms can be dissolved and converted into either of the other two forms. Thus all three are aggregates of the same needle-shaped monomer units (see Fig. 5.12C, D). The fundamental unit of collagen structure, 2800 Å long, is called *tropocollagen* (from the Greek meaning "turns into collagen"). It is the tropocollagen molecule that is composed of three helical peptide chains, that has a "head" and a "tail," and that is synthesized by fibroblasts.

This information suggests that the three forms of collagen are simply different arrangements of tropocollagen molecules. In the native collagen fibril, the molecules are lined up facing in the same direction and overlapping by one-fourth of their lengths (see Figs. 5.12D, 5.13C). It is the overlapping that creates the 700 Å periodicity. In the FLS form the molecules lie side by side but do not all face in the same direction and do not overlap (see Fig. 5.13E). Hence the major periodicity measures 2800 Å. The random positioning of "heads" and "tails" of adjacent molecules explains the symmetrical fine structure, even though the discontinuities, or "bumps," along the lengths of the individual molecules are asymmetrically spaced. In the SLS form the molecules are again nonoverlapping, but they all face the same way (see Fig. 5.13F). As a result, the characteristic distributions of various basic and acidic amino acid residues along the lengths of the individual molecules show up as an asymmetrical band pattern.

In summary, fibroblasts secrete collagen molecules into the amorphous intercellular matrix, and there, under the influence of the physiological ionic atmosphere, temperature, and perhaps certain rate-regulating factors, the molecules aggregate spontaneously to form fibrils. The part, if any, played in this process by the noncollagenous components of the matrix, such as mucopolysaccharides and other proteins, is still obscure. It is possible that these substances are important in the ordering of the fibrils into bundles. The properties of collagen obviously account for its crucial biological functions. We shall later learn how the properties of similar specifically structured fibrous molecules account

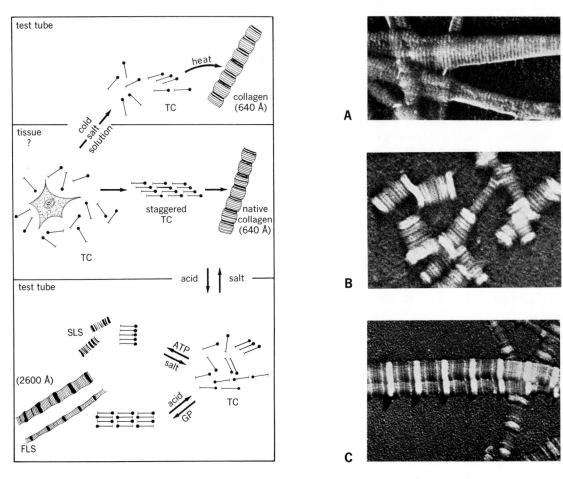

5.13 Tropocollagen and reconstitution of collagen showing different modes of tropocollagen aggregation. In the tissues, tropocollagen molecules synthesized by fibroblasts overlap in a staggered array to form native collagen (electron micrograph A). When salt, hydrogen ion, or ATP concentrations are varied in solutions of extracted tropocollagen, the reaggregated fibrils can demonstrate the normal band pattern (A), the SLS or "segment long spacing" pattern (electron micrograph B) or FLS or "fibrous long spacing" (electron micrograph C). The observed fine structure reflects the asymmetry of the tropocollagen molecule. Magnification of the electron micrographs, ×30,000. (From J. Gross. In *Comparative Biochemistry*, Vol. 5, Academic Press, N.Y. 1963.)

for such diverse phenomena as muscular contraction, bone formation, blood coagulation, and wound healing.

Intercellular substance. The matrix of connective tissue is that material that is intercellular and nonfibrillar. Chemically it is a complex mixture consisting of collagen, mucopolysaccharides (including hyaluronic acid and chondroitin), water, and soluble materials originating in the blood plasma. Its main functions are the structural stabilization of connective tissue and the maintenance of a medium for the transport of metabolites. It is the matrix as a whole that constitutes the actual environment of most body cells. Thus it is a channel through which materials pass on their way from blood to cell and back again.

MUSCLE TISSUE

The body moves by contracting its muscles, muscle tissue performing the necessary mechanical work by physically shortening its cells and molecular components. How the molecular machinery of muscle converts the energy of ATP into mechanical work will be discussed in Chapter 17. The answer is not yet fully known, but a host of brilliant investigators has obtained important insights from detailed studies of muscle structure.

There are two fundamental types of muscle: *smooth* muscle and *striated* muscle. Although this classification is made on the simple basis of microscopic appearance, it has a good deal of functional significance. Only striated muscle is under voluntary control. Smooth muscle operates involuntarily and much more slowly. *Cardiac* muscle is an involuntary form of striated muscle that contracts automatically and rhythmically.

Smooth muscle. Smooth muscle is found primarily in the internal organs. A smooth muscle cell is a long spindle-shaped body, somewhat thickened in the middle, with a centrally placed oval nucleus (Fig. 5.14A). Its length varies from 20 to several hundred μ, and its greatest diameter from 3 to 8 μ. Electron microscopy reveals many long thin filaments running parallel to the axis of the cell. These constitute the bulk of the cytoplasm.

In smooth muscle tissue the cells are generally parallel to one another with the thick portion of one cell opposite the thin ends of several adjacent cells (Fig. 5.14B). Small amounts of connective tissue lie between the cells, usually in dense networks of reticular and elastic fibers. The cells are most often gathered into sheets or bands, as in the tubular muscle coats of the intestines and the respiratory passages. In this form smooth muscle contracts in slow rhythmic waves, its movement being known as *peristalsis*. Smooth muscle also appears as small individual fibers, as in the walls of blood vessels and in certain parts of the eye. These contract more rapidly than muscle sheets.

Striated muscle. Voluntary muscle is called striated muscle because of its characteristic alternating dark and light cross stripes (Fig. 5.15). It is also spoken of as *skeletal* muscle because it is attached to bone. It represents almost half of the total body weight.

The components of striated muscle are shown in Fig. 5.16. It consists of long cylindrical multinucleated cells called *muscle fibers*, which appear striated under a light microscope. A muscle fiber may reach a length of 40 mm. Its cytoplasm is called *sarcoplasm*, and its cell membrane is called the *sarcolemma*. The many nuclei are flattened oval bodies lying just beneath the sarcolemma.

Together with nuclei and mitochondria, *myofibrils*, 1 to 2 μ thick, make up the muscle fiber. Isolated muscle fibers are readily separated from gross striated muscle, and individual myofibrils can be teased from them. Electron microscopy of a myofibril discloses a striking pattern of alternating light and dark bands in a repeating sequence (see Fig. 5.15). Each repeating segment constitutes a *sarcomere*, and narrow dense *Z lines* (from *Zwischenscheibe*, meaning "between discs") mark its limits (see Fig. 5.16D). The other bands are two less dense *isotropic* or *I bands*; a dense *anisotropic* or *A band*; a narrow light *H zone* (named for Victor Hensen, its dis-

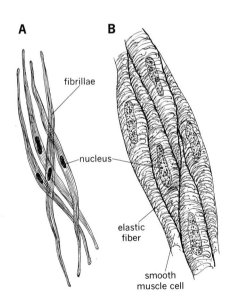

5.14 Smooth muscle: A, individual cells; B, arrangement of cells in smooth muscle of intestine. (From W. Bloom and D. W. Fawcett, *A Textbook of Histology*, 9th Ed., W. B. Saunders Co., Philadelphia, 1968.)

In the figure: A, B, fibrillae, nucleus, elastic fiber, smooth muscle cell.

5.15 Striated muscle, showing the band pattern (×19,200). Clearly visible are the dense A bands, bisected by H bands containing M lines; and the lighter I bands, bisected by Z lines. (Dr. Hugh E. Huxley)

coverer*) in the center of the A band; and a fine dense *M line* in the center of the H zone. A narrow light *pseudo H zone* immediately surrounds the M line.

Electron micrographs show that the band pattern is attributable (as in collagen) to the overlapping of filaments (see Fig. 5.16E). Myofibrils contain two types of filaments, thick and thin ones. Neither type changes in length during muscle contraction. The A band, made up mainly of thick filaments, is entirely anisotropic (or noncontracting). The I band, made up of

---

* Some authorities believe the H zone was named for *hell*, the German word for "light."

thin filaments, and the H zone do shorten in muscle contraction but not through a shortening of filaments. As we shall see in Chapter 17, muscle contraction is now believed to involve a *sliding* of thick and thin filaments over one another.

In addition to the myofibrils, Fig. 5.15 shows prominent endoplasmic reticulum in the interfibrillar sarcoplasm. This is often termed *sarcoplasmic reticulum*. At each Z line the sarcoplasmic reticulum is interrupted by a system of transverse tubules, called the *T system*, that are continuous with the sarcolemma. We shall later learn of the great importance of these tubules.

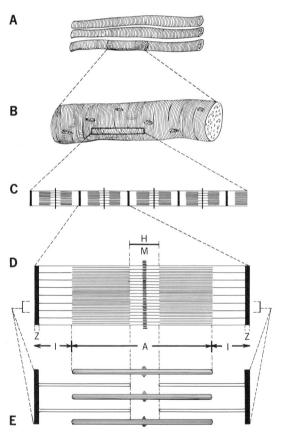

5.16 Schematic dissection of striated muscle: A, gross muscle, muscle fibers; B, single muscle fiber, showing nuclei; C, single myofibril; D, sarcomere of a myofibril, showing major bands and lines; E, thick and thin filaments, which account for the band pattern.

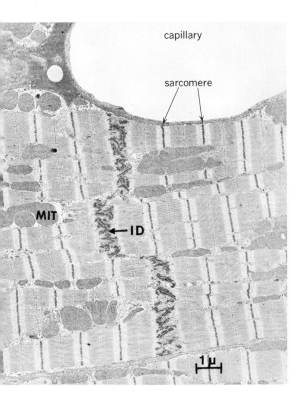

capillary

sarcomere

MIT

←ID

1 μ

5.17 Cardiac muscle. MIT, mitochondrion; ID, intercalated disc. (From Sonnenblick, Spotnitz, and Spiro, *Circ. Research, 15,* 11–70, 1964.)

Cardiac muscle. The muscle forming the bulk of the heart wall is intermediate in structure between smooth and striated muscle. Although its striations differ from those of striated muscle in detail only, its myofibrils, unlike those of striated muscle, may branch. The branching pattern enables cardiac muscle to function rapidly and concertedly in performing its unique task. It was once thought that cardiac muscle is a *syncytium* —i.e., a network in which cell borders are nonexistent or indistinct. Electron microscopy, however, has shown that discrete cells exist (Fig. 5.17). The cells are joined at the *intercalated discs*, elaborate structures containing desmosome-like junctions that link the elongated cells end to end.

In summary, cardiac muscle is distinguished by its involuntary mode of action, branching fibers, and intercalated discs. It is also noted for the size and number of its mitochondria. Presumably these are needed to satisfy its high ATP requirements.

NERVOUS TISSUE

Nerve cells, or *neurons*, the functional and structural units of the nervous system, are specialized to transmit impulses from one part of the body to another. They are extremely varied

5.18 Various types of neurons: A, unipolar neuron (spinal ganglion); B, multipolar neuron (anterior spinal cord); C, pyramidal cell (cerebral cortex); D, Purkinje cell (cerebellum); E, Golgi cell (spinal cord); F, fusiform cell (cerebral cortex); G, autonomic ganglion cell; a, axon; d, dendrites; c, collateral branch. (From Morris, *Human Anatomy,* 10th Ed., Blakiston, 1961.)

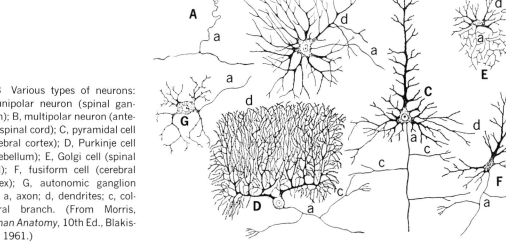

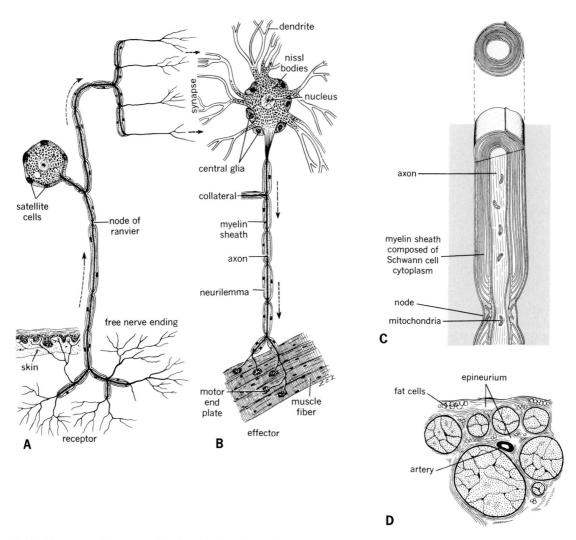

5.19 Two types of neurons, showing the directions of nerve impulses: A, sensory or afferent neuron; B, motor or efferent neuron; C, section of an axon, showing the arrangement of myelin in concentric lamellae (e.g., Figs. 12.3, 12.5) and mitochondria in the cytoplasm; D, transverse section of a nerve. (A and B from King and Showers, *Human Anatomy and Physiology*, 5th Ed., W. B. Saunders Co., Philadelphia, 1963. C and D from G. G. Simpson and W. S. Beck, *Life: An Introduction to Biology*, 2nd Ed., © 1957, 1965 by Harcourt Brace Jovanovich, Inc. and reproduced by permission.)

in form (Fig. 5.18), but all have one or more long cytoplasmic processes. The processes called *axons* often divide into many terminal branches and may be remarkably long. These transmit nerve impulses *away* from the cell body. The processes called *dendrites*, which generally branch freely close to the cell body, transmit nerve impulses *toward* the cell body. *Sensory* or *afferent* neurons transmit impulses from the

body tissues toward the central nervous system, and *motor* or *efferent* neurons transmit them from the central nervous system toward the muscles (Fig. 5.19).

Neurons may be classified by the number of cytoplasmic processes extending from the cell body (see Fig. 5.18). *Multipolar neurons* have many dendrites and one axon. *Bipolar neurons* have one dendrite and one axon. *Unipolar* neurons have but one axon, which divides some distance from the cell body into two long *collaterals*.

Neurons link together to form long pathways for the conduction of impulses. Two successive neurons make contact at a region called the *synapse*, which permits impulse conduction in one direction only (see Fig. 5.19). One-to-one junctions of neurons are unusual. More often, one neuron connects with many others—and vice versa. Many synapses occur directly on the surface of a cell body. Fig. 5.20 shows a large number of terminal branches from other neurons synapsing on the surface of a single neuron cell body.

A striking feature of neuron structure is the great distance that may be traversed by the cytoplasmic processes of a cell body. Even in the longest smooth muscle cells, cytoplasm does not extend farther than 250 μ from the nucleus. The axons of some neurons, however, extend several feet.

Neuron cell body. Regardless of external form, all neurons contain in their cell bodies and processes fine fibrils less than 1 μ in diameter. These are the *neurofibrils* (see Fig. 5.19B). Within the cytoplasm of all neurons and the dendrites of multipolar neurons are small, deeply basophilic particles that classic microscopists knew as *Nissl bodies*. Today we recognize that these are massive clusters of ribosomes that are unattached to membranes of endoplasmic reticulum. We shall deal with current concepts of their function in Chapter 12.

Nerve fiber and nerve. As shown in Fig. 5.19C, the central portion of the nerve fiber is the axon. The substance of the axon, the *axoplasm*, is strikingly free of ribosomes, though it contains a scattering of mitochondria. The axon is surrounded by (1) a thin sheath called the *axolemma*; (2) a lamellated sheath of lipid-containing material called *myelin* (lacking in certain nerves), which is arranged in concentric layers; and (3) an outer membrane called the *neurolemma*. At intervals along the neurolemma are *Schwann cells** and constrictions called *nodes of Ranvier*. Only one Schwann cell is present in each nodal segment.

A nerve is a bundle of many nerve fibers sheathed in connective tissue called *epineurium* (see Fig. 5.19D). Blood vessels penetrate the nerve, supplying it with food and oxygen.

Tissue Architecture

In this chapter we have surveyed the fundamental types of body cells and the elementary tissue patterns in which they are found. In later chapters we shall encounter far more complex patterns of tissue organization. For example, we shall see in the microscopic structures of liver, kidney, and pancreas architectural designs that are unique and that combine features of the basic tissue types.

It is not known how such organizational harmony develops, but it is obvious that it depends upon a highly ordered system of cellular interactions. Knowledge of these interactions, however, is also scanty. Presumably cells interact in two ways: either a cell elaborates a diffusible substance that affects another cell some distance

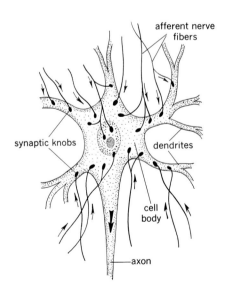

5.20 Terminal branches from many neurons synapsing on the surface of a single neuronal cell body and its dendrites.

afferent nerve fibers

synaptic knobs

dendrites

cell body

axon

---

* The process by which the Schwann cell gives rise to the multilayered myelin sheath will be described in Chapter 12.

away; or a cell affects another cell by direct contact. Interactions through diffusible agents (e.g., the hormones secreted by the endocrine glands) are much better understood than contact interactions—but both appear to control the evolution of tissue architecture.

One current approach to the problem of contact interactions involves the disintegration of a tissue into individual cells by destruction of all intercellular structures. Cells that have been separated this way, washed, and seeded out in a tissue culture frequently reaggregate into arrangements structurally and functionally similar to the original tissue. This suggests that the "blueprint" for construction of a cell community is somehow ingrained in each individual cell. Conceivably it is inherent in the molecular characteristics of the cell surface or of some substance secreted by the cell. If so, we have to suppose that each cell type has a distinctive surface or secretion that causes its cells to orient themselves in a specific three-dimensional pattern. Such thoughts are conjectural at present. But some mechanism is at work, and if we are to understand the body in mechanistic terms, we need to know what it is.

REFERENCES AND SUGGESTIONS FOR FURTHER READING

Arey, L. B., *Human Histology*, 2nd ed., Saunders, Philadelphia, 1963.

Bloom, W., and D. W. Fawcett, *A Textbook of Histology*, 8th ed., Saunders, Philadelphia, 1962.

Bonner, J. T., *Morphogenesis*, Princeton Univ. Press, Princeton, 1952.

Bourne, G. H., ed., *The Structure and Function of Muscle*, Academic Press, New York, 1960.

Brachet, J., *The Biochemistry of Development*, Pergamon Press, New York, 1960.

Copenhaver, W. M., *Bailey's Textbook of Histology*, 15th ed., Williams & Wilkins, Baltimore, 1964.

Curtis, A. S. G., "Cell Contact and Adhesion," *Biol. Rev.*, 37, 82 (1962).

Eccles, J., "The Synapse," *Sci. Am.*, 212, 56 (Jan., 1965).

Glimcher, M. J., "Molecular Biology of Mineralized Tissues with Particular Reference to Bone," in J. L. Oncley, ed., *Biophysical Science*, Wiley, New York, 1959.

Gross, J., "Comparative Biochemistry of Collagen," in M. Florkin and H. S. Mason, eds., *Comparative Biochemistry*, Academic Press, New York, 1963.

———, "Collagen," *Sci. Am.*, 204, 120 (May, 1961).

Gustavson, K. H., "On the Chemistry of Collagen," *Federation Proc.*, 23, 613 (1964).

Hall, M. C., *The Architecture of Bone*, Thomas, Springfield, Ill., 1966.

Hodge, A. J., and F. O. Schmitt, "The Tropocollagen Macromolecule and Its Properties of Ordered Interaction," in M. V. Edds, ed., *Macromolecular Complexes*, Ronald Press, New York, 1961.

Huxley, H. E., "Mechanism of Muscular Contraction," *Sci. Am.*, 213, 18 (Dec., 1965).

Levi-Montalcini, R., and P. U. Angeletti, "Growth and Differentiation," *Ann. Rev. Physiol.*, 24, 11 (1962).

Porter, K. R., and M. A. Bonneville, *An Introduction to the Fine Structure of Cells and Tissues*, Lea & Febiger, Philadelphia, 1963.

Schiller, S., "Connective and Supporting Tissues: Mucopolysaccharides of Connective Tissue," *Ann. Rev. Physiol.*, 28, 137 (1966).

Schmitt, F. O., "Interaction Properties of Elongate Protein Macromolecules with Particular Reference to Collagen (Tropocollagen)," in J. L. Oncley, ed., *Biophysical Science*, Wiley, New York, 1959.

Sussman, M., *Animal Growth and Development*, Prentice-Hall, Englewood Cliffs, N. J., 1960.

Verzar, F., "The Aging of Collagen," *Sci. Am.*, 208, 104 (Apr., 1963).

Weiss, P., "The Cell in Development," *Lab. Invest.*, 8, 415 (1959).

———, "Cell Contact," *Intern. Rev. Cytol.*, 7, 391 (1958).

PART TWO

Systems of the Body

His rise was neither insignificant nor inevitable. Man *did* originate after a tremendously long sequence of events in which both chance and orientation played a part. Not all the chance favored his appearance, none *might* have, but enough did. Not all the orientation was in his direction, it did not lead unerringly human-ward, but some of it came this way. The result *is* the most highly endowed organization of matter that has yet appeared on this earth. . . .

George Gaylord Simpson, in *The Meaning of Evolution*, 1949

6 Systems of the Body: A Prologue

The Human Organism

The human body contains about 100 trillion cells. We have spoken of the structural and functional attributes shared by *all* cells and those that permit the grouping of cells into different categories. Obviously, the differences among body cells account for the integrated nature of body organization. Specialized cells are woven into tissues, tissues into organs, and organs into functioning systems whose "purpose" is the preservation of a constant and ideal internal environment.

APPARENT PURPOSIVENESS OF BODY FUNCTIONS

A fundamental quality of the body's organizational hierarchy is the apparent purposiveness of its many activities. Another is the way in which this apparent end-serving or goal-seeking behavior integrates small local events into complex programs of self-regulation and self-reproduction. Body activities thus appear subservient to future needs. Reproduction serves the needs for continuation of the species; thirst serves the need for water; the synthesis of blood clotting factors serves the need for protection against fatal bleeding from minor wounds. All of these needs may be grouped under a general heading, the need for *self-preservation*. A striking illustration of the response of the body's systems to this need is the fact that in starvation, when the body must feed upon its own materials to survive, the order in which proteins break down is the reverse of the order of their physiological importance; hence the most vital tissues are preserved the longest.

We have modified the noun "purposiveness" by the adjective "apparent" because we cannot state on scientific grounds that the body contains certain components "for a reason or a purpose." This is a metaphysical idea beyond the scope of scientific inquiry. Nevertheless, modern biology considers the *apparently* purposive, goal-directed character of body functions an emergent property of material systems that are fully accessible to scientific analysis.

It is interesting to realize that the mechanisms of self-preservation are beyond our control. Hunger makes us seek food, fear makes us seek refuge, and cold makes us seek warmth. We are automatons kept alive by systems that allow us to exist in a wide range of environmental conditions. In the phrases quoted on p. 189, George Gaylord Simpson reminds us that the human body arose in organic evolution. What worked for survival survived; what worked against it perished. The emergence of an instinct for self-preservation and of the physiological devices for implementing it had obvious survival value. Therefore, despite the astonishing elegance with which body systems operate, we may regard their apparent purposiveness as a natural consequence of the body's hierarchical organization.

CONCEPT OF HOMEOSTASIS

Homeostasis is the word used by physiologists for the tendency of an organism to maintain a constant internal environment. We encountered the mechanisms by which the cell controls its internal environment in Chapter 4. We are here concerned with the control mechanisms of the body at large. Indeed, much of the subject matter of physiology deals with the manner in which each tissue, organ, and system promotes homeostasis.

If we consider that the body has two broad major functions, self-reproduction and self-regulation, we must conclude that processes such as respiration, digestion, metabolism, and excretion are but specifications of the latter. Through self-regulation, the self-perpetuation of structural integrity and functional stability—or, in a word, homeostasis—the body maintains itself, at least for a time, in the face of a universe of destructive forces.

Cold is warded off with heat generated in metabolism; depletion of available nutrients is remedied by the intake of food; physical injuries are repaired by complex means; foreign matter is walled off; bacteria are ingested by white blood cells or inactivated by antibodies; poisons are chemically detoxified; and external danger is avoided by physical movement, voluntary as in flight or involuntary as in the reflex action of rapid recoil. All factors, internal and external, tending to disturb the body's inner balance, however subtly, are resisted by systems that are triggered by the insult. Thus it is a physiological axiom that every manifestation of life is not an action but a reaction restoring the status quo.

The second law of thermodynamics* decrees the need for these defenses, which Walter Bradford Cannon was the first to call the mechanisms of homeostasis. They are the means by which stimulus begets response to restore equilibrium. We have discussed the receptor-conductor-effector sequence in individual cells. Now we must enlarge it to embrace the whole organism. Cannon recognized that the homeostatic state, whose existence we can observe, is itself prima-facie evidence that certain agencies are acting to maintain it, whether we can observe them or not. He pointed out that these agencies are largely automatic. To him they exemplified "the wisdom of the body."

Every function of the body fits into the framework of self-regulation. It follows then that a failure in self-regulation ultimately leads to death. The concept of homeostasis is truly one of the great central ideas of biology, for it interconnects all the aspects of biological organization. It is homeostasis that gives to the body functions the appearance of purposiveness.

CHARACTERISTICS OF CONTROL SYSTEMS

Control processes have been usefully considered in theoretical terms. What interests us in this body of ideas is the universal pattern of effect acting back upon cause to inform it of the consequence of its previous action and thereby to determine its future behavior. We have already noted how feedback mechanisms control the activity levels and rates of synthesis of en-

---

* The second law of thermodynamics holds that entropy (or disorder) in a closed system increases with time unless energy is put into the system (see p. 118).

zymes (see Chapter 4). A more familiar example is the ordinary room thermostat, which turns the heat off when the air temperature exceeds the thermostat setting. Another example is the steersman of a boat, who swings the rudder to windward when he sees his vessel moving too far to leeward. The function of the steersman (or the thermostat) is to hold the course (or the temperature) by directing the rudder (or the heater) so as to offset any deviation from that course (or that temperature).

This general process is called *negative feedback* because the response to the initiating stimulus is negative.* A rising end-product concentration *decreases* enzyme activity; a rising temperature *decreases* heat production. In all cases a *reflex system* is essential. Such a system contains three elements: a receptor, which receives the stimulus; a conductor; and an effector, which responds to the stimulus.

Frequently negative feedback systems demonstrate *oscillation*. This occurs when there is a delay in the transmission of signals from the receptor to the effector. The effector acts longer than it should, and the system "overshoots." This accounts for the zigzag course of the boat. Oscillation characterizes many normal physiological mechanisms, among them breathing, the heartbeat, and menstruation.

SYSTEMS OF THE BODY

Homeostasis depends on the feedback operations of the organs and systems of the body. An *organ* is an association of tissues especially adapted for the performance of a particular function. A *system* is an arrangement of organs, tissues, or mechanisms closely allied for the performance of a single function or of related functions. For example, the urinary system consists of the following organs: (1) kidneys, which pro-

duce urine by clearing the blood of waste materials; (2) ureters, which convey urine from the kidneys to the bladder; (3) the bladder, a urinary reservoir; and (4) the urethra, a tube through which urine is voided from the bladder. The labor-sharing character and the interdependence of these organs are evident.

An index of the body systems, their components, and their homeostatic functions appears in Table 6.1. In the following chapters we should bear in mind that each body system, in its way, acts to maintain a steady state. The common denominator in all is the phenomenon of control.

Gross Body Structure

What has been said of the reciprocal relationship between structure and function in the cell (see Chapter 3) is no less true in the whole body. Therefore, we need to consider the anatomy of the human body if we are to deal with its functions.

The chapters that follow contain summaries of the anatomy relevant to the topics under discussion.† In order to place the summaries in context, we shall here examine some generalizations about body structure. These should facilitate our understanding of the locations of various structures in the overall plan and acquaint us with the terminology of anatomical description.

STUDY OF ANATOMY

It is sometimes suggested that human anatomy includes only classic *gross* anatomy, whose study over the centuries has consisted mainly in the dissection of human cadavers. Gross anatomy does remain the foundation of all anatomical study. In the dissecting room a student usually investigates one body region at a time. For example, to study the neck, first he cuts the cadaver's skin; then he exposes the trachea and observes the appearance of the larynx, its surrounding structures, and the attached blood

---

* *Positive feedback* can occur but is undesirable. When an "enough" signal causes an effector to increase its output, instability is inevitable. Such a system is known as a *vicious cycle*. For example, the heart normally responds to blood loss by pumping more rapidly than usual. If blood loss is so severe that the heart is weakened as a pump, its output begins to decrease. The decrease in output weakens the heart further and further decreases its pumping efficiency. When the stimulus of decrease in output causes the output to decrease further, positive feedback exists. Unless the cycle is reversed, the result will be death.

---

† Standard reference works should be consulted for the details of body structure.

TABLE 6.1 BODY SYSTEMS

System	Components	General function	Chapter
Body water and salt	Intra- and extravascular fluids and dissolved salts	Solvent for body constituents and maintenance of proper osmotic pressure, pH, body temperature, etc.	7
Blood	Blood, bone marrow, spleen, lymphatic vessels, and reticuloendothelial cells	Transport of nutrients and wastes to and from cells and regulation and defense	8
Circulatory	Heart, arteries, veins, and capillaries	Delivery of blood to all regions of body	9
Urinary	Kidneys, ureters, bladder, and urethra	Elimination of soluble wastes from blood and maintenance of proper concentrations of essential solutes	
Respiratory	Nose, pharynx, larynx, trachea, bronchi, lungs, and diaphragm	Oxygenation of blood and elimination of carbon dioxide	11
Nervous	Brain, spinal cord, ganglia, nerve fibers, nerve endings, eyes, ears, and olfactory, taste, and tactile receptors	Regulation of other systems, movement, emotion, thought, and sensation	12
Digestive	Mouth, salivary glands, teeth, esophagus, stomach, intestines, pancreas, and liver	Digestion and absorption of food, storage of food reserves, and elimination of wastes	13
Metabolic	Intracellular enzymes of all tissues, liver, and muscles	Extraction of chemical energy from nutrient materials and biosynthesis of body materials	4, 14
Endocrine	Pituitary, thyroid, parathyroids, thymus, adrenals, testes, ovaries, and other ductless glands	Control of metabolism	15
Reproductive	Testes, ovaries, genital organs, and associated glands	Reproduction	16
Musculoskeletal	Voluntary muscles, bones, and ligaments	Movement and support	17
Integumentary	Skin, hair, and nails	Protection of body parts from desiccation, light, heat, cold, and infection	18
Immune	Mechanisms synthesizing antibodies	Defense against foreign substances	19

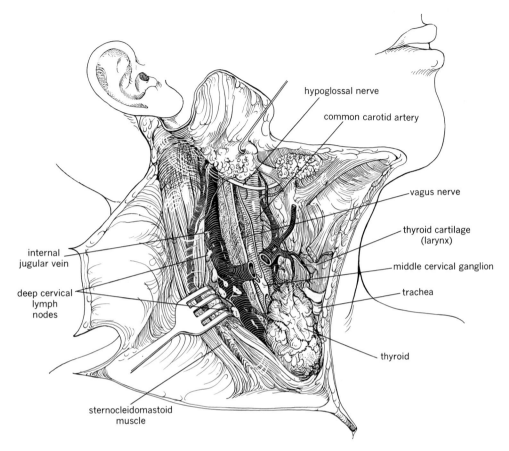

6.1 Dissection of the neck, showing relations of major muscles, nerves, blood vessels, lymphatic vessels, and glands.

The image includes the following labels: hypoglossal nerve, common carotid artery, vagus nerve, thyroid cartilage (larynx), middle cervical ganglion, trachea, thyroid, internal jugular vein, deep cervical lymph nodes, sternocleidomastoid muscle.

vessels and nerves; and finally he uncovers the thyroid and observes its appearance and its relations with surrounding structures. With care and diligence he obtains a preparation resembling the one in Fig. 6.1, where all structures are identified and the significant relationships are apparent. This diagram exemplifies both the style and the goal of classic gross anatomical dissection and offers evidence of the body's structural complexity.

Human anatomy, however, may have other forms and other goals. *Microscopic* anatomy, as we have noted, includes cytology and histology. *Developmental* anatomy (embryology) involves the study of carefully dissected or sliced embryos such as that shown in Fig. 6.2. *Pathological* anatomy involves the gross and microscopic analysis of specimens from a diseased body. The surgical removal of a small tissue sample from a living body for the express purpose of anatomical examination is called a *biopsy;* the systematic examination of a dead body is called an *autopsy.*

Still another approach to human anatomy is *topographical* anatomy, which involves the study of spatial relations among the body structures. These are of obvious importance to the surgeon, to whom body sections like those in Fig. 6.3 are indispensable. Much topographical anatomy can also be learned from x-rays or roentgenograms of a living body.

Last there is what might be termed *systematic* anatomy, in which the structural features of a single body system are considered, and more emphasis is placed upon the structural relations within the system than upon those between the

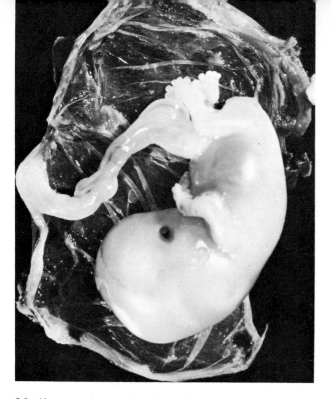

6.2 Human embryo at 8 weeks (×3).

system and its immediate surroundings. Anatomists interested particularly in system structure have developed many ingenious techniques. In one of the most impressive, they inject colored plastics into blood vessels, remove surrounding tissue by digestion, and embed the blood vessel cast in a permanent clear mounting. The preparation of kidney blood vessels shown in Fig. 6.4 surely gives us a keener appreciation of their structural complexity than could be obtained from gross dissection or microscopic analysis. To the systematic anatomist body structure is a composite of the structures of the circulatory system, the musculoskeletal system, and all the other systems. Therefore, since our anatomical discussions are meant to illuminate our understanding of physiology, we shall approach anatomy as systematists.

Terms used in anatomical descriptions. We should pay particular attention to anatomical terminology. In using it, one should always imagine that the body is in an upright position,

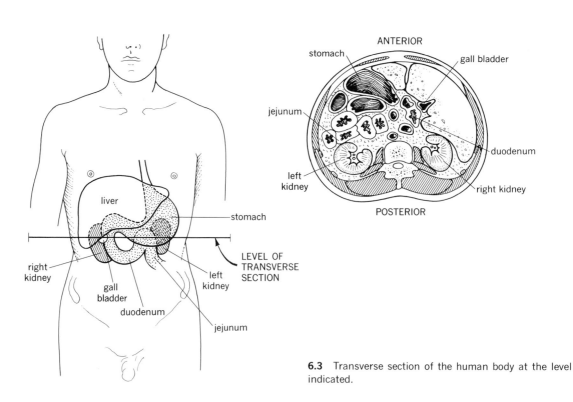

6.3 Transverse section of the human body at the level indicated.

6.4 Vascular system of the kidney (sheep), prepared by the injection technique. Arteries are red, veins blue, and urinary passages yellow (dark, medium, and light gray, respectively, here). (From Ward's Natural Science, Rochester, N.Y.)

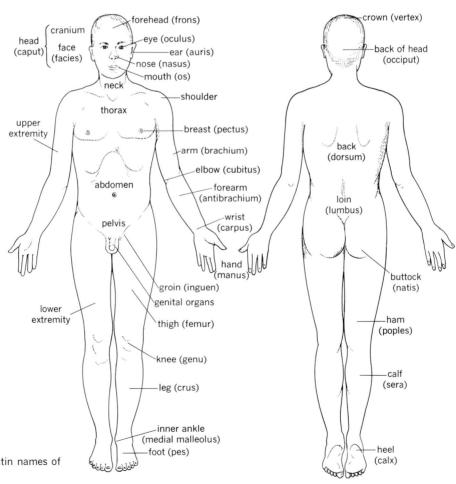

head (caput) { cranium / face (facies) }

forehead (frons)
eye (oculus)
ear (auris)
nose (nasus)
mouth (os)
neck
shoulder
thorax
breast (pectus)
upper extremity
arm (brachium)
elbow (cubitus)
forearm (antibrachium)
abdomen
wrist (carpus)
pelvis
hand (manus)
groin (inguen)
genital organs
thigh (femur)
lower extremity
knee (genu)
leg (crus)
inner ankle (medial malleolus)
foot (pes)

crown (vertex)
back of head (occiput)
back (dorsum)
loin (lumbus)
buttock (natis)
ham (poples)
calf (sera)
heel (calx)

6.5 English and Latin names of the body regions.

arms at the sides, palms of hands forward (as in Fig. 6.5).

The three fundamental planes of the body are the *sagittal, transverse,* and *frontal* planes. The vertical plane through the long axis dividing the body into right and left halves is the *median* or *midsagittal* plane; any plane parallel to this is a sagittal plane. Any vertical plane at right angles to any sagittal plane, and dividing the body into front and rear portions, is a frontal plane. Any horizontal plane at right angles to the sagittal and frontal planes is a transverse plane.

The front of the body is designated as *ventral* or *anterior;* the rear as *dorsal* or *posterior;* the upper part as *cranial* or *superior;* and the lower part as *caudal* or *inferior.* These terms are independent of body posture and therefore apply equally well to vertebrates with horizontal body axes.

Medial means nearer the median or midsagittal plane, and *lateral* farther from that plane. These terms should be carefully distinguished from *internal* (inner) and *external* (outer). Internal as now used means deeper, or nearer the central axis of the body or part, while *external* means more superficial. *Proximal* refers to a position nearer the trunk, while *distal* refers to a more peripheral position.

Functional terms are commonly employed in descriptive anatomy. Thus *pronation* is the act of rotating the forearm so that the palm of the hand points backward, and *supination* is the act resulting in the reverse position of the hand. *Adduct* means to draw toward the axis of the body, and *abduct* means to carry away from the axis. The act of extending a limb is known as *extension,* whereas the approximation of related parts, in bending a joint, is known as *flexion.* *Efferent* means conducting outward or centrifugally (thus a motor nerve is an efferent nerve), and *afferent* means opposed to efferent, or conducting inward or toward the center (thus a sensory nerve is an afferent nerve).

BODY PLAN

The primary divisions of the body are the *head, neck, trunk,* and *extremities.* The head includes the cranium and face. The neck connects the head and trunk. The trunk includes the thorax, abdomen, and pelvis. The upper ex-

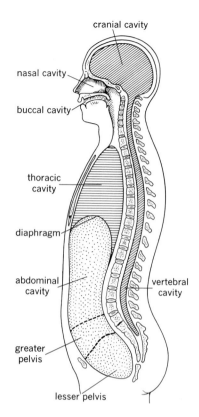

6.6 Midsagittal section of the body, showing various body cavities.

tremity includes the arm, forearm, and hand; and the lower extremity includes the thigh, leg, and foot. As shown in Fig. 6.5, each of these parts has further subdivisions. (The figure also gives the corresponding Latin terms for the various anatomical structures. These help us to remember the meanings of anatomical terms.)

A characteristic of all vertebrates is tubular structure. A tube called the *body wall* encloses a tube called the *viscera,* the cavity between the two tubes being the *coelom,* or *body cavity.* The body wall is composed of skin, connective tissue, muscles, a so-called *serous membrane,* and bone. In mammals the coelom becomes subdivided during embryonic life by a dome-shaped, musculomembranous partition, the *diaphragm,* into the *thoracic* and *abdominal cavities* (Fig. 6.6). The *pericardial cavity* also develops embryologically from the coelom.

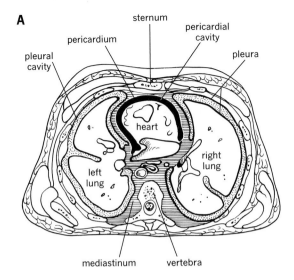

A

pleural cavity · pericardium · sternum · pericardial cavity · pleura · heart · right lung · left lung · mediastinum · vertebra

which divides it into right and left *pleural cavities,* each containing a lung. The other thoracic organs lie in the *mediastinum* between the pleural cavities (Fig. 6.7A).

The abdominal cavity contains the stomach, liver, gall bladder, pancreas, spleen, kidneys, and small and large intestines. It is lined with the *peritoneum* enclosing the *peritoneal cavity.* The kidneys are described as retroperitoneal, since they are behind the peritoneum (Fig. 6.7B). The pelvic cavity is that portion of the abdominal cavity below an imaginary line drawn across the prominent crests of the hip bones. It is more completely bounded by bone walls than the rest of the abdominal cavity. It is

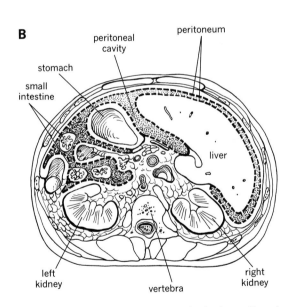

B

peritoneal cavity · peritoneum · stomach · small intestine · liver · left kidney · vertebra · right kidney

6.7 Transverse sections of the major body cavities: A, thoracic; B, abdominal.

The thoracic cavity contains the trachea, bronchi, lungs, esophagus, nerves, heart, and great blood and lymph vessels connected with the heart. It also contains lymph nodes and the thymus. The thoracic cavity is lined with *pleura,*

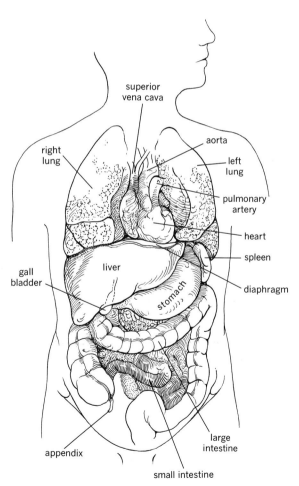

superior vena cava · right lung · aorta · left lung · pulmonary artery · heart · spleen · diaphragm · gall bladder · liver · stomach · large intestine · small intestine · appendix

6.8 Viscera of the thoracic and abdominal cavities.

divided by a narrow bony ring, the *pelvic inlet,* into the *greater,* or *false,* pelvis above and the *lesser,* or *true,* pelvis below. The greater, or false, pelvis is the lower part of the peritoneal cavity and contains parts of the organs of the abdominal cavity. The lesser, or true, pelvis contains the bladder, rectum, and reproductive organs. A frontal view of the viscera of the thorax and abdomen is seen in Fig. 6.8.

Within the skull is the *cranial cavity,* which constitutes the brain case, and a number of other special cavities: the *orbital cavities,* containing the eyes, optic nerves, ocular muscles, and lacrimal organs; the *nasal cavity;* and the *buccal cavity,* or mouth.

REFERENCES AND SUGGESTIONS FOR FURTHER READINGS

Anson, B. J., ed., *Morris' Human Anatomy,* 12th ed., McGraw-Hill, New York, 1966.

Arey, L. B., *Developmental Anatomy. A Textbook and Laboratory Manual of Embryology,* 7th ed., Saunders, Philadelphia, 1965.

Beck, W. S., *Modern Science and the Nature of Life,* Harcourt Brace Jovanovich, New York, 1957.

Cannon, W. B., *The Wisdom of the Body,* 2nd ed., Norton, New York, 1939.

Goss, C. M., ed., *Gray's Anatomy of the Human Body,* 27th ed., Lea & Febiger, Philadelphia, 1959.

Grant, J. C. B., *A Method of Anatomy,* 7th ed., Williams & Wilkins, Baltimore, 1965.

———, *Atlas of Anatomy,* 5th ed., Williams & Wilkins, Baltimore, 1962.

Patten, B. M., *Human Embryology,* 2nd ed., McGraw-Hill, New York, 1953.

Romer, A. S., *The Vertebrate Body,* 3rd ed., Saunders, Philadelphia, 1962.

Rugh, R., *Vertebrate Embryology: The Dynamics of Development,* Harcourt Brace Jovanovich, New York, 1964.

Simpson, G. G., *The Meaning of Evolution,* Yale Univ. Press, New Haven, Conn., 1949.

Simpson, G. G., and W. S. Beck, *Life: An Introduction to Biology,* 2nd ed., Harcourt Brace Jovanovich, New York, 1965.

Smith, H. W., *From Fish to Philosopher,* Little, Brown, Boston, 1953.

Wiener, N., *Cybernetics,* Wiley, New York, 1948.

Woodburne, R. T., *Essentials of Human Anatomy,* 3rd ed., Oxford Univ. Press, London, 1965.

The living organism does not really exist in the *milieu extérieur*—the atmosphere if it breathes, salt or fresh water if that is its element—but in the liquid *milieu intérieur* formed by the circulating organic liquid which surrounds and bathes all the tissue elements; this is the lymph or plasma which, in the higher animals, is diffused throughout the tissue and forms the ensemble of the intercellular liquids and is the basis of all local nutrition and the common factor of all elementary exchanges.

> Claude Bernard, in *Leçons sur les phénomènes de la vie communs aux animaux et aux végétaux*, 1878–79

7 Body Water and Salt

Water and Its Biological Significance

Water is one of the most important chemical substances. It is a major constituent of our environment and accounts for 60 to 90% of the bulk of all living organisms. Although much of the earth's water is in the liquid state, large portions of it are in the vapor and solid states. All three forms are biologically significant.

Despite the wide distribution of water, important aspects of its basic nature remain unknown. It is known, however, that water is a strikingly unusual substance with many distinctive properties.

UNIQUE PHYSICAL PROPERTIES OF WATER

The ancients believed that water is an element. In the late eighteenth century Henry Cavendish showed that water is formed when hydrogen is burned in air, and a short time later Lavoisier recognized that water is a compound of hydrogen and oxygen. It was then established that its formula is H_2O.*

We saw in Chapter 2 that water molecules can dissociate into hydrogen and hydroxide ions. But, in a vessel containing pure water, only 1×10^{-7} moles per liter do so. Thus the pH of pure water is 7. Since a mole of any substance contains 0.6024×10^{24} molecules (Avogadro's number) and since 1 liter of water contains 52.72 moles, 1 liter of water contains $31.758 \times$

---

* This is easily verified by weighing of the amounts of hydrogen and oxygen liberated from water by electrolysis.

10^{24} molecules. Of these, only 0.6024×10^{17}, or 1 in 527,200,000, undergo dissociation.

To the chemist, water presents many apparent anomalies of behavior. Lawrence Henderson in his classic monograph *The Fitness of the Environment* cogently pointed out that the unusual properties of water gave it a key role in guiding the course of evolution. Indeed, he concluded, life, as it developed on earth, would not have been possible were it not for these features.

Our understanding of the relationship between the structure and function of the water molecule dates back to a 1933 article by J. D. Bernal and R. H. Fowler, "A Theory of Water and Ionic Solutions, with Particular Reference to Hydrogen and Hydrogen Ions." This led to the realization that pure water is more than a collection of inert molecules having no interrelationships and to explanations of some of its properties.

Expansion on freezing. Most substances decrease in volume and thus increase in density as their temperature is lowered. Water is unusual in that at a certain temperature (4°C) its density is higher than at both lower *and* higher temperatures; both cooling and warming increase the volume and lower the density. This phenomenon is due to the ease with which individual water molecules join together through hydrogen bonding (see Chapter 2).

A water molecule has a strong tendency to form hydrogen bonds because it has two attached hydrogen atoms and two unshared electron pairs. Hence it can form four hydrogen bonds. It is known from x-ray diffraction studies that an ice crystal has a tetrahedral structure in which each oxygen atom is surrounded by four other oxygen atoms, one at each corner of the tetrahedron. This coincides with Pauling's picture of a water molecule with its four hydrogen bonds extending in four directions. The structure of an ice crystal, then, with each molecule having only four immediate neighbors, is usually open. Ice therefore has an abnormally low density. As ice melts, the tetrahedral arrangement is partially destroyed, and the molecules move closer together. Thus water has a greater density than ice.

Because water reaches a maximum density at 4°C instead of at the freezing point (0°C), water just above the freezing point is heavier than water at the freezing point and sinks. Since the warmer water moves toward the bottom, freezing begins at the surface, and the bottom is last to freeze. Organisms living at the bottoms of fresh-water lakes are thereby protected from freezing.*

---

* Organisms in the oceans have similar protection, though by a somewhat different mechanism. As ice freezes out of salt water to float on the surface, the warmer water under the ice tends to become more concentrated and dense and hence to sink. As in fresh water, therefore, the lower depths of the seas tend to escape freezing.

High surface tension. The tendency of a liquid surface to contract as much as possible causes it to resemble a stretched membrane. This property is called *surface tension*. Except for certain molten metals, water has the highest surface tension of any known liquid. One consequence is the unusually high level to which water can rise in a narrow capillary tube.

Some substances when dissolved in water decrease its surface tension by collecting at the interfaces between water and the solid and gaseous phases. This effect is significant in the functioning of many membranes (see Fig. 2.7).

High dielectric constant. The special property of water that perhaps more than any other accounts for its biological importance is its abnormally high *dielectric constant*. This is responsible for the striking ability of water to dissolve various substances. It is due to water's power to form hydrogen bonds.

We can explain the dielectric constant by picturing two metal plates connected to the positive and negative poles of a battery (as in Fig. 7.1). If we placed between them a bar that pivots on its center and bears a positive charge on one end and a negative charge on the other end, we would observe that the positive plate attracts the negative end of the bar and the negative plate attracts the positive end. One result would be an increase in what is termed the electrical *capacity* of the plates. Such a system is called a *condenser*.

Many molecules are like the pivoting bar. They have positive and negative charges at opposite ends so that they line up appropriately when placed between oppositely charged plates. Again the result is an increase in the electrical capacity of the plates. The ratio of the

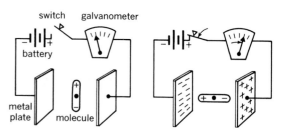

7.1 Action of an electric field on a polar molecule. When the switch is open, the metal plates are uncharged. With the switch closed, one plate develops a positive charge, and the other a negative charge, depending on the pole of the battery to which it is connected. The polar molecule then pivots so that its positively charged end faces the negative plate and its negatively charged end faces the positive plate.

plates' electrical capacity with molecules between them to their electrical capacity with a vacuum between them is the dielectric constant.

The greater the distance between the positive and negative ends of a molecule (i.e., the longer the bar), the greater the plates' electrical capacity, and the greater the dielectric constant. Since the formation of hydrogen bonds between groups increases molecular length and since water is a prolific hydrogen bond former, water has an exceedingly high dielectric constant. At room temperature its value is 80. This means that two oppositely charged groups (analogous to the electrically charged plates) attract each other in water with a force $\frac{1}{80}$ as great as that they exert in a vacuum. Thus, if a crystal of sodium chloride is placed in water, its Na^+ and Cl^- ions are far more likely to leave it than they are if it is placed in a vacuum because the ionic bonds or electrostatic forces tending to draw the ions back to the crystal are only $\frac{1}{80}$ as powerful in water as in a vacuum. The result is the gradual passage of all the Na^+ and Cl^- ions into the surrounding water. In other words, the water *dissolves* the salt. This example shows us how a uniquely high dielectric constant makes water a uniquely successful solvent.

SPECIAL THERMAL PROPERTIES OF WATER

Regulation of external and internal temperatures. Water has many properties that have made it an excellent medium for the origin, development, and survival of life. Some of these have conferred upon water a major role in keeping the earth's surface at a relatively constant temperature. Since organisms can survive only within a restricted temperature range, this is a matter of utmost importance.

The energy metabolism of a living organism, as we have seen, invariably yields heat and so raises the temperature. An average man weighing 70 k (or 154 lb) might produce in an average day 3000 Cal. In a closed system containing 70 k of water, this amount of heat would raise the water temperature more than 40°C. But a much greater temperature rise would occur with most other liquids.

Heat capacity. The *heat capacity* of a substance is the amount of heat required to raise the temperature of a unit quantity of the substance —1 mole or 1 g—1°C without changing its phase (i.e., without converting it from solid to liquid or liquid to gas). The heat capacity of liquid water is extremely high.* This means that, compared with most other substances, more calories are needed to increase the temperature of a given amount of water by a given number of degrees and that a smaller temperature rise is produced by a given amount of heat. Thus water acts as a temperature buffer, tending to maintain its temperature, despite shifting environmental temperature, more successfully than most other substances.

The huge quantity of water on the surface of the earth—about 1.25×10^{24} g, or enough to form a layer 1.56 miles deep if it were evenly distributed—tends also to prevent sudden sharp variations in atmospheric temperature, as, for example, between night and day. The heat capacity of the air is considerably lower than that of water. Therefore, the air is susceptible to much more abrupt temperature changes than are the oceans. Because the water vapor in the air is in equilibrium with the water in the oceans, however, the latter serves as a stabilizer of air temperature. This explains why the climates of regions next to large bodies of water are by far the most equable.

Since water is the principal component of living organisms, it tends to prevent too sudden temperature shifts within them as well. The body temperatures of fishes, amphibia, and reptiles change in accordance with the environmental temperature. These animals are referred to as *poikilotherms,* or cold-blooded animals. Birds and mammals, conversely, are referred to as *homotherms,* or warm-blooded animals. Their body temperatures are, within broad limits, independent of the outside temperature.† We

---

* Only one other liquid—liquid ammonia—surpasses water in this respect. If the heat capacity of water is taken as 1.0, that of liquid ammonia is 1.23. Other values are as follows: chloroform, 0.24; alcohol, 0.6; glass, 0.2; solid iron, 0.10.

† Incidentally, evolutionists have wondered whether the common warm-bloodedness of birds and mammals bespeaks common ancestry. Physiological and paleontological evidence, however, is overwhelmingly against this supposition, and it is now accepted that warm-bloodedness developed independently in two widely separated evolutionary lines. Warm-blooded animals, of course, cannot rely exclusively upon water for body temperature maintenance. Complex regulatory mechanisms are also necessary and will be discussed in Chapters 12 and 14.

earlier encountered some of the reasons why temperature control is physiologically significant (see Chapter 2). We shall merely note here that both poikilotherms and homotherms would exist with much more difficulty were it not for the high heat capacity of water. This may be considered the physical basis for all physiological temperature-regulating mechanisms.

Heat of vaporization. Vaporization and condensation are of particular importance in the maintenance of steady temperatures in systems that are not closed. In order to change 1 mole of liquid water to water vapor, it is necessary to invest approximately 9700 cal from the immediate environment. This is called the *heat of vaporization* of water. The conversion of the same quantity of water from vapor to liquid requires the transfer of an equal amount of heat from the water to the environment. The high heat of vaporization of water is due to the strong intermolecular forces resulting from hydrogen bonding.

Water's high heat of vaporization contributes to its role in the regulation of atmospheric temperature. Air above a body of water, for example, takes up more and more water vapor as the temperature rises. Since water enters the air by evaporation, and since its evaporation utilizes large amounts of heat because of its high heat of vaporization, water tends to prevent rapid rises in air temperature at times of sudden heat input (such as sunrise).

Vaporization from the skin and lungs is the primary regulatory mechanism in body temperature control, serving adequately to expel the heat produced by metabolic activity. Its effectiveness is obviously dependent on the magnitude of the heat of vaporization of water.

Heat of fusion. The *heat of fusion* is the amount of heat necessary to convert a given quantity of a substance from solid form to liquid form at the melting point (or from liquid to solid at the freezing point). To change 1 mole of liquid water to ice at the freezing point requires the withdrawal of 1435 cal from the liquid water. This is an exceptionally high value compared to the heats of fusion of other substances. It is explained by the tendency of water to form hydrogen bonds. If water temperature is to drop below the freezing point, then, a considerable amount of heat must be surrendered to the immediate environment. This keeps the air temperature above a large body of water from dropping below freezing as winter comes on. Correspondingly, the heat absorbed by melting ice slows the rate of temperature rise in the spring.

Water's high heat of fusion is biologically significant mainly because of its modulating influences on the environmental temperature. It also tends to protect a living organism against the adverse effects of freezing, though this is a rare phenomenon.

Physiology of the Body Fluids

BODY WATER

We may suppose that the organisms which arose in the waters of the primitive planet depended in some way upon the solvent power of water—perhaps materials dissolved in water constituted their only nutrients. Although living forms emerged at last from the sea onto dry land, they had first to develop means for maintaining a suitable watery internal environment. Thus the living cell continued to depend upon a fluid milieu.

The human body remains dependent upon internal water, and, as we shall presently see, a sizeable portion of the body's physiological machinery is devoted solely to the critical task of maintaining body water within precise limits.

Amount. The amount of water in the human body varies with size and age but in all individuals constitutes a high fraction of total body weight. Since the amount of body fat varies widely among individuals, the impression may be gained that the proportion of total body water to total body weight also varies widely. Actually, however, fat has a relatively low water content. If the amount of body water is related to the weight of fat-free tissue (the so-called *lean body mass*), it is always remarkably close to 70%. In an average adult body water is approximately

60% of body weight; in an early embryo approximately 97%; and in a newborn baby approximately 80%. The water content of the body generally decreases with age until adulthood, when it levels off for a time and then decreases at a slower rate than before.*

Distribution. Water permeates all of the tissues and organs of the body, but it can be shown that it divides itself into several natural "compartments." These have been traditionally shown as in Fig. 7.2A. There is the *intracellular* water, within the body's cells, and the *extracellular* water, outside the cells. The fluid fraction of blood is part of the latter. The extracellular water constitutes about 20% of the body weight, and a fourth of it is within the blood vessels. Thus it consists of *extravascular* and *intravascular* portions. The intracellular water makes up about 40% of the body weight.

Fig. 7.2B is an up-to-date version of Fig. 7.2A, incorporating recent discoveries and suggesting some of the serious difficulties encountered in the experimental measurement of the water compartments. These difficulties all relate to the fact that body water recognizes no anatomical boundaries. Rather it diffuses continuously throughout the body. The water of blood, for example, passes rapidly through the walls of the capillaries—a huge surface when considered in the aggregate. If we could label all of the water molecules in the blood, we would find only about half of them still there a minute later. Yet at any one moment about 8% of the total body water is present in the blood plasma. A third of the total body water lies in the interstitial spaces of tissue; and two-thirds lies within the cells.

We now know of a fraction of body water that in some ways behaves as though it were intracellular but in fact is extracellular. This is the so-called *transcellular* water, found in the cavities of hollow organs such as the intestine. Its functions include the lubrication of joints and other surfaces where organs rub and slide past each other. It acts protectively by equalizing mechanical pressures, as, for example, upon a developing embryo, or by maintaining normal pressure levels in hollow structures like the eye. *Ciscellular* water is that portion of the extracellular water which is unambiguously extracellular.

Measurement. The methods of measurement of body water are all variations of what is called the *dilution principle.* In each, the body is pictured as a container. If a known amount of a freely soluble and easily identifiable substance is added to an unknown volume of water and the substance is then distributed uniformly in the water, the total water volume is readily determined by measuring the concentration of the added substance in a small sample of the solution. Thus, if we added 1 ml of water containing 1000 μg of dye to the water in a vessel, mixed it well, drew off a sample, and found it to contain 1 μg of dye per milliliter, we would calculate that the vessel contained almost exactly 1000 ml (or 1 liter) of water.

The volume of body water is similarly computed by dividing the final concentration of a test substance into the amount added. To be useful in the measurement of body water, a test substance must be nontoxic, freely soluble in water, readily distinguishable from substances already in the body, and capable of diffusing quickly throughout all of the body water. In addition, it must not be destroyed by the body's metabolism and it should not become more concentrated in some parts of the body than in others. Otherwise, the sample removed for analysis will not be representative. Few substances have been discovered with all these properties. Urea, antipyrine and its derivatives, and water containing an atom of isotopic hydrogen (either deuterium or tritium) in place of one of its two hydrogen atoms have all been employed successfully. Within an hour after injection into the blood, each of these substances has diffused throughout the body water. Presumably it mixes completely with the body water so that its concentration in the blood is a reliable measure of total body water.

Problems arise when attempts are made to measure the water contained in any one of the several water compartments, again by the dilution principle. An ideal substance for measurement of the extracellular water compartment

---

* These figures are not firmly established norms differing little from individual to individual. On the contrary, it is surprising to realize (1) how wide is the range of observed values in a group of ostensibly normal subjects, (2) how few in number are the acceptable accounts of investigations in this area, and (3) how few in number are the subjects studied.

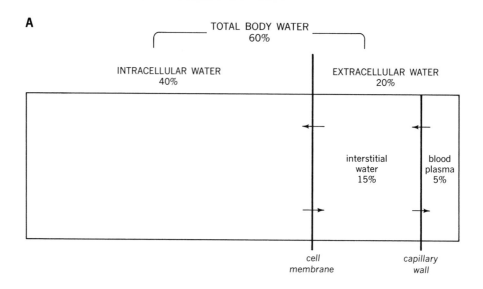

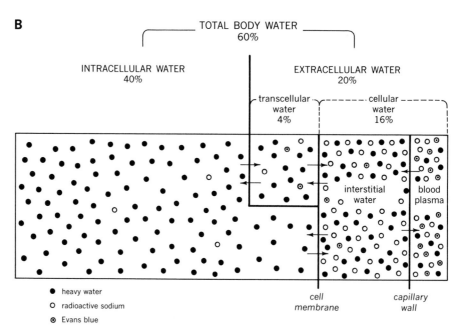

7.2 Distribution of the body water into compartments: A, traditional representation; B, modern representation, showing substances commonly used to study body water compartments.

should diffuse into and out of the circulatory system with rapidity and ease and should work its way into all the interstitial spaces while remaining entirely outside the cells. As yet, no known substance entirely fills these specifications. The substances most closely approaching the ideal are sucrose, thiocyanate, and radioactive chloride, sulfate, and sodium.

Fig. 7.2B illustrates how labeled water is distributed throughout all of the body water; radioactive sodium (^{24}Na) almost but not exclusively throughout the extracellular water; and the dye Evans blue almost but not exclusively throughout the intravascular water. The amount of intravascular water cannot be measured directly by any dilution method. Instead it is estimated from the difference between simultaneously measured total body water and extracellular water.

WATER BALANCE

Definition of balance. Physiological balance is defined as follows:

$$balance = gain - loss$$

By this definition an organism is "in balance" when it maintains a constant concentration of a specific body constituent. If gain exceeds loss, the balance is "positive"; and if loss exceeds gain, the balance is "negative." Momentary gains and losses may temporarily throw an organism out of balance, as immediately following a large drink of water. However, balance is usually viewed in terms of the average situation prevailing during a given period of time (e.g., a day). Consequently, the transient effects of intermittent intakes and outputs are neutralized.

Normally an adult man is in balance with regard to all body constituents. His body composition remains constant, for each day he loses exactly as much as he gains of carbon, nitrogen, oxygen, and so on.* He maintains his water balance each day by acquiring an amount of water just equal to the amount he loses. It should be noted that the percentage of the extracellular water entering and leaving each day is considerably higher in an infant than in an adult. As

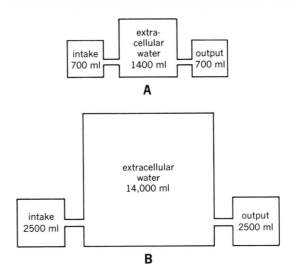

7.3 Average daily water exchange: A, of a normal infant weighing 15 lb (6.8 kg); B, of a normal adult weighing 150 lb (68 kg).

shown in Fig. 7.3, a 15 lb (6.8 kg) infant has an extracellular water volume of about 1400 ml, or 20.6% of his total body weight. He takes in and puts out 700 ml of water each day, or 50% of his extracellular water. An adult weighing 10 times as much has an extracellular water volume of 14,000 ml, but his daily intake and output approximates 2500 ml, or only 18% of his extracellular water. These figures clearly indicate why infants are more vulnerable to dehydration than adults. An infant who through water deprivation fails to obtain his normal water supply or because of fever loses more water than he should may quickly go into dangerous negative water balance.

Sources of body water. The body obtains water in two ways: (1) by drinking and by eating food containing "preformed" water (i.e., the water present in "solid" food); and (2) by forming water in the oxidative reactions of metabolism. As shown in Fig. 7.4, an average man in balance with respect to water who consumes and excretes about 2500 ml of water per day obtains 1200 ml from drink and 1000 ml from food.† The remainder is produced in the proc-

---

* A growing child, continuously adding to his body substance, is in positive balance with regard to most materials.

---

† From this it is evident that water is normally absorbed into the body from the intestine. It can also be introduced artificially (i.e., by subcutaneous or intravenous injection).

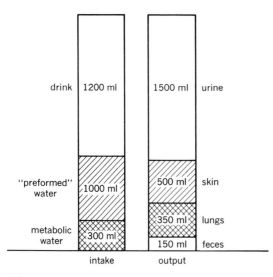

7.4 Sources of body water and routes of water loss.

esses of metabolism. The amount of water from all sources that must be taken in by an ordinary adult varies from 2200 to 2800 ml—approximately 1 ml per calorie of ingested food.

Routes of water loss. Water is eliminated from the body in the urine, feces, and saliva and by evaporation from the skin surface and lungs. According to Fig. 7.4, a man of average size doing light work in a moderate climate excretes each day about 1500 ml in the urine, 150 ml in the feces, 500 ml from the skin, and about 350 ml in expired air. The amount lost via the skin and lungs varies with the surrounding temperature and humidity and with the level of physical exertion. Much of the water lost from the skin diffuses through the skin cells. This should be distinguished from sweat, which is actively secreted by the sweat glands. In moderate or dry atmospheres the water lost by diffusion through the skin evaporates as quickly as it forms. Since it is therefore imperceptible, it is termed the *insensible perspiration*. In humid atmospheres this water evaporates poorly and thus becomes apparent. When heavy work is performed in a hot or humid environment, the amount of sweat secreted may rise to 5 liters. Moreover, during exercise water loss through the lungs increases in proportion to the respiratory rate.

Control of water balance. The water content of the body is remarkably constant. The fact that daily fluctuations in body weight may be no more than 150 g suggests that daily variations in the amount of body water are probably no greater. In percentages this means that body water content fluctuates only 0.036% (i.e., 150/42,000 for a 70 kg adult whose body water constitutes 60% of his total body weight). The variation is often even smaller. Such constancy implies the existence of an exquisitely precise control system.

This, like all other control systems, operates by negative feedback (see Chapter 2). When excessive water loss occurs, as during heavy exercise, several mechanisms act to reduce the deficit. Urine flow decreases (though, as we shall see, it does not cease entirely), and thirst develops so that water intake is increased. Conversely, when there is excessive intake of water, as in experimental forced drinking, the rate of urine formation increases so that elimination is accelerated. In both water excess and water deficiency, the rate at which the water level returns to normal is nearly proportional to the amount of water that the body must lose or gain.

It is quite difficult for a human being voluntarily to effect a large change in his body water content. Water deprivation causes thirst, which becomes progressively more intense and uncomfortable with time.* On the other hand, if he sets out deliberately to drink a massive quantity of water, he soon develops a strong distaste for it. Indeed, forced drinking, the so-called "water cure," was an ancient method of torture. The kidneys partially nullify the water excess by producing a flood of urine; but, if drinking continues, the condition called *water intoxication,* characterized by headache, nausea, irritability, convulsions, coma, and finally death, may ensue.

---

* It should be noted that the sensation of thirst, the conscious desire to drink, is a complex psychic phenomenon that may be associated with intracellular or extracellular dehydration. We shall learn in Chapter 12 that electrical stimulation of a discrete region in the brain causes an experimental animal to begin drinking within a few seconds. The relation of this "drinking center" to the normal thirst mechanism is unknown. Accounts of men adrift in lifeboats or lost in deserts who have had to witness others dying of dehydration indicate that the sensation of thirst abates completely just before death, however. This curious fact has not been satisfactorily explained.

The symptoms of water intoxication result primarily not from the water excess but from a relative salt deficiency, for the concentration of salt in body water diminishes as body water volume increases. Therefore, salt may relieve them just as water may relieve the thirst created by excessive salt intake.

SALTS OF EXTRACELLULAR AND INTRACELLULAR FLUID

Evolutionary significance. Some species of animals require no water other than the moisture in their solid food (preformed water) and that generated by oxidative metabolism (metabolic water). These include various desert reptiles and mammals and certain migratory birds. A spectacular example is the desert-dwelling kangaroo rat, which lives on air-dried seeds. It grows, reproduces, and nurses its young on a diet containing only 5 to 10% free water. Man, however, needs a great deal of water, and it is interesting to consider the reasons why.

A large number of organic and inorganic substances are dissolved in intracellular and extracellular water. Most are *nonelectrolytes* (i.e., they do not form ions). The *electrolytes* consist mainly of inorganic salts that break down into such ions as sodium (Na^+), potassium (K^+), magnesium (Mg^{++}), calcium (Ca^{++}), chloride (Cl^-), bicarbonate (HCO_3^-), and phosphate (HPO_4^{--}). We have seen that water molecules diffuse freely through the membranes of the body, the rate of flow being governed by the laws of osmosis.* The salts dissolved in the body water significantly affect its distribution by lowering the tendency of the water to diffuse out of the body.

According to current theories, the body fluids of the early salt-water organisms contained the same salts as the surrounding sea water. When these organisms attempted to migrate from salt water into fresh water, they presumably were challenged by the tendency of fresh water to pass inward through their external body membranes as a result of osmotic pressure produced by their high internal salt concentrations. We may assume that they did not accomplish the transition to fresh water until they had developed external waterproof armor. The massive influx of water that would have occurred otherwise would have inflated them into waterlogged gelatinous balloons.

It is important to our understanding of the human body that the later major evolutionary developments took place in fresh-water rather than salt-water organisms. As complex physiological devices such as fins, jaws, teeth, and sense organs gradually emerged, it became increasingly necessary, to ensure their smooth functioning, that body fluids have a precisely regulated composition. A controlling mechanism had to evolve that could provide the stability needed for the increasing complexity of function. It was the evolution of the kidney in the early vertebrates that made further progress possible. We shall discuss it in detail in Chapter 10.

Patterns of distribution. The main inorganic cations and anions of the body fluids have been noted. In addition, certain organic acids (e.g., lactate) and proteins are anions at physiological pH levels. It should be remembered that in any electrolytic solution the total number of positive charges equals the total number of negative charges. This is evident in Fig. 7.5, which depicts the cationic and anionic compositions of various fluids. Note that the cation and anion

---

* Although the membranes of the body are permeable to water molecules, they are not permeable to many of the molecules dissolved in water. Hence they are called *semipermeable membranes.* In general, the molecules that do not diffuse through the membranes are the larger ones (e.g., proteins). When the concentration of nondiffusible molecules is greater on one side of a semipermeable membrane than on the other, water passes through the membrane toward the side with the greater concentration, by osmosis (see p. 159). The *osmotic pressure* is the pressure that would have to be applied to prevent this transfer of water. It should be kept in mind that the osmotic pressure of a solution is determined by the *number,* not the kind, of particles dissolved in it. Nondiffusible ions cause osmosis and osmotic pressure in the same manner as nondiffusible molecules.

The numerical unit in which osmotic pressure is expressed is the *osmol.* We speak of osmols or milliosmols per liter. In calculating the osmolarity of a solution, we disregard ionic valence since osmotic pressure is determined only by the number, not the charge, of dissolved particles. Thus a bivalent ion such as Mg^{++} exerts no more osmotic pressure than a univalent ion such as Na^+.

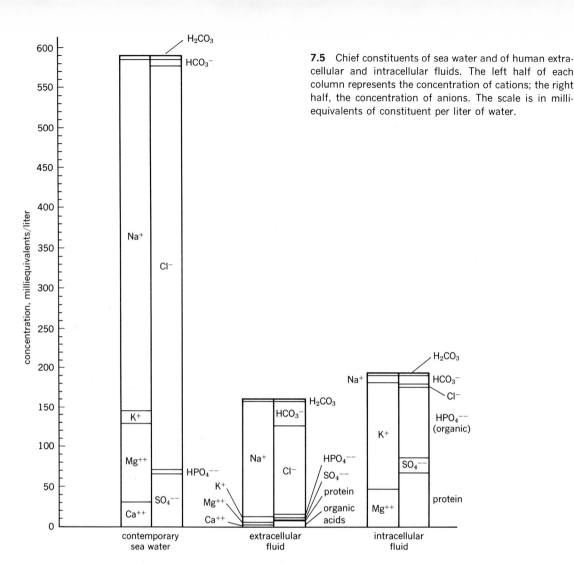

7.5 Chief constituents of sea water and of human extracellular and intracellular fluids. The left half of each column represents the concentration of cations; the right half, the concentration of anions. The scale is in milliequivalents of constituent per liter of water.

columns are the same height. Note also that the substances are represented quantitatively in milliequivalents per liter.* Concentrations could be expressed more conventionally in terms of weight per unit volume, but the use of equiv-

alents emphasizes the ionic charges and simplifies calculations, since electrical neutrality requires that the total number of milliequivalents of cations equals the total number of milliequivalents of anions.

Fig. 7.5 reveals some interesting properties of the body fluids. The most striking is the predominance of sodium in extracellular fluid and the predominance of potassium in intracellular fluid. We saw in Chapter 4 that this difference has not yet been explained. Although there is reason to believe that organisms derived their salt patterns from ancient oceans and brackish waters, what would account for the high-sodium–low-potassium pattern of extracellular

* From the footnote on p. 39, the number of milliequivalents of a substance in solution equals the concentration (in milligrams per liter) times the ionic charge divided by the atomic weight. For example, a solution containing 40 mg of Ca^{++} per liter contains $40 \times 2 \div 40 = 2$ meq per liter. Phosphate presents a special case. In plasma at pH 7.4, 20% of the phosphate is present as monovalent $H_2PO_4^-$ and 80% as divalent HPO_4^{--} (see p. 41). Thus the "virtual valence" is $(0.2 \times 1) + (0.8 \times 2) = 1.8$. The number of milliequivalents of phosphate phosphorus in solution, then, equals the concentration (in milligrams per liter) multiplied by 1.8 divided by 31.

fluid and the reverse pattern of intracellular fluid? Comparative studies show that high intracellular potassium is a characteristic of all cells, both plant and animal. The development of high-potassium protoplasm in cells existing in a high-sodium medium must, therefore, have occurred early in evolution. Wallace Fenn has pointed out an interesting parallel in the relation between earth and sea: "Potassium is of the soil not the sea; it is of the cell not the sap." The meaning of this distinction in the body and the mechanisms of its preservation are among physiology's deepest enigmas.

Water diffuses freely through a cell membrane, but the membrane is selectively permeable to positively charged ions. The mechanisms of selection are extremely efficient, for the disparities between extracellular and intracellular concentrations of sodium and potassium are very great. We shall return to this subject repeatedly in later chapters. We note here only that the cell membrane possesses active transport mechanisms that draw potassium into the cell—from a surrounding fluid low in potassium —and pump sodium out of the cell (see Fig. 4.26); whenever sodium leaks inward, it is actively ejected by a system commonly called the *sodium pump,** which operates with energy derived from cellular metabolism.

Fig. 7.5 also shows that high concentrations of chloride and bicarbonate and low concentrations of phosphate, sulfate, and magnesium are present in extracellular fluid, whereas the reverse is true in intracellular fluid. Presumably active transport systems operate for many of these ions, too. Perhaps the major function of the steep ionic gradients between cell and environment is the maintenance of an electrical gradient across the cell membrane (see Fig. 4.25D). With the death of the cell, the gradients are rapidly dissipated.

Intracellular fluid seems peculiarly suited for the many intracellular enzymatic processes. For example, magnesium is known to be involved in most of the enzymatic reactions of ATP, and potassium may contribute to the stability of organic components and to the regulation of enzyme activity.

As for extracellular fluid, its partial resemblance to sea water is apparent in Fig. 7.5. Both contain the same four cations, with the same dominance of sodium and chloride. Just as the seas of today are far saltier than those of early times, contemporary sea water is saltier than extracellular fluid.

The composition of extracellular fluid is readily determined by analysis of the fluid portion of blood. The absolute composition of the extravascular (interstitial) part of extracellular fluid is less easily determined, since it does not usually accumulate in large enough quantities for analysis by conventional methods. The values given in Fig. 7.5 are averages from the data of many workers; undoubtedly wide variations exist. The ionic pattern of intracellular fluid is also determined with difficulty. Muscle is perhaps the best-studied tissue in this regard, but, as indicated in the figure, the exact levels of sodium and chloride in intracellular fluid are still in question. Nevertheless, the view that sodium and chloride are found mainly in extracellular fluid serves as a useful generalization.

Salt balance. Maintenance of salt balance, like that of every other body constituent, requires that daily intake equal daily loss. It happens, however, that the daily intake of sodium chloride in the average American or European diet far exceeds the body's actual needs. Daily intake varies from 5 to 15 g, whereas only 0.25 to 0.75 g would be adequate. Does this mean that man has an appetite for salt that is analogous to his thirst for water? An appetite for salt is observed in other animals; herbivores, for example, have frequent recourse to salt licks.† Yet, despite the fact that salt-depleted human beings

---

* The sodium pump is not fully understood (see p. 160), and even less is known of the potassium pump, if indeed one exists. Whether the accumulation of potassium by the cell involves active potassium transport (or merely active sodium ejection) is not yet clear. Probably the two processes are coupled.

---

† The diets of herbivores contain approximately 20 times as much potassium as sodium; the sodium chloride obtained from a salt lick reduces this disparity. Carnivores, on the other hand, consume diets containing only five times as much potassium as sodium, and they do not seek out salt licks. The proportion of sodium to potassium in human diets ought to resemble that in carnivore diets, which apparently provide enough sodium to meet physiological needs. Instead human diets often contain huge quantities of sodium, equalling or exceeding their potassium contents.

may crave salt, it appears that under ordinary circumstances man's heavy salt consumption is more a matter of acquired taste than one of instinctual drive. Indeed, in some regions of the world, salt is considered distasteful and is never added to food.

ABNORMALITIES OF WATER AND SALT DISTRIBUTION

Although detailed study of the causes and consequences of physiological abnormalities belongs to the fields of medicine and pathology, much can be learned about normal body function from the investigation of abnormal states. From what we know of the distribution and composition of the body fluids, we might reasonably predict the existence of a number of abnormalities. The major manifestations of disturbances of body water and salt result directly from increases or decreases in the (1) volume, (2) osmotic pressure, (3) composition, and (4) pH of the body fluids. We shall here briefly mention all except those involving pH, which will be discussed in Chapters 10 and 11.

Water excess. It is helpful to follow changes in the volumes and osmotic pressures of the intracellular and extracellular fluid compartments by visualizing a stepwise sequence of events: (1) the normal resting state; (2) the situation immediately following an insult; and (3) the adjustments to the insult.

For example, when a substantial quantity of pure water is added to the extracellular fluid compartment, as by forced drinking or intravenous injection, the immediate result is a dilution of the extracellular fluid, which causes it to become hypotonic (i.e., low in osmotic pressure) with respect to the intracellular fluid. Large amounts of water then enter the cells by osmosis, and a new osmotic equilibrium is reached.

Fig. 7.6A illustrates the useful method of representing such a sequence devised by Darrow and Yannet. The extracellular and intracellular fluid compartments are rectangles plotted left and right, respectively, of a central vertical line. The *height* of a rectangle depicts the osmotic pressure (in milliosmols per liter), the *width* the volume (in liters), and the area the total number of milliosmols in the compartment. In the nor-

mal resting state, the osmolarity is 300 in both compartments, the extracellular fluid volume is 15 liters, and the intracellular fluid volume is 25 liters. Immediately after the addition of, say, 10 liters of water to the extracellular fluid, its volume increases to 25 liters. However, its osmolarity decreases proportionally, since the total number of milliosmols (originally $15 \times 300 = 4500$) is unchanged (now $25 \times 180 = 4500$). Since the osmolarities of extracellular fluid and intracellular fluid must be equal, water moves into the cells, and a new osmotic equilibrium is established. The volume of the extracellular fluid becomes 18.75 liters, that of the intracellular fluid becomes 31.25 liters, and the osmolarities of both become 240. Careful study of the diagram reveals how osmotic equilibration between fluid compartments tempers the effects of fluid abnormalities.

Salt excess and water deficiency. The concentration of salt (particularly sodium) increases in individuals who drink sea water or receive strong salt solutions intravenously and in a number of other situations to be discussed later. Fig. 7.6B shows the events following the intravenous injection of 2 liters of NaCl solution with an osmolarity of 1500. We might expect the immediate effects to be an increase in the volume of the extracellular fluid to 17 liters and an increase in the extracellular osmolarity from 4500 to 7500. However, since such a massive departure from the normal osmolarity of extracellular fluid would be lethal, the body immediately activates mechanisms that combat the rising salt concentration. These consist chiefly of an osmotic shift of water from the intracellular compartment to the extracellular compartment and an increased rate of salt excretion by the kidneys. Until the extra burden of salt has been eliminated in the urine, the water shift preserves osmotic equilibrium.

In certain diseases the urinary excretion of salt is impaired. As we shall learn in Chapters 9 and 10, salt retention results in the retention of water. This is the consequence of osmotic forces that maintain normal osmotic pressures in the body fluid compartments even when their volumes are abnormal. The accumulation of fluid in the extracellular compartment produces the condition called *edema*, a common complication

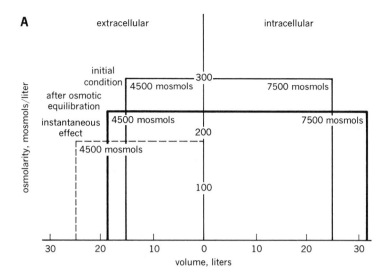

A

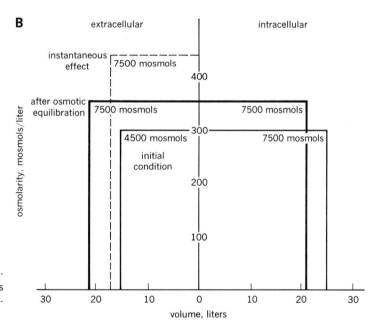

B

7.6 Darrow-Yannet diagrams demonstrating the effects of water excess (A) and salt excess (B) on the distribution of extracellular fluid.

of heart disease, kidney disease, and pregnancy.

Primary water deficits result simply from limited water intake. The extracellular fluid volume is maintained at the expense of the intracellular fluid, so that the blood pressure remains normal; urine volume decreases; and thirst is intense.

Salt deficiency. When there is pronounced water loss, as in excessive sweating, with simultaneous restriction of salt intake, the extracellular fluid volume decreases, and the blood pressure drops. Similar events accompany prolonged vomiting or diarrhea. This combination of salt and water deficiency, like the state of water excess, results in decreased salt concentrations and osmotic pressures in the extracellular fluid (see Fig. 7.6A). In water excess the body gradually restores the normal pattern of fluid distribution by excreting an excessively dilute urine. The compensatory reactions to salt deficiency are more complex because of electrolyte differences between extracellular fluid and intracellular fluid. For example, in a disorder such as diarrhea, in which fluid is lost mainly from the extracellular compartment, the major cation lost is sodium. If the only compensatory reaction were a shift of water from the intracellular compartment to the extracellular compartment, the intracellular potassium concentration would become too high, and the extracellular sodium concentration would decrease further. The principal compensatory reactions, therefore, are the retention of water and sodium by the kidneys* and the leakage of potassium from the intracellular compartment to the extracellular compartment. The latter reaction may be dangerous, for high potassium levels in the extracellular fluid have many adverse effects (e.g., the depression of nerve conduction and muscular contraction). The ideal remedy is the prompt ingestion or injection of sodium chloride and water to replace what has been lost.

A notable feature of medical practice in recent years has been the growing recognition of the importance of body water and salt. Not long ago the rule was that individuals recovering from surgery should receive nothing by mouth, even when fluid had been withheld for a substantial period before operation. This was en-

forced despite the intense suffering caused by thirst. It was eventually realized that significant amounts of fluid (sometimes in excess of 3 liters) were lost during surgery and must be replaced—either by mouth or by intravenous infusion—and today a large variety of sterile salt solutions is kept on hand in hospitals for the maintenance or restoration of salt and water balance in surgical patients and in victims of burns and other injuries or diseases.

REFERENCES AND SUGGESTIONS FOR FURTHER READING

Adolph, E. F., *Physiology of Man in the Desert*, Interscience, New York, 1947.

Bernal, J. D., and R. H. Fowler, "A Theory of Water and Ionic Solutions, with Particular Reference to Hydrogen and Hydrogen Ions," *J. Chem. Phys.*, **1**, 515 (1933).

Black, D. A. K., *The Essentials of Fluid Balance*, 2nd ed., Thomas, Springfield, Ill., 1960.

Darrow, D. C., and H. Yannet, "Changes in Distribution of Body Water Accompanying Increase and Decrease in Extracellular Electrolyte," *J. Clin. Invest.*, **14**, 266 (1935).

Elkinton, J. R., and T. S. Danowski, *The Body Fluids*, Williams & Wilkins, Baltimore, 1955.

Gamble, J. L., *Chemical Anatomy, Physiology and Pathology of Extracellular Fluid: A Lecture Syllabus*, 6th ed., Harvard Univ. Press, Cambridge, Mass., 1954.

Henderson, L., *The Fitness of the Environment*, Macmillan, New York, 1913.

Leaf, A., and L. H. Newburgh, *Significance of the Body Fluids in Clinical Medicine*, 2nd ed., Thomas, Springfield, Ill., 1955.

Robinson, J. R., "Metabolism of Intracellular Water," *Physiol. Rev.*, **40**, 112 (1960).

Steinbach, H. B., "The Prevalence of K," *Perspectives Biol. Med.*, **5**, 338 (1962).

---

* The mechanisms by which the kidneys perform this function involve several body systems, including the endocrine system. We shall consider them in detail later.

To be a hematologist nowadays is to inhabit a whole continent of knowledge and to have perhaps only a distant acquaintance with large areas of it. Some divisions of the subject have already become specialties in their own right, and the study of leukemia or blood coagulation is more than enough to occupy a man for life. Nevertheless, hematology remains an entity in so far as all its practitioners are interested in that quite peculiar fluid which circulates through the blood vessels. Moreover, whereas much medical research is applied research . . . this is not true of hematological research. It may also have an intrinsic value, due to its fundamental biological nature . . . . It is a story which can be read with equal interest by those who are engaged in the treatment of patients and those who enjoy seeing nature so brilliantly put to the question.

L. J. Witts

8 Blood

Introduction

HISTORICAL BACKGROUND

Beginnings of hematology. Since the dawn of history, men have recognized that blood is essential to life. The Biblical phrase "to shed blood" meant "to kill," and, in the Books of Moses, the word "blood" is synonymous with "soul." The earliest scientific studies of blood came in the seventeenth century with the improvement of the microscope, and red blood cells were first described by Jan Swammerdam in 1650 from examinations of frog blood. Human red blood cells were first described by van Leeuwenhoek in 1673.

Much of our early knowledge of blood emerged from observations, scientific and other-wise, of individuals with various blood diseases. For example, the importance of iron in the diet was first suspected when iron was found useful in the treatment of a common anemia, and, as has happened so often in the history of medicine, the treatment was discovered long before the disease was understood. The disease was chlorosis, which Johannes Lange in 1554 termed *morbus virgineus*, the disease of virgins. Chlorosis was widespread until recent years among young unmarried women who by custom led cloistered lives and for some reason ingested diets deficient in iron. Cures followed marriage and a freer existence with more attention to the preparation of food and less to the shape of the figure. The elders called chlorosis "love sickness," and, when the benefits of marriage could not be obtained for a victim of the disease, a substitute was sought in a remedy prepared with the sword of a handsome knight. The sword was immersed in water for several days, and the rusty water was administered orally. In this manner the necessary iron was supplied, and a genuine cure achieved. Today the sword and the rust are replaced by a simple iron salt, and the effects are even more striking.

Many such examples could be cited. Suffice it to say that the road to our modern understanding of blood has wound extensively through the provinces of clinical medicine; never has this been more evident than it is today.

The modern era. The first phase in the modern era of blood research began in 1879 with Paul Ehrlich's development of staining techniques for human blood preparations, methods that for the first time permitted careful microscopic study of the morphological characteristics of blood cells. The impact of these techniques was considerable. Blood cells could now be classified, and it became apparent that each type behaved differently in health and in disease. The subsequent staining of bone marrow led to an awareness of the important relationships between the cells of bone marrow and those of peripheral (i.e., circulating) blood.

A second phase in blood research began in the 1920's as interest grew in the physiological aspects of normal and abnormal blood. With morphological descriptions of blood cells completed, investigators began to inquire into the physiology of their production, their life span, their role in body function, and their fate. They found, for example, that the production and removal of both fluid and cellular blood elements depend upon various organs outside the vascular system: the gastrointestinal tract supplies the nutritional factors needed for the maturation of red blood cells; the reticuloendothelial system provides for the destruction of blood cells and the disposal of blood pigments; the liver is essential for the production of some plasma proteins and the lymphoid tissues for the production of others.

The third and latest phase in the study of blood centers on the biochemical, biophysical, and genetic mechanisms underlying both morphological and physiological phenomena. Today the science of blood, or *hematology* as it is called, is one of the most dynamic areas of physiology, for blood is an easily accessible tissue

and hence readily examined by the techniques of modern biology and biochemistry.

Vocabulary of hematology. The vocabulary of hematology consists largely of words having Greek or Latin roots. Therefore, familiarity with the commonly used roots simplifies the learning of new terms. The Greek word for blood, *haima*, gives rise to the prefixes *hemo-*, *hema-*, and *hemato-;* thus *hematology* is the science of the blood, and *hematopoiesis* the production of blood. It also gives rise to the suffix *-emia*, as in *anemia*, insufficient blood, *hypercalcemia*, excess calcium in the blood, and *septicemia*, sepsis or infection in the blood. Table 8.1 lists some of the combining forms frequently encountered.

Blood and Blood Formation

BIOLOGICAL SIGNIFICANCE OF BLOOD

Evolution of blood. Early unicellular organisms were directly exposed to the immediate environment. Simple exchanges of nutrients and waste products took place across the cell membranes separating the cell interiors from the surrounding sea water. Hence blood vascular systems were not needed.

We learned in Chapter 7 that the evolution of higher forms was possible only through the establishment of a stable, self-regulated, internal fluid milieu. As animals became increasingly complex, they achieved such internal constancy by the development of special canal systems that conveyed the sea water to the deeply placed portions of their bodies. In time the canal systems closed, and regulatory mechanisms gradually evolved for the maintenance of internal constancy, including elegantly discriminating kidneys and a heart whose rapid and uniform pumping action forced the fluid through the canal system and minimized local disturbances in composition by promoting mixing. These and the closed canals became the circulatory system, and the fluid within became the blood.

Functions of blood. Blood circulates through every tissue and participates in every major functional activity of the human body. If it can

TABLE 8.1   COMMON COMBINING FORMS IN THE VOCABULARY OF HEMATOLOGY

Combining form	Greek word	Examples
Erythro-	*Erythros*, red	Erythrocyte, red cell
		Erythrocythemia, abnormal increase of red cells in blood
Hema-, hemo-, hemato-*	*Haima*, blood	Hematology, science of blood
Leuko-	*Leukos*, white	Leukocyte, white cell
		Leukemia, "white blood," a disease sometimes characterized by an increase of white cells in blood
Myelo-	*Myelos*, marrow	Myeloma, tumor of bone marrow
Thrombo-	*Thrombos*, clot	Thrombocyte, "clotting cell," or platelet
-blast	*Blastos*, sprout	Erythroblast, early precursor of red cell
-crit	*Kritēs*, judge	Hematocrit, instrument for measuring blood cell volume; blood cell volume
-emia	*Haima*, blood	Anemia, deficiency in number of red cells in blood
		Septicemia, infection in blood
-oma	*Onkoma*, tumor	Lymphoma, tumor of lymphoid tissues
-penia	*Penia*, poverty	Thrombocytopenia, deficiency in number of platelets in blood
-phil	*Philein*, to love	Acidophil, white cell with affinity for acidic stain
		Basophil, white cell with affinity for basic stain
-poiesis	*Poiein*, to produce	Hematopoiesis, production of blood

* In Great Britain the conventional spellings are haema-, haemo-, and haemato-.

be called an organ, it is the body's most voluminous one.

Blood has three primary functions: (1) it is the body's main transport system; (2) it plays a leading role in the regulation of salt and water distribution, acid-base balance, and temperature; and (3) it is responsible for essential defense mechanisms.

The many substances transported in blood include nutrients, wastes, gases, regulatory agents such as hormones, and ingredients of the defense mechanisms such as antibodies, phagocytes, and clotting factors. Thus blood carries oxygen from lungs to tissues and carbon dioxide from tissues to lungs; it delivers lipids, carbohydrates, proteins, minerals, and vitamins from the digestive tract and storage depots to the tissues; it removes the waste products of tissue metabolism from the tissues to the excretory organs—the lungs, kidneys, skin, intestines, and liver; and finally, it distributes the hormonal regulators produced in the endocrine glands—indeed, the effective hormonal integration of body metabolism is entirely dependent on the transport function of blood.

Water distribution in the body is regulated chiefly by the osmotic pressure of proteins dissolved in blood plasma. Salt distribution and acid-base balance are governed in part by the water and electrolyte contents of blood and the states and capacities of its buffer systems. As noted in Chapter 7, an important consequence of proper water balance is temperature control. Metabolic processes continuously produce heat, but mammalian tissues function efficiently only within a narrow range of temperature. The constant flow of blood through the small blood vessels helps to maintain the organs within this range by minimizing minor variations in local temperature.

One of the most intricate of the body's defense mechanisms is the remarkable system that prevents fatal blood loss after blood vessel injury. This is the clotting (or coagulation) mechanism, which we shall discuss later. Blood also carries antibodies and the bacteria-ingesting white cells, both essential in defending the body against microbial infection.

PROPERTIES OF BLOOD

Physical characteristics. When blood is removed from the body, it congeals to form a clot, which, if undisturbed, eventually shrinks or contracts, leaving a clear yellow fluid, the *serum* (Fig. 8.1). Therefore, to examine whole blood in the fluid state, we must add a chemical substance that prevents coagulation.

The difference between blood serum and *plasma* should be clearly understood. We have seen that serum is the fluid remaining after clot formation. Plasma is the supernatant fluid remaining after whole blood containing an anticoagulant has been centrifuged. A clot results when the soluble plasma protein fibrinogen is converted to insoluble fibrin. Serum differs from plasma, then, chiefly in that it contains no fibrinogen. Serum cannot clot; plasma can because its fibrinogen is intact. Clotting does not normally occur within a blood vessel, however. Plasma within the body contains no antico-

8.1 Clotted blood, with a fully contracted clot surrounded by clear serum.

agulant, but its fibrinogen is prevented from becoming fibrin by other mechanisms.

Whole blood consists of a plasma suspension of three cell types: nonnucleated red cells, or *erythrocytes;* nucleated white cells, or *leukocytes;* and curious fragmentlike platelets, or *thrombocytes.* So unusual are the erythrocytes and thrombocytes that some have questioned whether they deserve the name cells. Perhaps that is why the evasive terms "corpuscle" and "formed element" are sometimes used. We shall call them cells, but we must keep in mind their special properties.

The *viscosity* of blood is due to the viscosity of plasma itself plus the friction between the cells and their surrounding plasma. The operation of the circulatory system is significantly affected by blood viscosity. In diseases increasing blood viscosity (e.g., conditions with abnormally increased numbers of red cells, such as polycythemia vera), blood flow to critical organs like the brain is greatly reduced. In severe anemia (in which the number of red cells is decreased), viscosity decreases, and tissue blood flow correspondingly increases. The viscosity of normal whole blood at 37°C is about four times that of water.

The *specific gravity* of blood is about 1.057 for men and 1.053 for women. The specific gravity of plasma is about 1.030, that of serum about 1.027, and that of red cells about 1.098. The relatively high specific gravity of the red cells is apparent in anticoagulant-treated blood standing in a test tube. The red cells fall to the bottom, and the plasma rises to the top. Obviously the specific gravity of blood is altered by any disturbance that shifts the ratio of red cell volume to plasma volume. For example, dehydration increases blood specific gravity, as well as blood viscosity, whereas anemia decreases it.

Sedimentation rate. The rate at which red cells of whole blood settle to the bottom of a container is normally quite slow, but, in a variety of abnormal conditions, it is rapid. Often the acceleration is related to the severity of the disease. Determination of the sedimentation rate is thus a helpful laboratory test for diagnosing a disease and following its course.

The sedimentation rate is a measure of the so-called *suspension stability* of blood. Studies have shown that sedimentation occurs in three phases. First, there is a brief period of red cell aggregation with little fall. The cells tend to come together with their flat surfaces in apposition, forming neat piles called *rouleaux* (Fig. 8.2). Second, the aggregates fall rapidly because the ratio of mass to surface area is higher than for solitary cells. Finally, the aggregates pack together at the bottom of the vessel until the upper plasma layer is clear.

Red cells normally remain in uniform suspension within the body. Only when the sedimentation rate outside the body is grossly abnormal does sedimentation occur inside the body—"sludging" of red cells in small capillaries, with resulting local obstructions in blood flow.

 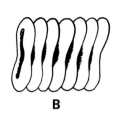 

8.2 Shapes of an individual red cell (A), a rouleau of red cells (B), and a clump of agglutinated red cells (C). (From Miale, *Laboratory Medicine–Hematology*, 2nd Ed., The C. V. Mosby Company, St. Louis, 1962.)

In normal blood the tendency of red cells to form rouleaux is slight. Since the number of large sedimenting particles is therefore low, sedimentation is slow. In diseases in which the tendency to form rouleaux is great, sedimentation is very rapid. Increased rouleau formation is due chiefly to increased levels of certain plasma proteins, particularly fibrinogen and globulin.

Sedimentation rates are expressed in millimeters of fall per unit of time (usually an hour), and normal values vary depending on the method of measurement. For example, in the Wintrobe method the normal sedimentation rate is 0 to 9 mm per hour for men and 0 to 15 mm per hour for women. Rates as high as 25 and 30 mm per hour, respectively, are common in disease. To the physician they are signals that further investigation is called for.

Agglutination. Rouleau formation, as we have just noted, is promoted by abnormalities in the plasma. Another type of red cell clumping, *true agglutination*, is a result of variations in the red cell surfaces (see Fig. 8.2C). It occurs, for example, when red cells of one blood group are mixed with serum of a different blood group. This *isoagglutination* is caused by the group antigen on the red cell surfaces and accounts for incompatible blood transfusion reactions.

Autoagglutination is an abnormal process in which an individual's red cells are agglutinated by his own serum. Frequently it takes place only when red cells and serum are chilled. In such cases it is called *cold agglutination*. Autoagglutination is caused by an abnormal agglutinin in the serum.

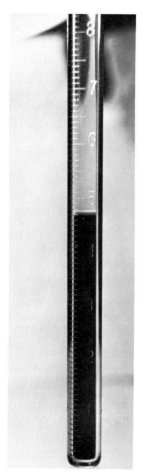

8.3 Hematocrit tube containing centrifuged blood with packed red cell volume of about 48%.

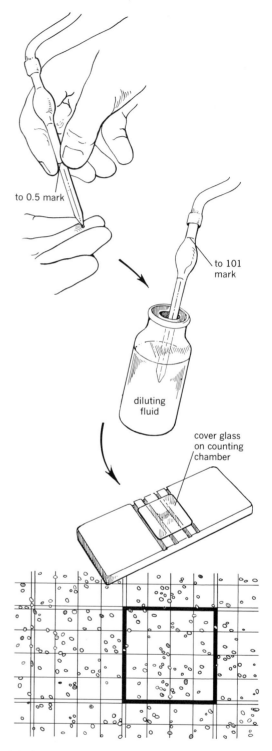

to 0.5 mark

to 101 mark

diluting fluid

cover glass on counting chamber

8.4 Method used in counting blood cells.

Recently it was found that certain viruses cause red cells to agglutinate in a test tube, the extent of agglutination being proportional to the number of virus particles. Because red cell agglutination is so easily detected, it constitutes a useful technique for the assay of these viruses.

Hematocrit. The word *hematocrit* is accurately applied to an instrument for determining the relative amounts of blood cells and plasma in whole blood. The quantity measured is the volume of packed cells (particularly red cells) or, more correctly, the volume of packed cells per 100 ml of blood. However, in common usage, hematocrit is synonymous with this percentage.

The hematocrit is determined by centrifugation of a blood specimen in a graduated tube

TABLE 8.2 NORMAL CELL COUNTS IN PERIPHERAL BLOOD

Cell type	Average count, cells/mm³	Range of count in 95% of normal population, cells/mm³
Red cells	5,400,000 (men)	4,500,000–6,500,000 (men)
	4,800,000 (women)	3,900,000–5,600,000 (women)
	4,500,000 (infants)	4,000,000–5,600,000 (infants)
White cells	7,500 (adults)	4,500–11,000 (adults)
	12,000 (infants)	10,000–25,000 (infants)
Platelets	250,000 (all ages)	140,000–440,000 (all ages)

until the red cells are firmly packed in the bottom of the tube. Above them lies a white layer of white cells and platelets, and above this lies clear cell-free plasma (Fig. 8.3). The volume of circulating red cells and the total volume are read from the graduations. The red cell hematocrit averages 45 (i.e., 45 ml of red cells per 100 ml of blood) in normal men and 42 in normal women.

The white cells and platelets settle on top of the red cells because their specific gravities are lower than that of red cells and higher than that of plasma. They form a little white cap on the column of red that is usually called the *buffy coat*.* It varies in depth with the number of circulating white cells and platelets.

Number of circulating blood cells. The methods for counting the blood cells in a given volume of blood are well known, but the large error inherent in these methods is not sufficiently appreciated. In practice, a carefully measured sample of blood and a carefully measured quantity of diluting fluid are drawn into a special diluting pipet (Fig. 8.4). The pipet is then shaken to ensure uniform mixing. A small portion of the diluted blood is introduced into a *hemocytometer*, a special counting chamber

made to accommodate a known volume; its bottom surface is engraved with a grid of microscopic dimensions for use as a guide in the counting operation. The chamber is placed under a microscope, and the cells are counted. A simple calculation correcting for the dilution gives the cell count per cubic millimeter of undiluted blood.

The chief sources of error in this procedure are inaccurate measurement of blood and diluting fluid, inadequate mixing, and nonuniform cell distribution in the hemocytometer. Moreover, the cell count in blood removed from a capillary differs slightly from the cell count in blood removed simultaneously from a vein, and many factors produce temporary rises and falls in the count. The normal diurnal variation in the red cell count, for example, may be as high as 1,000,000 cells per cubic millimeter.

Normal cell counts for the three types of blood cells are listed in Table 8.2. The preponderance of red cells is striking; there are approximately 1000 per white cell. Yet white cells are evidently larger than red cells since the volume of the buffy coat is about ⅟₅₀ that of the packed red cells.

PHYSIOLOGY OF BLOOD FORMATION

Hematopoiesis in the embryo. In the course of embryonic development, body cells differentiate into three types: the ectoderm, which ultimately covers the embryo, forming the skin, hair, enamel of the teeth, etc.; the endoderm,

---

* William B. Castle has pointed out that this is a misnomer. The term was first used to describe the whitish upper end of a blood clot that had formed after rapid red cell sedimentation. With the introduction of anticoagulants, physicians saw fewer samples of clotted blood and gradually applied the term to the white cell–platelet layer appearing in centrifuged blood.

which lines the primitive gut; and the meso-derm, which lies between the ectoderm and endoderm and gives rise to the circulatory apparatus, blood cells, muscles, bones, ligaments, and other connective tissues. In some regions of the embryo body, the mesoderm is represented at first by unorganized masses of actively migrating cells known collectively as mesenchyme. It is the mesenchyme that develops into the circulatory system and blood cells.

Late in the third week of embryonic life, certain cells of the mesenchyme begin to cluster into *blood islands*. The peripheral cells of these islands soon become flattened and join to form the tubes of a primitive vascular system. During this period the central cells of the blood islands differentiate into detached free elements carried along by a mounting stream of primitive plasma. Some of them turn yellow and gradually acquire *hemoglobin*. These, the *primitive erythroblasts*, are the first hemoglobin-synthesizing cells of the embryo. Unlike the erythroblasts of adult bone marrow, they do not mature into red cells. Instead they are believed to die out, to be replaced by erythroblasts of the type found in the adult body.

In the third month of embryonic life, the liver becomes the chief site of blood cell formation, with additional contributions by the *spleen*, *lymph nodes*, and *thymus*. It may continue in this role until after birth. However, bone marrow becomes an active hematopoietic site in midembryonic life, and, by the end of gestation, it is the major site. Blood cell production in the bone marrow is called *medullary hematopoiesis*. After birth it diminishes within the shafts of the long bones, so that in the adult it is limited to the ends of these bones and to the sternum, ribs, skull, vertebrae, and innominate bones. Although normally hematopoiesis in the adult occurs only in the bone marrow, in abnormal situations it may resume in those organs that were hematopoietically active in embryonic life, such as the liver and spleen. This is called *extramedullary hematopoiesis*.

Hematopoiesis in the adult. Details of the production of red cells (*erythropoiesis*), white cells (*leukopoiesis*), and platelets (*thrombopoiesis*) in the adult will be reviewed later. Here we shall make some general observations.

Red cells, certain white cells, and platelets arise in the bone marrow. It appears that the *reticulum*, or *stroma*, of bone marrow is the essential blood-forming tissue. The primitive *reticulum cell* of bone marrow is the closest approach in the adult to the primitive mesenchyme cell. The ability of reticulum to form blood cells depends upon its ability to form the blood cell precursors called *erythroblasts*, *myeloblasts*, and *megakaryoblasts*, which develop into red cells, certain white cells, and platelets, respectively.

Morphological hematologists have long searched for structural similarities and differences between cells that might reveal the early steps in the pathways of cell differentiation. Although they have contributed much basic knowledge, confusion has arisen from variations in terminology and from the inadequacy of the evidence upon which they have based some of their conclusions. In general, the various theories of hematopoiesis have attempted to resolve two major points: (1) the origin or origins of the mature blood cells—whether all are derived from one distinct precursor cell (the *monophyletic theory*) or whether red cell, white cell, and platelet stem from three distinct precursor cells (the *polyphyletic theory*); and (2) the nature of the crossover from one cell type to another under abnormal conditions. Though recent evidence obtained by the methods of tissue culture, biochemistry, and electron microscopy has failed to provide decisive solutions to these problems, it does suggest a single totipotential precursor of all blood cells differentiates into separate primitive red cell, white cell, and platelet precursors. Hence we shall assume that there is such a cell, the undifferentiated reticulum cell, or *stem cell*, from which all three cell types evolve.

As we have noted, in the adult the hematopoietically active bone marrow—known as the *red marrow*—is confined almost entirely to the flat bones (e.g., the sternum, ribs, skull, and vertebrae). There is ordinarily little or no red marrow in the long bones of the legs and arms, though small amounts may be found in the bone ends. The bone marrow in an adult weighs 1500 to 3000 g. It is therefore an organ of formidable size. About half its volume is red marrow; the remainder is fatty, inactive *yellow marrow*. Yellow marrow retains the essential reticular structure of hematopoietic tissue and is capable under appropriate stimulation of becoming red marrow. The transformation of yellow marrow to red marrow is one of the body's mechanisms for rapidly increasing blood cell production in times of emergency, such as during hemorrhage. Conversely, there are certain pathological conditions in which red marrow is gradually replaced by yellow marrow. The result is a potentially dangerous decrease in the number of circulating red cells, white cells, and platelets.

LYMPH AND THE LYMPHATIC SYSTEM

One type of white blood cell, the *lymphocyte,* is not formed in the bone marrow. So that we may discuss lymphocytes in context, we shall here introduce the lymphatic system at large.

Scheme of the lymphatic system. Tissues contain innumerable vessels that are separate and distinct from the capillaries which transport blood (Fig. 8.5). They collect *lymph,* the watery extravascular fluid of the tissue spaces, and convey it to the blood stream. Thus they help to maintain the constancy of the cellular environment and of the blood itself and also conduct a variety of substances from the cells to the blood. These *lymphatic capillaries* make up a complex network of fragile distensible vessels resembling veins. Since their function is drainage, not perfusion or circulation, they end blindly in the tissues. In other words, they form a closed system, with fluid entering through their walls.

The capillaries unite to form larger and larger lymphatic vessels, which finally converge in two main lymphatic channels, the *thoracic,* or *left lymphatic duct,* and the *right lymphatic duct.* The thoracic duct receives lymph from the left side of the head, neck, and chest and from all of the lower half of the body; lymph from the right side of the head and neck, the right arm, and the upper half of the trunk empties into the right lymphatic duct. The destination of the large lymphatic ducts is the great veins, into which they pour their lymph at the point where the veins enter the heart, carrying returning blood of the circulation (see Fig. 8.5A).

In general, the lymphatic vessels are in close juxtaposition to the veins, and, like the blood vessels, they are present in nearly every tissue and organ. One exception is the bone marrow, which contains no lymphatic vessels. Other exceptions are body structures like hair, nails, and cartilage, which lack both blood and lymphatic vessels. More will be said of the relationship of the two systems of vessels in Chapter 9.

It is of interest that the lymphatic system is an evolutionary newcomer encountered only in the higher vertebrates. As organisms grew more complex, hydrostatic pressures increased within the blood circulatory system. As a result, the small blood vessels in the tissues became leaky, and fluids seeped out of the blood stream. A drainage system was needed, and we assume that the lymphatic vessels evolved to meet this need.

Lymph nodes: structure. Lymph nodes (or "lymph glands," as they are sometimes improperly called) are small oval or bean-shaped bodies occurring at intervals along the larger lymphatic vessels (see Fig. 8.5A) and ranging from 1 mm in diameter to the size of a large grape. Despite an occasional solitary node, most nodes are arranged in groups or chains. Like the lymphatic vessels, lymph nodes have both superficial and deep locations. Typical superficial nodes are the *inguinal, axillary, supratrochlear,* and *cervical* groups. Typical deep nodes are the *mesenteric* group, which lies near the small intestine (see Fig. 8.5A). Each group drains lymph from a particular area.*

The internal structure of a typical lymph node is shown in Fig. 8.5B. The *hilus* is a slight depression through which both blood vessels and lymphatic vessels pass. The outer covering is a *capsule* of connective tissue from which fibrous bands, the *trabeculae,* proceed into the substance of the node, dividing it into irregular, freely communicating spaces. Suspended within this framework is the *reticular framework.* Its

---

* We cannot elaborate here the anatomical details of the many lymph node groups and their subdivisions. The individual who most needs such information is the surgeon, who is frequently asked to remove an enlarged node for microscopic examination. This procedure (biopsy) facilitates diagnosis of many benign and malignant diseases associated with lymph node enlargement. When he identifies a cancerous growth, he must find and remove each of the regional nodes, since a remaining node may contain a stray cancer cell that will give rise to a second, or *metastatic,* tumor growth. The surgeon must also deal occasionally with an obstruction to lymph flow caused by a benign or malignant disease of the nodes. Such an obstruction results in *lymphedema* (i.e., swelling in the drainage area due to collected lymph in the tissue spaces). The most spectacular form of lymphedema is *elephantiasis,* a chronic obstruction caused by invasion of the nodes by filarial worms.

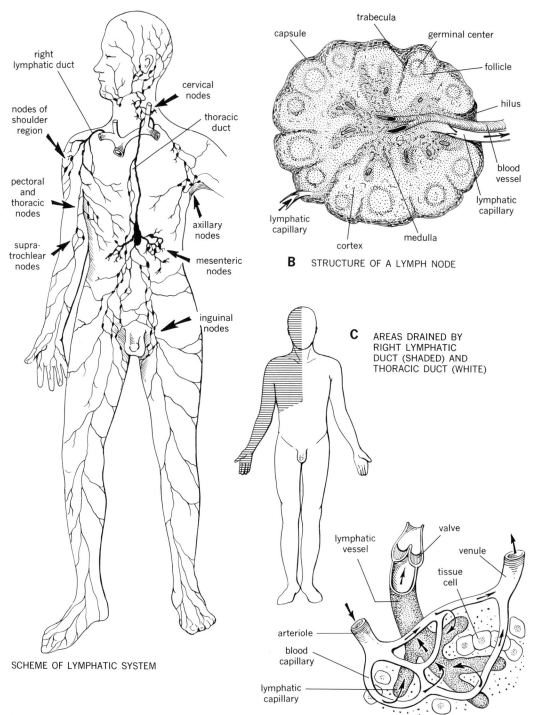

A SCHEME OF LYMPHATIC SYSTEM

right lymphatic duct

cervical nodes

nodes of shoulder region

thoracic duct

pectoral and thoracic nodes

axillary nodes

supra-trochlear nodes

mesenteric nodes

inguinal nodes

B STRUCTURE OF A LYMPH NODE

trabecula

capsule

germinal center

follicle

hilus

blood vessel

lymphatic capillary

lymphatic capillary

cortex

medulla

C AREAS DRAINED BY RIGHT LYMPHATIC DUCT (SHADED) AND THORACIC DUCT (WHITE)

D DETAIL OF BLOOD AND LYMPHATIC CAPILLARIES

lymphatic vessel

valve

venule

tissue cell

arteriole

blood capillary

lymphatic capillary

8.5 Lymphatic system. Note in D that plasma flows from blood capillaries into intercellular spaces and then into the lymphatic capillaries which end blindly.

most loosely meshed areas constitute the *sinuses* through which the lymph percolates. The sinuses are lined with reticulum cells. These cells are capable of ingesting (phagocytizing) foreign materials and therefore are part of the diverse group of cells throughout the body that is known collectively as the *reticuloendothelial system*, or *RES*. The RES includes those tissues and cells that are capable of phagocytizing bacteria and other foreign particles. They include: (1) the reticulum cells of the bone marrow, lymph nodes, spleen, and liver; (2) the circulating white blood cells; and (3) the tissue *macrophages*, wandering cells in the tissues in constant quest of foreign matter. All are closely related and have a common origin in the primitive reticulum cell.

Under low magnification a freshly cut section of a lymph node shows an outer *cortex* and an inner *medulla*. Both consist of *lymphoid tissue*. However, the cortex contains lymphatic *nodules*, or *secondary follicles*, temporary structures whose appearance varies with their state of activity.* They are made up of masses of lymphocytes of various ages arranged in a characteristic nodular pattern. When stimulated under conditions to be described later, the central portions of the nodules can produce new lymphocytes. Thus these areas are called *germinal centers*.

Lymph enters a lymph node from the *afferent* lymphatic vessels at the hilus, circulates through the sinuses, and leaves by the *efferent* vessels emerging from the hilus. In passing through the node, it picks up any newly formed lymphocytes that may have arisen in the germinal centers.

Lymph nodes: functions. The lymph nodes have two main functions. First, they remove foreign particles from lymph before it enters the blood stream. No lymph reaches the blood stream without passing through at least one node, and, in its tortuous course through the node, it is cleansed by what appears to be simple mechanical filtration and by the phagocytic cells in the node, which can ingest and destroy bacteria, dead tissue cells, and many foreign particles. This defensive function is exercised when lymph is infected. The infection is eliminated by the first node or group of nodes in

the pathway of the lymph, unless infection is severe. There is some indication that a mild degree of infection exists normally.

Second, the nodes are centers for the proliferation of lymphocytes and other antibody-manufacturing cells. When bacteria, viruses, or other antigens arrive at a node, they stimulate the production of these cells. This is one of the body's major defense mechanisms, and it will be described in detail in Chapter 19.

Spleen: structure. The spleen is a lymphoid organ roughly the size of a man's fist, situated directly beneath the diaphragm, behind and to the left of the stomach (see Fig. 6.8). It is covered by peritoneum and held in position by peritoneal folds.

The spleen comprises the largest single collection of lymphocytes and reticuloendothelial cells in the body. Like a lymph node, it has a connective tissue capsule from which trabeculae extend inward (Fig. 8.6). A few smooth muscle fibers are present in both capsule and trabeculae. These are less prominent in man, however, than in dogs and other animals, in which the capsule is capable of contracting and expelling splenic contents into the blood stream.

The spaces between trabeculae contain *splenic pulp*. In a cross section of spleen, three types of pulp—*white pulp, red pulp,* and *marginal zone*—can be distinguished with the naked eye. The white pulp is scattered throughout the spleen as tiny grayish islands, each less than 1 mm in diameter. These are the nodules. Microscopic examination reveals that most contain germinal centers. The marginal zone is a poorly defined region between the white pulp and the red pulp. It is made up of a reticular meshwork, blood vessels, and free cells and is notable for the fact that many arteries terminate within it. The red pulp consists primarily of *cords* separated by *sinuses*. Both constitute vascular spaces lined by reticuloendothelial cells arrayed upon a lattice-like reticular framework. They differ in that the sinuses are relatively broad, uninterrupted channels, whereas the cords are crossed by numerous incomplete septums of reticulum which divide them into small intercommunicating compartments. The cords lie between the sinuses, which have linings that are unusual compared to ordinary blood vessel linings (Chapter 9), in that the cells are not tightly joined to one another and the basement membrane is pierced with large holes. Thus free cells, as well as the blood issuing from the arteries ending in the cords, may pass directly into the sinuses.

Arterial blood entering the spleen follows a complex pathway before emerging as venous blood. An understanding of this pathway is essential to an understanding of splenic function. The main *splenic artery* has many branches, called *trabecular arteries*, and these

---

* As we shall see, similar nodules occur in the spleen, though not in the thymus. They are also scattered throughout the body in regions of *loose* or *diffuse* lymphoid tissue, such as in the walls of the intestines and respiratory passages. The nodules are sometimes called *Malpighian bodies* after the anatomist who first described them, but, since this term has been used in different ways by different histologists, it is not recommended.

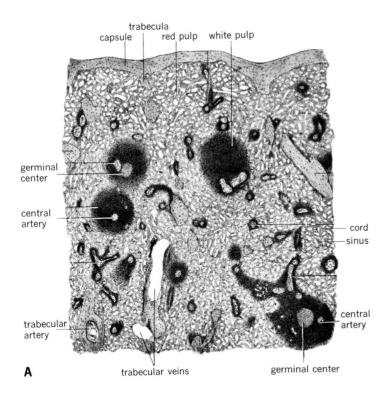

germinal center

central artery

trabecular artery

capsule

trabecula

red pulp

white pulp

cord

sinus

central artery

trabecular veins

germinal center

A

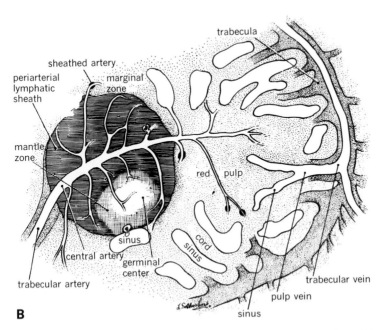

sheathed artery

periarterial lymphatic sheath

mantle zone

central artery

trabecular artery

marginal zone

trabecula

red pulp

sinus

germinal center

cord sinus

sinus

pulp vein

trabecular vein

B

8.6 Spleen: A, section perpendicular to capsule (×32); B, organization of blood vessels. (A from W. Bloom and D. W. Fawcett, *A Textbook of Histology*, 8th Ed., W. B. Saunders Co., Philadelphia, 1962. B from L. Weiss, in R. Greep, ed., *Histology*, McGraw-Hill, 1965. Used by permission.)

enter the white pulp as the *central arteries*. The central arteries also have many branches. Some branches terminate in the white pulp; later branches terminate in the marginal zone; and still later branches terminate in the red pulp. Before terminating in the white pulp, some acquire nodular sheaths composed of lymphocytes, reticular cells, and macrophages, and are therefore termed *sheathed arteries*. A few arteries communicate directly with sinuses.

Two main routes are taken by blood passing through the spleen. A small proportion of blood passes directly into the sinuses—or into the cords at a point which affords free and immediate transfer into the sinuses. Blood taking this pathway has a virtually unobstructed route to the venous collecting system and out of the spleen via the *splenic veins*. This is therefore the *rapid-transit pathway* or *closed circulation component* of the splenic blood flow. A larger proportion of the blood empties into the cords of the red pulp. (The substantial amount that empties into the white pulp and marginal zone finds its way into the cords.) Blood in the cords is obliged to navigate the circuitous and macrophage-lined cordal compartments before penetrating the narrow holes through which it gains access to the sinuses. This is the *slow-transit pathway* or *open circulation component* of the splenic blood flow. In traversing this pathway through the spleen, blood is brought into intimate contact with the phagocytic reticuloendothelial cells of its white pulp, marginal zone, and red pulp.

Spleen: functions. The spleen is not essential for life, and no serious disturbance follows its removal—though, as we shall see, a number of events occurring after splenectomy provide clues to its functions. We have already mentioned its role in the embryo (*hematopoietic function*). It is known to perform the following tasks in the adult.

1. It produces lymphocytes (*immunological function*). Likely in early life lymphocytes leave the thymus and colonize the spleen and lymph nodes. The lymphocytes found in the adult spleen are partly descendants of these original thymus lymphocytes and partly sequestered blood-borne lymphocytes. The spleen is rich in lymphocytes and reticuloendothelial cells and takes an active part in antibody synthesis and other defense mechanisms (Chapter 19).*

2. It exerts poorly understood controls over the number of red cells, white cells, and platelets in the blood (*endocrine function*). Some authorities have suggested that the spleen secretes one or more hormones that affect the rates of blood cell production in the bone marrow.

3. It destroys by phagocytosis aged or imperfect red blood cells (*culling function*). In this situation the spleen acts as an inspector on an assembly line, scrutinizing the circulating red cells and removing the few that do not, so to speak, meet specifications.† Such selective destruction is made possible by the unique splenic circulation. As blood passes through the white pulp, plasma tends to be skimmed off, and the cells concentrated. This phenomenon, coupled with the slow-transit open circulation pathway, leads to stasis of flow so that opportunities for phagocytosis of old or damaged red cells by the reticuloendothelial cells increase. Even after passage through the cords, the red cells may remain for some time in the sinuses. There their glucose supply is rapidly used up and the oldest cells fail to survive. The red cells are thereby subjected to additional culling.

4. It serves as a store or reservoir of platelets (*reservoir function*). In dogs, cats, and guinea pigs, it also serves as a reservoir of red blood cells, which tend to accumulate in the sinuses during sleep. The cells are ejected back into the blood under stress or under the influence of epinephrine. Recent studies have cast doubt on the importance of this function in normal man. A healthy adult human spleen contains only 20 to 30 ml of blood.

Thymus: structure. The thymus is an elongated bilobed structure, chiefly lymphoid tissue, lying in the upper part of the thorax above the heart and behind the top of the sternum (Fig. 8.7A).‡ It is largest relative to the body as a whole at birth, when it comprises 0.5 to 1% of the total body weight. For 8 to 10 years it grows

---

* Collections of lymphoid tissue containing germinal centers are found in locations other than the lymph nodes and spleen. Among the most important are the *tonsils* and the intestinal *Peyer's patches*, both of which are associated with the alimentary tract (Chapter 13). Their special immunological significance will be discussed later (Chapter 19).

---

† In this role it also performs what is called the *pitting function*. Occasional red cells contain granular deposits. Rather than destroy a cell, the spleen removes the granule by pinching off a small granule-containing portion of the cell as it squeezes through a narrow hole opening into a sinus. The red cell then reseals itself. When such a cell returns to the blood, it appears as though a bite had been taken from it.

‡ Calf thymus as a food is called *sweetbreads*. This term is also applied to calf pancreas.

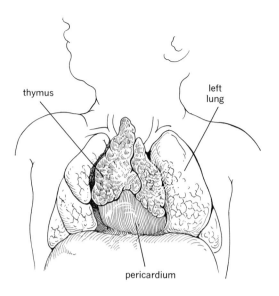

A LOCATION OF THYMUS IN CHILD

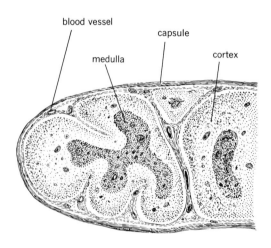

B STRUCTURE

8.7 Thymus.

slowly, its relative rate of weight increase lagging behind that of the body. It then regresses, its lymphoid tissue being largely replaced by adipose tissue, so that in the adult its substance may be difficult to distinguish from the fat in which it is embedded. Nevertheless, some thymus elements persist into old age.

As a lymphoid organ the thymus differs strikingly from lymph nodes and spleen in structure and function. In section each of its many lobules reveals a grossly visible central portion or medulla and a peripheral por-

tion or cortex (Fig. 8.7B). The two main cell types are *epithelial-reticular cells* (special reticular cells) and *thymus lymphocytes,* or *thymocytes* (which some authorities believe are identical to the small lymphocytes of blood). In the fully developed thymus, the lymphocytes are densely packed in the cortex, where they greatly outnumber the epithelial-reticular cells. Unlike the lymphocytes of lymph node and spleen, thymus lymphocytes are not arranged in nodules. They are separated from connective tissue and blood vessels by a basement membrane and a layer of epithelial-reticular cells. The epithelial-reticular cells, of ectodermal origin, are most prominent in the medulla.

The medulla also contains Hassall's corpuscles, rounded structures 30 to 100 μ in diameter, which are found only in the thymus (Fig. 8.7C). The corpuscles, scattered throughout the medulla, are composed of concentrically arranged cells that are believed to be derived from macrophages. Though the corpuscles are known to be phagocytic, their physiological significance is not known.

Thymus: functions. The functions of the thymus were long obscure, though, rightly or wrongly, the organ was frequently referred to as a gland. Recent studies have indicated that it does secrete one or more hormones and so indeed may be an endocrine gland. However, this is but one of several aspects of thymus function.

In the early 1960's the important observation was made in several species that the normal growth and maturation of lymph nodes and spleen ceases following experimental removal of the thymus just after birth (neonatal thymectomy). The affected animals also become immunologically unresponsive (e.g., incapable of forming antibodies). Subsequent research on the role of lymphocytes in the body's immune mechanisms have suggested that a major function of the thymus may be to supply immunologically competent lymphocytes to other lymphoid tissues during a critical period in early life.

In considering the activities of the thymus, we must distinguish prenatal events from postnatal events. During the first two months of embryonic life, the cells of the developing thymus are primarily epithelial. Thymus lymphocytes then appear for the first time, arising directly from epithelial cells, with the mesenchyme providing only the initial stimulus to differentiation and not contributing at all to the population.

If these lymphocytes arise by direct transformation of the epithelial cells, how do the rest of the body's lymphocytes arise? According to the work of R. Auerbach, lymphocytes appear first in the epithelial part of the thymus and later in the spleen, lymph nodes, blood, and other tissues—in some species they do not appear in nonthymic lymphoid tissue until the time of birth. In other words, the precursors of some of the nonthymic lymphoid elements originate in the thymus and migrate at various times before or after birth to sites where they establish germinal centers for the further development of the self-perpetuating lymphoid structures—the lymph nodes and spleen, the organs responsible for maintaining part of the body's immunological defenses. Interestingly, thymus lymphocytes are not considered responsible for the establishment of germinal centers in the tonsils and Peyer's patches.

A number of experiments have supported these conclusions, disclosing that (1) lymphocyte production within the thymus, in contrast to that in lymph nodes and spleen, is stimulated not by antigens but by agencies intrinsic to the thymus itself; (2) the epithelial-reticular cells are thymus-specific cells that somehow influence the mitotic behavior of neighboring thymus lymphocytes; (3) the rate of lymphocyte production remains higher despite advancing age in thymus cortex than in any other lymphoid tissue (seven times as great in the thymus as in the lymph nodes or spleen); and (4) the thymus sends to the lymph nodes and spleen a factor (or hormone), probably secreted by the epithelial-reticular cells, that prompts conversion of the seeded (and perhaps the resident) lymphocytes to cells capable of participating in immune reactions.

Recent studies have revealed a second source for the precursors of thymic lymphocytes. Certain hematopoietic cells in the bone marrow are capable of migrating to the thymus, there to be "instructed" in some way—that is, they come to resemble lymphocytes within the thymus. Their progeny then move into the lymph nodes and spleen. It appears that the number of resident precursor cells in the thymus capable of giving rise by differentiation or mitosis to thymus lymphocytes is too small to account for the high lymphocyte production rate after birth; therefore, the maintenance of this rate must depend on the importation of precursor cells from another location (i.e., the bone marrow). Before

birth the thymus is the major source of lymphocytes; in later life it is the precursor cells that migrate to the thymus from elsewhere.

Normally the thymus contributes few cells directly to the circulating pool of lymphocytes in adult blood and lymph; but it may play an essential part in rebuilding the pool when lymphocytes have been rapidly destroyed—as, for example by the effects of severe stress. Thus it is necessary not only for establishing an immunologically active lymphoid system during development but also for restoring the system after it has been damaged.

At times of antigenic stimulation, the nodules in the lymph nodes, spleen, and Peyer's patches may also contribute to the pool of circulating lymphocytes. Lymphoid tissues that were originally seeded by thymus lymphocytes (i.e., lymph nodes and spleen) produce distinctive small lymphocytes. Lymphocytes produced in Peyer's patches and similar lymphoid tissues are considerably larger. Under ordinary circumstances most circulating lymphocytes are not newly formed cells but repeatedly recirculating old ones. J. L. Gowans, using isotope-labeled cells, has demonstrated that long-lived lymphocytes continuously pass from the blood, through the lymph nodes, into the lymph, and back to the blood via the thoracic and right lymphatic ducts. The major migratory pathways of these cells do not travel to the bone marrow or thymus. What governs their homing behavior to peripheral lymphoid tissues is unknown. Current evidence suggests that specific disaccharides in the cell surfaces may guide these migrations, the functional significance of which is not yet known.

"Formed Elements" of the Blood

ERYTHROCYTES

Physical properties. We turn now to the blood cell type produced in greatest numbers—the red cell, or erythrocyte. Morphologically it is a biconcave disc 7 to 8 μ in diameter, varying in thickness from 1 μ in the center to 2.4 μ at the edge (see Fig. 8.2A). The volume of an average normal red cell is 90 μ^3 (range 78 to 94 μ^3). The so-called *mean corpuscular volume,* or *MCV,* is derived from the hematocrit value and the red cell count according to the following formula:

MCV (in cubic microns)

$$= \frac{\text{milliliters of packed red cells per liter of blood}}{\text{millions of red cells per cubic milliliter of blood}}$$

In unstained preparations the red cell is pale reddish yellow, but, since it absorbs acidic dyes, in stained films it is red with a slight pallor in the central area. About one-third of its substance is hemoglobin. The *mean corpuscular hemoglobin concentration,* or *MCHC,* is determined as follows:

MCHC (in percent)

$$= \frac{(\text{grams of hemoglobin per 100 ml of blood}) \times 100}{\text{milliliters of packed red cells per 100 ml of blood}}$$

Because this is a concentration, it is expressed as a percentage. In a normal individual it is about 33%. The amount of hemoglobin per red cell in units of weight is known as the *mean corpuscular hemoglobin,* or *MCH.*

MCH (in micromicrograms)

$$= \frac{\text{grams of hemoglobin per liter of blood}}{\text{millions of red cells per cubic millimeter of blood}}$$

The normal value is about 29 $\mu\mu$g (range 27 to 32 $\mu\mu$g).

Functions. The primary function of the red cells is to transport hemoglobin, which in turn transports oxygen and carbon dioxide. Why, one wonders, has evolution produced such a complex method of carrying hemoglobin from place to place? Could it not have permitted hemoglobin to dissolve freely in the plasma, thus eliminating the need for red cells?

The answer to the last question is a qualified "no." A pigment similar to hemoglobin occurs in solution in the plasma of crayfish and crabs, but, when free hemoglobin is released into human plasma, it is quickly oxidized to a form incapable of oxygen transport, some is bound to a protein in the plasma, some is ingested and biochemically altered by the reticuloendothelial cells, and some is excreted by the kidneys. As a result, most of it is rapidly lost.

Moreover, it is physically possible for much more hemoglobin to be carried within red cells than in solution, where it would increase blood viscosity prohibitively. Hemoglobin may not transport oxygen as rapidly in red cells as it would in solution, but in red cells its life span is the same as that of the cells—120 days—and there it is continuously protected against oxidation by red cell enzymes. Finally, red cells expedite gas transfer by causing turbulence in the blood vessels as pebbles do in a stream.

Transport of hemoglobin is not the only function of red cells. They significantly affect blood viscosity. They contain the enzyme carbonic anhydrase, which catalyzes the reversible conversion of carbon dioxide to bicarbonate. And they contain an active glycolytic system that generates the energy necessary for their several metabolic functions.

Osmotic fragility. Under certain conditions injury of the red cell membrane results in the release of hemoglobin. This phenomenon, known as *hemolysis,* occurs in a variety of normal and abnormal circumstances.

Lowering the osmotic pressure of the suspending fluid is the best-known method of hemolyzing red cells. The cells remain intact for hours in isotonic salt solution (of osmotic pressure equal to that within them) but are rapidly hemolyzed in distilled water or hypotonic salt solution; because the osmotic pressure is higher in the cells than in the suspending fluid, water passes into them. The cell volume increases, and the pores thought to exist in the cell membranes increase in diameter, so that the cells liberate their hemoglobin and other contents. Thus red cells are excellent *osmometers.*

Normal red cells do not undergo hemolysis until the osmotic pressure of the suspending fluid has dropped to a certain level. This behavior is the basis of the *osmotic fragility test* (Fig. 8.8). Red cells suspended in 0.85% sodium chloride solution act as they do in whole blood. When the salt concentration is lowered to $0.44 \pm 0.02\%$, hemolysis begins. It is complete when the salt concentration falls to $0.34 \pm 0.02\%$. Abnormally increased or decreased osmotic fragility is characteristic of a number of blood diseases.

Hemolysis has causes other than osmotic shock. Various synthetic chemicals (e.g., detergents) are potent hemolytic agents. So are various natural products (e.g., powerful en-

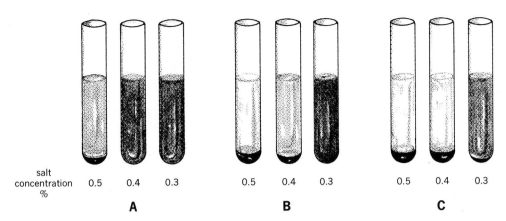

salt concentration %

0.5 0.4 0.3   0.5 0.4 0.3   0.5 0.4 0.3

A   B   C

8.8 Osmotic fragility of red cells: A, in hereditary spherocytosis; B, normal; C, in thalassemia. In the osmotic fragility test, the salt concentration is decreased 0.02% in successive tubes. Thus many more tubes are employed than are shown here. Normal hemolysis begins at a salt concentration of $0.44 \pm 0.02\%$ and is complete at a salt concentration of $0.34 \pm 0.02\%$. Hereditary spherocytosis is a genetically determined anemia associated with red cells of unusually great osmotic fragility (spherocytes). Thalassemia is a genetically determined anemia in which defective hemoglobin synthesis results in the production of target cells, with decreased osmotic fragility.

zymes present in snake venoms) known as *hemolysins*. Certain immunological reactions and physical stresses such as freeze-thawing and mechanical buffeting also result in hemolysis.

Stroma. The red cell is a remarkable example of biological engineering. Its biconcave shape provides a much larger surface for gas diffusion than would a spherical shape; therefore, if red cells were spherical, we would need a much larger number to distribute oxygen with the same efficiency.*

How and why the red cell maintains its biconcavity are subjects of great current interest. We assume that the shape of the cell is determined by a definite internal ultrastructure. Its nature is still unknown, however. When red cells are hemolyzed, colorless baglike structures remain that have been variously called *stroma, ghosts, membranes,* and *posthemolytic residues.*† Under suitable conditions they recover the biconcave configurations. It is evident, therefore, that their substance is the main framework of the cells.

Electron microscopy indicates that the stroma includes a typical cell membrane about 70 to 75 Å thick. The nature of the contact between stroma and hemoglobin is also uncertain. Although hemoglobin may be associated with stroma lipoprotein in the cell, it does not appear to be an essential structural element. Probably the physical relationship between stroma and hemoglobin is complex. ‡

Recent studies of the stroma have revealed its highly ordered molecular structure. Like other cell membranes (see Fig. 3.14), it is composed of proteins and lipids. The protein portions are parallel to the surface, and the lipid portions perpendicular to the surface with their nonpolar groups pointed outward (although the exact arrangement has not been established). The membrane is believed to contain many *pores* that are lined with polar groups (see Fig. 4.25). These are channels for cations and are only 7 Å in diameter.

---

* Dissenting opinions have been heard recently. Calculations of F. J. W. Roughton and others, which cover factors other than simple diffusion, indicate a smaller advantage of biconcave form over spherical form than has previously been assumed.

† The trend in the literature of hematology is to reserve the term "ghost" for the residue following hemolysis in hypotonic solution and the term "stroma" for the same structure in the intact cell.

---

‡ For example, when a red cell is cut in two, hemoglobin does not necessarily ooze away from the fragments. This experiment shows that hemoglobin is somehow bonded to the stroma. If the hemoglobin is to be released, the bonds must be broken, and their breakage depends on the properties of the suspending fluid and other conditions.

The major lipids of red cell stroma are phospholipids, free cholesterol, and glycolipids. Phosphatidyl serine, phosphatidyl choline, phosphatidyl ethanolamine, and sphingomyelin constitute 90% of the phospholipids (see Fig. 2.7). Lipid synthesis does not occur in the mature red cell, but the lipids of the membrane are in a dynamic equilibrium with those of the plasma, entering and leaving the membrane with surprising ease. Under abnormal circumstances, the membrane may acquire an excess or deficiency of lipid. In these instances, total surface area increases or decreases and cell shape changes accordingly.

The proteins of stroma have several distinctive properties. They contain many free —SH groups, which are essential to the structural and functional integrity of the membrane, and many enzymes (e.g., glycolytic enzymes, ATPase), which are components of the cation pumping mechanism of the membrane. They also include many glycoproteins. Some of these contain a component called sialic acid which is on the outside of the surface and accounts for the highly acidic character of the red cell surface. This means that the cell surface has a strong negative charge. Since red cells carry like surface charges, they repel one another in the blood with considerable force. The stroma continues to be actively investigated, for it is readily obtainable in essentially pure form and what is learned of its chemical composition and fine architecture may be valid for all cell membranes.

Production. Red cells are produced in the red bone marrow. Approximately one-fourth to one-third of the marrow cell population is normally devoted to red cell formation, or erythropoiesis, while the remainder is primarily concerned with white cell formation. The so-called *erythroid* component of marrow consists of cells in various stages of maturation.

The process of maturation has several remarkable features. The most primitive red cell precursor, the erythroblast, is an actively dividing cell that gives rise to the whole maturation sequence (Plate 1). Yet, when it divides into two new erythroblasts, both of them cannot very well mature, for, if they did, the marrow would become devoid of erythroblasts. On the contrary, the number of erythroblasts present in the marrow is constant. This must mean that one of the two erythroblasts resulting from a single division remains an immature stem cell capable of further division and that the other undergoes maturation (Fig. 8.9). Only in this way could there be continuing red cell production without depletion of the precursor cells.

The sequence of maturation is as follows:

erythroblast ⟶ pronormoblast ⟶ normoblast ⟶ reticulocyte ⟶ erythrocyte

During this sequence striking changes occur in the size, architecture, and biochemical characteristics of the cell.

The erythroblast derives from the undifferentiated stem cell. Both the early erythroblast and the *pronormoblast* are large nucleated cells with strongly basophilic cytoplasm. The *normoblast* is the earliest cell with recognizable hemoglobin. It has three developmental stages showing progressive decreases in size and cytoplasmic basophilia and a progressive increase in hemoglobin content: the *basophilic* normoblast; the *polychromatophilic* normoblast; and the *acidophilic* normoblast. As the nature of the cytoplasm alters, the nucleus changes dramatically from a large open structure to a small shrunken remnant and then disappears. The mechanism for its elimination is entirely unknown. It is of interest that the nucleus is not lost from the red cell in fishes, reptiles, birds, and other lower animals.

The *reticulocyte* is the earliest cell without a nucleus. In its small size, high hemoglobin content, and absence of a nucleus, it resembles a mature red cell. However, there are significant differences. The main characteristic of the reticulocyte is a blue-staining reticulum (i.e., network), revealed by staining of a fresh cell with brilliant cresyl blue.* The stained material is thought to be precipitated cytoplasmic RNA. The reticulocyte contains ribosomes (the source of the RNA) and mitochondria, and its metabolism is like that of its precursors. Hemoglobin synthesis continues, and the cell is actively engaged in oxidative phosphorylation. However, the reticulocyte rapidly develops into the adult erythrocyte and, in doing so, loses its distin-

---

* The procedure of applying stain to a wet film of freshly drawn blood is called *supravital staining.*

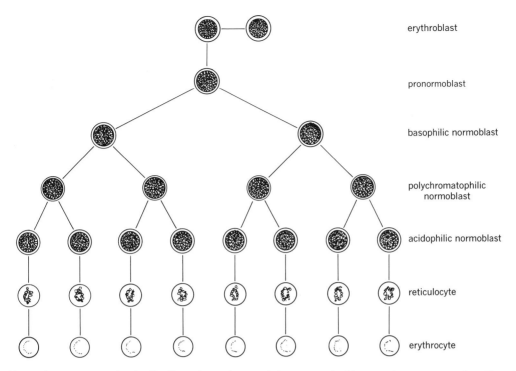

erythroblast

pronormoblast

basophilic normoblast

polychromatophilic normoblast

acidophilic normoblast

reticulocyte

erythrocyte

8.9 Maturation sequence of red cells. Note that only one of the two erythroblasts undergoes maturation; the other remains an erythroblast, capable of dividing again. (Adapted from A. J. Erslev, *Blood, 14,* 386, 1959. Grune & Stratton, Inc., New York.)

guishing features—its RNA, ribosomes, and mitochondria, and its capacity to synthesize protein and to utilize oxygen in metabolism.

Red cell maturation, then, may be regarded as a progressive and systematic loss of biological potentialities. Unlike its precursors, the mature red cell cannot synthesize protein and cannot divide. In a sense, this is the price of specialization. For, despite their many other capabilities, the red cell precursors cannot transport oxygen throughout the body with dispatch and efficiency.

Whether red cell formation takes place inside special capillaries within the marrow or outside these capillaries—i.e., extravascularly—has been debated for years. Electron microscopy suggests that it is extravascular and that, when the cells reach the reticulocyte stage, they are released into the marrow capillaries and thence into the blood stream. There is some evidence of a reticulocyte pool in the marrow, a reservoir of young cells awaiting the signal to enter the blood stream. The reticulocyte stage lasts about 48

hours, and it is possible that a reticulocyte spends half of this time in the pool. Normally 0.5 to 1.5% of the circulating red cells are reticulocytes. When red cell production is stimulated, an early and reliable sign is a sharp increase in this proportion.

Control of production: the erythron. In 1936 W. B. Castle and G. R. Minot introduced the term *erythron* to indicate the total mass of developing and mature red cells in the body—i.e., all the cells in the erythrocyte series, extravascular as well as intravascular. The value of the concept lies in its emphasis on the functional unity of the red cells and their precursors, whether in blood or bone marrow. The erythron remains constant for long periods of time but can change in an orderly way in response to internal and external stimuli. Therefore, mechanisms of regulation must exist.

These mechanisms have been partly clarified. Since the number of red cells in blood is stable, a steady state, in which rates of production and destruction are equal, was early apparent. Two factors are known to accelerate production: an abnormality, such as hemorrhage or anemia, that causes the erythron to shrink; and oxygen deficiency (anoxia) in the tissues. Thus individuals living at high altitudes, where oxygen concentrations are relatively low, have higher than normal red cell counts.* High red cell counts are also usual in individuals suffering from pulmonary disease or heart disease, in which tissue oxygenation is defective. The common denominator in anemic and anoxic situations is insufficient tissue oxygen, for a decrease in the cells' capacity for oxygen transport deprives the tissues of oxygen as effectively as primary anoxia. Since the life span and rate of destruction of the red cells are normal in both instances, it appears that in some way the tissue oxygen concentration influences the rate of red cell production in the marrow.

How is information on the tissue oxygen concentration communicated to the marrow? Scientists long sought some substance that might be released by tissues receiving too little oxygen and carried in the blood to the marrow, where it would stimulate red cell production. This would be a *humoral* agent since it would be delivered by the body fluids.

In general, their experiments consisted of transfusions of plasma from animals made anemic by bleeding or anoxic in some other manner into normal animals. If a soluble agent were present in the plasma of the abnormal animal, it might induce a spurt of red cell production in the normal animal. Such experiments were successful only after much difficulty and debate. Apparently red cell production is controlled by a hormone named *erythropoietin,* whose secretion increases as the level of oxygen in the tissues decreases. Erythropoietin appears to be a glycoprotein. It is synthesized in the kidneys, though not exclusively. Indeed, current evidence suggests that the kidney may secrete an enzyme called renal erythropoietic factor, or

REF, that converts a plasma globulin into erythropoietin. Indications are that erythropoietin acts on undifferentiated stem cells, accelerating their conversion to erythroblasts. Erythropoietin seems to be one of the hormones that stimulate certain genes (see p. 156), thereby promoting the synthesis of mRNA and thus of specific proteins. It is known to induce the formation of RNA in marrow cells, and that this RNA is mRNA is suggested by the ability of actinomycin D to prevent its synthesis following erythropoietin treatment.

Red cell production is also influenced by the thyroid, hypophysis, gonads, and adrenal cortex. Loss of any one of these glands results in anemia. Their mechanisms of action have not yet been thoroughly studied, but it has been shown that testosterone, secreted by the testes (Chapter 16), stimulates red cell production by causing the kidneys to elaborate erythropoietin.

Hemoglobin. Hemoglobin, fabricated in the immature red cells, is a member of a large class of conjugated proteins widely distributed in nature, the *hemoproteins.* The hemoproteins are a class of chromoproteins (see p. 33). They consist of protein molecules linked to one or another *porphin* derivative.

The structural formula of the parent porphin nucleus is shown in Fig. 8.10. It is a large ring made up of four smaller pyrrole rings united by methene ($=CH-$) "bridges." (We earlier encountered the cyclic tetrapyrrole structure in the cytochromes—see Fig. 4.7.) Substitution of various side groups for the hydrogen atoms yields the many porphin derivatives, or porphyrins (see p. 54). Porphyrins, always deeply colored, complex with a metal atom that serves as the active site in a variety of reactions, including the reversible binding of oxygen (in hemoglobin) and the oxidative transfer of electrons (in cytochromes). Hence hemoproteins are sometimes referred to as *respiratory pigments* and *metalloporphyrins.*

In hemoglobin the protein is *globin,* and the porphyrin group is *heme,* whose metal component is *iron.* Hemocyanin, in snails and crustaceans, contains copper rather than iron and is blue rather than red. Chlorophyll is a green hemoprotein whose metal is magnesium. The particular porphyrin found in heme is *protopor-*

---

* Recent studies have shown that Andean natives living at an altitude of 14,900 ft have an average red cell count of 6,150,000, an average hemoglobin level of 20.8 g per 100 ml, and an average hematocrit of 60%. At an altitude of 20,000 ft, the average red cell count is lower, indicating that severe anoxia ultimately interferes with hemoglobin production.

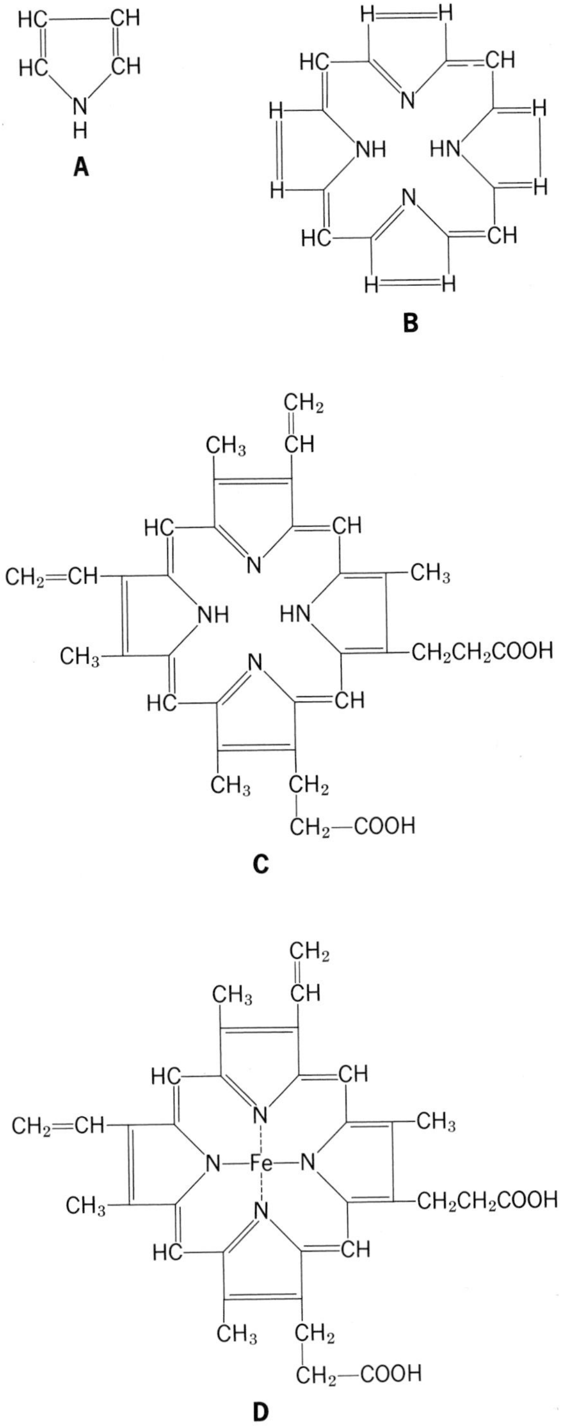

HC————CH
HC   CH
    N
    H
A

B

C

D

8.10 Structures of pyrrole and its derivatives: A, pyrrole; B, porphin; C, protoporphyrin; D, heme.

phyrin IX (see Fig. 8.10C), which has a distinctive arrangement of methyl, vinyl, and propionyl substituents. Protoporphyrin IX joins an atom of iron in the ferrous state (Fe^{++}) to form heme (see Fig. 8.10D). Heme then attaches to globin to form hemoglobin.

The hemoglobin molecule is an ellipsoidal structure of molecular weight about 67,000 (see Fig. 2.11). It contains four polypeptide chains and four heme groups, each of the latter being enfolded into a pocket in one of the polypeptide chains on the surface of the molecule. Normal adult hemoglobin has two identical alpha (α) chains, each containing 141 amino acid residues, and two identical beta (β) chains, each containing 146 amino acid residues.

Studies of hemoglobin in the 1950's led to two of the most important discoveries of modern biology. One will be mentioned presently. The other was the elucidation of the molecular structure of hemoglobin by M. F. Perutz and his associates, which paralleled the elucidation of the molecular structure of myoglobin by J. C. Kendrew (see Fig. 2.11). Using new techniques of x-ray crystallography, Perutz and his co-workers, after years of investigation, established the three-dimensional configuration of horse hemoglobin (which closely resembles human hemoglobin). Myoglobin is, in a sense, a quarter molecule of hemoglobin, being about one-fourth its size. It consists of a single polypeptide chain of 153 amino acid residues and a single heme group.

The iron atom in the center of each wafer-shaped heme group serves as the point of attachment of protoporphyrin to globin and of oxygen to hemoglobin. Hemoglobin actually binds oxygen reversibly—that is, it takes up oxygen when the surrounding partial pressure of oxygen is high and releases it when the surrounding partial pressure of oxygen is low. This behavior is apparent from the oxygen saturation curves in Fig. 8.11. It should be emphasized that the binding of oxygen involves no change in the valence of iron, which remains +2. Therefore, we must distinguish between the *oxygenation* of hemoglobin (which produces *oxyhemoglobin*) and its *oxidation*, wherein Fe^{++} is converted to Fe^{+++} to produce *methemoglobin*, a substance incapable of oxygen binding. Nevertheless, the oxygen-free product left when oxyhemoglobin releases

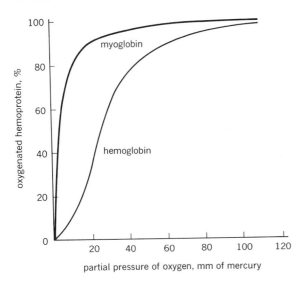

8.11 Oxygen saturation curves of hemoglobin and myoglobin.

its oxygen is sometimes called *reduced hemoglobin*—a misnomer, since to the chemist "reduced" means that electrons have been added. A better term for oxyhemoglobin minus its oxygen is *deoxyhemoglobin*.

When one, two, or three of the four heme groups have been oxygenated, the remaining groups are more easily oxygenated. Thus there are *heme-heme interactions,* implicit in the sigmoid shape of the oxygen saturation curve of hemoglobin. This phenomenon has been known for many years; indeed, the resemblance between the oxygen saturation curve of hemoglobin and the substrate saturation curve of an allosteric enzyme (see Fig. 4.21) contributed to the realization that interactions between hemes and those between catalytic and regulatory sites in an allosteric enzyme (see pp. 51 and 152) are alike in principle. Both depend upon conformational changes of polypeptide chains. Perutz discovered that oxygenation causes a striking displacement of the β chains in hemoglobin, decreasing the distance between their iron atoms from 40.3 to 33.4 Å. We have learned that an allosteric enzyme deprived of its regulatory site no longer has a sigmoid substrate saturation curve but instead has a classic hyperbolic one (see Fig. 4.21). Similarly, a compound with a single polypeptide chain and heme group (i.e., myoglobin), which cannot experience heme-heme interactions, has a simple hyperbolic oxy-

gen saturation curve (see Fig. 8.11). We shall discuss later the extraordinary physiological advantages of the sigmoid oxygen saturation curve of hemoglobin.

It is noteworthy that *free* heme cannot bind oxygen reversibly. Heme can perform this function only when it is combined with globin. Whereas free heme is relatively insoluble in water, hemoglobin is highly soluble. Clearly, high solubility is essential to its efficiency as an oxygen carrier. Nor can any of the several other hemoproteins containing heme—among them cytochromes, catalase, and peroxidase—bind oxygen reversibly. Although their heme is presumably right, their protein is wrong. Thus, if heme is to participate in oxygen transport, it needs globin; other proteins give it quite different properties.

Hemoglobin also plays a role in carbon dioxide transport, which we shall examine in Chapter 11.

Data on the normal distribution and composition of hemoglobin are given in Table 8.3. Despite the frequent necessity of assaying blood hemoglobin, none of the routine methods determines hemoglobin as such—chiefly because blood hemoglobin is a rapidly fluctuating mixture of oxyhemoglobin and deoxyhemoglobin. Since these two forms have different colors, the first step in the easiest accurate assay techniques, those employing spectrophotometry, is the conversion of hemoglobin to a chemical derivative of stable color. Some of the common assay methods are as follows: (1) determination of hemoglobin as *cyanmethemoglobin,* formed on the addition of an oxidant, which converts hemoglobin to methemoglobin, and cyanide, which converts methemoglobin to the desired product; (2) determination of hemoglobin as *hematin,* formed on the addition of acid or alkali to blood; (3) determination of *total blood iron,* more than 92% of which is present in the heme of hemoglobin; and (4) determination of blood *oxygen-binding capacity,** since 1 g of hemoglobin combines with exactly 1.34 ml of oxygen.

Methemoglobin and the hemoglobin spectrum. Iron-porphyrin compounds can be divided into three classes: (1) those in which iron is always reduced (Fe^{++}), such as hemoglobin; (2)

---

* This is thought to be the most precise of all the methods.

TABLE 8.3 QUANTITATIVE DATA ON NORMAL HUMAN HEMOGLOBIN*

Concentration in whole blood	
Newborn infant	19.5 g per 100 ml
1-yr-old child	11.8 g per 100 ml
Adult male	15.0 g per 100 ml
Adult female	14.0 g per 100 ml
Concentration in red cells	
Packed	33% wet weight; 95% dry weight; 0.34 g per 100 ml
Single ($90 \mu^3$)	$29 \mu\mu$g; 3.4×10^8 molecules
Iron content	3.34 mg per gram, or 50.1 mg per 15 g (i.e., per 100 ml of blood)
Oxygen-binding capacity	1.34 ml per gram, or 20.1 ml per 15 g (i.e., per 100 ml of blood)
Oxygen saturation in vivo	
Arterial blood	95%
Venous blood	70%

* Values given are averages; values in normal individuals deviate from the averages no more than 10%.

those in which iron is reversibly oxidized and reduced, such as the cytochromes; and (3) those in which iron is always oxidized (Fe^{+++}), such as catalase and peroxidase.* All three classes have the same basic iron-porphyrin structure, with the fundamental biochemical reactions in which they function centering about the iron atom. Nevertheless, the properties of the classes are radically different. Only the Fe^{++} compounds are capable of oxygen transport. Let us consider the interesting consequences of this fact.

The ferrous iron of hemoglobin, like that of any inorganic ferrous compound, is converted to ferric iron (Fe^{+++}) by an oxidizing agent. The product is methemoglobin, which is incapable of transporting oxygen. When methemoglobin is exposed to a reducing agent, ferrous hemoglobin is regenerated, and the capacity for oxygen transport is restored.

If an aqueous solution of pure hemoglobin is allowed to stand exposed to the air at room tem-

perature, almost all of the hemoglobin is rapidly converted to methemoglobin. In this case, the oxidizing agent is the oxygen of the air, and, because of its abundance and the absence of countervailing reducing agents, the oxidative reaction proceeds to completion. However, if a solution of pure hemoglobin in plasma is allowed to stand exposed to the air at room temperature, only about 50% of the hemoglobin is converted to methemoglobin. We must conclude, therefore, that something exists in plasma that opposes the complete oxidation of hemoglobin. Presumably this is one or more of the reducing agents normally found in plasma.

If hemoglobin still within the red cells is exposed to the air at room temperature, only 1 or 2% of it is oxidized. From this we must infer that reducing agents exist in red cells that are even more potent than the agent or agents in plasma. The reducing agents in red cells are the enzymes (or enzyme systems) *methemoglobin reductase* and *"diaphorase."*† Both can recon-

---

* Catalase and peroxidase are iron-porphyrin enzymes that catalyze the conversion of hydrogen peroxide (H_2O_2) to H_2O and O. The peroxidase reaction requires the presence of an oxidizable substrate (AH_2), thus:

$$AH_2 + H_2O_2 \longrightarrow A + 2H_2O \qquad (8.1)$$

Catalase can catalyze the same reaction, but under certain conditions a second molecule of H_2O_2 serves as the oxidizable substrate.

$$H_2O_2 + H_2O_2 \longrightarrow O_2 + 2H_2O \qquad (8.2)$$

† Diaphorases are properly defined as FAD-linked enzymes that catalyze the transfer of hydrogen from $NADH_2$ to a redox dye, such as 2,6-dichlorophenol indophenol or methylene blue. They occur in liver and other tissues, though their function is obscure. Although it has not been established that red cells contain such enzymes, the term diaphorase has been indiscriminately used to refer to a red cell enzyme that transfers hydrogen from $NADH_2$ to methemoglobin. To minimize confusion, we shall call the red cell enzyme "diaphorase" (in quotation marks), after the suggestion of E. R. Jaffé.

vert methemoglobin to hemoglobin and so are of the greatest biological significance. The observation that intact red cells contain reducing agents was first made in 1909 by Otto Warburg, who found that red cells reduce methemoglobin and that the reduction is dependent on the availability of glucose.

A second observation of historical importance was made late in the 1920's by E. S. Guzman Barron, who found that red cells—which, unlike most other cells, do not consume oxygen (though they do transport it)—utilize large quantities of oxygen in the presence of methylene blue. This phenomenon is also dependent on the availability of glucose.

We learned in Chapter 4 that the first step in glucose metabolism is the conversion of glucose to glucose 6-phosphate. As shown in Fig. 4.13, glucose 6-phosphate has at least two major pathways open to it. One is the Embden-Meyerhof glycolytic pathway, in which it is converted to two triose phosphates, which are then oxidized and eventually converted to pyruvate (see Fig. 4.3). In the oxidation of each triose phosphate, a molecule of NAD is converted to $NADH_2$, which acts as hydrogen donor in various reductions. The alternate glycolytic pathway for glucose 6-phosphate is the pentose phosphate shunt, in which NADP is reduced to $NADPH_2$. We now know that methemoglobin reductase catalyzes the rapid transfer of hydrogen from $NADPH_2$ to methemoglobin, but apparently only under artificial circumstances—that is, upon the addition of methylene blue. Methylene blue is a hydrogen acceptor when oxidized (MB) and a hydrogen donor when reduced (MBH_2). When MB is added to red cells, it promotes the reconversion of $NADPH_2$ to NADP by accepting hydrogens, thereby being itself converted to MBH_2. When methemoglobin is present, MBH_2 reduces it to hemoglobin. When methemoglobin is not present, MBH_2 transfers its hydrogens to the oxygen of the air.

$$MBH_2 + \tfrac{1}{2}O_2 \longrightarrow MB + H_2O \quad (8.3)$$

Thus, in the presence of methylene blue, O_2 can become the final hydrogen acceptor, whereas in its absence (for reasons that we shall not explore) O_2 is not reduced. The reduction of O_2 to H_2O is synonymous with O_2 *utilization*, in contrast to O_2 *transport*.

What, then, reduces methemoglobin to hemoglobin under normal conditions—that is, when methylene blue is not added? The answer is the "diaphorase" of red cells. This enzyme specifically catalyzes the transfer of the hydrogens of $NADH_2$ generated in triose phosphate oxidation to methemoglobin. A dye is not needed as an intermediary. Since $NADH_2$ is produced continuously in the glycolytic pathway of glucose metabolism (hence the glucose requirement in the experiments of Warburg), its reaction with "diaphorase" provides a mechanism whereby the tendency toward methemoglobin production can be continuously opposed. Since "diaphorase" is absent from plasma, this mechanism would explain the superiority of red cells over plasma in the prevention of hemoglobin oxidation.

Methemoglobin exceeds the normal level of 1 or 2% (that is, methemoglobinemia exists) in three situations. The first, which is very rare, occurs in individuals with a genetically determined defect in the methemoglobin-reducing system. It is the $NADH_2$-linked "diaphorase" (not the $NADPH_2$-linked methemoglobin reductase) that is deficient or absent in most of these cases. This suggests that "diaphorase" is physiologically the more important reducing agent in red cells. Lack of "diaphorase" accounts for the cardinal feature of the disorder: the oxidized state of a large proportion of the total hemoglobin—typically, about 40% of it is in the methemoglobin form. The fact that not all of the hemoglobin is oxidized suggests that the secondary reducing mechanisms are still operating, even though the major one is defective.

The second situation occurs in individuals with one of the genetically determined abnormal hemoglobin molecules—known collectively as hemoglobin M—which are characterized by an unusual sensitivity to oxidation. In the hemoglobin M's, of which five varieties are known, an amino acid substitution occurs in the α or β polypeptide chain at or near the locus of heme attachment. The abnormality favors oxidation of the iron atom and thus prevents its reversible oxygenation. Hemoglobin M is found only in heterozygotes; homozygosity would be incompatible with life.

The third occurs when normal red cells are exposed to environmental oxidizing agents (such

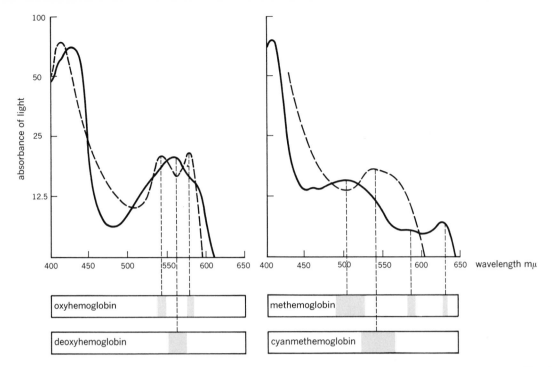

8.12 Absorption spectra of hemoglobin and its derivatives showing curves obtained with a spectrophotometer and band patterns obtained with a hand spectroscope. Dark bands correspond to absorption peaks.

as nitrites, aniline dyes, or any of a long list of drugs) in concentrations that overburden the normal reducing mechanisms. This type of methemoglobinemia is obviously acquired, not inherited. It is common, for example, in farm children who drink well water contaminated by nitrites from fertilizer. Individuals with a genetically determined deficiency of the enzyme glucose-6-phosphate dehydrogenase develop some methemoglobinemia when their abnormal red cells are exposed to environmental oxidizing agents in low concentrations, which would not disturb normal red cells.

The various forms of hemoglobin are identified by their visible absorption spectra (Fig. 8.12). The spectrum of deoxyhemoglobin has an absorption maximum (peak) at a wavelength of 555 mμ; that of oxyhemoglobin has two peaks, one at 540 mμ and the other at 577 mμ, showing why the oxygenation of deoxyhemoglobin alters its color. Oxidation to methemoglobin changes it further; the major methemoglobin peaks are at 500 and 630 mμ. Spectral curves are obtained with a spectrophotometer. A simple hand spec-

troscope yields similar information, producing dark bands in the zones of maximum absorption. The band identifying methemoglobin is the 630 mμ band. An abnormal compound, sulfhemoglobin, has an absorption band in the vicinity of 618 mμ and must be carefully distinguished from methemoglobin.*

When the blood has been deprived of oxygen, it contains an increased percentage of deoxyhemoglobin, which produces a bluish skin coloration called *cyanosis*. Normally a certain fraction of the total hemoglobin is in deoxygenated form. According to Table 8.3, this fraction averages 5% in arterial blood and 30% in venous blood. The size of the fraction varies with the partial pressure of oxygen in the environment, as is clearly shown in Fig. 8.11. The *percent* of reduced hemoglobin at a given oxygen saturation can be read from the curve; the actual *amount* of deoxyhemoglobin is calculated by multiplying the percent by the total hemoglobin concentration. If arterial blood is 95% saturated with oxygen and its hemoglobin content is 15 g per 100 ml,

---

* On the addition of cyanide to methemoglobin, cyanmethemoglobin forms, and the 630 mμ band is eliminated. Sulfhemoglobin is not affected by cyanide.

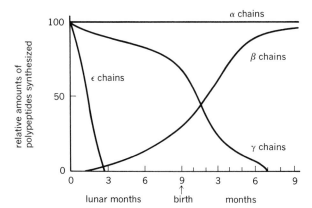

8.13 Hemoglobin polypeptide production before and after birth.

the amount of unsaturated or deoxyhemoglobin is $0.05 \times 15 = 0.75$ g per 100 ml. In contrast, the amount of deoxyhemoglobin in venous blood is

$$0.30 \times 15 = 4.5 \text{ g per 100 ml.}$$

When incoming arterial blood is insufficiently oxygenated—either because of a low atmospheric oxygen level or defective function of the heart and lungs—the normal oxygen uptake in the tissues augments the degree of unsaturation. Cyanosis occurs when the concentration of deoxyhemoglobin in the blood exceeds 5 g per 100 ml.

Embryonic and fetal hemoglobin. In early embryonic life, the embryo contains two types of distinctive hemoglobins with the awkward names *hemoglobin Gower 1* and *hemoglobin Gower 2*. These are replaced by the end of the third month of embryonic life by a new variety of hemoglobin, *fetal hemoglobin.* Fetal hemoglobin differs strikingly from the hemoglobin of adult blood. For example, although adult hemoglobin is rapidly denatured in alkaline solution, fetal hemoglobin is alkali-resistant. This difference is the basis of a useful assay for fetal hemoglobin. The important functional difference between fetal hemoglobin and adult hemoglobin is the greater affinity of fetal hemoglobin

for oxygen—though this is more difficult to demonstrate in man than in other mammals.*

Fetal hemoglobin is present in high concentrations (75 to 95%) in blood at birth but gradually disappears during the first year of life. However, it may persist into adulthood, so that 20 to 30% of the total hemoglobin is fetal. While this might appear to be a compensatory response to oxygen deficiency, it does not occur in individuals living at high altitudes. Current thought postulates a mutant "high fetal hemoglobin" gene.

Biochemical research has shown that the embryonic hemoglobin Gower 2 and fetal hemoglobin both contain α chains, but instead of β chains, each contains a unique polypeptide chain—ϵ in embryonic hemoglobin Gower 2 and γ in fetal hemoglobin. Thus, if normal adult hemoglobin is formulated $\alpha_2\beta_2$, fetal hemoglobin would be $\alpha_2\gamma_2$ and Gower 2 $\alpha_2\epsilon_2$. Gower 1 turns out to be ϵ_4. As shown in Fig. 8.13, α chain synthesis occurs from the beginning of embryonic life, ϵ chain synthesis ceases early, and γ chain synthesis stops during the first year of life, being replaced by β chain synthesis.

Abnormal hemoglobins. In 1910 James Herrick, while examining a Negro boy with an obscure ailment, discovered that many of the boy's red cells had bizarre crescent or sickle shapes (Fig. 8.14). Physicians later found many such cases of *sickle cell anemia.* It proved to be

8.14 Red cells in sickle cell anemia.

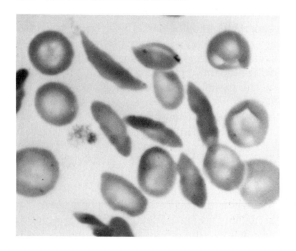

---

* Many factors promote a tendency toward oxygen deficiency in the embryo, among them being the several temporary openings between the arterial and venous blood streams. This situation undoubtedly explains the high hemoglobin concentration in the blood of a newborn infant (see Table 8.3) and may explain the existence of the special fetal hemoglobin, with superior oxygen-binding properties.

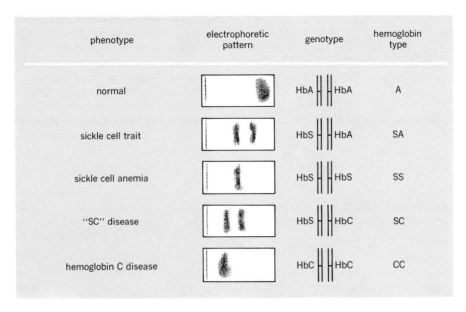

phenotype	electrophoretic pattern	genotype	hemoglobin type
normal		HbA ‖ ‖ HbA	A
sickle cell trait		HbS ‖ ‖ HbA	SA
sickle cell anemia		HbS ‖ ‖ HbS	SS
"SC" disease		HbS ‖ ‖ HbC	SC
hemoglobin C disease		HbC ‖ ‖ HbC	CC

8.15 Electrophoretic patterns and genetics of normal and abnormal hemoglobins.

a hereditary disorder, to which Negroes are particularly susceptible, associated with an abnormal hemoglobin molecule. This was named *hemoglobin S* to distinguish it from normal adult hemoglobin, or *hemoglobin A*, and fetal hemoglobin, or *hemoglobin F* (sometimes abbreviated Hb S, Hb A, and Hb F, respectively).*

The molecular abnormality produced by the mutant gene was obviously a subtle one, for, when the oxygen supply is adequate, red cell shape is normal but, when the partial pressure of oxygen is reduced, the sickle shape appears dramatically. Sickle cell anemia aroused much interest, for it involved a genetic trait manifested in the peculiar behavior of a specific molecule. It was thus the clearest illustration to date of the one gene–one enzyme theory of gene action (see Chapter 3) and the prototype of the so-called *molecular diseases*—indeed, this term was first used in referring to sickle cell anemia.

Investigators set about trying to determine the abnormality in hemoglobin S, and in 1949 Linus Pauling and Harvey Itano discovered that

hemoglobin S differs from hemoglobin A in its electrophoretic mobility, moving more slowly because it has two less electrical charges. They showed that in severe sickle cell anemia nearly all the hemoglobin is of the slow-moving variety, whereas in milder anemias the proportions are roughly half hemoglobin S and half hemoglobin A (Fig. 8.15). James Neel then demonstrated that the hemoglobin S molecule is transmitted as a simple Mendelian trait, with homozygotes (SS) exhibiting severe sickle cell anemia and heterozygotes (SA) a milder disorder called *sickle cell trait.*†

When electrophoresis became a widely used laboratory tool, many other abnormal hemoglobins were observed in various parts of the world. These were designated by the letters of the alphabet in the order of discovery until the let-

---

* About 2% of the hemoglobin in the normal adult is a variant called hemoglobin A_2. It, too, contains a unique polypeptide chain in place of the β chain—the δ chain; thus hemoglobin A_2 is $\alpha_2 \delta_2$.

---

† Whether the hemoglobin S trait is described as dominant, codominant, or recessive depends on one's point of view. If the phenotype is defined by the presence of sickling in the blood, then the hemoglobin S trait is dominant, since both homozygote and heterozygote have this property. If the phenotype is defined by the presence of abnormal hemoglobin in the blood, then the hemoglobin S and hemoglobin A traits are codominant, since each produces its characteristic product. Finally, if the phenotype is defined by the presence of sickle cell anemia in the blood, then the hemoglobin S trait is recessive, since only the homozygote has this disease.

ter Q was reached. With the end of the alphabet in sight, they received other designations: for example, *hemoglobin Barts* (Hb Barts) is named after the hospital in which it was first detected, St. Bartholomew's of London;* and we identify *hemoglobin Zürich*, *hemoglobins* M_{Boston},

---
* Hemoglobin Barts is formulated γ_4^F.

$M_{Saskatoon}$, $M_{Milwaukee}$, and so on. All are different in certain respects. They vary in mobility on paper or starch, in solubility, or in other characteristics, and many are associated with no recognizable clinical abnormalities.

The *charge* difference between hemoglobin A and hemoglobin S established that the *chemical* difference between them is small. However, a

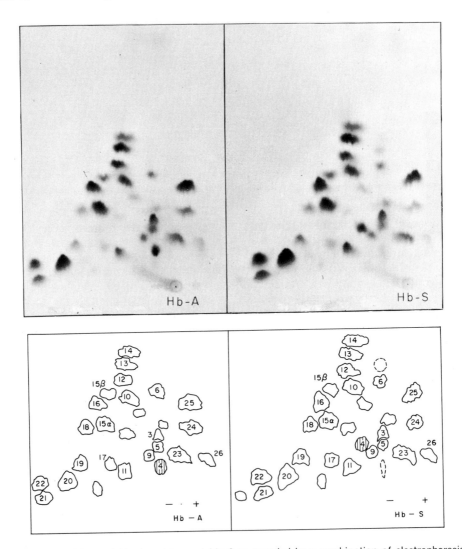

8.16 "Fingerprints" of hemoglobin A and hemoglobin S as revealed by a combination of electrophoresis and chromatography. Shading in the lower diagrams indicates differences. The hemoglobin sample was placed in the lower right corner of the papers. Then electrophoresis was conducted in the horizontal direction, followed by ascending paper chromatography (at right angles to the direction of electrophoresis). The peptides were made visible when the paper was sprayed with ninhydrin. (From V. M. Ingram, *The Hemoglobins in Genetics and Evolution*, Columbia University Press, 1963.)

Plate 1: Morphology and maturation sequences of the "formed elements" of blood and bone marrow. Cells have been stained with Wright's stain, except for the reticulocytes which have been stained with brilliant cresyl blue. The erythroblast and several other of the maturation stages mentioned in the text have been omitted. The morphology of stem cells (or undifferentiated reticulum cells) has not been established with certainty. The cells depicted in the center of the figure may or may not be accurate representations of stem cell morphology. Asterisk (*) denotes cells termed hemocytoblasts and hemohistioblasts by some authorities. (From "Maturation of Human Blood Cells: A Photographic Presentation of Hematopoiesis," by John J. Butler, M.D., Lynn C. Wall, and H. L. Gibson. Published by Radiography Markets Division, Eastman Kodak Company.)

TE SERIES

THROMBOCYTE SERIES

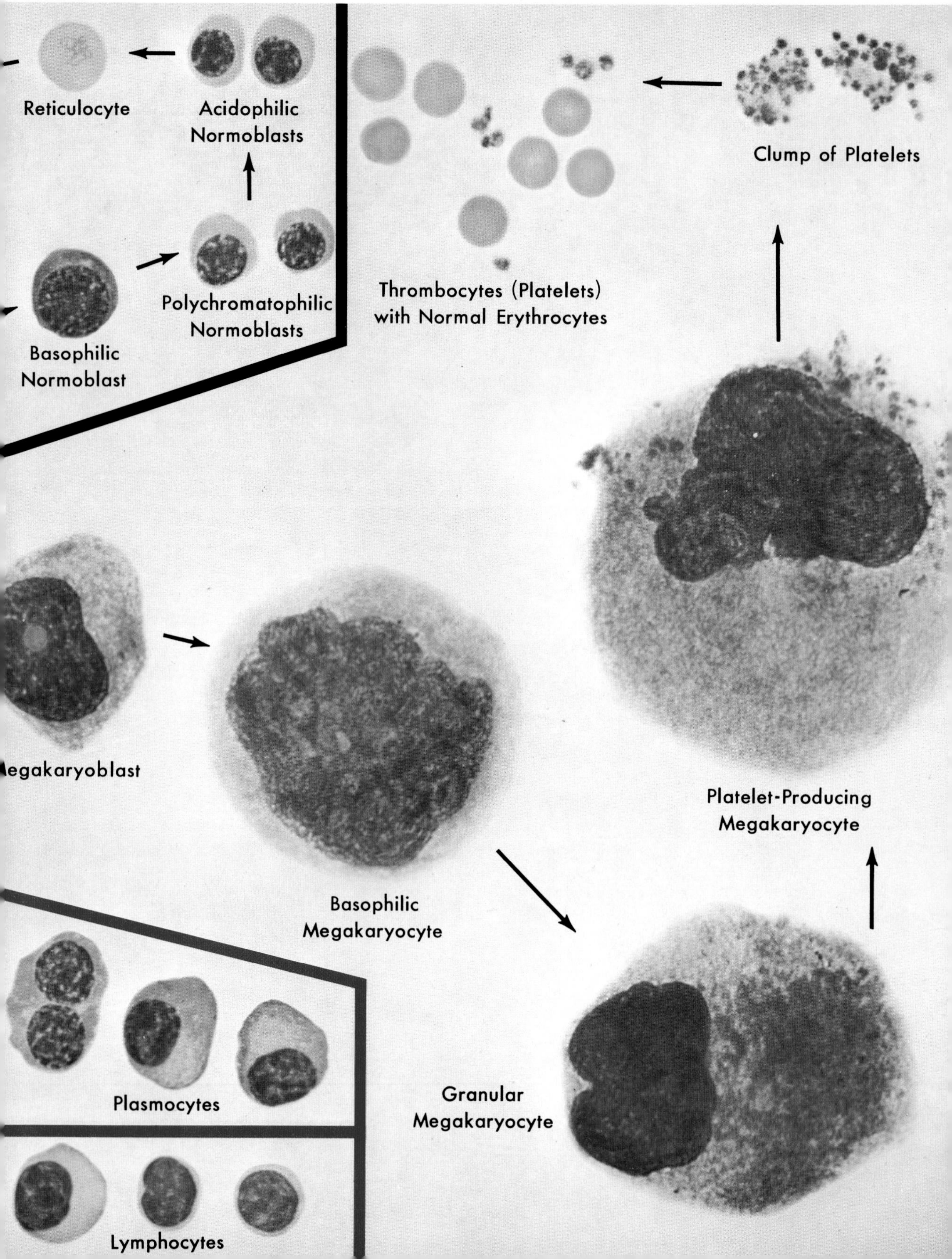
Reticulocyte
Acidophilic Normoblasts
Polychromatophilic Normoblasts
Basophilic Normoblast
Thrombocytes (Platelets) with Normal Erythrocytes
Clump of Platelets
Platelet-Producing Megakaryocyte
egakaryoblast
Basophilic Megakaryocyte
Granular Megakaryocyte
Plasmocytes
Lymphocytes

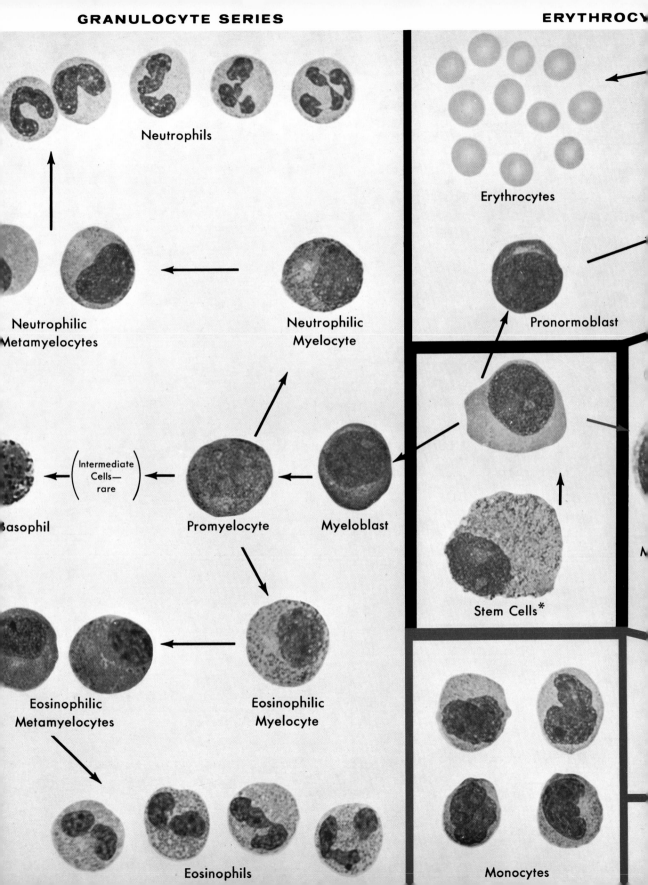

Neutrophils

Erythrocytes

Neutrophilic
Metamyelocytes

Neutrophilic
Myelocyte

Pronormoblast

$\left(\begin{array}{c}\text{Intermediate}\\\text{Cells—}\\\text{rare}\end{array}\right)$

Basophil

Promyelocyte

Myeloblast

Stem Cells*

Eosinophilic
Metamyelocytes

Eosinophilic
Myelocyte

Eosinophils

Monocytes

$$\text{hemoglobin A} \quad NH_2—\overset{+}{Val}—\overset{+}{His}—Leu—Thr—Pro—\overset{-}{Glu}—\overset{-}{Glu}—\overset{+-}{Lys}—\uparrow\ldots.$$

$$\text{hemoglobin S} \quad NH_2—\overset{+}{Val}—\overset{+}{His}—Leu—Thr—Pro—Val—\overset{-}{Glu}—\overset{+-}{Lys}—\uparrow\ldots.$$

$$\text{hemoglobin C} \quad NH_2—\overset{+}{Val}—\overset{+}{His}—Leu—Thr—Pro—\overset{+-}{Lys}\uparrow\ \overset{+-}{Glu}—\overset{+-}{Lys}—\uparrow\ldots.$$

8.17 Amino acid sequences of the βI peptides of hemoglobins A, S, and C. The charges on each residue are shown (see Table 2.5). Since valine is the N-terminal amino acid, its free —NH_2 (or —NH_3^+) group carries a positive charge. The arrows indicate where trypsin has cleaved the peptides. Since trypsin hydrolyzes a peptide bond to the right of a lysine residue, and since the abnormal amino acid in hemoglobin C is lysine, a hexapeptide and a dipeptide are obtained instead of an octapeptide.

whole hemoglobin molecule contains 574 amino acid residues, and the best methods of amino acid analysis were accurate only to within 3%. Hence single amino acid substitutions would not be apparent from analyses of whole molecules. The problem was ingeniously solved by Vernon Ingram in 1957. Each of the hemoglobin molecule's two identical halves yields 26 short peptides on digestion with trypsin. By a simple method that has come to be known as fingerprinting, Ingram found that 25 of these peptides are the same in hemoglobin A and hemoglobin S (Fig. 8.16). The difference in the remaining one, which contains only 8 amino acid residues, could be detected by electrophoresis: the mobility of the abnormal peptide, like that of its parent protein, is abnormal. Ingram thus localized the fundamental defect of the hemoglobin S molecule within a short amino acid sequence and correspondingly simplified the technique of identifying an abnormal amino acid.

Ingram then determined that seven of the eight amino acids are identical in the normal peptide and its abnormal counterpart. The other —the sixth—is glutamic acid in hemoglobin A peptide and valine in hemoglobin S peptide (Fig. 8.17).* Since a glutamic acid molecule carries one negative charge, the net effect of the substitution of valine for glutamic acid in hemoglobin is the deletion of two negative charges, one from each half of the hemoglobin molecule, precisely as predicted from electrophoretic behavior.†

Hemoglobin C (Hb C, formulated $\alpha_2^A\beta_2^C$ or $\alpha_2^A\beta_2^{6\text{Glu}\to\text{Lys}}$), another abnormal hemoglobin, has an electrophoretic mobility approximately half that of hemoglobin S. We surmise, then, that its net electric charge is more positive than that of hemoglobin S. In fact, its abnormality is a substitution of lysine for glutamic acid at the same locus as the hemoglobin S abnormality (see Fig. 8.17). Instead of being sickle-shaped, a cell carrying hemoglobin C is abnormally thin, with a central dot of hemoglobin, its appearance having suggested the name *target cell* (Fig. 8.18).

_____

† The genetic significance of this observation—that a single mutant gene causes a single amino acid substitution in a protein molecule—was discussed in Chapter 3.

8.18 Red cells in target cell anemia (thalassemia). Red cells carrying Hb C have a similar appearance. (From Miale, *Laboratory Medicine—Hematology*, 2nd Ed., The C. V. Mosby Company, St. Louis, 1962.)

_____

* We have noted that hemoglobin A contains two identical α chains and two identical β chains. To show that these are normal chains, the formulation may be written $\alpha_2^A\beta_2^A$. When an abnormality occurs in the β chain, as in hemoglobin S, the formulation is written $\alpha_2^A\beta_2^S$, to indicate that the β chain is abnormal. As it happens, the abnormal peptide is the first (peptide βI) of the β chain, at the N-terminus, and so the sixth amino acid in the peptide is the sixth amino acid in the β chain. The most accurate formulation for hemoglobin S is $\alpha_2^A\beta_2^{6\text{Glu}\to\text{Val}}$. This system of identification facilitates the formulation of the many other abnormal hemoglobins whose amino acid substitutions have since been worked out.

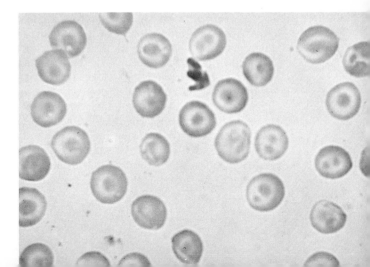

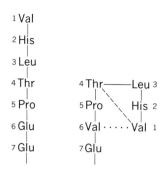

8.19 Theory explaining why replacement of glutamic acid by valine in position 6 of β chain leads to deformity of molecule and ultimately to tactoid formation and sickling. Hydrogen bond is denoted by dashed line; nonpolar interaction by dotted line (compare with Fig. 2.10). (Adapted from M. Murayama, *Nature*, 202, 258, 1964.)

Unlike a sickle cell, which is specific for hemoglobin S, a target cell may have many causes.

After years of uncertainty as to why sickling occurs only in cells containing hemoglobin S—or, indeed, why it occurs at all—it was found in 1950 by John Harris that, whereas oxyhemoglobin S has the same solubility as oxyhemoglobin A, deoxyhemoglobin S formed when the partial pressure of oxygen decreases precipitates rapidly, in both tissues and test tube. It aggregates into long *tactoids*, which distort the red cell into the sickle shape. An individual with sickle cell anemia is sick because distorted cells block the capillaries and are destroyed in massive numbers by the spleen, so that a severe hemolytic anemia results. Why the replacement of glutamic acid by valine causes the abnormal solubility behavior of deoxyhemoglobin S was given a plausible explanation by M. Murayama who discovered in 1964 that the normal N-terminal valine residue interacts by nonpolar attraction (see Fig. 2.10) with the abnormal valine residue in the 6 position. The resulting loop at the end of the β chain (Fig. 8.19), stabilized by hydrogen bonds, fits like a key into a crevice on the surface of the deoxygenated α chain. As a result, the hemoglobin molecule becomes distorted, forming the longitudinal aggregates called tactoids that deform the red cell.

It should be noted that the amino acid substitutions determined so far are in perfect harmony with the genetic code (see Table 3.1).

One of the triplet code words for glutamic acid (GAG) differs from one for valine (GUG) and one for lysine (AAG) by only a single nucleotide. Thus the mutation responsible for the replacement of hemoglobin A by hemoglobin S or hemoglobin C may be tentatively visualized as having altered only a single nucleotide in the DNA sequence comprising the hemoglobin A gene.

Over 100 abnormal hemoglobins have been discovered. Some have abnormalities in the α chain—e.g., *hemoglobin I* (Hb I, formulated $\alpha_2^I \beta_2^A$ or $\alpha_2^{16\text{Lys}\rightarrow\text{Asp}}\beta_2^A$)—and others in the β chain—e.g., *hemoglobin E* (Hb E, formulated $\alpha_2^A \beta_2^{26\text{Glu}\rightarrow\text{Lys}}$). Genetic studies have disclosed numerous gene combinations in afflicted families. The *SC* combination is illustrated in Fig. 8.15.

Nutritional and biochemical features of production. Biochemically speaking, the production of red cells has two major aspects, one concerned with the synthesis of hemoglobin and the other with cell division and maturation *per se*. The latter processes have nutritional and biochemical features that, in general, resemble those in other cells and upon which we shall not dwell here. However, certain features of red cell division and maturation merit attention. These relate to the particular roles of the vitamins.

Like almost all dividing cells, the red cell precursors require amino acids, minerals (among them iron, copper, and cobalt), and vitamins (including vitamin B_{12}, folic acid, riboflavin, nicotinic acid, pyridoxine, ascorbic acid, pantothenic acid, thiamine, choline, and biotin). Perhaps because red cells are produced so rapidly, they are one of the first of the body's cell lines to reflect deficiencies of particular vitamins. Each deficiency leads to anemia. It was through studies of deficiency anemias that some of the vitamins of human nutrition were discovered.

Vitamin B_{12} was found as a direct result of the pioneering work of George Minot, William Murphy, and William B. Castle in the late 1920's on human pernicious anemia, a previously fatal disease of unknown cause and mechanism. Minot and Murphy learned that remarkable and sustained improvement could be achieved by the feeding of large quantities of liver. Castle then postulated that red cell formation is arrested in pernicious anemia owing to a deficiency of an essential *extrinsic factor* in food whose absorption in the intestines is dependent upon prior binding to an *intrinsic factor* normally secreted in the stomach. The massive efforts of over 20 years to identify the extrinsic factor in liver culminated in 1948 with the isolation of vitamin B_{12}. Unusual because of its potency—1 ton of liver yielded only 250 mg of vitamin B_{12}, and pernicious anemia responds to only a few micrograms per day—and because of its curious porphyrin-like molecular structure (see Table 4.1), the vitamin was unequivocally demonstrated to be an active principle essential to red cell production.

Pernicious anemia develops when the mucous membrane of the stomach loses the ability to synthesize intrinsic factor, perhaps owing to a genetic defect. Because of the loss, vitamin B_{12} is usually administered by injection rather than by mouth. Only very large oral doses are effective—such as those given in massive liver feeding. The course of events following vitamin B_{12} injection is shown in Fig. 8.20A. When vitamin B_{12} is administered orally together with a source of intrinsic factor, a prompt response is also obtained (Fig. 8.20B). Chemically, intrinsic factor appears to be a low–molecular weight glycoprotein.

The role of vitamin B_{12} in cell division was not revealed until 1964. In vitamin B_{12}–requiring bacteria, a coenzyme form of the vitamin is essential for the action of ribonucleotide reductase, the enzyme that converts ribonucleotides to deoxyribonucleotides, the building blocks of DNA (see Chapter 4).* It is assumed, though proof is still lacking, that vitamin B_{12} coenzyme plays a similar role in human and other animal cells. If it does, vitamin B_{12} deficiency would block DNA synthesis in the red cell precursors,

---

* It also participates in the biosynthesis of methionine and the conversion of methylmalonyl CoA to succinyl CoA in the human body.

8.20 Treatment of pernicious anemia with vitamin B_{12}: A, administration by injection; B, oral administration with and without intrinsic factor. Solid lines represent reticulocytes, and dots total red cell counts.

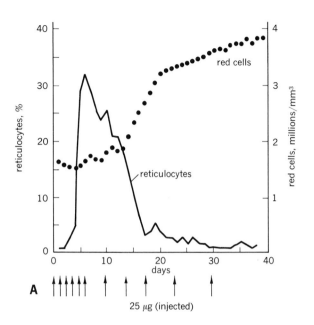

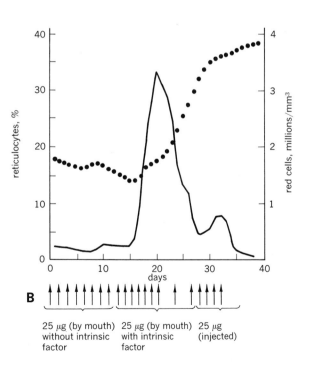

and therefore lead to abnormal cell division, or unbalanced growth. The cells become unusually large, with deeply basophilic cytoplasm. They have been given the name *megaloblasts*. The red cells that emerge from these abnormal precursors are themselves enlarged and are thus called *macrocytes*.

Folic acid is a vitamin essential for the synthesis of both nucleic acids (RNA and DNA) and of certain amino acids. Discovered a few years before vitamin B_{12}, it was thought for a brief time to be the long-sought extrinsic factor of liver, since treatment with it temporarily alleviates pernicious anemia. Folic acid is required for normal red cell production, and folic acid deficiency leads to megaloblastic bone marrow changes that are similar to those of vitamin B_{12} deficiency. In DNA synthesis folic acid participates in the methylation of deoxyuridylate to thymidylate (see Chapter 4.)

We turn now to a unique biochemical aspect of red cell production, hemoglobin synthesis. It involves four processes: (1) the synthesis of protoporphyrin III; (2) the insertion of iron in protoporphyrin III to form heme; (3) the synthesis of globin; and (4) the attachment of heme to globin.

Protoporphyrin III synthesis consists in the stepwise enzymatic conversion of simple precursor compounds, specifically succinic acid and glycine, to porphyrin. Mitochondria participate in the reactions in two ways: (1) the citric acid cycle, a mitochondrial enzyme system, produces succinyl CoA from succinic acid (see Fig. 4.4); and (2) δ-aminolevulinic acid (ALA) synthetase, which exists only in mitochondria, catalyzes the condensation of succinyl CoA with glycine of form ALA (Fig. 8.21). ALA dehydrase then catalyzes the combination of two molecules of ALA to form porphobilinogen, a monopyrrole. Four porphobilinogen molecules join to form porphyrin, which eventually becomes protoporphyrin III.

The incorporation of iron into protoporphyrin III by heme synthetase is the last step in heme synthesis. This, too, occurs in mitochondria.

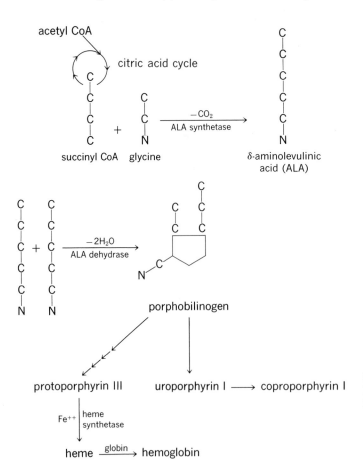

8.21 Schematic diagram of pathway of hemoglobin synthesis. Only carbon skeletons are shown.

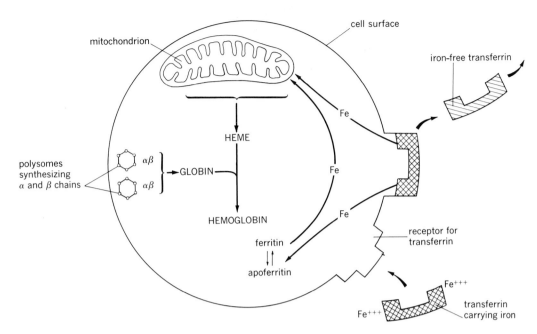

8.22 Diagram of the incorporation of iron and the synthesis of heme and globin in the immature red cell. The nucleus is not shown. A molecule of iron-laden transferrin is approaching a receptor on the cell surface; another molecule is adhering to a receptor and transferring its iron to the cell; a third molecule, free of iron, which it has transferred to the cell, is returning to the plasma. (Adapted from I. M. London, *The Biosynthesis of Hemoglobin and Its Control in Relation to Some Hypochromic Anemia in Man*, Series Hematologica, Vol. 2, p. 1, 1965.)

Iron metabolism will be discussed in Chapter 14. Here we shall merely note that iron is delivered to the bone marrow by an iron-carrying plasma protein called *transferrin*. Transferrin is preferentially bound to immature red cells (Fig. 8.22). Though it gives up its iron to them, it does not itself combine with them. The entry of iron into the cells is by micropinocytosis (see Chapter 4), involving invaginations only a few hundred Angstrom units in diameter. Within the cells most of the iron reacts with a protein called *apoferritin* to form small granules of *ferritin*, an Fe^{+++}-protein complex that is easily demonstrated by staining. The granules appear to be iron storage depots on which the cells draw as heme synthesis proceeds. At the completion of maturation, the iron has gone, leaving apoferritin. Cells containing ferritin are called *sideroblasts*. The absence of sideroblasts from bone marrow is a sign of iron deficiency.

Globin synthesis is a particular instance of protein synthesis—indeed, its study contributed significantly to present knowledge of protein synthesis in general. Globin molecules are assembled on the ribosomes, whose RNA gives the erythroblast cytoplasm its basophilic character. The α and β chains, produced independently, join to form α-β dimers. Two of these then join to form a complete tetramer ($\alpha_2\beta_2$). The precise stage at which heme is bound to globin has not been established. It appears to be added before chain synthesis is complete.

Although the syntheses of heme and globin are dissimilar processes taking place in different cell regions,[*] they are sensitively coordinated by some unknown mechanism. Heme synthesis is controlled by the feedback inhibition of ALA synthetase by heme. There is evidence that heme also stimulates globin synthesis, as does iron. These effects may provide a basis for a system capable of regulating the two channels of the assembly line with great precision.

_____

[*] The participation of mitochondria and ribosomes in hemoglobin synthesis makes it obvious why mature red cells, having lost both, synthesize no hemoglobin.

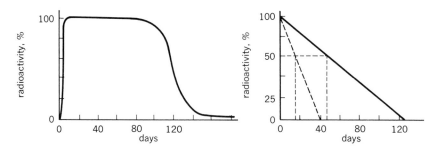

8.23 Curves showing red cell survival times, as measured by two methods: A, a single-age population of cells labeled in vivo with glycine-^{14}C: B, a mixed-age population of cells labeled in vitro with ^{51}Cr and reinjected before (dashed line) and after (solid line) splenectomy for hypersplenism.

Life span. If the blood volume is 5 liters, and the red cell count is 5,000,000 per cubic millimeter, the blood contains 25×10^{12} red cells. The life span of a normal red cell is 120 days. Therefore, $\frac{1}{120}$ the total red cell mass is produced each day, and an equal fraction is destroyed—a daily turnover of 2×10^{11} cells, whose packed volume is approximately 20 ml. It is no coincidence, then, that the total number of circulating reticulocytes is 2×10^{11}, for these are the visible traces of each day's new red cells.

The red cell life span can be directly determined by labeling the cells with a distinctive permanent tag whose rate of disappearance from the blood can be measured; if the cells are made radioactive, it can be presumed that they are extant as long as blood samples contain measurable radioactivity. Two methods are used: (1) labeling all of the new cells formed during a short time interval (and only those cells) and measuring the rate of disappearance from the blood of the radioactive single-age population (sometimes called a *cohort*); and (2) labeling all of the cells in the red cell population (or a representative sample) at a given time and measuring the rate of disappearance from the blood of the radioactive mixed-age population.

The first method is easily carried out after administration of a single dose of radioactive glycine, a heme precursor. All of the red cells developing for several hours will contain radioactive hemoglobin, and they will be "born" together and die together. This procedure yields the survival curve shown in Fig. 8.23A. The long flat portion of the curve attests the stability of the label (once inside the cell, hemoglobin stays until the cell is destroyed). It can be seen that the cells of the cohort begin to disappear from the blood at 105 days and are all gone at 140 days, the mean survival time being 120 days.

The second method is technically simpler. Red cells are removed from the body and mixed with sodium chromate containing a radioactive isotope of chromium, ^{51}Cr. This forms a firm bond with hemoglobin that remains largely intact for the life of the cells. The cells are then reinjected, and the survival time is calculated on the basis of repeated blood radioactivity determinations.* Since the labeled population includes all ages, a constant proportion ($\frac{1}{120}$) of the original number becomes senescent each day. When the percentages of remaining radioactivity are plotted on logarithmic graph paper, a straight line is obtained (Fig. 8.23B). For convenience this is usually used to estimate the half-life—i.e., the time required for 50% of the ^{51}Cr to disappear from the blood—rather than the total life span. Theoretically the half-life should be 60 days, but a number of factors (such as leakage of ^{51}Cr from its binding sites on hemoglobin) cause the observed half-life to approximate 30 to 32 days.

---

*In the older, classic method, red cells from a donor of a slightly different but compatible blood group are transfused into a recipient, and differential blood-grouping tests instead of radioactivity determinations are used to measure the rate of disappearance of the transfused cells. This procedure is called the *Ashby differential agglutination technique.*

Red cell life span is demonstrably shortened in the hemolytic anemias. Hemolysis is often associated with enlargement of the spleen, the locus of normal red cell destruction. Results of the ^{51}Cr test in individuals with this variety of hemolytic anemia are shown in Fig. 8.23B. The ^{51}Cr is doubly useful here, since careful scanning of the body surface reveals an accumulation of radioactivity in the splenic area. This abnormal red cell sequestration is the prelude to excessive red cell destruction.

A life span of 120 days is remarkably long for a cell that has been estimated to travel 200 miles before its demise, on a journey that subjects it to much physical wear and tear, swelling and shrinking, compression and decompression. We are not sure why the cell eventually breaks down—physical wear and tear seem to make little difference—but it somehow becomes mechanically fragile and chemically defective. The aged cell is recognized by the reticuloendothelial cells, and then the final act begins.

Destruction. The senescent red cell is destroyed by the reticuloendothelial system, many of whose cells are in the spleen. Hence the

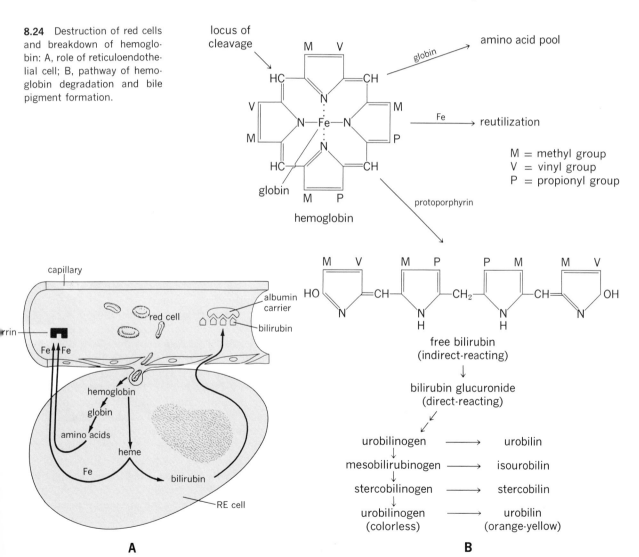

8.24 Destruction of red cells and breakdown of hemoglobin: A, role of reticuloendothelial cell; B, pathway of hemoglobin degradation and bile pigment formation.

M = methyl group
V = vinyl group
P = propionyl group

spleen may be considered the graveyard of the red cells. Reticuloendothelial cells phagocytize the red cells (Fig. 8.24A) and break them down within their cytoplasm.

When the red cell is destroyed, its hemoglobin is chemically degraded. Globin is broken down to its component amino acids, which rejoin the body's general amino acid pool. Iron is completely conserved, delivered from the reticuloendothelial cell to plasma transferrin, and reutilized in the synthesis of new hemoglobin. The fate of the remainder of the heme molecule is of particular interest.

The bulk of the protoporphyrin is chemically split into an open-chain pigmented tetrapyrrole compound called *bilirubin* (Fig. 8.24B). This is transported from the site of hemoglobin degradation by the plasma, in which it is bound to albumin. It is normally present in the plasma in concentrations of less than 1 mg per 100 ml. The bound bilirubin is insoluble in water and does not react in chemical tests with Ehrlich's diazo reagent until it is dissolved in alcohol; it is therefore termed *indirect-reacting* bilirubin. When it passes through the liver, it becomes attached to two molecules of glucuronic acid to form bilirubin diglucuronide, a water-soluble compound that reacts *directly* with Ehrlich's reagent. This is excreted from the liver in the bile. Upon reaching the intestine, it is converted into a mixture of pigments commonly referred to collectively as *urobilinogen*. These mechanisms are discussed in detail in Chapter 13.

Metabolism. Despite its relatively simple structure, the red cell contains complex metabolic machinery. Moreover, like other cells, it contains some enzymes, whose physiological functions are not yet known. For example, although no detectable breakdown of protein occurs in the mature red cell, several active proteolytic enzymes reside there.

Much of importance has been learned of red cell metabolism in recent years. Perhaps because red cells are readily available, many investigators have employed them as models for the study of metabolism in general. An instance of such research concerns the effects of certain nerve gases whose toxicity results from their inhibition of an enzyme of nerve cells, cholinesterase. This enzyme is also present, though of unknown physiological function, in red cells, where the enzyme-inhibitor interaction is more conveniently examined than in nerve cells.

We have already noted, in Chapter 4, that the red cell derives most of its energy from glycolysis, despite the availability of oxygen. As shown in Figs. 4.2 and 4.3, glycolysis converts one molecule of glucose to two molecules of lactic acid with the liberation of energy. In the red cell there is almost no subsequent oxidation of lactic acid to carbon dioxide and water in the citric acid cycle. Hence there is almost no oxygen utilization. Since oxidative metabolism extracts far more energy than glycolytic metabolism from a molecule of glucose, the processes of red cell energy metabolism must be deemed inefficient. Perhaps the red cell does not require a large fund of energy. However, it does require some energy, and this must be supplied by glycolysis.

Four main red cell functions depend upon the generation of energy.

(1) Energy in the form of ATP is necessary to initiate glycolysis. (2) Energy is necessary to maintain concentration gradients across the cell surface. A potassium concentration 25 times as great in the cell as in plasma and a sodium concentration lower in the cell than in plasma do not arise passively. They can be established only if the tendency of the ions to distribute themselves uniformly is overcome (see Chapter 7). These gradients are upset whenever glycolysis is disturbed. (3) Energy is necessary for the active transport of materials through the cell surface. The renewal of particular cell components continues throughout the life of the cell—although hemoglobin and probably stroma are not replenished. (4) Energy is necessary to produce $NADH_2$ or $NADPH_2$, which keeps hemoglobin in the reduced form. Without glycolysis it is gradually oxidized to methemoglobin.

Another facet of red cell metabolism was mentioned in Chapter 3: a new antimalarial drug, primaquine, was found to cause hemolysis in individuals whose red cells lack a single enzyme owing to a genetic defect. The enzyme, glucose 6-phosphate dehydrogenase, catalyzes the conversion of glucose 6-phosphate to 6-phosphogluconate (see Fig. 4.13). When the enzyme is absent, primaquine triggers a train of events that first oxidizes hemoglobin to methemoglobin and then denatures it. The damaged cell is rapidly destroyed in the spleen. Primaquine is an oxidizing agent, and many other oxidants have been shown to act in the same manner. Remarkably enough, the abnormal red cells behave quite normally in the absence of oxidants (which, in ordinary concentrations, are innocuous to normal red cells).

One more aspect of red cell metabolism attracts much current attention. Although red cells survive 120 days in the blood stream, blood bank blood—that is, refrigerated whole blood stored in anticoagulant solutions—does not last more than 3 weeks. Red cell glycolysis somehow becomes defective, and hemolysis follows. Because of the obvious need for preserving blood as long as possible, this phenomenon and its prevention have been extensively studied. Despite a number of useful maneuvers that have been devised, efforts to avoid the "storage lesion" have had limited success. The investigation continues, but the recent discovery of methods for preserving blood by freezing may ultimately circumvent the problem.

Diseases. Abnormalities of the red cells may be quantitative or qualitative. There may be too few red cells—*anemia*—or too many—*erythrocythemia*. Or red cells may be biochemically or structurally abnormal, as in sickle cell anemia or primaquine hemolysis; ordinarily, defective red cells are destroyed more rapidly than normal ones.

Anemia is a manifestation of disease, not a disease itself. Broadly speaking, it has two causes: inadequate red cell production and accelerated destruction or loss. The opposite is true of erythrocythemia. The common specific causes of anemia are outlined in Table 8.4.

LEUKOCYTES

Physical properties. White cells, or leukocytes, can be distinguished from red cells by the fact that they contain nuclei. Stained films create the illusion that white cells are spherical. Actually they are ameboid in shape; for, like amebae, they have the power of movement, a property that has earned them the name *wandering cells*. Migration through the walls of capillaries and surrounding tissues is called *diapedesis*. It is normal but under pathological conditions is greatly stimulated.

The ability to move is the second great difference between white cells and red cells. Red cells remain in the peripheral blood throughout their life span. In view of their role in gas transport, they would be useless otherwise. White cells, in contrast, appear in the peripheral blood not because their main function is there but because the blood is a convenient carrier for them. Their essential services are performed elsewhere. Unlike red cells, white cells are useless until they escape from the blood stream.

Classification: the differential count. There are several methods of classifying white cells, based on staining reactions, nuclear forms, and sites of origin. A common classification divides them into two major types: *nongranulocytes* and *granulocytes*. Nongranulocytes are further subdivided into *lymphocytes* and *monocytes*. Granulocytes are further subdivided into *neutrophils*, *eosinophils* (or *acidophils*), and *basophils*, depending on the staining properties of their cytoplasmic granules, which are specifically tinted with Wright's stain.* Cells whose granules take the acidic red stain, eosin, are eosinophils or acidophils; those whose granules take the basic blue stain are basophils; and those whose granules take a neutral mixture of both are neutrophils. All derive from myeloblasts.

Perhaps the simplest classification separates white cells into *mononuclear* and *polymorphonuclear* groups. A mononuclear cell has a single rounded or slightly indented nucleus that constitutes most of the cell volume. Lymphocytes and monocytes are mononuclear. A polymorphonuclear cell has a literally polymorphous nucleus, which assumes curious shapes with many filaments and lobules (see Plate 1). Granulocytes are polymorphonuclear.

Finally, lymphocytes and monocytes are produced in reticular tissues of the lymph nodes and spleen (and accordingly are referred to as *lymphoreticular* cells), and granulocytes are produced in the bone marrow (and accordingly are referred to as *myeloid* cells).

---

* Wright's stain is a popular blood stain prepared from eosin and methylene blue (see Chapter 3).

TABLE 8.4 CAUSES OF COMMON ANEMIAS

I. Inadequate red cell production
   A. Nutritional deficiencies
      1. Iron deficiency (small red cells, low MCH)
         a. Low intake, e.g., starvation
         b. Poor absorption in intestine, e.g., chronic diarrhea
         c. Poor utilization in tissues, e.g., lead poisoning
         d. Increased requirements, e.g., pregnancy
         e. Excessive loss, e.g., chronic bleeding
      2. Deficiency of hematopoietic vitamins (megaloblasts, macrocytes)
         a. Vitamin B_{12} deficiency
           (1) Low intake, e.g., meat-free diet
           (2) Lack of intrinsic factor, e.g., pernicious anemia
           (3) Vitamin B_{12}–consuming parasite in intestine, e.g., fish tapeworm
         b. Folic acid deficiency
           (1) Low intake, e.g., starvation
           (2) Increased requirement, e.g., pregnancy
           (3) Poor absorption in intestine, e.g., chronic diarrhea
      3. Miscellaneous deficiencies (pyridoxine, copper, riboflavin, etc.)
   B. Abnormal bone marrow function
      1. Injury by poisons, radiation, etc.
      2. Displacement by tumor cells, etc.
      3. Failure of unknown cause

II. Accelerated red cell destruction (hemolytic anemia)
   A. Intrinsic defectiveness of red cells
      1. Sickle cell anemia and other hemoglobinopathies
      2. Glucose-6-phosphate dehydrogenase deficiency and related defects
      3. Other inborn defects such as spherocytosis and elliptocytosis
   B. Increased activity of red cell–destroying mechanisms
      1. Acquired immunohemolytic anemia
   C. Increased mechanical buffeting of red cells
      1. Abnormal heart valve
      2. Narrowing of small blood vessels

The approximate percentage distribution of the cell types in normal adult blood is as follows:

Polymorphonuclear cells, %	
Granulocytes	60–70
Neutrophils	55–65
Eosinophils	2–3
Basophils	1–2
Mononuclear cells, %	
Lymphocytes	20–30
Monocytes	4–7

Children under 4 years of age normally have 50 to 60% lymphocytes and 25 to 40% granulocytes.

The determination of the percentage distribution of the cell types in an individual blood sample is called the *differential count*. Ordinarily it consists in careful classification of 100 randomly selected cells in a stained film.

Granulocytes: neutrophils. Neutrophils, the most numerous of the normal circulating white cells, show the following characteristics when treated with Wright's stain: diameter, 10 to 12 μ;

cytoplasm, delicate pink; granules, bluish lavender; nucleus, several deeply basophilic lobules connected by filament—that is, the nucleus is *segmented.*

The maturation of a neutrophil involves these distinctive stages (see Plate 1):

myeloblast ⟶ promyelocyte ⟶
    myelocyte ⟶ metamyelocyte ⟶
        band form ⟶ neutrophil

The *myeloblast,* with a large round nucleus containing fine chromatin strands and a prominent nucleolus, develops into the *promyelocyte,* with no nucleolus and cytoplasm displaying some poorly defined granules. The *myelocyte* stage is the first to have cytoplasmic granules with recognizable staining preferences. Thus we see neutrophilic, eosinophilic, and basophilic myelocytes. The *metamyelocyte* has a nucleus that is indented slightly into a kidney shape. In the *band form* the nucleus is horseshoe-shaped or elongated with parallel borders; the indentation that began in the metamyelocyte has grown deeper. Maturation is completed with further shrinkage and, finally, segmentation of the nucleus. The coarse chromatin clumps in the mature neutrophil stain more deeply than the lacy chromatin network in the myeloblast.

Granulocytes: eosinophils. Eosinophils are characterized by their red-staining cytoplasmic granules. These granules are larger than neutrophilic granules, so that, with experience, one can identify eosinophils by the granules without staining. Eosinophils undergo a maturation sequence similar to that of neutrophils (see Plate 1).

Despite extensive study, the function of eosinophils is poorly understood. *Eosinophilia,* an increase in the number of eosinophils in the blood, often accompanies immune or allergic reactions, but the significance of this phenomenon is obscure. The number of eosinophils in the blood is also influenced by certain hormones, as we shall see in Chapter 15.

Granulocytes: basophils. Basophils are easily recognized by their coarse, dark blue–staining cytoplasmic granules. Their maturation also parallels that of neutrophils (see Plate 1).

Though few in number, basophils are interesting cells whose physiology has only recently been explored. They are believed to be members of the mast cell family. We learned in Chapter 5 that tissue mast cells contain the anticoagulant heparin; histamine, which, when isolated and administered as a drug, powerfully dilates the small blood vessels; and hyaluronic acid, a high–molecular weight polysaccharide. It is clear that basophils synthesize these three substances; that two of them (heparin and hyaluronic acid) are strong acids may account for the basophilic nature of the cytoplasmic granules.

Lymphocytes. We earlier encountered lymphocytes in the interspaces of the lymphoid organs. Their cytoplasm contains no specific-staining granules (see Plate 1), and their maturation sequence is considerably simpler than that of granulocytes.

lymphoblast ⟶ prolymphocyte ⟶
                        lymphocyte

The *lymphoblast* closely resembles the myeloblast. Maturation consists in a progressive intensification in the staining of the chromatin and a coarsening of its texture, with both cell and nucleus becoming smaller. The *prolymphocyte* is merely the cell halfway between the lymphoblast and the lymphocyte.

We have already noted that lymphocyte maturation is a function of the lymph nodes, spleen, and scattered lymphoid tissues of the body, some of which find their way into bone marrow. About 5% of the cells in bone marrow smears are lymphocytes, and occasional lymphatic nodules are also seen in histological sections of marrow.

Investigations of the role of lymphocytes in antibody synthesis provide the basis for intriguing new theories of immunology. We shall review these in Chapter 19.

Monocytes. Monocytes are probably the circulating counterparts of fixed tissue macrophages (see Chapter 5). A monocyte is larger than a neutrophil and typically has an indented or kidney-shaped nucleus (see Plate 1). Its cytoplasm does not contain specific-staining granules. The monocyte maturation sequence is monoblast ⟶ promonocyte ⟶ monocyte

Mature monocytes are phagocytic and act to defend the body against infection.

Production and distribution. Some granulocytes are found outside the blood and bone marrow. It was once believed that many such cells are retained within certain organs for variable periods of time, that a given granulocyte probably enters and leaves the peripheral blood several times during its life, and that it is therefore more difficult experimentally to determine the life span and fate of white cells than of red cells. Investigators now agree that there is no appreciable pool of granulocytes outside the blood and bone marrow in normal subjects and that granulocytes which do enter the tissues do not return to the blood in significant numbers. Hence, the normal cell cycle represents a dynamic balance among formation of cells in the marrow, release of cells into the blood, circulation of cells throughout the body, the one-way egress from blood to tissues of cells chosen randomly (i.e., not on the basis of age), and peripheral destruction. The sum of these forces acting together maintains the granulocyte count normally between 4,000 and 8,000 per cubic millimeter.

Life span measurements with tagging techniques similar in principle to those used in red cell life span determinations have employed three labeled substances: radioactive phosphate (^{32}P) and radioactive thymidine (^{3}H-TdR), both of which are incorporated into granulocyte DNA, and radioactive diisopropylfluorophosphate (DF^{32}P), which binds irreversibly to a serine residue of a granulocyte protein. When administered to an intact subject, all three label cohorts of granulocyte precursors in vivo, though ^{32}P and ^{3}H-TdR best label the early DNA–forming granulocyte precursors (the myeloblasts, promyelocytes, and myelocytes), and DF^{32}P best labels mature cells and their more mature precursors (myelocytes, metamyelocytes, and band forms). When added to blood temporarily removed from the body, one of the three, DF^{32}P, labels in vitro all the granulocytes present, a mixed-age population.

Studies employing DF^{32}P as an in vitro granulocyte label revealed two interesting facts. Immediately on reinfusion of labeled cells, half of them "disappear" from the blood—that is, they cannot be found in a freshly withdrawn blood sample. This observation led to the discovery that the total number of granulocytes in the blood (the *total blood granulocyte pool,* or *TBGP*) consists of two compartments: the *marginal granulocyte pool,* or *MGP,* and the *circulating granulocyte pool,* or *CGP.* The MGP consists of cells which adhere to the endothelial linings of small capillaries and venules (i.e., they "marginate.") When blood is removed from the axial stream of a large vein, only the CGP is sampled. The two pools contain an equal number of cells and there is a continuous and rapid exchange of cells between them.

In vitro labeling with DF^{32}P also disclosed that reinjected cells are removed from the blood in random fashion and that their half-life in the circulation is about 7 hours (Fig. 8.25A). It may be calculated, then, that the CGP turns over 2.3 times per day and the average time that a granulocyte spends in the blood is 10.4 hours.

When granulocyte precursors were labeled in vivo, information was obtained on the rates of granulocyte production in the marrow, the identification of various pools of granulocyte precursors in the marrow, and the transit times of individual cells through the stages or pools of the maturation sequence. Collectively, this body of information is referred to as *granulocyte kinetics* or *leukokinetics.* When granulocyte DNA is labeled in vivo by ^{32}P or ^{3}H-TdR over a brief time interval, the amount of radioactivity in the peripheral blood in the ensuing days is shown in Fig. 8.25B. Essentially no radioactivity appears in the blood for 4 or 5 days after ^{32}P administration; this is presumably the time needed for complete myelocyte maturation. There then occurs a sharp upward surge of blood radioactivity indicating the release of labeled cells from bone marrow. Radioactivity begins to diminish in between 10 and 14 days. Comparable results are obtained when the later granulocyte precursors are labeled in vivo with DF^{32}P. The time-course of the changing specific radioactivity of blood granulocytes (Fig. 8.25C) shows that the most mature labeled marrow precursor cells rapidly enter the blood and disappear with a half-life of 7 hours. A plateau then follows which reflects the slow, steady maturation and release of labeled myelocytes, metamyelocytes, and band forms. A

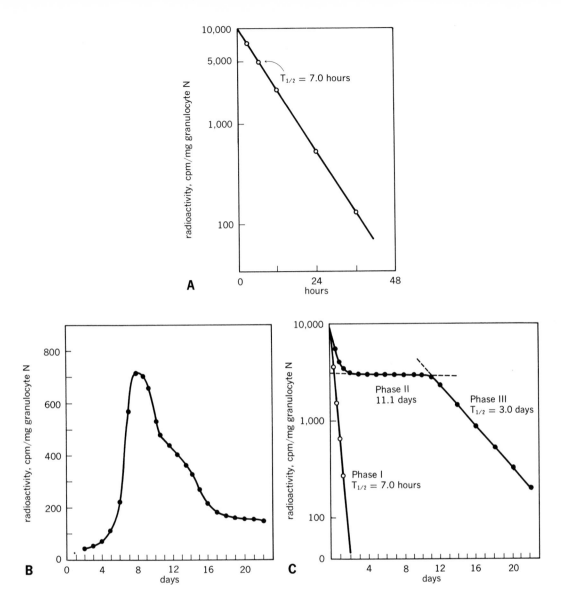

8.25 Time-course of blood granulocyte radioactivity after labeling by different methods. A, cells labeled in vitro with DF^{32}P and returned to circulation; B, cells labeled in vivo with ^{32}P; C, cells labeled in vivo with DF^{32}P. (A and C adapted from G. Cartwright *et al.*, *Blood, 24,* 780, 1964; B adapted from C. Craddock *et al.*, *Blood, 15,* 840, 1960, Grune & Stratton Inc., New York.)

third phase is then reached, in which an exponential decline in specific radioactivity reflects dilution of the label as a result of mitosis within the myelocyte pool (i.e., the continuous formation of new and unlabeled myelocytes). Such studies further indicate that of the total number of granulocytes and their precursors in the bone marrow—about 1860×10^7 per kg body weight —400×10^7 per kg are myelocytes and 1450×10^7 per kg are metamyelocytes or band forms. A maturing cell remains a myelocyte for 2.9 days and a metamyelocyte or band form for 8.5 days. As we have noted, the mature released granulocyte survives 10.4 hours. These figures agree satisfactorily with the conclusions of DNA-labeling studies, which suggest that the life span of the granulocyte precursor is 10 to 14 days.

In the normal adult, the TBGP contains about 70×10^7 cells per kg; the MGP and CGP each contain about 35×10^7 cells per kg. Cells in the MPG and CGP continuously exchange places, but certain factors (e.g., violent exercise, epinephrine) cause the cells of the MGP to move en masse into the CGP, thereby doubling the granulocyte count in a removed blood sample. Infection has a similar effect; it also causes the bone marrow to accelerate the rate of granulocyte production.

Although the life history of granulocytes is complicated, it is agreed that they are end cells without the capacity to divide. They are removed from the circulation and disintegrated by the lungs, intestines, liver, and spleen. The removal and breakdown mechanisms are poorly understood.

As we have learned (see p. 226), uncertainty also surrounds the production sites, turnover rates (i.e., rates of addition to and removal from the blood), and distribution patterns of lymphocytes. Since enough lymphocytes enter the blood via the thoracic and right lymphatic ducts alone to replace those circulating several times a day, the turnover is quite rapid, or the total body pool is enormous compared to the circulating pool, or both. Studies of retention of radioactive labels by the DNA of normal human blood lymphocytes have revealed two populations: one with an age of 2 to 3 days (20% of the total) and one with an age of 100 to 300 days (80% of the total).* These figures indicate a slow rate of lymphocyte production and a considerable recirculation of lymphocytes through the body.

There is much evidence that the morphologically defined lymphocyte does, in fact, comprise a large family of cells with different life spans, genealogies, and functions. We shall consider the significance of this evidence in Chapter 19.

Chemical composition and metabolism. A mature granulocyte plays a major role in *inflammation*, the body's normal response to injury, whose three components are localized increase in blood flow, passage of fluid from the capillaries into the tissues, and rapid localized accumulation of white cells. The function of a granulocyte in inflammation depends upon (1) its ameboid motility, which enables it to pass through capillary walls at the inflammation site, and (2) its power of phagocytosis. Both properties derive from the cell's special biochemical and enzymatic machinery.

Like red cells, granulocytes have an active glycolytic metabolism. However, they also have oxygen-utilizing metabolic systems, though quantitatively meager ones. The ability to develop energy without much oxygen may be an adaptive device allowing granulocytes to survive and do battle in tissues far from oxygen-supplying red cells.

Granulocytes contain an interesting enzyme called *alkaline phosphatase.* Although its metabolic function is unknown, its levels are dramatically altered in some diseases, providing physicians with a useful diagnostic tool and a provocative clue (so far unexploited) to the basic nature of the diseases.

In addition, granulocytes have certain constituents that sometimes seem to place them in a category of their own. The heparin, histamine, and hyaluronic acid content of basophils was mentioned earlier. Neutrophils contain massive quantities of a curious iron-porphyrin enzyme called *verdoperoxidase.* As its name implies, it is green (it accounts for the greenish tint of certain pus collections). Verdoperoxidase is a form of peroxidase. As noted on p. 235, peroxidase cleaves hydrogen peroxide (H_2O_2) into water and oxygen. However, the physiological significance of verdoperoxidase is unknown. No metabolic process of the neutrophil is known to produce peroxide in any quantity, although certain ingested bacteria are thought to release peroxide within the neutrophil. Even so, another iron-porphyrin enzyme, catalase, is present to eliminate it. Despite these uncertainties, the fact that verdoperoxidase is lacking from lymphocytes permits a "peroxidase test" for distinguishing primitive cells of the granulocyte series from those of the lymphocyte series.

Defense functions. Although many microscopists had previously observed that granulocytes

---

* Some results suggest that the life span of small lymphocytes may extend to 10 years.

literally devour solid objects such as red cell fragments and carbon particles, it was the Russian biologist Élie Metchnikoff who established the importance of phagocytosis in body defenses (see Chapter 4). In his first experiment in 1882, he placed a rose thorn beneath the skin of a starfish larva in an area possessing no blood vessels or nerve fibers. Immediately the thorn became surrounded by mobile cells. Metchnikoff's proposal that phagocytosis is a prime defense mechanism had a cool reception from those who believed in the supremacy of humoral defense mechanisms, i.e., antibodies. And to this day, debate continues over the relative significance of humoral and cellular defenses.

Phagocytic defense encompasses three phases: chemotaxis, phagocytosis *per se*, and digestion. Chemotaxis means directional movement of a cell or organism in response to a chemical substance in the environment. The ordinary ameboid locomotion of white cells is undirected, whereas their locomotion in chemotaxis is directed. When a plasma suspension of granulocytes is placed next to a clump of bacteria on a warm microscope stage, the granulocytes move at once toward the microbes, exhibiting *positive* chemotaxis. When certain substances (e.g., kaolin) are substituted for bacteria, the granulocytes move away, exhibiting *negative* chemotaxis. Neutrophils illustrate chemotaxis most strikingly. Eosinophils and monocytes react similarly, but lymphocytes do not, at least under laboratory conditions.

Chemotaxis is due to the release of a *chemotactic factor* as a result of the attack on foreign particles by antibody and serum proteins belonging to the *complement system* (see Chapter 19). The actual attractive forces are not known. A gradient of some kind exists, but it is not necessarily due to diffusion from the particles. For example, insoluble particles cause chemotaxis, and repeated washing does not diminish their effects. Perhaps particles remove material from the surrounding plasma to set up a concentration gradient with it. Alternatively, perhaps particles generate an electric field. At present, we are sure only that white cells "know" where trouble is beginning and, with unerring aim, go there.

Neutrophils actively ingest bacteria and other foreign particles in the body and in a test tube.

The most active phagocytosis is shown by young forms with two nuclear lobes. Their activity depends in part on the nature of the foreign particles. Some bacteria, for example, are protected against phagocytosis by their heavy polysaccharide capsules. Conversely, some plasma substances strongly stimulate phagocytosis. These include the agents called *opsonins*, which are apparently of immunological origin. These, too, are thought to arise from the action of antibodies and the complement system on foreign materials (see Chapter 19). Thus certain antibodies appear to operate against bacterial invaders by promoting phagocytosis.

After ingestion, bacteria may be destroyed by hydrolytic enzymes in the white cells. On the other hand, virulent bacteria may destroy the ingesting white cells; these bacteria secrete the toxin *leukocidin*, which destroys white cells by interfering with their metabolism. Time-lapse motion pictures of these processes are genuinely fascinating to witness. Such studies first showed *egestion* following ingestion. Bacteria sometimes emerge from a white cell, actively dividing and, like Jonah, apparently none the worse for their experience.

One interesting type of phagocytosis was discovered in 1948. It is the ingestion of one white cell by another, which occurs specifically in the disease *lupus erythematosus*, or *LE*. The so-called *LE cell phenomenon* is considered an important aid in the diagnosis of the disease. Abnormal antibodies developing in LE attack the nuclei or DNA of circulating white cells, and the resulting damaged cells, now behaving as foreign bodies, are ingested by normal white cells.

It should be emphasized that we are speaking here of phagocytosis in circulating white cells. The many fixed cells of the reticuloendothelial system are also phagocytic, though their investigation is rather more difficult technically than the investigation of circulating white cells. We may tentatively assume that the differences between the two cell groups are geographical rather than physiological.

Diseases. There may be too many circulating white cells—leukocytosis—or too few—leukopenia. Leukocytosis occurs in both pathological and physiological circumstances. Only rarely is it due to proportional increases in all types of white cells; commonly it is due to an increase in one type only. Therefore, neutrophilic leukocytosis (neutrophilia), eosinophilic leukocytosis (eosinophilia), basophilic leukocytosis, lymphocytic leukocytosis (lymphocytosis), and monocytosis may arise.

Neutrophilic leukocytosis is most frequently caused by bacterial infection. Though infection is a pathological or abnormal event, the body's response, of which leukocytosis is a part, is a physiological or normal one. Hence we refer to this condition as *physiological* leukocytosis.

Leukopenia is most frequently due to the depression of bone marrow function by toxic diseases and chemicals—e.g., in many instances it is a side effect of drug action. Severe and fatal leukopenias from drugs occur in a few susceptible individuals, though the basis for this unpredictable idiosyncracy is obscure.

The outstanding example of *pathological* leukocytosis is *leukemia*, a disease or class of diseases involving malignant proliferation of the white cell precursors. Any white cell type may be affected. There are granulocytic leukemias, monocytic leukemias, and lymphocytic leukemias. These vary in clinical course, but all are characterized by elevations of the white cell count, with one cell type predominating, enlargement of the hematopoietic tissue mass with marked swelling of the spleen and lymph nodes, replacement of the bone marrow cells by malignant leukemic cells, and invasion of various organs by these cells.

PLATELETS

Physical properties. Platelets, or thrombocytes, are perhaps the most unusual formed elements of blood. They were recognized as blood cells in the midnineteenth century only after debate concerning their very existence. Moreover, the controversy over their origin and function has not yet entirely subsided.

The differences between platelets and the other blood cells are striking (see Plate 1). They are much smaller than red cells, measuring only 2 to 4 μ in diameter, and they lack true nuclei. In stained films they appear singly or in clumps as rounded or frayed bluish bodies with purple central granules. Normally they number from 200,000 to 400,000 per cubic millimeter of blood. Despite their unusual structure, they have phagocytic ability.

Their clumping tendency led shrewd early observers to conclude that platelets play a role in blood clotting. This view was not universally accepted until quite recently, some investigators arguing as late as 1920 that platelets are nothing but artifacts, fragments of red or white cells with no physiological function whatever. Platelets have three remarkable properties that determine their part in clotting: (1) an endothelial supporting function; (2) the ability to form a plug at a bleeding point; and (3) the ability to release substances essential in the clotting reaction—*serotonin*, a vasoconstrictive product of tryptophan metabolism, *ADP, platelet fibrinogen*, and a lipoprotein called *platelet factor 3*. The endothelial supporting function (according to some authorities, but not all) results from the capacity of platelets to collect between capillary endothelial cells in gaps caused by minute local injuries and thereby to reinforce the capillary wall. When the number of circulating platelets is abnormally low, this function is impaired, and a hemorrhage occurs at each of the many sites of minor capillary injury. The ability of platelets to form a plug results from *adhesiveness* (their tendency to stick to foreign surfaces—especially of collagen fibers) and *aggregation* (their tendency to stick to each other or clump), which is greatly enhanced in vitro by released platelet ADP. We shall refer to these properties in discussing the clotting reaction.

Megakaryocytes and their functions. Platelets are derived from the curious polyploid cells called megakaryocytes, which are themselves derived from megakaryoblasts (see Plate 1). They are found mainly in bone marrow, though the capillaries of the lungs contain about one-tenth as many megakaryocytes as the marrow. These are formed in the marrow and delivered to the lungs in the blood. The outstanding characteristics of the megakaryocyte are enormous size (21 μ diameter) and a large irregular nucleus, almost filling the cell, which gives it its name. The chromosome number (i.e., ploidy) is 8 to 32 times the haploid number. Close examination reveals an irregular cell border and a granular cytoplasm. Platelets are fragments of this cytoplasm, which arise by budding.

Less than 0.1% of the bone marrow cells are megakaryocytes. A megakaryocyte produces 400 to 800 platelets per day. Each day about 35,000 platelets are produced per cubic millimeter of blood. Thus in an adult whose blood volume is 5 liters, about 2×10^{11} platelets (total volume, about 0.7 ml) are produced daily. The average life span of a platelet is 10 days. Like the other blood cells, platelets have an active glycolytic metabolism.

Diseases. The chief significance of *thrombocytopenia* (too few platelets) and *thrombocythemia* or *thrombocytosis* (too many platelets) lies in their effects on the clotting reaction. We mention them here simply to illustrate how platelet abnormalities parallel those of the other formed elements.

Secondary thrombocytopenia results from the disappearance of megakaryocytes from the bone marrow with a consequent depression of platelet production. Like leukopenia, it frequently is caused by drug toxicity. The causes of *primary* thrombocytopenia are: (1) excessive utilization of platelets by abnormal intravascular blood coagulation; (2) excessive destruction of platelets by an abnormal antibody; and (3) excessive pooling of platelets in an enlarged spleen. In primary thrombocytopenia, the bone marrow contains its normal complement of megakaryocytes. The main outcome of thrombocytopenia, whatever the cause, is a tendency to spontaneous bleeding.

Thrombocythemia usually occurs in association with proliferative disorders of the other marrow elements (erythrocythemia and leukemia), although it may occur alone. Oddly, the platelets of thrombocythemia are often functionally defective, producing a bleeding tendency similar to that of thrombocytopenia. Abnormal platelets may develop without a change in number. The resulting condition is *thrombocytopathia* ("sick" platelets) or *thrombasthenia* ("weak" platelets).

One of the four main functions of the spleen (see p. 225) has to do with control over the production or release of new red cells, white cells, and platelets. This function has been established by two observations: (1) subjects with enlarged spleens, irrespective of cause, frequently have low red cell, white cell, and/or platelet counts, which promptly return to normal after splenectomy; and (2) removal of a normal spleen often leads to a temporary increase in the blood cell counts.

One explanation for these observations suggests that the normal propensity for removing aged blood cells may be abnormally exaggerated in a large spleen. Another postulates a splenic humoral agent that acts to retard blood cell maturation or release in the bone marrow; oversecretion of this hypothetical hormone would account for "hypersplenism" (the term implies hyperactivity of the spleen in its phagocytic function). The evidence of recent years has tended increasingly to support the former, "graveyard" theory. Not only has research failed to disclose a splenic hormone, but also radioactive scanning has revealed excessive sequestration of blood cells in enlarged spleens.

Plasma and the Plasma Proteins

PLASMA

Chemical composition. Plasma's three categories of ingredients reflect its main functions. They are (1) water, the essential vehicle of transport; (2) the myriad substances being transported to and from the tissues; and (3) proteins, which, by maintaining the osmotic pressure, regulate the distribution of water between the intravascular and extravascular compartments. The proteins also contain the materials of the clotting reaction and the antibodies, both essential in body defense. Some of the major components of plasma are listed in Table 8.5.

Let us briefly consider the three main functions of plasma.

Transport. All plasma constituents are continuously transported. However, some are passengers, and others vehicles. As shown in Table 8.5, the former comprise the nutrients absorbed from the intestinal tract and produced in the liver, metabolic intermediates and waste products (collectively, metabolites), and an assortment of hormones.

The nutrients transported by plasma include amino acids from the digestion of dietary proteins, glucose from the digestion of higher carbohydrates, fatty acids from the digestion of lipids, and plasma proteins, which to a certain extent are utilized by the tissues as foods. More or less constant concentrations of nutrients are maintained through adjustments in the balance among storage in body depots, removal from storage, intestinal absorption, excretion, and metabolic utilization by body cells. Variations from normal concentrations generally occur only following meals. On these occasions, the concentrations rise to temporary peaks and then subside to their original levels.

Since plasma is a carrier of metabolites, it is the medium of exchange for the many biochemical intermediates to be studied in Chapter 14. Interestingly, some of these small molecules are free in plasma, while others are bound to pro-

TABLE 8.5 MAJOR COMPONENTS OF PLASMA, PER 100 ml

Electrolytes and metals

Bicarbonate	152–190 mg
Calcium	8.5–10.5 mg
Chloride	355–380 mg
Cobalt	0.4–0.7 μg
Copper	88–124 μg
Iodine	5–6 μg
Iron	50–150 μg
Magnesium	1.8–3.6 mg
Manganese	4–6 μg
Phosphate	3.0–4.5 mg
Potassium	14–20 mg
Sodium	315–330 mg
Sulfate	0.5–1.5 mg
Zinc	250–350 μg

Nutrients and vitamins, mg*

Amino acids	
Alanine	4.0
Arginine	2.3
Aspartic acid	1.0
Citrulline	0.5
Cystine	1.4
Glutamic acid	0.8
Glycine	1.8
Histidine	1.4
Isoleucine	1.6
Leucine	1.9
Lysine	3.0
Methionine	0.5
Phenylalanine	1.4
Proline	2.5
Threonine	2.0
Tryptophan	1.1
Tyrosine	1.5
Valine	2.8
Glucose	70–100

Vitamins

Ascorbic acid	0.4–1.5
Biotin	1.0–1.7 μg
Carotenoids	100–200 μg
Folic acid	0.6–2.5
Niacin	0.2–0.3 μg
Pantothenic acid	6–35 μg
Riboflavin (B_2)	2.6–3.7
Vitamin A	50–100
Vitamin B_{12}	20–50 μg
Vitamin E	0.6–1.6

Metabolic intermediates, mg

Bile acids	0.2–3.0
Choline	26–35
Citric acid	2.0–3.0
Lactic acid	6–16
Lipids	360–820
Cholesterol	150–280
Esterified	150–193
Free	26–106
Fatty acids	200–450
Phospholipids	135–170
Triglycerides	100–250
Pyruvic acid	1.0–2.0

Waste products, mg*

Acetone	0.3–2.0
Ammonia	40–70 μg
Bilirubin	
Direct-reacting	0.4
Indirect-reacting	0.7
Creatine	0.8–1.0
Creatinine	0.7–1.5
Glutamine	5–12
Urea	25–52
Uric acid	3–6

Hormones, traces

Adenohypophyseal hormones
  Adrenocorticotrophic hormone (ACTH)
  Follicle-stimulating hormone (FSH)
  Luteinizing hormone (LH)
  Luteotrophic hormone (LTH)
  Thyrotrophic hormone (TSH)
Adrenal cortical steroids
Adrenal medullary hormones
  Epinephrine
  Norepinephrine
Insulin
Neurohypophyseal hormones
  Vasopressin
  Oxytocin
Ovarian hormones
  Estrogens
  Progesterone
Parathyroid hormone
Testicular hormones
  Testosterone
Thyroid hormone

Proteins, g*

Albumin	4.6
Antibodies	1.5
Clotting factors	uncertain
Enzymes	uncertain
Fibrinogen	0.38
Globulins	2.6
α_1-globulin	0.3
α_2-globulin	0.6
β-globulin	0.8
γ-globulin	0.9
Metal-binding proteins	trace

Dissolved gases

See Chapter 11 (Table 11.5)

* Unless otherwise noted.

teins or other large molecules. We have earlier noted, for example, that bilirubin, the breakdown product of hemoglobin, travels attached to plasma albumin. Such an arrangement yields a high–molecular weight complex whose size prevents excretion by the kidneys and passage through the capillaries. Binding to the so-called *transport proteins* is best illustrated by the trace metals in plasma, among them iron and copper. Each is bound to a specific protein whose sole known function is the binding and transport of the metal. In this manner quantities of substances that would be toxic in the free state are moved in a state of physiological inertness. The iron-binding protein of plasma is *transferrin* (see p. 245)*; the copper-binding one is *ceruloplasmin*. Even vitamin B_{12} has two specific transport proteins, *transcobalamin I* and *II*. Like intrinsic factor (p. 243), they contain a large amount of carbohydrate (i.e., they are glycoproteins).

Plasma transports the waste products of metabolism to the chief excretory organs—the skin, intestines, liver, lungs, and kidneys. It carries the breakdown products of protein metabolism, such as urea, creatinine, uric acid, and phenol; the breakdown products of hormone and hemoglobin metabolism; and others. Many metabolic end products from one tissue —e.g., lactic acid, the product of glycolysis— may be utilized in other tissues and thus escape excretion.

Regulation of water distribution and acid-base balance. The importance of the osmotic pressure of plasma proteins was mentioned in Chapter 7. If plasma were separated from distilled water by a membrane permeable to water but impermeable to all plasma solutes (including electrolytes), an osmotic pressure of 7 atm (1 atm = 760 mm of mercury) would develop. Such a pressure does not arise in the capillaries because their walls are impermeable only to the

* It has also been called *siderophilin*. "Transferrin" combines Latin roots meaning "transfers iron"; "siderophilin" combines Greek roots meaning "loves iron."

large protein molecules dissolved in plasma (Table 8.6). These alone exert an osmotic force across the walls of the capillaries.

Plasma proteins are also polyvalent electrolytes capable of strong buffering action by virtue of their many dissociating amino and carboxyl side groups. Consequently, plasma functions in the maintenance of acid-base equilibrium.

Defense. The first major defense system of plasma is the clotting mechanism, an intricate device that, when triggered by injury, acts to seal the vascular system against an otherwise lethal blood loss. Since the body must also be protected against a lethal degree of intravascular clotting, a variety of circulating anticlotting factors are present to maintain blood fluidity within the vessels. Both the clotting and anticlotting factors are specific plasma proteins.

The second major defense system of plasma consists of the antibodies, also specific plasma proteins. Injection of an antigen into the body stimulates the formation of antibodies that antagonize the foreign substance. Their production continues long after the disappearance of the antigen. Hence they provide protection or *immunity* for years of a lifetime. The astonishing "memory" of the immune mechanism will be discussed in Chapter 19.

PLASMA PROTEINS

Fractionation. The classic technique for the separation and purification of plasma proteins, of prime importance until recently, was de-

TABLE 8.6 MOLECULAR DIMENSIONS OF CERTAIN PLASMA COMPONENTS

Component	Molecular weight, g	Length, mμ	Diameter, mμ
Na⁺	23	0.19	0.19
Cl⁻	35.5	0.36	0.35
Glucose	180	0.95	0.65
Albumin	69,000	15	3.8
Globulin	156,000	32	3.6
Fibrinogen	500,000	90	3.3
Capillary pore			2.8

veloped in the late nineteenth century. In this method individual proteins are precipitated from plasma solution on the addition of increasing quantities of a neutral salt such as ammonium sulfate. One species of protein molecule might precipitate with $2M$ ammonium sulfate, another with $3M$ ammonium sulfate, and so on. The *salting-out effect* depends on the size and charge of the protein molecule. Newer techniques of protein fractionation include low-temperature alcohol precipitation, heavy-metal precipitation, electrophoresis, ultracentrifugation, and chromatography (see Chapter 2).

During the early years of World War II, it became evident that large quantities of blood might be needed for the treatment of battle casualties. Because of the perishability of whole blood, the possibility was considered of using plasma or certain purified plasma proteins as a blood substitute. Since the main consequence of blood loss is shock due to decreasing osmotic pressure and shrinking blood volume, it seemed likely that infusions of osmotically active plasma protein(s) instead of whole plasma or blood would maintain blood volume.

The large-scale efforts initiated then to purify the plasma proteins led to much of our present knowledge. The bulk of the work was done in the laboratories of E. J. Cohn,* some of whose prewar albumin preparations were actually used to treat the wounded at Pearl Harbor. Under the demands of war, Cohn and his associates produced large quantities of plasma albumin and other proteins, such as the antibodies, which help to control infection.

The diversity of the plasma proteins is suggested by the electrophoretic patterns of normal plasma. Fig. 8.26 shows the Tiselius pattern of classic electrophoresis and the patterns of the more recent (and more convenient) techniques of paper and starch gel electrophoresis.

We recall that, in diagrams of the type in Fig. 8.26A, each protein or class of proteins of a singular electrical charge has a discrete peak. An ideally homogeneous protein, if it exists, has a peak of knifelike sharpness. The relative breadth of a plasma protein peak is a measure of heterogeneity. Each peak represents not an in-

---

* Cohn was also prominent in the long search for the active "anti–pernicious anemia principle" of liver, eventually identified as vitamin B_{12}.

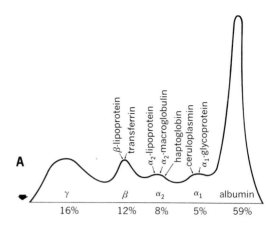
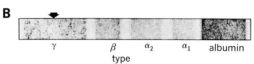
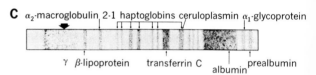

8.26 Electrophoretic patterns of normal serum: A, Tiselius; B, paper; C, starch gel. The broad vertical arrow represents the starting point. β_{2M}-globulin remains in the starting slot in starch gel electrophoresis but moves in the γ to β range in other methods. (From F. W. Putnam, in H. Neurath, ed., *The Proteins*, Vol. III, 2nd ed., Academic Press, New York, 1965.)

dividual protein but a class of proteins, the concentration of the class being proportional to the area under the peak. Thus plasma contains only six major electrophoretic groupings (under the usual conditions of analysis), although there are perhaps hundreds of individual plasma proteins.

The peaks (or spots) of the electrophoretic groupings appear in the following sequence in the electrophoretic pattern(s): albumin, α_1-globulin, α_2-globulin, β-globulin, and γ-globulin.† This, curiously, is the same as the sequence

---

† The α_1, α_2, β, and γ designations for the globulins are arbitrary names for the successive peaks in the electrophoretic pattern(s). We shall presently see that the globulins have other names and classifications.

order of their precipitation by increasing concentrations of salt. Concentrations of the plasma proteins are given in Table 8.5. The molecular dimensions of some of them are given in Table 8.6.

When mass and density are the bases of separation and classification of the plasma proteins, new major groupings emerge. Thus plasma contains several classes of proteins with distinctive sedimentation behavior in the ultracentrifuge (Fig. 8.27). Each class has a characteristic sedimentation constant (see p. 60). Though the proteins in such a grouping have similar masses and densities, they differ in electrical charge and hence in electrophoretic mobility.

Let us briefly consider the major electrophoretic groupings of plasma proteins. Albumin, which has the highest electrophoretic mobility, is also the most abundant. Accordingly, its peak is the largest. It constitutes 50 to 60% of the total plasma proteins and accounts for 80% of the plasma osmotic pressure. Therefore, it is critically important in the maintenance of blood volume. In diseases causing a decrease in plasma albumin (e.g., certain liver ailments), water moves from the blood into the tissues to produce severe edema. We have spoken of the role of albumin in the transport of bilirubin. It functions similarly in the transport of a number of otherwise poorly soluble compounds (e.g., drugs, dyes, and fatty acids).

Fibrinogen, a long threadlike molecule, constitutes only 4 to 6% of the total plasma proteins and may not be seen in the electrophoretic pattern(s) of normal plasma.* In the clotting reaction it is converted to fibrin, the tough insoluble substance of the clot. The dozen or more remaining clotting factors are distributed among the globulins.

The globulins comprise several groups of proteins of high molecular weight, among them the glycoproteins, the lipoproteins (carriers of plasma lipids that would be insoluble without them), the metal-binding proteins, and the antibodies. They make up about 30% of the total plasma proteins. Glycoproteins occur principally in the α_1-globulin and α_2-globulin peaks; trans-

* It migrates between β-globulin and γ-globulin. It is not seen in the electrophoretic pattern of serum because fibrinogen is not present in serum.

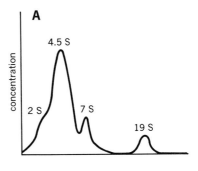

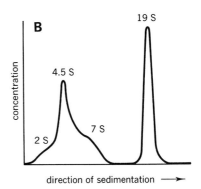

8.27 Ultracentrifuge patterns of normal serum (A) and serum from a patient with macroglobulinemia (B). In normal serum there are four main fractions: 2S, lipoprotein and β-globulin (about 4%); 4.5S, albumin and α_2-globulin (81%); 7S, γ-globulin (12%); and 19S, macroglobulin (3%). In the abnormal serum the 19S fraction is greatly increased.

ferrin in the β-globulin peak; antibodies in the slowest-moving globulin peak, that of γ-globulin; and lipoproteins in both the α- and β-globulin peaks (see Chapter 14).

Immunoglobulins. Much insight regarding the antibodies, or *immunoglobulins* as they have come to be known, has been gained recently— from new methods of separating and characterizing them (notably the technique of *immunoelectrophoresis*, Chapter 19), from studies of their tertiary structures and amino acid sequences, and from genetic studies of the differences between immunoglobulins of different individuals and of the genes controlling the synthesis of different parts of the immunoglobulin molecule.

TABLE 8.7 NORMAL IMMUNOGLOBULINS

Class	Synonyms	Sedimentation constant	Electrophoretic mobility*	Carbohydrate content, %	Heavy chain type	Light chain type	Approximate molecular weight, g	Relative abundance, % total immunoglobulin
Immunoglobulin G	IgG, γG, γ_2, 7Sγ	7S	γ to α_2	3	γ	κ or λ	150,000	80
Immunoglobulin A	IgA, γA, γ_{1A}, β_{2A}	7–14S	Slow β	10	α	κ or λ	50,000–450,000	15
Immunoglobulin M	IgM, γM, β_{2M}, 19Sγ	19S	Fast γ	10	μ	κ or λ	900,000	5
Immunoglobulin D	IgD, γD	7–8S	Slow β	†	δ	κ or λ	†	0.2
Immunoglobulin E	IgE, γE	8S	Slow β	†	ϵ	κ or λ	†	trace

* See Figs. 8.26 and 19.6.
† Not yet known.

Table 8.7 summarizes the names and properties of the five main classes of immunoglobulins. Those with high sedimentation constants (14S and 19S) are called *macroglobulins.* All classes contain carbohydrates, and the macroglobulins are especially rich in them. The most plentiful immunoglobulin is immunoglobulin G (IgG), the γ-globulin with a molecular weight of 150,000 and a sedimentation constant of 7S. It appears to be a symmetrical molecule consisting of two identical units of a small or light (*L*) chain covalently linked through a disulfide bond to a large or heavy (*H*) chain; the two units are themselves joined through a disulfide bond (Fig. 8.28A). The immunoglobulin subunits are folded into a roughly cylindrical form some 35 Å in diameter and 280 Å in length (Fig. 8.28B)—about one-millionth the volume of a typical bacterium. The end portions of each L and H chain have unique structural features. These confer the specificity upon an immunoglobulin, whereby it binds only the correct antigen. We shall examine the significance of this circumstance in Chapter 19.

As indicated in Fig. 8.28A, the immunoglobulin structure can be divided into several types of fragments by various treatments. L and H chains are readily dissociated by the use of a reducing agent which cleaves disulfide bonds (e.g., mercaptoethanol). Treatment of the molecule with the proteolytic enzymes papain or pepsin yields fragments that are named *Fab* (*an*tigen-*b*inding), *Fc* (*c*rystallizable), and *Fd*. For example, papain treatment yields an Fc fragment—the combined terminal portions of the H chains—and two Fab fragments—each an intact L chain plus the remaining portion of an H chain. This portion alone is an Fd fragment. Such techniques led to the discovery that each class of immunoglobulins has a distinctive H chain (as judged by immunological methods or by amino acid analysis). These are designated γ-, α-, μ-, δ-, and ϵ-type H chains, respectively, in IgG, IgA, IgM, IgD, and IgE (see Table 8.7). However, L chains are of only two types, designated κ and λ, and within each category of immunoglobulin in a single individual two-thirds of the molecules contain κ-type L chains and one-third contains λ-type L chains. Therefore, IgG molecules may be formulated $\gamma_2\kappa_2$ or $\gamma_2\lambda_2$; IgA, $\alpha_2\kappa_2$ or $\alpha_2\lambda_2$; IgM, $\mu_2\kappa_2$ or $\mu_2\lambda_2$; IgD, $\delta_2\kappa_2$ or $\delta_2\lambda_2$; and IgE, $\epsilon_2\kappa_2$ or $\epsilon_2\lambda_2$.

Finally, it has been shown that IgG and IgD molecules have a molecular weight of 150,000 and that IgM molecules and many (but not all) IgA molecules are polymers whose monomer unit resembles an IgG or IgD molecule (Fig. 8.28C). IgM molecules are pentamers, though some may be hexamers; hence their molecular weight is 5 or 6 × 150,000. IgA molecules may be monomers, dimers, or trimers; hence their molecular weight covers a wide range.* It is of interest that IgA is found, not only in plasma, but in the secretions of various glands (e.g., in saliva). *Secretory* IgA is a dimer that differs from *plasma*

---

* Polymeric immunoglobulins are formulated as follows: IgM's, $(\mu_2\kappa_2)_5$ or $(\mu_2\lambda_2)_5$; IgA's, $(\alpha_2\kappa_2)_{2 or 3}$ or $(\alpha_2\lambda_2)_{2 or 3}$; etc.

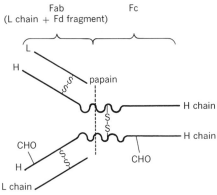

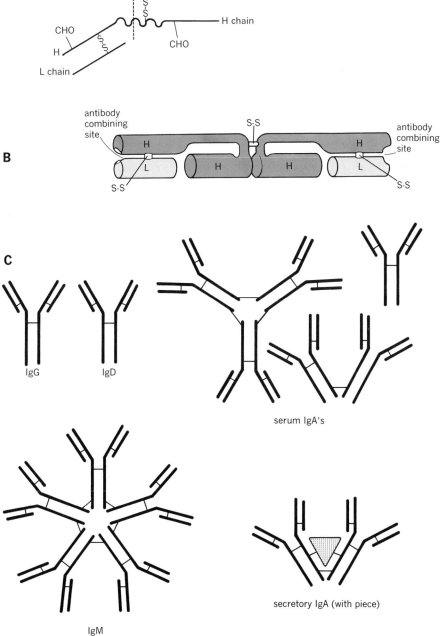

8.28 Structure of the immunoglobulin molecules. A, nomenclature and relationships of chains and fragments produced by papain (CHO denotes a carbohydrate moiety); B, theoretical three-dimensional model, showing the interchain disulfide bond and two antibody combining sites; C, schematic comparison of the several types of immune globulin molecules.

IgA in possessing an additional polypeptide chain (molecular weight, 50,000 to 60,000), the *secretory piece*, which enhances the stability of the molecule.

Production. The liver is the chief site of formation of the plasma proteins other than the immunoglobulins. This has been demonstrated experimentally by the comparison of blood from the vessels entering the liver with that from the vessels leaving the liver, by the study of animals whose livers have been removed, and by the study of liver sections. It has been found that incubated liver slices synthesize radioactive plasma proteins from radioactive amino acids.

The immunoglobulins are formed mainly in lymphoid tissue. The arrival of an antigen molecule is now known to stimulate certain lymphocytes in the tissue to differentiate and divide into *plasma cells* (see Plate I). These synthesize the immunoglobulins. As noted above, immunoglobulins are also synthesized in certain glands.

Diseases. Plasma proteins may vary quantitatively and qualitatively in disease. A decrease in the amount of plasma proteins is called *hypoproteinemia*, and an increase is called *hyperproteinemia*. The occurrence of abnormal plasma proteins is called *dysproteinemia*.

In the most interesting of the hypoproteinemias, a single plasma protein is missing as a result of defective gene action. Examples are genetically determined instances of single clotting factor deficiency with resulting abnormal bleeding—e.g., hypofibrinogenemia and hemophilia, in which the absent proteins are fibrinogen and antihemophilic globulin, respectively—and of inadequate antibody synthesis with resulting abnormal susceptibility to infection—e.g., hypogammaglobulinemia (or agammaglobulinemia). Similar disorders have nongenetic causes. Starvation and severe acquired liver disease, for example, produce nonheritable decreases in the synthesis of some or all of the plasma proteins.

In the most striking example of hyperproteinemia, *multiple myeloma*, there is a pathological increase in immunoglobulin G or immunoglobulin A due to malignant proliferation of the particular globulin-synthesizing plasma cells. The electrophoretic patterns in Fig. 8.29 are the diagnostic key to this disease. Many patients with multiple myeloma excrete in their urine large amounts of an unusual substance called *Bence-Jones protein* (after its discoverer, Sir Henry Bence-Jones). This has a curious property: it precipitates in urine heated to 60°C but dissolves with further heating. It is now known, from electrophoresis, that pure Bence-

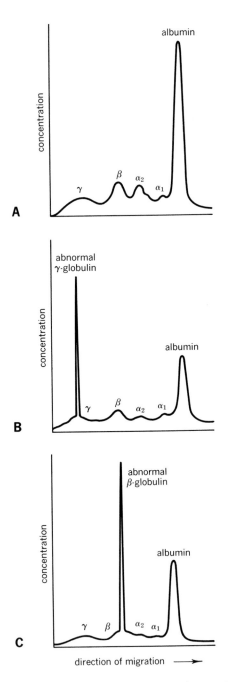

8.29 Tiselius electrophoretic patterns of normal (A) and multiple myeloma (B,C) sera. The abnormal protein may have differing mobilities in different patients, although its mobility in any individual patient remains constant. In B the abnormal protein is an immunoglobulin G; in C it is an immunoglobulin A.

Jones protein is a homogeneous collection of free L chains identical to those in the patients' abnormally plentiful immunoglobulin. Soon after Bence-Jones protein was identified, a case was found in which a myeloma-like hyperglobulinemia was attributed to excessive production of free Fc fragments. This situation is much rarer than multiple myeloma. Both disorders—in a sense experiments of nature—have greatly extended our knowledge of the immunoglobulins.

Abnormal proliferation of the plasma cells synthesizing immunoglobulin M occurs in an uncommon disease called *macroglobulinemia*. Its recognition depends upon ultracentrifugal or immunoelectrophoretic analysis of plasma or serum (see Fig. 8.27B).

Blood Coagulation

CLOTTING REACTION

Hemostasis. We have learned that blood remains fluid within the vessels but rapidly solidifies into a clot when shed. The solidification process, termed *clotting* or *coagulation*, is among the most complex and interesting self-regulating processes in physiology, one that manifestly had to emerge in evolution if species were to survive.

Blood coagulation results from an integrated sequence of vascular, extravascular, and intravascular events, which act together in *hemostasis*, or the prevention or arrest of bleeding. Following injury, a small blood vessel constricts locally near the site of injury. This reaction, the *vascular* or *vasoconstrictive* component of hemostasis, depends in part on nerve reflexes. It lowers the rate of blood loss but does not stop it completely. The smallest blood vessels do not contain contractile smooth muscle fibers and therefore cannot actively constrict. However, their cut surfaces close off by adhesion to each other.

Disruption of the endothelial lining of the injured blood vessel exposes a layer of loosely packed collagen fibers (Fig. 8.30A). When plate-

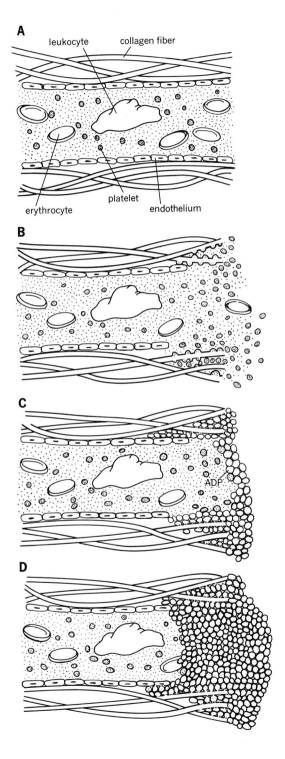

8.30 Primary hemostasis, showing formation of primary hemostatic plug. A, intact blood vessel; B, injury damages integrity of vessel, platelets begin to accumulate on exposed collagen; C, platelets discharge ADP and thereby promote platelet aggregation and loss of platelet granules; D, loose platelet aggregate plugs vessel.

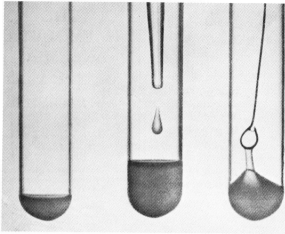

A   B

8.31 Two clotting tests. A, clotting time; the end point is reached when the tube can be safely inverted; B, prothrombin time: left, plasma; center, addition of tissue thromboplastin and Ca^{++}; right, final clot.

lets are exposed to collagen, they undergo a remarkable transformation. They are attracted to collagen to which they adhere. They then swell and release many of their intracellular constituents into the local area, so that they become degranulated (Fig. 8.30B). As we have learned, the released materials include serotonin, ADP, platelet factor 3, and platelet fibrinogen. Serotonin prolongs the local vasoconstrictive reaction and ADP aggregates the platelets, causing them to form a whitish friable plug, or thrombus, that temporarily seals the ruptured blood vessel (Fig. 8.30C). The lipoprotein, platelet factor 3, initiates the clotting reaction. The function of platelet fibrinogen will be discussed later.

Hemostasis, then, can be divided into two stages: *primary* hemostasis, in which the flow of blood is stanched by a combination of vasoconstriction and platelet disintegration and aggregation; and *secondary* hemostasis, in which the stanching is maintained until healing is complete. The clot is the bulwark of secondary hemostasis. The final clot is a tough, solid meshwork of fibrin strands and entrapped red cells.

Fibrin, which is insoluble, is not normally present in blood. It arises from plasma fibrinogen, a soluble protein. To explain the clotting reaction is to recount the series of events that begins with the release of platelet factor 3 and tissue juices (i.e., tissue fluid containing substances released from damaged tissue cells) and ends in the conversion of fibrinogen to fibrin.

Numerous obstacles have stood in the way of the elucidation of the clotting reaction. Some were methodological. For years, means for testing the various clotting factors, separating them, and determining their purity were lacking. Many false steps were taken before pitfalls were recognized. For example, it was not known until recently that the glass surface of a test tube artificially accelerates clotting and that this phenomenon is due to a specific clotting factor.

Much of our information has been gained from the study of hemorrhagic diseases, in which a deficiency of a single clotting factor hinders clotting. The best known of these, hemophilia, is still a common and difficult medical problem, but its treatment, as we shall see, has been drastically improved by modern developments.* We shall speak of these diseases not only for their intrinsic interest but because of their contributions to progress in clotting research. A separate and distinct genetically determined disease involving a lack of each known clotting factor has now been found. Indeed, most of the clotting factors were discovered in the first place when a clinical clotting abnormality was encountered which could not be ascribed to deficiency of a known factor.

Classic clotting tests. The competence of the immediate vascular and platelet reactions to injury may be estimated from measurement of the *bleeding time,* the time required for the cessation of bleeding from a small standard-sized puncture wound. Normal bleeding time is 3 to 6 minutes. There is little relationship between bleeding time and the ability of blood to clot. Bleeding time is prolonged by vascular and platelet abnormalities (i.e., abnormalities of primary hemostasis) and not by deficiencies of clotting factors (i.e., abnormalities of secondary hemostasis). With a clotting factor deficiency (e.g., in hemophilia), bleeding stops within the normal time interval as a result of vascular contraction and thrombus formation; but, because clotting is defective, it eventually resumes, often taking the form of oozing.

If normal whole blood is removed from a vein and placed in a clean test tube, the blood clots in 6 to 8 minutes. The time required for the formation of a firm clot is called the *clotting time* (Fig. 8.31A). This is perhaps the most familiar of the so-called *clotting tests.* It is also the least helpful, since severe deficiencies of a number of clotting factors can exist without materially lengthening clotting time. Nevertheless, certain deficiencies and certain anticoagulant drugs prolong it.

If normal whole blood is removed from a vein and mixed with an anticoagulant (e.g., oxalate, citrate, or versene) that acts by removing calcium ion (Ca^{++}), the blood does not clot. This circumstance indicates that Ca^{++} is essential for clotting. If the blood cells (including platelets) are removed from the unclotted, calcium-free blood sample by centrifugation, clear "decalcified" plasma is obtained. This contains all of the necessary clotting factors except Ca^{++} and platelets, whose function it is to liberate the substances that initiate clotting. If Ca^{++} and platelets are then added, the plasma clots.

Tissue juices and extracts of certain tissues (e.g., brain or lung) also contain substances capable of initiating clotting and can therefore substitute for platelets in this system. This means that care must be exercised in drawing blood for clotting tests. Careless use of the syringe may result in an admixture of tissue juices with blood that affects the clotting behavior of "recalcified" plasma.

Historical development of the theory of clotting. The broad outlines of the modern theory of clotting had already been formulated at the beginning of the twentieth century. These can be stated briefly: (1) under the catalytic influence of substances arising from platelets or tissue juices, the plasma protein *prothrombin* is converted to *thrombin;* (2) thrombin catalyzes the conversion of the soluble plasma protein *fibrinogen* to insoluble *fibrin,* which forms the meshwork of the clot.

These two clearcut steps were recognized half a century ago. Today we recognize a third step, the initial stage of clotting. The essential substances arising from platelets and tissue juices have both been named *thromboplastin.*† Thromboplastin (of either origin) and Ca^{++} are among the factors required for the conversion of prothrombin to thrombin. The initial stage of clotting encompasses the reactions that transform thromboplastin to *prothrombin activator,* the active prothrombin-converting substance.

---

* Hemophilia is the famous bleeding affliction suffered by the male offspring of Queen Victoria. However, the disease and its hereditary nature were first noted by the ancient Jews, who recorded instances of severe bleeding following ritual circumcision of the newborn. The Talmudic sages wrote that, if a child died in these circumstances, its later brothers would also die of hemorrhage. To forestall such disasters, they freed these brothers from the necessity of circumcision.

---

† Application of this term to both substances has caused some confusion. In fact, the substances are poorly defined and almost certainly not identical, although their activities are similar.

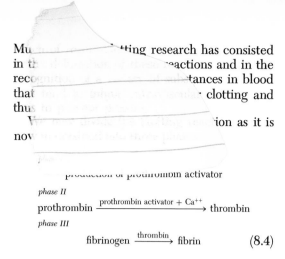

Mu... ...ting research has consisted
in t... ...actions and in the
rec... ...tances in blood
that... ...clotting and
thu...

...ion as it is
nov...

...production of prothrombin activator

phase II

$$\text{prothrombin} \xrightarrow{\text{prothrombin activator} + Ca^{++}} \text{thrombin}$$

phase III

$$\text{fibrinogen} \xrightarrow{\text{thrombin}} \text{fibrin} \qquad (8.4)$$

One notable feature of this scheme should be
emphasized. Certain of the reactants already
exist in plasma, among them Ca^{++}, prothrombin, and fibrinogen. The two proteins are actually precursors of the clotting factors thrombin
and fibrin. In the procedure in which Ca^{++} and
tissue thromboplastin are added to decalcified
plasma, the chief determinants of the *rate* of
clotting are the levels of prothrombin and fibrinogen in the plasma. If plasma's capacity to
activate added tissue thromboplastin is intact,
if inhibitors are relatively inactive, and if excess
Ca^{++} is present, only a low plasma prothrombin
or fibrinogen level can delay clotting time.
Usually the fibrinogen level is adequate. Therefore, clotting time in this test, first described in
1935 by Armand J. Quick, is generally taken as
a measure of the plasma prothrombin level and
is called the *prothrombin time* (Fig. 8.31B).

It was soon observed that, when plasma is
allowed to age by standing in a test tube, its
prothrombin time gradually increases. This phenomenon is reversed on the addition of a small
amount of fresh plasma. That the increasing
prothrombin time is not due to a spontaneous
disappearance of plasma prothrombin or fibrinogen was readily shown. Accordingly, it was concluded that there must exist a previously unknown plasma substance that is essential in the
clotting reaction, that is labile (since it vanishes
on standing), and that (if the clotting scheme is
correct) must function in phase I in the conversion of tissue thromboplastin to prothrombin
activator. This was termed the *labile factor*. By
1947 labile factor had been independently discovered and named in several laboratories

throughout the world. Because of the resulting
confusion, investigators agreed to use noncommittal Roman numerals to designate the known
clotting factors, as follows: Factor I, fibrinogen;
Factor II, prothrombin; Factor III, tissue thromboplastin; Factor IV, Ca^{++}; and Factor V, labile
factor.

Next it was established that prothrombin is
literally consumed in its conversion to thrombin,
only traces remaining after clot formation. Between 1947 and 1950 Quick pointed out that
the serum of hemophilic patients (i.e., the fluid
left after clotting, slow though it may be) contained a large quantity of prothrombin; in other
words, in hemophilia very little prothrombin is
converted to thrombin. It was then recalled that
in 1911 Thomas Addis had found that a globulin-
like fraction from normal plasma corrected the
prolonged clotting time of hemophilic plasma.
The concept thus arose that the basic abnormality in hemophilia is a deficiency of an "antihemophilic" factor. Since this factor is present
in normal plasma, presumably it is necessary for
the formation of prothrombin activator, and
lack of it results in poor conversion of prothrombin to thrombin—i.e., in poor prothrombin consumption and poor clotting.

We now employ the *prothrombin consumption test*, determining the prothrombin content
after clot formation and comparing it with the
prothrombin content of whole anticoagulant-
treated plasma. Normal plasma utilizes 95% of
its prothrombin in clot formation. Hemophilic
plasma utilizes very little. The missing factor in
hemophilia was designated *antihemophilic factor*, or *AHF*. It became Factor VIII.

Current views of the clotting mechanism. As
we have noted, added tissue thromboplastin
stimulates the formation of prothrombin activator. So does added platelet thromboplastin. The
two systems for elaborating the activator are the
extrinsic system, which begins with tissue thromboplastin, and the *intrinsic system*, which begins
with platelet thromboplastin. The extrinsic system operates in the test tube when platelets are
removed and tissue thromboplastin is introduced artificially. The intrinsic system operates
in the test tube when platelets are allowed to
remain. Both systems operate in the body—the
extrinsic system when tissue juices are liberated

TABLE 8.8 CLOTTING FACTORS

Number	Common synonyms (and abbreviations)	Required for clotting via Intrinsic system	Required for clotting via Extrinsic system
I	Fibrinogen	Yes	Yes
II	Prothrombin	Yes	Yes
III	Tissue thromboplastin (TPL)	No	Yes
IV	Ca^{++}	Yes	Yes
V	Labile factor, proaccelerin, accelerator globulin (AcG)	Yes	Yes
VI*			
VII	Proconvertin	No	Yes
VIII	Antihemophilic factor (AHF)	Yes	No
IX	Plasma thromboplastin component (PTC), Christmas factor	Yes	No
X	Stuart factor	Yes	Yes
XI	Plasma thromboplastin antecedent (PTA)	Yes	No
XII	Hageman factor (HF)	Yes	No
XIII	Fibrin-stabilizing factor (FSF)	Yes	Yes

* This number is no longer used. It was originally assigned to a factor later revealed to be activated Factor V.

by injury and the intrinsic system when activation is entirely dependent on factors within the blood stream.

A succession of discoveries since 1947 has shown that both systems require the participation of several specific clotting factors, as listed in Table 8.8. Some of these factors function in both the extrinsic and intrinsic systems, and some in only one system.

Although we shall not discuss the details surrounding the discoveries of all the clotting factors, we should emphasize that each discovery since 1947—like many earlier—came about from exhaustive study of an unusual patient or a peculiar clotting test pattern that did not fit the accepted theory of clotting. One technique that helped to validate the advances was the *cross-correction test*. For example, hemophilic (Factor VIII–deficient) blood and Factor V–deficient blood both have prolonged clotting times. However, a mixture of equal parts of the two has a normal clotting time because Factor VIII–deficient blood supplies Factor V and Factor V–deficient blood supplies Factor VIII. On the other hand, a mixture of equal parts of two Factor VIII–deficient bloods (or of two Factor V–deficient bloods) still has a prolonged clotting time because it still lacks an essential factor. The principle is simple and clear: when mixing two bloods with abnormal clotting behavior produces an improvement in this behavior, the two bloods lack different factors; when mixing them produces no improvement, they both lack the same factor. This applies even when the source of one of the bloods is unknown. Indeed, the procedure is a useful analytical tool.

Some of the unusual patients studied in the late 1940's and the 1950's had hemorrhagic manifestations and abnormal clotting behavior typical of hemophilia (e.g., poor prothrombin consumption). Yet, when their plasmas were mixed with certified hemophilic plasma, normal clotting behavior resulted. The inference was that their bloods contained Factor VIII but lacked some other factor. Whatever the missing factor was, it was apparently present in Factor VIII–deficient blood.

Benjamin Alexander examined a subject who lacked a factor that participates in the extrinsic system. This was later named Factor VII. The evidence that it does not participate was that the subject's prothrombin consumption was normal in spontaneously clotting blood but abnormal in platelet-free blood to which tissue thromboplastin had been added.

From a procession of such discoveries, today's clotting scheme has emerged. By analyzing it carefully, we see why individuals with Factor VIII deficiency have normal prothrombin times (which depend on the extrinsic system) and prolonged clotting times (which depend on the intrinsic system): Factor VIII is not needed in the extrinsic system. We also see why individuals with Factor V deficiency have prolonged prothrombin times *and* clotting times: Factor V functions in both systems. We could say, then, that the scheme helps us to understand test abnormalities due to factor deficiencies, but we should acknowledge that these test abnormalities gave rise to the scheme in the first place.

Factor XII, the *Hageman factor,* or *HF,* merits special comment. It is responsible for the rapid clotting of blood in glass tubes compared to that in plastic or surface-treated (siliconized) tubes. Oscar Ratnoff, studying a subject named Hageman in 1955, noted a normal prothrombin time (meaning intact extrinsic system) and a prolonged clotting time (meaning defective intrinsic system) without an overt bleeding tendency. The test abnormalities were corrected when Hageman blood was mixed with blood deficient in any known factor. The significant feature was that the abnormalities were observable *only in glass tubes!** Clotting behavior in

* Later work has shown that the phenomenon occurs with any charged wettable surface (e.g., glass or powdered diatomaceous earth).

plastic or surface-treated tubes—and in the subject—was indistinguishable from that of normal blood.

The order in which the various clotting factors combine in phase I to form prothrombin activator in the intrinsic system has only recently been elucidated (Fig. 8.32). The first step is the activation of Factor XII by a glass (or other charged wettable) surface. Active Factor XII then converts inactive Factor XI (PTA) to active Factor XI. Active Factor XI then converts inactive Factor IX (PTC) to active Factor IX. It, in turn, converts inactive Factor VIII (AHF) to active Factor VIII. Finally, active Factor VIII converts inactive Factor X to active Factor X, which has been identified as prothrombin activator. All but the first two steps require Ca^{++}.

R. G. Macfarlane has described this sequence of reactions as a "cascade of proenzyme-enzyme transformations." Such an arrangement has two properties of great biological importance. First, it amplifies small effects. Only a few molecules of active Factor XII are needed to set the whole sequence in motion. Second, it affords many opportunities for limitation of the scope of the clotting reaction once it has begun. This is achieved by rapid inactivation of the activated clotting factors. It appears that each of the activated clotting factors has its own specific inactivating system. In addition, certain of the activated clotting factors are removed from the blood as it passes through the liver by a mechanism that is unrelated to the reticuloendothelial system.

In the extrinsic system the activation of Factor X is more direct, requiring only Factors VII and V and Ca^{++}.

The chemical events in phase II, the conversion of prothrombin to thrombin, are incompletely understood. Probably the reaction is a proteolytic cleavage of prothrombin, for it liberates carbohydrate and peptide fragments. Indeed, one synonym for prothrombin activator (active Factor X) is *prothrombinase.* Factor V and Ca^{++} are involved, and the reaction proceeds slowly unless certain phospholipids are present. In the intrinsic system these come from platelets (the main one being platelet factor 3), and in the extrinsic system from tissue thromboplastin in the tissue juices.

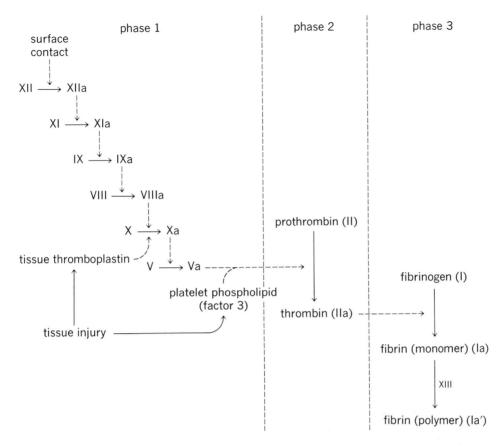

8.32 Scheme of the clotting reaction, showing cascade, or waterfall, theory of factor activation. Solid arrows (→) denote chemical transformations; dashed arrow (⇢) catalytic effects; the letter a denotes active form of the clotting factor.

Thrombin is a proteolytic enzyme, which in phase III splits from the end of the long fibrinogen molecule four short *fibrinopeptides** and some carbohydrate. What is left is a soluble fibrin *monomer*. In addition to its action on fibrinogen, thrombin powerfully affects platelets, causing them to release their constituents much as does collagen. Platelet fibrinogen released in the platelet plug helps to produce a fibrin meshwork that converts the primary hemostatic plug to the strong definitive hemostatic plug, the secondary, or final, hemostatic plug.

The final stable clot does not form until the fibrin monomers polymerize into long tough fibers (*fibrin polymer*). This step requires the participation of active Factor XIII (FSF). Interestingly, thrombin activates Factor XIII. The final opaque clot, consisting of fibers and enmeshed platelets and red cells, soon shrinks. The process, called *clot retraction,* begins with contraction of the trapped platelets† and ends with contraction of the fibers. It serves to pull together the edges of a small wound. Fig. 8.33 summarizes the main events of the normal hemostatic sequence.

---

* The peptides appear to promote the constriction of small blood vessels. Thus they may be helpful in the vasoconstrictive reaction of hemostasis.

† It has recently been found that a contractile protein, *thrombosthenin,* constitutes 15% of the total protein in platelets. It is remarkably similar to *actomyosin,* the contractile protein in muscle (Chapter 17).

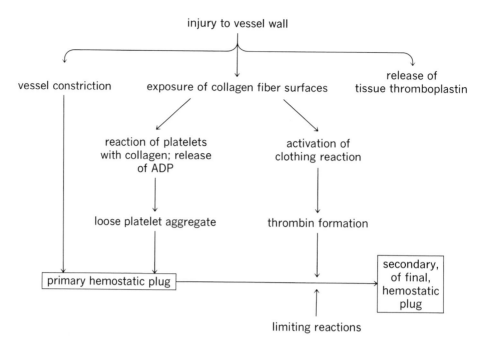

injury to vessel wall

vessel constriction → exposure of collagen fiber surfaces → release of tissue thromboplastin

reaction of platelets with collagen; release of ADP

activation of clothing reaction

loose platelet aggregate

thrombin formation

primary hemostatic plug

secondary, of final, hemostatic plug

limiting reactions

8.33 The hemostatic sequence. (Adapted from D. Deykin, *New Eng. J. Med., 276,* 622, 1967.)

Table 8.9 presents a classification and selected examples of the major bleeding disorders. Hemostasis may become defective at any point in the complex chain of events—though it is noteworthy that no hemorrhagic state has ever been attributed to calcium deficiency.

Fibrinolysin system. We have already considered three mechanisms that oppose the unchecked growth of a localized clot: (1) consumption or utilization of certain clotting factors in the clotting reaction; (2) specific inactivators or inhibitors for the various activated clotting factors; and (3) removal of activated clotting factors by the liver. A fourth mechanism pits in opposition to the clotting system the *fibrinolysin system,* which dissolves, or lyses, clots. Not only are both systems complex and dynamic; each is also responsive to the other's activities. If this were not so, *thrombosis* (formation of an intravascular clot) or hemorrhage would occur more often than it does.

The fibrinolysin system is not fully understood, nor has a choice been made between two conflicting nomenclatures. It is known that the proteolytic enzyme attacking the fibrin of the clot,* *fibrinolysin* (also called *plasmin*), does not appear in plasma until its inactive precursor *profibrinolysin* (or *plasminogen*) is activated by an *activator* (or *plasminogenase*), which itself comes from an inactive *proactivator.* Clearly, nature is ultracautious in this area. The several substances that convert proactivator to activator are referred to as *lysokinases.* Some lysokinases are present naturally in blood. Others have been found in urine (urokinase), in streptococci (streptokinase), and in extracts of various tissues.

Even after fibrinolysin is formed, the body may escape its influence through the action of the two (or more) fibrinolysin inactivators that exist in plasma. A summary of the system is shown in Fig. 8.34.

Fibrinolysis takes place in normal and pathological conditions. Clotted blood removed from

---

* Fibrinolysin also attacks fibrinogen, Factor V, and Factor VIII. The products released from fibrinogen by fibrinolysin are potent inhibitors of thrombin and of platelet aggregation reactions.

TABLE 8.9 CLASSIFICATION AND EXAMPLES OF BLEEDING DISORDERS

Disorders	Spontaneous bleeding	Platelet count	Bleeding time	Clotting time	Clotting factor and related tests	Other features
I. Defects of capillaries	Yes	Normal	Prolonged	Normal	Normal	Increased "capillary fragility"
A. Hereditary						Various genetic patterns

B. Acquired
  1. Metabolic disorders—scurvy
  2. Drug reactions, allergy						

II. Defects of platelets	Yes	Varies	Prolonged	Normal	Decreased prothrombin consumption; no factor deficiency	

A. Thrombocytopenia
  1. Primary
  2. Secondary
		Decreased				

B. Thrombocythemia		Increased				Platelets may be qualitatively abnormal
C. Thrombocytopathia		Normal				Platelets are abnormal
III. Clotting factor deficiencies	Varies	Usually normal	Usually normal	Varies	Specific deficiency	
A. Phase I	Varies	Normal	Normal	Prolonged	Decreased prothrombin consumption	Hereditary or acquired
1. AHF—"classic hemophilia"	Yes				Decreased Factor VIII	Sex-linked recessive only
2. PTC—"Christmas disease"*	Yes				Decreased Factor IX	Sex-linked recessive or acquired
3. PTA	Yes (mild)				Decreased Factor XI	Sex-linked dominant or acquired; rare
4. Stuart factor	Yes				Decreased Factor X	Autosomal recessive or acquired; rare
5. Hageman factor	No			Normal in plastic tubes	Decreased Factor XII	Genetics unknown; rare
B. Phase II	Yes	Normal	Normal	Prolonged	Increased prothrombin time	

1. Hypoprothrombinemia
    a. Congenital
    b. Acquired
						Genetics unknown
						Complication of liver disease and vitamin K deficiency

C. Phase III	Yes	Usually normal	Varies	Varies		
1. Hypofibrinogenemia				Prolonged	Decreased fibrinogen; normal prothrombin; increased prothrombin time	
a. Congenital		Occasionally decreased				Genetics unknown
b. Acquired						Complication of liver disease and pregnancy
2. Fibrinolysins				Usually normal	Increased fibrinolysins	See text
IV. Anticoagulant effects	Yes	Normal	Normal	See text	See text	

* Hemophilia-like disease named for the family in which it (and PTC) were discovered.

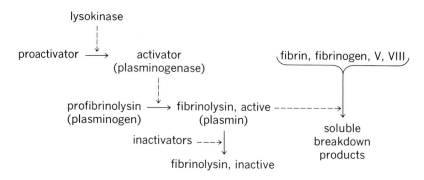

8.34 The fibrinolysis system. Solid arrows (→) denote chemical transformations; dashed arrows (--→) catalytic effects.

the body immediately after death rapidly becomes fluid because of the development of fibrinolysin. After hemorrhage into body cavities, blood may clot and then undergo fibrinolytic digestion to become fluid. The fluidity of menstrual blood is explained in this way. The main physiological role of fibrinolysis is in defense against thrombosis.

ANTICOAGULANTS

Physiological anticoagulants. Clotting is normally limited to a site of blood vessel injury. A number of substances in plasma act continuously to suppress Factor X activation and the several reactions catalyzed by thrombin in the large portion of fluid blood not in contact with the site of injury. These are essential for survival, since blood normally contains a huge excess of clotting factors.

Antithrombin inhibits thrombin by binding it in an inert complex that is incapable of catalyzing the conversion of fibrinogen to fibrin. Antithrombin occurs in the α-globulin fraction. A lipid *antithromboplastin* has been claimed to inhibit the intrinsic system, and there is some evidence that an abnormal increase in this substance contributes to the defective hemostasis of hemophilia. A second antithromboplastin inhibits the extrinsic system.

Anticoagulant drugs. A variety of agents can be used to retard blood clotting deliberately. We have already mentioned those that prevent clotting by removing Ca^{++} (oxalate, citrate, and versene). Obviously, these are applicable only to blood that has been removed from the body. The anticoagulant drugs are administered for the purpose of decreasing clotting *in the body*, e.g., in the treatment of thrombotic diseases. Two types of anticoagulant drugs are most commonly employed.

Heparin is a highly acidic polysaccharide with many sulfonic acid ($—SO_3H$) groups that is isolated from livers (hence its name) and lungs. It is produced in the tissue mast cells (see Chapter 5 and p. 251), though it exercises no known physiological function. Purified heparin, given intravenously, prolongs the clotting time by causing a decrease in thrombin production.[*] Its mechanism of action involves the formation of complexes, with clotting factors. Since its ability to form complexes is partially due to its acidic character, its anticoagulant effect is counteracted by the basic protein *protamine*.

Dicumarol is typical of the class of anticoagulant drugs called *coumarins*. It is given orally. Dicumarol is not an anticoagulant *per se;* that is, it does not prevent coagulation when added to blood in a test tube. Rather it inhibits the synthesis in the liver of clotting factors—notably prothrombin, Factor VII, Factor IX, and Factor X—whose production requires the presence of vitamin K. As we learned in Chapter 4, vitamin K and its simpler but equally effective analogue menadione (see Fig. 8.35) apparently participate in oxidative phosphorylation, perhaps shuttling electrons between NAD and cytochromes in the mitochondrial respiratory

---

[*] Heparin has another important property, apparently unrelated to its anticoagulant property. When it is given intravenously to an animal that has just had a fatty meal and whose serum has thus become turbid with small fat droplets, a visible clearing of the serum occurs within minutes. This effect implicates heparin in plasma fat transport, which will be discussed in Chapters 9 and 14.

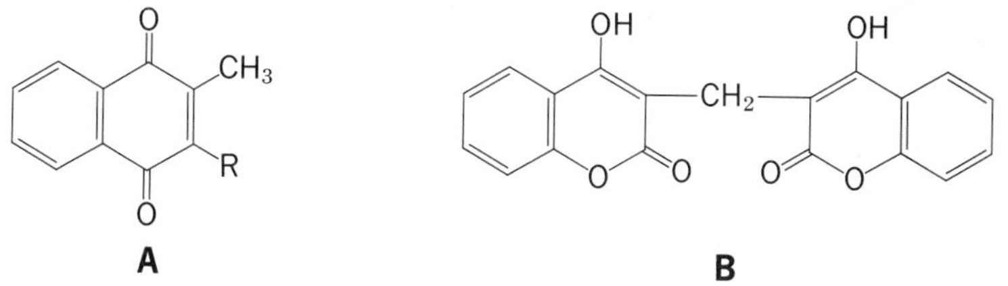

8.35 Structural formulas of vitamin K (A) and dicumarol (methylene-bis-3,4-hydroxycoumarin) (B). In vitamin K_1 (natural), $R = C_{20}H_{39}$; in vitamin K_3 (synthetic menadione), $R = H$.

chain.* Dicumarol and other coumarins are chemically similar to vitamin K (Fig. 8.35) and displace it from liver cells.† The result is *hypoprothrombinemia*, a deficiency of prothrombin, and deficiencies of other factors. As would be expected, the effect of such drugs is readily reversed on administration of vitamin K.

Pathological anticoagulants. At times powerful anticoagulant substances appear spontaneously in the blood. They are easily demonstrated, as suspect plasma prolongs the clotting time of recalcified normal plasma.

A heparin-like agent sometimes forms in the course of severe diseases and may cause serious bleeding. Another type of anticoagulant, which interferes with thromboplastin activation in an unknown manner, sometimes arises in hemophilic subjects who have received many transfusions, in normal women during the first weeks or months after childbirth, and in elderly individuals.

Blood Groups and Blood Transfusion

Blood typing, or the classification of human blood into groups, depends upon immunological reactions between antigens of the red cell surfaces and antibodies of testing sera. Over 60 years ago, while searching for possible differ-

ences among individual human bloods, Karl Landsteiner mixed serum from one normal individual with red cells from other normal individuals. He found that certain combinations of cells and serum showed marked clumping of the red cells, while other combinations showed no clumping. By means of such reactions, Landsteiner proved that human beings could be divided into blood groups. It was this discovery that made possible the safe transfusion of blood.

BLOOD GROUPS

Individuality of human blood. According to Landsteiner, the naturally occurring blood group antibodies or agglutinins‡ are of two chief types, which he designated *anti-A* and *anti-B*. The red cell surfaces contain two corresponding antigens or agglutinogens, which he designated *A* and *B*. If serum bearing anti-A agglutinin is mixed with a suspension of red cells bearing agglutinogen A, the cells clump together in large masses visible to the naked eye (Fig. 8.36). However, no clumping is observed in a mixture of anti-B serum with A cells or of anti-A serum with B cells.

This simple principle underlies all blood grouping tests, including those used to detect recently discovered blood groups for which testing sera are of nonhuman origin. There are four possible combinations of anti-A and anti-B sera with A and B cells, which comprise the four

---

* Although vitamin K appears to take part specifically in the hepatic synthesis of clotting factors, there being no decrease in the synthesis of other proteins in vitamin K deficiency, it is interesting that it is found in plants and bacteria (which do not have clotting systems) as well as in the heart and kidneys (which are not involved in the synthesis of clotting factors).

† Recent studies have shown that the coumarins competitively inhibit the enzyme vitamin K reductase.

‡ Antibodies are actually of many types. When the antibody-antigen reaction results in agglutination of cells, the antibody may be called an *agglutinin*, and its corresponding antigen an *agglutinogen*. When the antibody-antigen reaction results in precipitation, the antibody may be called a *precipitin*, and its antigen a *precipitinogen;* etc. (See Chapter 19.)

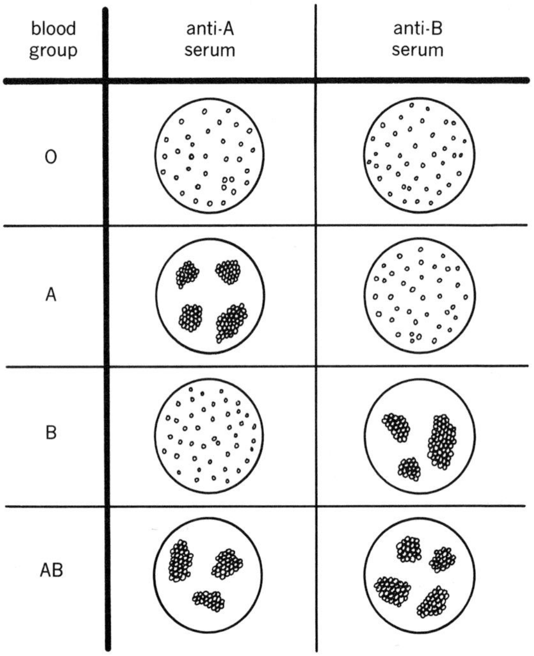

blood group	anti-A serum	anti-B serum
O		
A		
B		
AB		

8.36 Reactions of blood groups of the ABO system. Diagrams show microscopic appearance of red cells.

well-known blood groups of the *ABO system:* O, A, B, and AB. Fig. 8.36 shows the results of testing cells of each group with anti-A and anti-B sera. Testing of an unknown red cell suspension with both sera reveals its blood group.

Antibodies are usually looked upon as specific substances whose synthesis follows exposure to foreign antigens. From this viewpoint, the invariable presence of anti-A or anti-B agglutinin in human serum presents some interesting problems. As far as we know, an individual with anti-B in his serum has been exposed to only one type of red cell, his own, and these are A cells. Moreover, an individual with A cells always has anti-B in his serum. For these reasons the normal serum agglutinins of the ABO system have been termed *naturally occurring antibodies* in contrast to the conventional antibodies of immunological origin. Only the sketchiest of evidence suggests that blood group antibodies are uniquely naturally occuring, and in fact their source is unknown. It is clear that the agglutinins present in an individual's blood can be specific only for agglutinogens *not* present in his blood.

If blood groups are distinguished solely by their agglutination or nonagglutination upon the addition of testing sera, the possibility arises that novel testing sera of diverse origins may uncover hitherto undetected blood group substances. This possibility was realized in experiments. When washed human red cells were tested with sera from rabbits that had been immunized by earlier injections of human red cells, the rabbit antibodies distinguished two new human agglutinogens, which delineated three new blood groups. These were arbitrarily labeled *M, N,* and *MN.* They have little importance for blood transfusion compatibility because *anti-M* and *anti-N* rarely occur in human serum. But their simple hereditary patterns give them great value in the investigation of disputed parentage. Every individual must belong to one of three MN groups as well as one of four ABO groups. Thus for each individual there are $3 \times 4 = 12$ possible combinations of these groups.

Likewise, a completely different set of agglutinogens was found when human red cells were mixed with sera from guinea pigs or rabbits into which monkey (*Macacus rhesus*) red cells had previously been injected. By this means Landsteiner and Alexander Wiener discovered the *Rh* or *Rhesus* blood factor in 1940. The guinea pig or rabbit antibodies developed against the monkey red cells agglutinated not only the monkey red cells but also those in about 85% of the human blood samples tested from the white population of New York City. Therefore, human red cells can be categorized as Rh *positive* (Rh$^+$), containing the Rh factor, and Rh *negative* (Rh$^-$), lacking it. It was shown that the presence or absence of the Rh factor was in no way related to the presence or absence of antigens of the ABO system.

In summary, then, blood grouping involves the detection, differentiation, and classification of the various antigens of red cell surfaces. Owing to the many blood group systems now established, many thousands of blood groups can be distinguished. It is even possible that the complete pattern of blood group antigens in a given individual is as distinctive as his fingerprint.

ABO system. The four groups of the ABO system are these: (1) group O, in which red cells contain neither A nor B agglutinogen; (2) group A, in which red cells contain A but not B; (3) group B, in which red cells contain B but not A; and (4) group AB, in which cells contain both A and B. In the United States the frequencies of O, A, B, and AB among white people are 42.2%, 39.2%, 13.5%, and 5.1%, respectively.

It was discovered in 1911 that the sera of group B and group O individuals contain, in addition to anti-A antibody, which agglutinates all A and AB cells, another antibody that agglutinates only about 80% of the A and AB cells. The red cells attacked by this antibody were designated A_1 and A_1B and were considered to belong to blood subgroups A_1 and A_1B; the remainder were placed in subgroups A_2 and A_2B. With a specific anti-A_1 testing serum (which did not agglutinate *all* A cells), it was readily shown that, of the 39.2% of the population in group A, 30.4% are in subgroup A_1, and 8.8% are in subgroup A_2.

The classic laws of Mendelian genetics govern the inheritance of the ABO antigens; however, it is necessary to postulate three genes, for A, B, and O, rather than two, for A and B. Our earlier implication that O cells merely lack A and B antigens is a simplification suggested by the practical fact that O cells are not agglutinated by anti-A or anti-B antibody. In fact, O cells contain a special O agglutinogen for which there

exists a corresponding serum anti-O agglutinin. Anti-O agglutinin is rarely found and is therefore often ignored.

With three ABO genes, each capable of occupying the same locus on the chromosome, the possible gene combinations are those given in Table 8.10.

MN system. As we have noted, red cells contain M and N antigens that are separate and distinct from the ABO antigens. Hence a group O individual could be OM, ON, or OMN. In the general population the frequencies of M, N, and MN are 28%, 22%, and 50% respectively. Serum rarely contains antibodies for these groups. When they occur they are usually more active in the cold than at 37°.

It is now known that two genes govern the MN system. One is either *M* or *N*. The other is either *S* or *s*. Since both operate in all individuals, M includes *MS* and *Ms*; N includes *NS* and *Ns*; and MN includes *MNS* and *MNs*. In medicolegal and genetic studies, S and s are often ignored because of the scarcity of anti-S serum. Accordingly, we shall omit them from our discussion.

The MN system provides one of the clearest examples of Mendelian inheritance in man. Only three gene combinations are possible: *MM*, *NN*, and *MN*. These produce individuals of blood groups M, N, and MN, respectively. The offspring of two M's must be M, and those of two N's must be N. If parents are M and N, all children must be MN. Parents M and MN have children half of whom are M and half MN. Mating of two MN's produces 25% M, 50% MN, and 25% N offspring.

TABLE 8.10   POSSIBLE GENE COMBINATIONS IN THE ABO SYSTEM

Gene from one parent	Gene from other parent	Gene combination (genotype)	Serologically demonstrable blood group (phenotype)
A	A	AA	A
A	O	AO	A
B	B	BB	B
B	O	BO	B
A	B	AB	AB
O	O	OO	O

The MN test is applied as follows in cases of disputed paternity. If an N mother has an MN child, a man of group N could not be the father; only a man of group M or MN could be the father. The test does not constitute proof of paternity, since any man of group M or MN could be involved, but it does exonerate suspects. The ABO and Rh systems may be used in refining the analysis.

Rh system. Landsteiner and Wiener quickly realized that the first discovered Rh antigen is but one of a complex set, stronger than the antigens of other systems and of great clinical and genetic importance. Since human serum normally does not contain anti-Rh antibodies, detection of Rh antigens demands testing sera prepared from other animals.

Eight or more Rh antigens have been found so far. Of these, only six are common in Caucasian red cells. The burgeoning complexity of the Rh system led to a prolonged dispute among experts over nomenclature. Wiener urged the use of the names rh', Rh_o, rh'', hr', hr_o, and hr''. The British geneticists R. R. Race and R. A. Fisher simplified these terms to C, c, D, d, E, and e, labeling respective antibodies anti-C, anti-c, etc. Blood is Rh^+ if its red cells contain C, D, and/or E. It is Rh^- if its red cells contain c, d, and/or e.*

The red cells of every human being carry at least three Rh antigens; some carry more. Though the corresponding antibodies are normally lacking, they may develop in the serum of an individual receiving several transfusions of blood whose red cells contain a foreign antigen. Thus anti-C antibody may arise in the serum of a man with cDe red cells when he is repeatedly transfused with CDe blood.

The genetics of the Rh system are much more complex than those of the ABO system. At least eight genes are required to explain the many possible combinations. Nevertheless, Mendelian laws are followed.

Major and "private" blood groups. We have referred to three major blood grouping systems: ABO, MN, and Rh. A lengthening list of similar systems whose antigens are present or absent in all bloods according to genetic instructions has been discovered. These systems, with their antigen symbols in parentheses, include Lutheran (Lu), Kell (K, k), Duffy (Fy^a, Fy^b), Kidd (Jk^a, Jk^b), and Lewis (Le^a, Le^b).

Other antigens have been found that appear to exist only in certain families. These constitute what are referred to as "private" or "family" blood groups. Undoubtedly some of them are heralds of new major systems. Others may belong to systems already established. Proof that a newly discovered antigen is part of a new system rather than an old one is tedious to develop. It depends ultimately on the demonstration that the gene for the antigen is inherited independently of the genes for all known systems. The slow rate of human breeding is an obstacle in this type of study. Another difficulty is the ever-present possibility of illegitimacy. If the genes for a particular system were never found to segregate independently of the genes of the ABO system, we should conclude that the two systems are related. If a single exception were discovered, the possibility should at least be considered that illegitimacy is responsible and that the appearance of independence is a false one.

Blood group substances. The task of the biochemist studying human blood groups is to isolate and characterize a blood group antigen to determine which of its chemical configurations accounts for its immunological specificity. As yet, such work has been largely confined to the antigens of the ABO system. Interestingly, the materials used as sources of these antigens are usually not red cells, for ABO antigens in red cells are tightly bound to surface lipids and proteins and are therefore insoluble. Fortunately, high concentrations of A and B antigens that are water-soluble have been found in the saliva, gastric juice, and other tissue fluids of some individuals.† Careful examination has shown that they are complex glycoproteins whose specificities reside in their carbohydrate moieties (e.g., galactose is essential for B specificity) and in the nature of the combinations of these with each other and with the protein moieties.

Little is known of the basis for blood group antibody specificity beyond what is known of the basis for antibody specificity in general (see p. 262 and Chapter 19). Powerful agglutinins called *lectins* that mimic human anti-A, anti-B, and other antibodies are found in plant seeds.

---

* R. E. Rosenfield and F. H. Allen, Jr., designate the six common antigens Rh1, Rh2, Rh3, Rh4, Rh5, and Rh6.

---

† Individuals who secrete A and B antigens in their saliva are called *secretors*. The secretor state is hereditary, depending on a dominant gene.

For example, a potent and useful anti-A lectin occurs in the lima bean.

Blood groups in genetics and anthropology. We have seen the clear Mendelian pattern of ABO and MN genetics. The antigens of the ABO and MN systems and all the other antigens whose inheritance has been traced are uncomplicated dominant traits—no known blood group antigen is recessive.

Like other genes, blood group genes occasionally mutate, and mutation is presumably the basis of their great variety. Most of the mutations probably occurred in antiquity. If mutations occurred commonly, we should be extremely ill-advised to rely on blood groups in cases of disputed paternity. Nevertheless, recently developed methods apparently detect rare mutations of blood group genes in individual human red cells. This is not surprising in view of the rate of red cell production—perhaps 250,000,000,000 cells each day. Even one mutation per minute of, say, A to B would yield increasingly more foreign cells in the circulation. Though they could scarcely harm the host by inducing an incompatibility reaction, it is just possible that they might constitute the stimulus leading to the formation of anti-B antibody in the serum. If so, serum antibodies are not naturally occurring but rather immune antibodies like all other antibodies.

In some situations blood contains a mixture of groups—e.g., in cases of mismatched transfusion or escape of fetal blood into the maternal circulation. The most curious mixtures are those found in *chimeras*—dissimilar twins who shared a circulation in the uterus and thereby exchanged group-specific red cell precursors (Chapter 19).

Because of their hereditary nature, blood groups have become invaluable in anthropology. Unlike other phenotypic traits such as height and skin color, a blood group is totally uninfluenced by environment. Hence it reflects only genotype. When we study the geographical distribution of blood groups, we find that the human race falls into a number of large divisions, each occupying an area of subcontinental size and characterized by a different set of average frequencies of the various blood group genes. For example, Central and South Ameri-can Indians are exclusively O. More than half of all Africans are cDe, a proportion far higher than in any non-African population. There is a well-known gradient from high A in the south of England to high O in Scotland, thought to be due to the progressive retreat of high O peoples before the onslaught of high A immigrants from the Continent.

Statistical evidence suggests the association of certain ABO blood groups and certain diseases, particularly ulcer, pernicious anemia, and gastric cancer. If this evidence is valid, blood grouping may have been instrumental in the species' survival.

BLOOD TRANSFUSION

Choice of blood. The chief practical significance of blood groups is in the field of transfusion therapy. Although knowledge of the complete blood group profiles of donor and recipient is desirable in theory, it is not necessary in practice. The technician determines the ABO and Rh groups with testing antisera and selects the donor from the correct ABO and Rh categories. Then he directly mixes samples of blood from the intended donor and the recipient to discover whether any rare group incompatibilities exist.

Modern transfusion techniques have reduced accidental error to the vanishing point. One of today's elegant procedures is illustrated in Fig. 8.37. A plastic bag stores donor blood. Its plainly lettered label shows that the blood is B, Rh+. The attached plastic tube contains samples of both anticoagulant-treated blood and clotted blood in heat-sealed segments that can be cut apart. Since the bag need not be opened, bacterial contamination is prevented. Each segment supplies red cells and serum for major and minor cross-matching tests. Numbers indelibly printed on the segments provide for positive identification at all times.

The main consequence of incompatible transfusion is intravascular hemolysis of donor red cells by the antibody of recipient serum. If severe enough, grave kidney injury and death may ensue. A cells injected into B blood are destroyed by anti-A antibody, and B cells injected into A blood are destroyed by anti-B antibody. Lacking A or B antigens, O cells cannot

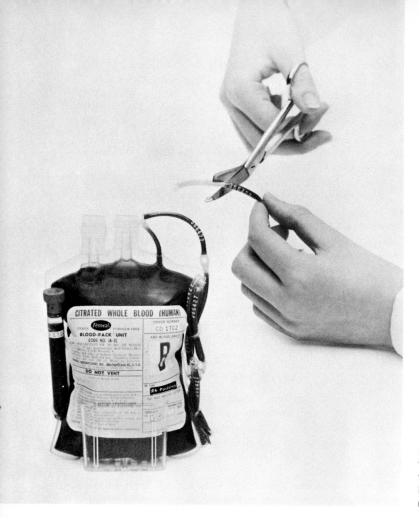

8.37 Use of a plastic bag and tubing for the handling of blood in modern transfusion therapy. (Courtesy of Fenwal Laboratories, Division of Travenol Laboratories, Inc.)

be attacked by recipient anti-A or anti-B antibody (and anti-O is rare); thus O blood can ordinarily be transfused into a member of any ABO group. For this reason O individuals are called *universal donors.* Occasionally donated O blood contains anti-A or anti-B antibody of sufficient potency to destroy recipient A or B red cells, the obverse of the usual transfusion reaction. Donors of such O blood are called *dangerous universal donors.*

Hemolytic disease of the newborn. In 1939 (before the Rh system had been discovered), Philip Levine and co-workers detected an unusual antibody in the serum of a woman who received a blood transfusion shortly after delivery of a stillborn infant. Her husband was the donor, and she suffered a severe transfusion reaction. With the discovery of the Rh system, the investigators postulated that the patient had Rh⁻ blood, that

her husband had Rh⁺ blood, and that the baby had inherited Rh⁺ blood from him. They concluded that the presence of this foreign factor in the mother's body caused the production of anti-Rh antibody in her serum —that she was, in fact, immunized by the Rh⁺ fetus in her uterus. Since antibodies readily cross the placental barrier, anti-Rh antibody would continuously attack the red cells of the fetus, with fatal results in this case. The patient's serum indeed contained anti-Rh antibody.

This relatively common sequence of events usually has less catastrophic effects. Typically, an infant is born who suffers from severe hemolytic anemia due to the action of anti-Rh antibody upon his red cells. His serum has a dangerously high bilirubin content due to excessive red cell destruction (see p. 248), and there are signs of intense compensatory red cell production in his bone marrow, liver, and spleen. The older name for this disorder is *erythroblastosis fetalis;* the newer name is *hemolytic disease of the newborn.* The studies of L. K. Diamond and others have shown that an infant with this disease can be saved by complete replacement of his blood, which removes the harmful maternal antibody. He is simultaneously bled and transfused until the antibody is diluted away.

An Rh⁻ mother may acquire anti-Rh antibody from previous transfusions with Rh⁺ blood or from previous pregnancies yielding Rh⁺ babies. Thus the amount of maternal anti-Rh antibody tends to increase with each pregnancy. Of course, the babies of an Rh⁻ mother have a 50% (or better, depending on the father's genetic pattern) chance of being Rh⁻ themselves, in which case they would be unaffected by high levels of maternal anti-Rh antibody.

Hemolytic disease of the newborn may develop similarly from fetal-maternal incompatibilities in the ABO system and other systems. In 1960, R. Finn observed that the presence of an ABO incompatibility between mother and fetus seemed to prevent the development of Rh incompatibility. This led to the important discovery that injections of γ-globulin, purified from pooled plasma from many donors, prevent Rh incompatibility with promising regularity. Presumably, the injected γ-globulin contains enough anti-Rh antibody to eliminate fetal Rh⁺ cells that leak through the placenta before the Rh⁻ mother can form a harmful amount of anti-Rh antibody.

REFERENCES AND SUGGESTIONS FOR FURTHER READING

General

DeGruchy, G. C., *Clinical Haematology in Medical Practice*, 2nd ed., Davis, Philadelphia, 1964.

Galton, D. A. G., ed., "Haematology," *Brit. Med. Bull.*, **15**, 1 (1959).

MacFarlane, R. G., and A. H. T. Robb-Smith, eds., *The Functions of the Blood*, Academic Press, New York, 1960.

Miale, J. G., *Laboratory Medicine–Hematology*, 3rd ed., Mosby, St. Louis, 1969.

Wintrobe, M., *Clinical Hematology*, 6th ed., Lea & Febiger, Philadelphia, 1967.

Wolstenholme, G. E. W., and M. O'Connor, eds., *Ciba Found. Symp., Haemopoiesis: Cell Prod. Regulation*, 1961.

"Formed Elements" of the Blood

Archer, R. K., *The Eosinophil Leucocytes*, Blackwell, Oxford, 1963.

Beck, W. S., "Deoxyribonucleotide Synthesis and the Role of Vitamin B₁₂ in Erythropoiesis," *Vitamins and Hormones*, **26**, 413, 1968.

————, "The Metabolic Functions of Vitamin B₁₂," *New Engl. J. Med.*, **266**, 708, 765, 814 (1962).

Bishop, C., and D. M. Surgenor, eds., *The Red Blood Cell*, Academic Press, New York, 1964.

Blaustein, A., ed., *The Spleen*, McGraw-Hill, New York, 1963.

Boggs, D. R., "Homeostatic Regulatory Mechanisms of Hematopoiesis," *Ann. Rev. Physiol.*, **28**, 39 (1966).

Bothwell, T. H., and C. A. Finch, *Iron Metabolism*, Little, Brown, Boston, 1962.

Caughey, W. S., "Porphyrin Proteins and Enzymes," *Ann. Rev. Biochem.*, **36**, 611 (1967).

Craddock, C. G., "The Production, Utilization, and Destruction of White Blood Cells," *Progr. Hematol.*, **3**, 92 (1962).

Dameshek, W., and F. Gunz, *Leukemia*, 2nd ed., Grune & Stratton, New York, 1964.

Galindo, B., and J. A. Freeman, "Fine Structure of Splenic Pulp," *Anat. Record*, **147**, 25 (1963).

Good, R. A., and A. E. Gabrielsen, eds., *The Thymus in Immunobiology*, Harper & Row, New York, 1964.

Gordon, A. S., ed., "Leukocytic Functions," *Ann. N. Y. Acad. Sci.*, **59**, 665 (1955).

Harris, J. W., *The Red Cell—Production, Metabolism, Destruction: Normal and Abnormal*, 2nd ed., Harvard Univ. Press, Cambridge, Mass., 1970.

Heller, J. H., ed., *Reticuloendothelial Structure and Function*, Ronald Press, New York, 1960.

Ingram, V. M., *The Hemoglobin in Genetics and Evolution*, Columbia Univ. Press, New York, 1963.

Jacobson, L. O., and M. Doyle, eds., *Erythropoiesis*, Grune & Stratton, New York, 1962.

Johnson, S. A., R. W. Monto, J. W. Rebuck, and R. C. Horn, Jr., eds., *The Platelets*, Little, Brown, Boston, 1960.

Johnson, S. A., D. L. Van Horn, H. J. Pederson, and J. Marr, "The Function of Platelets: A Review," *Transfusion*, **6**, 3 (1966).

Lehmann, H., and R. G. Huntsman, *Man's Hemoglobins*, Lippincott, Philadelphia, 1966.

MacDonald, G. A., T. C. Dodds, and B. Cruickshank, *Atlas of Hematology*, Williams & Wilkins, Baltimore, 1965.

Marcus, A. J., and M. B. Zucker, *The Physiology of Blood Platelets*, Grune & Stratton, New York, 1965.

Mayerson, H. S., "The Lymphatic System," *Sci. Am.*, **208**, 80 (June, 1963).

O'Brien, J. R., "Platelet Stickiness," *Ann Rev. Med.*, **17**, 275 (1966).

Pauling, L., H. A. Itano, S. J. Singer, and I. C. Wells, "Sickle Cell Anemia, A Molecular Disease," *Science*, **110**, 543 (1949).

Perutz, M. F., "The Hemoglobin Molecule," *Sci. Am.*, **211**, 80 (Nov., 1964).

Riley, J. F., *The Mast Cells*, Livingstone, Edinburgh, 1959.

Shemin, D., "The Biosynthesis of Porphyrins," *Harvey Lectures*, Ser. 50 (1954–55), 112 (1956).

Snook, T., "A Comparative Study of the Vascular Arrangements in Mammalian Spleens," *Am. J. Anat.*, **87**, 31 (1950).

Stohlman, F., Jr., ed., *The Kinetics of Cellular Proliferation*, Grune & Stratton, New York, 1959.

Weiss, L., "The Spleen," in R. Greep, ed., *Histology*, McGraw-Hill, New York, 1965.

———, "The White Pulp of the Spleen: The Relationships of Arterial Vessels, Reticulum and Free Cells in the Periarterial Lymphatic Sheath," *Bull. Johns Hopkins Hosp.*, **115**, 99 (1964).

Zuckerkandl, E., "The Evolution of Hemoglobin," *Sci. Am.*, **212**, 110 (May, 1965).

Plasma and the Plasma Proteins

Bernier, G. M., and F. W. Putnam, "Myeloma Proteins and Macroglobulins: Hallmarks of Disease and Models of Antibodies," *Progr. Hematol.*, **4**, 160 (1964).

Blood Coagulation

Biggs, R., and R. G. MacFarlane, *Human Blood Coagulation and its Disorders*, 3rd ed., Blackwell, Oxford, 1962.

Brinkhous, K., ed., *Hemophilia and Other Hemorrhagic States*, Univ. of North Carolina Press, Chapel Hill, N. C., 1959.

Feamley, G. R., *Fibrinolysis*, Williams & Wilkins, Baltimore, 1965.

Gitlin, D., "Current Aspects of the Structure, Function, and Genetics of the Immunoglobulins," *Ann. Rev. Med.*, **17**, 1 (1966).

Kline, D. L., "Blood Coagulation: Reactions Leading to Prothrombin Activation," *Ann. Rev. Physiol.*, **27**, 285 (1965).

———, "Chemistry and Biochemistry of the Fibrinolytic System," *Federation Proc.*, **25**, 31 (1966).

Laki, K., and J. A. Gladner, "Chemistry and Physiology of the Fibrinogen-Fibrin Transition," *Physiol. Rev.*, **44**, 127 (1964).

Lennox, E. S., and M. Cohn, "Immunoglobulins," *Ann. Rev. Biochem.*, **36**, 365 (1967).

MacFarlane, R. G., "An Enzyme Cascade in the Blood Clotting Mechanism, and Its Function as a Biochemical Amplifier," *Nature*, **202**, 495 (1964).

McKusick, V. A., "The Royal Hemophilia," *Sci. Am.*, **213**, 88 (Aug., 1965).

Blood Groups and Blood Transfusion

Boyd, W. D., *Genetics and the Races of Man*, Little, Brown, Boston, 1950.

Goldsmith, K. L. G., ed., "Blood Groups," *Brit. Med. Bull.*, **15**, 89 (1959).

Kabat, E. A., *Blood Group Substances*, Academic Press, New York, 1956.

McConnell, R. B., "The Prevention of Rh Haemolytic Disease," *Ann. Rev. Med.*, **17**, 291 (1966).

Mollison, P. L., *Blood Transfusion in Clinical Medicine*, 3rd ed., Blackwell, Oxford, 1962.

Race, R. R., and R. Sanger, *Blood Groups in Man*, 4th ed., Blackwell, Oxford, 1962.

Szulman, A. E., "Chemistry, Distribution, and Function of Blood Group Substances," *Ann. Rev. Med.*, **17**, 307 (1966).

Wiener, A. S., *An Rh-Hr Syllabus*, 2nd ed., Grune & Stratton, New York, 1960.

———, "The Blood Groups. Three Fundamental Problems—Serology, Genetics and Nomenclature," *Blood*, **27**, 110 (1966).

I began to think whether there might not be a motion, as it were, in a circle. Now this I afterwards found to be true . . . in all likelihood, does it come to pass in the body, through the motion of the blood; the various parts are nourished, cherished, quickened by the warmer, more perfect, vaporous, spirituous, and as I may say, alimentive blood; which, on the contrary, in contact with these parts becomes cooled, coagulated, and so to speak, effete; whence it returns to its sovereign the heart, as if to its source, or to the inmost home of the body, there to recover its state of excellence or perfection. Here it resumes its due fluidity and receives an infusion of natural heat . . . and thence it is again dispersed; and all this depends on the motion and action of the heart.

William Harvey, in *De Motu Cordis et Sanguinis*, 1628

9 Circulatory System

Anatomical Survey

INTRODUCTION

Functions of the circulatory system. Blood circulates continuously throughout the body in a network of blood vessels. It is driven through these vessels by the action of the heart, which is at the functional and anatomical center of the entire circulatory system. The *arteries* conduct blood away from the heart, and the *veins* return it to the heart for recirculation. Between the arteries and veins blood flows through the minute *capillaries*.

It is by means of the capillaries that blood makes contact with tissues, discharging their nutrients and receiving their wastes. Discussions of circulatory physiology have traditionally given greater emphasis to the heart than to the capillaries, doubtless because so much less is known of capillary properties. However, despite the functional importance of the heart—it is the prime source of pumping energy, and it maintains the pressure difference between arteries and veins that is essential for the circulation of the blood and the exchanges across capillary walls—capillaries are the *raison d'etre* of the whole circulatory enterprise, for at the capillary walls the circulatory system performs its ultimate functions.

Scheme of the adult circulation. Fig. 9.1 is a diagram of the adult circulatory system. Blood leaves the *left ventricle* of the heart through the body's greatest artery, the *aorta*. Branches of the

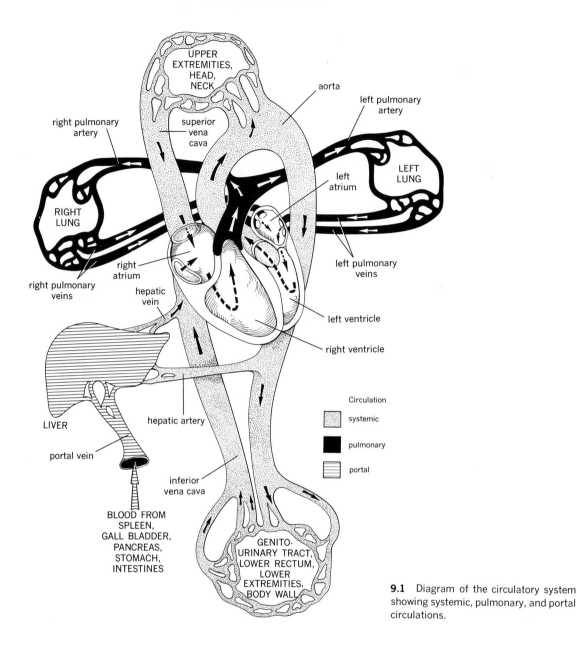

UPPER EXTREMITIES, HEAD, NECK

aorta

left pulmonary artery

right pulmonary artery

superior vena cava

left atrium

LEFT LUNG

RIGHT LUNG

right atrium

left pulmonary veins

right pulmonary veins

hepatic vein

left ventricle

right ventricle

LIVER

hepatic artery

Circulation

systemic

pulmonary

portal

portal vein

inferior vena cava

BLOOD FROM SPLEEN, GALL BLADDER, PANCREAS, STOMACH, INTESTINES

GENITO-URINARY TRACT, LOWER RECTUM, LOWER EXTREMITIES, BODY WALL

9.1 Diagram of the circulatory system showing systemic, pulmonary, and portal circulations.

aorta extend to the head, arms, internal organs, and legs. In the periphery, arterial blood moves into and through the capillaries, supplying the tissues with oxygen and emerging as venous blood, laden with carbon dioxide and other waste products. The veins from the lower portion of the body converge into the *inferior vena cava;* those from the head and upper extremities

converge into the *superior vena cava.* Both of these vessels empty into the *right atrium* of the heart. This completes the *systemic circulation.*

Although blood leaves the heart freshly oxygenated and returns to it exhausted of oxygen, nowhere in this circuit does it come in contact with the gas-exchanging surfaces of the lungs. It must traverse a second loop to do so—the *pul-*

monary circulation. In the pulmonary circulation, venous blood moves from the right atrium to the *right ventricle,* which pumps it via the *pulmonary arteries* (unique because they carry "venous" blood) into the lungs. After discharging carbon dioxide and taking up oxygen, the blood travels via the *pulmonary veins* (also unique because they carry "arterial" blood) to the *left atrium* and then to the left ventricle for another trip through the body. Anatomically there is a single heart (of two cylinders) between two lungs, but physiologically there is evidently one lung between two pumps (Fig. 9.2): the *right heart,* which receives venous blood from the body and pumps it to the lung; and the *left heart,* which receives arterial blood from the lung and pumps it to the body.

The *portal circulation,* which begins in the intestines and other digestive organs and ends in the liver, constitutes a third loop in the circulatory system. Note that the liver has a double circulation. The portal circulation serves a specific purpose: it transmits nutrients absorbed from the alimentary tract to the liver. However, the bulk of the liver substance receives arterial blood from the *hepatic artery,* a branch of the systemic circulation.* Both blood supplies drain into the *hepatic vein,* which connects with the inferior vena cava.

Later sections of this chapter will deal with the functions of the circulatory system. We should note here that structurally it comprises pumps and circular loops of tubing and that, among the many factors ensuring normal circulation, are the valves of the heart and veins, the elasticity of the arteries and arterioles, and the viscosity of the blood.

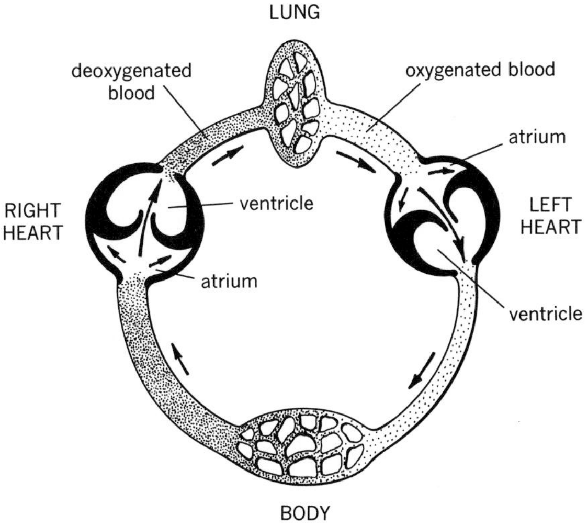

Evolution of the circulation. As we have seen, the heart consists of two pumps, side by side, beating in unison and thereby maintaining the two major circulations. It is instructive to consider the evolutionary origin of this arrangement. In lower orders, simpler hearts are adequate. For example, a fish has a single heart with

9.2   Functional relationships of systemic and pulmonary circulations.

two chambers, an atrium (auricle)† and a ventricle (Fig. 9.3A). A single stream of deoxygenated blood passes through the heart. The ventricle pumps the blood through the ventral aorta to the gills, where it is oxygenated. The oxygenated blood then passes through the dorsal aorta to the rest of the body. After discharging oxygen into the tissues, it returns through the veins to the atrium.

An amphibian has a more complicated pump. For example, a frog has two atria (Fig. 9.3B). Oxygenated blood returning from the lungs enters the left atrium, and deoxygenated blood returning from the body enters the right atrium, as in man. Here the resemblance ends, for in the frog both atria discharge their blood into a single ventricle, which pumps the mixture of aerated and nonaerated blood into a single aorta. The aorta branches in two beyond the heart, one branch going to the lungs and one to the rest of the body. This system is obviously inefficient since some of the blood entering the lungs has already been oxygenated and some of the blood reaching the tissues has already been deoxygenated.

---

* The unusual circulation through the liver is discussed in detail in Chapter 13. An understanding of liver function requires an understanding of the interrelations within the liver of the small branches of the portal vein and the hepatic artery.

---

† Although physiologists and physicians frequently use the words "auricle" and "atrium" interchangeably, in human anatomy the auricle is, strictly speaking, a small saclike appendage of the chamber properly called the atrium.

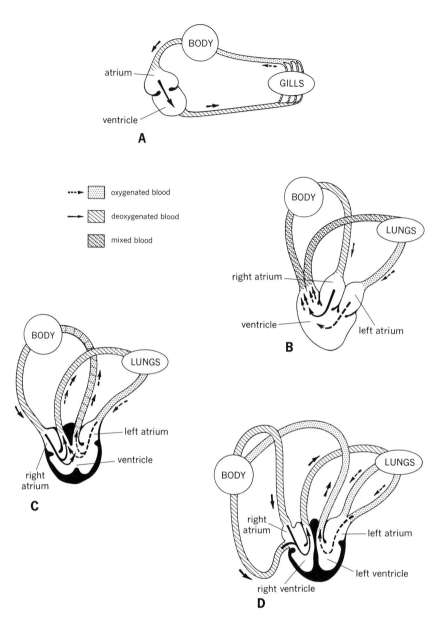

9.3 Diagrams of circulatory system of different species: A, fish; B, amphibian; C, reptile; D, mammal.

The reptile heart shows the beginning of a partition dividing the ventricle into two chambers (Fig. 9.3C). Mixing of oxygenated and deoxygenated blood still takes place, but to a lesser extent than in the frog. The partition is complete in birds and mammals (Fig. 9.3D), so that mix- ing is prevented. Later we shall learn how the embryonic development of the human heart retraces this evolutionary sequence and how a defect in embryonic development may result in mixing of the circulations with varying consequences depending on its locus.

HEART

Three components. The heart is a hollow, muscular organ, shaped like a blunt cone and about the size of a large fist, situated in the chest between the lungs and above the central depression of the diaphragm (see Figs. 6.7, 6.8). It is suspended by the great blood vessels with its broad end, or base, directed upward, backward, and to the right and its pointed end, or apex, directed downward, forward, and to the left.* Because its position is oblique, the terms "right heart" and "left heart" are more physiological than anatomical, for the right heart is almost in front of the left heart. In an adult the maximum impulse of the heart against the chest wall is felt in the space between the fifth and sixth ribs, a little below the left nipple, and about 8 cm to the left of the mid-line.

The heart has three main structural components: *pericardium, myocardium,* and *endocardium.*

The pericardium, which covers the heart, consists of an *external fibrous* membrane and an *internal serous* membrane (Fig. 9.4). The fibrous pericardium is attached above the heart to the large emerging blood vessels, covering them for about 1½ in. and finally blending with their sheaths; below the heart to the diaphragm; and anteriorly to the sternum. Posteriorly it is the sole barrier between the left atrium and the esophagus. The serous pericardium has a parietal layer and a visceral layer. The parietal serous pericardium lines the fibrous pericardium, and the visceral serous pericardium tightly invests the heart. The narrow space between the two layers is the *pericardial sac,* or *cavity.* In disease it may become distended with fluid or blood.

The myocardium is the muscular portion of the heart. It includes the muscles of the atria and the ventricles and the specialized tissue of the *conduction system.* It was noted in Chapter 5 that cardiac muscle is a particular form of muscle tissue, differing from skeletal muscle in the interlocking pattern of its fibers, which permits the rapid transmission of impulses from cell to cell (see Fig. 5.17). The muscular walls of the ventricles are much thicker that those of the atria, and the wall of the left ventricle is thicker than that of the right ventricle. Indeed, the walls of the atria and ventricles are not continuous with one another but are separated by a strong fibrous band around the "waist" of the heart. The only connection between the muscle tissue of the atria and that of the ventricles is a small bundle of specialized cells called the *atrioventricular bundle of His,* which is an essential part of the conduction system.

The endocardium lines the cavities of the heart, covering the valves and other internal structures. It is

---

* The words *base* and *apex* should be clearly understood in this context, for they are frequently used in descriptions of heart anatomy. We must remember that the base of the heart is closer to the head than the apex.

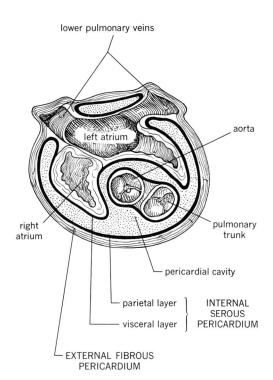

9.4 Cross section through atria, showing relations of heart, internal serous pericardium, external fibrous pericardium, and pericardial cavity. In actuality the two layers of the internal pericardium nearly touch each other. The diagram exaggerates the pericardial cavity.

continuous with the endothelial lining of the attached blood vessels.

Cavities. The mammalian heart has two atria and two ventricles (see Figs. 9.1, 9.3D). The ventricles are separated by a muscular partition, the *ventricular septum.* The *atrial septum,* between the atria, is similar but considerably thinner. The upper portion of the ventricular septum is also quite thin and is referred to as the *membranous septum.*

As shown in Fig. 9.5, muscles called *trabeculae carneae* extend inward from the inner surfaces of the ventricles. Some are attached along their entire lengths, forming ridges or columns; others are attached at their extremities but free in the middle; still others are the tent-shaped *papillary muscles* whose bases are continuous with the ventricular walls and whose apices are the fibrous cords, called *chordae tendineae,* that are attached to the cusps of the atrioventricular valves and control their movement.

Orifices and valves. The orifices of the heart are the passages between the atrium and ventricle on left and right and the openings into the large blood vessels con-

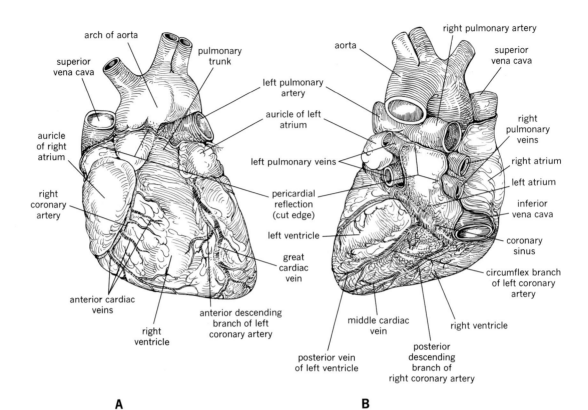

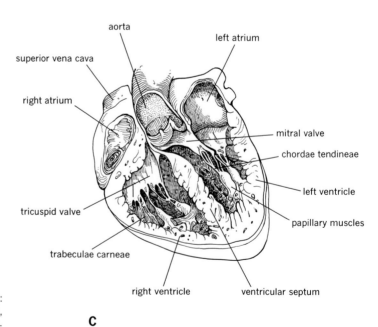

9.5 Gross anatomy of the heart: A, anterior view; B, posterior view; C, heart opened to show atria and ventricles.

nected with the heart. These are the superior and inferior vena cavae and pulmonary artery of the right heart and the four pulmonary veins and aorta of the left heart (Fig. 9.6A,B).

The atrioventricular orifices are somewhat constricted channels that are strengthened by fibrous rings. They and the orifices into the pulmonary artery and the aorta are guarded by valves—the *tricuspid, mitral, pulmonary,* and *aortic valves,* respectively. These four valves act to prevent the reversal of blood flow when the heart contracts. The orifices between the two great veins and the right atrium and between the left atrium and the four pulmonary veins are not guarded by valves.

The tricuspid valve lies between the right atrium and ventricle. It consists of three irregularly shaped flaps, or cusps (Fig. 9.6C,D), composed chiefly of fibrous tissue covered by endocardium. At their points of attachment, the cusps are continuous with one another, forming a ring-shaped membrane around the margin of the atrial opening. The cusps project into the ventricle and are supported by the chordae tendineae.

The mitral (or *bicuspid*) valve is the left atrioventricular valve. It consists of two cusps and is so named because of its resemblance to a bishop's mitre. Except for a somewhat stronger and thicker construction, it is generally like the tricuspid valve.

As shown in Fig. 9.6A,B, the tricuspid and mitral valves permit a free flow of blood from the atria into the ventricles because their free edges are directed into the ventricles. The pressure created by ventricular contraction forces the cusps together, completely closing the orifices and preventing any backward flow. The chordae tendineae limit the movement of the cusps so that they cannot be pushed back into the atria. The papillary muscles keep the chordae tendineae taut by contracting.

The orifice between the right ventricle and the pulmonary artery is guarded by the pulmonary valve; that between the left ventricle and the aorta by the aortic valve. Each of these consists of three half-moon–shaped cusps that have given rise to the name *semilunar valves* (see Fig. 9.6C). A small fibrous nodular body is attached to the center of the free edge of each cusp (Fig. 9.6E). At the roots of each great artery are three dilatations, one behind each cusp. These are the *sinuses of Valsalva.* But for the sinuses, the cusps would be thrown back upon the arterial wall when the valve is open.

The semilunar valves permit a free flow of blood from the heart into the arteries but form barriers against the passage of blood in the opposite direction. The nodular bodies assist in the closure of the valves, helping to make the barriers complete.

Conduction system. A special type of cardiac muscle is found in the tissue of the conduction system, one of whose components is the atrioventricular bundle of His. Since this tissue plays a fundamental role in the origin and propagation of the heartbeat, it is best understood in functional terms.

The heartbeat begins in the wall of the right atrium near the point of entrance of the superior vena cava. Here can be seen the first sign of muscular contraction. The contraction passes like a wave over the muscles of both atria simultaneously, driving blood into the ventricles. A fraction of a second later, the contraction spreads through the ventricles, ejecting blood into the arteries.

The principal features of this phenomenon are (1) the localized initiation of the contraction and (2) its wavelike progress over the heart. The site of the initial impulse is a small structure called the *sinoatrial,* or *S-A, node,* a mass of specialized cardiac muscle into which certain nerves lead. This is also referred to as the *cardiac pacemaker,* and the contractions beginning in it travel throughout the heart via strands of the specialized cardiac muscle.

Ordinary cardiac muscle conducts contractions, too, but not in such a controlled manner. A specialized conduction system makes possible the rapid and simultaneous spread of excitation waves through both ventricles. The result is a better synchronized and hence more efficient pumping action than would be produced if excitation waves wandered about randomly.

Atrial muscle is a continuum of interconnecting muscle cells. As we have mentioned, however, it is not continuous with ventricular muscle but is separated from it by a heavy fibrous band, the sole connecting pathway being the atrioventricular bundle of His. This bundle of specialized muscle cells begins in the *atrioventricular,* or *A-V, node,* a mass of specialized cardiac muscle low in the right atrium, and proceeds into the ventricular septum, where it divides into right and left branches, one running down each side of the septum just beneath the endocardium. Each branch divides further into smaller branches, the *Purkinje fibers,* which spread over the inner surfaces of the ventricles and ramify in the body of the musculature.

Blood supply of the heart wall: the coronary circulation. Like all active tissues, cardiac mus-

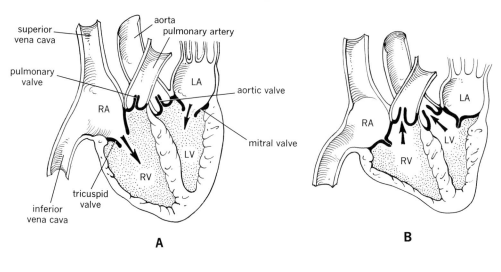

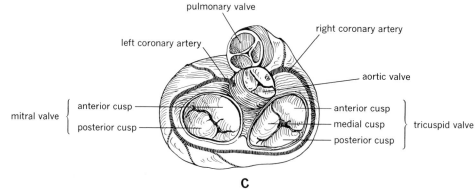

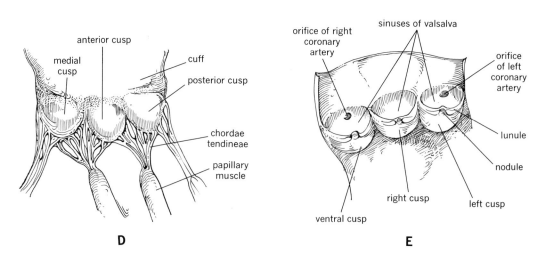

9.6 Chambers, orifices, and valves of the heart: A, when the ventricles relax, the pulmonary and aortic valves close, and blood flows from the atria through the open tricuspid and mitral valves; B, when the ventricles contract, the tricuspid and mitral valves close, and blood is forced through the open pulmonary and aortic valves; C, transverse section of valves (heart viewed from above after removal of atria); D, tricuspid valve (spread out); E, aortic valve (spread out).

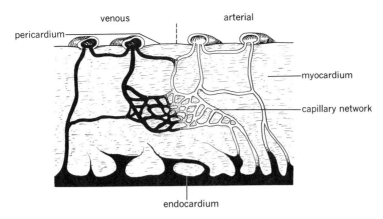

venous   arterial

pericardium

myocardium

capillary network

endocardium

9.7 Scheme of the myocardial circulation. (Adapted from J. C. B. Grant: *An Atlas of Anatomy,* 3rd ed., p. 508, The Williams & Wilkins Co., Baltimore, 1951.)

cle requires a constant blood supply. Though we might assume that the heart wall is nourished by the blood passing through the heart chambers, this is true of only the thin innermost layers. The wall is much too thick to derive its entire supply of oxygen and nutrients by diffusion from chamber blood.

Therefore, like other tissues, the heart has blood vessels of its own. The *right* and *left coronary arteries* arise from the *aortic sinus* just above the aortic valve, encircling the heart like a crown (see Figs. 9.5, 9.6E). The left coronary artery divides in two almost immediately. Blood is returned by two sets of veins: (1) those that empty into the *coronary sinus,* a wide venous channel flowing into the right atrium; and (2) three or four small *cardiac veins,* which bypass the sinus and enter the right atrium directly.

Within the heart wall the coronary arteries branch into an extraordinarily complex capillary network (Fig. 9.7), most easily explained in the light of its embryology. The primitive human heart (like the invertebrate heart) is nourished by chamber blood percolating through the spaces between adjacent trabeculae on the inside of the ventricular wall. Some of the spaces extend clear through the wall. Since the myocardium is therefore spongy, a coronary circulation is unnecessary. With later development coronary veins appear. They unite with some of the many spaces. Finally coronary arteries sprout from the future aorta, spread over the heart, and join the web already formed by the veins, and most of the spaces become conventional capillaries linking arteries with veins. A few remain as *Thebesian*

veins (for Thebesius, the anatomist who described them in 1708), opening directly into the chambers of the heart. Physiologists debate the significance of these vessels in the adult heart and the relative proportions of venous blood returning via them, the coronary sinus, and the cardiac veins.

One of the many safety factors built into the coronary circulatory system is the location of the orifices of the coronary arteries. Their positions tend to ensure that the blood being perfused through them is under high pressure. The architecture of the rest of the system affords a slight margin of safety in the capillary circulation between branches of the vessels on the right and left sides. Other body tissues use as little as 25% of the oxygen brought to them by the blood; cardiac muscle uses about 80%. Accordingly, there is little latitude for increased oxygen consumption on those occasions when increased activity raises the oxygen requirement. Moreover, there is little overlap in the territories served by the main arteries. If one is suddenly blocked, the affected portion of myocardium is unable to obtain blood except through the capillaries. The result is local death of muscle fiber, local cessation of muscular contraction, and—if life itself is not extinguished—local formation of scar tissue. For these reasons, the consequence of any interruption of coronary blood flow depends on the size of the occluded vessel.*

---

* The term for the muscle injury produced by sudden coronary obstruction is *myocardial infarction.* Its most common cause is *coronary thrombosis,* a hardening and narrowing of an artery (i.e., *atherosclerosis*) with the sudden development of a localized *thrombosis. Angina pectoris* is an old name for the severe chest pain produced by exertion in an affected individual. It is frequently due to a sudden temporary constriction or spasm of a coronary artery.

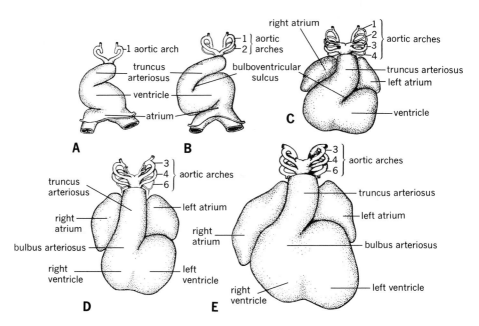

9.8 Development of the heart: A, in a 2.1 mm long embryo; B, at 3.0 mm; C, at 5.2 mm; D, at 6.0 mm; E, at 8.8 mm. (From B. M. Patten, *Human Embryology*, 2nd Ed., © 1953 by McGraw-Hill, Inc. Used by permission of McGraw-Hill Book Co., New York. After Kramer.)

Embryology of the heart and circulation. We now consider those aspects of embryonic development essential to an understanding of the adult circulatory system. The account parallels that of the evolution of the heart (see p. 285).

Most of the development of the heart and blood vessels occurs between the third and eighth weeks of embryonic life. In the early embryo the heart is a mere tube, which receives blood at its caudal end and discharges it at its cephalic end. This tubular heart expands into five sacculations, or bulges, as follows (reading from the caudal to the cephalic ends): *sinus venosus, atrium, ventricle, bulbus arteriosus,* and *truncus arteriosus* (Fig. 9.8A). There is no need for a pulmonary circulation at this stage, since the lungs are not in use.

As the tubular structure lengthens, it forms a loop so that the two caudal segments, the sinus venosus and the atrium, come to lie behind (i.e., dorsal to) and above the cephalic segments (Fig. 9.8B,C). By the fifth week of embryonic life, a septum separates the right and left atria and is growing within the ventricle. Next, the right atrium absorbs the sinus venosus. Then a septum appears in the truncus arteriosus, dividing it into

the pulmonary artery and the aorta. Curiously, this septum takes a spiral course, resulting in a crossover between the pulmonary artery and the aorta in the adult (see Fig. 9.6). In the eighth week of embryonic life, the ventricular septum is completed (Fig. 9.8D,E). Meanwhile, blood vessels are arising everywhere, lung buds are developing, and the liver and other organs are differentiating.

A characteristic of the chest contents of the embryo is their symmetry. In the adult thorax the aorta arches to the left and the heart lists to one side. The loss of bilateral symmetry is due to the disappearance of certain veins from the left side and of certain arteries from the right side. We shall review these changes in our discussion of the anatomy of the blood vessels.

Two facts underline the outstanding structural and functional differences between the prenatal (Fig. 9.9) and postnatal circulations: (1) the embryonic lungs contain no air; and (2) gases, nutrients, and waste products enter and leave the embryo via the *placenta,* where embryonic blood is in contact with maternal blood.

The embryonic heart is adapted structurally to the special requirements of embryonic life.

Since the lungs do not yet exchange gases, the blood flow to them is minimal. In fact, some venous blood is shunted directly from the right atrium to the left through a special opening in the atrial septum, the *foramen ovale*. The remainder of the blood in the right atrium follows the conventional route, through the right ventricle into the pulmonary artery. However, only a small amount of it goes to the airless lungs. Most of it traverses a second direct shunt from the pulmonary artery to the aorta, the *ductus arteriosus*. Clearly, this vessel has to disappear at birth; otherwise, little blood would manage to pass through the pulmonary circulation.

The blood reaching the left atrium (through the foramen ovale and the small pulmonary veins) proceeds to the left ventricle and the aorta, where it joins the blood from the ductus arteriosus. As might be expected, these bloods are poorly oxygenated.

How then does the embryo obtain its oxygen supply? Blood of the embryonic systemic circulation flows through the *hypogastric arteries* to the *umbilical arteries* and finally to the placenta, where waste products are transferred to the maternal blood and oxygen is taken up. Blood returns from the placenta through the *umbilical vein* toward the liver. For a time in early embryonic life, blood empties from the umbilical vein into the portal vein and thus passes through the liver on its way to the heart. However, the development of the *ductus venosus*, another shunt pathway, provides a shortcut to the inferior vena cava that bypasses the liver. In the embryo, then, oxygenated blood, admixed though it is with vena caval blood, enters the heart through the right atrium. In the adult the left atrium is the first heart chamber to receive oxygenated blood.

Inasmuch as both sides of the embryonic heart deliver blood to the body, the heart of the embryo resembles that of an amphibian or reptile (see Fig. 9.3B,C). Because in the embryo the right ventricle must pump blood through the collapsed lungs and the systemic circulation—whereas the left ventricle has only to pump part of the blood through the systemic circulation—the wall of the left ventricle is proportionately thinner than in the adult heart while that of the right ventricle is proportionately thicker. One consequence of the chronically deoxygenated state of the embryo is a relatively high level of circulating hemoglobin (see Table 8.3).

Transformations at birth. Dramatic and fundamental changes in the circulation occur at birth. With great suddenness the maternal oxygen supply is literally cut off. An immediate exchange of oxygen and carbon dioxide through previously collapsed lungs is therefore essential. The circulation must also be rebalanced. Instead of one circulation, two—the systemic and the pulmonary—must be established, and further mixing of oxygenated and deoxygenated blood must be prevented.

When the umbilical cord is tied off, circulation through the umbilical vessels and the ductus venosus ceases. In time the closed ductus veno-

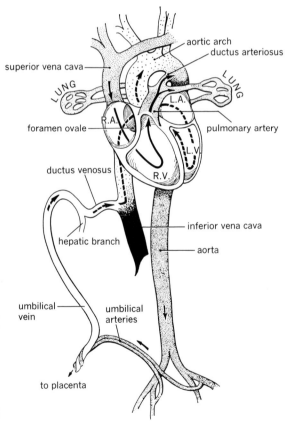

9.9 Diagram of the embryonic circulation. (From D. E. Reid: *Textbook of Obstetrics*, W. B. Saunders Co., Philadelphia, 1962.)

sus forms a fibrous ligament, the *ligamentum venosum*. With the first few breaths, the lungs expand, so that the pressure within them decreases and the flow of blood to them through the pulmonary artery increases. The volume of blood returned to the left atrium via the pulmonary veins increases correspondingly, raising the pressure in that atrium and closing the foramen ovale (though permanent closure may take a year).

A third passage must also close—the ductus arteriosus. With the onset of respiration, pressure in the pulmonary circulation drops below that in the systemic circulation, and the direction of flow through the ductus arteriosus is reversed. Soon the pressures in the two sides of the heart are equal, and flow through the ductus stops, partly owing to the contraction of muscles in its wall. Gradually the vessel becomes a fibrous cord, the *ligamentum arteriosum*, that remains through life.

Malformations of the heart result when development is arrested or defective at some point in this train of events. Such developmental errors may be genetic in origin or due to environmental assaults of one sort or another. For example, German measles in early pregnancy commonly causes congenital heart lesions; since they ordinarily affect only one portion of the heart, the remainder of the circulatory system usually develops normally. Other possible malformations include a single truncus arteriosus, due to faulty development of the septum between the aorta and the pulmonary artery; a single ventricle, due to faulty development of the ventricular septum; and a patent ductus arteriosus or foramen ovale, due to faulty closure. When we take up the physiology of the circulation, the functional consequences of such abnormalities will be obvious.

BLOOD VESSELS

Generally an artery, one or two veins, and a nerve occur together within a supporting sheath of connective tissue. The walls of the blood vessels reflect their different functions (Fig. 9.10).

Arteries and arterioles. The arteries carry blood from the heart to the capillaries. Although arterial structure changes with distance from the heart, every artery has three coats. The inner

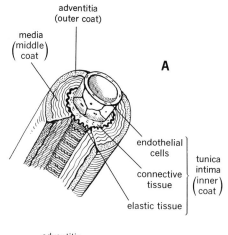

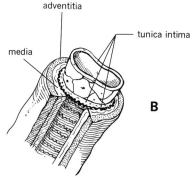

9.10 Major blood vessel types: A, artery; B, vein.

coat, or *tunica intima*, consists of three layers: a layer of endothelial cells; a layer of delicate connective tissue, which is most prominent in the larger vessels; and a layer of elastic fibers containing microscopic perforations. The middle coat, or *tunica media*, consists of elastic fibers and fine bundles of smooth muscle arranged in layers circularly disposed around the vessel. The outer coat, the *tunica externa* or *adventitia*, consists of loose connective tissue containing scattered smooth muscle cells. The blood that flows through an artery nourishes only the inner coat. The other coats are supplied through minute arteries, capillaries, and veins, called *vasa vasorum*, the "blood vessels of the blood vessels."

Because of the structure of the middle coat, arteries are both extensile and elastic. These properties enable them to receive the blood sent forth by each heartbeat, to withstand high pressure, to propel the blood onward, and to assume their original diameters in time to receive the blood leaving the heart with the next beat. The middle coat also gives form to an artery, so that an empty artery does not collapse and a cut artery remains open. However, as noted in Chapter 8, arteries in the vicinity of an injury contract somewhat to permit plugging of the wound by a blood clot. The constriction of a severed artery is an important factor in the arrest of hemorrhage.

The smallest arteries, the *arterioles*, contain more smooth muscle than elastic tissue. This reflects the great physiological importance of arteriolar vasoconstriction.

Capillaries. Of all the blood vessels, capillaries have the thinnest walls and the smallest diameters. They communicate freely with one another and form networks of variable shape and size that permeate all the tissues except those such as cartilage, hair, nails, cuticle, and cornea, which have almost no blood supplies. A capillary may range in diameter from 8 to 20 μ. This means that erythrocytes of 7 μ diameter must pass through some capillaries in single file and that a blood cell larger than a capillary must undergo distortion to negotiate passage.

The small sizes and large numbers of capillaries in the tissues result in some interesting statistics. For example, although 1 ml of blood takes 5 to 7 hours to traverse a single capillary, the entire blood volume circulates through the body in a few minutes. The total length of the capillaries is estimated to be 60,000 miles, and their total surface area to be almost 70,000 ft^2. Taken together, the capillaries constitute the body's largest organ, their total bulk being more than twice that of the liver.

Capillaries are most abundant in tissues that need blood for purposes other than routine local nutrition. In glandular organs capillaries provide the materials from which cells synthesize their secretions. Since the hormones produced by endocrine glands travel via the blood stream to distant sites of action, many endocrine glands possess rich capillary networks into which hormones are rapidly transferred. Similarly, in the alimentary tract rich capillary networks take up some elements of digested food; in the lungs they absorb oxygen and give up carbon dioxide;

and in the kidneys they participate in the elimination of waste products.

Veins and venules. From the capillaries blood flows into the *venules* and then the veins. The basic structures of veins and arteries are similar. Both have three coats—an inner endothelial lining, a middle coat of muscle, and an outer coat of loose connective tissue. The main differences between veins and arteries are as follows: (1) veins have relatively thinner walls, so that they collapse when empty; (2) veins have few elastic fibers in the middle coat; and (3) many veins have valves.

A venous valve consists of two or three crescent-shaped folds of the vein lining with their free edges pointed toward the heart. If for some reason blood is interrupted in its course toward the heart and is driven backward, the valve flaps are pressed together, the channel closes, and the retrograde flow is blocked.

Valves are most plentiful in the veins in which reflux is most likely, i.e., the veins of the extremities. Accordingly, more are found in the legs than in the arms. Valves are not present in the large veins of the trunk, in many small veins, and in veins not subjected to muscular pressure.

Topography of the vascular network. The arterial tree is a branching system whose offshoots grow progressively smaller and more numerous. The total cross-sectional area of the arteries at any level is always greater than that of the arteries from which they sprang. This means that as blood moves from the heart toward the capillaries, it flows in an ever-widening bed.

The smaller arteries relate to each other in several ways. When their distal ends unite at frequent intervals, a network called an *anastomosis* results. An anastomosis permits communication between blood currents and tends to obviate the effects of local interruption of blood flow and promote equality of distribution and pressure. The absence of extensive anastomoses between branches of the major coronary arteries is one reason why the myocardium is so defenseless against localized obstructions to blood flow. This lack is especially surprising in view of the extensive arrangements that circumvent local obstructions to circulation in the brain and the extremities. Anastomoses are of great impor-

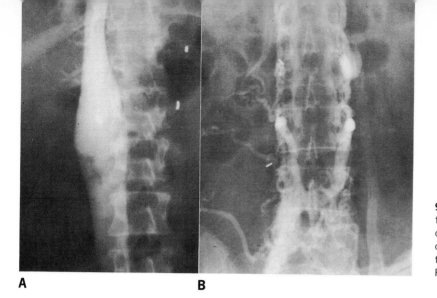

tance in surgery. When an artery is tied off, a new collateral circulation develops by extension of an existing anastomosis.

A *plexus* is a network formed by the union of a large number of veins in a limited area. Venous obstruction may be compensated for by the expansion of an existing plexus.

Until recently the only techniques for observing complex vascular networks were dissection—at either the operating table or the autopsy table—and the injection of colored plastics into organs taken from the body (see Fig. 6.4). Today investigators rely increasingly on the method of *angiography* (for arteries) or *venography* (for veins), in which blood vessels are x-rayed after injection of an indicator substance that is opaque to x-rays. The advantages of such an approach are evident. Fig. 9.11 illustrates the usefulness of venography in delineating an obstruction and the anastomotic collateral circulation in the inferior vena cava.

DIVISIONS OF THE VASCULAR SYSTEM

General plan. We must now extend our general conception of the circulation to include the principal divisions of the circulatory system (Fig. 9.12), which have deep significance for our later discussions of physiology.

Pulmonary circulation. The major vessels of the pulmonary circulation are the pulmonary artery, the capillaries within the lungs, and the pulmonary veins.

The main trunk of the pulmonary artery, about 2 in. long, arises from the right ventricle and passes upward, backward, and to the left. It then divides into right and left pulmonary arteries, which proceed to the right and left lungs, respectively, where they subdivide extensively and merge into a notably rich capillary network within the walls of the air spaces. The pulmonary veins, four short valveless veins, two from each lung, join to return the blood to the left atrium.

Systemic circulation. The systemic circulation includes the aorta and its many arterial branches, the capillaries, and the veins that ultimately drain into the superior and inferior vena cavae.

The aorta is the main trunk of the systemic arterial system. Emerging from the left ventricle, it extends upward and then gracefully arches toward the back and the left to begin its descent through the diaphragm and the abdomen, where it divides into the great *common iliac arteries* supplying the lower extremities. The aorta has different names throughout its length: the *ascending aorta*, the *aortic arch*, and the *descending aorta*. The part of the descending aorta above the diaphragm is called the *thoracic aorta*, and the part below the diaphragm, the *abdominal aorta*.

The only branches of the ascending aorta are the right and left coronary arteries (see Fig. 9.5). The aortic arch has three chief branches (Fig. 9.13): the *innominate artery*, which quickly divides into the *right subclavian* and *right common carotid arteries*; the *left subclavian artery*; and the *left common carotid artery*. The thoracic and abdominal portions of the descending aorta give off arteries supplying the chest and abdominal walls (the *parietal* groups) and the internal organs (the *visceral* groups). As already noted, the abdominal aorta ends by dividing into right and left common iliac arteries.

VEINS ARTERIES

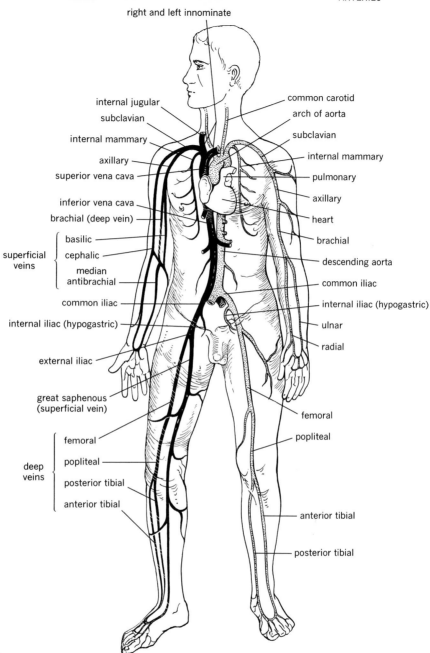

right and left innominate

internal jugular
subclavian
internal mammary
axillary
superior vena cava
inferior vena cava
brachial (deep vein)
superficial
veins
{ basilic
cephalic
median
antibrachial }
common iliac
internal iliac (hypogastric)
external iliac
great saphenous
(superficial vein)
deep
veins
{ femoral
popliteal
posterior tibial
anterior tibial }

common carotid
arch of aorta
subclavian
internal mammary
pulmonary
axillary
heart
brachial
descending aorta
common iliac
internal iliac (hypogastric)
ulnar
radial
femoral
popliteal
anterior tibial
posterior tibial

9.12 Circulatory system.

Regional arterial circulations. The pathway of blood from the aorta to individual regions of the body follows the branching of the arterial tree (see Figs. 9.12, 9.13).*

---

* Details of the regional circulations are considered in later chapters dealing with the systems of the body.

The arteries of the head and neck derive from the two common carotid arteries, each of which divides in the neck into an *external carotid artery* and an *internal carotid artery*. The former sends branches to the thyroid, throat, tongue, face, ears, and meninges, and the latter to the brain, eyes, forehead, and nose. A slight

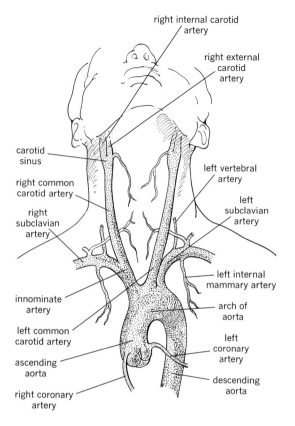

right internal carotid artery

right external carotid artery

carotid sinus

right common carotid artery

right subclavian artery

left vertebral artery

left subclavian artery

left internal mammary artery

innominate artery

arch of aorta

left common carotid artery

left coronary artery

ascending aorta

descending aorta

right coronary artery

9.13 Branches of the aortic arch.

bulbous dilatation occurs at the point where a common carotid artery bifurcates into external and internal carotid arteries. This is the *carotid sinus.*

The arteries of the upper extremities derive from the subclavian arteries. After sending branches into the chest, each subclavian artery continues into an arm to become an *axillary artery* and then a *brachial artery.* The latter divides into a *radial artery* and an *ulnar artery* at the level of the elbow.

The arteries of the chest arise directly from the aorta or its main branches. For example, those coming from each subclavian artery include the *vertebral artery,* which enters the skull to supply the brain, and the *internal mammary artery,* which descends along the undersurface of the anterior chest wall to supply the breast, diaphragm, and chest wall itself. The vessels arising directly from the thoracic aorta are generally named for the structures they supply. They include a visceral group (the *pericardial, bronchial, esophageal,* and *mediastinal arteries*) and a *parietal* group (the *intercostal, subcostal,* and *superior phrenic* arteries).

The arteries of the abdomen and pelvis also include a visceral group (the *celiac, superior mesenteric, renal,* and other *arteries*) and a parietal group.

The arteries of the lower extremities derive from the two common iliac arteries, each of which bifurcates into an *external iliac artery* and an *internal iliac artery.*

Embryology of the great arteries. We have mentioned the structural symmetry of the embryonic circulation and the structural asymmetry of the adult circulation. Our review of arterial anatomy accentuates this difference.

From Fig. 9.14A we see that the early embryo has six pairs of aortic arches, comparable, though not identical, with the gill (branchial) vessels of a fish (see Fig. 9.3A). They connect the two primitive ventral aortae (which fuse into one at the truncus arteriosus) with the corresponding primitive dorsal aortae (which fuse to form the descending aorta). Later the first, second, and fifth pairs of arches disappear (Fig. 9.14B,C; cf. Fig. 9.8C,D,E). Portions of the third pair develop into the proximal segments of the common and external carotid arteries. The changes in the two remaining pairs account for the loss of symmetry. The fourth pair develops into the great vessels of the systemic circulation. The ventral root of the right fourth arch becomes the innominate artery; the arch itself becomes the right subclavian artery. The left fourth arch becomes the aortic arch. The sixth pair develops into the pulmonary arteries. At first these might be designated right and left pulmonary arches (rather than aortic arches), because, when the truncus arteriosus splits into the ascending aorta and the pulmonary artery, the sixth pair (unlike the fourth) remains joined to the pulmonary artery, uniting it with the right and left dorsal aortae. The right sixth arch early loses its connection with the dorsal aorta; the left sixth arch retains its connection with the dorsal aorta as the ductus arteriosus. Not until this is broken are the pulmonary and systemic circulations completely separate.

In adulthood amphibians and reptiles have left and right aortic arches; birds have a right arch; and mammals have a left arch (Fig. 9.15). Rarely, owing to a developmental error, a human adult has both right and left arches or only a right arch. Similarly, both common carotid arteries may arise from the innominate artery; the right subclavian artery may arise from the descending aorta with no innominate artery intervening; etc. As long as arterial flow is adequate, most of these malformations are not injurious.

Regional venous circulations. Many veins resemble arteries in distribution and nomenclature. For example, the *common iliac veins* draining the lower extremities form from the convergence of the *external* and *internal iliac veins.* There are certain interesting exceptions, however.

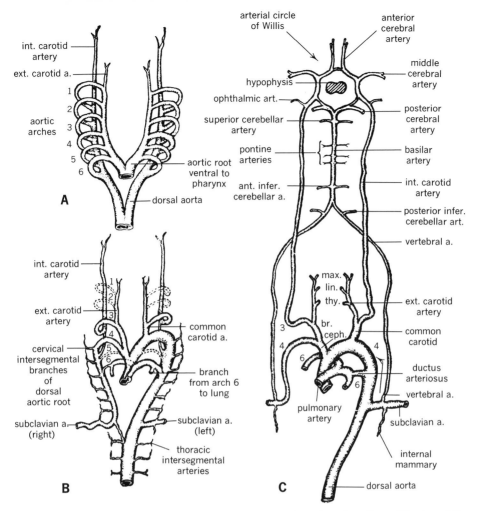

9.14 Development of the great arteries: A, plan of the complete set of aortic arches; B, modification of the arches in the embryo; C, derivatives of the arches in the adult. (From B. M. Patten, *Human Embryology*, 2nd Ed., © 1953 by McGraw-Hill, Inc. Used by permission of the McGraw-Hill Book Co., New York.)

One striking feature of the systemic venous circulation is the parallel arrangement of *superficial* and *deep veins*, connected by small *communicating veins* (see Fig. 9.12).* The superficial veins lie just beneath the skin in the superficial connective tissue and are sometimes called *cutaneous veins*. The deep veins generally travel with the arteries. A smaller artery (e.g., the brachial, ulnar, or radial artery) has a pair of accompanying veins, one on each side. These are called *companion veins*, or *venae comitantes*. A larger artery (e.g.,

---

*The systemic circulation contains yet another kind of vein, the *venous sinus*, typified by the sinuses in the skull that are formed from layers of dura mater, the fibrous sheath covering the brain (see Chapter 12).

the subclavian, axillary, or femoral artery) has only one accompanying vein. Other deep veins (e.g., the hepatic veins and the large veins carrying blood from the bones) have no arterial counterparts.

The veins of the lower extremities, pelvis, and abdomen empty into the inferior vena cava. Those of the head, neck, upper extremities, and thorax empty into the superior vena cava, as do the *azygos veins*, though they arise in the lower part of the body. The azygos veins constitute an important collateral channel for conveying blood from the lower part of the body to the superior vena cava when the inferior vena cava is obstructed (as in Fig. 9.11B,C).

The veins corresponding to the internal and external carotid arteries are the *internal* and *external jugular*

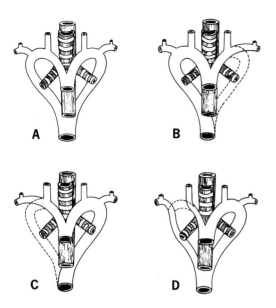

9.15 Aortic arches in various animals: A, amphibian; B, bird; C, man; D, man, abnormal. (Adapted from J. C. B. Grant: *An Atlas of Anatomy,* 3rd Ed., p. 522, The Williams & Wilkins Co., Baltimore.)

veins. The external jugular veins drain into the *sub-clavian veins.* In contrast to the arterial system, there are two innominate veins. The right internal jugular vein joins the right subclavian vein to form the *right innominate vein;* the left internal jugular vein joins the left subclavian vein to form the *left innominate vein.*

Embryology of the great veins. Like the arteries and other components of the early embryo, the veins display precise bilateral symmetry (Fig. 9.16). Three main pairs of embryonic veins enter the primitive sinus venosus—the *common cardinal veins,* the *umbilical veins,* and the *omphalomesenteric veins.* The common cardinal veins arise from the convergence of the *anterior* and *posterior cardinal veins,* which drain regions of the body later drained by the vena cavae. The umbilical veins come from the placenta (see Fig. 9.9). The omphalomesenteric veins pass to and through the liver and eventually unite to become the *portal vein* and the hepatic vein.

In development, symmetry is lost in all but the veins in the periphery of the body. Only two of the six great embryonic veins remain after birth: the right common cardinal vein becomes the terminal portion of the superior vena cava; and the segment of the right omphalomesenteric vein lying between the liver and the heart becomes the terminal portion of the inferior vena cava. The other four lose their connections with the heart, which are made unnecessary by new vessels like the innominate and *hemiazygos veins.* Since systemic

9.16 Development of superior and inferior vena cava: A, B, in a 4 week old embryo; C, at 5.5 weeks; D, E, at 6 weeks; F, at 7 weeks; G, H, at 8 weeks; I, at full term. (From B. M. Patten: *Human Embryology,* 2nd Ed., © 1953 by McGraw-Hill, Inc. Used by permission of the McGraw-Hill Book Co., New York.)

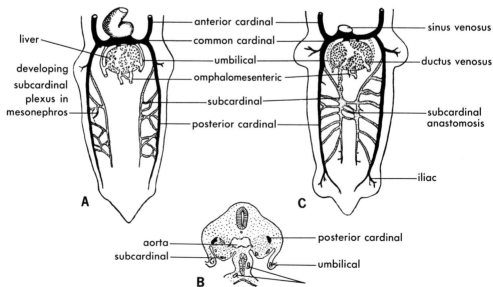

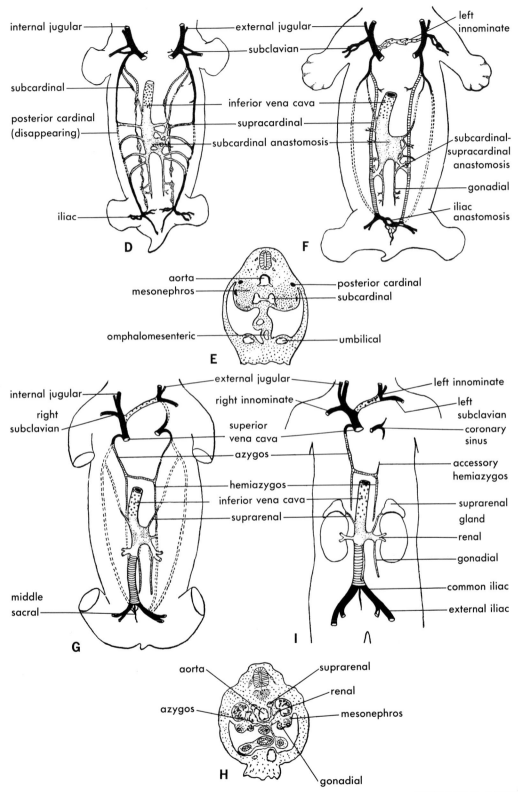

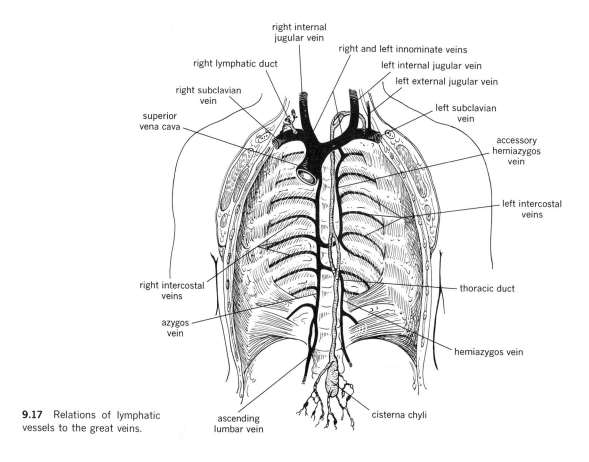

right internal
jugular vein

right and left innominate veins

right lymphatic duct

left internal jugular vein

right subclavian
vein

left external jugular vein

left subclavian
vein

superior
vena cava

accessory
hemiazygos
vein

left intercostal
veins

right intercostal
veins

thoracic duct

azygos
vein

hemiazygos vein

cisterna chyli

ascending
lumbar vein

9.17 Relations of lymphatic
vessels to the great veins.

venous blood can return to the heart through either the
superior vena cava or the inferior vena cava (both on
the right side of the body), there is no further need for
veins entering the sinus venosus from the left. They
disappear, leaving only the coronary sinus, a remnant
of the left common cardinal vein. The left common cardinal vein receives half of the venous blood from the
body. The coronary sinus has no tributaries from the
body, surviving simply as the vein into which the coronary veins empty.

The derivation of the inferior vena cava is complex,
involving the replacement of many channels by new
ones and the gradual development of various small local
networks into major ones. The *subcardinal veins*,
which appear early as channels parallel to the posterior
cardinal veins, eventually form the middle segment of
the inferior vena cava. The *supracardinal veins*, which
drain blood from the dorsal body wall into the posterior
cardinal veins near the heart, arise later. The affected
part of the right posterior cardinal vein joins the superior vena cava. That of the left posterior cardinal vein
vanishes. The portions of the supracardinal veins per-

sisting in the adult are the hemiazygos and azygos veins,
which empty into the superior vena cava.

LYMPHATIC VESSELS

General plan. The relations of the lymphatic vessels
to the great veins are shown in Figs. 8.5A and 9.17.
The thoracic, or left lymphatic, duct begins in the dilatation called the *cisterna chyli*, which is located in
front of the second lumbar vertebra. It ascends through
the aortic opening of the diaphragm, passes in front of
the thoracic vertebrae, and then curves over to the left
innominate vein at the angle between the left internal
jugular and left subclavian veins. It is about 15 in. long
and ³⁄₁₆ in. in diameter and has several valves. A bicuspid valve at the end excludes venous blood.

The right lymphatic duct is about ½ in. long. It is
formed by the convergence of three smaller vessels and
pours its contents into the right innominate vein at the
junction of the right internal jugular and right subclavian veins. Its orifice is also guarded by a bicuspid
valve.

Physiology

NATURE OF THE HEARTBEAT

Special properties of cardiac muscle fibers. To comprehend the remarkable mechanism of the heartbeat, we must first understand four fundamental properties of cardiac muscle fibers: *automaticity, conductivity, irritability,* and *contractility.*

Automaticity is the capacity to originate contractions spontaneously. As soon as the heart is formed, even before its muscle fibers acquire their typical histological pattern, it begins to contract rhythmically. Interestingly, the rhythm is more rapid in certain regions of the heart wall than in others. When the heart is removed from the body, it continues to beat for a time. The ventricles are first to stop beating, then the left atrium stops, and finally the right atrium stops. Even when contractions have ceased in both atria, careful examination may reveal a small area of contracting muscle where the superior vena cava enters the right atrium. This is the location of the sinoatrial node.

The activity of the isolated heart can be prolonged if it is continuously perfused with fresh blood or nutrient fluids. By this means mammalian hearts, including human ones, have been kept beating for several hours, amphibian hearts for several days, and chick hearts for several years. We must conclude, therefore, that the contraction of cardiac muscle is completely independent of nervous and humoral control.*

Conductivity, as noted earlier (see p. 290), is the capacity to transmit excitation as nerves transmit nerve impulses (see Chapter 4). A contraction impulse arising automatically in a circumscribed area of the heart spreads throughout the cardiac muscle like a ripple in a pool.

The irritability of cardiac muscle enables it to respond to external stimuli, such as certain

chemicals and electrical impulses, as well as to stimuli initiated by its own automaticity. The result of such stimulation is muscular contraction. When an activated muscle fiber contracts, it temporarily loses its irritability (Fig. 9.18). For a brief interval it does not respond to a second stimulus, no matter how intense. Then irritability gradually returns. During this period the fiber responds to very strong stimuli. These recovery phases are known, respectively, as the *absolute refractory period* and the *relative refractory*

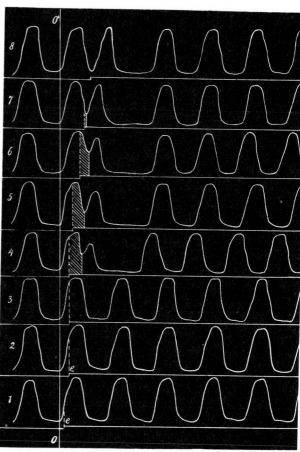

9.18 Effect of electrical stimulation on cardiac muscle (frog) during contraction and relaxation. In 1–3 the stimulus, e, is applied during the refractory period and elicits no response. In 4–8 the stimulus is applied later and elicits a premature ventricular contraction followed by a compensatory pause. The later the stimulus is applied, the shorter the time between stimulation and contraction. (From Howell, *Textbook of Physiology,* W. B. Saunders Co., Philadelphia, 1940.)

---

* The fundamental nature of the automaticity of cardiac muscle was dramatically demonstrated recently. Isolated single cells of mammalian cardiac muscle in vitro were seen to beat with a regular rhythm!

period. The contraction elicited during the relative refractory period has less amplitude than a normal contraction. Also, a *compensatory pause* follows an artificially stimulated premature beat, since cardiac muscle is still in the refractory state when the time comes for the next spontaneous beat.

The contractility of cardiac muscle has a unique feature that distinguishes it from that of skeletal muscle. Any stimulus of strength just sufficient to cause cardiac muscle to contract spreads over the entire muscle mass, producing a contraction as great as that produced by a maximum stimulus. This is a manifestation of a venerable law of physiology, the *all-or-none law of the heart*. In contrast, when a stimulus is applied to skeletal muscle, only the fibers near the stimulus respond; the more remote ones are not influenced. In other words, when the strength of a stimulus equals or exceeds the *threshold strength* required to evoke any response from cardiac muscle, the heart responds with a maximum contraction. Skeletal muscle responds to increasingly strong stimuli with increasingly strong contractions. The all-or-none law of the heart is explained by the fact that cardiac muscle fibers are so interconnected physically that any excitation is propagated throughout the whole heart.

Normal excitation of the heart: the pacemaker. The portion of the heart muscle that generates the most impulses per unit time imposes its own rhythm on the entire organ. Normally this is the S-A node, which is therefore designated the pacemaker of the heart.

It is relatively easy to demonstrate the pacemaker's action experimentally. If a test tube containing hot or cold water is placed near the S-A node, the heartbeat is altered. Cooling slows it, and warming accelerates it. Changing the temperature of other parts of the heart has no such effect. Moreover, if the S-A node is surgically removed or damaged by disease, the heartbeat is drastically slowed. The atrial muscle or the A-V node takes over as pacemaker, and the atria and ventricles go on beating simultaneously but at a slower rate. The normal rhythm set by the S-A node is termed the *sinus rhythm*. It is 75 to 80 impulses per minute. The atrial muscle and the A-V node can produce only 40 to 50 impulses per minute, but this is adequate to maintain the circulation. Many individuals have survived more than 25 years with the A-V node substituting as pacemaker for the S-A node. Impulse generators even lower in the conduction system can act as pacemakers if necessary.

What sort of mechanism is responsible for the emission of impulses with astonishing regularity some 110,000 times a day? Since an excised heart keeps beating when artificially supplied with fresh blood or nutrient solutions, these fluids presumably provide one or more ingredients essential for the heartbeat. In 1882 the English physiologist Sidney Ringer discovered that a solution containing sodium, potassium, and calcium ions in concentrations identical to those in whole blood sustains the beat of an excised frog or mammalian heart for some time. It was then found that in Ringer's solution with glucose added the heart performed for an even longer time. Thus scientists at the turn of the century knew that the heartbeat—and probably the pacemaking spark in the S-A node—depends upon the availability of (1) a balanced mixture of sodium, potassium, and calcium ions, (2) oxygen, and (3) an energy-yielding compound such as glucose.

Later research has extended our knowledge of the ways in which inorganic elements participate in the initiation and spread of impulses throughout cardiac muscle. As we learned in Chapters 4 and 7, a cell constantly expels sodium ions and accumulates potassium ions against sizeable concentration gradients. Consequently, sodium predominates in extracellular fluid and potassium in intracellular fluid. The result is a potential difference across the cell membrane of about 0.1 v, with the interior being negatively charged relative to the exterior. We shall see in Chapters 12 and 17 that this *resting potential* across the cell membrane has a role in nerve impulse transmission and in muscular activity. It is not known exactly how the cell membrane maintains the ionic segregation or the relative negativity of the cell interior. Perhaps the latter condition is due to the orientation of protein molecules within the membrane. If they were arranged in rows with their negatively charged ends ($R-COO^-$) pointed outward and their positively charged ends ($^+H_3N-R$) pointed inward, clusters of cations might be attracted to the cell surface. In any event the charge difference means that each cell membrane is a small storage battery.

At a given moment a cell membrane of the S-A node starts to leak slightly (Fig. 9.19A). Sodium ions begin to enter the cell, so that the potential difference between exterior and interior drops. When the difference has decreased by some critical amount—about 0.006 v—the minute pores in the cell membrane abruptly open. With the barriers now removed, sodium ions rush in, while potassium ions rush out. Owing to the sudden increase of membrane permeability, the inside becomes positive with respect to the outside. When this occurs, the membrane is said to be *depolarized*. The *action potential* created by depolarization in the S-A node initiates an impulse leading to a heartbeat. The excitation wave that subsequently spreads through the conduction system—called the *wave of accession*—is a progressive shift of the depolarized zone along the cell surfaces in the conducting fibers (Fig. 9.19B).

As soon as a contraction is over, the cell membrane *repolarizes;* that is, it reconstitutes itself as a storage battery. How this is accomplished is not certain, though it is known that glucose oxidation (or its equivalent) and intense metabolic activity fashion a metabolic pump that literally ejects the sodium ions which entered the cell at the beginning of the cycle. The resting potential of the membrane is restored, and the cell returns to the resting state. Before all the sodium ions are gone, some potassium ions leak out through the still permeable membrane. Physiologists believe that the restitution process thus overshoots and that the cell exterior inevitably becomes too positive. The cell's attempt to reverse this situation, to decrease the positive charge outside, reinitiates the inward leakage of sodium and hence the entire cycle.

Such spontaneous firing can be produced in any type of cell by alteration of the stability (or permeability) of the cell membrane (e.g., by the placing of appropriate molecules on the cell surface). The sense of smell involves such a mechanism. The molecules that we smell transform our nasal end organs into "pacemakers" that send fine nerve impulses to the olfactory centers of the brain. A similar mechanism is the basis of

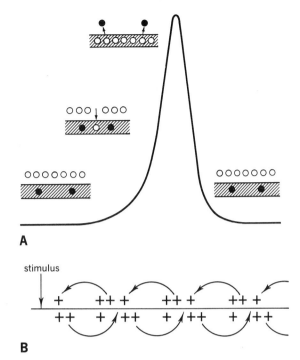

A

B

9.19 Electrophysiology of the conduction system: A. Generation of an electrical impulse in the S-A node. Diagrams are of cross sections of a cell. At bottom left positively charged potassium ions (black) are inside the cell, and a larger number of positively charged sodium ions (white) are outside it. Therefore, the inside of the cell is negatively charged with respect to the outside. At center left a sodium ion leaks across the cell membrane. At top left sodium ions rush into the cell, and potassium ions rush out, depolarizing the cell and producing an action potential (peak in curve). At bottom right the original situation is restored. B. Transmission of an impulse through the conduction system, not as an electric current but as an electrical reaction. When stimulated, a cell in the S-A node discharges and generates a local current, which causes the depolarization and discharge of adjacent cells. In effect a wave of + charge passes through the system, as indicated by the arrows.

action of many drugs, including some that affect the heartbeat.

The frequency with which a cardiac pacemaker cell discharges (i.e., with which its membrane depolarizes) depends upon at least two factors: (1) the rate at which sodium ions leak into it; and (2) the potential difference across the cell membrane necessary to discharge it completely. The rate of sodium entry is increased by warming and decreased by cooling. This accounts for the local effects described earlier (see p. 304) and perhaps also for the rapid heartbeat of an individual with fever. The potential difference required to discharge the cell depends on the characteristics of the membrane. The concentration of calcium ions is important here, for calcium promotes membrane stability. If its concentration falls below a certain critical value, the rate of discharge increases; if its concentration is too high, the rate decreases. The rate of discharge is also affected by nervous factors, blood oxygen and carbon dioxide content, blood pH, hormones and drugs, notably digitalis, etc.

ELECTRICAL ACTIVITY OF THE HEART

Transmission of impulses through the conduction system. The spark from the pacemaker is transmitted to the heart muscle via the special conduction system This does not mean that electricity flows as it does in copper wires. We may recall (see Chapter 4) that any irritable tissue develops at a site of stimulation a localized potential negative relative to the neighboring area. The negativity spreads to the neighboring area, meanwhile subsiding at its original locus. By progressively moving so as to neutralize adjacent positive areas, an area of depolarization is able to migrate (see Fig. 9.19B). When cardiac pacemaker cells discharge, therefore, they generate a localized electric current, which causes depolarization (localized surface negativity) and discharge of an adjacent group of cells and initiates the accession wave that passes through the system until the muscle is induced to contract.

This reaction can be demonstrated experimentally with electrodes and a simple galvanometer. If two electrodes are placed on the atrium, the one nearer the S-A node records a negative potential first. The times necessary for impulses to travel from the S-A node to various points in the heart are shown in Fig. 9.20. The node transmits an impulse over the conduction system of the heart as a series of localized electric currents, relaying it step by step over special tissue to the contracting cells. On arrival at these cells, the undiminished electrical charge triggers

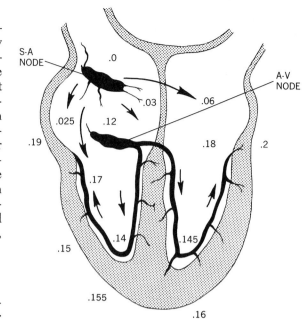

9.20 Transmission of impulses through the conduction system, showing the time (in seconds) required for passage from one area to another. (Adapted from A. A. Luisada: *Development and Structure of the Cardiovascular System.* McGraw-Hill, Inc. Used by permission of the McGraw-Hill Book Co., New York.)

the cellular metabolic process that releases the chemical energy for contraction.

Fig. 9.20 reveals that the time needed for an impulse to travel from the S-A node to the outermost portions of the atrium is brief—0.06 second at the most. Twice that time is needed for the impulse arising in the S-A node to emerge from the A-V node. This delayed conduction in the normal A-V node accounts for the sequential contractions of atria and ventricles. It is obvious that without a reasonable interval between the end of atrial activation and the onset of ventricular activation, the atria and ventricles would contract simultaneously. Since the atria could not then empty themselves, atrial function would be lost.

Electrocardiogram. An impulse originating in a living heart is also transmitted to neighboring structures of the body—being disseminated by the same direct pathways as from an isolated heart suspended in a body-sized volume of salt water. Electrical impulses at the body surface were first reported by A. D. Waller in 1879. Willem Einthoven later used a galvanometer and electrodes attached to any pair of points on the body surface to obtain a continuous graphic record of the electric currents emitted by the beating heart. This became known as the *electrocardiogram*, or *ECG*. A more outstanding development in the history of physiology is hard to imagine.

It soon became clear that the observed electrocardiographic pattern depends upon the locations of the positive and negative electrodes. If an excitation wave moves toward a positive electrode, the galvanometer needle is deflected in a positive direction; if the wave recedes from a positive electrode, the deflection is in a negative direction. For a complete picture of the heart's electric field, Einthoven decided to place the heart in the center of a large imaginary triangle, with electrodes at the three corners.

So that we may understand the physiological significance of the electrocardiogram, let us consider the *Einthoven triangle* (Fig. 9.21). The heart is regarded as a point source of electrical impulses at the center of an equilateral triangle, whose three corners are at the junctions of the arms to the torso, RA and LA, and in the pubic region, LL. Since the limbs do not affect electrical potentials, an electrode low on the right arm behaves exactly as it does at RA, and an electrode low on the left leg behaves exactly as it does at LL. Thus Einthoven chose, for convenience, the right arm, the left arm, and the left leg as sites for electrodes. He was able then to record in sequence the potential differences of three possible electrode combinations—each affording a different "view" of the electric field around the heart. A given electrode pair is called a *lead*. The three standard leads are as follows: I, right arm and left arm; II, right arm and left leg; and III, left arm and left leg. The arrows indicate the directions in which current must flow to cause a positive deflection of the galvanometer needle; current moving in the opposite direction causes a negative deflection.

Fig. 9.22A is an electrocardiogram of a normal heart in lead III. It shows a series of similar groupings of positive and negative deflections, with the intensity (voltage) of the deflections on the vertical axis plotted against time on the horizontal axis. Each grouping represents a single heartbeat. Hence the heart rate can be calculated. When a grouping is magnified (Fig. 9.22B), it demonstrates a small preliminary *P wave*, a brief but vigorous *QRS complex*, a low rolling *T wave*, and a small (and frequently absent) *U wave*.*

The P wave corresponds to the spread of excitation from the S-A node throughout the atria to the A-V node. The electrical impulse producing it causes atrial contraction. The PR interval is the time from origination of the impulse to the end of its passage through the A-V node. The QRS complex marks the beginning of the spread of excitation throughout the ventricles. The ST segment is the time from complete ventricular depolarization to the onset of repolarization. The ST interval is the time from complete ventricular depolarization to the completion of repolarization. The T wave results from ventricular repolarization. The meaning of the U wave is still under debate.

By studying electrocardiograms, physicians and physiologists can reach astonishingly accurate conclusions regarding the mechanism of myocardial activa-

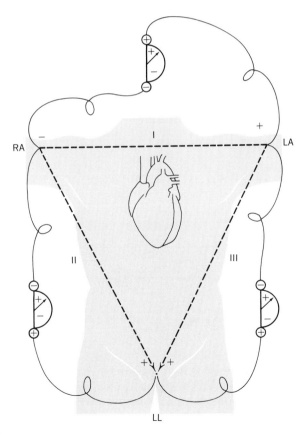

9.21 The Einthoven triangle.

---

* Einthoven proposed these names from the middle of the alphabet to avoid confusion with other waves known to physiologists—e.g., A (atrial) waves, C (carotid) waves, and V (venous) waves.

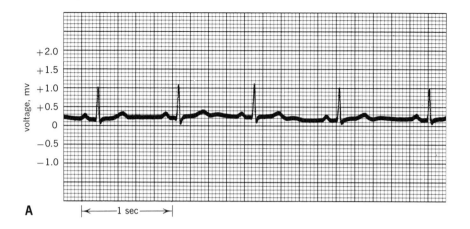

A |← 1 sec →|

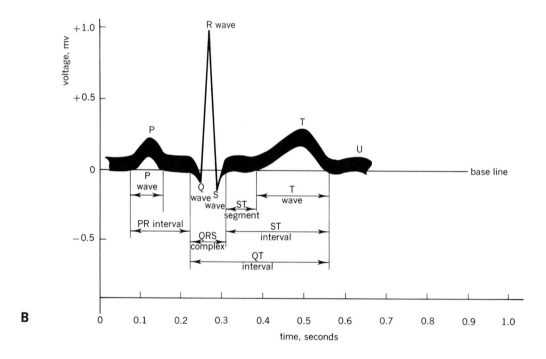

B

9.22 Electrocardiogram of a normal heart in lead III: A, total record traced in 4 seconds; B, single heartbeat magnified. (A, courtesy of Dr. Roman DeSanctis.)

tion; the locus of the pacemaker and the regularity of the heartbeat; and the size and integrity of the various chamber walls.

Defects of rhythm and conduction. Considering the complexity of the excitation mechanism, it is remarkable how seldom it becomes im-

paired. When it does go wrong, the physician must analyze the disturbance and then attempt to determine its cause. Many defects that lead to strikingly abnormal electrocardiograms do not seriously diminish the heart's functional efficiency. Others have catastrophic conse-

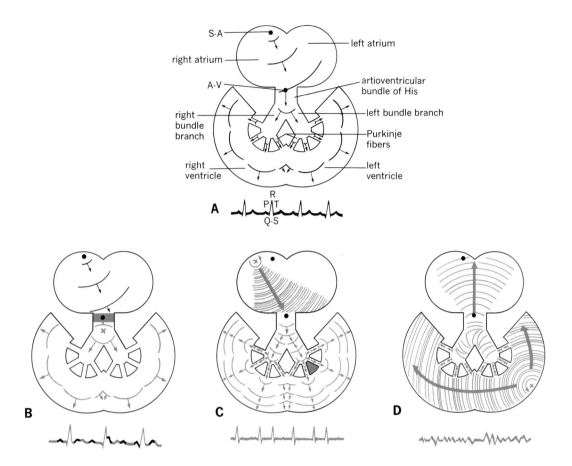

9.23 Diagrams illustrating impulse formation and conduction: A, normal sinus rhythm; B, complete atrioventricular block, with atrial rate 80 and ventricular rate 32; C, atrial fibrillation, with atrial rhythm chaotic and ventricular rate 160; D, ventricular fibrillation, with ventricular rhythm chaotic. The curved lines represent impulses; the arrows the directions in which they travel; and the spaces between the curved lines the time intervals between successive impulses (the wider the space, the slower the heartbeat). Beneath each diagram is an electrocardiogram. Solid lines indicate normal features, and screened lines abnormal features.

quences. In every case electrocardiography is a key to precise diagnosis.

A disorder may be characterized by: (1) a heart block—a delay or interruption in the conduction of excitation waves; (2) an abnormality of the pacemaker or its replacement by a pacemaker lower in the conduction system; or (3) the appearance in the myocardium of areas—referred to as *ectopic* or *irritable foci*—that initiate impulses in an irresponsible manner. When any of these disturbances alters the rhythm of the heartbeat, an *arrhythmia* is said to be present.

Although most arrhythmias are caused by disease, some arise physiologically.

Fig. 9.23 shows the sources of impulse formation, the pathways of impulse conduction, and the electrocardiograms of normally and abnormally functioning hearts. We have already traced impulses in the normal heart, where the S-A node is pacemaker and the rhythm is a normal sinus rhythm (see Fig. 9.23A). Many physiological factors may prevent the S-A node from dispatching impulses with perfect regularity. For example, inspiration and expiration may quicken and slow the heartbeat, producing *sinus arrhythmia*. Emotional tension and fever may speed the heartbeat, producing *sinus tachycardia*. Both these conditions are due to nervous influences upon the pacemaker.

A *premature beat* occurs when an irritable focus somewhere in the myocardium initiates an impulse that is not synchronized with the orderly rhythm of the S-A node. The result is called a *premature atrial contraction* or a *premature ventricular contraction*, depending on whether the abnormal focus is in an atrium or a ventricle. The P wave of a premature atrial contraction is usually inverted, since the excitation wave moves in an abnormal direction in the atria. However, the QRS complex is usually normal in appearance, since the ventricles cannot distinguish that the impulse arose ectopically. The QRS complex of a premature ventricular contraction is typically bizarre. Generally, premature beats cause only "flip-flop" sensations in the chest. They have no adverse affects.

When the S-A node is depressed, the A-V node may take over as pacemaker, producing an *A-V nodal rhythm*, characterized by slow ventricular contractions and inverted P waves due to conduction from the A-V node backward to and through the atria. Though this mechanism is not notably efficient, it does not necessarily weaken the pumping action of the heart.

A *complete atrioventricular block* results from physical destruction or functional depression of the atrioventricular bundle of His (see Fig. 9.23B). No impulses are transmitted from the atria to the ventricles, and the two pairs of chambers contract quite independently.*

Disorders in which the heartbeat is excessively fast are designated *tachycardias*, in contrast to *bradycardias*. These often occur in attacks or paroxysms, with the sudden appearance and disappearance of an ectopic focus causing rapid regular impulse formation for minutes, hours, or days. In *paroxysmal atrial tachycardia* an atrial pacemaker initiates impulses that stimulate the ventricles to contract about 180 times per minute. The P waves almost vanish in a forest of QRS complexes. The ventricles pump inefficiently, and the victim may become weak and faint. When atrial tachycardia is extreme (300 or more contractions per minute), the overwhelmed atrioventricular bundle of His unable to conduct all the impulses that it receives transmits to the ventricles every second, third, or fourth impulse. The

name *atrial flutter* is applied to this disorder because of the characteristic movements of the atria.

Atrial fibrillation (see Fig. 9.23C) is probably the most important of the common disturbances of cardiac rhythm. In all the cases discussed so far, excitation arises in an orderly manner. Here a disorderly flood of impulses is initiated in an ectopic atrial focus at the rate of 350 to 600 per minute. Because some atrial fibers are contracting while others are relaxing, the atria never actually complete a contraction. Rather they quiver wildly in the pattern called *fibrillation*. Only a small number of the impulses result in ventricular contractions, for the ventricles are in a refractory condition most of the time and unable to respond to the barrage of impulses coming down from the atria. If the ventricular contraction rate is not too fast, the circulation may be quite satisfactory.

The next two arrhythmias, in which impulses are initiated in the ventricular walls, have ominous significance. *Ventricular tachycardia* is characterized by rapid regular ventricular contractions. Though less serious than *ventricular fibrillation*, the most dangerous of the arrhythmias (see Fig. 9.23D), it is often a prelude to it. Ventricular fibrillation probably occurs as the terminal event in the course of death from any cause, irrespective of the primary event.

Sometimes the heart stops beating because of a temporary biochemical derangement not due to intrinsic disease of the heart itself. Such *cardiac arrest* may follow an overdosage of anesthetic, an acute oxygen deficiency, etc. The preservation of life depends on swift action to restore the heartbeat. The treatment consists of immediate external or internal cardiac massage. If the time elapsing from the moment of arrest to the moment of restoration exceeds 3½ minutes by even a few seconds, permanent damage is done to the nervous system. If the heartbeat is restored in time, there is often no further difficulty.

HEART AS A PUMP

William Harvey and the discovery of the circulation. The quotation on p. 283 commemorates William Harvey's discovery in 1628 of the circulation of the blood, one of the greatest events in the history of science. We esteem it so highly for a number of reasons. In the first place, the work of Harvey discredited the belief of fourteen centuries that the heart is not a muscular organ and that blood passes through the septum between the right and left ventricles. Second, by conceiving of the heart as a pump, Harvey introduced an entirely new view of the living organism. The organism was for the first time regarded as a material machine differing

---

* In atrioventricular block (or block elsewhere in the conduction system) the rate of ventricular contraction may be too slow to maintain adequate circulation. To combat this situation—particularly when there is reason to believe that it is temporary—an electronic cardiac pacemaker has been devised. Electrodes applied to the chest and back send out intermittent electrical impulses 70 times per minute that successfully replace the nonfunctioning S-A node. In many instances external cardiac pacemakers have been life-saving. More recently internal pacemakers, with direct connections to the heart, have been implanted surgically.

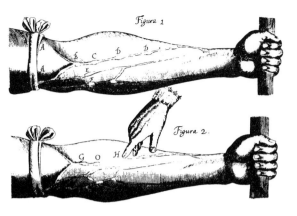

Figura 1

Figura 2

9.24 Illustrations from Harvey's presentation of experiments on venous flow. (From William Harvey: *De Motu Cordis et Sanguinis*, 1628.)

from man-made machines only in degree of complexity. Finally, he utilized the *scientific method* in arriving at his conclusions—its first important application in biology. In many ways, Harvey's discovery marked the dawn of modern biology.

Harvey observed that the beating heart expels the blood within it. He reasoned that, if the heart contains 2 oz of blood and beats 80 times per minute, then it must eject into the body 10 lb of blood per minute. It had previously been thought that blood derives from the food that is eaten. But the formation of 10 lb of blood each minute from the amount of food a man consumes seemed impossible. Thus Harvey postulated that blood expelled by the heart must circulate through the body and return to the heart. This was his scientific hypothesis. He next performed experiments to test the hypothesis.

Fig. 9.24 is a reproduction of a famous plate in Harvey's treatise *De Motu Cordis et Sanguinis in Animali* (On the Motion of the Heart and Blood in Animals). In this simple experiment Harvey showed that obstruction of a vein causes a pooling of blood on the side of the obstruction *away* from the heart. He also showed that bleeding arises from the cut end of a severed artery *nearest* the heart and from the cut end of a severed vein *farthest* from the heart. With elegant simplicity he demonstrated the function of the venous valves, concerning which he wrote that

"so provident a cause as nature had not so plac'd many valves without design." Harvey's conclusions are summarized in the quotation on p. 283.

We have discussed the anatomy and electrical activity of the heart. We shall now consider its mechanical function as a pump and see what

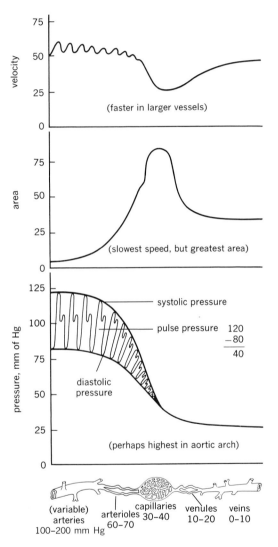

velocity

(faster in larger vessels)

area

(slowest speed, but greatest area)

pressure, mm of Hg

systolic pressure

pulse pressure   120
                 −80
                  40

diastolic pressure

(perhaps highest in aortic arch)

(variable)   arterioles   capillaries   venules   veins
arteries   30–40   10–20   0–10
100–200 mm Hg 60–70

9.25 Relationships among (A) rates at which blood moves through various blood vessels; (B) areas of various blood vessels; and (C) arterial, capillary, and venous blood pressures. Systolic pressure is the pressure during cardiac contraction, diastolic pressure is the pressure during cardiac relaxation, and pulse pressure is the difference between them.

modern physiology has added to Harvey's monumental contribution.

Elementary hemodynamics. Hydrodynamics is the study of physical laws governing the flow of water. *Hemodynamics* is the study of physical laws governing the flow of blood. The movement of fluid through a tube is controlled by the pressure exerted by the fluid against resistance. For example, the rate of flow of water from a faucet is determined by the pressure in the water pipe and the extent to which the faucet is opened. Here the resistance is due chiefly to the faucet but is also affected by the diameter of the pipe, the length of the pipe, and bends, obstructions, variations in diameter, and other sources of turbulence in the pipe. If the pipe were delivering blood instead of water, we should add viscosity to this list.

In the circulatory system a given amount of blood is transferred from the heart each minute into an elastic network of branching tubes, which become progressively larger in total cross-sectional area and progressively smaller in individual cross-sectional area (Fig. 9.25). The blood meets relatively little resistance in the large arteries, whose total wall surface is small. But it encounters much resistance in the many arterioles and capillaries, whose total wall surface area is enormous. If the large vessels were large water pipes, the arterioles and capillaries would constitute numerous small faucets turned off and on by contractions and relaxations.

The factors controlling fluid flow through small tubes are illustrated in Fig. 9.26, which depicts a reservoir connected to a horizontal tube from which lead several vertical tubes of the same diameter as the horizontal tube. The horizontal tube can be closed or opened to any extent by a pinch clamp. If the clamp is closed and the reservoir is filled, the columns of fluid in the reservoir and the vertical tubes all have the same height, and all exert the same pressure at the base against the horizontal tube. The height to which the fluid rises in a reservoir or vertical tube is a measure of the *hydrostatic pressure*.

If the reservoir is kept filled to a certain level and the clamp is opened completely, fluid flows from the end of the horizontal tube at a steady rate. If the horizontal tube had a large diameter and there were no impediment to outflow from

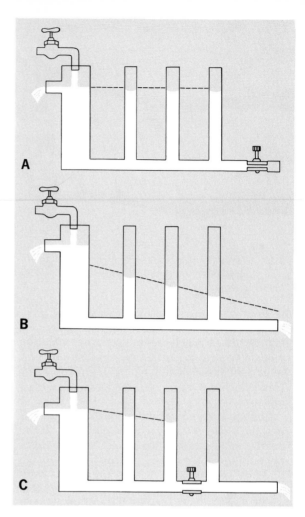

9.26 Relationships between pressure and flow in a small horizontal tube: A, when outflow is prevented, pressure is constant throughout the system; B, when constant pressure is applied, outflow is constant, but lateral pressure decreases progressively; C, when the tube is constricted locally, lateral pressure increases upstream and decreases downstream.

it in the form of friction, lateral pressure, or turbulence, fluid would not rise at all in the vertical tubes. Instead, it would pour from the horizontal tube as through a hole in the bottom of the reservoir. Since the horizontal tube has a small diameter, however, the vertical tubes show progressively lower fluid levels.

We conclude that the reservoir pressure is used up in overcoming the resistance to flow imposed by the horizontal tube. If the reservoir is made deeper (and hence the reservoir pressure increased), other factors remaining equal, the rate of flow through the horizontal tube is increased. If the horizontal tube is lengthened (and the reservoir pressure kept constant), the rate of flow is correspondingly decreased. To maintain a given flow rate through a longer tube —which therefore offers greater resistance—a corresponding increase in reservoir pressure is needed. Finally, if a section of the horizontal tube between two vertical tubes is narrowed, the pressure increases in the vertical tube nearer the reservoir and decreases in the vertical tube farther from the reservoir; and, of course the drop in the flow rate decreases.

These simple physical principles should help us understand the phenomena associated with blood flow in the heart and the vessels. In summary, the arteries are analogous to horizontal tubes, and the arterioles to pinch clamps. As Fig. 9.25 indicates, pressure throughout the arterial system gradually falls as a result of increasing "tube length" and dissipation of the propelling force of the heart's contraction, the greatest reduction in pressure occurring beyond the arterioles.

Modern techniques in the study of hemodynamics. We have already mentioned many methods for investigating the circulatory system, including such recently developed techniques as angiography or venography and the more classic techniques of dissection, electrocardiography, and injection of blood vessels with colored plastics. Since all scientific conclusions ultimately rest upon the methods used in gathering data, we must now inquire into the methodology of hemodynamics.

For many years investigators could study blood flow in a living man only by indirect methods. For example, they could measure pressure in the arteries with the conventional blood pressure apparatus, pressure in the veins by inserting a needle into a vein and observing the height reached by the blood in a vertical glass tube, and cardiac output by other means, to be outlined later. Though much was learned by these techniques, flow and pressure changes *within* the human heart could only be estimated. Resulting data were only approximations.

This situation has been changed by a number of recent discoveries. One can now, with small risk, insert a long tube into a blood vessel of the arm and thread it all the way into the heart. This procedure is called *cardiac catheterization.**

Cardiac catheterization was first performed in Germany in 1924 by W. Forssmann, who was his own experimental subject. In his first attempt, Forssmann, aided by another physician, introduced a long rubber catheter into an arm vein. Because his colleague considered the experiment too dangerous, Forssmann did not complete it, although he wrote, "I felt perfectly fine." A week later, he made another attempt, this time alone, and succeeded in passing the catheter into his right atrium, verifying its position on a fluoroscopic (x-ray) screen ("I observed the progress of the catheter in a mirror which was held by a nurse in front of the fluoroscopic screen."). Forssmann felt no unusual sensations and considered the procedure safe. For this and his later physiological work, he received the Nobel Prize in 1955.

It was now possible to study the circulation in a living man without extensive trauma or anesthesia. Hence cardiac catheterization has been widely employed for diagnostic and other purposes. The catheter contains at least two channels: one is a conduit for a sensitive pressure-recording element in the tip; the other permits the withdrawal of blood from various locations for oxygen assay. A saline solution of the anticoagulant heparin is continuously dripped through the catheter to prevent blood clotting during the procedure. The catheter is inserted and moved under direct observation on a fluoroscopic screen.

A venous catheter can enter the right heart only—unless there is a defect in the atrial or ventricular septum (and it should be noted that catheterization can be used to detect such a defect). A catheter in the right heart provides information on pulmonary and systemic blood flows, blood pressure in the pulmonary artery, and a variety of congenital and acquired abnormalities.

---

* A *catheter* is a hollow rubber or plastic tube designed to be inserted into a body cavity. Thus urethral catheters are inserted into the bladder through the urethra, eustachian catheters into the middle ear cavity, etc. *Catheterization* refers to the act of inserting or passing a catheter.

A similar approach to the left heart requires catheterization of an artery, but investigators found it all but impossible to pass a catheter from the aorta to the left atrium without running afoul of the cusps of the aortic valve, which point in the opposite direction. They finally developed a successful method of left heart catheterization in which a long needle inserted in the back enters the left atrium (Fig. 9.27A). A second such needle may simultaneously enter the left ventricle via the atrium. Later refinements allow the introduction of a catheter into the left ventricle through a fine needle in the chest (Fig. 9.27B) and into the left atrium from the right atrium through the thin atrial septum; this is the simplest and safest route.

With fine catheters in the left atrium and the left ventricle and another threaded into the aorta through an artery of the arm—usually the brachial artery—the sizes of the orifices of the heart and the flow rates through the mitral and aortic valves can be determined.* Fig. 9.27 shows *simultaneous* left and right heart catheterization. Usually it is performed when cardiac surgery is contemplated.

Cardiac cycle and its mechanical consequences. A single heartbeat—the *cardiac cycle* —consists of three phases: (1) a period of contraction, called *systole;* (2) a period of dilatation, called *diastole;* and (3) a period of rest. With a pulse rate of 70 to 72 heartbeats per minute, the time required for a cardiac cycle is 0.8 second; half of this, or 0.4 second, represents the quiescent phase. When the heart beats more rapidly, as it does after exercise, the rest period is shortened.

Curves depicting pressure changes in the left atrium and the left ventricle during a cardiac cycle reveal five stages or phases (Fig. 9.28).† Stage 1 is the *atrial contraction phase.* Since the mitral valve is open, there is only a small increase in atrial pressure, and almost no increase in ventricular pressure. As blood rushes into the ventricle, eddy currents begin to close the mitral valve.

At the start of stage 2, the *isometric contraction phase,* the ventricle contracts. Ventricular systole commences within 0.1 second after atrial systole. At first, ventricular volume remains constant while ventricular pressure rapidly rises. But, as the growing pressure completes the closure of the mitral valve, ventricular volume also rises. The aortic valve remains closed because pressure in the aorta still exceeds pressure in the ventricle. Meanwhile, pressure in the atrium declines, though the drop is interrupted by a brief upsurge due to a small thrust of pressure from ballooning of the mitral valve. When the atrium is relaxed, pressure in the ventricle is at its peak; with the mitral valve closed, no blood can return to the atrium and thereby relieve the pressure.

Stage 3, the *maximum ejection phase,* begins with the opening of the aortic valve. Pressure in the ventricle forces blood through the open valve into the aorta, tending to dilate it. Atrial pressure continues to fall; and, with the ejection of ventricular blood, the ventricle relaxes, and ventricular pressure also falls.

In stage 4, the *reduced ejection phase,* ventricular pressure is less than aortic pressure—so that the aortic valve closes—and greater than atrial pressure—so that the mitral valve stays closed. For this reason incoming blood builds up the atrial pressure. Until the ventricular pressure drops below the atrial pressure, accumulated atrial blood cannot enter the ventricle.

In stage 5, the *rapid filling phase,* the pressure reversal occurs, causing a rapid influx of blood into the ventricle without actual atrial contraction. The ventricle rapidly increases in volume, atrial pressure decreases, and all is in readiness for the next atrial systole, which starts a new cycle.

9.27 Simultaneous right and left heart catheterization: A, with two fine needles inserted in the back entering the left atrium and left ventricle, a catheter through an arm artery into the aorta, and a catheter through an arm vein into the pulmonary artery; B, with a fine needle inserted in the chest entering the left ventricle, a catheter through an arm artery into the aorta, and a catheter through an arm vein into the pulmonary artery. (Adapted from *Seminar Report.*)

---

* Through direct ventricular puncture the aortic valve can be examined independently of the mitral valve.

† The events in the left heart are typical of the events in both hearts.

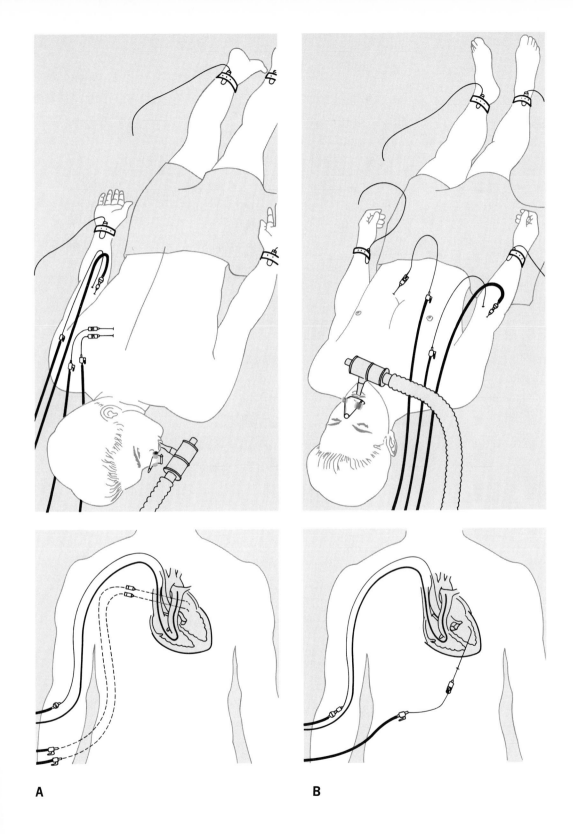

A

B

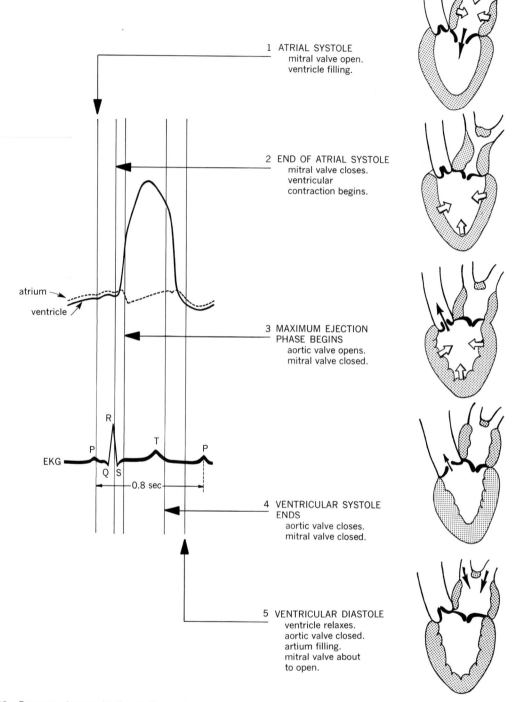

1 ATRIAL SYSTOLE
mitral valve open.
ventricle filling.

2 END OF ATRIAL SYSTOLE
mitral valve closes.
ventricular
contraction begins.

3 MAXIMUM EJECTION
PHASE BEGINS
aortic valve opens.
mitral valve closed.

atrium

ventricle

R

T

P   P

EKG

Q S

0.8 sec

4 VENTRICULAR SYSTOLE
ENDS
aortic valve closes.
mitral valve closed.

5 VENTRICULAR DIASTOLE
ventricle relaxes.
aortic valve closed.
artium filling.
mitral valve about
to open.

9.28   Pressure changes in the cardiac cycle.

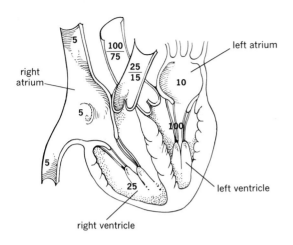

9.29 Normal intracardiac pressures, in millimeters of mercury, as determined by catheterization. The paired figures are systolic and diastolic pressures. A useful memory aid is the "rule of fives." Right atrial pressure is 5 or less. Left atrial pressure is 2 × 5. Note that ventricular pressures are approximately equal to the squares of the corresponding atrial pressures.

Pressure tracings like those shown in Fig. 9.28 are obtained by catheterization. The highest pressure in each chamber is reached during systole. This is the *systolic pressure*. The *diastolic pressure* is the minimum pressure achieved during diastole. Typical normal intracardiac pressures are given in Fig. 9.29.

Heart sounds. "With each beat of the heart," wrote Harvey, "when there is the delivery of a quantity of blood from the veins to the arteries, a pulse takes place and can be *heard* within the chest." Though the heart sounds are audible when we press an ear against a subject's chest, the stethoscope—invented in 1819 by René Laënnec—aids immeasurably in amplifying the sounds and eliminating adventitious noises.

As Laënnec recognized, two major heart sounds are associated with each cardiac cycle. The first is comparatively long and booming; the second, short and sharp. Together they resemble the spoken syllables *lubb dup*. The first sound is attributed to the contraction of ventricular muscle and to the snapping closure of the atrioventricular valves. The second is attributed chiefly to the sudden closure of the semilunar valves.

In certain heart diseases these sounds change, becoming muffled, prolonged, or roughened. Most such sounds are called *murmurs*. They are often due to (1) an improperly closing valve, which allows regurgitation of blood, and (2) a valvular lesion, which narrows an orifice and decreases the amount and the velocity of blood flowing through it. The former disorder is termed valvular *insufficiency*, and the latter, valvular *stenosis*. By careful analysis of the character and timing of murmurs, physicians can determine which valve is afflicted and how. For example, a murmur during ventricular diastole indicates stenosis of the mitral or tricuspid valve, for this is the phase of rapid ventricular filling, when no blood enters the great arteries. Mitral or tricuspid insufficiency creates a murmur only during the period when the valve should be closed, i.e., during ventricular systole.

As might be expected, a machine, the *phonocardiograph*, has been devised to record heart sounds. Fig. 9.30 shows tracings of the first sound associated with mitral valve closure in a normal heart, one with a leaky or insufficient valve, and one with a valve so stiffened by disease that it is both stenotic and leaky.

Starling's law of the heart. From 1912 to 1914 the distinguished British physiologist E. H. Starling performed a series of experiments with his famous *heart-lung preparation*, an isolated beating animal heart attached to outflow and inflow tubes in such a way that the investigator can control blood flow into the right atrium and mechanical resistance to blood flow out of the right ventricle. The right ventricle pumps blood to artificially ventilated lungs or to an oxygenator, either of which maintains constant oxygen and carbon dioxide concentrations in the blood. An essential feature of the arrangement is the elimination of the nerves to the heart, which makes possible examination of the responses of the heart alone, free of nervous influences.

In an experiment blood is allowed to flow into the right atrium from a reservoir, placed at any desired height above the atrium. The pressure at which atrial filling occurs is expressed as the vertical distance (in centimeters) from the atrium to the blood level in the reservoir. Ventricular volume is determined by enclosing the ventricles in an airtight chamber and measuring

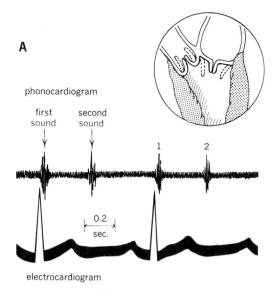

A

phonocardiogram

first
sound

second
sound

1   2

0.2
sec.

electrocardiogram

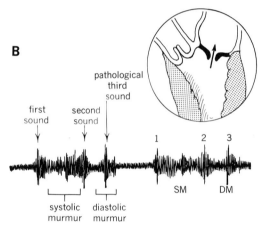

B

pathological
third
sound

first
sound

second
sound

1   2   3

SM   DM

systolic
murmur

diastolic
murmur

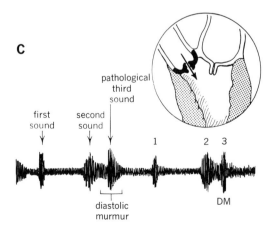

C

pathological
third
sound

first
sound

second
sound

1   2   3

DM

diastolic
murmur

the pressure in the space between them and the container during the cardiac cycle. The volume at the end of diastole represents the volume of the filled ventricles; this volume minus the volume at the end of systole represents the volume of blood expelled, or the *stroke volume* of the ventricles.

Starling carefully observed the effects of the variables that he could control: venous pressure, arterial pressure (i.e., peripheral resistance), and temperature. In studying the effect of increasing venous pressure, he recorded the ventricular volume and heart rate when the reservoir was at a certain height and then abruptly elevated it, say, 10 cm. The diastolic volume increased markedly, since more blood was fed in at the higher pressure, and simultaneously stroke volume increased. Starling therefore concluded that an increase in diastolic volume of the ventricles—which necessarily causes an increase in length of the cardiac muscle fibers—strengthens the contraction (so that more blood is expelled per stroke).

The heart rate did not change. Since the volume of blood pumped by the heart each minute is the product of this rate and the stroke volume, and since the latter increases while the former remains constant, cardiac output increases in direct proportion to stroke volume. Stated in general terms, the heart—in the absence of its nerve supply—responds to an increase in venous pressure with an increase in output. This serves to prevent blood from accumulating in the great veins.

If venous pressure is progressively increased, the ventricle progressively enlarges (or dilates), and the progressively lengthening cardiac muscle fibers produce stronger and stronger ventricular contractions. However, a point is reached at which an increase in venous pressure causes the ventricle to become so dilated that stroke volume, and thus cardiac output, begin to decrease

9.30 Mechanical basis of the heart sounds, demonstrated in the left side of the heart (with the aortic valve at left and the mitral valve at right): A, normal closing of the mitral and aortic valves; B, partial closing of a leaky (insufficient) mitral valve; C, partial closing of a leaky (insufficient) aortic valve. Phonocardiographic tracings show heart sounds and systolic and diastolic murmurs.

(Fig. 9.31). We see, then, that when certain stresses exceed a critical point the result can be heart failure.

In the venous pressure experiment, arterial pressure was kept constant. Conversely, venous pressure is kept constant in studies of the effect of increasing arterial pressure. As arterial pressure is increased, diastolic filling pressure increases secondarily, and cardiac output remains the same (see Fig. 9.31). When arterial pressure becomes excessive, the result is a decrease in cardiac output.

The heart-lung preparation illustrates an important means by which the heart maintains its output and balances the circulation. It does these remarkable things by itself, without nervous control. The important factor is always the relationship between the *length* of the cardiac muscle fibers and the *strength* of their contractions. As the former increases, so does the latter. This is *Starling's law of the heart*. By "obeying" this law, the heart is able to keep down venous congestion and keep up arterial flow. We shall better appreciate this singular capability when we discuss congestive heart failure.

Cardiac output and the work of the heart. Starling's law is one of the great principles of classic physiology. However, direct measurements of cardiac output in man (i.e., by catheterization) indicate that the artificial circumstances in dog heart-lung preparations have led to some overemphasis of the significance of diastolic

filling pressure as a primary determinant of cardiac output. Physiologists now agree on the importance of two regulatory factors in addition to venous inflow, which controls initial myocardial fiber length. These are arterial resistance, which influences the ventricular output, and the innate characteristics of the fibers, whose contractile power may be affected by coronary flow and nervous and chemical stimuli, including drugs.

The first useful method for measuring cardiac output was suggested in a brief note published by A. Fick in 1870. Fick reasoned that, if he knew (1) the difference between the exact oxygen contents of the freshly oxygenated blood in the arteries or left heart and the unoxygenated venous blood in the right heart (i.e., the *arteriovenous oxygen difference*) and (2) the total amount of oxygen consumed by the body each minute, he could readily calculate the total blood flow, since all oxygen utilized in the tissues is brought there by the blood. Thus, if arterial blood contains 19 ml of oxygen per 100 ml of blood and right heart blood contains 13 ml of oxygen per 100 ml of blood, then 6 ml of oxygen per 100 ml of blood is given up to the tissues. If the body requires 250 ml of oxygen per minute, then cardiac output is $(250/6) \times 100 = 4160$ ml per minute. The *Fick principle* is equally valid when another blood constituent, such as carbon dioxide, or a foreign gas is used as indicator.* In the normal circulation, cardiac output of the right heart equals cardiac output of the left heart.

A second method for measuring cardiac output involves the injection of a dye into the blood stream and determination of the rate at which it is diluted. If 12 mg of a dye (chosen for the properties listed on p. 203) is injected into a vein, serial samples of blood drawn from an artery show that the dye concentration reaches a peak, falls, and then rises again owing to recirculation. Calculations are based on the slope of the descending limb of the concentration curve. If the average concentration during the first 30 seconds (before recirculation begins) is 6 mg per liter, then the 12 mg has obviously been diluted

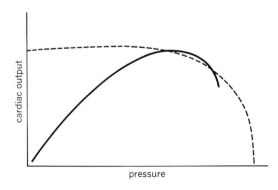

9.31 Effects of venous (———) and arterial (– – –) pressures on cardiac output.

_____

* If 206 ml of carbon dioxide is eliminated per minute and the arteriovenous carbon dioxide difference is 5 ml per 100 ml of blood, cardiac output is $(206/5) \times 100 = 4120$ ml per minute.

by 2 liters of blood in the 30 second interval, and the cardiac output is 4 liters per minute.

Such methods, despite the sources of error that precise work has revealed, have proved that cardiac output is closely regulated. With each systole a volume of blood estimated at around 70 ml in a man at rest is forced from the left ventricle into the aorta. As in heart-lung preparations, this is called the *stroke volume*. A similar volume is ejected from the right ventricle. With 70 heartbeats per minute, approximately $70 \times 70 = 4.9$ liters of blood leaves each ventricle each minute (the normal value ranges from 3 to 5 liters, depending on body size). This is called the *minute volume*.

Cardiac output can change from a resting figure of 5 liters per minute to 15 liters per minute in mild exercise and very much more in heavy exercise. The increases are brought about by increases in stroke volume and heart rate. When the heartbeat accelerates, the stroke volume can be kept constant only if the venous inflow is adequate. With a very rapid heartbeat, as in a paroxysmal tachycardia, the heart does not relax enough between beats to take in sufficient blood, and so the stroke volume is actually reduced. Hence the contribution of the heart rate to the minute volume is limited. With a constant stroke volume, minute volume can only triple. Yet in violent exercise the amount of oxygen consumed by the tissues may increase 12-fold. The need for extra oxygen is met by either an increased cardiac output or an increased utilization of blood oxygen. Since it has been shown that in exercise cardiac output increases as much as 9-fold, we conclude that in exercise stroke volume triples, even at a maximum heart rate, and that increased oxygen requirements are satisfied primarily by a rise in cardiac output. Thus exercise temporarily makes the heart larger. Other situations in which cardiac output increases include fever, anemia, and instances of abnormal shunts between large arteries and veins.

Although the two chief means by which cardiac output is increased are increases in stroke volume and heart rate, there is a third means—permanent enlargement of the heart muscle itself. When the left ventricle must continuously increase its stroke volume, it dilates, and its individual muscle fibers thicken and lengthen. The resulting condition is known as *muscular hypertrophy*. The extent to which any heart may exploit these measures is termed the *cardiac reserve*. The degree of hypertrophy is an index of encroachment upon the cardiac reserve; a grossly enlarged heart is therefore assumed to be closer to failure than a normal-sized one.

Nervous control of the heart rate. We have already observed that an impulse can be initiated in the S-A node in the total absence of nerve connections. The cardiac nerves act only as regulators of the rate of impulse formation, the rate of conduction, and the strength of contraction. Branches of the vagus nerves entering the S-A and A-V nodes act to slow the heartbeat, while the accelerator nerves act in the opposite manner. Which set dominates depends largely upon information reaching the central nervous system from various regions.

One of the most important of these regions is the carotid sinus (see p. 298). Its walls contain sensitive nerve endings (*pressoreceptors*) that are stimulated by the mechanical stretching caused by increased blood pressure. Through reflex reaction they induce the vagus nerve branches to slow the heart and dilate the arteries. The result is a fall in blood pressure, which shuts off the reaction.*

Congestive heart failure. When the limit of the cardiac reserve is reached, the heart fails as a pump. This means that the myocardium has been overwhelmed by its load and the various adaptive mechanisms—increases in stroke volume, heart rate, and fiber size—have been overextended. When the heartbeat becomes too rapid, diastolic filling is incomplete, and cardiac output drops. When the ventricle becomes too dilated, the elasticity of the fibers is lost, and cardiac output drops. When hypertrophy becomes extreme, the myocardium outgrows its blood supply, its contractions weaken, and cardiac output drops. As the output decreases, the vicious cycle of *congestive heart failure* begins.

Sir Thomas Lewis wrote, "Congestive failure is a condition in which the heart fails to discharge its con-

---

* The carotid sinus reflex is demonstrated when the neck in the region of the carotid sinus is massaged; the heartbeat slows noticeably. This is one of the methods used to terminate attacks of paroxysmal atrial tachycardia. On occasion, it produces temporary cardiac arrest, particularly in individuals with abnormally sensitive carotid sinus reflexes.

tents adequately." In this statement are direct explanations for two of the three major manifestations of heart failure. The word "congestive" here implies congestion of the blood vessels, particularly of the veins; the manifestations of *venous congestion* are increased volumes of lungs, liver, and veins, with resulting shortness of breath. A *decreasing tissue blood supply* produces such difficulties as cyanosis and other metabolic derangements due to the unmet demands of the tissues for oxygen. A third group of symptoms relates to *edema*—or *dropsy* as it was once known—the accumulation of large quantities of fluid in the tissues, due to a breakdown in the efficiency of the kidney as a regulator of body salt and water balance. The kidney fails when the heart fails. The result is marked salt and water retention.

Congestive heart failure is a failure of the same circulatory system that Harvey found was driven in "a motion, as it were, in a circle." We should naturally assume that, when a circulating system slows down, all its parts slow at once. Yet it has been debated for more than a century whether the chain of events in heart failure follows primarily from a decrease in cardiac output (i.e., a decrease in blood flow to the tissues downstream)—the *forward failure theory*—or from an increase in venous congestion upstream—the *backward failure theory*. Obviously decreased cardiac output eventually leads to venous congestion, and venous congestion beyond a certain point leads to decreased cardiac output (see Fig. 9.31). Modern studies with cardiac catheterization have reconciled the two theories, establishing that their application rests on the conditions under which failure develops.

We now recognize that, though common factors control the contractions of both ventricles, each ventricle, with its venous inflow and arterial outflow systems, operates in a distinct fashion. The right ventricle, with a smaller muscle mass and greater distensibility than the left ventricle, can accommodate increases in venous inflow with only small increases in diastolic pressure. Moreover, unlike the left ventricle, it pumps its blood into a circulation whose pressure (and hence resistance) is low (see Fig. 9.29) (pulmonary arterial pressure is low compared to aortic pressure). These differences mean simply that, when right heart failure occurs first, initiating a series of events culminating in left heart failure, venous congestion temporarily predominates over decreased cardiac output.

For example, an obstruction to pulmonary blood flow due to lung disease first affects the right ventricle. Too rapid adminstration of intravenous fluids, especially in the aged, increases the venous inflow, further straining the right heart. As the right ventricular output gradually decreases, the left ventricular output remains unchanged, so that blood continues to pour into the veins. This situation cannot continue indefinitely. Ulti-

mately the left ventricular output drops, coronary and renal blood flow decrease correspondingly, and forward failure sets in.

Conversely, the left heart fails first if the left ventricular wall is selectively injured by coronary thrombosis. As the left ventricular output decreases, the pulmonary vessels become congested because they still receive the full right ventricular output. Pressure builds up in them, finally increasing the load on the right ventricle until it, too, fails.

In both these examples, cardiac output declines in proportion to venous inflow in one ventricle first. Then the other fails as a direct consequence of the fundamental relationship between the length of a muscle fiber and the tension it exerts in contracting, as set forth in Starling's law of the heart.

The most common causes of congestive heart failure are myocardial infarction due to coronary thrombosis; developmental defects, such as massive cardiac enlargement (particularly when associated with extensive mixing of the venous and arterial blood owing to abnormal passageways between the right and left hearts); hypertension (high blood pressure); and valvular injury and scarring with resulting valvular insufficiency or stenosis, usually a late consequence of rheumatic fever.

The three best therapeutic measures are rest, decreased salt intake, and digitalis. Rest decreases the need for a high cardiac output. Decreased salt intake greatly diminishes water retention. Digitalis* increases the force of systole and the cardiac output and decreases heart size and rate. Though it has prolonged many lives its mechanism of action is still unknown. Recent work suggests that it affects the transport of sodium, potassium, and other ions in the membranes of cardiac muscle cells.

One of the remarkable advances of modern medicine is the development of surgical techniques for correcting valvular injuries. A surgeon can insert a finger or knife

---

* The story of the introduction of digitalis into medicine is fascinating. A pamphlet published in 1785 by William Withering, a physician of Birmingham, England, entitled *An Account of the Foxglove and Some of Its Medical Uses*, told of Withering's discovery of a secret remedy for dropsy (edema) that had been used for centuries by the Hutton family of Shropshire. The grand dame of the family, "the Old Woman of Shropshire," was said to treat dropsy successfully with a tea brewed of herbs. At first skeptical, Withering was convinced of the tea's efficacy when he learned that the dropsical Dr. Cawley, Dean of Brazen Nose College, Oxford, "had been cured . . . after some of the first physicians of the age had declared they could do no more for him." After some difficulty, Withering borrowed the recipe and showed that the active ingredient was the plant foxglove (*Digitalis purpurea*), whose leaves contain several of the chemical compounds now known collectively as digitalis.

into the beating heart and widen or replace constricted valves.*

Myocardial metabolism. The metabolic activities of cardiac muscle cells reflect their main function, the performance of mechanical work. Myocardial metabolism is chiefly concerned with the production of the ATP that is required in extraordinary quantities. So great, in fact, are the energy needs of the ceaselessly contracting fibers that their patterns of organization allow for almost no metabolic pathways except energy-yielding ones. Thus a cardiac muscle cell contains many mitochondria in clusters around the nuclei and beneath the sarcolemma. Compared to those of other tissues, these mitochondria are unusually large, with many cristae.

Much has been learned of myocardial metabolism from studies of differences in composition between fresh arterial blood and coronary sinus blood. Since the latter is returning from the myocardium, its composition reflects additions and deletions resulting exclusively from myocardial metabolism. Hence certain inferences are possible concerning the overall metabolic exchanges of the myocardium—if not the specific metabolic pathways underlying them. Such studies have shown that, though the major fuel of myocardial metabolism is glucose, the normal heart can utilize pyruvate, lactate, and a variety of noncarbohydrate materials including fatty acids and ketones (Chapter 14). The ATP generated in the catabolism of these substances is used principally in the processes of muscle contraction† but is also responsible for the electrical activity of heart tissue. The requirements of the myocardium for oxygen and oxidizable substrates such as glucose are stringent, and even brief deprivations of either may cause serious damage.

------

* For an operation in which the heart must be completely opened for an extended period, a heart-lung machine substitutes for the heart and lungs. Basically, it consists of four blood pumps, an artificial lung, and three sets of devices that detect and correct for changes in temperature, oxygenation, and pH. Recently a similar machine has been used to rest a heart that has been weakened by disease. In the future such artificial hearts may have wide applications in the treatment of congestive heart failure.

† We shall defer to Chapter 17 discussion of the special contractile proteins of cardiac muscle and the mechanisms of contraction.

BLOOD FLOW IN THE PERIPHERAL VESSELS

Arterial circulation and its control. The elasticity of the arteries tempers fluctuations in flow and pressure. Flow is more uniform near the periphery of the arterial tree than near the heart, since the larger vessels act as reservoirs of blood under mechanical pressure. The chief mechanism by which arterial flow is controlled is the contraction and relaxation of the arteriolar muscles. Even when these are surgically disconnected from incoming nerves, they remain in a state of partial contraction, which is called *tonus,* or *tone.*

One advantage of arteriolar tonus is that it permits local variations in resistance to flow that regulate the blood supplies of the organs, shunting blood from inactive to active ones—e.g., the digestive organs after a meal. As the total blood volume is too small to fill the entire vascular system, arteriolar tonus maintains a balance in flow that regulates the capacities of various regions of the system. Such regulation is impossible in an invertebrate, whose circulation is open and whose blood pressure is low. Although the blood of an invertebrate is efficient in transporting oxygen, it is grossly inefficient in distributing oxygen to the sites of greatest demand.

If the arterioles are to shunt blood from one region of the body to another, it follows that dilatation must occur in one area and constriction in another. The integration of these responses depends on *vasomotor nerves,* first demonstrated by Claude Bernard in 1852. Vasomotor control of arteriolar tonus is effected by two opposing systems: the *vasoconstrictor* nerves, which contract the blood vessels; and the *vasodilator* nerves, which widen them. Since arterioles are normally in a state of partial contraction, it appears that the vasoconstrictor nerves act continuously. Thus they contribute to the maintenance of arteriolar tonus, which in turn contributes to the maintenance of blood pressure.

The curious actions of the vasomotor nerves are familiar phenomena of everyday life. Vasoconstriction of the skin vessels in cold weather conserves heat in the central portions of the body. Vasodilatation of the skin vessels accounts for the flush of warmth on entrance into a hot room. Psychological events also powerfully influence vasomotor activity. Blushing is psychically stimulated, and acute coronary insufficiency with angina pectoris is frequently precipitated by emotion. Balanced vasodilatation and vasoconstriction is the essential cause of erection of the penis or clitoris following erotic stimulation.

Many drugs affect arteriolar tonus. We have already mentioned nitroglycerin, used to dilate the coronary arterioles in angina pectoris. Conversely, epinephrine (or adrenalin) constricts the arterioles when given in large doses. A group of epinephrine-like drugs, the amphetamines (which include benzedrine and dexedrine), resemble epinephrine in constricting the blood vessels and raising the blood pressure. Intense mental activity has a similar result. Perhaps this reflects increased blood flow through the brain at the expense of decreased blood flow elsewhere. Perhaps increased blood flow through the brain explains the mental stimulation produced by the amphetamines.

Arterial blood pressure. The arterial pressure curves in Fig. 9.25 illustrate the events of the cardiac cycle. The highest arterial pressure is the systolic pressure, and the lowest is the diastolic pressure. In a normal subject at rest, these are usually around 120 and 80 mm of mercury, respectively, written 120/80 and referred to verbally as "120 over 80." The difference between systolic and diastolic pressures is called the *pulse pressure.*

We have already spoken of the interaction of arterial elasticity and arterial resistance to flow. A large proportion of the cardiac output is stored in the large arteries just after systole and held there by the relatively great resistance in the arterioles beyond. At this moment the pressure within the arteries is systolic pressure. The recoil of the arterial walls causes the continued flow of blood during diastole. It is evident that the volume of the cardiac output, the resistance of the small arteries, and the elasticity of all the arteries are the primary factors determining systolic pressure, diastolic pressure, and pulse pressure.

Although a rough estimate of arterial blood pressure is possible from feeling an artery at the wrist, accurate measurements are made with the *sphygmomanometer,* which consists of a graduated column of mercury or an aneroid spring scale connected by rubber tubing to an elastic air bag in a fabric cuff. The air bag in turn is connected to a small hand pump. The cuff is wrapped snugly about the arm just above the elbow, over the brachial artery, and the bag is inflated until the pulse at the wrist disappears (the bag compresses the artery, obstructing blood flow). The bag is then slowly deflated until the pulse can just be felt again, or until the first pulsation in the brachial artery distal to the obstruction can be heard by stethoscope. The pressure in the bag at that time, as indicated by the manometer or scale reading, is equal to the systolic pressure. The bag is further deflated, and the pressure just before the last sound of pulsation in the brachial artery is equal to the diastolic pressure.

Systolic pressure is usually a little higher in men than in women, but it varies with exercise, emotion, digestion, and position. A systolic pressure that is consistently higher than 150 suggests *hypertension.*

Arterial pulse. As we have noted, when a finger is placed over an artery coursing near the surface of the body—particulary over a bone—a pulse, which is synchronized with the heartbeat, can be felt. In some locations the pulse can be seen.

Though present in all arteries, the pulse is best observed in the following ones: the *radial arteries* at the wrists, the *temporal arteries* at the temples, the *carotid arteries* in the neck, the *brachial arteries* along the inner sides of the upper arms, the *femoral arteries* over the pelvic bone, the *popliteal arteries* behind the knees, and the *dorsalis pedis arteries* over the insteps.

The pulse is produced by the rapid rise of arterial pressure from diastolic level to systolic level during systolic discharge. The *sphygmograph* records the movement of the arterial walls graphically and shows that the pulse has several components. The main thrust is due to ventricular systole, but a small secondary thrust, the *dicrotic wave,* results from closure of the aortic valve.

In certain diseases the shape of the sphygmographic curve and the feel of the pulse are grossly altered. When the aortic valve is functionally insufficient, the pulse drops rapidly after reaching the systolic peak as aortic blood is regurgitated into the ventricle. With stenosis of the aortic valve, systolic ejection of the blood is delayed, and the amplitude of the waves remains low.

Hypertension and atherosclerosis. Hypertension and atherosclerosis are perhaps the two most widespread human ailments. They illustrate in different ways the single feature of arterial disease from which come all adverse consequences: occlusion of arterial flow.

Hypertension in a sense owes its discovery to the invention of the sphygmomanometer, which makes possible its detection in even ostensibly normal individuals in the course of a routine physical examination. The common variety, *essential* hypertension, is of unknown cause. Over the years many forms of hypertension have been removed from this category as their causes have been found. One severe form is associated with epinephrine-secreting tumors called *pheochromocytomas* (Chapter 15). This can usually be distinguished from essential hypertension, and cure follows tumor removal.

Another form results from reduced blood flow to a kidney. Harry Goldblatt observed that when the blood flow in one renal artery of an experimental animal was partially obstructed a with clamp, severe hypertension developed. Further investigation disclosed that a kidney deprived of its normal blood supply releases a proteolytic enzyme, known as *renin*, which acts upon a blood α_2 globulin known as *angiotensinogen* to split off angiotensin I, a decapeptide. A converting enzyme in the blood removes the two terminal amino acids of angiotensin I to yield *angiotensin II*, an octapeptide. Angiotensin II raises the systolic and diastolic blood pressures—indeed it is the most potent pressor substance known. Perhaps this is an adaptive mechanism whereby the kidney seeks to obtain more blood. Extensive efforts to show that human essential hypertension is due to the "Goldblatt phenomenon" have been unsuccessful.

We shall encounter the renin-angiotensin system in Chapter 10, in connection with the renal regulation of body salt. A number of recent theories have been based on the interesting fact that restriction of dietary salt leads to a lowering of the blood pressure; apparently excess salt somehow exerts a vasoconstrictive effect.*

Atherosclerosis is the major consequence of cardiovascular aging in man. About two-thirds of the deaths in the United States are attributed to cardiovascular diseases, of which atherosclerosis is the most common. About one-third of the deaths occur in relatively young people in whom the aging process has somehow been accelerated in the blood vessels, the main sites of involvement being the coronary and cerebral arteries and the large peripheral arteries. The characteristics of atherosclerosis are as follows: (1) hardening of the arteries with deposition in their walls of lipid-containing plaques called *atheromata*; (2) elevation in the blood of the level of cholesterol, a lipid component of blood and body tissues; and (3) an ill-defined relationship to stress from the pace of modern living.† The incidence of atherosclerosis is higher in men than in women, and a tendency toward it may be inherited.

A fatty meal normally causes *lipemia*, a turbidity of the plasma due to insoluble fat droplets. It was recently observed that heparin injected after such a meal eliminates the lipemia. This *clearing reaction* suggests that heparin, in addition to its well-known function as an anticoagulant (see Chapter 8), may be implicated in plasma fat transport (Chapter 14). Heparin itself does not clear lipemic plasma in a test tube, but plasma from an animal previously injected with heparin does. Upon the administration of heparin, then, the blood acquires a *clearing factor* not previously present. The significance of this phenomenon in the genesis of atherosclerosis remains a riddle. The role of the diet in preventing atherosclerosis is discussed in Chapter 14.

Capillary circulation and its control. When blood leaves the arterioles and passes into the capillaries, it enters the microcirculation. If all the capillaries of the body were open at once, they would accommodate the entire blood volume. Since under normal circumstances they are not, we must inquire how the flow of blood through the capillaries is regulated so as to meet the changing needs of the different tissues without interfering with the efficiency of the circulatory system as a whole.

Unlike such organs as the muscles, liver, and kidneys, the interwoven capillary bed cannot be removed from the body and studied as an intact unit. It is possible, however, to examine part of it in a living animal by placing a thin sheet of living tissue under a microscope. The capillary bed is displayed in almost diagrammatic simplicity (Fig. 9.32).

---

* In addition to a salt-free diet, available remedies for hypertension include a long list of drugs that variously decrease vasoconstriction. In certain cases vasoconstrictor nerves have been cut surgically. The effectiveness of this procedure, *sympathectomy*, is unsettled at present.

† The importance of stress is almost impossible to prove or disprove. The discovery of atherosclerosis in Egyptian mummies does not shed light on the question. Neither does the startling announcement that autopsy records at the Philadelphia Zoo show a 10- to 20-fold increase in atherosclerosis in birds and mammals in the last 40 years, even though some authorities blame this on increased "social pressures" resulting from the increasing animal population in the zoo.

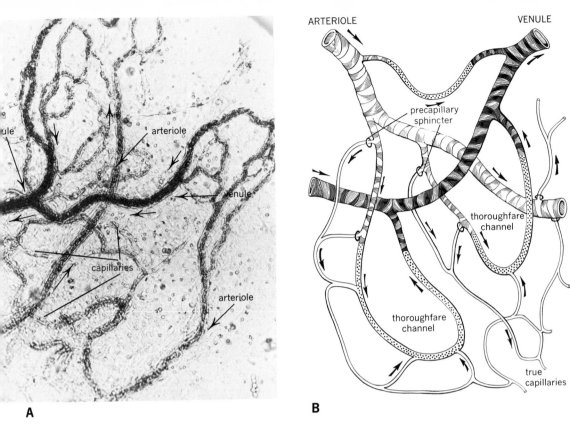

precapillary
sphincter

thoroughfare
channel

thoroughfare
channel

true
capillaries

A

B

9.32 Capillary bed. A, rabbit omentum (× 300); B, diagram of typical bed. (A, courtesy of Dr. B. W. Zweifach.)

The muscular sheath of a larger blood vessel does not continue into a capillary. Yet, as long ago as the late nineteenth century, physiologists saw that the capillaries change in diameter. Moreover, the pattern of flow through a living capillary bed constantly changes. At one moment flow proceeds through one part of the network, and a few minutes later flow shuts off in that part and shifts to another or reverses. Nevertheless, certain vessels of the capillary bed maintain a steady, unvarying flow.

If the capillaries contain no muscle fibers, how is flow controlled? We might suspect that the endothelium is contractile, for we know that in many lower animals blood vessels consisting only of endothelium do contract and relax rhythmically, although movements of this kind have not been observed in mammals. Another explanation was advanced by Charles Rouget, a French histologist. He had discovered peculiar star-shaped cells, each of which was wrapped around a capillary, and inferred that they were primitive muscle cells that open and close capillaries. However, he was unable to prove or disprove his theory. It is now possible to probe Rouget cells with extremely fine needles and electrodes, and this technique—*microsurgery*—has established that neither the endothelium nor the Rouget cells control the capillary circulation in a mammal by contraction.

On the other hand, microsurgery has established that not all vessels in the capillary bed are entirely lacking in muscle. Locally applied epinephrine causes some of them to narrow. Even without added stimulating substances, these vessels open and close with the ebb and flow of blood in the capillary bed. Moreover, these are the same vessels through which blood was seen to flow steadily from the arterial system to the venous system. Careful study showed that their walls contain muscle cells placed so far apart that their channels are structurally almost indistinguishable from *true capillaries*. At the point where a true capillary branches off from a *thoroughfare channel,* there is a prominent muscle, the *precapillary sphincter,* which acts as a floodgate, regulating the flow of blood from the thoroughfare channel to the capillary.

The capillary bed, therefore, is not a mere web of simple vessels between the arterial and venous systems. Rather, it is a physiological unit with two specialized components: (1) the thoroughfare channel, into which blood flows from the arterial system and which provides continuous muscular control; and (2) the true capillary, connected to the thoroughfare channel. The precapillary sphincters along a channel open and close periodically, irrigating first one part of the capillary bed and then another. When they are all closed, blood is restricted to the thoroughfare channel. Current research is aimed at determining the nature of the chemical and nervous mechanisms governing the sphincters. Among the known effective chemical agents are epinephrine, the adrenocortical hormones, acetylcholine, histamine, serotonin, and heparin, the last three perhaps arising from the mast cells.

Capillary permeability. Substances pass back and forth across a capillary membrane by the simultaneous processes of filtration and diffusion, which fulfill the function of the entire circulatory system, namely, the exchange of material between blood and tissue fluids.

A capillary membrane is freely permeable to water and dissolved substances of low molecular weight (gases, salts, sugars, amino acids, etc.). Molecules as large as proteins (molecular weights of 80,000 and higher) can penetrate it but do so to a relatively smaller extent. Thus solutions that seep through the membrane have, to some degree, the typical characteristics of a filtrate.*

The hydrostatic pressure necessary for filtration out of the capillaries results from the difference between the blood pressure and the pressure of the tissue fluids. A counterforce opposes the outward thrust of capillary hydrostatic pressure. This is the osmotic pressure, which is higher in the capillaries than in the tissue fluids

because of the osmotically active plasma proteins (we have learned that only those substances that fail to pass through the pores of a semipermeable membrane exert osmotic pressure). The *effective filtration pressure,* therefore, is the difference between the hydrostatic pressure (which tends to force fluids out of the capillaries into the tissues) and the osmotic pressure (which tends to retain fluids or to draw them from the tissues into the capillaries.)

It is instructive to consider these forces in quantitative terms. If the hydrostatic pressure of the blood at the arteriolar end of a capillary is 30 mm of mercury and the hydrostatic pressure of the surrounding tissue fluids is 10 mm, the effective hydrostatic pressure is $30 - 10 = 20$ mm. Conversely, if the osmotic pressure of the blood at that point is 25 mm of mercury and the osmotic pressure of the surrounding tissue fluids is 15 mm, the effective osmotic pressure is $25 - 15 = 10$ mm. The net result, the effective filtration pressure, is $20 - 10 = +10$ mm. Since this is positive, outward filtration is favored. Hydrostatic pressure, however, decreases toward the venous end of a capillary. There the hydrostatic pressure of the blood is about 15 mm of mercury, so that the effective hydrostatic pressure is $15 - 10 = 5$ mm. The effective osmotic pressure remains unchanged, and so effective filtration pressure is $5 - 10 = -5$ mm. Thus inward filtration is favored. This situation accounts for the uptake of tissue waste products by the capillaries.

Venous circulation. We have seen that the veins have thin walls containing only a few muscle fibers; hence they can contract, but they lack the vigorous vasomotor apparatus of the arterioles. Their hydrostatic pressure is relatively low. Typically, we measure venous pressure in millimeters of water rather than millimeters of mercury. Although it varies in different veins, a typical vein in the arm (held at heart level) has a pressure of 80 to 120 mm of water.

Pressure diminishes progressively from the capillaries to the right atrium. This pressure drop is the principal reason for the movement of blood from the capillaries to the heart. Without it blood would pool in the veins in the lowermost portions of the body. Valves in the larger veins of the lower body guard against an even tem-

---

* The transfer of leukocytes through the capillary membrane is another matter. These are not filtered passively through the pores; rather they migrate actively through temporary openings in the membrane by the process of diapedesis (see Chapter 8).

porary tendency in this direction. The main sources of the pressure drop are the pumping action of the heart and the negative pressure within the chest during breathing (Chapter 11). When either of these devices is disturbed, there is a rise in venous pressure. Congestion of the veins is one of the cardinal manifestations of heart failure.

Although the superficial veins do not have a pulse that can be felt with a probing finger— indeed, the ease with which a surface vein can be obstructed was shown by Harvey (see Fig. 9.24)—a pulse in veins near the heart can be recorded mechanically. This reflects not the force of blood propulsion but the changes in the cardiac cycle due to the intermittent sucking or negative pressure in the right atrium. In heart disease the venous pulse, particularly in the external jugular veins, is a useful clue to the diastolic filling pressure. Abnormalities of the pulse's amplitude are a sign of disease in the right heart valves.

The two most common disorders of the veins are *varicose veins* and *venous thrombosis*. A varicose vein is a vein that is abnormally dilated and tortuous. Sacculations due to patchy thinning of areas of the wall are usually present. The underlying cause of the condition is not known, but evidence mounts that it is a genetic defect ordinarily not manifesting itself until middle and later life. The dilatation of a vein causes a breakdown in competence of its valves, whose cusps can no longer meet and hence can no longer prevent blood from pooling under the pull of gravity.

Thromboses may form in the veins with or without associated inflammation. The former condition is called *phlebothrombosis;* the latter, *thrombophlebitis.* The danger of both is that a piece of clot may break off, be carried to the right heart by the blood stream, and be ejected into the pulmonary circulation, producing a *pulmonary embolism.*

REFERENCES AND SUGGESTIONS FOR FURTHER READING

Bellet, S., *Clinical Disorders of the Heart Beat,* 2nd ed., Lea & Febiger, Philadelphia, 1963.

Bing, R. J., "Cardiac Metabolism," Physiol. Rev., **45,** 171 (1965).

Braunwald, E., "Heart," *Ann. Rev. Physiol.,* **28,** 227 (1966).

Burch, G., and T. Winsor, *Primer of Electrocardiography,* 5th ed., Lea & Febiger, Philadelphia, 1966.

Burton, A. C., *Physiology and Biophysics of the Circulation,* Year Book, Chicago, 1965.

Guyton, A. C., "Peripheral Circulation," *Ann. Rev. Physiol.,* **21,** 239 (1959).

Harary, I., "Heart Cells *in Vitro,*" *Sci. Am.,* **206,** 41 (May, 1962).

Hecht, H. H., ed., "Electrophysiology of the Heart," *Ann. N. Y. Acad. Sci.,* **65,** 653 (1957).

Kossmann, C. E., *Advances in Electrocardiography,* Grune & Stratton, New York, 1958.

Krayer, O., *Pharmacology of Cardiac Function,* Macmillan, New York, 1965.

Levy, M. N., and R. M. Berne, "Heart," *Ann. Rev. Physiol.,* **32,** 373 (1970).

Luisada, S., *Cardiology: An Encyclopedia of the Cardiovascular System,* McGraw-Hill, New York, 1959.

Mellander, S., "Systemic Circulation: Local Control," *Ann. Rev. Physiol.,* **32,** 313 (1970).

Mommaerts, W. F. H. M., B. C. Abbott, and W. J. Whalen, "Selected Topics on the Physiology of the Heart," in G. H. Bourne, ed., *The Structure and Function of Muscle,* Vol. 1, Academic Press, New York, 1960.

Murray, J. F., "Systemic Circulation," *Ann. Rev. Physiol.,* **26,** 389 (1964).

Olson, R. E., ed., "Symposium on the Physiology of Cardiac Muscle," *Am. J. Med.,* **30,** 649 (1961).

Rushmer, R. F., "Anatomy and Physiology of Ventricular Function," *Physiol. Rev.,* **36,** 400 (1956).

Scher, A. M., "The Electrocardiogram," *Sci. Am.,* **205,** 132 (Nov., 1961).

Shepherd, J. T., *Physiology of the Circulation in Human Limbs in Health and Disease,* Saunders, Philadelphia, 1963.

Starling, E. H., *Law of the Heart Beat,* Longmans, Green, London, 1918.

Taccardi, B., and G. Marchetti, *Electrophysiology of the Heart,* Pergamon Press, New York, 1965.

Taussig, H., *Congenital Malformations of the Heart,* 2nd ed., Harvard Univ. Press, Cambridge, Mass., 1960.

Wilson, F. N., "The Distribution of Potential Differences Produced by the Heart Beat Within the Body and at Its Surfaces," *Am. Heart J.,* **5,** 599 (1930).

Zweifach, B. W., "The Microcirculation of the Blood," *Sci. Am.,* (Jan., 1959).

———, "General Principles Governing the Behavior of the Microcirculation," *Am. J. Med.,* **23,** 684 (1957).

The urine of man is one of the animal matters that have been the most examined by chemists, and of which the examination has at the same time furnished the most singular discoveries to chemistry, and the most useful application to physiology, as well as the art of healing. This liquid, which commonly inspires men only with contempt and disgust, which is generally ranked amongst vile and repulsive matters, has become, in the hands of chemists, a source of important discoveries, and is an object in the history of which we find the most singular disparity between the ideas which are generally formed of it in the world, and the valuable notion which the study of it affords to the physiologist, the physician, and philosopher.

A. F. Fourcroy, in *General System of Chemical Knowledge,* 1801

10 Urinary System

Kidneys

INTRODUCTION

Significance. We have seen that the body cells are immersed in a fluid of nearly fixed composition; that protein-free plasma and interstitial tissue fluid are essentially identical and are the two interchanging portions of this extracellular fluid; that the constancy of composition of this fluid is maintained by the kidneys; and that the walls of capillaries act as filters that permit the passage of all soluble blood components other than plasma proteins.

We have also seen evidence of the chemical complexity of plasma (see Table 8.5). Therefore, to say that the kidneys maintain the constancy of plasma composition is to say a very great deal.

How do the kidneys achieve such regulation in a body that takes in varying mixtures and quantities of food substances at different times of the day? The answer is that they first remove all plasma constituents including water and excepting only protein and then restore to the body, in correct proportions, the precise amounts of water and plasma ingredients needed for internal constancy. The kidneys then excrete the remaining watery solution of diverse waste materials as urine.

The functions of the normal kidney can be briefly summarized: (1) elimination of metabolic waste products; (2) regulation of plasma volume and body water content; (3) regulation of osmotic equilibrium and maintenance of optimal salt concentrations in body fluids; and (4) regulation of acid-base equilibrium. This list might be supplemented with (5) secretion of certain hormones, since the kidney secretes renin, which is concerned with the maintenance of blood pressure, and probably secretes erythropoietin, which controls erythrocyte production. This function, in which the kidney behaves as an endocrine gland, is apparently not directly related to its excretory functions.

Evolutionary and embryological development. In perhaps no other body system are the traces of the past more clearly written than in the kidney. To comprehend its ingenious design, we must review some of the landmarks of its evolutionary development.

We have noted (see Chapter 7) that early organisms, in migrating from salt to fresh water, took with them an "internal sea" in the form of salt-containing body fluids. We have also noted the fundamentally different challenges facing the salt-water organism and the fresh-water organism with regard to salt and water conservation.

An invertebrate that lives in the sea is *isosmotic* with its environment—that is, its body fluids have the same salt content, and hence the same osmotic pressure, as the surrounding sea water. For it the maintenance of salt and water balance is thus relatively simple. Excess salt is partly, if not entirely, excreted through the epithelium of the respiratory organs, and the problem of conserving water does not exist. The only other excretory function is the elimination of nitrogenous wastes. This is managed by a meager set of membranes and segmented tubules emptying into the main body cavity, the *coelom* (Fig. 10.1A). This type of primitive excretory equipment, strikingly similar to that of the early human embryo, deals mainly with waste products incapable of leaving the body by simple diffusion.

It has not been rigorously established whether the first vertebrates arose in salt water or fresh water, but the weight of evidence suggests the latter. Part of this evidence is found in the kidney, for a discrete kidney first appeared in vertebrates, the early fishes. Fresh-water fishes would have needed an excretory apparatus capable of excreting water and conserving salt, in addition to excreting metabolic waste products. Water, for the first time, would have needed to be eliminated, and the elaboration of dilute urine would have kept salt loss to a minimum. The kidney would have acquired major control of the composition of the internal environment, including water, salt, and metabolic products.

It is instructive to consider the parallel between the evolutionary development and the embryological development of the human kidney. We recall that the body cells of vertebrate embryos differentiate into three types: ectoderm, endoderm, and mesoderm. Mesoderm eventually gives rise to the muscles, nerves, bones, circulatory system, and kidneys, the kidneys being the only abdominal organs developing from mesoderm. Like the muscles, nerves, and spine, the excretory system begins as a *segmented* structure. From this fact and from observations of the adult kidney in primitive fishes, we infer (despite the lack of fossils) that the excretory organ of the early vertebrate was also segmented and that each body segment carried a pair of *nephric tubules** opening into the coelom via *nephrostomes*. There is evidence that the membrane lining the coelom once played a role in excretion and that the nephric tubules first formed by the multiplication of cells in this membrane. The main difference between the nephric tubules of embryology and those of evolution is that those of the embryo do not each open to the body surface as do those of the early segmented vertebrate. The embryonic tubules fuse together into two *collecting ducts*, one on either side of the coelom (Fig. 10.1B). These ducts empty, with the intestine, into a common posterior vent called the *cloaca* (Fig. 10.2). When evolution finally encased the vertebrate in "armor," it had to devise an internal conduit through which urine could leave the body, for the multiple external openings of the segmentally arranged tubules were now covered over.

The primitive nephric tubules probably also carried the eggs and sperm shed freely by the gonads into the coelom. Embryologically, the testes and ovaries are derived from the lining membrane adjacent to the tissue forming the nephric tubules, suggesting that reproduction and excretion have had a close affiliation throughout evolutionary history.

The segmented kidney—if we may so designate a dozen or more pairs of primitive nephric tubules—is typical of the segmented animal whose coelom participates in excretion, a feeble stream of coelomic fluid serving to wash the excretion from the tubules. With this meager equipment, the early fresh-water vertebrate had to prevent overdilution of its body fluids by an irreducible influx of salt-free water (much of it entering through a permeable body surface) by increasing the rate of water excretion. It accomplished this by means of a cardiac pump. Close to the open mouths of the existing nephric tubules, remarkable filtering devices, tufts of capillaries, developed (see Fig. 10.1A). Under the high hydrostatic pressure of blood, water filtered through these capillaries and drained away through the attached tubules, leaving blood cells and plasma proteins within the capillaries and the blood stream.

---

* *Nephric* and *renal* both mean "pertaining to the kidney." The former comes from the Greek word for kidney, *nephros*, and the latter from the Latin equivalent, *ren*.

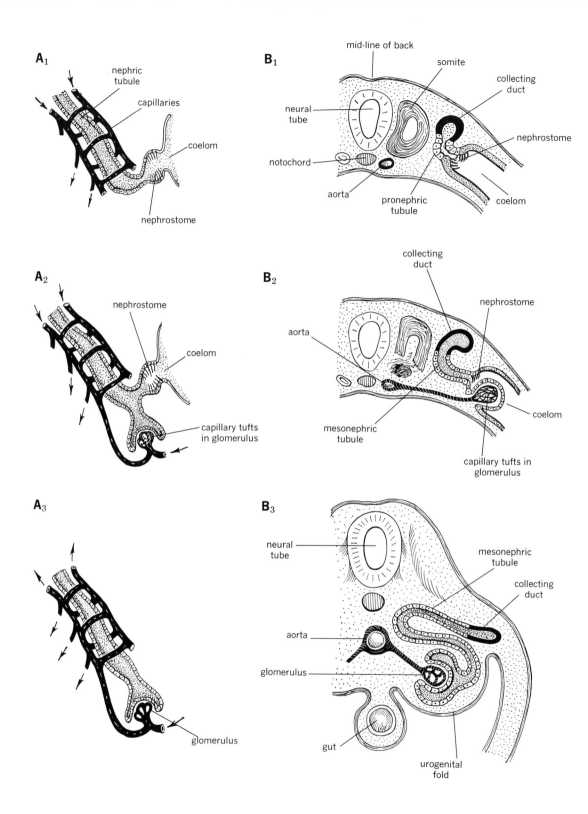

A₁

nephric tubule

capillaries

coelom

nephrostome

A₂

nephrostome

coelom

capillary tufts in glomerulus

A₃

glomerulus

B₁

mid-line of back

somite

collecting duct

neural tube

nephrostome

notochord

aorta

pronephric tubule

coelom

B₂

collecting duct

nephrostome

aorta

mesonephric tubule

coelom

capillary tufts in glomerulus

B₃

neural tube

mesonephric tubule

collecting duct

aorta

glomerulus

gut

urogenital fold

A similar construction appears in the embryo (see Fig. 10.1B). Drainage of the filtrate is made efficient by the insertion of the capillary tuft into a blind end of the nephric tubule. The result is the *glomerulus*,* which characterizes the adult kidney in all higher vertebrates. In the glomerulus the tubule is expanded into a double-walled sphere; the thin inner wall is closely applied to the lobulated tuft of capillaries, while the outer wall forms a capsule, known as *Bowman's capsule*, that collects the filtrate and directs it into the tubule without transit across the coelom.

Finally, in embryology and in evolution, the nephrostome closes. Henceforth filtration is glomerular, under the hydrostatic pressure of heart action; drainage into the tubules from the coelom is lost. The final unitary structure of glomerulus and tubule is the *nephron*, the basic functional unit of the kidney. With the evolution of the nephron, the primary function of the tubule became excretory and regulatory.†

With no opening into the coelom, the tubule could no longer carry eggs and sperm, and these had to find a new route to the exterior. When a suitable urinary duct system had evolved, the reproductive system invaded it. Thus in the male human the urinary and reproductive systems have common conduits.

With further development the embryo loses its conspicuously segmented character, so that its urinary system, which has come to consist of a pair of glomerular nephric tubules in each segment draining into collect-

---

* Since the glomerulus was first described by Malpighi in 1666, it is also known as a *Malpighian corpuscle*.

† These paragraphs outline current views of the evolution of the kidneys in land vertebrates. Present-day fresh-water and salt-water fishes retain kidneys that are believed to resemble those of the ancestors of land vertebrates. Elasmobranchs, fresh-water bony fishes, and amphibians have tubules with large glomeruli capable of excreting large quantities of dilute urine. Salt-water fishes, on the other hand, have small or no glomeruli because they must conserve water. In marine teleosts, some salt and nitrogenous waste is excreted directly from the gills. The problem of land vertebrates more closely resembles that of salt-water than fresh-water fishes. They live in a dry environment, water is continuously lost through the body surface, and conservation of water is necessary, particularly since there are no gills for the elimination of salt or wastes. Thus these vertebrates have glomeruli, though they are relatively small.

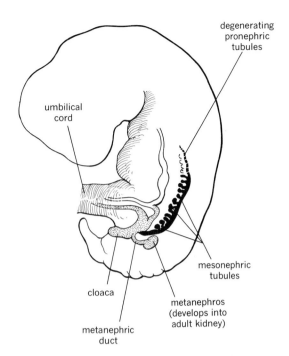

10.2 Urogenital system of the early human embryo.

ing ducts that run the length of the coelom, degenerates. In its place arise bilateral clusters of nephrons, localized in single tissue buds at the caudal end of the old embryonic duct system. These develop into adult kidneys (see Fig. 10.2).

We shall turn now to the structure of the adult kidneys. We shall see that the tubule attached to the glomerulus is far more than a simple duct for conveying glomerular filtrate to the exterior.

STRUCTURE

The urinary system consists of two nearly identical kidneys located at the back of the abdominal cavity. Each kidney, sheathed by a tough fibrous *external capsule*, lies within a large deposit of fat behind the peritoneum. These relations are shown in Fig. 6.7.

The characteristic indented shape of the kidney has given rise to the familiar descriptive phrase "kidney-

---

10.1 Parallels between the evolutionary and embryological development of the kidneys. A, nephrons of different species (adapted from H. W. Smith, *From Fish to Philosopher*, Little, Brown, Boston, 1953). A_1, nephric tubule of protovertebrate; A_2, tubule of early vertebrate, with glomerulus and persisting nephrostome; A_3, typical nephron of higher vertebrate. B, kidney region of human embryos at different stages (adapted from L. B. Arey, *Developmental Anatomy*, 4th ed., W. B. Saunders Co., Philadelphia, 1940). B_1, 2.5 mm embryo (3.5 weeks); B_2, 5 mm embryo (5 weeks); B_3, 8 mm embryo (8 weeks).

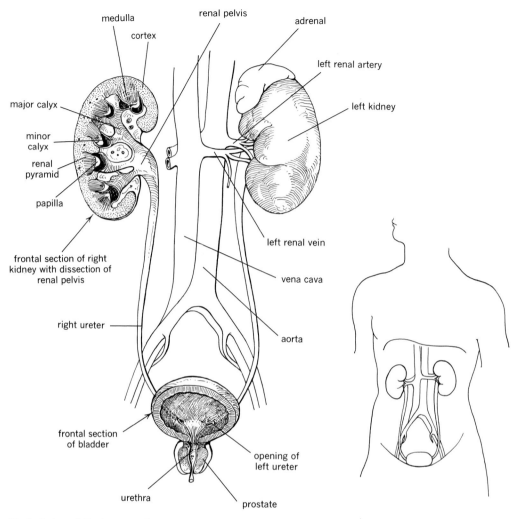

10.3 Ventral view of the human urinary system. Insert shows position in body.

shaped." In longitudinal section the exterior third of the kidney has a typical brownish red color. This is the *cortex*. A number of more or less separate red triangular masses, the *renal pyramids*, with apexes pointed away from the cortex, penetrate the interior two-thirds of the kidney, the *medulla*. Cortical substance dips deeply between adjacent pyramids.

Each kidney contains about a million nephrons. These drain into a treelike arrangement of *collecting tubules* that convey the urine to the *renal pelvis* and thence by way of the *ureters* to the *bladder*. Urine is ejected from the bladder through the *urethra* by the process called *micturition*, or *urination*. The gross structural outlines of the urinary system and its position in the body are given in Fig. 10.3.

The renal pelvis is funnel-shaped, with two (sometimes three) main divisions, the *major calcyes*, which drain the upper and lower halves of the kidney. These in turn divide into several smaller conduits, the *minor calcyes*, each of which terminates around the base of one or more *urinary papillae* in the form of a *calyx*, or cup, into which the urine is delivered. The major and minor calyces contain circular smooth muscle fibers whose contractions propel the urine into the ureter, whence further contractions force it into the bladder.

Nephrons and collecting tubules. A single nephron is shown in Fig. 10.4. Near the glomerulus the tubule is extensively convoluted and is called the *proximal convoluted tubule*. It descends in a fairly straight course

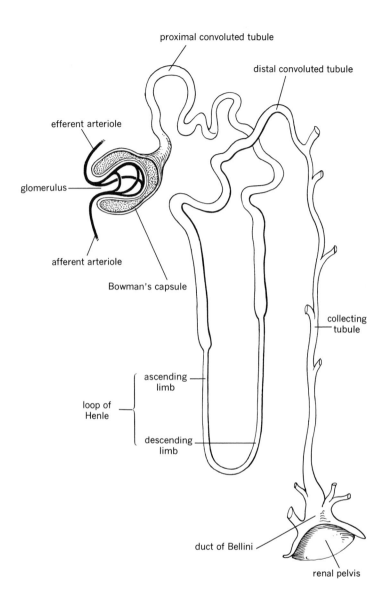

proximal convoluted tubule

distal convoluted tubule

efferent arteriole

glomerulus

afferent arteriole

Bowman's capsule

collecting tubule

ascending limb

loop of Henle

descending limb

duct of Bellini

renal pelvis

10.4 Nephron.

through the cortex toward a pyramid, where it makes a sharp hairpin turn, proceeding back in a straight course to the cortical area near its own glomerulus. The hairpin turn with its parallel descending and ascending limbs comprises the *loop of Henle.* After returning to the cortex, the tubule is further convoluted. This section, the *distal convoluted tubule,* then joins a collecting tubule. Adjacent convoluted tubules may be somewhat intertwined, but, since each tubule in the embryo elongates only after its proximal and distal ends are in fixed positions, no two nephrons can be anatomically crossed.

Throughout its length the tubule is constructed of a single layer of epithelial cells, photographs of which appear in Fig. 10.5. The proximal convoluted tubule is made up of thick cells with brush borders on the surfaces facing the tubular lumen. Thinner cells predominate in the thin portion of the descending limb of the loop of Henle. The cells are thicker again in the ascending limb of the loop of Henle and the distal convoluted tubule.

The functional implications of these structural differences are not yet known in detail. It is customary, therefore, to subdivide the renal tubule into three segments

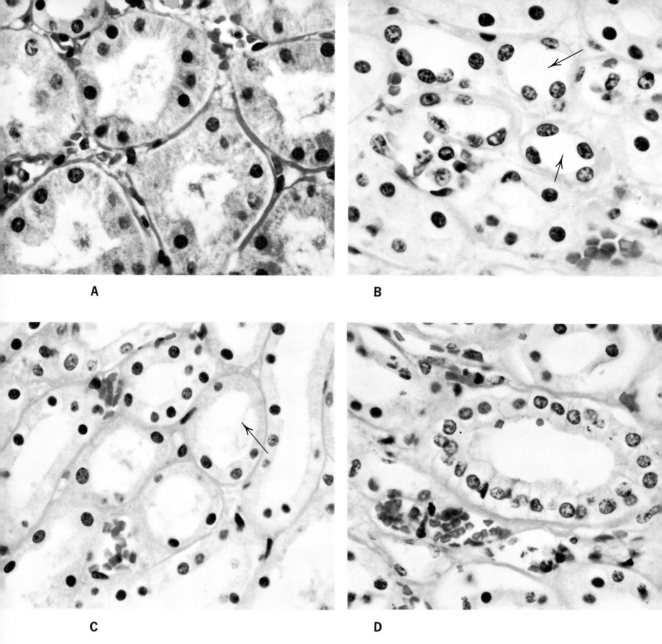

10.5 Cytological patterns of a nephron tubule (× 800): A, proximal convoluted tubule; B, descending limb of the loop of Henle; C, ascending limb of the loop of Henle; D, distal convoluted tubule. (Courtesy of Dr. S. Robboy.)

that have distinctive functional if not anatomical properties. They are (1) the *proximal segment,* which includes the proximal convoluted tubule and the thick part of the descending limb of the loop of Henle, all parts of which are essentially identical cytologically; (2) the *thin segment,* which is variably located in the descending limb of the loop of Henle; and (3) the *distal segment,* which includes the thick ascending limb of the loop of Henle and the distal convoluted tubule. These three segments plus the glomerulus comprise the nephron.

Functional adjuncts to the nephrons proper are the collecting tubules and the *papillary ducts of Bellini* (see Fig. 10.4). These have separate embryonic origins

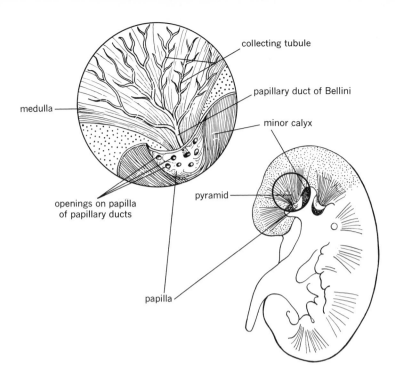

collecting tubule

papillary duct of Bellini

minor calyx

medulla

pyramid

openings on papilla
of papillary ducts

papilla

10.6 Convergence of the collecting tubules into papillary ducts, which open into a papilla.

from the nephrons. Each collecting tubule begins in the cortex and runs directly toward the medulla, where it is connected to a lateral, outwardly curved branch of a nephron. The collecting tubules converge to form the papillary ducts of Bellini (Fig. 10.6). A dozen or more such ducts open on each papilla. Two or more papillae discharge their contents into a minor calyx, and several minor calyces unite into a major calyx.

In recent years investigators have had notable success in isolating individual nephrons for anatomical and physiological study (Fig. 10.7). In such a preparation

10.7 Photographs of nephrons and collecting tubules isolated by the microdissection method of Oliver: A, cluster of nephrons, showing how they are assembled to form a lobule; B, individual nephron, showing connection to the collecting system. (Courtesy of Dr. Jean R. Oliver.)

B

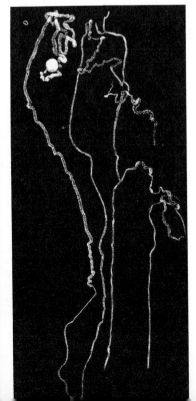

A

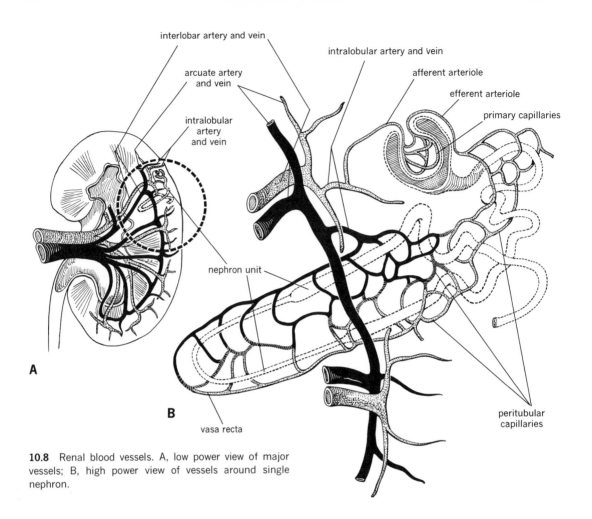

interlobar artery and vein

arcuate artery
and vein

intralobular
artery
and vein

intralobular artery and vein

afferent arteriole

efferent arteriole

primary capillaries

nephron unit

peritubular
capillaries

A

B

vasa recta

10.8 Renal blood vessels. A, low power view of major vessels; B, high power view of vessels around single nephron.

we can readily distinguish the glomerulus, the proximal convoluted tubule, the loop of Henle, the distal convoluted tubule, and the collecting tubule.

We can explain certain features of the cut surface of a kidney (see Fig. 10.3) with what we have learned of the arrangement of the nephrons within it. For example, the fine granular texture of the cortex is due to the irregular convolutions of the proximal and distal tubules, and the fingerlike projections pointing outward from the renal pyramids are due to extensions of the straight renal and collecting tubules into the cortex. These are called *medullary rays.*

Renal blood vessels. The richness of the renal circulation is clear from the injected preparation of Fig. 6.4. Each kidney receives blood from a single *renal artery* arising directly from the abdominal aorta (see Fig. 10.3). Within the renal pelvis each renal artery divides into seven to nine branches, which enter the solid substance of the kidney between adjacent renal pyramids (Fig. 10.8A). Because the vessels lie between the pyramids or lobes of the kidney, they are called *interlobar arteries* at this point. They curve laterally from the outer bases of the pyramids to form incomplete arterial arches; accordingly, they are called *arcuate arteries* at this point. These, in turn, give rise to *intralobular arteries,* which penetrate the cortex radially, running between subdivisions of the pyramids or lobules and giving off many short side branches. These are the *afferent arterioles,* each of which enters a glomerulus (Fig. 10.8B).

The glomerular tuft is supplied with blood by an afferent arteriole and is drained by an *efferent arteriole.* The two vessels are close together, for they penetrate Bowman's capsule and its edges are firmly sealed around them.

The afferent arteriole is a typical arteriole, but its short length and moderately large diameter permit

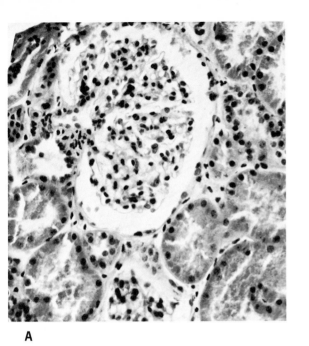

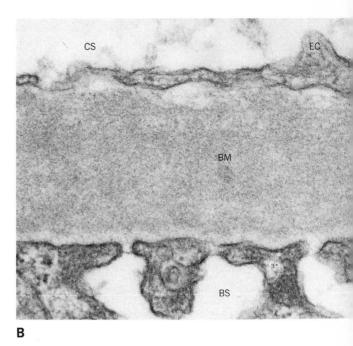

A

B

10.9 Glomerulus: A, section of kidney tissue (×250); B, wall of a glomerular capillary tuft (×90,000): lumen; EC, endothelial cytoplasm; BM, basement membrane; FP, foot processes of surrounding epithelial cells; CS, capillary space; BS, Bowman's space. (A, Courtesy of Dr. S. Robboy; B, courtesy of Dr. James Caulfield.)

blood to enter the glomerular capillaries at a relatively high pressure. Since the afferent arterioles are a more important determinant of the vascular resistance of the kidney than the larger renal arteries, they contribute significantly to the control of renal blood flow and glomerular hydrostatic pressure.

Inside of Bowman's capsule the afferent arteriole abruptly breaks up into four, six, or eight primary capillaries, which branch further. After pursuing intricate courses through the capsule, the capillary loops return to their points of origin, where they coalesce to form the efferent arteriole. The efferent arterioles constitute another component affecting renal vascular resistance.

After emerging from Bowman's capsule, the efferent arteriole breaks up into a second network of capillaries, the *peritubular capillaries,* around the tubular portions of the nephron. These empty into *intralobular veins,* whose arrangement parallels that of the intralobular arteries. Blood then passes through the *arcuate veins,* the *interlobar veins,* and finally, the major *renal veins,* which drain into the inferior vena cava. Unlike the arcuate arteries, the arcuate veins form closed arches and abundant anastomoses.

Glomeruli. It is increasingly difficult to review the structure of the glomerulus without dwelling on its functions. Thus does anatomy blend into physiology. Therefore, we shall merely note here the curious microscopic appearance of the glomerulus (Fig. 10.9). We see the lobulated tuft of capillaries within the spherical Bowman's capsule. Unlike the tubular epithelium, the cells lining the capsule are quite flat. With light microscopy the cells enveloping the capillary tuft are indistinguishable from the endothelial cells of the capillary walls. Electron microscopy, however, reveals interesting fine structure. The wall of the glomerular capillary tuft has three components: (1) a thin inner endothelial layer; (2) a basement membrane that is continuous with the basement membrane of the glomerular arterioles; and (3) a complex epithelial layer whose curious footlike projections make direct contact with the basement membrane. Since the endothelial layer has many *pores* or fenestrations, 400 to 1000 Å in diameter, the only continuous filter is the basement membrane. Although it has interspaces of 30 to 75 Å in its fibrillar structure, it has no actual pores. Thus it is a barrier to all large molecules.

Juxtaglomerular apparatus. At the point where the distal convoluted tubule touches the afferent arteriole, before it empties into the collecting tubule, there is a proliferation of cells, comprising the structure known as

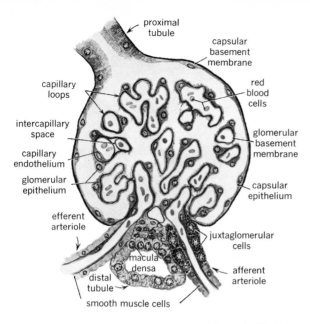

10.10 Juxtaglomerular apparatus. (From A. W. Ham, *Histology*, 5th ed., J. B. Lippincott Co., Philadelphia, 1965.)

Labels in figure: proximal tubule; capsular basement membrane; capillary loops; red blood cells; intercapillary space; glomerular basement membrane; capillary endothelium; glomerular epithelium; capsular epithelium; efferent arteriole; juxtaglomerular cells; macula densa; afferent arteriole; distal tubule; smooth muscle cells

the *juxtaglomerular apparatus* (Fig. 10.10). The cells between the distal tubule and the glomerulus form the *macula densa*. Macula densa cells are readily distinguishable from the cells of the distal tubule by electron microscopy.

The juxtaglomerular cells in the wall of the afferent arteriole are notable for their granular cytoplasm. Electron microscopy shows that the granules are larger than mitochondria and surrounded by membranes and that the cells are richly supplied with endoplasmic reticulum. Such structural features suggest that they have a secretory function, and they are now believed to secrete renin.

FORMATION OF URINE

Historical notes on the theory of renal function. In his classic paper (1842) describing the relation between glomerulus and renal tubule, William Bowman observed that the glomerular structure is ideally suited for the separation of water from plasma. He speculated that the glomerulus performs such a function and that all other urinary constituents are added by the tubule cells—that is, that the kidney is a true gland, and that the water from the glomerulus serves only to wash its secretions down the tubule. In 1844 Carl Ludwig proposed that urine

formation begins in the glomerulus with the separation from plasma of a protein-free ultrafiltrate sufficient in volume to contain all the urinary constituents and that the filtrate is progressively concentrated in the tubule by reabsorption of a large fraction of its water. Renal physiology was thus divided into two schools of thought, one identified with Bowman's secretion theory and the other with Ludwig's filtration theory. The difficulty was that conclusive proof of glomerular filtration was lacking.

In 1917 a monograph titled *The Secretion of the Urine* was published by Arthur Cushny. Modern renal physiology may be considered to have begun with this work. Interestingly, it contained few new facts. Rather, it extracted from the known facts a concise, if somewhat erroneous, new theory of renal function. Cushny rejected the secretion concept and accepted the concept of filtration of a large volume of fluid as the initial step in urine production yet problems arose when he interpreted the function of the tubules. He correctly believed that the formation of the glomerular filtrate is "due to a blind physical force" (i.e., hydrostatic pressure) but incorrectly assumed that the tubules reabsorb an "ideal fluid" of constant composition resembling protein-free plasma. His contribution was one of inspiration rather than experimental implementation. He proposed a theory but adduced no evidence to support it. This had to await the brilliant experiments of A. N. Richards and his associates, which began in 1923 and extended over many years.

Richards first developed methods of directly observing the glomeruli of a living frog. He then evolved a microdissection method that made it possible by a micropuncture technique to withdraw fluid from the Bowman's capsule of a single living nephron, together with special procedures for the chemical analysis of fluid samples smaller than 0.0005 ml. He quickly found that the fluid in a Bowman's capsule contained less than 30 mg of protein per 100 ml. Since his method did not permit the detection of protein below this concentration, he could not categorically state that protein was absent. He also found that glucose and chloride were present in the capsular fluid in concentrations similar to those in plasma although the urine contained no glucose and chloride. Clearly, this fact could be explained only on the basis of *glomerular filtration* (i.e., filtration of all plasma constituents except protein) and *tubular reabsorption* (i.e., reabsorption in the tubules of the ingredients of the glomerular filtrate that the body should conserve, among them glucose and chloride).

Glomerular filtration was therefore accepted as proved, as was tubular reabsorption. Thus the major tenets of Cushny's and Ludwig's theories were firmly established by a series of investigations emanating almost entirely from a single laboratory over a period of 20 years.

Important developments were still to come, some of them showing Cushny partly in error, from workers who felt that urine formation could not be entirely explained by filtration and reabsorption and who saw the need for quantitative measurement of these processes in living man.

Glomerular filtration and the concept of clearance. The methods by which Richards and his associates demonstrated glomerular filtration had obvious limitations. For example, capsular puncture was less useful in the mammal than in the amphibian, where the glomeruli lie just beneath the kidney surface. A new approach applicable to living man was necessary. It was forthcoming in the development of *clearance methods.*

The clearance concept had its origin in early observations of Van Slyke and others that the urine excreted by the kidneys each minute contains an amount of urea equivalent to that in 75 ml of blood. Since urea is a soluble end product of protein metabolism that is destined for elimination, it could be said that the normal kidney *clears* 75 ml of blood of its urea each minute. (This does not mean that only 75 ml of blood perfuses the kidney each minute. If, for example, 1300 ml of blood perfused the kidney each minute and gave up 6% of its urea, we could still say that the kidney cleared approximately 75 ml of blood of its urea each minute.)* For most purposes it is more accurate to speak of plasma clearance than blood clearance, since the kidney clears materials only from the fluid

compartment of blood. The blood cells are regarded as mere space-occupying inert bodies.

We can calculate a *clearance* (a virtual volume of plasma that is completely cleared each minute) for any substance in blood that finds its way into urine. If we compare the concentrations of plasma and urine creatinine, uric acid, and glucose, we find that of the 700 ml of plasma perfusing the kidney each minute, the volume cleared of creatinine is 175 ml; the volume cleared of uric acid is 20 ml; and the volume cleared of glucose is 0 ml. From these figures alone, we can draw two fundamental inferences concerning kidney function: (1) the mechanisms of excretion for these three substances differ; and, (2) since all three are filtered through the glomeruli at least one, glucose (and probably uric acid), must be reabsorbed in the tubules.

For these clearances to have concrete physiological meaning, the volume of filtrate formed in all the glomeruli each minute must be known. If the clearance of any substance is less than this volume, we can conclude that the substance is reabsorbed in the tubules. If the clearance is exactly the same as the volume of filtrate, we can conclude that none of the substance is reabsorbed in the tubules.

A rational search for an indicator substance that could be injected into the blood and that would be excreted by glomerular filtration without subsequent tubular reabsorption ended with the discovery that *inulin,* a polysaccharide from dahlias and artichokes is excreted in this manner. After intravenous injection, inulin passes into the glomerular filtrate. The amount appearing in the urine is exactly equal to the amount filtered. Since the concentration is necessarily the same in the filtrate as in the plasma, the amount in 1 minute's urine is thus exactly the same as that in the plasma filtered by the glomeruli in 1 minute. Knowing both the amount and the concentration of inulin in 1 minute's filtrate, we can estimate the volume of the filtrate or, in formal terminology, the *glomerular filtration rate.*

For example, if the plasma inulin concentration after an intravenous injection of inulin is 1 mg per milliliter and 130 mg of inulin per minute is excreted in the urine, each minute's urine contains the inulin from 130 ml of plasma (or filtrate). Therefore, 130 ml of plasma has

---

* It is of interest to compare the concept of clearance with the Fick principle employed in the measurement of cardiac output (see p. 319).

been filtered each minute, and the glomerular filtration rate is 130 ml per minute. This is a normal figure.

Development of the inulin method of measuring glomerular filtration rate was a major achievement; it is a simple matter to compute the amount of any substance delivered into the tubules with knowledge of this rate and the plasma concentration of the substance. If the plasma urea concentration is 20 mg per 100 ml and the glomerular filtration rate is 130 ml per minute, $20 \times 130/100 = 26$ mg of urea enters the tubules each minute. If the urine passing into the bladder in 1 minute contains only 11 mg of urea, we must assume that 15 mg has been reabsorbed in the tubules. Inulin clearance thus becomes a point of reference for other quantitative determinations.

Hemodynamic aspects of glomerular function. The production of 130 ml of glomerular filtrate each minute requires the passage of an extremely large volume of blood through the kidneys. This could have been anticipated from the fact that the kidneys are responsible for the minute to minute regulation of the composition of 45 liters of body fluids, of which 5 liters is blood. To accomplish their task, the kidneys must process the entire blood volume at frequent intervals. Of the 4 to 5 liters per minute of cardiac output, they receive about 1 liter, and the rest of the body receives the remainder. In other words, 20 to 25% of the cardiac output at rest goes to two organs making up less than 0.5% of the body weight. In 5 minutes or less, the total blood volume passes through the kidneys.

Because of their relatively small size and disproportionately large blood flow, the kidneys require a profuse arterial system (see p. 336). A glomerular capillary is unique in four ways: (1) it lies between two arterioles, in contrast to an ordinary capillary, which lies between an arteriole and a venule; (2) the pressure of the blood within it is 50 to 70 mm of mercury, roughly twice that in other capillaries; (3) it is triply invested—by vascular endothelium, basement membrane, and capsular epithelium—and this helps to prevent the escape of blood cells and protein into Bowman's capsule; and, (4) in contrast to ordinary capillaries, in which fluid filtration into the tissues is roughly balanced by

fluid reabsorption, it permits a bulk transfer of water and dissolved substances in an outward direction.

The energy for filtration derives from the hydrostatic pressure transmitted through the blood from the heart—which necessarily exceeds the osmotic pressure of the plasma proteins. Normally, the outpushing hydrostatic pressure within a glomerular capillary is about 65 mm of mercury (Fig. 10.11). The pressures opposing filtration include the osmotic pressure of blood, about 25 mm, a renal interstitial pressure, about 10 mm, and a renal intratubular pressure, about 10 mm, for a total of 45 mm of mercury. The difference between these pressures is the *filtration pressure*, normally $65 - 45 = 20$ mm of mercury.

The glomerular filtration rate tends to remain constant at about 130 ml per minute despite moderate variations in systemic arterial pressure and renal blood flow. This situation is possible because the arterioles, one on each side of the glomerular capillary bed, allow the renal blood flow and hydrostatic capillary pressure to change independently of one another. When systemic arterial pressure rises, the afferent arteriole constricts, so that the renal blood flow, capillary pressure, and glomerular filtration rate stay nearly the same. When systemic arterial pressure falls, the afferent arteriole dilates, and the efferent arteriole constricts, so that, although blood flow diminishes, the capillary pressure increases sufficiently to maintain the filtration rate. When systemic arterial pressure drops to a point (below 70) where it equals the combined osmotic and renal tissue pressures, filtration stops. The result is the cessation of urine formation, or *anuria*. Anuria may occur despite maximal dilatation of the afferent arterioles and extreme constriction of the efferent arterioles. When normal pressure is restored, filtration resumes if the tissues have not been damaged by anoxia.

When systemic arterial pressure is constant and osmotic pressure or renal tissue pressure increases, capillary pressure and filtration rate decrease. Conversely, with a decrease in osmotic pressure, as in diseases associated with hypoproteinemia, capillary pressure and filtration rate increase.

The filtration rate is also influenced by the renal blood flow. With an increase in blood flow,

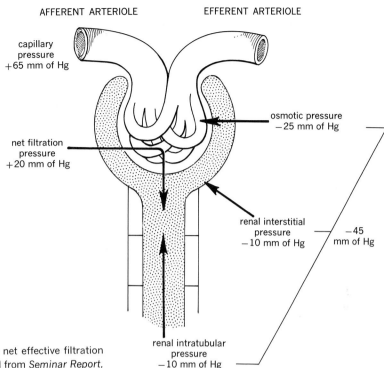

AFFERENT ARTERIOLE   EFFERENT ARTERIOLE

capillary
pressure
+65 mm of Hg

osmotic pressure
−25 mm of Hg

net filtration
pressure
+20 mm of Hg

renal interstitial
pressure
−10 mm of Hg

−45
mm of Hg

renal intratubular
pressure
−10 mm of Hg

10.11 Pressures determining the net effective filtration pressure in a glomerulus. (Adapted from *Seminar Report,* 1947.)

the filtration rate increases, even when capillary, osmotic, and renal tissue pressures remain the same. Similarly, with a decrease in blood flow, the filtration rate decreases. Hence impaired excretory function is common in individuals with congestive heart failure and diminished cardiac output. Since changes in osmotic pressure and renal tissue pressure occur slowly, filtration rate from moment to moment depends primarily on the balance between renal blood flow and capillary pressure.

Renal tubular excretion. We have seen that glomerular filtrate is identical to plasma except that it lacks protein and protein-bound substances. If the enormous volume of filtrate entered the collecting tubules without further modification, the extracellular spaces would be rapidly drained of fluid. Instead of this hypothetical calamity, materials needed by the body are restored to the blood, and unnecessary or harmful ones are excreted in the urine. The disparity between what appears in the filtrate (e.g.,

glucose, chloride, and urea) and what is finally excreted in the urine represents tubular activity.

A careful inspection of Table 10.1 shows that the clearances of some substances are higher than the clearance of inulin. Thus 180 ml of plasma per minute is cleared of creatinine whereas only 130 ml is cleared of inulin. When the clearance of any substance is greater than the filtration rate, it must be excreted (i.e., secreted) by the tubule cells in addition to being filtered in the glomeruli. When a constituent of urine is absent from the glomerular filtrate (e.g., ammonia), it must be derived entirely from the tubule cells. Tubule cells, then, either transfer substances from the surrounding capillaries to the renal tubules (e.g., potassium and creatinine) or synthesize substances *de novo* (e.g., ammonia).

The examples of tubular activity given earlier (i.e., the tubular handling of glucose and urea) involve reabsorption by the tubule cells. In tubular excretion substances move in the opposite direction, toward the tubular lumen. The fact

TABLE 10.1 RENAL CLEARANCES*

Component	Plasma concentration, mg†/100 ml	Amount filtered, mg†/min	Amount reabsorbed in tubules, mg†/min	Amount excreted by tubules, mg†/min	Amount excreted in urine, mg†/min	Plasma clearance, ml/min	Remarks
PAH‡	2	2.0§	0	12	14	700	Clearance equals *renal plasma flow* (see Fig. 10.12).
	50	52	0	80	132	264	Tubular excretion equals *maximum tubular excretory capacity* (T_m) (see Fig. 10.12).
Phenol red‡	5	1.0§	0	14	15	300	
Creatinine	1	1.3	0	0.5	1.8	180	
Inulin‡	20	26	0	0	26	130	Clearance equals *glomerular filtration rate* (see Fig. 10.12).
Urea	20	26	11	0	15	75	Urine flow $\geq$ 2 ml/min
	20	26	15	0	11	55	Urine flow $\leq$ 2 ml/min
Glucose	100	130	130	0	0	0	
	400	520	360	0	160	40	
Phosphate	0.2 meq	0.26 meq	0.21 meq	0	0.05 meq	25	
Potassium	0.5 meq	0.65 meq	0.65 meq	0.05 meq	0.05 meq	10	
Water	100,000	130,000	129,300	0	700	0.7	
Sodium	14.2 meq	18.5 meq	18.0 meq	0	0.5 meq	0.3	
Protein	7,000	10	10	0	0	0	
Ammonia	0	0	0	0.03 meq	0.03 meq		

* Typical values in a normal adult on a normal diet excreting 1500 ml of urine per 24 hours.
† Unless otherwise noted.
‡ An indicator substance injected into the body.
§ The amount filtered is limited by the binding of the component to plasma proteins.

that a substance is excreted by the tubules does not mean that it is not also filtered by the glomeruli. Except in the unusual case of ammonia (which arises in tubule cells), tubular excretion supplements glomerular filtration. Yet the aglomerular kidney of a fish (see Fig. 10.1A) is perfectly competent to excrete most of the ordinary urinary constituents—creatine, creatinine, uric acid, potassium, chloride, and sulfate—as well as a variety of foreign substances—thiosulfate, *p*-aminohippuric acid, and phenol red.

It is instructive to consider the importance of knowledge of the glomerular filtration rate in any attempt to determine whether a given substance is excreted by the tubule cells. The micropuncture method of Richards cannot decide this question. As filtrate passes down a tubule, a filtered substance, such as creatinine, urea, and phosphate, becomes more concentrated, owing either to tubular reabsorption of water or to tubular excretion of additional quantities of the same substance. Which of these two factors is responsible for the concentration can be decided only by comparing it with the concentration of a reference substance (e.g., inulin) that is known to undergo no tubular additions or subtractions on its way to the bladder.

Despite the necessity of such clearance data, E. K. Marshall, Jr., shrewdly inferred the tubular excretion of the dye phenol red before the discovery of quantitative methods for measuring filtration rate. In 1923 Marshall intravenously injected phenol red into dogs whose urine excretion he had stopped by reducing their blood pressure to 40 mm or below. He showed that phenol red nevertheless accumulated in high concentrations within the proximal convoluted tubules. He also showed that phenol red excretion in a normal animal does not increase in direct proportion to plasma phenol red concentration, as it would if the substance were excreted solely by glomerular filtration. At high plasma phenol red concentrations, the excretion rate levels off, approaching a constant value. He interpreted this to mean that the tubule cells participate in the excretion and that the tubule cell mechanisms for transporting phenol red from capillary to tubule eventually become saturated.

In the context of our discussion, the most interesting of Marshall's observations was that as much as 70% of the phenol red could be removed from the blood perfusing the kidney during a single circulation through the kidney. Since only 130 ml of glomerular filtrate is removed from the 700 ml of plasma flowing through the kidney each minute, and since the fraction of any substance removed can never exceed that of the plasma water filtered through the glomeruli—in this case, $(130/700) \times 100 = 18.6\%$ —less than a fifth of the dye could be removed by filtration. The conclusion is inescapable that many times more phenol red appears in the urine than could possibly be accounted for by filtration alone.

Renal plasma flow. The list of substances excreted by tubule cells includes the organic compounds diodrast and *p*-aminohippuric acid (abbreviated PAH). The clearances of these two compounds, like the clearance of inulin, have a special significance because they are almost completely removed from the plasma in one circulation through the kidney. Their clearances may therefore be taken as approximate measures of *renal plasma flow*.

If we inject PAH into the blood at such a rate that the plasma PAH level is kept at 2 mg per 100 ml and if 14 mg of PAH appears in the urine produced each minute, each minute's urine contains the PAH of 700 ml of plasma (Fig. 10.12). Since PAH removal is virtually complete, this volume of plasma must be the total volume of plasma passing the tubules each minute. Hence we conclude that renal plasma flow is 700 ml per minute (corresponding to a *renal blood flow* of about 1400 ml per minute). This is in the normal range.

The volume of plasma cleared of PAH per minute is limited only by the amount of blood flowing through the kidney. In fact, a small amount (perhaps 10%) of the renal blood flow is not cleared of PAH. This supplies the nonexcretory portions of the kidney—the external capsule, the pelvis, and the rest. Blood returning from these structures still contains its full PAH complement. In any event, PAH clearance indicates a normal renal plasma flow of 600 to 700 ml per minute (a normal renal blood flow of 1200 to 1400 ml per minute, or 20 to 25% of the cardiac output).

Tubular excretory capacity. The amount of PAH or similarly excreted material appearing in the urine is the amount filtered through the glomeruli plus the amount excreted by the tubules. If the plasma concentration of such a substance is increased until it exceeds the capacity of the tubules to transfer it into the tubular fluid, a further increase in concentration then only increases the amount filtered. The amount excreted is therefore maximal and a measure of the *maximum tubular excretory capacity*, or T_m. In the case of PAH, this is 75 to 85 mg per minute in normal individuals (see Fig. 10.12). Tubular excretory capacity is useful in assessing the total mass of normally functioning renal tissue.

Significance of plasma protein binding. The relative importance of glomerular filtration and tubular excretion of a given compound depends entirely upon (1) the extent to which it is filtered from the plasma, (2) the extent to which the tubule cells excrete it, and (3) the glomerular filtration rate and the renal plasma flow. For example, at plasma concentrations below that saturating the tubular excretory mechanisms, 7%

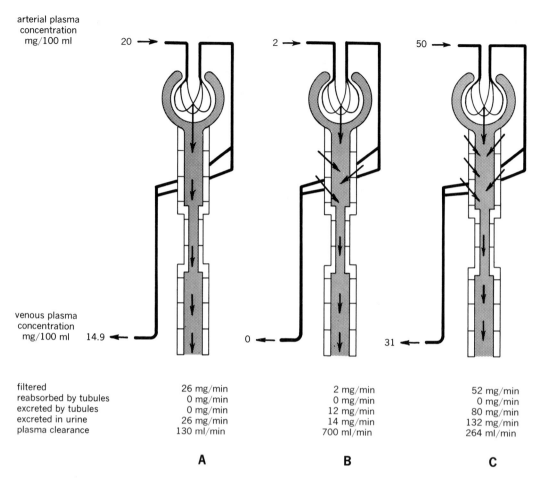

arterial plasma
concentration
mg/100 ml

	A	B	C
arterial plasma concentration mg/100 ml	20 →	2 →	50 →
venous plasma concentration mg/100 ml	14.9 ←	0 ←	31 ←
filtered	26 mg/min	2 mg/min	52 mg/min
reabsorbed by tubules	0 mg/min	0 mg/min	0 mg/min
excreted by tubules	0 mg/min	12 mg/min	80 mg/min
excreted in urine	26 mg/min	14 mg/min	132 mg/min
plasma clearance	130 ml/min	700 ml/min	264 ml/min

10.12 Clearances for the measurement of renal function (see Table 10.1): A, inulin; B, PAH; C, PAH at maximum tubular excretory capacity. (Adapted from *Seminar Report,* 1947.)

of urinary phenol red comes from glomerular filtration, and 93% from tubular excretion. For PAH, the comparable values are 14 and 86%, respectively.

The difference between these two compounds derives from the fact that only 20% of plasma phenol red is filterable whereas 80% of plasma PAH is filterable. The remainder is bound to plasma protein and hence cannot pass through the glomerular filter. The extent to which a substance is protein-bound in plasma is therefore a determinant of the manner of its excretion. Protein binding minimizes the loss through the kidneys of many trace materials, among them iron, copper, thyroxine, vitamin B_{12}, and various peptide hormones.

Tubular reabsorption of glucose: a model system. Richards' discovery that the glucose concentration in glomerular filtrate is similar to that in plasma disclosed a filtrate component that is completely reabsorbed by the tubule cells. Glucose must be completely reabsorbed in order to be absent from the urine.

The handling of glucose by the kidneys typifies the process by which the reabsorptive tubule cells regulate the body content of any essential substance. If the plasma glucose level is normal, say, 100 mg per 100 ml, the 130 ml of glomerular filtrate delivers into the tubules 130 mg of glucose per minute (Fig. 10.13A). Normally all of this is absorbed as it passes down the proximal convoluted tubules.

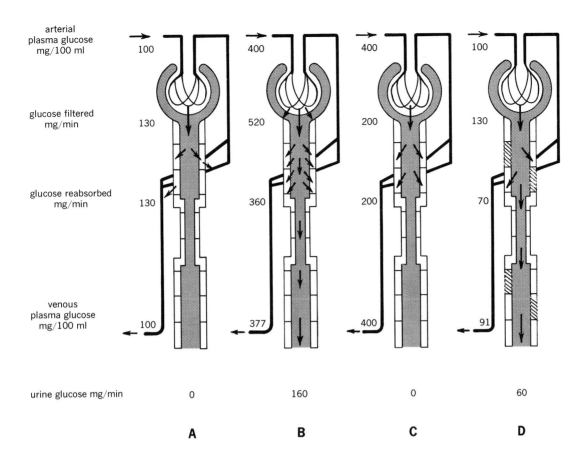

	A	B	C	D
arterial plasma glucose mg/100 ml	100	400	400	100
glucose filtered mg/min	130	520	200	130
glucose reabsorbed mg/min	130	360	200	70
venous plasma glucose mg/100 ml	100	377	400	91
urine glucose mg/min	0	160	0	60

10.13 Renal handling of glucose: A, normal; B, in glycosuria due to hyperglycemia; C, in glycosuria due to hyperglycemia alleviated by decreased filtration (high threshold); D, in glycosuria due to renal tubule damage (renal glycosuria).

If the plasma glucose level increases—a condition known as *hyperglycemia*—additional glucose is filtered and presented to the tubules for reabsorption (Fig. 10.13B). Accordingly, additional glucose is reabsorbed. However, as with PAH, a point is reached at which the active glucose-transporting mechanisms of the tubule cells are saturated. More glucose cannot then be reabsorbed. The unabsorbed glucose stays in the tubular fluid, where, because of its osmotic activity, a portion of the tubular water that would normally have been reabsorbed also remains. The glucose is excreted in the urine, carrying with it the increased volume of water. This condition—called *glycosuria*—is an indication that more glucose is present in the glomerular filtrate than the tubule cells can reabsorb. We can measure the reabsorptive capacity of the glucose system merely by raising the blood glucose level and determining the difference between the amount of glucose filtered and the amount appearing in the urine each minute. The *maximum tubular capacity for glucose reabsorption* averages about 360 mg per minute.

As shown in Fig. 10.13C, glycosuria due to hyperglycemia is alleviated or abolished when the glomerular filtration rate decreases. With less glucose in the glomerular filtrate, the challenge presented to the tubular reabsorptive mechanism diminishes. The disease known as *diabetes mellitus* is the most common cause of hyperglycemia and hence of glycosuria. However, diabetics frequently have atherosclerotic or other changes in their renal arteries that tend to lower the glomerular filtration rate and, in effect, raise the threshold that blood glucose must cross before glucose appears in the urine.

Glycosuria may also follow renal tubule damage (Fig. 10.13D). This may be induced experimentally by the injection of a compound such as *phlorizin*—which affects almost specifically the tubular mechanism of glucose reabsorption —or it may result from disease. If the tubular capacity for glucose reabsorption drops to 70 mg per minute while the glomerular filtration rate remains at 130 mg per minute, 60 mg of glucose appears in the urine each minute. Since most kidney diseases damage the glomeruli and the tubules simultaneously, a decrease in tubular reabsorptive capacity is often accompanied by a proportionate decrease in glomerular filtration rate and accordingly does not manifest itself.

We have considered glycosuria in some detail not only because it occurs commonly but also because the kidney controls plasma levels of many other substances—among them ascorbic acid and other vitamins and most of the blood amino acids—as it does the plasma level of glucose. They are retained by the body and not excreted when the amounts filtered do not exceed the reabsorptive capacities of the tubules. Most reabsorptive mechanisms are highly specific in that they function with blood constituents of definite chemical structures. Thus the reabsorption of a variety of filtrate components simultaneously demands different tubular mechanisms each proceeding independently. All of these depend upon the energy generated in tubule cell metabolism, which therefore becomes a limiting factor.

Nitrogen excretion. The major nitrogen-containing constituents of all living organisms are amino acids and proteins. Metabolism in the animal body never converts them completely to free nitrogen but instead yields a number of end products that contain nitrogen in chemical combination. The three main end products are ammonia, urea, and uric acid.

$$NH_3$$
ammonia

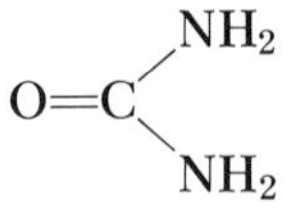

$$O{=}C{\raise1ex\hbox{$\diagup$}\lower1ex\hbox{$\diagdown$}}{\begin{array}{l}NH_2\\ NH_2\end{array}}$$

urea

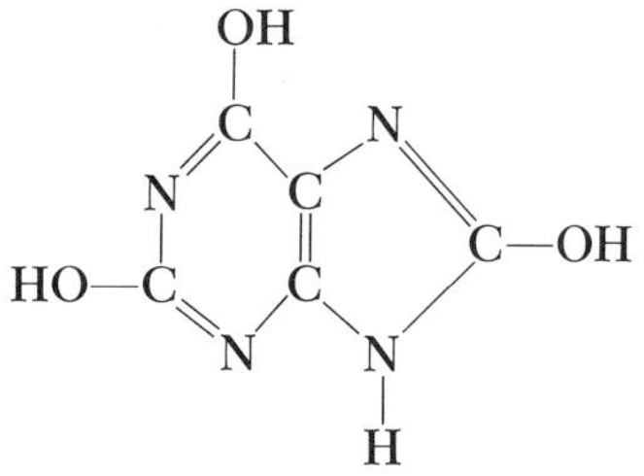

uric acid

Free ammonia is extremely soluble in water, where it forms ammonium ions (NH_4^+). There is almost no ammonia in the blood, and injected ammonia is highly toxic. By rapidly excreting ammonia arising in the body into the surrounding water, however, many aquatic animals prevent its internal accumulation in toxic concentrations. Terrestrial animals, which do not have an unlimited water supply, have evolved powerful enzymatic mechanisms for converting newly formed ammonia into more complex nontoxic compounds, which are excreted. These mechanisms (in the liver in man) rapidly eliminate ammonia.

The relatively harmless substance to which ammonia is converted in mammals, amphibians, and fishes is *urea;*[*] that in birds, most reptiles, terrestrial snails, and insects is *uric acid*. Since urea is freely soluble in water, it is excreted in a watery solution. Indeed, the osmotic activity of urea in the tubules is such that a good deal of water must be excreted along with it. Animals in which such a depletion of water reserves would be a major handicap survive by excreting uric acid. Uric acid is insoluble. When urine becomes concentrated, uric acid readily crystallizes out of solution, and water is reabsorbed. The droppings of birds, for example, consist of brownish fecal matter and a whitish semisolid substance, which in fact is urine containing white uric acid crystals.

Interestingly, all animals in which uric acid is an end product of amino acid metabolism reproduce by egg laying. Clearly the embryo, which must live exclusively on the limited water supply within the egg, needs to conserve water. Since ammonia would be far too toxic for the embryo, and since means do not exist for the

---

[*] The metabolic production of urea is discussed in Chapter 4.

production of a copious watery urine containing urea, the excretion of uric acid solves the problem admirably. In mammals, embryonic wastes are readily removed by the maternal circulation, hence urea can be readily eliminated. A small amount of uric acid forms in man, but it derives chiefly from the breakdown of nucleic acids rather than proteins (Chapter 14). Quantitatively, urea is the principal vehicle of nitrogen elimination.

The fact that almost no ammonia appears in plasma must mean that none is present in the glomerular filtrate. However, Table 10.2 indicates a substantial amount in the urine, a pattern which exactly reverses that of glucose. The source of urinary ammonia and its significance will be dealt with later.

Salt and water excretion. The fundamental mechanisms of salt and water excretion are not yet known. Yet the reabsorption of water is quantitatively the foremost function of the tubules, and sodium reabsorption is a close second. Moreover, these are among the most precisely regulated of all physiological processes, being exquisitely balanced against the body's requirements and intake.

For example, some 600 g of sodium is filtered from the plasma every 24 hours, though the average dietary intake and excretion of sodium is only 4 g (see Table 10.2). Thus the tubules must reabsorb 596 g from the glomerular filtrate. When the dietary intake is deliberately decreased to 0.2 g per day, excretion drops accordingly, so that sodium balance is maintained. Negative sodium balance results when the daily intake falls below this level. At the other extreme, the kidneys are able to excrete as much as 10 g of sodium a day when the intake is high. However, at best, they can excrete only about 2 g of sodium chloride (or 0.78 g of sodium) per 100 ml of urine, a 2% salt solution. This explains

TABLE 10.2   COMPOSITION OF URINE*

Specific gravity	1.020(1.002–1.030)
pH	5.5(4.5–7.8)
Volume, ml	1200(600–2500)
Titratable acidity, meq	30(20–40)
Total solids, g	50(30–70)
Osmolality, mosmols/liter	960(50–1400)

COMPONENT

Inorganic	meq	g	Organic, g	
Chlorine (as chloride)	170(85–380)	10.0 (5.0–16.0)	Nitrogen compounds	30(25–35)
			Urea	25 (20–30) (half total solids)
Sodium	170(85–300)	4.0 (2.0–7.0)	Creatinine	1.4 (1.0–1.8)
Potassium	50(40–80)	2.0 (1.5–3.0)	Uric acid	0.7 (0.5–0.8)
Sulfur (as sulfate)	125(45–22)	2.0 (0.7–3.5)	Creatine	0.10 (0.06–0.15)
Phosphorus (as phosphate)	90(60–120)	1.5 (1.0–2.0)	Amino acids, ascorbic acid, copro- porphyrins, hor- mones, oxalic acid, purines, and pyrimidines	Trace
Ammonia	40(17–60)	0.7 (0.3–1.0)		
Calcium	7(5–10)	0.15 (0.1–0.2)		
Magnesium	12(4–16)	0.15 (0.05–0.2)		

*Per 24 hours in a normal adult on an average diet. Average normal data (with ranges of normal in parentheses).

the lethal consequences of drinking sea water (see p. 210, Fig. 7.6), a 3.5% salt solution. To eliminate the excess salt, the kidneys must excrete up to 1.5 liters of urine for every liter of sea water ingested. The difference is made up by body water, and the inevitable result is dehydration.

Because of the limit in man's capacity to excrete urine with high salt concentrations, physiologists have inquired into the means by which birds that drink sea water eliminate excess salt. Such birds have well–regulated internal environments with salt concentrations considerably lower than that of sea water. Moreover, their kidneys are even less well adapted to salt excretion than ours; for example, a sea gull would have to produce more than 2 liters of urine to dispose of the salt in a liter of sea water. This enigma has only lately been solved. Marine birds and reptiles possess paired *salt glands* just above their beaks or nostrils. These glands produce an external secretion that is almost pure 5% sodium chloride—many times as salty as tears and nearly twice as salty as sea water.

Some mammalian kidneys function at this high level. Those of the kangaroo rat, which thrives in the desert on dried foods and no free water and in the laboratory on dried soybeans and sea water, can produce urine twice as salty as the sea (see Chapter 7).

The problems of water excretion parallel those of salt excretion. The normal rate of glomerular filtration is about 130 ml per minute, or 187 liters (49.3 gal) per 24 hours. The readily exchangeable extracellular fluid in an average man is 11 liters. Consequently, this volume of fluid is reworked by glomerular filtration and tubular reabsorption some 16 times a day. Plasma itself is reworked 58 times a day, or once every 25 minutes. Of the 130 ml filtered each minute, approximately 106 ml is reabsorbed in the proximal tubules, leaving a flow of about 24 ml per minute at their lower ends. If reabsorption ended there, the daily urine volume would be about 35 liters (or 9.2 gal)!

It is important to remember that the glomerular filtrate is isosmotic with the plasma (see Chapter 7); both contain about 290 mosmols per kilogram of water. Any hypothesis that seeks to explain how the tubules reabsorb salt and water must account for the kidneys' ability under conditions of dehydration and salt excess to excrete a highly concentrated urine, containing 1000 to 1400 mosmols per kilogram of water, and under conditions of overhydration to excrete a large volume of *dilute* urine, containing as little as 50 mosmols per kilogram of water.

Specific gravity is a simple measure of the extent to which the glomerular filtrate is concentrated or diluted in its passage through the tubules. Like plasma, the filtrate has a normal specific gravity of 1.010 (compared to 1.000 for water). When the need arises, however, normal kidneys may excrete urine of specific gravity as high as 1.034 or as low as 1.002, resulting from concentration (preponderant water removal) or dilution (preponderant solute removal), respectively. This adaptability is one of the casualties of kidney disease.

It should be understood that the osmolarity of a solution cannot be determined from its specific gravity.* The reason is that osmolarity is a function of the total number of molecules of all solutes in a given quantity of solution irrespective of molecular size whereas specific gravity is a measure of the density of the solution in com-

---

* The concentration of total solutes in the urine expressed as specific gravity and as milliosmols per kilogram of water is as follows:

Sp gr

| 1.005 | 1.007 | 1.010 | 1.015 | 1.020 | 1.025 | 1.030 | 1.035 |

Mosmols per kg of water

| 200 | 300 | 400 | 600 | 800 | 1000 | 1200 | 1400 |

Although the concentrations of osmotically active substances can be measured directly in an osmometer, indirect methods of measurement are more convenient. The freezing point depression technique (*cryoscopy*) is usually used with body fluids. According to the *law of Blagden*, the depression in freezing point of a given solvent (e.g., water) is proportional to the concentration of solute; 1 g-mole of solute lowers the freezing point 1.86°C. A 0.1M (0.1 osmol per liter) aqueous solution of a nonionizing substance freezes at −0.186°C, corresponding to an osmotic pressure of 1900 mm of mercury. Blood has a freezing point of −0.56°C, corresponding to an osmotic pressure of about 5510 mm of mercury.

The osmotic concentration measured by freezing point depression is properly termed an *osmolal* concentration—milliosmols of solute per kilogram of solvent—in contrast to an *osmolar* concentration—milliosmols of solute per liter of solution. The two values differ only slightly in ordinary dilute solutions, however, since moles per kilogram of solvent is practically identical with moles per liter of solution. In this discussion osmolarity is used except when reference is made to data obtained by freezing point depression.

parison to that of pure water and hence depends upon the molecular sizes and masses of the solutes. An electrolyte dissociating into multiple osmotically active ions raises the osmolarity more than a nondissociating organic solute in equivalent molar concentration; for example, $1M$ sodium chloride contains 2000 mosmols per liter, and $1M$ urea contains 1000 mosmols per liter. Consequently, a solution of high specific gravity may have a lower osmolarity than one of low specific gravity.

Large fractions of the osmotically active constituents of glomerular filtrate are reabsorbed in the proximal tubules, together with an osmotically equivalent quantity of water. This phase of water reabsorption is therefore governed solely by the osmotic activity of the reabsorbed solutes.* If the urine in the bladder consisted entirely of the fluid emerging from the proximal tubules, then, it would be isosmotic with the plasma. Since, to maintain internal constancy, the kidneys must be able to produce urine of higher or lower osmotic activity than that of plasma (depending on the conditions of the moment), special mechanisms must modify the fluid issuing from the proximal tubules by adjusting the further reabsorption of salt and water in accord with the needs of the body. Moreover, they must handle salt and water independently, for homeostasis would be impossible if water reabsorption were controlled only by the osmotic activity of salt. This later phase of water reabsorption, in the distal tubules, is termed *facultative* rather than *obligatory*. To use an analogy, if we call obligatory reabsorption involuntary, we can call facultative reabsorption voluntary, that is, fine adjustments made to regulate the osmolarity and volume of body fluids. Facultative reabsorption is the operative limb of a feedback control system, regulating body salt and water levels by regulating their excretion. We shall presently consider the control mechanisms that produce dilute or concentrated urine according to need.

Loci of tubular activities. Until recently, the segments of the renal tubule concerned with specific reabsorption and excretion systems were identifiable only from studies in lower animals and various indirect data. Micropuncture in frogs, rats, and guinea pigs revealed that the main site of glucose reabsorption is the proximal segment; the distal segment is incapable of carrying out this operation. On this evidence, it was established that in man glucose is exclusively reabsorbed by the proximal segment and that, at normal plasma glucose levels, reabsorption is completed within the first, or convoluted, portion of this segment. That phenol red excretion occurs only in the proximal segment in man was inferred from the fact that phenol red excretion occurs in aglomerular fishes, which have only proximal segments.

The *stop-flow technique* is a new and powerful method for detecting the locus of a tubular function. A catheter is tied tightly into the renal pelvis of an anesthetized animal and clamped shut at a given moment to arrest urine flow. During the stop, intratubular pressure rises, and glomerular filtration ceases. The continuing activity of tubule cells in contact with a stationary column of tubular fluid produces exaggerated changes in the local composition of the fluid. Small specimens of urine are then released from the clamped catheter every few seconds. Analysis of these produces a pattern reflecting tubule cell functions at different levels of the nephron. The first urine emerging after the clamp is released is from the collecting tubules and distal parts of the nephron, that emerging later is from the proximal parts of the nephron, etc. Various "markers" disclose the sources of the urine samples: high PAH and low glucose concentrations indicate proximal fluid,† and inulin, injected just prior to release of the clamp, indicates glomerular fluid.

Among the significant observations made since the introduction of the stop-flow technique are these: the zones of PAH excretion and glu-

---

* If 1200 ml of urine of specific gravity 1.020 (800 mosmols per kilogram of water) were formed per day, the total solute output would be 960 mosmols per day. If 960 mosmols of solute were excreted per day in urine whose concentration was 1400 mosmols per kilogram of water (the maximum achievable), the total urine volume would have to be 686 ml per day. This obligatory volume is determined by the osmotic activity of unreabsorbed solutes.

---

† The fact that all proximal fluid must subsequently pass through the distal segment, in which it may undergo considerable modification, limits the usefulness of this method in characterizing events in the proximal segment. It is surprising that concentrations in the fluid rushing through the nephron and pelvis after release of the clamp are not distorted by mixing more than they are.

cose reabsorption overlap and are used to define proximal fluid; sodium and chloride are reabsorbed in both proximal and distal segments; bicarbonate may be strongly reabsorbed at a distal site, though some may be reabsorbed proximally; and potassium and ammonium are excreted distally, as are the bulk of urinary hydrogen ions.

Three types of reabsorptive processes. Thus far we have spoken of two distinct categories of reabsorptive processes. The first and simplest is *diffusion.* Its prototype is the reabsorption of urea. Except for water, oxygen, and carbon dioxide, urea is the most diffusible body constituent. At the beginning of the proximal tubules, no difference exists between the urea concentrations of tubular fluid and plasma, but a gradient develops as water is reabsorbed. If 85% of the water is reabsorbed in the proximal tubules, the ratio of tubular urea to plasma urea is 6:1 or 7:1 at the end of the proximal tubules and 100:1 in the distal tubules. The higher the gradient, the greater the tendency for diffusion to force urea back into the plasma. High urine flow tends to decrease the concentration gradient and hence reduce back-diffusion. Approximately 58% of the filtered urea is reabsorbed at a urine flow of 1 ml per minute, 42% at a urine flow of 2 ml per minute, and only 15 to 20% at a urine flow of 20 ml per minute. In sum, urea reabsorption is believed to occur passively, without a tubular transport system.*

The second category of reabsorptive processes involves active tubular transport systems with *limited* transfer capacities. Its prototype is glucose excretion. When the tubular reabsorptive system is overloaded, a portion of the filtered glucose escapes into the urine.

A third category of reabsorptive processes is concerned with salvaging water and major electrolytes from the tubules. Such processes are distinguished by their complexity, by their localization in proximal *and* distal tubules, by their subservience to physical osmotic forces *and* specific hormones, and finally by their importance in the regulation of acid-base balance.

In recent years understanding of these proc-

---
* Certain features of urea excretion have yet to be explained satisfactorily. These will be mentioned later.

esses has reached a notable turning point. The earlier position was roughly as follows. A massive amount of sodium (accompanied mainly by Cl^- and HCO_3^-) is delivered to the tubules in the glomerular filtrate. Of this, less than 1% is excreted in the urine (see Table 10.1). Although some may be passively reabsorbed along with reabsorbed anions, specific tubular mechanisms actively reabsorb the rest. About seven-eighths of the filtered sodium is reabsorbed in the proximal tubules, and about one-eighth in the distal tubules.

The excretion of potassium is especially peculiar. As noted in Table 10.1, its clearance is less than 10% of the glomerular filtration rate. It was always assumed, therefore, that over 90% of the filtered potassium is reabsorbed and the rest excreted. However, careful study showed that, when the plasma potassium concentration is elevated, the potassium clearance may rise to twice that of inulin. This must be attributed to tubular excretion. It was then found that all filtered potassium is reabsorbed proximally and that the potassium in urine is excreted distally. The demonstration in 1949 that potassium can be both reabsorbed and excreted by the tubules seemed to complicate overwhelmingly the interpretation of clearance data. We shall see in a moment that clarity may again be at hand.

Countercurrent exchange. Claude Bernard suggested many years ago that veins lying next to arteries take up heat from the arteries, thus preventing some body heat from reaching the extremities. Recent measurements have confirmed his theory. This type of heat exchange is minor in man compared to that in animals adapted to severe cold. Such animals have special networks of interweaving arteries and veins, called *rete mirabile* ("wonderful net"), that serve as heat traps, transferring heat from arteries to veins before the arteries enter the extremities.

This mechanism of heat conservation involves *countercurrent exchange.* Let us suppose that water at 30°C flows through an insulated pipe (Fig. 10.14A) at a rate of 10 ml per minute. It passes through a heating unit that puts in 100 cal per minute and leaves the unit opposite its point of entry. Since 100 cal per minute heats 10 ml of water per minute, the temperature of the water

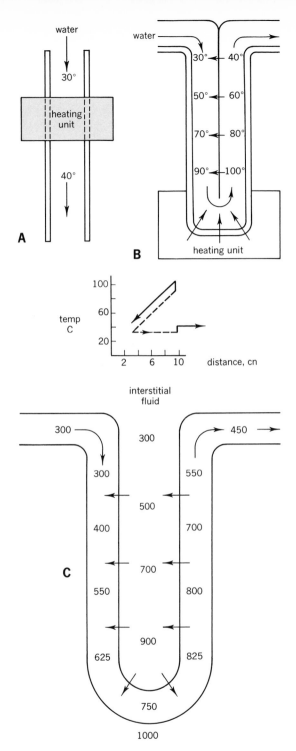

10.14 Countercurrent exchange. Figures are explained in the text. (Adapted from R. W. Berliner *et al.*, *Am. J. Med.*, 24, 730, 1958.)

leaving the heating unit is 10° higher than that of the water entering, or 40°C.

Let us introduce the factor of countercurrent flow by having the pipe double back upon itself (Fig. 10.14B) after emerging from the heating unit, with its two limbs close enough for heat to be exchanged between them, although still insulated from the rest of the surroundings. Water coming out of the heating unit loses some of its heat to the water coming in, warming it before it even enters the unit. Since heat production remains constant within the unit, the temperature of the water leaving it rises above the 40°C of the through-flow system. The temperature increase continues until a steady state is reached, giving a temperature pattern similar to that in the graph. If we look upon the water flow as a device intended to cool the heating unit, it is clear that, with the same rate of water flow, the system with through flow is much more effective than the one with countercurrent flow. In other words, a *countercurrent exchanger* reduces the effective flow.

We have here used heat input as a means of explaining the principle of countercurrent exchange. Identical considerations apply to the addition of a solute (such as sodium chloride) instead of heat, if the solute is permitted to diffuse passively from the outflowing pipe to the inflowing pipe (Fig. 10.14C). To G. Hargitay and W. Kuhn, it seemed that the limbs of the loop of Henle (accompanied by blood capillaries with similar loops) could form a countercurrent system (Fig. 10.15), particularly since, as Marshall had pointed out, only an animal with a loop of Henle in each nephron—namely, a mammal or a bird—can produce concentrated urine. They reasoned that the hairpin countercurrent flow of the tubular fluid and of blood in the capillaries might be responsible for the concentration of urine.

The experiments of H. Wirz in 1953 gave their hypothesis decisive support. Using the freezing point depression technique to determine salt concentration, Wirz discovered that the salt concentration of blood drawn carefully from the capillaries around the loop of Henle is up to three times that of blood from the renal cortex and other parts of the body. He then found that osmolality increases progressively from cortex to papillae, being identical in all

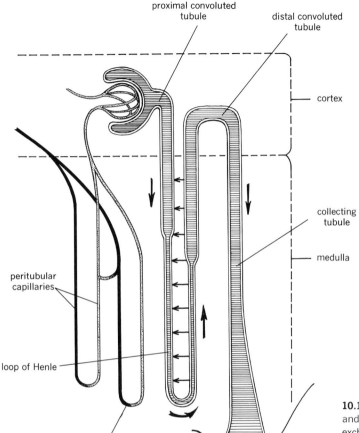

proximal convoluted
tubule

distal convoluted
tubule

cortex

collecting
tubule

medulla

peritubular
capillaries

loop of Henle

vasa recta

papilla

10.15 Nephron, peritubular capillaries, and collecting tubule as a countercurrent exchange system.

structures at any level—capillaries, descending and ascending limbs of the loop of Henle, collecting tubules, and interstitial spaces of tissue.*

A countercurrent exchanger such as the heat-transfer system described is limited by the passive nature of the transfer. It can only conserve inequalities of temperature or osmolarity; it cannot develop them. To do this, a similar hairpin arrangement is needed in which transfer between the two limbs is active. Such an arrangement is a *countercurrent multiplier.* As shown in Fig. 10.16, Wirz concluded that sodium is

actively extruded from the thick part of the ascending limb of the loop of Henle into the interstitial spaces. This operation initiates a continuous series of alternating processes. From the interstitial spaces sodium escapes passively into the adjacent blood vessels and the descending limb of the loop of Henle, according to the concentration gradient in the tissue. The fluid in the descending limb of the loop thereby becomes hypertonic with respect to plasma. Near the bend of the loop, the descending limb and the ascending limb contain hypertonic fluids that for the moment are osmotically identical. Beyond the bend the further active excretion of sodium by cells of the ascending limb results in a new osmotic difference, so that the fluid in the descending limb becomes even more hypertonic, and so on. A countercurrent multiplier system is thus established.

---

* Direct sampling demonstrated that the fluid in the bend of the loop of Henle, the blood in the bends of the adjacent capillary loops, and the interstitial fluid in the immediate vicinity have virtually identical osmolalities several times that of ordinary plasma (as determined by the freezing point depression technique).

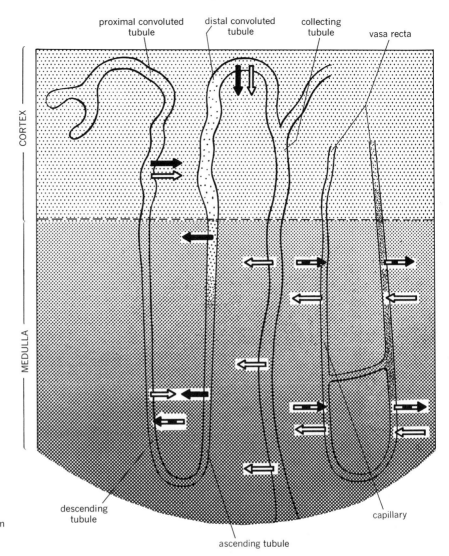

10.16 Loop of Henle as a countercurrent multiplier. Deepening shading denotes increasing osmolarity.

⟹ passive transport of water

⟹ passive transport of sodium

⟹ active transport of sodium

proximal convoluted tubule

distal convoluted tubule

collecting tubule

vasa recta

CORTEX

MEDULLA

descending tubule

ascending tubule

capillary

Since sodium is not excreted or reabsorbed actively by either limb of a *capillary* loop, the capillaries serve simply as countercurrent exchangers, promoting the overall efficiency of the multiplier system and carrying away water removed from the tubules. Later workers have suggested that the peritubular capillaries, not the loops of Henle, might be the prime movers of countercurrent multiplication, the motive force being the hydrostatic pressure difference between their apposed limbs.

The ascending limb of the loop of Henle delivers a hypotonic fluid to the distal convoluted tubule, where, under hormonal influences, water is reabsorbed. Hence the fluid leaving the distal tubule and passing to the collecting tubule is isotonic. The collecting tubule traverses the medulla through strata of successively higher osmolarities that withdraw more water, rendering the urine hypertonic:*

The demonstration that the osmolality of the renal medulla greatly exceeds that prevailing in the body fluids generally is strong evidence that medullary hyperosmolarity is a key part of the mechanism for concentrating urine. Comparative anatomy also offers strong circumstantial evidence for this conclusion. For example, desert rodents noted for their ability to conserve water by excreting concentrated urine (see p. 207) have unusually deep medullas and long papillae. Indeed, the capacity to concentrate urine is directly proportional to the thickness of the medulla relative to that of the cortex.

The preceding paragraphs account for the excretion of concentrated urine. When the body's needs are served by the excretion of dilute urine, the major change is a lowering of the permeability to water of the distal convoluted tubule and the collecting tubule. How does this occur?

Water diuresis, antidiuresis, and the antidiuretic hormone. Water balance is maintained by the renal excretion of dilute urine during overhydration and of concentrated urine during dehydration. The two processes are called *water diuresis* and *antidiuresis*, respectively.†

When a man drinks a large amount of water (1 to 2 liters), diuresis begins in less than 30 minutes, reaches its peak in 1 to 2 hours, and sub-

sides in 3 to 5 hours. It results from dilution of the blood, a slight and short-lived phenomenon that inhibits the release of a neurohypophyseal hormone, *antidiuretic hormone* or *ADH* (also called *vasopressin*), which directly affects the kidney, causing it to excrete a more concentrated urine. Suppression of ADH secretion therefore leads to the excretion of a dilute urine. Increasing plasma osmolarity indirectly stimulates the neurohypophysis to release ADH. Sensitive nerve endings in a brain center—the *hypothalamus*—respond to small changes in plasma osmolarity. These *osmoreceptors* in turn set up nerve impulses that increase or decrease the secretion of ADH by the hypothalamus. The hormone is transferred to the neighboring neurohypophysis, where it is stored until released into the blood. We shall learn more in Chapter 15 of the special significance of the anatomical and functional proximity of the neurohypophysis and the central nervous system. ADH is transported to the kidneys by the blood stream. The result of its action is *increased water retention* by the kidneys and hence *reduced urine volume*.

Remarkable recent experiments have shown that addition of ADH to a living membrane in vitro stimulates the active transport of sodium and increases the bulk movement of water across the membrane. The water movement is passive, ADH apparently acting by enlarging the "pores" or channels in the membrane.‡ ADH thus may be considered to promote the removal of water from the distal convoluted tubule and collecting tubule by directly increasing the bulk flow of water across the tubular membrane into the surrounding interstitial spaces. In the presence of ADH, passive water reabsorption in the distal segment occurs in proportion to the ADH level in body fluids and the solute load. Maximum ADH stimulation in man can lead to a urinary concentration of 1400 mosmols per liter, about 4 times that of body fluids. In the absence of ADH, the distal segment remains relatively impermeable to water, and a dilute urine is excreted whose concentration may be as low as 50 mosmols per liter and whose volume is 7 to 10 times the normal volume. ADH, then, can

---

* The countercurrent hypothesis has led to a reexamination of many excretory processes. The following mechanism of urea excretion has recently been proposed (cf. p. 350). After filtration, the fluid traverses the proximal tubule, now viewed as impermeable to urea. At the bend of the loop of Henle, urea is actively transported against a relatively small concentration gradient from the ascending limb of the loop across into the descending limb. By means of the countercurrent multiplier system, a high concentration of urea is thus built up in the tip of the papilla, the maximum concentration being virtually identical with that in the urine. Hence urea could pass from this region into the collecting tubule by diffusion. In fact, experiments on rat kidney slices show that the urea concentration in interstitial fluid of the papilla tip is about 1000 mmoles per liter (almost as high as that in the urine) while that in interstitial fluid of the outer medulla is 140 mmoles per liter and that in interstitial fluid of the outer cortex is only 40 mmoles per liter. The urea in interstitial spaces of the kidney, however, may reflect local urea synthesis rather than concentration differences related to a countercurrent system.

† *Diuresis* is defined as an increase in the production of urine. A *diuretic* is a drug or other agent that promotes diuresis.

---

‡ ADH also increases membrane permeability to certain other uncharged molecules (e.g., compounds possessing an amide group, typified by urea).

alter the rate of water excretion over a 28-fold range. It is supposed that at normal blood osmolarity there is steady, moderate ADH secretion.* The exquisite sensitivity of the ADH release mechanism is noteworthy: a change of no more than 2% in plasma osmolarity activates it. Its regulation of urine volume combined with the accurate regulation of water intake by thirst maintains a normal body water content.

Removal of an animal's neurohypophysis leads to the production of copious dilute urine (*polyuria*) and continuous severe thirst with prodigious water intake (*polydipsia*). Injection of concentrated salt solutions into the blood of an animal so deprived fails to induce its kidneys to conserve water. The identical picture is seen in the human disease *diabetes insipidus*,† which results from destruction of the neurohypophysis or hypothalamus. This condition is benefited by the administration of ADH.

Adrenocortical hormones: aldosterone. The osmoreceptor–antidiuretic hormone system regulates the overall concentration of osmotically active body solutes. Other hormones powerfully influence the renal excretion of specific ions. Since sodium constitutes approximately 90% of the extracellular cations, the control of sodium excretion is of first importance in determining body fluid composition. Responsibility for this control rests with the *adrenal cortex*, the outer layer of the adrenal lying over each kidney (see Fig. 10.3).‡ When adrenocortical hormone is deficient (as in Addison's disease or after adrenalectomy), an excess of sodium is lost from the body with an osmotic equivalent of water. When adrenocortical hormone is overabundant—due to overproduction by an abnormal gland or overdosage by mouth or injection—an excess of sodium is retained in the body with an osmotic equivalent of water.

The adrenal cortex secretes three types of hormones: *mineralocorticoids, glucocorticoids,* and *sex hormones.* Structurally, all are members of the steroid family, a category of organic compounds of wide biological distribution (see Chapter 2). We are here concerned with the mineralocorticoids, 95% or more of whose activity is attributable to *aldosterone.*

Aldosterone acts specifically upon the renal tubule to increase the reabsorption and retention of sodium (and thus of water, chloride, and bicarbonate) and the excretion of potassium (Fig. 10.17). There is good evidence that it primarily stimulates sodium reabsorption in the distal convoluted tubule, probably by facilitating the entry of sodium into the transport mechanisms of the tubule cells.§ Evidence has been obtained that protein synthesis in the tubule cells is an essential step in aldosterone action (see p. 156) and that the sodium transport system stimulated is dependent on oxidative metabolism. Unlike aldosterone, ADH directly increases the permeability of the distal segment to water. However, it may secondarily increase the amount of sodium entering the tubule cells, thereby making more sodium available to the active transport mechanisms. Thus, although aldosterone and ADH can act independently, they can also act in concert to promote the retention of sodium and water.

Stop-flow experiments confirm that the locus of aldosterone action is the distal tubule. As

---

* Assay methods are not yet sensitive enough to detect ADH in the blood stream. A new assay technique, however, has recently disclosed that in normal hydrated subjects ADH in the urine totals less than 10 milliunits per day and that in dehydrated subjects 10 to 12 milliunits of ADH per hour is excreted in the urine.

† The word *diabetes* identifies a disease marked by the excretion of large quantities of urine. In diabetes mellitus the large urine volume is due to the osmotic effect of massive glycosuria. "Mellitus" means "sweet" or "honeylike," and, indeed, the urine of diabetes mellitus is sweet to the taste. In diabetes insipidus the large urine volume is due to an ADH deficiency. The urine resembles pure water and hence is tasteless or insipid, unlike the urine of diabetes mellitus. A rare form of diabetes insipidus presumably due to an inherited primary defect of the renal tubule cells has been described. In these cases ADH treatment has no effect.

‡ The thyroid and parathyroids affect calcium and phosphate excretion. These and other endocrine glands will be discussed in Chapter 15.

§ It is not known how aldosterone promotes potassium excretion. It is known that in the distal tubule some potassium ions are excreted in exchange for absorbed sodium ions, that potassium excretion increases as the sodium load increases, that potassium excretion requires intact adrenals, and that aldosterone administration does not result in increased potassium excretion in individuals on a low sodium diet—presumably because insufficient sodium reaches the distal tubule to permit a sodium-potassium exchange. Conceivably, the exchange is mediated by a mechanism such as that described in Fig. 4.26. In any event, it is apparently enhanced by aldosterone.

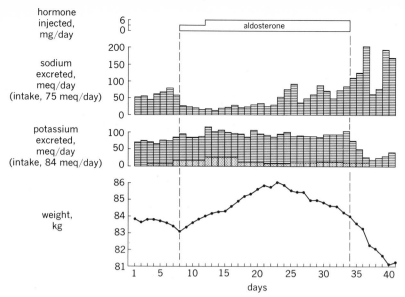

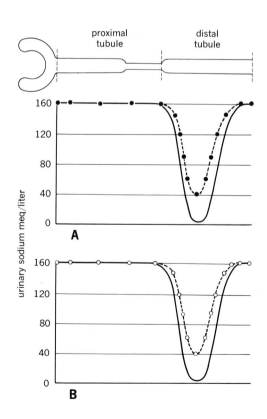

10.17 Effects of aldosterone on sodium and potassium excretion. Comparison of the daily excretion with the daily intake shows the state of balance. The weight increase is due to water retention. The curves reveal that the effects diminish—in an "escape" phenomenon—after prolonged hormone administration. (From J. T. August, D. H. Nelson, and G. W. Thorn, *J. Clin. Invest.*, 37, 1549, 1958.)

10.18 Determination of the locus of aldosterone action in the renal tubule by stop-flow experiments. A. ——, intact dog; ----, adrenalectomized dog. B. ——, intact dog; ----, intact dog administered an antialdosterone.

shown in Fig. 10.18A, adrenalectomy decreases the tubule's capacity to reabsorb sodium. It is restored on the administration of aldosterone. Administration to a normal animal of an *antialdosterone* (or aldosterone antagonist)—a steroid similar in structure to aldosterone but an inhibitor of aldosterone action—converts a normal stop-flow pattern to one resembling that observed after adrenalectomy (Fig. 10.18B).

The weight curve in Fig. 10.17 indicates that aldosterone causes some water retention. This is secondary to sodium retention, not a primary effect of the type produced by ADH. We must remember that, despite the obligatory osmotic interconnection between water excretion and sodium excretion, ultimate control of the two processes involves different facultative mechanisms. Perhaps water balance is maintained separately from sodium balance because the dietary intake of sodium fluctuates more widely than that of water. As noted in Chapter 7, man, unlike a herbivorous animal, does not have a craving for salt that corresponds to his thirst for water.* It is logical, therefore, that the hormonal mechanisms governing salt and water excretion should be different and highly specific. ADH primarily causes water retention; lack of ADH causes water loss; and aldosterone primarily causes sodium retention and potassium loss.

---

* However, salt-depleted individuals often develop strong desires for salty foods.

Like ADH, aldosterone is secreted in response to changes in the volume and osmotic pressure of the body fluids. It was early apparent that the kidneys produce an agent stimulating aldosterone secretion. For example, the outpouring of aldosterone after hemorrhage in normal dogs does not occur in nephrectomized dogs. Recent findings indicate that a decrease in renal arterial pressure (and/or renal blood flow) signals sensitive volume receptors in the afferent arterioles of the kidney, causing the juxtaglomerular cells to release renin (see p. 337), the same substance that elevates blood pressure (see p. 324). As we have seen, renin enters the blood, where it converts a specific globulin to angiotensin I. A plasma enzyme then converts angiotensin I to angiotensin II, which constricts arterioles and thereby raises blood pressure *and also* acts directly on the adrenal cortex to promote aldosterone production. Both effects in turn increase renal arterial pressure and blood flow and through negative feedback shut off the secretion of renin by the juxtaglomerular cells.

Biochemical mechanisms of renal tubular transport. Although much work is being devoted to the biochemical events underlying the active transfer of ions and compounds in the renal tubules, these events remain major unsolved problems of kidney physiology. Moreover, the essential problem is obviously not unique to the kidneys, for all living cells have the capacity to establish and maintain concentration gradients across their membranes. However, the manner in which metabolic energy directs the orderly movement of molecules across these membranes still eludes understanding.

Research in this area consists in attempts to link specific enzyme systems with the transport of specific substances. The most carefully studied systems have been those for the reabsorption of glucose and the secretion of PAH. In part, this work has been a search for agents that inhibit specific tubular processes. Many have been found. We have already mentioned the glycosuria induced by phlorizin. Though this inhibition has obvious theoretical interest, it has a practical aspect as well. For instance, the antibiotic penicillin is excreted in the same manner as PAH or diodrast—by massive tubular excretion plus glomerular filtration. The observation that PAH excretion inhibits simultaneous penicillin excretion led to the discovery that the two compounds are excreted by the same tubular enzyme system. When the transport system is saturated with PAH, it can no longer transport penicillin. Still another compound, probenecid, completely blocks PAH and penicillin excretion by combining so tightly with the transport enzymes that it even inhibits its own excretion. We shall later speak in detail of the diuretics, a group of agents that includes many tubule cell–inhibitors.

ACID-BASE BALANCE

We have dealt thus far with the mechanisms controlling (1) the overall concentration of osmotically active solutes in body fluids and (2) the concentrations of certain individual anions and cations. We must now consider the control of a critically significant and unique cation, the *hydrogen ion*. The term *acid-base balance* refers to the vast apparatus with which the body controls the concentration of this one ion in its fluids.

We recall from Chapter 2 that, according to the theory of Brønsted, an acid is any compound that can donate a hydrogen ion (or proton). Thus HCl is an acid because it dissociates in water into H^+ and Cl^-.

Role of bicarbonate ion. The pH of body fluids is controlled by buffers. One of the most important consists of HCO_3^-, the bicarbonate ion, and H_2CO_3 (or $H-HCO_3$), carbonic acid. Normally the pH of blood is 7.4. Since the pK of H_2CO_3 is 6.1, we can easily calculate the ratio of $[HCO_3^-]$ to $[H_2CO_3]$ that must exist in blood if this pH is to be maintained.

$$pH = 7.4 = 6.1 + \log \frac{[HCO_3^-]}{[H_2CO_3]}$$

$$1.3 = \log \frac{[HCO_3^-]}{[H_2CO_3]}$$

$$\text{antilog } 1.3 = 20$$

$$\frac{[HCO_3^-]}{[H_2CO_3]} = 20 \qquad (10.1)$$

Normally the respective concentrations are 26 and 1.3 meq per liter.

There are two reasons for the particular importance of this buffer system. The first is that

TABLE 10.3 DISTURBANCES OF ACID-BASE EQUILIBRIUM

Disorder	Typical causes	Effect
Acidosis, respiratory	inadequate elimination of CO_2, as in pneumonia or poliomyelitis	compensated: $[H_2CO_3]$ increase balanced by $[HCO_3^-]$ increase; normal pH uncompensated: $[H_2CO_3]$ increase without proportionate $[HCO_3^-]$ increase; lowered pH
Acidosis, metabolic	renal failure, diabetes, NH_4Cl ingestion	compensated: $[HCO_3^-]$ decrease balanced by $[H_2CO_3]$ decrease; normal pH uncompensated: $[HCO_3^-]$ decrease without proportionate $[H_2CO_3]$ decrease; lowered pH
Alkalosis, respiratory	rapid breathing and over-elimination of CO_2	compensated: $[H_2CO_3]$ decrease balanced by $[HCO_3^-]$ decrease; normal pH uncompensated: $[H_2CO_3]$ decrease without proportionate $[HCO_3^-]$ decrease; elevated pH
Alkalosis, metabolic	vomiting with HCl loss, $NaHCO_3$ ingestion	compensated: $[HCO_3^-]$ increase balanced by $[H_2CO_3]$ increase; normal pH uncompensated: $[HCO_3^-]$ increase without proportionate $[H_2CO_3]$ increase; elevated pH

CO_2 is constantly produced in metabolism (see Chapter 4). The second is that this system, unlike others (e.g., the HPO_4^{--}–$H_2PO_4^-$ system), has two components whose concentrations are readily regulated. An enzyme, *carbonic anhydrase*, catalyzes the reaction

$$H_2CO_3 \rightleftharpoons CO_2 + H_2O \qquad (10.2)$$

This means that H_2CO_3 is in constant equilibrium with CO_2. Since CO_2 is a gas that can be expired in the lungs, the pulmonary system provides one means of controlling the concentration of the H_2CO_3 portion of the buffer; by decreasing or increasing the ventilation rate, the lungs retain or discharge more of the CO_2 formed from H_2CO_3 by carbonic anhydrase. The kidneys sensitively regulate the HCO_3^- level. Because of the easy variability of both components of the buffer system, the blood pH can be shifted up or down by the actions of the lungs and kidneys.

The interaction of the HCO_3^-–H_2CO_3 buffer system and pulmonary CO_2 elimination is apparent from the events following the entrance into the blood stream of extra acid or extra base. When acid is added to the blood, H^+ is removed by combination with HCO_3^-, and the H_2CO_3 formed is split into H_2O and CO_2.

$$H^+ + HCO_3^- \rightleftharpoons H_2CO_3 \rightleftharpoons$$
$$H_2O + CO_2 \qquad (10.3)$$

The CO_2 is expired by the lungs, so that the $[HCO_3^-]/[H_2CO_3]$ ratio is kept at 20. Much plasma bicarbonate can be utilized in this manner in resisting a decrease in pH. When base is added to the blood, $[HCO_3^-]$ increases.

$$OH^- + H_2CO_3 \rightleftharpoons$$
$$HCO_3^- + H_2O \qquad (10.4)$$

The rise in $[HCO_3^-]$ is partially offset by a fall in pulmonary CO_2 elimination. This results in an increase in $[H_2CO_3]$ tending to restore the normal $[HCO_3^-]/[H_2CO_3]$ ratio. Mechanisms are therefore set in motion to readjust the $[HCO_3^-]/[H_2CO_3]$ ratio to 20 after either acid or base addition.

Acid-base disturbances. When the $[HCO_3^-]/[H_2CO_3]$ ratio in the blood decreases, blood pH drops until and unless compensatory reactions occur. The condition is called *acidosis*. The opposite condition is called *alkalosis*. When buffer adjustments successfully restore the pH to normal, the acidosis or alkalosis is *compensated*. When they do not, the acidosis or alkalosis is *uncompensated*.

It is convenient to categorize acidosis and alkalosis according to origin—*respiratory* or *metabolic*—as in Table 10.3. This classification should be considered in the light of Eq. 10.3. It is evident that defects in pulmonary function causing retention or overelimination of CO_2 cause corresponding changes in $[H_2CO_3]$—*respiratory acidosis* or *respiratory alkalosis*—and that retention or loss of H^+ secondarily affects the $[HCO_3^-]/[H_2CO_3]$ ratio to produce *metabolic acidosis* or *metabolic alkalosis*. Note that all the reactions in Eq. 10.3 are reversible. This means that any primary change in the pulmonary ventilation of CO_2 results in a secondary change in $[HCO_3^-]$. Conversely, every H^+ ion entering the system converts an HCO_3^- ion to CO_2, which is expelled in the lungs.

Role of the kidneys. We saw in Table 10.2 that the average pH of urine is between 5 and 6 and that the extremes of urinary pH are approximately 4.5 and 7.8. A pH of 4.5 signifies 5.01×10^{-3} moles of H^+ per liter. The kidneys, then, can excrete a solution whose hydrogen ion concentration is over 1000 times that of plasma.

The kidneys' ability to regulate the pH of urine is a physiological asset of the greatest importance. Food intake varies, and the metabolism of foodstuffs inside cells gives rise to variable quantities of *fixed acids* (HA).* The hydrogen ions released by these acids react with HCO_3^- in the extracellular fluid to yield the salts of the acids, plus H_2O and CO_2. The homeostatic task of the kidneys is to regenerate the HCO_3^- consumed in the reactions with endogenously produced acids. This is accomplished by the tubular excretion of acid—i.e., by the production of an acidic urine.

When the kidneys must excrete a highly acidic urine, they must promote the excretion of

---

* These result mainly from (1) the oxidation of organic sulfur (e.g., in methionine) to sulfate, (2) the formation of unmetabolized organic acids (e.g., uric acid), and (3) the hydrolysis of phosphoesters (e.g., nucleotides).

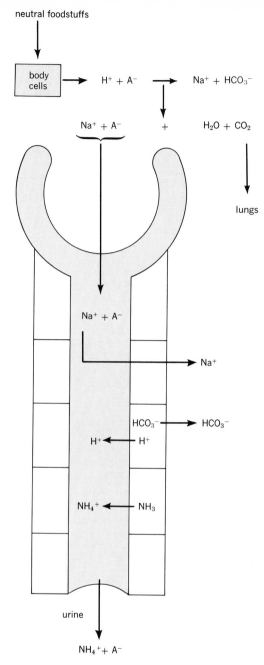

neutral foodstuffs

body cells → $H^+ + A^-$ → $Na^+ + HCO_3^-$

$Na^+ + A^-$ + $H_2O + CO_2$

lungs

$Na^+ + A^-$

→ Na^+

HCO_3^- → HCO_3^-

H^+ ← H^+

NH_4^+ ← NH_3

urine

$NH_4^+ + A^-$

10.19 Role of the kidney in the regulation of acid-base balance. (From A. S. Relman, *Advan. Internal Med.*, 12, 295, 1964.)

hydrogen ions and other acidic groups (i.e., potential hydrogen ions) while simultaneously conserving essential, or fixed, cations, such as Na^+, K^+, Ca^{++}, and Mg^{++}. If anions were excreted with fixed cations instead of with H^+, urinary $[H^+]$ would not increase, and plasma cations would be squandered. As illustrated in Fig. 10.19, the kidneys acidify the urine by promoting three closely related phenomena: (1) the replacement of some Na^+ in the tubular fluid by H^+ arising within the tubule cells; (2) the replacement of some Na^+ in the tubular fluid by NH_4^+ derived from the tubule cells; and (3) the conservation of HCO_3^-. In a sense the renal tubule can thus be regarded as a special kind of gland that secretes acid into the tubular fluid and base into the blood.

The fluid in the cavity of a Bowman's capsule is an ultrafiltrate of plasma and is therefore of the same acid-base composition. We know that tubule cells acidify this fluid because the concentration of H^+ is higher in urine than in glomerular filtrate. Theoretically, the acidification could result from the tubular excretion of acid or the tubular reabsorption of base or both. In fact, the primary acidification mechanism is the addition of H^+ to tubular fluid in exchange for Na^+. H_2CO_3 is synthesized in the tubule cells from H_2O and CO_2 (derived from blood and tubule cell metabolism) under the influence of carbonic anhydrase (see Fig. 10.19).* The H_2CO_3 dissociates into H^+ and HCO_3^-, and the H^+ passes into the tubules. This intracellular process, which generates and conserves a bicarbonate ion, requires a large expenditure of energy by the tubule cells. When one H^+ ion is excreted into the tubules, one Na^+ (or K^+ or Ca^{++} or Mg^{++}) ion is reabsorbed from the tubular fluid by the tubule cells. This exchange has three interesting consequences. First, filtered Na^+ is conserved, but filtered HCO_3^- is destroyed.

$$NaHCO_3 \quad + \quad H^+ \quad \longrightarrow$$
(in tubular fluid)   (from tubule
pH 7.4   cells)

$$Na^+ \quad + \quad H_2CO_3 \qquad (10.5)$$
(reabsorbed)   (split to CO_2 and
                            H_2O; CO_2 reabsorbed)
                            pH 4.5

---

* Any of the circumstances in Table 10.3 that elevate blood $[H_2CO_3]$ proportionately elevate tubular $[H^+]$.

Note that the fate of the filtered HCO_3^- contrasts with that of the HCO_3^- generated within the cell from H_2O and CO_2.

Second, the excess H^+ produced by the tubule cells (i.e., the H^+ remaining after the available HCO_3^- is "used up") converts HPO_4^{--} in the tubular fluid to the more acidic $H_2PO_4^-$. This also promotes the replacement of Na^+ by H^+.

$$Na_2HPO_4 \quad + \quad H^+ \quad \longrightarrow$$

(in tubular fluid)   (from tubule
$\quad$ **pH 7.4** $\qquad$ cells)

$$Na^+ \quad + \quad NaH_2PO_4 \qquad (10.6)$$

(reabsorbed)   (excreted in
$\qquad\qquad\qquad$ urine)
$\qquad\qquad\qquad$ **pH 4.5**

Third, NH_4^+ is produced and eliminated. The replacement of tubular cation by NH_4^+ is particularly interesting. Although the toxic cation NH_4^+ is absent from blood and hence from glomerular filtrate, according to Table 10.2 it is a prominent urinary constituent. This situation suggests that the tubule cells excrete NH_3, which reacts with H^+ in the tubular fluid to form NH_4^+. The NH_3 is a product of the breakdown of glutamine and other amino acids by glutaminase and other enzymes.* Since the tubule cells are freely permeable to NH_3 but not to NH_4^+, NH_4^+ is not reabsorbed. The NH_4^+ content of urine is negligible until the pH falls below 6. Then NH_4^+ excretion increases linearly as urinary pH decreases. NH_4^+ thus plays a vital and unique role in conserving fixed cations. Its production can proceed at very high rates, and its excretion, unlike that of phosphate, requires no simultaneous excretion of fixed cations.

In summary, the chief function of the kidneys in acid-base regulation is to counterbalance the body's excess acid production (which normally averages 50 to 100 meq per day). They do this by regenerating bicarbonate and by excreting acid into the tubular fluid. The acid appears in the urine free as H^+ and in combination as $H_2PO_4^-$ and NH_4^+. Since NH_4^+ is a cation, it substitutes

for and thereby conserves Na^+ and other fixed cations.

The urinary excretion of NH_4^+ in a normal individual on an average diet is 30 to 50 meq per day. The amount of cations conserved by the addition of H^+ to tubular fluid can be determined from the amount of base needed to bring the urine pH to 7.4. This quantity is referred to as the *titratable acidity* of urine. It is normally 20 to 40 meq per day. The sum of titratable acidity and urinary ammonium (minus whatever small amount of urinary bicarbonate may be present) is referred to as the *total renal defense* of the plasma bicarbonate buffer system. As long as acid production in the body does not exceed the capacity of these mechanisms, the plasma $[HCO_3^-]$ remains normal. Otherwise, uncompensated metabolic acidosis results.

The excretion of alkaline urine is perhaps more easily explained. It occurs whenever the plasma $[HCO_3^-]$ exceeds the normal level of 26 to 28 meq per liter. Below this level all bicarbonate in the tubular fluid is reabsorbed. Above it $NaHCO_3$ is rapidly lost in the urine (see Eq. 10.5), which is alkaline, and the alkalosis is gradually corrected.

Diuretics and edema. We noted in Chapters 7 and 9 that in certain abnormal conditions, principally congestive heart failure, the kidneys fail to excrete enough sodium. Sodium retention is accompanied by water retention, since normal osmolarity must be maintained in the body fluids, and its chief manifestation is edema. In such instances an imbalance exists between glomerular function and tubular function. A decrease in glomerular filtration rate, whatever the cause, without a proportional reduction in tubular reabsorptive activity inevitably leads to sodium and water retention. An increase in tubular reabsorptive activity without a corresponding increase in glomerular filtration rate produces the same result. The former circumstances arise mainly from a decrease in cardiac output, with a consequent decrease in blood flow to the kidney. The latter circumstances arise mainly from an excessive secretion of aldosterone.

In treating edema, the physician first attempts to eliminate the primary disorder. The salt and water retention of heart failure, for instance, is completely reversed only when the heart failure is reversed. Since the glomerular filtration rate depends upon the energy imparted by cardiac systole, we can relieve edema merely by strengthening the heart muscle with digitalis, limiting the dietary sodium intake, and decreasing the demands upon the heart with rest. However, we can sup-

---

* The conversion of glutamine to NH_3 (and its place in the overall pattern of amino acid metabolism) is shown in Fig. 4.18.

TABLE 10.4 DIURETICS

Type and effect	Examples
PHYSIOLOGICAL Increasing glomerular filtration of sodium and/or chloride	Agents increasing cardiac output: digitalis Agents dilating afferent glomerular arterioles: xanthine derivatives
Decreasing tubular reabsorption of sodium and/or chloride	Agents inhibiting aldosterone secretion: amphenones Agents inhibiting aldosterone action: spirolactones Competitive osmotic agents: urea
PHARMACOLOGICAL Increasing tubular reabsorption of sodium and/or chloride	Organic mercurials: meralluride Chlorothiazide Xanthine derivatives
Interfering with sodium-hydrogen exchange	Carbonic anhydrase inhibitors: acetazolamide Potassium salts: potassium chloride

plement these measures with diuretics. Their study tells us much of renal physiology.

Diuretics are commonly defined as agents that increase the volume of urine by increasing the difference between the volume of glomerular filtrate formed and the volume reabsorbed in the tubules. R. F. Pitts more precisely defines diuretics as agents that promote the urinary excretion of sodium and either chloride or bicarbonate—the major ionic components of extracellular fluid (see Fig. 7.6). This definition properly emphasizes that the excretion of these ions is primary, that they are drawn from extracellular rather than cellular stores, and that increased urine volume and lost body weight are proportional to and osmotic consequences of the ionic losses.

Table 10.4 indicates that diuresis is induced by two kinds of substances: (1) *physiological* diuretics, which act indirectly on the kidneys, reversing a glomerular-tubular imbalance by either elevating depressed glomerular filtration or depressing elevated tubular reabsorption; and (2) *pharmacological* diuretics, which act directly on the kidneys, inhibiting tubular reabsorption of sodium or sodium-hydrogen exchange. The two categories overlap to some extent in that some diuretics affect the kidneys both directly and indirectly.

The diuretic effect of digitalis occurs only in individuals with failing hearts and depends entirely upon an increase in cardiac output. The xanthine derivatives (such as theophylline and theobromine), chemical relatives of the purines, briefly and feebly increase glomerular filtration (physiologically) and decrease tubular reabsorption of sodium (pharmacologically).* These

* They also act as mental stimulants.

compounds occur naturally throughout the world in various plants, whose aqueous extracts natives have long used as beverages. Theophylline is found in tea; theobromine in cocoa; and their close chemical relative caffeine in coffee. The diuretic effects of tea and coffee are well known.

The antialdosterones are of two general types: those blocking the secretion of aldosterone by the adrenal cortex, the amphenones (Chapter 15); and those blocking the action of aldosterone on the renal tubules, the spirolactones. We noted earlier (see p. 356) the ability of an antialdosterone to neutralize the effect of aldosterone in stimulating sodium reabsorption in the distal convoluted tubule. Antialdosterones promote sodium and water excretion by blocking both sodium reabsorption and potassium excretion in the distal tubule.

If a large quantity of urea is administered by mouth, massive diuresis results. Urea is one of a group of diuretics that are filtered by the glomeruli and poorly reabsorbed by the tubules. Hence its osmotic equivalent of water is lost.

The most important pharmacological diuretics are the mercurials—i.e., mercury compounds. Calomel (mercurous chloride, $HgCl$) was used as a diuretic by Paracelsus in the fifteenth century. However, its powerful laxative action has led to its replacement by organic mercurials. These are perhaps the most effective diuretics known. They act directly upon the kidneys. Thus experimental injection of a mercurial into the left renal artery produces diuresis only from the left kidney. Current evidence indicates that mercurials inhibit tubule cell enzymes concerned with the reabsorption of sodium (and chloride) in the proximal convoluted tubule. The pharmacological diuretic effect of chlorothiazide, first synthesized in 1957, is due to inhibition of sodium reabsorption.

Any inhibitor of carbonic anhydrase will interfere with the sodium–hydrogen exchanges of Eqs. 10.5 and 10.6, since these depend on the cleavage and synthesis of H_2CO_3 by carbonic anhydrase. It was discovered in 1949 that sulfanilamide promotes diuresis in subjects with congestive heart failure. This was attributed to sulfanilamide inhibition of tubular carbonic anhydrase. Other carbonic anhydrase inhibitors have since been prepared (e.g., acetazolamide, or Diamox) and have proved useful in the treatment of mild edema.*

Maximum diuresis may require the simultaneous administration of several diuretics with different sites of action. Interestingly, a combination of diuretics produces a synergistic rather than a simple additive effect.

Research in the field of diuretics is extremely active. Beyond its immediate clinical value, it furnishes much-needed information on the nature of renal transport mechanisms, and, as more is learned about these mechanisms, improved methods for alleviating acid–base and salt–water imbalances will surely follow.

PHYSIOLOGICAL IMPLICATIONS OF KIDNEY DISEASE

The pathology of the kidney is uncommonly interesting and complex. Here we shall discuss those matters that will further our understanding of renal physiology.

Significance of proteinuria. The historic demonstration by A. N. Richards that glomerular filtrate contains less than 30 mg of protein per 100 ml involved the direct sampling and chemical analysis of fluid from a Bowman's capsule (see p. 338). Since Richards' method could not detect proteins in lower concentrations, and since the volume of glomerular filtrate is 180 liters per 24 hours, as much as 54,000 mg of protein could be filtered in 24 hours and escape detection. In later years more sensitive methods were developed which revealed that glomerular filtrate is, in fact, *not* protein free but contains about 10 mg of protein per 100 ml. Again reckoning from the known volume of filtrate per 24 hours, we calculate that 18 g of protein is actually filtered each day.

Yet urine contains very little protein—normally less than 200 mg per 24 hours—and most of this is due to cells sloughing from the walls of the lower urinary tract. We conclude that the 18 g of protein filtered daily is reabsorbed in the renal tubules (Fig. 10.20)—the reabsorption, like that of glucose, being complete.

Proteinuria, the appearance of protein in the urine, is due to (1) the faulty reabsorption of protein by a damaged or saturated tubule or (2) the leakage of an excessive amount of protein through a damaged glomerulus (the usual cause). It is therefore a presumptive indication of kidney damage.† The rate of protein excretion in proteinuria roughly parallels the plasma protein level and the glomerular filtration rate. Ordinarily the chief protein in urine is albumin, although normal and abnormal globulins are found in disorders characterized by abnormal protein synthesis (e.g., multiple myeloma; see p. 264).

Urinalysis. Examination of the urine is an ancient means of studying human disease, *urinalysis* (urine analysis) having been widely employed by medieval physicians. Though their observations were unscientific, they were correct in assuming that the urine provides valuable information on the state of the body and the kidneys.

Urinalysis requires careful and systematic study of the physical, chemical, and microscopic properties of the urine. Normal urine varies in color from pale yellow to deep amber, depending on its degree of concentration. The color is due to a mixture of urinary bile pigments. An abnormal red color may result from the presence of hemoglobin (*hemoglobinuria*), erythrocytes (*hematuria*), or porphyrins (*porphyrinuria*). Normally, freshly voided urine is clear, but turbidity may develop owing to the precipitation of calcium phosphate in alkaline urine and of urates in strongly acid urine. Normal urine has a mild, faintly pungent or aromatic odor, due in part to minute amounts of certain volatile organic acids. Decomposing urine has a penetrating ammonia odor. A fruity odor occurs in diabetes mellitus because of the presence of acetone. A high specific gravity—1.025 or more—in a specimen of urine free from protein and glucose is good evidence of normal renal function.

According to Table 10.2, the urine excreted in 24 hours contains about 50 g of solids in a volume of 600 to 2500 ml. Half of the solids are organic, and half are inorganic. Most of the organic solid matter is urea. The inorganic matter is chiefly chlorides, phosphates, sulfates, potassium, and sodium. With an ordinary diet half of it is sodium and chloride. The appearance of more than a trace of glucose suggests diabetes mellitus with hyperglycemia. The usual chemical test for glucose is a nonspecific one for reducing agents.‡

---

* Both sulfanilamide and acetazoleamide contain sulfonamide groups ($-SO_2NH_2$).

---

† An exception is *physiological* or *postural proteinuria,* an uncommon situation in which intermittent proteinuria from otherwise normal kidneys.

‡ The familiar Benedict test is based on the reduction of cupric ions in alkaline copper sulfate. The resulting cuprous ions produce a reddish color whose intensity is a measure of the amount of reducing agent present. Rarely, other sugars (e.g., fructose and galactose) that also actively reduce Benedict's solution occur in urine. On those occasions more specific assay methods may be necessary.

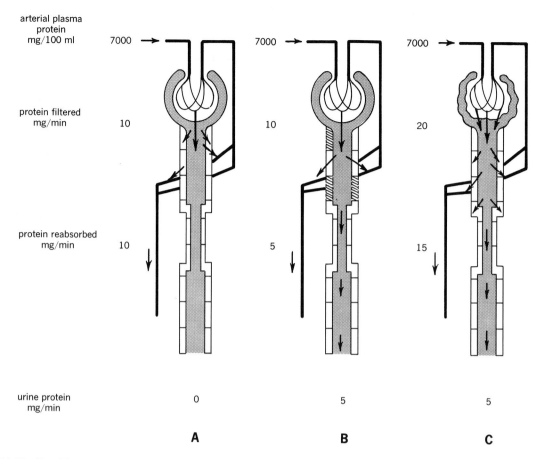

10.20 Renal handling of protein: A, normal slow filtration and complete reabsorption; B, slow filtration and incomplete reabsorption due to tubular injury; C, fast filtration and incomplete reabsorption due to glomerular injury. (Adapted from *Seminar Report*, 1947.)

A centrifuged sample of urine yields a small pellet of solid matter, whose microscopic examination is one of the most important aspects of urinalysis. The main components of abnormal urinary sediment are *casts*, erythrocytes, leukocytes, and epithelial cells. Casts are cylindrical bodies so named because they are in fact molds or casts of the tubular lumen. They may be formed anywhere along the nephron by the precipitation of protein (as the tubular fluid becomes concentrated) or by the conglutination of any insoluble materials in the lumen.

Fig. 10.21 shows the common types of casts. *Hyaline* casts are clear structures formed by the precipitation of protein. If they encase erythrocytes, we must conclude that the cells come from the kidney and not the lower urinary tract. *Epithelial* casts consist of clumped cells from the walls of damaged tubules. *Blood* casts are actual blood clots They imply severe glomerular damage

with bleeding. *Broad* casts, also called *renal failure* casts, are formed in the wide collecting tubules. They have grave significance, for they appear only when entire groups of nephrons have ceased functioning.

Erythrocytes in the urine (hematuria) indicate bleeding in the urinary tract. *Gross* hematuria follows hemorrhage within the kidneys or bladder. With greatly elevated blood pressure, kidney hemorrhage may occur in the same manner as a spontaneous nosebleed. More frequent is *microscopic* hematuria, in which a small number of erythrocytes are detected microscopically in the sediment. Of the other elements in the sediment, leukocytes are a sign of infection; epithelial cells slough from tubules and membranes lining the urinary tract; and inorganic crystals precipitate from the urine.

Blood chemistry. When renal disease is extensive and glomerular destruction has permanently decreased the

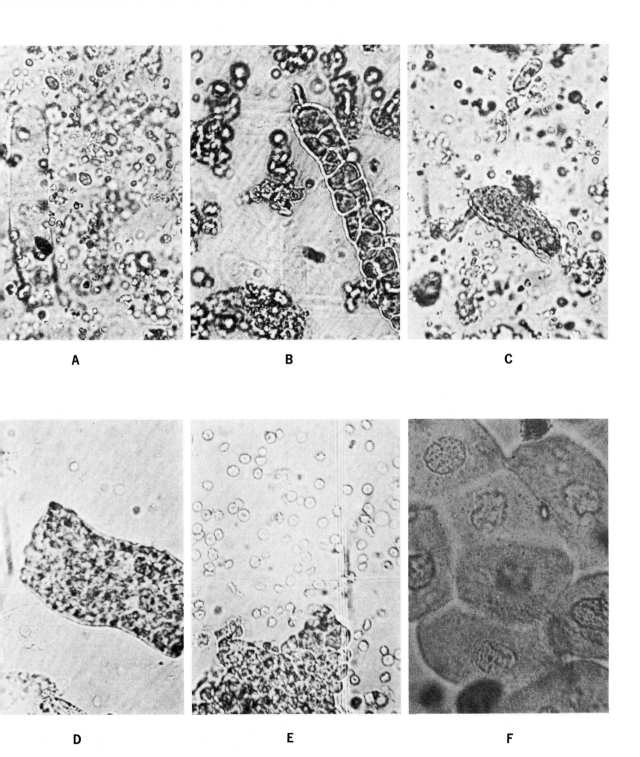

10.21 Urinary sediment (×500): A, hyaline cast; B, epithelial cast; C, blood cast; D, broad cast; E, erythrocytes; F, bladder epithelial cells. (Dr. R. Lippman)

filtration rate, nitrogenous substances that are normally excreted by filtration (urea, creatinine, uric acid, etc.) accumulate in the body until equilibrium between rate of production and rate of excretion is reached. This elevation in the concentration of plasma nonprotein nitrogenous constituents is known as *azotemia*.

Because of the relatively large quantities of urea produced in protein metabolism, its retention is most marked in azotemia. That of creatinine is next. Their importance lies in the fact that they are retained in the blood in approximate proportion to the degree of renal involvement. The nature of this relationship is simple: as glomerular function is reduced by the obliteration of glomeruli, the plasma urea concentration increases until the rate of urea filtration through the remaining glomeruli becomes equal to the rate of urea production —i.e., until production and excretion return to a balanced state. The same considerations apply for plasma levels of creatinine, uric acid, etc., as well as phosphate, sulfate, and organic anions.

Certain renal functions may be judged from the plasma bicarbonate level. The metabolic acidosis of renal insufficiency (see Table 10.3) is due to impairment of the renal mechanisms for conserving bicarbonate and acidifying the urine.

Consequences of renal insufficiency. Uremia means "urine in the blood." The term was first used in 1840 for the intoxication due to defective purification of the blood by the kidneys. Though correct in essence, such a definition fails to give an adequate picture of the complex constellation of biochemical and physiological derangements characterizing advanced renal insufficiency.

Uremia is a condition that occurs when the kidneys are unable to maintain a normal internal environment, when there is an imbalance between the rates of intake or production in the body of chemical substances and their rates of excretion. This comprehensive definition emphasizes three important points: (1) there is no one outstanding manifestation of renal failure; (2) uremia arises from an imbalance between intake or production and renal function; and (3) uremia includes manifestations arising from the loss of essential substances as well as those arising from the retention of waste products.

The biochemical features of uremia result from the progressive destruction of nephrons and, to some extent, from the predominant type of injury—glomerular or tubular. As the nephron population falls below 50%, there is decreasing flexibility in the excretory and regulatory functions of the kidneys. The ability of the remaining nephrons to respond to excesses or deficits of any substance is greatly restricted, although each must carry an increased excretory load if an internal environment capable of supporting life is to be maintained; in other words, the remaining nephrons must excrete the same amount of solutes excreted by normal kidneys.

Progressively this goal is less successfully approached.

There is progressive reduction in the capacity of the kidneys to dilute or concentrate the urine according to need so that excreted urine maintains the "fixed" specific gravity of the body fluids, 1.010. There is impaired conservation of sodium and chloride attributable to a sustained relative water diuresis in each nephron, specific defects in the active transport system responsible for the tubular reabsorption of sodium and chloride, and defective production of H^+ and NH_4^+ ions.

As the nephron population drops further, uremia becomes disabling. The victim displays loss of appetite and weight loss, disordered heart function with frequent arrhythmia and heart failure, bone marrow intoxication with resulting anemia, and progressive acidosis. When a critical point is passed, he dies.

The artificial kidney. Since most diseases causing chronic renal insufficiency are irreversible, little can be done therapeutically to halt the disease process. Rather, it is the physician's grave responsibility to assist in promoting homeostasis, supplementing or substituting for the imperfect performance of a damaged kidney whenever possible. He must attempt to achieve and maintain normal body composition, and, to do this, he must closely observe body metabolism and clearly understand its physiological significance.

The so-called *artificial kidney* is a device that removes blood from the artery of an arm, pumps it through a dializing membrane that permits accumulated toxic materials to pass into a surrounding bath, and returns it to a vein. The procedure is called *hemodialysis*. Although it is reasonably formidable, technical advances are making it more and more accessible to those requiring it. Its greatest usefulness is in the treatment of acute renal failure of the type that is reversible if life can be sustained for 10 to 14 days. Here hemodialysis can be life saving. Repeated long-term use of the artificial kidney also sustains the lives of a small but increasing number of individuals with chronic irreversible renal insufficiency.

Kidney transplantation. If it were possible to replace a damaged kidney with a normal one, renal disease would obviously not be the difficult problem that it is. Aggressive efforts have been made in recent years to accomplish this feat, and a number of promising results have been obtained.

We shall not here enter into the nature of the immune responses that cause an individual to reject a tissue transplant except to note that at present grafts are possible only between genetically identical (or closely related) individuals (e.g., identical twins) and in subjects whose graft-rejection mechanisms have been pharmacologically suppressed. A normal kidney has been successfully transplanted into a victim of advanced kidney disease in hundreds of cases. More transplants have been unsuccessful, however. Nevertheless, these dramatic advances, to be discussed in Chapter 19, hold promise of important future developments.

Excretion of Urine

ANATOMY OF THE LOWER URINARY TRACT

Ureters. The two ureters convey the urine from the kidneys to the bladder (see Fig. 10.3). Each is a tubule 10 to 12 in. long and ¼ in. in diameter. The upper half is in the abdomen, and the lower half in the pelvis.

The wall of a ureter has three layers: an outer *fibrous* coat, a *muscular* coat, and an inner *mucosal* (mucous membrane) lining made up of transitional epithelium. The contractions of the muscular coat produce peristaltic waves, beginning at the upper end and moving progressively downward.

Bladder. The bladder is a muscular receptacle situated behind the pubic bone, in front of the rectum in the male (Fig. 10.22) and in front of the vagina and neck of the uterus in the female (Fig. 10.23). It is a freely movable organ but is held in position by folds of peritoneum and fascia. When empty or slightly distended, it is round and rests deep in the pelvis. When greatly distended, it is ovoid and rises to a considerable height in the abdominal cavity. Its capacity varies. It holds about a pint of urine when moderately distended.

The bladder wall has four coats: (1) the *serous* coat is continuous with the peritoneum and covers only the upper and the upper lateral surfaces; (2) the *muscular* coat has three layers—inner longitudinal, middle circular, and outer longitudinal—whose fibers form a *sphincter* muscle at the opening of the bladder into the urethra that is normally contracted and that relaxes only when the bladder is being emptied; (3) the *submucosal* coat consists of loose connective tissue; (4) the *mucosal* lining is of transitional epithelium, like the lining of the ureters and the urethra, and is thrown into deep folds when the bladder is empty.

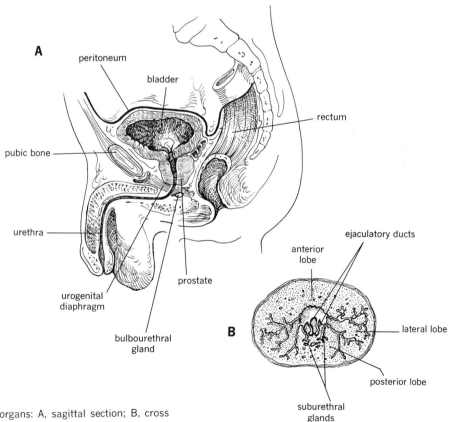

10.22 Male pelvic organs: A, sagittal section; B, cross section of the prostate.

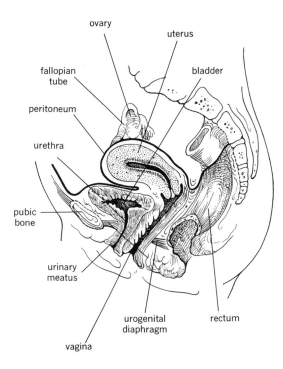

ovary
uterus
fallopian tube
bladder
peritoneum
urethra
pubic bone
urinary meatus
urogenital diaphragm
rectum
vagina

10.23 Female pelvic organs in sagittal section.

There are three openings into the bladder. The two ureters enter the lower part about ½ in. from the mid-line, taking an oblique course, downward and inward, through the bladder wall. The urethra leads from the bladder mid-line below and in front of the ureter openings.

Male urethra and prostate. The male urethra is a fibroelastic tube about 8 in. long (see Fig. 10.22). It is divided into three portions: (1) the *prostatic* urethra, through the *prostate;* (2) the *membranous* urethra, through the urogenital diaphragm; and (3) the *spongy* or penile urethra, from the membranous urethra to the *urinary meatus,** external orifice of the penis.

The wall of the male urethra is composed of a *mucosal* lining, which is continuous with that of the bladder and extends into the ducts of glands opening into the urethra, and a *submucosal* coat, which connects it with the underlying structures.

The prostate is the most important accessory sex gland in the male. Although we shall discuss its physiology in Chapter 16, we are compelled by its location to consider its anatomy here. As shown in Fig. 10.22, the prostate is situated below the bladder, above the urogenital diaphragm, behind the pubic bone, and in front

---

* A meatus is a passage or opening.

of the rectum. Because it strategically surrounds the urethra and its opening into the bladder, diseases of the gland obstruct the lower urinary tract; and the prostate is notorious for the frequency with which it becomes infected, enlarged, or malignant.

The adult prostate weighs about 20 g. Its major mass consists of right and left lateral lobes and a middle lobe. The glandular alveoli are lined with columnar epithelium embedded within a thick fibromuscular *stroma.* The secretions of the prostate are concerned with the maintenance and activation of the spermatozoa.

Female urethra. The female urethra courses obliquely downward and forward from the bladder to the external orifice, the *urinary meatus* (see Fig. 10.23).* It is located behind the pubic bone and is embedded in the anterior wall of the vagina. Its diameter, when undilated, is ¼ in.; its length is about 1½ in. The meatus, its narrowest part, lies between the clitoris and the opening of the vagina.

The wall of the female urethra has three layers: (1) an outer *muscular* coat, continuous with that of the bladder; (2) a thin *spongy* coat, containing a plexus of veins; and (3) a *mucosal* coat, continuous internally with that lining the bladder and externally with that of the vulva.

The female urethra is surrounded by numerous small glands that many believe to be homologues of the male prostate.

PHYSIOLOGY OF MICTURITION

Function of the ureters and bladder. Urine is excreted continuously by the kidneys. It is propelled down the ureters to the bladder, where it is temporarily stored, and intermittently expelled through the urethra by the process of micturition.

The ureter is not a passive conduit but an active organ. Urine normally enters the bladder in spurts. To be sure, it may issue continuously from a ureter that for some reason has lost its contractility, but the flow is less than with normal peristalsis. During the period of storage in the bladder, urine is prevented from regurgitating into the ureters by the anatomical arrangement of the ureter openings, which are easily compressed by the contraction of the bladder wall. The openings therefore serve as valves.

Bladder muscle exhibits two types of activity: a continuous sustained contraction and intermittent contractions. Since micturition requires coordination of the common smooth muscle responses with the movements of the sphincter, integrated functioning of the nervous system is essential. We shall deal with details of neural physiology in Chapter 12. Suffice it to say here that micturition is normally voluntary. Normal man first feels the need to void when the bladder contains between 100 and 150 ml of urine, and he becomes uncomfortably aware of distension when the bladder volume reaches 350 to 400 ml. The sensations of bladder filling and of painful distension and the conscious perception of the desire to urinate are dependent on afferent impulses carried by the pelvic nerves.

Voluntary micturition involves the abrupt cessation of the continuous nerve impulses that maintain the sphincter in a state of contraction. The normal bladder empties completely. The bladder neck opens more easily in the female than in the male. Moreover, the entire circumference of the orifice relaxes in the female whereas chiefly the posterior portion relaxes in the male. This difference between the sexes is associated with the more rigid nature of the structures around the bladder neck in the male and explains, in part, the higher incidence of bladder neck obstruction in the male.

REFERENCES AND SUGGESTIONS FOR FURTHER READING

Addis, T., *Glomerular Nephritis, Diagnosis and Treatment*, Macmillan, New York, 1949.

Berliner, R. W., "Renal Mechanism for Potassium Excretion," *Harvey Lectures Ser. 55 (1959–60)*, 141 (1961).

Cushny, A. R., *The Secretion of Urine*, 2nd ed., Longmans, Green, London, 1926.

Forster, R. P., "Kidney, Water, and Electrolytes, *Ann. Rev. Physiol.*, **27**, 183 (1965).

———, "Kidney Cells," in J. Brachet and A. E. Mirsky, eds., *The Cell*, Vol. 5, Academic Press, New York, 1961.

Giebisch, G., "Kidney, Water, and Electrolytes," *Ann. Rev. Physiol.*, **24**, 357 (1962).

Hartroft, P. M., "The Juxtaglomerular Complex," *Ann. Rev. Med.*, **17**, 113 (1966).

Leaf, A., "Kidney, Water, and Electrolytes," *Ann. Rev. Physiol.*, **22**, 111 (1960).

———, and R. M. Hays, "The Effects of Neurohypophyseal Hormone on Permeability and Transport in a Living Membrane," in G. Pincus, ed., *Recent Progress in Hormone Research*, Vol. 17, Academic Press, New York, 1961.

Lotspeich, W. D., *Metabolic Aspects of Renal Function*, Thomas, Springfield, Ill., 1959.

Luetscher, J. A., Jr., "Studies of Aldosterone in Relation to Water and Electrolyte Balance in Man," in G. Pincus, ed., *Recent Progress in Hormone Research*, Vol. 12, Academic Press, New York, 1956.

Orloff, J., and J. S. Handler, "The Cellular Mode of Action of Antidiuretic Hormone," *Am. J. Med.*, **36**, 686 (1964).

Peart, W. S., "The Renin-Angiotensin System," *Pharmacol. Rev.*, **17**, 143 (1965).

Pitts, R. F., *The Physiological Basis of Diuretic Therapy*, Thomas, Springfield, Ill., 1959.

Relman, A. S. "Renal Acidosis and Renal Excretion of Acid in Health and Disease," *Advan. Internal Med.*, **12**, 295 (1964).

Richards, A. M. "Urine Formation in the Amphibian Kidney," *Harvey Lectures Ser. 20 (1934–35)*, 93 (1936).

Schmidt-Nielsen, B., and D. F. Laws, "Invertebrate Mechanisms for Diluting and Concentrating the Urine," *Ann. Rev. Physiol.*, **25**, 631 (1963).

Schmidt-Nielsen, K., "Salt Glands," *Sci. Am.*, **200**, 109 (Jan., 1959).

Scholander, P. F., "The Wonderful Net," *Sci. Am.*, **196**, 96 (April, 1957).

Sharp, G. W. G., and A. Leaf, "Mechanism of Action of Aldosterone," *Physiol. Rev.*, **46**, 539 (1966).

Smith, H. W., *From Fish to Philosopher*, Little, Brown, Boston, 1953.

———, *Principles of Renal Physiology*, Oxford Univ. Press, London, 1956.

———, *The Kidney: Structure and Function in Health and Disease*, Oxford Univ. Press, London, 1952.

Stanbury, S. W., "Some Aspects of Disordered Renal Tubular Function," *Advan. Internal Med.*, **9**, 231 (1958).

Ullrich, K. J., and D. J. Marsh, "Kidney, Water, and Electrolyte Metabolism," *Ann. Rev. Physiol.*, **25**, 91 (1963).

Wesson, L. G., Jr., "Hormonal Influences on Renal Function," *Ann. Rev. Med.*, **12**, 77 (1961).

Wolf, A. V., *The Urinary Function of the Kidney*, Grune & Stratton, New York, 1959.

Certainly the first step in the direction of land life was the respiration of air instead of, or in addition to, water. This important step was taken by the fishes long before fins were converted into large legs with which to crawl about on land. Some of the earliest fishes, living far back in the Devonian period, were air-breathers. They had a lung, of sorts, into which they swallowed air and they were enabled by this unique organ to live in pools and rivers when the dry season came and the water disappeared or became foul with rotting animals and plants. They were able to survive when the other fishes were driven out, driven possibly back into the sea, or into the Beyond. . . . Here is really your connecting link in a bigger sense: a fine new organ, a new power that broadened the organism's way of living, increased its physical freedom, widened its margin of safety. The evolution of the air-breathing lungs marked an elevation in the stream of life; it had gotten over an obstacle, surmounted it by finding a new way of living.

Homer W. Smith, in *Kamongo,* 1932.

11 *Respiratory System*

Anatomy of the Respiratory Organs

INTRODUCTION

Functional guide to respiratory anatomy. The term *respiration* refers to all the processes involved in satisfying the body's oxygen needs. Since carbon dioxide is usually produced as oxygen is consumed in animal metabolism, respiration is also defined as the *gas exchange* between an organism and its environment. (This definition is applicable to plants, too, which take in carbon dioxide and give off oxygen.) Of all the factors concerned in the maintenance of normal function, respiration has perhaps the greatest immediate significance. Every activity of the body, every manifestation of life, requires constant utilization and liberation of energy. This energy is ultimately derived from the potential energy foodstuffs by the biochemical processes of *oxidation* (see Chapters 2, 4).

In air-breathing animals respiration involves three separate but related phenomena: (1) *external respiration*, which includes the withdrawal of O_2 from the air, the excretion of CO_2 into the air, and the nervous and mechanical devices controlling these processes; (2) *oxygen and carbon dioxide transport*, which includes the chemical and physical processes of gas exchange between air and blood in the lungs and between blood and cells in the tissues; and (3) *internal respiration*, which includes the physicochemical and biochemical processes of cellular metabolism in which O_2 is utilized and CO_2 is produced.

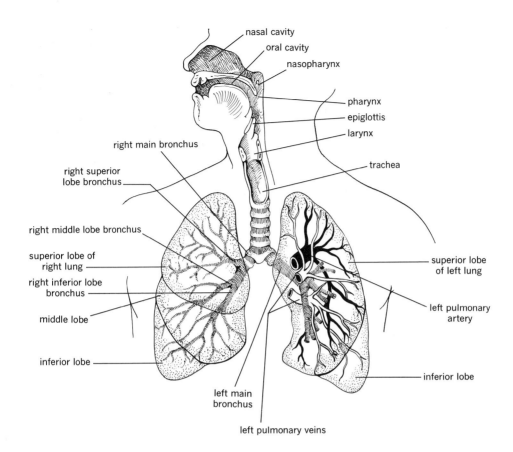

nasal cavity
oral cavity
nasopharynx
pharynx
epiglottis
larynx
trachea
right main bronchus
right superior
lobe bronchus
right middle lobe bronchus
superior lobe of
right lung
right inferior lobe
bronchus
middle lobe
inferior lobe
left main
bronchus
left pulmonary veins
superior lobe
of left lung
left pulmonary
artery
inferior lobe

11.1   Respiratory system.

The first two aspects of respiration constitute a coordinated system that furnishes the body cells with the O_2 they need and removes the CO_2 they produce. As we shall see, the system incidentally aids in the elimination of excess heat and water.

In a unicellular organism the gas exchange that takes place readily across the cell membrane —in effect, the body surface—sustains life. A multicellular animal, however, cannot survive on the small amount of O_2 diffusing through its relatively small body surface. Therefore, in the course of evolution, animals have acquired special respiratory organs providing larger surfaces through which greater quantities of O_2 can pass. When the respiratory organ extends outward from the body to form an appendage, we call it a *gill*. When it turns inward to form a cavity, we call it a *lung*. Thus lungs are devices in which

the blood stream is brought in close contact with a continuously renewed supply of fresh air across a thin membrane of extremely large area.

Successful gas exchange in the lungs depends on the frequent periodic inflow and outflow of gases occurring in the involuntary act of *breathing*. Since the lungs themselves have no muscles, their filling and emptying require a bellows-like motion of the chest wall. Lung expansion and contraction, then, result from movements of the ribs, which form a rigid cage around the lungs; the diaphragm, the muscular partition separating the chest cavity from the abdominal cavity (see Fig. 6.6); and the other respiratory muscles. Later we shall consider the mechanics of respiration, pulmonary gas exchange, and the control of respiration. But first we must become acquainted with the anatomy of the respiratory organs.

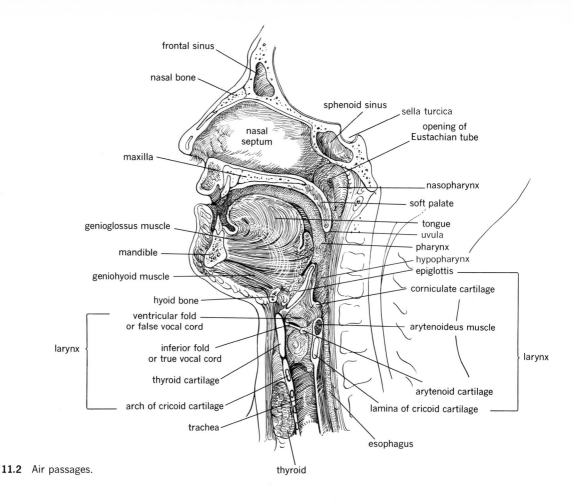

frontal sinus

nasal bone

sphenoid sinus

sella turcica

opening of
Eustachian tube

nasal
septum

maxilla

nasopharynx

soft palate

tongue

genioglossus muscle

uvula

pharynx

mandible

hypopharynx

epiglottis

geniohyoid muscle

corniculate cartilage

hyoid bone

ventricular fold
or false vocal cord

arytenoideus muscle

larynx

inferior fold
or true vocal cord

thyroid cartilage

larynx

arch of cricoid cartilage

arytenoid cartilage

lamina of cricoid cartilage

trachea

esophagus

thyroid

11.2 Air passages.

General plan of the respiratory system. The complex of organs and structures referred to as the *respiratory system* has three components: (1) the *air passages*, which inspired air traverses on its way to the lungs and which include the *nose, pharynx, larynx, trachea,* and *bronchi;* (2) the *lungs*, which include the *bronchioles, alveoli* (*air sacs*), and the veins, arteries, and capillaries of the pulmonary circulation; and (3) the *skeletal* and *muscular respiratory apparatus*, which includes the *ribs, diaphragm, intercostal muscles,* and other *accessory muscles.*

The arrangement of the air passages and lungs in the body is shown in Fig. 11.1 (see also Figs. 6.6, 6.7, 6.8).

AIR PASSAGES

Nose and nasal cavities. The nose is the special organ of the sense of smell. It also serves as a passageway in which air en route to and from the lungs is filtered, warmed, and moistened.

The external nose, a triangular framework of bone and cartilage covered by skin, encloses the *nasal cavities* (Fig. 11.2). The prominent upper bone, the *nasal bone*, is easily broken. Because of its strategic location near the brain, fractures of this bone are always potentially serious.

The two wedge-shaped nasal cavities are separated from one another by a vertical partition called the *nasal septum*. The soft front portion of the septum contains much of the cartilage of the nose. The hard remaining part of the septum is an intricate combination of delicate bones that will be described in Chapter 17. Each nasal cavity is incompletely divided into four regions by the *nasal conchae*, longitudinal mounds in the lateral wall (see Fig. 11.3). The furrows between adjacent conchae are the *superior, middle,* and *inferior meatuses* (or passages). The superior meatus is frequently missing. The *palate* and *maxillae* separate the nasal cavities from the mouth, and the horizontal plate of the *ethmoid bone* separates them from the cranial cavity.

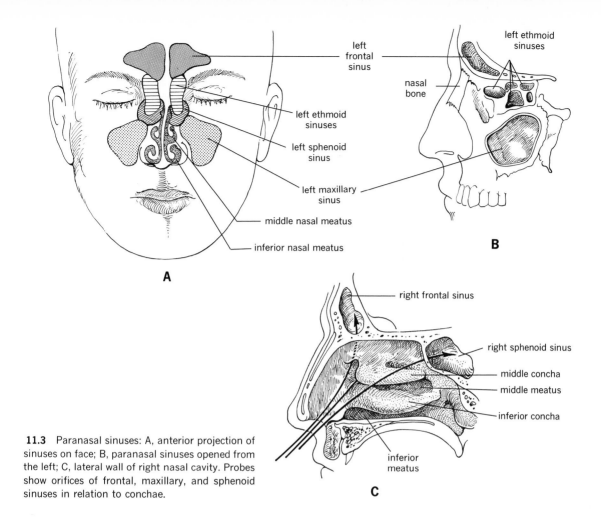

left frontal sinus

left ethmoid sinuses

left sphenoid sinus

left maxillary sinus

middle nasal meatus

inferior nasal meatus

A

left ethmoid sinuses

nasal bone

B

right frontal sinus

right sphenoid sinus

middle concha

middle meatus

inferior concha

inferior meatus

C

11.3 Paranasal sinuses: A, anterior projection of sinuses on face; B, paranasal sinuses opened from the left; C, lateral wall of right nasal cavity. Probes show orifices of frontal, maxillary, and sphenoid sinuses in relation to conchae.

The nasal cavities communicate with the exterior through the two oval *anterior nares,* or *nostrils.* The two *posterior nares* lead into the *nasopharynx,* the space between the nasal cavities and the pharynx proper. The nasal cavities are lined with thick, highly vascular mucosa, consisting chiefly of ciliated columnar epithelium (see Fig. 5.2C). At the anterior nares it contains a number of coarse hairs, and in the upper, or *olfactory,* area the olfactory cells, receptors for the sense of smell (Chapter 12). It is continuous externally with the skin and internally with the mucous membrane lining the sinuses and other structures connected with the nasal cavities. When inflammatory conditions cause it to swell, the air passages may be obstructed.

Blood is supplied to the external nose by branches of the external and internal maxillary arteries, which are derived from the external carotid arteries, and to the walls of the nasal cavities by branches of the ethmoidal arteries, which are derived from the internal carotid arteries.

Normally, air is inhaled through the nose rather than the mouth. This is advantageous for several reasons. The hairs of the nostrils and the cilia of the lining epithelium prevent much dust and other foreign matter from entering the nasal cavities. The arrangement of the conchae makes the upper cavities very narrow, thus providing further filtration. The thick lining and abundant blood supply of the cavities tend to keep the temperature relatively high, thus warming and moistening the air before it reaches the lungs.

The mouth may serve as a passageway for air, however, and the pharynx (throat) transmits air from the nose or mouth to the larynx. The mouth and pharynx will be described in Chapter 13.

Paranasal sinuses. In Chapter 17 we shall learn that the 23 bones of the head—which constitute the *skull,* or *cranium*—are the most intricately sculptured and interrelated components of the human skeleton. We shall defer until then a detailed discussion of the topography of the sinuses, including here only general remarks on their location and structure.

The *paranasal sinuses* are mucosa-lined chambers in the bones adjacent to or forming the walls of each nasal cavity. The sinuses connect with the nasal cavity via narrow channels through which probes can be passed. Each sinus is named for the bone in which it occurs (Fig. 11.3). The *frontal* sinus, between the outer and inner layers of the frontal bone, opens into the middle meatus of the nasal cavity. The *ethmoid* sinus, in the ethmoid bone, opens into the superior and middle meatuses of the nasal cavity. Structurally it may be likened to a collection of variously inflated small balloons packed into an oblong box. Indeed, one or more "balloons" may burst through the lid of the "box" and invade the neighboring frontal bone. The *sphenoid* sinus, behind the ethmoid sinus in the sphenoid bone, opens into the nasal cavity above the superior nasal concha. The *maxillary* sinus, the largest of the paranasal sinuses, a pyramid-shaped space in the maxilla, opens into the middle meatus of the nasal cavity.

The sinuses are well known for their tendency to become inflamed in the course of upper respiratory infections. The maxillary sinuses are also frequently involved in infections spreading from the molar teeth, whose roots extend close to the floors of the sinuses.

Larynx. The larynx, the organ of voice, is located in the upper and front part of the neck between the root of the tongue and the trachea (see Figs. 11.1, 11.2). Above and behind it lies the pharynx, which opens downward into the *esophagus;* and on either side of it lie the great blood vessels of the neck. The adult larynx is shaped somewhat like a triangular box, with a broad base above, flat sides, a prominent front ridge, and a narrow, rounded bottom, where it joins the trachea. It is constructed of nine discrete cartilages—the large unpaired *epiglottis* and *thyroid* and *cricoid* cartilages and the small paired *arytenoid, corniculate,* and *cuneiform* cartilages—linked by extrinsic and intrinsic ligaments and moved by numerous muscles.

The thyroid cartilage is a large shieldlike structure that is easily felt in the neck. It rests upon the ring-shaped cricoid cartilage and consists essentially of two square plates connected at an acute angle in the front mid-line. The union forms the laryngeal prominence called the *Adam's apple.*

The epiglottis is shaped like a leaf, with its stem inserted in the notch between the two thyroid plates. It is critically situated between the beginning of the trachea and the beginning of the esophagus.

Like the rest of the air passages, the larynx is lined with mucous membrane, which is continuous above with the lining of the pharynx and below with that of the trachea. The cavity of the larynx is divided in two by two folds of mucous membrane stretching from front to back and just failing to meet in the mid-line. The elongated fissure between them is the *glottis,* the narrowest segment of the air passages. The glottis is protected above by the epiglottis.

Embedded in the mucous membrane at the edges of the glottis are fibrous and elastic ligaments, which strengthen the edges and give them elasticity. These ligaments, firmly attached at both ends to the cartilages of the larynx, are called the *inferior* or *true vocal folds* because they function in sound production, or *phonation.* They are also referred to as *vocal cords,* but this term disregards their shelf-like architecture. Above the true vocal folds are two other folds that do not function in sound production. These are called the *ventricular* or *false vocal folds.*

A complex set of highly specialized muscles tightens and slackens both the true and false vocal folds. In effect, it is a sphincter, opening and closing the glottis. The muscles are under neural control and can operate voluntarily or involuntarily. The sphincter shuts off the respiratory tract during swallowing and breath holding and on many other occasions. Its action is separate and distinct from the production of sound and from the production of speech, which is clearly under the control of the higher nerve centers.

The anatomy and physiology of the larynx are especially rich in implications for biology, evolution, endocrinology, and psychology. The ability to produce sound was a late evolutionary development, and the ability to produce speech was a very late one. The familiar differences between the voices of men and the voices of women or boys are all explained by anatomical changes secondary to hormonal influences.

Trachea and bronchi. The trachea, below the larynx, is a membraneous and cartilaginous tube about 1 in. in diameter. C-shaped segments of cartilage embedded in fibrous tissue and completed behind by bands of muscular and elastic tissue stiffen its walls. The bands form the flattened surface that comes in contact with the esophagus (see Fig. 11.2). The ciliated epithelium of the mucosal lining of the trachea keeps it free from dust particles, the movements of the cilia continually sweeping dust-laden mucus upward in the general direction of the pharynx.

Soon after entering the chest, the trachea divides into a *right bronchus* and a *left bronchus.* These are air conduits to the respective lungs. After entering the lungs, they branch repeatedly into bronchioles (see Fig. 11.1). The bronchi resemble the trachea in structure; but, as they divide and subdivide, their walls become thinner, the rings of cartilage appear less frequently and then not at all, and the fibrous tissue disappears, until only a thin layer of muscular and elastic tissue lined by ciliated epithelium remains in the narrow *terminal* bronchioles (Fig. 11.4).

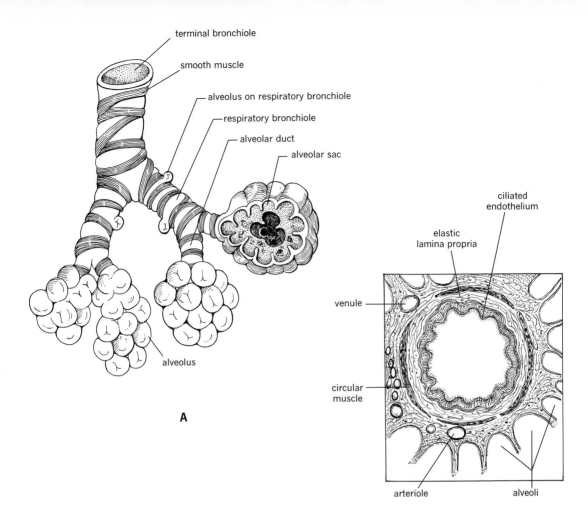

terminal bronchiole

smooth muscle

alveolus on respiratory bronchiole

respiratory bronchiole

alveolar duct

alveolar sac

ciliated
endothelium

elastic
lamina propria

venule

circular
muscle

alveolus

arteriole

alveoli

A

B

11.4   Trachea and bronchi: A, reconstruction of a primary lung lobule; B, cross section of a terminal bronchiole.

In recent years a technique called *bronchography* has been developed, which permits x-ray examination of the bronchial tree. A medium opaque to x-ray is instilled in the bronchi, and a *bronchogram* is obtained. A bronchogram and an ordinary chest film are compared in Fig. 11.5.

LUNGS

Gross anatomy. The lungs are cone-shaped organs that occupy the two lateral chambers of the thoracic cavity and are separated from each other by the heart and other components of the mediastinum (see Fig. 11.1). Each lung lies freely in its chamber, within a closed membranous sac. The lungs are connected to the heart and trachea by clusters of tubes and vessels, including the bronchi, the pulmonary arteries and veins,

the bronchial arteries and veins, lymphatic vessels and lymph nodes, and nerve plexuses, all in a meshwork of loose connective tissue. Together these constitute the pleura-covered "roots" of the lungs, their only points of attachment. On the medial surface of each lung is a notch, the *hilus*, through which the root enters the lung substance.

The right lung is larger and broader than the left, owing to the leftward inclination of the heart. It is divided by *interlobar fissures*, which extend obliquely downward and forward, into three *lobes:* the *superior, middle,* and *inferior* lobes. The left lung is smaller, narrower, and longer than the right. It is divided into two lobes: the *superior* and *inferior* lobes. All lobes except the right superior lobe touch the upper surface of the diaphragm. The projections of the lungs upon the body surface are shown in Fig. 11.6.

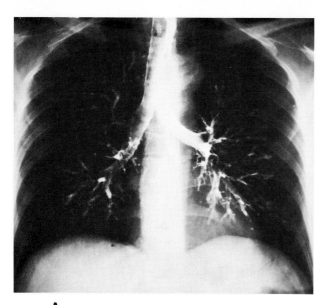
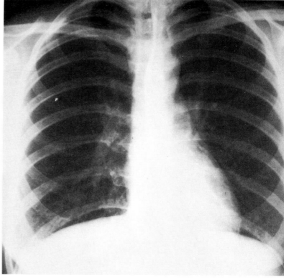

A **B**

11.5 Comparison of a bronchogram (A) and a normal chest x-ray (B). (Dr. Murray Janower.)

A curious black mottling is usually visible beneath the glistening pleural surface of a lung observed grossly at surgery or autopsy. This is due to inhaled carbon particles that over the years have managed to escape the filtering and cleansing mechanisms of the air passages. They enter the lungs where they are phagocytized and either permanently walled off in the lung substance or deposited in the lymph from which they are removed by regional lymph nodes. This condition is pronounced in coal miners and, to a lesser degree, in city dwellers exposed to a smoky, dusty atmosphere. Since carbon particles are chemically inert, the only result of their entrance into the lungs is tissue pigmentation. Other dusts, however, may be extremely harmful.

A functioning lung is full of air. This is obvious when lung tissue is manually squeezed, for it emits a characteristic crackling sound known as *crepitation*. Moreover, a lung removed from a body floats in water. This fact is used in medicolegal investigations to test whether the lungs of a dead newborn infant contain air. If they do not, it can be concluded that the child never breathed after birth.

The high air content of the lungs is also apparent from a chest x-ray, for the x-ray beam passes more easily through air than through solid tissue. The air-containing substance of the lung is textured with a delicate webbing produced by the small vessels and bronchioles.

Microscopic anatomy. We have seen that the bronchial tree is a system of passages dividing dichotomously many times and eventually ending blindly. The tubing from the trachea to the terminal bronchiole serves only as an air conduit, having no other respiratory function. Beyond the terminal bronchiole are the "leaves" of the bronchial tree.

Fig. 11.7 shows these fine structures in varying degrees of magnification and schematization. The terminal bronchiole is the last of the air passages to retain a complete wall. Beyond it are the *respiratory bronchioles*. These may be sorted into three general types, called A, B, and C (see Fig. 11.7B), depending on how much of the wall is typically bronchiolar and how much is alveolar. Entirely alveolar walls characterize the *alveolar ducts*. These divide into blind *alveolar sacs*. The outpouchings in the walls of the alveolar ducts and sacs are individual alveoli themselves. Each air passage, then, is a closed tube whose limits are the alveolar sacs. Adjacent passages are sometimes separated only by two layers of ultrathin alveolar epithelium and a little connective tissue containing capillaries and elastic tissue, and they may intercommunicate through defects or pores in the alveolar walls. A film of lipoprotein about 50 Å thick lines the alveoli, lowering the surface tension of the small amount of moisture present and thereby helping to keep the alveoli open. It contains phospholipids that originate in special granular cytoplasmic bodies in the *pneumonocytes*, the large flat cells lining the alveoli, pass through the cell membrane, and spread upon the alveolar surface. The fine structure of an alveolar wall is shown in Fig. 11.7C. The fragility of the membranes between capillary blood and alveolar air is noteworthy.

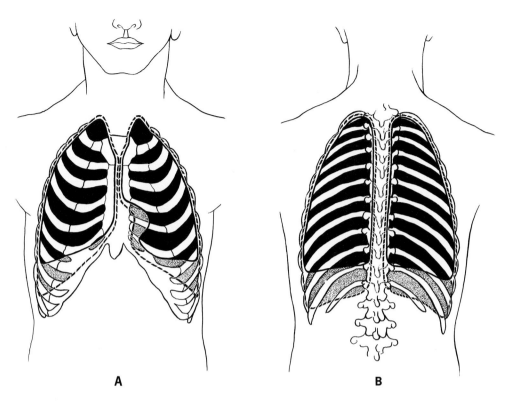

11.6 Projections of the lungs on the thorax, with shading showing their movements in respiration: A, ventral view; B, dorsal view.

The cluster of structures at the end of a terminal bronchiole is regarded as the fundamental pulmonary unit, the *primary lobule* of the lung. Like the nephron of a kidney, it ultimately performs the physiological functions of its parent organ. Each lung lobe has many lobules, and the two lungs have 300,000,000 or more alveoli. The lung surface exposed to air is thus enormous, totaling about 70 m² or more than 45 times the skin surface of the adult body. Of this about 55 m² is involved in respiratory gas exchange.

Pulmonary blood vessels. We learned in Chapter 9 that two sets of arteries go to the lungs: (1) branches of the pulmonary arteries, which carry blood from the right ventricle; and (2) branches of the bronchial arteries, which carry blood from the aorta (Fig. 11.8).

The branches of the pulmonary arteries accompany the branches of the bronchial tree and form the capillary plexi around the alveoli through which respiratory gas exchange takes place. Approximately 99% of the output of the right heart is constantly flowing through these arteries so that venous blood may be converted to arterial blood. We have earlier noted that the blood

pressure in the pulmonary arteries is 20 to 22 mm of mercury, low compared to that in the systemic arteries. However, the capillaries of the systemic circulation are surrounded by tissue fluid that exerts pressure against them, whereas the capillaries of the alveolar walls are surrounded only by gases. This means that hydrostatic pressure must be low in the pulmonary circuit to prevent disturbances of fluid exchange. It is remarkable indeed that the cardiac output—even when it rises drastically in exercise—hurtles through this delicate system without appreciable leakage of fluid into the alveolar spaces.

The bronchial arteries supply fresh aortic blood to the bronchi, bronchioles, blood vessel walls, lymph nodes, and pleura. The pulmonary arteries cannot fulfill this function because, until they perfuse the alveoli, they actually carry venous blood. The flow through the bronchial arteries is less than 1% of that through the pulmonary arteries. Lately it has become evident that some blood is exchanged in the bronchial tree between the systemic circulation and the pulmonary circulation. The anastomosis probably occurs in the walls of the smaller bronchioles where the two systems come in

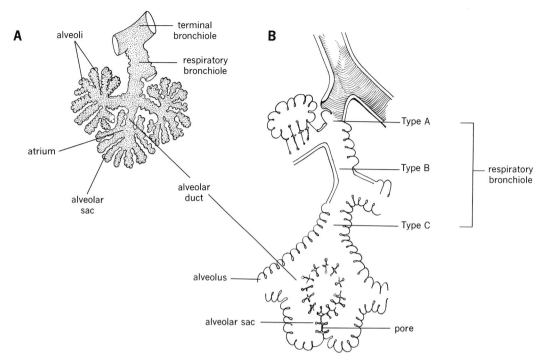

A   alveoli
terminal
bronchiole

respiratory
bronchiole

atrium

alveolar
sac

alveolar
duct

B

Type A

Type B

Type C

respiratory
bronchiole

alveolus

alveolar sac

pore

11.7 Structure of the bronchial tree: A, primary lobule; B, respiratory bronchiole; C, electron micrograph of alveolar wall (×4,400). Alv, alveolar space; GP, granular pneumonocyte; Mp, membranous pneumonocyte; Col, collagen fibers; Os, osmiophilic lamellar bodies; BM, basement membrane; Mac, macrophage lying in alveolar space; M, mitochondrion; Cap, capillary; Sep, septal cell. (A, after W. S. Miller, *The Lung*, Thomas, Springfield, 1947; B, after K. McLean, *Am. J. Med.*, 25, 62, 1958; C, from Dr. S. Recavarren.)

close contact. It constitutes a shunt by which blood may bypass the alveoli and escape respiratory gas exchange. We shall find that this phenomenon is important in disease.

There are both pulmonary and bronchial veins. The pulmonary veins begin in the pulmonary capillaries, coalesce to form larger vessels, and accompany the pulmonary arteries and bronchi to the hilus, whence they travel to the left atrium. The bronchial veins correspond to the bronchial arteries. The right bronchial vein ends in the azygos vein, and the left bronchial vein in the highest intercostal, or the hemiazygos, vein (see Fig. 9.17).

Pleura and mediastinum. The fibrous sac around a lung is the pleura. Like the pericardium, it consists of a *visceral layer*, which adheres to the surface of the lung, and a *parietal layer*, which adheres to the wall of the chest and the diaphragm (see Fig. 6.7A). Since these two moist layers are normally in close contact, the so-called *pleural cavity*, like the pericardial cavity, is a potential cavity rather than an actual one. A thin film of lubricating fluid permits the pleural layers to move easily upon one another with each respiration. Only a small amount of this fluid is normally secreted, and its absorption by the lymphatic vessels keeps pace with its secretion, so that the amount between the layers re-

C

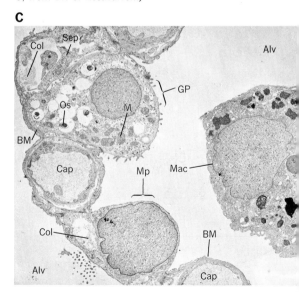

mains small. However, when the pleura is inflamed, as in *pleurisy*, or *pleuritis*, the amount of fluid may increase sufficiently to separate the two layers, producing a real pleural cavity. This condition is known as *pleurisy with effusion*.

Between the two pleural compartments is the mediastinum, or interpleural space. It extends from the *sternum* to the spinal column and is entirely filled with the following thoracic viscera: thymus, trachea, esophagus, heart, aorta (and its branches, pulmonary arteries and veins, venae cavae, azygos vein, thoracic duct, nerves, lymph nodes, and lymphatic vessels, all lying in a matrix of connective tissue.

Physiology of the Respiratory Organs

MECHANICS OF BREATHING

Respiratory movements. Although the alveolar surface is in contact with the air, diffusion alone cannot provide the necessary degree of gas exchange. Accordingly, a bellows-like mechanism draws in fresh air and expels impure air 16 to 20 times per minute.

Mechanically, the act of breathing depends entirely upon the fact that the thoracic cavity is in effect a *closed* compartment, whose only opening to the exterior is the trachea. Therefore, when the total volume of the cavity increases, negative pressure in the cavity causes air to be drawn inward through the trachea; when the volume decreases, positive pressure in the cavity causes air to be expelled. Breathing consists simply in periodic expansions and contractions of the thoracic cavity, brought about by intermittent contractions of the respiratory muscles and passive recoils of the elastic lungs. Both these factors are essential for respiration. The movements of the respiratory muscles are controlled by the nervous system. To obtain a clear picture of thoracic action in breathing, we must consider the anatomy of the thoracic bones and muscles.

The thoracic "bellows." The spinal column, which we shall describe in detail in Chapter 17, contains 33 (or 34) separate vertebrae: 7 *cervical* vertebrae, 12 *thoracic* vertebrae, 5 *lumbar* vertebrae, 5 *sacral* vertebrae, and 4 (or 5) *coccygeal*

vertebrae.* The spinal column lengthens slightly during inspiration, but we may regard it as a rigid structure for purposes of discussion.

The bony structures of the thorax are the thoracic vertebrae, the ribs with their cartilages, and the sternum (Fig. 11.9A). A pair of ribs is connected to each thoracic vertebra. The first seven pairs progressively increase in length and are attached to the sternum by cartilages. The eighth, ninth, and tenth pairs progressively decrease in length and are attached to the sternum by strong cartilages that join the cartilages of the seventh pair. The anterior ends of the last two pairs are unattached; hence they are sometimes called *floating ribs*. Functionally they are part

---

* In the adult, the sacral and coccygeal vertebrae are fused into two bones, the sacrum and coccyx (Chapter 17).

11.8 Pulmonary blood vessels, with branches of pulmonary artery black, of pulmonary vein white and of bronchial artery stippled gray.

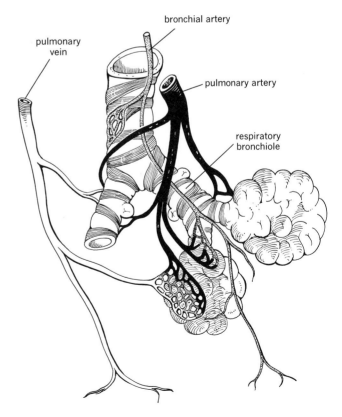

pulmonary vein

bronchial artery

pulmonary artery

respiratory bronchiole

of the abdominal wall and do not participate actively in respiration. As shown in Fig. 11.9B, rib and vertebra join (articulate) at two points. For this reason a rib has only one motion, rotation around an axis passing through the centers of both joints. Fig. 11.9C shows how the ribs change position during inspiration and expiration. The sternum moves upward and forward with them during inspiration. The result is an increase in the volume of the thoracic cavity.

Respiratory muscles. Contractions of various muscle groups known collectively as *respiratory muscles* are responsible for the rib movements. These may be classified as *inspiratory* and *expiratory* respiratory muscles. The inspiratory muscles run from the head, neck, and arms to the ribs and from rib to rib (the *external* intercostal muscles; see Fig. 11.9A). When they contract, they raise the ribs, expanding the thoracic cavity. Lowering the floor of the thoracic cavity also induces inspiration. This action is accomplished by the most important of the respiratory muscles, the diaphragm. Additional muscles of the trunk, larynx, pharynx, and face function in forced inspiration.

Normal expiration is primarily a passive act due to the elastic recoils of the lungs. But in forced expiration an assortment of expiratory muscles contracts to depress the ribs and diminish thoracic volume. In addition, muscles of the abdominal wall contract, pressing upon the viscera and forcing the diaphragm up into the thoracic cavity. Thus the diaphragm is "pushed" up by increased intraabdominal pressure and "sucked" up by decreased intrathoracic pressure.

Diaphragm. That the diaphragm is the principal inspiratory muscle (see Fig. 11.9C) is implicit in the fact that it can maintain adequate respiration when all the other respiratory muscles are paralyzed. Though it is often thought of merely as a musculomembranous partition separating the thoracic and abdominal cavities, it is

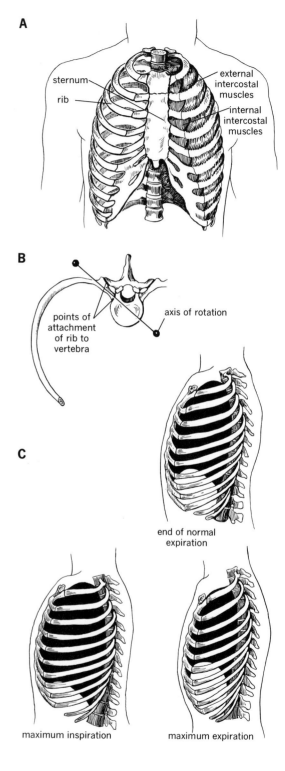

11.9 Mechanics of breathing: A, thoracic bellows; B, articulation of the sixth rib with the vertebra; C, rib and diaphragm positions during respiration.

an organ of remarkable design and prodigious activity (Fig. 11.10); it is essential for respiration and important for circulation, urination, defecation, coughing, sneezing, yawning, sobbing, straining, singing, and speaking.

The peripheral or muscular portion of the diaphragm is attached to the sternum, ribs, and vertebrae. The muscle fibers are inserted in a *central tendon.* The inspiratory depression of the diaphragm is brought about by contraction of these fibers, which draws the central tendon downward, thereby increasing the vertical dimension of the chest. The characteristic dome shape is due in part to negative intrathoracic pressure and positive intraabdominal pressure. In quiet breathing the diaphragmatic dome descends only slightly. Since the area of the diaphragm is about 250 cm^2, a descent of only 1 cm would increase thoracic volume by 250 cm^3 if the action of the diaphragm resembled that of a piston. Of course, it does not, but this is a reasonable approximation. The model in Fig. 11.11 illustrates diaphragm action in inspiration.

At the end of inspiration, when the lungs are momentarily quiet, and at the end of expiration, when they are quiet again, the forces tending to stretch the lungs and the forces tending to collapse them are equal. At the end of inspiration, when intraalveolar pressure equals atmospheric pressure (1 atm = 760 mm of mercury), the tendency of the lungs to recoil (approximately 9 mm) causes the intrapleural pressure to decrease to 751 mm of mercury.* At the end of expiration, the tendency of the lungs to recoil (about 6 mm) produces an intrapleural pressure of 754 mm.

The relationships are summarized in the pressure-volume curve of natural breathing (Fig. 11.12A). The curve is a small ellipse whose height indicates the volume of air exhaled or inhaled and whose width indicates the pressure differential in the alveoli. We see that the pressure changes during quiet breathing are indeed rather small. Nevertheless, they are sufficient to move air in and out. When the air passages are obstructed, the pressure changes are greater,

---

* It is often incorrectly stated that there is negative intrapleural pressure. A negative pressure is less than zero. Intrapleural pressure is slightly less than 760 mm of mercury, and so a better descriptive term is *subatmospheric.*

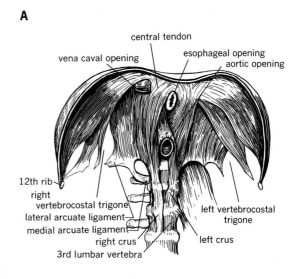

A

central tendon
vena caval opening
esophageal opening
aortic opening
12th rib
right vertebrocostal trigone
lateral arcuate ligament
medial arcuate ligament
right crus
3rd lumbar vertebra
left vertebrocostal trigone
left crus

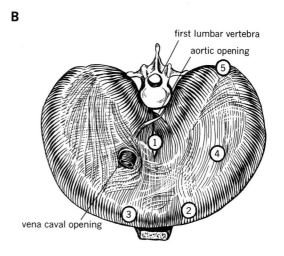

B

first lumbar vertebra
aortic opening
vena caval opening

11.10 Diaphragm: A, ventral view; B, superior view. Circles indicate common sites of herniation, with numbers showing order of frequency.

and the pressure-volume ellipse is wider (Fig. 11.12B).

We should note that, though the diaphragm serves to separate the thoracic and abdominal cavities, it connects them as well. It is pierced with three openings that permit passage of the aorta, vena cava, and esophagus (see Fig. 11.10). Occasionally an abnormal opening develops, and a *diaphragmatic hernia* results. If the hernia

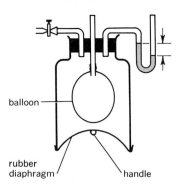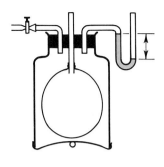

balloon

rubber
diaphragm   handle

11.11 Model showing the action of the diaphragm in breathing.

is large, the relatively high intraabdominal pressure may force part of the abdominal contents into the thoracic cavity, usually the left chest.

Pneumothorax. When air gains entrance to a pleural cavity for any reason, the lungs decrease in size in proportion to the volume of air introduced. This state is called *pneumothorax* (Fig. 11.13). Both lungs are involved because the mediastinum shifts toward the normal side.

Pneumothorax may be induced deliberately by the insertion of a needle into the pleural cavity and the injection of air; it may occur spontaneously through rupture of the lung surface; or it may follow penetrating injury of the chest wall. When the pleural cavity contains air but is occluded from the atmosphere, the pneumothorax is said to be closed.

Pneumothorax is frequently induced as part of the treatment of pulmonary disease, especially tuberculosis. It permits relaxation of the diseased lung, thus promoting healing and scar formation. Air introduced around a lung in this manner disappears in the course of time. The lung slowly expands and in a few weeks again adheres to the chest wall. If the lung requires more rest, a refill of air is required. Obviously the reabsorption of air from the pleural cavity provides important protection for the breathing apparatus. Without it, any air bubble, however small, would grow rather than shrink, owing to the subatmospheric intrapleural pressure, widening the gap between lung and chest wall. The mechanism of this gas reabsorption will be given later.

Spontaneous pneumothorax is common, even in the apparently healthy. Frequently it results from the rupture of a small bleb on the lung surface. The lung collapses rapidly, and the pleural cavity fills with air. Usually the rupture heals, the air is slowly reabsorbed, the lung reexpands, and little harm is done. More frequently spontaneous pneumothorax results from penetration of the pleura by some disease process within the lung.

Traumatic pneumothorax may occur during the induction of therapeutic pneumothorax. It may also follow a rib fracture or a gunshot or stab injury. A particularly dangerous form is associated with open chest wounds. Here a pneumothorax is established in which the intrapleural pressure equals atmospheric pressure. During inspiration air freely enters through the damaged chest wall, competing with air entering the trachea. If the opening is large enough to allow into the pleural cavity more air than the larynx can accommodate, respiratory movements result only in the passage of air from one lung to the other in a useless exchange. Unless the chest opening is quickly closed, there is increasing air hunger, with eventual suffocation.

Artificial respiration. When the act of breathing has ceased for any of a variety of reasons, among them drowning, bulbar poliomyelitis, or electrocution, or when there is danger of respiratory failure, it is possible to renew the air in the lungs by means of artificial respiration. In artifical respiration the thoracic bellows is manipulated from the outside, the normal rhythmic respiratory motion having ceased.

The many methods of artificial respiration can be classified as mechanical or manual. The two types of mechanical techniques are so-called *resuscitators* and *body respirators* ("iron lungs"). A resuscitator ventilates the lung by rhythmically applying positive pressure (or alternating positive and negative pressures) through a face mask or tracheal catheter. This device is often

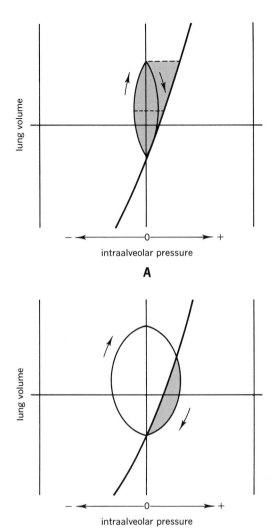

11.12 Pressure-volume curves. A. Natural breathing: the shaded area between the inspiratory side (ascending arrow) of the breathing loop and the pressure-volume curve of the resting lung represents inspiratory work; the triangular area to the right of zero represents elastic work expanding the lungs and chest. B. Deep breathing: the breathing loop overlaps the pressure-volume curve of the resting lung; the shaded area of the overlap represents expiratory work.

employed by firemen and rescue workers who must deal with acute cases of respiratory failure in locations remote from a hospital.

An "iron lung" differs from a resuscitator in that it operates the thoracic bellows, not by applying pressure through the nose and mouth, but by directly applying alternating positive and negative pressures to the entire thorax. The subject is placed in a cylindrical sealed pressure chamber with only his head protruding through a tightly fitted rubber collar. The pressure within the chamber is alternately increased and decreased by an electric pump. The alternating pressures expand and compress the thorax, substituting successfully for normal respiratory movements. This type of apparatus has kept many individuals alive for years after the onset of respiratory failure.

Manual techniques of artificial respiration are used only in emergencies. Many are available. In the *rocking* method the subject is placed in a prone position on a trestle that can be tilted to elevate head or feet. Rocking about 10 times a minute changes the thoracic volume rhythmically by shifting the weight of the abdominal viscera upon the diaphragm. Other well-known techniques include the *back-or-chest pressure* method, involving periodic compression of the chest by the hands of the operator, and the *arm-lift* method.

Careful studies of manual techniques have shown that they are inadequate in most situa-

11.13 X-ray showing pneumothorax. Arrow shows edge of partially collapsed lung. Compare with Fig. 11.5B. (Dr. Murray Janower.)

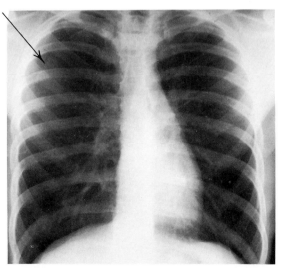

tions—even when administered by experts. Authorities now advise against them and recommend instead a *mouth-to-mouth* procedure in which the operator places his mouth directly over that of the victim and alternately blows air into and sucks air out of the lungs of the victim. In this way the operator himself becomes a resuscitator.

SURVEY OF THE FUNDAMENTAL GAS LAWS

Physical principles governing gas exchange. The exchange of gases through the thin membrane between an alveolar space and a pulmonary capillary is governed by the classic physical laws of gas behavior, which are common to all gases. Although hemoglobin in the erythrocytes stores and transports O_2 (as plasma bicarbonate does CO_2), the actual interchange of these gases in the lungs depends on their pressures (or tensions) in lung air and in plasma and extracellular fluid.

We shall here briefly summarize the laws governing gas transport and exchange. Further details are available in standard textbooks of physics and physical chemistry. Our discussion will not rely on mathematical analysis. However, it will be useful to understand the following simple relationships in quantitative terms.

Kinetic theory of gases. Any quantity of gas placed in a container of any size expands in volume to the limit set by the walls of the container. This is because gas molecules are in continuous motion. The concept that views a gas as a collection of molecules in motion and that explains all gas behavior in relation to these movements is called the *kinetic theory of gases.* Individual molecules are visualized as careening through space at high velocity, changing their courses only after colliding with other gas molecules or with the walls of the container.

Relationship of pressure to volume: Boyle's law. The steady bombardment of the walls of a container by confined gas molecules accounts for the gas *pressure.* The greater the number of molecules within the space, the higher the rate of bombardment, and the higher the pressure. When the space is made smaller, the molecules are brought more closely together, the rate of bombardment increases, and the pressure rises.

Boyle's law states this relationship quantitatively: the temperature remaining constant, the pressure of a gas varies inversely with its volume. If p_1 and p_2 are the pressures corresponding to volumes v_1 and v_2, then

$$\frac{p_1}{p_2} = \frac{v_2}{v_1} \qquad (11.1)$$

We see from this expression that the product of the pressure and volume of a gas has a constant value; that is, $p_1v_1 = p_2v_2 = p_nv_n$, where n denotes any condition.

Relationship of temperature to volume: Charles' law. The second fundamental gas law considers the influence of temperature. A rise in temperature increases the velocity of movement of the individual gas molecules, thus increasing their rate of bombardment of the vessel walls and the force of their impact. The temperature of a gas, in fact, is ultimately defined in terms of the velocity of motion of the individual molecules. *Charles' law* states that a gas kept at constant pressure expands by $\frac{1}{273}$ of its volume at $0\,^\circ C$ with each $1\,^\circ C$ rise in temperature. Thus the volume of a gas at constant pressure is directly proportional to its absolute temperature (absolute zero being $-273\,^\circ C$).

$$\frac{v_1}{v_2} = \frac{T_1}{T_2} \qquad (11.2)$$

Note that an interdependence of pressure and volume is also implied in this expression. The volume remaining constant, the pressure of a gas increases by $\frac{1}{273}$ of its pressure at $0\,^\circ C$ with each $1\,^\circ C$ rise in temperature. This follows from Boyle's law.

Diffusion and Dalton's law of partial pressure. The individual molecules of all confined gases tend to distribute themselves evenly throughout their container. Hence the pressure is the same on all parts of the confining walls. If, for example, a gas were concentrated in one portion of an enclosed space—say, near the floor in a sealed room—molecular movements would soon distribute it uniformly throughout the room and equalize the pressure on floor, walls, and ceiling. This tendency of gases to fill spaces uniformly is called *diffusion.*

In a mixture of two or more different gases, each diffuses throughout the container *as though the others were absent.* The significant result (for our discussion of pulmonary gas exchange) is that the pressure of each individual gas in the mixture depends solely on the concentration of that gas and is independent of the concentrations of other gases in the mixture. In other words, the pressure exerted by a gas depends on its percentage in the mixture. In essence, this is *Dalton's law of partial pressures.*

The pressure exerted by each gas in a mixture is called the *partial pressure* of the gas. With p representing partial pressure, pO_2 refers to the partial pressure of O_2, pCO_2 to that of CO_2, etc. According to Dalton's law, the sum of the partial pressures in a mixture of gases A, B, and C equals the total pressure of the gas mixture, P.

$$pA + pB + pC = P \qquad (11.3)$$

For example, in a mixture of 20% O_2 and 80% N_2 at a pressure of 760 mm of mercury, pO_2 is $760 \times 20/100 = 152$ mm, and pN_2 is $760 \times 80/100 = 608$ mm. The sum of the two partial pressures ($152 + 608$) is 760, the total pressure of the mixture. If the concentration of CO_2 in air is 5.5% (which it is in alveolar air), pCO_2 is $760 \times 5.5/100 = 41.8$ mm.

Any determination of the partial pressure of a gas in a mixture must take into account *all* the gases in the mixture. Unless the mixture is rigorously dried, it is likely to contain water vapor. Air in contact with water continually receives water molecules from the surface of the liquid, and these obey all the rules of gas behavior. Water vapor is, in fact, a gas, which exerts a partial pressure in any gas mixture in proportion to its concentration. The higher the temperature, the more water a gas mixture can hold before becoming saturated (at which point water vapor condenses into droplets of liquid) and hence the greater the partial pressure of water vapor. Typical partial pressures of water vapor in air at different temperatures are given in Table 11.1.

TABLE 11.1 PARTIAL PRESSURES OF WATER VAPOR IN MOISTURE-SATURATED AIR AT DIFFERENT TEMPERATURES

Temperature, °C	Partial pressure, mm of Hg
0	4.6
10	9.2
20	17.5
30	31.8
37	47.0
40	55.3
50	92.5

The contribution of water vapor is particularly important in calculations of the partial pressures of gases in alveolar air, which is saturated with water vapor. From Table 11.1 at 37°C water vapor has a partial pressure of 47 mm of mercury. This value must be subtracted from the total gas pressure before computation of the partial pressures of CO_2 and O_2 in a sample of alveolar air containing 5.3% CO_2 and 14% O_2 at a pressure of 756 mm. Since $756 - 47 = 709$, $pCO_2 = 709 \times 5.3/100 = 37.6$ mm, and $pO_2 = 709 \times 14/100 = 99.3$ mm.

It is clear that the temperature of the gas mixture must also be considered. The temperature of air is 37°C in the alveoli but falls to room temperature on leaving them. Since cooler air holds less water vapor, some condensation occurs, so that the partial pressure of water vapor becomes correspondingly smaller. Table 11.1 shows that the partial pressure of water vapor in room air (20°C) is only 17.5 mm of mercury when the air is fully saturated with water (i.e., when the humidity is 100%). Room air is usually far from saturated, and the partial pressure of water vapor in it is ordinarily only 4 or 5 mm.

Solubility of gases in liquids: Henry's law. When a gas (or a gas mixture) is in contact with the surface of a liquid, the gas molecules enter the liquid and become dissolved in it until the pressures of the gas within and without the liquid are equal. The gas is then said to be in equilibrium with the liquid.

It is important to comprehend the difference between a *solution* of a gas in a liquid and a *combination* of a gas with a liquid. For example, gaseous O_2 passing through a certain liquid may oxidize the liquid or its solutes. In this case the gas combines chemically with the liquid and is no longer subject to the laws of gas behavior. On the other hand, CO_2 may be dissolved in a liquid under high pressure as it is in soda water, a solution of a gas in a liquid. Continued high pressure is necessary to keep the CO_2 in solution. This is why a carbonated beverage must be sealed with an airtight cap. When the cap is removed, the pressure drops, and CO_2 is released as a free gas until its pressure in the liquid equals its pressure in the atmosphere. This explains the effervescence in a newly opened bottle of soda.

These relationships are summarized in *Henry's law*, which states that at constant temperature the quantity of a gas going into solution in a liquid is proportional to the partial pressure of that gas. The *volume* of a gas absorbed by a liquid at a given temperature is constant whatever the pressure. If a liquid in a dish inside a chamber whose volume can be compressed with a plunger absorbs 2 ml of the surrounding gas when chamber volume is, say, 500 ml, it still absorbs 2 ml when the chamber volume is 250 ml, because the pressure has doubled. However, it now absorbs twice the number of molecules of the gas.

When a gas combines chemically with a liquid, Henry's law is inapplicable. However, the chemical combination may be affected by the partial pressure of the gas in solution. Since the concentration of a dissolved gas depends upon its partial pressure—at least initially—the extent of its participation in a chemical reaction has the same relation to its partial pressure as does that of any reactant in a chemical reaction to its concentration. In the combination of O_2 with hemoglobin, for example, the amount of oxyhemo-globin formed is determined by the partial pressure of dissolved O_2.

The maximum quantity, in milliliters, of a particular gas dissolving in 1 ml of a particular liquid at 1 atm of pressure and a given temperature is indicated by the *absorption coefficient*. The absorption coefficient of O_2 in water as $0°C$ is 0.049 (i.e., 0.049 ml of O_2 dissolves in 1 ml of water) and that of CO_2 in water at $0°C$ is 1.71. The coefficient varies inversely with the temperature (Table 11.2).

TABLE 11.2 ABSORPTION COEFFICIENTS OF VARIOUS GASES IN WATER AT DIFFERENT TEMPERATURES

Temperature, $°C$	O_2	CO_2	CO	N_2
0	0.049	1.71	0.035	0.024
20	0.031	0.87	0.023	0.016
40	0.023	0.53	0.018	0.012

PULMONARY GAS EXCHANGE

Basic functions of the lungs. The primary function of the lungs is the aeration of blood.* Under normal circumstances this includes the addition of O_2 to venous blood and the withdrawal of excess CO_2 from venous blood. Under unusual circumstances it includes other operations. For example, the lungs may be called upon to take up anesthetic gases, volatile drugs, or even toxic gases such as carbon monoxide or "nerve gases." We speak of these functions collectively as *pulmonary gas exchange*.

Pulmonary gas exchange is divided conveniently into three separate processes (Fig. 11.14): (1) *ventilation*, the flow of fresh air into the lungs and its distribution to all 300,000,000 alveoli; (2) *diffusion*, the transfer of gases across the alveolar-capillary membranes; and (3) *pulmo-nary circulation*, the flow of blood essential for the transport of gases to and from the lungs.

Lung volumes and approved nomenclature. To facilitate discussion of pulmonary gas exchange, a precise nomenclature for the various lung volumes is necessary. The terminology had developed rather chaotically in the writings of various authorities, with resulting confusion. A committee of American physiologists therefore met in 1950 and recommended certain standard definitions and symbols. Their proposals were widely approved and will be given here.

Fig. 11.15 reproduces the tracing of respiratory movements that was published by the committee. The vertical dimension represents volume. Thus the *total lung capacity* for air is the distance between the bottom and top lines. The other lung volumes are shown in between. The oscillating curves depict respiratory sequences, the upward strokes indicating inspiration and the downward strokes expiration. Such a tracing is obtained with a *spirometer*, which consists of a cylindrical bell, closed at the upper end, that is immersed in a tank of water and counterbalanced. When air is blown into the bell through a pipe passing through the water, the bell rises. When air is removed from the bell through the pipe, the bell sinks. The distances traversed are proportional to the volumes of air introduced and withdrawn. These values can be read on a scale and recorded continuously.

Resting tidal volume is the volume of air inspired and expired in each respiratory cycle during quiet breathing. In an adult it averages 500 ml, varying from 350 to 600 ml. The resting tidal volume multiplied by the number of respirations per minute is the total volume of air breathed per minute, or the *respiratory minute volume*. At rest this is normally 5 to 8 liters.

The total lung capacity is about 5 liters. The resting tidal volume is about 10% of this total and about 12.5% of the *vital capacity*, which is the maximal volume of air that can be expired after a maximal inspiration. The vital capacity averages about 4 liters in men—though exceptionally it may reach 6 liters—and 3 liters in women. *Inspiratory capacity* is the maximal volume of air that can be inspired after a normal expiration. In the average man it is about 3 liters. *Inspiratory reserve volume* is the maximal vol-

---

* The lungs also perform certain subsidiary functions. One is the elimination of about 350 ml of water (see p. 206) and 300 cal of heat per day in exhaled air. The quantities depend on the volume of pulmonary ventilation, the metabolic rate, the atmospheric temperature, and the humidity. The lungs also act as filters, removing from the general circulation small clots that may arise in capillary and venous channels.

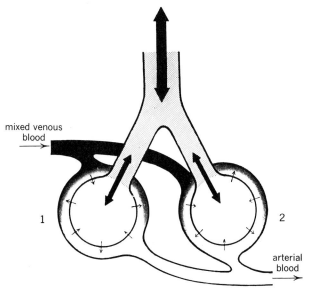

11.14 Schema of the lungs and pulmonary circulation. Round areas 1 and 2 represent alveoli; the shaded tubes leading to them represent all of the conducting airways. Mixed venous blood (dark) flows through vessels in intimate contact with ventilated alveoli and becomes arterial blood (light). The large arrow represents air flow to the lungs, and the medium-sized arrows represent distribution of the air to the alveoli. The small arrows represent the passage of O_2 out of the alveoli and that of CO_2 into the alveoli. (From J. H. Comroe, Jr., *Physiology of Respiration: An Introductory Text*, Year Book Medical Publishers, Inc., Chicago, © 1965. Used by permission.)

ume of air that can be inspired after a normal inspiration. *Expiratory reserve volume* is the maximal volume of air that can be expired after a normal expiration. Vital capacity is the sum of the inspiratory capacity and expiratory reserve volume.

Residual volume is the volume of air remaining in the lungs after a maximal expiration. *Functional residual capacity* is the volume of air remaining in the lungs after a normal expiration. It is the sum of expiratory reserve volume and residual volume.

The air expelled into the vital capacity apparatus is not the total volume of air in the lungs, for a forced expiration leaves a 1500 ml residual volume, which can be expelled only if the chest wall is opened surgically and the lungs are allowed to collapse. A small volume of air is trapped even in collapsed lungs; it gives them their buoyancy in water.

Dead space. In addition to the volumes just defined, another important volume of air is present within the respiratory organ complex. It occupies those parts of the respiratory system—the trachea, bronchi, and bronchioles—that act as mere conduits for air passing to and from lung tissue. Since no exchange of gases occurs between the air and the blood stream in these structures, they are referred to as the *anatomical*

dead space. During quiet breathing the volume of the dead space is between 100 and 200 ml. This means that approximately 150 ml of the 500 ml of freshly inspired tidal air is used to fill the dead space, the air previously in the dead space at the end of an expiration joining the remaining 350 ml of freshly inspired air in entering the lungs. Air entering the lungs is therefore a mixture of freshly inspired air and expired air from the dead space.

The volume of the dead space cannot be measured directly in a living subject, but it can be calculated. Table 11.3 shows that the average CO_2 content of *alveolar air* is 5.3% whereas that of *expired air* is 4.0%. If the resting tidal volume is 500 ml, then the total CO_2 expired in each breath is $500 \times 4.0/100 = 20.0$ ml. This quantity of CO_2 is contained in $20 \times 100/5.3 = 377$ ml of alveolar air. Hence 500 ml of expired air contains the same amount of CO_2 as 377 ml of alveolar air. This indicates that 377 ml of alveolar air has been added to 123 ml of dead space air (which contains no CO_2) to produce the 500 ml of expired air. The volume of the anatomical dead space varies, not only among different individuals but also in the same individual, depending on posture and other factors.

The *anatomical* dead space is sometimes overshadowed by the *physiological* dead space, which is the total space receiving air but not

TABLE 11.3 COMPOSITION OF RESPIRATORY GAS AT STANDARD ATMOSPHERIC PRESSURE

Gas	Inspired air % (wet)	Inspired air Partial pressure, mm of Hg	Expired air % (wet)	Expired air Partial pressure, mm of Hg	Alveolar air % (wet)	Alveolar air Partial pressure, mm of Hg
O_2	20.71	157	14.6	111	13.2	100
CO_2	0.04	0.3	4.0	30	5.3	40
H_2O	1.25	9.5	5.9	45	5.9	45
N_2	78.00	593	75.5	574	75.6	574

receiving pulmonary arterial blood. If alveoli are inadequately perfused with this blood they are useless in gas exchange and thus part of the physiological dead space.

Ventilation. The physiology of ventilation has two important aspects: volume and distribution. The volume of fresh air that enters the nose and mouth and is destined for the alveoli is, as we have already noted, about 350 ml per inspira-tion—500 ml resting tidal volume minus the 150 ml in the anatomical dead space. It is essen-tial, of course, that this volume be large enough to provide the O_2 required by the body.

In normal ventilation air is more or less uni-formly distributed among the millions of alveoli, owing to an ingenious structural arrangement within the lungs (Fig. 11.16A,B). Where dis-tances from the trachea to the alveoli are short, the bronchioles are narrow in diameter, and vice versa. As a result, all the alveoli, regardless

11.15 Lung volumes and capacities. (Adapted from J. Pappenheimer, et al., Fed. Proc. 9, 602, 1950.)

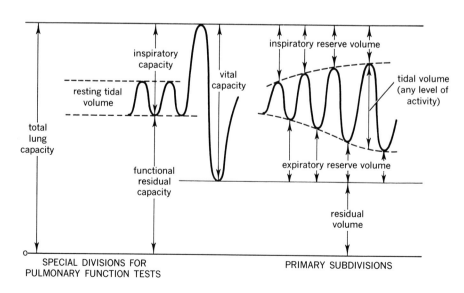

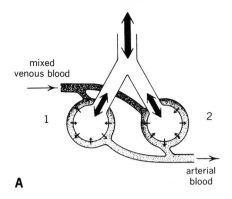

A

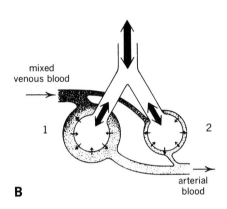

B

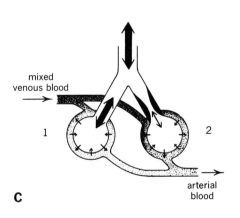

C

11.16 Relations of alveolar ventilation and capillary blood flow. Diagrams are based on schema in Fig. 11.14: A, uniform ventilation and uniform blood flow; B, uniform ventilation and nonuniform blood flow; C, nonuniform ventilation and uniform blood flow. (From J. H. Comroe, Jr., *Physiology of Respiration: An Introductory Text,* Year Book Medical Publishers, Inc., Chicago, © 1965. Used by permission.)

of their locations, tend to expand simultaneously. Nevertheless, in many circumstances certain alveoli receive different quantities or mixtures of air from other alveoli. Nonuniform air distribution is diagramed in Fig. 11.16C.

Uniform distribution connotes that during each inspiration every alveolus receives at the same time the same volume of gas of the same chemical composition and that this gas mixes almost instantaneously with the residual gas in the alveolus. That this ideal state is fairly well approached can be shown with a simple test (Fig. 11.17). The lungs of a normal subject breathing room air contain the same proportion of N_2 (80%) as does room air. If, after an expiration, there is a single deep inspiration of pure O_2 (say, 2000 ml), the N_2 level in the lungs will drop to, say, 40%. If distribution is uniform, the 40% mixture will be present throughout the lungs. If the N_2 in successive small volumes of expired air is measured instantaneously by a continuous gas analyzer, the levels will be found to increase rapidly and then remain constant. In a number of lung diseases—particularly those marked by a loss of elasticity of lung tissue, bronchial obstruction, or other localized disturbances—they will be found to vary, being low in the air expired first and high in that expired last.

The "single breath analysis" in Fig. 11.17 reveals a rapidly rising N_2 level in the 2000 ml or so of expired air. Such a record permits determination of the volume of the dead space. This is proportional to the length of time after expiration begins that the curve indicates no N_2. Clearly, this time represents the volume of pure O_2 that had remained in the nose, mouth, pharynx, trachea, bronchi, and bronchioles and thus had not mixed with alveolar gas (which contains 80% N_2 when the subject is breathing room air).

Such a record also permits estimation of the lung volume. In 1799 Sir Humphrey Davy, having prepared pure H_2 gas, inhaled two breaths of it and computed his own residual volume by analyzing the H_2 content of his expired air. He wrote, "It is more than probable that gas inspired into the lungs, from being placed in contact with the residual gas on such an extensive surface, must instantly mingle with it. Hence possibly deep inspiration and compleat expiration of the whole of a quantity of hydrogene will be sufficient to determine the capacity of

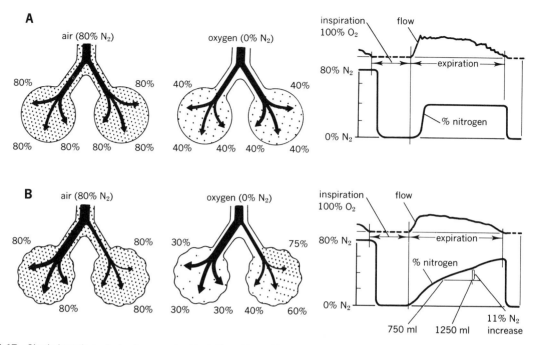

11.17 Single breath analysis: A, normal subject; B, subject with pulmonary emphysema. (From J. H. Comroe, Jr., *et al., The Lung: Clinical Physiology and Pulmonary Function Tests,* 2nd ed., Year Book Medical Publishers, Inc., Chicago, © 1962. Used by permission.)

the lungs . . . and the nature of the residual air." Precisely the same principle is involved in the single breath analysis of N_2.

Diffusion. Air must diffuse across the alveolar-capillary membranes if it is to combine with hemoglobin for distribution throughout the body. Just as measurement of total ventilation does not measure alveolar distribution, so measurement of alveolar distribution measures neither diffusion nor blood flow.

A normal alveolar-capillary membrane is less than 1 μ thick, so that circulating blood is in almost direct contact with alveolar gas. Fig. 11.18 shows arrangements for maximal as well as abnormal diffusion.

The characteristics of the alveolar-capillary membranes—area, thickness, and overall competence—can be summarized in one value, the so-called *diffusing capacity* of the lungs. There are several methods of measuring this. The simplest requires the use of CO in a variation of the single breath test. A normally breathing subject inhales a single breath of a nontoxic concentration of CO, holds it for 10 seconds, and then

exhales. The expired air is collected and analyzed for CO. The difference between the quantity of CO inspired and the quantity of CO expired is the amount that diffused across the alveolar-capillary membranes during the time interval. It is thus an indicator of CO uptake by blood, which is limited only by its rate of diffusion across the membranes.

Pulmonary circulation. Obviously pulmonary circulation is the major factor controlling the supply of O_2 to the tissues just as ventilation is the major factor regulating the elimination of CO_2 from the tissues. Although ventilation brings O_2 from the atmosphere to the alveoli, blood is the carrier that ultimately delivers it to the tissues. Likewise, although circulation brings CO_2 from the tissues to the pulmonary capillaries, ventilation is the mechanism that ultimately delivers it to the atmosphere.

The techniques that measure cardiac output also measure the volume of blood in the pulmonary circulation (see Chapter 9). These include catheterization and techniques based upon the Fick principle involving the uptake of an inert

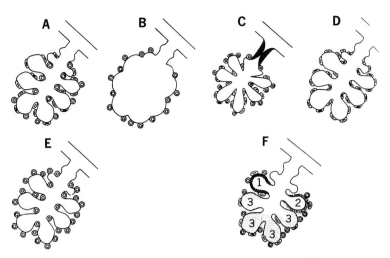

11.18 Arrangements of alveoli and capillaries: A, normal, with about half the capillaries open and about half closed; B, with alveolar septa destroyed and about half the total available number of capillaries; C, with the bronchiole obstructed but no decrease in the potential alveolar-capillary contact surface; D, with the pulmonary circulation obstructed and no alveolar-capillary blood flow; E, with an increase in the number of open capillaries as might occur in exercise; F, with (1) the alveolar epithelium thickened, (2) tissue separating the alveolar epithelium from the capillaries, and (3) pulmonary edema. (From J. H. Comroe, Jr., *Physiology of Respiration: An Introductory Text,* Year Book Medical Publishers, Inc., Chicago, ©1965. Used by permission.)

gas such as nitrous oxide or acetylene. The gas methods, of course, require analyses only of inspired and expired gases and not of arterial and mixed venous bloods as does catheterization.

We recall that the blood flow through the lungs averages 3 to 5 liters (depending on body size) per minute in a normal man at rest. The extraordinarily low pressures in the pulmonary blood vessels (see pp. 317, 377) signify their large cross-sectional area and relatively low resistance, qualities that are essential if the lungs are to maintain adequate ventilation.

In a normal man, during severe physical exertion, nearly 90 liters of air is inhaled, uniformly distributed to the alveoli, and exhaled each minute. At the same time some 25 liters of blood—nearly 7 gallons per minute or 1 pint per second—courses through the pulmonary vessels. It is a remarkable feat of engineering that the lungs function without interference from this massive quantity of fluid. Direct measurements show that the large increase in the volume of blood in the pulmonary capillaries during exercise is *not* accompanied by an increase in pulmonary blood pressure, however. This presumably means that exercise widens the pulmonary vessels. In an abnormal situation characterized by increased pul-

monary blood volume (e.g., severe congestive heart failure), pulmonary blood pressure strikingly increases during exercise, rising sometimes to over 100 mm of mercury. This explains the rigidity of congested lungs.

Interestingly, the composition of the alveolar air and the composition of the capillary blood leaving the alveoli depend not only on the volumes of ventilation and pulmonary circulation but also on their distribution. Just as inspired air may ventilate the alveoli nonuniformly, blood may flow at different rates in different areas of the lungs.

These all-important determinants of gas exchange are summarized conveniently in the *ventilation-to-perfusion ratio,* the ratio of air flow to blood flow. Normally this ratio is a little less than 1.0—that is, the alveoli receive slightly less air than blood per minute. For example, if 4 liters of air per minute ventilates the alveoli and 5 liters of blood per minute perfuses the pulmonary capillaries, the ratio is 0.8. If the ratio is decreased (i.e., if air flow is reduced out of proportion to blood flow), the result is a disproportionate increase in the amount of "venous" blood relative to that of "arterial" blood returning from the lungs to the heart. If the ratio is increased (i.e., if blood flow is reduced out of proportion to air flow), the result is in effect an increase in the physiological dead space because ventilated alveoli are not perfused and are therefore physiologically useless. Moderate exercise produces proportionate increases in ventilation and perfusion, so that the ratio remains normal. Hence there is no change in the gaseous contents of air and blood. Cardiac output increases, but respiratory rate and volume increase proportionately. Conversely, pneumothorax produces proportionate decreases in ventilation and perfusion. Fig. 11.19 presents in essence all of the foregoing concepts.

We shall now see why nonuniform distribution of inspired air interferes with normal gas exchange. It is illogical to assume that alveoli receiving too much air serve as counterbalances for those receiving too little. If, in an extreme hypothetical case, all inspired air entered only a single portion of the lungs—say, shunt X in Fig. 11.19—from which blood was totally excluded, the subject would suffocate. Although ventilation, cardiac output, and circulation might be otherwise normal, the inspired air would not come into contact with blood. In less extreme situations nonuniform distribution of inspired air leads to low blood O_2 and CO_2 retention.

Significance of foreign gas in pulmonary function tests. It is instructive to note why one gas is used in the measurement of pulmonary ventilation (nitrogen), another in the measurement of pulmonary circulation (nitrous oxide), and a third in the measurement of the diffusing capacity of the lungs (carbon monoxide).

For the determination of pulmonary ventilation, a gas that is relatively insoluble in water, so that very little "leaks out" of the alveoli into the blood, is necessary. Nitrogen fills this prescription admirably because it is physiologically inert and quite insoluble.

For the determination of pulmonary circulation, a more soluble gas (nitrous oxide), which readily "leaks out" of the alveoli into the blood but which has limited solubility in the blood, is necessary. As venous blood flows through the pulmonary capillaries, it becomes saturated with the gas almost immediately—perhaps a tenth of the way along the capillaries—the actual volume uptake depending upon the properties and

pressure of the gas. More gas is absorbed only if the blood flow is increased.

Carbon monoxide uptake is diffusion-limited, not flow-limited. The tremendous capacity of hemoglobin for CO makes CO exceedingly useful in physiological research. Though its ability to diffuse through the alveolar-capillary membranes is limited, its solubility in the blood (i.e., its ease of combination with hemoglobin) is so great that it would cross the membranes and be taken up by hemoglobin without saturating it even if the pulmonary circulation stopped for a brief period. We shall comment again later on the special significance of the reaction of CO with hemoglobin.

Mechanical factors. We have already spoken of the mechanics of breathing and the remarkable accomplishments of the lungs during exercise, when massive quantities of air and blood pass through them each minute. The thoracic bellows always operate against certain resistances. When these are unusually great, the act of breathing is less efficient than usual, and pulmonary ventilation suffers. The additional work required for breathing against these resistances in disease play a considerable part in the genesis of *dyspnea,* or difficult breathing.

Both elastic and nonelastic forces oppose the inspiratory effort. We have learned that the expansion of the chest in inspiration is pitted against the elastic recoil of the lung tissue. In other words, work must be expended to stretch the lungs. It has been shown, in fact, that the elastic recoil of the lungs is proportional to their increase in volume. The work necessary to overcome this elastic resistance comprises 65 to 70% of the total work of quiet breathing.

There is a strict relationship between the size of a lung and its capacity to be stretched. Physiologists call this the *mechanical compliance* or, more simply, the *compliance* of the tissue; it is defined as the volume change per unit of pressure change. For a given change in intrapleural pressure, a small lung changes volume less than a large lung—that is, a small lung is less distensible than a large lung. For example, normally the resistance to stretching is greater in an infant lung than in an adult lung. In effect, then, a smaller lung is a "stiffer" lung, resembling in some ways the relatively inelastic lung of congestive heart failure (see p. 391).

Nonelastic resistances to breathing include frictional resistance, to the flow of air through the air passages and to the movements of the lung tissue, and turbulence, due largely to sharp angles at the bifurcation of the trachea and the origins of the smaller bronchioles.

Work, in the physical sense, is force times distance or pressure times volume. Therefore, the product of pressure and volume of air moved at each instant equals work. The *total work of breathing* can be deduced from a pressure-volume curve. Normally the area between the inspiratory side of the breathing loop and the pressure-volume curve of quiet breathing represents the total work of one breath (see Fig. 11.12A). In deep breathing the breathing loop overlaps the pressure-volume curve of quiet breathing (see Fig. 11.12B); the area of the overlap is a measure of the work done by expiratory muscles. Obviously in normal breathing the muscles do much less work than they are capable of doing.

For a given rate of ventilation and a given anatomical dead space, the rate and depth of quiet breathing are so well adjusted that a minimal amount of work is necessary to overcome the various resistances. Slow deep breathing requires the least work against nonelastic resistances but the most work against elastic resistances. Rapid shallow breathing requires the least work against elastic resistances but excess work against nonelastic resistances; it also involves a great deal of useless ventilation of the dead space. An individual automatically selects an optimum rate and depth of breathing, which require the least work. This is true in health and in disease. An individual with chronic heart failure and lung congestion breathes rapidly and shallowly in order to avoid elastic work, whereas an individual with an obstructed air passage breathes slowly and deeply in order to minimize the effects of increased nonelastic resistance.

A dyspneic individual is able to exert as much effort in expanding his thorax as a normal individual, but the volumes of gas inhaled and exhaled are smaller than normal for comparable pressure swings in the thoracic bellows. He may therefore work as hard to obtain a ventilation of 40 liters per minute as a normal individual does to obtain one of 120 liters per minute. The heavy work with its poor return explains the distress and discomfort of labored breathing.

Composition of respiratory gas. Table 11.3 gives typical percentages and partial pressures of O_2, CO_2, H_2O vapor, and N_2 in the three chief categories of respiratory gases—*inspired, expired,* and *alveolar* air. The table reveals the

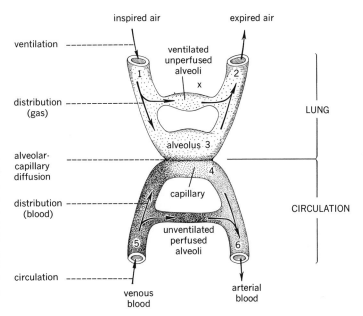

11.19 Respiratory gas exchange diagram showing partial pressure gradients associated with the five major functions. The net effects of the functions may be expressed quantitatively as follows: ventilation gradient, pO_2 at 1 minus pO_2 at 2; gas distribution gradient, pO_2 at 2 minus pO_2 at 3; diffusion gradient, pO_2 at 3 minus pO_2 at 4; blood distribution gradient, pO_2 at 4 minus pO_2 at 6; circulation gradient, pO_2 at 6 minus pO_2 at 5. Shunt X is explained on p. 392.

expected drop in pO_2 and rise in pCO_2 from inspired air to alveolar air.

The significance of the composition of alveolar air is apparent in Fig. 11.20, which shows the changes in pO_2 and pCO_2 in alveolar air and blood as it passes through the lungs. A sample of venous blood entering the lungs via the pulmonary artery contains CO_2 at a partial pressure of 46 mm of mercury (Table 11.4). As the sample circulates through the pulmonary capillaries, CO_2 diffuses into the alveoli until the pCO_2 of both alveolar air and arterial blood leaving the lungs becomes 40 mm. Simultaneously the pO_2 of blood rises from 38 to 100 mm as blood is exposed to alveolar air with a pO_2 of 100 mm.

In sum, expired air contains about 15% O_2 and 4% CO_2, while alveolar air contains about 13% O_2 and 5% CO_2. However, these figures vary from subject to subject. To obtain true percentages of O_2 and CO_2 in alveolar air, one needs a weighted mean value, which would reflect the compositions of all alveolar airs and thus overcome possible nonuniformity of distribution. The ingenious techniques providing such data are beyond the scope of our discussion.

GAS TRANSPORT

Transport of oxygen by blood. Oxygen is carried in the blood in two ways: (1) in physical solution; and (2) in chemical combination with deoxyhemoglobin in the erythrocytes. First it diffuses into the pulmonary capillaries and dissolves in the blood plasma, where it remains chemically uncombined. Then most of it diffuses further into the erythrocytes.

The amount of O_2 remaining in solution in the plasma is determined by Henry's law and is therefore proportional to its partial pressure. As noted earlier (see p. 386), the absorption coefficient of a gas is the volume that dissolves in 1 ml of liquid at 1 atm of pressure. Since the absorption coefficient of O_2 in plasma at 40°C is approximately 0.023 (see Table 11.2), at an arterial blood pO_2 of 100 mm of mercury, the amount of O_2 in 1 ml of plasma is $0.023 \times 100/760 = 0.003$ ml, or 0.3 ml per 100 ml of plasma.

In contrast, some 20 ml of O_2 per 100 ml of blood is transported by hemoglobin. We recall that hemoglobin's capacity to bind O_2 is based upon the reversible formation of oxyhemoglobin ($HHbO_2$), which gives the distinctive red color to arterial blood (see Chapter 8), 1 g of deoxyhemoglobin combining with 1.34 ml of O_2. Hence 15 g of hemoglobin—the normal amount per 100 ml of blood—can combine with up to $15 \times 1.34 = 20.1$ ml of O_2 (Table 11.5). This maximum amount of O_2 that can be bound by hemoglobin, expressed in milliliters per 100 ml of blood, or volumes percent, is termed the *oxygen capacity* of the blood. In fact, arterial blood carries about 5% less O_2 than it is capable of carrying, or about 19 ml per 100 ml. Thus we speak of the *saturation* of arterial blood with O_2 (O_2 content of arterial blood divided by O_2 capacity of whole blood). Saturation is normally 95 to 97%. The O_2 content of venous blood depends on body activity. At rest it is about 15 ml per 100 ml of blood ($pO_2 = 40$ mm of mercury), and during strenuous exercise it is about 4 ml per 100 ml of blood ($pO_2 = 20$ mm of mercury or less).

Transport of carbon dioxide by blood. Carbon dioxide is carried in the blood in three forms: (1) as dissolved CO_2; (2) as HCO_3^-; and (3) as *carbaminohemoglobin* ($HHbCO_2$),* a com-

---

* All proteins with free amino groups, including plasma proteins, form carbamino derivatives. However, the carbamino derivatives of the plasma proteins play a minor role in CO_2 transport.

TABLE 11.4 GAS PRESSURES IN ARTERIAL AND VENOUS BLOOD

Gas	Partial pressure	
	Arterial blood, mm of Hg	Venous blood, mm of Hg
O_2	100	38
CO_2	40	46
H_2O	46	46
N_2	574	574
Total	760	704

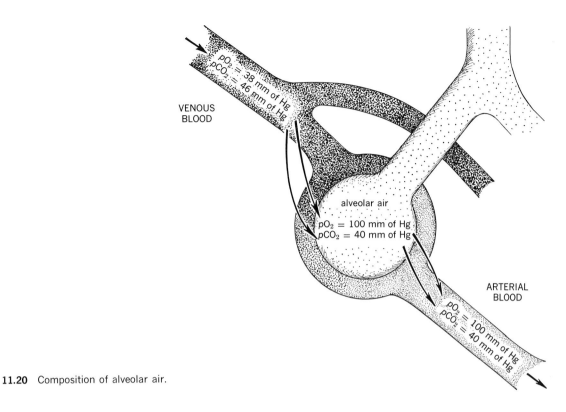

VENOUS
BLOOD

$pO_2 = 38$ mm of Hg
$pCO_2 = 46$ mm of Hg

alveolar air
$pO_2 = 100$ mm of Hg
$pCO_2 = 40$ mm of Hg

ARTERIAL
BLOOD

$pO_2 = 100$ mm of Hg
$pCO_2 = 40$ mm of Hg

11.20   Composition of alveolar air.

pound resulting from its combination with hemoglobin. Table 11.6 outlines the distribution of CO_2 in arterial and venous blood.

The 200 ml or so of CO_2 that is produced each minute in the tissues diffuses freely into the plasma. The amount capable of being dissolved in the plasma depends, like that of O_2, on partial pressure and solubility coefficient. However, although 100 ml of water at body temperature can dissolve 53 ml of CO_2 without chemical combination (see Table 11.2), only 1.8 ml of the total 52.1 ml of venous blood CO_2 or $(1.8/52.1) \times 100 = 3.4\%$ of the total remains dissolved. The rest is transferred or chemically converted.

Only a small amount of the CO_2 dissolved in the plasma participates in the following hydration reaction (the same as Eq. 10.3), because plasma contains little carbonic anhydrase (see Chapter 10).

TABLE 11.5   NORMAL VALUES FOR BLOOD GASES, ml per 100 ml of blood

O_2 capacity, whole blood	18–21 (g of Hb per 100 ml of blood $\times$ 1.34)
O_2 content, arterial blood	17–20 (saturation, 95–97%)
O_2 content, venous blood	10–16
CO_2 content, arterial blood	40–55
CO_2 content, venous blood	45–60

TABLE 11.6 TRANSPORT OF CO_2 BY BLOOD, ml per 100 ml of whole blood

	Arterial blood			Venous blood		
	Whole	*Cells*	*Plasma*	*Whole*	*Cells*	*Plasma*
CO_2 in solution	2.4	0.8	1.6	2.7	0.9	1.8
CO_2 as HCO_3^-	42.9	9.8	33.1	45.7	10.5	35.2
CO_2 as $HHbCO_2$	3.0	2.0	1.0	3.7	2.6	1.1
Total combined CO_2	45.9	11.8	34.1	49.4	13.1	36.3
Total CO_2 in all forms	48.3	12.6	35.7	52.1	14.0	38.1

$$CO_2 + H_2O \rightleftharpoons H_2CO_3 \rightleftharpoons$$
$$H^+ + HCO_3^- \quad (11.4)$$

We have noted that plasma $[HCO_3^-]$ is normally 20 times plasma $[H_2CO_3]$. Moreover, Table 11.6 indicates that $(35.2/52.1) \times 100 = 67.6\%$ of the total CO_2 of venous blood is accounted for by plasma HCO_3^-. As we shall presently see, most of this HCO_3^- comes from the erythrocytes.

Diffusion into the erythrocytes is the fate of the bulk of the CO_2 entering the plasma in the tissues. Some of this CO_2 remains in solution within the cells. Most of it is transformed by hydration to H_2CO_3 under the catalytic influence of carbonic anhydrase. The H_2CO_3 then dissociates into H^+ and HCO_3^-.

Some of the free CO_2 in the erythrocytes combines with the free uncharged amino groups ($-NH_2$) of hemoglobin—presumably the amino groups of 4 N-terminal valines—to form carbaminohemoglobin.

$$R-NH_2 + CO_2 \rightleftharpoons R-NHCOOH \rightleftharpoons$$
$$R-NHCOO^- + H^+ \quad (11.5)$$

Although only a small amount of the total CO_2 in the blood is carried as carbaminohemoglobin, it accounts for 18 to 25% of the difference in CO_2 content between arterial blood and venous blood.

It is interesting that deoxyhemoglobin (HHb) is about three times as effective as oxyhemoglobin ($HHbO_2$) in forming carbaminohemoglobin. This was first observed in 1914 by J. S.

Haldane, and so the phenomenon is known as the *Haldane effect*. It means that venous blood is a more competent carrier of CO_2 than arterial blood, whose $HHbO_2$ level is relatively higher. The CO_2 dissociation curves in Fig. 11.21 relate the amount of CO_2 carried by blood to the pCO_2 to which blood is exposed. Perhaps the Haldane effect is an evolutionary adaptive mechanism to promote the return of CO_2 from the tissues to the lungs. As $HHbO_2$ sheds its O_2, it becomes a preferred vehicle for CO_2 transport.

In summary, of the 3.8 ml difference in CO_2 content between whole arterial and whole venous blood, about $(0.3/3.8) \times 100 = 7.9\%$ reflects increased dissolved CO_2, $(2.8/3.8) \times 100 = 73.5\%$ increased HCO_3^-, and $(0.7/3.8) \times 100 = 18.4\%$ increased carbaminohemoglobin. Of the total venous blood CO_2, almost 88% is transported as HCO_3^-; of this about 77% is carried in the plasma. The large fraction of blood CO_2 carried in the plasma is in marked contrast to the small fraction of blood O_2 carried in the plasma.

Significance of shapes of gas dissociation curves. We referred to the curious S shape of the O_2 dissociation curve on pp. 152 and 234, noting that the oxygenation of hemoglobin is facilitated at high O_2 pressures. We shall now compare the O_2 dissociation curve with the CO_2 dissociation curve (see Fig. 11.21). Their differences have much significance.

The CO_2 dissociation curve is almost straight within the range of CO_2 pressures occurring physiologically. Changes in CO_2 pressure pro-

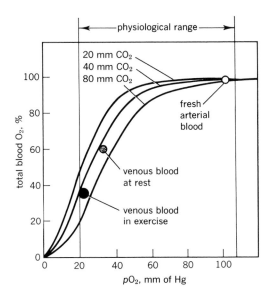
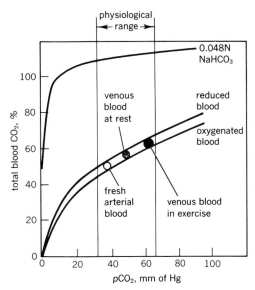

11.21 O_2 and CO_2 dissociation curves.

duce nearly proportional changes in total blood CO_2. The O_2 dissociation curve has a steep portion and a flat portion in the physiological range. Small changes in O_2 pressure produce large changes in total blood O_2 on the steep portion, and large changes in O_2 pressure produce small changes in total blood O_2 on the flat portion.

The flat portion of the O_2 curve above a pO_2 of 70 mm of mercury ensures a relatively constant arterial blood O_2 content even though alveolar air pO_2 may drop from 100 to 70 mm. Hence man can live at reasonably high altitudes without much reduction in the amount of O_2 carried by the blood.

The steep portion of the curve between pO_2 values of 10 and 60 mm ensures ready delivery of O_2 to the tissues. Metabolically active tissues have pO_2's in this range. Thus $HHbO_2$ in the capillary blood serving such tissues surrenders much of its O_2, the tissues receiving a large delivery of O_2 for a relatively modest decrease in pO_2. At a pO_2 of 40 mm, for example, hemoglobin gives up 25% of its O_2 to a resting tissue; at a pO_2 of 20 mm, it gives up 65%; and at a pO_2 of 10 mm (as in hard exercise), it gives up 87%.

O_2 dissociates from hemoglobin most readily in the acidic environments produced by rising CO_2 pressures and lactic acid concentrations.

With decreasing pH, the O_2 dissociation curve shifts to the right. In exercise, when CO_2 is released rapidly, the resulting local acidity sensitively influences oxyhemoglobin to alter its dissociative behavior in the right direction.

An understanding of the shapes of the dissociation curves is helpful in the interpretation of abnormalities. If we begin with normal ventilation and the arterial blood O_2 and CO_2 pressures given in Table 11.4—100 and 40 of mercury, respectively—we shall find that reducing the ventilation by one-half will result in an increase of pCO_2 to 80 mm and a decrease of pO_2 to 50 mm (with about 80% saturation) and that reducing ventilation by one-fourth will result in an increase of pCO_2 to 60 mm and a decrease of pO_2 to 80 mm (with about 94% saturation). In other words, decreases in ventilation produce proportional increases in CO_2 pressure and proportional decreases in O_2 pressure but relatively small decreases in O_2 saturation; a drop in O_2 pressure from 100 to 80 mm is associated with a drop in O_2 saturation from 98 to 94%—a change almost too small to be measured accurately.

The significance of these facts for a victim of pulmonary disease is that a large drop in O_2 pressure must occur before appreciably less O_2 is carried away from the lungs by the arterial blood. This situation is disadvantageous to physiologists, who would like to assess pulmonary function from the arterial blood O_2 content. Since a small drop in O_2 saturation signifies a large drop in O_2 pressure, visible cyanosis signifies a massive drop in O_2 pressure and a severe physiological defect.

The shapes of the curves indicate the large decrease in total gas pressure from arterial blood to venous blood (see Table 11.4). A 62 mm decrease in O_2 pressure minus a 6 mm increase in CO_2 pressure gives a net loss of 56 mm in total gas pressure. This explains why gas does not diffuse from venous blood into the pleural cavity. For the same reason gas introduced into body spaces is always absorbed.

Henderson-Hasselbach equation revisited. We have earlier suggested that hemoglobin plays an important role in acid-base balance, pointing out that it dissociates into Hb^- and H^+ (the extent of dissociation decreasing with decreasing pH) and that its O_2 dissociation curve shifts to the right when the pH is lowered or the CO_2 pressure is raised. This shift of the O_2 dissociation curve is called the *Bohr effect*, after one of its discoverers, Christian Bohr, father of the physicist Niels Bohr. We have seen that the Bohr effect facilitates the delivery of O_2 to the tissues and the removal of CO_2 from them.

The basis of the effect is the difference between the buffering capacities of HHb and $HHbO_2$. In the physiological pH range, $HHbO_2$ is a stronger acid than HHb (i.e., $HHbO_2$ dissociates to H^+ to a slightly greater extent, so that its solutions have lower pH's). If 1 mmole of $HHbO_2$ at pH 7.40 is deoxygenated to HHb, the pH rises to 7.67, and a substantial amount (0.7 mmole) of H^+ is needed to bring it back to 7.40. This means that the two major changes in the tissue capillaries—release of O_2 and uptake of CO_2—facilitate one another. The uptake of CO_2 increases pCO_2 and indirectly elevates $[H^+]$, as shown in Eq. 11.4. A rise in $[H^+]$ stimulates the deoxygenation of $HHbO_2$ (the Bohr effect). The HHb formed is a weaker acid than $HHbO_2$. Therefore, it is a better buffer—indeed the buffer pair $[Hb^-]/[HHb]$ assists in the regulation of blood pH. Since HHb is more effective than $HHbO_2$ in transporting CO_2 as carbaminohemoglobin (the Haldane effect), a rise in $[H^+]$ that

increases the release of O_2 from $HHbO_2$ results in improved CO_2 transport and pH control. Thus the reversible dissociation of O_2 from $HHbO_2$ becomes a factor in acid-base balance, for here is a weak acid whose strength is fixed by its pK (6.68) but whose pK rises (to 7.93) and hence whose acidic strength falls with deoxygenation.*

When CO_2 enters an erythrocyte, H_2CO_3 forms, as shown in Eq. 11.4. Since H_2CO_3 is an acid, the intracellular $[H^+]$ tends to increase. If this tendency were not curbed by an adequate buffering system, impaired pH control would result from absorption of CO_2 by a cell. The $[Hb^-]/[HHb]$ buffer pair acts in this instance.

$$Hb^- + H_2CO_3 \Longrightarrow$$
$$HHb + HCO_3^- \quad (11.6)$$

In effect, the conversion of $HHbO_2$ to HHb and O_2 supplies the very buffer needed to cover the CO_2 taken up by the cell.

Practically all of the H_2CO_3 in the erythrocyte eventually becomes HCO_3^-. An anion must, of course, be associated with a cation, and the chief cation in the cell is K^+. When H^+ combines with Hb^- to form HHb (Eq. 11.6), the released K^+ is free to cover the newly formed HCO_3^-.

Most of the bicarbonate produced in the erythrocyte must diffuse out into the plasma to eliminate the concentration gradient across the cell membrane. This creates a problem, for the cell membrane is rapidly permeable to anions (such as HCO_3^- and Cl^-) but only slowly permeable to cations (such as K^+ and Na^+). The problem is solved by the tendency of anions of any kind to move into the cell from the plasma. Since Cl^- ions are the most abundant anions in the plasma, a *chloride-bicarbonate shift* takes place. When HCO_3^- unaccompanied by cations diffuses out into the plasma, anions, particularly Cl^-, are drawn into the red blood cell (RBC) until the following equilibrium is reached:

$$\frac{[HCO_3^-]_{RBC}}{[HCO_3^-]_{plasma}} = \frac{[Cl^-]_{RBC}}{[Cl^-]_{plasma}} \quad (11.7)$$

Every Cl^- ion entering the cell replaces an

---

* The changes in acidic strength caused by oxygenation and deoxygenation are due to conformational changes of the hemoglobin molecule—the same changes that account for the sigmoid shape of the O_2 dissociation curve (see p. 234).

HCO_3^- ion. Just the reverse reaction occurs in the pulmonary capillaries, where the blood loses CO_2 and gains O_2.

These considerations show why plasma is a better carrier of CO_2 when it is in contact with cells than when it is separated from them. Plasma alone has a dissociation curve resembling that of a $NaHCO_3$ solution (see Fig. 11.21), whose shape is poor from a physiological viewpoint. The chloride-bicarbonate shift and the buffering capacities of $[Hb^-]/[HHb]$ and $[HCO_3^-]/[H_2CO_3]$ keep the blood pH remarkably constant and make gas exchange possible.

ABNORMALITIES OF GAS TRANSPORT

Introduction and terminology. Man has almost no oxygen reserve. Hence the elaborate mechanisms providing the tissues with oxygen must operate properly. When they do not, when the cells do not receive or cannot use enough oxygen for normal functioning, *anoxia,*

an extremely hazardous condition, exists. "Anoxia" means an insufficiency of oxygen in the organs and tissues. *Anoxemia* is a more restricted term meaning a subnormal amount or partial pressure of O_2 in arterial blood.*

Hypercapnia means an increase, and *hypocapnia* a decrease, in the CO_2 pressure of the blood. *Asphyxia* denotes simultaneous decreases in the intake of O_2 and the elimination of CO_2—i.e., anoxia plus hypercapnia.† Though these two disorders are usually associated, anoxia is by far the more damaging.

Table 11.7 summarizes the main types of anoxia and their causes. *Stagnant* anoxia may or may not be accompanied by anoxemia, since the fundamental defect is a subnormal flow of oxyhemoglobin molecules through the capillaries. The individual who stands erect too quickly after bending over and "blacks out" temporarily suffers from stagnant anoxia of the brain due to a momentary inability of the circulation to compensate for the forces of gravity.

The effects of anoxemia, or *anoxic* anoxia, are general. In anemia the arterial blood contains less hemoglobin and consequently less O_2 than normally. Yet its O_2 pressure is normal because it equilibrates with the normal O_2 pressure of alveolar air. As the blood flows through the tissues, the usual amount of O_2 is removed. Since the total amount of O_2 in the blood is low, the percent removed is greater than usual, and so the O_2 pressure of venous blood is below normal. If the O_2 pressure of the blood in the pulmonary capillaries cannot equilibrate with that of alveolar air because of a thickened alveolar membrane or if, as in certain congenital abnormalities of the circulatory system, unoxygenated blood from the systemic veins mixes with oxygenated blood from the lungs, the O_2 pressure of arterial blood also drops. We have already discussed one chemical alteration of the hemoglobin molecule that renders it incapable of transporting O_2. In methemoglobinemia, both congenital and acquired (see p. 236), hemoglobin cannot combine with O_2 because its iron is in the oxidized, or ferric (Fe^{+++}), state.

Another interesting condition is *carbon monoxide poisoning.* We have seen that CO has a strong affinity for hemoglobin; hence it is valuable in pulmonary function tests. However, in such tests only minute doses are used. CO reacts with hemoglobin to form *carboxy-*

TABLE 11.7 TYPES OF ANOXIA

Type	Typical causes
Stagnant anoxia	Decreased blood flow
General	Shock, heart failure
Local	Arterial or venous obstruction
Anoxemia, or anoxic anoxia	Decreased arterial O_2
	Low hemoglobin level
	Low atmospheric O_2—high altitudes
	Defective pulmonary function — respiratory obstruction
	Defective hemoglobin —methemoglobinemia, CO poisoning
Overutilization anoxia	Increased need for O_2 requirements
	Extreme muscular activity, convulsions
Histotoxic anoxia	Poisoned tissue enzymes Cyanide poisoning

---

* *Hypoxia* and *hypoxemia* have been proposed as substitutes for "anoxia" and "anoxemia" because the former words signify a decrease in O_2, which is a more accurate statement of the physiological picture than the latter words, which signify a lack of O_2. Most physiologists, however, take the perverse position that if it is acceptable to use "anemia," which means lack of blood, it is acceptable to use "anoxia."

† It should be remembered that asphyxia always implies both O_2 deficiency and CO_2 excess, whereas anoxia conveys no implications regarding CO_2.

Gas	Amount, liters	Site and availability
O_2	1	Blood: bound by hemoglobin;* readily available
CO_2	17	Soft tissues: bound as HCO_3^- and H_2CO_3; readily available
	100	Bone: bound as carbonates; slowly available
N_2	1	Fat and soft tissues: dissolved; readily available

* A small quantity of O_2 is bound by myoglobin, a hemoglobin-like hemoprotein pigment of muscle. This is not available for respiratory gas exchange unless the blood O_2 pressure falls excessively.

hemoglobin, thereby blocking the combination of O_2 with hemoglobin. Moreover, carboxyhemoglobin is stable. Herein is the danger of CO poisoning: the O_2-carrying capacity of the blood is reduced just as though the hemoglobin level were low. Moreover, the reduction is rapid, unlike that in true anemia, to which the body can adapt. The severity of the anoxia is in direct proportion to the amount of CO inhaled. Since the gas has no odor, an individual may become unconscious without having been aware of his exposure to it. Death may follow.

Overutilization anoxia is the physiological consequence of intense muscular activity. We shall later discuss some of its metabolic effects.

When the utilization of O_2 by the cells is impaired, *histotoxic* anoxia results. Alcohol, narcotics, and certain poisons such as cyanide inhibit the oxidative processes of cellular metabolism even though the supply of O_2 is entirely adequate. Actually, the O_2 content of venous blood is higher than normal in this situation because the arterial blood cannot unload its O_2 in the capillaries.

Effects of oxygen deficiency. In general, the consequences of anoxia depend more on the rapidity with which it develops than on its severity. The symptoms are due entirely to the malfunction of O_2-deprived body cells. When the O_2 pressure of the blood is lowered at a moderate rate of speed (as during an ascent to high altitudes in an airplane), the symptoms

resemble those produced by alcohol. As the higher functions of the brain deteriorate, there is a sense of well-being. Therefore, a pilot at high altitudes should not await the symptoms of O_2 lack before he begins breathing pure O_2. If he does not take this precaution, he will gradually lose his judgment, muscular coordination, vision, and memory. Collapse and unconsciousness will follow, with convulsions and death soon after.

When the onset of anoxia is so slow that compensatory mechanisms keep pace, symptoms are few, the chief ones being fatigue and loss of psychological inhibitions. Even acclimatized dwellers at high altitudes may occasionally suffer from "mountain sickness." We recall that one important compensatory mechanism in such individuals is an increase in the erythrocyte count (see Chapter 8). Another is a high vital capacity. A striking manifestation of anoxemia is cyanosis, the bluish skin color due to the circulation of large amounts of deoxyhemoglobin.

Total body gas reserves. Quantities of the respiratory gases are stored—either in solution or chemical combination—in various locations in the body outside the lungs, from which they are available in greater or lesser degree for use in emergencies. The sizes of the gas reserves, the tissues in which they occur, and the forms in which they exist determine their availability for respiratory gas exchange (Table 11.8).

O_2 stores are regulated chiefly by cardiac output, and CO_2 stores by ventilation. When ventilation decreases, blood O_2 decreases, and soft tissue CO_2 increases; when ventilation increases, the reverse happens. Though only slowly available, the large CO_2 reserve in the carbonates constituting a small part of the solid matter of bone is also depleted when ventilation increases for a long period, with resulting decalcification of the bone.

CONTROL OF RESPIRATION

Nature of respiratory control. A remarkable complex of regulatory mechanisms controls respiration. The rhythm and depth of breathing are determined, not in the lungs or the muscles of the chest and diaphragm, but in a nerve center of the brain whose behavior, in turn, is determined by integrated nervous, chemical, and muscular influences.

An arresting feature of the respiratory control machinery is that it includes both voluntary and involuntary mechanisms. We may assume that a pacemaker within the nervous system produces the rhythmic pattern of involuntary breathing. But, unlike the involuntary cardiac pacemaker, this pacemaker may be superseded by voluntary mechanisms. One cannot will his heart to beat more slowly, but he can, for a time at least, adjust the rate of his breathing.

We shall discuss anatomical details of the nerves and their central attachments that are responsible for breathing control in Chapter 12. Here we shall concentrate on certain functions of the incredibly efficient message center that continuously receives information concerning the state of inflation of the lungs, the levels of O_2 and CO_2 in the plasma, the blood pH, and the blood pressure—as well as communications advising it that swallowing is taking place (for we do not breath while swallowing), that sharp pains have occurred or offensive odors have entered the nose (for then we catch our breath), or that the air passages are irritated (for then we cough or sneeze). Having sorted these messages, the center sends forth nerve impulses that act on the breathing muscles.

Respiratory centers. In the early nineteenth century, physiologists discovered a small region of the brain whose destruction in an experimental animal caused cessation of breathing and hence death. They concluded that the region regulates breathing, and it became known as the *respiratory center* of the brain. Later investigations disclosed that this vital center crosses the mid-line, extending to each side. The concept of one small, bilateral, inherently rhythmic center of respiratory events has been attractive. We now know, however, that several respiratory centers exist. They not only govern the sequence of inspiration and expiration; they must be connected to many areas of the brain in order to integrate the complex patterns of breathing and other respiratory movements.

It appears that the respiratory centers discharge rhythmically in somewhat the same manner as the cardiac pacemaker, though their intrinsic rhythmicity is the subject of controversy among neurophysiologists. Rather than review the competing theories, we shall note only that the centers do have a pacemaking function but that a number of special arrangements stimulate them to alter their rhythm in response to changes in lung inflation and in blood O_2, CO_2, and/or pH. Thus respiration may be considered to be controlled *primarily* by the intrinsic periodicity of the respiratory centers and *secondar-*

ily by nervous and chemical devices. Like a thermostat, the centers receive signals informing them of conditions throughout the body and refine their control of ventilation accordingly.

Nervous control of breathing. Nerve impulses reaching the respiratory centers are of several types. Some come from the lungs. During quiet breathing the depth of respiration depends upon *stretch receptors* in the lungs. Impulses from the inspiratory section of a respiratory center initiate inspiration. As the lung tissue stretches, nerve impulses generated within the stretch receptors ascend to the expiratory section of the center and, in effect, notify it that the lungs are filling. The expiratory section sends forth inhibitory nerve impulses that stop the action of the inspiratory section and initiate expiration. As the lungs deflate, nerve impulses from the stretch receptors ascend to the inspiratory section, which sends forth impulses inhibiting the expiratory section, and the cycle is repeated. The impulses from the stretch receptors are named *Hering-Breuer reflexes,* after the two physiologists who, in 1868, demonstrated that breathing could temporarily be halted if the lungs were forcibly inflated.

Functioning along with the Hering-Breuer reflexes to inhibit the inspiratory section of a center is a higher brain center, the *pneumotaxic* center, which is thought to exert some complex integrative or modulating influence over the inherent rhythm of the lower respiratory centers.

Chemical control of breathing. Respiration is controlled by blood pH, CO_2, and O_2 by direct and indirect means. Arterial blood CO_2, the most important chemical determinant of ventilation, acts directly on the respiratory centers, the respiratory rate rising as the CO_2 level rises. The nerve cells of the respiratory centers are acutely sensitive to changes in arterial blood pCO_2; an increase of only 2.5 mm of mercury doubles the resting ventilation rate. It is the powerful effect of CO_2 that keeps one from successfully holding his breath voluntarily. The pH is also significant in this situation, but to a lesser degree.

It is interesting that CO_2, a toxic waste product on its way to elimination, is so critical in a life-supporting physiological system. Though it is a poison, it is also a messenger from the body at large to the respiratory centers, and it triggers the very devices that eliminate it. We shall later encounter circumstances in which an excess of CO_2 denies this function, however.

Arterial blood O_2 acts indirectly on the respiratory centers (it acts directly, too, but much less effectively). Next to the carotid sinuses in the forks of the carotid bifurcations are small, highly specialized masses of neurovascular tissue called *carotid bodies* (Fig. 11.22). Similar structures called *aortic bodies* are located near the aortic arch. Both contain special cells called *chemoreceptors*, which detect changes in arterial blood pO_2. When it falls considerably below normal, the chemoreceptors stimulate the respiratory centers to increase the rate and depth of respiration.

The carotid body chemoreceptors also cause an increase in the rate and depth of respiration when the pCO_2 rises considerably above normal. However, CO_2 directly stimulates the respiratory centers long before this point is reached, and special experimental maneuvers are required to demonstrate the sensitivity of the chemoreceptors to CO_2.

Abnormal respiration. The complex interactions of the respiratory centers and the reflexes based on chemical and mechanical stimuli account for both normal and modified patterns of respiration. Perhaps the most familiar modified pattern is the heavy breathing that follows exercise. Respiration of this type is called *hyperpnea*—more air than usual is breathed per minute as a result of an increase in the rate or depth of respiration or both. Many nervous and chemical factors contribute to the hyperpnea of exercise, and the phenomenon has not yet been thoroughly explained. Anoxemia ranks next to exercise as a cause of hyperpnea, the chemoreceptors of the carotid and aortic bodies probably bearing the primary responsibility for an increased depth of respiration.

Exercise and moderate anoxemia may be regarded as normal physiological states, for they occur in all normal individuals from time to time. A number of breathing patterns are usually associated with abnormal states. These include *tachypnea*, rapid, shallow breathing; *apnea*, suspended breathing; and *dyspnea*, difficult or labored breathing (see p. 393). In this lexicon *eupnea* means normal breathing.

Tachypnea accompanies high fever, prolonged severe anoxemia, and lung collapse. When ordinary reflex mechanisms stimulate respiration, they usually increase the depth of respiration before its rate. In tachypnea rate increases, and depth decreases. This may be due to fatigue of the respiratory center, or, in the case of lung collapse, to impaired Hering-Breuer reflexes. Since a collapsed lung is never distended, messages would not go from the stretch receptors to the respiratory center switching off the inspiratory effort.

"Apnea" literally means absence of breathing but generally denotes a temporary cessation of breathing, as when the breath is held voluntarily, when sharp rises in blood pressure stimulate the carotid sinuses, when the respiratory center is depressed by toxic substances (e.g., narcotics, certain anesthetics, and poisons), or when the arterial CO_2 pressure falls too low. An individual who breathes deeply and rapidly for a few minutes (*hyperventilates*) exhales an excessive amount of CO_2. The result is a temporary loss of the desire to breathe; this impressively demonstrates the importance of CO_2 as a respiratory stimulant.

"Dyspnea" implies a consciousness of breathing difficulty, a sensation of discomfort or breathlessness (taken literally, this would indicate that dyspnea cannot occur in unconsciousness), which can also arise in hyperpnea and tachypnea. In dyspnea the neurochemical demand for breathing is inappropriate to the existing O_2 needs (as at high altitudes), the neuromuscular respiratory effort is inappropriate to the breathing achieved (as in obstruction of the bronchioles), or the neural effort is

11.22 Carotid and aortic bodies.

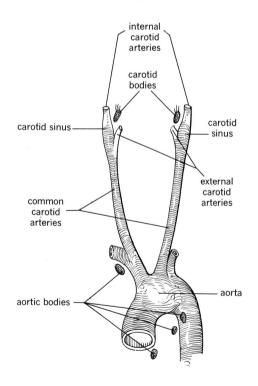

internal carotid arteries

carotid bodies

carotid sinus

carotid sinus

external carotid arteries

common carotid arteries

aortic bodies

aorta

inappropriate to the muscular actions achieved (as in paralysis of the diaphragm or respiratory muscles).

Respiratory control of blood pH. We have seen that abnormalities in body chemistry can alter the pattern of respiration. Table 10.3 showed that a decrease or increase in $[HCO_3^-]$ is compensated for by a corresponding decrease or increase in $[H_2CO_3]$. The compensatory decreases and increases in $[H_2CO_3]$ are achieved by increases and decreases, respectively, in the amount of CO_2 eliminated from the lungs.

Table 10.3 also showed that, conversely, respiratory abnormalities result in overretention of CO_2 (respiratory acidosis) or overelimination of CO_2 (respiratory alkalosis). Here compensation consists in increased or decreased excretion of HCO_3^- by the kidneys.

Physiology of the bronchioles. The muscles of the bronchial walls (see p. 375) relax with inspiration and contract with expiration. As a result of these changes, the air passages tend to enlarge during inspiration.

This muscle system is controlled by two sets of nerves, one producing *bronchodilatation* and the other producing *bronchoconstriction.* Anatomical studies indicate that the muscle fibers surrounding the small bronchioles just as they enter the alveoli are unusually highly developed. By relaxing thoroughly during inspiration, they greatly facilitate the inspiratory effort.

Frequently the muscles of the bronchioles contract excessively, causing a *bronchospasm.* Because of their strategic position, this constitutes a serious obstruction to air flow. Bronchospasm is the outstanding feature of *asthma,* one of the common causes of severe dyspnea in which the neuromuscular effort is inappropriate to the breathing achieved. A number of chemicals are powerful bronchodilators, and these are used in the treatment of asthma. The best known are epinephrine, ephedrine, and related compounds. Conversely, many are powerful bronchoconstrictors—among them morphine, histamine, and agents mimicking the actions of certain nerves. These may initiate asthmatic attacks.

Coughing, sneezing, and other curious phenomena. The body possesses a number of remarkable devices for eliminating foreign materials from the air passages. We have already commented on the cleansing action of the whip-like cilia of the lining epithelium (see Fig. 5.3). These microscopic fibers beat in rhythmic waves to push fluid or mucus up toward the glottis with astonishing efficacy. In addition, the muscles of the bronchioles exhibit a peristaltic motion that extrudes foreign matter from the bronchioles. More dramatic, however, are the protective reflexes, the *cough* and the *sneeze.* Both are coordinated, purposeful responses to mechanical or chemical irritations of the nerve endings in the respiratory mucous membrane.

A cough is elicited by irritation of the mucous membrane of the air passages below the nose; a sneeze may be described as an upper respiratory cough. Both involve a sudden preliminary deep inspiration and a violent expiration, whose force is proportional to the volume of inspired air. The air expelled under such great pressure carries with it the foreign matter that initiated the reflex. Interestingly, a violent contraction of the expiratory muscles is harder to avoid after a deep inspiration than after a shallow inspiration. Perhaps this situation is due to the Hering-Breuer reflexes. It is noteworthy that, although both coughing and sneezing can occur involuntarily only coughing can be induced voluntarily.

These normal responses to irritants are considerably modified under anesthesia or in other forms of unconsciousness—with danger to the individual. Deprived of his protective reflexes, he may suffer respiratory obstruction from a foreign body or secretion that a cough or sneeze would normally dislodge.

Other strange respiratory movements are rather poorly understood. These include *laughing, crying, yawning, hiccoughing,* and *snoring.* Of the nonlachrymal aspects of laughing and crying—the sounds, grimaces, and so on—we can say only that human beings laugh when they are amused and cry when they are sad. These curious behaviorisms remain unexplained and peculiar physiological events. The lachrymal aspect serves a useful purpose, as we shall see in Chapter 12.

Yawning is ordinarily an involuntary act accomplished by the rapid inhalation of large quantities of air with the mouth widely opened. Its causes range from reflexes to psychic phe-

11.23 Deep diver with self-contained underwater breathing apparatus.

may last from a few minutes to months and years. The consequences of prolonged hiccoughing are serious because sleep and nourishment are impaired. Some obstinate cases have yielded only to surgical interruption of the respiratory nerves. On occasion even this has been unsuccessful. The common effect of the many familiar remedies for ordinary hiccoughing—including holding the breath and either drinking a glass of water or counting—is a temporary increase of CO_2 in the arterial blood. This also follows brief inhalation of air containing 5 to 7% CO_2. Why this effect stops hiccoughing is unknown.

Snoring is noisy breathing through the mouth in sleep, caused by vibrations of the relaxed soft palate and the thin edge of the posterior tonsillar pillars (see Chapter 13). It is not abnormal breathing but normal breathing through the wrong passageway.

PROBLEMS OF COMPRESSION AND DECOMPRESSION

Deep diving. Interest in deep diving has grown in recent years as a result of the development of a compact "self-contained underwater breathing apparatus," or *scuba* (Fig. 11.23). The scuba diver receives his air supply from a tank that he carries with him and he can thus swim about under water independently.

The physiology of deep diving rests upon the fundamental gas laws. We recall that air is compressible and water is not. Fig. 11.24 shows that a weight acts on a cylinder containing water and an air-filled balloon to compress the air in the balloon but not the water, which retains its original volume. An identical weight acts on an attached cylinder containing air to compress the air and restore the balloon to its original size, the pressure within the balloon again becoming equal to that of the water outside (Fig. 11.24C).

These relationships mean that the pressure within the air-containing cavities of the body of a diver must be raised to equal that of the surrounding water. If the pressure within the lungs, paranasal sinuses, and middle ears—or even beneath faulty dental fillings or inlays—is not equalized with that of the surrounding water, a vacuum will exist. Of course, the lungs, like the balloons in the figure, would collapse without difficulty until the air in them attained the same

nomena. Since it commonly occurs after meals, one of its causes may be the pooling of blood in the abdominal organs. During yawning intraabdominal pressure markedly increases while intrathoracic pressure decreases. The pressure differential helps to return blood from the abdominal cavity to the thoracic cavity and thence to the heart.

A hiccough represents a sudden violent and purposeless contraction of the diaphragm. The first phase is a deep inspiration, which is abruptly terminated. A characteristic sound results. Hiccoughing is frequently due to reflexes originating in local irritations of the respiratory tract, diaphragm, or alimentary tract (this is the cause of alcoholic hiccoughing). It may also be associated with emotional or toxic disorders (e.g., uremia), which probably involve abnormal stimulation of the respiratory center. Episodes of hiccoughing

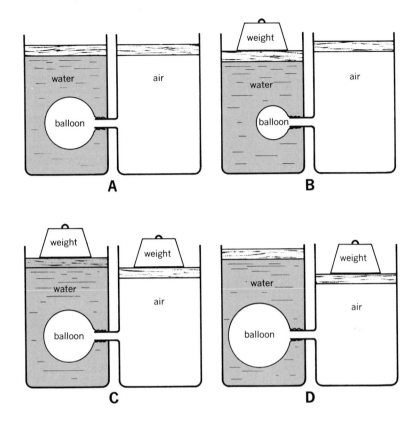

11.24 Demonstration of the compressibility of air and the incompressibility of water. Diagrams are explained in the text. (Adapted from *The Ciba Collection of Medical Illustrations* by Frank H. Netter, M.D. Copyright CIBA.)

pressure as the surrounding water. However, the sinuses, which have rigid walls, might painfully fill with fluid or blood. Obviously, the greater his depth under water, the greater the pressure of air that must be supplied a diver.

Water pressure increases about 23 mm of mercury with every foot of depth. Hence, at a depth of 33 ft, a diver must withstand twice the pressure at the surface. When the pressure in the balloon of Fig. 11.24 (as controlled by the pressure on the air-containing cylinder) remains unchanged, the volume of air in the balloon varies inversely with the pressure of the water surrounding it (Boyle's law). When the pressure on the water-containing cylinder decreases, the volume of air in the balloon increases. Therefore, a lung half inflated at a depth of 33 ft is fully inflated at the surface, and a lung fully inflated at a depth of 33 ft may rupture at the surface unless it expels air on its way up.

Dalton's law of partial pressures and Henry's law also have special relevance to diving. With increasing depth pO_2 and pN_2 increase proportionately. At the surface pO_2 is 152 mm of mercury. At a depth of 125 ft, it is 727 mm,* approximately the same as the pressure of pure O_2 at the surface. At depths greatly in excess of this (e.g., 270 ft), pO_2 may be so high as to cause convulsions. The partial pressures of atmospheric contaminants, such as CO_2, are also increased at great depths, thus augmenting their toxic effects.

From Henry's law pN_2 at a depth of 33 ft, where the pressure is 2 atm, is twice pN_2 at the surface. Therefore, twice the normal amount of N_2 is dissolved in the blood and tissue fluids of a diver at 33 ft. Conversely, only half the amount

---

* Air is approximately 20% O_2. Accordingly, $125 \times 23 \times 20/100 = 575$ mm; $575 + 152 = 727$ mm.

of N_2 dissolved in the blood and tissue fluids at a depth of 33 ft remains in solution at the surface. As a diver rises to the surface, excess N_2 tends to come out of solution in his blood and tissue fluids as bubbles of gas, as when a cap is removed from a bottle of soda.

Anyone under water faces the dangers of asphyxia or drowning.* Asphyxia can occur if CO_2 is not eliminated from the air by adequate ventilation or properly operating absorbents. Divers exhaling into the water do not face this particular risk. It is limited to those using closed-circuit devices in which exhaled air is freed of CO_2 and reinhaled. For convenience, we shall divide the other hazards of diving into those encountered in descent, those encountered at the bottom, and those encountered in ascent and consider them separately.

Problems during descent. The pressure on a diver is 1910 mm of mercury at a depth of 50 ft and 3060 mm, or more than four times that to which man is accustomed, at a depth of 100 ft. Man can withstand this increased pressure quite well. Indeed, he has descended to 600 ft without any untoward effects due to compression per se; at this depth the pressure on him is more than 19 times that at the surface. However, if the increased pressure cannot be transmitted to closed body cavities, the result is the same as if a vacuum were to develop in the cavity at the surface. This is why divers occasionally experience pains in their sinuses, middle ears, and teeth.

Problems at the bottom. The special diving hazards mainly encountered at the bottom are *nitrogen narcosis* and *oxygen poisoning*. The extra N_2 that dissolves in the blood and tissue fluids at great depths produces a dizziness similar to that following the consumption of alcohol. Some individuals are unusually sensitive to this "rapture of the deep." Its exact cause is unknown. However, N_2 is highly soluble in fat, and, according to current theories, any substance having this characteristic may act as an anesthetic.

Despite intensive investigation the reason life-supporting O_2 causes convulsions and other nervous aberrations when present in the body in excess is also unknown. Only about 0.3 ml of O_2 is dissolved in 100

ml of plasma at the surface, in contrast to about 6.5 ml at a depth of 300 ft. (This is in addition to the 19.5 ml in erythrocytes per 100 ml of blood.) Since the brain utilizes only 6 to 6.5 ml of O_2 per 100 ml per minute, at a depth of 300 ft, its O_2 requirement is largely satisfied by plasma. As blood leaves the brain, then, it carries with it only a small amount of hemoglobin and a large amount of oxyhemoglobin. Because of the Bohr effect (see p. 398), oxyhemoglobin has a relatively poor capacity for CO_2 transport. Hence plasma pCO_2 increases, the increase amounting to about 4 mm of mercury at 300 ft. As in respiratory acidosis, the result is stimulation of the respiratory center, which leads to an increase in ventilation and a decrease in arterial blood pCO_2 of about 6 mm of mercury.

Cerebral blood flow varies directly with plasma pCO_2. Consequently, hyperventilation, which decreases pCO_2, is probably a safety mechanism by which the brain is able to resist the toxic effects of excessive O_2. When the removal of CO_2 from the respiratory mixture is impaired (by the failure of diving equipment, for example), plasma pCO_2—and thus cerebral blood flow—increases in spite of hyperventilation. The net result is a further increase in O_2 in the brain and an increased tendency to O_2 poisoning.

Problems during ascent. On ascent, two major problems arise. The first is due to the expansion of gas as pressure decreases. The second is due to the escape of N_2 from solution in the form of bubbles.

Although swallowed air or sinus cavity air expands painfully during an ascent from the depths, it does not cause serious injury. The most dangerous result of pressure reduction, or decompression, is the expansion of lung air. A diver must exhale continuously during a rapid ascent. If he does not, the alveoli progressively distend, and some may rupture. The consequent introduction of air into the blood stream is called *air embolism*. Air embolism to the brain may be fatal.

The escape of gaseous N_2 during decompression is well known as "the bends." The appearance of symptoms depends upon the depth and duration of immersion and the rapidity of ascent. The greater the depth, the greater the pressure on N_2 to go into solution; and the longer the immersion, the greater the opportunity for it to do so. Depths to 30 ft can be tolerated indefinitely without decompression difficulties. Those below 60 ft are safe for progressively shorter periods of time. More than 30 minutes at 100 ft permits the absorption of so much N_2 that slow decompression is imperative. The greater the pressure difference between the dissolved N_2 and the atmosphere (that is, the greater the depth), the larger and more numerous the bubbles of N_2 escaping during decompression and the greater the possibility of symptoms.

Once gas bubbles have formed, the severity of the

---

* Drowning occurs when the lungs fill with water. If the water contained enough dissolved O_2 (and could be pumped in and out rapidly enough), air-breathing animals could get O_2 from water as fish do with their gills. This has been successfully demonstrated. When O_2 has been dissolved in water under pressure and adequate turbulence has been maintained, mice and dogs have lived under water and returned to air breathing.

symptoms depends upon where they lodge. Pain, the chief symptom, is largely centered around the joints and such poorly distensible tissues as tendons and ligaments. Symptoms appear in 85% of "the bends" cases within 4 to 6 hours and in the remaining 15% within 12 to 24 hours.

The only method of relieving decompression pain is *recompression.* The victim is placed in a special chamber in which air pressure can be maintained at a desired level. With increased pressure, N_2 is forced back into solution, and pain is alleviated. Pressure is then reduced gradually so that excess gas may be excreted slowly. Curiously, there is no correlation between the depth to which a diver has descended and the amount of recompression required. One diver who has been down to 100 ft may require pressure equivalent to that at a depth of 165 ft, whereas another may require much less. The reason for this discrepancy is not known.

PHYSIOLOGY OF THE LARYNX

Evolution of the larynx. We can best understand the physiology of the larynx by considering its evolutionary origin. The most primitive larynx is found in the lungfish, the creature introduced in the quotation on p. 370. The lungfish, which inhabited rivers that periodically became dry, survived by developing lungs and the ability to breathe air. With the emergence of lungs, however, new problems arose, for it was essential that the trachea be protected from food during eating and from water during drinking and submersion.

The indispensable protective mechanism that appeared in the upper end of the trachea of the lungfish was a simple circular group of muscle fibers constituting a sphincter. When the sphincter was closed, the lower respiratory passages were isolated from the upper ones. This larynx contained only constricting muscles and dilated on their relaxation. Higher animals developed special dilating muscles. In man these remain as the *posterior cricoarytenoid muscles,* which rotate the arytenoid cartilages outward, thereby separating the vocal cords.

There is no evidence that the primitive larynx evolved for the purpose of sound production. Yet it produced sound, we imagine, through the same physical principles that apply when we constrict the neck of a balloon and allow air to escape under pressure. Thus, though presumably not needed, an apparatus for sound production had come into existence. Still the sphincter function of the larynx was physiologically more significant than the sound production function.

The sphincter came to serve a number of useful purposes besides lung protection. In man, closure of the glottis permits intrathoracic pressure to build up for coughing. In the deer, whose sense of smell must be highly efficient to protect him against enemies, the epiglottis can actually meet the soft palate. The continuous separate air passage thus formed from the nose to the lungs permits respiration to continue during swallowing and directs inspired air past the nasal olfactory centers. A similar air passage characterizes every animal with a well-developed sense of smell. It no longer exists in man, whose sense of smell is inferior to that of many other animals.

Man's loss of the nasotracheal airway is not due to any foreshortening of the epiglottis or the palate but rather to the descent of the larynx from a position high up under the base of the skull, where it occurs in a lower animal and in the human embryo, to the mid-neck region during growth to adulthood. From the standpoint of olfaction, the descent is an evolutionary backward step.

Phonation and speech. "Phonation" means the production of vocal sounds. "Speech" means the shaping or articulation of vocal sounds into patterns that man and custom have made meaningful. This astonishing function involves the brain centers that control the tongue, jaw, and other structures organizing sound into speech.

The anatomy of the laryngeal muscles is extremely complex. They widen the opening in the glottis during respiration and narrow it during phonation. Then air driven by contractions of the thorax causes a vibration of the vocal folds, which results in sound. The larynx cannot produce sound unless both vocal folds are vibrating. In this regard it differs from a flute, a violin, a piano, and a bugle, in which sound is produced by single vibrating elements. The larynx is not the organ of speech. The sound emanating from it is not the sound that emanates from the mouth. Spoken sound is colored, textured, and improved by the resonating chambers of the upper respiratory passages. The larynx merely produces the sound, which is modified into the voice and articulated into speech by structures outside the larynx. The loudness or intensity of the voice depends upon the volume and force of the expired air. The pitch, or frequency, depends upon the length, thickness, and elasticity of the vocal folds and the tension with which they are held.

With practice and discipline some individuals acquire the ability to speak while inspiring air. This is the basis of *ventriloquy*. Other sounds produced during inspiration are sighing in man and neighing in horses.

The endocrine glands play an important role in determining the character of the voice. Under endocrine stimulation the larynx becomes relatively larger in a male at puberty than in a female at puberty. It is evident as the Adam's apple. The vocal folds may lengthen so fast that the ability to control them expertly cannot keep pace. This situation accounts for the squeaking and breaking voice that embarrasses adolescent boys. At the other end of the scale, a senescent larynx atrophies and loses its elasticity, so that the voice becomes high-pitched and often tremulous. These changes usually occur after the age of 70.

REFERENCES AND SUGGESTIONS FOR FURTHER READING

Andersen, H. T., "Physiological Adaptations in Diving Vertebrates," *Physiol. Rev.,* **46**, 212 (1966).

Aviado, D. M., *The Lung Circulation,* Vols. 1–2, Pergamon, New York, 1965.

Bates, D. V., and R. V. Christie, *Respiratory Function in Disease,* Saunders, Philadelphia, 1964.

Caro, C., ed., *Advances in Respiratory Physiology,* Williams & Wilkins, Baltimore, 1966.

Comroe, J. H., Jr., *Physiology of Respiration: An Introductory Text,* Year Book Medical, Chicago, 1965.

———, "The Functions of the Lung," *Harvey Lectures Ser. 48 (1952–53),* 110 (1954).

———, and others, *The Lung: Clinical Physiology and Pulmonary Function Tests,* 2nd ed., Year Book Medical, Chicago, 1962.

Cunningham, D. J. C., and B. B. Lloyd, eds., *The Regulation of Human Respiration,* Blackwell Scientific, Oxford, 1963.

Fenn, W. O., "The Mechanism of Breathing," *Sci. Am.,* **202**, 138 (Jan., 1960).

———, and H. Rahn, ed., *Handbook of Physiology: Sec. 3. Respiration,* Vols. 1–2, Williams & Wilkins, Baltimore, 1964–65.

Heinemann, H. O., and A. P. Fishman, "Nonrespiratory Functions of Mammalian Lung," *Physiol. Rev.,* **49**, 1 (1969).

Knowles, J. H., *Respiratory Physiology and Its Clinical Application,* Harvard Univ. Press, Cambridge, Mass., 1959.

Miller, W. S., *The Lung,* Thomas, Springfield, Ill., 1947.

Mitchell, R. A., "Respiration," *Ann. Rev. Physiol.,* **32**, 415 (1970).

Nahas, G. G., ed., "Regulation of Respiration," *Ann. N. Y. Acad. Sci.,* **109**, 411 (1963).

Pappenheimer, J., and others, "Standardization of Definitions and Symbols in Respiratory Physiology," *Federation Proc.,* **9**, 602 (1950).

Pattle, R. E., "Surface Lining of Lung Alveoli," *Physiol. Rev.,* **45**, 48 (1965).

Permutt, S., "Respiration," *Ann. Rev. Physiol.,* **28**, 177 (1966).

Pitts, R. F., "Organization of the Respiratory Center," *Physiol. Rev.,* **26**, 606 (1946).

Pressman, J. J., and G. Keleman, "Physiology of the Larynx," *Physiol. Rev.,* **35**, 506 (1955).

Riley, R. L., "Pulmonary Gas Exchange," *Am. J. Med.,* **10**, 210 (1951).

Rossier, P. H., and A. Buhlmann, "The Respiratory Dead Space," *Physiol. Rev.,* **35**, 860 (1955).

Staub, N. C., "Respiration," *Ann. Rev. Physiol.,* **31**, 173 (1969).

Could we but look altogether naively at the question of a seat of the mind within the body we might suppose the mind diffused, not confined to any one part. An individual, one's dog, one's self, is a mass of microscopic lives, each one self-centered. It might then perhaps be that our mind, at least so far as sentience, would extend through all our parts. That is not found to be so.

The finite mind of an individual as to "place" is related with one only of the systems of the body. That system is the opposite of diffused. The behavior of the complex animal, such as the cephalopod, the insect and the vertebrate is largely an expression of interaction in the clumped cell masses of the nervous system. The largest of these masses is the brain. It is with that that recognizable mind correlates. Much as one special organ, the heart, maintains the flow of nutriment throughout the body; so one organ, the brain, is provider of mind for the whole individual.

Sir Charles Sherrington in *Man on His Nature*, 1951

12 Nervous System

Introduction

BIOLOGICAL SIGNIFICANCE OF NEURAL INTEGRATION

To survive, an organism must be capable of responding to changes in the environment. To respond quickly and appropriately, its several parts must be in direct or indirect communication with one another and with the outside world. This is true of a small unicellular organism or of a large multicellular one. We learned in Chapter 4 that an individual cell displays irritability. A *stimulus* activates a *receptor,* which initiates the conduction to another region of the cell of a zone of depolarization, which triggers an *effector.* In a single cell the surface area is large in relation to the total volume, and most internal structures are close to one another and to the surface. Communication over the short intracellular distances therefore depends on simple biophysical and biochemical mechanisms and does not require a specialized network.

A large multicellular organism maintains a high ratio of surface area to volume only by drastically modifying its form as it grows. A plant develops huge spreads of foliage and intricately branching systems of roots and root hairs. An animal develops such devices as intestinal villi and respiratory alveoli.

Compared to an animal, a plant has a notably small and unsophisticated repertoire of responses to environmental change; it can turn to the sun, but it cannot flee from danger. Accordingly, it too lacks a specialized communication network. Despite adaptations in form, however, an animal body requires a system by which information is transmitted inward and sorted and appropriate coordinated responses are effected. The nervous system makes possible these complex functions. Together with the endocrine system, it controls and integrates almost all body functions.

MODEL NERVOUS SYSTEMS

If we regard the nervous system as an enlarged version of an individual cell's primitive *receptor-conductor-effector* system, we can devise an instructive theoretical model of the simplest of all possible nervous systems (Fig. 12.1A). We can then build upon this model in stepwise fashion to obtain a diagram of the fully developed nervous system (Fig. 12.1A–E).

In the simplest possible nervous system, a fiber extending directly from a receptor, or sensory, cell makes direct contact with an effector cell. Thus the stimulated receptor cell itself transmits an impulse to the effector cell, which responds appropriately (for example, a muscle cell contracts). Systems of this type occur in lower animals.

At the next stage of complexity (Fig. 12.1B), a receptor cell initiates an impulse in a *separate* nerve cell. This is a conductor, conveying the impulse to an effector cell. Systems of this type also occur in lower animals.

In both of these models, reaction patterns are necessarily simple and entirely inflexible. When the stimulus is of adequate intensity, the receptor cell is "turned on" like a switch, and the impulse is invariably conducted to one particular effector, whose predictable response can vary only in duration.

In higher animals receptor cells rarely make direct contact with effector cells. In vertebrates they never do. Instead, vertebrate nervous systems introduce another complication. There is almost always more than one nerve cell between the receptor and effector cells (Fig. 12.1C). In the lower vertebrates a separate *sensory neuron* carries an impulse from a receptor cell and passes it on to a separate *motor neuron* across a *synapse* (see p. 185). The motor neuron in turn conducts the impulse to an effector cell.

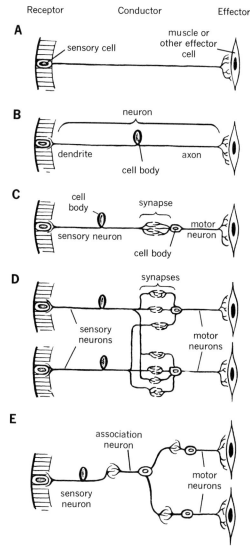

12.1 Model nervous systems. (From G. G. Simpson and W. S. Beck, *Life: An Introduction to Biology,* 2nd ed., © 1957, 1965, by Harcourt Brace Jovanovich, Inc. and reproduced by permission.)

Though this system is still a simple one, the chain provides for greater effectiveness and flexibility of reaction than does a direct connection between receptor and effector. Since a sensory neuron can stimulate more than one motor neuron (Fig. 12.1D), a single stimulus can arouse an extensive response, and continuous stimulation can involve more and more effectors.

Each of the models discussed so far constitutes a *reflex arc*, the essence of which is a direct or simple connection between receptor and effector. Clearly, each permits little control or modification of behavior.

In the arrangement characteristic of higher vertebrates, still other nerve cells are interposed between the sensory and motor neurons (Fig. 12.1E). These are called *association neurons.* Through them, a single impulse may be passed on selectively to one or more effectors, or alternatively impulses from several different receptors may be brought together and routed to one or several different effectors. Simple reflexes can still occur, but the possibility of greater complexity and flexibility of reaction is greatly enhanced.

EVOLUTION OF THE NERVOUS SYSTEM

We believe that the models of Fig. 12.1 essentially recapitulate the early evolution of the nervous system. We cannot state dogmatically how it evolved, for nervous tissue is never preserved in fossils and, furthermore, many steps in evolution are missing from the fossil record. Rather we judge the pathway of development from studies of extant lower animals. By comparing them with animals that have changed extensively, we reach a tentative formulation of the course of evolution.

The early unicellular animals, or Protozoa, displayed primitive irritability. When evolution gave rise to multicellular animals, or Metazoa, cells adapted specifically for intercellular conduction appeared for the first time. Sponges were perhaps the only major group of multicellular animals lacking them, each body cell continuing to transmit impulses within itself. The special cells were responsible for the irritability of the organisms at large. Except for the new phenomenon of cell acting upon cell, nothing fundamentally new was involved in this forward step.

Sea anemones, for example, developed receptor-effector systems similar to those in Fig. 12.1A,B. In these and other coelenterates, nerves joined into diffuse *nerve nets.* A local stimulus anywhere eventually spread throughout the net, moving in either direction. Because of the numerous intersections, a stimulus had to be strong and prolonged to spread widely. Hence conduction was slow. The chief function of such a nervous system was to coordinate local stimuli and disseminate simple responses. Nerve nets, we shall find, are still concerned with rhythmic movements of certain smooth muscles (e.g., in the human intestinal wall).

In the evolution of the vertebrate nervous system, there arose an elaborate communication network, the *peripheral nervous system,* extending to all portions of the body, and a coordinating apparatus, the *central nervous system.* Lower animals such as the flatworms developed a central nervous system consisting of a nerve cord running the length of the body with a large nerve center, or *brain,* containing association neurons in the anterior end. Undoubtedly this was a precursor of the vertebrate brain and *spinal cord.*

The central nervous system is, in essence, a complex of association neurons resembling that in Fig. 12.1E. Lying between receptors and effectors, it evidently derived from the primitive nerve net. A related evolutionary trend was the appearance of distinctive *afferent* and *efferent* nerves in the peripheral nervous system. The former carry sensory impulses *to* the central nervous system; the latter carry motor impulses *from* the central nervous system.

As animals increased in complexity, they acquired *distance receptors*—eyes, nose, and ears—at the anterior ends of their bodies. In contemporary higher animals these sense organs are connected directly to the brain, but at one time their reflex centers in the enlarged anterior end of the spinal cord consisted of massive collections of cell bodies, or *ganglia.** Thus a special "eye brain," "nose brain," and "ear brain" supplemented the older ganglia representing the "visceral brain." Because of the linear arrangement and segmental distribution of the older, master ganglia, their part of the brain is generally designated as the "brain stem" (Fig. 12.2). Because they were well developed in the early fishes, it is also sometimes termed the "old brain."

As the early vertebrates began crawling and swimming, another ganglion arose to form the

---

* Most, though not all, nerve cell bodies occur in the clusters called ganglia.

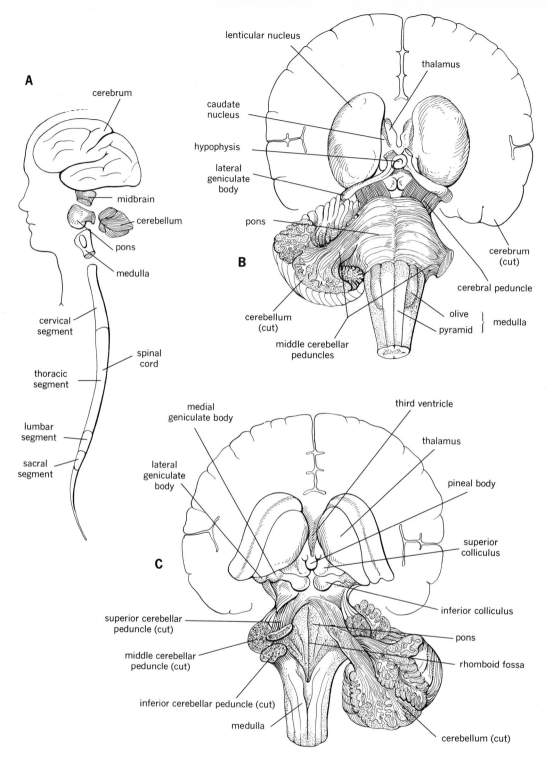

12.2 Central nervous system. A, relation of brain and spinal cord, shown with divisions artificially separated from one another; B, ventral aspect of brain stem and cerebellum; C, dorsal aspect of brain stem and cerebellum.

cerebellum, a part of the brain concerned chiefly with motor coordination and the maintenance of body position in space. The front part of the brain, the *cerebrum*, appeared very early, even in primitive fishes, as small swellings. Yet it is the portion of the brain that expanded most dramatically in evolution (Fig. 12.3).* It is concerned with higher functions such as mental activity and memory.

In all this multiplying of nerve cells and connections, no new functional features were added to the basic devices that went back to the dawn of evolution. Greater complexity resulted merely from the superimposition of new structures upon old: brain stem was added to spinal cord, and cerebellum and cerebrum to brain stem. The resulting hierarchical system operates as it does today because the new was always capable of dominating the old.

The evolution of the nervous system bought new flexibility and adaptability into the old stimulus-response pattern. And, as the number of cells making up the organism grew, so did the number of environmental stresses and stimuli capable of disturbing the organism's equanimity.

Neurons and Neuronal Circuits

EXCITATION AND SYNAPTIC TRANSMISSION

For all its complications and functional variations (including sensation, memory, and consciousness), the nervous system is made up of cells—neurons—all of which perform in essentially the same manner. Neuron function consists in the transmission of impulses from one

---

* The progressive enlargement of the central nervous system in the anterior end of the body is called *encephalization* (the development of a brain in the head). The brain itself is sometimes called the *encephalon*.

12.3 Evolution of the cerebrum. (Adapted from A. S. Romer, *The Vertebrate Body*, 2nd ed., W. B. Saunders Co., Philadelphia, 1955.)

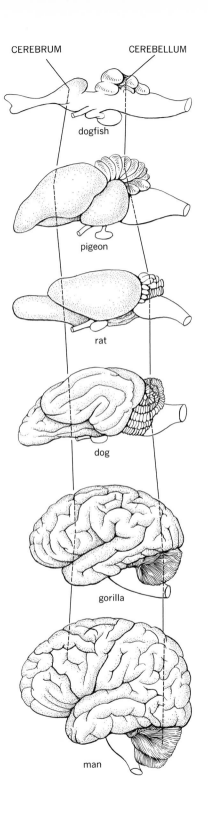

CEREBRUM   CEREBELLUM

dogfish

pigeon

rat

dog

gorilla

man

end of the cell to the other and from one cell to another; the impulses are electrical disturbances, fundamentally similar in all nerves.

It is not the separate neurons themselves but their varying arrangements, connections, and effectors that determine the qualitative differences in their actions. Thus, though we may consider neurophysiology at two levels—that of the individual neuron and its synapses and that of the entire nervous system—we could not possibly deduce knowledge of the latter from knowledge of the former unless the anatomical distribution of the cells was fully specified. Even with this information, we could not always predict the behavior of the whole from the behavior of its parts.

FINE STRUCTURE

Introduction. We recall from Chapter 5 that a neuron contains a *cell body,* a single long *axon,* and numerous short *dendrites.* Generally the dendrites bring impulses in on one side of the cell body, and the axon carries them away on the other side toward a synapse or an effector. The neurons of higher animals can conduct impulses in either direction, but synapses can conduct them in only one direction. Hence, neurons in a chain have a functional polarity.

The axolemma, myelin, and neurolemma sheaths enclosing an axon (see p. 185) are not essential to impulse conduction, which proceeds in an unsheathed axon as well. Before dealing with the nature of a nerve impulse, we shall briefly review certain aspects of the fine structure of a neuron.

Schwann cells and the myelin sheath. Fig. 5.19 revealed curious cells lying at intervals along the axon sheath. They are named for their discoverer, the same Theodor Schwann who with Schleiden was cofounder of the cell theory (see p. 67). The cytoplasm of the Schwann cells enfolds both myelinated and unmyelinated axons, although each myelinated axon has its own Schwann cell (Fig. 12.4A), whereas groups of unmyelinated axons share the same cell (Fig. 12.4B).

It has been established by tissue culture experiments that Schwann cell activity is necessary

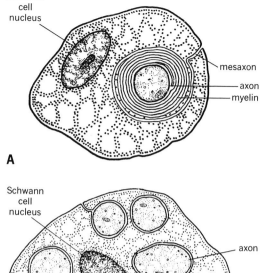

12.4 Relationships of Schwann cells to myelinated and unmyelinated nerve fibers. A, axon within its own Schwann cell; B, several axons within a single Schwann cell. (From G. M. Wyburn, *The Nervous System,* Academic Press, New York, 1960.)

for the deposition of myelin around the axon. Electron micrographs reveal that myelin exists in concentric layers. The question of how a Schwann cell produces these layers was recently given an ingenious answer. As shown in Fig. 12.5, the Schwann cell envelops the growing axon and, by continuously infolding its cell membrane, forms the layered structure. The connecting structure along the root of invagination, consisting of two fused portions of Schwann cell membrane, is called the *mesaxon.*

Although the physiological significance of Schwann cells and myelin is not yet known, one of their functions has been discovered. When a nerve of the peripheral nervous system is cut, all of the fibers distal to the cut (i.e., detached from the parent cell bodies) degenerate.* If the two cut ends are placed close together, the fibers that are still connected to their cell bodies grow into the detached fibers and replace the degenerating ones. During the degeneration of fibers and their myelin sheaths, Schwann cells multiply as if to form cellular tubes through which regenerating fibers may extend at the rate of a few millimeters per day.† Interestingly, regeneration does not take place when a nerve of the central nervous system is cut.

NERVE IMPULSES

Propagation and transmission. The mechanism by which signals are rapidly transmitted over long distances in the body has puzzled biologists for over a hundred years. Since the nineteenth century it has been known that nerves not only are excited by external electric currents; they also produce electricity in the course of their activity. Investigation of the nature of this activity has been marked by unusual and instructive difficulty. Until recently a nerve impulse could be studied only by observation of its consequence, a response in an effector. The main obstacles to more direct investigation were (1) the minute size of an individual nerve fiber, (2) the brief duration of electrical activity during impulse transmission (no more than a few thousandths of a second), and (3) a lack of promising ideas on how an electrical impulse could be generated by biochemical means.

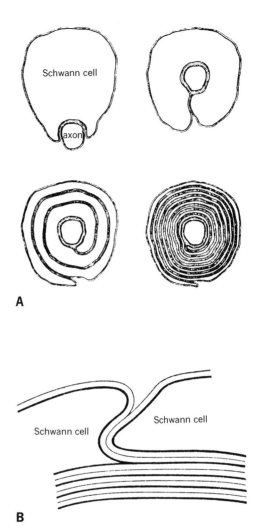

12.5 Membrane theory of myelin origin. A, four stages in myelin formation; B, infolding of the Schwann cell membrane to form layers. (From F. O. Schmitt, "Molecular Organization of Nerve Fiber," in L. J. Oncley, ed. *Biological Science,* John Wiley & Sons, Inc., New York, 1959.)

---

* It is important to draw a distinction between a *nerve* (which is a grossly visible structure) and a *nerve fiber* (which is an axon of a single neuron visible only under a microscope). A nerve is a large bundle of fibers of as many different neurons, each of which has its own cell body. As already noted, most neuron cell bodies are in ganglia near the brain or spinal cord. Some individual nerve fibers run through nerves for distances up to several feet (e.g., from the mid-back to the foot). Since it is extremely difficult in the laboratory to follow a single fiber through its entire course, early investigators could not bring themselves to believe that a fiber can actually be part of a cell whose nucleus is several feet away.

† The mass of protoplasm in a long fiber may be thousands of times that in a cell body. To accomplish its huge task of biosynthesis, the neuron is well equipped with endoplasmic reticulum and a rich supply of ribosomes (see pp. 73, 185). This apparatus is strikingly similar to that in highly active secretory gland cells.

The essential character of a nerve impulse is already familiar to us. We have traced the movement of a zone of depolarization through an individual cell (Fig. 4.28) and through the cardiac conduction system (Fig. 9.19B). The electrical events at the cell surface in both instances are identical to those concerned with the transmission of a nerve impulse. The impulse is a zone of change of electrical charges in the fiber that moves with small purely local currents from the fiber through the fluids immediately around it.

Systematic study of the electrical basis of nerve impulses dates back to the beginning of the twentieth century. Despite the relative insensitivity of their instruments, workers of that day demonstrated a potential difference across the membrane of a resting neuron, the inside being more negatively charged than the outside. Impulse propagation was associated with a wave of negativity along the external membrane surface. In 1902 the German physiologist Julius Bernstein theorized that, if the membrane of the resting neuron were selectively permeable to K^+ alone and impermeable to Na^+, Cl^-, and other ions, the interior would be electrically negative relative to the exterior because K^+ concentrated inside the cell would diffuse outward and no Na^+ or other positive ions would enter to correct the resulting deficit of positive charges. The internal negativity would tend to prevent an excessive outflow of K^+. Bernstein's theory offered an attractive explanation for the differing ionic patterns as well as for the potential difference across the membrane of the resting neuron.

Bernstein went one step further and proposed that membrane permeability suddenly increases during the passage of a nerve impulse so that Na^+ rushes into the fiber from the extracellular fluid, the potential difference drops to zero, and electric currents are generated that depolarize the adjacent membrane. Hence the impulse migrates.

Regardless of the meagerness of supporting evidence, Bernstein's theory was accepted for many years on the basis of its plausibility. Not until 1938 was it technically possible to test the theory experimentally. Then K. S. Cole and H. J. Curtis mounted one of the easily isolated giant axons of a squid as one arm in a Wheatstone bridge, a standard device for determining the electrical properties of conductors, and found (by measuring resistance and impedance across the membrane) that membrane permeability does indeed increase during impulse transmission.

In 1939 an even more critical test of the Bernstein theory yielded a disconcerting fact. Cole, Curtis, A. L. Hodgkin, and A. F. Huxley inserted a micropipet into an axon (Fig. 12.6)

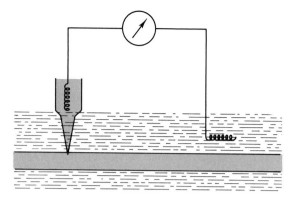

12.6 Method of measuring the internal potential of a nerve fiber. A micropipet only 0.0005 mm across at the tip is thrust through the cell membrane.

and to their surprise discovered that the potential difference did not fall to zero during impulse transmission but reversed. For a brief moment the interior was positive in relation to the exterior. This situation negated the view that membrane permeability simply increases during impulse transmission, so that the membrane is equally permeable to all ions. If all ions traversed it freely, charges on the two sides should become equalized, and the potential difference should become zero.

The dilemma may have been resolved by the later work of Hodgkin and Bernard Katz, who concluded that the membrane does not fully lose its ionic selectivity during impulse transmission, as Bernstein supposed, but rather becomes highly and specifically permeable to Na^+ for about half a millisecond. Hence Na^+ rushes inward, creating a temporary relative positivity within.*

The next task of the physiologist was to find the means by which Na^+ is extruded when the impulse has passed—the process that moves Na^+ ions "uphill" and outward as fast as they pour inward, impelled by the "downhill" concentration gradient. At this point the sodium pump hypothesis was introduced (see p. 160, p. 209). The pump is an ATP-dependent mecha-

---

*We should note the similarity of these events to the ionic fluxes accompanying the activity of the cardiac pacemaker (see Chapter 9).

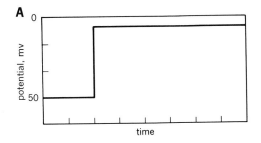

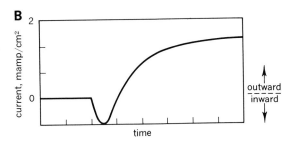

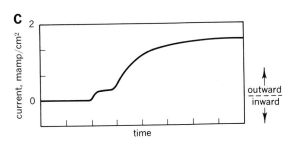

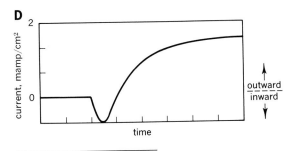

12.7 Records of a voltage-clamp experiment. A, change in internal potential; B, axon in salt water; C, axon in sodium-free medium; D, axon returned to salt water.

nism in the cell membrane that drives sodium from the fiber in exchange for potassium and thereby restores the original potential difference.

Rigorous testing of some of these proposals was made possible by two important technical advances: (1) the development of the so-called *voltage-clamp* technique; and (2) the introduction into experimental biology of radioactive sodium and potassium. The voltage-clamp is an arrangement that records potential differences across a membrane while maintaining the internal potential at any desired level (that is, it "clamps" the internal voltage). Current flowing across the membrane can be measured the instant after the operator has established a new internal potential. Typical results of a voltage-clamp experiment (Fig. 12.7) show that abrupt alteration of the internal potential of an axon in salt water produces first a brief inward current and then a prolonged outward current. In sodium-free fluid the initial inward current is lacking, an indication that it is initiated by Na$^+$. Proof that the prolonged outward current is induced by K$^+$ was obtained with radioactive potassium. A summary of the reactions associated with impulse transmission appears in Fig. 12.8, which contrasts the *resting potential* with the *action potential* at the locus of the impulse.*

At present we understand little of the molecular mechanisms underlying the permeability changes in the nerve membrane. Presumably charged groups of unknown type within the membrane structure are rearranged.† Certainly no known chemical discriminates between sodium and potassium as effectively as the membrane. We can infer, however, that conduction mechanisms in vertebrate and invertebrate nerves are fundamentally similar and that conduction mechanisms in vertebrate nerves are

---

* An action potential may arise from electrical, chemical, or mechanical stimulation.

† There is general agreement that excitable membranes consist of a bimolecular layer of lipid and phospholipid coated with protein on both sides. The protein seems less important for electrical activity than the lipid-phospholipid layer. According to one theory, calcium combines with membrane elements at specific sites and thereby controls the availability of ion-passing channels. Some of these are permeable to sodium when they are not occupied by calcium. However, they may change their configuration owing to collisions and become permeable to potassium.

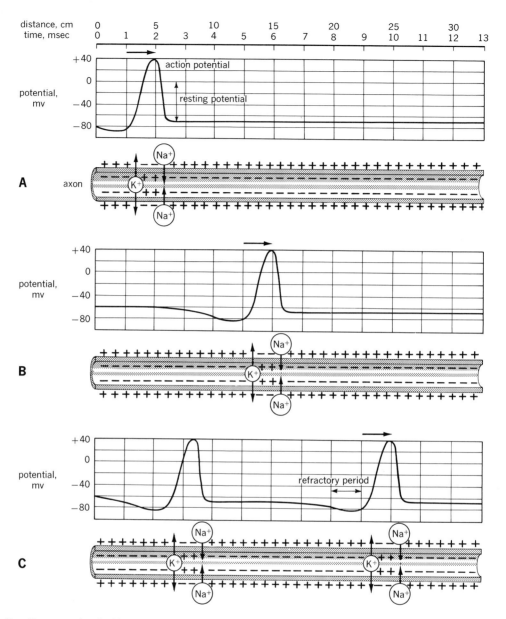

12.8 Reactions associated with nerve impulse transmission. A, B. When a nerve impulse arises, a "gate" opens and lets Na⁺ from the fluid outside pour into the axon in advance of the impulse, making the axon interior locally positive; in the wake of the impulse, the sodium gate closes, and a potassium gate opens, which lets K⁺ flow out, restoring the normal negative potential. C. As the impulse moves along the axon, it leaves the axon in a refractory state briefly, after which a second impulse can follow.

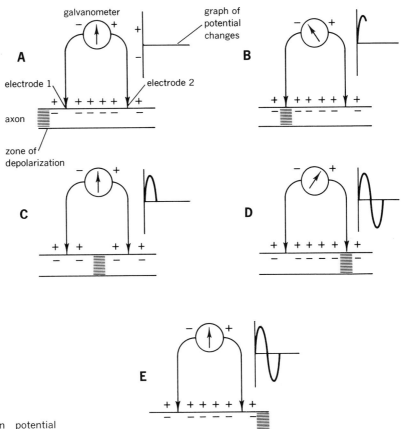

12.9 Changes in the action potential during impulse transmission.

similar to those in other kinds of excitable vertebrate tissue (e.g., the cardiac pacemaker).

Functional characteristics. Certain characteristics of the action potential can be demonstrated with an experiment similar to that performed on the heart during impulse transmission (see p. 307). When we place the two electrodes of a galvanometer on the surface of an intact resting nerve fiber (Fig. 12.9), we observe a potential difference of zero. As an impulse passes by, we see in sequence (1) a negative deflection of the galvanometer needle when the impulse reaches electrode 1, because the potential there is negative relative to that at electrode 2; (2) electrical neutrality while the impulse is between the electrodes; (3) a positive deflection of the needle when the impulse reaches electrode 2; and (4) a return to resting position. This diphasic series of potential changes clearly indicates that ordinary nerve impulse transmission is associated with local depolarization.

Interesting and perhaps predictable events occur when electrodes are applied to the fiber surface for the purpose of introducing electric currents. Currents too weak for stimulation make the exterior in the region of the cathode (negative electrode) more negative and thus more like the interior, so that it is abnormally sensitive to subsequent stimulation. Conversely, they artificially enhance the potential difference across the membrane in the region of the anode (positive electrode), so that it requires an abnormally large stimulus for depolarization. These consequences of the application of constant subthreshold currents are called *electrotonus*. The cathodal effect, increasing the sensitivity of the membrane, is *catelectrotonus;* the anodal effect, decreasing its sensitivity, is *anelectrotonus.*

The speed of an impulse in nervous tissue is similar to that in cardiac tissue (see Fig. 9.20). It is considerably slower than the speed of electricity in a wire because of time-consuming membrane changes and ionic migrations and is different in different nerves and different species. In some human fibers it is about 100 m per second (220 miles per hour), and in certain invertebrate fibers it is only 5 cm per second (about 1/10 mile per hour).

Another parallel between cardiac conduction and nervous conduction exists in the all-or-none phenomenon (see p. 304). The strength of a nerve impulse does not vary as does that of electricity in a wire. An impulse starts and remains at full strength. Its passage resembles the ignition of a sprinkled line of gunpowder more than the activation of an electric wire. When the powder is ignited at one end of the line, a flash travels to the other end with a strength independent of the heat of the igniting match and of path length. The strength of a nerve impulse likewise depends solely upon energy generated by the nerve itself and on local events.

A stimulus must be of sufficient intensity to create a zone of depolarization, but, once depolarization begins, it goes to completion, producing a full-sized action potential that propagates itself along the membrane surface. If the stimulus is too weak, it evokes only a local subliminal response, which quickly extinguishes itself.

Although an individual nerve impulse has a fixed strength, a given stimulus can produce effects of varying intensities. Most external stimuli (unless they are unusually brief) initiate many impulses in succession. Strengthening the stimulus, then, increases the frequency of impulses, so that more of them arrive at the effector per unit time, thus enhancing the response. The nerves therefore constitute an FM (frequency modulation) system rather than an AM (amplitude modulation) system.

The *refractory period,* as with the cardiac conduction system (Fig. 9.18), is the period immediately after stimulation or impulse transmission during which nervous tissue is first inexcitable—the *absolute* refractory period—

and then less excitable than normal—the *relative* refractory period.* Before returning to its resting state, the tissue often also passes through a *supernormal phase* of excitability. The refractory period is assumed to represent restitution of the normal Na^+ and K^+ orientation, and the supernormal phase to reflect a slight overshooting by the two species of ions as they rush back to their proper places (see Chapter 9).

Synapses. Synapses are so variable that we cannot speak of a typical one. Yet we can say that within every synapse is a discontinuity. No fibers cross the gap from one cell to another; hence, if a nerve impulse is to continue beyond the synapse, it must be regenerated afresh on the other side. This traditional statement has until recently been based upon evidence more physiological than anatomical. However, electron micrography has clearly confirmed it. Interestingly, pictures of synapses between axons and dendrites show previously unsuspected collections of mitochondria and curious vesicles within the ends of the axons. We shall refer to these later.

The fact that impulses are not conducted from one neuron to another through intervening fibers has important consequences. Synaptic transmission is therefore more complex than nerve transmission, and synapses have the capacity to alter impulses and to determine which ones get through and which do not. Isolated or infrequent impulses may arrive at a synapse and fail to cross it. Conversely, a suitable volley of impulses may arrive and effect a crossing. If the impulses arrive simultaneously in many presynaptic terminals, transmission is enhanced through *spatial summation.* If they arrive in rapid succession in a single presynaptic terminal, it is enhanced through *temporal summation.* If through summation a given volley crosses many synapses in a nerve network, the effector response may be strong and widespread, especially as association neurons become involved.

---

* The refractory period accounts for the fact that nerve fibers normally carry impulses in only one direction. Impulses traveling in opposite directions would stop and be extinguished at their meeting point, for each would have left in its wake a refractory region that could not be traversed by its advancing opponent.

Other features of synaptic transmission should be briefly noted. As we have learned, impulses are conducted through synapses only from presynaptic terminals to the successive postsynaptic neurons and never in the reverse direction, according to the *Bell-Magendie law*. More time is consumed in the transmission of an impulse across a synapse than can be accounted for by conduction alone. The extra time is called *synaptic delay*. When a presynaptic terminal is continually and repetitively stimulated, the number of impulses transmitted by the postsynaptic neuron progressively decreases, owing to *fatigue* of synaptic transmission. The opposite situation also occurs. When a series of impulses has crossed a synapse, subsequent impulses cross more readily. For example, although a long volley of closely spaced impulses may be required to cross a given synapse, a short volley or even a single impulse will follow it with ease. This effect is known as *facilitation*. If additional impulses are not forthcoming, the effect diminishes rapidly; if they continue to arrive, it remains operative and even increases. Continual crossing of a particular synapse at appropriate intervals can thus maintain its facilitated condition for a lifetime. There is evidence that certain established pathways in the nervous system are determined by facilitated synapses. Impulses move readily and rapidly along these routes because they have been traveled before. Clearly a mechanism like facilitation may underlie habit, training, memory, and learning (see p. 459).

Chemical transmission. Since the synaptic gap is a physical reality, we must inquire how nerve impulses jump across it.

In 1920 Otto Loewi, while studying the behavior of isolated frog hearts, passed saline solution through one beating heart into another. When he applied electrical stimulation to the vagus nerve of the first heart, the usual slowing of the heartbeat took place (see p. 320). But surprisingly the second heart, which had earlier been detached from its nerves, soon began beating more slowly, too. The salt water flowing from the first heart to the second apparently carried an inhibitory substance of some kind whenever the first heart's vagus nerve was stimulated. Loewi called the substance *Vagusstoff* and concluded that it was liberated by the vagus nerve

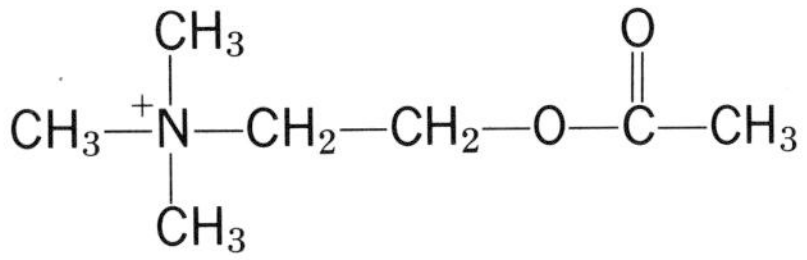

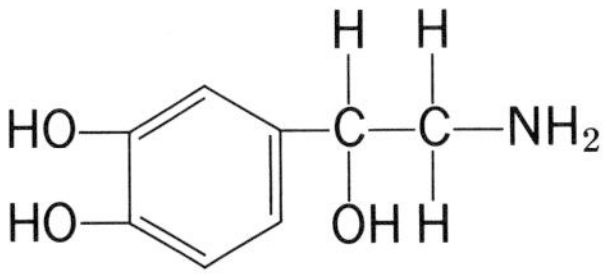

12.10 Chemical mediators of synaptic transmission.

endings in the heart wall. As a result of earlier pharmacological studies by Sir Henry Dale, *Vagusstoff* was soon shown to be *acetylcholine* (Fig. 12.10).

Presumably two basically different modes of synaptic transmission are possible: *electrical* and *chemical*. Although purely electrical transmission, in which electric current from one cell stimulates the next, has been demonstrated in lower animals, a large volume of work indicates that the electrical link is almost always broken at the synapse and impulses are transmitted by a locally secreted chemical substance. In many synapses (but not all) this substance is acetylcholine.

The specific chemical stimulant is stored (and probably synthesized) inside the presynaptic terminal in small *synaptic vesicles* (Fig. 12.11). Between the *synaptic knob* and the *subsynaptic membrane* of the postsynaptic neuron is a space of about 200 Å, the *synaptic cleft*, near which the vesicles are concentrated. It is believed that arriving impulses somehow cause them to discharge their contents into the cleft,* with the released substance attaching itself to special receptor sites on the subsynaptic membrane of

---

* Some authorities contend that one vesicle is ruptured for every impulse transmitted. This would require rapid replacement of the vesicles and hence their actual formation within the presynaptic terminal. Such a process could be dependent on enzymes and other materials (possibly messenger RNA) that are synthesized in the cell body and that flow down the axon.

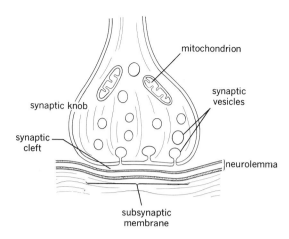

mitochondrion

synaptic vesicles

synaptic knob

synaptic cleft

neurolemma

subsynaptic membrane

12.11 Structure of the synapse.

the postsynaptic neuron.* This hypothesis is supported by the discovery that the transmitter is released in "packages" of a few thousand molecules.

The transmitter produces local depolarization in the postsynaptic neuron. In most cases this is inadequate to initiate a nerve impulse. Rather, it has only a local electrotonic effect. Through summation the electrotonic effects from the many dendrites of a single neuron together reach the depolarizing threshold and recreate a typical all-or-none impulse in the initial segment of the axon.

An enzyme within the synapse, *cholinesterase*, destroys acetylcholine during and after its action, thereby curbing and controlling its stimulatory effect. If cholinesterase or some similar device did not perform this function, a single impulse would be enormously amplified within a synapse, and the nervous system would lose the capacity to regulate. Just such nervous "storms" occur when large quantities of acetylcholine are injected into the blood or when certain cholinesterase-inhibiting chemicals are administered.

Nerve impulses are transmitted across some synapses by norepinephrine, a substance that is

also secreted (along with its close analogue epinephrine) by the adrenal medulla (see Fig. 12.11). The presynaptic terminal of such a synapse contains granules varying in diameter from 300 to 1000 Å. These store (and presumably synthesize) norepinephrine and are responsible for its release into the synaptic cleft, where, like acetylcholine, it depolarizes the postsynaptic neuron. In sum, most synaptic transmission is mediated by either acetylcholine or norepinephrine.† Therefore, we classify synapses as *cholinergic* and *adrenergic*.‡

Chemical synaptic transmission has several important implications. As we have learned, the synapses determine the direction of impulse transmission in neuronal circuits. Presumably they do so because only one-half of a synapse can synthesize acetylcholine (or norepinephrine) and only one-half can be stimulated by it. Presumably the energy-requiring processes of acetylcholine (or norepinephrine) synthesis, storage, and release account for the many mitochondria in the presynaptic terminals. These processes may be considered to constitute a special case of *neurosecretion*.

Excitatory and inhibitory postsynaptic potentials. Two types of synapses have been identified in the nervous system: *excitatory* and *inhibitory*. Although they are morphologically similar, only the latter type secretes an inhibitory transmitter. However, it inhibits only when the postsynaptic neuron is being excited by the excitatory transmitter.

An impulse crossing an excitatory synapse generates an *excitatory postsynaptic potential* (EPSP) in the postsynaptic neuron. An oscillo-

† This generalization derives mainly from studies of synapses outside the central nervous system. It is probable that yet another transmitter operates within it. A leading candidate for such a role is γ-*aminobutyric acid* (also called *gamma-aminobutyric acid* or *GABA*), a compound arising from glutamic acid. It now appears that γ-aminobutyric acid is a transmitter in crustaceans.

‡ The term "adrenergic" derives from "adrenalin," a frequently used synonym for "epinephrine." It is of interest that adrenergic synapses actually contain some acetylcholine. Some investigators have suggested therefore that in adrenergic synapses the arriving impulse causes a release of acetylcholine into the synaptic cleft, which then, through effects on calcium and the ionic equilibrium, causes release of norepinephrine.

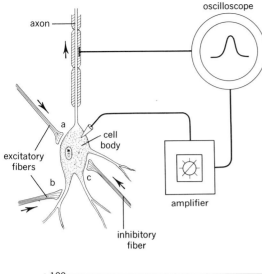

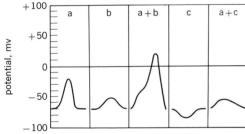

12.12 Response of a neuron to incoming impulses, as registered by an oscilloscope. (From R. Galambos, *Nerves and Muscles*, ©️ 1962 by Educational Services Inc. (Science Study Series). Reprinted by permission of Doubleday & Co., Inc.)

scope connected to an intracellular microelectrode indicates the EPSP produced. Fig. 12.12 shows a spinal cord neuron upon which three different presynaptic terminals (*a*, *b*, and *c*) converge. Let us consider that a volley of impulses from *a* provokes an EPSP that rapidly dies away and that one from *b* provokes an EPSP that is smaller and somewhat more prolonged. Neither alone causes complete depolarization in the postsynaptic neuron, but both together cause it to fire so that a "spike" appears in the oscilloscope, showing that a nerve impulse has been generated.

The significance of this cumulative effect is easily illustrated. Let us suppose, for example, that the postsynaptic neuron goes to a muscle fiber that helps pull the arm away from a hot object and that *a* comes from a skin receptor in the finger that sends an impulse toward the spinal cord when the finger touches heat. If *b* arises within the brain, it is active only when the subject is awake and alert. Ordinarily, then, a message from the heated skin causes arm withdrawal because impulses from *a* and *b* simultaneously provoke EPSP's. With the body anesthetized, there would be no impulse from *b*, and the EPSP from *a* alone would not activate the postsynaptic neuron.

The withdrawal of a finger from a hot object is a reflex response involving a chain of neurons (Fig. 12.13). This would be a reflex arc within the definition given on p. 411. Sensory nerves run from the finger to the spinal cord. Impulses aroused in the finger traverse them and are distributed to many spinal cord neurons, where they produce EPSP's. Depolarized under this

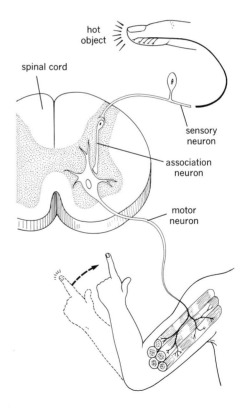

12.13 Impulse transmission involved in touching a hot object and jerking the arm away.

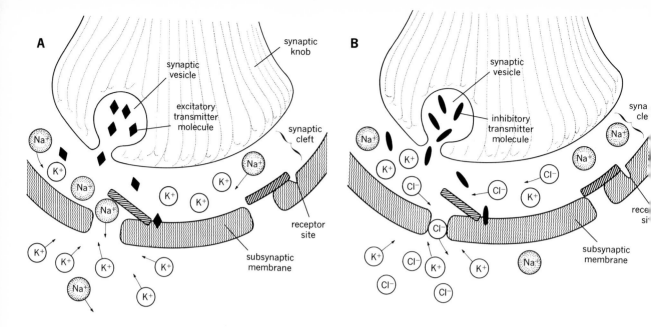

12.14 Effects of excitatory transmitter (A) and inhibitory transmitter (B) on the conductivity of a subsynaptic membrane.

barrage of EPSP's are the motor nerves. Their impulses pass outward to the arm muscle and cause it to jerk the arm away.

The complete reflex arc here includes an interposed association neuron—as does the model nervous system of Fig. 12.1E. Association neurons, unlike sensory and motor neurons, which exist partly inside and partly outside the spinal cord, start and end inside it. Their messages therefore originate only in prior neuron activity and result only in more neuron activity. As mediators between incoming and outgoing impulses, association neurons perform valuable modulating and regulating functions.

We mention association neurons at this time in order to bring out a particular aspect of postsynaptic potentials. Let us review again the painful experience of touching a hot object. It is not absolutely necessary that one jerk his arm away by reflex. He can know in advance that the object is hot, deliberately touch it, and keep his finger on it despite his discomfort. In other words, one can prevent or inhibit a reflex response if one desires. In view of what we have learned about what causes a reflex in the first place, we must now find an explanation for our ability to prevent or inhibit a reflex.

A relatively simple mechanism—the generation of an *inhibitory postsynaptic potential* (IPSP) at an inhibitory synapse—accounts for this phenomenon. When an inhibitory fiber (*c* in Fig. 12.13) delivers its impulse at the synapse, it creates in the postsynaptic neuron an event (IPSP) whose electrical sign is opposite to that of the event (EPSP) created by an excitatory fiber. IPSP's oppose EPSP's; when an IPSP occurs along with an EPSP, the two cancel out, and the postsynaptic neuron shows no response at all. Thus a reflex response, such as withdrawal from a painful stimulus, is inhibited when a counterbarrage of IPSP's descends upon the motor nerves and cancels out the EPSP's.

Investigators have begun to understand how EPSP's and IPSP's develop in the postsynaptic neuron. The ionic flow causing an EPSP is ascribed to a local increase in membrane conductivity due to the chemical transmitter (Fig. 12.14). The excitatory transmitter (e.g., acetylcholine) is believed to open large channels in the subsynaptic membrane that permit Na^+ to enter and K^+ to leave (see Fig. 12.8). The resulting EPSP is related to the potential difference between the ions.

The production of an IPSP is not so clear. For

example, the inhibitory transmitter has not been identified with certainty. There is evidence, however, that it opens channels in the subsynaptic membrane that permit Cl^- to enter and K^+ to leave but that are too small to let Na^+ enter. This might explain why an IPSP makes a postsynaptic neuron's internal voltage more negative than normally (see Fig. 12.12). Perhaps the response elicited depends upon the subsynaptic membrane rather than on the transmitter. If so, the specificity may reside in the molecules serving as receptors for the excitatory transmitter or in the channels permitting the passage of different ions into and out of the neuron.

Extremities of Neuronal Circuits

RECEPTOR ORGANS: BIOLOGICAL TRANSDUCERS

Varieties of receptors. The foregoing discussion dealt with the *conduction* of nerve impulses within and between neurons. We shall now consider the *initiation* of nerve impulses at the receptor end of the receptor-conductor-effector pathway.

TABLE 12.1 RECEPTORS

Classification according to stimulus		
Stimulus	*Name of receptor*	*Location of example*
Mechanical energy		
Touch	Mechanoreceptor	Skin (p. 427)
Pressure	Mechanoreceptor	Skin (p. 427)
Gravity, motion	Statoreceptor	Vestibular organ (p. 500)
Sound	Phonoreceptor	Organ of Corti (p. 498)
Blood pressure	Pressoreceptor	Carotid sinus (p. 320)
Heat		
Environmental	Thermoreceptor	Skin (p. 427)
Internal	Thermoreceptor	Hypothalamus (p. 463)
Light		
Refracted	Photoreceptor	Retina of eye (p. 483)
Chemical substances		
Volatile	Chemoreceptor (smell)	Olfactory epithelium (p. 505)
In solution (except O_2)	Chemoreceptor (taste)	Taste bud (p. 503)
O_2 in solution	Chemoreceptor	Carotid body (p. 402)

Classification according to source of stimulus		
*Source of stimulus**	*Name of receptors*	*Location of receptors*
Position of body in space	Proprioceptors	Muscles, tendons, joints
Immediate environment	Exteroceptors	Skin
Viscera	Interoceptors	Viscera
Events at a distance	Teleceptors	Eyes, ears, nose

*All sensory reception does not lead directly to conscious sensation. Thus proprioceptors, which initiate reflexes in response to changes in the positions of the limbs, the posture, etc., and interoceptors, which initiate reflexes in response to changes in blood pressure, blood gas content, etc., transmit information to levels of the central nervous system far below that of the conscious mind.

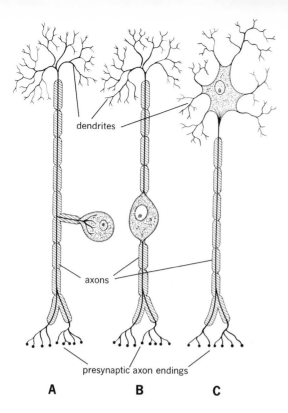

dendrites

axons

presynaptic axon endings

A　　　B　　　C

12.15 Typical neurons: A, monopolar sensory neuron; B, bipolar sensory neuron; C, motor neuron.

A receptor is a specialized neuron (or outgrowth of a neuron) that initiates nerve impulses within itself when stimulated by external physical and chemical events such as physical contact, mechanical pressure, light, heat, sound, and chemical substances.* The impulses are conducted off through the attached nerve fiber to a

synapse where they meet one of the succeeding neurons shown in Fig. 12.1.

Biologists have long wondered how an external physical or chemical event is converted into a physiological event *of proportional intensity* within the organism. The proportionality between the stimulus intensity and the frequency of the resulting nerve impulses means that receptors resemble the physicist's transducers in converting one form of energy into another.† Despite their sensitivity and efficiency, however, biological transducers have not yet been discovered for many stimuli to which physical transducers respond. Radio waves, for example, are not converted into electrical energy in known living organisms. Still the list of stimuli provoking responses in them grows. Recent additions to it include polarized light, wind, gyroscopic deflections, infrared radiation, and ultrasound.‡ As Table 12.1 shows, the human nervous system is entirely or relatively insensitive to many of these stimuli.

Principles of function. Current thought suggests that common mechanisms are at work in all receptors. We may well have found models in the dendrites of postsynaptic neurons. The commonest receptors in vertebrates are monopolar, though bipolar ones do occur. In Fig. 12.15 the dendrites through which impulses enter the cell bodies of motor or association neurons are compared with those of sensory neurons. In each type of cell, messages are ultimately transmitted through the axon as all-or-none impulses at frequencies varying from 4 or 5 per second to perhaps 500 per second. A striking difference between the dendrites and axon of a motor or association neuron is that the dendrites do not respond to incoming impulses in an all-or-none

---

* From antiquity there has existed a popular misconception that man has but five senses: sight, hearing, smell, taste, and touch. It is true that the receptors of four of these are localized in the head and that we consider them separately as "the special senses." But, curiously, we have never included among them another sense organ in the head, the organ of equilibrium in the inner ear. Innumerable small, anatomically simple receptors are scattered throughout the body. They vary and intergrade so much in structure, in stimulus, and in sensation that it is difficult to say how many senses they represent. Undoubtedly many receptors are adapted to receive specific stimuli. This is easily demonstrated with a pin and small hot and cold rods. Yet it appears that certain sensations arise not because specific receptors are activated but because different stimuli affect the same fibers in different ways. Moreover, some sensations apparently arise in more than one type of receptor. For example, the sensation of wetness may result from stimulation of a complex combination of receptors.

---

† In physical transducers, too, output energy is proportional to input energy. Physical transducers are essential in most communication or signaling systems. Those whose output is in the form of electrical energy usually feed into wires ("nerves"), which lead to recording or controlling devices that operate on feedback principles. For example, transducers in the telephone convert sound waves to electrical energy and vice versa.

‡ Responses not shared by man include olfactory distinction of isomers by insects, "radar"-directed movement by bats, auditory amplitude modulation frequency analysis by crickets, and orientation to turbulence and the movement of other organisms in water by fishes.

manner but rather with graded electrotonic effects that spread and summate to produce all-or-none impulses in the initial segment of the axon.

In 1950 Katz discovered that the physical stretching of a receptor in striated muscle, a *neuromuscular spindle*, generates a local electric current (in less than 0.001 second). When the current reaches a certain intensity, it triggers the firing of an impulse to the central nervous system; it itself does not travel along the nerve fiber. Accordingly, it was given the name *generator current*. Significantly, the intensity of the generator current varies in proportion to the intensity of the stimulus.

The receptor in Katz' experiment was a *mechanoreceptor* (see Table 12.1). It happens that much of our knowledge of receptor action was acquired in experiments on mechanoreceptors, particularly those of the skin. The known mechanoreceptors of skin are shown in Fig. 12.16: the *Pacinian corpuscle, Meissner's corpuscle, Golgi-Mazzoni corpuscle,* and *tactile disc of*

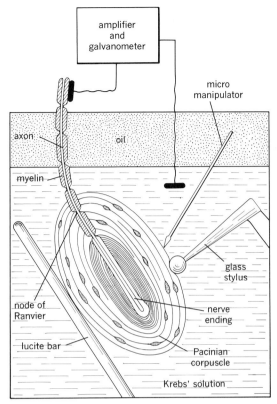

12.17 Mode of stimulation of an isolated Pacinian corpuscle. A rod attached to a vibrating phonograph crystal induces an impulse, which is picked up by a pair of electrodes. (Adapted from W. R. Loewenstein and R. Rathkamp, *J. Gen. Physiol., 41,* 1245, 1958, by permission of the authors.)

Merkel. The best studied of these is the Pacinian corpuscle, known for its large size and onionlike architecture. Many concentric lamellae enclose the nerve fiber ending.

Experiments with isolated Pacinian corpuscles (Fig. 12.17) have revealed that 99.9% of the lamellae can be peeled away without impairing transducer function (Fig. 12.18). Short of injuring the nerve fiber ending, the only maneuver that prevents a mechanically induced generator current from initiating an impulse in a corpuscle is placing a block at the first *node of Ranvier* (see Fig. 5.21). This circumstance suggests that the first node of Ranvier corresponds functionally to the initial segment of the axon of a motor neuron.

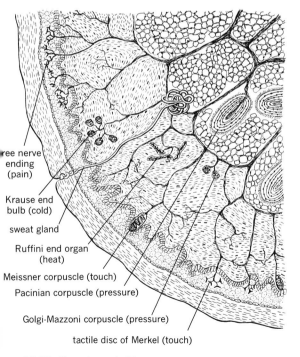

12.16 Receptors of skin.

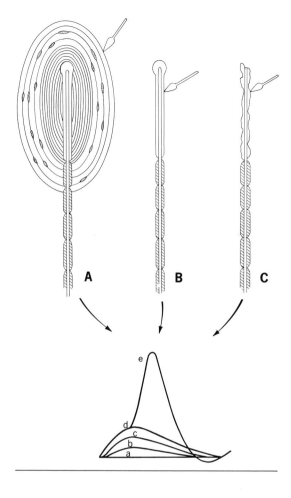

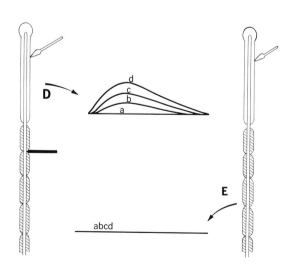

Thus sensory impulse initiation is a two-step process. Although there is the usual potential difference across the membrane of a corpuscle, this membrane differs from those of other neurons in its ability to produce a generator current after mechanical deformation. The current flows to the first node of Ranvier, where it triggers all-or-none depolarization and the propagation of nerve impulses at a frequency proportional to the intensity of the generator current. It is not known exactly why the intensity of the generator current is proportional to that of the stimulus, though recent evidence suggests that many local all-or-none generator currents may summate to produce a single large generator current. If local all-or-none generator currents summate (instead of local electrotonic effects as in a postsynaptic neuron), we might picture the corpuscle membrane as containing many tiny holes, too small in the resting state for ions to pass. If a mechanical deformation somehow opened (excited) a certain number of holes, the resulting ion flow would create a generator current whose intensity would rise as the stimulus increased in strength, opening more and more holes. Some authorities, however, believe that the summations occurring in receptors and in postsynaptic neurons are precisely analogous. Further research is needed to clarify this issue.

Still another factor ensures that the intensities of stimulus and sensation are proportional. Skin contains many corpuscles in close proximity. A weight pressing on the skin stimulates them—the greater the weight, the larger the number—to dispatch impulses to higher nerve centers. Moreover, since several receptors are generally the twigs of a single nerve fiber, the impulses converge. When a strong stimulus increases the

12.18 Effects of stimulation of a Pacinian corpuscle. Stimulation of the corpuscle when intact (A), with the outer layers removed (B), or after partial destruction of the core sheath (C) produces the same response: a, b, c, d, progressively stronger generator currents with stronger subthreshold stimuli; e, an all-or-none impulse with a threshold stimulus. A block at the first mode of Ranvier (D) prevents initiation of an all-or-none impulse, and damage to the axon ending (E) prevents any response at all. (Adapted from W. R. Loewenstein and R. Rathkamp, *J. Gen. Physiol.*, *41*, 1245, 1958, by permission of the authors.)

number of activated receptors, the frequency of the impulses correspondingly increases.

The proportionality between the intensity of the stimulus and the intensity of the generator current has been demonstrated in many receptors besides mechanoreceptors and has been postulated for the rest. In some cases the initial response may take place in a specialized cell other than the receptor, with a series of generator currents triggering it. Later we shall point out the critical events in other types of receptors believed to correspond to mechanical deformation in mechanoreceptors.

Pain. For many reasons pain is an unusually complicated problem for physiologists. We are hampered, to begin with, by its emotional components: it is unpleasant and a warning of existing or impending trouble. Powerful emotion or mental effort may diminish it (as in athletes in violent action or some women in childbirth) or enhance it (as in fearful children or the seriously ill). *Superficial* pain (arising in the skin) may or may not have features in common with *deep* pain (arising in the muscles, tendons, and joints) and *visceral* pain (arising in the internal organs). It is often more severe because the skin is more richly supplied with receptors than the deeper tissues. Indeed, many deep tissues can be cut, crushed, or burned with no sensation whatever. Certain injuries are painless; the impact of a bullet, for example, may temporarily cause no sensation. On the other hand, sustained muscle contraction and visceral distention may be acutely painful. Finally, testing for pain is notably difficult, since tests depend upon subjective responses.

For many years the nature of pain has been debated by advocates of the *specificity theory*, which holds that pain is a specific sense like vision or hearing,* and advocates of the *pattern theory*, which holds that the nerve impulse pattern for pain arises through intense stimulation of nonspecific receptors. It was once believed that the fine bare-ended nerves (see Fig. 12.13) are responsible for pain and that other sensations

---

* Portions of the evidence for this view follow: (1) pain alone is evoked by stimulation of certain skin areas; (2) other skin areas are pain-free; (3) transection of certain spinal cord tracts suppresses pain without affecting other sensations; (4) certain analgesics suppress pain without affecting other sensations.

depend on nerves with encapsulated endings. Undeniably the distribution of naked nerve endings correlates roughly with the distribution of pain sensitivity (skin areas lacking them are insensitive to pain). However, it is now known that the bare-ended nerves respond to a great variety of stimuli, some being highly selective and others less so.

In one recent attempt to reconcile these facts and theories, it was noted that the spinal cord ordinarily receives a continual bombardment of incoming nervous impulses—even in the absence of obvious stimulation—and that it somehow monitors this input, acting as a gate controlling what is passed to the brain via special pathways. A painful stimulus is thought to produce a barrage of impulses that is spatially and temporally summated in the monitoring cells in the cord. When a critical threshold is reached, pain results. According to this concept, the threshold can be raised and lowered by the central nervous system. This would account for the ability of emotion, diversion, and memories of prior experience to modify pain. Thus the sensation of pain may depend upon the central classification, selection, and interpretation of diverse patterns of incoming nerve impulses from many types of receptors.

CONDUCTOR-EFFECTOR JUNCTION

Structure. We have now considered the initiation of nerve impulses in receptors and their conduction through neurons and synapses. From the models of Fig. 12.1D,E, we conclude that two areas remain for discussion. One is the junction of conductor and effector, the *neuroeffector junction;* and the other is the association neuron.

The junction between a motor neuron and striated muscle, called a *motor endplate,* is the best-studied of the neuroeffector junctions (Fig. 12.19), and we regard it as a model of such structures. Unlike junctions associated with glands and smooth muscle, it is notable for the rapidity with which impulses cross it, to cause muscular contraction. Both structurally and functionally, it has much in common with an ordinary synapse.

The axon of the motor neuron, which has no myelin sheath near its end, enters the muscle fiber, expanding and ramifying beneath the sar-

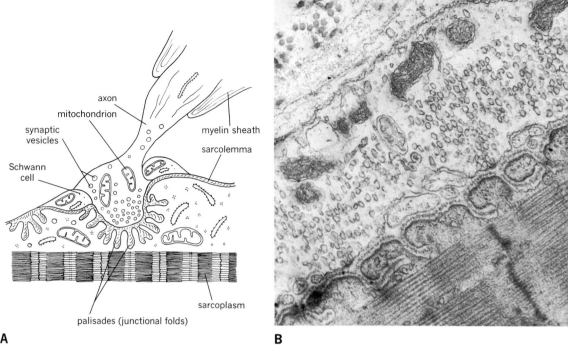

axon
mitochondrion
synaptic
vesicles
myelin sheath
sarcolemma
Schwann
cell
sarcoplasm
palisades (junctional folds)

A

B

12.19   Motor endplate. A, diagram; B, electron micrograph (× 53,000). (B from Dr. Bernard Katz.)

colemma over a disclike endplate of muscle substance containing many muscle cell nuclei. The
Schwann cell membrane continues over the endplate and keeps open a narrow space between
the axon ending and the muscle. This open zone
has a characteristic striated border on the muscle
side, the *palisades*, or *junctional folds*, that electron microscopy has shown to be merely folds of
muscle membrane in cross section. As in an ordinary synapse, the end of the axon contains many
small vesicles and mitochondria.

Junctional impulse transmission. What was
said about cholinergic synaptic transmission is
generally true of impulse transmission across a
neuromuscular junction.* Impulses arrive at the
axon ending; acetylcholine is discharged in the
space between it and the endplate—probably
from the vesicles in the axon ending; and the
acetylcholine attaches itself to receptor molecules in the endplate membrane and initiates depolarization. The action potential thus produced

---
*Neuromuscular junctions are cholinergic; their chemical
transmitter is acetylcholine. Examples of adrenergic neuroeffector junctions will be described later.

inside the muscle fiber causes it to contract.

A profound question remains for future research. How does the metabolic machinery of
the cell maintain and swiftly repair the unstable
membrane structures responsible for ion distributions in the resting state and ion movements
in the active state? We must anticipate that one
day we shall find a common basis for metabolic
and electrical events in structural and organizational changes at the molecular level as the cell
goes from rest to activity and back again.

Spinal Cord
and Spinal Nerves

STRUCTURE

Embryological development. The central nervous
system begins as a tube and remains a tube throughout
life. The first sign of it in an embryo is a groove, the
neural groove, along the mid-line of the back (Fig.
12.20). By a process of folding, so typical of early

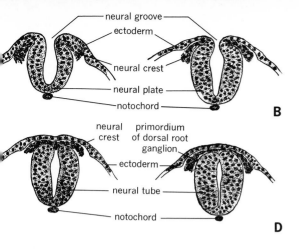

12.20 Closure of the neural tube and formation of the neural crests in the pig: A, at the 8-somite stage; B, at the 10-somite stage; C, at the 11-somite stage; D, at the 13-somite stage. (From B. M. Patten, *Human Embryology*, 3rd ed., © 1968. McGraw-Hill Inc. Used by permission of the McGraw-Hill Book Co., New York.)

embryogenesis, the groove deepens, and the edges thicken into *neural folds*, which fuse together to form a long *neural tube*. Undifferentiated ectoderm that was continuous with the neural folds separates from them and becomes an external covering over the submerged tube. The cephalic end of the tube enlarges into the primitive *brain vesicle*, whose further development we shall describe later, and the rest of the tube evolves into the spinal cord.

As the neural tube forms, a column of cells appears on each side of it in the angle between the neural fold and the ectoderm. When the tube closes, these *neural crests* remain, at first continuous along the length of the tube but eventually breaking into clumps, one pair for each body segment. They ultimately become the *dorsal root ganglia* of the *spinal nerves*.

Thirty-one pairs of spinal nerves emerge segmentally from the tube, which lies within a bony *vertebral canal* made up of 33 vertebra. Until the third month of embryonic life, the tube extends the length of the canal. Then the canal elongates rapidly, so that the nerves from the lower tube must travel some distance before making their exits. This bundle of nerves is named for an object it closely resembles, *cauda equina* (horsetail).

The simple columnar epithelium lining the neural groove is indistinguishable from the adjoining ectoderm. But, with closure of the neural tube, it proliferates into three distinct layers (Fig. 12.21): a thin outer *marginal layer*, which becomes the *white matter* of the spinal cord; a thick middle *mantle layer*, which becomes the *gray matter* of the cord; and a thin inner *ependymal layer*, which lines the small central canal of the cord (Fig. 12.22). *Ependymal cells* have only a supporting and limiting function; they do not transmit nerve impulses.

Adult spinal cord. The adult spinal cord extends from the base of the skull to the lumbar region. It is about 18 in. long and ½ in. wide, with enlargements in the cervi-

12.21 Five stages in the development of the spinal cord in human embryos: A, 10 mm; B, 15.5 mm; C, 30 mm; D, 45 mm; E, 80 mm. (Schematized from Streeter, in Kiebel and Mall, vol. II, J. P. Lippincott Company, Philadelphia, 1912.)

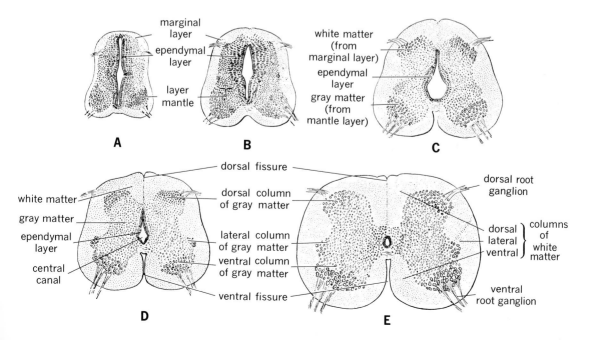

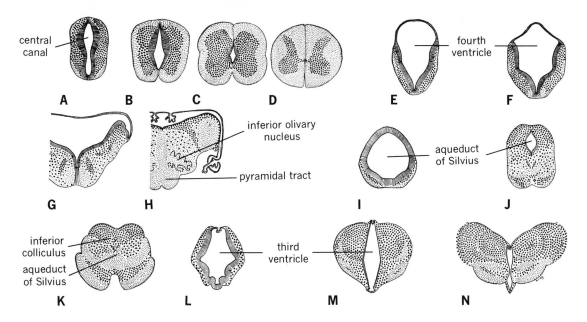

central canal

fourth ventricle

A   B   C   D   E   F

inferior olivary nucleus

pyramidal tract

aqueduct of Silvius

G   H   I   J

inferior colliculus
aqueduct of Silvius

third ventricle

K   L   M   N

12.22 Cross section of different levels of the central nervous system (arbitrarily enlarged to comparable sizes): A–D, spinal cord; E–H, medulla and pons; I–K, midbrain; L–N, thalamus. Striations represent the ependymal layer; large dots the gray matter; and small dots the white matter. (From W. Krieg, *Functional Neuroanatomy*, 2nd ed., © 1953 by McGraw-Hill, Inc. Used by permission of the McGraw-Hill Book Co., New York.)

cal and lumbar regions, which supply nerves to the arms and legs (Fig. 12.23).

Slung within the vertebral canal (Fig. 12.24), the cord is encased in three membranes, the *meninges:* a thick outer *dura mater;* a sheath lining the dura mater whose spiderweb thinness accounts for its name, the *arachnoid;* and a thin sheath lying directly over the cord and its emerging nerves, the *pia mater.* The meninges are continuous with those covering the brain. The dura mater and the adherent arachnoid are attached by loose connective tissue to the periosteum of the vertebral canal. The wide space between the arachnoid and the pia mater, the *subarachnoid space,* contains *cerebrospinal fluid.* Since the dura mater and arachnoid continue well below the end of the spinal cord, so does a sizeable section of subarachnoid space.* Thus the cord is protected by the bony walls of the vertebral canal and a shock-absorbing fluid.

The essential nature of the spinal cord is shown in the middle portion of Fig. 12.1D. The cord receives sensory fibers and sends out motor fibers. Within its

substance are (1) *cell bodies* of the motor neurons; (2) *synapses;* and (3) *ascending* and *descending* fibers providing connections between the cord and higher centers and between various levels of the cord (Fig. 12.25). The cord substance is divided into right and left halves by the *ventral,* or *anterior, fissure* and the *dorsal,* or *posterior, fissure.*† The striking dark central area in the form of a letter H is gray matter, consisting of cell bodies, their dendrites, unmyelinated initial segments of axons, and a supporting network of nonneural connective tissue cells called *neuroglia.* Around the gray core is white matter, consisting exclusively of longitudinal columns of the myelinated segments of axons in a framework of neuroglia. The white color is due to the myelin.

The two parallel limbs of the H are the *ventral* and *dorsal columns* of the gray matter.‡ The small protuber-

---

* It is into this cord-free region between the lumbar and sacral vertebrae that the physician can safely insert a needle to obtain samples of cerebrospinal fluid or to inject anesthetic agents.

---

† In the literature of neuroanatomy, *ventral* and *dorsal* are frequently used as if they were synonyms of *anterior* and *posterior.* Although strictly speaking the two sets of terms have different meanings (see p. 196), we shall use them interchangeably in light of man's erect posture.

‡ The columns of gray matter are sometimes called *horns* because of their appearance in cross section. This term is useful but fails to emphasize their longitudinal continuity.

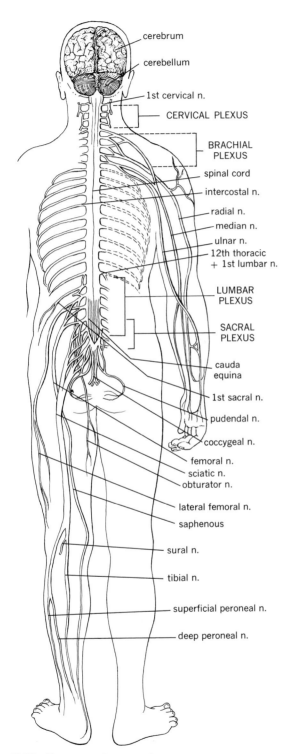

12.23 Nervous system.

Labels (top to bottom):
cerebrum
cerebellum
1st cervical n.
CERVICAL PLEXUS
BRACHIAL PLEXUS
spinal cord
intercostal n.
radial n.
median n.
ulnar n.
12th thoracic + 1st lumbar n.
LUMBAR PLEXUS
SACRAL PLEXUS
cauda equina
1st sacral n.
pudendal n.
coccygeal n.
femoral n.
sciatic n.
obturator n.
lateral femoral n.
saphenous
sural n.
tibial n.
superficial peroneal n.
deep peroneal n.

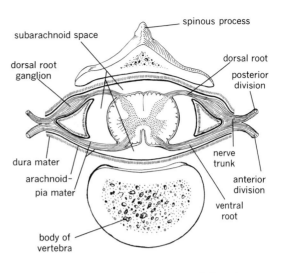

Labels: spinous process; subarachnoid space; dorsal root ganglion; dorsal root; posterior division; dura mater; arachnoid; pia mater; nerve trunk; anterior division; ventral root; body of vertebra

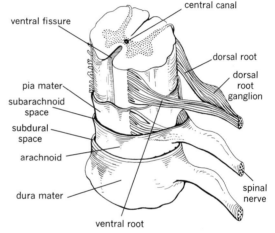

Labels: central canal; ventral fissure; dorsal root; dorsal root ganglion; pia mater; subarachnoid space; subdural space; arachnoid; dura mater; spinal nerve; ventral root

12.24 Spinal cord and associated structures.

ances along their sides are the *lateral columns.* The horizontal bar of the H is the *gray commissure.* The small *central canal* divides it into *ventral* and *dorsal* gray commissures, which contain fibers passing from one side of the cord to the other. The gray matter roughly divides the white matter into *ventral, dorsal,* and *lateral columns,* or *funiculi* (see Fig. 12.21). The relative quantities of gray matter and white matter vary in different parts of the cord (see Fig. 12.22).

As might be expected, the afferent tracts increase in size from caudal end to cephalic end of the cord, since new afferent fibers join them at each segment. Conversely, the efferent tracts become slimmer as fibers from the higher centers terminate by synapsing with effector neurons at different levels.

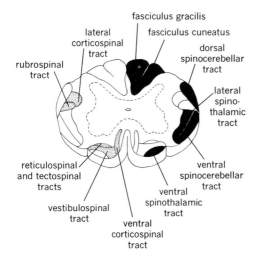

12.25 Ascending and descending fiber tracts of the spinal cord. All tracts shown are actually bilateral.

Spinal nerves and their cord connections. There are 31 pairs of spinal nerves, corresponding to the embryonic segments of the spinal cord, as follows: 8 cervical, 12 thoracic, 5 lumbar, 5 sacral, and 1 coccygeal. As shown in Fig. 12.24, a spinal nerve is formed by the junction of *dorsal* and *ventral nerve roots.* Each of these in turn is formed by the convergence of many dorsal or ventral nerve fibers emerging separately from the cord. Since a dorsal root is entirely sensory in function and a ventral root entirely motor, it is clear that a spinal nerve contains both sensory and motor fibers. It proceeds for only a few millimeters before dividing into anterior and posterior branches.

Let us trace a sensory impulse that has reached a spinal nerve through the axon of a receptor. The sensory neuron of the spinal nerve is monopolar (see Fig. 12.15), and its cell body lies within a swelling or nodule of the dorsal root, the dorsal root ganglion (see p. 431 and Fig. 12.24). Its axon traverses the remaining length of the dorsal root and enters the cord near the tip of the dorsal gray column. Here it may take one of several pathways to make synaptic contact with the cell body of a motor neuron in the ventral gray column (or it may give off an ascending branch that rises to a higher center; we shall consider this possibility later). The pathway from sensory neuron cell body to motor neuron cell body may be *intrasegmental* or *intersegmental.* In the former case, the axon travels to the ventral gray column at the same level and there synapses directly with a motor neuron, or it relays its message through an *internuncial neuron* at the same level (Fig. 12.26). In the latter case, it descends one or more levels before synapsing, or it terminates in the dorsal gray column

and there synapses with a *second-order sensory neuron,* which ascends or descends before synapsing with a motor neuron.

In any event, the impulse moves from a sensory neuron in the dorsal gray column to a motor neuron in the ventral gray column. The axons of ventral column neurons, which constitute the ventral root, proceed out through the spinal nerves to supply the muscles and other body structures with motor innervation.

Peripheral nerves and plexuses. In general, the smaller posterior divisions of a spinal nerve supply the muscles and skin of the back, and the larger anterior divisions supply the muscles and skin of the extremities and the remainder of the trunk. In all areas except the thoracic region, the anterior divisions interlace to form nerve networks called *plexuses,* which give off branches to the various body parts (see Fig. 12.23). The branches are the *peripheral nerves,* and each has a special name.

The first four cervical spinal nerves form the *cervical plexus,* which innervates the neck region. One important branch is the *phrenic nerve,* which supplies the diaphragm. The *brachial plexus,* formed from the fifth, sixth, seventh, and eighth cervical spinal nerves

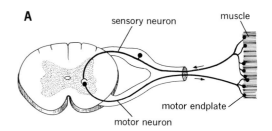

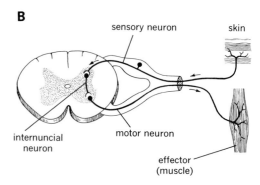

12.26 Intrasegmental connections between sensory and motor neurons of a spinal nerve: A, direct; B, through an internuncial neuron.

and the first thoracic spinal nerve, supplies the arm. Its major branches are the *radial, median,* and *ulnar nerves.* The *lumbar* and *sacral plexuses,* formed from the last thoracic and all the lumbar and sacral spinal nerves, serves the leg. Its great branches are the *obturator, femoral,* and *sciatic nerves.*

The rest of the thoracic spinal nerves do not form a plexus but pass out in the intercostal spaces as the *intercostal nerves.* They supply the intercostal muscles, the upper abdominal muscles, and the skin of the chest and abdomen.

Longitudinal fiber tracts of the spinal cord. We have noted that the white matter of the spinal cord consists of longitudinal fibers linking various levels of the cord with one another and with the higher centers. They are conveniently classified as short-distance and long-distance fibers. The short-distance fibers include association fibers between nearby levels of the cord and commissural fibers that cross the commissures between opposite sides of the cord. The long-distance fibers are those maintaining two-way communication between the cord and the brain. It is through them that sensory impulses reach the brain and that motor impulses travel from motor centers in the brain to the ventral gray columns of the cord and ultimately to the muscles and other effectors.

The longitudinal fibers are arranged in bundles that are structurally and functionally distinct from each other although indistinguishable in cross section on either gross or microscopic inspection. A *fiber tract* is defined as a bundle of fibers having the same origin, termination, and function.* Figure 12.25 shows that the white matter is composed of many such tracts in specific locations. Their positions and functions have been painstakingly mapped over the years in careful animal experiments in which small portions of the cords were cut and the functional consequences were observed. In addition, much has been learned from the correlation of neurological malfunctions in man with anatomical lesions found at autopsy. Since nerve fibers degenerate when separated from their cell bodies, investigators have also traced fiber tracts by cutting parts of a cord and then examining serial sections for evidence of fiber degeneration. This technique has helped them to determine whether a tract continues on the same side of the cord throughout its course or crosses to the opposite side. It has also helped them to determine whether the pathway between two levels of the nervous system consists of a single neuron or a chain of neurons.

Figures 12.25 and 12.27 illustrate some of the principal ascending and descending fiber tracts of the spinal

cord.† The *fasciculus gracilis* and *fasciculus cuneatus* (whose names violate the conventions of tract nomenclature) are the two main tracts of each dorsal white column. They are the ascending fibers of the primary sensory neurons of position, movement, touch, and pressure. They synapse in the *nucleus gracilis* and *nucleus cuneatus* on the same side. Hence they are uncrossed.

The *ventral spinothalamic tract,* ascending to the *thalamus* in each ventral white column, contains the fibers of second-order sensory neurons. Incoming axons of primary sensory neurons synapse in the dorsal gray column with the second-order neurons, whose axons cross to the opposite side of the cord before turning upward. The fact that impulses of position, movement, touch, and pressure rise to the brain through the *crossed* ventral spinothalamic tracts and the *uncrossed* fasciculi gracilis and fasciculi cuneatus explains why these sensations are not lost completely when only one side of the cord is transected.

The *dorsal spinocerebellar tract,* which terminates in the cerebellum, also contains the fibers of second-order sensory neurons. The primary sensory neurons originate in the neuromuscular spindles. The impulses transmitted are those responsible for muscular coordination.

The *lateral spinothalamic tract* contains the fibers of second-order sensory neurons of the pain and temperature pathways. Axons of the primary sensory neurons synapse in the dorsal gray column with the second-order neurons, whose axons cross the cord before ascending to the thalamus. Therefore, transection of one side of the cord results in loss of pain and temperature sense on the opposite side of the body below the level of the section.

The major descending fiber tracts originate in the higher brain centers and end in the ventral gray columns. Frequently the pathway to a motor neuron in a ventral gray column includes several internuncial neurons. More than one higher center may link with a single motor neuron. For this reason, the motor neuron whose axon emerges from the cord in the ventral root of a spinal nerve is often called the *final common pathway* of the outgoing messages of the central nervous system. We shall see that various kinds of "instructions" reach the final common pathway from higher centers.

---

* The *origin* of the tract is the locus of the cell bodies giving rise to the fibers forming the tract. The *termination* is the point at which the fibers synapse with the cell bodies of the next neurons in the chain.

† In general, a tract name is sufficiently descriptive to indicate the column in which it occurs (i.e., ventral, dorsal, or lateral) and the general locations of its origin and termination. Thus the name *lateral spinothalamic tract* reveals that the tract lies in the *lateral* column, originates in the *spinal* cord, and terminates in the *thalamus.* Since its route is from spinal cord to thalamus, it is an ascending, or afferent, tract. Similarly, the *ventral corticospinal tract* lies in the ventral column, originates in the (cerebral) cortex, and terminates in the spinal cord. It is a descending, or efferent, tract.

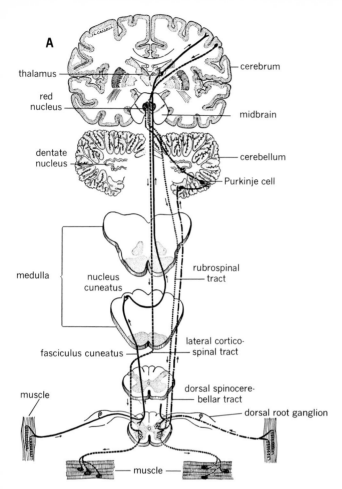

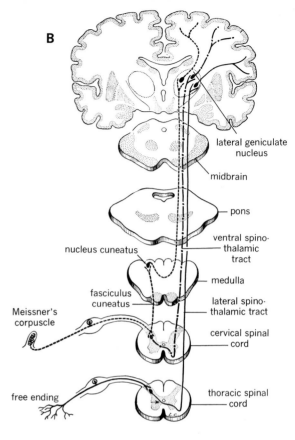

12.27 Sensory pathways of the spinal cord. A. ——, – –, conscious muscle sense; – · –, · · · , unconscious muscle sense. B. ——, pain and temperature; – – –, – · –, light touch. (From King and Showers, *Human Anatomy*, 5th ed., W. B. Saunders Co., Philadelphia, 1963.)

The first of the two main classes of descending fiber tracts consists of the great motor pathways from the *cerebral cortex*. They carry the impulses of voluntary movement, particularly those requiring the refined participation of small numbers of muscle groups. About four-fifths of the fibers cross to the opposite side in the *medulla** and descend in the cord as *lateral cortico-spinal tracts*. Because of their pyramidal shapes, these are known collectively as the *pyramidal tracts* (see Fig. 12.22H).† The remaining, uncrossed fibers descend as *ventral corticospinal tracts*. Both lateral and ventral

---

* Such a crossover is called a *decussation*.

tracts terminate in the ventral gray columns, either directly or through short internuncial neurons.

The second main class of descending fiber tracts is known as the *extrapyramidal system*. It includes the *reticulospinal, tectospinal, vestibulospinal,* and *rubro-spinal tracts*, among others.‡ Since these tracts arise from cell bodies placed considerably lower in the brain than those beginning the pyramidal tracts, extrapyramidal messages must come from the highest levels through a chain of neurons. The indirectness of this route is consonant with the view that the extrapyramidal system accounts for muscular movements accompanying discrete voluntary acts under pyramidal control. For example, the pyramidal tracts activate the finger muscles to pick up a pencil, but the extrapyramidal system produces the associated motions of the head, eyes, arms, and body.

---

† Some authorities attribute this name to the fact that the tracts arise in the pyramidal cells of the cerebral cortex.

‡ Respectively, these originate in the *reticular formation*, the *superior colliculi* of the *tectum*, the *vestibular nuclei*, and the *red nuclei*.

FUNCTIONS

Upper and lower motor neurons. The central nervous system controls all forms of motor activity—from voluntary actions stemming from conscious decisions to involuntary actions such as breathing. Between these is a large group of learned and associated functional patterns such as walking and speech that, though voluntary, are effortless. We have just seen that the neurons whose axons make up the descending fiber tracts and the neurons of the final common pathways supply motor impulses to the muscles. Thus there are two sets of motor neurons, *upper* and *lower.*

Upper motor neurons (whose cell bodies are in the brain and whose axons form the descending tracts) initiate motor activity. In addition, they exert important modifying influences upon lower motor neurons, as is demonstrated when the descending tracts are cut at different levels. In general, upper motor neurons originating high in the brain *inhibit* lower motor neuron function, whereas those originating lower in the brain deliver a constant low-frequency train of impulses to the lower motor neurons that *facilitates* their function, keeping them in a state of partial excitation ready to respond to any spatially or temporally more concentrated stimulus.

When the spinal cord is suddenly transected (as in a "spinal animal"), most cord functions are immediately depressed. This reaction, called *spinal shock,* is due to the sudden loss of the continuous facilitatory discharges from higher levels, particularly those transmitted through the pyramidal tracts and the vestibulospinal and reticulospinal extrapyramidal tracts. After a few days or weeks of spinal shock, the lower motor neurons gradually regain their excitability. Like other neurons anywhere in the nervous system, they make up for the missing source of impulses by increasing their own natural sensitivity. Their activity may then become exaggerated because of the absence of inhibitory effects from higher levels.

When only the highest brain centers are detached from the rest of the nervous system (as in a "decerebrate animal"), the flow of impulses is unaltered, and only the inhibitory effects are eliminated. The result is *muscular spasticity,* a state of greatly exaggerated activity of the lower motor neurons.

Simple spinal cord reflexes. The routine work of the nervous system is conducted principally by means of reflexes. In lower animals many important basic reflexes are mediated by the spinal cord alone. But, further up the phylogenetic scale, spinal cord reflexes are progressively obscured by the superimposition of upper motor neuron control. For this reason they are usually studied after spinal cord transection or decerebration.

The simplest spinal cord reflex, the *stretch,* or *myotatic reflex,* occurs when muscle fibers are stretched. Nerve impulses are initiated in spindles, mechanoreceptors in the muscle, transmitted to the spinal cord, and reflexly returned to the motor endplates of the same muscle, causing it to contract automatically following sudden stretching. This is a simple reflex because (1) it involves only two neurons; (2) it is always much the same; and (3) it is confined to a single level of the spinal cord. An example of a stretch reflex is the familiar *knee jerk.* When the tendon below the kneecap is tapped sharply, the leg kicks forward owing to the contraction of the "stretched" extensor muscle of the knee joint. Stretch reflexes serve to oppose changes in muscle length and hence to prevent jerkiness in muscle operation.

Physicians examine the stretch reflexes, in part to assess the facilitation or inhibition of lower motor neurons. When facilitation is depressed, the muscle jerk is weak or absent. When inhibition is depressed by disease in the higher centers or when excessive numbers of facilitatory impulses are transmitted from higher centers to the spinal cord, the muscle jerk is strong. Under certain conditions stretch stimulus produces an oscillating response, or *clonus,* in which the muscle contracts repeatedly. Clonus typically accompanies overfacilitation.

Another spinal cord reflex is the *flexor reflex.* If the skin on the limb of a spinal or decerebrate animal is painfully stimulated, the limb muscles flex, drawing the limb away from the stimulus. The reaction is much swifter than after a conscious decision, and its protective function is apparent. The reflex arc involves an internuncial neuron. Frequently, in fact, incoming sensory signals activate a whole pool of internuncial neurons, many of which branch extensively and ascend or descend in the cord. Results are (1) spreading of the reflex to the motor neurons of associated muscles at higher and lower levels and (2) inhibition of muscles that would oppose the reflex.

Brain and Cranial Nerves

COMPONENTS OF THE BRAIN

Introduction. The brain is the large organizing, integrating, and monitoring apparatus that controls all body activities. It contains (1) many *centers,* or groupings of special neurons that collectively regulate complex functions such as respiration and circulation; (2) *nuclei,* or clusters of cell bodies whose axons constitute the descending fiber tracts of the spinal cord and the 12 pairs of *cranial nerves,** the brain's counterpart of the spinal nerves; (3) special *motor areas* that determine the body's motor responses to stimuli arriving through the sensory cranial nerves and the ascending fiber tracts of the spinal cord; (4) special *sensory areas* that interpret sensory information reaching the conscious level; and (5) *association areas* concerned specifically with mental activity, memory, emotion, and learning.

---

* Some of the cranial nerves are exclusively motor, some are exclusively sensory, and some contain motor and sensory elements. The mixed cranial nerves have separate motor and sensory nuclei in the brain.

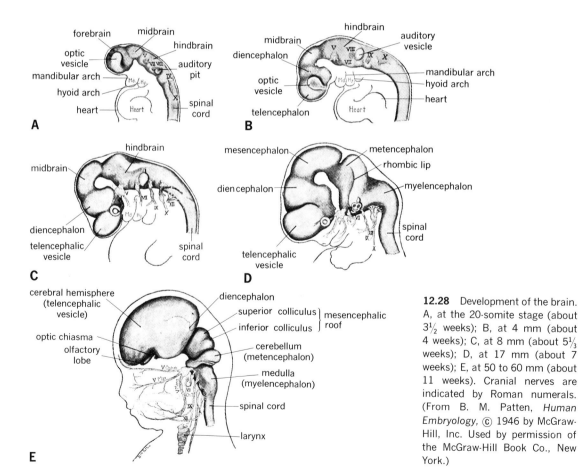

12.28 Development of the brain. A, at the 20-somite stage (about $3\frac{1}{2}$ weeks); B, at 4 mm (about 4 weeks); C, at 8 mm (about $5\frac{1}{3}$ weeks); D, at 17 mm (about 7 weeks); E, at 50 to 60 mm (about 11 weeks). Cranial nerves are indicated by Roman numerals. (From B. M. Patten, *Human Embryology,* © 1946 by McGraw-Hill, Inc. Used by permission of the McGraw-Hill Book Co., New York.)

TABLE 12.2 DIVISIONS OF THE BRAIN

Primary	Secondary	Components
Forebrain	Telencephalon	Olfactory bulbs Cerebrum Lateral ventricles
	Diencephalon	Thalamus Hypothalamus Third ventricle
Midbrain	Mesencephalon	Corpora quadrigemina Cerebral peduncles Red nucleus Aqueduct of Sylvius
Hindbrain	Metencephalon	Cerebellum Pons Fourth ventricle
	Myelencephalon	Medulla

No structure of the human body can compare in intricacy of design and virtuosity of function with "the great ravelled knot," as Sir Charles Sherrington aptly named the brain of man. It is certainly the least-understood facet of physiology or biology. Indeed, it has been suggested that man's brain is incapable of unraveling and comprehending its own complexity.

Embryological development. The embryological development of the primitive brain vesicle was described on p. 431. Almost from its first appearance, it shows signs of regional differentiation (Fig. 12.28A), and transverse grooves early divide it into *forebrain, midbrain,* and *hindbrain.* The forebrain is the largest of these three zones, partly because of the *optic vesicles,* which arise as outgrowths of its lateral walls and are forerunners of the eyes.

Soon after closure of the brain vesicle, both the forebrain and the hindbrain divide into two subsegments (Fig. 12.28B–D), making five secondary divisions in all, as presented in Table 12.2. The secondary divisions are precursors of the major subdivisions of the adult brain (Fig. 12.28E).

The brain then enters a period of growth so rapid that parts become folded upon themselves several times. Nevertheless, the fundamental relations among the five divisions remain unchanged.

Meninges. The meninges covering the brain are continuous with the dura mater, arachnoid, and pia mater covering the spinal cord (see Fig. 12.24). Although the brain may be defined as that portion of the nervous system within the cranial cavity, its several divisions occupy different compartments of the cavity. It is significant that the boundaries of these compartments are partly determined by meningeal folds.

The dura mater adheres to the inner surface of the cranium. Two prominent dural folds, the *falx cerebri* and the *tentorium cerebelli,* project inward (Fig. 12.29). The falx cerebri is a vertical sickle-shaped partition inside the fissure between the cerebral hemispheres. The tentorium cerebelli is a horizontal tentlike roof over the cerebellum, separating it from the posterior part of the cerebrum.

Between the layers of the dura mater are blood-filled endothelium-lined spaces that connect with veins. These are the *dural venous sinuses,* unique channels (see p. 299) through which blood returning from the brain reaches the internal jugular veins. The principal venous sinuses include the *superior sagittal sinus* on the superior border of the falx cerebri, the *inferior sagittal sinus* on the inferior border, the *straight sinus* where the falx cerebri joins the tentorium cerebelli, the *transverse sinus* along the occipital and temporal bones (Chapter 17), and the *cavernous sinus* in the region of the *sella turcica.*

The arachnoid sends many fingerlike projections, the *arachnoidal villi,* or *pacchionian bodies,* through the walls of the venous sinuses (see Fig. 12.29A). The subarachnoid space, like that in the spinal cord, contains cerebrospinal fluid.

Brain stem. The term "brain stem" refers to the hindbrain and midbrain exclusive of the cerebellum (see Figs. 12.2, 12.30). Despite its small size, about half the time available to teachers of neuroanatomy is usually allotted to its study. The reasons for such emphasis are apparent: (1) the brain stem contains the roots and nuclei of all the cranial nerves except pair I; (2) it is traversed by sensory fiber tracts ascending from the spinal cord and motor fiber tracts descending to the cord; (3) all of the special senses except smell operate through it; and (4) many important body functions are reflexly controlled in its centers. We may in fact regard the brain stem as a self-contained mechanism for regulation of the nervous system onto which cerebral and cerebellar controls were lately superimposed in the course of evolution. In spite of its complex topography, its general design is relatively simple.

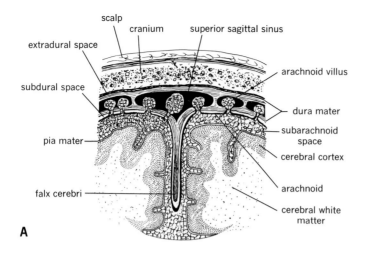

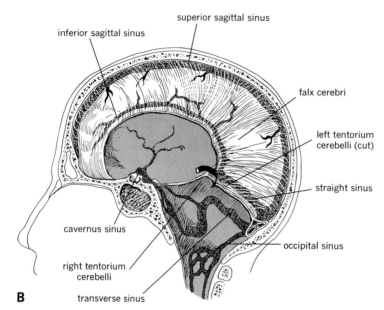

12.29 Meninges. A, frontal section through the scalp, cranium, meninges, and brain; B, sagittal section showing right half of skull as viewed from the left.

Early anatomists such as Vesalius could readily observe that 12 pairs of cranial nerves emerge from the undersurface of the brain. But, as the brain substance was little more than an amorphous pulp to them, they could describe and classify these nerves only on a purely anatomical basis. When the functions of individual neurons were finally appreciated, it became obvious that a single cranial nerve could contain both sensory and motor fibers, from cell bodies having different locations in the brain, and that, conversely, a single brain center could receive sensory impulses from more than one cranial nerve.*

The major differences between the brain stem and the spinal cord are as follows: (1) the gatherings of cell bodies called nuclei are of limited length in the brain stem, so that they

---

* For example, the cell bodies of the sensory neurons of taste are situated in one particular center in the brain stem. Yet the taste buds of the tongue and pharynx send impulses to the center through three different cranial nerves (VII, IX, and X). The taste fibers of these three nerves converge in the single taste center, which in turn sends a single taste tract to a higher center.

occur at specific levels, whereas those in the spinal cord (i.e., the gray matter) form long continuous columns; and (2) the brain stem is grossly flexed.

Medulla and pons. The lowermost portion of the stalklike hindbrain is the medulla, or *medulla oblongata* (see Figs. 12.2, 12.30). This small structure, only 3 cm long, is continuous with, and in many ways similar to, the spinal cord. All ascending and descending fiber tracts of the cord are found in the medulla. Some pass directly through it, some cross over it, and some—the fasciculi gracilis and fasciculi cuneatus—terminate in it (see Fig. 12.27).

The medulla also contains the nuclei of the last five cranial nerves—VIII, IX, X, XI, and XII—and the vital cardiac, vasomotor, and respiratory centers. Finally, a part of the *reticular formation* occupies a large portion of the medulla. The functional significance of this curious structure will be discussed later.

As shown in Fig. 12.22E–H, the narrow central canal of the spinal cord widens dorsally in the medulla into the large *fourth ventricle*. The *pons* forms its floor, and the cerebellum its roof (see Fig. 12.30).

The pons lies above the medulla and in front of the cerebellum (see Figs. 12.2, 12.30). It is essentially a bridge of fibers between the two halves of the cerebellum and between the medulla and the midbrain. As we shall presently see it is an important link in the complex chain of communication from cerebral hemispheres to cerebellum and back again. In addition the pons also contains the nuclei of cranial nerves V, VI, and VII.

Midbrain. The midbrain is the inch-long segment that connects the hindbrain and forebrain (see Figs. 12.2, 12.30). It consists of a dorsal part, the *tectum*, and a ventral part, the *cerebral peduncles* (Fig. 12.31). Running through it is the narrow *aqueduct of Sylvius* joining the fourth ventricle and the *third ventricle*. As shown in Fig. 12.31, the tectum contains two pairs of rounded elevations, the *corpora quadrigemina*, or the *colliculi*. The *superior colliculi* are concerned with vision, and the *inferior colliculi* with hearing. The cerebral peduncles are two thick stalks of white matter, bundles of fiber tracts to and from each half of the cerebrum.

The midbrain contains the nuclei of cranial nerves III and IV, some of the nuclei of pair II, and the so-called *red nuclei*, two masses of gray matter in which the extrapyramidal rubrospinal tract originates (see Fig. 12.27A).

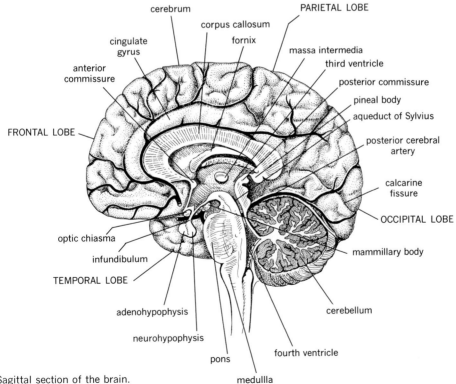

12.30 Sagittal section of the brain.

cerebrum

corpus callosum

fornix

PARIETAL LOBE

cingulate gyrus

massa intermedia

third ventricle

anterior commissure

posterior commissure

pineal body

aqueduct of Sylvius

FRONTAL LOBE

posterior cerebral artery

calcarine fissure

OCCIPITAL LOBE

optic chiasma

mammillary body

infundibulum

TEMPORAL LOBE

adenohypophysis

neurohypophysis

cerebellum

pons

fourth ventricle

medullla

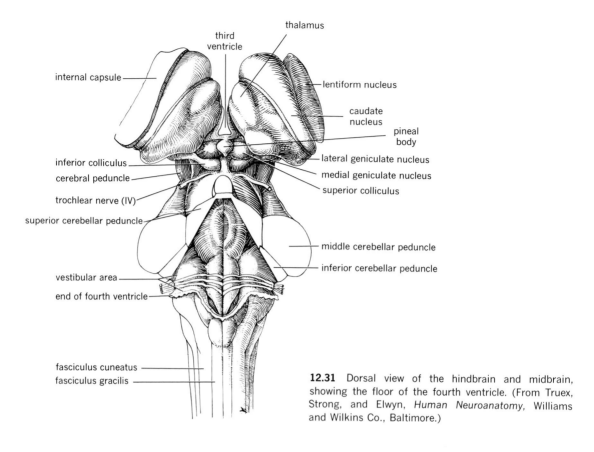

12.31 Dorsal view of the hindbrain and midbrain, showing the floor of the fourth ventricle. (From Truex, Strong, and Elwyn, *Human Neuroanatomy*, Williams and Wilkins Co., Baltimore.)

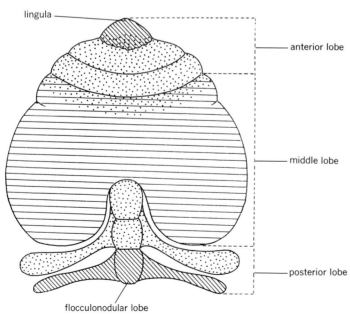

12.32 Cerebellar lobes: ▨, vestibular area; ▦, spinal area; ▤, cerebral area. (From G. M. Wyburn, *The Nervous System*, Academic Press, New York, 1960.)

Cerebellum. The cerebellum lies in the back of the cranial cavity, separated from the cerebrum by the tentorium cerebelli (see Figs. 12.2, 12.3, 12.30). It consists of a middle portion, the *vermis*, and two *lateral hemispheres*, although neurophysiologists and evolutionists find it more helpful to consider these as three *lobes*—anterior, middle, and posterior (Fig. 12.32). Three heavy bands of fibers, the *cerebellar peduncles*, connect the cerebellum directly to the cerebrum, pons, and medulla (see Fig. 12.31).

The wrinkled and creased appearance of the cerebellum is striking. Although it constitutes only 10% of the brain's total mass, its surface area is 75% that of the much larger cerebral hemispheres. If one dissects it along the midline, he observes a pattern of folding that resembles a tree in full leaf (see Fig. 12.30). Medieval anatomists called this the *arbor vitae*, the tree of life, and declared it to be the seat of the soul.

The cerebellum functions at a subconscious level. It is concerned with equilibrium, posture, and movement and, in the terminology of engineering, operates as the control box of a feedback circuit (Fig. 12.33). Such a control box does not originate the commands governing a system; it is informed of them by a command center. Like a room thermostat whose command center is a dial setting, it receives a feedback of data on existing conditions (room temperature in the thermostat; balance and position in the cerebellum) and orders the corrections necessary to bring conditions into closer accord with command decisions.

The form and relative size of the cerebellar lobes in a lower animal are related to the animal's posture, muscle distribution, and mode of locomotion (see Fig. 12.3). The small cerebellum of a primitive fish, for example, is merely an extension of the *vestibular apparatus*, the organ of equilibrium. In man, fibers from the vestibular apparatus enter the *flocculonodular lobe*—the back part of the posterior lobe—and the *lingula* —the front part of the anterior lobe (see Fig. 12.32). Early evolution also linked the *proprioceptors* (see Table 12.1) to the cerebellum via the spinocerebellar tracts, which terminate in the anterior and posterior lobes (see p. 435). Hence these lobes are termed the *old cerebellum*, or *paleocerebellum*. The upright position

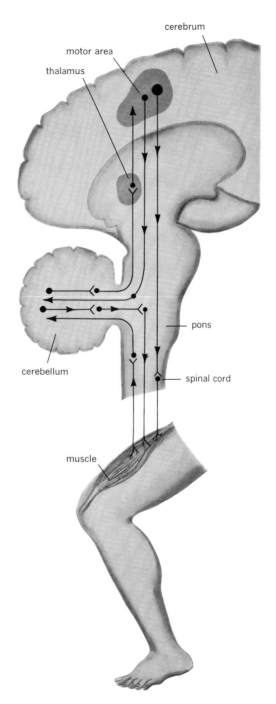

12.33 Circuits linking the cerebellum to the cerebrum, receptors, and effectors.

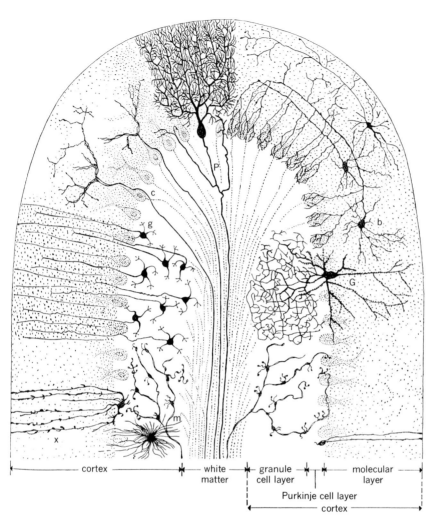

12.34 Cross section of a cerebellar fold: b, basket cell; c, climbing fiber; G, Golgi type II cell; g, granule cell; m, mossy terminal; P, Purkinje cell; x, y, cells of unknown function. (From R. S. Snider, "The Cerebellum," *Scientific American, 199,* 87 Aug. 1958. Copyright © 1958 by Scientific American, Inc. All rights reserved.)

cortex — white matter — granule cell layer — molecular layer

Purkinje cell layer

cortex

and independent use of arms and legs in recently evolved primates resulted in great enlargement of the middle lobe, which is accordingly called the *neocerebellum*. The neocerebellum marked the beginning of a two-way connection between the cerebellum and the cerebrum, the cerebrum being the great command center whose extensive development is unique to higher animals.

The cerebellum receives impulses from the proprioceptors, the skin, the vestibular and special sense organs, and the cerebrum. It sends messages, directly or indirectly, to the motor units of the spinal cord (via extrapyramidal tracts), to the brain stem, and to the cerebrum. All these connections to other areas of the nervous system admirably suit it for its function as a control box.*

How this function is performed is not known in detail, but the architectural plan revealed by microscopic examination of the cerebellum provides considerable insight into the mechanism. It is seen that the wrinkled outermost layer of the cerebellum, the *cerebellar cortex,* consists of gray matter containing three sublayers (Fig. 12.34): an outer *molecular layer*, a middle layer of large *Purkinje cells*, and an inner layer of small

---

* Similar feedback loops can be demonstrated between the cerebellum and three structures to be discussed later—the reticular formation, the thalamus, and the basal ganglia.

granule cells. A granule cell receives impulses from the central white matter through four to six short dendrites and sends them forth through a single axon that crosses the Purkinje cell layer and enters the molecular layer, where it connects with hundreds of branching dendrites of the Purkinje cells. Messages gathered by these dendrites are then borne to the key Purkinje cell bodies, which transmit them out of the cerebellum through long, threadlike axons that pass out with the central white matter.

It is noteworthy that impulses from a single granule cell axon reach many Purkinje cells. The process of amplification is furthered by the *basket cells* of the molecular layer. A basket cell also receives impulses from a granule cell and conveys them directly to Purkinje cell bodies via a long axon encompassing 6 to 12 Purkinje cells in a "basket." Thus information brought in by a single impulse spreads widely through the cerebellar cortex from a small number of granule cells to hundreds of Purkinje cells or, via the basket cells, thousands of them. The Purkinje cells then fire a huge volley of outgoing impulses—the original impulse many times multiplied—back into the white matter and thence to the brain stem.

Diencephalon. The forebrain has a central part, the *diencephalon,* and two large lateral expansions, the *cerebral hemispheres.* The main cavity of the diencephalon is the third ventricle, which arises from the widening of the aqueduct of Sylvius as it emerges from the midbrain. The third ventricle also communicates with the two *lateral ventricles,* the cavities of the cerebral hemispheres. The diencephalon includes two deep-lying areas of gray matter, the *thalamus* and *hypothalamus;* both contain many important nuclei that serve as relay centers for impulses coursing to and from the cerebral hemispheres.

The thalamus is a spheroidal structure (see Figs. 12.2, 12.30) almost completely divided by the third ventricle into right and left halves that protrude into the floors of the lateral ventricles. The two halves are connected across the mid-line by a bridge of gray matter called the *massa intermedia.* The groups of nuclei within the thalamus may be described topographically as medial, lateral, anterior, posterior, and ventral; or they may be described according to their afferent and efferent connections and functional significance. The principal nuclei are the *ventral nucleus,* the relay center for proprioceptive, tactile, pain, and temperature impulses; the *lateral geniculate nucleus* (or *body*),

which relays visual impulses to the cerebrum; and the *medial geniculate nucleus* (or *body*), which relays auditory impulses to the cerebrum.

The thalamus is the major relay station for all sensory impulses ascending to the cerebrum. In many lower animals it is the highest center for coordination of all the sensory input except for smell. In vertebrates, particularly mammals, the cerebral hemispheres have taken over coordination of the nonolfactory senses. The specific nature of the relationship between thalamus and cerebrum is still imperfectly understood.

The hypothalamus lies below (i.e., ventral to) the thalamus and forms the floor and part of the lateral wall of the third ventricle (see Fig. 12.30). Its several parts include the *optic chiasma,* where the nerves from the eyes meet and partially cross; the *infundibulum,* a slender stalk connecting it to the *hypophysis* (see Fig. 12.29B); and a pair of small rounded structures termed *mammillary bodies.* The nuclei of the hypothalamus are primarily concerned with the regulation of visceral activities. Among other things, they control the heart rate, blood pressure, and body temperature, acting mainly through the *autonomic nervous system.* We shall discuss the role played by the hypothalamus in the endocrine system in Chapter 15.

On the roof of the diencephalon immediately in front of the midbrain is the small *pineal body* (see Figs. 12.30, 12.31), once regarded as the seat of the soul. It functions in fishes, amphibians, and some reptiles as a median or third eye. In these species it contains distinctive receptor cells that appear to convert light of certain wavelengths into nerve impulses. In mammals it contains no receptor cells. Rather, its characteristic cells, *pinealocytes,* are secretory. The function of the pineal body in man has only recently been elucidated. It is apparently another neuroendocrine transducer—that is, it converts neural information into endocrine information (Chapter 15).

Cerebrum. By far the largest part of the mammalian brain is the cerebrum. Dividing it into right and left hemispheres is a deep cleft, the *great longitudinal fissure,* at the bottom of which are bands of fibers, the *corpus callosum,* bridging the two hemispheres (see Fig. 12.30). Human cerebral hemispheres are larger and more complex than those of any other animal.

The cerebrum is a mass of white matter covered by a layer of gray matter, the cerebral cortex. In lower animals the cortex is unwrinkled (see Fig. 12.3), but in mammals it becomes heavily folded during embryonic life. Only in this manner can its great surface area (estimated at 220,000 mm^2 or 340 $in.^2$ in man) fit into the limited confines of the cranial vault.

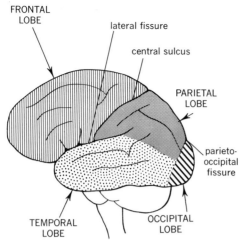

FRONTAL LOBE

lateral fissure

central sulcus

PARIETAL LOBE

parieto-occipital fissure

TEMPORAL LOBE

OCCIPITAL LOBE

12.35 Lobes of a cerebral hemisphere. (Adapted from Ranson and Clark, *Anatomy of the Nervous System*, 10th ed., W. B. Saunders Co., Philadelphia, 1959.)

The deepest depressions separating the many elevations, or *convolutions*, of the cerebral surface are called *fissures*, and the other depressions are called *sulci*. Although they are similarly arranged in all human brains, they vary in detail from brain to brain. Description is facilitated if the cerebrum is arbitrarily divided into *frontal, parietal, temporal*, and *occipital* lobes (see Figs. 12.30, 12.35). The *central sulcus* (or fissure of Rolando) separates the frontal and parietal lobes. The *lateral fissure* (or fissure of Sylvius) separates the temporal lobe below from the frontal and parietal lobes above. A fifth lobe, the *insula*, lies buried within the lateral fissure and cannot be seen unless the frontal and temporal lobes are parted (Fig. 12.36). The *parieto-occipital* fissure marks the boundary of the occipital lobe. The *calcarine fissure*, an important landmark on the

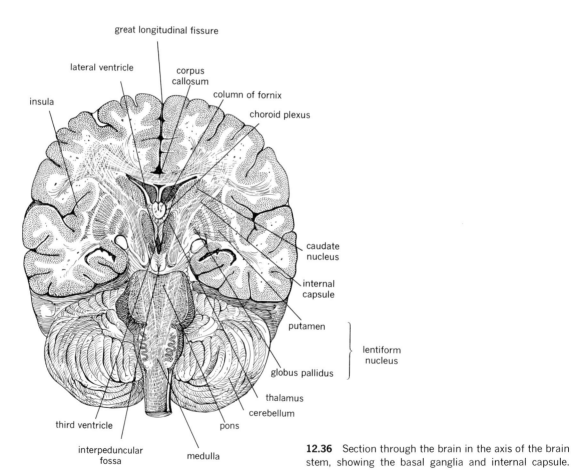

great longitudinal fissure

lateral ventricle

corpus callosum

column of fornix

choroid plexus

insula

caudate nucleus

internal capsule

putamen

lentiform nucleus

globus pallidus

thalamus

cerebellum

third ventricle

pons

interpeduncular fossa

medulla

12.36 Section through the brain in the axis of the brain stem, showing the basal ganglia and internal capsule.

medial surface of the occipital lobe, meets the parieto-occipital fissure anteriorly.

Just as a paleocerebellum and a neocerebellum exist, so do a *paleocortex* and a *neocortex*, the oldness and newness of which are determined by both evolutionary and embryological time scales. Embryologically development of the cerebrum is more complex than that of the other brain segments, owing to the ballooning of the lateral walls of the anterior end of the neural tube into the huge cerebral hemispheres and to the markedly unequal development of various parts of the walls. As shown in Fig. 12.37, an early cerebral hemisphere contains five areas that are distinguishable only by their different destinies: the *striatum*, the *septum*, the *hippocampus*, the *pyriform cortex*, and the *neocortex*. The first four are part of the "smell brain," or *rhinencephalon*, the deeply buried core of the paleocortex. The striatum and its overlying cortex become the insula, and the septum, hippocampus, and pyriform cortex become involved in the sense of smell and, in man, in other functions. Since man has a relatively poor olfactory sense, these ancient structures remain small in the human brain. Because other functions, including consciousness, sleep, memory, and basic control of emotions, have been attributed to the rhinencephalon, the more noncommittal name *limbic-midbrain system* is now generally used to denote its major functional elements and their outside connections, though this name too leaves something to be desired.

The neocortex, also known as the *general cortex*, expanded later in evolution than the paleocortex, and much more dramatically, to form the surface of most of the adult brain.* Its explosive growth produced the fissures, sulci, and convolutions of the cerebrum that made possible the fundamental advances of the mammalian nervous system. The neocortex becomes involved in the special senses other than smell.

In considering the complexities of the cere-

<hr>

* In man and other higher animals, large areas of neocortex may be surgically removed with little apparent effect on behavior. Partly for this reason and partly because of the important progress made in the early 1950's in researches on the brain stem, the rhinencephalon and other inaccessible regions of the cerebral hemispheres were long neglected by neurophysiologists.

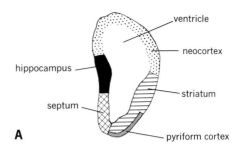

A

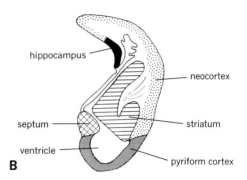

B

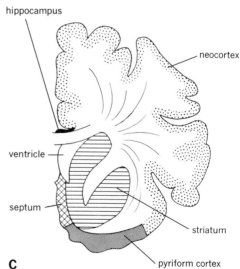

C

12.37 Development of the five areas of the cerebral hemisphere. A, 15 mm embryo; B, 50 mm embryo; C, adult (with the hippocampus shifted backward). (Adapted from W. Krieg, *Functional Neuroanatomy*, © 1942 by McGraw-Hill, Inc. Used by permission of the McGraw-Hill Book Co., New York.)

bral cortex, we must remember that the nervous system was competent to meet the needs of lower animals long before the neocortex expanded in evolution. We must regard it therefore as a recently added mechanism of analysis and control.

The cortex has been described as a holding corporation formed to integrate and extend the services of a number of older companies housed in the brain stem. The older companies act as useful adjuncts of the more modern organization —for example, they regulate such functions as breathing and digestion and keep things running in emergencies—but such functions as inquiry and planning are executed by the higher holding corporation. What goes on in this network of cells more than anything else distinguishes man from lower animals.

It was early discovered that specific areas of the cerebral cortex are concerned with specific functions in different regions of the body. The search for cortical areas began as an attempt to relate certain brain areas with particular talents and moral qualities. However, in 1861 a young French surgeon, Paul Broca, reported the case of an old man who had lost his speech but understood all that was said to him. When the man died, Broca found a lesion in the left third frontal convolution of the cerebrum. Thus was born the dogma that the speech function resides in this area, now known as *Broca's area* (Fig. 12.38A).

Experiments on the exposed brains of men* and lower animals showed that electrical or chemical stimulation of certain cortical areas produced discrete and highly localized responses, such as sensations or movements of small muscle groups, whereas it seemed to have no effect on other areas. These data, together with those collected following surgical removal of specific cortical areas, yielded a detailed functional map of the cerebral cortex (see Fig. 12.38).

The map reveals that the cortex contains specific motor and sensory areas. In addition,

---

* There is a fascinating literature on the early experiments in this field. It is said, for example, that two Prussian medical officers, Fritsch and Hitzig, first demonstrated that electrical stimulation of the brain causes muscular movements on the opposite side of the body in 1870 by using the exposed brains of war casualties on the battlefield of Sedan. Another early worker, Bartholow, reported similar reactions in experiments performed in 1874 upon his servant girl, who had a huge scalp ulcer that exposed her brain. After communicating his observations to the medical society, he was promptly expelled for unjustifiable experimentation on a human being.

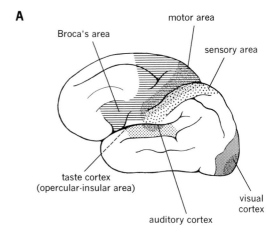

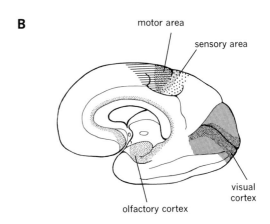

12.38 Localized functional areas of the cerebral cortex: A, lateral view; B, medial view. (Adapted from Ranson and Clark, *Anatomy of the Nervous System*, 10th ed., W. B. Saunders Co., Philadelphia, 1959.)

there are many large and ill-defined areas concerned with memory, reasoning, judgment, and other integrative processes. These are usually called *association areas*. This term implies that the general function of these areas resembles that of the association neuron in Fig. 12.1E. Since this may or may not be true,* some investigators have recommended the less precise term *intrinsic sectors*, which implies only that these areas lack major direct connections with peripheral structures.

The *motor area* of the cerebral cortex lies just in front of the central sulcus. Here are most of the large pyramid-shaped *Betz cells*, which give rise to most fibers of the corticospinal tracts (see Fig. 12.27A). Motor responses of individual muscles or small muscle groups to localized electrical stimulation show that the Betz cells are arranged along the convolution in an inverted order, beginning with those for movement of the toes in the upper part by the great longitudinal fissure and ending with those for movement of the jaws, palate, and face in the lower part by the lateral fissure. The crossing of pyramidal tracts accounts for the fact that stimulation of cells in one hemisphere causes movement of the muscles on the opposite side of the body.

The *sensory*, or *somesthetic, area* immediately posterior to the central sulcus receives information concerning touch, pressure, temperature, position, and pain. The distribution pattern along the convolution closely resembles that in the adjacent motor area. Thus sensations from the toes on the opposite side of the body are interpreted in the upper part of the convolution, near the great longitudinal fissure, whereas those from the trunk, arm, and face are interpreted progressively nearer the lateral fissure. When a specific region of the sensory area is stimulated electrically during brain surgery, a sensation occurs that seems to have arisen in the appropriate region of the body.

Lying between sensory and motor pathways, association areas integrate activities by connecting diverse sensory and motor centers—for example, the sense of vision controls fine finger movements through an association area—and somehow elaborate the higher intellectual functions of the brain.

Much that we know of these higher functions comes from studies of the consequences of removal of association areas either experimentally (*ablation*) or by injury or disease. We have learned that the anterior portion of the frontal lobe, sometimes called the *prefrontal lobe*, has a significant role in governing behavior, though it is without effect on reflexes, posture, and discrete movements. Removal of the prefrontal lobe (or interruption of its connections with the rest of the brain) from an experimental animal causes a characteristic alteration in behavior, with periods of hyperactivity involving aimless pacing, delayed reactions, and easy distractibility. However, removal of the prefrontal lobe from an experimental animal that through conditioning has developed behavioral abnormalities resembling human psychoneurosis results in improved behavior, with mistakes and failures, for example, no longer producing severe emotional outbursts.

Observations such as these led in 1935 to the first surgical prefrontal lobotomy in an emotionally disabled human being. The operation has now been performed many times. Although its evaluation is exceedingly difficult, it appears that, in certain disorders characterized by extreme anxiety, lobotomy causes a behavioral change that represents a net improvement. Unfortunately, it may also bring about a loss of certain higher sensibilities with a marked personality transformation.† In any event, large portions of the cortex are clearly expendable.

Impulses from the laryngeal region of the motor area produce sound, but the articulation of sound into syllables and words requires the associated control of Broca's area. Normally it is found on only one side of the brain, in the left hemisphere of a right-handed person or in the right hemisphere of a left-handed person. Although we shall not deal with the areas of the cerebral cortex concerned with vision and

---

* Association areas may also be *projection areas*, connecting directly with lower regions of the central nervous system. Hence they are not exclusively involved in the association of sensory and motor activities.

---

† Perhaps the best-known example of this phenomenon was the famous "crowbar" case of Phineas P. Gage in 1848. According to the description that appeared in the *Boston Medical and Surgical Journal*, "Phineas P. Gage, an efficient and capable foreman, was injured on September 13, 1848, when a 'tamping iron' was blown through the frontal region of his brain. He suffered the following change in personality according to the physician, J. M. Harlow, who attended him. 'He is fitful, irreverent, indulging at times in the grossest profanity (which was not previously his custom), manifesting but little deference to his fellows, impatient of restraint or advice when it conflicts with his desires, at times pertinaciously obstinate yet capricious and vacillating, devising many plans for future operation which are no sooner arranged than they are abandoned in turn for others appearing more feasible . . . . His mind was radically changed, so that his friends and acquaintances said he was no longer Gage.'" Gage's skull, tamping iron and all, is now in the museum of the Harvard Medical School.

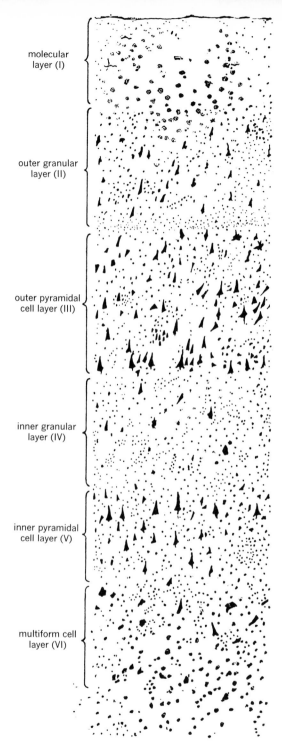

molecular layer (I)

outer granular layer (II)

outer pyramidal cell layer (III)

inner granular layer (IV)

inner pyramidal cell layer (V)

multiform cell layer (VI)

12.39 Cerebral cortex. (From B. M. Patten, *Human Embryology*, © 1946 by McGraw-Hill, Inc. Used by permission of the McGraw-Hill Book Co., New York.)

hearing until later, we shall locate them on the map now. The *visual cortex* is in the occipital lobe, in the calcarine fissure. The *auditory cortex* is in the temporal lobe, just below the lateral fissure. These areas interpret messages arriving from the eyes and ears. Two closely related areas are the *visual word center*, essential for the comprehension of written words, and the *auditory word center*, essential for the comprehension of spoken words.

Histologists have found almost 100 different cellular patterns in the cerebral cortex. Nevertheless, they are unable to account for the existence of specialized sensory and motor areas on the basis of cellular architecture.

All regions of the cerebral cortex have six basic layers (Fig. 12.39), designated by Roman numerals: I, a surface *molecular layer* (resembling that of the cerebellum —see Fig. 12.34) of small horizontal cells; II, an *outer granule cell layer* of small round cells; III, an *outer pyramidal cell layer;* IV, a closely packed *inner granule cell layer;* V, an *inner pyramidal cell layer,* with more and larger cells than the third layer; and VI, a bottom *multiform cell layer* containing spindle-like cells of various sizes and shapes. In general, sensory areas have fewer and smaller pyramidal cells and heavier collections of granule cells than motor areas, but otherwise the cortex displays a monotonously even appearance. In sum, its architecture is more homogeneous than regionally distinctive.

The white matter making up the noncortical component of the cerebrum consists of three kinds of fiber tracts: *association tracts, commissural tracts,* and *projection tracts.* Association tracts interconnect the lobes and convolutions of the same hemisphere. Commissural tracts join the two hemispheres and correlate their actions. For example, when the two legs are used together, as in walking, the main commissural tract, the corpus callosum, coordinates their movements. Projection tracts link specific regions of the cerebral cortex, the projection *areas,* to lower portions of the central nervous system. Visual radiations, auditory radiations, and pyramidal tracts are projection tracts.

Embedded deep within the white matter of the cerebrum are large masses of gray matter called *basal ganglia.* The older anatomists regarded the thalamus as one of the basal ganglia, but today the term is usually limited to the *caudate nucleus,* the *putamen,* and the *globus pallidus,* the last two being referred to collectively as the *lentiform nucleus* (see Fig. 12.36). The caudate nucleus and lentiform nucleus are separated by the *internal capsule,* a major band of projection fibers.

It is not yet possible to give a comprehensive statement of the functions of the basal ganglia. Even their

TABLE 12.3 CRANIAL NERVES

Number	Name	Type*	Origin of sensory fibers	Effectors innervated by motor fibers
I	Olfactory	Sensory	Olfactory receptors in nasal cavity	
II	Optic	Sensory	Photoreceptors in retina of eye	
III	Oculomotor	Motor		Four of the six extrinsic muscles of eye; ciliary muscle and pupillary sphincters of iris
IV	Trochlear	Motor		Superior oblique muscle of eye
V	Trigeminal	Mixed	Receptors in facial skin and teeth	Jaw muscles used in chewing
VI	Abducens	Motor		Lateral rectus muscle of eye
VII	Facial	Mixed	Taste buds of anterior two-thirds of tongue	Muscles of face; sublingual and submaxillary salivary glands
VIII	Auditory	Sensory	Phonoreceptors in cochlea and statoreceptors in vestibular apparatus	
IX	Glossopharyngeal	Mixed	Taste buds of posterior third of tongue and pharynx	Parotid gland; pharyngeal muscles used in swallowing
X	Vagus	Mixed	Receptors in many internal organs	Parasympathetic fibers to many internal organs
XI	Accessory	Motor		Muscles of shoulder
XII	Hypoglossal	Motor		Muscles of tongue

* Nerves classified as purely motor are believed to carry sensory fibers originating in the proprioceptors of the innervated muscles. To this extent they are mixed.

connections within the brain are disputed. It is generally believed that they coordinate voluntary movements, but recent experimental evidence suggests an even wider sphere of influence. Though they are closely interrelated with the cerebral cortex, they continue to function after its removal.

Cranial nerves. Fig. 12.40 shows the 12 pairs of cranial nerves emerging from the base of the brain. These nerves differ from spinal nerves in that (1) they are not attached by regular motor and sensory roots and (2) each pair has its own characteristics. Table 12.3 lists the cranial nerves and summarizes their functions.

It emphasizes that some are purely motor or sensory, whereas others are mixed motor and sensory. The motor fibers of cranial nerves III, VII, IX, and X are components of the *parasympathetic* division of the autonomic nervous system. The arrangement of motor fibers in the other cranial nerves more closely resembles that in the ventral root of a typical spinal nerve (see p. 434).

Three of the cranial nerves—I, II, and VIII—are associated with the special senses. Another three—III, IV, and VI—supply the muscles controlling the eyeball, pupil, and lens. The four largest—V, VII, IX, and X—have a number of important interconnections.

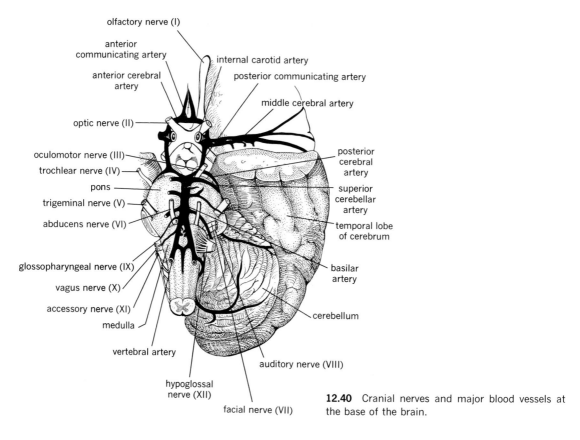

olfactory nerve (I)

anterior communicating artery

internal carotid artery

anterior cerebral artery

posterior communicating artery

middle cerebral artery

optic nerve (II)

oculomotor nerve (III)

trochlear nerve (IV)

pons

trigeminal nerve (V)

abducens nerve (VI)

glossopharyngeal nerve (IX)

vagus nerve (X)

accessory nerve (XI)

medulla

vertebral artery

hypoglossal nerve (XII)

facial nerve (VII)

posterior cerebral artery

superior cerebellar artery

temporal lobe of cerebrum

basilar artery

cerebellum

auditory nerve (VIII)

12.40 Cranial nerves and major blood vessels at the base of the brain.

The *trigeminal nerve* (V), the main nerve of the mouth region, has three widely distributed branches: the *ophthalmic, maxillary,* and *mandibular nerves.* Its motor nucleus is in the pons, and its sensory nuclei are in the large *semilunar* (or *Gasserian*) *ganglion* outside the brain stem (Fig. 12.41). The *lingual nerve,* a branch of the mandibular nerve, carries fibers of the *chorda tympani,* a branch of nerve VII.

The *facial nerve* (VII) has a motor nucleus in the lower part of the pons and a sensory nucleus in the *tractus solitarius* of the medulla. Taste receptors of the anterior two-thirds of the tongue deliver impulses through the chorda tympani to the *geniculate ganglion* and on to the medulla via the *nervus intermedius.*

The *glossopharyngeal nerve* (IX) carries motor fibers arising in the *nucleus ambiguus* of the medulla. Its sensory nuclei are in the *petrous ganglion* and *superior ganglion.* Impulses travel to them through the *sinus nerve* from chemoreceptors in the carotid sinus (see p. 320) and carotid bodies (see p. 402).

The *vagus nerve* (X) is the longest and most versatile cranial nerve, running from the medulla through the neck and chest to the abdomen. We shall discuss its functions later.

Ventricular system. The ventricles of the adult brain develop through local modifications of the neural tube (see Figs. 12.22, 12.42). The lumen becomes reduced in the spinal cord to form the narrow central canal, and it evolves in the brain into (proceeding anteriorly) the wide fourth ventricle, whose *lateral recesses (of Luschka)* open into the subarachnoid space, the narrow aqueduct of Sylvius, and the slitlike third ventricle, whose large lateral cavities, the right and left lateral ventricles, are also spoken of as the first and second ventricles (see pp. 441, 445). The passages between the third ventricle and the lateral ventricles are the *interventricular foramina,* or the *foramina of Monro.*

Choroid plexuses. In certain regions the walls of the ventricles contain specialized tufts of capillaries arranged in rich plexuses along the ependymal layer. These are the *choroid plexuses.* They are especially profusely distributed in the roofs of the fourth, third, and lateral ventricles

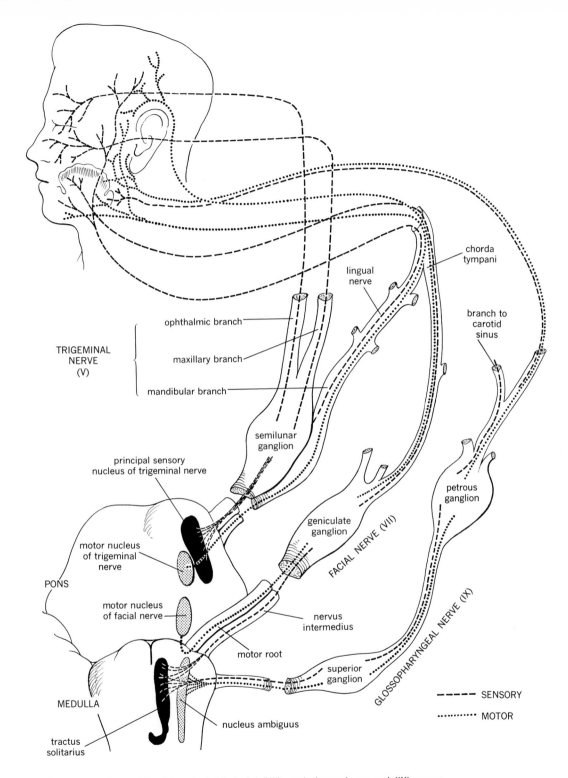

12.41 Interconnections of the trigeminal (V), facial (VII), and glossopharyngeal (IX) nerves.

Labels in figure:

TRIGEMINAL NERVE (V)

ophthalmic branch

maxillary branch

mandibular branch

semilunar ganglion

lingual nerve

chorda tympani

branch to carotid sinus

principal sensory nucleus of trigeminal nerve

motor nucleus of trigeminal nerve

PONS

motor nucleus of facial nerve

geniculate ganglion

FACIAL NERVE (VII)

petrous ganglion

nervus intermedius

motor root

superior ganglion

GLOSSOPHARYNGEAL NERVE (IX)

MEDULLA

tractus solitarius

nucleus ambiguus

- - - - SENSORY

· · · · · · · · · MOTOR

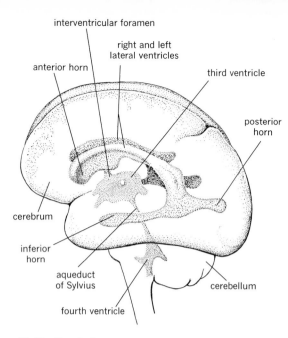

interventricular foramen

right and left
lateral ventricles

anterior horn

third ventricle

posterior
horn

cerebrum

inferior
horn

aqueduct
of Sylvius

cerebellum

fourth ventricle

12.42   Ventricular system.

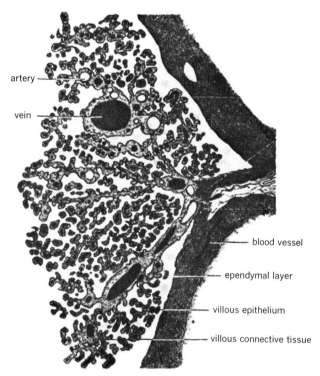

artery

vein

blood vessel

ependymal layer

villous epithelium

villous connective tissue

12.43   Choroid plexus. (From H. Davson, *Textbook of General Physiology*, 2nd ed., Little, Brown & Co., Boston, 1959.)

(see Fig. 12.36), areas in which neither fiber tracts nor nuclei appear.

Examination of the structure of a choroid plexus (Fig. 12.43) reveals that its capillaries have pushed the ependymal layer into villous tufts and folds that bulge into the ventricular cavity. A plexus can therefore be defined as an outpouching of the capillaries of the ependymal layer lining a ventricle—in many ways resembling a renal glomerulus (see p. 337).

Cerebrospinal fluid. Within the ventricles and the subarachnoid space is cerebrospinal fluid. Although it has many of the characteristics of an ultrafiltrate of plasma, it differs significantly from such a fluid in its formation, composition, circulation, and absorption.

As shown in Table 12.4, cerebrospinal fluid contains considerably lower levels of potassium, calcium, phosphate, and glucose than ordinary extracellular fluid. It also contains less cholesterol and no bilirubin. Conversely, its chloride level exceeds those of extracellular fluid and plasma by 20%. It can then be concluded that cerebrospinal fluid is not a simple filtrate of plasma but rather a *secretion* whose production involves active processes.

These processes are believed to take place in the epithelial cells covering the choroid plexuses of the lateral, third, and fourth ventricles. According to one theory, these cells secrete sodium and perhaps chloride. As a result, the osmotic pressure of the ventricular fluid exceeds that of plasma by approximately 160 mm of mercury, thus causing large quantities of water and solutes to move passively through the choroidal membranes into the ventricles. Blood proteins are almost entirely filtered out. This theory is based chiefly on the observable fact that the osmotic pressure of the substances in cerebrospinal fluid is 9 mosmols greater than that of the substances in plasma. This difference exerts an osmotic force of 160 mm of mercury.

Part of the protein and the occasional cells of cerebrospinal fluid enter through the ependymal layer. Radioactive albumin given intravenously passes more readily from the blood stream to the fluid of the subarachnoid space than to that of the ventricles. This explains why more protein is present in fluid from the lumbar region than in fluid from the brain.

12.44 Circulation of the cerebrospinal fluid.

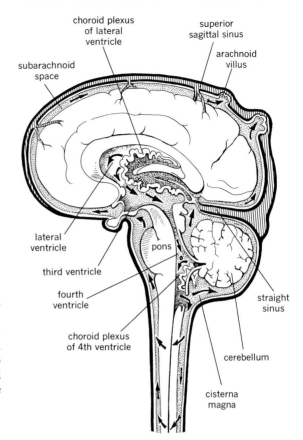

The total volume of cerebrospinal fluid in a normal adult is 100 to 150 ml. It is formed continuously, its daily production being estimated at 45 to 130 ml. An equal quantity must be disposed of each day.

Cerebrospinal fluid flows out of the fourth ventricle through its lateral recesses into the *cisterna magna*, a widening of the subarachnoid space between the cerebellum and the posterior surface of the medulla (Fig. 12.44). From there it spreads throughout the subarachnoid space.

Almost all of the cerebrospinal fluid is eventually reabsorbed into the blood through the arachnoidal villi, projecting from the subarachnoid space into the venous sinuses of the brain and the veins of the spinal cord (see Fig. 12.29). The mechanism of absorption is unknown. Conceivably the arachnoidal villi transfer the small amounts of protein in the cerebrospinal fluid to the veins, carrying with them the water needed to maintain the balance between hydrostatic and osmotic forces.

TABLE 12.4   COMPARISON OF COMPOSITIONS OF CEREBROSPINAL FLUID (CSF) AND ORDINARY EXTRACELLULAR FLUID (ECF)

Substance	CSF level, mg/100 ml	CSF level relative to ECF level (100%), % difference	ECF level relative to plasma level* (100%), % difference
Sodium	315–330	+5	−5
Potassium	12–16	−48†	−4
Calcium	4.0–6.0	−67†	−35
Magnesium	2.8–3.6	−20	−20
Chloride	445–460	+20	0
Bicarbonate	150–190	0	0
Phosphate	1.2–2.0	−66†	−4
Glucose	40–70	−36†	−3
Protein	20–40	−1	−99
Urea	10–35	−15	0

* See Table 8.5.
† Significantly different.

As cerebrospinal fluid passes from the ventricles to the subarachnoid space, the hydrostatic pressure is so great that an obstruction may cause an accumulation of the fluid under pressure. This condition, *hydrocephalus,* expands the ventricles. Hydrocephalus in newborn infants whose soft cranial bones have not yet knit together may result in grotesque enlargements of the head.

Cerebrospinal fluid provides mechanical and thermal protection to the brain and spinal cord, which in effect float suspended in a water jacket. The fluid is not known to have any metabolic functions, although it may aid in the removal of waste products from the brain and spinal cord surfaces.

BLOOD SUPPLY OF THE BRAIN

Major blood vessels. Arterial blood flows to the brain through an unusual traffic circle named for Thomas Willis, the English anatomist who first described it in 1664. The *circle of Willis* at the base of the brain receives arterial blood through two pairs of arteries, the internal carotids and the vertebrals (see p. 297, Fig. 9.14C). As also shown in Fig. 12.40, the vertebral arteries join to form the *basilar artery,* which enters the posterior portion of the circle of Willis. The circle sends three pairs of arterial feeder lines into the brain substance: the *anterior, middle,* and *posterior cerebral arteries.* The various *cerebellar arteries* stem directly from the vertebral arteries. We shall not consider them in detail.

The major arteries run in the subarachnoid space and give off *cortical* branches, which supply the cortex, and *central* branches, which supply the deeper structures including the basal ganglia and thalamus (see Fig. 12.36). Since the central branches do not communicate with one another, they are end arteries whose obstruction leads to death of the dependent brain tissue. In contrast, cortical arteries freely intercommunicate.

Within the brain the arteries break up into capillary networks. In general, gray matter receives more blood than white matter. Since the brain has a notably high oxygen requirement, interference with the blood supply causes damage and loss of function more rapidly in it than in any other organ.*

Venous blood leaves the brain through superficial and deep veins and drains into the dural venous sinuses (see p. 439). These in turn empty into the major extracranial veins, chief of which are the internal jugular veins in the neck (p. 299).

Cerebral circulation. The normal blood flow through the brain is 50 to 55 ml of blood per minute per 100 g of brain tissue. In the average-sized adult brain, this totals 750 ml of blood per minute, or 15% of the cardiac output under resting conditions. There is much current interest in the exquisitely precise control mechanisms that maintain cerebral blood flow within its narrow limits. Part of the interest arises from the fact that for some reason cerebral blood flow decreases with aging. The resulting morbidity is a primary cause of disability and death.

The most common method of measuring cerebral blood flow is the Kety-Schmidt technique, which employs the Fick principle (see p. 319) with a nitrous oxide indicator. Despite its complexity, it has been widely and profitably used. It has been found, for example, that CO_2 pressure is the main regulator of blood flow in brain tissue. A local increase in blood flow occurs in a specific brain area that has been "exercised"— so that each portion of tissue seems to govern its own blood supply according to need. The development of simpler procedures for determining total and local blood flow will surely disclose many subtleties of critical physiological importance.

Blood-brain barrier. Many native and foreign substances of both low and high molecular weights do not pass from the blood into brain tissue, although they pass readily into other tissues. Ehrlich observed in 1885 that certain ani-

---

* Mechanical occlusion of both internal carotid arteries in a healthy young person may cause unconsciousness but usually produces no lasting damage since arterial blood still enters through the vertebral arteries. The protection given these by the cervical vertebrae makes it impossible to occlude them from outside the body. This immunity to compression from an attack such as neck biting has evident advantages for a vertebrate whose head is connected to the trunk by a tenuous supply line.

line dyes injected into the blood stream stained all body tissues but those of the central nervous system. Similar observations with many other substances suggested a *blood-brain barrier*. Yet even electron microscopy has failed to reveal or identify any physical barrier.

A conclusion that no blood-brain barrier exists would be justified if it could be shown that the rates at which brain tissue and other tissues consume or metabolize substances differ. A relatively low concentration of a substance in brain tissue could reflect slow uptake or rapid destruction of the substance. Though the rate differences have not been established, another indication that the blood-brain barrier is illusory is the absence of lymphatic vessels and extracellular spaces in brain tissue.* This means that water and solutes can leave a capillary in the brain substance only by crossing brain cell membranes and is thought to explain the apparently low porosity of brain capillaries. Undeniably, absence of the extracellular spaces necessary for transcapillary exchange protects the brain from noxious substances in the plasma, but it also causes rapid starvation of the brain cells when their blood supply is temporarily interrupted.

PATTERNS OF BRAIN FUNCTION

Introduction. The neurophysiologist is a mechanist who implicitly believes that all behavior can be explained in terms of physicochemical events in a nerve network that, complex though it may be, is decipherable. Indeed, this outlook has been well justified by the many regularities already discovered in the spinal cord and brain stem. Even certain functions of the cerebrum in receiving incoming information and initiating motor actions can be interpreted in terms of various input-output sequences that are as fundamental as the spinal cord reflexes.

Accordingly, the nervous system could be described as a series of superimposed reflex arcs of increasing complexity. This hierarchical pattern has not been introduced into the models of Fig. 12.1. To illustrate it requires a diagram like

---

* Lymphatic vessels derive from mesoderm and remain confined to tissues of mesodermal origin. The nervous system derives from ectoderm.

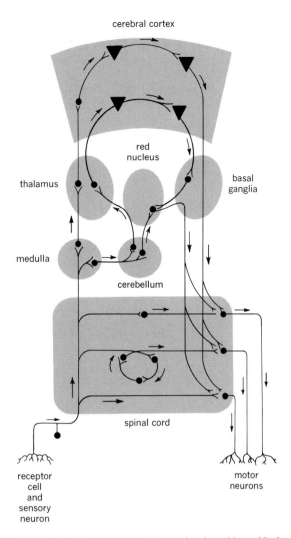

12.45 Model nervous system, showing hierarchical series of reflex arcs. Each relay station between the initial spinal segment and the cerebral cortex has access to a pathway capable of influencing the final motor neuron.

Fig. 12.45, which shows the several relay stations between the spinal cord and the cerebral cortex, the multiple circuits that directly and indirectly influence the motor neurons of the final common pathway, and the association neurons between the motor and sensory areas of the cortex. Our ignorance of much that takes place in this network, including the mechanisms that fuse the

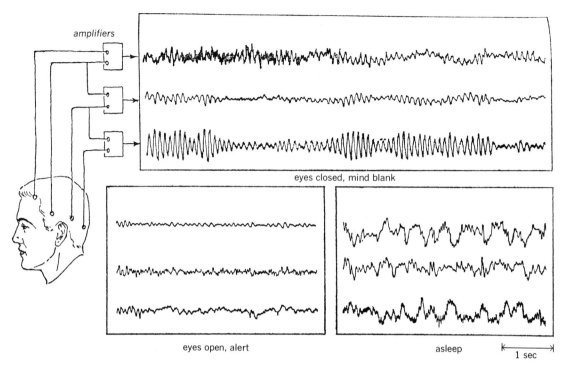

amplifiers

eyes closed, mind blank

eyes open, alert

asleep

1 sec

12.46 Electroencephalogram. Brain waves are recorded simultaneously from the front, middle, and back of the scalp. (From R. Galambos, *Nerves and Muscles,* ⓒ 1962 by Educational Services Inc. (Science Study Series). Reprinted by permission of Doubleday & Co., Inc., New York.)

activities of individual circuits into unified behavior, must be attributed in part to the enormous number of cells and connections and in part to experimental obstacles. For example, it is increasingly evident that a good deal of central integration involves information arising not from peripheral receptors but from the depths of the brain gray matter—the reticular formation, the cerebellum, the thalamus, and the limbic-midbrain system. Certain impulses dispatched from these centers evoke cortical responses that act back upon the centers.* Since such circuits are closed—that is, since they do not connect directly with peripheral receptors or effectors—and since their functions are so in-

timately woven into fundamental patterns of behavior, it is profoundly difficult for an experimenter to define their properties and sites of origin. Similar difficulties surround the study of the brain functions underlying phenomena such as consciousness, learning, memory, language, and symbolic processes—functions usually ascribed to the association areas.

Electrical activity. Electrodes placed directly on the scalp reveal continuous rhythmic electrical activity within the brain. Graphic records of the fluctuating potentials are called *brain waves.* A systematic recording of potential differences between specific scalp areas and a standard electrode is called an *electroencephalogram,* or *EEG* (Fig. 12.46). Voltages produced by the brain are so feeble—in normal adults they range from 0 to 300 μv—that extremely sensitive amplifying equipment is required.

By convention wave patterns are described

---

* Systems of this type have been termed *intrinsic systems* (cf. intrinsic sectors, p. 448). An example is a recently recognized set of fibers running from an area of the thalamus with no known afferent connections to an association area of the cerebral cortex.

in terms of frequency (i.e., number of waves per second). The character of the patterns depends upon the level and type of cerebral cortical activity, changing notably between wakefulness and sleep. Much of the time EEG's are asynchronous blends of frequencies with no sensible patterns. However, distinctive patterns occasionally appear, and these have been classified on the basis of frequency and amplitude.* *Alpha* waves, predominant in normal adults, are slow, fairly irregular waves with frequencies of 8 to 12 per second. *Beta* waves are faster, 18 to 32 per second, occurring in bursts most noticeably in the frontal and parietal lobes and usually in association with alertness, problem-solving, and the like. *Delta* waves are slower than 8 per second.

The fact that the human brain contains 10,000,000,000 neurons interconnected by complex fiber nets probably explains the incessant nature of its electrical activity. The spontaneous firing of one cell affects hundreds of others so that recorded waves represent the summed effects of many cortical units. The apparent synchrony of the classic wave patterns presumably reflects the close interrelationships.

Electroencephalography is useful in the detection of localized brain lesions, such as brain tumors, and in the diagnosis of *epilepsy*. Epilepsy is a poorly understood disorder characterized by episodic storms of neural activity either throughout the entire central nervous system or in a specific region. These are accompanied by seizures of greater or lesser degree—the generalized convulsions of *grand mal* or the momentary lapses of consciousness of *petit mal*. During the attacks distinctive and often bizarre EEG's are observed.

---

* Automatic frequency-analysis techniques have disclosed many interesting facts about brain waves. It has been shown, for example, that the dominant frequency—that is, the one most often seen in the waking brain wave of a person at rest—depends very much on age, being 3 to 5 per second in small children and 8 to 12 in adults. The EEG is first obtainable about 3 months before birth, with very low-voltage waves of frequencies of 1 to 3 per second. A month or so later it closely resembles that of a newborn infant, with large waves of frequencies of 2 to 5 per second. With increasing age it becomes progressively more like that of an adult, which appears at about 17 years. Many attempts have been made to correlate the rise in dominant frequency with changes in behavior, interests, and intellectual capacity.

In recent years electroencephalography has entered a dramatic new realm. It is now possible to implant a minute recording microelectrode within an animal's brain and leave it in place indefinitely while the animal lives a normal existence. Moreover, the electrode may be used to stimulate the brain electrically for experimental purposes.

Conditioned reflexes, learning, and memory. In a simple spinal cord reflex (see p. 437), the response to a stimulus is in no way influenced by the subject's experience or training. The knee is tapped, and it jerks, independent of mind or goal. The mouth is chemically stimulated by weak acid, and it salivates, a reaction that is automatic and unlearned. Such responses are known as *unconditioned reflexes*.

The classic experiments of the Russian physiologist Ivan Pavlov showed that experience or *conditioning* can influence otherwise unconditioned reflexes. If some stimulus—such as a bright light—that has no effect on the action of the salivary glands is regularly shone into a dog's eyes just before acid is placed upon his tongue, soon light alone will elicit salivation, without acid stimulation. This demonstrates a *conditioned reflex*. Higher-order conditioning can also be established. Thus, if a bell is rung each time the conditioned dog is stimulated by light alone, he will soon begin to salivate to the bell alone—and so on. A familiar instance of a conditioned reflex in man is salivation following the sound of a dinner bell in the total absence of the sight and smell of food.

The newly developed implanted-electrode technique of electroencephalography has revealed interesting changes in electrical activity in specific brain components during a conditioning period. For example, recordings from electrodes implanted in certain cortical areas of a cat indicate little response to a rhythmic clicking noise. If, however, such a noise is regularly associated with an unpleasant event (such as the blowing of air in the animal's face), large electrical disturbances occur when the clicking is later offered alone.

In many ways a conditioned reflex resembles learning. It is adaptive; it generalizes to new stimulus situations; and it disappears when no longer reinforced by experience. It differs from learning in that it lacks awareness, control, intent, and motivation.

The nature of the learning process is among the most challenging problems confronting modern neurophysiology. What parts of the nervous system participate in this process? What is its physiological mechanism? We can only give passing notice to these profound questions. Clearly an analysis of the neurophysiology of learning must explain how information is introduced into the brain, how it is stored, and how it is transmitted from place to place.

At present it is not known whether learning occurs diffusely in the entire central nervous system, in localized cortical areas, or in combinations of lower and cortical areas. One view derives from the work of Karl Lashley, who showed that maze-learning ability in rats depends on the amount of cerebral tissue destroyed and not on the anatomical area excised. This holds that all areas of the cortex are involved equally in the learning process and that learning may take place wherever two neurons meet. When an impulse from neuron A successfully fires neuron B, an increment of facilitation arises, so that an impulse from neuron A is more likely to fire neuron B in the future (see p. 421). Another view is that of Wilder Penfield, who believes that information comes via receptors to the neocortex, where an apparatus functioning like a tape recorder takes down the stream of impressions as a long coded strand holding the imprint of experience from birth to death; this then is deposited in some more or less discrete location. Between these two theories are others that localize the learning process in greater or lesser degree.

It is true that certain regions receive information more readily than others. Hence we might say that there are "learning" and "nonlearning" neural structures. The former are exemplified by the brains of vertebrates and the cerebral ganglia of invertebrates. Each of these—even the cerebral ganglia of the earthworm, which contain only a few hundred cells— has a layered cortex-like arrangement of neurons offering ample opportunity for dendritic overlap and direct contact between adjacent cells. The majority of such systems also exhibit wavelike patterns of electrical activity. In contrast, nonlearning structures are exemplified by spinal cords, all efforts to "educate" cords surgically isolated from higher centers having been unsuccessful. A spinal cord lacks both characteristics of a learning structure, a cortex and an undulating pattern of electrical activity. Conceivably these differences mean that a peculiar spatial orientation of cortical cells is essential for the acquisition of information.

The search for the mechanism of learning must also deal with the problem of information storage or *memory*. Presumably memory involves events at the molecular or submolecular level of organization. Although investigation is handicapped by obvious methodological difficulties, the consensus is that memory probably depends on molecular rearrangements—that is, that a specific molecule or class of molecules in the neuron serves as the *engram*, or permanent *memory trace*. The strongest suggestion of such a mechanism has come from experiments showing that a significant change in the purine and pyrimidine compositions of neuronal RNA occurs when a rat learns a balancing task. Administered RNA has also been claimed to improve the memory of elderly human subjects. However, rigorous experimental criteria have not yet established any one molecular species as a memory trace. RNA may be active in memory, but other compounds, such as lipids and proteins, may also have roles.

Memory could also depend on the electrical functions of whole neurons, the relations between cell body and synapse, and the relations between neuron and neuron. If memory were due to patterns of electrical activity, they would have to be sufficiently stable to withstand such holocausts as deep anesthesia, electric shock, and severe concussion—all of which would doubtless interfere with mechanisms requiring continuous electrical activity. Whatever the memory mechanism proves to be, it will be the job of the neurophysiologist to explain how the brain scans its file of memory traces, selects relevant ones, and then calls them into consciousness.

Consciousness, wakefulness, and sleep. It has been written that undefinable but intuitively recognized entities are a necessary foundation for most sciences. In neurophysiology, consciousness may be such an entity, for we can describe it only as the awareness of self and surroundings. Consciousness can only be experi-

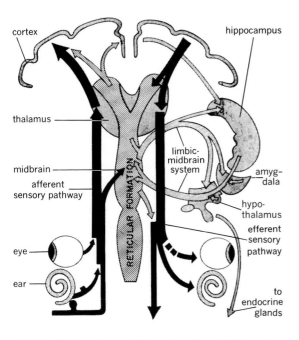

cortex

hippocampus

thalamus

limbic-
midbrain
system

midbrain

amyg-
dala

afferent
sensory pathway

RETICULAR FORMATION

hypo-
thalamus

efferent
sensory
pathway

eye

ear

to
endocrine
glands

12.47 Major systems of the brain. (From R. Galambos, "Neurophysical Studies on Learning and Motivation," *Fed. Proc. 20,* 603, 1961.)

enced, but loss of consciousness can be observed. The physician's well-known methods of assessing the level of consciousness include tests of the ability to recount past events, to respond to commands, and to avoid noxious stimuli. Consciousness is lost during sleep.

The unique value of consciousness is that it carries a residue of neural activity from one instant to the next, giving a semblance of continuity to what, in actual fact, may be extremely brief and isolated neural events. In this sense, it may be likened to a television tube that glows for a fraction of a second after it has been electrically excited and thus affords a continuous rather than a flickering image. Consciousness is not and never has been essential for the functioning of the nervous system. Rather, it is a supplement to the operations of the spinal cord, the brain stem, and the autonomic nervous system—all of which can and do function without it.

Investigation of the specific neural mechanisms of consciousness, wakefulness, and sleep has advanced rapidly in recent years. The progress stems directly from new insights into the function of the reticular formation, also known

as the *reticular core* (Fig. 12.47; see also p. 441). Structurally the reticular formation is a diffuse aggregate of cells and fibers running from the spinal cord junction through the brain stem to the upper end of the thalamus. It consists of short-branched interlacing neurons forming a matrix of fine fibers that is traversed by many larger fibers carrying impulses to and from higher and lower centers.*

It now appears that the reticular formation is one of the most important of the brain's integrating systems. Recent reviewers have gone so far as to suggest that it is the master control mechanism of the nervous system, the cortex being merely an amplifying and encoding device that simplifies patterns and feeds them back to this older but higher integrating apparatus. Whether the cortex should yet be downgraded remains to be seen. But it is clear that the reticular formation has at least three major functions: (1) it is responsible for wakefulness; (2) it modifies the reception, conduction, and integration of all incoming sensory messages in such a manner that some are selected for perception and others are rejected; and (3) it is critically involved in the maintenance of muscular tone.

Following the discovery that electroencephalography distinguishes between the sleeping and waking states (see p. 458), it was found that surgical separation of the cerebral hemispheres from the brain stem and spinal cord produces perpetual "electroencephalographic sleep" in the cortex. This suggests that, to maintain its waking state, the cortex must receive a constant inflow of sensory information through ascending fiber tracts. Evidence then appeared indicating the existence of a diffuse system arranged in parallel with and partaking of the same informational content as the classic sensory paths but perhaps projecting to the cortex by a different route. The brilliant studies of G. Moruzzi and H. W. Magoun in 1949 demonstrated unequivocally that stimulation of the reticular formation causes a sleeping animal to awaken instantly and a "sleeping" electroencephalogram to revert to a waking pattern. This arousal re-

---

* Such a mixture of cells and fibers with little apparent organization represents a phylogenetically ancient pattern, and indeed the reticular formation can be considered the oldest part of the brain.

sponse occurs even after interruption of the classic sensory paths. It was also recognized that individuals with lesions in the reticular formation suffer loss of consciousness. It was concluded therefore that wakefulness results from enhanced activity of the brain and sleep results from a deficiency of adequate sensory stimulation of the brain.*

This conception focuses attention on the reticular formation as a filter for sensory information. From its structure and strategic location, it is apparent that a continuous flow of impulses from all ascending sensory and descending motor channels pours through it. Anatomical study reveals that axons of reticular cells reach to many distant centers, upstream and down, through complexly branched networks. The targets include motor and sensory centers in the spinal cord, cranial nerve nuclei in the brain stem, and cerebral cortical areas. Even receptors such as the photoreceptors of the eye, the phonoreceptors of the ear, and the proprioceptors of muscles, tendons, and joints are under the direct modulating control of the reticular formation, which performs as a subtle manipulator of the perceived world, filtering and censoring incoming data.

By selective facilitation or inhibition, the reticular formation can focus attention on particular channels of sensory or motor activity, preventing the intrusion into consciousness of irrelevant information. This is the hallmark of alertness or attention—the ability to "look" and "listen." Apparently the reticular formation works in concert with another system, the *efferent sensory pathway* (see Fig. 12.47), a collection of descending fibers that may suppress the inflow of unwanted sensory data from the eyes, ears, and other receptors. We begin to see that the nervous system is designed to decide for itself what aspects of the immediate environment will enter for processing. Man's capacity to select one of several possible courses of action

may depend to a large extent on what sensory inputs are emphasized at the expense of others.

Emotion. In 1937 James W. Papez published a paper speculating that the limbic-midbrain system has a paramount role in the genesis of emotion. His much-quoted theory is one of the few positive contributions to this difficult area.

We recall from Fig. 12.47 that the limbic-midbrain system consists, in essence, of a loop of fibers extending from certain nuclei in the forebrain to the *hippocampus, fornix,* and *amygdala* in the rhinencephalon to the *mammillary bodies* in the hypothalamus to the part of the reticular formation in the midbrain and back again. Both limbs of the loop contain many synapses, particularly in the hypothalamus; hence there are many opportunities for the integration and modification of impulses. It should be noted that the function of the old "smell brain" also depends upon the reticular formation.

Papez held that the limbic-midbrain system is a "reverberating" circuit connecting the hypothalamus with the cerebral cortex. His conclusion that the processes of emotion originate in it was supported, but not proven, by ablation, stimulation, and electroencephalographic experiments. It is generally agreed that the hypothalamus is associated with certain overt manifestations of intense emotion—trembling, blushing, elevation of blood pressure, etc.—whereas the cortex accounts for the conscious experience of emotion.

In one series of experiments, J. Olds and N. E. Miller, working independently, implanted an electrode in a specific region of the limbic-midbrain system in an animal and placed within easy reach a lever that, when pressed, delivered an electric shock to the brain area surrounding the electrode. Surprisingly, animals of every species that they tested refused to stop pressing the lever once they had found it and, for the pleasure of stimulating the special *reward center,* would endure hardships and punishments that they would not undergo to obtain food. They manipulated the lever rapidly for days at a time, forgoing food, drink, and sleep. These experiments evidently illustrate a form of motivated behavior or emotion under the control of a sharply localized brain site.

---

* When we consider the significance of sleep, the necessity of spending one-third of our lives at it, and its curious cyclical nature, it is astonishing how little we know of it. It is easy to say that sleep is brought on by "metabolic exhaustion" of the neurons, but we have no idea what this means. The only obvious mechanism that might become exhausted is the sodium pump (see p. 416), but why it should take many hours for it to do so and many more for it to be regenerated is an unsolved riddle.

Current thought on the nature of emotion has been influenced by the study of drugs that affect consciousness and emotionality. The best known of these are the *tranquilizers* reserpine and chlorpromazine; *sedatives* such as the barbiturates; the *stimulants* caffeine, amphetamine, and their relatives; *depressants* like morphine; and *hallucinogens* such as lysergic acid (LSD), mescaline, and marijuana. Some of these produce symptoms mimicking those of psychiatric illnesses, whereas others reverse them. Although much information exists on the actions of these drugs on the brain and its components, they are not yet explainable in concrete terms.

Hypothalamic regulation of thirst and body temperature. We shall find in discussing the autonomic nervous system that the hypothalamus participates in the regulation of many visceral functions. Presumably it is responsible for the visceral manifestations of intense emotion. Two recently discovered hypothalamic centers deserve mention both for their intrinsic interest and because they illustrate the importance of specialized brain areas.

The *thirst center* of the hypothalamus is such a structure. We learned in Chapter 7 that animals drink the exact amounts of water needed to compensate for losses. In man the level of total body water fluctuates no more than 0.036%. It has been shown that drinking behavior can be totally disrupted by electrical stimulation of the hypothalamus. A goat with an electrode implanted in its hypothalamus drinks an amount of water equal to 40% of its body weight in a few hours. Apparently the thirst center normally "monitors" body water content, "decides" when drinking behavior is called for, and "orders" an animal to take appropriate action.

The other hypothalamic center serves as the body thermostat, precisely measuring body temperature and, when necessary, activating heat-dissipating mechanisms that maintain it within a fraction of $1°F$.* The problem of how the body keeps its temperature constant long puzzled investigators, although they knew that sweating and the dilation of peripheral blood vessels are the effectors for dissipating excess heat. As long ago as 1884, they found that body temperature rises after experimental puncture of the forebrain and concluded that a temperature control center exists in the hypothalamus, consisting of a collection of cell bodies that convert incoming sensory impulses (from temperature-sensitive receptors elsewhere in the body) into motor impulses to the heat-dissipating effectors.

The nerve endings of the skin are the sensory organs for another system, which bypasses the hypothalamic center and initiates voluntary corrective behavior. To sensations of heat or cold reported by the skin, the body reacts by deliberately seeking a cooler or a warmer environment. Despite their mastery over external circumstances, however, the skin receptors cannot regulate internal temperature with accuracy. For this purpose the hypothalamic center is required.†

Autonomic Nervous System

STRUCTURE

Introduction. The portion of the nervous system that remains to be described controls a large group of involuntary body functions—the so-called visceral functions—such as arterial blood pressure, respiration, gastrointestinal motility and secretion, urination, sweating, body temperature, and pupillary movements in the eye. Their involuntary character is suggested by the name "autonomic nervous system."‡

We must clearly distinguish the anatomical pattern of the autonomic nervous system from that of the voluntary nervous system. In the classic spinal cord reflex arc (see Fig. 12.26), a sensory neuron whose cell body lies in the dorsal root ganglion connects directly (or via an inter-

---

* The ultimate sources of body heat are the metabolic processes that generate it as a by-product (see Chapters 4, 14).

† The feedback system that *dissipates* excess heat has thus been elucidated, but not the one that increases metabolic heat production to keep body temperature from falling below the optimum level. The two systems apparently operate quite differently.

‡ *Involuntary nervous system, visceral nervous system,* and *vegetative nervous system* are sometimes used as synonyms for the autonomic nervous system. In contrast, the *voluntary nervous system* is called *somatic* and *nonvegetative.*

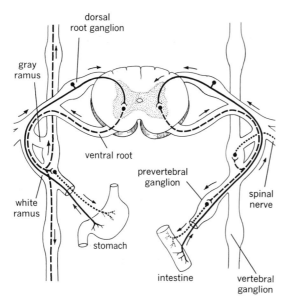

labels on figure:
dorsal root ganglion

gray ramus

ventral root

prevertebral ganglion

white ramus

stomach

intestine

spinal nerve

vertebral ganglion

12.48 Diagram of reflex arcs in the autonomic nervous system. Compare with Fig. 12.26.

nuncial neuron) with a motor neuron cell body in the ventral gray column of the cord. The motor neuron axon emerges in the ventral root and travels directly to the effector. Fig. 12.48 shows the arrangement of the reflex arcs in the autonomic nervous system. The sensory neuron is identical to its counterpart in the classic spinal cord reflex arc except that it arises in the smooth muscle of the viscera (e.g., intestinal wall or arteriole wall). Its axon also connects with a motor neuron cell body that is situated in the gray matter of the spinal cord. Here the similarity ends.

In the autonomic nervous system motor impulses from the spinal cord to an effector are conveyed along a *two-neuron pathway.* The synapse between the two neurons is in a special *ganglion* outside the spinal cord. The cell body of the first of the two neurons, the *preganglionic neuron,* is situated in the middle of the lateral column of gray matter. Its axon extends to the outside autonomic ganglion, where it synapses with the cell body of the second neuron, the *postganglionic neuron.* The axon emerging from the ganglion innervates the smooth muscle of the intestinal wall, arteriole wall, or other viscera.

The ganglia containing the synapses between preganglionic and postganglionic neurons and the cell bodies of postganglionic neurons can be classed in three groups depending upon their general locations (see Figs. 12.48, 12.49): (1) the *vertebral ganglia* lie on either side of the vertebral column; (2) the *prevertebral ganglia* lie in front of the vertebral column; and (3) the *peripheral ganglia* lie directly on or in the walls of the organs that they innervate. Accordingly, the axons issuing from them to the viscera are very long, intermediate in length, and extremely short, respectively.

The autonomic nervous system has two major anatomical and functional components, the *sympathetic* and *parasympathetic* divisions. Cell bodies of the parasympathetic preganglionic neurons are located in the *cranial* and *sacral* regions of the central nervous system—specifically, in the nuclei of the midbrain, pons, and medulla giving rise to cranial nerves III, VII, IX, and X and in the gray matter of the spinal cord at the level of the second, third, and fourth sacral nerves (see Fig. 12.49). For this reason the parasympathetic division is sometimes called the *craniosacral division.*

Cell bodies of the sympathetic preganglionic neurons are located in the gray matter of the *thoracic* and *lumbar* regions of the spinal cord, with axons emerging from each spinal cord segment from the first thoracic to the second lumbar. Hence the sympathetic division is sometimes designated the *thoracolumbar* division.

We see that sympathetic preganglionic neurons synapse with postganglionic neurons in vertebral or prevertebral ganglia, whereas parasympathetic preganglionic neurons synapse with postganglionic neurons in peripheral ganglia. Let us trace the various routes followed by these two-neuron chains.

Sympathetic preganglionic and postganglionic neurons. Axons of the sympathetic preganglionic neurons emerge from the spinal cord in the ventral root, as do axons of the corresponding voluntary neurons, but soon branch off toward an autonomic ganglion. Since most of these axons are myelinated and therefore white, the bundle that they form is called the *white branch,* or *white ramus communicans.*

The white ramus enters a vertebral ganglion, where a few of its axons synapse with postganglionic neurons.

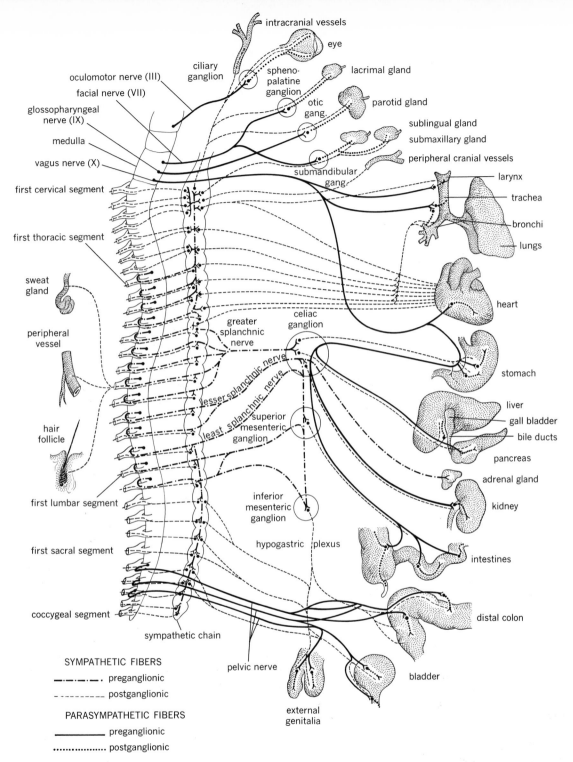

12.49 Autonomic nervous system. (Adapted from *The Ciba Collection of Medical Illustrations* by Frank H. Netter, M.D. Copyright CIBA.)

Each axon connects with several dozen cell bodies. This accounts for the prompt and widespread character of sympathetic responses. The vertebral ganglia, which appear as a series of 22 swellings joined into a long beaded cord down each side of the spinal column (see Fig. 12.49), form the *gangliated sympathetic chain*.

The axon of a postganglionic neuron whose cell body lies in a vertebral ganglion may take one of two courses: it may extend directly to the viscera via a visceral nerve branch and there terminate; or it may turn toward the spinal nerve through a short root called the *gray ramus communicans* (because its axons are unmyelinated) and terminate in an involuntary effector (e.g., smooth muscle in an arterial wall or sweat gland). Hence some sympathetic postganglionic fibers travel to the periphery of the body in the voluntary nerves. Like the voluntary nerve outflow, the sympathetic postganglionic outflow at each level innervates a segmental area of the body surface.

Most of the axons of a white ramus pass up or down the gangliated chain to enter vertebral ganglia and synapse with postganglionic neurons at higher or lower levels*—or they pass through vertebral ganglia en route to prevertebral ganglia in which they synapse with postganglionic neurons. The prevertebral ganglia are somewhat larger than the vertebral ganglia. Three are named for large neighboring arteries: the *celiac, superior mesenteric,* and *inferior mesenteric ganglia* (see Figs. 12.48, 12.49). Axons passing from the spinal cord to prevertebral ganglia are gathered on each side into three well-defined strands: the *greater, lesser,* and *least splanchnic nerves.* Axons emerging from the great prevertebral ganglia form massive plexuses whose branches supply the blood vessels and viscera.

The largest plexuses of the autonomic system are the *cardiac, celiac* (or *solar*), and *hypogastric plexuses.* Although they are essentially sympathetic, they contain a scattering of axons from the parasympathetic division. The cardiac plexus lies under the arch of the aorta just above the heart. It receives sympathetic branches from the cervical ganglia and parasympathetic branches from the right and left vagus nerves (X). All act in concert to regulate heart action (see p. 320). The celiac plexus, the largest of the autonomic system, lies behind the stomach and is associated with the aorta and the celiac arteries. It receives the sympathetic splanchnic nerves and parasympathetic branches of the vagus nerves. The hypogastric plexus begins in front of the fifth lumbar vertebra and continues downward in front of the sacrum to form the *right* and *left pelvic plexuses,* which supply the organs and blood vessels of the pelvis.

It connects the celiac plexus with the two pelvic plexuses.

Parasympathetic preganglionic and postganglionic neurons. We have seen that axons of parasympathetic preganglionic neurons travel without interruption to peripheral ganglia, on or in the walls of the organs being innervated. There they synapse with postganglionic neurons, whose axons reach no more than a few millimeters before terminating in the substance of the organs.

The absence of a clear segmental pattern such as that of the sympathetic division makes it difficult to generalize the structure of the parasympathetic division. As shown in Figs. 12.48 and 12.49, axons of parasympathetic preganglionic neurons leave the central nervous system in certain cranial and sacral nerves, approximately 80% in the huge vagus nerves (X), which travel to the thoracic and abdominal regions.† Those from the *oculomotor nerve* (III) extend via the *ciliary ganglion* to the pupillary sphincters and ciliary muscles of the eye; those from the facial nerve (VII) travel to the lacrimal, submaxillary, and nasal glands; and those from the glossopharyngeal nerve (IX) pass to the parotid gland.

The axons of sacral parasympathetic preganglionic neurons are collected into two *nervi erigentes* and distributed to the peripheral ganglia of the descending colon, rectum, bladder, lower ureters, and external genitalia.

FUNCTIONS

Sympathetic-parasympathetic relationships. We might now reasonably ask what common features justify combining the widely separated and structurally diverse cranial and sacral regions into the single parasympathetic division. Aside from the fact that both regions employ peripheral ganglia, the answer lies in their *functional* organization.

The parasympathetic division is primarily concerned with conservative and restorative processes such as slowing the heart rate, contracting the pupils to protect the eyes from light, and inhibiting the utilization of liver glycogen (Table 12.5). It has a restricted distribution with more or less local functions. In contrast, the sym-

---

* Since they leave the cord only in the thoracic and upper lumbar regions, they must course up or down the gangliated chain to innervate the cervical, lower lumbar, and sacral vertebral ganglia.

---

† The right and left vagus nerves send parasympathetic branches to the *cardiac* and *pulmonary* plexuses. Above the stomach these branches unite to form the *esophageal plexus*. Below it they supply the stomach, small intestine and ascending colon, pancreas, and gall bladder.

TABLE 12.5 FUNCTIONS OF THE AUTONOMIC NERVOUS SYSTEM

Tissue	Location	Effect of parasympathetic stimulation	Effect of sympathetic stimulation
Smooth muscle	Iris	Contraction of circular fibers, constricting pupil	Contraction of radial fibers, dilating pupil
	Ciliary body	Contraction, making lens thinner	Relaxation, making lens thicker
	Stomach wall	Contraction and increased motility	Inhibition of contraction
	Pyloric sphincter	Inhibition of contraction	Contraction
	Intestinal wall	Increase in tone and motility	Inhibition of motility
	Anal sphincter	Inhibition of contraction	Contraction
	Bladder wall	Contraction	Inhibition of contraction
	Bladder sphincter	Inhibition of contraction	Contraction
	Uterus (pregnant)		Contraction
	Uterus (nonpregnant		Inhibition of contraction
	Bronchioles	Constriction	Dilatation
	Hair follicle		Contraction
Gland	Eye (lacrimal)	Secretion	
	Mouth (salivary)	Secretion of copious thin saliva	Secretion of scanty thick saliva
	Stomach (gastric)	Secretion	Inhibition of secretion
	Liver	Inhibition of glycogen breakdown	Increase in glycogen breakdown
	Pancreas	Increase in enzyme and hormone secretion	Inhibition of enzyme and hormone secretion
	Adrenal medulla		Secretion
	Skin (sweat)		Secretion
Blood vessel	Cerebrum (arteries)	Dilatation	Constriction
	Heart (arteries)	Constriction	Dilatation
	External genitalia	Dilatation	Constriction
	Skin		Constriction
Heart	Heart	Deceleration of rate	Acceleration of rate

pathetic division is concerned with the expenditure of energy and defense in emergencies, widespread activities requiring diffuse distribution. Its functions, if discharged en masse, would be useful to man or a lower animal in an attitude of "fight or flight": acceleration of the heart rate, elevation of the blood pressure, stimulation of the breakdown of liver glycogen, and dilatation of the bronchioles.

In general, organs innervated by fibers from both divisions respond to them in opposite directions.* For example, sympathetic stimulation accelerates the heart rate, and parasympathetic

---

* Reciprocal relationships are not invariable, however. Many organs are controlled dominantly by one of the two divisions or similarly by both, so that active opposition does not occur. For example, parasympathetic stimulation excites the ciliary muscles of the eyes, whereas sympathetic stimulation is without effect, and both divisions excite the salivary glands. Unfortunately, no generalizations are available to permit us to predict whether sympathetic or parasympathetic stimulation will excite or inhibit a particular organ.

stimulation slows it. Through such balanced opposition the two divisions coordinate responses to widely varying internal and external conditions. Accordingly, the autonomic nervous system is essential for the preservation of internal constancy.

Chemical transmission. One of the considerations that originally led J. N. Langley to postulate two divisions within the autonomic nervous system was the discovery of characteristic patterns in the responses to certain drugs. The differences arise because acetylcholine (see p. 421) mediates impulse transmission in the neuroeffector junctions of the parasympathetic division and norepinephrine (see p. 422) mediates it in the neuroeffector junctions of the sympathetic division.* Thus the parasympathetic division is *cholinergic,* and the sympathetic division is *adrenergic.* The transmitters are different only in the junctions between postganglionic neurons and effectors. Acetylcholine mediates impulse transmission in the synapses between preganglionic and postganglionic neurons in *both* divisions.

The cholinergic nature of their neuroeffector junctions completes the justification for grouping cranial and sacral autonomic nerves into a single division. Likewise, the adrenergic nature of their neuroeffector junctions unites the thoracic and lumbar autonomic nerves. Figs. 12.48 and 12.49 reveal one striking exception to what has just been stated. The adrenal gland, whose medulla secretes norepinephrine and epinephrine directly into the blood (Chapter 15), receives special fibers that proceed straight from sympathetic preganglionic neurons in the spinal cord without synapsing. Moreover, these fibers terminate in the adrenal medulla with cholinergic endings. Perhaps this situation reflects the fact that the special secretory cells of the medulla are themselves derived embryologically from nervous tissue and therefore are, in a sense, analogous to postganglionic neurons;

indeed, they even have rudimentary fibers. Since the adrenal medulla secretes norepinephrine and epinephrine in conditions of stress, thereby activating all adrenergic synapses, we should properly consider it the equivalent of a ganglion and a part of the sympathetic division.

A large number of chemical agents have been discovered that, when administered as drugs, either mimic or totally block the actions of a division of the autonomic nervous system. Those causing behavior resembling that following injection or endogenous formation of acetylcholine—that is, enhanced activity of the parasympathetic division—are *parasympathomimetic;* those mimicking the actions of the sympathetic division are *sympathomimetic.* Agents inhibiting the actions of the parasympathetic and sympathetic divisions are, respectively, *parasympatholytic* and *sympatholytic.*†

In general, chemical agents function at one of three possible groups of sites: (1) in the autonomic ganglia, both sympathetic and parasympathetic, where acetylcholine is the transmitter; (2) in the adrenergic junctions between axons of sympathetic postganglionic neurons and effectors; and (3) in the cholinergic junctions between axons of parasympathetic postganglionic neurons and effectors, where the agents act differently from the way they do in the ganglia, although acetylcholine is the transmitter in both instances. At each group of sites, some agents stimulate, and others inhibit, impulse transmission.

Chemical agents excite the cholinergic system by (1) inhibiting the normal transmitter-destroying enzyme, thus causing an accumulation of transmitter (e.g., physostigmine is an anticholinesterase both in the ganglia and at the parasympathetic effectors); and (2) by reinforcing or duplicating normal transmitter action (e.g., pilocarpine directly stimulates parasympathetic effectors). They excite the adrenergic system by (1) combining directly with subsynaptic receptor sites and thus duplicating normal transmitter action (e.g., phenylephrine); and (2) by causing the release of stored nor-

---

* For many years it was uncertain whether the adrenergic transmitter was norepinephrine or epinephrine. We now know it to be norepinephrine, although an injection of either substance produces a mass adrenergic response. (For an instructive review, see the articles by Dale and Euler listed on p. 508).

---

† Although "cholinergic" and "adrenergic" were originally applied only to nerve action, they and *anticholinergic* and *adrenolytic* are now frequently used in referring to drug action.

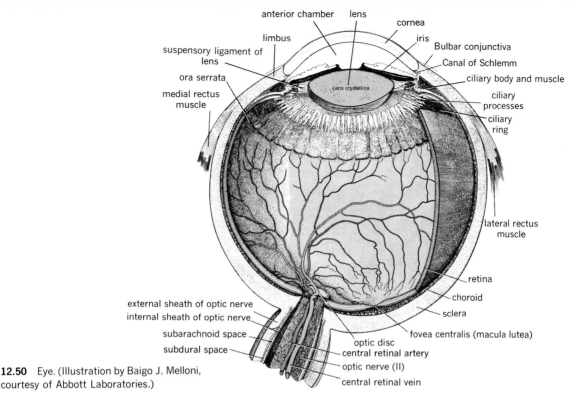

anterior chamber   lens
cornea
limbus   iris
Bulbar conjunctiva
suspensory ligament of lens
Canal of Schlemm
ciliary body and muscle
ora serrata
ciliary processes
medial rectus muscle
Lens crystallina
ciliary ring

lateral rectus muscle

retina
choroid
external sheath of optic nerve
sclera
internal sheath of optic nerve
fovea centralis (macula lutea)
subarachnoid space
optic disc
subdural space
central retinal artery
optic nerve (II)
central retinal vein

12.50 Eye. (Illustration by Baigo J. Melloni, courtesy of Abbott Laboratories.)

epinephrine (e.g., ephedrine and tyramine). Agents that prevent the inactivation of norepinephrine have not yet been found.

Chemical agents inhibit the autonomic nervous system by combining with the transmitter receptor sites (e.g., tetraethylammonium derivatives block transmission in the ganglia, atropine paralyzes parasympathetic effectors, and phentolamine paralyzes sympathetic effectors) and by suppressing the synthesis and release of transmitter (e.g., reserpine blocks norepinephrine production). Each of these neuropharmacological phenomena is of fundamental importance in the study of depolarization, synaptic transmission, and conduction.

Vision

ANATOMY OF THE EYE

Introduction. The *eyeball* is a nearly spherical structure, with a diameter of about 24 mm in an average adult. Anteriorly the transparent *cornea* forms a small convex protrusion upon the surface like the crystal of a

watch (Fig. 12.50). The junction of eyeball and cornea is called the *limbus*.

The wall of the eyeball has three layers. The external layer, the *fibrous tunic*, consists of the *sclera* and cornea. The middle layer, the pigmented *vascular tunic*, includes the *choroid, ciliary body,* and *iris*. The inner layer, the *nervous tunic*, is the *retina*. The retina contains light-sensitive neural elements that serve as receptors of the visual sense. By means of an optical system consisting of refracting media—*cornea, aqueous humor, lens,* and *vitreous body*—and a diaphragm—iris—images of the outside world form upon the receptors. They transform the images into nerve impulses, which are transmitted through special nerves to the visual cortex. How the receptors convert light energy into nerve impulse patterns and how the brain interprets the patterns are among physiology's most perplexing problems.

Fibrous tunic. The sclera is the tough white fibrous coat that covers the eyeball, except at the cornea, and gives it its protective semirigidity. The *ocular conjunctiva,* a transparent mucous membrane continuous with the inner surfaces of the *eyelids,* extends over it anteriorly. The small blood vessels visible through the conjunctiva in a living eye lie in the outermost part of the sclera, the *episcleral tissue.* The sclera itself is not abundantly supplied with blood vessels.

The cornea is the transparent anterior portion of the fibrous tunic, composed of a specialized form of fibrous tissue. Its remarkable transparency is due in part to the total absence of capillaries. It is nourished by aqueous humor on the inside and by *perilimbal capillaries* around its border.

The outer surface of the cornea bears a thin layer of epithelium that is continuous with the epithelium of the conjunctiva,* and its inner surface bears a layer of endothelium. These layers control the water content of the cornea. If either is damaged, the cornea takes up water and loses its transparency. Injuries to the fibrous portion of the cornea heal slowly and leave an opaque scar. The cornea receives branches from the *ciliary nerves* and is extremely sensitive to touch.

Vascular tunic. The choroid is a thin dark-brown membrane lining the sclera. Anteriorly it joins the ciliary body at the *ora serrata*, and posteriorly it contains an opening for the optic nerve (II). The choroid consists almost entirely of blood vessels and provides nutrition to the retinal receptors. It also contains numerous irregularly shaped pigment cells that contribute to the dark color of the interior of the eye.†

The ciliary body has a triangular shape in section, with one side lying against the sclera. Its three components are the *ciliary ring*, the *ciliary processes*, and the *ciliary muscle*. The ciliary ring is the posterior part of the ciliary body, continuous with the choroid. The ciliary processes, some 70 outpouchings of the ciliary body, are arranged meridionally, radiating from behind the iris and tapering toward the ciliary ring. They occupy the anterior third of the ciliary body. Their surfaces are covered by two layers of epithelium (one pigmented and one unpigmented) that secrete aqueous humor in the same way that the choroid plexuses secrete cerebrospinal fluid (see p. 454). For this reason the ciliary processes are sometimes referred to as the *ciliary gland*. They also provide attachment points for the *suspensory ligaments of the lens*.

The ciliary muscle is composed largely of meridional fibers, with a small band of circular fibers. Contraction pulls the ciliary processes forward, slackening the suspensory ligaments so that the lens becomes thicker and more convex than usual. In this way it *accommodates* for near vision.

The iris is the most anterior portion of the vascular tunic. Drawn snugly around the outer front margin of the lens, it is the colored adjustable diaphragm that controls the size of the lens opening, or *pupil*, and therefore the amount of light entering the eye. The black appearance of the pupil is due to the dark interior cavity of the eyeball seen through it. The position of the iris depends on two sets of smooth muscles, the circular pupillary *sphincters* and the radial pupillary *dilators*. The dilators receive sympathetic preganglionic fibers from the first and second thoracic ganglia via the first cervical ganglion and sympathetic postganglionic fibers from the opthalmic branch of the trigeminal nerve (V) via the *long ciliary nerve*. The sphincters receive parasympathetic fibers from the oculomotor nerve (III). Preganglionic fibers arise in the midbrain and extend to the *ciliary ganglion* behind the eyeball; postganglionic fibers enter in the *short ciliary nerve* and extend to the ciliary muscle and the sphincters.

The pigment, or lack of pigment, in the iris determines eye color. Although its genetics are complex, suggesting that more than one gene is involved, blue eye color generally behaves as a recessive trait, and brown eye color as a dominant trait. Albinism, in which the iris lacks all pigment and appears pink, also behaves as a recessive trait. The iris has two layers, which derive from retina and choroid. The inner (retinal) layer is deeply pigmented. The outer (choroidal) layer is a loose connective tissue mass with a scattering of pigment cells. When the outer layer is thin and its cells contain little or no pigment, the iris appears blue. With a greater amount of pigment in the outer layer, the iris appears gray, greenish, or brown, in the order of increasing pigmentation.

Nervous tunic. The retina is a highly differentiated sensory apparatus comprising receptors and nerve fibers that conduct impulses toward the brain. It is most sensitive in the posterior wall of the eyeball. It continues forward in front of the ora serrata but becomes gradually thinner and more insensitive to light.

Though the light-sensitive portion of the retina is only 0.4 mm thick, it consists of a complex arrangement of 10 layers (Fig. 12.51): (1) an outer layer of *pigment epithelium*, low columnar epithelium containing a pigment that prevents diffusion of stray light; (2) a layer of specialized photosensitive receptors called *rod* and *cone cells* because of their shapes; (3) and *external limiting membrane*; (4) an *outer nuclear layer*, the nuclei of the cones being generally in contact with the external limiting membrane and those of the rods next to them; (5) an *outer fiber layer* made up of the axons of the rods and cones; (6) an *inner nuclear layer* containing the nuclei of bipolar and of association (*horizontal* and *amacrine*) neurons; (7) an *inner fiber layer* made up of the axons of these cells; (8) a layer of large neurons called *ganglion cells;* (9) a *nerve fiber layer* made up

---

* Actually, the anterior layer of the cornea should be regarded as part of the conjunctiva.

† In many vertebrates the choroid contains a light-reflecting device, the *tapetum lucidum*, a highly colored iridescent layer that may aid vision in nocturnal animals by conserving scarce light rays, turning them back to the retina like a mirror. It is familiar to us from the shining eyes of cats seen in the glare of automobile headlights.

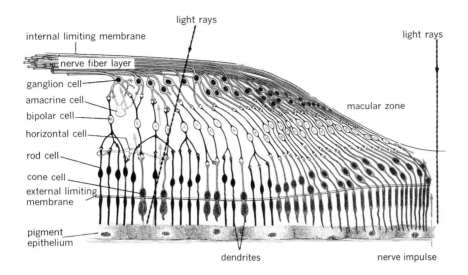

A

internal limiting membrane

light rays

light rays

nerve fiber layer

ganglion cell

amacrine cell

bipolar cell

horizontal cell

rod cell

cone cell

external limiting
membrane

pigment
epithelium

macular zone

dendrites

nerve impulse

B

retina

choroid

internal limiting membrane

nerve fiber layer

ganglion cell layer

inner fiber layer

inner nuclear layer

outer fiber layer

outer nuclear layer

external limiting membrane

rod and cone layer

pigment epithelium

choriocapillary layer

12.51 Retina: A, diagram; B, photograph. (A, illustration by Baigo J. Melloni, courtesy of Abbott Laboratories. B, from S. Polyak, *The Retina*, University of Chicago Press, 1941.)

of their axons, which eventually form the optic nerve (II); and finally (10) an *internal limiting membrane*.

The rods and cones are located in the posterior part of the retina facing *away* from the light source. We might expect that, in a "logically" constructed retina, the tips of the receptor cell bodies would point toward the light source and the fibers conducting impulses to the brain would be more deeply placed. Although some invertebrate eyes are so organized, vertebrate eyes are not. This anomalous situation is usually attributed to the fact that the vertebrate retina develops from the lateral wall of the brain. In an early embryo the neural groove infolds to form the neural tube, which then expands to form the brain. A local invagination occurs in the eye region, so that the receptors are located at the back of the retina. A vertebrate retina is therefore termed an *inverted* retina.

Certain macroscopic retinal landmarks are of physiological interest. If we take the midpoint of the cornea as the anterior pole of the eyeball and the diametrically opposite point as the posterior pole, we find a small bowl-shaped depression, the *fovea centralis*, situated slightly temporally* in the retina (see Fig. 12.50). Visual acuity is sharpest in the fovea; hence it may be regarded as the *functional* posterior of the eye. In the fovea region the retina contains little more than cones—as if its other layers have been deliberately swept aside to improve retinal sensitivity. Because of the resulting yellowish tinge, the area surrounding the fovea is known as the *macula lutea* (yellow spot). Toward the equator of the eyeball, the retina is thicker, and visual acuity is poorer.

The ganglion cells of the retina produce the long fibers that converge in the optic nerve (II). Since these arise on the internal side of the retina, they must turn and travel through the retina to reach a common junction. There they combine and plunge through the retinal substance, emerging from the eyeball posteriorly. At the point where they emerge, the *optic disc*, there is no room for rods and cones. Thus this small area is insensitive to light and creates a *blind spot* in the visual field.†

---

*In descriptions of the eye, the terms *temporal* and *nasal* refer, respectively, to the lateral and medial halves of the eye. The temporal portion of the retina perceives the nasal visual field, and the nasal portion of the retina perceives the temporal visual field.

† A simple experiment demonstrates the existence of the blind spot. Draw a small circle and a small cross 2 in. to its right on a white card. Close the left eye, and hold the card about 10 in. away from the right eye. While staring at the circle, slowly move the card toward the eye. When the card is about 6 in. from the eye, the cross will disappear from view. When the card is a little closer to the eye, the cross will reappear. The image disappears when it falls upon the blind spot.

The *central retinal artery* enters the eyeball at the center of the optic disc and then divides into four main branches, which with numerous smaller branches cover the retina (see Fig. 12.50). The small retinal veins converge in the *central retinal vein*, which leaves the eyeball at the center of the optic disc. The vessels are visible in the living eye through an ophthalmoscope. Since they are the most easily and directly observable blood vessels of the body, their inspection is an important part of a physician's examination.

Eye cavities and intraocular fluids. Much of the eyeball consists of cavities that function merely as conduits for light rays—and that therefore are filled with suitably transparent substances. The principal cavity of the eyeball, between the lens and the retina, contains the *vitreous humor*, a thick jellylike material within the fibrillar network of the vitreous body. The aqueous humor, a clear, free-flowing, watery fluid, occupies the space between the cornea and the lens. This space is divided into a large *anterior chamber* between the cornea and the iris and a small *posterior chamber* between the iris and the lens.

In terms of fluid and electrolyte turnover, the vitreous humor is almost inert. The aqueous humor, on the other hand, is formed and reabsorbed at an average rate of $2\ \mu l$ per minute. Obviously the total volume of aqueous humor and the *intraocular pressure* depend upon the balance between formation and reabsorption. As noted earlier (see p. 470), aqueous humor is formed by the ciliary processes.‡ From them it flows through the pupil into the anterior chamber, then into the angle between the cornea and the iris, and finally into the *canal of Schlemm*, a special thin-walled circumferential vein that drains into extraocular veins (Fig. 12.52A).

Intraocular pressure, normally about 15 mm of mercury, remains remarkably constant. Abnormal intraocular pressure results chiefly from *dehydration*, which leads to a drop in pressure and a softening of the eye (see p. 210), and *glaucoma*, in which fluid outflow is obstructed and pressure rises painfully (Fig. 12.52B). Glaucoma is a common cause of blindness.

Lens. The lens is a transparent, solid body suspended behind the iris between the anterior chamber and the vitreous body (see Fig. 12.50). It is *biconvex*, the posterior side being somewhat more convex than the an-

---

‡ The epithelium is believed to secrete sodium ions, and possibly chloride and bicarbonate ions. As osmolarity increases, large quantities of water enter the anterior and posterior chambers (see p. 454). Chemically, aqueous humor is nearer to ordinary extracellular fluid than to cerebrospinal fluid (see Table 12.4).

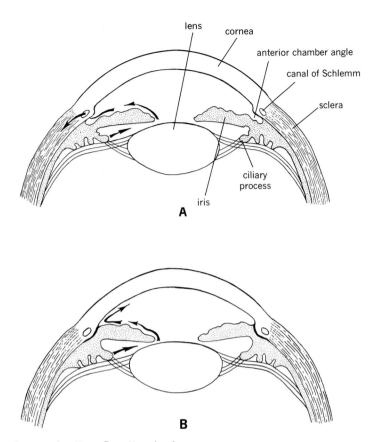

lens
cornea
anterior chamber angle
canal of Schlemm
sclera
ciliary process
iris

A

B

12.52 Circulation of the aqueous humor: A, normal pattern; B, pattern in glaucoma.

terior side, and measures 9 mm in diameter and 4 mm thick in a normal adult. It is held in place by suspensory ligaments that are attached radially to the ciliary processes. Normally the lens is composed of long, slender, transparent yellowish *lens fibers,* which in cross section resemble flattened hexagons. They are arranged in concentric lamellae like onion tissue within a clear, elastic external capsule.

If the lens were free of its suspensory ligaments, the elastic capsule would cause it to assume a spherical shape. However, its edges are normally pulled toward the ciliary body by the ligaments. As mentioned earlier, when the ciliary muscle contracts, the ligaments relax, and the lens becomes more nearly spherical.

Embryological development. We have learned that the eye originates as an outgrowth of the primitive brain (see Fig. 12.28), thus differing from all the other peripheral sense organs. Fig. 12.53 shows the early development of the *optic vesicle* and its subsequent conver-

sion into an *optic cup* lined with the retinal precursors. The optic nerve (II) arises as a stalklike extension from brain to optic cup.

The lens, in contrast to the optic cup and the optic nerve, develops as an ingrowth of superficial ectoderm. Interestingly, it is quite unlike other ectodermal structures, in which old cells on top are continuously replaced by new ones from beneath. The oldest lens fibers are the deepest, in the *embryonal nucleus,* and the new ones are added in overlying concentric cortical lamellae. The layering is visible in the oblique beam of a *slit lamp.* An important cause of the disorder known as *cataract,* in which the lens becomes opaque and milky in appearance and vision becomes progressively impaired, is maldevelopment of the lens.* An observer

---

* Another is excessive radiation. With many types of cataracts, a delicate operation to remove the lens may restore sight. A strong lens in the eyeglasses compensates for the lost natural lens. Since the artificial lens cannot alter its shape like the natural lens, however, visual adaptability is greatly diminished.

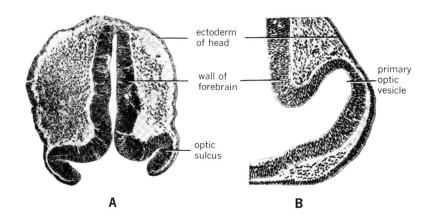

ectoderm
of head

wall of
forebrain

optic
sulcus

primary
optic
vesicle

A

B

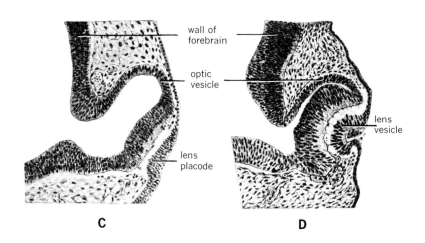

wall of
forebrain

optic
vesicle

lens
placode

lens
vesicle

C

D

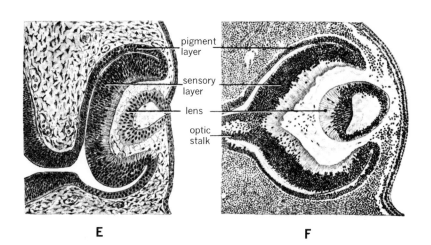

pigment
layer

sensory
layer

lens

optic
stalk

E

F

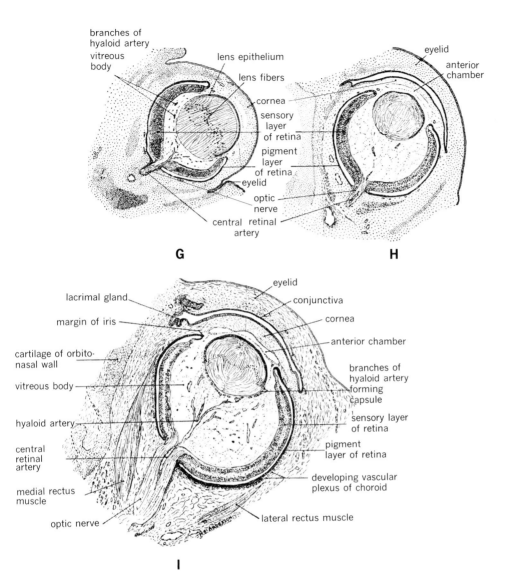

branches of
hyaloid artery
vitreous
body
lens epithelium
lens fibers
cornea
sensory
layer
of retina
pigment
layer
of retina
eyelid
optic
nerve
central retinal
artery

G

eyelid
anterior
chamber
pigment
layer of retina

H

lacrimal gland
margin of iris
cartilage of orbito-
nasal wall
vitreous body
hyaloid artery
central
retinal
artery
medial rectus
muscle
optic nerve

eyelid
conjunctiva
cornea
anterior chamber
branches of
hyaloid artery
forming
capsule
sensory layer
of retina
pigment
layer of retina
developing vascular
plexus of choroid
lateral rectus muscle

I

12.53 Development of the eye: A, at the 14-somite stage; B, at 4 mm; C, at 4.5 mm; D, at 5 mm; E, at 7 mm; F, at 10 mm; G, at 17 mm; H, at 33 mm; I, at 45 mm. (From B. M. Patten, *Human Embryology*, © 1946 McGraw-Hill, Inc. Used by permission of the McGraw-Hill Book Co., New York.)

can determine at what stage opaque fibers have formed by measuring how deeply within the lens they lie.

The sclera and other supporting structures of the eye derive from the mesenchyme surrounding the optic cup, as do the blood vessels in and about the eye. While the future refracting media are still opaque, a large artery, the *hyaloid artery,* runs directly from the optic disc to the lens through the very middle of the eyeball.

The cavity destined to contain vitreous humor becomes filled with a delicate network of fibers of glasslike transparency, the vitreous body.* A depression in the front of the vitreous body accommodates the lens, and a per-

* It is believed that the fibers of the vitreous body derive from the ectoderm lining the optic cup. This origin and their general arrangement would place them in the same category as the neuroglia of the central nervous system (see p. 432).

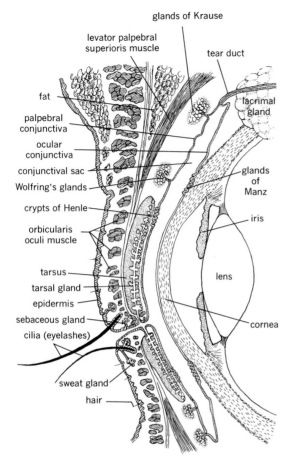

glands of Krause

levator palpebral
superioris muscle

tear duct

fat

lacrimal
gland

palpebral
conjunctiva

ocular
conjunctiva

conjunctival sac

glands
of
Manz

Wolfring's glands

crypts of Henle

orbicularis
oculi muscle

iris

tarsus

lens

tarsal gland

epidermis

sebaceous gland

cilia (eyelashes)

cornea

sweat gland

hair

12.54 Eyelids.

oculi muscles close the eyelids, and the *levator palpe-brae superioris* muscles raise the upper eyelid.

Lacrimal apparatus. The anterior surface of the eyeball is bathed by a constant stream of *tears*, secretions of the *lacrimal gland*. The gland is an oblong lobed structure that reposes in a bony depression above the outer (temporal) angle of the eye (Fig. 12.55). Two sets of accessory glands, the *glands of Krause* and W*olfring's glands*, also secrete tears (see Fig. 12.54). Tear secretion is increased by parasympathetic stimulation.*

Tears are discharged onto the posterior surface of the upper lid through a number of ducts. Much of the fluid is lost by evaporation. The remainder flows toward the inner angle of the eye, where it enters a small conjunctival space, the *lacrimal lake*. From this collecting point it passes by capillary action through two minute

---

* Parasympathetic preganglionic fibers leave the *lacrimal nucleus* in the pons in the facial nerve (VII), pass through the geniculate ganglion, and synapse in the *sphenopalatine ganglion* with postganglionic neurons. Fibers of these travel to the lacrimal gland in the *lacrimal nerve*, a branch of the maxillary division of the trigeminal nerve (V). Some sympathetic fibers, from the *superior cervical ganglion*, also reach the lacrimal gland. Their role is disputed.

manent canal through its substance marks the course of the defunct hyaloid artery.

Accessory structures. The eyeball lies in the pyramid-shaped *orbit*, a bony cavity of the skull, well cushioned with fat, that shields it on all sides except the front. The *eyebrow* helps to protect it from above. The eyelids form a protective curtain over the exposed anterior surface of the eyeball, (Fig. 12.54). The upper eyelid is larger and more freely movable than the lower one. The lids are composed of muscle tissue attached to dense fibrous plates, the *tarsi*, for shape. They are covered externally by skin and internally by the *palpebral conjunctiva*, a thin mucous membrane continuous with the ocular conjunctiva on the anterior surface of the eyeball. Posterior to the bases of the *cilia* or *eyelashes* are the *tarsal*, or *Meibomian*, *glands*. These secrete an oily substance called *sebum*. The *orbicularis*

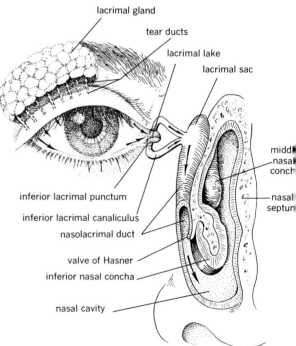

lacrimal gland

tear ducts

lacrimal lake

lacrimal sac

middl
nasal
conch

nasal
septum

inferior lacrimal punctum

inferior lacrimal canaliculus

nasolacrimal duct

valve of Hasner

inferior nasal concha

nasal cavity

12.55 Lacrimal apparatus.

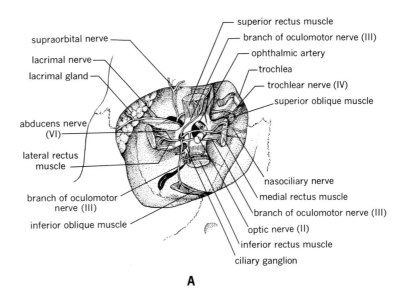

superior rectus muscle
branch of oculomotor nerve (III)
ophthalmic artery
trochlea
trochlear nerve (IV)
superior oblique muscle
supraorbital nerve
lacrimal nerve
lacrimal gland
abducens nerve (VI)
lateral rectus muscle
nasociliary nerve
medial rectus muscle
branch of oculomotor nerve (III)
branch of oculomotor nerve (III)
optic nerve (II)
inferior oblique muscle
inferior rectus muscle
ciliary ganglion

A

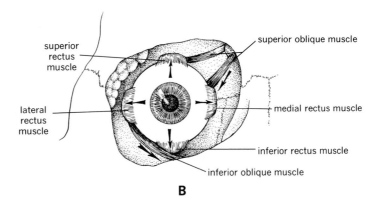

superior rectus muscle
superior oblique muscle
lateral rectus muscle
medial rectus muscle
inferior rectus muscle
inferior oblique muscle

B

12.56 Extrinsic muscles of the eye: A, origins; B, insertions (with arrows indicating the direction).

duct openings, the *superior* and *inferior lacrimal puncta,* into *superior* and *inferior lacrimal canaliculi,* which carry it into the *lacrimal sac.* It continues through the *nasolacrimal duct* and via the *valve of Hasner* into the nasal passage.*

Tear fluid contains dissolved salts, a small amount of mucus, and a highly bactericidal enzyme named *lysozyme,* which gives it antiseptic properties. Any irritation of conjunctival or nasal membranes increases the

flow of tears. So do certain emotional states. Presumably, irritant lacrimation involves local reflexes whereas emotional lacrimation originates in the higher centers of the nervous system. Man is the only creature that weeps in emotional distress. Other animals cry tearlessly like very young human infants, whose lacrimal glands are not yet fully functional. Weeping conceivably originated in evolution as an adaptive trait in species whose young continued to cry for unusually long periods of time. Since crying dehydrates the mucous membranes, the ability to weep prevents mucosal damage.

A mucoid secretion also bathes the eye. This comes from the *crypts of Henle* and the *glands of Manz* (see Fig. 12.54).

Extrinsic muscles. Six separate muscles are attached to the outside of the eyeball (Fig. 12.56). Every eyeball

---

* In lizards and certain snakes, the nasolacrimal duct opens into the mouth, an arrangement suggesting that tears developed to moisten the mouth as well as the eye and the nose. The histological similarities between the salivary glands and the lacrimal glands are well known, and it is of interest that many stimuli (e.g., spicy foods) excite both and some diseases dry up both.

movement involves several of them, but, for movement in each of the main directions, one is predominant.

Four of the six muscles, the *recti* (or straight) muscles, arise from a ring-shaped tendon at the back of the orbit. They insert on the eyeball in positions corresponding to their names. The *superior rectus muscle* attaches to the upper side and turns the eyeball upward. The *inferior rectus muscle* attaches to the underside and turns the eyeball downward. The *lateral* and *medial rectus muscles* attach to the temporal side and the nasal side, respectively, and turn the eyeball accordingly.

The remaining two muscles, the *oblique* muscles, aid in rotation of the eyeball. The *inferior oblique muscle,* shortest of the extrinsic muscles, arises anteriorly from the floor of the orbit near the lacrimal sac, passes backward laterally, and inserts posteriorly on the eyeball under the lateral rectus muscle. The *superior oblique muscle,* longest of the extrinsic muscles, arises posteriorly from the roof of the orbit above the medial rectus muscle, passes forward through a loop of cartilage known as the *trochlea,* turns downward and backward, and inserts posteriorly on the eyeball. The oblique muscles and rectus muscles act together to roll the eyeball.

Man looks directly at objects because visual acuity is greatest at the fovea centralis. As shown in Table 12.3, eye movement is controlled by cranial nerves III, IV, and VI.* The *oculomotor nerve* (III) innervates all of the rectus muscles except the lateral one, which the *abducens nerve* (VI) innervates. The oculomotor nerve also innervates the inferior oblique muscle and carries parasympathetic fibers to the sphincters of the iris and the ciliary muscle. The *trochlear nerve* (IV) innervates the superior oblique muscle.

Simultaneous movement of both eyes in the same direction is called *conjugate movement.* Its coordination in the central nervous system is essential for normal vision but is not fully understood. Lack of coordination, called *strabismus,* has many causes. If prolonged, it results in *double vision* or suppression of the vision of one eye.

OPTICAL ASPECTS

Physical principles of optics. So that we may comprehend the optical aspects of vision, we shall briefly review the physical basis of refraction and focusing. Light travels more slowly through solids and liquids than through gases or a vacuum. If a beam of light rays traveling through air strikes the surface of a second substance (e.g., glass) perpendicularly, the light rays continue on their course, slowed but undeviating (Fig. 12.57A). On the other hand, if the beam strikes the surface of the second substance at an angle (the *angle of incidence*), it is bent toward the portion of the surface that it hits first and is slowed.

The bending of light rays at such an interface is called *refraction.* The *refractive index* of a substance is inversely proportional to the speed at which light travels through it. The refractive index of air is arbitrarily set at 1.0. Thus, since light travels through glass at two-thirds of its speed through air, the refractive index of glass is $1.0 \div \frac{2}{3} = 1.50$. The degree of refraction at the interface between two light-transmitting media of different refractive indices is proportional (1) to the difference between the refractive indices of the two substances and (2) to the angle between the light beam front and the interface.

Fig. 12.57B illustrates the refraction of light by *convex* and *concave* lenses. Rays passing through the center of a lens are not refracted because there the angle of incidence is zero. They are refracted most at the edges. A convex lens causes parallel light rays to converge at a single *focal point* some distance from the lens, whereas a concave lens causes them to diverge.

The distance between a convex lens and the point at which the light rays converge is the *focal length* of the lens. Focal length depends upon whether the incident light rays are parallel or diverging from a *point source* (Fig. 12.57C). The stronger the lens, the greater the refraction, and therefore the shorter the focal length.

Fig. 12.57D shows a convex lens with two point sources of light. Each focal point is on an extension of a straight line running between the point and the center of the lens. Since a large object is a mosaic of point sources, it produces a corresponding mosaic of focal points in a plane on the other side of the lens. If a white surface were placed exactly in this plane, a sharply defined upside-down image of the object would appear on it. If the surface were moved nearer to or farther from the lens, the image would no longer be "in focus." We precisely determine this plane when we focus a camera. Parallel light

---

* Nerves III and IV arise from nuclei below the aqueduct of Sylvius and emerge near the anterior border of the pons. Nerve VI arises from a nucleus in the lower part of the pons, beneath the fourth ventricle, and emerges at the posterior border of the pons in the fissure between the pons and the medulla (see Fig. 12.40).

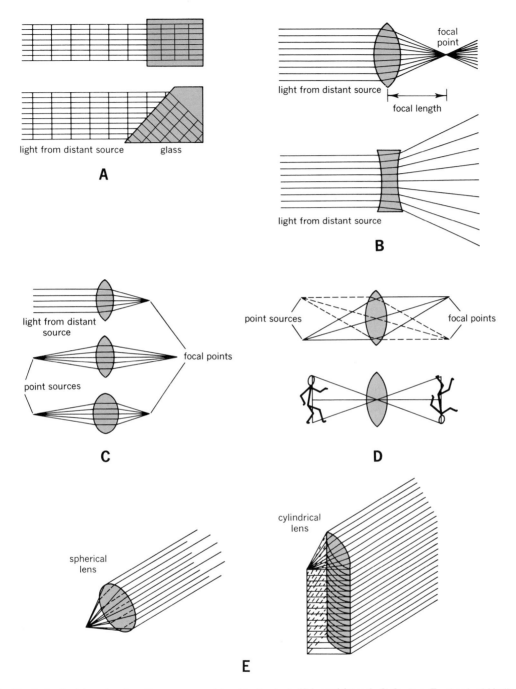

12.57 Physical principles of optics. Figures are explained in the text. (Adapted from A. C. Guyton, *Textbook of Medical Physiology,* 2nd ed., W. B. Saunders Co., Philadelphia, 1961.)

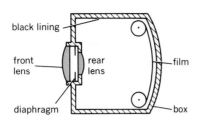 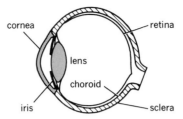

12.58 Comparison of the eye and a camera.

rays entering a convex cylindrical lens come to a focus in a line rather than a point (Fig. 12.57E).

Accommodation. The eye is like a camera (Fig. 12.58) in that a convex lens brings an inverted image to focus upon a light-sensitive surface, a diaphragm controls the amount of light admitted, and a dark interior absorbs stray light that would otherwise be reflected back and forth, obscuring the image. However, a camera ordinarily contains both a front lens and a rear lens, whereas other structures function as a second lens in the eye.

In order to reach the retina, light rays must pass successively through air, cornea, aqueous humor, lens, and vitreous body, whose refractive indices are, respectively, 1.00, 1.38, 1.33, 1.40, and 1.34. Since the degree of refraction depends upon the difference between the refractive indices of adjacent media, the lens has much less refractive power in the eye (where it is surrounded by fluids of comparable refractive indices) than it would have in the air. The greatest refraction occurs at the convex surface of the cornea, whose refractive power is about twice that of the lens. Hence light rays converge extensively before reaching the lens. Although the posterior surface of the cornea is concave— and therefore acts as a weak concave lens— its effect is minimized by the slightness of the difference between the refractive indices of it and aqueous humor.

Thus we see that the refracting system of the eye, like that of a camera, consists of more than one element. Nevertheless, it is the ability of the lens to change shape that permits the eye to accommodate to visual objects at different distances—in contrast to the camera, which accommodates only by changing the distance between lens and film. The eye lens becomes increasingly convex to focus the divergent rays of near ob-

jects upon the retina.* It does so through the *accommodation reflex*. When the image of a near object is not focused sharply upon the retina, sensory impulses pass through the optic nerve (II) to the brain, and motor impulses pass through parasympathetic fibers in the oculomotor nerve (III) to the ciliary muscle. When this contracts, the suspensory ligaments relax, so that the lens becomes more convex and focuses the image more sharply than before. With advancing age the elasticity of the lens and its capacity to accommodate progressively decrease. The result is a condition called *presbyopia*. Special glasses for near vision or distance vision, and frequently *bifocal* glasses for both, are essential.

Convergence. When two eyes gazing straight ahead at a distant object are confronted with a near object, they *converge* and appear "crossed" to an observer. When the near object is held at arm's length and then slowly brought toward the face, a point is reached beyond which a sharp focus is no longer possible. The *near point* is defined as the closest point at which an object can be clearly seen. There the eyes are fully accommodated and highly converged. Convergence is a necessary corollary of accommodation, since it gives the retinal images in the two eyes corresponding locations.

Action of the iris. Light entering the eye initiates the *pupillary light reflex*, whereby the

---

* In practice, light rays coming from objects more than 20 ft away from the eye are considered to be parallel, since they diverge so little. Light rays coming from objects less than 20 ft away diverge significantly. We can easily demonstrate accommodation. If, while gazing at a distant object, we place a finger about 12 in. from our eyes, we shall see the finger, but its image will be blurred. If we then gaze directly at the finger, we shall see it clearly, but the image of the distant object will be blurred.

sphincters of the iris constrict the pupil. The neuronal pathway of this reflex, like that of the accommodation reflex, begins in the retina. Light-induced sensory impulses pass through the optic nerve (II) to nuclei in the pons—specifically, the *pretectal nuclei* and the *Edinger-Westphal nucleus*. Motor impulses return through the parasympathetic fibers of the oculomotor nerve (III) via the ciliary ganglion to the sphincters (see Table 12.5).*

Sympathetic fibers arising in the gangliated sympathetic chain cause pupillary dilatation. This is apparent whenever stress or adrenergic drugs produce massive sympathetic discharges. Dilatation of the pupils in fear or pain is a familiar phenomenon.

Binocular vision, stereopsis, and depth perception. A one-eyed man judges distance or depth chiefly from the relative sizes and positions of objects. For example, he estimates the distance of a plant known to be 6 ft tall from the size of the plant's image on his retina and concludes that an object whose image partly covers the plant's image is nearer to him than the plant is. When he turns his head, the image of the near object moves across his retina more rapidly than that of the distant object. This *moving parallax* is a powerful depth cue in *monocular vision.*

In *binocular* vision (i.e., vision with both eyes) depth perception is improved by *stereopsis*, or stereoscopic vision. When both eyes gaze upon a reasonably near object, each sees the object from a different angle. Hence the two retinal images differ slightly. But, when superimposed and interpreted in the visual cortex, they are perceived as a single image that has depth. If the two retinal images are too discordant (as in strabismus), they do not fuse, and the cortex either perceives both simultaneously (double vision) or oscillates from one to the other, perceiving one while repressing its rival. In such an event stereopsis does not occur. Obviously stereopsis is less useful for distances beyond 20 ft than for those under 20 ft, since convergence is insufficient at the greater distances to cause a disparity of retinal images.

* The pupil also constricts during accommodation for near vision. Thus pupillary constriction and accommodation are associated in much the same way as convergence and accommodation.

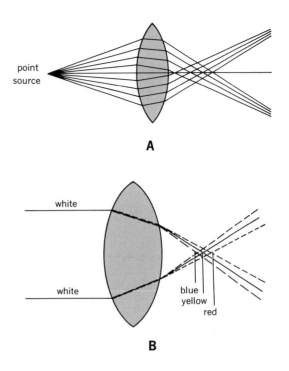

12.59 Aberrations of vision: A, spherical; B, chromatic.

An illusion of depth is easily produced in flat photographs by means of a *stereoscope.* When two pictures taken from slightly different angles, with the difference equalling the angle of eye convergence, are viewed through this device (which places one image upon each retina), they fuse to form an image having depth and reality.

"Normal" aberrations of vision. Lens aberrations create special problems for eye and camera. In lenses with convex spherical surfaces, light rays passing through the edges are focused sooner than those passing through the center (Fig. 12.59A). In a fine camera this *spherical aberration* is corrected by precise modifications of the lenses. The eye's relative freedom from spherical aberration is attributable to the decreasing curvature at the margins of the cornea, its principal refracting medium. Also, the lens is denser at its center, and accordingly refracts light more strongly there, than at its margins. What spherical aberration exists accounts for the loss of image sharpness when the pupil is widely dilated in dim light.

Another major lens aberration, *chromatic aberration,* or color error, is not as well corrected in the eye as spherical aberration (Fig. 12.59B). Every lens refracts light rays of short wavelengths more strongly than those of long wavelengths. Therefore, light rays of different colors are focused at different distances behind it. The result is a blurred image fringed with color. A camera corrects this error by using two different types of glass, whose chromatic aberrations neutralize one another, in the two lenses. The eye lacks such an arrangement but does have properties that partially oppose chromatic aberration. First is a slight yellow color of the lens, which screens out light rays of wavelengths shorter than 365 mμ, the region of the spectrum in which chromatic aberration is greatest. This action of the lens prevents vision in the near ultraviolet region of the spectrum. An individual whose lens has been removed for cataract may consequently have excellent vision in ultraviolet light. Second is the yellow color of the macula lutea, which filters light over the receptors of the fovea centralis, the area of the retina least protected by the filtering effect of the lens color. The filtering action of the macula lutea attenuates vision in the violet and blue regions of the spectrum, in which chromatic aberration is high.

Errors of refraction. An eye is considered normal, or *emmetropic,* if parallel light rays from a distant object focus sharply on the retina when the ciliary muscle is completely relaxed (Fig. 12.60A).*

Hypermetropia (also called *hyperopia,* or farsightedness) is due either to an eyeball that is too short from cornea to retina or to a refracting system that is too weak when the ciliary muscle is completely relaxed. Parallel light rays from a distant object focus *behind* the retina unless the eye accommodates by increasing the lens thick-

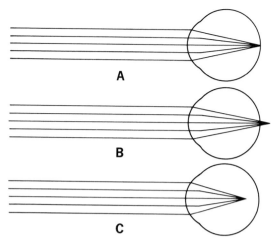

12.60 Normal refraction and errors of refraction: A, emmetropia; B, hypermetropia; C, myopia. (Adapted from A. C. Guyton, *Textbook of Medical Physiology,* 2nd ed., W. B. Saunders Co., Philadelphia, 1961.)

ness (Fig. 12.60B). With accommodation a distant object is seen clearly. With or without accommodation light rays from a near object focus only with difficulty. The result is blurred vision and eyestrain from close work. Correction requires glasses containing the proper convex lenses.

In *myopia,* or nearsightedness, either the eyeball is too long from cornea to retina, or the refractive power of the eye is too great. Parallel light rays from a distant object focus *in front of* the retina, so that vision is blurred (Fig. 12.60C). Since the lens cannot become flatter than it is when the ciliary muscle is completely relaxed, glasses containing concave lenses are the only means of correction. Though a myopic eye views distant objects with difficulty, it sees near objects clearly.

Astigmatism is usually due to imperfect curvature of the cornea. If the cornea curves identically in all directions (vertically, horizontally, etc.), light rays passing through it form a cone of light behind it focusing at a single point. If, for example, horizontal curvature is greater than vertical curvature (in which case the surface resembles the side of a football instead of the side of a basketball), light rays from the greater curvature focus at one point, and those from the lesser curvature focus at another point, farther

---

* In tests for visual acuity, a chart consisting of rows of letters in decreasing type sizes is usually placed 20 ft from the eyes. Each row of letters is marked by a number indicating its readability by a single eye with normal visual acuity. A rating of 20/20 means that the subject at 20 ft can read the same type that a normal eye can read at 20 ft. A rating of 20/100 means that the subject at 20 ft can read only the type that a normal eye can read at 100 ft. A rating of 20/15 means that the subject at 20 ft can read the type that a normal eye can read only at 15 ft.

back. Vision is therefore blurred. Correction requires glasses containing cylindrical or other lenses designed to compensate for the abnormal corneal curvature.

PHOTORECEPTOR FUNCTION OF THE RETINA

Introduction. Both the film of a camera and the retina of the eye contain photosensitive substances that are chemically transformed by light. In the retina these substances are within the rods and cones, the photoreceptors of the retina. Light induces reversible reactions that somehow generate nerve impulses, which are transmitted to the brain. We might say that the photochemical effect of light on the retinal photoreceptors is the key event in the visual process. Its recent elucidation has surely been a major contribution to biological science.

The retinal photoreceptor is a biological transducer par excellence, one whose properties have significance for our understanding of other sensory transducers. Its external physical stimulus or input is a specific form of electromagnetic radiation. In a sense the eye more resembles a television camera than an ordinary camera, for it not only takes a picture but also transmits it via the optic nerve (II). Its output is nerve impulses. Let us consider the chain of events between the arrival of the light image and the departure of the nerve impulses.

Rods and cones. An image projected by the lens of the eye falls upon a complex mosaic of rods and cones—about 125,000,000 per retina. Since only about 1,000,000 nerve fibers leave the retina, we conclude that many rods and cones share fibers (see Fig. 12.51A). Nevertheless, each rod or cone operates as a discrete unit in the process of photoreception.

Structurally a photoreceptor consists of an *inner segment* much like that of an ordinary neuron and a roughly rod- or cone-shaped *outer segment* that is sensitive to light (see Fig. 12.51B). The inner segment is filled with mitochondria—there are more in a cone than in a rod—and is therefore probably a locus of intense metabolic activity. The outer segment contains many membranous discs arranged like a stack of coins (Fig. 12.61). In a rod each disc is bordered by a rigid rim structure. In a cone

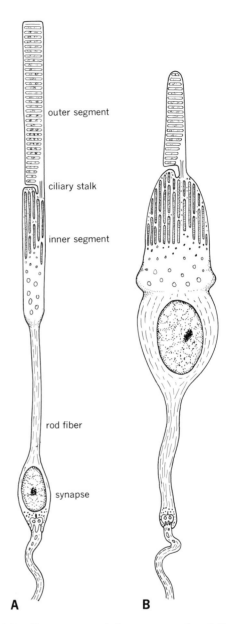

12.61 Fine structure of photoreceptors: A, rod; B, cone.

each disc is formed by an infolding of the cell membrane on the side opposite the *ciliary stalk*. In both rods and cones, the lamellar membranes are 40 to 50 Å thick. We shall later speak of their possible role in photoreception. The ciliary stalk connecting the outer and inner segments of a rod or a cone has the $9 + 2$ structure typical of all cilia (see p. 79).

Rods and cones are distributed unevenly throughout the retina. The number of cones decreases rapidly with increasing distance from the fovea centralis, and the number of rods increases and then decreases. As noted earlier, the fovea itself consists almost exclusively of densely packed cones; the peripheral retina contains many more rods than cones.

Rods and cones differ in their reactions to light in three ways: (1) cones require more light for stimulation than rods; (2) cones are most sensitive to longer wavelengths, whereas rods are most sensitive to shorter wavelengths; and (3) cones can distinguish different wavelengths, whereas rods can distinguish only different light intensities. Thus cones are the organs of vision in bright light and of color vision and rods are the organs of vision in dim light, although their excitation yields only neutral gray sensations.* The relatively insensitive cones are not stimulated at all until the light intensity is about a thousand times as great as the lowest intensity to which the eye responds—that is, until dilute sensations of color arise. Over an intermediate range of intensities, rods and cones function together; but, as brightness increases, the cones come to predominate.

As Fig. 12.51A shows, some cones have individual nerve fibers. Other cones and all rods share nerve fibers with other cones and rods. Thus cones can act individually, whereas rods must act in large groups; rod vision is correspondingly coarse. This means that the change from cone vision to rod vision, like that from slow film to fast film, involves a change from a fine-grained image to a coarse-grained one.

Photochemistry of the retina. Both rods and cones contain light-sensitive pigments, each consisting of a specific colorless protein, called an *opsin*, and a chromophore (i.e., colored group) derived from vitamin A (see Table 4.1).† The isolated and purified rod pigment is a reddish material named *rhodopsin (visual purple)*. The cone pigment has not yet been isolated from the human retina, but the cone pigments of lower animals differ from their rhodopsins only in their opsins.‡

Fig. 12.62 summarizes the chemical changes that occur in rods after light strikes them. Rhodopsin is transformed by light into the unstable compound *prelumirhodopsin*, which spontaneously becomes *lumirhodopsin*. The initial reaction is the only truly photochemical one in visual photoreception; the reactions following it can take place in the dark. Within a fraction of a second, lumirhodopsin decays into *metarhodopsin I* and *metarhodopsin II*, which in turn split into *retinal (visual yellow;* also known as *retinene* or *retinaldehyde)* and an opsin. Because this process converts reddish rhodopsin to yellow products, it is referred to as the *bleaching* of rhodopsin. (Iodopsin is bleached in similar stages.) During bleaching the opsin component undergoes progressive conformational changes, which tend to promote loss of organization and loosening of structure. Retinal and the opsin react slowly with one another to regenerate rhodopsin. Although its regeneration is not nearly so rapid as its original decomposition, it is completed in a few minutes.

All of these pigments are *carotenoid* proteins —that is, they bear carotenoid chromophores to which they owe their color and sensitivity to light. The term "carotenoid" indicates chemical similarity to β-carotene, a red hydrocarbon ($C_{40}H_{56}$) made up of two six-membered rings joined by a long unsaturated chain. Vitamin A (*retinol,* $C_{20}H_{29}OH$) consists of one-half of a β-carotene molecule (one ring and half of the chain) with a hydrogen and a hydroxyl group added at the broken chain end. Retinal is vitamin A aldehyde.

Only three of the 11 major animal phyla have well-developed image-resolving eyes: the arthropods (e.g., crab, spider, grasshopper), mollusks (e.g., squid, octopus), and chordates (e.g., fish, man). Since they evolved separately, we may conclude that the basic image-forming

---

* Accordingly, the retinas of nocturnal animals, which need mainly to discern moving shapes in the darkness, contain many more rods than cones, whereas cones are dominant in the retinas of animals that go to shelter at sundown.

† We have previously encountered such chromophore-bearing proteins. For example, the hemoproteins (see pp. 33, 232) contain protein and porphin.

---

‡ The cone pigment extracted from chicken retinas by George Wald is *iodopsin*, a yellow-green pigment. The grey squirrel's pure-cone retina has yielded a pigment whose light-absorbing properties are indistinguishable from those of the pigment of the hamster's pure-rod retina.

eye has appeared independently at least three times. And yet eyes from these three phyla contain almost identical visual pigments. Moreover, since no animals can synthesize carotene or vitamin A—they arise only in plants—all vision must depend upon dietary sources.

Why the carotenoid chromophores have such a special significance in the biology of photo-reception has been suggested by Wald and his associates. Wald has observed that the long carotenoid chain with its many double bonds is peculiarly capable of existing in many spatial arrangements. Organic chemistry tells us that, whenever two carbon atoms are joined by a double bond, their positions are fixed with respect to each other. If another carbon atom is

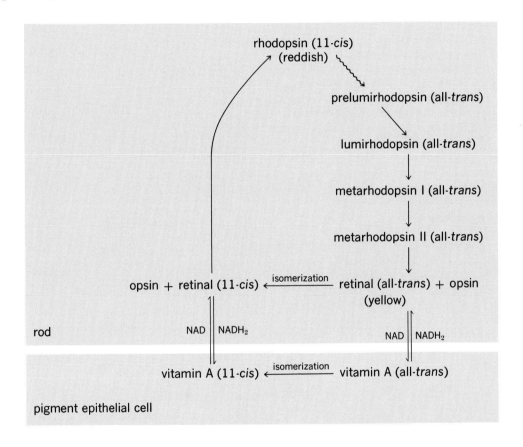

12.62 Rhodopsin cycle. Wavy lines denote photochemical reactions, and straight lines thermal (dark) reactions.

joined to each of the first two, the two new carbons may be on the same side of the double bond (the *cis* position) or on opposite sides of the double bond (the *trans* position). It is obvious from Fig. 12.63 that retinal and vitamin A have large numbers of *cis-trans* isomers, each with a characteristic shape. The all-*trans* molecule is relatively straight; a *cis* linkage at any point bends it. The available evidence indicates that

carotenoid chromophores are uniquely adapted to their visual function by virtue of their capacity for geometrical isomerization.

Only the 11-*cis* isomer of retinal combines with an opsin to regenerate rhodopsin. The 11-*cis* rhodopsin is stable until light acts upon it and isomerizes it to the all-*trans* form, prelumi-rhodopsin. The only thing that light does in any visual system is to initiate this isomerization.

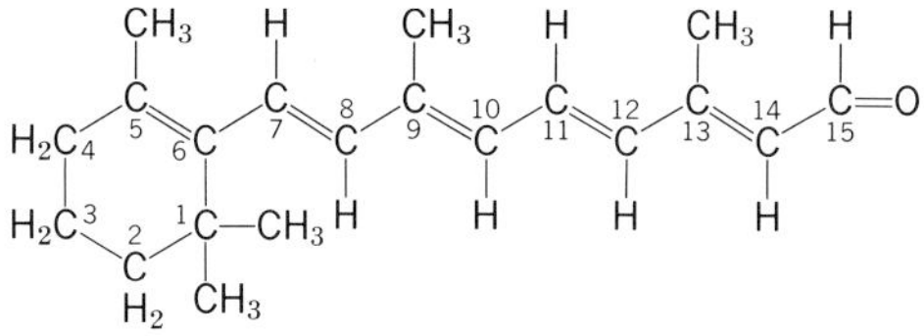

all-*trans* retinal ($C_{19}H_{27}CHO$)

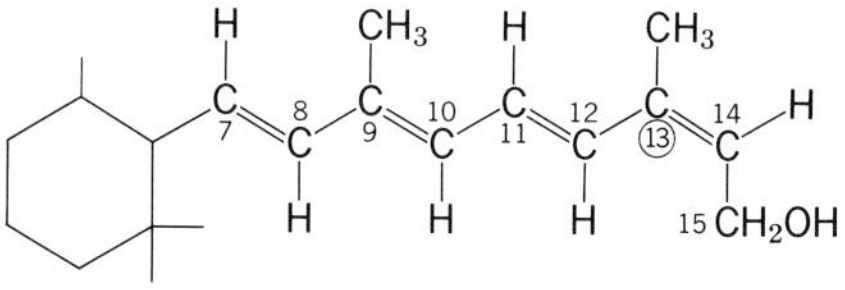

13-*cis* vitamin A

12.63 Retinal, vitamin A (retinol), and some geometrical isomers.

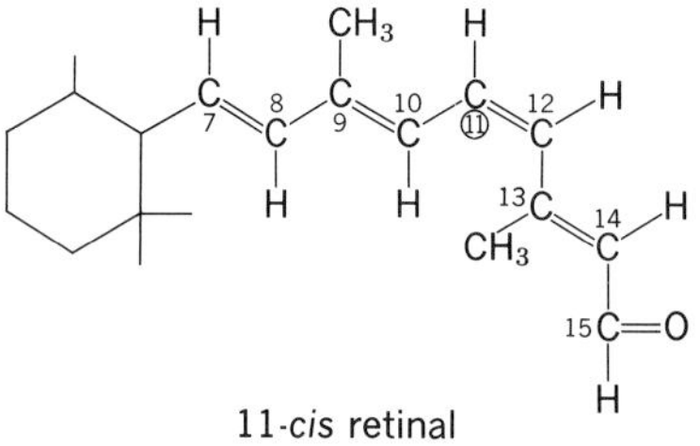

11-*cis* retinal

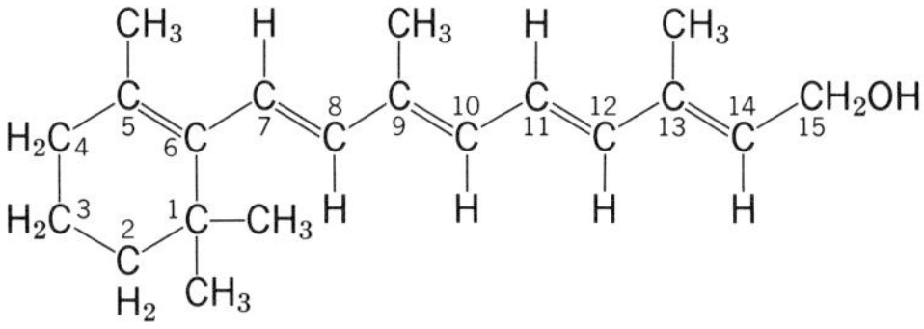

all-*trans* vitamin A ($C_{19}H_{27}CH_2OH$)

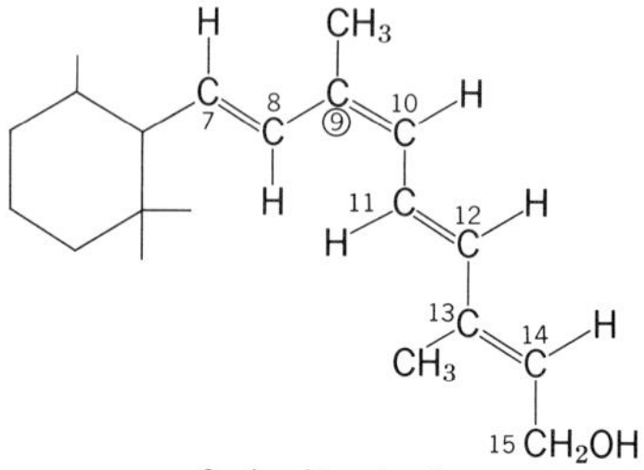

9-*cis* vitamin A

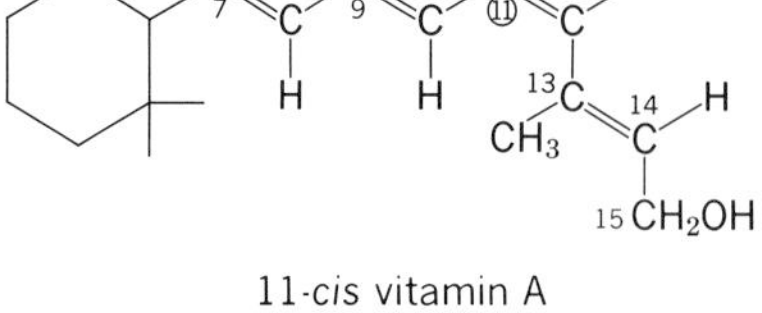

11-*cis* vitamin A

Retinal is believed to be an allosteric effector (see p. 152), which upon isomerization induces conformational changes in its opsin. As opsin is successively modified in the dark reactions that follow, the chromophore gradually ceases to fit the opsin and becomes detached. The liberated all-*trans* retinal is inactive in rhodopsin regeneration. Only when the enzyme *retinal isomerase* catalyzes its reconversion to the active *cis* isomer can it recombine with opsin.

Dark and light adaptation. The principal problem in vision relates to the range of light intensities that the eye must perceive—how to see by starlight and by the light of the noonday sun, brighter by 10 billion times. The versatility of the eye is vastly extended by *dark* and *light adaptation*.

Dark adaptation occurs gradually as the eye accommodates to darkness. The pupil dilates, and the rods become increasingly sensitive. The entire process requires about half an hour. Light adaptation occurs as the eye accommodates to normal or bright light. It is best demonstrated when the light is abruptly turned on in a dark room. For a moment the light appears uncomfortably bright, but in a few minutes the eye adjusts to the new conditions.

These adaptations result in part from changing rhodopsin levels in the rods, to which rod sensitivity is directly proportional. Light adaptation consists in an extensive breakdown of rhodopsin to retinal and the gradual conversion of the retinal to vitamin A, in a simple reversible enzyme-catalyzed oxidation-reduction reaction (see Fig. 12.62). These changes produce a marked drop in the concentration of photosensitive materials and a decreased sensitivity of the eye to light. The slow buildup of rhodopsin during dark adaptation draws upon stores of vitamin A in the pigment epithelium. A fully dark-adapted rod can respond to the absorption of a single photon of visible light.

Some individuals have abnormally poor vision in dim light. This defect is called *night blindness*. When it is due to a dietary deficiency of vitamin A, it is rapidly reversed if vitamin A is added to the diet. When vitamin A starvation is too prolonged, however, supplemental vitamin A will not reinstitute rhodopsin synthesis. Relatively little is known of the quantitative relationship between sensitivity and pigment concentration in cone vision.

Negative and positive afterimages. When an eye gazes steadily at a scene for a period of time, the bright portions of the image cause local light adaptation in the retina, and the dark portions of the image cause local dark adaptation. Hence the retina is "imprinted" with a pattern of insensitive and highly sensitive areas that corresponds to the image.

If the eye is then turned toward a uniform bright surface, the previous image is briefly seen in reverse, as a *negative afterimage*. This is due to the fact that dark-adapted areas of the retina are temporarily hypersensitive and light-adapted areas are temporarily insensitive. A *positive afterimage* has the same color and light-dark values as the preceding image. It persists beyond the actual exposure apparently because nerve impulses continue to develop in the retina for a brief time after the initial stimulus has passed. This phenomenon is exploited in motion pictures. The afterimage remains just long enough for the projector to substitute the succeeding picture in a series. Individual pictures are thus fused and appear to have movement.

Color vision and color blindness. To appreciate the physiology of color vision, we should briefly consider certain physical aspects of color. When Isaac Newton passed a narrow beam of sunlight through a glass prism, he discovered that it fanned out into the band of colors we know as the visible spectrum: red, orange, yellow, green, blue, indigo, and violet. When he reversed the process, gathering the colored beams together with a second prism, color vanished, and white light reappeared. When it was later found that light consists of waves,

these results were readily explained. The order of the colors in the spectrum follows the wavelength, the longest wavelengths falling at the red end of the spectrum and the shortest at the violet end (see footnote, p. 54). A pure color has a single wavelength. A compound color, such as brown, purple, or olive, does not appear in the spectrum and can be obtained only by the mixing of pure colors.

The classic experiments of James Clerk Maxwell and Hermann L. Helmholtz showed that mixtures of three pure colors—red, green, and blue—match all the other colors that can be seen. Accordingly, these three colors—in the red, green, and blue portions of the spectrum, respectively—are termed the *primary colors*. On the basis of this evidence, a *three-color,* or *trichromatic, theory* of color vision was proposed, which held that the eye responds to three different wavelengths of light and that all color sensation is the result of stimulation to varying degrees by these wavelengths. In this view the color seen at any point depends upon the wavelengths issuing from that point and their relative intensities or strengths. Which wavelengths issue from a point, of course, depend upon which wavelengths have been absorbed—since only nonabsorbed light is reflected, or transmitted to the eye.* Color sensation, then, is the brain's interpretation of the wavelengths impinging upon the retina. In this respect, color exists only in the mind.

The three-color theory of color vision implied that there exist three distinctive types of cones, each containing a pigment responding specifically to red, green, or blue light, and that color is uniquely specified by the relative absorptions of the three wavelengths of light by the three cone pigments. In 1963 the three pigments were demonstrated in the goldfish retina by W. B. Marks and in the human retina by Paul K. Brown and George Wald, by recording the distinctive absorption spectra of single cones. They were named *erythrolabe* (red-catching), *chlorolabe*

---

* For example, a ripe tomato in sunlight absorbs most of the wavelengths *except* those of around 600 mμ, namely, those corresponding to red light; hence what is reflected from the tomato onto the retina is seen as red. A "white" object reflects all wavelengths; a "black" object absorbs all wavelengths. The wavelengths absorbed by a substance depend upon its molecular structure.

(green-catching), and *cyanolabe* (blue-catching). Their absorption maxima are, respectively, near 570, 540, and 435 mμ.

In addition to receptors and pigments, color vision demands a nervous system than can extract the requisite information from the receptors. If the photoreceptors containing the three photopigments all fed the same kind of output into the same channel, an organism would be unable to discriminate wavelengths. A neural process is needed that can monitor the output of a receptor (assuming that its output is proportional to the light absorbed by its pigment) and that can transmit data to the visual cortex without confusing it with that from other neural units.

In perceiving color, the eye makes color matches—that is, it is concerned with whether two colors match and ignores the fact that a color pair has different appearances under different circumstances. We learned earlier that the rods adapt to the average brightness of the environment. For this reason it is difficult for the eye to judge the absolute level of light intensity. We estimate the brightness of individual objects in a scene in relation to the mean brightness of the whole scene.* It is obvious that we make color judgments in the same way: we estimate the colors of parts of a scene in relation to the mean wavelength of the whole scene. Color matches depend simply on the absorption spectra of the cone pigments and the wavelengths and intensities of the lights striking them; thus they are quite stable. Appearances vary, however, because they are subject to a whole complex of nervous interactions, not only among cones in the retina but also between sensations and preconceptions in the mind.

Color blindness is defective color vision. It has several forms. Most colorblind people have two-color vision, or *dichromia,* and are therefore called *dichromats.*† Dichromats were early classified on empirical grounds as *protanopes,*

deuteranopes, and *tritanopes.* Recent work supports this classification. Protanopes lack erythrolabe and are red-blind. About 1% of all men and 0.02% of women are protanopes. Deuteranopes lack chlorolabe and are green-blind. About 2% of men and 0.01% of women fit in this category. Protanopes and deuteranopes confuse red and green. Tritanopes lack cyanolabe and are blue-blind. They are quite rare—only one in 20,000 persons, male and female. Wald has proposed more rational names for the three types of dichromia—*anerythropia, achloropia,* and *acyanopia,* respectively.

The fact that the three color-vision pigments contain three different opsins is evidence for the operation of three different genes. Two loci on the X chromosome apparently specify the opsins of erythrolabe and chlorolabe. Red-blindness and green-blindness are inherited as sex-linked recessive traits, with these opsins lacking.‡ Thus, among the offspring of a woman (XX) carrying a gene for red- or green-blindness in one X chromosome and a man (XY) with normal color vision, half the male children will be colorblind, half the female children will be carriers, and the other children will have normal color vision. Since women have two X chromosomes and men have only one, these forms of color blindness are far commoner in men than in women. Blue-blindness is not sex-linked and is a dominant trait.

NEUROPHYSIOLOGY

Neural excitation in the retina. Bleaching—or, more accurately, *cis-trans* isomerization—of the visual pigments by light initiates nerve impulses in the rods and cones. Though it is not known at which point in the bleaching process and by what mechanism the impulses are generated, indications are that neural excitation in the rods and cones involves the same funda-

---

* Although the actual intensity of the light reflected from black paper in bright sunlight is greater than that of the light reflected from white paper at twilight, the former looks black, and the latter looks white.

† *Anomalous trichromats* also exist. They mix the three primary colors to match all the colors they see, but they mix them abnormally.

---

‡ Presumably the genes concerned with color vision determine the amino acid sequences of the three opsins. Thus, mutations causing defective color vision could be of the nonsense variety (see p. 112)—in which an amino acid is deleted, opsin synthesis is blocked, and color blindness results—or of the missense variety—in which an amino acid substitution occurs, a modified opsin is synthesized, and anomalous color vision (e.g., anomalous trichomia) results. In both cases, the other visual pigments would remain unaffected.

mental transducer phenomenon as in other receptors. The absorption of light leads to the production of a graded generator current, which causes the discharge of all-or-none nerve impulses. The challenge has been to explain how a completely dark-adapted rod emits nerve impulses on the absorption of a single photon of light, which can bleach no more than one molecule of rhodopsin. Clearly amplification by a factor of thousands or millions must intervene between the photochemical event and the first neural response.

The neural responses of the retina occur almost immediately after the entrance of light. Hence they appear to be triggered by one of the early steps in the bleaching of rhodopsin, perhaps the formation of lumirhodopsin. It has also been suggested that allosteric transformations of the opsin structure initiate the responses.

Among other hypotheses that have been offered to explain how a small photochemical reaction produces a large depolarization is one that likens the visual excitation system to the proenzyme-enzyme cascade in blood clotting (see Fig. 8.32). Like clotting, visual excitation takes place on surfaces—the membranes forming the discs in the outer segments of the rods and cones (see Fig. 12.61). The thickness of a disc membrane equals the diameter of a rhodopsin molecule (if it is spherical). Thus the membranes may be single layers of visual pigment in contact with soluble components. Light could, in effect, knock a hole in one membrane, thereby initiating a self-propagating depolarization process that would draw large returns from a minimal input.

The brilliant studies of H. K. Hartline imply that light generates all-or-none nerve impulses varying in frequency with the light intensity. Since recording the neural activity of an individual rod or cone in the vertebrate retina is technically difficult, Hartline used the eye of an invertebrate, the horseshoe crab *Limulus*, in his investigations. The production of a graded generator current following light stimulation was revealed by microelectrodes inserted directly into the retinal cells.

Retinal potential and the electroretinogram. When one electrode is placed on the cornea of the eye and another on the back of the eyeball or

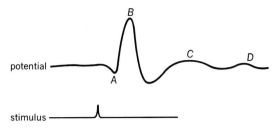

12.64 Typical normal electroretinogram, showing the names of the waves. (Adapted from B. Ziv, *New Eng. J. Med., 264,* 544, 1961.)

even the back of the head, a potential difference of approximately 1 mv is recorded. This is the *retinal potential.* It is present even when the eye is not stimulated by light.

When a light is suddenly shone into the eye, a characteristic series of changes in the retinal potential occurs, which is recorded as an *electroretinogram* (Fig. 12.64). Both the retinal potential and the retinogram result from potentials generated in the rods and cones and the neurons of the inner nuclear layer of the retina.

Interactions among retinal receptors. The retina is considerably more than a mere collection of discrete receptor-transducers of light energy with passive connections to the central nervous system. It is also a system that sorts and processes data in such a way that significant features are stressed and useful interpretations are made. The retina is itself a highly organized structure containing elements that participate in many integrative functions—summation, inhibition, and temporal, spatial, and chromatic interactions—even before transmitting visual information to the brain. The fact that developmentally the retina is part of the brain leads us to expect that it includes many neural cross-connections and integrating devices.

The most recently discovered of these is an arrangement of the processes of bipolar, amacrine, and ganglion cells (see Fig. 12.51A) whereby stimulation of an amacrine cell by a bipolar process causes an inhibitory feedback onto the bipolar cell by an amacrine process. This arrangement, common in the brain, reduces the sensitivity, or gain, of the bipolar cell in proportion to the level of its excitatory activity, and provides a second mechanism for dark and light adaptation.

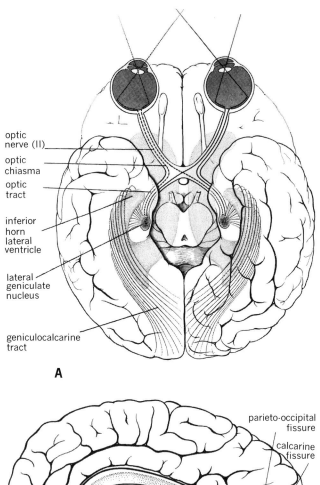

optic
nerve (II)

optic
chiasma

optic
tract

inferior
horn
lateral
ventricle

lateral
geniculate
nucleus

geniculocalcarine
tract

A

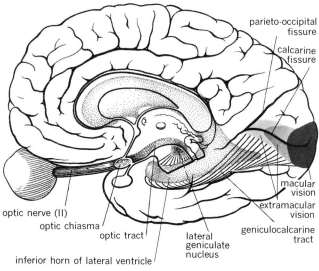

parieto-occipital
fissure

calcarine
fissure

macular
vision

extramacular
vision

geniculocalcarine
tract

lateral
geniculate
nucleus

inferior horn of lateral ventricle

optic tract

optic chiasma

optic nerve (II)

B

12.65 Pathway of nerve impulses from the retina to the visual cortex. A, view from below; B, lateral view. (Illustrations by Baigo J. Melloni, courtesy of Abbott Laboratories.)

The first step in the processing of visual data occurs in the instant when an image spreads across the mosaic of retinal receptors. Precise analysis shows that stimulation of a receptor causes inhibition of its immediate neighbors. Thus the response of a single receptor is influenced both by the stimulatory light shining on it and by the inhibitory activity of its neighbors. A brightly lighted receptor inhibits a dimly lighted one more than a dimly lighted one inhibits a brightly lighted one. For this reason an image is distorted in the nerve fibers emerging from the retina. The distortion serves a useful purpose, however, since it emphasizes *contrast*. And we detect an image only by virtue of the contrast between its light and dark areas.

Transfer of visual information to the brain. The pathway of nerve impulses from the eye to the visual cortex is illustrated in Fig. 12.65. After leaving the retina, impulses pass backward through the optic nerve (II) to the lateral geniculate nucleus in the thalamus. From there they are conveyed by the second-order neurons of the *geniculocalcarine tract* to the visual cortex, in the calcarine fissure of the occipital lobe of the cerebrum.

The optic nerves from the two eyes converge in the X-shaped optic chiasma, but only the fibers from the nasal halves of the two retinas cross to the opposite sides. Fibers from the temporal halves do not cross. It is important, therefore, to distinguish the optic *nerves*, which enter the chiasma, from the optic *tracts*, which leave it and which contain fibers from both retinas. The right optic tract carries left nasal and right temporal fibers, and the left optic tract carries right nasal and left temporal fibers. Obviously, if the left optic *nerve* is cut, the left eye is totally blinded. If the left optic *tract* is cut, the right half of the visual field in both eyes is eliminated through loss of the rightward-looking right nasal and left temporal retinal receptors. In other words, all fibers from the left-hand sides of both retinas (the rightward-looking sides) are bundled together and sent to the left side of the brain, and those from the right-hand sides are sent to the right side of the brain. Similarly, fibers from the upper and lower halves of the retinas travel to the corresponding upper and lower halves of the visual cortex. The chiasmic crossing results in good coordination between eyes and brain, since images from both retinas are sent to each side of the brain and each retina has connections to both sides of the brain.

Image representation in the visual cortex. Images from the macula lutea project upon the occipital pole of the visual cortex, and those from the peripheral portions of the retina project in concentric circles farther and farther forward from the occipital pole.

Evidence suggesting that each fiber in the optic nerve eventually connects to a specific location in the visual cortex has emerged from recent experiments. It was first observed that a severed optic nerve in a frog or newt grows together again. This discovery prompted workers to attempt certain anatomical rearrangements after severing the optic nerve, such as turning the eyeball upside down and reconnecting the nerve or connecting the nerve directly to the optic tract of the same side, thus eliminating the chiasmic crossing. They found that, after such manipulations and regrowth, (1) clear vision returns—hence the correct spatial relations of the nerve fibers must be reestablished; and (2) an image is perceived in a predictable manner. For example, a frog whose right optic nerve is grafted to the optic tract of the same side strikes to the right when an insect is placed at his left. Moreover, he never corrects his error. The results lead to two important conclusions. One is that individual optic nerve fibers travel to specific central points and that some sort of matching process occurs to rejoin the correct fiber ends after nerve section. The other is that the perception of distances and patterns is an inherent property of the visual cortex and is not a product of learning or experience, as was once believed.

Hearing and Equilibrium

ANATOMY OF THE EAR

Introduction. Hearing is a specialized sense whose external stimulus is the vibratory energy of sound waves. The ear receives sound waves, distinguishes their frequencies, translates this information into nerve impulses, and dispatches the impulses to the central nervous system.

The auditory apparatus consists of the ear, the pathway to the central nervous system—the *auditory nerve* (VIII)—and the interpreting center in the temporal lobe of the cerebrum. The ear itself is conveniently divisible into three parts: the *external ear,* the *middle ear,* and the *inner ear.*

External ear. The external ear comprises the *auricle* and the *external auditory canal,* or *meatus* (Fig. 12.66). The auricle collects sound waves and guides them into the canal. In lower animals the auricle is highly mobile and can be directed to the source of sound. The human auricle is composed of elastic cartilage covered by skin.

The canal, approximately 1¼ in. long, leads inward to the *eardrum,* or *tympanic membrane.* Slightly more than half of the angulated canal has a bony wall, and the skin covering the auricle extends into it. Close to the external opening are stiff protective hairs and modified apocrine sweat glands that produce *cerumen* (wax).

The eardrum, which makes an angle of about 50 degrees with the floor of the canal, completely separates the external ear and the middle ear. It is a thin fibrous membrane covered with skin externally and mucous membrane internally. All but the small upper *flaccid segment* is a taut vibrating diaphragm. The drum is easily seen when the auricle is pulled upward and backward to straighten the canal.

Middle ear. A small space in the petrous portion of the temporal bone houses the middle ear, whose cavity connects with the pharynx through the *Eustachian tube* (see Fig. 12.66).

Within the middle ear cavity is a chain of three remarkable small bones, or *ossicles:* the *malleus,* the *incus,* and the *stapes.* The malleus resembles a small mallet with its handle attached to the center of the inner surface of the eardrum and its rounded head tightly fitted into a depression in the head of the incus, where it is held in place by ligaments. When the malleus moves, the incus moves with it. The opposite end of the incus joins the head of the stapes, a stirrup-shaped bone whose footplate fits into the *oval window,* through which sound waves are transmitted to the inner ear (Fig. 12.67). The articulation of the incus with the stapes is freely movable, so that the stapes rocks every time the malleus is set in motion by the eardrum.

The two small muscles of the middle ear are the *tensor tympani,* which attaches by a tendon to the handle of the malleus and keeps the tympanic membrane tensed, and the *stapedius,* tiniest of the body's skeletal muscles, which attaches to the head of the stapes and opposes its inward movement.

The Eustachian tube connecting the middle ear with

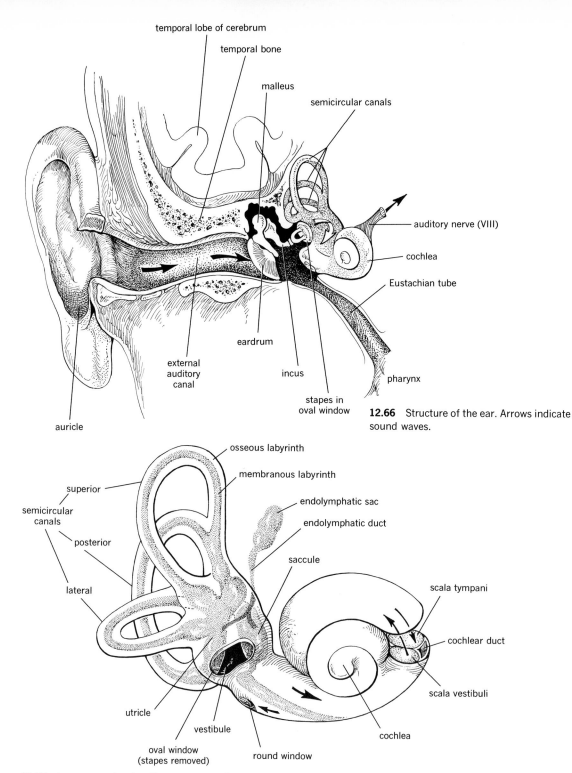

temporal lobe of cerebrum

temporal bone

malleus

semicircular canals

auditory nerve (VIII)

cochlea

Eustachian tube

eardrum

external
auditory
canal

incus

pharynx

stapes in
oval window

auricle

12.66 Structure of the ear. Arrows indicate sound waves.

osseous labyrinth

membranous labyrinth

endolymphatic sac

endolymphatic duct

superior

semicircular
canals

posterior

saccule

scala tympani

lateral

cochlear duct

scala vestibuli

utricle

vestibule

oval window
(stapes removed)

round window

cochlea

12.67 Inner ear, showing the osseous and membranous labyrinths. The stapes has been removed from the oval window, and the passage of sound is represented by black arrows.

the pharynx is about 1½ in. long and ⅛ in. in diameter at its narrowest point. By permitting air to enter the middle ear behind the eardrum, the Eustachian tube equalizes the pressure on the drum.*

Inner ear. The inner ear, or *labyrinth,* has two major intercommunicating divisions: the *anterior labyrinth,* which consists of the *cochlea,* the receptor organ of hearing; and the *posterior labyrinth,* which contains the *semicircular canals, utricle,* and *saccule,* the receptor organs of equilibrium (see Figs. 12.66, 12.67). The

---

* One commonly feels the effects of changing air pressure on the eardrum during rapid ascent or descent in an elevator or airplane. Since the lower pharyngeal opening of the Eustachian tube is closed except during swallowing or yawning, he can deliberately hasten pressure equalization by swallowing or yawning. If for some reason the Eustachian tube is obstructed, equalization does not occur, and the eardrum bulges inward or outward with resulting discomfort and impairment of hearing.

entire apparatus is deeply embedded in the petrous portion of the temporal bone.

The outer, bony wall of the inner ear, the *osseous labyrinth,* encloses a thin-walled *membranous labyrinth* (see Fig. 12.67). A fluid called *perilymph* separates them, and one called *endolymph* fills the membranous labyrinth.

The *vestibule,* a cavity of the osseous labyrinth, is directly medial to the oval window, which opens into it, as do the semicircular canals and the cochlea. The membranous labyrinth generally conforms in shape to the osseous labyrinth, except in the vestibule. There the membranous labyrinth has two separate chambers. The larger, the utricle, communicates with the membranous labyrinth of the semicircular canals, and the other, the saccule, communicates with the membranous labyrinth of the cochlea. Branches from the utricle and saccule merge as the *endolymphatic duct.* This begins as a sinus, narrows, and enlarges again as it terminates in the *endolymphatic sac.*

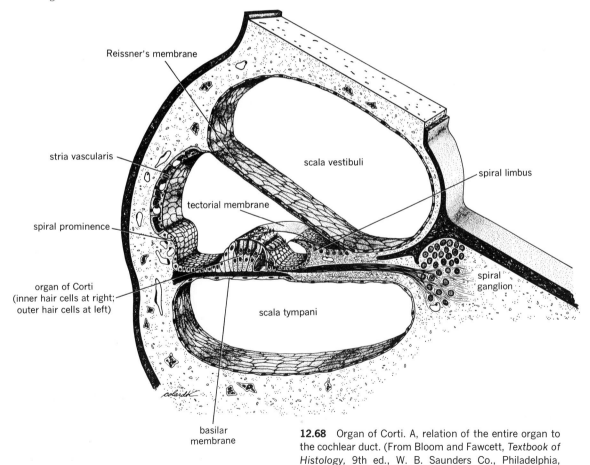

Reissner's membrane

stria vascularis

tectorial membrane

spiral prominence

organ of Corti
(inner hair cells at right;
outer hair cells at left)

scala vestibuli

spiral limbus

spiral ganglion

scala tympani

basilar membrane

A

12.68 Organ of Corti. A, relation of the entire organ to the cochlear duct. (From Bloom and Fawcett, *Textbook of Histology,* 9th ed., W. B. Saunders Co., Philadelphia, 1968.)

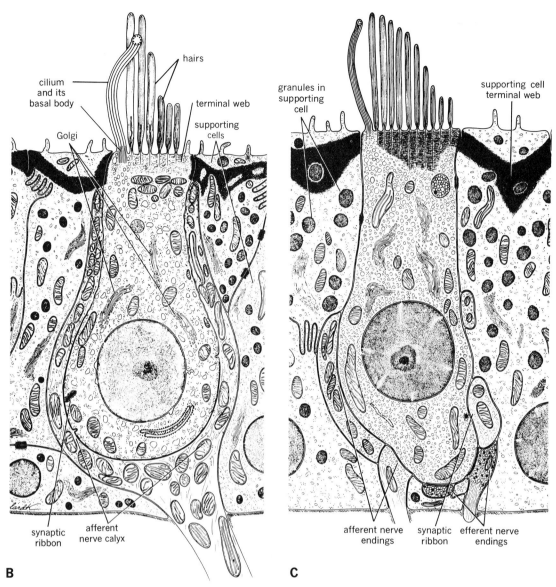

cilium
and its
basal body

hairs

terminal web

supporting
cells

Golgi

granules in
supporting
cell

supporting cell
terminal web

synaptic
ribbon

afferent
nerve calyx

afferent nerve
endings

synaptic
ribbon

efferent nerve
endings

B

C

12.68 (continued)   B, hair cell, Type I; C, hair cell, Type II. These are vestibular hair cells. Cochlear hair cells are structurally similar. (From W. Bloom and D. W. Fawcett, *Textbook of Histology,* 9th ed., W. B. Saunders Co., Philadelphia, 1968.)

The snail-like cochlea arises from the vestibule and spirals 2.75 times around a conical axis whose apex points anteriorly and laterally. Its osseous labyrinth is divided into two compartments: the *scala vestibuli,* which runs from the oval window to the apex; and the *scala tympani,* which runs from the *round window,* just below the oval window, to the apex. There the two passageways converge in the *helicotrema.*

The membranous labyrinth of the cochlea consists of the *cochlear duct,* or *scala media.* This structure, triangular in cross section, is separated from the scala vestibuli above by the thin *vestibular membrane (of Reissner)* and from the scala tympani below by the *basilar membrane* (Fig. 12.68A). Thus the basilar membrane forms the floor of the cochlear duct, extending from a thin bony shelf, the *spiral lamina,* and paradoxically widening as it approaches the blind end of the duct at the apex of the cochlea.

Along the basilar membrane is the *organ of Corti,* which contains rows of flask-shaped *hair cells,* the phonoreceptors. There are two types of hair cells: *inner* hair cells (Type I), numbering about 3,500, in a single row; and *outer* hair cells (Type II), numbering about 20,000, in three rows. These cells have hairlike cilia, that pass through openings in an overlying, supporting *reticular membrane* into the endolymph of the cochlear duct, projecting upward to a specialized fibrogelatinous structure, the *tectorial membrane.* A rich nerve supply, partly afferent and partly efferent, makes profuse synaptic contact with the bases of the hair cells. The network of afferent fibers leads to the *spiral ganglion* in the central portion of the cochlea and on to the *cochlear nerve* and the auditory nerve (VIII) (Fig. 12.69). Sound waves are transmitted from the perilymph to the endolymph and in turn to the hair cells.

PHYSIOLOGY OF HEARING

Physical principles of sound. Sound can be defined both objectively and subjectively. Objectively it is a particular type of wave motion generated in physical matter—whether gas, liquid, or solid—by a specific disturbance. Subjectively it is a perceptual experience of the nervous system.

When a gun is fired, sound waves consisting of alternating condensations and rarefactions of the air result. These waves are usually diagramed as simple curved lines tracing series of hills and valleys. On the other hand, the sound waves produced by most musical instruments consist of vibrations with *overtones,* so that their wave diagrams are complex. Equally complex wave diagrams arise from dissonance and noise.

It is well known that sound travels much more slowly than light. We see lightning many seconds before we hear thunder. Under standard conditions the speed of sound through air is 1,087 ft per second, whereas that of light through air is 186,000 miles per second.

Sound is received by any transducer capable of absorbing the vibrations and converting them to another form of energy or motion. Sound waves striking an inelastic substance like felt are converted into a minute quantity of heat, the

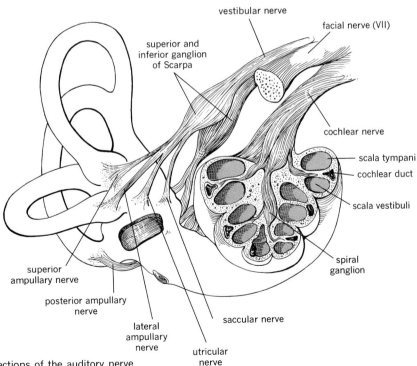

12.69 Branches and connections of the auditory nerve (VIII).

lowest form of energy. Sound waves striking a diaphragm or stretched membrane like the eardrum cause it to vibrate with a comparable frequency and force.

Physiologically the most important aspects of sound are *intensity, pitch,* and *timbre.* Sound intensity is commonly confused with loudness, although the two are not identical. Intensity is the physical energy content of transmitted sound. Loudness is a subjective psychological experience depending upon the auditory mechanisms of the ear and brain. Sound intensity is usually expressed quantitatively in *bels,* the logarithm of intensity. A 10-fold increase in intensity is represented by 1 bel, and a 1.26-fold increase in intensity by 1 *decibel.* As it happens, 1 decibel is the smallest change in intensity that the ear can detect.

Pitch depends upon the frequency of sound vibrations—that is, the number of vibrations per second. The lowest tones that can be perceived have frequencies of about 16 cycles per second, and the highest have frequencies of about 25,000 cycles per second. The highest note commonly employed in orchestral music (disregarding overtones) is the high D of the piccolo, with 4,702 cycles per second. The lowest is the E of the double bass, with 41 cycles per second. Frequencies too high or too low to be heard by the human ear are sometimes audible to dogs and other animals.

Timbre is the peculiar quality of sound that enables us to distinguish one musical instrument or voice from others producing the same note. The difference lies in the overtones. These are partial vibrations resulting from the combination of certain frequencies. Overtones always have frequencies higher than the principal frequency upon which they are imposed.

Physical mechanisms in hearing. Sound is transmitted to the phonosensitive apparatus of the inner ear through a long sequence of vibrating components. As noted, vibrations impinging upon the eardrum cause it to vibrate with the same frequency. The vibrations from the drum travel through the middle ear via the chain of ossicles. The footplate of the stapes rocks in the oval window of the inner ear, creating intermittent pressure upon the perilymph of the cochlea. This forces a thin membrane covering the round window, at the far end of the cochlear duct, to bulge in and out, thus setting up additional small fluctuations in the perilymph that ultimately stimulate the phonosensitive elements of the organ of Corti.

Sound is transmitted to the inner ear with remarkable fidelity. The amplitude of movement of the stapes footplate exactly equals that of the handle of the malleus. Hence the ossicle lever system has no amplifying function. However, the area of the eardrum is 22 times as great as that of the stapes footplate. Therefore, the pressure exerted upon the cochlear fluids is 22 times as great as that of the stapes footplate and accordingly 22 times as great as that exerted upon the eardrum by sound. Since a fluid has greater inertia than air, such pressure magnification is necessary to cause the same degree of vibration in the fluids as in air.

When extremely intense sounds reach the ossicles, a reflex contraction of the stapedius and tensor tympani muscles occurs after a latent period of only 10 mseconds. The opposing actions of the two muscles tend to stiffen the ossicle system, reducing the intensity of the sound transmitted to the cochlea by as much as 30 decibels. This *attenuation reflex* resembles the pupillary reflex of the eye in that it allows the ear to adapt to sounds of different intensities and protects the cochlea from damaging vibrations.

Because the cochlea is embedded in bone, vibrations of the entire skull can produce vibrations in the cochlear fluids. However, when vibrations are identical in the scala vestibuli and the scala tympani, the phonosensitive mechanism is not stimulated. Thus the cochlea's design minimizes the effects of bone conduction.

Function of the cochlea. Fig. 12.70 shows the cochlea as an uncoiled structure. The figure omits the vestibular membrane, which is so thin and flexible that it has no effect on the transmission of sound waves,* and represents the scala vestibuli and the cochlear duct as one element separated from the scala tympani by the basilar membrane.

---

* The only known function of the vestibular membrane is to surround the auditory receptors with a special fluid that may be essential for their function.

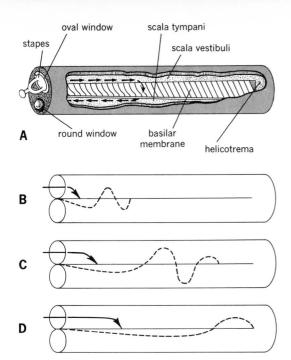

12.70 Sound wave transmission through the cochlea, shown uncoiled. A, movement of fluid following a forward thrust of the stapes; B, C, D, movement of traveling waves along the basilar membrane with high-, medium-, and low-frequency sounds.

Vibrations enter the scala vestibuli through the oval window, creating a hydrostatic pressure that causes the round window to bulge outward. Though the distal ends of the scala vestibuli and the scala tympani connect in the helicotrema, there is not enough time for fluid to flow from one scala to the other during a rapid vibratory sequence. Instead, the fluid wave takes a shortcut through the basilar membrane, forcing it to move back and forth.

How does the basilar membrane function? Does this remarkable structure enable the ear to discriminate sound frequencies? Perhaps the two most notable attempts to answer these questions are *On the Sensations of Tone,* written in 1877 by Helmholtz, and *Experiments in Hearing,* a republication in 1960 of the major papers of George von Békésy. We owe much of what we know of auditory physiology to these two works. Helmholtz compared the crosswise *basilar fibers* of the basilar membrane to the strings of a piano. The shortest fibers (0.04 mm) are at the base of the cochlea, and the longest

(0.5 mm) are at its apex. He recognized that stretched strings, such as those of the piano, vibrate sympathetically with the sound waves that strike them. Hence he suggested that the short fibers of the basilar membrane vibrate in sympathy with sound waves of high frequency and the long fibers in sympathy with sound waves of low frequency. Subsequent research by von Békésy and others showed that Helmholtz' hypothesis is correct—although the phenomenon is somewhat more complicated.

The basilar membrane consists of about 20,000 fibers projecting from the bony center of the cochlea to its outer wall. They are stiff, elastic structures relatively free to vibrate in the manner proposed by Helmholtz. In addition to the progressive increase in fiber length, however, there is also a progressive increase in the "load" of cochlear fluid upon the fibers. When a fiber vibrates, all the fluid between it and the oval and round windows must vibrate simultaneously. The total "load" of fluid on a fiber vibrating near the base of the cochlea is therefore slight compared to that on a fiber vibrating near the helicotrema. This difference is a second factor favoring high-frequency vibration near the windows and low-frequency vibration near the cochlear apex.*

According to current views, inward movement of the stapes causes the basilar membrane at the very base of the cochlea to bulge in the direction of the round window, and the tension built up in the basilar fibers as they bend toward the round window initiates a wave that travels along the membrane (see Fig. 12.70B). The traveling-wave theory implies that vibrations of any frequency ultimately affect the entire basilar membrane. All waves are weak at first, but they become strong when they reach that portion of the membrane whose natural frequency (as determined by fiber length and fluid load) equals the sound frequency. There the basilar membrane vibrates with such ease that the energy of the waves is completely dissipated. Thus the area of maximal vibration in the basilar membrane depends on sound frequency.

---

* The product of the length and the load is called the *volume elasticity* of the fibers. The volume elasticity of the fibers near the helicotrema is 100 times as great as that of the fibers near the stapes. This difference corresponds to a frequency difference of about 7 octaves.

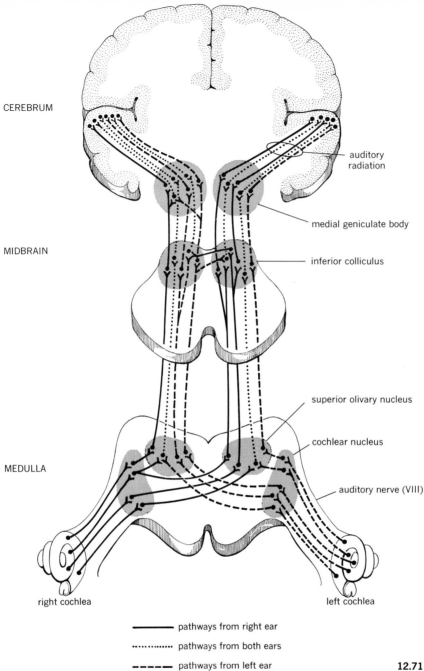

CEREBRUM

auditory
radiation

medial geniculate body

MIDBRAIN

inferior colliculus

superior olivary nucleus

cochlear nucleus

MEDULLA

auditory nerve (VIII)

right cochlea

left cochlea

——————— pathways from right ear

··············· pathways from both ears

– – – – – pathways from left ear

12.71 Auditory nerve pathways.

Mechanism of phonosensitivity. How the movements of the basilar membrane are communicated to the phonoreceptors, how the movements of the receptors trigger physicochemical processes that initiate nerve impulses, and how the auditory information contained in a stimulus is coded in a pattern of nerve impulses in the auditory nerve (VIII) are questions at the core of the "theory of hearing." There is reason to believe that the following events occur. The bending of the cilia of the hair cells causes mechanical deformations of critical areas of the cell surfaces.* The deformations create alternating potentials across the surfaces, which generate alternating receptor potentials, called *cochlear microphonic potentials*. These in turn lead to liberation of a chemical transmitter that stimulates the nerve fibers synapsing at the bases of the hair cells (see Fig. 12.68B,C). The stimuli summate to excite all-or-none impulses in the afferent fibers.

We noted earlier that the cochlear duct is filled with *endolymph*, not perilymph. Endolymph is high in potassium and low in sodium, the opposite of perilymph. Perhaps for this reason there is a continuous potential difference of 80 mv between endolymph and perilymph, with the potential inside the duct positive and that outside the duct negative. This difference is the *endocochlear potential*. It may result from the secretion of K^+ into the duct from the *stria vascularis*, a highly vascular area on the duct's outer wall (see Fig. 12.68A).

The endocochlear potential is important because the tops of the hair cells project into endolymph while perilymph bathes their bodies.† A hair cell has an intracellular potential of -70 mv relative to the perilymph potential. Therefore, the total potential difference between endolymph and intracellular fluid is $70 + 80 = 150$ mv. Obviously this high potential greatly sensitizes the cell and correspondingly increases its ability to respond to slight movements of its cilia.

Auditory neural pathways. Since the pathways of nerve impulses to and from the cochlea are complicated, we shall examine the anatomy of the nerve fibers and their relations to the operation of the phonoreceptors in some detail.

Pitch discrimination is currently explained by the *duplex theory*— a combination of the *place theory*, which holds that the hair cells immediately above the vibrating region of the basilar membrane are selectively stimulated and that they dispatch nerve impulses to a specific brain area where they are interpreted as sound of a given pitch, and the *frequency theory*, which holds that the cochlea, not the brain, distinguishes pitch by transmitting to the brain nerve impulses whose frequencies are exactly the same as those of the original sound waves. Elements of both theories are undoubtedly correct. It has been shown, for example, that the frequencies of nerve impulses carried by an individual nerve fiber have an upper limit and that this upper limit is related to the fiber's position along the cochlea. The specificity for frequencies of the individual fibers greatly exceeds what would be expected from the less specific pattern of basilar membrane activity. It appears, therefore, that there is an interplay between mechanical vibration frequency and neuronal impulse frequency in the discrimination of pitch. von Bèkésy has called the narrowing of the broad membrane response in the neuron response *funneling*.

The main auditory pathways to the central nervous system are shown in Fig. 12.71. Nerve fibers from the spiral ganglion (see Fig. 12.69) enter the *cochlear nucleus* in the medulla, where they synapse with second-order neurons. Most of the fibers of the second-order neurons cross to the *superior olivary nucleus* on the opposite side, and the rest go to the nucleus on the same side, where they synapse with third-order neurons. Fibers from the superior olivary nucleus travel upward to the inferior colliculus, where most of them synapse with fourth-order neurons. Others cross to the inferior colliculus on the opposite side before synapsing. Fibers from the inferior colliculus proceed to the medial geniculate nucleus, where they synapse with fifth-order neurons, the *auditory radiations*, which ascend to the auditory cortex in the temporal lobe of the cerebrum.‡

Nerve impulses from each ear reach the audi-

---

* Because the hair cells of the organ of Corti are stimulated by sound waves, we have classified them as phonoceptors (see Table 12.1). However, they may also be considered special types of mechanoceptors, for, like the Pacinian corpuscles, muscle spindles, and the stretch receptors of the carotid sinus, they are deformed by mechanical forces.

† It is of interest that the fluid in the *tunnel of Corti* (between the outer and inner hair cells) resembles perilymph rather than endolymph in its electrolyte composition.

‡ Pathways to other major brain areas also exist—for example, from the auditory pathways to the reticular formation and from the cochlear nucleus, the inferior colliculus, the reticular formation, and the auditory cortex to the cerebellum. The latter pathways instantaneously activate the cerebellum when a sudden noise enters the ear. There is also an efferent cochlear nerve pathway, the *olivocochlear bundle (of Rasmussen)*. Its function has not been established.

tory pathways on both sides of the brain, with only a slight preponderance of transmission to the opposite side. Since the pathways from the cochlea to the cerebral cortex may consist of from five to seven neurons, some tracts are more direct than others. This probably accounts for the arrival of some impulses at the cortex before others that originated simultaneously.

Little is known of the functions of the various nuclei of the auditory pathways. A lower animal can hear even when its cerebral cortex has been removed; hence the nuclei in the medulla and midbrain presumably operate successfully without it. This situation may reflect the fact that some neurons of the auditory pathways terminate at each station whereas others synapse and still others continue uninterrupted.

Auditory localization. Hearing with two ears, *binaural hearing,* has the same importance for depth perception as binocular vision (see p. 481). Sounds heard through one ear only are almost impossible to localize. Two separated ears receive slightly different sound patterns from a given source, and somehow the brain utilizes the difference in fixing the position of the source.

However, we have just seen that messages from both ears reach both sides of the brain; in fact, most impulses from each ear cross to the opposite side of the brain. A recent experiment has directly attacked this apparent anomaly. Tiny electrodes inserted at various points in the auditory pathways of an anesthetized animal were attached to an amplifier and recorders. Accurately timed simultaneous records were made from both sides while the ears were separately stimulated with clicking sounds. As expected, at every level of the auditory pathways, an input to the ear of the opposite side elicited greater electrical response than one to the ear of the same side. Removal of the auditory cortex immediately eliminated the ability to locate a sound source.

Presumably, then, when the auditory cortex of one side of the brain is stimulated, the sound is interpreted (i. e., heard) as coming from the other side. Since the pathways from the two ears overlap at all but the lowest levels, nerve impulses from one ear have an increasing chance of encountering impulses from the other as they approach the cortex. Depending on the relative positions of listener and sound source, the converging impulses make some groups of cells more active and others less so. The different activity patterns projected upon the auditory cortex are correlated with different locations of sound sources.

PHYSIOLOGY OF EQUILIBRIUM

Organ of equilibrium. The saccule and utricle and the membranous labyrinth of the semicircular canals are special mechanoreceptors, *statoreceptors,* stimulated by changes in head position. They constitute the *statokinetic labyrinth.*

Each of the semicircular canals is oriented in a plane at an approximately right angle to the other two. Thus the canals form a three-coordinate system ensuring that movement of the head around any axis produces sensory stimulation, and thus they are designated *lateral, superior,* and *posterior* (see Fig. 12.67). The osseous labyrinth of the canals has a diameter of about 1 mm, whereas their membranous labyrinth has a diameter of less than ¼ mm. As we have noted, endolymph fills the membranous labyrinth. The canals communicate with the vestibule by five rather than six openings because the medial extremities of the vertical superior and posterior canals converge and enter the vestibule as one channel.

Near the end of each canal is an expansion, the *ampulla.* Fibers of the *vestibular nerve,* the nonauditory branch of nerve VIII, pierce the membranous labyrinth of each ampulla and terminate in a tuft of hair cells, the *ampullary crest* (Fig. 12.72A). Rising above the crest is a wedge-shaped structure, fixed at the base and free at the apex, known as the *cupula.* The hair cells in the ampullary crest send processes upward into the *gelatinous substance* of the cupula. Bending of the cupula in response to movements of the endolymph stimulates the hair cells and initiates impulses in the vestibular nerve (Fig. 12.72B).

As we have seen, the membranous labyrinth does not follow the osseous labyrinth in the vestibule but divides into the utricle and saccule. They communicate directly with the semicircular canals and indirectly with each other through a fine duct. Within the utricle is a thickened oval area, the *macula,* which consists of supporting cells and hair cells (Fig. 12.72C). It is covered by a gelatinous layer containing minute crystals of bonelike material called *otoconia.* They resemble the larger *otoliths* of certain invertebrates and fishes. The weight of the otoconia upon the hair cell processes accentuates the mechanical effects of movements of the endolymph. The saccule has a similar macula. The cells of both maculae give rise to fibers that join to form the vestibular nerve.

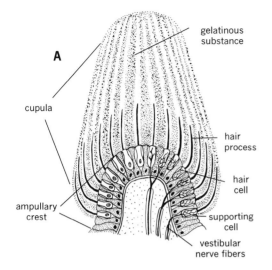

A

gelatinous substance

cupula

hair process

hair cell

ampullary crest

supporting cell

vestibular nerve fibers

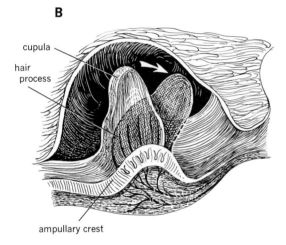

B

cupula

hair process

ampullary crest

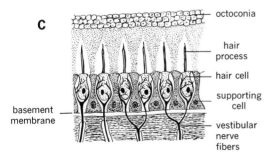

C

octoconia

hair process

hair cell

supporting cell

basement membrane

vestibular nerve fibers

12.72 Receptors of the vestibular apparatus. A, ampullary crest; B, response of the cupula to movement of the endoymph; C, macula. (A and C adapted from G. M. Wyburn, *The Nervous System*, Academic Press, New York, 1960.)

Vestibular neural pathways. The vestibular nerve fibers travel in the auditory nerve (VIII) to the *superior, lateral, medial,* and *spinal vestibular nuclei* at the junction of the medulla and pons (Fig. 12.73). Some of the fibers pass through these nuclei to the flocculonodular lobe of the cerebellum (see Fig. 12.32), and others synapse with second-order neurons, whose fibers also proceed to the flocculonodular lobe. In a sense this lobe is an extension of the vestibular nuclei. The vestibular nuclei also connect with the nuclei of the oculomotor (III), trochlear (IV), and abducens (VI) nerves and with the lower motor neurons of the spinal cord through the extrapyramidal vestibulospinal tract (see p. 436).

Through this complex network the vestibular apparatus plays an essential part in the equilibrium sense, which is closely integrated with the visual and proprioceptive senses. Equilibrium is maintained by the shifting of the body forward, backward, or sideways in accordance with information reaching the vestibular nuclei from the vestibular apparatus, eyes, and proprioceptors.

Vestibular function. The semicircular canals are stimulated by acceleration or deceleration, whether movement is in a straight line or rotational. A familiar reflex response to acceleration in a straight line is the *placing reaction* of a cat: its forelegs extend, and its toes spread to ensure a safe landing when it jumps down from a high place, even if it is blindfolded.

Angular or rotational acceleration elicits a number of interesting responses. An individual being rapidly rotated in a swivel chair displays characteristic involuntary back-and-forth eye movements. This sequence occurs repeatedly: first a slow movement of the eyes in a direction opposite to that of the turning as they attempt to remain fixed on some object; then a quick movement of the eyes in the direction of the turning to focus on a new object. This pattern, called *nystagmus,* is due to both acceleration and rotational motion. When the rotation is stopped suddenly, a postrotatory nystagmus occurs in which the slow movement is in the direction of the rotation and the quick movement is in the direction opposite to that of the rotation.

If the head is erect during rotation, the horizontal semicircular canals are primarily affected,

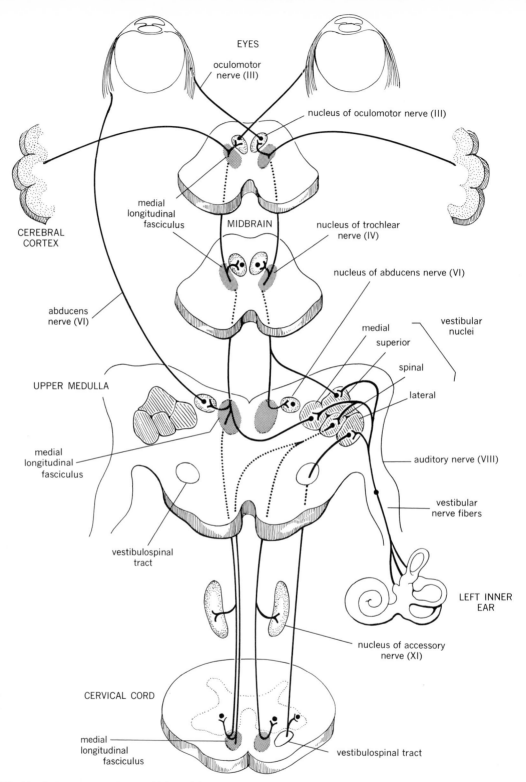

12.73 Vestibular nerve pathways. (Adapted from *The Ciba Collection of Medical Illustrations* by Frank H. Netter, M.D. Copyright CIBA.)

and nystagmus is horizontal. If the head rests on the shoulder in such a way that the anterior and posterior canals lie in the plane of rotation, the nystagmus is vertical, and the eyes move up and down. Nystagmus is a reflex beginning in the vestibular apparatus and traveling from the vestibular nuclei to the nuclei of the oculomotor (III), trochlear (IV), and abducens (VI) nerves, the motor nerves of the extrinsic eye muscles (see p. 478).* Other effects of rotation—nausea, pallor, and sweating—may be due to reflexes involving the autonomic nervous system.

Whereas acceleration and deceleration affect the semicircular canals, the utricle is chiefly affected by gravity. Changes of position involving tilting away from the horizontal plane affect the macula of the utricle. Motion sickness, including seasickness, is thought to be due to its continuous stimulation. The differences between the activities of the saccule and utricle, if any, are not known.

Much contemporary research concerns vestibular function in aviation and space flight. Acceleration is probably the most important of the several physical stresses imposed upon a spacecraft pilot (others being heat, vibration, radiation, weightlessness, and noise). Aside from its direct effects upon the body at large, it gives rise to many sensory illusions. For example, a pilot who is deprived of visual data may have false sensations of climbing or falling during acceleration and deceleration.

Chemical Senses

TASTE

Introduction. Taste and smell are considered chemical senses because their primary stimuli are chemical substances in the mouth and nose.

---

* A similar reflex is involved in *past pointing*. After being rotated, a blindfolded subject is unable to bring his arm straight down in a vertical plane. Instead he points to one side, in the direction that he was rotated—in other words, in the same direction as the slow movement of the eyes in postrotatory nystagmus. In vestibular reflexes of this type, widespread motor stimuli activate the muscles of the neck, trunk, and appendages that attempt to maintain balance during rotation or to regain a normal position after rotation.

They have specialized receptors that differ from the visceral chemoreceptors (e. g., the aortic and carotid bodies) in that they evoke conscious sensations.

What is commonly called taste is a complex sensory pattern made up of the gustatory sense as well as the tactile, thermal, and olfactory senses. We are here concerned only with the gustatory sense. Its significance lies in the role it plays in the selection and enjoyment of food.

Primary taste sensations. When smell is eliminated by blocking of the nasal passages, four primary taste sensations can be demonstrated: *sweet, sour, salty,* and *bitter.* The hundreds of tastes that we perceive are but combinations of the four primary sensations.

Certain areas of the tongue react more strongly to one primary taste sensation than to the others (Fig. 12.74A). The posterior part of the tongue responds to bitter substances (quinine, alkaloids, etc.); the edges of the tongue respond to sour substances (all acids) and salt (sodium chloride, potassium chloride, etc.); and the tip of the tongue responds to sweet substances (sugars and other alcohols). The center of the tongue contains few taste receptors.

To stimulate taste receptors, chemical substances must be present in sufficient concentrations. Some *threshold concentrations* follow: hydrochloric acid, about $0.0009N$ for stimulation of the sour taste; sodium chloride, about $0.01M$ for stimulation of the salt taste; sucrose, about $0.01M$ for stimulation of the sweet taste; and quinine, about $0.000008M$ for stimulation of the bitter taste. Since the bitter taste protects the body against many toxic agents, it is logical that it is the most sensitive of the four primary sensations.

Recent evidence indicates that the ability to taste certain substances is under genetic control. The substance phenylthiocarbamide is extremely bitter to about 65% of a tested population, whereas it is either tasteless or not bitter to the remainder. This variation is one of many that have little apparent adaptive significance and that together account for human individuality.

Taste receptors. The existence of different primary taste sensations suggests physiological differences among the receptors in the various

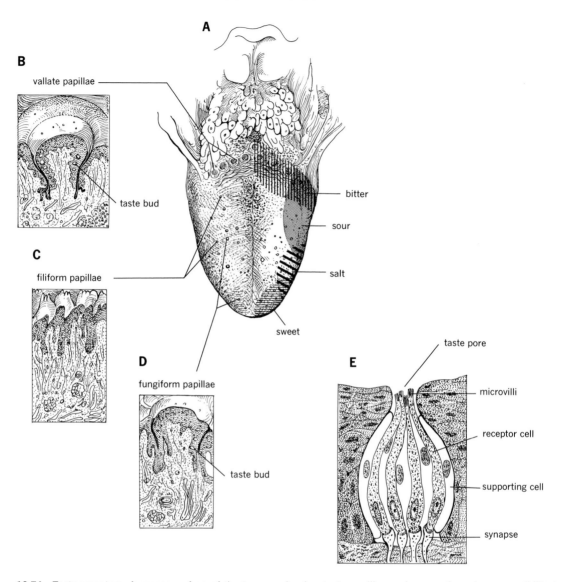

12.74 Taste receptors. A, upper surface of the tongue, showing taste papillae and areas of maximum sensibility to the four primary taste sensations; B, C, D, longitudinal sections of three types of papillae; E, longitudinal section of a taste bud.

tongue areas. Nevertheless, there is only one anatomical type of taste receptor, or *taste bud* (Fig. 12.74E). Furthermore, individual taste buds cannot be assigned to the primary taste sensations since many respond to several of the primary taste stimuli.

Taste buds are minute oval bodies composed of modified epithelial cells, some supporting and others specialized elongated receptors. The receptor cells have small hairlike processes, *microvilli*, that converge toward a small cavity, the *taste pore*, which chemical stimuli enter. Interwoven among the receptor and supporting cells is a branching nerve network. The mechanism of taste bud stimulation is not known. Presumably a stimulant combines chemically with a substance on the surface of a receptor cell, thereby altering the permeability of the surface and generating a volley of nerve impulses.

The taste buds are located in small, rounded elevations called *papillae*, particularly the large *vallate* papillae, which form a V-shaped row at the back of the tongue, and the *fungiform* papillae, which are scattered sparsely over the rest of the tongue (Fig. 12.74B,D). The slender *filiform* papillae, generally distributed on the tongue, contain no taste buds. Although most taste buds are situated on the top of the tongue, many also occur over the palate, the epiglottis, and the tonsillar region. An adult has about 10,000 taste buds, but after age 45 the number gradually decreases.

A taste bud is very sensitive to many stimuli. Yet it is seldom damaged permanently. Studies with labeled thymidine show that individual taste bud cells live about 250 hours. They appear to die from indiscriminate injury, not from age. When they die, they are replaced.

Neural pathways of taste. Two cranial nerves transmit impulses from the taste buds on the tongue. A branch of the *facial nerve* (VII) serves those of the anterior two-thirds of the tongue, and branches of the *glossopharyngeal nerve* (IX) serve those of the posterior one-third. The *vagus nerve* (X) transmits impulses from scattered taste buds in the pharynx (see also the footnote on p. 440).

Fibers from all primary neurons terminate in the gray matter of the pons and medulla (in the tractus solitarius), where they synapse with second-order neurons (see Fig. 12.41). Their fibers proceed to the thalamus, where they synapse with third-order neurons. Their fibers finally lead to the parietal *opercular-insular area* of the cerebral cortex (see Fig. 12.38), which is thought to be closely associated with the sensory information–interpreting area for the face as a whole.

SMELL

Introduction. Smell is the least understood sense. In lower animals the smell sense provides a reliable method of identifying mate, friend, foe, and food. In man it is vestigial and difficult to study.

Olfactory receptors. Smell is a chemical sense whose stimuli are volatile substances entering the nostrils. The *olfactory membrane* is an area of about a square inch covering the upper part of the superior nasal concha and the nasal septum just opposite it high in each nostril (Fig. 12.75).* It is sufficiently far above the ordinary airstream to require sniffing for optimal stimulation.

The olfactory receptor is a primitive type of bipolar neuron derived originally from the central nervous system. It projects into the nasal cavity, where its exposed end forms a *knot* bearing a tuft of minute *olfactory hairs* resembling cilia. The fiber extends inward as part of the *olfactory nerve* (II).

Mechanism of smell. Although generator potentials have been demonstrated in chemically stimulated olfactory receptors—*electroolfactograms* have been recorded—chemoreceptor activity in smell is as yet unexplained. There is no adequate description of the critical features of olfactory stimuli or of the transducer for either quality or intensity of olfactory sensations.

Presumably a chemical substance alters the permeability of an olfactory receptor surface and thereby initiates impulses. The surface probably contains specific chemically sensitive molecules, or "traps," which trigger or inhibit an action current. Some authorities believe that individual receptors respond selectively to large groups of chemical substances (e.g., benzene-like, acetone-like, etc.). In any event, odor-specific elements are clearly analogous to frequency-specific elements in the cochlea and color-specific elements in the retina.† The olfactory membrane has a yellow color, and the discovery that it contains vitamin A and other carotenoids has suggested to some that olfaction, like vision, may involve a *cis-trans* isomerization of a carotenoid chromophore.

---

* The anatomy of the nose is described in Chapter 11.

† The olfactory disorder corresponding to color blindness is *partial anosmia*, or smell blindness. An individual with this disorder is able to detect the almond odor from, say, nitrobenzene but unable to detect the same type of odor from, say, hydrogen cyanide. This condition furnishes evidence for the existence of traps on the receptor surface—certain ones would be missing in partial anosmia.

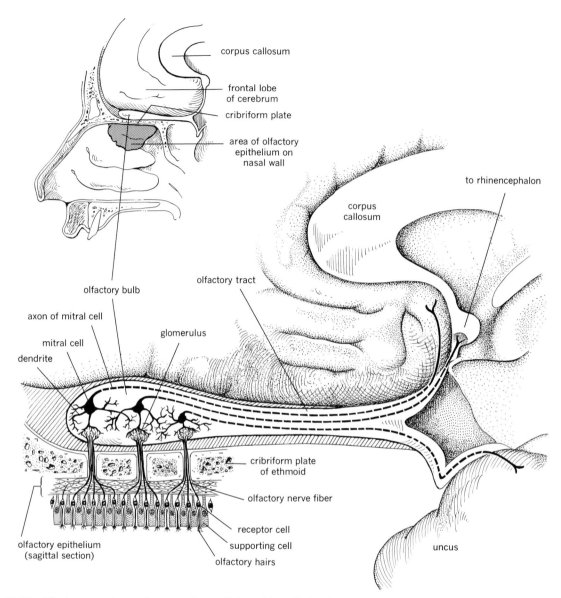

12.75 Olfactory receptors and nerve pathway. (Adapted from D. Krech and R. S. Crutchfield, *Elements of Psychology,* A. A. Knopf, Inc., New York, 1969. Used by permission of the authors.)

There are no primary odors as there are primary tastes. To be smellable, a substance must be volatile at ordinary temperatures and soluble in water and organic solvents. All known odorous substances are either gases or solids or liquids with high vapor pressures. Most inorganic substances have low vapor pressures and hence no discernible odors. Others, such as chlorine and iodine, appear to have odors but actually have only irritant effects.*

* Some authorities hold that such chemical irritation is attributable to a third chemical sense, the *common chemical sense* —a primitive sense from which taste and smell evolved. In lower animals receptors for chemical irritation are spread over the skin; in man only the mucous membranes are sensitive. Its impulses are transmitted to the brain from the membranes of the head via the trigeminal nerves (V). The resulting sensation is distinct from pain, touch, taste, and smell.

It is a familiar experience that olfactory receptors rapidly adapt to a strong odor. On entering a room where someone is smoking or cooking, we are aware of the odor but after a brief time become insensitive to it. This phenomenon is not understood. It is not due to general fatigue of the receptor mechanism, for we readily smell new odors at these times. The rate at which such adaptation occurs increases with the intensity of the stimulus. Another curious phenomenon concerns the nonadditive character of certain odors. The results of mixing them are unpredictable—indeed this is the basis of perfume making (and using). The situation has parallels in the variations, noted earlier, in appearances of objects in different surroundings.

Olfactory neural pathway. Fibers of the olfactory receptors, which constitute the *olfactory nerve* (I), pass through an opening in the bony root of the nose (the cribriform plate of the ethmoid; see Fig. 11.3) to an *olfactory bulb* on the undersurface of the frontal lobe of the cerebrum. There they synapse with second-order neurons called *mitral cells* in elaborate structures called *glomeruli* (see Fig. 12.75). These provide for complex interactions among the fibers within the bulb. Fibers from functionally similar receptors, if such exist, may converge on particular glomeruli. The glomeruli are, in fact, shunting stations that, while relaying incoming impulses to the brain, can also return them to primary neurons. Thus a second signal can be sent to the brain following a single stimulus. This reverberation phenomenon may explain why olfactory receptors are 10,000 times as sensitive as taste receptors.

Axons of the mitral cells join in an *olfactory tract*, which leads inward to the rhinencephalon and terminates in the anterior portion of the pyriform cortex, especially the *uncus* and *amygdala*. From these areas, fibers of third- and higher-order neurons travel to the hippocampus, septum, prefrontal lobe of the cerebrum, caudate nucleus, and mammillary bodies. Despite, or perhaps because of, these many connections, knowledge of olfactory perception is meager.

Perhaps even more than the other senses, smell has powerful psychological associations. Its sensations are so woven into our memories that they frequently recall an introspective panorama of long-past scenes and events.

REFERENCES AND SUGGESTIONS FOR FURTHER READING

General References

American Physiological Society, *Handbook of Physiology. Section 1. Neurophysiology*, Williams & Wilkins, Baltimore, 1959–60.

Ashby, W. R., *Design for a Brain*, Wiley, New York, 1953.

Bishop, G. H., "My Life Among the Axons," *Ann. Rev. Physiol.*, **27**, 1 (1965).

Campbell, H. J., *Correlative Physiology of the Nervous System*, Academic Press, New York, 1966.

Lorente de No, R., "A Study of Nerve Physiology," *Studies Rockefeller Inst. Med. Research*, **131** (1947).

Sherrington, C. S., *The Integrative Action of the Nervous System*, Scribner's, New York, 1906.

Wyburn, G. M., *The Nervous System*, Academic Press, New York, 1960.

Neurons and Neuronal Circuits

Axelrod, J., "Metabolism of Epinephrine and Other Sympathomimetic Amines," *Physiol. Rev.*, **39**, 751 (1959).

Baker, P. F., "The Sodium Pump," *Endeavour*, **25**, 166 (1966).

Davis, H., "Some Principles of Sensory Receptor Action," *Physiol. Rev.*, **41**, 391 (1961).

Eccles, J. C., "The Synapse," *Sci. Am.*, **210**, 56 (Jan., 1965).

——, *The Physiology of Synapses*, Academic Press, New York, 1964.

——, *The Physiology of Nerve Cells*, Johns Hopkins Press, Baltimore, 1957.

Ferry, C. B., "Cholinergic Link Hypothesis in Adrenergic Neuroeffector Transmission," *Physiol. Rev.*, **46**, 420 (1966).

Geren, B. B., "Structural Studies of the Formation of the Myelin Sheath in Peripheral Nerve Fibers," *Symp. Soc. Study Develop. Growth*, **14** (1955), 213 (1956).

Hebb, C., "CNS at the Cellular Level: Identity of Transmitter Agents," *Ann. Rev. Physiol.*, **32**, 165 (1970).

Hodgkin, A. L., *The Conduction of the Nervous Impulse*, Thomas, Springfield, Ill., 1964.

——, "The Croonian Lecture: Ionic Movements and Electrical Activity in Giant Nerve Fibres," *Proc. Roy. Soc. (London), Ser. B*, **148**, 4 (1958).

——, and A. F. Huxley, "Movement of Sodium and Potassium Ions During Nervous Activity," *Cold Spring Harbor Symp. Quant. Biol.*, **17**, 43 (1952).

Hydén, H., "The Neuron," in J. Brachet and A. E. Mirsky, eds., *The Cell*, Vol. 4, Academic Press, New York, 1960.

Katz, B., "How Cells Communicate," *Sci. Am.,* **205,** 209 (Sept., 1961).

Keynes, R. D., "The Nerve Impulse and the Squid," *Sci. Am.,* **199,** 83 (Dec., 1958).

Krnjević, K., "Chemical Transmission in the Central Nervous System," *Endeavour,* **25,** 8 (1966).

Kuffler, S. W., "Excitation and Inhibition in Single Nerve Cells," *Harvey Lectures, Ser. 54 (1958–59),* 176 (1960).

Lim, R. K. S., "Pain," Ann. Rev. Physiol., **32,** 269 (1970).

Loewenstein, W. D., "Facets of a Transducer Process," *Cold Spring Harbor Symp. Quant. Biol.,* **30,** 29 (1965).

———, "Biological Transducers," *Sci. Am.,* **203,** 98 (Aug. 1960).

———, "The Generation of Electric Activity in a Nerve Ending," *Ann. N. Y. Acad. Sci.,* **81,** 367 (1959).

McLennan, H., *Synaptic Transmission,* Saunders, Philadelphia, 1963.

Miller, W. H., F. Ratliff, and H. K. Hartline, "How Cells Receive Stimuli," *Sci. Am.,* **205,** 222 (Sept., 1961).

Nachmansohn, D., *Chemical and Molecular Basis of Nerve Activity,* Academic Press, New York, 1959.

Rosenblith, W. A., ed., *Sensory Communication,* Wiley, New York, 1962.

Schmitt, F. O., "Molecular Organization of the Nerve Fiber," in J. L. Oncley, ed., *Biophysical Science—A Study Program,* Wiley, New York, 1959.

"Sensory Receptors," *Cold Spring Harbor Symp. Quant. Biol.,* **30** (1965).

Strumwasser, F., "Nervous Function at the Cellular Level," *Ann. Rev. Physiol.,* **27,** 451 (1965).

Wilson, V. J., "Inhibition in the Central Nervous System," *Sci. Am.,* **214,** 102 (May, 1966).

Peripheral, Central, and Autonomic Nervous Systems

Bailey, P., "The Seat of the Soul," *Perspectives Biol. Med.,* **2,** 417 (1959).

Benzinger, T. H., "The Human Thermostat," *Sci. Am.,* **204,** 134 (Jan., 1961).

Brazier, M. A. B., "The Analysis of Brain Waves," *Sci. Am.,* **206,** 142 (June, 1962).

Crombie, A. C., "Early Concepts of the Senses and the Mind," *Sci. Am.,* **210,** 108 (May, 1964).

Dale, H. H., "Opening Address," *Ciba Found. Symp. Adrenergic Mechanisms,* 1961.

Dingman, W., and M. B. Sporn, "Molecular Theories of Memory," *Science,* **144,** 26 (1964).

Dobbing, J., "The Blood-Brain Barrier," *Physiol. Rev.,* **41,** 130 (1961).

Eccles, J. C., *The Neurophysiological Basis of Mind,* Clarendon Press, Oxford, 1953.

Euler, U. S., "Twenty Years of Noradrenaline," *Pharmacol. Rev.,* **18,** 29 (1966).

French, J. D., "The Reticular Formation," *Sci. Am.,* **196,** 54 (May, 1957).

Green, J. D., "The Hippocampus," *Physiol. Rev.,* **44,** 561 (1964).

Hardy, J. D., "Physiology of Temperature Regulations," *Physiol. Rev.,* **41,** 130 (1961).

Hartman–von Monakow, K., ed., *The Autonomic Nervous System,* Vols. 1–2, S. Karger, Basel, 1966.

Hydén, H., "Satellite Cells in the Nervous System," *Sci. Am.,* **205,** 62 (Dec., 1961).

Lloyd, D. P. C., "Spinal Mechanisms Involved in Somatic Activities," in American Physiological Society, *Handbook of Physiology. Section 1. Neurophysiology,* Vol. 2, Williams & Wilkins, Baltimore, 1960.

Magoun, H. W., *The Waking Brain,* Thomas, Springfield, Ill., 1958.

Morrell, F., "Electrophysiological Contributions to the Neural Basis of Learning, *Physiol. Rev.,* **41,** 443 (1961).

Triggle, D. J., *Chemical Aspects of the Autonomic Nervous System,* Academic Press, New York, 1966.

Underwood, B. J., "Forgetting," *Sci. Am.,* **210,** 91 (March, 1964).

Wolstenholme, G. E. W., and M. O'Connor, eds., *Ciba Found. Symp. Nature Sleep,* 1961.

Special Senses

Adey, W. R., "The Sense of Smell," in American Physiological Society, *Handbook of Physiology. Section 1. Neurophysiology,* Vol. 1, Williams & Wilkins, Baltimore, 1959.

Amoore, J. E., J. W. Johnston, Jr., and M. Rubin, "The Stereochemical Theory of Odor," *Sci. Am.,* **210,** 42 (Feb., 1964).

Armington, J. C., "Vision," *Ann. Rev. Physiol.,* **27,** 163 (1965).

"Biological Receptor Mechanisms," *Symp. Soc. Exptl. Biol.,* **16,** (1962).

Botelho, S. Y., "Tears and the Lacrimal Gland," *Sci. Am.,* **210,** 78 (Oct., 1964).

Brindley, G. S., "Central Pathways of Vision," *Ann. Rev. Physiol.,* **32,** 259 (1970).

Brown, P. K., and G. Wald, "Visual Pigments in Single Rods and Cones of the Human Retina," *Science,* **144,** 45 (1964).

Crescitelli, F., "Physiology of Vision," *Ann. Rev. Physiol.,* **22,** 525 (1960).

Dartnall, H. J. A., *The Visual Pigments,* Methuen, London, 1957.

———, and K. Tansley, "Physiology of Vision: Retinal

Structure and Visual Pigments, *Ann. Rev. Physiol.,* **25,** 433 (1963).

Diamond, I. T., and W. C. Hall, "Evolution of Neocortex," *Science,* **164,** 251 (1969).

Gernandt, B. E., "Vestibular Mechanisms," in American Physiological Society, *Handbook of Physiology. Section 1. Neurophysiology,* Vol. 1, Williams & Wilkins, Baltimore, 1959.

Graham, C. H., and others, "Symposium on Mechanisms of Color Vision," *Proc. Natl. Acad. Sci.* U.S., **55,** 1311 (1966).

Hartline, H. K., "Visual Receptors and Retinal Interaction," *Science,* **164,** 270 (1969).

Hawkins, J. E., Jr., "Hearing," *Ann. Rev. Physiol.,* **26,** 453 (1964).

Hubbard, R., D. Bownds, and T. Yoshizawa, "The Chemistry of Visual Photoreception," *Cold Spring Harbor Symp. Quant. Biol.,* **30,** 301 (1961).

Hubel, D. H., "The Visual Cortex of the Brain," *Sci. Am.,* **209,** 54 (Nov., 1963).

Katsuki, Y., "Comparative Neurophysiology of Hearing," *Physiol. Rev.,* **45,** 380 (1965).

Land, E. H., "Experiment in Color Vision," *Sci. Am.,* **200,** 84 (May, 1959).

Moulton, D. G., and L. M. Beidler, "Structure and Function in the Peripheral Olfactory System," *Physiol. Rev.,* **47,** 1 (1967).

Oakley, B., and R. M. Benjamin, "Neural Mechanisms of Taste," *Physiol. Rev.,* **46,** 173 (1966).

Pfaffmann, C., "The Sense of Taste," in American Physiological Society, *Handbook of Physiology. Section* Physiological Society, *Handbook of Physiology. Section 1. Neurophysiology,* Vol. 1, Williams & Wilkins, Baltimore, 1959.

Ruston, W. A. H., "Visual Pigments in Man," *Sci. Am.,* **207,** 120 (Nov., 1962).

Wald, G., "The Molecular Organization of Visual Systems," *Symp. Light Life, Baltimore, 1960* (1961).

Wenzel, B. M., and M. H. Sieck, "Olfaction," *Ann. Rev. Physiol.,* **28,** 381 (1966).

Wever, E. G., "Electrical Potentials of the Cochlea," *Physiol. Rev.,* **46,** 102 (1966).

Wolken, J. J., *Vision: Biophysics and Biochemistry of Retinal Photoreceptors,* Thomas, Springfield, Ill., 1966.

To understand digestion we need first a gross, qualitative description of ingesta: what is secreted in response to the meal, how ingesta are transformed by reaction with the secretions and by the propulsive mechanism, what is absorbed and what rejected. This should be followed by an exact quantitative and temporal description in the style of chemical kinetics in which net changes are reported in terms of two-way fluxes. Coupled with this should be a detailed account of the physiological and morphological basis of secretion, transformation, and absorption reaching down to the deepest levels at which anatomy, biochemistry, and biophysics become indistinguishable from one another. Finally, we require a description complete at all levels of controlling processes. Then we can put the fragments of our knowledge together to characterize what happens in the gut. . .

Gastroenterological physiology has its peculiar trouble. What is ingested may range from a ten course dinner served at the Duchess of Guermantes' to repulsive mixtures of mud and mush appropriately called "test meals." Consequently, when we describe what occurs during digestion, we must specify what meal was eaten by what animal under what circumstances.

Horace W. Davenport in Annual Review of Physiology, 1959.

13 Digestive System

Introduction

BIOLOGY OF DIGESTION

All living organisms require food. However, most foodstuffs are useless without preliminary *digestion*. By digestion we mean the hydrolytic breakdown of the relatively large molecules of food materials into simpler compounds that can be absorbed into the body for use as energy sources, tissue components, and food reserves.

In animals digestion is accomplished in two ways: (1) food is ingested phagocytically by individual cells and then digested *intracellularly;* or (2) food is digested *extracellularly* before being absorbed. Since phagocytosis is possible only with particles up to a certain size, we may assume that extracellular digestion arose as an adaptation to the need for breaking up large food masses.

In many lower animals, notably among the protozoa and sponges, intracellular digestion is the only mechanism available for the assimilation of food, but in higher phyla it is not uncommon to find intracellular digestion and extracellular digestion taking place simultaneously. If an animal is carnivorous (e.g., a coelenterate like the hydra), its extracellular digestive enzymes chiefly attack proteins; nonprotein materials are digested intracellularly. If it is herbivorous (e.g., a lamellibranch mollusk), the extracellular enzymes are carbohydrate-splitting. In such vertebrates, as in man, the general disintegration of the food mass by extracellular enzymes is facilitated by mechanical movements of the walls of the digestive tract. Thus the food mass is eventually reduced to particles small enough to be taken by phagocytosis into the cells, where intracellular enzymes complete its digestion.

In animals as complex as man, digestion is entirely extracellular. Indeed, the only remnant of a phagocytic system is found in the fixed and wandering scavenger cells of the reticuloendothelial system (see pp. 223, 254).

Evolution has introduced one other digestive device that is unique to human beings: *cooking*. Though many foods can be digested without cooking, the cooking process performs an initial digestive function without which the human body would find certain foods relatively indigestible. Cooking denatures food proteins and bursts the granules of natural starch, thus facilitating attack by digestive enzymes. Without prior cooking less than half of the raw starch of, say, potatoes or cereals is assimilated by the body, whereas 90% of cooked starch is digested. The only naturally starchy food that does not require cooking is the banana; for this reason it is the first solid food given to babies

UNIFORMITIES OF DIGESTIVE FUNCTION

The chemical reactions involved in digestion are substantially the same in all animals, whether digestion is intracellular or extracellular. They constitute enzyme-catalyzed *hydrolyses*, digestive enzymes all being classifiable as *hydrolases*. As food passes along the digestive tract, it is acted upon by a series of hydrolytic enzymes, each of which catalyzes one particular digestive step or small group of steps. However, the digestive process is not divisible into clearly

TABLE 13.1 MAJOR DIGESTIVE ENZYMES

Source	Name	Optimum pH	Substrates	Products
Salivary glands	Salivary amylase	6.6–6.8	Cooked starches	Disaccharides (maltose)
Gastric glands	Pepsin	1.0–2.0	Proteins	Peptides
	Rennin	4.0	Milk casein	Clotted casein (curd)
Pancreas	Trypsin	7.9	Proteins, peptides	Amino acids
	Chymotrypsin	8.0	Proteins, peptides	Amino acids
	Carboxypeptidase	7.5	Peptides	Amino acids
	Pancreatic amylase	7.0	Starches	Maltose
	Lipase	8.0	Triglycerides (lipids)	Fatty acids, monoglycerides, glycerol
	Ribonuclease	7.0–8.0	RNA	Ribonucleotides
	Deoxyribonuclease	7.0–8.0	DNA	Deoxyribonucleotides
Small intestine	Aminopeptidase	7.8–9.3	Peptides	Amino acids
	Dipeptidase	7.5–9.0	Dipeptides	Amino acids
	Tripeptidase	7.5–9.0	Tripeptides	Amino acids
	Sucrase	5.0–7.0	Sucrose	Fructose, glucose
	Maltase	5.8–6.2	Maltose	Glucose
	Lactase	5.4–6.0	Lactose	Glucose, galactose

defined discrete steps, each yielding specific products. We frequently read that pepsin digests proteins to a point, that trypsin digests the products to a farther point, and so on. Though each enzyme, taken by itself, does carry out certain well-defined operations (Table 13.1), food in the digestive tract is exposed to many enzymes at once. Digestion is a continuous process under the control of a *system*, not a collection, of enzymes, so ordered as to produce a long and complex chain of reactions.

Man has a digestive apparatus that is equipped in its different parts to (1) receive solid food, (2) mince it mechanically into small fragments, (3) secrete required digestive enzymes from the intestines and accessory glands, (4) transport absorbed material into the blood stream for distribution to other parts of the body, (5) remove most of the water from the food residue, and finally (6) eliminate the solid remnant, freed of its useful nutrients. We shall see in the human digestive system how ingeniously nature has designed for the execution of these many functions.

OUTLINE OF DIGESTIVE SYSTEM STRUCTURE

In essence, the digestive system consists of a *tube* that receives, digests, and absorbs food and its nutrients and *accessory glands* that discharge essential enzymes and other digestive agents into the tube. The tube is called the *alimentary tract* or *canal*.* A less elegant but perfectly acceptable term is *gut*.

The major components of the digestive system are shown in Fig. 13.1. The alimentary tract includes the *mouth, pharynx, esophagus, stomach, small intestine* (which comprises the *duodenum, jejunum,* and *ileum*), and *large intestine*. To facilitate discussion, we divide the tract into *upper* and *lower* segments, the point of division being the juncture of the stomach and small intestine. The chief accessory glands are the *salivary glands, pancreas,* and *liver*. Each of these pours its secretions into the main tract through a special duct. In addition, the mucous membrane lining the tract is studded with numerous tiny glandular invaginations that pour important digestive fluids directly into the stream.

---

* *Alimentation* is the act of obtaining nourishment.

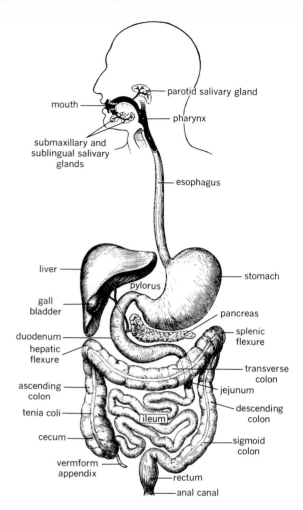

13.1 Digestive system. (From *The Human Body* by Logan Clendening. Illustration by W. C. Shepard and Dale Beronius. © 1927 and renewed 1945 by Alfred A. Knopf, Inc. Reprinted by permission of the publisher.)

The relations of structures and functions in the digestive system, as in other systems, are striking. Though the alimentary tract is basically a tube, it is capable of motility, of propelling its contents forward and, on occasion, backward. Though many glands open into it, it is capable of timing the entry of glandular secretions. Many other examples of such adaptive harmony will become apparent in the following pages.

EMBRYOLOGY OF THE DIGESTIVE SYSTEM

Early stages in the embryological development of the digestive system are shown in Fig. 13.2. It begins to form even before the body of

SAGITTAL SECTIONS

TRANSVERSE SECTIONS

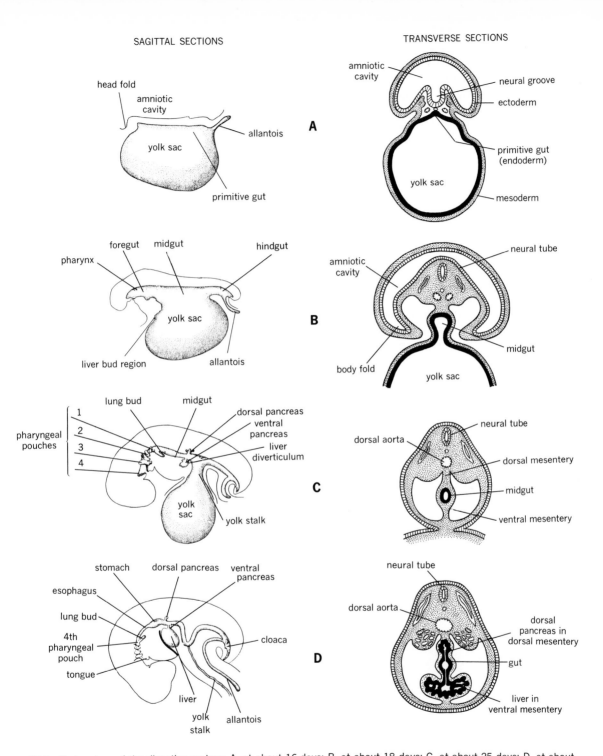

13.2 Embryology of the digestive system: A, at about 16 days; B, at about 18 days; C, at about 25 days; D, at about 4½ weeks. (From B. M. Patten, *Human Embryology*, 3rd ed., © 1968 McGraw-Hill, Inc. Used by permission of the McGraw-Hill Book Co., New York.)

the embryo takes shape. The *primitive gut,* arising from endoderm as an outgrowth of the yolk sac, ultimately becomes the epithelial lining of the adult alimentary tract and its glands. The surrounding tissues, arising from mesoderm, eventually differentiate into the muscular and connective tissue elements of the wall of the adult alimentary tract.

The embryonic body is undercut and delimited until the only communication with the yolk sac and other extraembryonic structures is the yolk stalk. The first portions of the embryonic gut to be definitely incorporated within the body proper are the *foregut* and *hindgut.* The foregut develops areas of dilatation and constriction that correspond to the pharynx, esophagus, and stomach. In the pharynx area four saclike *pharyngeal pouches* push out laterally as though attempting to form gills. The floor of the pharynx near the most posterior of the pharyngeal pouches bulges outward in the *lung bud,* which gives rise to the larynx, trachea, bronchi, and lungs.

Immediately caudal to the stomach are outgrowths that constitute the primitive pancreas and liver. A diverticulum of the main duct from the liver, the *common bile duct,* enlarges into the *gall bladder.*

The elongation resulting in the characteristic intestinal coiling begins in the sixth week of embryonic life and continues as the yolk stalk progressively shrivels. *Mesenteries,* which are folds of peritoneum, support the intestines. The *ventral* mesentery is soon lost, but the *dorsal* mesentery remains as their chief supporting structure in the adult.

The dilated area at the caudal end of the hindgut is the *cloaca* (see p. 329 and Fig. 10.2). An opening from the cloaca to the outside serves temporarily as posterior vent for both the digestive system and the urogenital system. By the eighth week, however, the cloaca has two parts, each opening separately to the outside. The dorsal part is the rectum, and the anterior part remains as a common orifice for the urogenital system for some time.

UNIFORMITIES OF ALIMENTARY TRACT STRUCTURE

The walls of the adult alimentary tract from esophagus to rectum have a generally uniform

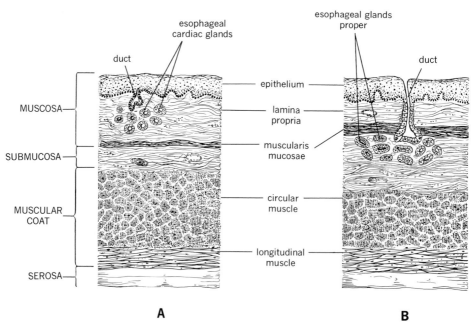

13.3 Wall of the alimentary tract as illustrated by wall of esophagus: A, high esophagus; B, low esophagus.

structure of four coats, only the character of their layers varying in different portions of the tract (Fig. 13.3). The inner coat is the *mucous membrane*, or *mucosa*, composed of specialized glandular epithelium, an underlying basement membrane, and supporting connective tissue, the *lamina propria*. This is surrounded in some places by a thin layer of muscle cells, the *muscularis mucosae*.

The second coat is the *submucosa*, a loose connective tissue sheath containing the larger blood vessels and lymphatic vessels. Next is the *muscular coat*. Most of the tract is covered with two layers of smooth muscle, with fibers of the inner layer arranged circularly around the tract and those of the outer layer arranged longitudinally. The coordinated contractions of these fibers produce *peristalsis*, the characteristic movement that propels contents along the tract. The fourth coat is the external fibrous *serosa*. In the portion of the alimentary tract below the diaphragm, the outer layer of serosa is the peritoneum.

At several points the circular layer of muscle fibers thickens into heavy bands. These *sphincters*, by relaxing or contracting, control the passage of food. They are located at the junctures of the esophagus and stomach, the stomach and duodenum, and the small intestine and large intestine, and, finally, at the anus.

Anatomy of the Upper Alimentary Tract

ORAL CAVITY

Mouth. The *mouth*, or *buccal cavity*, extends from the lips and cheeks in front and at the sides to the pharynx in back. The roof is formed by the *hard* and *soft palates*, and the greater part of the floor by the tongue, sublingual region (i.e., region beneath the tongue), and lower jaw. The space bounded externally by the lips and cheeks and internally by the gums and teeth is called the *vestibule*. Behind this is the mouth proper.

As shown in Fig. 11.2, the hard palate consists of a bony plate covered by mucous membrane. Suspended from its posterior border is the soft palate, a movable fold of mucous membrane, enclosing muscle fibers, blood vessels, nerves, lymphoid tissue, and mucous

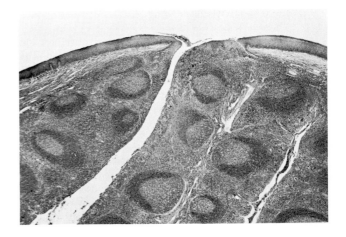

13.4 Section of a palatine tonsil, showing tonsillar crypts. (From *Atlas of Descriptive Histology*, by E. J. Reith and M. H. Ross. Hoeber Medical Division, Harper & Row, Publishers, New York, 1967.)

glands. Dangling from the middle of the lower edge of the soft palate is the cone-shaped *uvula*.

The *fauces* is the space between the mouth and pharynx. At the base of the uvula on either side is a curved fold of muscular tissue covered by mucous membrane, which splits into two *pillars* shortly after leaving the uvula; one runs downward, laterally, and forward to the side of the base of the tongue; the other runs downward, laterally, and backward to the side of the pharynx. The arches formed by the two pillars viewed through the open mouth are known respectively as the *glossopalatine arch* (anterior pillars of the fauces) and the *pharyngopalatine arch* (posterior pillars of the fauces).

Tonsils and adenoids. Between the anterior and posterior pillars, an oval mass bulges into the pharynx. This is a *tonsil*—specifically, one of the paired *palatine* tonsils.

Figure 13.4 shows that the surface of a palatine tonsil is stratified squamous epithelium pierced with numerous deep openings. These are *tonsillar crypts*, communicating with channels that course through the substance of the tonsil, which consists of lymphoid tissue.

Two other sets of tonsils, also lymphoid tissue, are found in the pharynx. At the base of the tongue are the *lingual tonsils*, and high in the pharynx are the *pharyngeal tonsils*, also called the *adenoids*. The pharyngeal tonsils, in a portion of the pharynx that functions only as an air passage, are not, strictly speaking, part of the digestive system. They are mentioned here only because of the frequency with which they and the palatine tonsils are infected together and removed together—with or without justification. The function of the tonsils is similar to that of the lymph nodes and other lymphoid structures (see p. 223). Typically, they rapidly increase in size during the first few years of life. However, before age 10 they begin to shrink, and at puberty they are considerably atrophied. If the palatine tonsils are much enlarged, they may fill the throat cavity and interfere with the passage of air to the lungs. Inflammation—i.e., *tonsillitis*—is not a prerequisite for such enlargement. Yet it seems more than coincidental that the peak in tonsillar size occurs at the age when infections are most frequent and that a decrease in tonsillar size accompanies a decline in these infections. This association was once considered evidence that large tonsils are a cause of trouble and should therefore be removed. Today it is believed that tonsillar tissue is significantly concerned with the establishment of immunity toward infections and that they recede when this task has been accomplished (Chapter 19).

Many of the reasons advanced for tonsillectomy are more valid for adenoidectomy. When the pharyngeal tonsils enlarge, they do so in a relatively small space, thus easily obstructing the posterior nares and the openings of the Eustachian tubes. Blockage of the nasal passage results in mouth breathing, nasal speech, snoring, and the characteristic "adenoidal" expression

of some boys and girls. Blockage of the Eustachian tube orifices commonly leads to potentially serious ear infection. Consequently, adenoidectomy is necessary more often than tonsillectomy, though both operations have been performed infrequently since the advent of antibiotics.

Tongue. The tongue assists in chewing, swallowing, and digestion by helping to move the food and to keep it between the teeth. Glands of the tongue secrete mucus that lubricates the food and further facilitates swallowing. The tongue also carries the taste receptors (taste buds) (see p. 504).

The top of the tongue is shown in Fig. 12.74. The numerous elevations are the filiform, fungiform, and vallate papillae. Taste buds are distributed over the surface of the tongue but are most conspicuous on the sides of the vallate papillae.

A sagittal section of the tongue appears in Fig. 11.2. The anterior portion is mobile, and the posterior portion is actually part of the anterior wall of the pharynx. The muscles of the tongue elevate, depress, protrude, and retract it. In profound anesthesia or coma, relaxation of the tongue muscles permits the organ to fall backward and obstruct the larynx. The proximity of the tongue to the epiglottis is evident in Fig. 11.2.

Salivary glands. The mucous membrane lining the mouth contains many minute *buccal glands*, which pour their secretions into the mouth. The bulk of the

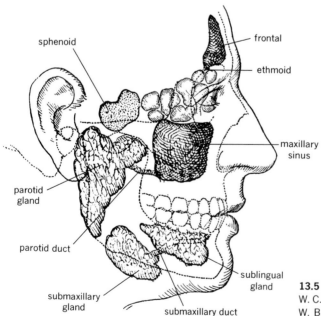

13.5 Salivary glands. (From T. Jones and W. C. Shepard, *Manual of Surgical Anatomy,* W. B. Saunders, Philadelphia, 1945.)

mouth secretions, however, is supplied by three pairs of compound alveolar salivary glands: the *parotid, submaxillary,* and *sublingual glands.*

A parotid gland is located just under and in front of each ear (Fig. 13.5). Its duct, the *parotid (Stensen's) duct,* opens onto the inner, upper surface of the cheek. The submaxillary and sublingual glands lie under the jaw and the tongue, respectively, the submaxillary behind the sublingual. A *submaxillary (Wharton's) duct* from the submaxillary gland and a number of small ducts from the sublingual gland open onto the floor of the mouth. The secretions of the salivary glands and buccal glands make up *saliva.*

Teeth. The upper and lower jaws contain the *alveoli,* or *sockets,* that hold the teeth. The gums, or *gingivae,* consist of dense connective tissue covered by smooth mucous membrane. They lie over the jawbones and extend a short distance into each alveolus. The alveoli are lined with connective tissue that is attached to the gums and that serves to fix teeth in position.

The three parts of a tooth are as follows: the *root,* of which there are one to three; the *crown,* which rises above the gum line; and the *neck,* the constricted portion between root and crown (Fig. 13.6A). A longitudinal section shows a tooth to be composed principally of an ivorylike substance called *dentine,* which gives it shape, and a central *pulp cavity.* The dentine of the crown is capped by a dense layer of *enamel;* that of the root, by *cement.* These three substances—dentine, enamel, and cement—are all harder than bone, enamel being the hardest substance found in the body. The pulp cavity is continuous with a canal through the center of each root and open at its tip. It is filled with dental *pulp,* a soft, sensitive connective tissue mass containing blood vessels and nerves.

Two sets of teeth develop during life: the *temporary (deciduous,* or *milk)* teeth, which begin to appear at about 6 months; and the *permanent* teeth, which begin to displace the temporary teeth in the seventh year and remain, in the fortunate, until old age. The temporary teeth are 20 in number, 10 in each jaw. There are 32 permanent teeth, 16 in each jaw.

According to their shape and use, teeth are classified as *incisors, canines, premolars* (or *bicuspids*), and *molars* (Fig. 13.6B,C). The incisors are the four front teeth of each jaw. They have sharp cutting edges and are well adapted for biting. Next are the canines, two in each jaw. They have sharp points, are longer than the incisors, and serve for biting and tearing. Next are the *premolars,* four in each jaw (two behind each canine). They are broad, each having two points, or cusps, on its crown and only one root, which is more or less completely divided in two. Their function is to grind food. There are no premolars in the temporary teeth. The *molars,* three on each side in each jaw, have

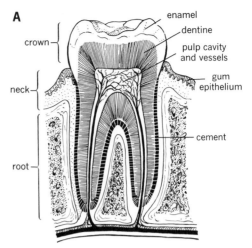

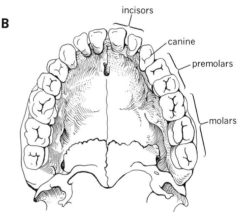

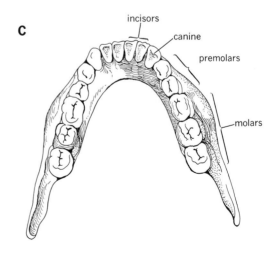

13.6 Teeth: A, section of a molar in its alveolus; B, upper jaw; C, lower jaw.

broad crowns with small, pointed projections that fit them for crushing food. Each upper molar has three roots, and each lower molar has two. The molars do not replace temporary teeth (although there are eight temporary molars). Rather, they are gradually added as the jaws grow. Since the hindmost molars may not appear until age 25, they are termed *late teeth*, or *wisdom teeth*.

PHARYNX AND ESOPHAGUS

Pharynx. We have already spoken of the pharynx, particularly its upper portion, the nasopharynx, which is above the soft palate. The two other portions of the pharynx are the *oral pharynx* and the *laryngeal pharynx* (or *hypopharynx*) (see Fig. 11.2). The oral pharynx extends from the soft palate to the epiglottis. In it the air and food passages cross one another. The laryngeal pharynx lies behind the larynx and reaches to the point where the esophagus begins.

The pharynx transmits air from the nose or mouth to the larynx and serves as a resonating cavity in sound production. It also transmits food from the mouth to the esophagus, its highly specialized musculature functioning in the act of swallowing.

The anatomy of the muscles surrounding the pharynx is quite complex. Much of the framework of the lateral and posterior walls of the pharynx comprises two muscle layers. The outer layer consists largely of the *superior, middle,* and *inferior constrictor muscles,* which are arranged circularly and overlap each other somewhat. The muscles of the inner layer are more nearly longitudinal. When swallowing is about to be performed, the longitudinal muscles contract, drawing the pharynx upward and dilating it to receive food. Then they relax, letting the pharynx drop, and the constrictor muscles contract in sequence, grasping the mass (or *bolus*) of food and passing it onward into the esophagus.

Esophagus. The esophagus is a muscular tube about 10 in. long. It begins at the lower end of the pharynx behind the trachea, descends in front of the vertebral column, passes through the diaphragm, and terminates at the lower end of the sternum (see Figs. 11.2, 13.1).

The structure of its walls is shown in Fig. 13.3. Its mucosa consists of thick stratified squamous epithelium containing two kinds of glands: *esophageal glands proper,* randomly distributed small compound tubular glands with many mucous glands (see Fig. 5.6); and *cardiac glands,* short simple tubular glands, mainly in the lower esophagus, whose deep terminal portions are moderately branched. In the lower esophagus the muscularis mucosae is unusually thick. It is noteworthy that the muscular layers in the top third of the esophagus are exclusively striated, in the bottom third, exclusively smooth, and in the middle third, a mixture of the two.

STOMACH

Gross structure. The esophagus ends in the stomach, a collapsible baglike dilatation of the alimentary tract that serves as a temporary receptacle for food. It has two openings and two so-called *borders,* or *curvatures.* The opening by which the esophagus enters is the *esophageal,* or *cardiac, orifice* (since it is just below the heart). The opening communicating with the duodenum is the *pyloric orifice* (since it opens into the *pylorus,* the juncture between stomach and duodenum). Both openings are guarded by special sphincters, whose contractions close them. Food is kept in the stomach until it is ready for intestinal digestion, and then the pyloric sphincter relaxes.

Figure 13.7 illustrates the regions of the stomach. An oblique line drawn between the esophageal orifice and the greater curvature separates the *fundus* from the *body.* An oblique line drawn between the greater curvature and an indentation (the *incisura angularis*) on the lesser curvature separates the body from the pyloric region. Just proximal is the *pyloric antrum;* just distal is the *pyloric canal.* The actual position of the stomach in the abdomen varies from person to person, and from time to time in a given individual. The greater curvature, for example, is lowest when the body is erect, rising several inches when the body is prone.

Microscopic structure. The microscopic structure of the gastric (stomach) mucosa is extraordinarily interesting. As shown in Fig. 13.7, there are three types of gastric glands, each indigenous to one area of the stomach—*cardiac, fundic,* and *pyloric* glands. All are short simple tubular structures. The cardiac glands of the stomach resemble those of the esophagus, the fundic glands are more branched, and the pyloric glands are more numerous and more branched still.

The principal digestive agents secreted by the stomach, hydrochloric acid and pepsin, originate mainly in the fundus and body. A narrow band of mucosa near the esophageal orifice secretes only mucus.

Four distinctive types of cells are found in the gastric mucosa. The surface and the *gastric crypts* opening onto it are composed of columnar *epithelial cells.* These contain *mucinogen* granules and secrete the visible mucus of the *gastric juice.* The gastric glands proper open into the bases of the crypts. In the necks of the glands are *neck chief cells.* These occur in all three types of gastric glands and are the only cells present in the pyloric glands and in the narrow mucus-secreting band near the esophageal orifice. They secrete

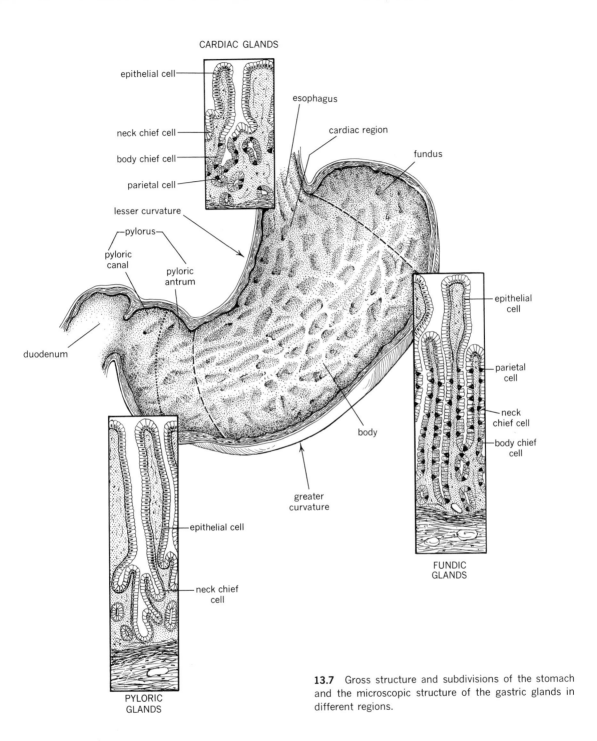

CARDIAC GLANDS

epithelial cell

neck chief cell

body chief cell

parietal cell

lesser curvature

pylorus

pyloric canal

pyloric antrum

duodenum

esophagus

cardiac region

fundus

body

greater curvature

epithelial cell

parietal cell

neck chief cell

body chief cell

FUNDIC GLANDS

epithelial cell

neck chief cell

PYLORIC GLANDS

13.7 Gross structure and subdivisions of the stomach and the microscopic structure of the gastric glands in different regions.

a mucus that is less viscous and more soluble than that of the surface epithelium and a moderate amount of pepsin. In the bodies of all but the pyloric glands are

body chief cells, which secrete more pepsin and less mucus, and *parietal* (or *oxyntic*) *cells,* which alone secrete hydrochloric acid (Fig. 13.8). These curious cells

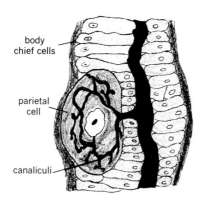

13.8 Relationship between a parietal cell and the lumen of a gastric gland. (From A. C. Guyton, *Textbook of Medical Physiology*, 2nd ed., W. B. Saunders, Philadelphia, 1961.)

stud the external walls of a gland, lying between the chief cells and the basement membrane without making direct contact with the lumen—hence the name parietal. The secretion of each parietal cell is collected by intracellular canaliculi, which converge to form a small intercellular canaliculus passing between adjacent chief cells into the gland lumen.

The surface epithelium is constantly being renewed. Neck chief cells reproduce themselves and also metamorphose to replace body chief cells and parietal cells, which almost never divide. Therefore, to the other functions of the neck chief cells must be added that of originating the major exocrine cell types of the gastric mucosa.

Physiology of the Upper Alimentary Tract

PRELIMINARY PHASES OF DIGESTION

Chewing and swallowing. Chewing, or *mastication,* is a voluntary act that mixes ingested food particles with saliva and reduces them to a size convenient for swallowing.* It also partici-

---

* It has been claimed that prolonged chewing improves digestion. This is unlikely, for the digestive enzymes of saliva continue to act in the stomach for about 30 minutes; each bolus of food entering the stomach tends to lie in the center of the preceding one and does not reach the outside layer, where the acidic gastric secretions inactivate the salivary enzymes, for about 30 minutes.

pates in the stimulation of the salivary glands. Chewing is less important in dogs, cats, and other carnivores, which bolt their food in large lumps, than in man.)

Swallowing, or *deglutition,* is a reflex action that usually begins involuntarily but may be initiated voluntarily. We have already spoken of the muscles responsible for swallowing. In the *first stage* of swallowing, the bolus of food passes through the fauces into the pharynx. The chewed food is shaped into a bolus by the movement of the tongue against the palate; it is then pushed toward the back of the mouth by elevation of the front of the tongue. At the same time, it is lubricated by saliva.

The *second stage* of swallowing is an involuntary process in which the bolus passes through the pharynx into the esophagus. Food in the pharynx can exit in four directions, only one of which is desirable. Ordinarily it is forced into the esophagus by the reflex closing of the other three routes: the mouth is shut off by the elevation of the tongue against the hard palate; the opening into the nasal cavity is closed by the elevation of the soft palate; and the opening into the larynx is closed by the vocal cords and by the elevation of the entire larynx—the familiar "bobbing of the Adam's apple."† If the elevation of the larynx does not take place, swallowing becomes impossible.

In the *third stage* of swallowing, the bolus passes along the esophagus and through the esophageal sphincter into the stomach. The movement of the food through the esophagus varies with the physical character of the swallowed material. Liquids are shot through by the force of the first stage of swallowing. Semisolids are propelled by peristalsis. When the bolus arrives at the esophageal orifice, the sphincter relaxes and permits it to enter the stomach.

Salivation and saliva. The sight, smell, and thought of food elicit salivary (and gastric) secretions through reflex control. Saliva assists in the process of chewing (1) by dissolving readily soluble food components; (2) by partly digesting a portion of starch; (3) by softening the food

---

† Respiration stops temporarily during swallowing so that food or drink is not drawn into the larynx by inhalation.

mass; and, as already noted, (4) by covering the bolus with a lubricant fluid that facilitates its movement in the mouth.

The composition of saliva is different under different conditions. In fact, at least two distinct kinds of saliva are secreted by the three pairs of salivary glands. One kind is watery; the other is rich in mucus. Mucus is the glycoprotein (see p. 33) that gives saliva its lubricating properties. The main digestive enzyme in saliva is *salivary amylase;** it rapidly hydrolyzes starch into disaccharides, principally maltose, and a scattering of small polysaccharide fragments called *dextrins*. As we have mentioned, salivary amylase continues to act until stopped by the stomach acids. As a result, the starch of potatoes or bread may be three-quarters digested by it.†

Dental health is undoubtedly influenced by certain properties of saliva. One theory on the cause of dental decay holds that tooth enamel is eroded by acids formed by the bacterial decomposition of food residues that adhere to the teeth. Saliva contains a number of substances that suppress bacterial growth and consequently may suppress decay.

GASTRIC DIGESTION

Functions of the stomach. When the semiliquid bolus of food reaches the stomach, it is converted into *chyme,* a thick grayish liquid. The stomach (1) serves as a temporary storage reservoir, which retains food until it has been reduced to chyme, a form in which it is acceptable to the small intestine; (2) digests some protein, though almost all of the digestive functions of the stomach can be carried out elsewhere if necessary;‡ and (3) exerts a bactericidal influence through the acidity of its contents. Al-

though the stomach is not essential to survival, its surgical removal greatly alters digestive efficiency.

Methods of studying gastric secretion. Strictly speaking, gastric juice is the digestive secretion produced by the mucous membrane of the whole stomach. However, because the cells and the secretion of the pyloric region differ from those of the rest of the stomach, it has come to be regarded as a separate and distinct secretory organ. Hence gastric juice ordinarily refers to the secretion of the fundus and body of the stomach.

A remarkable chapter in the history of physiology concerns the evolution of methods for the collection of gastric juice and for the study of digestive processes in the stomach. Investigators seeking to determine whether the chief function of the stomach is to soften food mechanically or to attack it chemically early observed that gastric juice seemed to melt food away. The ingenious Lazzaro Spallanzani (1729–1799) swallowed small perforated wooden tubes containing food in small linen bags. He recovered these tubes intact from his feces and found that the food was considerably digested despite the fact that the tubes were not crushed. Since the food was protected from mechanical manipulation, chemical attack must have occurred.

A classic contribution, the first involving direct, systematic observations of the human stomach, was made by William Beaumont, a United States Army surgeon. In 1822, Alexis St. Martin, a Canadian voyageur, was accidentally injured by a shotgun discharge in a frontier trading post at Mackinac. Beaumont, called to attend him, found that the skin and muscles of the upper abdomen had been torn away, leaving a gaping wound in the outer wall of the stomach. He stitched the edges of the open stomach to the skin. St. Martin survived, but the opening into his stomach remained. Beaumont recognized a unique opportunity for investigating stomach physiology and arranged to keep St. Martin in his employ so that he could carry out a series of studies. Peering through the opening, he watched the stomach move. He inserted a string with a piece of meat attached to its end and withdrew it 2 hours later to find that the meat had dissolved. He learned that gastric juice

---

* This was named *ptyalin* before adoption of the *-ase* ending for an enzyme.

† A simple experiment readily demonstrates the action of salivary amylase. Starch plus iodine has a blue color; free sugar plus iodine does not. When 1 ml of saliva is added to 10 ml of dilute boiled starch in a test tube, a sample is taken from the tube each minute, and a drop of iodine is added to each sample, each sample is less able than the preceding one to produce a blue color with iodine.

‡ Perhaps the single exception is the secretion of intrinsic factor, the substance essential to the later absorption of vitamin B_{12} (see p. 243).

appeared only when food entered the stomach or while it was being chewed in the mouth. He withdrew pure gastric juice and, with the aid of Professor Dunglison of the University of Virginia, discovered that it contained hydrochloric acid.*

Physiologists have devised clever methods for obtaining pure gastric juice from experimental animals. In order to collect gastric juice uncontaminated with food and other secretions, Heidenhain surgically isolated part of the stomach as a pouch (the *Heidenhain pouch*) that drained directly to the outside through an opening in the abdominal wall. The Heidenhain pouch and its later modifications have contributed much to our understanding of gastric function.

Composition of gastric juice. The three most important components of gastric juice are hydrochloric acid, pepsin, and mucus. In addition,

---

* Two other subjects have served for extensive studies of human gastric function. In 1916 A. J. Carlson reported on "Mr. V," whose esophagus had been sealed off by accidentally ingested strong alkali, making necessary the introduction of food through a surgically created opening directly into the stomach (gastrostomy). In 1943 S. Wolf and H. G. Wolff reported on "Tom," whose esophagus had been permanently closed by extremely hot clam chowder when he was 9 years old. When their investigations began, Tom had been feeding himself through a gastrostomy for 47 years. Carlson paid particular attention to the secretory and motor activities of the stomach, and Wolf and Wolff to the effects of emotional states on the stomach—for example, gastric secretion, motility, and vascularity were all found to increase during periods of anger and other emotional stresses.

intrinsic factor and a scattering of other digestive enzymes are present. Table 13.2 lists these, together with some of the characteristics of gastric juice.

The main evidence implicating the parietal cells in the secretion of gastric hydrochloric acid is data correlating the number of parietal cells in a given area with the total acid and chloride content in the same area. The mechanism by which these cells elaborate a fluid whose hydrogen ion concentration is 4,000,000 times that of the surrounding blood has intrigued physiologists for more than two centuries. We shall consider it later.

The hydrogen ion concentration of blood is $4 \times 10^{-8}N$ (i.e., 4×10^{-5} meq per liter). The hydrogen ion concentration of the secretions of parietal cells is about $0.16N$ (i.e., 160 meq per liter). However, this is partly neutralized by the more alkaline secretions of the nonparietal cells. Since gastric juice is a mixture of secretory products, the acidity at any given time depends upon the relative proportions of parietal and nonparietal secretions.

Pepsin, the principal enzyme of gastric juice, is stored in the chief cells as its inactive precursor, *pepsinogen*. Acid solutions (pH below 6) convert pepsinogen to pepsin, conversion occurring most rapidly at pH 1.5. This reaction proceeds autocatalytically; that is, free pepsin promotes the further transformation of pepsinogen to free pepsin.

Pepsin acts in digestion by splitting protein molecules. It is the first of several proteolytic

---

TABLE 13.2 COMPOSITION OF FASTING GASTRIC JUICE

Acids	Solids	Enzymes and protein factors
HCl, free, 0.40–0.50%*	Organic, 0.42–0.46%*	Pepsin
Total, 0.45–0.60%*	Inorganic, 0.13–0.14%*	Rennin
pH, 0.9–1.0	Sp gr, 1.002–1.004	Cathepsin
		Lipase
		Lysozyme
		Mucus
		Intrinsic factor

* By weight; 0.1N HCl is 0.35% HCl by weight.

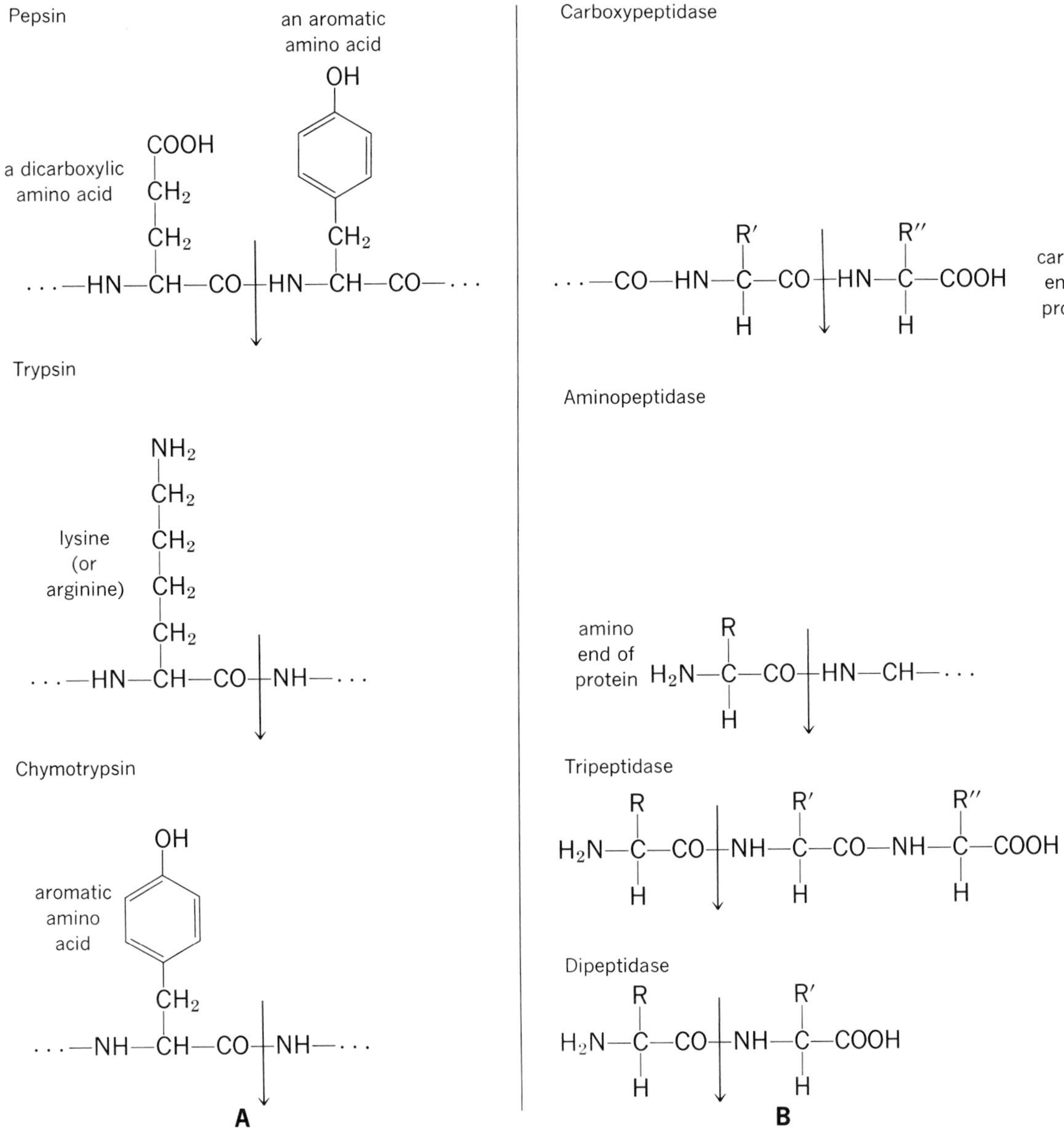

13.9 Specificities of the proteolytic enzymes of the stomach and small intestine. Vertical arrows show which peptide bonds are cleaved specifically by each proteolytic enzyme. A, Endopeptidases, which cleave a peptide bond regardless of its distance from the end of the protein molecule; B, exopeptidases, which cleave a peptide bond only when it connects a terminal amino acid to the rest of the protein molecule.

enzymes that food encounters in its passage through the alimentary tract. We recall that proteins are large molecules composed of many amino acids linked together by their amino (—NH$_2$) and carboxyl (—COOH) groups (see p. 30), which form peptide bonds (—CO—NH—). The proteolytic effect of pepsin ordinarily involves an attack on those peptide bonds whose amino groups are contributed by aromatic amino acids and whose carboxyl groups are contributed by dicarboxylic amino acids (Fig. 13.9). Hence pepsin is an *endopeptidase*—

i.e., it cleaves specific peptide bonds *within* a protein molecule, to yield peptides of varying lengths. An *exopeptidase,* on the other hand, cleaves only those peptide bonds attached to *terminal* amino acid residues, clipping amino acids one at a time from the ends of a polypeptide chain.

Pepsinogen is secreted not only *externally* into the lumen of the stomach; approximately 1% of the total amount is secreted *internally* into the blood stream. This appears in the urine as *uropepsinogen,* which becomes *uropepsin* in acid solutions. Consequently, the urinary uropepsinogen level is a convenient index of pepsin secretion in the stomach.

Rennin is the milk-curdling enzyme, produced by the chief cells of the fundic glands.* The clotting of milk improves its utilization in infants by slowing its passage through the alimentary tract and thus keeping it in the stomach long enough for pepsin to act upon it.

The mucus of gastric juice consists of at least two glycoproteins, differing in sites of origin, physicochemical properties, and secretory response to stimuli. One of these substances, visible mucus, has a gelatinous consistency and forms a white coagulum in hydrochloric acid. It is secreted by the surface epithelium. The other, soluble or dissolved mucus, does not coagulate visibly in hydrochloric acid. It is a product of the neck chief cells. Little is known of the chemical identities of these glycoproteins except that they contain the polysaccharide mucoitin sulfate (the sulfuric acid ester of hyaluronic acid).

The mucus that coats the stomach lining neutralizes some of the free hydrochloric acid and also tends to inhibit the action of pepsin, thereby safeguarding the gastric mucosa against erosion by its own digestive secretions. Under normal conditions the mucosa of the stomach and duodenum is resistant to the enzymes, but, when a local area undergoes trauma, hemorrhage, or weakening, cellular breakdown occurs, particularly when acid and

---

* *Rennet* is a commercial preparation of rennin from calf stomach. It reacts with milk to yield a clot called *junket.* When junket is allowed to stand, it separates into a solid portion (curd) and a liquid portion (whey). Curd prepared in this manner has been used for centuries as the starting material in the manufacture of cheese.

pepsin production concomitantly increase. The result is a *peptic ulcer,* in which devitalized cells are digested away. The incidence of peptic ulcers is highest in those portions of the alimentary tract that are in closest contact with gastric juice, the upper duodenum being first and the stomach itself second. Although the primary cause of peptic ulceration has not been established, it is frequently associated with emotional and nervous tension. Those who subscribe to the neurogenic theory regard an imbalance in the autonomic nervous system as the leading factor. This imbalance is often accompanied by an overproduction of acid and pepsin.

Other constituents of gastric juice include cathepsin, a proteolytic enzyme of as yet undetermined significance; lipase, a weak lipid-splitting enzyme of little importance in lipid digestion; urea, amino acids, histamine, sodium, potassium, calcium, magnesium, bicarbonate, phosphate, and sulfate.

Mechanism of hydrochloric acid secretion. The capacity of the parietal cells of the gastric glands to secrete strong hydrochloric acid continues to evade explanation. Careful chemical analysis shows that the composition of gastric juice depends upon its rate of formation. When this is slow, the composition approaches that of isotonic NaCl. When it is fast, the composition approaches that of isotonic HCl.

How the gastric glands form HCl and why the rate of formation of gastric juice determines its composition are questions that have generated much controversy. The central problem is the secretion of H^+ in a concentration millions of times as high as that of blood. Evidently the secretory machinery transports H^+ from the tissue fluids surrounding the gastric glands to the stomach lumen. In other words, a parietal cell performs as a hydrogen pump, somehow transferring H^+ from the blood to the canaliculi and gland lumen. The efficiency of the pump depends upon the energy produced by the cell.

Chloride ions also appear in gastric juice. These, too, derive from the tissue fluids and must therefore be actively transported across the gastric epithelium. However, the difference in $[Cl^-]$ between tissue fluids and stomach lumen is less spectacular than that in $[H^+]$, increasing only from 110 to 160 meq per liter. Nevertheless,

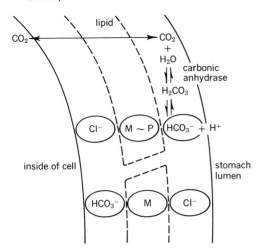

critical part of cell membrane

13.10 Proposed method of HCl formation. (From R. P. Durbin and D. K. Kaskebar, *Fed. Proc.*, 24, 1377, 1965.)

it is significant, for it reflects a movement of Cl^- against a concentration gradient. Thus a chloride pump exists as well as a hydrogen pump. Their mode of operation is now under investigation.

The demonstration that carbonic anhydrase (see pp. 228, 359, 394) is present in abundance in the parietal cells has implicated it in the formation of HCl. Administration of sufficient carbonic anhydrase inhibitor, such as acetazolamide (see p. 363), substantially curtails HCl secretion.[*] According to one recent proposal (Fig. 13.10), the luminal side of the parietal cell membrane contains a carrier, M. First, M is phosphorylated by ATP to form a complex that binds intracellular Cl^- at the inside of the cell membrane.

$$M + ATP + Cl^- \rightleftharpoons$$
$$M{\sim}P{-}Cl + ATP \quad (13.1)$$

(This reaction gives ATP a role in HCl secretion.) Next, HCO_3^-, supplied by the carbonic anhydrase–catalyzed hydration of metabolic CO_2, attaches to the phosphorylated carrier at the outside of the cell membrane. The two

---

[*] No such effect occurs until the inhibitor concentration is quite high ($10^{-3}M$). The maximum effect occurs with $10^{-2}M$ inhibitor.

anions, Cl^- and HCO_3^-, then exchange places through physical rotation of the carrier, and the anions and inorganic phosphate (P_i) are released, HCO_3^- and P_i within the secreting cell and Cl^- just outside.

$$M{\sim}P{-}Cl + HCO_3^- \longrightarrow$$
$$M + P_i + HCO_3^- + Cl^- \quad (13.2)$$

The Cl^- thus transported joins the H^+ remaining from the ionization of H_2CO_3 to form HCl. P_i is utilized to regenerate ATP, and HCO_3^- is taken up in the body fluids. Though this scheme seems plausible, more work is needed to test its validity.

A temporary rise in urinary alkalinity often follows a meal. This has been called the *alkaline tide*. It is generally attributed to an increase in blood alkalinity resulting from the temporary loss of HCl into the gastric juice. Whether the tide occurs depends on the diet, the rates of HCl secretion in the stomach and reabsorption in the intestines, and other factors.

Control of gastric secretion. Since the work of Pavlov and Beaumont, it has been recognized that both gastric secretion and gastric movement are intermittent. Since they often coincide, they may be under the same nervous or hormonal control.

The secretion of gastric juice in the absence of food or the sight, smell, or thought of food is referred to as *interdigestive* secretion. It is often blamed for indigestion. It includes the so-called *continuous* secretion—a misnomer since the secretion is actually intermittent—and the *emotogenic* (i.e., produced by emotion) secretion. Interdigestive secretion has been reported throughout a 40-day fast (though one wonders whether in this case it would not be due in part to the thought of food). The mechanism responsible for it is unknown.

The secretion of gastric juice directly related to food is referred to as *digestive* secretion and consists of three phases. First is the *cephalic*, or *psychic, phase*. Stimuli acting upon the brain initiate the discharge of impulses via the vagus nerves (X), which specifically trigger parietal cell secretion of water and HCl. Vagal impulses arise from unconditioned reflexes, as to "sham feeding" (in which food is not allowed to reach the stomach) in an experimental animal whose higher brain centers have been removed surgically, and from conditioned reflexes, as to the thought, sight, smell, or taste of food. The most important manifestation of the cephalic stage in man is the copious flow of gastric juice during the chewing of appetizing food. This juice comprises almost half the output of the gastric glands during the digestive period. It is rich in all components and contributes to both the initiation and the subsequent efficiency of gastric digestion. The cephalic phase is abolished if the vagus nerves are severed. This indicates that the secretory stimulus is entirely nervous in origin.

The second or *gastric phase* is so named because its stimuli originate in the stomach. They may be mechanical or chemical. The only mechanical stimulus capable of promoting secretion is the gastric distention following food intake. Chemical stimuli are substances called *secretagogues,* which are present in certain foods and are released in the process of digestion. The high secretagogue content of meat, particularly liver and fish, and its extractives (as in bouillon) explain their exclusion from the diet in acute cases of peptic ulcer. The chemical natures of the secretagogues have not yet been established.

A hormone known as *gastrin* is secreted by the mucosa of the pyloric antrum in response to distention or secretagogues. Gastrin, also called "the antral hormone," circulates in the blood (as do all hormones) and eventually returns to stimulate the parietal cells to secrete HCl. In 1964 R. A. Gregory and H. J. Tracy isolated from the antral mucosa two polypeptides, which they termed *gastrin I* and *gastrin II*. Each is physiologically potent (Fig. 13.11), and each contains 17 amino acids. Analysis of the amino acid sequences revealed that gastrin I differs from gastrin II only in lacking a single sulfate ester group. These workers have synthesized gastrin I and gastrin II and have thereby discovered several peptide fragments containing only 5 amino acids, which have the full physiological activity of gastrin.*

Both inhibition and stimulation of gastric juice secretion occur as a result of stimuli originating in the small intestine; this consti-

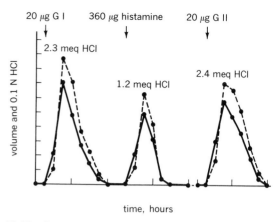

13.11 Comparative effects on HCl secretion of gastrin I, gastrin II, and histamine, given subcutaneously to a dog with a Heidenhain pouch. (From R. A. Gregory and H. J. Tracy, *Gut,* 5, 103, 1964.)

tutes the third or *intestinal phase* of digestion. Various substances—among them alcohol, histamine, and various meat extractives—are absorbed into the circulation by the small intestine and return to act directly on the parietal cells to stimulate HCl secretion. Of the three phases of digestive secretion, the intestinal phase is least important.

By the time an appreciable portion of the gastric contents has been delivered into the intestine, regulatory mechanisms are in operation to terminate digestive secretion. The filling of the stomach and the beginning of absorption of the products of intestinal digestion lead to *satiety* and with it the cessation of eating and the withdrawal of cephalic stimuli.

A pH of 1.5 or less within the stomach inhibits the release of gastrin by the antral mucosa and, together with the progressive emptying of the stomach, causes withdrawal of the gastric stimuli. Other inhibitory influences, applied indirectly through the vagus nerve, result from the presence of the food mass in the upper intestine. These are hormonal (release of the gastric inhibitory agent *enterogastrone*), mechanical (distention), chemical (acidity), and physical (osmolarity).

The quantity of gastric juice secreted varies. According to the usually quoted figures, the total amount secreted by a normal man after an average meal of meat, bread, vegetable, coffee,

---

* Administered histamine has a stimulatory effect on HCl secretion like that of gastrin, but on a molar basis gastrin is 500 times more potent. It is not known whether histamine normally present in the body has a similar effect.

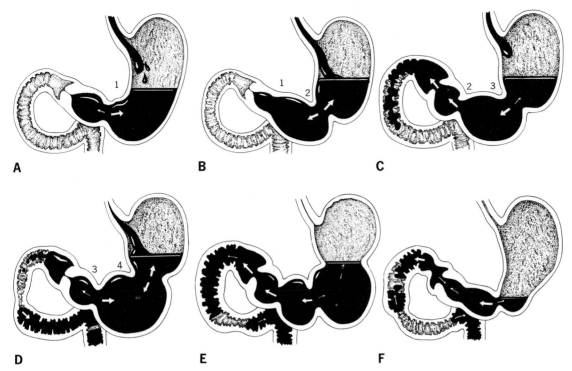

13.12 Movements of the stomach and duodenum during gastric filling and emptying. A. The stomach is filling, and a mild peristalsic wave (1) is pushing its contents in both directions. B. The first wave fades as the pyloric sphincter remains closed, and a stronger wave (2) is pushing the contents in both directions. C. As the second wave approaches, the sphincter opens, and some contents pass into the duodenum; a third wave (3) is pushing the contents in both directions. D. The sphincter is again closed, although the stomach has not been evacuated; a fourth wave (4) is pushing the stomach contents in both directions, and a wave pushes the contents of the second portion of the duodenum forward. E. Peristalsic waves originating higher and higher in the body of the stomach push the contents forward intermittently through the sphincter and the first portion of the duodenum. F. The stomach is nearly empty 3 or 4 hours later; a mild peristalsic wave is emptying the first portion of the duodenum, with some reflux into the stomach, and peristalsis continues throughout the duodenum.

and milk is about 700 ml—200 ml in the first hour, 150 ml in the second hour, and 350 ml in the next three hours. The capacity of the average stomach is about 5000 ml.

Movements of the stomach. The motor activity of the stomach is related to the thickness of the muscle layers. In the fundus, where the musculature is thinnest and the contractions are least vigorous, *receptive relaxation,* in which stomach capacity increases without a corresponding increase in intragastric pressure, occurs during gastric filling. The stomach is thus able to function as a reservoir for enough food to provide nourishment for several hours.

In the pyloric region, where the musculature is thicker, churning reduces the stomach contents to chyme. The distal and thickest muscular structure, the pyloric sphincter, controls the discharge of the stomach contents so that they are retained long enough for maximum gastric digestion and then passed into the small intestine at a rate that prevents excessive distention and favors maximum intestinal digestion and absorption. The movements associated with gastric filling and emptying are shown in Fig. 13.12.

The stomach is not quiescent even when empty. By the method illustrated in Fig. 13.13, changes have been recorded in intragastric pres-

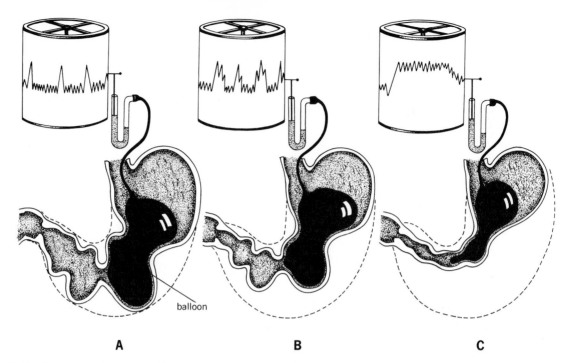

13.13 Movements of the stomach when empty: A, rhythmic (Type I) contractions; B, tonus (Type II) rhythm; C, tonic (Type III) contractions.

sure in an empty stomach that are sufficiently consistent from individual to individual to permit classification. Periods of activity occur at regular intervals, beginning with mild *rhythmic contractions,* also designated *Type I contractions,* which register as shallow undulating waves, three per minute. After a while the elevations in pressure become more marked and more frequent, registering as sharp peaks of about 30 seconds duration recurring in close succession. Carlson termed this pattern *tonus rhythm,* assuming that it reflects changes in the tonus of the gastric musculature. Its waves represent contractions known as *Type II contractions,* originating at about the level of the incisura and passing downward to the pylorus. As the empty stomach grows yet more active, the pressure elevations are sustained, and waves are superimposed upon previous waves. Ordinarily activity terminates after a series of Type II contractions. However, when the gastric musculature is exceptionally irritable or during prolonged fasting, even more rapid contractions, called *tonic* or *Type III contractions*—or *gastric*

tetany—may follow. These characterize a stomach in a state of hypertonus.

The cause of motor activity in an empty stomach is not known. The fact that it persists, although weakened, in a completely denervated stomach has led to the conclusion that the gastric musculature is autonomous, that it is set in motion by its own pacemaker. The fact that a piece of fundus grafted beneath the skin continues to contract almost synchronously with the stomach suggests a humoral mechanism.

Nervous controls. The two main functions of the stomach, secretion and motor activity, are under nervous control. The brain area regulating them is just in front of and behind the central sulcus. It connects with the stomach via the thalamus, hypothalamus, and motor nucleus of the vagus nerve (X).

The parasympathetic nerve supply to the stomach also arrives through the vagus nerve from the motor nucleus. The sympathetic nerve supply originates in the fifth to twelfth segments of the spinal cord and proceeds through the cor-

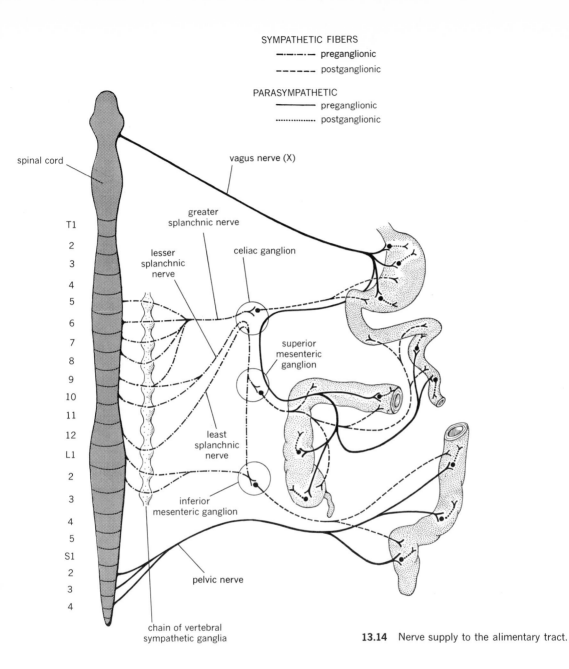

SYMPATHETIC FIBERS
—·—·—·— preganglionic
— — — — — postganglionic

PARASYMPATHETIC
——————— preganglionic
·············· postganglionic

spinal cord

vagus nerve (X)

greater
splanchnic nerve

T1
2
3
4
5
6
7
8
9
10
11
12
L1
2
3
4
5
S1
2
3
4

lesser
splanchnic
nerve

celiac ganglion

superior
mesenteric
ganglion

least
splanchnic
nerve

inferior
mesenteric ganglion

pelvic nerve

chain of vertebral
sympathetic ganglia

13.14 Nerve supply to the alimentary tract.

responding vertebral sympathetic ganglia to form the greater splanchnic nerves, which reach the stomach by way of the celiac ganglia (Fig. 13.14; see also Fig. 12.51). Both the vagus and splanchnic nerves, the *extrinsic nerves* of the alimentary tract, are distributed in the tract walls as *intrinsic plexuses,* through which the nerve endings are brought into intimate contact with the smooth muscle and the gland cells.

An intrinsic plexus is composed of numerous groups of ganglion cells interwoven in a mesh of fibers—postganglionic sympathetic fibers, preganglionic and postganglionic parasympathetic fibers, and afferent fibers. A plexus lying be-

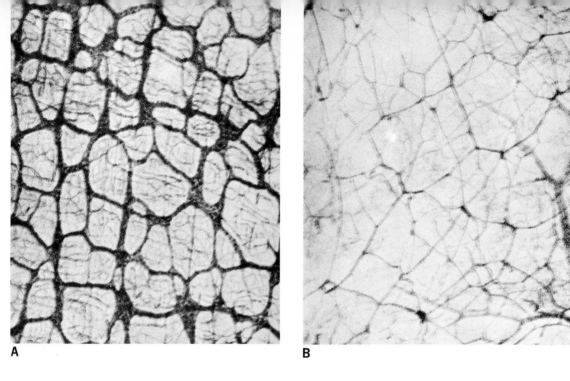

A   B

13.15 Microscopic appearance of intrinsic plexuses in the alimentary tract: A, myenteric (Auerbach's) plexus; B, submucous (Meissner's) plexus. (Copyright *The CIBA Collection of Medical Illustrations* by Frank H. Netter, M.D.)

tween layers of the muscular coat is termed a *myenteric (Auerbach's) plexus,* and one in the submucosa is termed a *submucosal (Meissner's) plexus* (Fig. 13.15) The distribution of the plexuses varies in different parts of the alimentary tract, both types being more plentiful in the stomach and large intestine than in the esophagus and small intestine. The plexuses undoubtedly account for the continuing muscular movement of isolated pieces of gut, although extrinsic nerves are still essential for muscular coordination.

The vagus nerve has both stimulatory and inhibitory effects. If it is cut, gastric tonus, motility, and secretion are all permanently reduced. The splanchnic nerves have relatively minor effects, principally inhibitory of motility. Cutting them does not essentially alter stomach activity. The stomach continues to function adequately after complete removal of the extrinsic nerves by virtue of the autonomy of the intrinsic plexuses.

It has been suggested that the central nervous system influences gastric secretion via a neurohormonal pathway, involving successively the hypothalamus, the adenohypophysis, and the

adrenal cortex. This hypothesis awaits experimental support.

Consequences of gastrectomy. Both human beings and lower animals can survive complete gastrectomy.* Hence the stomach is not indispensable. However, serious physiological dislocations follow. Loss of the stomach's reservoir function deprives a subject of the capacity to hold a meal of normal size, necessitating frequent small meals, although compensatory dilatation of the upper intestine eventually may permit the resumption of normal eating habits.

More important is the loss of the normal churning action of the stomach muscles, which converts particles of solid food into semiliquid chyme and for which the small intestine cannot compensate. Only chewing remains as a means of breaking up solid food.

Another important loss is the mechanism that gears the emptying of the stomach to the readiness of the intestine to receive its contents. The accelerated passage of food through the intestine

---

* When the stomach is removed surgically, a connection is made between the lower esophagus and the upper small intestine.

results in impaired digestion and nutrition. For example, both the acid of the normal stomach and its controlled emptying favor the intestinal absorption of iron; without them, iron deficiency is likely. Also, as has been noted, loss of the intrinsic factor ensures the eventual development of vitamin B_{12} deficiency. Both deficiencies can be prevented by the administration of adequate amounts of iron and vitamin B_{12}.

HUNGER AND APPETITE

Definitions and concepts. Although familiar words in everyday speech, *hunger* and *appetite* are notoriously inexact in physiological parlance. Many workers have attempted to clarify their meanings with arbitrary definitions. Others have invented new terms without contributing new insights, thus adding to the confusion.

For many years the dominant conception of hunger was a mechanical one. Although hunger is a subjective sensation, Walter Cannon in 1912 showed, in studies on his students, that hunger pangs occurred simultaneously with gastric contractions. In 1916 Carlson, in *The Control of Hunger in Health and Disease,* started with a broader view of hunger as a "biological condition which leads to the taking of food," which he ultimately reduced to the narrow one of pain in the upper abdomen. To these investigators hunger was the result of autonomous periodic gastric contractions.

Our concept of hunger has been extended beyond the gastric contraction theory by recent studies of the behavior of gastrectomized men and other animals, the normal eating patterns of men and other animals whose stomachs have been completely denervated, and the observed effects of various drugs on food intake.

This has led to the modern conception of hunger as an element in the nervous regulation of food intake. Hunger, the bodily state arising from deprivation of food, manifests itself in *hunger behavior* and *hunger sensations.* Hunger behavior consists of a general restlessness and increased motor activity, which is most obvious in human infants and other animals. Hunger sensations, the mental adjuncts of hunger, comprise feelings of generalized weakness, fatigue, irritability, emptiness, and tension and cramps and pain in the upper abdomen.

With learning or conditioning, man or another animal associates the ingestion of food with a decrease in hunger sensations and behavior so that hunger sensations give rise to a desire to eat —i.e., appetite—and hunger behavior becomes incorporated into the learned activity called *appetitive behavior.* The ingestion of food gradually abolishes the desire to eat, leading to satiety, or physiological loss of appetite (anorexia).*

Nausea and vomiting. It is well known that appetite can be dissipated rapidly by emotions such as anxiety, grief, erotic feelings, or revulsion. These, together with a variety of physical disorders, can produce *nausea,* an unpleasant sensation that is accompanied by excessive autonomic activity: salivation, swallowing, pupillary dilatation, sweating, and constriction of the skin capillaries with resulting pallor. Nausea frequently, though not always, culminates in *vomiting.*

Vomiting is a reflex reaction, dependent on a *vomiting center* in the medulla. Impulses from the abdominal cavity—as might result from overdistention of the stomach—traverse both the vagal and sympathetic pathways. Stimuli from the pharynx—such as tickling of the throat —traverse the trigeminal (V) and glossopharyngeal (IX) nerves. The efferent impulses are carried by the vagus nerves (X) to the stomach, by the phrenic nerves to the diaphragm, and by the various spinal nerves to the abdominal muscles. All act in concert to expel the gastric contents.

Anatomy of the Lower Alimentary Tract

SMALL INTESTINE

Gross structure. Stomach contents passing through the pyloric sphincter enter the lower alimentary tract, which begins with the small intestine, a thin-walled

---

* Under abnormal circumstances hunger and appetite may be dissociated. Hunger may exist without the desire to eat, as in serious illness; and appetite may exist without a need for food, as in neurosis or neurological disease.

tube 20 to 25 ft long arranged in serpentine coils (see Fig. 13.1). As mentioned earlier, its three consecutive portions are the duodenum, jejunum, and ileum.

The duodenum is about 10 in. long. Unlike the jejunum and ileum, it is attached directly to the posterior abdominal wall, following a C-shaped path as it winds around the head of the pancreas. The jejunum is the next 6 to 8 ft. The ileum comprises the remainder of the small intestine and its narrowest portion. It joins the large intestine at a right angle. The *ileocecal valve*, a valvelike sphincter at the orifice, prevents the return of material from the large to the small intestine. There are no sharp division points or major structural differences between successive portions of the small intestine.

The four coats of the alimentary tract are perhaps best represented in the small intestine (Fig. 13.16). Its features of special interest include *circular folds, villi, special glands, lymphoid nodules.*

Circular folds begin projecting into the lumen of the small intestine 1 or 2 in. beyond the pylorus. Some extend all the way around the intestine, and others only part of the way. The folds are most frequent and pronounced in the duodenum and upper jejunum, becoming smaller in the lower jejunum and almost

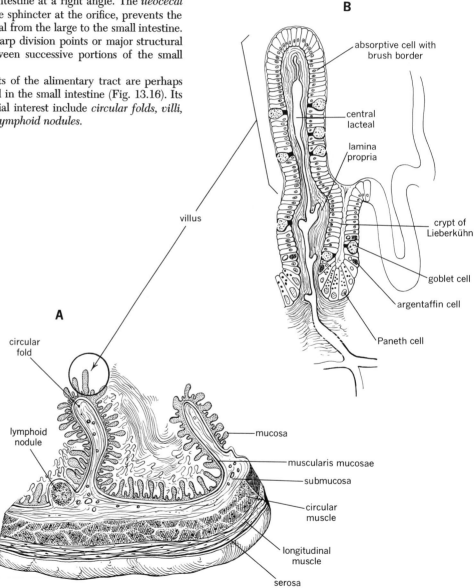

13.16 Structure of the small intestine. A, longitudinal section through wall; B, enlargement of villus and crypt of Lieberkühn. (See also Fig. 5.5B.)

invisible in the lower ileum. They serve to increase the area of the surface that secretes digestive juices and absorbs digested food and tend to slow the passage of the intestinal contents. Unlike folds in the stomach walls, the circular folds do not disappear with distention.

Villi are minute projections from the mucosal surface of the small intestine (see Fig. 13.16). They too serve to increase the absorptive area. The dense forests of these remarkable little structures give the mucosa its characteristic velvety appearance. Each villus is covered with simple columnar epithelium and contains a rich capillary network and a central, blind-ending lymphatic vessel called a *lacteal*. The continuous swaying movements of the villi aid absorption by creating local turbulence. The villi vary in shape in different regions of the small intestine. In the duodenum they are low, broad, and leaflike; in the jejunum they are slender and fingerlike; and in the ileum they are clublike.

Simple tubular *intestinal glands*, or *crypts of Lieberkühn*, are found throughout the small intestine. Economically placed between adjacent villi, they extend to the muscularis mucosae but do not penetrate it. *Duodenal*, or *Brunner's, glands* penetrate the submucosa. They are compound tubuloalveolar glands with much branching.

Lymphoid tissue in the form of nodules is present throughout the small intestine. The nodules appear separately or in aggregations known as *Peyer's patches*. The scattered solitary nodules are larger and more numerous in the lower portions of the small intestine than in the higher portions, and Peyer's patches occur chiefly in the ileum.

Microscopic structure. Fig. 13.16 shows the mucosa of the small intestine as revealed by light microscopy. The epithelium of the intestinal glands is continuous with that of the villi. In a gland four types of epithelial cells can be distinguished: (1) *Paneth cells,* which line the gland base and contain large eosinophilic cytoplasmic granules; (2) *goblet cells* (or *mucous glands*), each of which resembles a goblet when its cytoplasm is filled with mucus and its nucleus is pushed basally; (3) *undifferentiated crypt cells,* many in mitosis; and (4) *enterochromaffin,* or *argentaffin, cells,* which require special stains for histologic identification.

In contrast, only two types of epithelial cells cover a villus: *absorptive cells,* with obvious brush borders and columnar shapes, and goblet cells, which may be filled with mucus but may also be in various stages of mucus synthesis or discharge. An epithelial cell divides in a gland, and its daughter cells migrate up the sides of a villus to be extruded from its tip into the intestinal lumen at the completion of their lifespans. Thus a major function of a gland is to provide absorptive and

goblet cells to replace those that are continuously lost from the villi. Studies with radioactive thymidine have shown that most of these are renewed every 1.3 to 1.9 days. The Paneth and enterochromaffin cells have low renewal rates and demonstrate no mitotic activity. Their function is uncertain, although their structure suggests that they synthesize and secrete proteins and carbohydrates, not digestive enzymes. It has been postulated that their secretions nourish the epithelial cells covering the villi.

The main function of the epithelium covering a villus is to absorb material from the intestinal lumen. Fig. 13.17 is a diagram of an absorptive cell as revealed by electron microscopy. The cell is columnar in shape, and its nucleus is located in its lower half. The brush border at the cell apex consists of many long, slender microvilli (see also Fig. 5.3). This adaptation of the cell membrane lies at the critical interphase between the external environment, as represented by the intestinal contents, and the internal environment, as represented

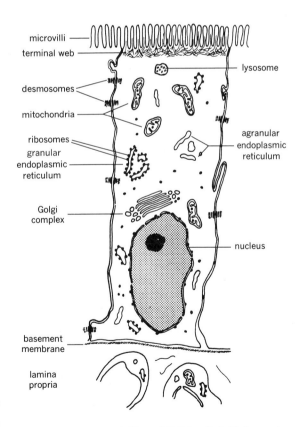

13.17 Absorptive cell. (From J. S. Trier, P. C. Phelps, and C. E. Rubin, *J. Am. Med. Assoc.,* 183, 768, 1963.)

by the cell cytoplasm. The great increase in surface area that it provides undoubtedly facilitates absorption of the intestinal contents. Immediately beneath the microvilli is the *terminal web,* a meshwork of closely packed fine filaments, parallel to and presumably stiffening and stabilizing the apical surface. Some filaments enter the cores of the microvilli. The cell cytoplasm contains mitochondria, endoplasmic reticulum (both agranular and granular), ribosomes, lysosomes, and desmosomes, regions of attachment to neighboring cells.

PERITONEUM AND MESENTERIES

We learned earlier that the abdominal cavity is lined by a fibrous sheath, the peritoneum (see p. 197). Like the sheaths lining the pleural and pericardial cavities, the peritoneum has two layers, a parietal layer over the abdominal walls and a visceral layer over the surfaces of the organs within the cavity (Fig. 13.18). The peritoneum prevents friction between contiguous abdominal organs and affords a large surface area, as great as that of the skin, through which fluid may pass in edema (see p. 210).

It must be emphasized that the organs of the abdominal cavity are all "outside" the peritoneum—just as the heart is "outside" the pericardium (see Fig. 9.4). We have seen how, in the course of embryological development, the intestine wanders some distance from the body wall toward the center of the abdominal cavity, dragging along a fold of peritoneum that later serves as its sole attachment to the body wall (see Fig. 13.2). This is the mesentery.*

The abdomen may be said to exist in order to accommodate two sets of organs: (1) the *gastrointestinal tract* and its derivatives, the liver and pancreas (with which the spleen is associated); and (2) the *urogenital system,* including the kidneys, ureters, and testes or ovaries. All the components of the first set are unpaired, whereas those of the second set are paired.

During early embryonic life the paired urogenital structures lie on either side of the aorta, covered with the peritoneum of the posterior abdominal wall (see Fig. 13.2). The gut, a straight tube of uniform caliber, is slung from the front of the vertebral column by the primitive dorsal mesentery, which is pierced by the large unpaired branches of the aorta: the celiac, superior mesenteric, and inferior mesenteric arteries. As the gut lengthens, it ceases to be a straight tube, and its mesentery and its major blood vessels follow it in its

---

* The prefix *mes-* means supporting membrane. Hence *mesentery* is the supporting membrane of the intestine; *mesogastrium* is the supporting membrane of the stomach; *mesocolon* is the supporting membrane of the colon; etc.

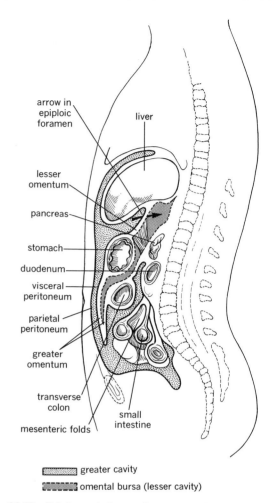

13.18 Relation of the peritoneum to the abdominal structures.

convolutions. All remain ventral to the urogenital structures behind the peritoneum, which could be viewed as the hard cover of an open book. The mesentery would be its pages.

We see, then, the relations of the adult peritoneum and the mesentery attaching the small intestine to the posterior abdominal wall. Only the duodenum, which lies directly upon this wall, has no mesentery.

One set of peritoneal folds merits special attention. Extending from the undersurface of the liver to the upper lesser curvature of the stomach is the *lesser omentum;* extending from the lower greater curvature of the stomach to the transverse colon and spreading out like an apron over the small intestine is the *greater omentum,* most conspicuous of all the peritoneal folds.

These effectively divide the entire abdominal cavity in two: the greater cavity; and the lesser cavity, or *omental bursa*, behind the stomach. The small opening between the greater and lesser cavities is the *epiploic foramen*.

LARGE INTESTINE

Gross structure. The last 5 ft of the alimentary tract is the large intestine. It begins at the ileocecal valve, ends at the *anus*, and consists of the *vermiform appendix, cecum, colon, rectum, anal canal,* and anus (see Fig. 13.1).

The cecum is a sac-like blind pouch about 2½ in. long and 3 in. wide just below the point of attachment of the ileum. From its lower end emerges the worm-shaped tubular vermiform appendix. This little organ, often the locus of inflammation necessitating surgical treatment, varies in length from 3 to 8 in. It is lined by mucous membrane and has circular and longitudinal coats of smooth muscle. The triangular *mesoappendix* attaches the appendix to the posterior abdominal wall.

The colon is divided into *ascending, transverse, descending,* and *sigmoid* portions, which roughly form three and a half sides of a square framing the convolutions of the small intestine. The ascending colon proceeds upward along the right side of the abdomen to the undersurface of the liver, where it bends in the *hepatic flexure*. The transverse colon continues across the abdominal cavity from right to left below the stomach to the undersurface of the spleen, where it bends downward in the *splenic flexure*. It is attached to the posterior abdominal wall by the *transverse mesocolon*. The descending colon extends downward along the left side of the abdomen to the brim of the pelvis. From there the sigmoid colon pursues an S-shaped course, ending in the rectum, whose lower portion, the anal canal, passes through the floor of the pelvis into the anus.

The walls of the large intestine generally have the four coats typical of the alimentary tract. The mucosa resembles that of the small intestine except that it has no circular folds or villi. Intestinal glands are present, but their chief secretion is mucus. Lymphoid tissue is present in solitary nodules but not in aggregated nodules. The muscular coat, however, differs strikingly from that in the rest of the tract. The longitudinal layer does not completely surround the intestine but instead is confined to three narrow bands called *teniae coli*, nearly equidistant from each other (see Fig. 13.1) The teniae are shorter than the intestine and consequently gather it into sacculations. In addition, the serosa bears rows of fat-containing pockets of peritoneum known as *epiploic appendages*.

The transverse and sigmoid colons retain mesenteric attachments to the abdominal wall. The ascending and descending colons lose theirs in varying degrees, but often the loss is complete, so that peritoneum covers all but their posterior surfaces.

The point of transition from descending colon to sigmoid colon is indefinite. The sigmoid colon is usually considered to be the part of the large intestine between the descending colon and the rectum that, because of its mesentery, is freely movable. Its average length in adults is about 40 cm, although sigmoid colons 84 cm long have been reported. The shapes of the sigmoid colon and its mesentery are subject to great variation.

The walls in the most distal portion of the sigmoid colon are distinctive. The epiploic appendages diminish in number and size, and the three teniae spread to form an encircling longitudinal muscle layer at the junction with the rectum. In the same region the circular muscle thickens until it structurally resembles a sphincter.

The terminal part of the intestine comprises the rectum and the anal canal (Fig. 13.19). The rectum extends 10 to 15 cm from the rectosigmoid junction, but only its uppermost portion is completely enveloped by peritoneum. A very small *mesorectum* may be present near the rectosigmoid junction. In a male the peritoneum is reflected from the upper anterior rectal surface into the space between the rectum and the bladder, forming the *rectovesical recess* or *pouch*. In a female the peritoneum is reflected to the uterus, forming the *rectouterine recess*. The anal canal is 2 to 3 cm long, ending in the anus, or anal orifice.

Figure 13.19 shows the several *rectal valves* (which are in fact mucosal folds running halfway round the rectum); the anal canal; and the mucosal and muscular apparatus that constitutes the anal sphincters.* The mucosa of the rectum near the anal canal is thicker and more vascular than that of the sigmoid colon. Simple columnar epithelium resembling that in the colon lines most of the rectum, becoming simple cuboidal epithelium toward the rectoanal junction. The lining gradually changes into stratified squamocuboidal epithelium in the upper anal canal and then into simple squamous epithelium, which eventually merges into true skin.

The *superior, middle,* and *inferior hemorrhoidal arteries* supply the rectum and anal canal, and the *superior, middle,* and *inferior hemorrhoidal veins* drain them. The inferior hemorrhoidal vein spreads into rich *internal* and *external hemorrhoidal plexuses*. These

---

* The figure reveals several structures that have no obvious functions. For example, at the rectoanal junction there is a corrugated *dentate*, or *pectinate*, *line*, made up of permanent mucosal folds known as *rectal columns*. Their lower ends are united by smaller folds known as *anal valves*. Between the columns lie valleylike *rectal sinuses*, which extend behind the anal valves to form small pockets, *anal crypts*. Opening into the crypts are the ducts of small *perianal glands*.

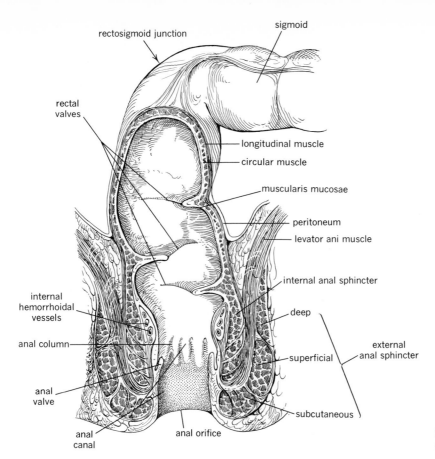

sigmoid

rectosigmoid junction

rectal valves

longitudinal muscle

circular muscle

muscularis mucosae

peritoneum

levator ani muscle

internal anal sphincter

internal hemorrhoidal vessels

deep

anal column

external anal sphincter

anal valve

superficial

anal canal

anal orifice

subcutaneous

anal canal

13.19   Rectum   and   anal canal.

venous networks tend to develop varicosities similar to those in the veins of the lower extremities Such varicosities, called *hemorrhoids,* are common. The predisposing factors are insufficient anatomical support of the thin-walled hemorrhoidal veins and gravity, which favors venous stasis and dilatation.

The musculature of the anal region is unique and complex—a combination of typical alimentary tract smooth muscle and striated muscle. It appears in cross section in Fig. 13.19.

The main sphincters surrounding the anus are the *internal* and *external anal sphincters,* the latter with *subcutaneous, superficial,* and *deep* subdivisions. The internal anal sphincter represents a gradual enlargement of the inner circular smooth muscle, ordinarily innervated by autonomic nerves. However, the sphincter is intimately associated with the striated muscles of the region and may respond reflexly to stimuli from peripheral nerves. The superficial subdivision of the external anal sphincter is its largest and strongest portion. It arises from the tip of the coccyx and curves around the side of the anus, inserting in the central point of the *perineum.*

The external anal sphincter anchors the anus but does not support it well either in front or in back. Keeping it in position is the muscle that fixes the pelvic floor, the *levator ani muscle.* It is shaped somewhat like a funnel, with its lowest point marked by the rectoanal junction. The levator ani muscle acts as a base of resistance against increased abdominal pressures during lifting, coughing, and defecation.

Nervous controls. Like the upper alimentary tract, the lower alimentary tract is provided with a web of autonomic nerves that coordinates and regulates motor and secretory functions and carries most of the afferent impulses. Again intrinsic plexuses modify the effects of autonomic stimuli.

The sympathetic and parasympathetic pathways to the intestine are diagrammed in Fig. 13.14 (see also Fig. 12.49). The hypothalamus connects with both efferent and afferent fibers and thereby is a most important relay station. Parasympathetic efferent fibers descend from the hypothalamus to synapse with cells in the motor nuclei of the vagus nerves (X) and in the sacrum. The axons of these cells are the preganglionic parasym-

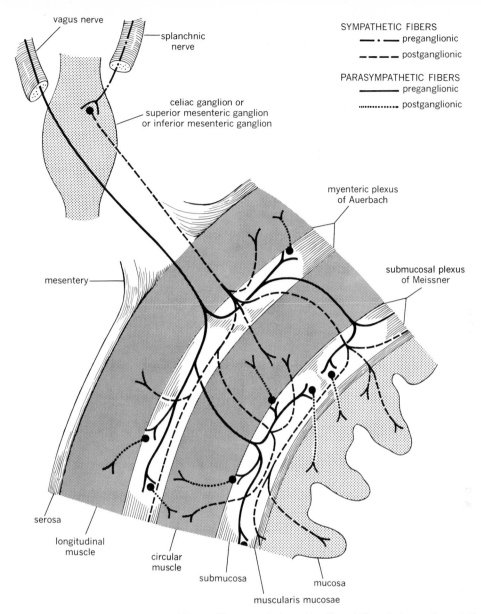

vagus nerve

splanchnic nerve

celiac ganglion or
superior mesenteric ganglion
or inferior mesenteric ganglion

myenteric plexus
of Auerbach

mesentery

submucosal plexus
of Meissner

serosa

longitudinal
muscle

circular
muscle

submucosa

muscularis mucosae

mucosa

13.20 Efferent nerves in the intestinal wall. Afferent fibers travel alongside efferent fibers but pass through the celiac ganglion without synapsing. Fine structure of intrinsic plexuses is shown in Fig. 13.15. (Adapted from *The CIBA Collection of Medical Illustrations* by Frank H. Netter, M.D. Copyright CIBA.)

pathetic efferent fibers in, respectively, the vagus nerves and the pelvic splanchnic nerves. These nerves together serve almost the entire alimentary tract. The vagus nerves are distributed to the parts derived from the foregut and midgut (pharynx to end of transverse colon); the pelvic splanchnic nerves are distributed to the parts derived from the hindgut (end of transverse

colon to anus). The fibers all end and synapse with postganglionic neurons in the intrinsic plexuses (Fig. 13.20). The postganglionic parasympathetic fibers, with the corresponding sympathetic fibers, innervate the smooth muscle of the intestinal wall, the wall capillaries, and the glands.

The sympathetic efferent fibers descend from the

hypothalamus to synapse with cells in the lower thoracic and first two lumbar segments of the spinal cord. The axons of these cells are preganglionic sympathetic fibers, which pass through or synapse in the sympathetic ganglia. The fibers leaving these ganglia form the greater, lesser, and least splanchnic nerves. Their fibers synapse with neurons in the superior and inferior mesenteric ganglia and the superior and inferior hypogastric plexuses. The emerging postganglionic sympathetic fibers accompany various arteries directly to the intestines.

Nerves for the striated muscles of the rectum and anal canal arise in the motor area of the cerebral cortex, as do those for other striated muscles.

The intestines are insensitive to ordinary touch, pain, and temperature, stimuli. Therefore, it is not surprising that specialized nerve endings resembling those in the skin are absent. The exact mode of termination of the afferent fibers is unclear, as is their pathway in the central nervous system. The main afferent and efferent fibers travel in the same nerves, however.

MAJOR ACCESSORY GLANDS

Pancreas. The liver and pancreas are the major accessory glands of the lower alimentary tract (see Fig. 13.1). The pancreas is a long slender organ with a distinct head and tail that give it somewhat the appearance of a tadpole. The head points to the right within the loop of the duodenum. Like the duodenum, it lies behind the peritoneum, tightly fixed to the posterior abdominal wall (see Fig. 13.18). The upper anterior portion of the pancreas is covered with peritoneum of the lesser omentum, and the lower anterior portion by peritoneum of the greater omentum. *Pancreatic juice*

is discharged into a long *pancreatic duct,* which runs from tail to head and empties into the duodenum through an orifice common to it and the common bile duct in the ampulla of Vater (Fig. 13.21). The normal human pancreas weighs 50 to 75 g.

The microscopic structure of the pancreas illustrates a novel principle of glandular architecture (Fig. 13.22). The bulk of the tissue superficially resembles that of the salivary glands, and so the pancreas has been fittingly called "the salivary gland of the abdomen." As in the salivary glands, there are many alveoli whose enzyme-producing cells secrete digestive juice into small ducts for transmission to the main duct.* However, islands of distinctive pale-staining cells that are evidently not part of the main glandular apparatus also appear throughout the pancreas, clearly distinguishing it from the salivary glands. These are the *islets of Langerhans.* They are about 1,000,000 in number and constitute about 1 g of tissue (i. e., about 1.5% of the weight of the pancreas). The islets are made up of several cell types—as is suggested by the diverse nuclear forms in the figure—one of which secretes the hormone insulin. Thus insulin is not produced by the same cells that secrete pancreatic juice, nor is it discharged into the duodenum through the pancreatic duct. Rather, it passes directly into the blood capillaries traversing the islets.

We must conclude, therefore, that the pancreas is both an exocrine gland (one that discharges its secretions into the immediate vicinity through ducts) and an endocrine gland (one that deposits its secretions into the blood for distribution throughout the body), though it is dispersed. The endocrine function will be discussed later.

---

* Table 13.1 notes that both the pancreas and the salivary glands secrete amylase.

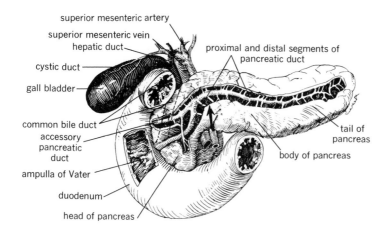

13.21 Relations of the pancreas to the duodenum and gall bladder. (From C. L. Callander, *Surgical Anatomy,* 2nd ed., W. B. Saunders Co., Philadelphia, 1939.)

superior mesenteric artery
superior mesenteric vein
hepatic duct
cystic duct
gall bladder
common bile duct
accessory pancreatic duct
ampulla of Vater
duodenum
head of pancreas
proximal and distal segments of pancreatic duct
tail of pancreas
body of pancreas

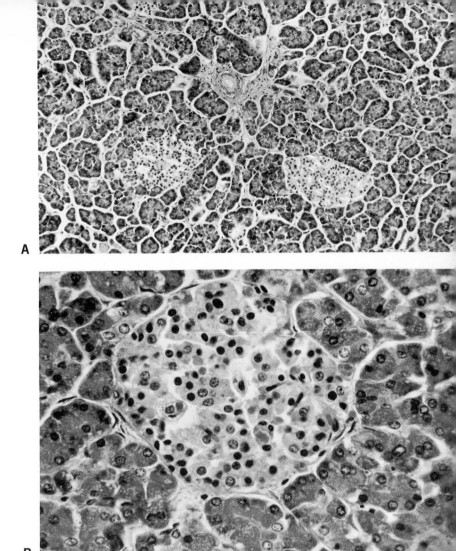

A

B

13.22 Microscopic sections of a normal human pancreas, showing enzyme-producing cells and islets of Langerhans: A, low power; B, high power. (From E. J. Reith and M. H. Ross, *Atlas of Descriptive Histology*, Hoeber Medical Division, Harper & Row, Publishers, 1967.)

Liver. The liver is the largest organ in the body, weighing 1500 g in an adult male and a bit less in an adult female. From the front it is triangular. Situated primarily on the upper right side of the abdominal cavity (see Fig. 13.1), it is sheltered by the lower right ribs and molded to the bottom of the diaphragm, whose central tendon alone separates it from the heart and pericardium.

The broad, concave undersurface of the liver is scored by a deep H-shaped fissure that divides the organ into *right, left, quadrate,* and *caudate lobes.* The line of demarcation between the right and left lobes is indicated on the upper surface by the *falciform ligament,* which connects the liver to the diaphragm and the anterior abdominal wall and conveys in its free border the *round ligament,* a fibrous remnant of the occluded umbilical vein (see p. 300). Except for a bare area on its posterior surface that is attached directly to the diaphragm, the liver is covered with peritoneum (see Fig. 13.18). Beneath the peritoneum a dense fibrous layer, the *capsule of Glisson,* spreads over the entire surface and continues into the liver substance at the hilus as a sleeve enveloping the *portal vein, hepatic artery,* and *hepatic duct.*

A section of liver viewed grossly shows no structural organization. However, the microscope reveals long interconnecting rows, one cell wide, of characteristic, variously shaped *parenchymal liver cells* (Fig. 13.23). Under low magnification the cell rows seem to radiate around a small *central vein* (Fig. 13.24A).

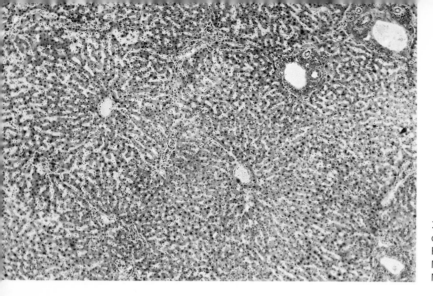

13.23 Microscopic appearance (low power) of a human liver. (From E. J. Reith and M. H. Ross, *Atlas of Descriptive Histology,* Hoeber Medical Division, Harper & Row, Publishers, New York, 1967.)

The interpretation of these observations has been modified somewhat in recent years. According to classic histology, the microscopic pattern is a cross section of the idealized structure in Fig. 13.24B, the *liver lobule.* The liver cells are arranged in *cords,* two cells thick, extending from the central vein to the periphery of the lobule. Between adjacent cords lie blood-filled vascular channels called *sinusoids.* The endothelium lining the sinusoids and wrapping the cords is studded with occasional cells of a special type known as *Kupffer cells.* These and the other endothelial cells are actively phagocytic and together constitute a large portion of the reticuloendothelial system (see p. 223). Between the two rows of liver cells making up a cord runs a narrow tubule, the *bile canaliculus,* which collects and transports *bile,* the dark green fluid secreted by the cells.

In this traditional theory of liver structure, the fundamental architectural unit is viewed as an elongated lobule having five, six, or seven sides.* At its corners are *portal areas.* Each portal area is bounded by a sleevelike limiting plate of liver cells and contains a branch of the portal vein, a branch of the hepatic artery, a network of bile canaliculi, and a network of lymphatic vessels. From the branch of the portal vein, small venules penetrate the limiting plate and empty into the sinusoids. Encircling each lobule are *interlobular veins* (fine branches of the portal vein), *interlobular arteries* (fine branches of the hepatic artery), and *interlobular bile ducts.* The interlobular veins drain into the sinusoids at the periphery of the lobule; the sinusoids drain into the central vein; and the central vein drains into the *sublobular vein,* a tributary of the hepatic vein. The interlobular arteries also drain into

---

* It is important to realize that the classic liver lobule was more postulated than observed. A lobule could not be dissected free as could an individual renal tubule (see Fig. 10.7).

the sinusoids. Hence sinusoidal blood is derived from two circulations. In a cross section of a limiting plate, the bile canaliculi appear as polygons around the cells. In a longitudinal section, they appear as dots among the cells. They begin in the Golgi apparatuses of the cells and form loops that lead into *bile ductules,* which lead in turn into the interlobular bile ducts

In recent years histologists have reasoned that, if liver cells are arranged in cords like those in Fig. 13.24, some liver sections should show cell groupings representing cross sections of cords. Since they have seen no such cross sections, they have proposed that the liver is a spongelike mass of one-cell–thick plates tunnelled by lacunae—in other words, that the liver is a biliovascular tree with the spaces between the branches filled with plates and lacunae. The lobules are viewed as ill-defined areas surrounding central veins and continuous with one another. We see, then, that classic histology regarded the liver as a meshwork of cylinders of liver cells surrounded by blood and that modern histology regards it as a meshwork of cylinders of blood-filled sinusoids surrounded by liver cells.

Portal and hepatic circulations. The liver has a unique circulatory feature: it receives blood from two sources (see Chapter 9). One, the portal vein, carries a large volume under low pressure; the other, the hepatic artery, brings in only about 30% of the liver's blood supply under high pressure. All blood leaves the liver through the hepatic vein.

The portal vein drains most of the blood from the capillary beds of stomach and intestines, spleen, and pancreas. It is the great vein that completes the circulatory loop from the splanchnic arteries supplying these organs. Blood passing from the intestines to the liver through the portal vein is laden with nutrient substances resulting from intestinal digestion and absorp-

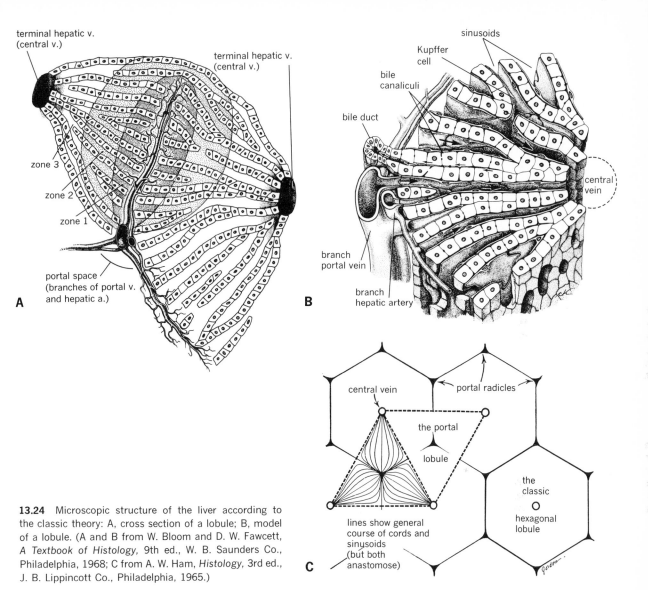

A

zone 3

zone 2

zone 1

terminal hepatic v.
(central v.)

terminal hepatic v.
(central v.)

portal space
(branches of portal v.
and hepatic a.)

B

sinusoids

Kupffer
cell

bile
canaliculi

bile duct

central
vein

branch
portal vein

branch
hepatic artery

C

central vein

portal radicles

the portal

lobule

the
classic

hexagonal
lobule

lines show general
course of cords and
sinusoids
(but both
anastomose)

13.24 Microscopic structure of the liver according to the classic theory: A, cross section of a lobule; B, model of a lobule. (A and B from W. Bloom and D. W. Fawcett, *A Textbook of Histology*, 9th ed., W. B. Saunders Co., Philadelphia, 1968; C from A. W. Ham, *Histology*, 3rd ed., J. B. Lippincott Co., Philadelphia, 1965.)

tion. In fact, the portal vein alone conserves these substances for the body, for if they were to enter the systemic circulation directly, most would be rapidly eliminated by the kidneys. The existing arrangement permits the liver to store and transform them.

The flow of blood in the portal vein is streamlined. The three major veins converging to form the portal vein are the *splenic vein* (from the spleen) and the *superior* and *inferior mesenteric veins* (from the intestines) (see Fig. 9.1). The portal vein ascends to the liver, where it divides into right and left branches serving the corre-

sponding liver lobes. India ink injected into the superior mesenteric vein is deposited only in the right lobe; India ink injected into the splenic vein is deposited only in the left lobe. Thus, though they join in the portal vein, the incoming streams remain relatively unmixed.

The substance of the liver, its covering capsule, and the bile ducts receive arterial blood from the hepatic artery. As we have seen, the two blood supplies meet within the sinusoids, and blood leaves the liver by a single channel, the hepatic vein. In some ways the situation parallels that in the pulmonary and bronchial arteries, in which the main business of gas exchange is conducted by the pulmonary arteries. The crucial difference is that each artery entering the lung transfers its blood to a corresponding vein, though there is minor mixing (see p. 377). Two vessels enter the liver, but only one leaves it.

In a resting adult the combined blood flow to the liver is approximately 1400 ml per minute, about 20% of the cardiac output.* Variations in blood flow through the liver result from vasomotor changes in the intrahepatic vessels. When the liver is well filled with blood, its size may increase appreciably. This occurs to some extent following meals and, of course, in congestive heart failure.

Gall bladder and external bile passages. One function of the liver is the continuous secretion of bile. Bile discharged deep within the liver enters the minute bile canaliculi and travels to the large interlobular bile ducts, which merge into *right* and *left hepatic ducts.* These unite on leaving the liver into the single large hepatic duct, which courses toward the duodenum.

Fresh bile may not enter the intestine immediately. It may turn instead into the *cystic duct* and enter the gall bladder, a pear-shaped sac attached to the bottom of the liver (see Figs. 13.1, 13.21). The gall bladder drains into the cystic duct, which joins with the hepatic duct to form the common bile duct. This pierces the duodenal wall about 10 cm below the pylorus. Within the wall it connects with the pancreatic duct in a short and narrow passage, the ampulla of Vater. The ampulla, which is surrounded by muscle fibers of the *sphincter of Oddi,* opens into the duodenum at the summit of the *duodenal papilla.*

Physiology of the Lower Alimentary Tract

COMPLETION OF DIGESTION

Introduction. We interrupted our discussion of the digestive process as the semifluid chyme was passing from the stomach into the duodenum. At this point (1) the food mass had been broken up mechanically by chewing; (2) it had been liquefied by saliva and gastric juice; (3) starch and protein digestion had been initiated by salivary amylase and gastric pepsin, respec-

tively; and (4) the pH of the food mixture had been lowered by the addition of HCl in the stomach.

In the small intestine digestion proceeds in earnest, continuing until complex food elements are sufficiently simplified to be absorbed. The most important digestive agents secreted into the small intestine are *pancreatic juice, bile,* and *intestinal juice.*

Composition and function of pancreatic juice. Pancreatic juice, the secretion of the pancreas, is the main digestive juice. The volume produced in 24 hours varies from 700 to 2000 ml. The juice is a colorless liquid with a pH of about 8, consisting of water, electrolytes (including bicarbonate, which is responsible for its alkalinity; Table 13.3), mucus, and digestive enzymes of four major types: *proteolytic* (protein-splitting, proteases), *amylolytic* (carbohydrate-splitting, carbohydrases), *lipolytic* (lipid-splitting, lipases), and *nucleolytic* (nucleic acid-splitting, nucleases). These enzymes are virtually capable of digesting all food without the aid of any other digestive agents. This explains the serious defects in digestion resulting from reduced pancreatic secretion in disease.

The principal proteolytic enzymes are *trypsin, chymotrypsin,* and *carboxypeptidase.* Like gastric pepsin, pancreatic trypsin is secreted as an enzymatically inert precursor. The trypsin precursor is *trypsinogen.* Upon entering the small intestine via the pancreatic duct,

	TABLE 13.3 CONCENTRATIONS OF ELECTROLYTES IN PANCREATIC JUICE AFTER STIMULATION

Electrolyte	Concentration, meq/liter
Sodium	138–143
Potassium	5–9
Bicarbonate	90–140*
Chloride	15–65*

* The sum of the bicarbonate and chloride concentrations is approximately 155 meq/liter.

---

* We recall that a similar fraction of the cardiac output continuously perfuses the kidney (see p. 340).

trypsinogen is converted to trypsin by *enterokinase,* an enzyme present in the intestinal juice. The conversion is attended by the removal of a large polypeptide (molecular weight 5000), which may be regarded as a "masking substance." Once some trypsin is formed, it can itself activate trypsinogen to trypsin, though in this case the masking substance removed has a molecular weight of only 700. Therefore, trypsinogen is converted completely to trypsin via two pathways.

$$\text{trypsinogen} \xrightarrow{\text{enterokinase}} \text{trypsin}$$
$$\text{trypsinogen} \xrightarrow{\text{trypsin}} \text{trypsin} \quad (13.3)$$

Chymotrypsin and carboxypeptidase (actually a family of enzymes) are also secreted as their precursors, *chymotrypsinogen* and *procarboxypeptidase,* respectively. In each case conversion of precursor to active enzyme is catalyzed by trypsin but not by enterokinase. Chymotrypsin itself also catalyzes the conversion of chymotrypsinogen to chymotrypsin.

$$\text{chymotrypsinogen} \xrightarrow[\text{chymotrypsin}]{\text{trypsin or}} \text{chymotrypsin} \quad (13.4)$$
$$\text{procarboxypeptidase} \xrightarrow{\text{trypsin}} \text{carboxypeptidase} \quad (13.5)$$

Trypsin, once believed to be the sole proteolytic enzyme in pancreatic juice, has been crystallized and shown to be an endopeptidase (see Fig. 13.9A). It specifically hydrolyzes a peptide bond to which arginine or lysine contributes the carboxyl group. Chymotrypsin has also been crystallized and shown to be an endopeptidase. Like pepsin, it hydrolyzes a peptide bond involving an aromatic amino acid (e.g., tyrosine or phenylalanine), but pepsin attacks on the amino side of the acid whereas chymotrypsin attacks on the carboxyl side. Carboxypeptidase has been purified and found to contain one atom of zinc per molecule. It and intestinal *aminopeptidase* are exopeptidases (see Fig. 13.9B). Carboxypeptidase attacks the carboxyl end of a protein molecule, and aminopeptidase the amino end, until only a dipeptide or tripeptide remains. Thus they complete the digestion of proteins begun by the endopeptidases.

We see, then, that the three principal proteolytic enzymes of pancreatic juice break down large proteins to tripeptides and dipeptides and many free amino acids. Because they attack different peptide bonds, extensive cleavage of the diverse dietary proteins can take place.*

Pancreatic amylases break down large starch molecules to dextrins and disaccharides, as does salivary amylase, but they are far more powerful. Indeed, starch is not effectively digested until it is exposed to them.

The disaccharide maltose must be cleaved into its two component glucose molecules to be absorbed. Pancreatic juice contains a weak enzyme, *maltase,* that splits a small fraction of the maltose, but the rest requires intestinal juice enzymes.

Pancreatic lipase breaks down lipids (i.e., triglycerides) to varying degrees, the major products being free fatty acids (and glycerol) and monoglycerides. Its activity depends on the presence of Ca^{++} and bile. Pancreatic lipase is extremely important in fat digestion. Though other digestive juices contain proteases and carbohydrases, the pancreas appears to be the only source of a lipase that can function in the conditions prevailing in the alimentary tract. Consequently, defective fat digestion is an early sign of pancreatic insufficiency.

Pancreatic ribonuclease and deoxyribonuclease cleave RNA and DNA, respectively, to low molecular weight ribo- and deoxyribo-mononucleotides and oligonucleotides. It is interesting and significant that the pH at which all the pancreatic enzymes function best is between 7 and 9. The alkaline character of the pancreatic secretion is therefore appropriate. As Table 13.3 indicates, the maximum bicarbonate concentration in pancreatic juice is three to four times that in plasma. Pancreatic HCO_3^- is derived from blood HCO_3^- and metabolic CO_2. The mechanism of its secretion by the pancreas has not yet been established, but one theory suggests that the carbonic anhydrase in pancreatic tissue participates. It appears, for example, that

---

* The proteolytic enzymes of the human alimentary tract do not digest all proteins. For example, they have no effect on the fibrous protein that composes wool. (Proteinases in the alimentary tract of the clothes moth, however, attack wool readily.) Moreover, they do not attack living tissue except in abnormal situations such as peptic ulcer.

the carbonic anhydrase inhibitor acetazolamide partially inhibits bicarbonate secretion (see p. 525). Conceivably, a HCO_3^-–Cl^- exchange, opposite to that postulated in gastric parietal cells, occurs.

Control of pancreatic secretion. The secretion of pancreatic juice is continuous and greatly influenced by humoral and neural factors, though their relative effects have not yet been defined.

It was early shown that HCl introduced into the duodenum promotes the secretion of pancreatic juice—and, surprisingly, it was found that removal of the nerves to the pancreas did not interfere with this reaction. It remained for Sir William Bayliss and Starling (see p. 317) to demonstrate in 1902 that the effect of HCl is mediated by a blood-borne hormone manufactured in the duodenal mucosa; an injection of the aqueous extract of the duodenal wall of one animal into the blood of another animal causes a profuse secretion of pancreatic juice. Bayliss and Starling named the hormone *secretin.* It was the first hormone to be recognized.

The pancreatic juice resulting from secretin stimulation is rich in bicarbonate and poor in digestive enzymes; 1 mg of secretin stimulates the secretion of 12 liters of pancreatic juice that is about $0.13N$ in HCO_3^-. We may therefore regard the secretin mechanism as an elegant regulatory device. Strongly acid chyme enters the duodenum and induces the duodenal mucosa to produce secretin, which enters the blood stream and eventually induces the pancreas to secrete a more alkaline juice. The outcome is neutralization of the duodenal contents and hence elimination of the stimulus for further pancreatic secretion.

A second hormone, *pancreozymin,* released by the duodenal mucosa also affects pancreatic secretion. Whereas secretin chiefly stimulates the secretion of water and bicarbonate, pancreozymin stimulates the secretion of digestive enzymes. The modes of action of the hormones upon the pancreatic cells are unknown.

Nerve impulses traveling via the vagus nerves (X) and cholinergic drugs stimulate the secretion of digestive enzymes by the pancreas and increase the volume of pancreatic juice. The role of the sphanchnic nerves in pancreatic secretion has not been established.

Composition and function of bile. The complexity of the liver cells is apparent in Table 13.4, which divides their functions into six categories. Those not concerned with digestion are considered elsewhere. The primary digestive function is bile secretion.

The composition of typical human bile is

TABLE 13.4 FUNCTIONS OF THE LIVER

Digestive	Secretion of bile salts into bile
Excretory	Secretion of products formed in liver cells (e.g., bile salts, bilirubin conjugates, cholesterol) into bile
	Excretion in bile of substances cleared from blood by liver cells (e.g., drugs, bromsulphalein)
Detoxicant and protective	Detoxication by conjugation, methylation, oxidation, etc.
	Removal of NH_3 from blood
	Phagocytosis of foreign particles by Kupffer cells
Metabolic	Carbohydrate, protein, lipid, mineral, and vitamin metabolism
Circulatory	Transfer of blood from portal circulation to systemic circulation
	Storage of blood
Hematological	Formation of blood in embryo
	Formation of clotting factors and other plasma proteins

TABLE 13.5 COMPOSITION OF BILE

Substance	Amount in hepatic duct bile, %	Amount in gall bladder bile, %
Water	97.48	83.98
Bile salts	0.93	8.70
Inorganic salts	0.83	1.02
Mucus and pigments	0.53	4.44
Fatty acids (soaps)	0.12	0.85
Cholesterol	0.06	0.87
Lecithin	0.02	0.14

given in Table 13.5. The principal components are water, bile salts, inorganic salts, mucus, pigments, fatty acids, cholesterol, and lecithin. The bile salts are most important in digestion.

Bile salts are the sodium salts of *glycocholic* and *taurocholic acids* (Fig. 13.25). These acids are formed by conjugation of the amino acids *taurine* and *glycine,* respectively, with cholic acid, a product of cholesterol breakdown. Bile salts behave as natural detergents to emulsify lipids, which otherwise would remain inaccessible to lipase action. Emulsification of a lipid in this case consists in its dispersal throughout an aqueous medium, with which it is immiscible. Detergents emulsify by decreasing the surface tension of the medium. In addition, their *hydrotropic effect* (i.e., their ability to form water-

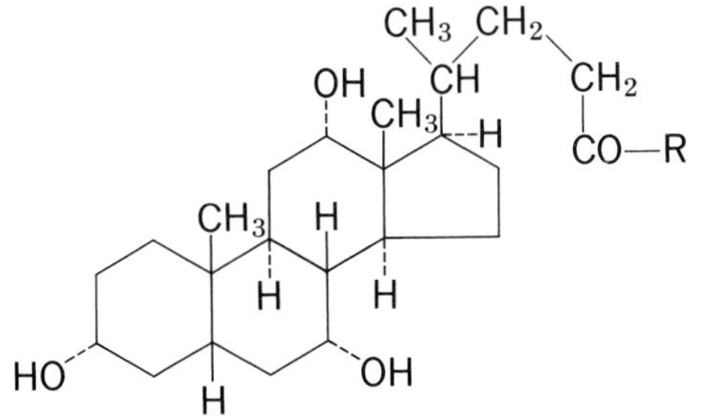

soluble complexes, or *micelles**) promotes the solution of certain fatty substances (e.g., fatty acids and fat-soluble vitamins A, D, E, and K†).

Since bile salts are products of cholesterol breakdown, they may also be regarded as excretions of the liver (see Table 13.4). Interestingly, only a small fraction of the bile salts excreted into the small intestine is found in the feces. The rest is reabsorbed farther along the alimentary tract, returned to the liver via the portal vein, and again secreted in the bile. Thus bile salts given by mouth can be almost completely recovered in the bile stream. This *enterohepatic* circulation represents considerable economy for the body.

Storage and evacuation of bile. A normal adult male secretes 15 ml of bile per kilogram of body weight per day—about a liter per 70 kg. The secretion of bile is a continuous process, distinct from its intermittent evacuation into the small intestine. Normally, when the pressure within the single hepatic duct reaches 50 mm of mercury, bile begins to flow through the cystic duct into the gall bladder, whose capacity is about 50 ml. There bile loses salts and water and receives gall bladder mucus. Table 13.5 illustrates the consequences of this concentration.

The arrival of chyme in the duodenum provokes relaxation of the sphincter of Oddi and contraction of the gall bladder, which force bile into the intestine at the time of greatest need. Then the sphincter contracts, the gall bladder relaxes, and the evacuation of bile ceases. These activities alternate during the early hours of digestion.

One abnormal consequence of the concentration of bile in the gall bladder is the formation of a precipitate of one or more bile components. This is particularly common in a poorly functioning or infected gall bladder. The particles of

13.25 Bile salts: glycocholic acid, R = NHCH$_2$COOH; taurocholic acid, R = NHCH$_2$CH$_2$SO$_3$H (cholic acid, R = OH).

---

* A micelle is a particle—40Å in diameter according to recent data—in which the bile salts and glycerides are arranged with their polar groups facing outward, so that it "looks" polar to a solvent (i.e., water), even though its interior with its long hydrocarbon chains is nonpolar. Micelles dissolve in water to form a *micellar solution,* although their insides can be regarded as a separate phase.

† Vitamin K is essential for the synthesis of clotting factors (see p. 274). It follows that defective lipid digestion and absorption lead eventually to defective blood coagulation.

precipitate grow with time into *gallstones,* ranging in size from sandlike grains to large masses. An average diameter is ½ in. Stones are precipitates of cholesterol, pigments, calcium carbonate, or, most frequently, combinations of these substances.

Intestinal juice. Intestinal juice is the secretion of the duodenal glands.* The secretory processes respond to at least three rather poorly understood stimuli: (1) the normal intestinal contents; (2) humoral factors such as the intestinal hormone *enterocrinin;* and (3) mechanical factors such as a balloon distending a segment of intestine or a bolus of some other inert material (e.g., paraffin or cotton). The response of an isolated intestinal segment is undiminished when the extrinsic nerves are cut; therefore, secretion may be mediated by the intrinsic plexuses.

Not until recently have accurate measurements of the pH of intestinal juice been possible. Employing a sensitive glass electrode on the end of a long cable inserted in the intestine through the mouth, investigators have found mean pH values of 5.66 in the duodenum, 5.83 in the jejunum, and 6.40 in the ileum. Despite the alkaline character of pure pancreatic juice, then, the intestinal contents are not neutralized. After an acid drink, the pH transitorily declines in the upper portions of the small intestine, though, surprisingly, it often rises simultaneously in the lower portions.

Intestinal juice contains several significant enzymes, some contributed by the shed epithelial cells (and bacteria) with which it is mixed. They include enterokinase, which activates pancreatic trypsinogen; the specific disaccharidases lactase, maltase, and sucrase (see Table 13.1); an aminopeptidase and specific tri- and dipeptidases, which cleave the tri- and dipeptides remaining from pancreatic proteolysis; and a variety of nucleases, which cleave both types of nucleic acid and their several low-molecular weight derivatives.

° It was once believed that a second intestinal juice, *succus entericus,* was secreted by the intestinal glands (see p. 533). It is now known that the enzymes attributed to the succus entericus in fact come from sloughing absorptive and goblet cells.

ABSORPTION AND TRANSPORT

Movements of the small intestine. Three kinds of movements are observed in the small intestine: (1) rhythmic *segmentation* movements; (2) *peristalsis;* and (3) *pendular* movements. In concert, these movements mix and knead the intestinal contents, rub them across absorbing surfaces, and propel them toward the large intestine.

Walter Cannon regarded rhythmic segmentation movements as the most constant and fundamental intestinal motions. In 1911 he wrote in his historic monograph *The Mechanical Factors of Digestion:* "A mass of food is seen lying quietly in one of the . . . loops. Suddenly . . . constrictions at regular intervals along its length cut it into little ovoid pieces. A moment later each of these segments is divided into two particles, and immediately after the division, neighboring particles rush together . . . and merge to form new segments. The next moment these new segments are divided, and neighboring particles unite to make a third series and so on." (Fig. 13.26.)

This pattern is repeated 6 to 10 times per minute. Segmentation movements blend the intestinal contents but do not push them along. Since the movements occur in isolated strips of intestinal muscle, we must conclude that they arise within the muscle itself.

Although the term peristalsis is often carelessly used for any form of motor activity in the alimentary tract, it actually denotes the special wavelike movements that propel the intestinal

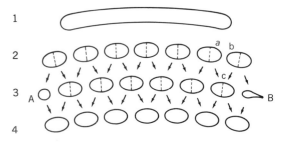

13.26 Diagram of rhythmic segmentation: 1–4, sequence of segments formed in a single intestinal loop. (From W. B. Cannon, *The Mechanical Factors of Digestion,* Longmans, Green & Co., New York, 1911.)

contents forward. It involves progressive contractions above the contents and relaxations below them, in either slow, steady *pulses* or rapid, abrupt spurts that traverse the whole small intestine without pause.

Peristalsis in the small intestine depends upon the intrinsic plexuses and continues even when all connections to the central nervous system are cut. (In contrast, the so-called peristalsis in the upper esophagus—whose intrinsic innervation is sparse—depends entirely upon the central nervous system.) The small intestine is "polarized" in that it transmits waves forward more easily than backward.

Pendular movements are due to annular constrictions that travel back and forth along a short length of the small intestine. They are not well understood.

Significance of intestinal absorption. The small intestine is by far the major locus of absorption; only a few highly diffusible substances (e.g., alcohols, free sugars, and amino acids) are absorbed in the stomach. We have studied the anatomical arrangements that maximize the surface area of the small intestinal mucosa. This membrane contains complex active transport systems whose selectivity recalls that in the renal tubular epithelium.

In physical terms, intestinal absorption involves the active transport of substances from the immediate vicinity of the luminal side of an absorptive cell (such as that in Fig. 13.17) (1) across the microvillous surface, (2) some distance into the cytoplasm, (3) out through the cell membrane, (4) across the epithelial basement membrane; (5) across the extracellular space, and finally (6) through the basement membrane and endothelial cell of a capillary or lacteal.*

This transfer proceeds against a concentration gradient and without leakage or back-diffusion—at least the rate of entry exceeds the rate of leakage. Thus it requires the performance

of electrochemical work. Undoubtedly different mechanisms operate for water, electrolytes, small molecules, and large molecules, but at present we have no precise explanation of any of them. We may have to distinguish between the processes transferring substances from lumen into cell and those transferring them from cell into blood. The absorption of certain substances (e.g., cholesterol, vitamin B_{12}, and iron) apparently involves a considerable time interval between the two events. This may signify storage within the cell, with passage into the blood stream being regulated in some way by the body's needs.

Absorption of sugars. We have seen that salivary and pancreatic amylases convert starches to disaccharides, which intestinal enzymes split into monosaccharides such as glucose, fructose, and galactose. It has been convincingly shown that disaccharides can not be absorbed as such in the intestine.

The absorption rates of the different monosaccharides vary much more than might be expected from their sizes and structures (Table 13.6). Therefore, absorption cannot be attributed to diffusion alone. If only diffusion occurred, pentoses would be absorbed more rapidly than the larger hexoses. In fact, the opposite is true, and many pentoses are absorbed at the same rate as indifferent substances such as sodium sulfate. The absorption of certain pentoses (e.g., xylose) is more rapid than that of indifferent substances. This is termed *facilitated diffusion.*

The notable rapidity of galactose and glucose absorption suggests a special mechanism in their case. The early conclusion that this is phosphorylation is known to be incorrect. However, the view that movement against a concentration gradient requires the formation of a complex that dissociates within the cell accords with current thought on renal tubular transport (see p. 357). Since phlorizin abolishes glucose absorption in the small intestine as well as in the kidney (see p. 346), we may conclude that intestine and renal tubule absorb glucose by similar means.

Absorption of protein digestion products. What we have learned of protein digestion implies that the intestine absorbs small water-

---

* It is important to understand the meaning of active absorption. Dissolved substances naturally tend to diffuse from a place of high concentration to one of low concentration. When the concentration of glucose in the intestine is high, it may enter the blood, where its concentration is only 0.1%, by diffusion. But its complete absorption in the intestine cannot be explained by diffusion alone, for glucose molecules removed by diffusion would tend to leak back by diffusion (see p. 159).

TABLE 13.6   RELATIVE RATES OF ABSORPTION OF SUGARS FROM THE SMALL INTESTINE

| | | Relative absorption rate* | |
Sugar	Number of carbons	Data of Cori and Cori	Data of Wilbrandt and Laszt
Galactose	6	110	115
Glucose	6	100	100
Fructose	6	43	44
Mannose	6	19	33
Xylose	5	15	30
Arabinose	5	9	29

* The rate of glucose absorption is arbitrarily set at 100.

soluble amino acids, not intact proteins. Indeed, nutrition can be maintained with pure amino acid mixtures in place of proteins. The teleologist might argue, in addition, that animals would not have evolved a complex enzyme system for splitting proteins into amino acids if they retained the capacity to absorb proteins intact.

The nature of the mechanisms responsible for amino acid absorption is unknown, though there are many resemblances between amino acid transport into intestinal mucosal cells and amino acid transport into other tissue cells. Presumably intestinal cell membranes contain carriers whose functioning requires metabolic energy. Studies of transport kinetics and of the competition among various amino acids for transport indicate that different groups of amino acids are transported by different carriers. The transport of L-stereoisomeric forms is strongly favored over that of D-stereoisomeric forms.

Despite the generalization that the intestine absorbs amino acids, not proteins, evidence exists that protein breakdown is not essential for absorption in at least two situations.* First, newborn mammals are temporarily able to absorb intact proteins in moderate quantities. Since antibody globulin is deficient in their blood, they would be susceptible to infection until

* This is not surprising. If passage across a cell membrane is considered in statistical terms (as it should be), it follows that a sufficiently sensitive method of detection will show that *any* substance crosses to *some* extent.

globulin synthesis got under way if maternal globulins, plentiful in the first milk, did not pass through their intestinal walls intact in the first days of life. Hydrolysis of antibody globulin would destroy its specificity. In this case, absorption is unidirectional (excessive leakage of plasma protein is thus avoided) and selective (foreign proteins are absorbed less well than maternal globulins) and takes place not by diffusion but by pinocytosis.

Second, quantitatively small "statistical" absorption of intact proteins occurs in adults. It is potentially dangerous, for foreign proteins within the body may provoke antibody formation. Antibody responses, or allergies, to proteins such as egg white may be due to this absorption.

Absorption of lipids. About 95% of food lipids is absorbed in the intestine. The problem of explaining the absorption process, however, has been complicated by the insolubility of lipids and of the fatty acids and monoglycerides liberated by the action of pancreatic lipase.

As noted earlier, the two chief functions of bile salts are (1) the emulsification of lipids and (2) the solubilization of fatty substances by the formation of micelles. Emulsified lipid droplets and micelles are readily absorbed by the intestinal mucosal cells—though it is still not clear whether the micelles, with their bile salt and lipid moieties, penetrate a cell as such or whether fatty acids and monoglycerides are released from the micellar state at the microvilli of the cell to enter in some other physical form.

A scheme of lipid absorption is shown in Fig. 13.27. After entry into the cell, both fatty acids and monoglycerides are esterified to triglycerides. One of the two major metabolic routes for esterification involves the acylation of fatty acids to acyl CoA thioesters, which react with glycerophosphate* to yield phosphatidic acid and then diglycerides. The other involves the direct interaction of monoglycerides with fatty acyl CoA molecules to yield diglycerides. Diglycerides arising by either pathway are readily converted to triglycerides. We see then that in *digestion* triglycerides are broken down to fatty acids and monoglycerides whereas in *absorption* an obligatory resynthesis of triglycerides takes place within the cell.

For many years some physiologists believed that lipid particles as large as 0.5 μ in diameter are absorbed. In part, this view was based on the observation that thoracic duct lymph becomes extremely turbid after a fatty meal, owing to minute droplets of undigested fat named *chylomicrons*. We now know that chylomicrons are lipoproteins produced from triglycerides (abbreviated TG in Fig. 13.27) and proteins

synthesized within the absorptive epithelial cell of the intestinal mucosa (see Fig. 13.27).†

The absorption processes just described are well established for derivatives of fatty acids with chains of 12 or more carbon atoms (see Table 2.4). Interestingly, triglycerides containing fatty acids with chains of 8 or 10 carbon atoms may enter the mucosal cell as such and be split there by specific enzymes. The liberated fatty acids are not esterified but pass out of the cell intact and enter the portal rather than the lymphatic system.

Absorption of water and salts. It is difficult to distinguish experimentally between the active transport of water from the blood to the intestine and the active absorption of salts from the intestine into the blood. Free inorganic salts, such as sodium chloride, can be absorbed without preliminary digestion. Inorganic salts combined with organic materials are often liberated in the course of digestion. It was once believed that they had to be so combined in order to be absorbed, but it is now known that free inorganic salts are absorbed as well as or better than bound forms.

---

* Glycerol arising from the breakdown of glycerides in the intestine may serve as precursor for glycerophosphate within the mucosal cell. Glucose can also be converted to glycerophosphate.

† In the hereditary disorder known as *congenital β-lipoprotein deficiency*, lipid transport into the lymphatic channels is markedly impaired. Triglycerides accumulate in the mucosal cells as a result of a deficiency of lipoprotein synthesis. An almost identical situation is induced in a rat by administration of an inhibitor of protein synthesis (e.g., puromycin).

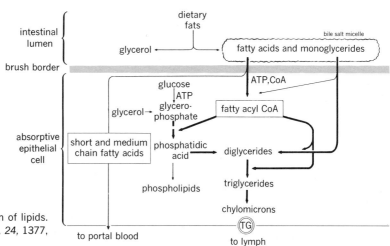

13.27 Pathways in the absorption of lipids. (From K. J. Isselbacher, *Fed. Proc.*, 24, 1377, 1965.)

The rates of absorption of electrolytes vary considerably. Sodium, potassium, and chloride are taken up rapidly; phosphate and magnesium, slowly.* In some cases the absorption rate is affected by the oxidation state. For example, reduced (ferrous) iron is more readily absorbed than oxidized (ferric) iron. The absorption rate may also be influenced by the local concentration of another element. A large excess of phosphate ion inhibits the absorption of calcium ion by forming insoluble calcium phosphate.

ELIMINATION OF WASTE

Functions of the large intestine. The time required for the passage of food from stomach to cecum varies widely, averaging 3½ hours. Material from the small intestine proceeds through the ileocecal valve into the large intestine in spurts. The large intestine is distensible and can retain its contents for long periods. It (1) conserves water, (2) temporarily stores the waste products of digestion, and (3) acts as an incubator for a variety of bacteria that serve the body's nutritional needs. The colon is the most important locus of these functions. The large intestine has no important digestive functions, for they have all been completed.

The semiliquid mass entering the large intestine consists mainly of the residues of digestive secretions (gastric, pancreatic, and intestinal juices, bile, and mucus), epithelial cells, and leukocytes. About 550 ml of water passes into the cecum per 24 hours. The feces normally contain only about 150 ml of water and 35 g of solids for the same period. The large intestine absorbs the remaining 400 ml of water and thereby converts semiliquid intestinal contents to solid feces. Water is absorbed principally in the cecum and ascending colon, though the rest of the large intestine is capable of absorbing it—a fact that is exploited when a physician administers fluids through a rectal tube.

The secretions of the mucosa of the large intestine are scanty. A thick fluid rich in mucus and of pH about 8.4 is elicited on local irritation, but it contains no digestive enzymes and acts simply as a lubricant to facilitate the progress of the intestinal contents. A herbivore, which subsists on bulky cellulose-containing vegetable fibers that are digested in part by colonic bacteria, has an unusually long and capacious colon. A carnivore has a short, narrow colon and almost no cecum. Man is between the two. Although his large intestine does not participate in digestion, it is the part of the alimentary tract in which bacteria grow most actively.

Composition of feces. We have indicated that in man, and carnivores, practically all of the food mass is absorbed in the small intestine.† Feces, then, are more a product of the alimentary tract than a residue of digested food. This is shown by (1) the absence from feces of soluble carbohydrates or proteins; (2) the absence from feces of muscle fibers after a meat meal; (3) the lack of effect of the nature of the diet on the chemical composition of feces; (4) the continued production of feces of normal chemical composition after fasting; and (5) the slow production of a semisolid substance resembling feces in a length of large intestine closed off from the rest of the alimentary tract.

As long as excessive quantities of vegetables or cereals are excluded from the diet, feces are composed of water (65 to 75% by weight), nitrogen (3 to 6% by weight), salts (11 to 20% by weight), lipids (12 to 18% by weight), large numbers of bacteria, and the products of bile pigment disintegration.

Defecation. Usually feces are stored in the sigmoid colon until discharged. At the rectosigmoid junction is a narrow portion of colon that operates as a sphincter (see p. 535). Only when peristalsis forces the feces past this obstacle do they enter and distend the rectum, creating a sensation that constitutes the desire to defecate.

Defecation, like other complex neuromuscular phenomena (e.g., swallowing, micturition, and respiration), involves both voluntary and involuntary responses. Distention of the rectum involuntarily provokes strong peristalsis, followed by contraction of the rectum and relaxa-

---

* The laxative action of magnesium sulfate (Epsom salts) depends upon its low rate of absorption and its resulting ability to draw water into the intestine by osmosis.

† In herbivores and vegetarians a large amount of unabsorbed food increases the bulk of the material in the colon.

tion of the internal and external anal sphincters. The emptying process is aided by voluntary contractions of the diaphragm and abdominal muscles, which increase intraabdominal pressure. That defecation is essentially a reflex mechanism is suggested by the fact that it is not affected when the spinal cord is cut above the lumbosacral level. Destruction of the lumbosacral region of the cord results in loss of the ability to defecate and fecal incontinence, though a degree of control sometimes returns.

EXCRETORY FUNCTION OF THE LIVER

Unusual aspects of certain bile components. Although uncertainties of liver anatomy have delayed our understanding of bile formation (see p. 544), we know that it is an unusual process. In its origin bile resembles urine—which derives from a sequence of filtration, reabsorption, and secretion—*and* milk, saliva, and other fluids—which derive directly from the secretory cells of glands.

Three groups of ingredients are found in bile. Group 1 includes substances whose concentrations in hepatic duct bile and blood are equal: sodium, potassium, chlorine, and glucose. Group 2 includes substances that are less concentrated in bile than in blood: cholesterol and alkaline phosphatase. Group 3 includes substances that are more concentrated in bile than in blood: bile salts, bile pigments, and a variety of foreign materials whose elimination from the body is a function of the liver—among them dyes that have long been employed in testing the excretory function of the liver, as phenol red is employed in testing the excretory function of the kidneys (see Table 10.1 and p. 344). The best known of these is *bromsulphalein*, or *BSP*. The high-concentrations of group 3 substances in bile leave no doubt that only *active excretion* could account for their transfer from blood.

The substances of group 3 have certain chemical characteristics in common: they are free organic acids (or their salts); their molecular weights are 300 or more; and they tend to form complexes with plasma proteins, particularly albumin (see pp. 248, 261).

When 5 mg of BSP per kg of body weight is injected into a vein, less than 6% remains in the serum after 45 minutes. As phenol red is cleared from blood by the kidneys, BSP is cleared by the liver. Although the concentration of BSP is far greater in bile than in blood, there is a point beyond which increases in dose or infusion rate cause no increase in excretion rate. The excretory mechanism appears then to be saturated. Other substances have similar effects; in fact, their competition for uptake (e.g., between bilirubin and BSP and between bilirubin and bile salts) by the liver is reminiscent of the competition between penicillin and PAH for renal tubular excretion (see p. 357).

The biochemical pathways between blood and bile are extremely interesting. Studies of the transfer from blood to bile of BSP labeled with radioactive sulfur reveal temporary storage or delay in the liver cells.* To approach recent information on BSP excretion in an orderly manner, we must consider first a class of compounds whose excretion resembles that of BSP.

Bile pigment metabolism. We first encountered bile pigments in our discussion of hemoglobin degradation, and we spoke of their behavior in the Ehrlich reaction (see Chapter 8). Much of our present knowledge of bile pigment metabolism rests on this chemical test. Ehrlich observed in 1880 that, when a mixture of sodium nitrite, hydrochloric acid, and sulfanilic acid (now termed *diazo reagent*) was added to the urine of an individual with an elevated serum bilirubin level (jaundice), the solution often became reddish purple. Ehrlich found that the addition of alcohol ensured appearance of the color. He offered the procedure as a qualitative test for bilirubin.

Some 30 years later A. A. Hijmans van den Bergh attempted to adapt the Ehrlich reaction to the quantitative determination of bilirubin. He carefully followed the prescribed method, alcohol and all. However, even when he omitted alcohol, the color developed. Intrigued, he studied the reaction with and without alcohol and discovered that serum from individuals with one type of jaundice developed color promptly without alcohol, whereas serum from

---

* Rat liver cells take up BSP, but, instead of excreting it into the bile as such, they degrade it chemically and excrete the degradation products.

individuals with another type of jaundice developed color only with alcohol. The first reaction became known as the *direct van den Bergh test,* and the second as the *indirect van den Bergh test.*

Despite efforts over the years to identify the seemingly different bilirubins, direct-reacting bilirubin could never be isolated. Confusion was compounded when direct-reacting fluids (e.g., bile itself) regularly yielded crystalline indirect-reacting bilirubin when extracted with appropriate solvents. Not until 1956 was the mystery solved. Then the British workers B. H. Billing, P. G. Cole, and G. H. Lathe established that direct-reacting bilirubin is a water-soluble diglucuronide conjugate of indirect-reacting bilirubin.

The conjugation of bilirubin with glucuronic acid takes place in the liver cells (Fig. 13.28).* Bilirubin is produced from hemoglobin. At the end of their life spans, red cells are destroyed in the reticuloendothelial system (see Fig. 8.24), and the liberated hemoglobin is broken down in the reticuloendothelial cells by a process involving the stripping of the porphyrin heme from globin, the opening of the porphyrin ring, and the expulsion of iron. The product is free, or indirect-reacting, bilirubin (a group 3 substance). It enters the blood, attaches loosely to albumin, and journeys around the body, finally arriving at the liver. The liver cells actively transport bilirubin from blood to bile. In the course of the transfer, it is converted to direct-reacting bilirubin diglucuronide. This reaches the intestine via the common bile duct, and there it is transformed by bacterial action into a group of colorless substances collectively designated *urobilinogen.* A portion of the urobilinogen is absorbed into the portal circulation, whence it returns to the liver. There is an enterohepatic circulation of urobilinogen similar to that of the bile salts (see p. 545). A small amount—normally equivalent to no more than 1% of the daily output of bilirubin—escapes clearance by the liver cells, passes into the systemic circulation, and is excreted by the kidneys. The bulk is excreted in the feces. Some is converted by bacteria to *urobilin,* which accounts in part for the brown color of normal feces.†

The immature liver of a newborn infant is deficient in its ability to conjugate bilirubin with glucuronic acid. Hence the bilirubin released from hemoglobin is not excreted, and the serum bilirubin level rises, the result being jaundice. Within 3 to 5 days the liver "learns" to conjugate bilirubin, and the jaundice disappears. With a hemolytic process as in Rh incompatibility (see p. 280), tremendous elevations of serum bilirubin may occur. Free bilirubin in toxic concentrations damages the brain and other tissues.

Jaundice. Jaundice is not a disease but a symptom of disease. It is a yellow coloration of the skin and mucous membranes due to an abnormally high serum bilirubin level. The outline of bilirubin metabolism provides a simple key to understanding its causes (see Fig. 13.28). The primary causes are overproduction of bilirubin (*prehepatic* jaundice), liver cell damage (*hepatic* jaundice), and obstruction of the extrahepatic bile passages (*posthepatic* jaundice).

In prehepatic jaundice more bilirubin is produced than the normally functioning liver cells can excrete. Accordingly, free bilirubin leaks into the systemic circulation, so that the serum level increases, it is absent from the urine (the kidneys excrete only bilirubin glucuronide), urine and fecal urobilinogen levels rise, and other signs of excessive hemolysis appear (anemia, active bone marrow, etc.). In hepatic jaundice a normal amount of free bilirubin is formed, but its excretion is impaired because of defective liver cells. The bile canaliculi may also be damaged, so that a degree of ob-

---

† In a normal adult with a blood volume of 5 liters and a hemoglobin concentration of 15 g per 100 ml, about 1% of the total hemoglobin, or $0.01 \times 750 = 7.5$ g, is broken down each day. Since 1 g of hemoglobin yields 35 mg of bilirubin or its derivatives, approximately 263 mg of bilirubin is produced daily. Most of this is recoverable as fecal urobilinogen or urobilin.

---

* The discovery that direct-reacting bilirubin is bilirubin diglucuronide explained the role of alcohol in the van den Bergh tests. Direct-reacting bilirubin is soluble and reacts promptly. Indirect-reacting bilirubin is insoluble and cannot react. Alcohol solubilizes it, thus permitting it to react. Unwittingly chemists had always employed conditions that hydrolyzed bilirubin diglucuronide to bilirubin in trying to isolate direct-reacting bilirubin.

13.28 Bile pigment metabolism: A, normal; B, in prehepatic jaundice; C, in hepatic jaundice; D, in posthepatic jaundice; E, with defective conjugation. (Adapted from L. Page and P. J. Culver, *Syllabus of Laboratory Examinations in Clinical Diagnosis,* Harvard University Press, 1960.)

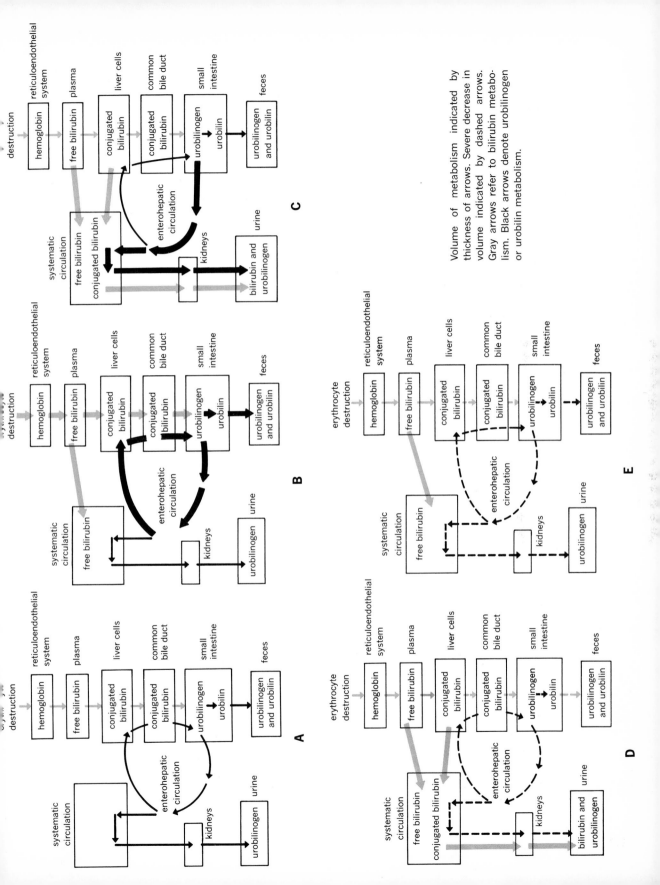

Volume of metabolism indicated by thickness of arrows. Severe decrease in volume indicated by dashed arrows. Gray arrows refer to bilirubin metabolism. Black arrows denote urobilinogen or urobilin metabolism.

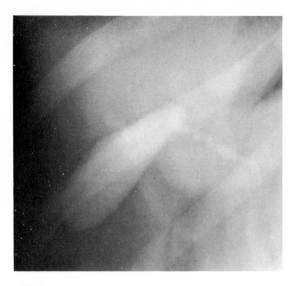

13.29 X-ray of a gall bladder. The subject has ingested diodrast, an organic iodide. (Dr. Murray Janower)

struction is present. Thus hepatic and posthepatic jaundice often occur together. Posthepatic jaundice is due particularly to mechanical blockage of the common bile duct. Bilirubin glucuronide therefore leaks into the systemic circulation and passes out through the kidneys. Since the feces receive no bile pigments, they are light in color. Frequently the liver is damaged secondarily, with a concomitant further increase in serum bilirubin. A rare congenital disorder has recently been recognized in which the liver never develops the ability to conjugate bilirubin with glucuronic acid. Its victims remain jaundiced.

Significance of conjugation in hepatic excretion. The liver excretes many substances, including most of those of group 3, exactly as it does bilirubin. Furthermore, in a victim of the congenital jaundice just mentioned, the liver is incapable of conjugating any of them with glucuronic acid.

Conjugation reactions are the means by which the liver eliminates steroids, poisons, drugs, and bile salts (products of the conjugation of cholic acid with taurine and glycine). Conjugation results in an increase in solubility that favors excretion and, in the case of substances like drugs, a decrease in pharmacological activity. It appears to be one of the body's major detoxification mechanisms. The liver's ability to excrete organic iodides has practical applications. These

compounds are opaque to x-ray, and, when they pass from the liver into the gall bladder, they render that organ visible on x-ray examination (Fig. 13.29).

The new knowledge of the importance of conjugation in liver physiology prompted a re-examination of BSP excretion. As anticipated, it was found that BSP is conjugated during its sojourn in the liver cells, the conjugate consisting of a complex of BSP and the tripeptide glutathione. Evidently BSP combines with glutathione in much the same way that bilirubin combines with glucuronic acid; hence the competition for excretion between BSP and bilirubin that we noted earlier.

REFERENCES AND SUGGESTIONS FOR FURTHER READING

Bender, M. L., and F. J. Kézdy, "Mechanism of Action of Proteolytic Enzymes," *Ann. Rev. Biochem.,* **34,** 49 (1965).

Benson, J. A., Jr., and A. J. Rampone, "Gastrointestinal Absorption," *Ann. Rev. Physiol.,* **28,** 201 (1966).

Billing, B. H., P. G. Cole, and G. H. Lathe, "The Excretion of Bilirubin as a Diglucuronide Giving the Direct van den Bergh Reaction," *Biochem. J.,* **65,** 774 (1957).

Bradley, S., "The Excretory Function of the Liver," *Harvey Lectures, Ser. 54 (1958–59),* 131 (1960).

Brauer, R. W., "Liver Circulation and Function," *Physiol. Rev.,* **43,** 115 (1963).

Crane, R. K., "Intestinal Absorption of Sugars," *Physiol. Rev.,* **40,** 789 (1960).

Davenport, H. W., "The Digestive System," *Ann. Rev. Physiol.,* **21,** 183 (1959).

Elias, H., "The Liver Cord Concept One Hundred Years After," *Science,* **110,** 470 (1949).

Farrar, G. E., and R. J. Bower, "Gastric Juice and Secretion: Physiology and Variations in Disease," *Ann. Rev. Physiol.,* **29,** 141 (1967).

Frazer, A. C., "Fat Absorption and Its Relation to Fat Metabolism," *Physiol. Rev.,* **20,** 561 (1940).

Gregory, R. A., "Secretory Mechanisms of the Digestive Tract," *Ann. Rev. Physiol.,* **27,** 395 (1965).

Grossman, M. I., "The Glands of Brunner," *Physiol. Rev.,* **38,** 675 (1958).

Haslewood, G. A. D., "Recent Developments in Our Knowledge of Bile Salts," *Physiol. Rev.,* **35,** 178 (1955).

Hendrix, T. R., and T. M. Bayless, "Digestion: Intestinal Secretion," *Ann. Rev. Physiol.,* **32,** 139 (1970).

Hightower, N. C., Jr., "The Digestive System," *Ann. Rev. Physiol.,* **24,** 109 (1962).

Hirshowitz, B. I., "Pepsinogen: Its Origins, Secretion and Excretion," *Physiol. Rev.,* **37,** 475 (1957).

Hogben, C. A. M., "The Alimentary Tract," *Ann. Rev. Physiol.,* **22,** 381 (1960).

Hollander, F., "Current Views on the Physiology of the Gastric Secretions," *Am. J. Med.,* **13,** 453 (1952).

Hunt, J. N., "Gastric Emptying and the Secretion in Man," *Physiol. Rev.,* **39,** 491 (1959).

Ingelfinger, F. J., "Esophageal Motility," *Physiol. Rev.,* **38,** 533 (1958).

Isselbacher, K. J., "Metabolism and Transport of Lipid by Intestinal Mucosa," *Federation Proc.,* **24,** 16 (1965).

James, A. H., *The Physiology of Gastric Digestion,* Arnold, London, 1957.

Janowitz, H. D., "Quantitative Tests of Gastrointestinal Function," *Am. J. Med.,* **13,** 465 (1952).

Matthews, D. M., and L. Laster, "Absorption of Protein Digestion Products: A Review," *Gut,* **6,** 411 (1965).

Obrink, K. J., "Digestion," *Ann. Rev. Physiol.,* **20,** 377 (1958).

Saunders, S. J., and K. J. Isselbacher, "Intestinal Absorption of Amino Acids," *Gastroenterology,* **50,** 586 (1966).

Senior, J. R., "Intestinal Absorption of Fats," *J. Lipid Res.,* **5,** 495 (1964).

Thomas, J. E., "Mechanics and Regulation of Gastric Emptying," *Physiol. Rev.,* **37,** 453 (1957).

Turner, D. A., *Intestinal Absorption,* Blackwell, Oxford, 1959.

A preoccupation with eating, the getting of nourishment, is a characteristic of all living things. . . . Eating may be postponed, but only for relatively short periods of time, and this denial soon leaves its stamp in debilitation, weakness, lethargy, and finally death. As Erasmus Darwin was prompted to say, "Eat, or be eaten!"

It comes as no surprise, therefore, that in human affairs this unending compulsion to be forever busy eating should be revealed in some of man's deepest concerns. For early man, eating—or not eating—was a matter never completely out of his mind. The first gregarious tendencies in man probably had their reward in the increased efficiency of the clan or tribe in coping with the problem of getting food. In his bewildering adjustments to a hostile world, prehistoric man certainly found one of his greatest satisfactions in a stomach momentarily distended with an abundant meal. . . . Is it any wonder, therefore, that man, in his immemorial concern with the problem of food, should have turned finally to a scientific inquiry about these substances that he finds he is compelled to put into his mouth, chew, and swallow?

Howard A. Schneider, *Perspectives Biol. Med.*, 1958.

14 Nutrition and Metabolism

Quantitative Aspects of Total Body Metabolism

Once food has been digested and absorbed, it has entered the body to be used as a source of immediate or reserve energy or of structural building blocks. The metabolic pathways responsible for its transformations in body cells were described in Chapter 4.

Here we shall consider the problems of whole body nutrition and metabolism. The metabolism of the whole body is simply the total of all the metabolic processes in all the cells of the body. Since the nutrition of an individual cell depends ultimately upon the food taken in through the digestive system, it is evident that whole body

food requirements can be interpreted only in the light of quantitative studies of whole body metabolism.

We learned in Chapter 4 that only a portion—slightly more than half—of the potential energy of foodstuffs is transformed into the energy-rich bonds of ATP. The remainder is dissipated as heat during ATP formation. Still more energy is dissipated as heat during the utilization of ATP in such processes as glandular secretion, nerve conduction, muscular contraction, and biosynthesis—so that actually only about a quarter of the initial food energy participates directly in body functions.

However, with one exception, even this fraction is eventually converted into heat. Each of the many endergonic, or energy-requiring, physiological processes using ATP is balanced by a simultaneous exergonic, or heat-producing, process. For example, ATP energy is consumed in protein synthesis, but concomitant protein breakdown releases as heat the energy stored in peptide linkages. Similarly, ATP energy is consumed in muscular contraction, largely in overcoming the viscosity of muscle components, but the friction involved in this activity generates heat. The single exception arises when the muscles perform *external* work—creating either potential energy by lifting a mass against gravity or kinetic energy by operating a tool or turning a wheel. Since during rest practically all the energy of ingested foodstuffs appears as heat, the rate at which heat is liberated by the body is a measure of the overall *metabolic rate*.

METABOLIC RATE

Caloric values of foodstuffs. We noted on p. 118 that a *calorie* (cal) is the amount of heat needed to raise the temperature of 1 g of water 1°C. In discussing body metabolism, we shall speak only of *Calories* (Cal), or *kilocalories* (kcal), one of which equals 1000 calories.

We mentioned earlier the bomb calorimeter (see p. 120), a device that accurately measures the heat liberated during the total combustion of a food sample. The calorimeter consists essentially of a heavy-walled steel cylinder with a screw-on cap containing two electrodes. A weighed sample is placed in the cylinder, the cap is affixed, and oxygen is introduced until the internal pressure is several atmospheres. The bomb is then immersed in a larger insulated vessel containing a weighed amount of water. A delicate thermometer registers the temperature of the water, and a stirrer agitates it for uniformity. When equilibrium has been reached, an electric current is passed through a fine iron wire joining the two electrodes. The wire, above the sample, melts and drops on the sample, causing it to burn. The heat is transmitted to the surrounding water and recorded by the thermometer. The end products of the combustion of carbohydrate or lipid in the bomb are CO_2 and H_2O, the same as those of the oxidation of carbohydrate or lipid in the body. Therefore, the heat produced in the bomb is equivalent to the energy yield in the body.

The situation is not as simple for protein, whose end products in the bomb are NO_2, CO_2, and H_2O. The body is incapable of oxidizing protein nitrogen completely to NO_2. Rather it breaks it down to urea, $(NH_2)_2C{=}O$. Hence the energy available from protein equals the heat produced in the bomb minus the heat that would be evolved if the body's nitrogenous end products were completely oxidized.

When a food's carbohydrate, lipid, and protein contents are known, its available energy is calculated from the mean bomb calorimeter values for the three constituents. These are as follows: carbohydrate, 4.1 Cal per gram; protein, 4.1 Cal per gram; lipid, 9.3 Cal per gram. With allowance made for the possibility of incomplete digestion or absorption, these values are sometimes rounded off, respectively, to 4, 4, and 9 Cal per gram.

Direct calorimetry. When the body is performing no external work, the total metabolic rate can be determined from the rate of heat liberation by *direct calorimetry.*

Although measuring body heat production by recording the temperature change in a known weight of water is theoretically simple, in practice it presents a number of problems. The expensive apparatus consists of an insulated chamber large enough to accommodate the whole body. The heat generated is transmitted either to water circulating in copper coils in the chamber or to the air (including its water vapor) circulating in the chamber. Both methods permit accurate calculation of the total heat loss in calories.

From measurements of this type, the average total heat production of a postabsorptive (i.e., after food absorption in the small intestine has ceased) adult resting male has been found to be between 1500 and 1800 Cal per day. We shall presently note the effects upon heat loss of activities such as exercise and food consumption.

Respiratory quotient. The processes of metabolism are associated with the consumption of O_2 and the production of CO_2 in the combustion of energy-yielding foodstuffs.

It is obvious that the magnitude and nature of this gaseous exchange vary with the type of foodstuff being oxidized, since the relative proportions of carbon and oxygen differ in carbohydrates and lipids. The relationship between O_2 consumed and CO_2 produced is easily calculated from the familiar equations describing the overall oxidation of a carbohydrate or lipid.

For example, the complete oxidation of glucose, a typical carbohydrate, was represented in Chapter 4 as

$$C_6H_{12}O_6 + 6O_2 \longrightarrow$$
$$6CO_2 + 6H_2O + \text{energy} \quad (14.1)$$

We see that the molar ratio of CO_2 produced to O_2 consumed—called the *respiratory quotient* (abbreviated R.Q.) in a living organism*—is 1.0. For any carbohydrate the R.Q. is 1.0.

The oxidation of the typical triglyceride tripalmitin occurs as follows:

$$2C_{51}H_{98}O_6 + 145O_2 \longrightarrow$$
$$102CO_2 + 98H_2O \quad (14.2)$$

Hence the R.Q. is $102 \div 145 = 0.703$. An R.Q. below 1 reflects a low proportion of oxygen to carbon in the molecule being oxidized. This is characteristic of lipids and means that more oxygen must be consumed in lipid oxidation than in carbohydrate oxidation. Although other triglycerides yield R.Q. values slightly different from that for tripalmitin (e.g., triolein, 0.713; tristearin, 0.699), the average R.Q. for mixed dietary or body lipids is 0.71.

Since the chemical compositions of most proteins cannot be stated precisely, a simple equation for their oxidation is not possible. However, the average R.Q. for proteins is about 0.8, according to indirect evidence.[†]

The amount of protein broken down can be determined from the rate of urinary N excretion. Since the average protein is 16% N, each gram

of urinary N must come from 6.25 g of protein $(100 \div 16 = 6.25)$, and oxidation of this amount of protein (from Loewy's data) involves the consumption of 5.94 liters of O_2 and the production of 4.76 liters of CO_2. The differences between the total amounts of O_2 consumed and CO_2 produced and the amounts computed from the urinary N excretion provide a ratio termed the *nonprotein respiratory quotient*. This is used in calculations of the amounts of lipid and carbohydrate undergoing combustion per liter of O_2.

The proximity of the R.Q. to 1 at any time during metabolism is a measure of the degree to which carbohydrate predominates in the mixture being oxidized.[‡] Generally the R.Q. for an ordinary mixed diet is between 0.80 and 0.87, indicating that a mixture of foodstuffs is being oxidized. During fasting the R.Q. approaches 0.71, indicating that energy production is almost entirely at the expense of the lipid reserves. Under certain circumstances the nonprotein R.Q. may be outside the range 0.7 to 1.0. For example, during the force-feeding of carbohydrate (as in the fattening of geese), it may be 1.2 to 1.5. Under these conditions an oxygen-rich foodstuff (carbohydrate) is converted into an oxygen-poor one (lipid), and a portion of the O_2 for the oxidation of the lipid comes from the carbohydrate instead of from extraneous sources. Thus CO_2 production exceeds O_2 consumption.

Indirect calorimetry. The size and cost of direct calorimeters prevents their widespread use. Instead, heat loss is usually estimated from O_2 consumption and CO_2 production. Two types of apparatus are employed. In one, the

---

* Alternatively, the R.Q. is calculated from the *volumes* of CO_2 produced and O_2 consumed; under standard conditions the volume of a mole of each gas is 22.4 liters.

† Calculation of a theoretical R.Q. for protein is complicated by the facts that the oxidation of protein does not proceed completely to CO_2 and H_2O and that its nitrogenous end products are for the most part excreted in the urine. However, studies by A. Loewy in 1910 provided a basis for establishing a protein R.Q. He found that 100 g of meat protein contained 52.38 g (4.36 moles) of C, 7.27 g (3.64 moles) of H, 22.68 g (0.709 mole) of O, 16.65 g of N, and 1.02 g of S and that after digestion the urine (and feces) contained 10.88 g (0.91 mole) of C, 2.88 g (1.44 moles) of H, 14.99 g (0.468 mole) of O, 16.65 g of N, and 1.02 g of S. Hence 3.45 moles of C and 2.20 moles of H were oxidized. Oxidation of these quantities

would require $3.45 + \frac{1}{2}(2.20) = 4.55$ moles of O_2. Deducting $0.709 - 0.468 = 0.241$ moles of O_2 that combined with the protein left $4.55 - 0.241 = 4.31$ moles of O_2 that must have come from extraneous sources. Oxidation of the protein yielded 3.46 moles of CO_2. Therefore, the protein R.Q. was $3.46 \div 4.31 = 0.80$. If the average N content of proteins is assumed to be 16% and if correction is made for the N lost in the feces, it may be calculated that on the average 5.94 liters of O_2 is consumed and 4.76 liters of CO_2 is produced per gram of urinary N excreted.

‡ The interpretation of the R.Q. is subject to limitations, particularly for short periods of observation. An intermediate of carbohydrate metabolism, for example, may enter many metabolic pathways, with a different respiratory quotient for each pathway.

closed-circuit type, the subject inspires air that is enriched with oxygen and recirculated. In the other, the *open-circuit type,* he inspires air from the outside atmosphere. Expired air is collected in an appropriate container for a definite period of time, at the end of which it is measured and analyzed for O_2 and CO_2.

The amount of energy liberated can be expressed in terms of the amount of O_2 consumed. However, the heat produced per liter of O_2 consumed varies with the food being oxidized. For example, a liter of O_2 liberates more heat in the combustion of a carbohydrate (5.047 Cal) than in the combustion of a lipid (4.686 Cal) or a protein (4.485 Cal). The figures reveal that the amounts of energy liberated per liter of O_2 in the oxidative metabolism of carbohydrates, lipids, and proteins are not grossly different. Since the average diet is a mixture of foodstuffs, the *average* amount of energy liberated per liter of O_2 consumed is about 4.825 Cal.

A more precise determination of heat production from O_2 consumption can be made if the R.Q. is known, since this indicates what foods are being oxidized and permits a reasonably accurate correction for the volumes of O_2 and CO_2 exchanged in protein catabolism.

A simple example illustrates these principles. Let us assume that the following figures apply to a normal subject during a 24-hour period: O_2 consumed, 400 liters; CO_2 produced, 350 liters; urinary N, 8.0 g.

Since 5.94 liters of O_2 is consumed and 4.76 liters of CO_2 is produced for each gram of urinary N, 8.0 g of urinary N corresponds to $8.0 \times 5.94 = 47.52$ liters of O_2 and $8.0 \times 4.76 = 38.08$ liters of CO_2. Thus

$$\text{nonprotein R.Q.} = \frac{\text{total } CO_2 - \text{protein } CO_2}{\text{total } O_2 - \text{protein } O_2}$$

$$= \frac{350 - 38.08}{400 - 47.52} = \frac{311.92}{352.48} = 0.88$$

Hence 352.48 liters of O_2 is consumed and 311.92 liters of CO_2 is produced in the oxidation of carbohydrate and lipid.

Since the R.Q. of carbohydrate is 1.0 and the R.Q. of lipid is 0.71, a nonprotein R.Q. of 0.88 means that 59% (or 208 liters) of the O_2 consumed in nonprotein oxidation is used in carbohydrate catabolism and 41% (or 144.5 liters) is used in lipid catabolism; therefore, $5.047 \times 208 = 1050$ Cal is derived from carbohydrate, and $4.686 \times 144.5 = 677$ Cal from lipid. In nonprotein oxidation, then, 1 liter of O_2 is equivalent to $(1050 + 677) \div 352.48 = 4.889$ Cal, and the total

heat liberated is $352.48 \times 4.889 = 1727$ Cal (i.e., $1050 + 677$ Cal). In summary, the amount of each foodstuff oxidized can be calculated from its known caloric value (see p. 557).

Foodstuff	Amount oxidized, g	Energy liberated, Cal
Carbohydrate	$1050 \div 4.1 = 256$	1050
Lipid	$677 \div 9.3 = 22.8$	677
Protein	$8 \times 6.25 = 50.0$	$50 \times 4.1 = 205$
		1932

In terms of total heat, $(1050 \times 100) \div 1932 = 54.3\%$ is contributed by carbohydrate; $(677 \times 100) \div 1932 = 35.0\%$ is contributed by lipid; and $(205 \times 100) \div 1932 = 10.7\%$ is contributed by protein.

Indirect calorimetry is preferable to direct calorimetry not only because of its convenience but also because it measures total energy expenditure. Since the energy utilized in the performance of external work is not translated into heat, direct calorimetry overlooks it.

Basal metabolic rate. Ordinarily the techniques of indirect calorimetry are standardized to facilitate meaningful comparisons of different individuals.

We have seen that the energy derived from food may be (1) converted to heat, (2) utilized to perform work, and (3) stored in the form of synthesized cellular constituents. It is technically difficult at any given time to assess the relative significance of heat energy, metabolic energy, or stored energy. We minimize the variables by deliberately curtailing food ingestion and work, measuring energy exchange in a postabsorptive period during rest in a room maintained at 20°C. When the subject has fasted for several hours and is lying quietly on a cot, his only work is associated with breathing and heart action.

Under these conditions—i.e., in the *basal state*—energy is liberated from the body solely as heat, and stored energy is its only possible source. Since energy cannot be created or destroyed, the decrease in stored energy equals the heat loss. The heat loss is determined by direct or indirect calorimetry and is termed the *basal metabolic rate,* or *BMR.* The BMR is the metabolic rate of the basic cellular processes. The total energy requirement of these processes

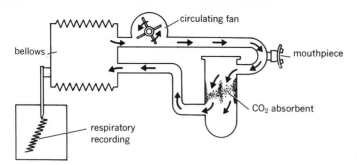

circulating fan

bellows

mouthpiece

CO_2 absorbent

respiratory
recording

14.1 Basal metabolism apparatus.

is equal to about half the total energy expenditure during a day of ordinary activity.

Indirect calorimetry is the usual method for assessing the BMR. Because of the standardized conditions, the BMR may be computed with sufficient precision for most purposes from the O_2 consumption alone. Since in the basal state the R.Q. is within 2% of 0.82, 1 liter of O_2 liberates 4.825 Cal.

The equipment generally employed is the Benedict-Roth apparatus or a modification thereof (Fig. 14.1). This is a closed-circuit instrument. The subject, who has not eaten for at least 12 hours, lies on his back. A rubber mouthpiece is inserted into his mouth, and his nostrils are closed with a clamp so that all breathing is done through the mouth. The inhaled gas is O_2, from the bellows. The exhaled gas passes over soda lime, which removes CO_2. The subject breathes freely for 10 to 15 minutes. As the body takes up O_2, the stylus traces a curve on graph paper. The slope of the curve indicates the volume of O_2 consumed. This value multiplied by 4.825 equals the total quantity of heat produced during the test.

Basal metabolism is usually expressed in Calories per hour, and BMR in Calories per hour per square meter of body surface. Physicians usually express the BMR as a percentage above or below an accepted normal value.

The BMR is quite constant in a given individual and among similar individuals of the same species. Many factors, however, affect it—among them body size, age, sex, climate, physical training, diet, drugs, and endocrine status. As it deviates from the normal in a variety of abnormal conditions, it is sometimes useful diagnostically.

The four major factors governing heat loss from the body are (1) the temperature differ-

ence between the environment and the organism, (2) the nature of the surface radiating the heat, (3) the area of the radiating surface, and (4) the thermal conductance of the environment. In the conditions of the basal metabolism test, body surface area is the most important of these factors. Although the *total quantity of heat produced* varies widely among various species and is inversely proportional to body weight, the quantity produced per square meter of body surface per 24 hours is essentially the same for all species (e.g., 948 Cal in the horse, 1042 Cal in man, 969 Cal in the goose, 917 Cal in the rabbit without ears, and 776 in the rabbit with ears*). The relationship of basal metabolism to surface area was first demonstrated by M. Rubner in 1883 and is often referred to as *Rubner's law.*

The earliest determinations of total surface area in man were made by E. F. Du Bois and D. Du Bois in the 1920's. They took measurements from flexible paper molds that they had fitted around their subjects. The difficulties of this method led to the development of a mathematical formula relating surface area to height and weight.

$$\log A = 0.425 \log W + 0.725 \log H + 1.8564 \tag{14.3}$$

where A is the surface area in square centimeters, W is the weight in kilograms, and H is the height in centimeters. A chart giving approximate surface areas for known heights and weights has been constructed from this equation (Fig. 14.2).

The BMR increases rapidly up to maturity and then slowly declines with age. It is somewhat

---

* The large area of rabbit ears accounts for this exception.

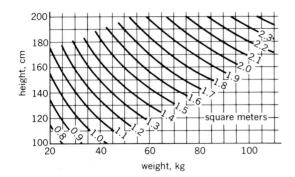

14.2 Du Bois's chart for determining the surface area of an individual from height and weight. (From E. F. Du Bois, *Basal Metabolism in Health and Disease*, 3rd ed., Lea & Febiger, Philadelphia, 1936.)

higher in a male than in a female of the same age and in an athlete than in a sedentary individual. It is lower in tropical climates than in temperate or cold climates. It rises with fever, with hyperactivity of some endocrine glands, especially the thyroid, and with some drugs (e.g., epinephrine and benzedrine). It falls with hypoactivity of these glands and with fasting. F. G. Benedict found that it dropped to 23% of the normal value in a man deprived of food for 21 days.

TOTAL DAILY ENERGY PRODUCTION

We have learned that the BMR is determined under conditions that minimize the variable effects of food and voluntary muscular activity. However, the human body is rarely in a basal state, and we must now consider the extent to which food and work elevate the total daily energy output above the basal level.

Effect of food. When an individual in the basal state ingests food, his heat production increases almost immediately. Part of the increase is due to the digestive and absorptive work of the alimentary tract. The rest is due to the metabolic transformations of the products of digestion and absorption. This component also follows the intravenous injection of glucose or amino acids.

Rubner called the stimulating effect of food on heat production the *specific dynamic action* of food (abbreviated SDA). The extra heat

liberated by the body amounts to approximately 30% of the energy value of ingested protein, 6% of the energy value of ingested carbohydrate, and 4% of the energy value of ingested lipid. Thus a meal consisting simply of 50 g of protein, equivalent to about 200 Cal, causes the production of 60 Cal over and above the BMR. The metabolic rate begins rising within 2 hours of the meal and remains elevated for up to 12 hours.

The high SDA of protein is undoubtedly due to the many energy-releasing reactions involved in amino acid catabolism. In the dog, for example, removal of the liver leads to significant decreases in the SDA's of individual amino acids. The extra heat resulting from the SDA must arise in cellular metabolism and must be at the expense of nutrient materials—from the diet, from tissue reserves, or from tissue protoplasm.

Effect of work. Because of the large proportion of body tissue that is muscle, muscular activity accounts for a substantial fraction of total metabolism. Voluntary muscles do *internal* or *external* work.* Since only 1 Cal is needed for 427 kg-m of work, it may be shown, for example, that the energy equivalent of shoveling 20 tons of coal into a bin 3 ft above ground is but 40 Cal. A laborer who uses 100 Cal for external work in the course of a day may convert 4000 Cal from the potential to the kinetic state in 24 hours. Subtracting 1600 Cal from this for the BMR, another 200 Cal for the SDA of food, and another 100 Cal for the external work leaves 2100 Cal. With the efficiency of external work assumed to be 25%, 300 Cal more must be produced to perform the work. Therefore, 2100 − 300 = 1800 Cal is expended in the remaining daily activities.

A 70-kg man who climbs a mountain 6000 ft high uses a considerable amount of energy in internal work related to the physical difficulties of climbing and loses another 900 Cal as heat. Yet, despite his exhausting effort and the correspondingly great total transformation of energy, the energy equivalent of the external work of

---

* Internal work includes the work necessary to overcome the viscosity of the muscles themselves. External work is the work necessary to produce potential or kinetic energy outside the body.

TABLE 14.1 ENERGY EXPENDITURE DURING MUSCULAR ACTIVITY

Activity	Energy expended by average (70 kg) man, Cal/hr
Sleeping	65
Lying still	90
Sitting, at rest	100
Sitting, playing a horn	120
Standing, relaxed	100
Standing, at attention	115
Standing, shaving	150
Standing, conducting an orchestra	165
Typing	170
Peeling potatoes	175
Driving a car	185
Making beds	200
Dancing, fox trot	275
Walking on a level (3.5 miles/hr)	350
Dancing, rumba	560
Swimming, breast stroke	600
Walking upstairs	1100

lifting his body through the vertical distance is only 300 Cal—less than 10% of the total expenditure. Table 14.1 lists caloric requirements of other muscular activities. These figures amusingly illustrate the completeness of the data of R. Passmore and J. V. G. A. Durnin.

Nutritionists have recently made calculations that add new meaning to such data. Table 14.2 lists the caloric values of certain common foods and also presents the *energy-use equivalents* of these foods. The figures are based on experiments showing these energy costs for an average (70 kg) man: reclining, 1.3 Cal per minute; walking (at 3.5 miles per hour), 5.2 Cal per minute; riding a bicycle, 8.2 Cal per minute; swimming, 11.2 Cal per minute; and running, 19.4 Cal per minute. We see, for example, that no less than 19 minutes of running or 290 minutes of reclining is necessary to "work off" the calories in a slice of apple pie.

In contrast to muscular activity, mental ac-tivity has little influence on metabolism. Benedict reported that the extra energy required for one hour of intense mental effort could be supplied entirely by one oyster cracker or half a salted peanut!

Daily caloric requirements. It is apparent that the total energy output in any 24-hour period is the sum of basal metabolism, the SDA of food, and a highly variable component related to muscular activity. It is also apparent that the total number of food calories needed per day varies considerably with age, sex, diet, activity, etc. It is impossible, therefore, to do more than indicate a range of daily caloric requirements. Table 14.1 permits the following estimate for an average (70 kg) man.

8 hr sleep at 65 Cal/hr	520
2 hr light work at 150 Cal/hr	300
8 hr moderate work at 200 Cal/hr	1600
2 hr evening chores at 170 Cal/hr	340
4 hr sitting at 100 Cal/hr	400
Total	3160 Cal/day

Men in sedentary occupations may need less than 3000 Cal per day, while those doing heavy muscular work may need much more. Moderately active women need about 2500 Cal. A growing child or adolescent needs more calories per pound of body weight than an adult engaged in the same activity. One reason for this difference is the generally higher BMR of children, due to a greater surface area per pound. A second reason is the energy demands of growth, a process involving the synthesis of new tissue. For the same reasons pregnancy and lactation increase caloric requirements.

We shall presently see that the relationship between caloric intake and energy expenditure is the prime factor determining whether the body gains or loses weight over a period of time.

Principles of Nutrition

Preceding discussions have dealt chiefly with quantitative aspects of nutrition. Energy production and tissue synthesis depend upon a sufficiency of nutrient materials. But the prob-

TABLE 14.2   CALORIC VALUES AND ENERGY-USE EQUIVALENTS OF COMMON FOODS*

Food	Energy, Cal	Time required for average (70 kg) man to consume calories in food, min				
		Reclining	Walking	Cycling	Swimming	Running
Apple, large	101	78	19	12	9	5
Bacon, 2 strips	96	74	18	12	9	5
Beer, 1 glass	114	88	22	14	10	6
Bread and butter, 1 serving	78	60	15	10	7	4
Cake, 2-layer, 1 serving	356	274	68	43	32	18
Carrot, raw, 1	42	32	8	5	4	2
Cheese, cottage, 1 tbsp	27	21	5	3	2	1
Chicken, fried, ½ breast	232	178	45	28	21	12
Cookie, plain, 1	15	12	3	2	1	1
Doughnut, 1	151	116	29	18	13	8
Egg, fried, 1	110	85	21	13	10	6
French dressing, 1 tbsp	59	45	11	7	5	3
Ham, 2 slices	167	128	32	20	15	9
Hamburger sandwich, 1	350	269	67	43	31	18
Ice cream, ⅙ qt	193	148	37	24	17	10
Malted milk shake	502	386	97	61	45	26
Milk, 1 glass	166	128	32	20	15	9
Orange juice, 1 glass	120	92	23	15	11	6
Pancake with syrup, 1	124	95	24	15	11	6
Peas, green, ½ cup	56	43	11	7	5	3
Pie, apple, 1 serving	377	290	73	46	34	19
Pork chop, loin, 1	314	242	60	38	28	16
Potato chips, 1 serving	108	83	21	13	10	6
Shrimp, French fried, 1 serving	180	138	35	22	16	9
Spaghetti, 1 serving	396	305	76	48	35	20
Strawberry shortcake, 1 serving	400	308	77	49	36	21

* Adapted from F. Konishi, *J. Am. Dietet. Assoc.*, **46**, 186 (1965).

lems of nutrition also have a qualitative side. The body's food supply must do more than provide calories. It must also provide substances such as phosphorus, calcium, iron, zinc, sulfur, and vitamins.

NUTRITIONAL BALANCE

In discussing body water (see p. 205), we noted that the balance of the body for a given constituent is the gain minus the loss of the constituent over a given time interval. When in balance, gain equals loss, and the quantity of the constituent in the body remains constant. When in positive balance, the body gains more of the constituent than it loses; when in negative balance, the opposite is true.

Metabolic balance studies. In investigations of metabolic balances, always elaborate and complex procedures, subjects are fed chemically analyzed diets for weeks or months, all urine and feces are collected and chemically analyzed, body weights are regularly recorded, and precise accounts of the intake and output of each body constituent are kept. Although data of this type do not directly reveal the events of metabolism, they do permit a number of significant generalizations and predictions.

For example, the nitrogen (N), potassium (K), sulfur (S), phosphorus (P), and calcium (Ca) balances indicate the type of body tissue being synthesized or destroyed. Muscle, the largest single tissue of the body, contains these elements in the following weight ratios:

$$N : K : S : P : Ca$$
$$100 : 8.4 : 6.9 : 6.7 : 0.0$$

Other tissues (such as liver and skin) contain them in similar ratios. Thus a convenient common denominator exists for the portion of the body we may call "protoplasm." Other major components of the body contributing to its weight—water, bone, fat, and glycogen—differ significantly from protoplasm in this respect.

In one of Fuller Albright's well-known studies, a subject in positive N balance (owing to the administration of testosterone) gained (or retained) 211 g of N, 18.2 g of K, 12.5 g of S, 12.2 g of P, and 1.48 g of Ca over a 70-day period. Most of the body Ca is in bone. Since the Ca:P ratio of bone is 2.23:1, it can be assumed that $(1 \div 2.23) \times 1.48 = 0.66$ g of P was incorporated into bone. Therefore, $12.2 - 0.66 = 11.54$ g of P was incorporated into tissue other than bone. With retained N set equal to 100 (after correction for the elements entering bone), calculation shows

that the other elements were retained in the following ratios:

$$N : K : S : P : Ca$$
$$100 : 8.6 : 5.8 : 5.4 : 0.0$$

Their correspondence with the ratios of protoplasm is sufficiently close to suggest that the bulk of the weight gain is attributable to an increase in protoplasm rather than to an increase in water, fat, or glycogen.

In 1915 Benedict gathered balance data on an adult male undergoing a 31-day fast (Table 14.3). The N lost in the urine during the fast totaled 276.4 g. The net losses of the 31-day period expressed in ratios by Albright's method are as follows:

$$N : K : S : P : Ca$$
$$100 : 8.4 : 6.4 : 7.7 : 0.0$$

Again it appears that the balance data reflect changes in protoplasm. Since protein is 16% N, a net loss of 276.4 g of N reflects a loss of $6.25 \times 276.4 = 1728$ g of protein. Benedict calculated that the average daily total energy expenditure of his fasting man was 1395 Cal. The possible sources of this energy were body fat, protein, and carbohydrate. However, the body stores of carbohydrate (i.e., glycogen) are rapidly exhausted during fasting. Hence only lipid and protein were available. The subject lost 1728 g of protein, an average of 55.7 g per day. Since oxidation of protein liberates 4.1 Cal per gram, he obtained $55.7 \times 4.1 = 228$ Cal per day from this source—only 16.3% of his energy output. The remaining 1167 Cal must have come from the oxidation of lipid. Since it liberates 9.3 Cal per gram, $1167 \div 9.3 = 125.4$ g of lipid must

TABLE 14.3   BENEDICT'S FASTING MAN

Days	Weight at beginning of each period, kg	Urine nitrogen, g/day	Energy, Cal/day	Cal from protein, %
3 preliminary		14.0		
1–7	60.6	9.9	1650	15
8–14	55.1	10.3	1450	18
15–21	52.8	8.4	1290	17
22–28	50.1	7.8	1250	16
29–31	48.1	7.2	1260	15
3 following*	47.4	3.8		

* The first 3 days after resumption of a normal diet.

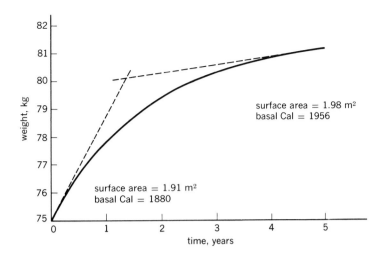

14.3 Diminishing weight gain with an initial excess of 80 Cal per day. (From J. W. Conn, "Obesity. II. Etiological Aspects," *Physiol. Rev., 24,* 31, 1944.)

have been oxidized. Only 10% of the weight of adipose tissue is water; in contrast, about 80% of the weight of nonadipose nonbony tissue (i.e., protoplasm) is water (cf. pp. 14, 26). Thus the loss of 55.7 g of protein and 125.4 g of lipid involved water losses of 222.8 g and 13.9 g, respectively. Summing up the *calculated* daily losses, we have

Protein loss	55.7 g
Associated water loss	222.8 g
Lipid loss	125.4 g
Associated water loss	13.9 g
Total	417.8 g

The *observed* weight loss during the 31-day fast was 13,200 g—or 425.8 g per day. The remarkable agreement between observed and calculated results validates the assumptions upon which the calculations are based.

Body weight. Body weight is determined ultimately by the balance between food income and energy expenditure. When the balance is negative—either in calories or in essential nutrients such as minerals and vitamins—the body feeds upon itself, attacking first its energy reserves and eventually its tissues. The consequence, as we have seen in Benedict's fasting man (who lost 13,200 g, or 29.1 lb, in 31 days), is weight loss.

When more food is ingested than is required, weight increases. The continued deposition of adipose tissue leads to *obesity.* It was stated some years ago that a normal adult ingesting only 80 Cal per day—the equivalent of one pat of butter—in excess of his energy requirements would double his weight in 20 years. This con-

clusion was based upon the assumption that an excess of 80 Cal per day means an excess of 29,200 Cal per year, which represents 3.46 kg (7.6 lb)* of adipose tissue per year or 69 kg (152 lb) of adipose tissue in 20 years.

The error of this conclusion is revealed in Fig. 14.3. Excess calories lead to early weight gain, but body surface area increases with weight, and basal metabolism increases with body surface area. Accordingly, a weight increase leads to an increase in basal energy expenditure, and the extra energy from the single pat of butter is invested in basal energy production rather than fat deposition. Similarly, basal metabolism decreases as weight decreases, and the rate of weight loss slackens as energy requirements lessen. Benedict's fasting man showed a progressive drop in daily energy expenditure and hence in weight loss.

The changing sources of body energy during the course of prolonged starvation are summarized in Fig. 14.4. The carbohydrate reserve is consumed almost entirely within the first 24 hours (Table 14.4). Then the fat stores are exhausted. Meanwhile, relatively little protein is utilized. In the subject described in Fig. 14.4, fat initially comprised 15% of the body weight. Thus 5 to 6 weeks was needed to deplete it. At

---

* This is computed as follows: 29,200 Cal is equivalent to 29,200 ÷ 9.3 = 3141 g of fat or 3455 g of adipose tissue, since in the tissue fat is associated with 10% of its weight in water. These figures indicate incidentally that 1 lb (454 g) of adipose tissue is equivalent to about 3800 Cal.

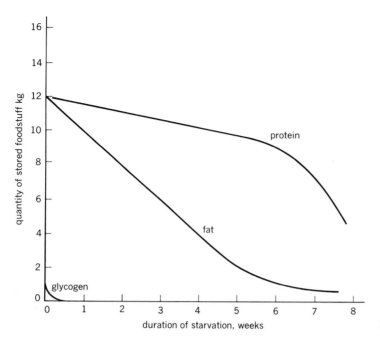

14.4 Effect of starvation on the body's reserves of foodstuffs.

the end of this period, essentially nothing remained but protein. Even the protein compartment sheds its most expendable portions first. For example, the liver, which is thought to contain a certain amount of "storage protein," loses 40% of its protein before muscle, skin, and skeleton lose 10% of theirs. Continued underfeeding at this stage threatens the continued functioning of the cells and, if carried on too long, causes death.

Much of the weight lost during undernutrition is body water. Since water loss depends upon renal salt and water excretion and subsequent osmotic adjustments, it may not occur in tempo with tissue destruction due to caloric deficits. L. H. Newburgh demonstrated many years ago that this discrepancy often accounts for the temporary failure of some individuals on reducing diets to lose weight (Fig. 14.5). Though such water retention may mask tissue destruction for a time, diuresis eventually brings the actual weight and theoretical weight curves together.

In planning a reducing diet for an ordinary case of mild to moderate obesity, we can conveniently calculate the desired daily caloric deficit on the basis that 1 lb of body fat is equivalent to approximately 3800 Cal (see footnote, p. 565). Thus a daily deficit of 500 Cal will produce over a long period an average weight

TABLE 14.4 CARBOHYDRATE CONTENT OF NORMAL MAN*

Component	Amount	
	%	g
Muscle glycogen	0.70	245
Liver glycogen	6.00	108
Blood and extracellular fluid glucose	0.08	17
Total		370
Caloric equivalent	370 × 4.1 = 1517 Cal†	

* Body weight, 70 kg; liver weight, 1800 g; muscle weight, 35 kg; volume of blood and extracellular fluid, 21 liters.
† In an individual requiring 2800 Cal per day (116.7 Cal per hour), the total body carbohydrate would supply caloric needs for 1517 ÷ 116.7 = 13 hours.

loss of almost 1 lb per week, a daily deficit of 1000 Cal an average weight loss of almost 2 lb per week, etc. It is useful to increase the daily caloric requirement by adding, for example, an hour of walking—though, as shown in Table 14.1, an hour's walk increases energy expenditure by only 350 Cal. Rigorous restriction of the caloric intake, which drastically increases the daily deficit, accelerates weight loss. However, most individuals find it difficult to maintain a daily intake of less than 1500 Cal for a long period.

Overnutrition is far more prevalent than undernutrition in developed countries of the world, despite the fact that obesity decreases life expectancy. Obesity is ultimately traceable to a long-term caloric surplus, but the reasons why individuals habitually eat too much are complicated and controversial. We spoke of hunger and appetite in Chapter 13, and abnormalities of these mechanisms are doubtless involved, but whether they are genetic, emotional, cultural, or neurological in origin is unclear.* We do know that the primitive satisfactions of eating mentioned by Schneider in the quotation on p. 556 are deeply rooted in human emotions.

Composition of the diet. Four factors must be considered in the planning of an adequate diet: (1) total caloric content; (2) proportions of the major foodstuffs—protein, lipid, and carbohydrate; (3) vitamin content; and (4) mineral content.

It is obvious that satisfactory diets can vary widely in composition; indeed, they vary widely among individuals and among geographical regions. The average American, for example, receives 12 to 15% of his energy from protein, 37 to 43% from lipid, and 45 to 48% from carbohydrate. In parts of the world where rice and

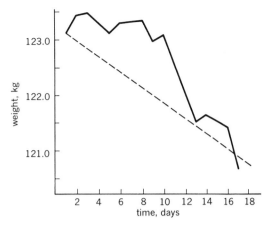

14.5 Effect of water retention in temporarily delaying weight loss during undernutrition: ——, actual weight; ----, predicted weight. Intake is 1600 Cal, output 2800 Cal, and deficit 1200 Cal per day.

other cereals are dietary staples, the quantity of energy derived from carbohydrate far exceeds that derived from both protein and lipid. Nevertheless, the diet must contain adequate amounts of all three components.

Protein is necessary in the diet to replace the proteins constantly being degraded into amino acids in the cells. We noted in Chapter 4 that some amino acids are deaminated and converted to intermediates capable of being oxidized in the citric acid cycle. For this reason cells must form new proteins to replace those destroyed. An average (70 kg) man can maintain his body protein stores provided that his daily protein intake is approximately 45 g. This is a *minimal* requirement for nitrogen balance and may be insufficient for general fitness and well-being. A rule of thumb holds that the optimal protein intake is 1 g per kilogram of body weight. Persons doing heavy work or living in cold climates need more than this for ideal performance.

We recall from Chapter 2 that fatty acids may be saturated or unsaturated. The human body is capable of desaturating fatty acids only to a very slight extent. For this reason unsaturated fatty acids such as arachidonic, linoleic, and linolenic acids are essential in the diet. As Mongolians exist on diets with lipid contents as low as 10%, presumably only small quantities

---

* Any attempt to establish the extent to which obesity is genetically determined encounters obvious difficulties (for example, one cannot do breeding experiments with human beings). Nevertheless, a recent study has ingeniously shown that the weight patterns of natural children correlate closely with those of their parents, whereas the weight patterns of adopted children do not correlate with those of their foster parents. This suggests (but does not prove) that obesity is a genotypic as well as a phenotypic trait.

of lipid are needed to supply the requisite un-saturated acids, and additional lipid consumed is either utilized immediately in energy production or stored in adipose tissue. An average diet might contain 135 g of lipid.

Carbohydrate is required in the diet for energy. We saw in Chapter 4 that the body metabolizes lipid when carbohydrate is not available in sufficient quantities. One reason for carbohydrate in the diet is to minimize this ten-dency, for exaggerated lipid catabolism pro-duces a disorder called *ketosis*. Since carbo-hydrate is also oxidized in preference to protein, it is a protein-sparing agent. An average Amer-ican or European diet might contain 380 g of carbohydrate.

Standard dietetic manuals may be consulted for information on the protein, lipid, and carbo-hydrate contents of foods. Most meat products contain high proportions of protein and lipid, whereas most vegetables contain a high propor-tion of carbohydrate.*

Special importance of dietary protein. The amount of protein in the diet is merely one aspect of protein nutrition. Proteins differ greatly in amino acid com-position and physical structure—i.e., in what is termed *biological value.* When a protein is relatively indigest-ible or lacks certain amino acids, its biological value is low.

In general, animal proteins are more digestible than vegetable proteins. As regards amino acid composition, T. B. Osborne and L. B. Mendel found in feeding ex-periments with purified proteins that some proteins maintain growth, some maintain body weight but not growth, and some maintain neither. Those that do not maintain growth are called *incomplete* proteins. For example, gelatin lacks tryptophan, cystine, valine, and phenylalanine. When these amino acids are added to a diet whose only protein is gelatin, growth is normal. Gelatin without these supplements cannot sustain life.

Since one incomplete protein may compensate for another, the diet should contain several proteins, par-ticularly if it is a vegetarian diet. When only a small amount of protein can be given, it should be an animal protein of high biological value (e.g., milk or whole egg). An ideal diet contains the proteins of eggs, milk, meat, potatoes, and other vegetables.

VITAMINS
AND VITAMIN REQUIREMENTS

The brief time between the identification of vitamin C in 1933 and the identification of vitamin B_{12} in 1955 witnessed the discovery of all the known vitamins of human nutrition. Even a casual backward glance at this era dis-closes that, with few exceptions, an orthodox sequence of events preceded the final identifica-tion of each vitamin. First it was observed that a pathological state in an individual could be re-versed by a food or food product. Then the active principle in the food or food product was isolated from natural sources by methods that, in many cases, developed from extraordinarily crude and tedious assays based on the responses of the deficient individual. Finally the purified vitamin was chemically characterized (and in some cases synthesized), and investigations of its mechanism of action were initiated.

Inquiry into the mechanism of action of a vitamin always begins in an untidy miscellany of data on functional, anatomical, and chemical features of the deficiency syndrome. The bio-chemist seeks to understand vitamin function at the molecular level. Although the vitamins are chemically diverse, most of those whose func-tions are known enter into metabolic reactions only after conversion to the coenzyme forms listed in Table 4.1. The biochemist's goal is therefore (1) to identify the coenzyme form of the vitamin, (2) to identify the enzyme in a test-tube system whose function depends on the presence of the coenzyme, (3) to determine the mechanism of coenzyme function (e.g., donor-acceptor of acyl groups, one-carbon groups, or hydrogen), (4) to establish that it is the enzyme in question whose function is impaired in a vitamin-deficient organism and that the impair-ment accounts for the deficiency syndrome, and (5) to elucidate the pathways of vitamin syn-thesis in lower organisms and to determine why the human body cannot accomplish the same synthesis.

The biologist is interested in the singular fact

---

* Lipid can be a deceptive ingredient of food. Since it is usually associated with little water and frequently occurs in pure form, and because it contains more than twice the calories per gram of protein and carbohydrate, it may represent less than 25% of the food weight but contribute 40% of the food energy. One pat of butter may contain as many calories as a potato. This is why lipids are scrupulously avoided in reducing diets.

that certain essential organic molecules are vitamins because some organisms can synthesize them and others cannot. If all organisms could synthesize them, they would not be vitamins; if none could, life would have evolved differently or not at all. Survival of nonsynthesizers has demanded the presence of synthesizers—at least before synthesis in the laboratory became possible. The relevance of such interrelations to ecological and evolutionary thought is obvious.

The physiologist and nutritionist wish to know daily dietary requirements of vitamins, natural sources, mechanisms of absorption and storage, and functional consequences of deficiency. It is to these aspects of vitaminology that we shall direct our attention.

For historical reasons, vitamin nomenclature is often confusing. To begin with, the name "vitamin" is not strictly correct, since all vitamins are not amines. The letter designations A, B, C, etc., arose before chemical structures were known. They are still used, but chemical names, such as ascorbic acid and thiamine, are generally preferred. However, even when a vitamin's chemical structure is known, a convenient trivia name may not be in general usage (e.g., vitamin D).

Vitamins are usually classified as *water-soluble* or *lipid-soluble*. The first group includes ascorbic acid (vitamin C) and a large number of compounds termed collectively the "vitamin B complex." The second group includes vitamins A, D, E, and K. The intestinal absorption of these vitamins, like that of lipid, is stimulated by bile. The intestinal bacteria may significantly affect vitamin nutrition. Some vitamins (e.g., vitamin K) are supplied almost entirely by them, and in some circumstances ingested vitamins are destroyed by them.

The usually recommended daily requirements of vitamins range from a few micrograms to a few milligrams. The requirements of some have not been established with precision. Requirements are influenced by many factors. All increase with (1) large body size, (2) active growth, (3) continuous active exercise, (4) disease and fever, and (5) pregnancy and lactation. Those of specific vitamins rise under certain conditions: examples are the thiamine requirement when greater than normal quantities of carbohydrate are metabolized and the vitamin D requirement during the period of greatest growth in children.

Vitamins are stored to a degree in all cells, but the main storage site of many is the liver. The vitamin A and vitamin B_{12} in the liver are sufficient to maintain a person without any intake of either for 2 to 3 years; the vitamin D in the liver is sufficient for 3 to 5 months. Very little is known of the storage of vitamin E, but the storage of vitamin K and most of the water-soluble vitamins—especially the vitamin B complex (except for vitamin B_{12})—is relatively slight. Since the diet is usually low in these vitamins, signs of deficiency may appear within a few days.

Lipid-soluble vitamins. Vitamin A exists in two forms—as vitamin A_1, whose formula is shown in Fig. 12.63, and as vitamin A_2, which is identical to vitamin A_1 except for an additional double bond in the ring. Each vitamin A consists chemically of one-half of a β-carotene molecule; hence it is a carotenoid (see p. 484). Carotenoids are synthesized only in plants, particularly green leafy and yellow vegetables. The body converts the yellow and red carotenoid pigments into colorless vitamin A. Since other animal bodies also carry out this conversion, foods of animal origin can provide vitamin A.

A prime function of vitamin A is in the photochemistry of the retinal pigments; deficiency of the vitamin leads to night blindness (see pp. 484, 487). In addition, vitamin A is essential for normal cell growth, especially that of epithelial cells. In vitamin A deficiency the epithelial structures of the body tend to become stratified and keratinized, with resulting scaliness of the skin and corneal opacity and blindness. Vitamin A–deficient lower animals may also show failure of growth and failure of reproduction due to atrophy of the germinal epithelium of the ovaries and testes.

Vitamin D promotes the intestinal absorption of calcium. We shall discuss it in detail in Chapter 15.

Vitamin E is poorly understood. It consists of a group of compounds, the *tocopherols*, which are found principally in wheat and other grains, meat, and milk. Vitamin E deficiency is most evident in experimental animals. Consequences of its deficiency include (1) degeneration of the germinal epithelium of the testes—hence male sterility, (2) death and absorption of embryos, (3) muscular weakness and paralysis, and (4) hemolytic anemia. The anemia is due to the damaging effects of peroxides upon the unsaturated fatty acids in the erythrocyte membrane. Vitamin E serves as a lipid antioxidant. Its principal role may be to ensure the stability and integrity of biological membranes.

This function probably accounts in large part for its effects on reproduction and muscle activity.

The fourth main lipid-soluble vitamin is also a group of compounds, natural and synthetic, designated *vitamin K* (the *Koagulations* vitamin) by its discoverer, Henrich Dam. We noted the chemical formula of vitamin K in Fig. 8.35 and the antivitamin K activity of the coumarin derivatives shown on p. 274. Natural vitamin K compounds are synthesized by colonic bacteria. Therefore, a dietary source is usually unnecessary. The vitamin is scarce in the normal diet. Consequently, since large quantities of antibiotic drugs may destroy the colonic bacteria, vitamin K deficiency may develop rapidly during treatment with the drugs. Vitamin K is necessary for the synthesis of prothrombin, Factor VII, and certain other essential clotting factors in the liver, possibly acting as an allosteric effector for a regulatory protein controlling the transcription of genetic information from DNA to mRNA (see p. 154). In vitamin K deficiency clotting is impaired.

Vitamins of the B complex. Of the water-soluble vitamins, we encountered *thiamine* (vitamin B_1) earlier in thiamine pyrophosphate, an essential coenzyme of pyruvate and α-ketoglutarate metabolism (see Chapter 4). Thiamine is synthesized by plants and is found primarily in beans, peas, roots, and enriched flour. Good animal sources are eggs and lean meat.

The first member of the vitamin B complex to be discovered, thiamine is the vitamin whose deficiency in man produces the disease called *beri-beri*. Thiamine deficiency leads to a decrease in the conversion of pyruvate to acetyl CoA in the tissues. Hence carbohydrates do not enter the citric acid cycle (see Fig. 4.4) and lipid utilization increases.

The decrease in the utilization of carbohydrates in the tissues accounts for much of the debility of thiamine deficiency. The central nervous system depends almost entirely on carbohydrates for its energy. In severe thiamine deficiency its consumption of glucose may drop as much as 60%. Thus its function is seriously impaired. Thiamine deficiency also causes degeneration of the myelin sheaths of peripheral and central nerve fibers. Lesions in the peripheral nerves frequently result in severe *polyneuritis*. Thiamine deficiency also causes weakening of the muscles, including the myocardium, and congestive heart failure is common in extreme thiamine deficiency. Usually the right side of the heart is affected first.*

---

* The heart failure of thiamine deficiency is associated with an *increased* return of blood to the heart and an *increased* cardiac output, approaching three times the normal output. Apparently thiamine deficiency causes peripheral vasodilatation, due perhaps to weakness of the smooth muscle of the vascular system.

Riboflavin (vitamin B_2), as we learned in Chapter 4, functions biochemically in the oxidative flavoprotein coenzyme flavin adenine dinucleotide (FAD) (see p. 132). Although riboflavin is essential for growth and well-being and its mode of action is known, it has not been possible to define a characteristic riboflavin-deficiency syndrome (such as beri-beri in thiamine deficiency, for example) in man and other higher animals. In lower animals riboflavin deficiency causes widespread debility arising from general depression of oxidative metabolism. In man severe riboflavin deficiency is rare, probably because riboflavin is present in a variety of foods (e.g., liver, eggs, milk, and leafy vegetables). The signs of mild riboflavin deficiency include nonspecific digestive disturbances, burning sensations of the skin and eyes, cracking at the corners of the mouth, and mental depression.

Nicotinic acid, or *niacin*, is found in liver and other meats, wheat germ, and eggs. It is converted in the body to the hydrogen donor-acceptor coenzymes NAD and NADP (see p. 53). In niacin deficiency, therefore, oxidative metabolism is depressed. Symptoms include diarrhea and other gastrointestinal disturbances, mental derangement and other neurological disorders, and dermatitis—a complex of symptoms known as *pellagra*. Pellagra occurs predominantly among poor peoples whose diets consist mainly of cornmeal, white flour, polished rice, and sugar.

Pyridoxine (vitamin B_6) is widely distributed in foods, good sources being meat, especially liver, vegetables, and cereals. In the form of the coenzyme pyridoxal phosphate, it participates in many types of biochemical reactions (see Table 4.1). In general, these relate to amino acid and protein metabolism. Dietary lack of pyridoxine in lower animals causes dermatitis, retardation of growth, anemia, and mental deterioration. Since pyridoxine deficiency without simultaneous deficiency of other vitamins is rare in man, the precise consequences of human pyridoxine deficiency, other than the depression of hemoglobin synthesis, are not clear.

Pantothenic acid, present in liver, eggs, vegetables, and many other foods, is incorporated into coenzyme A (see p. 126). Thus in pantothenic acid deficiency the metabolism of both carbohydrates and lipids is depressed. In lower animals pantothenic acid deficiency causes retardation of growth, failure of reproduction, graying of the hair, dermatitis, fat deposition in the liver, and numerous other abnormalities. In man no definite deficiency syndrome has been established; however, experimental evidence suggests that deficiency leads to malfunction of the adrenal cortex.

Biotin, among whose richest sources are liver, egg yolk, and yeast, is required in carboxylations and transcarboxylations, which are associated with fatty acid oxidation and synthesis (see pp. 126, 136) and with certain reactions of the citric acid cycle. There is an interesting relationship between biotin and *avidin,* a protein in egg white. Avidin combines chemically with biotin, rendering it metabolically inert. Accordingly, purified avidin is useful in testing enzymes for biotin dependence. In man biotin deficiency occurs only when one-third or more of the caloric intake comes from egg white.

Vitamin B$_{12}$ and *folic acid* were discussed in Chapter 8. In addition to other functions, they are essential in DNA synthesis and hence in cell division (see Table 4.1). Therefore, the earliest signs of deficiencies appear in the body's most rapidly dividing cells, the blood cell precursors and the mucosal cells of the alimentary tract.

Vitamin B$_{12}$ is synthesized only by bacteria and is thus unique among the vitamins. Wherever it is found in nature, it can be traced back to bacteria or other microorganisms growing in soil, sewage, or intestine. Its chief sources in the human diet are meat and dairy products. Folic acid occurs in plants (whence its name) but is also plentifully supplied by meat, eggs, and milk.

Table 4.1 shows the interesting structure of the cobalt-containing porphyrin-like vitamin B$_{12}$ molecule. Because of its unusual potency,* an extraordinarily small quantity, about 2 μg, is needed daily in the diet. The liver, the main storage depot, contains less than 100 μg per 100 g wet weight.

Ascorbic acid. Scurvy, the disease produced by a deficiency of *ascorbic acid* (vitamin C) has a prominent place in the history of war and oceanic exploration. In the centuries when sailors and soldiers subsisted for long periods on easily stored dried meats, grains, and beans, this scourge influenced the outcome of many military campaigns. With the discovery of the antiscorbutic (i.e., antiscurvy) effect of limes, lemons, and other citrus fruits and fresh vegetables, these foods became requirements in the diets of all military personnel. For this reason a British sailor is known as a "limey" and the active vitamin as "ascorbic" acid.

Interestingly, most animals are able to synthesize ascorbic acid and hence do not require it in their diets. The exceptions are guinea pigs and primates. The major dietary sources are fruit and vegetables, but prolonged cooking and oxidation may rapidly destroy their ascorbic acid contents.

Two characteristics of scurvy—excessive capillary bleeding and defective wound healing—provide the only clues to the metabolic role of ascorbic acid. Presumably the vitamin is essential for the maintenance or synthesis of intercellular substances throughout the body. This includes the formation of collagen (see p. 178), intercellular cement (see p. 80), bone matrix

---

* The marine chrysomonad *Monochrysis lutheri* requires only 40 molecules of vitamin B$_{12}$ per cell for maximal growth.

(Chapter 17), and dentine (see p. 517). Defective wound healing in scurvy is attributable to the inability of cells to deposit collagen fibrils and intercellular cement. Capillary bleeding is due to the extreme fragility of blood vessel walls resulting from abnormality of the cement between endothelial cells. Similarly, lack of ascorbic acid in childhood causes cessation of bone growth and failure of fractured bones to heal.

The biochemical mechanism of these effects is unknown. Ascorbic acid is a powerful reducing agent that probably can be reversibly oxidized and reduced within the body. Presumably it functions either as a reducing agent in specific metabolic processes or as a general-purpose donor-acceptor of electrons in diverse oxidation-reduction systems. The processes and systems in which it appears to act are the oxidation of tyrosine and phenylalanine, the intestinal absorption of iron and the removal of iron from cellular ferritin (both of which increase the concentration of iron in body fluids), and those dependent on vitamin B$_{12}$ and folic acid (ascorbic acid somehow enhances their nutritional effectiveness).

MINERAL REQUIREMENTS

The inorganic, or mineral, components of the diet are just as essential as the organic components. We saw in Table 2.1 that body tissues contain a large number of inorganic substances. Traditionally investigators have measured these in the laboratory by incinerating tissue—thereby volatilizing carbon, hydrogen, oxygen, and nitrogen—and analyzing the remaining ash. Today they assay inorganic substances in body fluids by sensitive spectrophotometric and spectroscopic techniques.

According to Table 2.1, body minerals can be divided roughly into two categories based upon the amounts present in the body and needed in the diet: *macronutrients,* including sodium, potassium, calcium, and phosphorus; and *micronutrients,* including magnesium, iron, copper, zinc, and other trace elements. Indeed, most of the elements of the periodic table have been found in living tissue, but this fact does not prove that they are all essential in normal metabolism.

We have already encountered many of the body minerals. The problems of sodium and potassium balance were considered in Chapter 7. Calcium and phosphorus, required for the growth and repair of bones and teeth and in larger quantities for the milk of nursing mothers, will be surveyed in Chapter 15. We shall note here only that calcium, which constitutes almost 2% of the body weight (the highest percentage of any inorganic substance), is supplied in the form of salts such as calcium phosphate, calcium sulfate, and calcium carbonate.

Iodine, needed only in minute amounts, is essential to the normal functioning of the thyroid, whose hormones are iodine-containing amino acids. Iodine is present in traces in drinking water and in vegetables grown where iodine is available in the soil and in abundance in sea foods. Iodized salt provides adequate amounts of iodine in parts of the world that are deficient in natural sources.

Most of the trace elements necessary in the diet are metals that act through their associations with enzymes. An enzyme that contains a metal atom as an integral component is called

a *metalloenzyme.* Dietary deficiency of the metal leads to loss of metalloenzyme activity. The essential trace metals include magnesium, whose balance with calcium is critical in muscle and nerve function; manganese and copper, components of various metalloenzymes (e.g., copper-containing enzymes participate in tyrosine metabolism and heme synthesis); chromium, necessary for insulin action and optimal glucose metabolism; arsenic and selenium, promoters of growth (when present in low concentrations); and zinc, a component of the metalloenzymes carbonic anhydrase and alcohol dehydrogenase —the latter of which by oxidation eliminates ingested alcohol from the body.

Iron metabolism. A micronutrient of particular importance is iron. We recall from Chapter 8

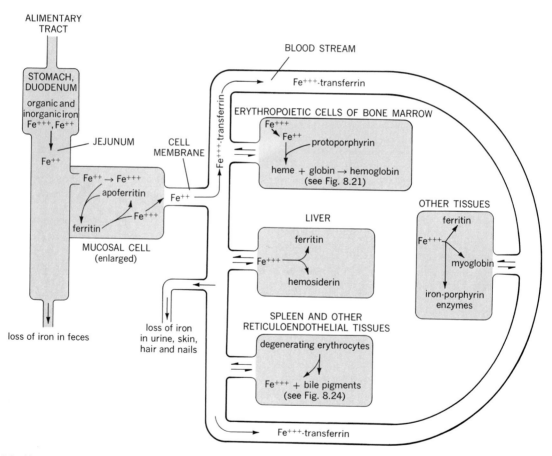

14.6 Normal iron metabolism.

that body iron is rigorously conserved, with the iron released during hemoglobin breakdown being reutilized again and again. The minuteness of daily iron losses, revealed in the classic studies of McCance and Widdowson in 1936 (see p. 14) and confirmed by balance studies using radioactive iron, ^{59}Fe, implies that daily iron intake must be equally minute—otherwise, the body would accumulate iron. Since the average diet contains much more iron than is needed each day, body iron levels can be kept constant only through the regulation of intestinal absorption. The mechanism by which the intestine controls the amount of iron absorbed is not yet understood. It is known that ferrous iron (Fe^{++}) in the intestinal lumen enters readily into the mucosal cells of the duodenum and upper jejunum (Fig. 14.6). There it is oxidized to ferric iron (Fe^{+++}), some or all of which complexes with the protein apoferritin to yield ferritin. The mucosal cells, which are continuously replaced (see p. 533), retain most of the iron in ferritin form; and, when they are sloughed away at the end of their 1 or 2 day life span, it goes with them. However, a small quantity of iron passes through the cells into the mucosal capillaries. Somehow the rate of this transfer is governed by the body's iron requirements. In iron deficiency more iron than usual leaves the cells to enter the body; when iron deficiency is corrected, the rate returns to normal. The agency that transmits to the cells information on the body's iron stores is unknown.

Despite these uncertainties, most of the basic features of iron metabolism and balance are well established. The total iron of the adult body, between 4.0 and 4.5 g, is distributed in three main compartments. Approximately 3.0 g is in hemoglobin, from 1.0 to 1.5 g is storage iron (chiefly ferritin and hemosiderin), and from 0.1 to 0.2 g is in myoglobin, transport iron in the plasma, and iron-containing enzymes (cytochromes, peroxidase, catalase, etc.). These compounds can be classified in a number of ways: as porphyrin compounds (hemoglobin, myoglobin, catalase, etc.) and nonporphyrin compounds (ferritin, hemosiderin, etc.); as compounds functional in H or O_2 transport (hemoglobin, cytochromes, etc.) and inert storage compounds; etc.

All body iron is combined with protein of one type or another—including plasma iron, which is transported in association with the specific iron-binding β_1-globulin transferrin. Inorganic iron itself is extremely toxic. If more than 10 mg, an amount that just exceeds the binding capacity of transferrin, is injected intravenously, serious poisoning results. This reaction undoubtedly accounts for the widespread existence in nature of iron-protein conjugates and for the limitations on intestinal iron absorption.

Possibly because of the restrictions on iron absorption, iron deficiency is one of the most common human ailments, even though iron is the second most abundant element on earth and the poorest of diets usually contains about 10 times the amount of iron needed per day. Normally a diet provides 10 to 20 mg of iron per day, of which only 1 to 2 mg is absorbed. The chief dietary sources are eggs, meat, and leafy vegetables. The amount of hemoglobin normally synthesized per day requires between 20 and 25 mg of iron. Most of this comes from the erythrocytes that break down each day—about 1% of the total number in the circulation.

Metabolism of Major Body Constituents

We discussed the principal metabolic pathways in Chapter 4 in connection with the physiology of the individual cell. Now we shall consider those aspects of metabolism of particular importance in the context of body physiology and nutrition, emphasizing the body's arrangements for supplying the nutrients necessary for cell metabolism.

It is, of course, only by convention that we treat the body constituents one at a time. Since the discovery of the citric acid cycle, it has been recognized that carbohydrates, lipids, and proteins—the three major foodstuffs—proceed along independent metabolic pathways only until they reach the keto acid stage (pyruvate, α-ketoglutarate, acetoacetate, etc.). Then they enter a final common pathway, each yielding energy and short carbon chains for the synthesis of new carbohydrates, lipids, proteins, and nucleotides (see Fig. 4.10). We must therefore bear in mind their interconvertibility.

Blood glucose. Glucose is the chief, and in a practical sense the only, transport form of carbohydrate. It is the main carbohydrate absorbed in the small intestine (see Chapter 13), and it is the carbohydrate furnished by the liver to the other tissues of the body.

The concentration of glucose in the blood and extracellular fluids normally ranges from 70 to 100 mg per 100 ml (see Table 8.5). Except for the period immediately after a meal, it remains remarkably constant. This regularity elegantly illustrates the integrated functioning of body systems in homeostasis.

In a fasting animal the liver is virtually the sole source of blood glucose.* The other body tissues continuously draw upon the blood glucose, utilizing it as their principal metabolic fuel. Since blood glucose levels stay constant through long periods of fasting, the glucose added to the blood by the liver must come from noncarbohydrate precursors and/or stored carbohydrate. The noncarbohydrate precursors are lipid and protein, and the stored carbohydrate is glycogen (see p. 138), first identified by Claude Bernard.

Bernard's career as a physiologist began in 1841, when morphological study was universally preferred to physiology. However, Bernard pursued his own interests and early set himself the task of tracing the transformations undergone by foodstuffs in the body. He undertook first to follow the course of glucose, through the alimentary tract to the blood stream, through the portal vein to the liver, from the liver to the right side of the heart, the lungs, and the left side of the heart, and then to the body tissues. "I shall find," he wrote, "that the dextrose (glucose) disappears, is destroyed, or is in some way or other changed."

He fed a dog a diet rich in sugar and found abundant sugar in the hepatic veins. He then fed the dog meat only and again found abundant sugar in the hepatic veins. "The liver had produced sugar!" he exclaimed. After repeated experimentation he concluded that the liver forms glucose by an act analogous to glandular secretion. He demonstrated the presence in the liver of a "glycogenic substance" (i.e., glycogen) that by a process of "fermentation" was readily converted to glucose. He also drew an accurate analogy between liver glycogen and plant starch.

Proof that the liver is the source of blood glucose was eventually obtained from liverless animals. F. C. Mann developed techniques for removing the entire liver of a living animal and keeping the animal alive for up to 30 hours afterward. The effects of hepatectomy were revealing. In every instance the blood glucose level fell, the result being hypoglycemia. The drop took place at a relatively constant rate and was accompanied by increasing restlessness, muscular weakness, loss of sight and hearing, muscle paralysis, convulsions, and coma. The symptoms were relieved when glucose was injected intravenously. Amino acids were useless. The symptoms of hypoglycemia reflect the fact that glucose is essential for the functioning of the nervous system and hence the life of the organism. Hypoglycemia after hepatectomy was not prevented by the breakdown (*glycogenolysis*) of the large quantities of glycogen present in the muscles. The liver alone (and perhaps the kidneys) contains a glucose 6-phosphatase, which catalyzes the conversion of glucose 6-phosphate to glucose (Fig. 14.7; see also Fig. 4.12). Since the muscles lack this specific enzyme, their glycogen reserves are available only for their own metabolic needs.

Regulation of blood glucose. The glucose concentration of blood is kept within the normal range by the interaction of two groups of mechanisms: those that prevent an abnormal increase in blood glucose (hyperglycemia) and those that prevent hypoglycemia. When the blood glucose level falls below normal, the sympathetic division of the autonomic nervous system is stimulated, presumably through some direct effect of hypoglycemia upon the hypothalamic centers. Sympathetic impulses cause the adrenal medulla to release large amounts of norepinephrine and epinephrine, which pass to the liver via the blood stream. Epinephrine to a greater extent and norepinephrine to a lesser extent activate the normally inactive phosphorylase in the liver, which makes possible the breakdown of glycogen and the production of glucose (see Fig. 4.12, p. 151). The glucose enters the blood

---

* The kidneys may contribute some, but the amounts are insignificant in relation to the total carbohydrate requirement of the body.

stream, elevating the blood glucose level, which often rises as rapidly as 15 mg per 100 ml per minute. When it becomes normal, sympathetic excitation and further glycogen breakdown cease. The feedback nature of this control mechanism is apparent.

When the blood glucose level rises above normal—as it tends to do after a heavy meal—the islets of Langerhans release insulin (see p. 538). The rate of release is approximately proportional to the need—that is, the more glucose that enters the blood per unit of time, the more insulin that is secreted in response. When the blood glucose level returns to normal, so does the insulin output—another feedback mechanism. In disorders associated with deficient insulin production (e.g., diabetes mellitus), the blood glucose level rises, and glycosuria develops (see p. 344).

We shall speak of insulin in detail later, but we may note here that it is believed to lower the blood glucose level by promoting the transfer of glucose into the cells. In the liver this leads to glycogen formation (*glycogenesis*). Thus the liver serves as a buffer in the maintenance of the blood glucose level. In severe hyperglycemia approximately two-thirds of the excess glucose enters the liver, where it is rapidly converted to glycogen, while the remaining one-third enters other body cells. Glycogen is released from the liver into the blood as glucose when needed.

A constant blood glucose level is important for two reasons. First, the rate of glucose transport into brain and heart tissue is highly dependent on the glucose concentration; in hypoglycemia brain and heart functions are impaired. Second, it is mainly as glucose that carbohydrates travel from one part of the body to another. When active muscle cells begin to remove glucose from the blood, the liver begins to release glucose into the blood. Liver glycogen is therefore eventually delivered to the muscles.

Under certain circumstances a sizeable portion of carbohydrate is transferred from the muscles to the liver, but not in the form of glucose. Once glucose has entered the peripheral tissues, whether to be stored as glycogen or used, it cannot return to the blood stream as glucose. During periods of intense muscular exertion, the amount of oxygen that the blood can supply to the muscle cells is limited. Con-sequently, the oxidation of carbohydrate in the muscles may be incomplete. With such an *oxygen debt*, a large quantity of lactic acid may be discharged from the muscles into the blood. The lactic acid is carried to the liver and converted to liver glycogen, and it ultimately reappears as blood glucose. Although this *lactic acid cycle* is an indirect source of blood glucose under extreme conditions, it is probably of minor significance under ordinary conditions.

Formation of carbohydrates from proteins and lipids. The production of blood glucose—and body carbohydrate metabolism in general—centers on glycogenesis and glycogenolysis in the liver. However, when body carbohydrate reserves are so depleted that the blood glucose level starts to fall, moderate amounts of glucose can be formed from certain amino acids, the glycerol of triglycerides, and, as we have just learned, lactic acid. In these cases we speak of *gluconeogenesis*.

For example, the blood glucose of a starving hepatectomized 10-kg dog with no accessible carbohydrate stores can be held at the normal level if about 60 g of glucose is injected per day. It follows that a starving normal dog of similar weight and normal blood glucose level produces *at least* 60 g of glucose per day. This glucose must arise through gluconeogenesis.

Any glycolytic intermediate can be converted into glycogen; so, of course, can any compound capable of being converted to a glycolytic intermediate. The glycerol of lipids can be phosphorylated to glycerophosphate, which can be oxidized to the glycolytic intermediate dihydroxyacetone phosphate. Similarly, approximately 60% of the amino acids of body proteins can be converted into keto acids capable of conversion to carbohydrates (see Fig. 4.10).

Low carbohydrate levels in the tissues and low blood glucose levels are the basic stimuli of gluconeogenesis. They cause reversal of the glycolytic sequence, thereby promoting the conversion of amino acids and glycerol into carbohydrates. Several hormones, including ACTH and thyroxin, also stimulate gluconeogenesis. ACTH stimulates the adrenal cortex to produce a hormone that mobilizes proteins from all body cells, making them available as amino acid donors. Most of the amino acids are quickly deaminated in the liver, leaving many substances capable of conversion into glucose.

Ketogenesis and hyperketonemia. When carbohydrate is ingested in sufficient quantity, the normal body converts somewhat more than half of it to energy, 30 to 40% to lipid, and 5% to glycogen. As suggested in Table 14.4, the body carbohydrate reserve is less than 400 g, the equivalent of only 1500 Cal. In contrast, the lipid reserve may vary widely, being equivalent to 90,000 Cal in the average well-nourished body. In the absence of an adequate diet, the carbohydrate reserve is depleted, and the body draws upon its lipid reserve to meet its energy needs. Simultaneously, gluconeogenesis furnishes carbohydrate for the maintenance of the blood glucose level.

Ketogenesis, the production of the so-called *ketone bodies* (an old term for ketones and reduced ketones), is a normal function of the liver. The normal fasting blood ketone level is 5 mg per 100 ml. A ketone is an organic compound containing a keto (or carbonyl) group (see Table 2.2), and the major ketone bodies are acetoacetic acid (CH_3COCH_2COOH), β-hydroxybutyric acid ($CH_3CHOHCH_2COOH$), and acetone (CH_3COCH_3). Since β-hydroxybutyric acid and acetone are derived from acetoacetic acid in the following reactions,

$$CH_3-\overset{\overset{O}{\|}}{C}-CH_2-\overset{\overset{O}{\|}}{C}-OH \xrightarrow{+2H}$$

acetoacetic acid

$$CH_3-\overset{\overset{OH}{|}}{CH}-CH_2-\overset{\overset{O}{\|}}{C}-OH \quad (14.4)$$

β-hydroxybutyric acid

$$CH_3-\overset{\overset{O}{\|}}{C}-CH_2-\overset{\overset{O}{\|}}{C}-OH \xrightarrow{-CO_2}$$

acetoacetic acid

$$CH_3-\overset{\overset{O}{\|}}{C}-CH_3 \quad (14.5)$$

acetone

we shall confine our discussion to factors influencing the synthesis of acetoacetic acid.

Ketogenesis depends upon acetyl CoA. As shown in Figs. 4.4 and 14.7, acetyl CoA arises from pyruvate oxidation and the β-oxidation of the long-chain fatty acids. The fate of acetyl CoA depends upon metabolic circumstances. If enough oxaloacetate is present, acetyl CoA can be converted into citrate, which is further oxidized in the citric acid cycle. Acetyl CoA can also participate in other acetylation reactions (see Figs. 4.5, 14.7). Finally acetyl CoA molecules can combine to form acetoacetyl CoA.

$$2 \text{ acetyl CoA} \longrightarrow$$
$$\text{acetoacetyl CoA} + \text{CoA} \quad (14.6)$$

Acetoacetyl CoA, which also arises directly from fatty acid breakdown, has several possible fates: (1) it can be cleaved to yield acetyl CoA; (2) it can be reconverted into long-chain fatty acids; and (3) it can condense with acetyl CoA to yield β-hydroxy-β-methylglutaryl CoA, a precursor of acetoacetic acid and of cholesterol. It should be noted that, although ketones are normally produced in the liver, they are normally oxidized in extrahepatic tissues.

Ketogenesis increases as the carbohydrate supply decreases. An excess of ketones in the blood, called *hyperketonemia* or *ketosis,* occurs chiefly with general starvation, carbohydrate deprivation (or diabetes mellitus, a disease of insulin deficiency), and excessive fat intake. A theory proposing the primary event of ketosis in each of these situations has been offered.

The first theory emphasizes an increase in gluconeogenesis. The second points out that the production of oxaloacetate, necessary to take acetyl CoA into the citric acid cycle, decreases as glycolysis decreases. Ordinarily we think of oxaloacetate as a substance generated in the citric acid cycle. However, as it is rather unstable, a certain amount is continuously being broken down. This is usually replaced by oxaloacetate formed from the glycolytic intermediate phosphoenolpyruvate or from pyruvate by way of malate; the pyruvate must also arise through glycolysis, since acetyl CoA cannot be reconverted to pyruvate. When glycolysis is impaired, so is the ability of the citric acid cycle to accommodate the available acetyl CoA molecules. The overflow is shunted to the acetoacetate pathway. The third theory stresses the exceedingly rapid production of fatty acids from adipose tissue,

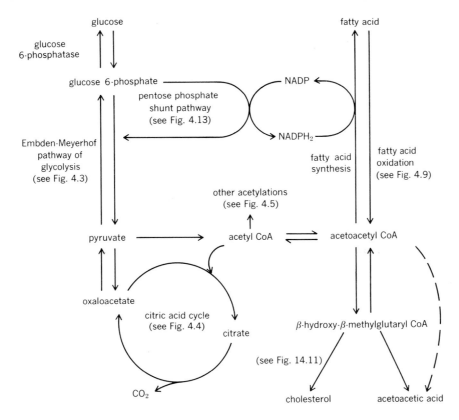

14.7 Some interrelationships between glucose metabolism and fatty acid metabolism in the normal liver.

which leads to a plethora of the acids. As a consequence, the liver removes increasing quantities of them from the blood, converting them to acetyl CoA that cannot enter the citric acid cycle.

After 5 days of fasting, ketones may account for 50% of the total energy production. The danger of such severe ketosis lies in the fact that the major ketone bodies are organic acids. They must be neutralized with equivalent amounts of base if normal pH's of blood and tissue fluids are to be maintained. As indicated in Table 10.3, the neutralization causes a severe metabolic acidosis, which, if uncorrected, may be fatal.

Insulin. We earlier noted the chemical structure of insulin and identified the islets of Langerhans as its source. We learned that insulin is a protein composed of two polypeptide chains, one of 21 amino acids (the A chain) and the other of 30 (the B chain), joined by two disulfide

bonds (see Fig. 2.8). In addition, an intrachain disulfide bond forms a loop on the A chain. The insulin molecule contains no metal, although it readily associates with certain metals, particularly zinc.

Special staining techniques distinguish three cell types in the islets of Langerhans. These have been termed α, β, and δ *cells.* Normally 60 to 90% of the islet cells is β cells, 1 or 2% δ cells, and the remainder α cells. Insulin is secreted only by the β cells (see Fig. 13.22), which are easily recognized with electron microscopy by their granular elements and rich endoplasmic reticulum (Fig. 14.8). Insulin is stored in the β cells in a polymerized form having a molecular weight of 24,000 to 48,000, from which units of molecular weight 6,000 are released into the blood. Degranulation of the β cells occurs when insulin is released, as after hyperglycemia.

The synthesis of insulin in the laboratory was

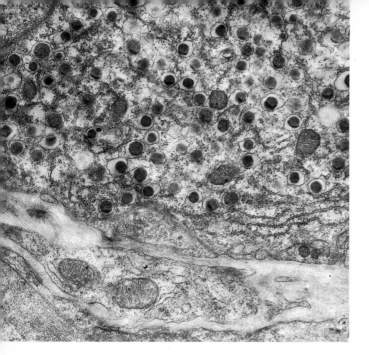

14.8 Electron micrograph of β cell. (Paul E. Lacy.)

achieved simultaneously in 1963 by P. G. Katsoyannis and J. Meienhofer. In β cells A and B chains are synthesized separately—so at least two structural genes are involved—and insulin results from their spontaneous combination.* Insulins from the pig, dog, and sperm whale are identical. Those from the horse, sheep, and pig differ from each other only in the three adjacent amino acid residues beneath the disulfide loop on the A chain. Those from the pig and man differ only in one amino acid residue in the B chain. These small structural differences have no detectable influence on the biological activities of the various insulins; all appear to have the same effect.

Despite intensive study in recent years, we do not yet know the precise mechanism of insulin action. Nevertheless, we can state some consequences of its actions and describe their roles in carbohydrate metabolism. Insulin facilitates the transfer of sugars across cell membranes in extrahepatic tissues and thus enhances their uptake by these tissues. It seems likely that glucose entry into the cell is the rate-limiting

---

*The analogy is obvious between the synthesis of insulin with its A and B chains and that of immunoglobulin with its H and L chains (see p. 262).

step in glucose metabolism. It apparently involves a transport system that operates before hexokinase on sugars that cannot be further metabolized inside the cell.

That insulin acts at this point is suggested by ingenious experiments employing nonutilizable sugars to distinguish the effect of insulin on sugar entry into the cell from its subsequent effect on sugar metabolism. For example, galactose, an isomer of glucose (see Fig. 2.4), is utilized primarily by the liver, kidneys, and intestine. Galactose injected into a dog whose liver, kidneys, and intestine have been removed behaves like nonutilizable urea rather than utilizable glucose (Fig. 14.9A). Its concentration in the blood (normally zero) declines rapidly at first and then steadies (excretion is impossible because the kidneys are gone). When galactose and insulin are injected simultaneously (Fig. 14.9B), blood galactose concentration falls even more rapidly than before and levels off at a lower value. This reaction implies that insulin enables galactose to penetrate to a previously inaccessible intracellular region. When insulin is injected after the galactose concentration has reached equilibrium (Fig. 14.9C), the concentration drops immediately and levels off at a value near that for the simultaneous injection.

Fig. 14.10 shows that insulin has similar effects on D-xylose and L-arabinose and no such effects on D-mannose, D-fructose, L-sorbose, D-sorbitol, and D-arabinose. The sugars responsive to insulin have common chemical configurations around the first, second, and third carbon atoms.

These phenomena are the basis for the *transmembrane transport theory* of insulin action. Since sugars whose entry into cells is accelerated by insulin have similar structures, the inference is that, somewhere in the transfer process, some sort of chemical union occurs between the sugar and a specific receptor or carrier substance. Evidence of such a union is the competition for transfer among several sugars whose entry is enhanced by insulin. Some sugars completely block others. A recent proposal is that insulin may be one of several hormones whose prime function is to set the redox potential (see p. 42) gradient across membranes and other interfaces. According to this view, changes in redox potential on the two sides of the membrane underlie changes in transport behavior.

Whether other effects of insulin on metabolism are primary effects or results of a primary effect on glucose transport has not been established. For example, insulin promotes glycogenesis, apparently by stimulating the enzymes that synthesize glycogen. It also appears to promote lipogenesis, in adipose tissue at least, by selectively stimulating the pentose phosphate shunt of glycolysis, in which $NADPH_2$ is produced. This in turn

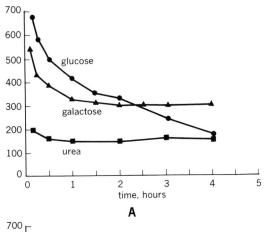

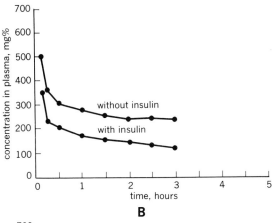

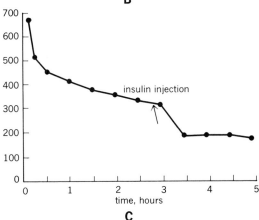

14.9 Body utilization of galactose: A, fates of injected utilizable and nonutilizable substances in eviscerated nephrectomized animal; B, effect of insulin on fate of galactose; C, effect of insulin injected after the galactose concentration has reached equilibrium. (Adapted from R. Levine, and M. S. Goldstein, *Recent Prog. Hormone Res.*, *11*, 343, 1955.)

is believed to stimulate the pathway of fatty acid synthesis (see Fig. 14.7). The effect of insulin on protein anabolism, as on lipogenesis, is probably secondary to its effect on glycogenesis. In the absence of insulin, protein loss stems from either excessive catabolism, as the body "attempts" to derive the glucose that it needs from amino acids, or defective anabolism. Insulin spares protein by furnishing glucose for energy and favors protein synthesis.

Perhaps the greatest difficulty in defining the site of insulin action is the marked difference in insulin effects between peripheral tissues and the liver. Since muscle tissue (such as the diaphragm) in vitro responds to insulin by increasing its glucose uptake and glycogen deposition and liver tissue in vitro does not, the conviction has been growing that insulin acts purely or at least primarily upon peripheral tissues. We recall that insulin in sufficient dosage lowers the blood glucose level—i.e., accelerates hypoglycemia—in a hepatectomized animal. Evidently the liver is not essential for an insulin-induced increase in glucose uptake in extrahepatic tissues but is essential for hyperglycemia in the absence of insulin.

Brief mention should be made here of *glucagon*, a second pancreatic hormone, which probably arises in the α cells of the islets of Langerhans.* It was early observed that certain insulin-containing pancreatic extracts caused an initial rise in the blood glucose level instead of the expected fall. At first, this was attributed to an incidental liberation of epinephrine, but in time glucagon was isolated and identified as a straight-chain peptide of molecular weight 3450 containing 29 amino acids. In many respects the action of glucagon resembles that of epinephrine and is opposite to that of insulin. It stimulates hepatic glycogenolysis (by activating hepatic phosphorylase), raises the blood glucose level, and inhibits the utilization of glucose in peripheral tissues. It is believed to play an essential role in the maintenance of blood glucose levels. A diabetes-like disorder is associated with excessive glucagon secretion.

Transport of lipids. After absorption in the small intestine, dietary lipid is transported in several forms by several routes (see Chapter 13). Most of the triglycerides reaggregate as lipoproteins in chylomicrons, which enter the lymphatic vessels draining the intestine and proceed to the

---

* The function of the islet δ cells is unknown.

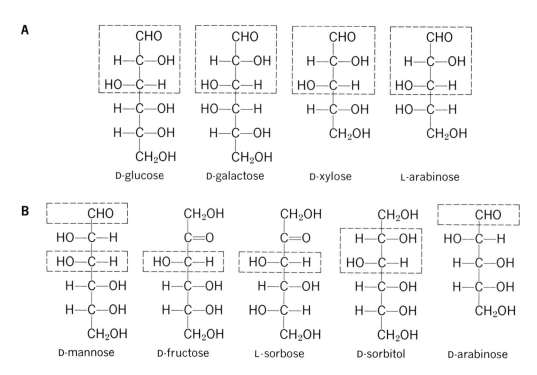

A

| | CHO | CHO | CHO | CHO |

D-glucose   D-galactose   D-xylose   L-arabinose

B

D-mannose   D-fructose   L-sorbose   D-sorbitol   D-arabinose

14.10   Structures of sugars responsive (A) and unresponsive (B) to insulin.

blood, where they remain for variable lengths of time. Only free medium- and short-chain fatty acids escape inclusion in the chylomicrons. They are absorbed directly into the portal vein blood (see p. 549).

We know that the chylomicrons are removed from the blood because they appear in it only after meals, as milky lipemia (see pp. 274, 324, 549). Disruption of chylomicrons and hydrolysis of their triglycerides is catalyzed in the blood and in the liver by the *clearing factor,* a lipoprotein lipase. The freed fatty acids either remain in the blood or are utilized in the liver. Heparin seemingly acts as a cofactor of lipoprotein lipase and clears the plasma of chylomicrons. Since heparin is highly acidic, it may alter the surface charges of the chylomicrons.

Chylomicrons are absent from fasting plasma. All its lipids are in the form of soluble lipoproteins. Of the total plasma proteins, about 3% is lipoproteins that migrate with α_1-globulins, and about 5% is lipoproteins that migrate with β_1-globulins. As shown in Table 14.5, α_1-lipopro-

tein is denser than β_1-lipoprotein because its lipid content (19% cholesterol and 26% phospholipid) is lower than that of β_1-lipoprotein (47% cholesterol and 23% phospholipid). When sodium chloride is added to plasma and the mixture is ultracentrifuged, β_1-lipoprotein rises to the top. Significantly, the fat-soluble vitamins and carotenoids are in this fraction. A scattering of lipid-protein complexes migrates with α_2-globulins. The density of α_2-lipoprotein is still lower than that of β_1-lipoprotein.

Presumably the plasma lipoproteins provide a mechanism for the transport of lipids from one part of the body to another. Even during periods of active conversion of the lipid of adipose tissue to carbohydrate, the main plasma lipid is lipoprotein. Neither cholesterol nor phospholipid is known to be transported in any other form. Indeed, none of the plasma lipids is sufficiently polar to circulate free in solution. Their ability to dissolve in plasma depends on their interactions with proteins to form the macromolecules called lipoproteins. The nature of the

TABLE 14.5 LIPOPROTEINS

Ultracentrifugal fraction	Density	Composition, %				Electrophoretic fraction
		Triglyceride	Phospholipid	Cholesterol	Protein	
Very low-density lipoproteins	<1.019	52	18	22	7	α_2-globulin
Low-density lipoproteins	1.019–1.063	9	23	47	21	β_1-globulin
High-density lipoproteins	1.063–1.210	8	26	19	46	α_1-globulin

bonds between lipid and protein in the lipoproteins is unknown.

As noted in Chapter 2, cholesterol belongs to a class of compounds called sterols or steroids. Almost all its physical properties are identical to those of other lipids, and it forms esters with fatty acids (approximately 70% of plasma cholesterol is esterified; see Table 8.5).

Cholesterol is obtained from the diet (exogenous cholesterol); it is also synthesized within the body (endogenous cholesterol), largely from the same products of fatty acid breakdown that serve as precursors of fatty acid synthesis. As shown in Fig. 14.11, two molecules of acetyl CoA join to form acetoacetyl CoA, which accepts another acetyl CoA to form β-hydroxy-β-methylglutaryl CoA (see Fig. 14.7). This compound, with the participation of $NADH_2$, is converted to *mevalonic acid,* which is converted to its pyrophosphate derivative and then to isopentenyl pyrophosphate. Three molecules of isopentenyl pyrophosphate combine to form *farnesyl pyrophosphate,* two molecules of which combine to form *squalene.* In a series of oxidations, squalene is converted to *lanosterol* and then *cholesterol.* The entire sequence from acetyl CoA involves about 26 steps.

The considerable interest in this pathway derives from the seemingly significant correlation between the concentration of plasma cholesterol—particularly β_1-lipoprotein cholesterol—and the incidence of atherosclerosis (one of whose manifestations is coronary heart disease). The idea has evolved that Western man is widely afflicted with excessively high levels of circulating cholesterol and other lipids because of his habitual consumption of a diet rich in lipids, particularly lipids of animal origin. Although modification of the diet—for example, the substitution of unsaturated fatty acids for saturated fatty acids—leads to a reduction in plasma cholesterol, the ultimate effect on the risk of atherosclerosis (and serious heart disease) is still in question (see p. 324).

Investigators are also examining the possibilities of blocking the pathway in Fig. 14.11 with antimetabolites or other inhibitors. Their results to date have been disappointing.

NITROGENOUS COMPOUNDS

We have already considered the mechanisms of protein synthesis, protein breakdown, nucleotide synthesis, nitrogen balance, and protein nutrition. We shall now deal with problems related to the synthesis and breakdown of each of these important nitrogenous body constituents.

Metabolic relations of the plasma proteins. We discussed the plasma proteins, whose major components are albumin, the globulins, and fibrinogen, in Chapter 8. Our intention here is to emphasize their metabolic roles.

Essentially all of the plasma albumin and fibrinogen is synthesized in the liver, and essentially all of the plasma globulins in the lymphoid tissues and reticuloendothelial system. When tissues are depleted of their proteins, plasma proteins can serve as replacements. It appears,

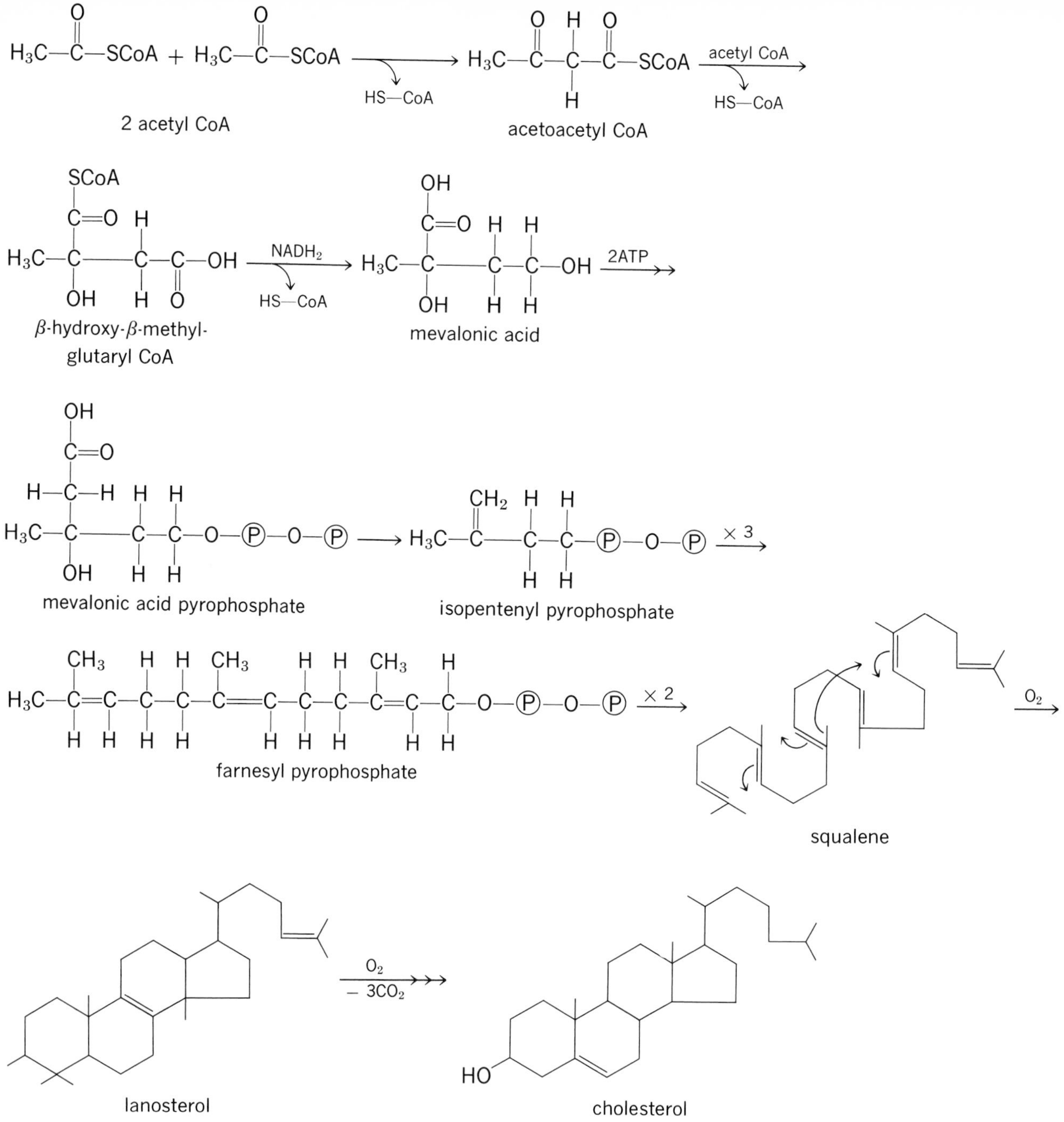

14.11 Cholesterol synthesis.

in fact, that an equilibrium exists between tissue and plasma proteins. Even during starvation or severe debilitating disease, the ratio of tissue protein to plasma protein remains relatively constant at about 33:1. Whole protein molecules are ingested intact by a tissue cell, and, once inside the cell, are split into peptides or amino acids and rebuilt into other proteins.

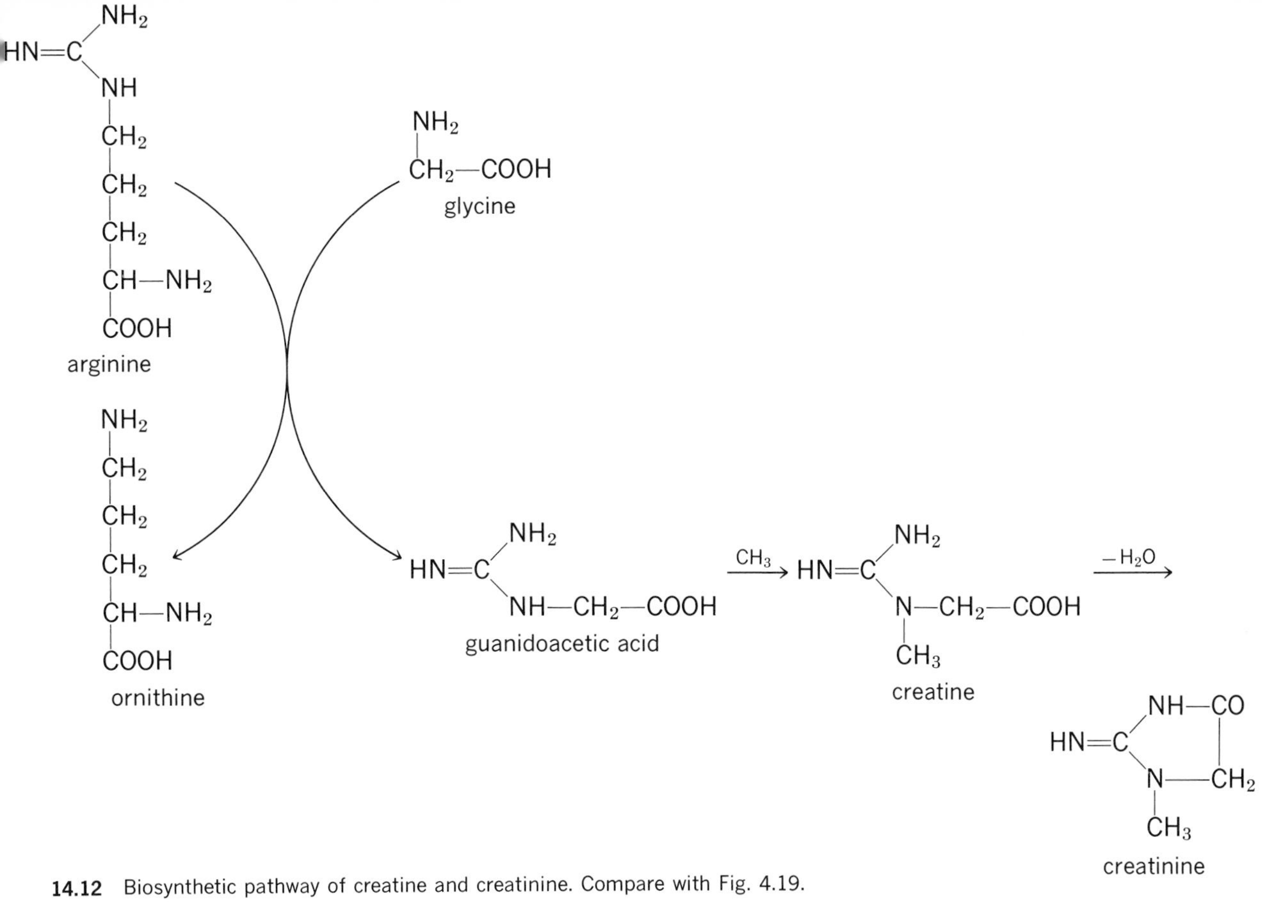

14.12 Biosynthetic pathway of creatine and creatinine. Compare with Fig. 4.19.

Creatine and creatinine. The term *nonprotein nitrogen* refers to nitrogenous substances that are not precipitated by the usual protein-precipitating reagents. It covers a heterogenous mixture of compounds, including some listed as waste products in Table 8.5—ammonia, creatinine, urea, and uric acid. All are of low molecular weight, and most are readily diffusible. To a large extent they are catabolic products en route to the kidneys, which eliminate them (see Chapter 10). Several merit brief comment in this chapter.

Creatine is an ingredient of muscle tissue. It is produced when *guanidoacetic acid,* formed from arginine and glycine, is methylated (Fig. 14.12). Creatine is an amino acid, though it is not of the α-amino type and is not present in protein. Creatine occurs in muscle both in the free state and as the phosphorylated derivative *creatine phosphate.* Creatine phosphate is an important reservoir of energy-rich phosphate

bonds. Because it promotes the synthesis of ATP, it is known as a *phosphagen.* Interestingly, creatine phosphate is found only in vertebrates. The corresponding phosphagen in invertebrates is arginine phosphate.

Creatine undergoes slow conversion to its anhydride, creatinine, which is a normal component of both blood and urine (see Tables 8.5, 10.2). Creatine itself appears in the urine only in certain diseases, notably muscular dystrophy, or in other conditions in which a large mass of muscle is lost (e.g., after leg amputation). Presumably this means that creatine is synthesized outside muscle tissue and excreted when muscle tissue cannot make use of it.

Catabolism of amino acids, purines, and pyrimidines. We discussed the pathways of purine and pyrimidine synthesis in Chapter 4. The nucleic acids synthesized in all cells break down at the end of the cells' life span. The main product

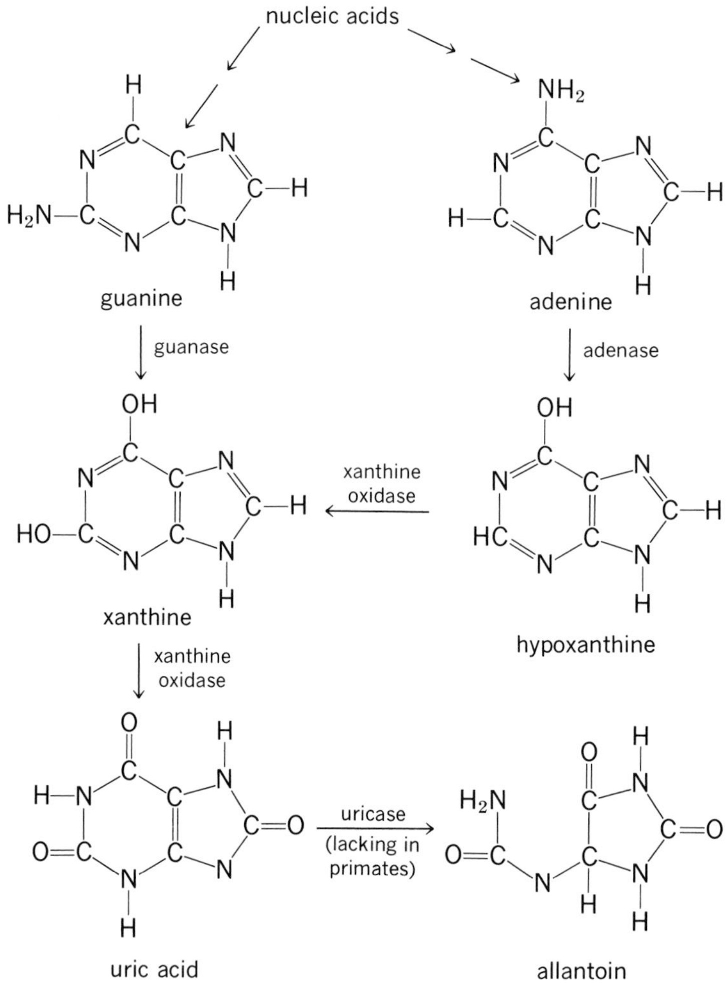

14.13 Principal steps in the degradation of purines in mammals.

of purine catabolism is uric acid. The major purines, adenine and guanine, are converted into uric acid via the series of oxidative reactions outlined in Fig. 14.13. Pyrimidine catabolism occurs chiefly in the liver, and the main products are β-alanine (from cytosine and uracil) and β-aminoisobutyric acid (from thymine).

It is of interest to consider the biological significance of uric acid excretion. The adult human on an adequate diet is in nitrogen balance: his body does not store excess nitrogenous compounds (e.g., amino acids) in the sense that it stores excess carbohydrate as glycogen and excess lipid as triglycerides. Nitrogenous matter absorbed in excess of what is needed to replace proteins, purines, pyrimidines, and other nitrogenous compounds lost through wear and tear must be eliminated as a nitrogenous waste product of one kind or another.

The arrangements for maintaining nitrogen balance vary greatly in different phyla. The simplest scheme, followed by a few primitive organisms, is to eliminate excess amino acids intact. Most organisms, however, convert amino acids (ingested in excess or arising from protein catabolism) to useful substances, such as α-keto acids and thence to lipid and carbohydrate (see Fig. 4.18) or purines (see Fig. 4.14), or to waste products. In man the principal waste product is urea (see Fig. 4.19)—which accounts for more than two-thirds of the nitrogen excreted (see p. 346). In certain birds, reptiles, and insects, it is uric acid. In aquatic invertebrates, amphibia, aquatic reptiles, and other forms, it is ammonia. It is possible, therefore, to divide animals into three groups according to the chief nitrogenous excretory product: *ammonotelic, uricotelic,* and *ureotelic* species. The nature of the end product is clearly dependent on the animals' natural habitat. Aquatic invertebrates are all ammonotelic. The development of systems for converting toxic ammonium ions to innocuous materials was an essential evolutionary adaptation to an inadequate water supply. Uricotelism, characteristic of many reptiles and birds, is apparently a mechanism for taking advantage of the relative insolubility of uric acid. Since reptile and bird embryos live for a time within sealed eggs, the deposition of solid uric acid offers an ideal solution to the problem of nitrogen excretion. A mammlian embryo has no such problem, for it is in close contact with the maternal circulation, which removes urea, a highly diffusible waste product. Thus mammals are ureotelic.*

Although urea is their principal nitrogenous excretory product, mammals do retain vestiges of ammonotelism and uricotelism. As shown in Table 10.2, man excretes small quantities of ammonia and uric acid. The renal excretion of ammonia facilitates the elimination of excess hydrogen ion (see p. 361); that of uric acid removes excess amounts of the main product of purine breakdown. In mammals other than man and the higher apes, an enzyme, *uricase*, oxidizes uric acid to *allantoin* (see Fig. 14.13), which appears in the urine in place of uric acid.

In the human disorder known as *gout*, the concentration of uric acid in the plasma is abnormally high. Theoretically gout may be due to either overproduction or underexcretion of uric acid, though recent studies indicate that both mechanisms may operate simultaneously. The agents used in the treatment of gout include diuretics, which promote uric acid excretion, and *xanthine oxidase* inhibitors, which prevent its synthesis (see Fig. 14.13).

---

* If a human subject excreting 15 g of urinary N a day as urea (32 g of urea) were obliged to excrete the same amount as uric acid, he would have to excrete 50 g of uric acid daily. This quantity of uric acid would form insoluble precipitates and clog the urinary tract (see Table 10.2).

REFERENCES AND SUGGESTIONS FOR FURTHER READING

Ashmore, J., and G. Weber, "The Role of Hepatic Glucose 6-Phosphatase in the Regulation of Carbohydrate Metabolism," *Vitamins Hormones*, 17, 92 (1959).

Awapara, J., and J. W. Simpson, "Comparative Physiology: Metabolism," *Ann. Rev. Physiol.*, 29, 87 (1967).

Beck, W. S., "The Metabolic Functions of Vitamin B$_{12}$," *New Engl. J. Med.*, 266, 708, 765, 814 (1962).

Benedict, F. G., and C. G. Benedict, "The Energy Requirements of Intense Mental Effort," *Science*, 71, 567 (1950).

Broom, W. A., and F. W. Wolff, *The Mechanism of Action of Insulin*, Blackwell, Oxford, 1960.

DuBois, E. F., *Basal Metabolism in Health and Disease*, 3rd ed., Lea & Febiger, Philadelphia, 1936.

Frederickson, D. S., R. I. Levy, and R. S. Lees, "Fat Transport in Lipoproteins—An Integrated Approach to Mechanisms and Disorders," *New Eng. J. Med.*, 276, 32, 94, 148, 215, 273 (1967).

Frohman, L. A., "The Endocrine Function of the Pancreas," *Ann. Rev. Physiol.*, 31, 353 (1969).

Gross, F., ed, *Iron Metabolism: An International Symposium*, Springer-Verlag, Berlin, 1964.

Jones, M. E., "Amino Acid Metabolism," *Ann. Rev. Biochem.*, 34, 381 (1965).

Kleiber, M., "Body Size and Metabolic Rate," *Physiol. Rev.*, 27, 511 (1947).

———, and T. A. Rogers, "Energy Metabolism," *Ann. Rev. Physiol.*, 23, 15 (1961).

Konishi, F., "Food Energy Equivalents of Various Activities," *J. Am. Dietet. Assoc.*, 46, 186 (1965).

Lands, W. E. M., "Lipid Metabolism," *Ann. Rev. Biochem.*, 34, 313 (1965).

Levine, R., "Cell Membrane as a Primary Site of Insulin Action," *Federation Proc.*, 24, 1071 (1965).

Mayer, J., "Some Aspects of the Problem of Regulation of Food Intake and Obesity," *New Engl. J. Med.*, 274, 610, 662, 722 (1966).

Moore, C. V., "Iron Metabolism and Nutrition," *Harvey Lectures, Ser. 55 (1959–60)*, 67, 1961.

Newburgh, L. H., "Obesity. I. Energy Metabolism," *Physiol. Rev.*, 24, 18 (1944).

Passmore, R., and J. V. G. A. Durnin, "Human Energy Expenditure," *Physiol. Rev.*, 35, 801 (1955).

Pearson, W. N., and W. J. Darby, "Protein Nutrition," *Ann. Rev. Biochem.*, 30, 325 (1961).

Pett, L. B., "Vitamin Requirements of Human Beings," *Vitamins Hormones*, 13, 214 (1955).

Richardson, H. B., "The Respiratory Quotient," *Physiol. Rev.*, 9, 61 (1929).

Schneider, H. A., "What Has Happened to Nutrition?" *Perspectives Biol. Med.*, 1, 278 (1958).

Tepperman, J., and J. R. Brobeck, eds., "Symposium on Energy Balance," *Am. J. Clin. Nutr.*, 8, 527 (1960).

Wohl, M. G., and R. S. Goodhart, *Modern Nutrition in Health and Disease*, 3rd ed., Lea & Febiger, Philadelphia, 1964.

With the accelerating increase of our knowledge of the biochemistry of the hormones, any summary assessment of their role as biologically active substances is fated to be overmiserly in description and hazardous in extrapolation. The description is overmiserly not merely because of the necessity for condensation, but also because the acquaintance of any single individual with the diverse and heterogeneous fields of investigation is bound to be limited. The extrapolation is hazardous because one is in the position of a mathematician attempting to write the equation of an S-shaped curve which has not yet exhibited its inflection point.

Gregory Pincus, in *The Hormones*, 1956

15 Endocrine System

Introduction

The word "hormone" was first used by Starling in his Croonian lecture to the Royal College of Physicians in 1905. In describing his discovery with Bayliss of the intestinal hormone secretin (see p. 544), Starling generalized the idea of humoral control of body function with these words, "These chemical messengers, or hormones as we may call them, ..." We may therefore consider secretin the first of many hormones to be discovered.

Although claims were made in the late nineteenth century for the existence of what we now call hormones—chemical messengers whose lack can lead to malfunction or death—the nervous system was generally thought to be the sole coordinator of body functions. As a result, almost every proposed endocrine effect was questioned on grounds that observed consequences of endocrine gland removal could be due to incidental nervous system damage. It was difficult for the workers of that era to believe that the chemical products of a minute organ could determine the health or survival of a whole complex organism. With the discovery of secretin, however, the idea of chemical regulation and coordination was finally accepted, and the science of endocrinology was launched. Now the concept of humoral control even encompasses the central nervous system, for there is good evidence that the hypothalamus secretes hormones.

It appears, then, that two systems are responsible for the integration of body functions: the nervous system and the endocrine system. They share certain features. In both systems the process being controlled is located some distance from a control center. To a large extent both systems operate according to classic feedback principles (see Chapter 6): a message dispatched from a control center causes the target organ to increase or decrease its activity, and the intensity of the controlling stimulus fluctuates in accordance with incoming information on the results being achieved.

The two systems also differ. The messages of one are efferent nerve impulses that traverse anatomically defined cables and activate a limited number of cells near their peripheral endings. The messages of the other are chemical substances—hormones—that are deposited in the blood stream by endocrine organs and that spread widely, influencing cells in various parts of the body.

In general, nerve impulses control rapidly changing activities such as striated muscle movements, whereas hormones alter the rate behavior of cellular metabolic systems. Some hormonal effects are complete within seconds (for example, the increase in glycogen breakdown that follows an abrupt increase in epinephrine secretion); however, many others continue for days, months, and years (for example, the regulation of body growth). From an evolutionary viewpoint it appears that the endocrine system could not have reached its advanced stage of development until after an efficient circulatory system had been established.* It appears also that some ancient and inefficient hormonal systems were replaced in evolution by neural systems. Visceral activity, for instance, is now largely governed by the autonomic nervous system.†

ORGANIZATION OF THE ENDOCRINE SYSTEM

Although we have discussed one endocrine gland and its hormone in some detail (i.e., the islets of Langerhans and insulin; see pp. 538, 577), most of the glands we have studied have

---

* Nevertheless, hormones are secreted in lower forms that lack well-developed circulatory systems. An example is the juvenile hormone of the developing silkworm.

† A number of significant relationships exist between the nervous and endocrine systems. For example, the adrenal medulla secretes its hormones only in response to neural stimulation.

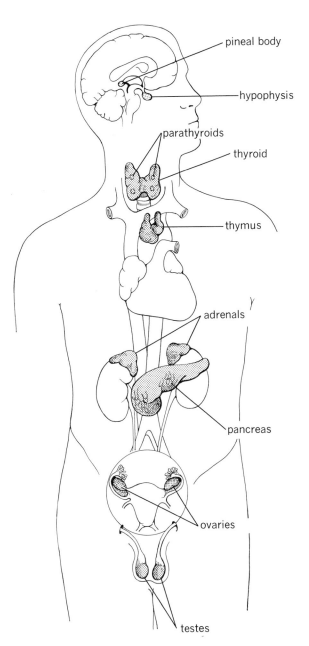

15.1 Major endocrine glands.

been exocrine glands, whose secretions are conveyed to specific neighboring areas through ducts.

The endocrine glands—the glands of internal secretion—have no ducts. Their products are deposited directly into the blood stream. The major endocrine glands are shown in Fig. 15.1.

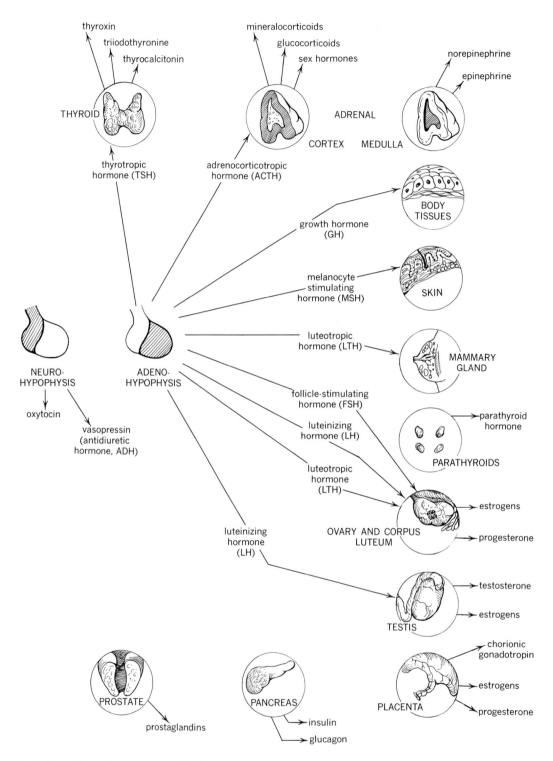

15.2 Outline of the endocrine system, showing functions of the tropic hormones of the adenohypophysis (and placenta).

It is evident that they form no anatomically coordinated system. On the contrary, they are scattered about the body and depend on the circulatory system for the transport of the hormones that they produce. Their microscopic architecture varies widely, as does the chemical nature of their hormones. Some hormones are proteins; others are steroids or amines. Some glands, such as the pancreas, testes, and ovaries, produce both exocrine and endocrine secretions.

The major endocrine glands and their major hormonal secretions are as follows (Fig. 15.2):

1. *Hypophysis.* The *hypophysis cerebri,* or *pituitary,** lying at the base of the brain, has two fundamental subdivisions, the *adenohypophysis* (or anterior pituitary) and the *neurohypophysis* (or posterior pituitary). The adenohypophysis occupies a special place in the endocrine system, for its hormones act not only upon various body tissues but also upon the other endocrine glands. The adenohypophyseal hormones include *adrenocorticotropic hormone, thyrotropic hormone, growth hormone,* the *melanocyte-stimulating hormones,* and two *gonadotropic hormones—follicle-stimulating hormone* and *luteinizing hormone* (it is not certain that a separate *luteotropic hormone* occurs in man). The principal neurohypophyseal hormones are *vasopressin,* or *antidiuretic hormone,* and *oxytocin.*

2. *Adrenals.* The two adrenals, resting atop the kidneys (and hence occasionally called *suprarenals*), contain an inner *medulla* and an outer *cortex.* The two regions constitute essentially different and unrelated endocrine glands. The adrenal medulla secretes *epinephrine* and *norepinephrine.* The adrenal cortex secretes a large number of steroid hormones, or *corticosteroids,* that may be grouped broadly into *mineralocorticoids, glucocorticoids,* and *sex hormones.* Synthesis of the corticosteroids, especially the glucocorticoids, is stimulated by adrenocorticotropic hormone.

3. *Thyroid.* The thyroid, in the anterior part of the neck, secretes *thyroxin, triiodothyronine,* and *thyrocalcitonin.* Synthesis of the first two is stimulated mainly by thyrotropic hormone.

4. *Parathyroids.* The small parathyroids, two to six (usually four) in number, lie behind (and sometimes within) the thyroid—hence their name. They secrete *parathyroid hormone* and apparently are not under hypophyseal control.

5. *Gonads.* The gonads, the *testes* in the male and the *ovaries* in the female, secrete several hormones important in determining the functions of other sex organs. These hormones include *testosterone* and *estrogens* in the male and *progesterone* and *estrogens* in the female.

6. *Pancreas.* The endocrine portions of the pancreas are the *islets of Langerhans* (see Chapters 5, 13). Their hormones are *insulin* and *glucagon.* As noted in Fig. 15.2, the pancreas is not directly controlled by the adenohypophysis.

7. Other endocrine glands include the relatively poorly understood *thymus* and *pineal body,* which secretes *melatonin,* a hormone affecting skin pigmentation; the mucosa of the pyloric antrum, which secretes *gastrin,* the hormone stimulating parietal cell secretion; the epithelial structures of the small intestine that secrete *secretin* and *pancreozymin,* the hormones initiating pancreatic exocrine secretion, and *enterocrinin,* a hormone governing intestinal gland secretion; the juxtaglomerular apparatuses of the kidneys, which apparently secrete *erythropoietin,* a hormone controlling the rate of erythocyte production, and *renin,* a hormone that elevates blood pressure and promotes aldosterone production in the adrenal cortex; and the prostate, which secretes the *prostaglandins,* a class of hormones that powerfully affect the contractility of smooth muscles. We shall omit discussion of the digestive, renal, prostatic, and sex hormones from this chapter.

INVESTIGATION OF THE ENDOCRINE SYSTEM

The study of hormones has passed through one historical phase and is now in a second. The first phase encompassed the discovery of the hormones, the identification of their glandular sources, and the description of their effects. The second phase is concerned with the chemical natures of the hormones and their mechanisms of action.

The investigative procedures of classic endocrinology still provide the basis for hormone

---

* The word "pituitary" is generally disapproved, since it originally referred to nasal mucus or phlegm ("pituita"). The word "hypophysis" indicates the location and development of the gland under the brain and is the favored term.

study. In general, these comprise the following sequence. First, an endocrine gland is removed surgically from an experimental animal (alternatively its action is suppressed by drugs), and the physiological effects of removal (or suppression) are observed. Second, a gland extract is prepared and administered to the animal. Proof that the extract contains the active principle is obtained when it restores normal physiological function. The active principle must also be demonstrated in the blood leaving a normal gland. Similar experiments are performed on human beings with disorders involving hypersecretion or hyposecretion of the gland. The active principle is then isolated and identified chemically. The next steps—production of the active principle in the laboratory and determination of its mode of action—belong to the second phase.

MECHANISMS OF HORMONE ACTION

We learned in Chapter 4 that cells have various mechanisms for regulating their own metabolic processes. These include devices that control the rate of enzyme activity (i.e., feedback inhibition) and the rate of enzyme synthesis (i.e., repression and induction). Hormones are agents that arise in locations remote from the cells and that act upon them by altering either the rates of activity of critical enzymes or enzyme systems or the permeabilities of cell membranes. In this way hormones influence cell metabolism to serve the interests of the body at large.

We are only now beginning to understand how hormones perform this function. Part of the difficulty has been the inability of investigators to produce a hormonal effect in vitro—that is, to demonstrate the stimulation of a specific enzyme or metabolic pathway by adding hormone to a system in a test tube, obviously a prerequisite to any elucidation of the mechanisms of hormone action. Even when an effect can be demonstrated in vitro, it remains to be proved that it is responsible for physiological effects in vivo. It also remains to be established precisely how the hormone affects the system.

Any explanation of the mechanisms of hormone action must answer two questions: (1)

How can trace amounts of hormone produce profound effects in target cells without contributing significant amounts of either energy or matter to the system? (2) How can the specificity of hormone action be accounted for? Hormones are now known to influence cell metabolism in a number of ways: some behave like an enzyme's natural coenzyme; some act directly upon the physical state of an enzyme protein, causing a polymer, for example, to disaggregate into catalytically inactive monomers; some stimulate enzyme synthesis by activating structural genes; and some regulate the permeabilities of certain membranes to specific substances. Examples of these mechanisms will be described later.

Hypophysis

The hypophysis, shown in Figs. 15.1 and 15.3, is an exceedingly small gland, only about 1 cm in length and 500 to 600 mg in weight (see Figs. 12.30, 12.65). It is well protected in a depression of the sphenoid bone, the *sella turcica* (see Fig. 11.2). The two most prominent parts of the hypophysis are the adenohypophysis and the neurohypophysis.

The adenohypophysis is divided into three parts: the *pars tuberalis*, a collar-like sheath around the stalk; the *pars distalis*, the anterior lobe of older terminology; and the *pars intermedia*, a small collection of cells between the adenohypophysis and neurohypophysis. The pars intermedia is much larger in lower animals than in man, forming a sizeable *intermediate lobe*. Consequently, this term is sometimes used for the human pars intermedia.

The neurohypophysis consists of two parts: the *infundibular stem*, a downward projection from the *median eminence* of the *tuber cinereum*, the portion of the brain underlying the third ventricle; and the *infundibular process*, the posterior lobe of older nomenclature. The median eminence and infundibular

15.3 Hypophysis (A) and hypothalamic-hypophyseal portal system (B). (Adapted from *The CIBA Collection of Medical Illustrations* by Frank H. Netter, M.D. Copyright CIBA.)

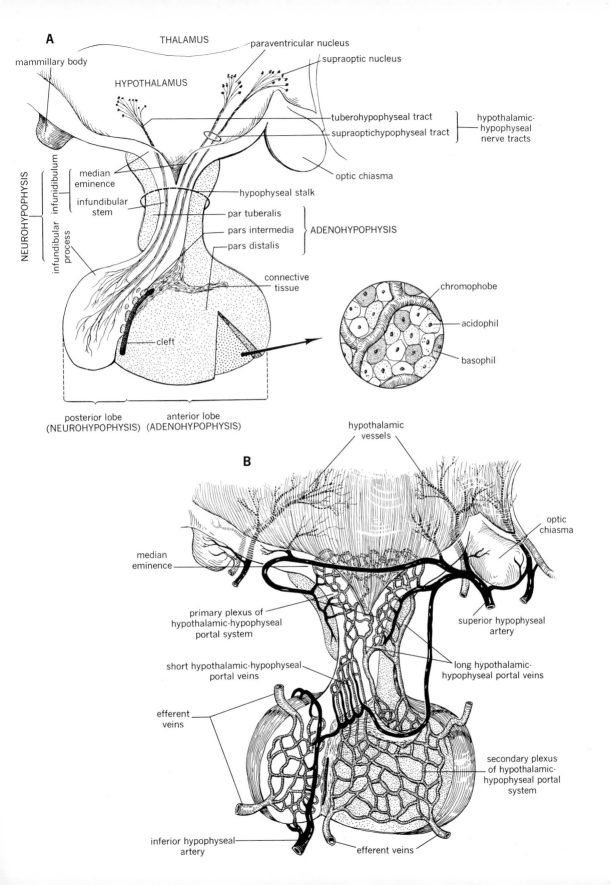

stem together form the *infundibulum* (or *neural stalk*). The infundibulum and its surrounding pars tuberalis together form the *hypophyseal stalk*.

The adenohypophysis and neurohypophysis have different embryonic origins. The adenohypophysis (i.e., glandular portion of the hypophysis) arises as a diverticulum of Rathke's pocket, a backward and upward extension of the embryonic oral epithelium (see Fig. 13.2). Therefore, the secretory cells of the adult adenohypophysis are epithelial in character. The neurohypophysis arises as a downgrowth of the part of the diencephalon that becomes the floor of the third ventricle. Therefore, it contains cells resembling neuroglia as well as numerous nerve fibers; the cells are called *pituicytes*.

The base of the hypophyseal stalk, which attaches the hypophysis to the brain, is located behind the optic chiasma. As shown in Figs. 12.30 and 15.3, short nerve tracts run in the infundibulum between the rest of the neurohypophysis and the hypothalamic nuclei (particularly the *paraventricular* and *supraoptic* nuclei).

It is now thought that the hormones of the neurohypophysis are actually produced by secretory cells in the hypothalamus and transmitted through axons to the neurohypophysis, in which they are merely stored until signals from the hypothalamus trigger their release. This view is based on the morphological difference between pituicytes and other glandular cells and the presence of neurohypophyseal hormones in the hypothalamic nuclei. Moreover, cutting the infundibulum does not necessarily interfere with neurohypophyseal secretion.

NEUROHYPOPHYSIS

It has long been known that crude extracts prepared from the neurohypophyses of slaughterhouse animals and injected into man have these effects: (1) reduction of water loss through the kidneys (an *antidiuretic* effect); (2) elevation of blood pressure (a *pressor* effect); (3) contraction of the uterine muscle (an *oxytocic* effect); and (4) stimulation of the smooth muscles in the breasts that eject milk (a *lactagogic* effect). These four phenomena are caused by two distinct neurohypophyseal hormones: vasopressin (antidiuretic hormone, ADH) and oxytocin.

Vasopressin. We previously discussed vasopressin as ADH in connection with the excretion of water (see p. 354). ADH promotes water retention by increasing the permeability of the distal convoluted and collecting tubules to

water.* As a result, body fluids are diluted. The osmoreceptors that control ADH secretion are apparently situated in the supraoptic hypothalamic nuclei.

The name "vasopressin" suggests a pressor effect. Vasopressin raises the blood pressure of an anesthetized dog or cat but not the blood pressure of man. In fact, human blood pressure does not change significantly even when the neurohypophysis is removed.

Oxytocin. In the early 1940's investigators isolated what appeared to be a pure protein of high molecular weight having antidiuretic, pressor, and oxytocic properties. When they subjected it to mild degradative treatment, however, they isolated two separate peptides, one having antidiuretic and pressor activity and the other oxytocic activity. They called the latter "oxytocin." When injected it aids in initiating or accelerating labor. Though removal of the neurohypophysis prolongs childbirth, the precise physiological role of oxytocin is unclear.

Just as vasopressin has a secondary pressor effect, oxytocin has a secondary lactagogic effect: it promotes the ejection of milk from the breasts by causing contraction of the *myoepithelial cells* (Chapter 16). To a degree, the actions of vasopressin and oxytocin overlap. We shall now learn why.

Structure-activity relationships. The observation that antidiuretic-pressor and oxytocic-lactagogic effects are produced by two separate peptides led to an epochal series of studies by Vincent du Vigneaud and his associates. By diligently purifying natural extracts—relying heavily on the technique of countercurrent distribution (see Chapter 2)—they isolated and analyzed vasopressin and oxytocin. Each contains only nine amino acids, and the amino acid sequences differ in only two residues (Table

---

* In the absence of ADH, the kidneys excrete extremely dilute urine while retaining electrolytes. In the presence of ADH, they excrete concentrated urine and conserve water. As noted on p. 355, removal of the neurohypophysis without injury to the hypothalamus causes polyuria. Diabetes insipidus is a similar pathological condition. It is a manifestation of neurohypophyseal deficiency and can be alleviated by the continuous administration of ADH. Injury to the hypothalamus or to the nerve tracts in the infundibulum also causes polyuria.

TABLE 15.1   STRUCTURES OF OXYTOCIN, VASOPRESSIN, AND SOME SYNTHETIC ANALOGUES*

Name	Structure†	Relative oxytocic activity	Relative pressor activity
Oxytocin§	$\overset{1}{\text{Cys}}$-$\overset{2}{\text{Tyr}}$-$\overset{3}{\text{Ileu}}$-$\overset{4}{\text{Glu}}$-$\overset{5}{\text{Asp}}$-$\overset{6}{\text{Cys}}$-$\overset{7}{\text{Pro}}$-$\overset{8}{\text{Leu}}$-$\overset{9}{\text{Gly}}$ (S—S; NH₂ NH₂; NH₂)	500	7
	Cys-**Phe**-Ileu-Glu-Asp-Cys-Pro-Leu-Gly (S—S; NH₂ NH₂; NH₂)	31	
Oxypressin	Cys-Tyr-**Phe**-Glu-Asp-Cys-Pro-Leu-Gly (S—S; NH₂ NH₂; NH₂)	20	3
Vasotocin	Cys-Tyr-Ileu-Glu-Asp-Cys-Pro-**Arg**-Gly (S—S; NH₂ NH₂; NH₂)	75	125
Arginine-vasopressin§	Cys-Tyr-**Phe**-Glu-Asp-Cys-Pro-**Arg**-Gly (S—S; NH₂ NH₂; NH₂)	30	600
Lysine-vasopressin§	Cys-Tyr-**Phe**-Glu-Asp-Cys-Pro-**Lys**-Gly (S—S; NH₂ NH₂; NH₂)	5	300
Histidine-vasopressin	Cys-Tyr-**Phe**-Glu-Asp-Cys-Pro-**His**-Gly (S—S; NH₂ NH₂; NH₂)	1	1

* The key to the amino acid abbreviations appears in Table 2.5.

† Residues in bold-face type differ from those in oxytocin.

§ Naturally occurring in mammals. *Arginine-vasopressin* occurs in man, cow, horse, sheep, and many other mammals. *Lysine-vasopressin* is known to occur only in the pig. Vasotocin occurs in amphibians.

15.1). Using special procedures, du Vigneaud was able to synthesize both peptides.

The difference between the structures of vasopressin and oxytocin is that phenylalanine and arginine in vasopressin are replaced by isoleucine and leucine in oxytocin. Oxytocin has a small pressor effect, and vasopressin has a small oxytocic effect. Having developed methods for synthesizing the two hormones, du Vigneaud had little difficulty in preparing "abnormal" peptides, among them hybrids of vasopressin and oxytocin, with various amino acid replacements. In this way he was able to assess the importance of each amino acid residue in each hormone. Substitution of a single residue in the oxytocin molecule—especially in

TABLE 15.2 ADENOHYPOPHYSEAL HORMONES

Name	Abbreviation	Synonyms	Chemical nature	Target organ
Adrenocorticotropic hormone	ACTH	Adrenocorticotropin, corticotropin	Peptide, 39 amino acids; mol wt, 4567	Adrenal cortex
Thyrotropic hormone	TSH	Thyrotropin, thyroid-stimulating hormone	Glycoprotein; mol wt, 26,000	Thyroid
Growth hormone	GH	Somatotropin, somatotropic hormone (STH)	Peptide, 188 amino acids; mol wt, 21,500	Body tissues
Gonadotropic hormones Follicle-stimulating hormone	FSH		Glycoprotein; mol wt, 41,000	Ovaries
Luteinizing hormone	LH	Interstitial cell–stimulating hormone (ICSH)	Glycoprotein; mol wt, 30,000	Ovaries, interstitial cells of testes
Luteotropic hormone*	LTH	Luteotropin, lactogenic hormone, prolactin	Protein	Ovaries, mammary glands
Intermedins Melanocyte–stimulating hormones	MSH		Peptide; α-MSH, 13; β-MSH, 18 amino acids	Skin (melanocytes)

* Luteotropic hormone occurs in a number of species—including the mouse, rat, and sheep. Whether it occurs in man and other primates has not yet been established (Chapter 16).

the ring (which bears an interesting resemblance to part of the insulin molecule; see Fig. 2.8A)—greatly diminishes oxytocic activity. A strongly basic amino acid in the eighth position seems essential for pressor activity.

The hormones are synthesized in the hypothalamus and transported along axons to the neurohypophysis (Fig. 15.3A). Vasopressin appears to act by altering the permeability of cell membranes to water. It is of interest that insulin, with a somewhat similar peptide structure, appears to act by altering the permeability of cell membranes to glucose.

ADENOHYPOPHYSIS

The anterior lobe of the hypophysis contains two main groups of cells (see Fig. 15.3A): *chromophobes,* small poorly staining cells with nongranular cytoplasm; and *chromophils,* larger readily staining cells with granular cytoplasm. Chromophils can be subdivided into *acidophils* and *basophils,* and, owing to the development of increasingly sophisticated histochemical techniques, these two categories can also be subdivided. Basophils, for example, may be classified as β^1, β^2, Δ^1, and Δ^2 cells.

About a quarter of the adenohypophyseal cells are chromophobes. Since they have an immature appearance, they may simply be resting cells or primordial precursors of chromophils. Acidophils constitute about half of all the cells, and basophils about a quarter. Acidophils and basophils are the sources of the adenohypophyseal hormones. The association of specific cells with specific hormones has not been universally accepted. However, some conclusions are possible, which we shall set forth in the following discussion.

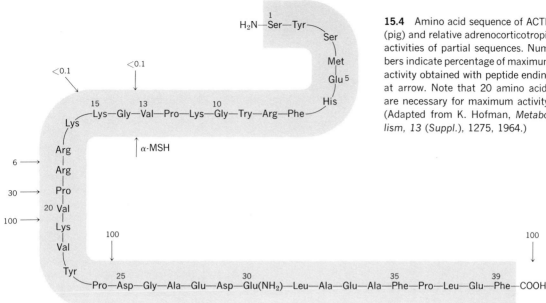

15.4 Amino acid sequence of ACTH (pig) and relative adrenocorticotropic activities of partial sequences. Numbers indicate percentage of maximum activity obtained with peptide ending at arrow. Note that 20 amino acids are necessary for maximum activity. (Adapted from K. Hofman, *Metabolism, 13* (Suppl.), 1275, 1964.)

Hypothalamic-hypophyseal portal system. We have seen that special nerve tracts connect the hypothalamus with the neurohypophysis (see Fig. 15.3A). A vascular link also connects the hypothalamus with the neurohypophysis and the adenohypophysis. As shown in Fig. 15.3B, small *superior hypophyseal arteries* supply the median eminence of the hypothalamus and the optic chiasma. Capillaries extend from these vessels into the substance of the median eminence, unite to form the *primary plexus* of the *hypothalamic-hypophyseal portal system,* and return to the surface, where they coalesce into *long* and *short hypothalamic-hypophyseal portal veins.* These spread downward around the hypophyseal stalk and form a rich *secondary plexus* of venous sinuses that permeates the pars distalis. Thus blood entering the secondary plexus has first passed through the primary plexus. The adenohypophysis receives no arterial blood.

The hypothalamic-hypophyseal portal system has an important role in the hypothalamic control of the adenohypophysis. Nerve endings in the median eminence secrete several neurohumoral *releasing factors*—specific low molecular weight peptides that are themselves hormones—that are transported via the system to the adenohypophysis, where they stimulate the

secretion of specific adenohypophyseal hormones. The median eminence, therefore, must be viewed as a neuroendocrine portion of the brain, the function of which is a control of the adenohypophysis.

Structures of adenohypophyseal hormones. As noted earlier, the adenohypophysis produces at least seven (and possibly more) separate hormones. Table 15.2 lists their names and properties. The table reveals that (1) all the adenohypophyseal hormones are proteins or peptides and (2) all but one act by stimulating another gland to act. The exception is growth hormone, which acts by stimulating the metabolism of the body tissues. We shall briefly survey the adenohypophyseal hormones in these two contexts.

When adrenocorticotropic hormone (ACTH) was first discovered, it was thought to be a protein of high molecular weight. It was then found that at least eight equally active low–molecular weight peptides could be isolated by countercurrent distribution and other methods from pig adenohypophyses, the major component being the one termed β-ACTH. On purification, β-ACTH was revealed to be a single-chain peptide with a molecular weight of 4567 containing 39 amino acids. The amino acid sequence was determined (Fig. 15.4), and in 1963 it was validated by R. Schwyzer and P. Sieber, who achieved the complete organic synthesis of β-ACTH. ACTH thus became the first adenohypophyseal hormone to be isolated, purified, "sequenced," and synthesized artificially.*

Studies of the peptides with ACTH activity in various species have shown that each contains 39 amino acids. Those of the pig, sheep, and cow are identical in the first 24, which constitute the portion of the peptide structure indispensable for hormonal action. Species specificity resides in the sequence from amino acid 25 to amino acid 32—a portion of the structure not essential for hormonal action.

	25	26	27	28	29	30	31	32
Pig	-Asp-	Gly-	Ala-	Glu-	Asp-	Glu-	Leu-	Ala-

NH₂

Sheep	-Ala-Gly-Glu-Asp-Asp-Glu-Ala-Ser-
Cow	-Asp-Gly-Glu-Ala-Glu-Asp-Ser-Ala-
Man	-Asp-Ala-Gly-Glu-Asp-Glu-Ser-Ala-

NH₂

This "variable" sequence is rich in acidic amino acids (see Table 2.5). ACTH is believed to be synthesized in the β^1 basophils and in certain large chromophobes of the adenohypophysis.

Thyrotropic hormone (TSH) is a glycoprotein of molecular weight 26,000 that is probably synthesized in β^2 basophils. It has not yet been completely purified, but preparations are available that are thousands of times as potent as crude adenohypophyseal extracts. Growth hormone (GH) is a protein of molecular weight 21,500. The structure of completely purified human GH was determined in 1966 by C. H. Li.† It contains 188 amino acids, arranged in a single chain with two loops, a large one caused by a disulfide bond between amino acids 68 and 162 and a small one caused by a disulfide bond between amino acids 179 and 186. The gonadotropic hormones are proteins that have not yet been completely purified. We shall deal with their functions in Chapter 16.

We shall consider the structures of the melanocyte-stimulating hormones later.

Tropic actions of adenohypophyseal hormones. We have noted that all the adenohypophyseal hormones except GH (and possibly the melanocyte-stimulating hormones) influence other endocrine glands and that GH influences the body tissues at large. The property is called *tropism.* A tropic‡ hormone acts on a target organ to maintain and stimulate its function. The ultimate physiological effects of tropic hormones upon the body, therefore, are those of the target organs themselves.

Removal of the hypophysis results in diminution of the activities of the target organs and a variety of degenerative changes but not necessarily death. Most of the effects that follow total hypophysectomy follow removal of the adenohypophysis alone. They include retardation of growth and sexual development, depression of the metabolic rate with consequent fat accumulation, and abnormalities in the thyroid and the adrenal cortex. Many diseases may cause adenohypophyseal hypofunction.

Treatment of normal young animals with adenohypophyseal extracts leads to acceleration of growth and early sexual maturity, enlargement of the external genitalia, and enlargement of the thyroid and adrenal cortex with evidence of hypersecretion. These effects are also symptoms of pathological adenohypophyseal hyperfunction, most frequently due to tumors made up of hypophyseal cells. The association of oversecretion of one adenohypophyseal hormone with abnormal proliferation of a single cell type (for example, oversecretion of GH with acidophil tumors) is important (if indirect) evidence that specific cell types are the sources of specific hormones. Adenohypophyseal hyperfunction may also result from disturbances in the regulatory hypothalamic centers.

Since the sole function of all but one of the adenohypophyseal hormones is the stimulation of other glands or skin, we shall defer further discussion of their physiology until we have reviewed the physiology of the target organs. We shall, however, deal with GH here.

---

*The two hormones of the pars intermedia have now been through these steps, and, but for the synthesis, so has growth hormone.

† Over 5000 human hypophyses, gathered at autopsies, were needed to produce the 5 g of GH used in Li's studies.

‡ *Trophic* is frequently used instead of *tropic.* Tropic derives from *tropos,* meaning "changing." Trophic derives from *trophe,* meaning "to nourish." Since there is no evidence that adenohypophyseal hormones nourish their target organs, and since the hormones do change activity levels when the target organs are glands, tropic is preferred.

Growth hormone. Human GH is secreted by adenohypophyseal acidophils. GH constitutes no less than 5 to 10% of the dry weight of the hypophysis, whereas hormones such as ACTH constitute only 0.2%. Secretion is regulated by the *GH-releasing factor* of the hypothalamus.

As its names indicate (see Table 15.2), GH is a tropic hormone to the whole body. Growth, fundamentally definable as a net increase in protoplasm, is a complex of processes that are dependent upon cellular and body metabolism and genetic and nutritional factors. Hormones can only regulate or modify the growth rate; they are not responsible for the growth processes.

Three basic manifestations of growth are (1) positive nitrogen balance, denoting a high rate of protein synthesis and accumulation; (2) decreased carbohydrate utilization; and (3) increased lipid utilization. Although GH apparently enhances all of these, its primary effect is probably on protein metabolism.* GH injected into experimental animals causes retention of nitrogen—thereby favoring the synthesis of protein and protoplasm (Fig. 15.5A)—conservation of carbohydrate, and consumption of lipid. At the same time, in contrast to insulin, it causes a gain in lean body mass. When a normal animal is starved, GH promotes conservation of protein and consumption of lipid. When a hypophysectomized animal is starved, its protein reserves, unprotected by GH, are rapidly depleted. In vitro GH selectively accelerates the transport of specific amino acids into cells.

GH is secreted continuously throughout life. However, growth in height ceases when the epiphyses of the long bones unite. In infants and children, adenohypophyseal hypofunction results in a form of dwarfism in which body proportions, facial features, and intelligence remain normal. Purified preparations of GH have been successfully used in the treatment of this pituitary dwarfism (Fig. 15.5B)—and of some cases of short stature without hypopituitarism—but only human GH is effective in man. Since the supply of human hypophyses is necessarily limited, the amount of human GH available for therapy is extremely small.

We noted in Chapter 14 that pancreatectomy leads

---

* GH also causes retention of potassium, phosphorus, and sodium, in a pattern reflecting the synthesis of protoplasm (see p. 564).

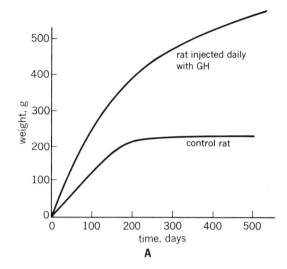

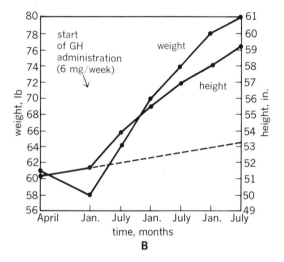

15.5 Effects of growth hormone on body weight of a rat (A) and on body weight and height of a pituitary dwarf (B). (Adapted from E. B. Astwood, (ed.), *Clinical Endocrinology*, Vol. I, Grune & Stratton, New York, 1960.)

to the condition known as diabetes mellitus because the source of insulin has been removed. In a classic experiment Bernardo Houssay found that the diabetes of a pancreatectomized animal is alleviated by hypophysectomy. His discovery prompted a prolonged search for a "diabetogenic" hypophyseal hormone, one that would cause diabetes if it were not for the countereffect of insulin. Although a number of hormones tend to raise the blood glucose level, diabetogenic action is largely attributable to GH. An animal given frequent,

15.6 Progression of acromegaly. A, normal, age 9 years; B, age 16 years with possible early coarsening of features; C, age 33 years, well-established acromegaly; D, age 52 years, end stage acromegaly with gross disfigurement. (Clinical Pathological Conference, *Am. J. Med. 20*:133, 1956.)

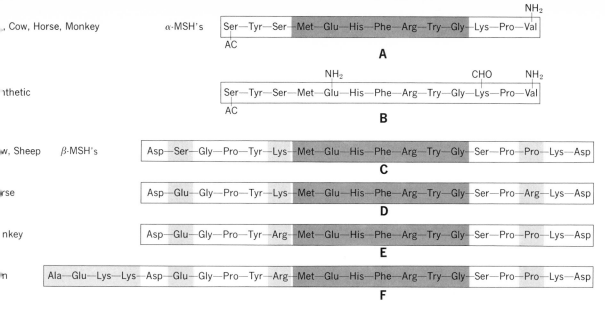

15.7 Amino acid sequences of the melanocyte-stimulating hormones: A, α-MSH; B, synthetic α-MSH; C, β-MSH of sheep, cow; D, β-MSH of horse; E, β-MSH of pig, sheep, monkey; F, β-MSH of man. The dark-shaded partial amino acid sequence is the same in α-MSH and β-MSH of several species. Light-shaded amino acid residues vary in the different species of β-MSH.

massive injections of GH eventually develops diabetes mellitus. If the injections are stopped, the diabetes may disappear, but, if they are continued long enough, it becomes permanent, the β cells of the islets of Langerhans eventually ceasing to function. In an animal whose pancreas has been partially or completely removed, the amounts of GH needed to produce or worsen diabetes are much smaller than in a normal animal. It is evident that normal carbohydrate metabolism depends upon a balance between GH activity and insulin activity. Thus naturally occurring diabetes may result not only from an absolute lack of insulin; it may result also from a relative lack of insulin—in which circumstance it is termed *pituitary diabetes.*

When the GH-secreting acidophils of the adenohypophysis are overactive or form a tumor, so that the body receives an excess of the hormone, two disorders are possible. In *gigantism* the bones of the arms and legs grow so fast that a height of 7 or 8 ft may be reached. For gigantism to develop, hyperfunction of the adenohypophysis must occur before growth would normally cease in adolescence. Hyperfunction in an adult, after growth is complete, gives rise to *acromegaly*—the result of exaggeration of all the GH effects. Excessive tissue formation in bone, cartilage, skin, and the various organs, especially in the brow ridges, nose, lips, lower jaw, hands, and feet (Fig. 15.6), deprives the face of intelligent expression and alters body proportions to cause a freakish appearance. Pituitary diabetes frequently accompanies gigantism and acromegaly.

Intermedins. Over 50 years ago P. E. Smith and B. M. Allen discovered independently that hypophysectomy brought about distinct lightening of the skin in frogs and other animals, whereas injected hypophyseal extracts darkened it. They also found that injected pineal extracts lightened the skin. Further study revealed that the changes in pigmentation were caused by a hormonally induced clumping or dispersion of dark *melanin* granules in special skin cells called *melanocytes.* We shall have more to say about skin pigmentation and its hormonal control in Chapter 18. Here we shall only note the hypophyseal and pineal hormones involved.

Investigators have isolated from hypophyseal extracts at least two varieties of peptide hormones that stimulate melanocytes in such a manner as to darken the skin. They named them *melanocyte-stimulating hormones* (MSH's), with the two varieties designated α-MSH and β-MSH. Since both are secreted by the pars intermedia, they are termed *intermedins.* Biological and chemical examination of the intermedins yielded some surprising facts.

To begin with, there is only one α-MSH, but there are many β-MSH's (Fig. 15.7). Apparently every species produces both varieties. The α hormone is a straight-chain peptide composed of

13 amino acids, in a sequence that is the same for pig, cow, horse, monkey, and probably man and essentially identical to that of the first 13 amino acids in ACTH (see Fig. 15.4). ACTH also darkens skin, but with only ⅟₃₀ the potency of α-MSH. On the other hand, α-MSH has no adrenocorticotropic activity. Two subtle differences do exist between the 13 amino acids of α-MSH and the first 13 amino acids of ACTH. The *N*-terminal amino acid, serine, is acetylated in α-MSH but not in ACTH,* and the *C*-terminal amino acid, valine, is an amide in α-MSH but is joined to the next amino acid, glycine, in ACTH (see Figs. 15.4, 15.7). When the *N*-terminal serine of ACTH is artificially acetylated, its skin-darkening capacity is increased sixfold. Several workers have synthesized α-MSH. Exactly how it acts upon the melanocytes is not yet known, however.

As Fig. 15.7 indicates, β-MSH's of different species, like their ACTH's, vary at several points. A β-MSH ordinarily has 18 amino acids, although human β-MSH has 22. Nevertheless, the β-MSH's shown and α-MSH share a common sequence of 7 amino acids. Presumably it (or a portion of it) is the active center of the molecule. Furthermore, all β-MSH's have free *N*-terminal and *C*-terminal amino acids. In general, β-MSH's are less active biologically than α-MSH: pig β-MSH's have about one-third the potency of pig α-MSH. Human β-MSH's and α-MSH have not been compared.

Adrenals

As noted on p. 589, each adrenal is actually two glands in one. That structures analogous to the adrenal cortex and the adrenal medulla are entirely separate in sharks and other fishes is evidence of their different origins and functions. Since the words "cortex" and "medulla" denote outer and inner portions of a layered organ, they are therefore inappropriate in such species.

Moreover, the two glands are not always "ad-renal" (i.e., next to the kidney) in these species. Instead, a single *interrenal* gland lying between the kidneys may correspond to the cortex, and a double row of small *chromaffin bodies* near the chains of sympathetic ganglia may correspond to the medulla.

In the human embryo the adrenal cortex derives from the cephalic end of the mesodermal region on each side that gives rise to the gonad and kidney. Small buds form an *adrenal ridge* that projects into the coelom between the mesonephros and the root of the mesentery (see Figs. 10.1, 10.2). As the buds develop into tissues with their own blood supply, they are infiltrated by ectodermal cells that originated in the prevertebral sympathetic ganglia but later detached themselves. Because these invading strands of cells are stained dark brown by chromic acid, they are called *chromaffin cells.* Eventually they compose the entire central or medullary portion of the adrenal. Histochemical analysis discloses two types of chromaffin cells. Presumably each type produces a different hormone. Preganglionic fibers reach the medullary tissue by way of the splanchnic nerves. There are no true postganglionic fibers; as we learned in Chapter 12, the medullary cells themselves fill this role.

As shown in Fig. 15.1, the adrenals lie on top of the kidneys in the adult. In a cross section of a fresh gland, the yellow cortex is readily distinguishable from the brownish or grayish medulla (Fig. 15.8). The whole gland is enclosed by a capsule of fibrous connective tissue. The cortex is differentiated into three zones: (1) the outer *zona glomerulosa,* (2) the middle *zona fasciculata,* and (3) the inner *zona reticularis.* These are among the body's most metabolically active tissues. Since the cortex and medulla secrete different hormones, we shall discuss them separately.

ADRENAL CORTEX

The effects of the several hormones of the adrenal cortex are so diverse that it is frequently difficult to tell whether they are primary or secondary. For example, if adrenocortical secretion were to fail, the functions of kidneys, liver, muscle, and circulatory system would be profoundly altered. Important changes would also occur in the metabolism of carbohydrates, proteins, lipids, amino acids, electrolytes, and water. As we shall see, the adrenocortical hormones are essential in the body's complex pattern of resistance to stress, a pattern that appears to involve every part of the body.

---

* This is the first instance of a hormone with an acetylated *N*-terminal amino acid, although this type of end group has been observed in certain virus proteins. An α-carboxyl amide end group has been demonstrated in oxytocin and vasopressin (see Table 15.1).

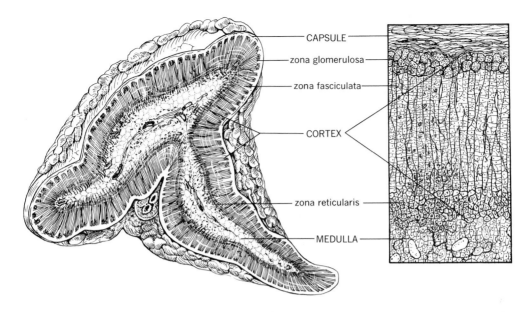

15.8 Gross and microscopic structure of an adrenal.

We can make three generalizations concerning the adrenocortical hormones: (1) many compounds—over 50 are now known*—can be isolated from the adrenal cortex; (2) all are steroid derivatives (see p. 28) and can therefore be designated corticosteroids (or corticoids); and (3) they can be grouped into three functional categories—mineralocorticoids, glucocorticoids, and sex hormones. Mineralocorticoids act upon the electrolytes of the extracellular fluids, particularly sodium, potassium and chloride. Glucocorticoids influence the metabolism of carbohydrates, proteins, and lipids. The relatively few sex hormones include androgens, which mimic the testicular hormone testosterone, progesterones, and estrogens, which mimic the ovarian hormones. We shall study adrenal sex hormones together with gonadal sex hormones in Chapter 16.

Among the numerous mineralocorticoids and glucocorticoids, three are of particular importance: *aldosterone, cortisol,* and *corticosterone* (Table 15.3). Aldosterone (see p. 355) accounts for at least 95% of all mineralocorticoid activity, and cortisol, for about 90% of all glucocorticoid activity. Corticosterone accounts for a small percentage of both activities.

Mineralocorticoid effects. We shall review the major mineralocorticoid effects briefly, since we considered them in Chapter 10. Deficiency of aldosterone and other mineralocorticoids causes excessive renal excretion of sodium—and with it an osmotically equivalent amount of water†—and excessive renal retention of potassium. On administration of the hormones, the kidneys retain sodium and water and eliminate potassium and hydrogen, presumably because the hormones stimulate sodium-potassium and sodium-hydrogen exchanges in the distal renal tubules.

These primary effects lead to many secondary effects. For example, large doses of aldosterone result in increased extracellular fluid volume, edema, increased cardiac output, and elevated blood pressure. The fact that the last two effects are observed even after the kidneys have been removed suggests that aldosterone acts on blood vessels and perhaps other body cells as well as those of the renal tubules.

---

* Of these, only a few are secreted into the blood stream. The rest are biosynthetic intermediates.

† This water loss is not a direct effect of aldosterone deficiency. A good part of it is due to ADH suppression by dilution and to osmotic diuresis (see p. 354).

TABLE 15.3 THREE MAJOR CORTICOSTEROIDS

Name	Structure	Properties
Cortisol (hydrocortisone, Kendall's compound F)	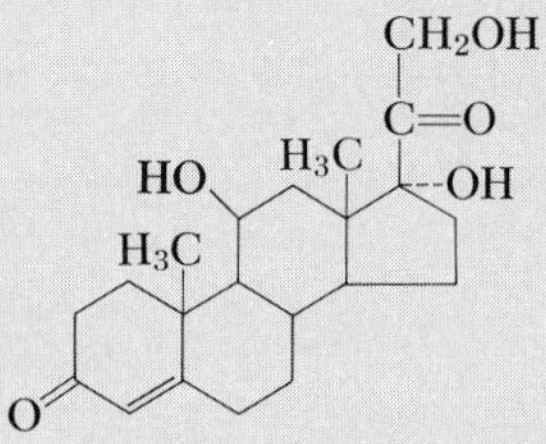 CH₂OH, C=O, HO, H₃C, –OH, H₃C, O	Main human glucocorticoid, produced in zona fasciculata; daily output, 10 to 25 mg; mineralocorticoid activity, <5% that of 11-desoxycorticosterone*; normal plasma level, 10 to 20 μg per 100 ml; production stimulated by ACTH
Corticosterone (Reichstein's compound H, Kendall's compound B)	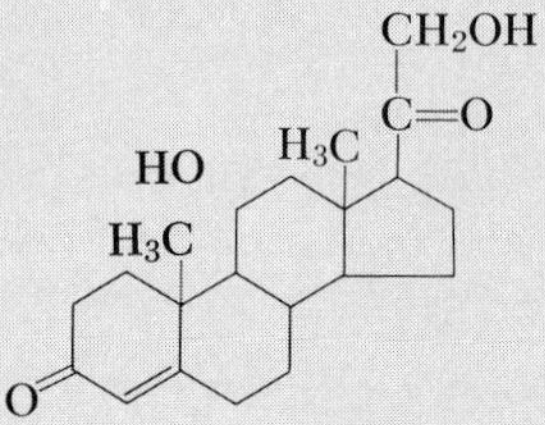 CH₂OH, C=O, HO, H₃C, H₃C, O	Produced in zona fasciculata; daily output, 2 to 5 mg; ratio of cortisol to corticosterone, secretion 7:1; glucocorticoid activity, 30 to 50% that of cortisol; mineralocorticoid activity, 10% that of 11-desoxycorticosterone; normal plasma level, 0.4 to 2.0 μg per 100 ml; production stimulated by ACTH
Aldosterone	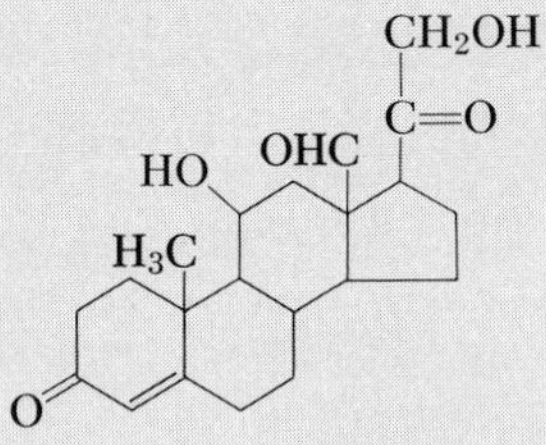 CH₂OH, C=O, HO, OHC, H₃C, O	Main human mineralocorticoid, produced in zona glomerulosa; daily output, 50 to 200 μg; normal plasma level, 0.005 to 0.015 μg per 100 ml; glucocorticoid activity, insignificant; mineralocorticoid activity, 30 to 100 times that of 11-desoxycorticosterone; production stimulated by renin-angiotensin system (see p. 357)

*11-Desoxycorticosterone (DOC) was the first corticosteroid available in large quantities. As its name implies, this compound lacks an oxygen on carbon 11. Since it is a mineralocorticoid, not a glucocorticoid, extensive efforts were made to transform it to a glucocorticoid of the 11-oxy (i.e., 11-keto or 11-hydroxy) type. Sarrett was finally successful, but his method was difficult and commercially unfeasible. The 11-oxy steroids became plentiful only when naturally occurring plant sources (such as the Mexican yam) were discovered and utilized.

Glucocorticoid effects. The primary effects of glucocorticoids have not yet been distinguished from their secondary effects. C. N. H. Long proposed that the primary effect is an acceleration of protein catabolism, which provides new sources of carbohydrate (i.e., gluconeogenesis), but it now appears that the effects upon protein and carbohydrate metabolism are not necessarily associated.

Glucocorticoid effects are as follows: (1) increased protein catabolism, deamination of amino acids, and gluconeogenesis with negative nitrogen balance; (2) increased deposition of liver glycogen; (3) increased blood glucose; (4) increased blood amino acids; (5) increased total body fat; (6) increased erythrocyte production in the bone marrow; (7) removal or lysis of blood and tissue lymphocytes and eosinophils; (8) decreased antibody formation; and (9) an antiinflammatory effect, with decreased migration of leukocytes from bloodstream to inflammatory sites.

Glucocorticoids have some interesting relations with other endocrine glands. They depress thyroid function by suppressing TSH secretion; they suppress ACTH secretion (it is noteworthy that TSH and ACTH are both apparently synthesized in β basophils of the adenohypophysis); and they suppress MSH secretion, thus indirectly affecting skin pigmentation. The mineralocorticoids do not suppress the production of these hypophyseal hormones.

Glucocorticoid activity increases greatly in response to nonspecific stress—trauma, extreme temperatures, surgical procedures, and so forth. Presumably this response is part of the body's pattern of resistance to stress, since an adrenalectomized individual is incapable of withstanding stress. Precisely how glucocorticoids operate in stress is not clear. They may furnish additional substrates for energy metabolism.

Functional significance of the cortical zones. The zones of the adrenal cortex exhibit several interesting biochemical features. Some are characteristic of all three zones—for example, the large lipid droplets and the ascorbic acid and cholesterol contents of the resting cells all disappear during active secretion stimulated by ACTH.

Although cortical hormones are synthesized by enzymatic processes, only the zona fasciculata is rich in glucose-6-phosphate dehydrogenase. In view of the probable role of NADP in the introduction of hydroxyl groups (hydroxylation) into specific positions in the steroid nucleus, it is significant that its distribution, like that of glucose-6-phosphate dehydrogenase (which converts NADP to $NADPH_2$; see Fig. 4.13), corresponds to the distribution of steroid hydroxylation reactions in the adrenal cortex. The zona fasciculata and the zona reticularis produce glucocorticoids under the direct control of ACTH. The zona glomerulosa produces mineralocorticoids in a process not controlled by hypophyseal hormones.

Structure and nomenclature of the corticosteroids. We learned in Chapter 2 that the steroids compose a large and biologically diverse group of lipid-soluble organic compounds. Among the steroids of physiological importance are the adrenocortical hormones, the bile salts, cholesterol, the vitamin D compounds, the gonadal hormones, and the steroid portions of the digitalis compounds. A hormonal steroid is, in general, an 18- to 21-carbon compound containing three six-membered rings and one five-membered ring. The numbering of the carbons is as follows:

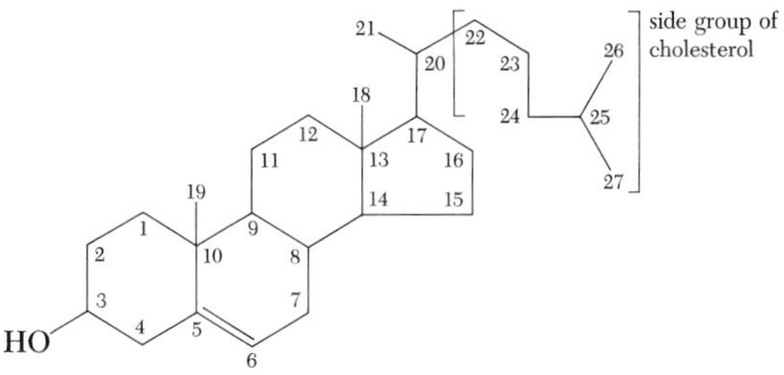

The side-chain containing carbons 20 and 21 continues to carbon 27 in cholesterol (see Fig. 14.11).

Since every carbon atom has four combining sites, many spatial arrangements of attached substituent groups, and therefore many opportunities for stereoisomerism, exist. Only one of a pair of isomers may possess biological activity. With the steroid molecule considered to lie in the plane of the printed page, carbons 18 and 19 project upward. All groups projecting in the same direction are termed *cis* groups. All groups projecting in the opposite direction are termed *trans* groups.* In naming a compound, it is customary to specify how an attached substituent group is oriented; the prefixes α (for *trans*) and β (for *cis*) serve this purpose. In a structural formula a solid line represents a *cis* group, and a dashed line a *trans* group.

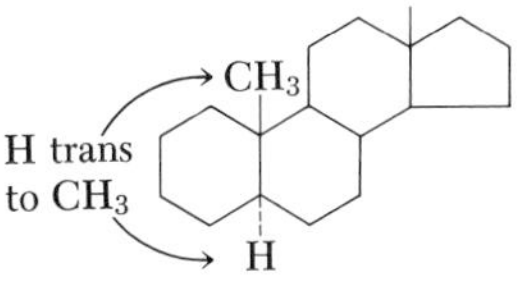

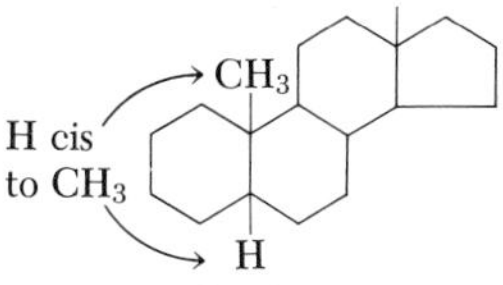

The number of the carbon to which a substituent group is attached is followed by the name of the group, with the steric symbol when necessary (e.g., 6-methyl- or 6α-methyl-).† A

---

* We encountered the *cis*- and *trans*- prefixes in discussions of the structures of vitamin A, retinal, and its isomers (see Fig. 12.63).

† In a corticosteroid the side-chain (carbons 20 and 21) at carbon 17 is oriented in the *cis* (β) position. A corticosteroid may also have a hydroxyl group at carbon 17. As the only position available for this substituent group is the *trans* (α) position, the α is frequently omitted from the name.

double bond in a ring is indicated with a Δ (delta) prefix and a superscript numeral for the lowest-numbered carbon to which it is connected (e.g., Δ^4-).

Corticosteroid activity of any kind requires four structural features in the steroid nucleus: (1) a 3-keto group; (2) a double bond between carbons 4 and 5; (3) a two-carbon chain attached to carbon 17 (i.e., carbons 20 and 21); and (4) a 20-keto group. Whether a corticosteroid is a mineralocorticoid or a glucocorticoid depends upon these additional structural features: (1) an 11-keto or 11β-hydroxyl group is essential for glucocorticoid activity but not for mineralocorticoid activity; (2) a 17α-hydroxyl group, though not essential for glucocorticoid activity, enhances it and diminishes mineralocorticoid activity; and (3) a 21-hydroxyl group greatly enhances both glucocorticoid and mineralocorticoid activity (see Table 15.3).

Production and clinical use of the corticosteroids. Corticosterone, cortisol, and the other corticosteroids were first isolated from adrenal cortical tissue by E. C. Kendall in the United States and by O. Wintersteiner and T. Reichstein in Switzerland over three decades ago. Each investigator designated the various corticosteroids with letter names in the order of discovery (compounds A, B, C, D, E, etc.). Since the order was different in the different laboratories, the nomenclature was considerably confused.

Although it was recognized that the compounds differed in biological activity, the significance of the large number of compounds within the adrenal cortex remained obscure until O. Hechter, G. Pincus, R. Dorfman, and others, largely on the basis of isotopic experiments, demonstrated an orderly metabolic pathway within the adrenal cortex that culminated in the biosynthesis of cortisol and the other corticosteroids (Fig. 15.9). We discussed the biosynthesis of cholesterol in Chapter 14. With cholesterol as the initial precursor, specific hydroxylases and other enzymes in the adrenal cortex catalyze a series of hydroxylations and other transformations through Δ^5-pregnenolone, progesterone, and 11-desoxycorticosterone (DOC) to corticosterone and aldosterone and through Δ^5-pregnenolone, progesterone, and 17α-hydroxyprogesterone to cortisol and the adrenal androgen Δ^4-androstenedione. We conclude, then, that many of the compounds isolated from the adrenal cortex are biosynthetic intermediates.*

During World War II, interest in the isolation and

---

* Recently, direct analysis has confirmed the presence of corticosterone, aldosterone, cortisol, and traces of Δ^4-androstenedione in adrenal vein blood.

synthesis of adrenal steroids was aroused by stories that Reichstein's Compound E, a glucocorticoid also known as *cortisone* and almost identical to cortisol, was being given to German pilots to reduce their fatigue. Although the stories were false, it was not long before synthetic glucocorticoids were available for clinical experimentation. In 1949 Philip Hench reported on the remarkable antiinflammatory effects of cortisone in rheumatoid arthritis. ACTH brought about similar effects by stimulating the adrenal cortex to produce excessive amounts of glucocorticoids. Successes were claimed for cortisone in other diseases, and it was soon in wide use as an antiinflammatory agent of unknown mechanism. In its applications it was administered in much larger amounts than those produced naturally by normal glands. Indeed, it soon became obvious that intensification of the normal effects—particularly the mineralocorticoid effect, with salt and water retention—led to undesirable side effects. In other words, prolonged treatment with glucocorticoids had troublesome consequences.

In attempts to eliminate the undesirable side effects, workers have synthesized many corticosteroid analogues (Table 15.4). With ingenious chemical modifications they have to some extent dissociated the antiinflammatory and the unwanted mineralocorticoid effects. For example, introduction of a double bond between carbons 1 and 2 of cortisol and cortisone increases antiinflammatory effects while lessening mineralocorticoid effects. Attachment of methyl groups to carbons 6 and 16 dissociates the wanted and unwanted effects even better. Interestingly, attachment of a halogen atom to carbon 9 greatly enhances antiinflammatory effects, but some compounds so formed (e.g., 9α-fluorocortisol) also have enhanced mineralocorticoid effects. The search for a purely antiinflammatory corticosteroid continues.

Hypo- and hyperadrenocorticalism. Early studies of adrenal hormone action consisted in observations of the physiological changes occurring after removal of the adrenals from an experimental animal. After such an operation a dog or cat seems normal for several days but then exhibits loss of appetite and muscular weakness. Blood pressure and body temperature begin to fall, and the animal suffers extreme prostration, with symptoms of shock, and finally dies. Most of these effects are due to an increase in the renal excretion of sodium and chloride and a resulting decrease in their concentrations in the extracellular fluids. Adrenalectomized subjects can be kept alive if large amounts of sodium chloride or mineralocorticoids are administered, but this procedure does not correct other injurious effects of adrenocortical insufficiency. Nevertheless, mineralocorticoid treatment—specifically DOC treatment—was for many years a routine life-saving measure in this situation.

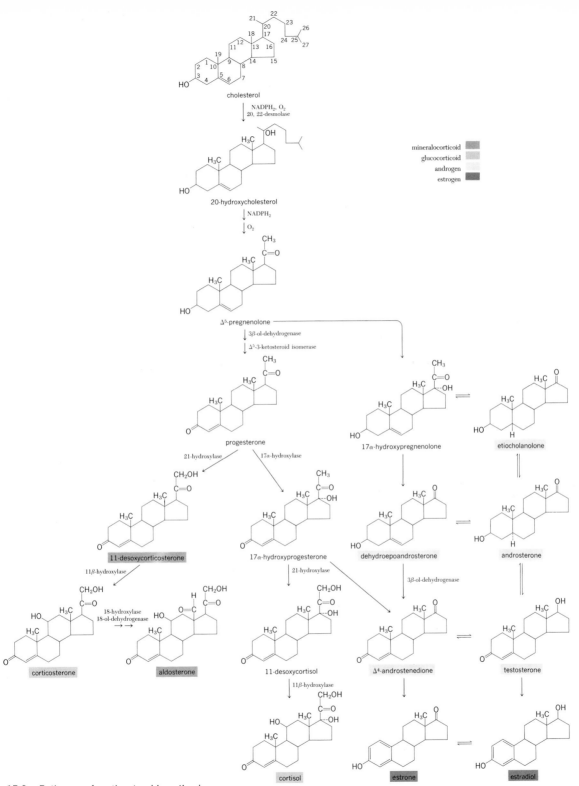

mineralocorticoid
glucocorticoid
androgen
estrogen

15.9 Pathways of corticosteroid synthesis.

TABLE 15.4   COMPARISON OF CORTICOSTEROID ANALOGUES AND ALDOSTERONE

Compound	Structure	Glycogen deposition (Cortisol = 1)	Antiinflammatory effect (Cortisol = 1)	Sodium retention (11-Desoxycorticosterone =
Prednisolone (Δ^1-cortisol)		3	4	<0.01
Fluorocortisol (9α-fluorocortisol)		8	12	5
Methylprednisolone (6α-methyl-Δ^1-cortisol)		10	6	0
Triamcinolone (9α-fluoro-16α-hydroxy-Δ^1-cortisol)		34	6	0
Dexamethasone (9α-fluoro-16α-methyl-Δ^1-cortisol)		17	100	0
Aldosterone		0.1	0	30–100

The defects that are not reversed by DOC and other mineralocorticoids are those due to glucocorticoid deficiency. Glucocorticoids are required for the body's resistance to stress. Glucocorticoid deficiency results in abnormal susceptibility to stress, as well as to infections and fatigue. The blood glucose level drops, the liver glycogen store is depleted, and nitrogen is retained. Glucocorticoids in large doses reverse these effects.

The major form of human *hypoadrenocorticalism* is *Addison's disease*. In 1855 the English physician Thomas Addison described a condition associated with the destruction of the adrenal cortex (frequently by tuberculosis) and characterized by extreme muscular weakness, low blood pressure, and weight loss. Especially characteristic is a peculiar increase in skin pigmentation caused by an increase in MSH production. Until recently Addison's disease was usually fatal. With salt and corticosteroid therapy, the outlook is now vastly improved.

Hyperadrenocorticalism takes several forms. One, already mentioned, is due to the administration of corticosteroids in large quantities. Others are due to oversecretion of ACTH. Still others are due to oversecretion by the adrenal cortex itself, which may lead to mineralocorticoid hyperactivity (as in *hyperaldosteronism*), to glucocorticoid hyperactivity (as in *Cushing's disease*), or, rarely, to adrenal sex hormone hyperactivity (as in the *adrenogenital* or *virilizing syndrome*). Hyperaldosteronism is manifested chiefly by sodium and water retention with consequent potassium loss—and thus muscular weakness—and elevated aldosterone levels in blood and urine. Cushing's disease is manifested by an excess of all the glucocorticoid effects (and frequently the mineralocorticoid effects as well), including a high blood glucose level ("steroid diabetes"), increased breakdown of tissue proteins, increased numbers of erythrocytes and leukocytes in the peripheral blood, and electrolyte abnormalities. The adrenogenital or virilizing syndrome is manifested by precocious sexual development in infancy and childhood.

Illuminating combinations of hypo- and hyperadrenocorticalism are seen in the recently discovered phenomenon of defective corticosteroid synthesis. In these cases a genetically determined defect in one of the synthetic enzymes—e.g., 21-hydroxylase, 18-hydroxylase, etc. (see Fig. 15.4)—causes (1) a block in the affected pathway with deficient production of the end product of that pathway; (2) increased ACTH release; (3) excessive stimulation of the normal pathways of corticosteroid synthesis; and (4) endocrine imbalance due to underproduction of certain corticosteroids and overproduction of others.

Control of adrenocortical activity. As we have seen, ACTH stimulates the adrenal cortex to increase the production of all corticosteroids. It is the main stimulus for glucocorticoid and adrenal sex hormone production. The main stimulus for mineralocorticoid production is the renin-angiotensin system.*

Within minutes of the secretion of ACTH by the adenohypophysis—or of its administration—increased glucocorticoid secretion can be demonstrated. The molecular basis of the ACTH effect upon the adrenal cortex is incompletely understood, but it is established that ACTH induces the consumption of cholesterol (in the synthesis of corticosteroids) and ascorbic acid† (the role of ascorbic acid, a potent reductant, is unknown).

The several theories that seek to explain the mechanism of ACTH action propose that in the biosynthetic pathway from cholesterol to corticosteroid the rate of a specific reduction is determined by the availability of $NADPH_2$; $NADPH_2$ is a key reductant in the pathway, particularly in the hydroxylation steps. One theory proposes that ACTH somehow provokes the accumulation in cortical tissue of a cyclic nucleotide, 3′,5′-adenylic acid, which stimulates phosphorylase to accelerate the conversion of glycogen to glucose 6-phosphate, which in turn stimulates the production of $NADPH_2$ in the pentose phosphate shunt of carbohydrate metabolism (see Fig. 4.13). Thus it is asserted, though not yet proved, that ACTH stimulates steroid synthesis by promoting $NADPH_2$-linked hydroxylation reactions.

It has also been observed that ACTH stimulates protein synthesis in the adrenal cortex and that continuous ACTH stimulation causes growth of the cortex. Conceivably this effect is related to the ACTH stimulation of the synthesis of specific mRNA's.

It has been suggested that the explanation for the ability of ACTH to stimulate aldosterone secretion only briefly is that an oxidation to

---

* Hypophysectomy practically abolishes cortisol secretion but reduces aldosterone secretion by only 60 to 80%. Administered ACTH doubles aldosterone secretion, but a decrease in blood volume raises it 30-fold. Clearly aldosterone serves primarily as a volume regulator. The second strongest stimulus of aldosterone secretion is an increase in the potassium level, possibly within the adrenal itself.

† Measurements of the extents of cholesterol and ascorbic acid depletion are well-known methods of assaying the potency of an ACTH preparation.

18-aldehyde must follow the special hydroxylation at carbon 18 (see Fig. 15.9). If the oxidation were extremely slow, it might limit the rate of aldosterone synthesis no matter how much ACTH enhanced the preceding hydroxylation. In this view aldosterone production would be influenced only by stimuli capable of altering the rate of 18-aldehyde formation. Inhibition of the oxidation accounts in part for the action of amphenones in blocking aldosterone production (see p. 362).

The major factors controlling ACTH secretion appear to be two (possibly more) *corticotropin-releasing factors* (CRF's), α-CRF and β-CRF, that are secreted by neurons in the median eminence and transported to the adenohypophysis via the hypothalamic-hypophyseal portal system. Interestingly, α-CRF is a small peptide containing the same amino acids as α-MSH—plus threonine, alanine, and leucine. β-CRF, which is more powerful, is not fully characterized, but its amino acid composition closely resembles that of vasopressin. A noteworthy feedback system apparently regulates CRF production—and consequently ACTH and glucocorticoid secretion. It is stimulated by stresses, such as trauma, intense heat, and intense cold, and by injected epinephrine and other sympathomimetic agents. It is inhibited by a rising plasma cortisol level.* Therefore, the plasma cortisol level is kept within narrow limits except in stressful situations. The mechanism by which cortisol influences CRF production is not

---

* ACTH secretion that is shut off as a result of a rising plasma cortisol level resumes when CRF is administered.

known. However, the secretion of other hypophyseal hormones (e.g., TSH and MSH) as well as ACTH is suppressed when the plasma cortisol level rises and may be enhanced when it falls.

Catabolism of the corticosteroids. Corticosteroids are transported in plasma bound to at least two separate proteins: (1) albumin, which has an affinity for all steroid hormones—the greatest for estrogens, next for progesterone, and least for corticosteroids; and (2) an α-globulin known as *transcortin,* or *corticosteroid-binding globulin* (CBG), which is relatively specific for corticosteroids. Transcortin maintains a hormone in ready supply in the circulation, protects it from inactivation, and solubilizes it.

Corticosteroids are ultimately inactivated by chemical transformation in the liver and in the blood. The main inactivating reactions are (1) reduction of the Δ^4-3-ketone followed by conjugation to glucuronide, sulfate, or phosphate (see p. 554); and (2) reduction of the 20-keto group. Inactive metabolites are excreted in the urine. Quantitative analysis of the urine therefore provides valuable information on the identities and activities of the steroid-secreting endocrine glands.

ADRENAL MEDULLA

It was known before 1900 that injected extracts of adrenal medulla cause a remarkable rise in blood pressure. In 1902 J. J. Abel isolated

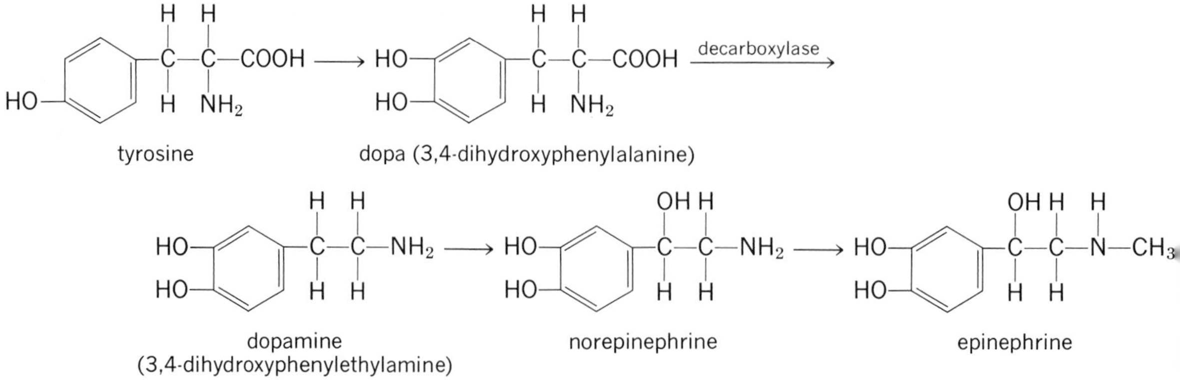

15.10 Synthesis of norepinephrine and epinephrine.

epinephrine, an active crystalline compound, from the adrenal medulla. It was the first hormone to be isolated in crystalline form.

We now know that the adrenal medulla secretes epinephrine and norepinephrine (Fig. 15.10).* These two compounds differ structurally in only one respect: a methyl group is present in epinephrine. Both are derivatives of 1,2-dihydroxybenzene, or catechol, $\underset{HO}{\overset{HO}{}}\!\!\!\diagdown\!\!\!\bigcirc$, and each has a two-carbon side-chain bearing an amine. Accordingly, they belong to the large category of *catecholamines*, which comprises many related compounds. Epinephrine arises metabolically from norepinephrine. Both are synthesized from the aromatic acid tyrosine.

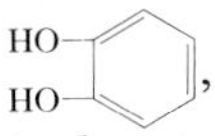

Epinephrine and norepinephrine are probably secreted by different types of chromaffin cells (see p. 600). Norepinephrine is also secreted in the sympathetic ganglia of the autonomic nervous system and in sympathetic postganglionic neurons. It is generally accepted as the chief humoral transmitter at sympathetic neuroeffector junctions (see pp. 422, 468).

The human adrenal medulla ordinarily contains about 2 mg of epinephrine and 1 mg of norepinephrine per gram of tissue.† These hormones are bound in granules, as they are in sympathetic neuroeffector junctions. In the embryo only norepinephrine is present; epinephrine appears later.

Under normal conditions only a small portion of the epinephrine and norepinephrine secreted by the adrenal medulla is excreted as such in the urine. Less than 2% of an administered dose is excreted in its original form. The conclusion is that the hormones undergo metabolic transformation. This has been confirmed by the recent discovery of the metabolic pathways outlined in Fig. 15.11. The main pathway includes a meth-ylation of one of the hydroxyl groups—a so-called O-methylation.‡ The methoxyhydroxy derivatives are converted ultimately to 3-methoxy-4-hydroxymandelic acid, a stable compound that appears in the urine in substantial quantities, an adult normally excreting 3 to 5 mg per day.

Physiological role. The physiology of the adrenal medulla is essentially a study of the effects of epinephrine and norepinephrine—and the special role of norepinephrine in sympathetic nerve function is of central significance. The medulla is not indispensable; totally adrenalectomized animals and humans with Addison's disease thrive when given sufficient adrenocortical hormones. Despite the loss of the medulla, the chief source of epinephrine, sympathetic nerves under proper excitation continue to liberate norepinephrine.

The effects of epinephrine and norepinephrine mimic those produced by stimulation of the sympathetic nervous system, summarized in Table 12.5. Among them are blood pressure elevation by arteriolar constriction, pupillary dilatation, bronchiolar dilatation, and inhibition of intestinal peristalsis. This identity of hormonal and sympathetic effects should perhaps be expected in light of the fact that adrenal medullary cells are modified sympathetic ganglion cells.§ Although injected epinephrine has striking effects, the function of secreted epinephrine in a normal unstressed individual is in doubt. Probably it acts directly in those stressful situations, normal and pathological, that elicit its secretion in massive amounts—e.g., after excitation of the hypothalamus or of the sympathetic fibers to the endocrine glands, in emotional states such as fear and rage, during pronounced muscular activity, upon sudden exposure to temperature or atmospheric extremes, and in hypoglycemia or severe hemorrhage. Since an adrenalectomized animal receiving adrenocortical hormones can survive these situations, the sympathetic system may assume the

---

* Other names in common use are *adrenalin* and *adrenin* (epinephrine) and *noradrenalin* and *arterenol* (norepinephrine).

† The ratio of epinephrine to norepinephrine in the adrenal medulla varies widely in different species, from almost exclusively epinephrine in the rabbit to almost exclusively norepinephrine in the whale. It has been suggested that aggressive animals have much norepinephrine in their adrenals while their prey, prepared to flee, have much epinephrine. Man apparently prefers "flight" to "fight."

‡ An enzyme, catechol-O-methyl transferase, catalyzes the methylation. This enzyme, distributed in all tissues, is believed to be the enzyme that inactivates norepinephrine in sympathetic neuroeffector junctions.

§ Thus epinephrine and its analogues are described as sympathomimetic (see p. 468).

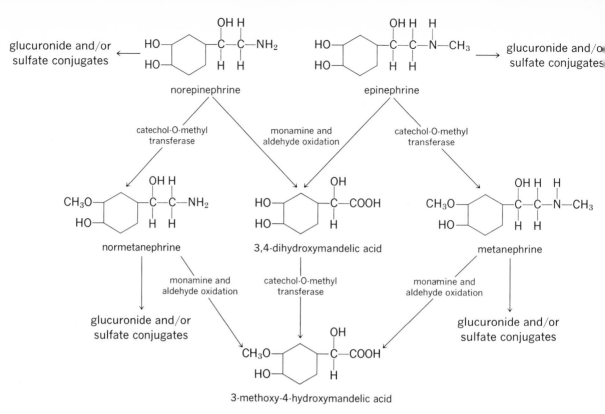

15.11 Metabolic pathways of norepinephrine and epinephrine.

functions of the absent adrenal medulla. Other tissues, such as the *paraganglia*, small bodies associated with sympathetic ganglia, may secrete epinephrine, but there is no direct evidence that they do.

Specific effects of epinephrine and norepinephrine. When either epinephrine or norepinephrine is injected into the veins, a sudden brief rise in blood pressure occurs. With epinephrine this effect is due primarily to increased cardiac output, with vasoconstriction limited to certain regions. With norepinephrine it is due primarily to generalized vasoconstriction.

Epinephrine brings about increases in the BMR, the rate of glycogen breakdown in liver and muscle, and the blood glucose level. The last effect is attributable to the activation of phosphorylase. Norepinephrine has the same effects to a lesser degree. Both hormones also promote the release of fatty acids from fat stores, which raises the blood fatty acid level.

Epinephrine and norepinephrine differ in the ability to stimulate the hypothalamus and adenohypophysis to secrete ACTH and TSH. Injected epinephrine has a stimulatory effect, but injected norepinephrine does not. The acceleration of ACTH production in stress (see p. 608) is therefore probably a result of the stress-induced upsurge in epinephrine secretion. It is not connected with norepinephrine secretion.

Excessive secretion of medullary hormones. The adrenal medulla is very stable in its function, and abnormalities are rare. Occasionally a *chromaffin tumor*, or *pheochromocytoma* (in the adrenal medulla or in the chromaffin tissue of the paraganglia), produces excessive amounts of both epinephrine and norepinephrine or of norepinephrine alone. Such a tumor typically causes paroxysmal elevations in blood pressure, which is normal between attacks. One method for distinguishing this disorder from ordinary hypertension is urinalysis; with a chromaffin tumor the concentrations of catecholamines in the urine are high.

Parallels between catecholamines and indolamines. Although 5-hydroxytryptamine, or *serotonin*, is not an

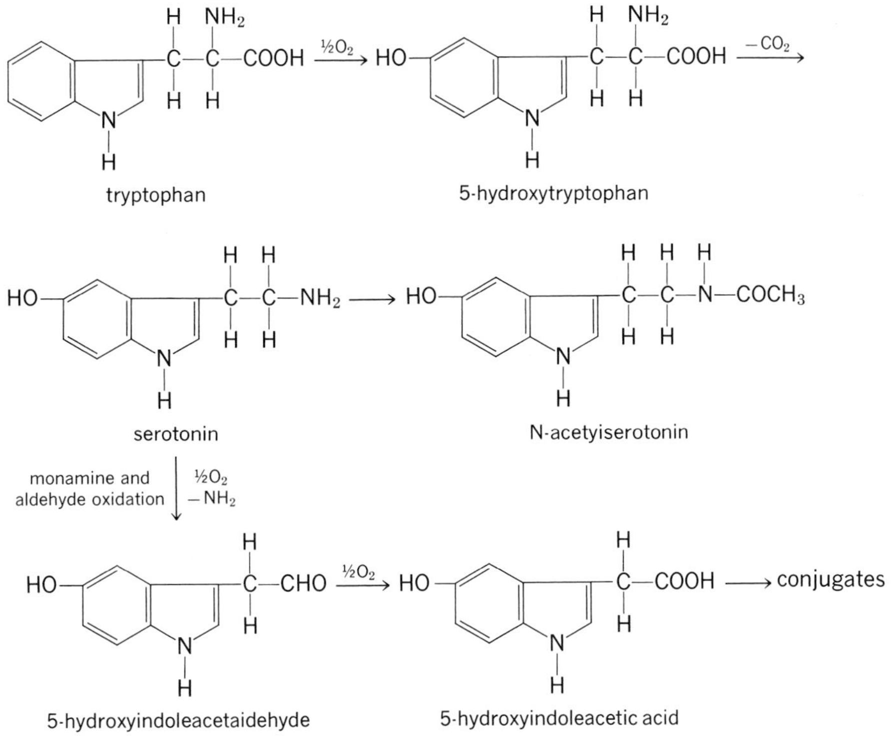

15.12 Synthesis and breakdown of serotonin.

adrenal hormone—and may not deserve to be called a hormone at all—we shall introduce it here because of certain parallels between its biochemistry and biology and those of the catecholamines.

Like epinephrine and norepinephrine, serotonin is an amine derivative of a ring structure—in this instance *indole;* hence it is classed as an *indolamine.* As shown in Fig. 15.12, it arises from the amino acid tryptophan. Serotonin is found chiefly in the gastrointestinal mucosa, platelets, and brain. We learned in Chapter 8 that platelet serotonin contracts the blood vessels in the early stages of hemostasis. This property accounts for its name.

Gastrointestinal serotonin is secreted by specific mucosal cells called *enterochromaffin cells* (i.e., chromaffin cells of the intestine) or *carcinoid cells.* When these cells proliferate into tumors that secrete massive quantities of serotonin, its pharmacological effects are most pronounced. They are vasodilatation with flushing and blood pressure lowering, hyperperistalsis, and bronchoconstriction. In such cases the urine contains a high concentration of 5-hydroxyindoleacetic acid, analogous to the 3-methoxy-4-hydroxymandelic acid of catecholamine metabolism.

The significance of brain serotonin is not known.

However, mood-altering drugs such as reserpine (see Chapter 12) promote the release of large amounts of serotonin in the brain. Interestingly, some of these drugs (e.g., the euphoria-producing agent iproniazid) also provoke a rise in the level of norepinephrine in the brain. Conceivably serotonin and norepinephrine have some part in producing the psychic drug effects. Lysergic acid, which causes active hallucinations, has an indole ring structure similar to that of serotonin. Serotonin and the enzymes involved in its synthesis are present in highest concentrations in the hypothalamus.

One more parallel between indolamines and catecholamines is that bananas contain both in high concentrations. Following the ingestion of bananas, both types of amines appear in the urine in levels far above normal.

Thyroid

The thyroid, under the primary control of the hypothalamic-hypophyseal portal system, manufactures, stores, and liberates iodine-containing hormones, which are transported by circulating

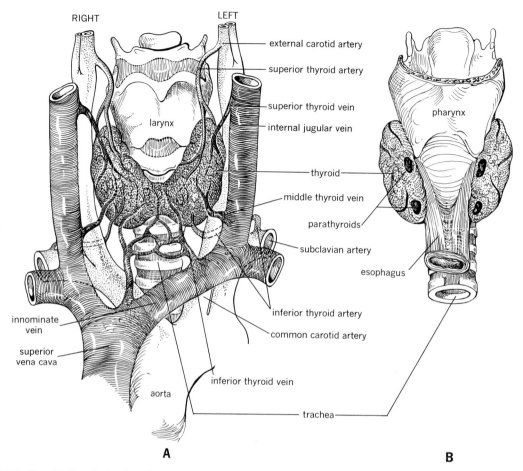

RIGHT   LEFT

external carotid artery

superior thyroid artery

superior thyroid vein

internal jugular vein

larynx

pharynx

thyroid

middle thyroid vein

parathyroids

subclavian artery

esophagus

inferior thyroid artery

common carotid artery

innominate vein

superior vena cava

inferior thyroid vein

aorta

trachea

A   **B**

15.13 Thyroid: A, anterior view; B, posterior view. Posterior view shows parathyroids.

proteins to the tissues, where they alter the rate of practically every fundamental process of intracellular metabolism. This effect is evident from the fact that lack of thyroid hormone (*hypothyroidism*) may cause the BMR to fall to as low as −50, whereas excess of thyroid hormone (*hyperthyroidism*) may cause the BMR to rise as high as + 100.

Except for the gonads, the thyroid was the first endocrine gland to appear in evolution and is the first to appear in the embryo. Originating as an outpocketing from the floor of the pharynx, it soon loses its connection with the pharynx and migrates from a location corresponding to the base of the tongue to a location in the neck (see Figs. 15.1, 15.13). It early develops its H-shaped bilobed structure, with the narrow isthmus between the lobes running anteriorly across the trachea

below the larynx. It grows during childhood and reaches its normal adult size at puberty. Although a normal thyroid weighs only 15 to 25 g, it is the largest of the glands that are entirely endocrine in function. Its blood supply, disproportionately rich for a tissue constituting such a small fraction of the body weight, derives from the external carotid arteries via the *superior thyroid arteries* and from the subclavian arteries via the *inferior thyroid arteries*.

Microscopically thyroid tissue is unusually interesting. As shown in Fig. 15.14, it is composed of many spherical, grapelike sacs or *follicles*. These are the secretory units of the gland. Although the follicles have no external openings, they are well provided with minute blood and lymphatic vessels. The walls of a follicle consist of a single layer of low cuboidal epithelial cells, and the cavity contains a viscous homogeneous fluid, known as *colloid* to the histologist and *thyroglobulin* to the biochemist.

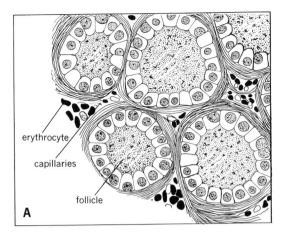

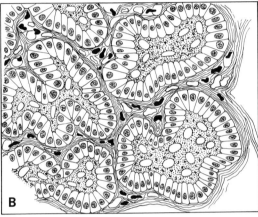

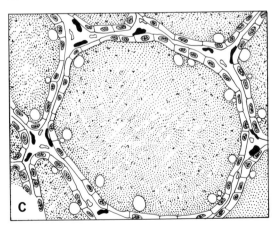

15.14 Microscopic structure of the thyroid: A, normal, with spherical follicles and cuboidal cells; B, in hyperthyroidism, with folded follicles and columnar cells; C, in hypothyroidism, with distended follicles and flattened cells.

It is convenient to consider thyroid physiology according to the scheme in Fig. 15.15, with ten discrete steps or processes, as follows:*

1. Intestinal absorption of dietary iodide into the blood; transport of plasma iodide to the thyroid; concentration of plasma iodide by thyroid tissue.

2. Enzymatic conversion of accumulated or "trapped" inorganic iodide to organic iodine derivatives of tyrosine, including the major thyroid hormone, thyroxine; retention of these compounds in the follicles within the peptide structure of thyroglobulin.

3. Release of iodine-containing thyroid hormones into the blood.

4. Binding of iodine-containing thyroid hormones to certain plasma proteins to form *protein-bound iodine* (PBI).

5. Transport of protein-bound thyroid hormones to peripheral tissues; entry of the hormones into the tissues.

6. Metabolic action of thyroid hormones in cells of peripheral tissues, possibly a transformation to more active forms; degradation of thyroid hormones.

7. Hypothalamic stimulation of adenohypophyseal secretion of TSH.

8. Stimulation of thyroid hormone synthesis by TSH.

9. Control of TSH secretion by circulating thyroid hormones.

10. Excretion of thyroid hormone degradation products by liver and kidneys.

PRODUCTION OF THYROID HORMONES

Iodide transport and the iodide pool. A unique feature of the thyroid hormones is their high iodine content. Indeed, the thyroid contains the bulk of the total body iodine, and iodine is not known to be utilized by any other tissue. It is not surprising, therefore, that the thyroid possesses a special mechanism for trapping iodide from the blood flowing through it. The study of thyroid physiology has been greatly facilitated in recent years by the use of ^{131}I, a radioactive

---

* These steps are concerned only with the production of the iodine-containing thyroid hormones. Thyrocalcitonin, a recently discovered thyroid hormone that does not contain iodine, will be discussed later.

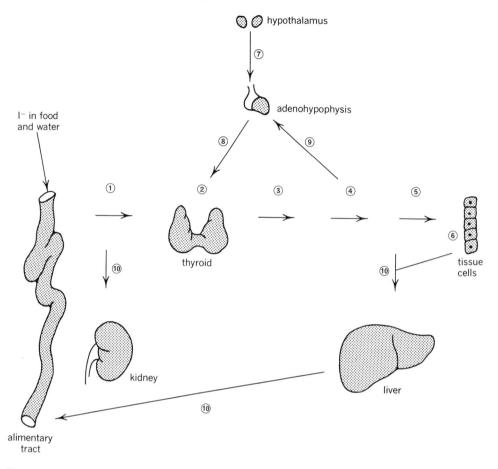

15.15 Phases of thyroid physiology.

isotope of iodine. The availability of ^{131}I has permitted evaluation of many aspects of iodide metabolism (Table 15.5). In particular, ^{131}I has shed light upon the body's *iodide pool*—i.e., its content of inorganic iodide.

The body obtains most of its inorganic iodine in the form of iodide in food and water. Except after meals, the plasma iodide concentration is less than 1 μg per 100 ml. To support normal thyroid function, about 1 mg of iodide is needed in the diet each week. Since many regions of the world are iodine-deficient, common table salt (NaCl) is now often supplemented with NaI, but to varying extents in different countries (e.g., 1 part NaI per 10,000 parts NaCl in the United States; 1 part NaI per 100,000 parts NaCl in Switzerland). Iodide ingested orally is absorbed

in the small intestine and circulates in the extracellular fluid in essentially the same manner as chloride. In addition to the iodide from dietary sources, the iodide pool receives iodide liberated in the metabolic degradation of iodine-containing hormones. This is subsequently reutilized.

Despite many similarities, iodide metabolism and chloride metabolism differ in several significant ways. For example, ^{131}I studies have shown that iodide does not remain long in the extracellular fluid. Within a few minutes of an intravenous injection, it is distributed throughout a space equal to 15% of the body weight and somewhat smaller than the total extracellular fluid compartment (see Figs. 7.2, 7.3). However, several hours later it is distributed throughout a space equal to 35% of the body weight—

TABLE 15.5 QUANTITATIVE ASPECTS OF IODIDE METABOLISM

Parameter	Estimated value
Distribution space of injected ^{131}I, % body weight	
After 2 min	15
After 30 min	26
At equilibrium	35
Thyroid clearance, ml/min	
Normal	17 (3–45)
Hyperthyroid	70–1000
Renal clearance, ml/min	35
Iodide uptake of thyroid, μg	
Normal	75
Hyperthyroid	890

considerably larger than the extracellular fluid compartment. Its spread is attributable to its passage into erythrocytes (and other body cells), saliva, and gastric juice. Within a few days two-thirds of the injected iodide is excreted in the urine. The remaining third is trapped by the thyroid and utilized in hormone synthesis. Thus the two modes of iodide removal from the pool are *thyroid clearance* and *renal clearance.*

Clearance is expressed as the volume of plasma cleared of iodide per minute. As indicated in Table 15.5, in a normal individual about 17 ml of plasma is cleared of iodide each minute by the thyroid, and about 35 ml by the kidneys. This totals 52 ml per minute, or $60 \times 52 = 3120$ ml per hour. Since iodide is ultimately distributed through fluids equivalent to 35% of the body weight (approximately 25 liters in a 70-kg man), then $3.12 \div 25 = 12\%$ of the total iodide pool turns over each hour.

We conclude that thyroid tissue contains a mechanism for concentrating iodide simply because the iodide level in the gland is more than 30 times that in the blood passing through it. That the concentration mechanism is separate and distinct from the process of hormone synthesis is demonstrated in several ways. In the iodine-containing fractions of thyroid tissue, in-

organic iodide is dialyzable,* whereas organically bound iodine is not. The reason is that organically bound iodine has entered the pathway of hormone synthesis. A remarkable discovery a few years ago showed that the drug thiouracil totally blocks the synthesis of thyroid hormone but does not affect the concentration of iodide.

Though the transport of iodide into the thyroid is the first step in thyroid hormone formation, the cellular mechanism and locus of iodide transport are unknown. Active transport across cell membranes is suggested by the fact that maximum transport occurs when the extracellular fluid is saturated with iodide. Transport is competitively inhibited by at least two groups of anions, of which thiocyanate (SCN^-) and perchlorate (ClO_4^-) are representatives. Transport requires oxidative metabolism and energy expenditure. Finally, transport is stimulated by adenohypophyseal TSH. Since the rate of iodide transport may be a factor governing the rate of hormone synthesis, TSH may thus stimulate thyroid secretion. This remains to be proved, however. It is of interest that, although various other tissues take up iodide (e.g., salivary glands, stomach, and mammary glands), TSH stimulates only thyroid iodide transport.

Synthesis and storage of thyroid hormones. The events following the entrance of iodide into the thyroid are summarized in Fig. 15.16. At least six enzyme systems are involved in the synthesis and liberation of thyroid hormone. After inorganic iodide is withdrawn from the plasma, it remains momentarily within the gland as free dialyzable inorganic iodide. The first step in hormone synthesis is the conversion of ionic iodide to elemental iodine—an oxidative reaction depending upon the presence of a tissue *peroxidase.* The next step is the combination of elemental iodine with tyrosine under the influence of *tyrosine iodinase.* About two-thirds of the tyrosine molecules acquire only one atom of iodine, producing *monoiodotyrosine* (MIT). The rest acquire two atoms, producing *diiodotyrosine* (DIT). Whether the two iodotyrosines are synthesized simultaneously or in series, with MIT a precursor of DIT, has not been estab-

---

* Dialysis is explained in Chapter 2.

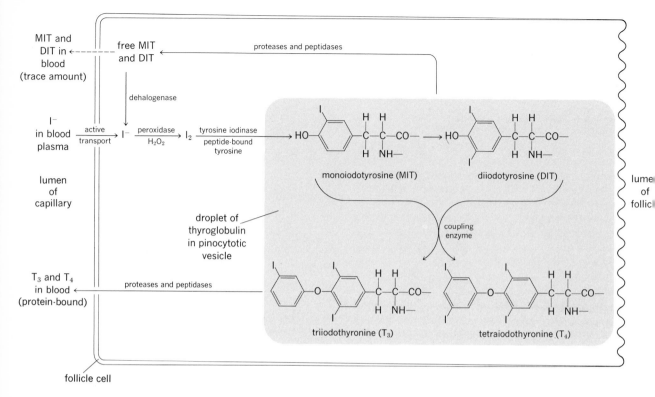

15.16 Thyroid hormone synthesis.

lished. Next the iodinated tyrosines couple with each other to form the *iodothyronines*. This process is catalyzed by a *coupling enzyme* and stimulated by TSH. The iodothyronines, which have two, three, or four atoms of iodine, are conveniently designated T_2, T_3, and T_4. The best-known and most abundant is thyroxine, or *tetraiodothyronine*, T_4; the most recently discovered is *triiodothyronine*, T_3.* Both are hormonally active.

An unusual feature characterizes the iodination and coupling reactions. The tyrosine mole-

cules entering the sequence are residues within the peptide structure of the large (molecular weight 650,000) glycoprotein thyroglobulin in the follicles (see Fig. 15.16). Thyroglobulin contains some 120 tyrosine residues, and about one-fifth of these are iodinated. The resulting iodotyrosines remain in thyroglobulin during their subsequent coupling. The iodothyronines also remain as residues in thyroglobulin.† Since the thyroglobulin molecule is too large to be transported across the follicle walls, they are thus temporarily unable to escape into the blood. The loci of the synthesis and storage of thyroid hormones are shown in Fig. 15.17.

---

* The structure of *thyronine* is

$$HO - \overset{3'}{\underset{5'}{\bigcirc}} - O - \overset{3}{\underset{5}{\bigcirc}} - \overset{H}{\underset{H}{\overset{|}{C}}} - \overset{H}{\underset{NH_2}{\overset{|}{C}}} - COOH$$

Thyroxine is 3,5,3′,5′-tetraiodothyronine; triiodothyronine is 3,5,3′-triiodothyronine.

---

† Varying numbers of the tyrosine residues are iodinated and coupled at a given moment. Since the iodotyrosines and iodothyronines formed are integral parts of the thyroglobulin molecule, their diversity in type and content provides an interesting exception to current views on specificity of protein structure (see p. 30).

Release and transport of thyroid hormones.
Thyroid hormone secretion—i.e., release of the hormones into the blood—cannot occur until after the separation of T_4 and T_3 from thyroglobulin (see Fig. 15.16). This reaction, catalyzed by a series of proteases and peptidases, occurs when lysosomes coalesce with droplets of colloid (thyroglobulin) that have entered the cells by pinocytosis.* It is stimulated by TSH. Over 80% of the thyroid hormones released into the blood is T_4. The remainder is T_3. The splitting also liberates small amounts of MIT and DIT. These are hormonally inactive, however, and a deiodinating enzyme, *dehalogenase*, releases iodine from them as iodides while having no effect on T_4 and T_3. The released iodide is available for further use. The importance of dehalogenase is seen in the fact that individuals lacking it excrete large amounts of MIT and DIT in the urine and develop hypothyroidism.

Most of the T_4 and T_3 is transported to the body cells bound to a carrier protein in plasma, *thyroxine-binding globulin (TBG)*, whose electrophoretic mobility is between that of α_1-globulin and that of α_2-globulin. In addition, small amounts of T_4 are bound to albumin and to a minor protein of mobility greater than that of albumin, *thyroxine-binding prealbumin (TBPA)*, and a trace is unbound. The binding to proteins prevents T_4 and T_3 from being excreted by the kidneys. The protein-bound hormones comprise the PBI of blood. They are readily precipitated from plasma, and their measurement is a reliable index of the quantity of thyroid hormones in the circulation. Normally the concentration of PBI is 4 to 8 μg per 100 ml of plasma. In hypothyroidism it may be less than 2 μg; in hyperthyroidism it may be 2 or 3 times the normal value.

Little is known of the mechanism of entry of the thyroid hormones into the tissues. Presumably unbound circulating hormones diffuse across cell membranes and reach binding sites within the cells. The largest portions go to the liver, where they produce many metabolic effects and are themselves metabolized. They are broken down in the liver and other body

---

* The last tyrosine residues to be iodinated are the first to be released by proteolytic enzymes. This has been called the "last in, first out" principle of thyroid iodine metabolism.

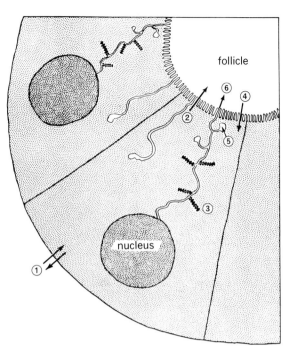

15.17 Sites of I⁻ transport and thyroglobulin iodination and transport. I⁻ is actively transported into the follicle cells (1) and concentrated in the follicle (2). Thyroglobulin is produced on ribosomes (3) and transferred as droplets into small vesicles (5) on the follicular sides of the cells. There it reacts with I⁻. Iodinated thyroglobulin is secreted into the follicle (6). Under the influence of TSH it re-enters the cell by pinocytosis (4) where lysosomal proteases release its hormones. (From Means, DeGroot, and Stanbury, *Thyroid and Its Disease,* 3rd ed., Copyright © 1963 by McGraw-Hill, Inc. Used by permission of McGraw-Hill Book Company, New York.)

tissues through deiodination, deamination, and decarboxylation. They also conjugate with glucuronic acid and sulfate in the liver, and sizeable amounts of inactivated hormones are excreted in the bile. The possible significance of these conversions in promoting the actions of thyroid hormones will be discussed in the following section.

ACTIONS OF THYROID HORMONES

Thyroid hormones stimulate almost every metabolic process. Therefore, the list of their actions (Table 15.6) is a long one that encom-

TABLE 15.6 ACTIONS OF THYROID HORMONES

ON METABOLISM

Maintenance of all tissues except those of the brain, gonads, spleen, lymph nodes, and smooth muscles

ON SPECIFIC METABOLIC PROCESSES

Lipids: decreases in liver and serum cholesterol concentrations

Carbohydrates: increase in utilization; mobilization of glycogen stores

Proteins: acceleration of synthesis and breakdown and of gluconeogenesis

Vitamins and minerals: increases in requirements

ON BLOOD AND THE CIRCULATORY SYSTEM

Increases in cardiac output, heart rate, and blood pressure, stimulation of myocardial metabolism and maintenance of hematopoiesis

ON THE NERVOUS SYSTEM

Enhancement of mental acuity and electrical activity

ON THE REPRODUCTIVE SYSTEM

Maintenance of fertility in ovulating females, of the uterine contents in pregnancy, and of lactation

ON GROWTH AND DEVELOPMENT

Maintenance of growth and development in the young (and of metamorphosis in lower vertebrates, e.g., frogs)

ON OTHER ENDOCRINE GLANDS

Augmentation of the effects of epinephrine and norepinephrine and acceleration of the hepatic inactivation of adrenocortical and gonadal steroid hormones

passes all the body systems. In examining this list, we should remember that in the normal body thyroid hormones *maintain* metabolic activity at a normal level. With the thyroid hormones, as with other hormones, we seek evidence of specific effects either by administering excessive amounts of hormone to a normal individual and looking for abnormal manifestations or by producing subnormal activity levels through removal of the gland and studying the normalizing properties of administered hormone. Table 15.6 has been compiled by the latter method.

Effects on oxidative phosphorylation. A list such as Table 15.6 clearly calls for a more fundamental theory of thyroid hormone action than has yet been advanced. The most promising recent hypothesis is that thyroid hormones play a leading role in the energy-yielding metabolic process of oxidative phosphorylation (see pp. 130 ff.). Considerable direct evidence exists that thyroid hormones affect the rate and efficiency of oxygen utilization and ATP formation. We recall that oxidative phosphorylation accounts for the production of most of the ATP of cell metabolism, that three molecules of ATP are normally generated for every atom of oxygen utilized (i.e., the P:O ratio is 3), and that the process occurs in the mitochondria. It has long been known that large quantities of thyroid hormones uncouple oxidative phosphorylation —that is, oxygen consumption is increased, but ATP production is decreased. This is easily demonstrated when the hormones are added to active mitochondrial preparations in vitro.

We noted in Chapter 4 that mitochondria shrink and swell under varying conditions. Thyroid hormones actively induce swelling, which presumably reflects an increase in permeability to water. It apparently is closely connected with the basic thyroid hormone effect, since other agents such as dinitrophenol uncouple oxidative phosphorylation but cause no mitochondrial swelling. It is not known whether thyroid hormones influence oxidative phosphorylation by producing swelling and thus altering the physical relations of adjacent enzymes secondarily or by altering the organizational states of specific enzymes and thus producing swelling secondarily. A. L. Lehninger has suggested that these are but two aspects of a single hormonal action upon what he terms mitochondrial "mechanoenzymes"—enzymes that change shape during catalysis and inhibition. Whatever the mechanism, it appears that thyroid hormones modify the mitochondria. The low concentrations of thyroid hormones normally present presumably change them only enough to increase the rate of oxygen uptake without disturbing the rate of ATP synthesis. Higher amounts so change them that phosphorylative energy transfer is reduced, even while hyperoxidation continues. The resulting ATP deficiency impairs energy-utilizing functions such as nerve impulse conduction, muscular contraction, intestinal absorption, and protein, lipid, and carbohydrate synthesis. It tentatively accounts for the many manifestations of hyperthyroidism.

Alterations of thyroid hormones in the tissues. It is now believed that the fate of the thyroid hormones in the tissues is in part conditioned by their interactions with the proteins of the extracellular fluid, which regulate the rate of delivery of the hormones to peripheral tissues. In the cells the hormones undergo at least three types of metabolic transformation: deiodination, conjugation of the phenolic group, oxidative deamination and decarboxylation of the side chain, and possibly *O*-methylation. Deiodination seems to be the major metabolic pathway, but its products have not been identified.* Although the conversion of T_4 to T_3 or its derivatives has been described, whether it is a general reaction has not been established. Thyroid hormones conjugate with glucuronic acid in the liver, and the resulting glucuronides are secreted in the bile. The liver is not essential for this process, however, since glucuronides also form in hepatectomized dogs. In these circumstances the conjugates appear in the urine and the plasma. Thyroid hormones can also esterify with sulfate. The importance of oxidative deamination and decarboxylation as a metabolic pathway for thyroid hormones and the physiological role of the derivatives are unknown.

The possibility exists that thyroid hormones themselves are not the substances that ultimately act on the machinery of the cells. Rather, they may be the parent molecules of the active principle, with the final conversion taking place in the peripheral tissues. To an extent this hypothesis is supported by the fact that several hours intervene before the metabolic effects of T_4 are observed and that a shorter latent period occurs with an analogue containing acetic acid in place of alanine.† Still, there is no direct evidence that another compound is the physiologically effective one.

Control of thyroid function. The activity of the thyroid is normally controlled by the adenohypophysis, which secretes TSH. Removal of the hypophysis leads to reduction but not elimination of thyroid function. Although, as we have mentioned, TSH stimulates both iodide transport and hormone synthesis in the thyroid, the actual mode of its action is not fully understood. Apparently, however, it regulates thyroid function in a purely quantitative manner.

TSH secretion by the adenohypophysis depends upon the hypothalamic secretion of a *thyrotropin-releasing factor (TRF)* into the hypothalamic-hypophyseal portal system. TRF is a tripeptide with the structure Glu-His-Pro(NH_2). When the hypophyseal stalk is cut, TSH secretion drops. Hypothalamic control of TSH secretion accounts for the stimulation of thyroid hormone secretion by intense emotion and cold. TSH secretion is sensitively regulated by the levels of T_4 and T_3 in the plasma (see Fig. 15.15): an increase in their concentrations causes a decrease in TSH secretion, and vice versa. The effect of TRF on the hypophysis is inhibited by T_4 and T_3. The thyroid hormones in the plasma also exert a measure of control directly upon the thyroid: the amounts administered govern thyroid size and function in hypophysectomized rats maintained on constant doses of TSH.

Abnormalities of thyroid function. Studies of abnormal thyroid function have greatly illuminated the investigation of normal thyroid physiology. Indeed, the study of various genetically determined thyroid disorders has led to the discovery of many of the enzymes involved in thyroid hormone synthesis. Similarly, the study of acquired thyroid disorders has deepened our

---

* Because of the almost exclusive use of ^{131}I for labeling the hormones, the complete elucidation of the metabolic pathways and their products has been difficult; after deiodination, a molecule can no longer be traced by isotopic methods.

† Many thyroid hormone analogues have been prepared and tested in attempts to find the hypothetical "active iodothyronine." These studies have shown that activity is markedly affected by minor changes in molecular structure.

A

B

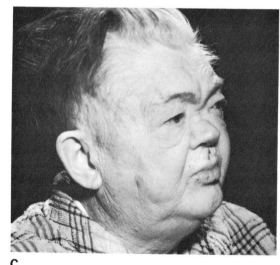

C

15.18 Manifestations of hypothyroidism. A, endemic goiter; B, cretinism (with endemic goiter); C, myxedema. (Dr. John B. Stanbury.)

understanding of the significance of iodine nutrition and of many inhibitory chemicals.

Several preliminary facts about thyroid disease should be noted: (1) thyroid disease may be associated with oversecretion of thyroid hormones (hyperthyroidism), undersecretion of thyroid hormones (hypothyroidism), or normal secretion of thyroid hormones (euthyroidism); (2) thyroid disease may or may not be associated with the disfiguring enlargement of the gland called *goiter* (Figs. 15.18, 15.19); (3) thyroid disease may or may not be genetically determined; and (4) each of these parameters may or may not be combined with any of the others.

For example, hypothyroidism may or may not be genetically determined, and it may or may not be associated with goiter. Acquired hypothyroidism with goiter occurs in two main situations. It results from dietary iodine deficiency and, in the years before iodization of salt, was common in iodine-deficient areas of the world, usually remote from the iodine-rich oceans (e.g., the Great Lakes region of the United States and the Swiss Alps). The records of some of these areas reveal that goiters were once so commonplace the individuals without them were considered abnormal and unattractive.

A second cause of goitrous hypothyroidism is certain *goitrogenic substances* that, when ingested, block the synthesis of thyroid hormones. These include perchlorate, thiocyanate, and thiouracil.* Many vegetables—particularly cauliflower, cabbage, kale, and Brussels sprouts—contain high concentrations of goitrogenic substances. Although it would be unusual for anyone to eat enough of these foods to develop goiter, a remarkable outbreak of goiter did take place in Tasmania following the initiation of a government free milk plan in 1950. It was found that, to satisfy the greatly increased demand for milk, farmers had begun feeding their herds a forage crop related to kale. A goitrogenic substance in the milk was proved to be responsible for the outbreak.

Genetically determined hypothyroidism with goiter has been investigated intensively only since 1950. It is due to deletions of specific enzymes essential in thyroid hormone synthesis, with defects having been demonstrated at almost every step in the synthetic pathway, including iodine-trapping (see Fig. 15.16).

Hypothyroidism without goiter appears when embryonic development of the thyroid has been faulty, after surgical removal of the thyroid in the treatment of thyroid tumor, with adenohypophyseal hypofunc-

---

* An interesting type of goiter arises with prolonged administration of large quantities of iodide. It evidently also involves the inhibition of thyroid hormone synthesis—possibly at the step in which iodine is converted to the organic form.

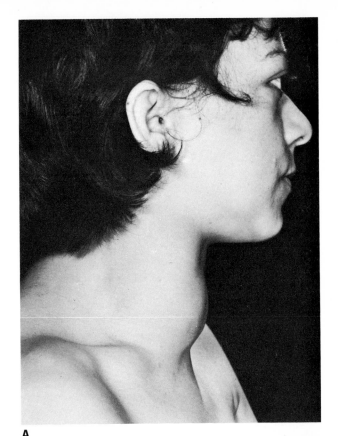

A

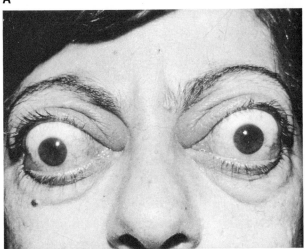

B

15.19 Manifestations of hyperthyroidism: A, enlarged thyroid and exophthalmos; B, close up of exophthalmos. (A, photo by Dr. Sidney H. Ingbar from R. H. Williams, *Textbook of Endocrinology*, W. B. Saunders Co., Philadelphia, 1968. B, Dr. Heskel M. Haddad, Beth Israel Medical Center, New York.)

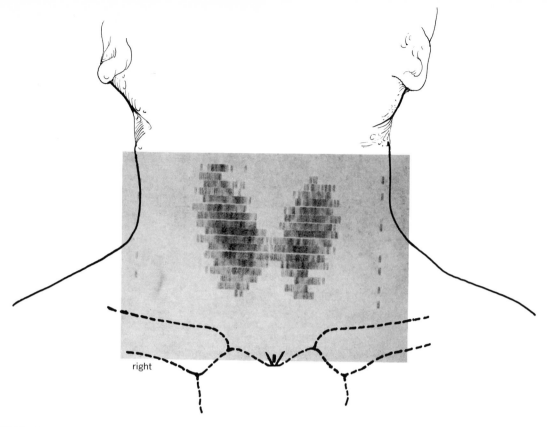

right

15.20 Scan of the thyroid area after administration of ^{131}I. (Dr. Farahe Maloof.)

tion, and perhaps most frequently after destruction of the thyroid by inflammation.

In general, goitrous enlargement of the thyroid is due to proliferation of thyroid tissue and gross distention of the thyroid follicles (see Fig. 15.14), with flattening of the epithelium and retention of a colloid that is poor in iodinated compounds. Thus both the "factory" and the "warehouse" increase in size. Presumably these changes are the compensatory reactions of a starved or blocked thyroid that is unable to supply the body with needed thyroid hormones.

Whether or not goiter is present, all types of hypothyroidism result in depressed metabolism, slowed circulation, anemia, delayed reflexes, and physical and mental sluggishness. Hypothyroidism has its most devastating effects in infancy and childhood. *Cretinism*, the severe hypothyroidism of infants and children, is characterized by failure of normal growth and development, mental deficiency, and a peculiar facial appearance (see Fig. 15.18B). *Myxedema*, adult hypothyroidism, is associated with extreme somnolence, feelings of coldness, coarsening of hair, skin, and voice, and alterations of many other body functions (see Fig. 15.18C).

Thyroid hormones administered orally dramatically reverse these manifestations. However, they will not remedy cretinism.

Probably the most widespread and most serious thyroid disorder is hyperthyroidism. Its symptoms consist of intensifications of all the effects in Table 15.6—among them intolerance to heat, weight loss, muscular weakness, high BMR, and low serum cholesterol.

Hyperthyroidism accompanied by goiter* is known as *Graves' disease*, after the Irish physician Robert Graves, who first described it in 1835. It is also called *thyrotoxicosis* in recognition of the profoundly toxic consequences of increased thyroid hormone secretion.

A manifestation of thyrotoxicosis of particular interest is *exophthalmos*, a characteristic bulging of the eyes due to accumulations of fluid, fat, mucopolysaccharides, and lymphoid tissue behind the eyes (see

---

* The goiter of hyperthyroidism is two or three times the size of a normal gland. In contrast to the goiter of hypothyroidism, its follicles contain little colloid. Their walls are folded inward, and the cells are increased in number and columnar in shape.

Fig. 15.19B). Usually the extrinsic muscles of the eyes are also weakened. Unlike the other signs of hyperthyroidism, exophthalmos is not produced in a normal individual by large doses of thyroid hormones. For this reason it has been proposed that it is caused by excessive adenohypophyseal secretion of TSH—or of a specific "exophthalmos-producing factor" that may be secreted in the adenohypophysis along with TSH. According to this theory, the primary difficulty in hyperthyroidism is excessive secretion of TSH (or a TSH-like substance) or perhaps excessive secretion of hypothalamic thyrotropin-releasing factor.

We have already spoken of the ability of the thyroid to concentrate iodide, and Table 15.5 shows that the rate of thyroid iodide accumulation is massively increased in hyperthyroidism. This information has given rise to a useful test of thyroid function and an important treatment for hyperfunction. The test consists in measurement of the percentage of a dose of ^{131}I that is retained by the body. In hyperthyroidism abnormal concentration of radioactivity in the gland is revealed when the neck is scanned by a radiation detector (an ingenious device that records local radioactivity as it scans a body region; Fig. 15.20). The detector provides a map that locates hyperactive tissues.

Since 80 to 90% of an injected dose of ^{131}I—an amount sufficiently radioactive to destroy many hyperactive cells—is absorbed by an overactive gland, injection of ^{131}I also constitutes effective treatment for hyperthyroidism. In cases responding favorably, a single dose of ^{131}I causes a gradual return to normal over a period of several months. Cases not responding, and others in which ^{131}I therapy may be inadvisable, are treated with drugs or surgery.

The drugs used in the treatment of hyperthyroidism include most of the known inhibitors of thyroid hormone secretion—e.g., potassium perchlorate and potassium thiocyanate, given to prevent thyroid iodide accumulation; and thiouracil and iodine, given to prevent thyroid hormone synthesis. Sedatives are also given to depress the actions of thyroid hormones in peripheral tissues.

Parathyroids

The parathyroids are composed of cords of secretory epithelial cells separated by fat cells (Fig. 15.21).* The epithelial cells are of two types: *chief cells* and *oxyphil cells*. A chief cell is

---
* There is considerable individual variation in location and number of the glands. About a quarter of the population has 4; the range is 2 to 6.

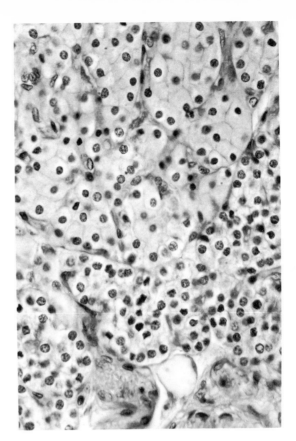

15.21 Microscopic structure of a parathyroid gland. (From E. J. Reith and M. H. Ross, *Atlas of Descriptive Histology,* Hoeber Medical Division Harper & Row, Publishers, New York, 1967.)

6 to 8 μ in diameter and has a large nucleus and pink-staining cytoplasm that contains secretory granules. An oxyphil cell is 11 to 14 μ in diameter and has dark red-staining cytoplasm that contains many mitochondria. Oxyphils and fat cells do not appear until after puberty, and so it is evident that parathyroid hormone is secreted by the chief cells.

The parathyroids are separate and distinct from the thyroid in structure and function, if not in origin. Embryologically they arise, like the thyroid, from the pharynx. Structurally they are the smallest of the known endocrine glands, each being only about 5 mm long, 3 mm wide, and 2 mm thick.

The parathyroids were discovered over a century ago by the British anatomist Richard Owen during a dissection of an Indian rhinoceros. In 1894 G. Vassales and F. Generali removed the parathyroids from an experimental animal and observed that it exhibited severe and progressive tetany—i.e., muscular tremors, twitching, and rigidity—with death ensuing. In 1909 W. MacCallum and Carl Voegtlin demonstrated that the muscular abnormalities were attributable to a sharp drop in the calcium level of the blood. Later it was found that this was followed by a rise in the inorganic phosphate level of the blood. It was also found that tetany was alleviated upon injection of an extract of parathyroid tissue or administration of calcium or certain drugs. It was concluded, therefore, that parathyroid hormone is concerned with the metabolism of calcium and phosphorus.

CALCIUM AND PHOSPHORUS METABOLISM

The normal concentration of calcium in the plasma is 10.0 mg per 100 ml. The constancy of this level is astonishing. Among the factors involved in its maintenance are the intestinal absorption of calcium, the transport and excretion of calcium, bone metabolism, and finally the action of parathyroid hormone. As we saw in Chapter 9, an optimal concentration of calcium is essential for the transmission of impulses through nerves and cardiac muscle. Hence there is adequate reason for its precise control.

Intestinal absorption. In the United States, Canada, and Northern Europe, the principal sources of calcium in the diet are milk and milk products.* These are also important sources of phosphorus, but phophorus is present in other foods as well. Calcium is relatively poorly absorbed from the small intestine because of the insolubility of its compounds. (In fact, all bivalent cations are poorly absorbed.) In contrast, phosphorus in the form of phosphate is well absorbed except when the diet contains enough excess calcium to form insoluble calcium phosphate in the intestine. Thus the problem of calcium and phosphate absorption is really the problem of calcium absorption, for, if calcium is absorbed, phosphate is absorbed.

Approximately 750 mg (500 to 1000 mg) of calcium is ingested daily by a normal adult. Larger quantities are required by growing children and during pregnancy and lactation. In the normal adult 75 to 90% of the calcium ingested is excreted in the feces and 10 to 25% in the urine. The net amount absorbed by the intestine each day is 100 to 200 mg.† Variations in the amount ingested have little effect on the amount absorbed, which rises only slightly with an intake as high as 3 g a day. This fact suggests that the intestinal absorption of calcium is geared somehow to body needs.

It has been known for some time that vitamin D has an important role in the metabolism of calcium and phosphorus. Early experimenters noted that, when animals were deprived of vitamin D, the concentrations of calcium and phosphorus in the plasma and in ashed bones decreased, while that of calcium in the feces increased. They concluded that vitamin D enhances calcium absorption in the intestine.

This hypothesis was strongly supported by later studies employing radioactive calcium. It is now believed that vitamin D facilitates the transfer of calcium across the intestinal cell membranes. Calcium absorption by the intestine is an active process that is markedly enhanced by prior vitamin D administration. Since this action of vitamin D is blocked by administration of actinomycin D, an antibiotic that prevents transcription of DNA (see p. 156), the expression of vitamin D action must involve both transcription of DNA into mRNA and protein synthesis. At least one if not the only site of its action is the brush border or microvilli of intestine, where it enhances the rate of calcium uptake into the mucosal cell. A *calcium-binding protein* has been isolated from the cytoplasmic fraction of intestine that appears in response to vitamin D. In addition, a vitamin-D-stimulated, calcium-dependent ATPase has been demonstrated in the brush borders of small intestine. Explorations of these two systems will undoubtedly cast much light on the mechanism of vitamin-D-induced calcium transport in the intestine.

In sum, one role of vitamin D is to maintain the body's calcium supply by promoting calcium absorption in the intestine. Its second role is promoting bone mineralization.

---

* In Latin America and the Orient, almost no milk is consumed. Nevertheless, the diet provides a fairly good calcium supply.

† The intestinal mucosa secretes about 600 mg of calcium into the intestinal lumen each day, since the calcium concentration in its various secretions equals that in plasma. However, most of this is reabsorbed.

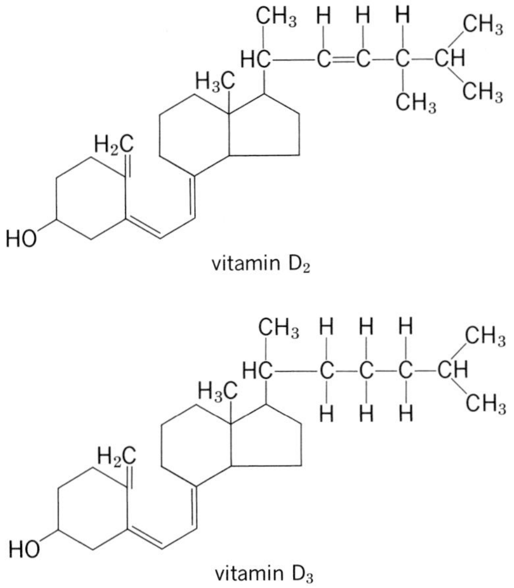

CH3 H H H CH3
HC—C=C—C—CH
H3C   CH3 CH3
H2C
HO
vitamin D2

CH3 H H H CH3
HC—C—C—C—CH
H3C   H H H CH3
H2C
HO
vitamin D3

15.22 Two of the D vitamins.

Vitamin D activity actually resides in several different sterols closely related to cholesterol. The two best-known are *vitamin D₂*, or *ergocalciferol* (usually called *calciferol*), and *vitamin D₃*, or *cholecalciferol* (Fig. 15.22, see Table 4.1). Vitamin D_3 is synthesized in the skin as a result of irradiation by the ultraviolet rays of sunlight.* Therefore, vitamin D_3 can be classified as a hormone: it is manufactured in one organ, the skin, and acts upon other organs, intestine and bone. However its production occurs only when the skin is exposed to sunlight. Since man gave up living in the nude, a substance that was once partially provided by internal synthesis—that is, a hormone—has become an essential factor in the diet—that is, a vitamin.

Studies with highly radioactive vitamin D preparations revealed that the vitamin is converted to at least three biologically active metabolites. Of major importance is a polar metabolite of vitamin D_3 now known as *25-hydroxycholecalciferol* (25-HCC) that was first described in 1966 and later isolated, identified, and synthesized. It acts much more rapidly than the parent compound vitamin D_3 in the cure of rickets and in the stimulation of intesti-

nal calcium transport. Experiments have shown that in isolated cultures 25-HCC in small amounts stimulates intestinal transport of calcium and resorption of bone, whereas huge doses of vitamin D_3 are ineffective. Thus it appears that 25-HCC is the metabolically active form of vitamin D_3. Recently, 25-hydroxyergocalciferol (25-HEC) has also been isolated and identified in the plasma of pigs given vitamin D_2. This compound probably represents the metabolically active form of vitamin D_2 as well.

The liver is the apparent locus of the hydroxylation of vitamin D_3 to 25-HCC. An enzyme system that carries out the 25-hydroxylation of vitamin D_3 has been found in liver homogenates. Conceivably, the pathologic state of vitamin D resistance may result from insufficiency of the hydroxylating enzyme.

Calcium and phosphate in body fluids. Calcium is present in the serum in three different forms. Half of it is bound to plasma proteins and hence is nondiffusible through capillary membranes. Most of this is bound to albumin, but small amounts are bound to γ-globulin and β-globulin. Another 5% of the serum calcium is bound to plasma substances but is diffusible through capillary membranes. The remaining 45% is freely diffusible ionic calcium (Ca^{++}). It is this fraction that is essential to the calcium-requiring systems of the body—heart, muscles, blood, nerves, and bones. A dynamic equilibrium exists between protein-bound calcium and Ca^{++}. If Ca^{++} is lost, some of the protein-bound calcium replaces it.

When the concentration of calcium in the plasma is increased by the addition of calcium, either in vivo or in vitro, the ratio of protein-bound calcium to Ca^{++} remains constant until the calcium level reaches a critical point, whereupon a calcium-phosphate complex forms, and the percentage of Ca^{++} drops sharply. In normal plasma the critical level is 20 mg per 100 ml. That the critical level is lower when the plasma phosphate level is high indicates a specific relationship between plasma phosphate and plasma calcium concentrations.†

Inorganic phosphate is present in the plasma

---

* A narrow range of ultraviolet light (maximum 2540 Å) is necessary to activate vitamin D precursors. This is why sunlight is more effective in summer than in winter and in rural smog-free areas than in cities.

---

† The product of the calcium and phosphate concentrations expressed in milligrams per 100 ml remains virtually constant —30 to 40 for adults and 40 to 55 for growing children.

as HPO_4^{--} and $H_2PO_4^-$, in proportions varying with the pH (see Chapter 2). The total phosphate concentration is normally 3.5 mg per 100 ml in an adult and slightly more in a child.

How the kidneys excrete calcium is still not fully understood, but they excrete from 50 to 65% of an infused dose within 12 hours. That it is cleared from the plasma at a constant rate implies a specific renal excretory mechanism. The Ca^{++} concentration in the glomerular filtrate equals that in plasma; therefore, the 180 liters of daily filtrate contains about 9 g of Ca^{++}. Most of the filtered Ca^{++} is reabsorbed by the tubules. Only about 1%—or 105 mg—appears in normal urine. Even when the amount filtered substantially increases, the amount in urine rarely exceeds 1000 mg.

Bone metabolism. We have considered the intestinal absorption and the transport and excretion of calcium. Far more important in the maintenance of the plasma calcium level is bone metabolism.

Until recently bone was regarded as an inert tissue, but isotopic studies have revealed that it is constantly being remodeled. We learned in Chapter 5 that the inorganic portion of bone is mainly crystals of hydroxyapatite, $Ca_{10}(PO_4)_6(OH)_2$. Nevertheless, the first calcium salts produced in the process of bone mineralization constitute a compartment termed *exchangeable bone* because they are readily soluble in body fluids unsaturated in calcium and phosphate. In an adult 99% of the calcium salts is in the form of *nonexchangeable bone*—a solid mass that does not liberate its ions readily.

In a normal adult enough bone is broken down each day by cells called *osteoclasts* to release about 500 mg of calcium into the body fluids. This is replaced by an equivalent amount of new bone salts—that is, 500 mg of calcium is deposited in the skeleton each day. Since the adult skeleton contains a total of 1 kg of calcium, this means an overall renewal rate of 18% per year.

Calcium ions of extracellular fluid are in direct equilibrium with those of exchangeable bone, the rate of bone mineralization being directly dependent upon the concentrations of calcium and phosphate. When the product of their concentrations at a bone surface reaches a critical point, salts are deposited, and new bone is formed. The concentrations of calcium and phosphate in plasma are too low to permit bone salts to precipitate, and specific mechanisms operate at the bone surface to achieve saturation. We shall discuss these mechanisms in Chapter 17.

PARATHYROID HORMONE

Purification and properties. Parathyroid hormone was isolated by G. D. Aurbach in 1959. H. Rasmussen and L. C. Craig then found that it is a peptide (molecular weight 8500) and determined its amino acid composition.* Studies reported in 1965 by J. T. Potts, Jr., and co-workers indicated that biological activity resides in approximately 25% of the molecule, a sequence 20 amino acids long at the *C*-terminal end (Fig. 15.23). Hence parathyroid hormone shares with α-MSH and ACTH the interesting feature that only a portion of its amino acid sequence is necessary for biological function.

Locus of action. We have already seen in the results of parathyroidectomy that parathyroid hormone affects the plasma calcium level. It was early believed that the critical function of the hormone is to maintain a normal plasma calcium level and that it does so by acting directly upon bone, accelerating bone resorption. It was then found that the hormone also lowers the serum phosphate level, and in 1929 Albright proposed that it acts primarily on the kidneys, promoting phosphate excretion. According to his view, the loss of phosphate through the kidneys leads to a decrease in plasma phosphate, and bone, exposed to extracellular fluid undersaturated in phosphate, releases bone salts that incidentally elevate the plasma calcium level.

Recently research has shown that parathyroid hormone acts directly on both bone and kidneys —but that its direct effect on bone is responsible for its critical function, maintenance of a normal plasma calcium level. These discoveries were

---

* Parathyroid hormone is not stored in the parathyroids but is synthesized and continuously secreted. To obtain a few milligrams of hormone, these investigators had to process fresh glands from hundreds of cattle. Their procedure culminated in repeated countercurrent distributions (see Chapter 2), the final step involving 2600 transfers.

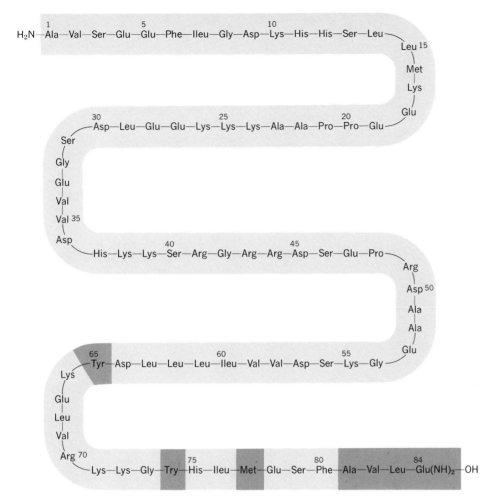

15.23 Structure of parathyroid hormone. The 34-residue sequence at the N-terminal end is biologically and immunologically active. Alteration of methionine, tryptophan, or tyrosine or the four residues at the C-terminal end reduces biological activity.

made possible by the purification of parathyroid hormone. Although the two major effects of parathyroid hormone, elevation of the plasma calcium level by a direct action on bone and depression of the plasma phosphate level by a direct action on the renal tubules, appear superficially to involve different molecular mechanisms, they are produced by a single hormone. The purity of the preparation used in reaching this conclusion precludes doubt that either effect could be due to a second hormone contaminating it.

The direct action of parathyroid hormone on bone was implied by the demonstration that the hormone can raise the plasma calcium level after removal of the kidneys from an experimental animal. It was confirmed by a series of ingenious tissue culture experiments in which thin sections of mouse bone were cultivated in vitro in the presence and absence of parathyroid hormone. The hormone caused bone resorption by intensifying osteoclast activity.

While this work was in progress, several laboratories established that parathyroid hormone acts directly upon the kidneys. When a solution of pure parathyroid hormone was infused into one of a dog's two renal arteries, a significant increase in the amount of phosphate in the urine from the infused side and a smaller increase in the urine from the other side were observed.

How parathyroid hormone alters the behavior of osteoclasts and decreases the tubular reabsorption of phosphate is not known. Recent data suggest that it acts by stimulating adenyl cyclase to form increased amounts of cyclic 3′, 5′-adenylic acid (see p. 607). There is reason to believe that these effects are mediated through induction of new enzyme formation (see p. 156).

A fall in the plasma calcium level stimulates the secretory activity of the parathyroids. Parathyroid hormone is transported in the blood to the bone cells, which it converts to osteoclasts that break down nonexchangeable bone and liberate calcium and phosphate, and to the renal tubule cells, which it induces to retain calcium and eliminate phosphate. These effects lead to an increase in the concentration of calcium in the blood and extracellular fluid. If it were not for the renal effect, calcium and phosphate levels would both rise, favoring crystal (hydroxyapatite) formation, which would lower the calcium level. The renal effect lowers the phosphate level, more than offsetting the increase in phosphate from bone resorption, while maintaining the calcium level. The net result is an increase in the amount of calcium and a decrease in the amount of phosphate in the blood and extracellular fluid. When the calcium level is restored to normal, the parathyroids reduce their output of hormone.

Relations of parathyroid hormone and vitamin D. We have seen that, unless the plasma calcium and phosphate concentrations are kept at certain levels, bone formation stops. The mechanisms for maintaining these levels are different in a growing child and an adult. During active growth the maintenance of an adequate calcium supply is primarily the function of vitamin D. It promotes calcium and phosphate retention in the body, thus reducing their excretion.

When vitamin D is lacking, total body calcium and phosphate decrease, bone mineralization ceases, and the bones stop growing, bend, and break. The condition is known as *rickets*. Obviously vitamin D cannot substitute for calcium in the diet. It can, however, cause maximum retention of whatever calcium is available and, by also causing phosphate retention, favor maintenance of the two ions at concentrations high enough for bone formation. Once the skeleton has matured, the relative importance of vitamin D diminishes, and maintenance of the plasma calcium level is primarily the task of the parathyroids.*

Whether vitamin D is necessary for parathyroid hormone action within the cells is unclear. There is evidence that both vitamin D and parathyroid hormone have qualitatively similar effects on bone, kidney, and intestine, although their relative potency differs at each site. Active calcium transport by intestine is impaired or absent in vitamin D deficiency, but once vitamin D intake is adequate, calcium transport may be affected by diet, age, or pregnancy, and these changes are independent of the vitamin. Similarly, as we have noted, vitamin D is required for bone resorption; in its absence plasma calcium decreases in a manner unresponsive to administered parathyroid hormone. These findings indicate a permissive role for vitamin D in calcium transport. Both vitamin D and parathyroid hormone stimulate the release of calcium from mitochondria in vitro. However, although vitamin D stimulates the release from mitochondria from parathyroidectomized animals, parathyroid hormone does not stimulate the release from mitochondria from vitamin D–deficient animals unless vitamin D is present. If vitamin D–deficient animals are less sensitive than normal animals to parathyroid hormone, it is not clear whether their poor resonse is due to a skeletal abnormality or to a specific requirement of vitamin D by the cells.

Thyrocalcitonin. In 1962 D. H. Copp and co-workers observed that perfusion of a dog's thyroid and parathyroids with blood containing a high concentration of calcium led to a more rapid fall in the plasma calcium level than could be accounted for simply by inhibition of parathyroid hormone secretion. They concluded that a hormone which lowers the plasma calcium level is liberated by parathyroids perfused with high-calcium blood. They called the hormone *calcitonin.* Later investigation revealed that a similar hormone is liberated by the thyroid, specifically by cells adjacent to the follicle walls, the *parafollicular cells.* This was called *thyrocalcitonin.* It now appears that calcitonin and thyrocalcitonin are identical and that the substance originates exclusively in the thyroid.

---

* In addition to stimulating the intestinal absorption of calcium, vitamin D acts similarly to parathyroid hormone in mobilizing calcium from bone. Because these effects tend to elevate the serum calcium level, vitamin D in large doses is frequently used in the treatment of parathyroid deficiency. Although vitamin D and parathyroid hormone are the two principal participants in this system, other hormones are also involved. These include the adrenocortical hormones, growth hormone, insulin, thyroid hormone, and the sex hormones. All influence one or more aspects of bone growth and resorption or calcium and phosphate metabolism. Ultimately they can also influence parathyroid function.

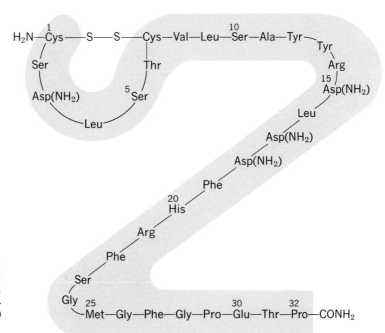

15.24 Structure of thyrocalcitonin. (Adapted from Potts, J. T., Jr., Niall, H. D., Keutmann, H. T., Brewer, H. B., Jr., and Deftos, L. J., *Proc. Nat. Acad. Sci. U.S.*, *59*, 1321, 1968.)

In 1968 the structure of thyrocalcitonin was elucidated by J. T. Potts and the hormone was synthesized by G. W. Anderson and P. H. Bell. It is a low–molecular weight single-chain peptide of 32 amino acids (Fig. 15.24) that contains no iodine and is unrelated to thyroglobulin. Unlike parathyroid hormone, it lowers both plasma calcium and plasma phosphate levels, probably by reducing bone resorption. Its physiological significance remains to be established in man. It may play a minor role in normal calcium homeostatis and may be of value mainly at times of hypercalcemia, pregnancy, and lactation.

Primary and secondary parathyroid dysfunction. The complex interrelationships of the several organ systems controlling calcium and phosphate metabolism offer many opportunities for functional abnormalities. As might be expected, when primary disease occurs in any one of the systems—bones, small intestine, kidneys, or parathyroids—secondary changes occur in all the others. Often the secondary phenomena are of a compensatory nature in that they tend to reverse the consequences of the primary defect.

For example, *hyperparathyroidism* usually arises from benign tumors of the parathyroids. It is charac-

terized by striking modifications in the functions of the bones, kidneys, and other body systems, due to a high plasma calcium level—i.e., hypercalcemia—maintained in part by calcium withdrawn from the bones. The skeleton may be greatly weakened through loss of calcium. The bones may fracture spontaneously and are generally deformed by cysts and fibrous tissue. Excess calcium may be deposited in other tissues, particularly the kidneys (as kidney stones). Loss of muscle tone and extreme muscular weakness—the opposite of tetany—are common. Surgical removal of excess parathyroid tissue is the best treatment.

Hypoparathyroidism has no apparent cause in some cases. In others it results from inadvertent removal of the parathyroids during thyroid surgery. Its major manifestation is increased neuromuscular activity, due to a low plasma calcium level. The muscles of the face contract or twitch when the facial nerve is tapped lightly. The fingers fold together in a characteristic fashion when pressure is applied to the arm. Severe hypoparathyroidism culminates in tetany and death. Although parathyroid hormone would seem to be the logical therapeutic agent for the disorder, since it raises the serum calcium level and lowers the serum phosphate level, it is not available in purified form in large quantities. Therefore, treatment usually consists of increasing the calcium and decreasing the phosphorus in the diet and administering massive doses of vitamin D_2 (calciferol).

Poorly Understood Endocrine Glands

Two additional structures shown in Fig. 15.1 must be mentioned, though their physiology is poorly understood. Indeed, it is not yet certain that they have endocrine functions. They are the thymus and the pineal body.

THYMUS

We discussed the thymus in Chapter 8, and we shall further examine it as part of the body's immune mechanisms in Chapter 19. Before recognition of its immunological function, many attempts were made to discover a hormonal secretion of the thymus. All were unsuccessful. Results of administration of thymus extracts and of thymectomy were inconclusive. Nevertheless, one or more thymus hormones may exist. Albert Szent-Györgyi reported in 1962 that he had finally isolated from calf thymus minute amounts of two chemicals. One, a potent promoter of growth, was named *promine;* the other, an equally potent inhibitor of growth, was named *retine.* Whereas promine stimulated the growth of malignant tumors in animals, retine inhibited their growth.

Retine was found not only in thymus but also in other organs. Hence it must be a general tissue constituent and not a specific thymus product. On the other hand, promine was found in no other organs. Hence it may be a thymus hormone. In view of promine's postulated role in growth, it is noteworthy that the thymus reaches its peak of development when the body is growing fastest. The physiological significance of promine and retine has not been assessed.

Abraham White reported in 1968 that thymus may produce yet another hormone. Thymus extracts contain a substance, which White named *thymosin,* that stimulates the proliferation of lymphoid tissue, enhances immune mechanisms, and promotes the survival and growth of thymectomized newborn mice. Thymosin is unrelated to promine and retine.

PINEAL BODY

The small dorsal outgrowth from the roof of the diencephalon (see Figs. 12.30, 12.31) is termed the pineal body or gland because it resembles a pine cone. Although it has an interesting evolutionary history as a photoreceptor (see p. 445), it has long been considered a vestige in man. Moreover, it has been considered unimportant because it calcifies in half of all adults.

Despite these views, its epithelioid structure has suggested that the pineal body is capable of hormone synthesis, and investigators have at last isolated a pineal hormone, *melatonin,* which affects skin pigmentation (see p. 599). It is now believed that light impinging on the retina generates impulses that travel via the sympathetic nerves to the superior cervical ganglia and thence via postganglionic fibers to the pineal body, where they cause the cells to secrete melatonin. The pineal body, then, is a neuroendocrine transducer—similar to the hypothalamus, whose cells respond to nervous input by secreting hormones that regulate hypophyseal function, and the adrenal medulla, whose cells respond to nervous input by secreting epinephrine and norepinephrine. Melatonin will be treated in detail in Chapter 18.

REFERENCES AND SUGGESTIONS FOR FURTHER READING

General References and Advanced Textbooks

Astwood, E. B., ed., *Clinical Endocrinology,* Grune & Stratton, New York, Vol. 1, 1960, Vol. 2, 1968.

Dorfman, R. I., ed., *Methods in Hormone Research,* Vols. 1–2, Academic Press, New York, 1962.

Litwack, G., and D. Kritchevsky, eds., *Mechanisms of Hormone Action,* Wiley, New York, 1964.

Pincus, G., K. V. Thimann, and E. B. Astwood, *The Hormones: Physiology, Chemistry, and Applications,* Vols. 1–5, Academic Press, New York, 1956–1964.

Tomkins, G., "Mechanism of Action of Hormones," *Ann. Rev. Biochem.,* **32,** 677 (1963).

Von Euler, U. S., and H. Heller, *Comparative Endocrinology,* Vols. 1–2, Academic Press, New York, 1963.

Williams, R., *Textbook of Endocrinology,* Saunders, Philadelphia, 1962.

Neurohypophysis

Du Vigneaud, V., "Hormones of the Posterior Pituitary Gland: Oxytocin and Vasopressin," *Harvey Lectures, Ser. 50 (1954–55)*, 1, 1956.

Farrell, G., L. F. Fabre, and E. W. Rauschkolb, "The Neurohypophysis," *Ann. Rev. Physiol.*, **30**, 557 (1968).

Heller, H., ed., *The Neurohypophysis*, Academic Press, New York, 1957.

Pinkerton, J. H. M., ed., *Advances in Oxytocin Research*, Pergamon Press, New York, 1965.

Sawyer, W. H., "Comparative Physiology and Pharmacology of the Neurohypophysis," *Recent Progr. Hormone Res.*, **17**, 437 (1961).

Adenohypophysis

Burgus, K., and R. Guillemin, "Hypothalamic Releasing Factors," *Ann. Rev. Biochem.*, **39**, 499 (1970).

Friesen, H., and E. B. Astwood, "Hormones of the Anterior Pituitary Body," *New Engl. J. Med.*, **272**, 1216, 1272, 1328 (1965).

Harris, G. W., *Hypothalamic-Hypophyseal Interrelationships*, Thomas, Springfield, Ill., 1956.

————, and B. T. Donovan, eds., *The Pituitary Gland*, Vols. 1–3, Univ. of Calif. Press, Berkeley, 1966.

Hofmann, K., "Chemistry and Function of Polypeptide Hormones," *Ann. Rev. Biochem.*, **31**, 213 (1962).

Knobil, E., and J. Hotchkiss, "Growth Hormone," *Ann. Rev. Physiol.*, **26**, 47 (1964).

Lerner, A. B., and J. S. McGuire, "Melanocyte-Stimulating Hormone and Adrenocorticotropic Hormone. Their Relation to Pigmentation," *New Engl. J. Med.*, **270**, 539 (1964).

Li, C. H., "The ACTH Molecule," *Sci. Am.*, **209**, 46 (July, 1963).

McCann, S. McD., A. P. S. Dhariwal, and J. C. Porter, "Regulation of the Adenohypophysis," *Ann. Rev. Physiol.* **30**, 589 (1968).

Reichlin, S., "Neuroendocrinology," *New Engl. J. Med.*, **269**, 1182, 1246, 1296 (1963).

Venning, E. H., "Adenohypophysis and Adrenal Cortex," *Ann. Rev. Physiol.*, **27**, 107 (1965).

Adrenal Cortex

Addison, T., *Disease of Suprarenal Capsules*, Highley, London, 1855.

Bransome, E. D., Jr., "Adrenal Cortex," *Ann. Rev. Physiol.*, **30**, 171 (1968).

Currie, A. R., T. Symington, and J. K. Grant, eds., *The Human Adrenal Cortex*, Williams & Wilkins, Baltimore, 1962.

Fieser, L. F., and M. Fieser, *Steroids*, Reinhold, New York, 1959.

McKerns, K. W., ed., *Functions of the Adrenal Cortex*, Vols. 1–2, Appleton-Century-Crofts, New York, 1968.

Moon, H. D., ed., *Adrenal Cortex*, Hoeber, New York, 1961.

Reichstein, T., and C. W. Shoppee, "The Hormones of the Adrenal Cortex," *Vitamins Hormones*, **1**, 346 (1943).

Venning, E. H., "Adenohypophysis and Adrenal Cortex," *Ann Rev. Physiol.*, **27**, 107 (1965).

Adrenal Medulla

Abel, J. J., "Methods of Preparing Epinephrine and Its Compounds," *Bull. Johns Hopkins Hosp.*, **13**, 29 (1902).

Axelrod, J., "Metabolism of Epinephrine and Other Sympathomimetic Amines," *Physiol. Rev.*, **39**, 751 (1959).

Blaschko, H., "Development of Current Concepts of Catecholamine Formation," *Pharmacol. Rev.*, **11**, 307 (1959).

Coupland, R. E., *The Natural History of the Chromaffin Cell*, Longmans, Green, London, 1965.

Krayer, O., ed., "Symposium on Catecholamines," *Pharmacol. Rev.*, **11**, 233 (1959).

Malmejac, J., "Activity of the Adrenal Medulla and Its Regulation," *Physiol. Rev.*, **44**, 186 (1964).

Thyroid

Barker, S. B., "Peripheral Action of Thyroid Hormones," *Federation Proc.*, **21**, 635 (1962).

Bogdanove, E. M., "Regulation of TSH Secretion," *Federation Proc.*, **21**, 623 (1962).

Care, A. D., "Secretion of Thyrocalcitonin," *Nature*, **205**, 1289 (1965).

Cassano, C., and M. Andreoli, eds., *Current Topics in Thyroid Research*, Academic Press, New York, 1965.

De Groot, L. J., "Current Views on Formation of Thyroid Hormones," *New Engl. J. Med.*, **272**, 243, 297, 355 (1965).

Fraser, R., "The Thyroid," *Ann. Rev. Med.*, **11**, 171 (1960).

Halmi, N. S., "Thyroidal Iodide Transport," *Vitamins Hormones*, **19**, 133 (1961).

Hazard, J. B., and D. E. Smith, *The Thyroid*, Williams & Wilkins, Baltimore, 1964.

Keating, F. R., Jr., ed., "Symposium on Thyroxine," *Proc. Staff Meetings Mayo Clinic*, **39**, 545 (1964).

Maloof, F., and M. Soodak, "Intermediary Metabolism of Thyroid Tissue and Action of Drugs," *Pharmacol. Rev.*, **15**, 43 (1963).

Means, J. H., L. J. De Groot, and J. B. Stanbury, *The Thyroid and Its Diseases*, 3rd ed., McGraw-Hill, New York, 1963.

Pitt-Rivers, R., and W. R. Trotter, eds., *The Thyroid Gland*, Butterworths, London, 1964.

Rosenberg, I. N., and C. H. Bastomsky, "The Thyroid," *Ann. Rev. Physiol.*, **27**, 71 (1965).

Solomon, D. H., and J. T. Dowling, "The Thyroid," *Ann. Rev. Physiol.*, **22**, 615 (1960).

Werner, S. C., and J. A. Nauman, "The Thyroid," *Ann. Rev. Physiol.*, **30**, 213 (1968).

Wolff, J., "Transport of Iodide and Other Anions in the Thyroid Gland," *Physiol. Rev.*, **44**, 45 (1964).

Parathyroids

Albright, F., and E. C. Reifenstein, Jr., *Parathyroid Glands and Metabolic Bone Disease*, Williams & Wilkins, Baltimore, 1948.

Aurbach, G., "The Parathyroids," *Advan. Metab. Disorders*, **1**, 45 (1964).

Behrens, O. K., and E. L. Grinnan, "Polypeptide Hormones," *Ann. Rev. Biochem.*, **38**, 83 (1969).

Cope, O., "The Story of Hyperparathyroidism at the Massachusetts General Hospital," *New Engl. J. Med.*, **274**, 1174 (1966).

Copp, D. H., "Endocrine Regulation of Calcium Metabolism," *Ann. Rev. Physiol.*, **32**, 61 (1970).

Munson, P. L., P. F. Hirsch, and A. H. Tashjian, Jr., "Parathyroid Gland," *Ann. Rev. Biochem.*, **25**, 324 (1953).

Potts, J. T., Jr., G. D. Aurbach, L. M. Sherwood, and A. Sandoval, "Structural Basis of Biological and Immunological Activity of Parathyroid Hormone," *Proc. Natl. Acad. Sci. U. S.*, **54**, 1743 (1965).

Rasmussen, H., and A. Tenenhouse, "Cyclic Adenosine Monophosphate, Ca^{++}, and Membranes," *Proc. Natl. Acad. Sci.*, **59**, 1364 (1969).

Thymus and Pineal Body

Relkin, R., "The Pineal Gland," *New Engl. J. Med.*, **274**, 994 (1966).

Szent-Györgyi, A., A. Hegyeli, and J. A. McLaughlin, "Constituents of the Thymus Gland and Their Relation to Growth, Fertility, Muscle, and Cancer," *Proc. Natl. Acad. Sci. U. S.*, **48**, 1439 (1962).

Wurtman, R., and V. Axelrod, "The Pineal Gland," *Sci. Am.*, **213**, 50 (July, 1965).

The most vaporous of women owes her sheerest femininity to a certain complex alcohol or sterol which can, among its other capacities, change the plumage of the capon and swell the uterus of the mouse. As for man, he is forced to admit that he gets his proud virility from another sterol (incidentally, only slightly different from the first) which also acts to darken the sparrow's beak and the thumbs of the frog. And these two principles—estrogen and testosterone, so powerfully and diversely morphogenic—are not restricted to working upon the body: they affect the instincts, the tendencies, the desires. . . .

Whether one likes or not, and whatever the idealism one may subscribe to, the whole edifice of human love—with all the word implies of animality and sublimation, of rage and sacrifice; with all that is frivolous, touching, or terrible in its meaning—is constructed upon the minimal molecular differences among a few derivatives of phenanthrene.

Jean Rostand, in *The Substance of Man* (1962)

16 Reproductive System

Introduction

HISTORICAL NOTES

Though the heart has been known for centuries to be a mechanical pump, it remains the symbolic seat of the emotions. In a similar way the uterus retains much of its ancient symbolic meaning, despite advances in our knowledge of biology and endocrinology. The Greek word for uterus, *hystera*, is the root of the word *hysteria;* and, according to Greek medical writings, "the uterus causes a thousand evils," and it is "the cause of all diseases." Presumably such ideas arose from the unsophisticated view that the uterus, the source of life, is a living being itself, endowed with the power of moving about the body. Indeed, a number of cultures believed that it could leave the body and walk cross-country.

It is fascinating to examine these old superstitions, whose distortions of anatomy and physiology obviously mirrored current beliefs in magic and mysticism. Many early drawings, for example, suggested that the semen of the male is deposited directly in the uterus, the vagina and uterus being depicted as a single canal of uniform diameter. Many anatomists thought that the embryo is nourished by suckling in the uterus, since a newborn mammal knows how to suckle. Hippocrates held that retained menstrual blood is the food of the embryo and that after delivery this blood travels through a direct canal to the breasts, where it is converted to milk. One of Leonardo da Vinci's drawings actually shows the vessel through which the blood was supposed to pass. Many similar concepts could be quoted.

These ancient errors were corrected following the dawn of modern science, and today the study of reproduction is the province of many disciplines: anatomy, endocrinology, genetics, obstetrics and gynecology, biochemistry, and zoology. We shall find in this chapter, however, that numerous problems are still unsolved. Though we have put down the burden of the centuries attributing mystical properties to the uterus, many mysteries still surround the organ of which John Wilmot, Earl of Rochester, said, "On this soft anvil all mankind is made."

BIOLOGY OF REPRODUCTION

We mentioned the probable origins of sexual reproduction in Chapter 3. Unicellular organisms, the first to appear in evolution, multiplied by the asexual process of division and were thus limited in their ability to vary genetically. Sex both facilitated the reproduction of multicellular organisms and provided a method by which two cells of differing genetic constitutions fuse to form one. The result was an immense increase in the frequency and ease of genetic variation.

Though we do not know exactly how sexual reproduction began, we may surmise that it developed from a process resembling phagocytosis. One cell may have engulfed another, and the two sets of genetic materials may have combined to yield a new genetic pattern. In time, certain cells may have come to resemble spermatazoa, and others to resemble ova, so that the act of fusion was regularized. We have already noted that sexual conjugation occurs between certain "male" and "female" bacteria. And it has been recently shown by E. N. Willmer that a certain protozoan can be made to appear sluggish, ameboid, and "ovalike" under one set of environmental conditions and swift-moving, flagellated, and "spermlike" under another. These may be clues to the mechanism by which sex emerged in early evolution.

However it emerged, it has been supremely successful in introducing genetic variation and thereby promoting evolutionary progress. Sexual reproduction is universal among vertebrates. In certain invertebrates, hermaphroditism—in which the same individual functions as male and female—is common. In vertebrates male and female are invariably separate individuals.

We shall presently speak of the differentiation of ova and sperm within the human embryo. It is evident that the cells destined to produce the next generation arise early in embryogenesis. They come to cluster in the region of the primitive excretory apparatus (see Fig. 10.2), where they develop into gonads—ovaries or testes—and become associated with the tubes and ducts of the female or male reproductive system.

Female Reproductive System

STRUCTURE AND DEVELOPMENT

The principal organs of the female reproductive system are shown in Fig. 16.1. The internal organs are the *ovaries, Fallopian tubes, uterus,* and *vagina.* Reproduction begins with the development of the ova in the ovaries. Each month a single ovum matures in an ovary and is expelled from the ovarian surface into the abdominal cavity in a process known as *ovulation.* The ovum then passes down one of the Fallopian tubes to the uterus. If it is fertilized by a spermatozoon during its passage through the tube, it becomes implanted in the lining of the uterus, where it develops into an embryo, with membranes and a placenta. An embryo from about the beginning of the third month, when it is recognizably human, until birth is usually called a *fetus.* Accordingly, we refer to fetal membranes, fetal hemoglobin (see Chapter 8), etc. If the ovum remains unfertilized, it breaks down and is lost in mucous secretions.

Ovaries. The paired ovaries lie on either side of the uterus below the Fallopian tubes. They are oblong bodies 2½ to 4 cm in length and 1½ to 2 cm from front to back. They lie behind the *broad ligaments,* support-

16.1 Uterus: A, Anterior view; B, Sagittal section through female pelvis; C, Posterior view of uterus and adnexae; D, Structure of wall. (From R. W. Kistner, *Gynecology: Principles and Practice.* Copyright © 1964, Year Book Medical Publishers. Used by permission.)

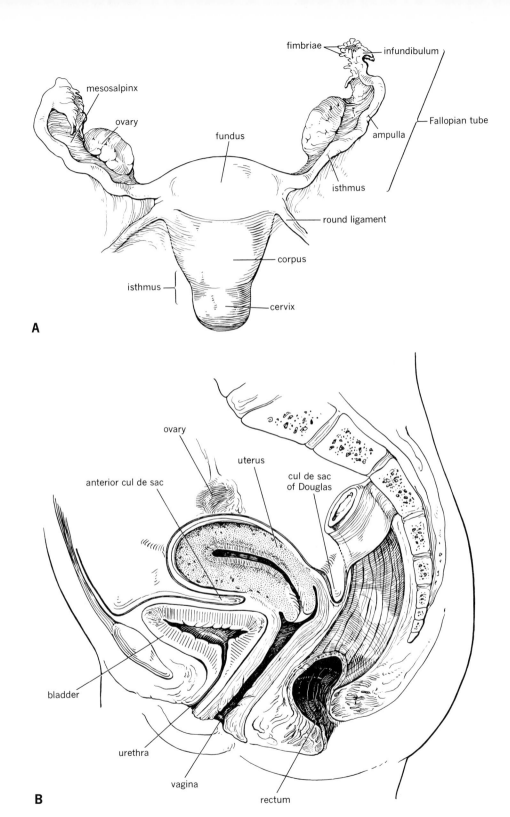

A

fimbriae — infundibulum

mesosalpinx

ovary

fundus

Fallopian tube

ampulla

isthmus

round ligament

corpus

isthmus

cervix

B

ovary

anterior cul de sac

uterus

cul de sac
of Douglas

bladder

urethra

vagina

rectum

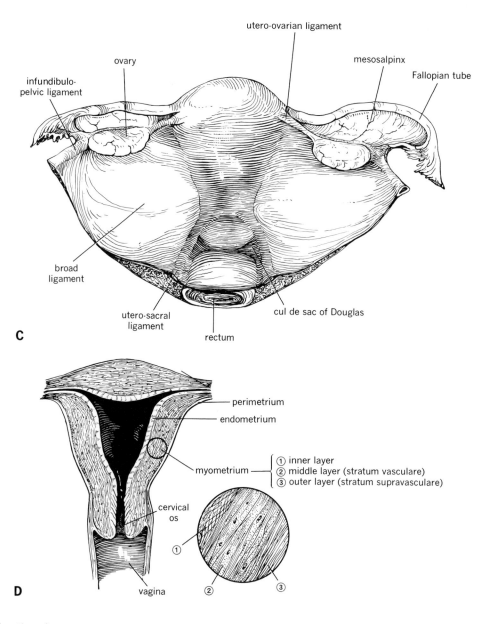

utero-ovarian ligament

ovary

infundibulo-
pelvic ligament

mesosalpinx

Fallopian tube

broad
ligament

utero-sacral
ligament

rectum

cul de sac of Douglas

C

perimetrium

endometrium

myometrium ⎰ ① inner layer
⎱ ② middle layer (stratum vasculare)
  ③ outer layer (stratum supravasculare)

cervical
os

vagina

①

②

③

D

16.1 *(continued)*

ing folds of the peritoneum, and are attached to the uterus by *ovarian ligaments*.

Internally an ovary consists of a *medulla* containing muscle cells and incoming blood vessels and nerves and a *cortex* containing numerous ova and follicles in various stages of development. The cortex is concerned mainly with the maturation of ova and the secretion of

ovarian hormones (Fig. 16.2). The surface of the ovary is covered by a delicate, single-layered membrane of cuboidal or low-columnar epithelium called the *germinal epithelium*. This is broken only when a follicle ruptures at ovulation.

The developmental history of the cellular components of the ovary has been debated for over 50 years.

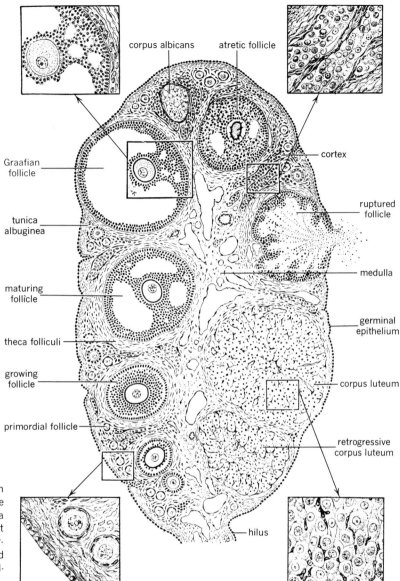

16.2 Composite diagram of an adult ovary, showing progressive stages in the differentiation of a follicle. Cycle starts at lower left part of diagram. (From C. D. Turner, *General Endocrinology*, 3rd ed., W. B. Saunders Co., Philadelphia, 1960.)

(Labels in figure:) corpus albicans · atretic follicle · Graafian follicle · tunica albuginea · maturing follicle · theca folliculi · growing follicle · primordial follicle · cortex · ruptured follicle · medulla · germinal epithelium · corpus luteum · retrogressive corpus luteum · hilus

The current theory is that ovarian development occurs in four major phases. In the first phase *primordial germ cells*, segregated very early, migrate from their sites of origin to settle in a localized thickening of the coelomic epithelium on either side of the dorsal mesentery (see Figs. 10.1, 10.2). This thickening appears in the fifth to sixth week of embryonic life. It divides lengthwise into a lateral *mesonephric ridge* and a medial *genital ridge* (Fig. 16.3).

In the second phase both germ and nongerm cells proliferate, forming distinct gonadal primordia known as *indifferent gonads* because they are identical in the two sexes. At the end of the sixth week, the third phase begins, with a peripheral cortex and an underlying central medulla becoming evident. The germ cells stay mainly in the cortex just beneath the epithelium. In the fourth and final phase, sex differentiation takes place. In a female the cortex proliferates, and the medulla involutes, forming an ovary; in a male the medulla proliferates, and the cortex involutes, forming a testis.

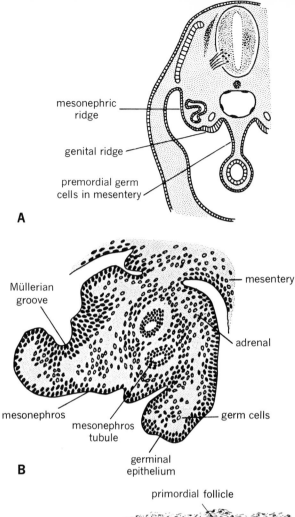

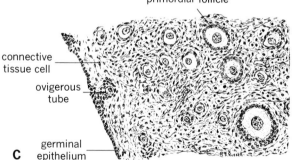

A

mesonephric ridge

genital ridge

premordial germ cells in mesentery

Müllerian groove

mesentery

adrenal

mesonephros

germ cells

mesonephros tubule

germinal epithelium

B

primordial follicle

connective tissue cell

ovigerous tube

germinal epithelium

C

16.3 Embryological development of the ovary: A, in a 7-mm embryo; B, in a 12-mm embryo; C, in a newborn infant. (A and B from *J. of Amer. Med. Assoc., 174:*1316 (1960). Reproduced by permission of the *J. Am. Med. Assoc.* and the author, Milton L. Bassis. C, from B. M. Patten, *Human Embryology*, 3rd ed., © 1968 by McGraw-Hill, Inc. Used with permission of McGraw-Hill Book Co., New York.)

Whereas the male germ cells are thus drawn into the medulla, the female germ cells remain in the cortex. It appears likely that certain cells of the indifferent gonads produce substances (hormones or "sex inductors") that determine whether the primordial germ cells are to become ova or spermatozoa. Although this has not been proved, it appears certain that the factors causing sex differentiation are intraembryonic.

During embryonic life, and to some extent during childhood, the female germ cells organize into *cords* and develop into larger *primordial ova*. Most of these migrate to the periphery of the cortex, and cells from the surface invest each with an envelope, thereby converting it into a *primordial follicle* (see Fig. 16.2). The investing cells are termed *granulosa cells*. The primordial follicles compose a thick layer just beneath the *tunica albuginea*, a dense connective tissue layer lying under the germinal epithelium.

The two ovaries contain a total of about 6×10^5 follicles in a 2-month-old embryo and a total of about 7×10^6 follicles in a 5-month-old embryo. Though new follicles continue to form, some of the formed ones undergo *atresia* (i.e., degeneration), so that only 2×10^6 are present at birth, half of which are undergoing atresia. At birth the formation of new follicles ceases, and at 7 years only 3×10^5 follicles remain. Only about 400 ever mature and expel their ova. The rest undergo atresia at various stages in their development so that only a few are still in the ovaries at the end of active sexual life.

Fallopian tubes and uterus. In both sexes a *Müllerian groove* indents each genital ridge (see Fig. 16.3). The groove advances caudally, closes into a tube, and swings medially to the midline near the cloaca where it fuses with its partner into a *genital cord* (Fig. 16.4). When the genital organs develop, the cord becomes the upper vagina and uterus, while the separate ducts become the Fallopian tubes. These tubes, which carry ova from the ovaries to the uterus, are referred to as *oviducts* in lower animals. In man, as shown in Fig. 16.1, they are slender structures lying in a horizontal plane above the ovaries. The ends near the ovaries flare out in funnel-like openings bearing fringed processes (*fimbriae*) or cystlike enlargements (*hydatids*) for guiding the ova into the tubes. Ciliated epithelium lines the tubes, and the beating of the cilia, together with the contractions of the smooth muscles in the walls of the tubes, propels the ova toward the uterus.

The uterus is a thick-walled muscular organ in the upper pelvis, usually tipped forward over the bladder (see Fig. 16.1) but varying widely in position. It is supported on each side by a broad ligament and a *round ligament*. Its cylindrical lowermost portion, the *cervix*, extends into the upper part of the vagina. The orifice of the cervical canal is the *cervical os*.

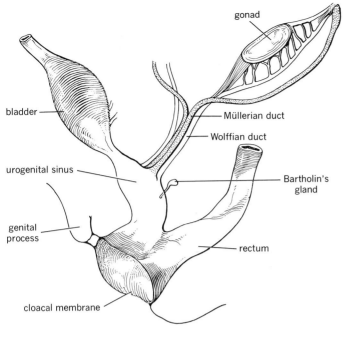

A

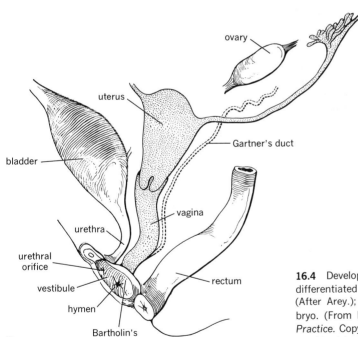

B

16.4 Development of female reproductive system: A, undifferentiated genital system of the 12-14 mm embryo. (After Arey.); B, differentiated system in 150 mm embryo. (From R. W. Kistner, *Gynecology: Principles and Practice.* Copyright © 1964, Year Book Medical Publishers. Used by permission.)

Though small during childhood, the uterus grows during puberty to become 3 in. long, nearly 2 in. wide, and about 1 in. thick. It is capable of great enlargement during pregnancy. Its muscular wall is called the *myometrium*, and its soft, vascular epithelial lining is called the *endometrium*.

Study of the uteri of lower animals provides interesting homologies that illuminate the evolution of the

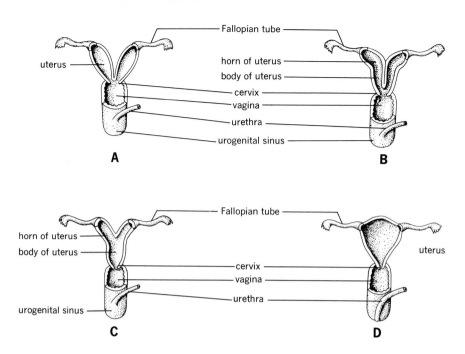

16.5 Progressive fusion of the oviducts in placental mammals: A, duplex uterus; B, bipartite uterus; C, bicornuate uterus; D, simplex uterus. (From A. S. Romer, *The Vertebrate Body,* 2nd ed., W. B. Saunders Co., Philadelphia, 1960.)

reproductive system. As shown in Fig. 16.5, mammals display a progressive tendency for the oviducts to fuse together, so that there may be a *duplex* uterus containing two separate chambers, with the fused segment constituting the vagina (as in marsupials); a *bipartite* or *bicornuate* uterus containing a small central chamber with two lateral horns (as in cattle); or a *simplex* uterus with complete fusion (as in man and other primates). A duplex, bipartite, or bicornuate uterus has ample room for the development of several embryos at once, a satisfactory arrangement for terrestrial animals that encounter little danger on the ground. For tree-dwelling primates and man, natural selection has favored a simplex uterus in which only a single embryo ordinarily develops per pregnancy. It has gone even further by evolving an endocrine system that allows, as a general rule, only one ovulation at a time.

Vagina. The muscular canal extending 7 to 10 cm between the cervix and the external genitalia is the vagina (see Fig. 16.1B). It lies almost perpendicularly to the plane of the uterus. The cervix projects a short distance into the vagina, creating pockets, or *fornices*, between it and the vaginal walls. The posterior fornix is deeper than the lateral and anterior fornices. In cross section the vagina resembles the letter H (Fig. 16.6A). It stretches to form the birth canal at childbirth.

The external opening of the vagina is partially covered by a membranous fold, the *hymen*. Depending upon age, activity, and previous childbirth, the hymen varies considerably in size, shape, and extensibility. Ordinarily it is no obstacle to sexual intercourse, or *coitus*, either because it is sufficiently flexible or because previous athletic or other activity has ruptured it.

Microscopic study of the vaginal walls reveals certain unique features. For the most part they consist of thick stratified squamous epithelium lacking any glands, but at the cervix they change abruptly to simple columnar epithelium. Vaginal epithelium has three main layers (Fig. 16.6B): a *basal* layer of several rows of small, darkly staining oval to round cells; an *intermediate* layer of larger, flatter cells with nuclei; and an inconstant *superficial* layer of flat, wide cells in various stages of cornification. The sex hormones exert a marked influence on this pattern, which we shall describe in connection with estrogen effects.

Vaginal epithelium, particularly the intermediate and superficial layers, is rich in glycogen. Indeed, the amount of glycogen in it during the reproductive years is second only to that in the liver, averaging 1½ to 3%. This too is affected by the sex hormones. Vaginal glycogen leads to active lactic acid production. As a result, the vaginal secretions are notably acidic, with the pH averaging between 4 and 5.

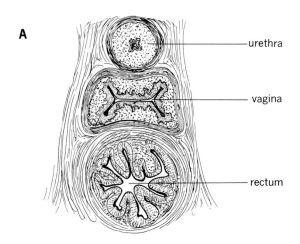

A

urethra

vagina

rectum

B

keratinized type   cornified type

superficial
layer

intermediate
layer

basal
layer

16.6 Structure of the vagina: A, cross section, showing its proximity to the urethra and rectum; B, layers of epithelium.

External genitalia. The external genitalia include the *mons pubis, labia majora, labia minora, vestibule,* and *clitoris* (Fig. 16.7), collectively known as the *vulva.* The mons pubis is an eminence of fat over the pubic bones, covered with pubic hair. The labia majora are two prominent fleshy folds continuous anteriorly with the mons pubis, in the *anterior commissure* of the vulva, and also covered with pubic hair. Posteriorly they blend with the *perineum* anteriorly to the anus to form the *posterior commissure* of the vulva. The labia minora are two membranous folds underneath and medial to the labia majora. They are reddish in color and devoid of fat and hair. Anteriorly they extend around the clitoris, a structure homologous to the penis, composed of erectile tissue, blood vessels, and nerves. The clitoris is about 2 cm long but is largely embedded in the surrounding tissues, with only its tip *(glans)* protruding. The glans contains a dense network of neural receptors and is an *erogenous zone,* a region whose stimulation leads to erotic sensations and sexual arousal. The vestibule is the space bounded by the labia minora and the clitoris. It evolves from the *urogenital sinus* of the

embryo and encompasses the external orifices of the urethra and the vagina. Vestibular glands, chief of which are *Bartholin's glands,* empty into it.

OVARIAN CYCLE

The two main functions of the ovaries are the production of ova and the secretion of hormones. The physiology of the ovaries is characterized by the cyclic nature of the processes. Their rhythmic pattern is referred to as the *ovarian,* or *sexual, cycle.* It includes the *uterine,* or *menstrual, cycle,* and its average period is 28 days.

The ovarian cycle has two significant results. First, it causes a single mature ovum to be released from one ovary each month. Second, it effectively prepares the uterine endometrium for implantation of the ovum, should it be fertilized. Thus the ovarian cycle is concerned with the well-being of the product of concep-

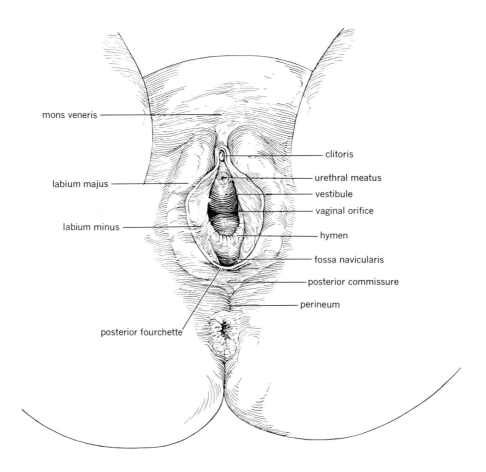

mons veneris

labium majus

labium minus

posterior fourchette

clitoris

urethral meatus

vestibule

vaginal orifice

hymen

fossa navicularis

posterior commissure

perineum

16.7 Female external genitalia. (From R. W. Kistner, *Gynecology: Principles and Practice.* Copyright © 1964, Year Book Medical Publishers. Used by permission.)

tion. When a fertilized ovum is not available, the endometrium breaks down, to be discharged in *menstruation.*

It is interesting to note that ovarian activity is controlled almost entirely by events outside the ovaries. In man the cycle depends directly on the production of gonadotropic hormones by the adenohypophysis (see Chapter 15). Under their periodic influence an ovarian follicle matures, and ovulation takes place. In mammals that breed seasonally, ovulation occurs only during the period called *estrus.* In some, ovulation occurs only after mating.

So that we may comprehend the ovarian cycle and its culminating events, ovulation and menstruation, we shall consider the gonadotropic hormones in detail.

Gonadotropic hormones. As shown in Table 15.2, the adenohypophysis secretes hormones that act upon the gonads. For convenience we shall assume that they are separate and distinct hormones. They are the follicle-stimulating hormone (FSH), luteinizing hormone (LH), and luteotropic hormone (LTH). In fact, it is not yet certain that the physiological effects attributed to FSH and LH are due to two hormones, though it was reported in 1966 that FSH had been purified until it was free of LH activity. Although LTH has been identified as a separate hormone in the rat, indications are that the effects attributed to it and GH are due to only GH in man. We should remember these qualifications in our discussions of FSH, LH, and LTH.

Without stimulation by the gonadotropic hormones, the ovaries are completely inactive. Such is the case until age 7 or 8, when the hypophysis slowly begins to secrete the hormones. After age 11 to 13, or *puberty*, their secretion proceeds as in an adult, initiating the ovarian cycles. The rates of production of the hormones are judged from measurements of their concentrations in urine. We shall mention one assay technique later.

Further evidence of the role of the gonadotropic hormones is seen following removal of the hypophysis. In a young animal ovarian development is retarded, and sexual infantilism persists. In an adult animal the ovaries regress or atrophy. Administration of potent adenohypophyseal extracts leads to enlargement of the ovaries in both normal and hypophysectomized animals.

Follicular phase and ovulation. Fig. 16.2 depicts the stages in the differentiation of a primordial follicle into a mature ovum, the process called *oögenesis*. We recall that a primordial follicle consists of an ovum surrounded by a layer of granulosa cells. The first stage is enlargement of the ovum itself. This is followed by proliferation of the granulosa cells into many layers and the formation of a clear liquid, the *liquor folliculi*, between the layers. Concurrently an external capsule, or *theca*, develops from the ovarian cortical tissue. It has two coats or *tunics*, the *theca externa* and the *theca interna*.

The liquor folliculi continues to collect, creating cavities within the follicle that eventually unite in one large fluid-filled *antrum*. The ovum is pushed to one side, in a hillock of granulosa cells, and keeps growing until it is the largest cell in the body, measuring 120 μ in diameter. Meanwhile, the follicle moves to the surface of the ovary, forming a vesicle with a thin, nearly bloodless surface. A follicle at least 10 mm in diameter is frequently termed a *Graafian follicle*, after the Dutch anatomist Regnier de Graaf, who first described it in 1672.

As continued secretion of liquor folliculi increases the intrafollicular pressure, a small rupture appears near the peak of the protruding follicle, liquor slowly oozes out, and finally the ovum, together with its surrounding cells, is extruded into the peritoneal cavity and received into the fimbriated open end of the adjacent Fallopian tube.

Luteal phase. After the follicle ruptures, its walls gradually collapse. It is then rapidly *luteinized*—that is, filled with cells containing yellow pigment—to form a *corpus luteum* (see Fig. 16.2). Although the moment of ovulation is arbitrarily taken as the beginning of the luteal phase, alterations in the follicle walls before ovulation are considered the first steps in luteinization.

The corpus luteum increases in size for 10 to 15 days. If conception has not occurred, it then undergoes retrogressive changes and is converted to an inert, white, scarlike *corpus albicans*. Under these circumstances the corpus luteum is often called a *false corpus luteum*, or *corpus luteum of menstruation*. If the ovum has been fertilized, the life of the corpus luteum is prolonged for many months. It continues to increase in size by multiplication of the lutein cells during the first few months of pregnancy and does not undergo retrogressive changes until the sixth month or later. It is therefore much larger than in menstruation and is called a *true corpus luteum*, or *corpus luteum of pregnancy*. There is no essential difference in structure between a false corpus luteum and a true one, although the latter body lasts longer and attains greater size. We shall deal with the functions of the corpus luteum presently.

Gonadotropic stimulation of the ovaries. FSH and LH are both necessary for the maturation of an ovum and the subsequent formation of a corpus luteum. FSH stimulates the ripening of the follicle, the development of the theca, and, the secretion of estrogens by the ovaries. LH stimulates ovulation and the construction of the corpus luteum. Even if large quantities of FSH were available, if LH were absent, the follicle would grow only slightly beyond the antrum stage before disintegrating. On the other hand, pure FSH appears to work about as well as LH as the final stimulus of ovulation.

As shown in Fig. 16.8, FSH production is predominant during the follicular phase of the ovarian cycle. As ovulation approaches, LH production rises suddenly. After ovulation FSH production falls to a low level, whereas LH pro-

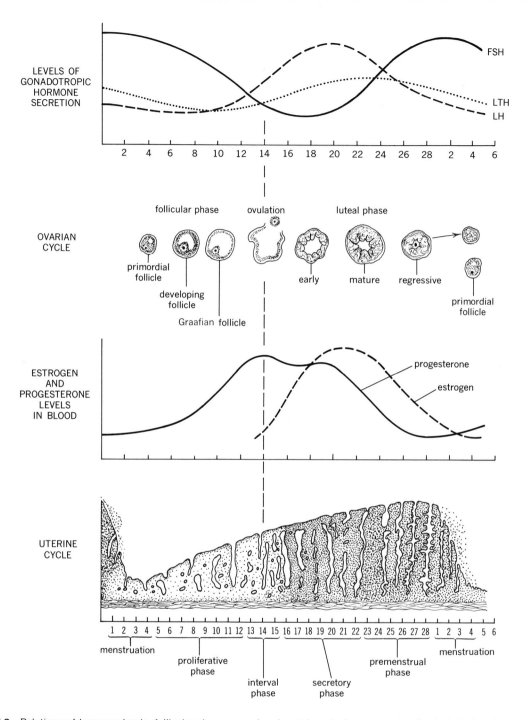

LEVELS OF
GONADOTROPIC
HORMONE
SECRETION

FSH

LTH

LH

2 4 6 8 10 12 14 16 18 20 22 24 26 28 2 4 6

OVARIAN
CYCLE

follicular phase ovulation luteal phase

primordial
follicle

developing
follicle

Graafian follicle

early mature regressive

primordial
follicle

ESTROGEN
AND
PROGESTERONE
LEVELS
IN BLOOD

progesterone

estrogen

UTERINE
CYCLE

1 2 3 4 5 6 7 8 9 10 11 12 13 14 15 16 17 18 19 20 21 22 23 24 25 26 27 28 1 2 3 4 5 6

menstruation

proliferative
phase

interval
phase

secretory
phase

premenstrual
phase

menstruation

16.8 Relations of hormone levels, follicular changes, and endometrium during an ovarian (and uterine) cycle.

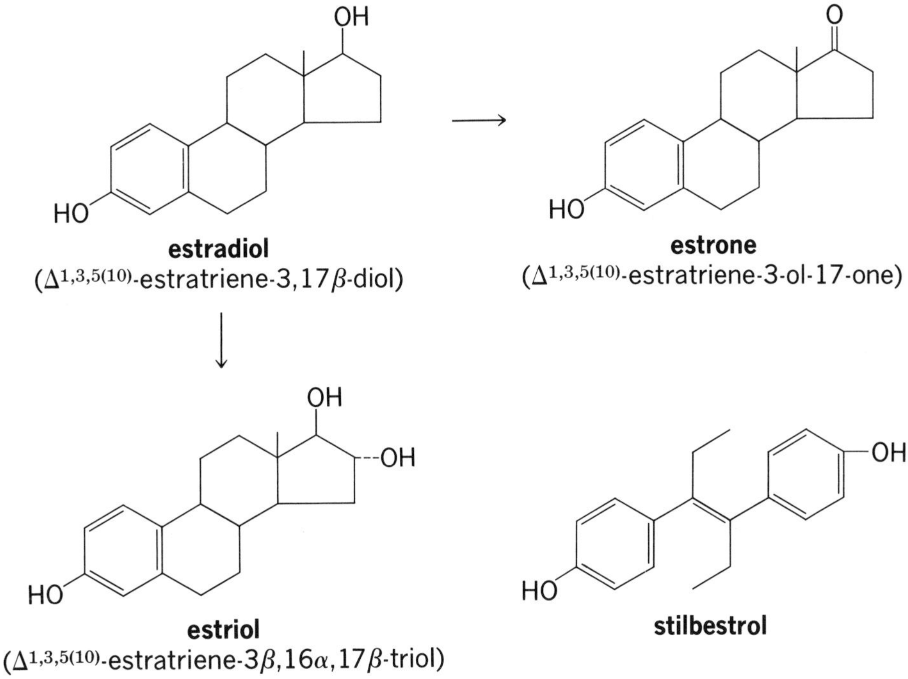

estradiol
($\Delta^{1,3,5(10)}$-estratriene-3,17β-diol)

estrone
($\Delta^{1,3,5(10)}$-estratriene-3-ol-17-one)

estriol
($\Delta^{1,3,5(10)}$-estratriene-3β,16α,17β-triol)

stilbestrol

16.9 Structures of the principal natural estrogens and stilbestrol, a synthetic estrogen.

duction continues unabated into the premenstrual phase.

In discussing the functions of LTH, we must turn our attention to the endocrine activities of the ovaries. The ovaries are responsible for the elaboration of *estrogens* and *progesterone.** Exactly where in the ovaries estrogens are secreted is not known, but progesterone is secreted by the corpus luteum. Indeed, the corpus luteum is a transitory endocrine gland whose primary function is the secretion of progesterone. LTH induces this secretion. LTH also induces lactation after childbirth; in this role it is usually designated *lactogenic hormone* or *prolactin*.

Sex hormones: estrogens and progesterone. The estrogens and progesterone control the development and function of the reproductive organs and the development of the secondary sexual characteristics in the female. As we have noted, their main sources in the nonpregnant female are the ovaries.† In pregnancy enormous amounts of both types of hormones are secreted by the placenta, the nourishing organ that provides for the exchange of nutrients, respiratory gases, and waste products between embryonic blood and maternal blood. In addition, the placenta produces a gonadotropic hormone called *chorionic gonadotropin*—the *chorion* is the membrane that forms the embryonic portion of the placenta—and a lactogenic hormone called *placental lactogen*.

The natural estrogens (that is, those secreted by the endocrine glands) comprise a group of 18-carbon steroids, many having a phenol-like arrangement of three double bonds and a hydroxyl group in the first ring (Fig. 16.9; cf. Fig. 15.9). At least six different estrogens have been isolated from the plasma of the human female, but only three are there in significant quantities: *estradiol* (also known as *17β-estradiol*), *estrone*, and *estriol*. Estradiol contains two hydroxyl groups, estrone one, and estriol three.‡ In terms

---

* Hormones of these types are also secreted by the placenta and adrenal cortex in the female—and, in fact, by the testes and adrenal cortex in the male (see Chapter 15).

† We shall consider a third ovarian hormone, *relaxin*, later.
‡ Estrone has a keto group on carbon 17 and is therefore a 17-ketosteroid.

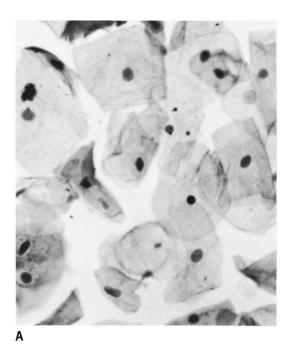 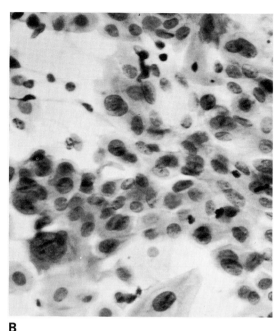

A

B

16.10 Smears showing vaginal epithelial cells under different degrees of estrogen stimulation (see Fig. 16.6B): A, with no estrogen—the smaller cells are from the basal layer, and the larger ones are from the intermediate layer; B, with estrogens—most of the cells are from the superficial layer. (Dr. Priscilla D. Taft.)

of estrogenic effects, estradiol has 12 times the potency of estrone and 80 times that of estriol. Estradiol and some estrone are present in ovarian venous blood, and estradiol is the major ovarian secretion. Estriol appears to be an oxidative product derived from the other two, with the conversion occurring mainly in the liver. The liver inactivates estrogens by conjugating them with glucuronic acid and sulfates (see Chapter 13). The inactive estrogens are then excreted in the bile and urine. From 5 to 10% of the active estrogens produced each day is also excreted in the urine. Hence determination of urinary estrogenic activity furnishes a useful index of estrogen production.

The physiological effects of the estrogens are of three types: (1) direct effects upon the female reproductive organs; (2) effects upon the adenohypophysis; and (3) general metabolic effects. One may summarize the effects of the first type by stating that estrogens function as specific growth hormones for the internal and external female reproductive organs. They are respon-

sible for the changes of puberty—the development of the genitalia, breasts, pubic and axillary hair, and the feminine body contour with broadening of the pelvis. The cyclic production of estrogens causes the proliferative phase of the uterine cycle. Estrogens also have interesting effects upon the mucus secreted by the glands of the cervix, increasing its amount, lowering its viscosity, and raising its pH, and thus greatly favoring the migration, motility, and longevity of spermatozoa.

We have already spoken of the effects of estrogens upon the vaginal epithelium—general thickening, increased glycogen content, and proliferation, enlargement, and cornification of the superficial cells. As estrogenic activity wanes, the vaginal epithelium thins. So distinctive are these changes that the examination of a vaginal smear is a simple means of estimating estrogen secretion (Fig. 16.10). Such an examination is frequently made in conjunction with studies of dried cervical mucus.

The principal effect of estrogens upon the

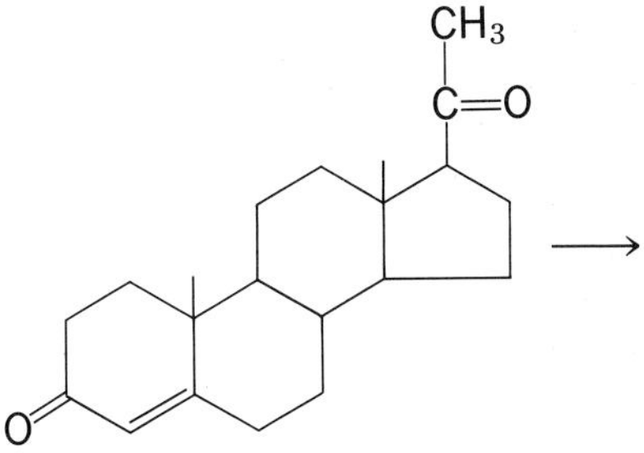

16.11 Structures of progesterone, its major metabolic breakdown product, and three synthetic progestational hormones.

progesterone
(Δ⁴-pregnene-3,20-dione)

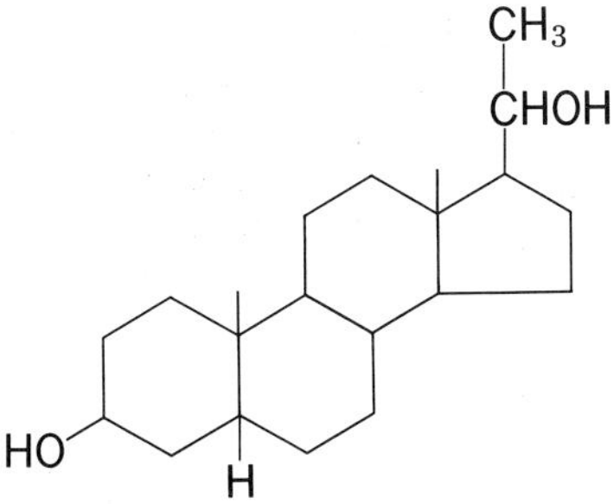

pregnanediol
(pregnane-3,20-diol)

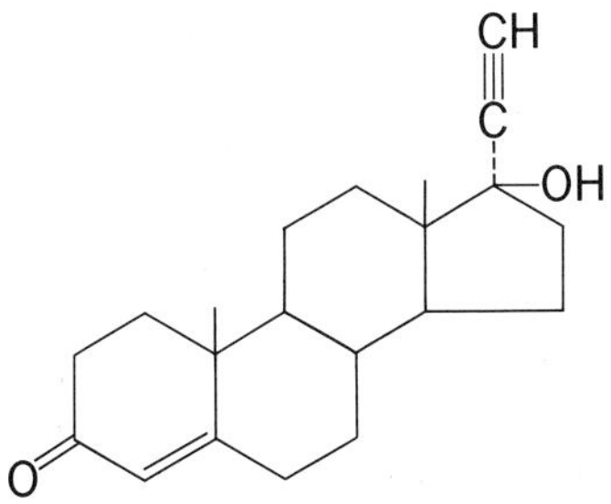

ethisterone
(Δ⁴-17α-pregnene-20-yn-3-one-17-ol)

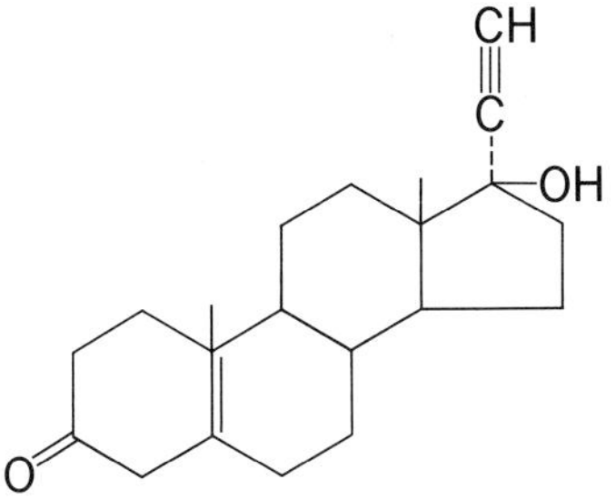

norethynodrel
(enovid; Δ⁵⁽¹⁰⁾-17α-pregnene-20-yn-3-one-17-ol-19-nor)

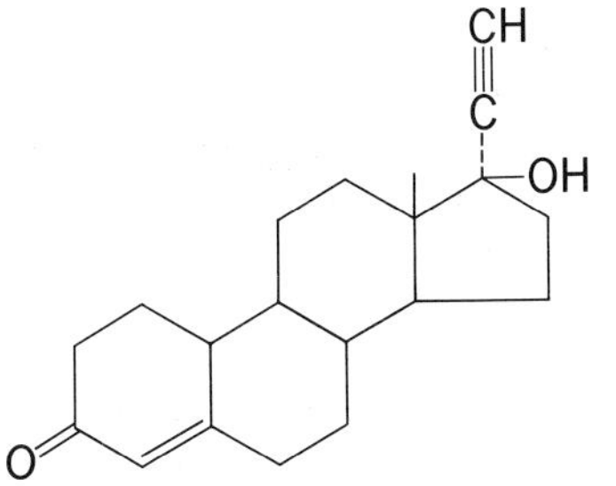

norethindrone
(Δ⁴-17α-pregnene-20-yn-3-one-17-ol-19-nor)

adenohypophysis (or upon the hypothalamic centers that control the adenohypophysis*) are (1) suppression of FSH secretion and (2) promotion of LH secretion (see Fig. 16.8). The effects upon general metabolism are slight in comparison to those of male sex hormones, consisting only in small increases in BMR and in protein anabolism, a tendency to renal sodium and water retention, and an acceleration of the processes of bone formation.

Natural progesterone, the progestational hormone (or *progestogen*) secreted by both the corpus luteum of menstruation and that of pregnancy,† is a steroid whose structure resembles that of a corticosteroid (Fig. 16.11; cf. Fig. 15.9). Indeed, it is a key intermediate in the biosynthesis of corticosteroids and testosterone. Progesterone has 21 carbons and a double bond between carbons 4 and 5. The only difference between it and the adrenal steroid desoxycorticosterone is that it lacks a hydroxyl group at carbon 21.

The main progesterone effects are the preparation of the uterus for pregnancy and the maintenance of conditions favorable for the continuance of pregnancy. Specifically, it (1) produces a secretory type of endometrium,

* Hypothalamic control over the adenohypophyseal secretion of gonadotropic hormones is well established. An FSH-releasing factor (FSH-RF) and an LH-releasing factor (LH-RF)—each analogous to CRF, the hypothalamic corticotropin-releasing factor (see Chapter 15)—have been identified in hypothalamic extracts, and it is assumed that they mediate the hypothalamic stimulation of FSH and LH secretion in vivo. The hypothalamus appears to inhibit LTH secretion.

† A progestational hormone prepares the endometrium for the implantation of a fertilized ovum and pregnancy (gestation). It is interesting that progesterone has been identified as an intermediate in steroid hormone biosynthesis in many phyla (see Fig. 15.9) but as a hormone only in higher vertebrates—in which the young develop in a uterus.

(2) diminishes muscular contractions of the Fallopian tubes and uterus, and (3) promotes glandular development in the breasts. Most of these effects occur only after prior estrogen action. Progesterone is also believed to inhibit LH secretion and to stimulate somehow FSH secretion in the next cycle. As we shall see, progesterone in high concentrations also inhibits ovulation. Thus its biological effects are evident chiefly in the reproductive tract and breasts; it appears to have no role in development of the secondary sexual characteristics.

Investigation of the metabolism of progesterone has been handicapped for want of a sensitive analytical method. The major pathway of breakdown involves reduction to an inactive compound, *pregnanediol* (see Fig. 16.10). This substance and its glucuronide are excreted in the urine in proportion to the amount of progesterone secreted;* consequently, measurement of their amounts in the urine is an assay of progesterone production. Males also excrete pregnanediol or a closely related substance, as a result of progesterone secretion in the testes and adrenal cortex.

In recent years the mechanism of action of the sex hormones has provoked much controversy, with different groups of investigators reaching different conclusions. First P. Talalay and H. G. Williams-Ashman produced evidence suggesting that the estrogens act primarily in the transport of hydrogen or electrons between the coenzymes NAD and NADP. These workers partially purified a 17 β-hydroxysteroid dehydrogenase that reacts with both NAD and NADP.

$$
\begin{array}{c}
\text{estradiol (R—OH) + NAD} \rightleftharpoons \text{estrone (R=O) + NADH}_2 \\
\text{estrone (R=O) + NADPH}_2 \rightleftharpoons \text{estradiol (R—OH) + NADP} \\
\hline
\text{NAD + NADPH}_2 \rightleftharpoons \text{NADH}_2 + \text{NADP}
\end{array}
$$

(16.1)

The equation indicates that, during the course of hydrogen transfer between the two coenzymes, estradiol undergoes reversible oxidation and reduction. Since it is regenerated in the reaction sequence, it can serve in catalytic quantities as a coenzyme-like cofactor. The net result, therefore, is a *transhydrogenation*, which could clearly determine the balance of the oxidized and reduced forms of the coenzymes and

thereby control metabolic pathways dependent upon them (see Chapter 4). C. Villee and D. D. Hagerman claimed that not one but three separate enzymes catalyze the reactions in Eq. 16.1—an NAD-linked dehydrogenase, an NADP-linked dehydrogenase, and a transhydrogenase. Subsequent studies yielded a highly purified, apparently single enzyme—a steroid-mediated transhydrogenase—that catalyzes the reactions, confirming the original conclusion. However, it is still difficult to trace the chain of events from the interconversion of NADPH$_2$ and NADP to the final physiological effects by which we recognize estrogens. G. Mueller has recently found that estrogens also specifically promote mRNA, protein, and phospholipid synthesis. Hence they seem to provoke a cell to transmit a portion of its genetic information by stimulating selective gene transcription (see p. 156). Mueller's observation, then, may cast doubt on the view that transhydrogenation is the major molecular mechanism by which estrogens control the growth and function of the tissues of the female reproductive system.

The mechanism of action of progesterone is not

TABLE 16.1   UTERINE CYCLE

Days	Events
1–4	Menstruation
5–12	Proliferative phase Preovulatory period Development of follicle (follicular phase) Growth of endometrium High estrogen level
13–15	Interval phase Ovulation
16–22	Secretory phase Migration and breakdown of unfertilized ovum Development of corpus luteum (luteal phase) High estrogen and progesterone levels
23–28	Premenstrual phase Regression of corpus luteum Fall of estrogen and progesterone levels Deterioration of endometrium

---

* Other progesterone metabolites in human urine include allopregnanediol, epipregnanolone, and epiallopregnanolone.

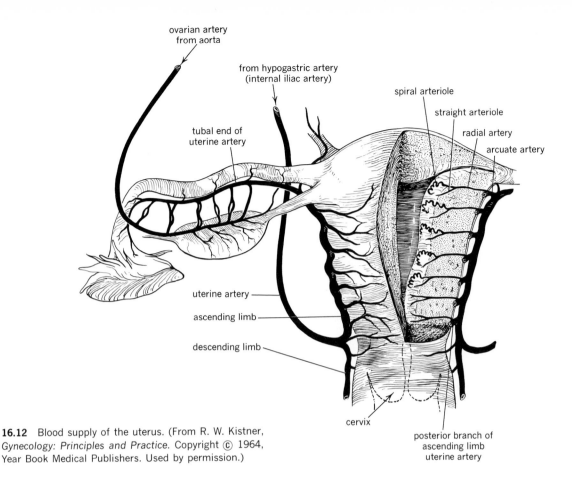

16.12 Blood supply of the uterus. (From R. W. Kistner, *Gynecology: Principles and Practice.* Copyright © 1964, Year Book Medical Publishers. Used by permission.)

known. One of the obstacles to its understanding is that, for some reason, progesterone acts only after the estrogens have acted, so that its biochemical effects may be confused with theirs. Progesterone stimulates the activity of uterine lactate dehydrogenase, which catalyzes the interconversion of NAD and $NADH_2$ (see Fig. 4.3). This effect upon NAD and $NADH_2$ availabilities may be the basis for its relationship with the estrogens. Progesterone also stimulates mitochondrial ATPase activity, inhibits cytochrome *c* reduction, and accelerates ascorbic acid oxidation.

A large number of unnatural and synthetic steroids and nonsteroids exhibit estrogenic or progestational activity. For example, the *stilbenes*, typified by stilbestrol (see Fig. 16.9), are similar enough to natural estrogens to be moderately active. Some of these compounds are widely used medically. Artificial progestational substances include the nonsteroid amphenone B and the steroids ethisterone and the recently

discovered 19-nor compounds,* norethynodrel and norethindrone (see Fig. 16.11).

Uterine cycle. The ovarian cycle terminates with the uterine cycle. Table 16.1 recapitulates the events shown in Fig. 16.8.

The period of menstruation is usually taken as the beginning of the uterine cycle. So that we can understand the nature of the menstrual process, we must study the structure of the endometrium. As shown schematically in Fig. 16.12, the uterus is generously endowed with blood vessels. The major arteries, the *uterine arteries,* give rise to branches, the *arcuate arteries,* just beneath the external myometrial surface. At intervals the arcuate arteries encircling the uterus give off *radial arteries,* which pass directly

---

* The *nor-* prefix indicates the lack of a methyl group (see Fig. 15.10).

inward. Before entering the endometrium, the radial arteries divide into two types of arterioles. The short *straight* arterioles supply only the deepest third of the endometrium, ending in a more or less horizontal meshwork of twigs. These vessels (and their portion of endometrium) are unaffected by cyclic hormonal changes. The *spiral* arterioles reach the endometrial surface and are notably affected by cyclic hormonal changes. Their branches supply the endometrial stroma, forming the *stromal* and *subepithelial capillary plexuses,* and the *endometrial glands,* forming the *glandular capillary plexus*. They also invest implanted ova. During the proliferative phase the spiral arterioles lengthen as the endometrium thickens under the influence of a high estrogen level.

After the interval phase, in which ovulation* takes place, the progesterone level rises, and the secretory phase commences. Endometrial growth continues, the spiral arterioles become distended, and the mucus-secreting glands become enlarged and active. In the premenstrual phase the estrogen and progesterone levels fall, and the endometrium starts to regress. As the tissue shrinks, the spiral arterioles fold or buckle, with resulting stasis of blood flow. Moreover, they become constricted, so that the capillary circulation is diminished. Leukocytes begin to migrate into the area, and the superficial tissue begins to break down. Bleeding occurs from degenerating capillaries, and the menstrual period follows, with sloughing of the outer two-thirds of the endometrial mass. The blood and discarded tissue are discharged into the vagina through the cervical canal.

The discomfort associated with the onset of menstruation is probably due to temporary hormonal imbalance. Since the corpus luteum has regressed and ceased secreting progesterone, the level of progesterone in the blood is low; and, since the ovaries secrete estrogens actively only after menstruation, the estrogen level is also low. Before the progesterone and estrogen levels fall, however, the hormones may cause the retention of sodium, with a consequent retention of water and edema. This is commonly given as the explanation for the premenstrual tension experienced by many women.

Menstruation is restricted to primates. The fundamental reason for it is unclear. It may be that preparation of the endometrium for implantation of a fertilized ovum takes place with such intensity that only tissue destruction and shedding will permit involution to the unprepared state if the fertilized ovum fails to arrive. The coiling of the spiral arterioles is probably an adaption that permits their extension in pregnancy, when uterine enlargement by growth gives way to enlargement by stretching. When the coiled arterioles do not extend, blood stagnation and tissue destruction are inevitable.

Menopause. The ovarian and uterine cycles end with the termination of ovulation and menstruation. The period of termination, the *menopause,* usually occurs between the ages of 45 and 50 years.† Emotional and other disturbances may arise for a time until the adjustment is completed.

The menopause signifies a "burning out" of ovarian activity. Throughout a woman's reproductive years, ova mature within their follicles. A small number are released in ovulation; the rest undergo atresia. Thus aging of the ovary begins early in life and progresses in an orderly fashion. By age 45 only a few primordial follicles remain to be stimulated by FSH and then LH. As a result, estrogen secretion drops (Fig. 16.13A). When the level falls below a critical value, estrogens no longer influence FSH and LH production sufficiently to create an oscillatory pattern. From then on, the gonadotropic hormones are continuously released in large quantities (Fig. 16.13B). Estrogen production decreases to zero, nevertheless, because the ovaries are thickened and scarred; the remaining primordial follicles become atretic; and reproductive life is over. Physicians occasionally administer estrogens to women in menopause in an attempt to mitigate some of the vasomotor and psychological consequences of the abrupt cessation of estrogen secretion.

CONCEPTION AND PREGNANCY

The union of spermatozoon and ovum is a remarkable phenomenon, and nature has gone to great lengths to ensure its occurrence. The means of promoting union varies widely among the species. Fishes, for example, generally produce millions of ova and millions of spermatozoa.

---

* Ordinarily outward signs of ovulation are meager, and so the exact time of its occurrence is difficult to determine. In many individuals it is accompanied by low fever and abdominal pain (*Mittelschmerz*).

---

† The oldest authenticated age of a woman giving birth is 52.

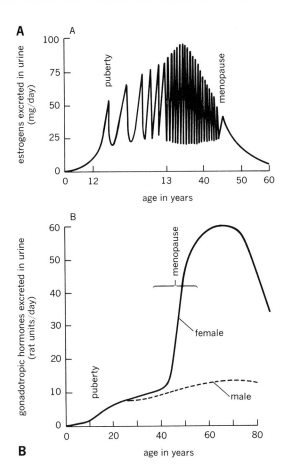

16.13 Changes in estrogen (A) and gonadotropic hormone (B) secretion during life.

apparently to make good use of this relatively small number of ova.

Sexual intercourse. Performance of the sex act by both male and female is profoundly affected by psychological, neurological, and physical factors. Both sexes possess a so-called "sex drive," but cultural and social factors significantly influence the manner in which this drive expresses itself. The desire to gratify the sex drive is an important psychological force in the lives of adult men and women that may or may not be directly associated with the desire to produce offspring. Suffice it to say that the prerequisites of successful sexual union, whatever its motivation, include a particular state of health, mind, and emotion and a complex pattern of preliminary tactile stimulations in the body's erogenous zones.

Once these have combined to cause sexual arousal in male and female, a number of reflex events occur that facilitate the sex act. Those in the female include the secretion of quantities of vaginal fluid of somewhat higher pH than usual.* This functions as a lubricant without which the elicited sensations would be painful or disagreeable. The intense stimulation ultimately initiates the reflexes constituting the *orgasm,* or *climax.* In the man this is accompanied by an ejaculation of *semen,* the spermatozoa-carrying fluid. In the woman it is less clearly defined, though it is believed to involve rhythmic contractions of the perineal muscles, uterus, and Fallopian tubes.

The disposition of the spermatozoa from the time they leave the penis until one penetrates an ovum is poorly understood. It is conjectured that movements of the penis spread much of the cervical mucus onto the vaginal walls. If so, the cervical os is partially cleared of the secretions that usually occlude it, and the impact of the first ejaculated mass of semen against it mixes the semen with the cervical mucus. It is likely that the sexual sensations during the female orgasm promote fertilization, perhaps by drawing into

The female lays her eggs and, probably by chemotaxis, attracts the male to the area. He showers sperm over the eggs, which thus are fertilized in a somewhat haphazard manner. Both eggs and young are easy prey, but enough manage to survive to perpetuate the species. With many amphibia, the male clings to the back of the female and fertilizes the ova as they emerge from the cloaca. In reptiles, birds, and mammals, fertilization is internal; therefore, the likelihood that a spermatozoon will meet an ovum is increased.

Despite these evolutionary advances a human male produces about 500,000,000 spermatozoa per ejaculation, and, as we have noted, a female has about 500,000 ova in her ovaries at birth. Yet, during a full reproductive life, she sheds only about 400. The purpose of sexual union is

---

* Over the years authorities have considered the cervix and Bartholin's glands the main sources of this vaginal fluid. Recent observations, however, indicate that it appears in minute coalescing droplets over the entire vaginal mucosa, in a process somewhat resembling sweating. Presumably it results from extreme local vascular congestion.

the cervical canal the mixture of fluids deposited in the cervical os. With the normal anteflexion of the uterus (see p. 16.1B), the cervix inclines posteriorly in the vagina and hence directly into the pool of secretions and semen.

Fertilization of the ovum. The vaginal secretions are normally acid (see p. 640). Semen is normally alkaline. Although the vaginal pH rises during and after sexual intercourse, it drops in several hours to its usual level, and spermatozoa left in the vagina lose their motility and die. Of the 500,000,000 spermatozoa present in the fluid spread over the vaginal walls on withdrawal of the penis, only a few hundred thousand find their way into the cervical canal before they succumb.

The mixture of semen and cervical mucus is continuous with the narrow column of clear fluid in the cervical canal, and this in turn with the thin layer of mucus over the endometrium. We learned (see p. 646) that the watery cervical secretion at ovulation is a favorable medium for spermatozoa. During ovulation many normally motile spermatozoa may be observed in cervical mucus even 48 hours after ejaculation. At other times in the ovarian cycle, the cervical mucus is viscous, and spermatozoa may or may not enter the cervical canal.

We may assume that some highly active spermatozoa aimlessly travel from the cervical canal into the uterus, where they then deploy in the broad, shallow lake of fluid lying between the anterior and posterior walls. At the lateral shores of this triangular lake are the two minute inlets of the Fallopian tubes. Individual spermatozoa move sideways along these shores, in one direction or the other, and some by chance swim into the tubes. The forward progress of a spermatozoon in cervical mucus has been measured at about 1 mm per minute. The direct distance from the cervical os to the opening of a tube is about 120 mm—a 2-hour trip if the spermatozoon moves only under its own power. In fact, spermatozoa may be found near the ovaries within 5 minutes of ejaculation. Thus uterine or tubal contractions must propel them. Several tens of thousands of spermatozoa reach the uterus, but only a few thousand at most enter the tubes.

In the Fallopian tubes the spermatozoa swim against the weak current established by the cilia of the tubal mucosa and by the downward

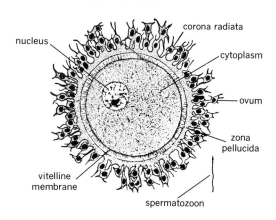

16.14 Structure of a mature ovum, showing its size relative to that of a spermatozoon. (From L. B. Arey, *Developmental Anatomy*, 4th ed., W. B. Saunders Co., Philadelphia, 1940.)

peristalsis of the tubal fluid. The force of peristalsis increases during ovulation. Of the few thousand spermatozoa from a single ejaculate that may enter the tubes, then, probably not more than a few hundred—approximately one-millionth of the spermatozoa in the ejaculate—reach the outer ends of the tubes. As the inert ovum, still in its ragged sheath of granulosa cells, is rolled and pushed down a tube, it probably, but not inevitably, encounters a few of these few hundred spermatozoa thrashing about in the current.

The mature ovum* is surrounded by a thin inner *vitelline membrane,* a thick, tough, lattice-like *zona pellucida,* and a *corona radiata* of granulosa cells (Fig. 16.14). The cells of the corona radiata are held together by hyaluronic acid, a polysaccharide common to many other cell aggregates and tissues of the body. This is broken down by an enzyme of semen, hyaluronidase. The corona radiata begins to disperse while the second meiotic division is being completed and finally leaves the zona pellucida exposed. If the zona were not so denuded, the spermatozoon would have difficulty reaching it.

---

* Only one ovum attains maturity. In the first meiotic division, a *primary oöcyte* yields a large *secondary oöcyte* and a small *first polar body.* This is the reductional division, in which the chromosome number is reduced from 46 to 23. In the second meiotic division, the secondary oöcyte yields a mature ovum and a small *second polar body.* During the second division, the first polar body divides into two second polar bodies. All the polar bodies are discarded.

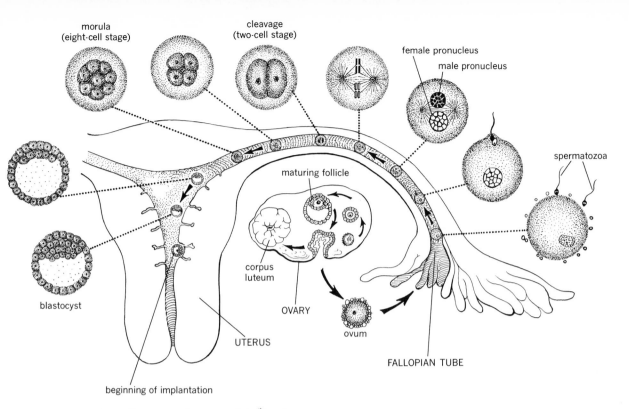

16.15 Fertilization and passage of an ovum into the uterus.

Little is known of the precise mechanism by which a spermatozoon penetrates the zona pellucida (Fig. 16.15). Remarkably, only one gains entry into the ovum. It is believed that, once the vitelline membrane is punctured, an unidentified substance diffuses out of the ovum and closes the latticework of the zona. Microscopic studies of fertilizations in vitro show that spermatozoa continue to invade the zona pellucida but are trapped and inactivated there, so that polyspermy is prevented.

When the spermatozoon is inside the ovum, its head swells to form a *male pronucleus*, which joins the *female pronucleus* already present. Finally the chromosomes of the male and female pronuclei combine. The sex of the embryo has already been established (see Chapter 3).

Cleavage of the fertilized ovum starts shortly after fertilization with the zona pellucida persisting for a time. At the eight-cell, or *morula*, stage, the embryo resembles a little ball of cells. Then certain peripheral cells begin to divide more rapidly than the cells beneath them, and a fluid-filled cavity, the *blastocoele*, appears.

The entire mass with its eccentrically placed inner cell grouping is called a *blastocyst*, and its outer wall is called the *trophoblast* (Fig. 16.16). All of these events are concluded before the

16.16 Structure of a blastocyst. (From Carnegie Collection, No. 8663.)

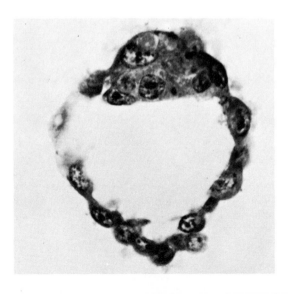

653

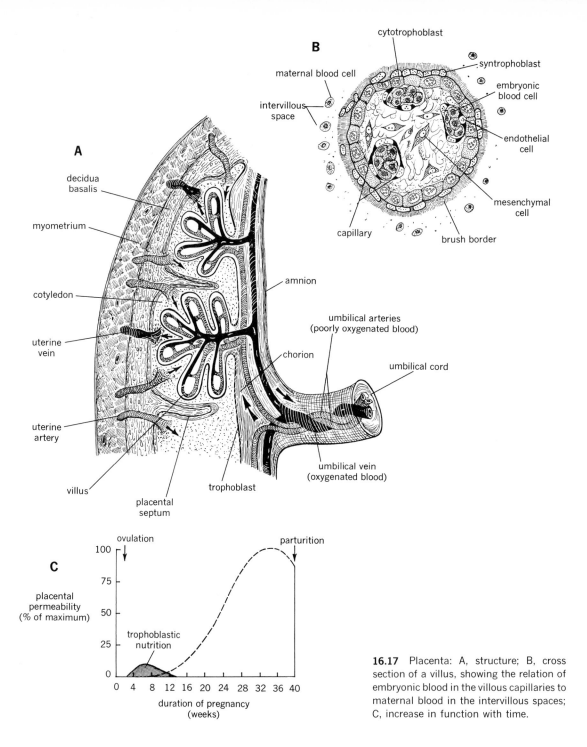

A

decidua basalis

myometrium

cotyledon

uterine vein

uterine artery

villus

placental septum

amnion

umbilical arteries (poorly oxygenated blood)

chorion

umbilical cord

umbilical vein (oxygenated blood)

trophoblast

B

cytotrophoblast

syntrophoblast

maternal blood cell

embryonic blood cell

intervillous space

endothelial cell

mesenchymal cell

capillary

brush border

C

ovulation

parturition

placental permeability (% of maximum)

100

75

50

25

0

trophoblastic nutrition

0   4   8   12 16 20 24 28 32 36 40

duration of pregnancy (weeks)

16.17 Placenta: A, structure; B, cross section of a villus, showing the relation of embryonic blood in the villous capillaries to maternal blood in the intervillous spaces; C, increase in function with time.

ovum completes its journey down the Fallopian tube. After the blastocyst reaches the uterus, it remains in its lumen for 24 to 48 hours before beginning its attachment to the uterine wall.

Implantation of the fertilized ovum. Hormones, whose role is so vital in conception, also control the preimplantation processes. Progesterone is required for survival of the blastocyst and for preparation of the endometrium for its receipt. It tends to convert endometrial cells into large swollen *decidual cells* that are rich in proteins, glycogen, and lipids.* These are present in large numbers in the hours before implantation, or *nidation,* and the ovum implants into them.

The trophoblastic cells invade the decidua, digesting it by proteolytic activity. During the first week of implantation, the embryo uses the stored nutrients in the decidua. As the proteolytic activity of the trophoblast increases, the local blood supply also increases. Implantation is considered to be accomplished when erosion of a maternal blood vessel occurs.

Structure and function of the placenta. The next stage after implantation is the development of the placenta, which gradually takes over the nutritive function from the disappearing decidua (Fig. 16.17). As we have observed (see p. 2), the placenta is an extremely important organ for the biological success of the most advanced group of mammals, the Eutheria, or so-called "true mammals." The structure of the placenta and its mode of attachment to the uterus vary in different species. O. Grosser has classified placentas as *epitheliochorial, syndesmochorial, endotheliochorial, hemochorial,* and *hemoendothelial,* and Fig. 16.18 illustrates the progressive elimination of a barrier between maternal and embryonic circulations. The human placenta is hemochorial, displaying a deep penetration of the maternal tissues by the fetal membranes.† Although our understanding of the various placen-

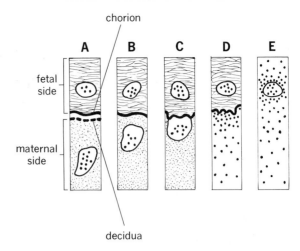

16.18   Types of placentas: A, epitheliochorial, with the decidua persisting and the chorion merely lying in contact with it (pig, horse); B, syndesmochorial, with the decidua gone and the chorion in contact with the maternal connective tissue (cow); C, endotheliochorial, with the chorion in contact with the endothelium of the maternal vessels (cat, dog); D, hemochorial, with the chorion in contact with the maternal blood (monkey, man); E, hemoendothelial, with the trophoblast gone and only embryonic endothelium between the embryonic and maternal blood streams (rat, rabbit, guinea pig).

tal types is limited, the number of layers separating the maternal and embryonic blood streams is apparently unrelated to functional efficiency.

The chorion is the outermost of the several membranes around the embryo and therefore the one in direct contact with the uterine wall. It consists of several layers of epithelial cells supported by mesoderm. It gives rise to the bulk of the placenta and is attached to the embryo by a body stalk that is later incorporated into the *umbilical cord* (Fig. 16.19).

The chorionic epithelium is still referred to as trophoblast because of its nutritive functions. While it is implanting in the endometrium, some of its cells form cords of connective tissue, capillaries, and even erythrocytes. The vascular system that develops in this manner then joins the vascular system of the embryo, and, by the sixteenth day after fertilization, blood begins to flow. Simultaneously, blood sinuses supplied with maternal blood appear between the cords. The trophoblast sends out more and more projections, which become the *chorionic villi* (see Figs. 16.17, 16.19).

The relationship between embryonic blood of a villus and maternal blood is shown in Fig. 16.17. A villus of a fully developed placenta consists of a thin outer layer of *syncytial trophoblastic cells,* the *syntrophoblast,* around mesenchymal tissue containing capillaries with

---

* Estrogens are secreted in increased quantities at this time. One of their actions is to release histamine at the locus of implantation, from the connective tissue mast cells of the endometrium (see Chapter 5). Almost all the mast cells degenerate before the blastocyst is embedded. The histamine transforms the endometrium into a decidual nest. When small amounts of an antihistaminic agent are instilled into the uterus, implantation is blocked, and pregnancy is terminated.

† There are interesting variations among placental mammals in gross, as well as microscopic, anatomy of the placenta. The placenta may be diffuse and cover the whole chorion; it may be a zonary band; it may be several discrete tufts; or it may be two discs, as in many monkeys, or one disc, as in man. Rarely, a human placenta has a shape and vascular supply resembling those of one of the evolutionary forms. Indeed, during its development the human placenta passes through a diffuse phase, becoming discoidal only at about the sixteenth week. Perhaps it is not fanciful to see during implantation a rapid transition through epitheliochorial, syndesmochorial, and endotheliochorial stages to the final hemochorial stage.

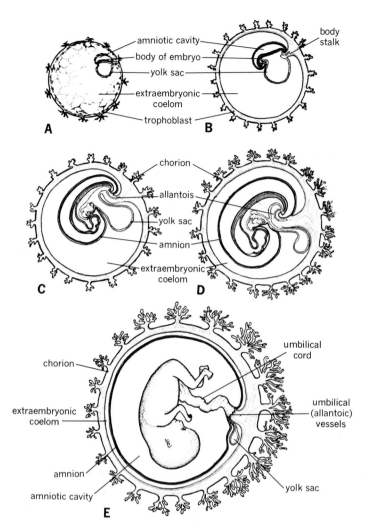

amniotic cavity
body of embryo
yolk sac
extraembryonic coelom
trophoblast

body stalk

A   B

chorion
allantois
yolk sac
amnion
extraembryonic coelom

C   D

chorion

extraembryonic coelom

amnion

amniotic cavity

E

umbilical cord

umbilical (allantoic) vessels

yolk sac

16.19 Development of the fetal membranes. (From B. M. Patten, *Human Embryology,* 3rd ed., ©️ 1968 by McGraw-Hill, Inc. Used by permission of McGraw-Hill Book Co., New York.)

extremely thin endothelial linings. During the first 16 weeks of pregnancy, an additional layer, the *cytotrophoblast,* composed of cuboidal cells known as *cytotrophoblastic,* or *Langhans',* *cells,* arises just beneath the syntrophoblast. The three layers that separate embryonic blood from maternal blood—mesenchyme, cytotrophoblast, and syntrophoblast—together constitute the *placental membrane.*

The territory of each enlarging major villous tree is marked off by *placental septa* into a distinctive lobule called a *cotyledon;* 15 to 20 cotyledons make up the embryonic portion of the placenta. The *decidua basalis,* the maternal contribution to the placenta, has a spongy layer, the *stratum spongiosum,* and a compact layer, the *stratum compactum,* both derived from endometrium.

In the fully developed placenta, embryonic blood flows through the *umbilical arteries* to the capillaries of the villi and then back through the *umbilical veins.* Maternal blood flows from the *uterine arteries* into the large blood sinuses surrounding the villi and then back into the *uterine veins.*

The mature placenta attains a diameter of 15 to 20 cm and a weight of 500 g. The placental membrane has a total surface area of approximately 10 m^2—about one-tenth the total surface area of the pulmonary alveolar membrane. However, the placental membrane is at least three cells thick (see Fig. 16.17), and so the distance between maternal blood and embryonic blood is several times the distance across the pulmonary alveolar membrane. Nevertheless, nutrients, gases, and wastes pass through the placental membrane by the same process of diffusion as through membranes elsewhere in the body.

Since the chief function of the placenta is to allow diffusion of foodstuffs from maternal blood into embryonic blood and diffusion of excretory products in the reverse direction, the permeability of the placental membrane is of special significance. As indicated in Fig. 16.17C, it is slight in the early months of pregnancy. This is true for two reasons: the total surface area of the placental membrane is still small; and the thickness of the placental membrane is still very great. As the placenta becomes older, permeability increases steadily until the last month of pregnancy, when it begins to decrease. The increase is due to enlargement of the surface area and thinning of the villi.* Occasionally, breaks occur in the placental membrane that permit embryonic blood cells to pass into the mother or, more rarely, maternal blood cells to pass into the embryo. Indeed, there are instances in which the embryo bleeds profusely into the maternal circulation because of a ruptured placental membrane.

The mean pO_2 in maternal blood in the placental sinuses is approximately 65 mm of mercury toward the end of pregnancy; the mean pO_2 in embryonic blood in the villi is approximately 50 mm of mercury. Thus a pressure gradient of at least 15 mm of mercury facilitates the diffusion of O_2 through the placental membrane. Yet, even though blood flows through the sinuses rapidly, O_2 diffuses through the membrane slowly enough that it never completely saturates the fetal hemoglobin. In the middle of pregnancy, fetal hemoglobin may become 90% saturated, but in the last month this figure falls below 80%, and immediately before birth to about 65%. We spoke of the implications of this relative anoxia for the embryonic circulatory system in Chapter 9.

Because CO_2 is an excretory product, the pCO_2 in embryonic blood is greater than the pCO_2 in maternal blood. However, not nearly so much difference exists between the CO_2 pressures as between the O_2 pressures.

Placental hormones. The placenta serves not only as the organ nourishing the embryo but also as a temporary endocrine gland secreting large quantities of estrogens, progesterone, chorionic gonadotropin, and placental lactogen. The syntrophoblast secretes both estrogens and progesterone, whereas the cytotrophoblast secretes chorionic gonadotropin.

Fig. 16.20 shows the daily production of estrogens, progesterone, and chorionic gonadotropin during pregnancy. In the early weeks estrogens and progesterone come mainly from the ovaries, but after the second month placental secretion may increase the daily estrogen output by 50 times and the daily progesterone output by 20 times. During pregnancy estrogens inhibit FSH secretion, thus suppressing follicle maturation, enlarge the uterus, and enlarge the breasts; and progesterone maintains the uterine lining in a condition favorable to the developing embryo, inhibits uterine movement, and aids in preparing the breasts for lactation.

Chorionic gonadotropin, a glycoprotein of molecular weight 30,000, is a luteinizing hormone devoid of follicle-stimulating activity. Hence its effects mimic those of LH. It preserves the corpus luteum of pregnancy, causing it to increase in size and activity, and prevents the sloughing of the endometrium that normally occurs approximately 14 days after ovulation. Menstruation after implantation of the fertilized ovum would terminate pregnancy.

Chorionic gonadotropin is released into the maternal body at a high rate between the sixth and sixteenth weeks of pregnancy. Under its influence the ovaries pour out estrogens and progesterone, which induce the endometrium to grow rather than degenerate, to store nutrients, and to form decidual cells. During the third or fourth month, secretion of chorionic gonadotropin begins to decline, and the corpus luteum begins to deteriorate. By that time the placenta itself is producing adequate amounts of estrogens and progesterone for development of the embryo.

Chorionic gonadotropin is present in both placental tissue and maternal urine. Injected into an immature mouse, rat, or rabbit, it causes

---

* Another mechanism promoting diffusion—one that is unrelated to the placenta itself—is provided by differing concentrations of binding proteins in maternal and embryonic sera. For example, permeability to thyroxine increases markedly during pregnancy, owing to a rising level of thyroxine-binding protein in embryonic serum.

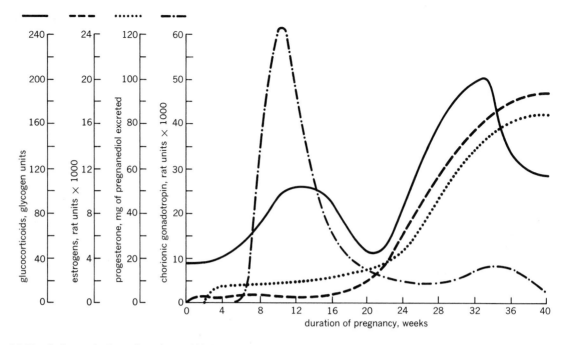

16.20 Daily production of corticosteroids, estrogens, progesterone, and chorionic gonadotropin during pregnancy.

dramatic changes in the ovaries. Some mature follicles erupt, and hemorrhagic spots appear in other follicles, which then become atretic corpora lutea. These changes are the basis of various tests for pregnancy, often applicable as early as 2 weeks after fertilization.*

Placental lactogen will be discussed later.

Anovulatory infertility. We have seen that the physiology of reproduction has two aspects: (1) mechanisms concerned with the maturation of the germ cells and (2) mechanisms that safeguard the mature germ cells and the embryo. Each of these aspects is governed by specific hormones. Although ova may develop to the brink of maturation independent of adenohypophyseal hormones, the growth and rupture of their follicles depend on them. Moreover, they are essential for the secretion of hormones by the ovaries. These hormones in turn regulate the growth and functioning of the accessory organs of the reproductive system and determine the orderly progress of a fertilized ovum through a Fallopian tube, its implantation in the uterus, the

maintenance of the embryo, and its expulsion at parturition. Because they influence tubal contractions, they also affect sperm transport in the female reproductive tract.

Normal fertility thus comprises a complex of processes that are subject at almost every step to a balanced program of hormonal controls. Infertility may arise from a defect in any of the steps. It may be spontaneous or artificially induced in a deliberate attempt to prevent conception. The natural or spontaneous causes of infertility include all manner of endocrine malfunctions and pelvic diseases. Artificially induced infertility has received a great deal of attention in recent years because of mounting concern over the accelerating population increase.†

---

* Tests requiring only a few hours can be performed with certain species of African and South American toads. Furthermore, the animals need not be sacrificed. When injected with urine containing chorionic gonadotropin, female toads extrude their eggs, and male toads discharge semen.

† Despite success in controlling his environment, man has thus far failed to control his reproductive rate. In a famous essay, Malthus in 1780 forecast the fate of nations whose rate of increase in population exceeds that in resources. It is war, pestilence, and famine. The validity of this doctrine is still generally accepted, and one of the major problems facing the human race is its growing population in areas where cultural patterns prohibit voluntary contraception. The world population, now over 2,700,000,000, is increasing by 40,000,000 per year—or doubling every 40 years. The implications of these figures are obvious, and they explain the seriousness of the current search for inexpensive and reliable means of regulating human fertility.

The most promising method of inducing infertility involves the continuous administration of progestational hormones. The result is an artificial duplication of the infertile conditions existing during the late luteal phase of the ovarian cycle (see Fig. 16.8) and during early pregnancy (see Fig. 16.20). At these times natural estrogen and progesterone levels are high, and consequently ovulation-stimulating FSH and LH levels are low. Ovulation does not occur; hence there is temporary, or *anovulatory, infertility.**

Such a "pseudopregnant" state was first observed in experimental animals to which progestational hormones were administered before expected ovulation. Normal ovulation was invariably restored when the hormones were withdrawn. Pincus found in 1953 that gonadotropic hormones administered to rabbits prevented from ovulating by administered progesterone rapidly brought about ovulation, mimicking the effects of naturally released endogenous gonadotropic hormones.

The data were first applied to human beings in the 1950's—particularly after discovery of the progestationally potent and orally effective synthetic 19-nor compounds norethynodrel and norethindrone (see Fig. 16.11). Usually one of these is given daily with a small amount of an estrogen, so that the hormonal pattern of early pregnancy is simulated. This regimen employs the so-called "combination pill." Alternatively, estrogens are given daily and progestational agents are given for only 3 to 5 days each month in order to produce cyclic menstrual bleeding. This has been called the "sequential" regimen. Large-scale field studies begun in 1955 have established the efficacy of both approaches to contraception, though more time will be needed to prove that they are free of long-term or late ill effects. Work is now in progress on newer progestational hormones.

Physiology of pregnancy. Implantation of the fertilized ovum in the uterine mucosa marks the beginning of pregnancy, or gestation. Fetal membranes form; the embryo grows, its inner cell mass differentiating into ectoderm, mesoderm, and endoderm (see Fig. 13.2); and, by processes of enlargement, constriction, evagination, and invagination, the various tissues, organs, and organ systems gradually develop.

The duration of a normal pregnancy is 280 days. Fig. 16.21 illustrates the growth of the embryo and its positions in the uterus at 8 weeks and at full term. At full term the embryo is fully

flexed with its head down and the umbilical cord coiling around its body. (This is the position of about 95% of all human embryos at full term.) It pushes the intestines upward toward the diaphragm and exerts extreme pressure upon the transverse colon, rectum, and bladder. Other implantation sites than that shown are possible. However, with one too near the cervical os, premature placental separation may occur as the cervix dilates.

The membranes enveloping the embryo appear early in pregnancy. Immediately surrounding it is the *amnion* (see Fig. 16.19); the closed space within the amnion is filled with *amniotic fluid.* As soon as the hindgut of the embryo is delineated, the diverticulum known as the *allantois* arises. It is rudimentary in man as compared to reptile or bird.

The amniotic fluid acts as a protective cushion permitting some embryonic movement. Normally, its volume is between 500 and 1000 ml. Isotopic studies of its rate of formation show that, on the average, it is completely replaced every 3 hours. However, its sources and mechanisms of reabsorption are unknown. Death of the embryo in the uterus only partially decreases the turnover rate.

Many metabolic changes take place in the mother during pregnancy. We have already spoken of the large quantities of estrogens, progesterone, and chorionic gonadotropin produced by the placenta and the consequent suppression of FSH and LH secretion. Other hormonal events in pregnancy include increased secretion of glucocorticoids (see Fig. 16.20), thyroxine, and parathyroid hormone.

The hormone *relaxin*, which is entirely independent of other female sex hormones, is secreted by the ovaries, primarily by the corpora lutea, of rats, rabbits, and guinea pigs during pregnancy. Its secretion in human females has not yet been established. Relaxin is a watersoluble peptide having a molecular weight of 10,000 to 12,000. It (1) relaxes the pelvic ligaments and the cervix, thereby dilating the birth canal; (2) inhibits uterine movement; and (3) stimulates the breasts.

The presence of a growing embryo in the uterus adds additional functional burdens to most of the body systems. First, it has many primary effects upon the sex organs. The uterus in-

---

* Presumably the inhibition of ovulation by progesterone accounts for the nonoccurrence of *superfetation* (i.e., conception during an established pregnancy).

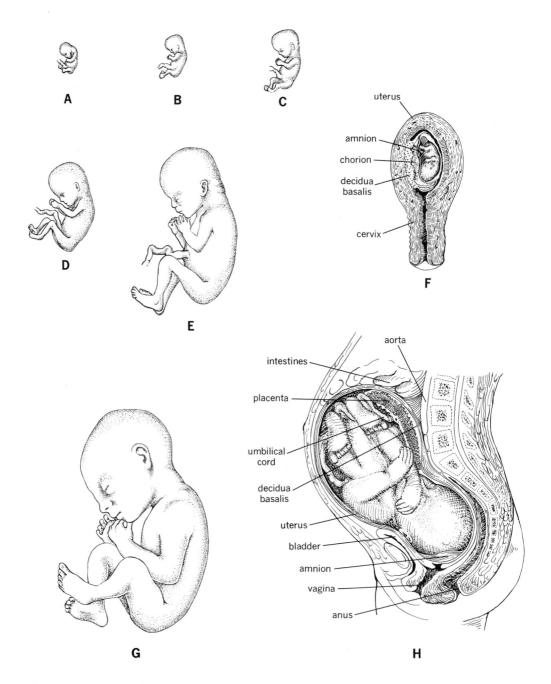

16.21 Development of an embryo: A, at 8 weeks; B, at 9 weeks; C, at 10 weeks; D, at 3 months; E, at 4 months; F, sagittal section of a pregnant uterus at 8 weeks; G, at 6 months; H, sagittal section of a pregnant uterus at full term. (A–G, one-half actual size; H, approx. one-fifth actual size.)

creases in weight from 30 to about 700 g; the breasts approximately double in size; the vagina enlarges; and the vaginal orifice widens. The various hormones may also produce facial changes, sometimes causing edema, acne, and skin pigmentation.

The flow of blood through the placenta (about 750 ml per minute) decreases the total peripheral resistance of the mother's circulatory system and consequently increases the cardiac output in the same manner that arteriovenous shunts do. Owing to this factor and the general increase in metabolic activity, the cardiac output rises to 30 to 40% above normal by the twenty-seventh week of pregnancy. The maternal blood volume shortly before term is approximately 30% above normal.

During the first months of pregnancy, possibly as a result of nausea, the mother ordinarily loses a few pounds, but over the entire period she gains an average of 24 lb, most of it in the last 6 months. Approximately 7 lb is baby, and approximately 4 lb is fetal membranes and amniotic fluid. The remaining 13 lb belongs to the mother herself—the uterus gains about 2 lb, the breasts about 3 lb, and the rest of the body about 8 lb.

As noted in Chapter 14, pregnancy increases the body's caloric requirements. Supplemental foods needed to supply the embryo and its membranes include minerals, vitamins, and proteins. The embryo assumes priority in regard to many of the nutritional elements in the mother's body fluids and ordinarily continues to grow even though the mother's diet is inadequate.

Parturition. The act of giving birth is called *parturition.* As pregnancy approaches its end, the uterus becomes progressively more excitable until finally *labor,* the series of strong rhythmic contractions whose force expels the baby, begins. The transition from pregnant uterus to parturient uterus is a gradual one, with the onset of labor its dramatic component.

Although the precise mechanism triggering parturition is unknown, both hormonal and mechanical changes lead up to the culminating contractions. Experimental studies indicate that the myometrium is "progesterone-dominated" until just before term, when it becomes "estrogen-dominated." During the period when pro-

gesterone effects exceed estrogen effects, uterine contractility is suppressed. This effect, known as the *progesterone block,* prevents expulsion of the embryo and thus maintains pregnancy. From the seventh month on, estrogen secretion increases more rapidly than progesterone secretion, and immediately before term relatively large quantities of estrogens appear in the extracellular fluid. They promote myometrial contractility and irritability. Apparently, therefore, the rising ratio of estrogens to progesterone is responsible, at least in part, for the increased uterine contractility at term.

Oxytocin, one of the hormones secreted by the neurohypophysis, specifically causes uterine contractions (see Chapter 15). There are four reasons for believing that it may be particularly important in increasing the contractility of the uterus at term: (1) the responsiveness of the uterus to a given dose of oxytocin is magnified by 10 times in late pregnancy; (2) oxytocin secretion appears considerably accelerated at the time of labor; (3) though hypophysectomized females can still deliver their young, their labor is prolonged; and (4) irritation or stretching of the uterus generates a reflex that accelerates oxytocin secretion.

It is well known that simply stretching smooth muscles usually increases their contractility.* Stretching or irritation of the cervix elicits uterine contractions. The obstetrician frequently induces labor by dilating the cervix or rupturing the membranes so that the head of the baby will dilate it more forcibly than usual or irritate it in some other way.

The following sequence summarizes the events that probably lead to the onset of labor. About halfway through pregnancy uterine irritability starts to intensify. As the active patches of the uterine wall expand, the amniotic fluid pressure rises cyclically—though the damping effect of the nonactive patches is sufficient to keep it from rising excessively. The mother is unaware of these changes. As pregnancy proceeds, however, the damping effect diminishes, and uterine activity increases until it crosses

---

* The intermittent stretching of the uterus by movements of the embryo probably increases uterine excitability. Significantly, twins are born an average of 19 days sooner than single offspring.

the threshold of perception and later that of pain. A few weeks before term, the mother experiences periodic brief episodes of weak rhythmic contractions, the *Braxton-Hicks contractions*. These become progressively stronger as gradually the cervix is prepared for dilatation. When the amniotic pressure reaches a certain critical level, regular labor pains begin.

When the uterine contractions are strong and painful, reflexes from the birth canal, or perhaps even from the uterus, to the spinal cord and thence to the abdominal muscles cause intense abdominal contractions. These add greatly to the force that eventually expels the baby. Most often the head emerges first; next most often the buttocks emerge first.

The first obstruction to be passed is the cervix. The *first stage of labor* is the period of cervical dilatation, which continues until the opening is as large as the baby's head—generally 8 to 12 hours in a first pregnancy. In the *second stage of labor*—generally 1 or 2 hours long—the head moves rapidly into the birth canal, wedging its way through until it emerges.

Immediately after delivery, the uterus shrinks. The result is a shearing effect between its walls and the placenta, which separates the placenta from its implantation site, opening the sinuses and causing a moderate amount of bleeding. However, since the smooth muscles of the uterine walls are arranged as figures of 8 around their blood vessels, the contraction of the uterus constricts the vessels, so that bleeding is soon stopped.

The uterus involutes within 4 or 5 weeks. In 1 week its weight drops to less than one-half its immediate postpartum weight. The endometrium at the implantation site autolyzes, producing a vaginal discharge named *lochia*, which is first bloody and then serous, but in about 10 days the endometrium is restored and ready for a normal uterine cycle.

BREASTS AND LACTATION

Anatomy of the breasts. Mammary glands, called breasts in man and some other animals, are distinguishing features of mammals. They are modified skin glands that develop from two rows of differentiated ectoderm, the *milk lines*, extending from the axillae to the inguinal regions in the embryo. In an animal having large litters, each milk line gives rise to a row of mammary glands. In man each milk line normally gives rise to only one breast, though remnants are frequently seen as accessory nipples or breasts, sometimes as far down as the groin.

The two breasts are located in the pectoral region, one on each side at a level between the second and sixth ribs (Fig. 16.22). They are present in both sexes, and their development is under the influence of female sex hormones. The female breasts remain relatively small until puberty, when the accumulation of fat adds to their size.* At the apex of each breast is a nipple containing 12 to 20 depressions, each at the site of an individual duct opening. Around the nipple is a pigmented circular area, the *areola*, whose color varies with complexion. The minute elevations in the areola are the loci of underlying sebaceous glands.

The glandular portion of the breast consists of 12 to 20 irregular *lobes*, arranged radially around the nipple. Each lobe is subdivided into lobules, and each lobule, in turn, into small glandular alveoli. Each lobe has a duct opening into the nipple, and ducts from the lobules empty into the larger duct. The ducts from the lobules contain sphincters under nervous control. The lobules regress late in life, the alveoli largely disappearing. The tubular structure remains, however, supported by connective and adipose tissue.

The breasts undergo additional growth and development during pregnancy, owing to the influence of placental estrogens and progesterone. Estrogens cause the duct system to extend and branch and increase the amount of connective tissue. Progesterone causes the alveolar system to expand, an effect recalling the secretory effect of progesterone upon the endometrium. Presumably these changes sensitize breast tissue to the ensuing lactogenic actions of glucocorticoids and LTH,† for only after they have occurred are the breasts capable of producing milk.

Initiation and suppression of lactation. At the end of pregnancy, the breasts are fully de-

---

* The male breasts enlarge, acquiring the characteristics of the female breasts, in *gynecomastia*, a pathological condition associated with an elevated estrogen level in the body.

† In discussions of breast physiology and lactation, LTH is usually referred to as lactogenic hormone or prolactin (see Table 15.2). As mentioned earlier, it is probably identical with GH in man. Some of the evidence for this conclusion is that (1) all preparations of LTH have GH activity, and vice versa; (2) the GH level rises during lactation; and (3) individuals with acromegaly (see p. 599) often lactate spontaneously.

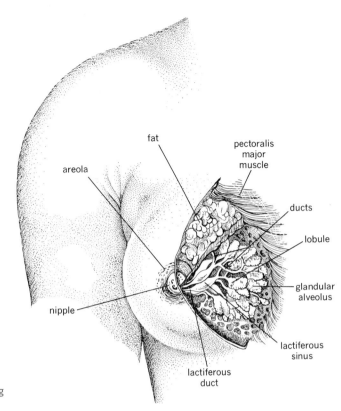

16.22 Sagittal section of a breast, showing the duct system.

veloped, but they discharge only a little fluid each day. This thin yellowish substance, *colostrum*, contains the same quantities of protein and carbohydrate as milk, but almost no lipid, and its maximum rate of production is about 1% the subsequent maximum rate of production of milk. The absence of lactation during pregnancy is probably due to the inhibitory effects of estrogens and progesterone on the breasts. The abrupt loss of placental estrogens and progesterone immediately after parturition removes these effects, and the milking or suckling stimulus generates the secretion of LTH by the adenohypophysis (Fig. 16.23). ACTH secretion and hence glucocorticoid secretion also increase. Within 2 or 3 days, the breasts are engorged with milk in place of colostrum.

The effect of LTH on the breasts is comparable to its effect on the corpus luteum. It induces the continuous secretion of milk into the alveoli of the breasts. Apparently LTH synthesis and release are normally suppressed by the hypothalamus, which secretes a *prolactin-inhibiting factor*, or PIF. Thus LTH is secreted when hypothalamic influence is removed. In contrast, the synthesis and release of other adenohypophyseal hormones usually require hypothalamic stimulation. Preliminary studies have shown that suckling initiates a reflex that decreases PIF secretion.

The effects of placental lactogen duplicate those of LTH. In a sense, the situation parallels that with LTH (from the adenohypophysis) and chorionic gonadotropin (from the placenta), two distinct proteins with overlapping functions. The function of placental lactogen remains to be elucidated.

Initiation of milk production does not mean that thenceforth milk passes freely and continuously from the alveoli into the ducts and out through the nipples. It must be ejected from the alveoli into the ducts before the nursing baby can obtain it. The ejection (also called "let down") of milk is the result of a combined

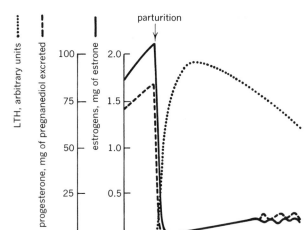

16.23 Effect of parturition on estrogen, progesterone, and LTH secretion.

neural and hormonal reflex involving oxytocin and its lactogogic action.

When a breast is suckled, impulses transmitted via the spinal cord to the hypothalamus stimulate the secretion of oxytocin and, to a lesser extent, vasopressin. These hormones cause the *myoepithelial cells* that surround the alveoli to contract, thereby forcing milk into the ducts, so that within a minute milk flows.* Oxytocin is five times as effective as vasopressin in promoting milk ejection.

When the suckling stimulus is absent, secretion of LTH, oxytocin, and vasopressin stops. Therefore, if milk is not regularly removed from the breasts, the capacity for milk production is lost in 1 or 2 weeks. Lactation, then, is ultimately controlled by the demand for milk. LTH secretion declines gradually over a period of 7 to 9 months despite continued milking, and milk production eventually terminates. Lactation may be deliberately arrested by the administration of estrogens, since these inhibit LTH secretion.

After parturition the uterus involutes more rapidly in women who lactate than in those who do not. The difference is probably due to re-

duced estrogen secretion during lactation, for estrogens enlarge the uterus. The uterus of a lactating mother usually becomes smaller than it was before pregnancy, whereas the uterus of a nonlactating mother often remains considerably larger.

Lactation prevents resumption of the ovarian cycle for a few months—as if the adenohypophysis were so preoccupied with LTH secretion that it neglected the secretion of the other gonadotropic hormones. However, after a time the adenohypophysis again produces enough FSH to reinstate the cycle. Since the rhythmic interplay of the ovarian and adenohypophyseal hormones during the cycle does not involve LTH to a significant degree, milk production continues.

Composition of milk. Table 16.2 compares the composition of human milk with that of cow's milk. The concentration of lactose (the main milk sugar) in human milk is approximately one and one-half times that in cow's milk, but the concentration of protein in cow's milk is two or more times that in human milk, and the mineral content of cow's milk is higher than that of human milk. The lipid contents of both vary during a single milking from 1.2% initially to 4.4% at the end.†

The rate of lactation in a mother subjected to optimal suckling generally increases in the first few months after parturition and thereafter decreases, dwindling to virtually nothing at the end of 7 months to a year. At the height of lactation, 1½ liters of milk may be formed each day. This requires large quantities of nutrients from the mother, including approximately 50 g of lipid, 100 g of lactose, and 2 to 3 g of calcium phosphate each day. Unless she drinks much milk and has an adequate supply of vitamin D, the output of calcium and phosphate by the lactating breasts will be greater than the intake of these substances. Indeed, the danger of decalcification of the mother's bones ordinarily is not very great during pregnancy but is extreme during lactation.

---

* In light of this mechanism, it is not surprising that suckling at one breast causes milk to flow from both breasts.

† This change is due to the clustering of milk globules within the alveoli. The few free (unclustered) globules and "milk serum," the nonsolid portion of milk, leave the alveoli first. The large lipid-rich clusters leave last.

TABLE 16.2 COMPOSITION OF MILK

Component	Approx. conc. in human milk, g per 100 ml	Approx. conc. in cow's milk, g per 100 ml
Water	88.5	87.0
Lactose (milk sugar)	7.0	4.8
Lipid	3.3	3.5
Casein	0.9	2.7
Lactalbumin and other proteins	0.4	0.7
Ash	0.2	0.7
Potassium	0.041	0.150
Calcium	0.030	0.120
Phosphorus	0.013	0.095
Sodium	0.011	0.050

Male Reproductive System

The major structural components of the male reproductive system (Fig. 16.24) are the *penis, testes, epididymides, vasa deferentia, seminal vesicles,* and *prostate.* The chief functions of the male reproductive system are (1) spermatogenesis, (2) participation in sexual intercourse, and (3) the production of male sex hormones, or *androgens.*

STRUCTURE AND DEVELOPMENT

Testes, epididymides, and *scrotum.* Spermatogenesis, as we learned in Chapter 3, takes place in the two testes. Each testis descends from an abdominal position before birth—actually it is drawn downward by a cord called the *gubernaculum*—and comes to lie in a sac, the *scrotum* (Fig. 16.25).* The scrotum has two com-

---

* Occasionally the testes fail to descend into the scrotum—a condition known as *cryptorchidism.* Although spermatogenesis is almost invariably defective in undescended testes, they do produce androgens. Frequently in later life the inguinal canal (through which a testis once descended) reopens, establishing communication between the abdominal cavity and the scrotum and giving rise to *inguinal hernia,* in which a loop of intestine or other viscera extends into the scrotum.

partments, one for each testis, with a medial seam (or raphe) where the embryonic scrotal folds have grown together.† The scrotal skin is wrinkled into transverse ridges by the contraction of cutaneous and subcutaneous muscle fibers.

A testis is essentially ovoid but a little flattened laterally and is about 2 in. long, 1 in. wide, and 1 in. thick. Its lateral border is free (Fig. 16.26), whereas its medial border is attached to a crescent-shaped body, the epididymis. The combination of testis and epididymis is usually referred to as the *testicle.*

Both testis and epididymis have complex duct systems. Internally the testis consists of about 250 lobules. Each lobule contains from one to three convoluted *seminiferous tubules,* which join to form a *straight tubule.* The straight tubules drain into a central network of fine tubules, the *rete testis,* which in turn drains into *efferent ducts.* These empty into the tightly coiled convoluted duct, only 0.4 mm in diameter but 18 to 20 ft in length, of the epididymis. The epididymis receives the efferent ducts at its upper pole.

The testis and epididymis are enveloped by the *tunica vaginalis,* a double-layered membrane whose embryonic derivation from peritoneum is shown in Fig. 16.25. Their outer covering, the *tunica albuginea,* is a tough white fibrous sheath resembling in appearance the sclera of the eye. Numerous septa pass inward from it to the connective tissue mass at the hilus of the testis.

Vasa deferentia and spermatic cords. At the lower pole of the epididymis, its duct is continuous with the *vas (ductus) deferens,* which extends upward from the testis through the *spermatic cord.‡* The cord, comprising the vas deferens, arteries, a rich venous plexus (the *pampiniform plexus*), lymphatic vessels, and nerves inside a common connective tissue sheath, runs upward through the inguinal canal and over the pubic arch, backward over the bladder, and downward to terminate in an *ejaculatory duct.* The course of the cord is that followed by the testis as it descended from its original location in the abdomen, carrying with it epididymis, vas deferens, blood and lymphatic vessels, and nerves.

Accessory glands. The accessory reproductive glands of the male are the *seminal vesicles, prostate,* and *bulbourethral (Cowper's) glands* (see Fig. 16.24). The two seminal vesicles lie above the prostate and behind the bladder. The duct from a vesicle joins the vas deferens to form an ejaculatory duct, of smaller diameter

---

† In the female the folds remain discrete as the labia majora.

‡ The vasa deferentia are unusually rigid and can be easily felt in the upper lateral regions of the scrotum (see Fig. 16.24).

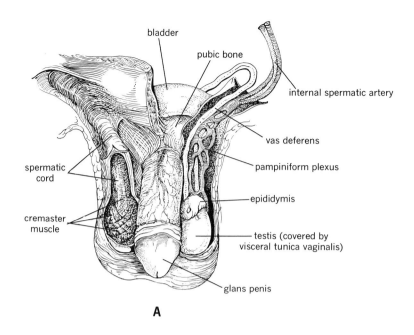

bladder

pubic bone

internal spermatic artery

vas deferens

pampiniform plexus

epididymis

testis (covered by
visceral tunica vaginalis)

spermatic
cord

cremaster
muscle

glans penis

A

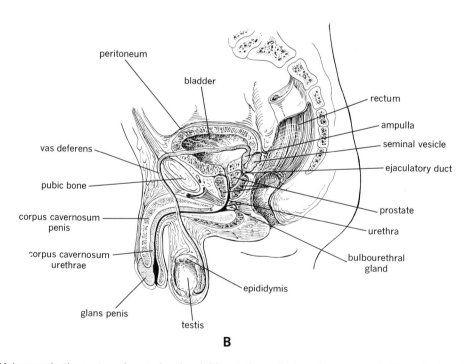

peritoneum

bladder

rectum

ampulla

seminal vesicle

ejaculatory duct

vas deferens

pubic bone

prostate

corpus cavernosum
penis

urethra

corpus cavernosum
urethrae

bulbourethral
gland

glans penis

epididymis

testis

B

16.24   Male reproductive system: A, anterior view (with anterior wall of scrotum removed); B, sagittal section.

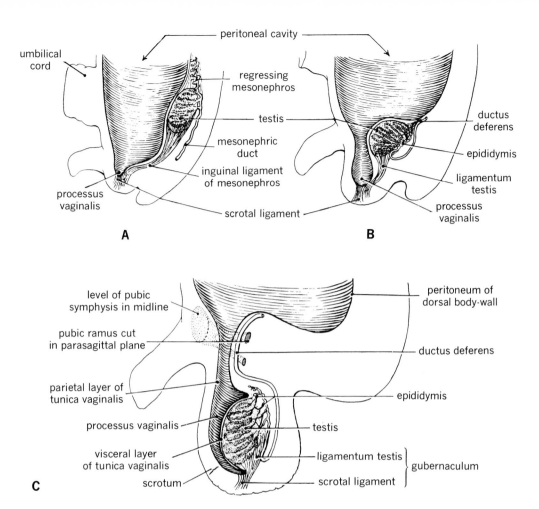

16.25 Descent of the testes: A, 10 week old embryo; B, at 6–7 months; C, just before birth. (From B. M. Patten, *Human Embryology*, 3rd ed., © 1968 by McGraw-Hill, Inc. Used with permission of McGraw-Hill Book Co., New York.)

than the vas deferens and only 2 cm long. The right and left ejaculatory ducts penetrate the tissue of the prostate and open into the urethra within it.

We described the prostate in Chapter 10. Its lobules discharge their secretion through 20 to 30 small ducts draining into the urethra through minute pores.

The bulbourethral glands are two yellow glands about the size of peas situated below the prostate, one on either side of the base of the penis. They empty into the urethra from below.

Penis. The penis is the male copulatory organ. Its body consists of three longitudinal columns of *erectile tissue* termed the *corpora cavernosa* (see Figs. 16.24, 16.27). Two of these, the *corpora cavernosa penis,* are in the dorsolateral part of the penis; the third, the

corpus cavernosum urethrae, is midventral and contains the urethra. Erectile tissue is full of vascular spaces. Sexual excitement causes blood to pour into these spaces from the arteries faster than it is drawn out by the veins. As a result the tissue becomes distended, and the penis, ordinarily soft and flaccid, becomes hard and erect. In this condition it can be inserted into the vagina in the act of sexual intercourse. After sexual excitement has passed, blood flows out of the erectile tissue, and the penis becomes soft again. (As we have noted, erectile tissue is also present in the clitoris.)

The corpus cavernosum urethrae is reflected back over the end of the penis like a cap and contains the slitlike external orifice of the urethra. The smooth tip of the penis, the *glans penis,* is sheathed by a loose

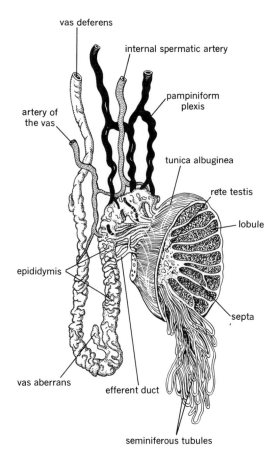

vas deferens

internal spermatic artery

pampiniform plexis

artery of the vas

tunica albuginea

rete testis

lobule

epididymis

septa

vas aberrans

efferent duct

seminiferous tubules

16.26 Sagittal section of the testis.

sleeve of skin, the *foreskin*, or *prepuce*. Circumcision, surgical removal of the foreskin, is frequently performed shortly after birth. This operation promotes cleanliness, for the area posterior to the glans in an

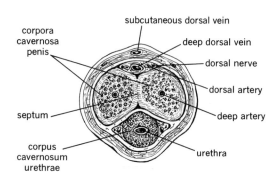

corpora cavernosa penis

subcutaneous dorsal vein

deep dorsal vein

dorsal nerve

dorsal artery

septum

deep artery

corpus cavernosum urethrae

urethra

16.27 Cross section of the penis.

uncircumcised penis accumulates secretions called *smegma*, which decompose and cause irritation. Circumcision has greatly reduced the incidence of cancer of the penis.

SPERMATOGENESIS

Spermatogenesis occurs in all the seminiferous tubules. It begins at about age 12 under the influence of adenohypophyseal hormones and continues throughout life.

Function of the seminiferous tubules. As shown in Figs. 16.28 and 16.29, the testis contains three main types of functioning cells: (1) the *germinal epithelial* cells of the seminiferous tubules; (2) the *Sertoli*, or *sustentacular*, cells, large cells extending from the base of the epithelium to the deep interior of the tubules; and (3) the *Leydig*, or *interstitial*, cells lying between adjacent tubules.

Until the time of puberty, the testis appears much the same as in early infancy. The seminiferous tubules are small cords lacking lumens, with a scattering of germ cells interspersed with more numerous undifferentiated Sertoli cells. The intertubular areas contain connective tissue and mesenchymal cells, precursors of the Leydig cells.*

The first evidence of puberal change in the testis is the differentiation of the Leydig cells. Almost simultaneously, the cells of the seminiferous tubules start to develop. Initially, the most conspicuous change is in the Sertoli cells, which elongate and exhibit the characteristic Sertoli cell nuclei. Not until the Leydig and Sertoli cells are differentiated does spermatogenesis commence.

Spermatogenesis consists in the differentiation of primordial cells, *spermatogonia*, into spermatozoa. The spermatogonia are arranged in two or three layers next to the connective tissue wall of the seminiferous tubule. They continuously divide by mitosis, but, as with the blood cell precursors (see Fig. 8.9), some of the

---

* For several months before birth, the intertubular areas contain many Leydig cells (see Fig. 16.28A). These secrete appreciable quantities of androgens as a result of stimulation by chorionic gonadotropin from the placenta. Shortly after birth, however, the Leydig cells disappear and are not seen again until puberty.

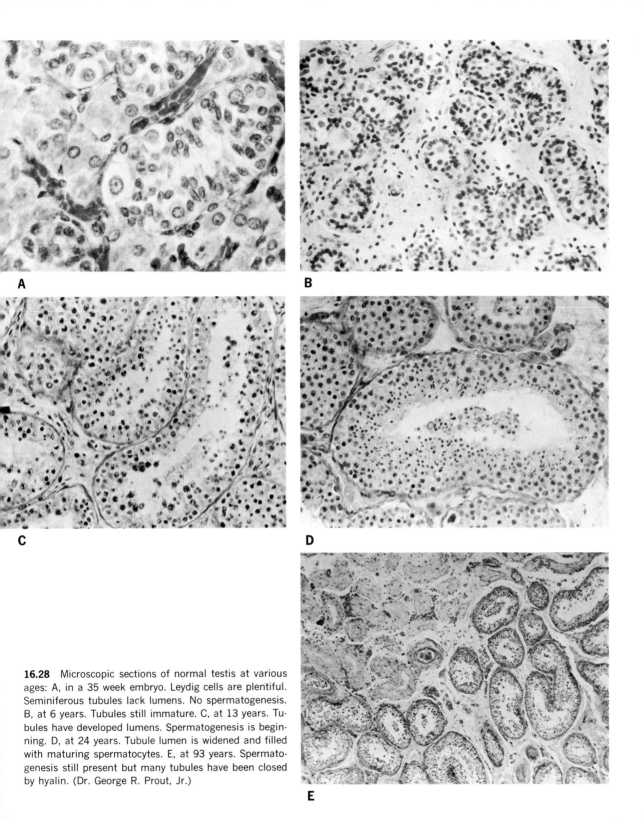

16.28 Microscopic sections of normal testis at various ages: A, in a 35 week embryo. Leydig cells are plentiful. Seminiferous tubules lack lumens. No spermatogenesis. B, at 6 years. Tubules still immature. C, at 13 years. Tubules have developed lumens. Spermatogenesis is beginning. D, at 24 years. Tubule lumen is widened and filled with maturing spermatocytes. E, at 93 years. Spermatogenesis still present but many tubules have been closed by hyalin. (Dr. George R. Prout, Jr.)

daughter cells enter the maturation process—which adapts them for their intended function but deprives them of the capacity for further mitotic division. Maturation includes the meiotic divisions that reduce the 46 spermatogonial chromosomes to 23 spermatozoal chromosomes (see Fig. 3.23).* It takes about 2 weeks.

The first stage in spermatogenesis is the growth of the spermatogonium into an enlarged *primary spermatocyte.* Each of the 46 chromosomes in the primary spermatocyte synthesizes a new, identical chromosome to which it remains attached at the centromere. The primary spermatocyte then undergoes meiosis. First, it divides into two *secondary spermatocytes,* each containing 23 paired chromosomes. Then each secondary spermatocyte divides into two *spermatids,* each containing 23 chromosomes, none of them paired. Hence is a haploid cell containing one of each maternal-paternal pair of chromosomes.

Each spermatid develops into a single spermatozoon.† A newly formed spermatid has the characteristics of an ordinary epithelial cell but soon grows a long flagellum. The cytoplasm of the cell contracts around the nucleus, and apparently some of the cytoplasmic material is actually cast away. Simultaneously the nuclear material is rearranged into a compact mass constituting most of the head of the spermatozoon (see Figs. 3.4, 16.29). An *acrosomal cap,* which may serve as a perforating device, covers the head. Though little cytoplasm remains in the middle piece of the body, a thin layer of cytoplasm and a cell membrane surround the flagellum, which becomes the motile tail. The tail has an elaborate connecting piece that contains mitochondria and an *axial filament* that extends its length. The axial filament protrudes from the tip of the tail as an end piece.

The most mature spermatozoa lie with their heads pointed toward the periphery of the tubule and their tails toward the lumen. Usually they are close to a Sertoli cell. The Sertoli cells

have a supporting function and also promote maturation. Spermatids attach themselves to Sertoli cells, which appear to provide essential nutrients or hormones. It has been demonstrated, for example, that, in conditions that interfere with spermatogenesis, lipids accumulate in the Sertoli cells.

Spermatozoa and semen. Following spermatogenesis in the seminiferous tubules of the testis, the spermatozoa pass into the connecting epididymis. Spermatozoa removed experimentally from the seminiferous tubules are completely nonmotile and therefore incapable of fertilizing an ovum. However, after 18 hours in the epididymis, they have the power to move and to fertilize. Likely they merely age or mature in the epididymis.

A small number of spermatozoa can be stored in the epididymis, but most are stored in the vas deferens. They are relatively dormant in storage, probably because their own metabolism releases considerable CO_2 into the surrounding fluid, the semen, lowering the pH to a level inhibitory to their activity. Apparently they can be stored without loss of fertility as long as 42 days.

As we noted earlier (see p. 652), motile and fertile spermatozoa can travel through fluid at a rate of about 1 mm per minute, somewhat faster in neutral or slightly alkaline media and somewhat slower in acid media. Their activity increases with rising temperature, but so does their rate of metabolism, so that their life span is shortened.‡

Early anatomists erroneously believed that spermatozoa were stored in the seminal vesicles

---

* See the footnote on p. 652 concerning maturation of the ovum.

† Since only one of each pair of the original 46 chromosomes is also present in each ovum, combination of a spermatozoon with an ovum reestablishes the original complement of 46 chromosomes (see Chapter 3).

---

‡ Elevating the temperature of the testes inhibits spermatogenesis, probably by increasing the rate of metabolism of the tubular epithelium until the regenerative spermatogonia "burn out." Immersion of the testes in hot water for prolonged periods of time may result in temporary or even permanent sterility. It is usually stated that the testicles were located in the dangling scrotum so that they could be maintained at a temperature below that of the body. Cold causes the scrotal muscles to contract, pulling the testicles closer to the body, whereas warmth causes them to relax, pushing the testicles farther from the body. Also, the scrotum is well supplied with sweat glands, which presumably aid in keeping the testicles cool. Thus the scrotum is apparently designed to act as a cooling mechanism without which spermatogenesis would be deficient.

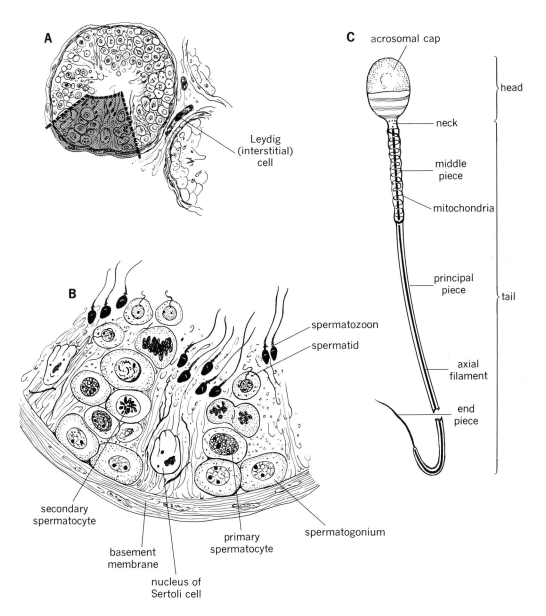

16.29 Spermatogenesis: A, seminiferous tubule; B, detail of area outlined in A; C, mature spermatozoon.

—whence the name "vesicles," which means fluid-containing sacs. It is now known that the vesicles are glands and not storage areas. They are lined with epithelium that secretes a mucoid material containing an abundance of fructose and smaller amounts of ascorbic acid, inositol, ergothioneine, amino acids, and phosphoryl-choline.* During ejaculation the vesicles empty their contents into the urethra at the same time that the vasa deferentia empty theirs into it. The

* When appropriately treated, phosphorylcholine liberates choline. This reaction is the basis for one of the older tests for semen in forensic medicine.

volume of semen is thereby increased. The fructose and other substances in the seminal fluid have nutritive and protective value for the spermatozoa.

The prostate secretes a thin, milky, alkaline fluid containing citric acid, acid phosphatase, fibrinolysin, spermine, and the prostaglandins, a group of 15 or more fatty acids with powerful vasodepressor effects and the capacity in low concentrations to cause contraction of uterine smooth muscle.* During ejaculation the capsule of the prostate contracts simultaneously with the vasa deferentia and seminal vesicles so that the volume of semen is further increased. Prostatic fluid may be essential for successful fertilization, neutralizing the fluid from the vasa deferentia and the vaginal fluid, which are relatively acid.

In summary, ejaculated semen is composed of fluids from the vasa deferentia, seminal vesicles, prostate, and mucous glands, especially the bulbourethral glands. Its average pH is 6.5. Prostatic fluid gives it a milky appearance, whereas the fluids from the seminal vesicles and the mucous glands give it a mucoid consistency. Within 30 minutes after ejaculation, the mucoid consistency is gone owing to the action of proteolytic enzymes in semen. Immediately after ejaculation the spermatozoa are relatively nonmotile, possibly because of the viscosity of the mucus. However, after the mucus thins, they are highly motile. Semen may be stored for several days at low temperatures, and spermatozoa of some animals have been found viable after storage for over a year at temperatures below $-100°C$.

SEX ACT

Stimuli. The principal receptors of stimuli initiating the male sex act are the sensory end organs in the penis, especially the glans, and the other erogenous zones. These transmit a distinctive sexual sensation, which passes through the *pudendal nerves,* thence through the sacral

plexus into the sacral region of the spinal cord, and finally up the cord to undefined areas of the cerebral cortex. Mental stimuli, such as sexual thoughts or dreams, can also initiate sexual excitement that may culminate in ejaculation. Indeed, nocturnal ejaculations during dreams occur in many males, especially during adolescence.

Though psychological factors usually play an important role in the male sex act, they are not essential. Appropriate genital stimulation causes ejaculation in a paraplegic man or in an experimental animal whose spinal cord has been cut above the lumbar region. It may be postulated, then, that the male sex act results from inherent reflex mechanisms generated by either mental or direct physical stimulation and integrated in the sacral and lumbar regions of the spinal cord.

Sequence of events. Erection is the first effect of male sexual stimulation. Parasympathetic impulses traveling from the sacral region of the spinal cord through the nervi erigentes to the penis dilate the arteries (and to a lesser extent constrict the veins) of the penis, admitting arterial blood under high pressure into the erectile tissue.

In addition to promoting erection, the impulses cause the bulbourethral glands to secrete fluid. This flows through and out of the urethra during intercourse and aids in lubrication. As stimulation intensifies, the reflex centers of the spinal cord begin to emit rhythmic sympathetic impulses. These leave the cord via the first two lumbar nerves and proceed to the genital organs, where they produce peristaltic contractions in the testes, epididymides, and vasa deferentia that expel spermatozoa into the urethra; peristaltic contractions in the seminal vesicles and the prostate that expel seminal fluid and prostatic fluid into the urethra; and the discharge of additional fluid from the bulbourethral glands into the urethra. All of the fluids combine to form semen. Strictly speaking, the process to this point is *emission*. Then rhythmic impulses pass from the cord over the pudendal nerves to skeletal muscles encasing the base of the erectile tissue, producing contractions that propel the semen from the urethra to the exterior. The complete process is ejaculation.

---

* The prostaglandins thus are hormones that are synthesized by the prostate. Their physiological role is not well understood. In 1968, five of them were artificially synthesized by E. J. Corey and coworkers. The hyaluronidase found in semen is believed to be secreted by the spermatozoa themselves.

SEX HORMONES

Gonadotropic hormones and androgens. The testes are influenced by two adenohypophyseal gonadotropic hormones, follicle-stimulating hormone (FSH) and interstitial cell–stimulating hormone (ICSH).* These are essential to stimulate, respectively, the seminiferous tubules to produce spermatozoa and the Leydig (or interstitial) cells to produce androgens. The gonadotropic hormone, if any, necessary for Sertoli cell function is not known.

The term "androgen" refers not only to a hormone secreted by the testes but also to any other hormone with similar physiological effects, wherever it arises in the body. An androgen, then, is the male counterpart of an estrogen. Among androgenic effects are maintenance of the structures and functions of the prostate and seminal vesicles and production of the pattern of body hair growth characteristic of the male. Many synthetic steroids have these effects (and others to be mentioned), too, and are therefore androgens.

Two different androgens have been isolated from venous blood draining from the testes, *testosterone* (Δ^4-androstene-17β-ol-3-one) and *androstenedione* (Δ^4-androstene-3,17-dione) (see Fig. 15.9). Both are secreted by the Leydig cells. Since testosterone is predominant by far, the consensus is that it is the principal basis for the endocrine effects of testicular activity.

The ovaries secrete a small amount of testosterone; and the adrenal cortex in both sexes secretes at least five androgens, though in quantities normally so slight that their masculinizing effects are insignificant even in females. However, an adrenocortical tumor may secrete enough androgens to produce prematurely or exaggeratedly all the usual secondary male sexual characteristics in a male and a virilizing syndrome including abnormal enlargement of the clitoris, male distribution of body hair, and deepening of the voice in a female.

Interestingly, all androgens and their breakdown products have a keto group at carbon 17. Hence the total amount of 17-ketosteroids in the urine is a useful index of androgen secretion.† In normal urine the chief 17-ketosteroids derived from androgens are *androsterone* and *etiocholanolone* (see Fig. 15.9). The normal daily output is 10 to 20 mg in the male and 6 to 14 mg in the female. The difference between the male and female values is the contribution of the testes.

Small amounts of estrogens are also produced in the testes, but by what cells is unknown. There is some evidence that the Sertoli cells secrete estrogens, and some that the Leydig cells secrete estrogens as well as androgens.

The plasma testosterone level, normally 0.3 to 1.0 μg per 100 ml, is 5 to 10 times as high in the male as in the female. The level of circulating androgens affects the secretion of gonadotropic hormones by the adenohypophysis. Even small additional quantities of androgens inhibit ICSH secretion. When an androgen other than testosterone is administered, Leydig cell function may be completely suppressed, and the plasma testosterone level may fall 80%. Since the secretion of adenohypophyseal hormones is under hypothalamic control, it may be that circulating androgens exert their suppressive effect on the hypothalamus.‡

In general, androgens are responsible for the primary and secondary sexual characteristics. Much of our knowledge of the action of these hormones has come from observations of castrated (i.e., gonadectomized) animals. It has been known for centuries that castration makes meat animals docile and easy to fatten. Castration causes a young cockerel to become larger than it would normally because closure of the zones of growth in the long bones is delayed. The resulting capon is desirable for its extra meat. The loss of androgens is especially apparent in the head, which lacks the tall red comb

---

* ICSH in the male is identical to LH in the female (see Table 15.2).

---

† As noted on p. 645, the estrogen estrone is also a 17-ketosteroid. Before the urine is assayed for androgens, estrone is eliminated by a simple extraction that takes advantage of its phenolic first ring (see Figs. 15.9, 16.9).

‡ The fact that the hypothalamus controls adenohypophyseal gonadotropic hormone secretion implies that mental stimuli can increase or decrease it. This effect is demonstrated in certain animals during the mating season, when mental or other neural stimuli influence the fertility of the males. For instance, transporting a bull under uncomfortable conditions may make him temporarily sterile.

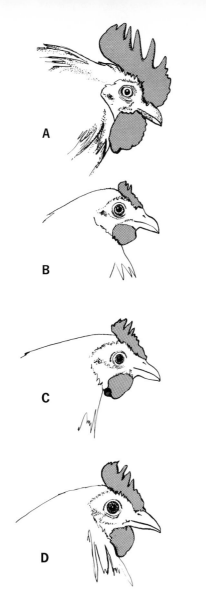

16.30 Effect of castration on a cockerel: A, normal rooster; B, capon, with fully regressed comb and wattles; C, same capon 3 days after an androgen injection; D, same capon 6 days after the androgen injection.

and wattles of a rooster and resembles the head of a female with a low pale comb (Fig. 16.30).

Testosterone is evidently the first hormone elaborated by the male embryo, appearing as early as the second month of embryonic life. Indeed, some workers believe that the major functional difference between male and female sex chromosomes is that the male chromosome promotes the secretion of testosterone by the genital ridge whereas the female chromosome promotes the secretion of estrogens. If testosterone is secreted by the genital ridge, it probably accounts, at least in part, for the embryonic development of male (instead of female) sexual characteristics—i.e., a penis and scrotum (instead of a clitoris and labia) and a prostate, seminal vesicles, and associated ducts. It also probably accounts for the descent of the testes into the scrotum.

Most of the events of puberty in the male are directly due to testosterone, whose secretion is initiated by a sudden unexplained increase in adenohypophyseal gonadotropic hormone production (Fig. 16.31), conceivably an expression of a maturation process in the adenohypophysis. The increase in adenohypophyseal activity is followed by an acceleration of body growth and muscular development.

In summary, testosterone and other androgens have the following specific effects: (1) they cause the development of primary and secondary sexual characteristics, including growth of the genital organs, distribution of the body hair in a male pattern and occasional baldness, deepening of the voice, and alteration of the skin with darkening of the pigment and frequently acne as a result of overactivity of the sebaceous glands; (2) they promote long bone growth and calcium retention; (3) they stimulate protein anabolism with increased gluconeogenesis, acting somewhat similarly to GH (see Chapter 15); (4) they increase the BMR; (5) they promote sodium and water retention to a small extent; and (6) they stimulate erythrocyte production in the bone marrow. Recent biochemical evidence suggests that androgens, like estrogens perhaps, act by promoting selective gene transcription in the cells of target organs.

Male climacteric. The adenohypophysis secretes gonadotropic hormones from puberty until old age, and spermatogenesis ordinarily continues until death. Many men, however, begin to decrease their sexual activity in the late 40's or 50's much as women do at the menopause. This situation in the male is known as the *male climacteric.*

Undoubtedly many of the features of the male climacteric are psychological in origin. However, among its objective events are a waning of androgen secretion, a consequent rise in FSH secretion, and a

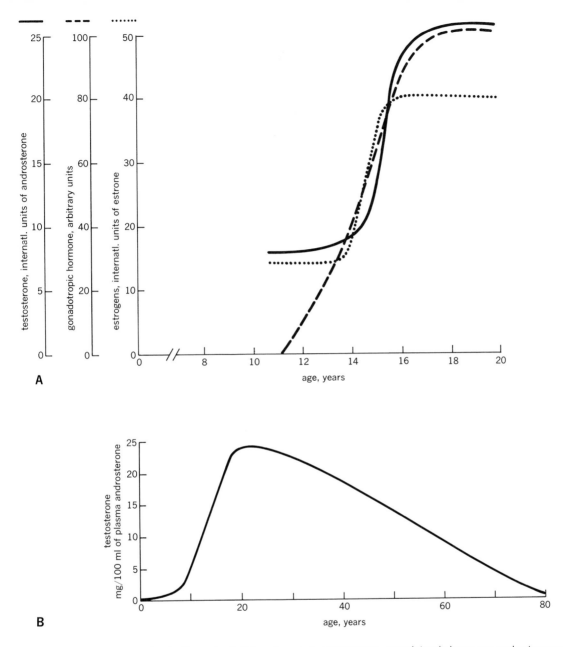

16.31 Sex hormone production in the male at different ages: A, testosterone, gonadotropic hormones and estrogens just before and after puberty; B, testosterone throughout life.

deterioration of the testicular architecture (see Fig. 16.28). Some men are benefited by administered testosterone.

Hypogonadism and infertility. Hypogonadism afflicts more than 0.13% of all males. Defects may arise at virtually every step in the complex chain of events leading to the production of normal spermatozoa and androgens, with consequent failure of either spermatozoa production or androgen production or both. Classification depends upon (1) whether the defect occurs before or after puberty; (2) the histological ap-

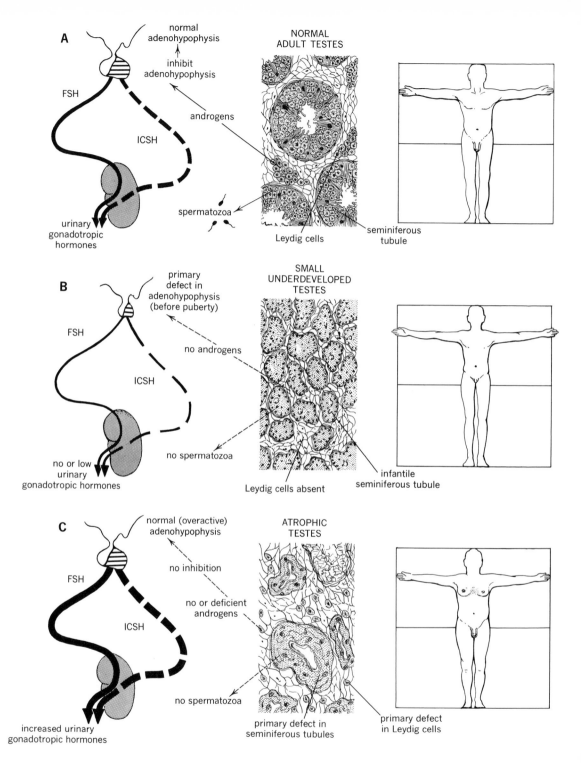

16.32 Main types of male hypogonadism: A, normal male; B, eunuchoidism, due to primary failure of FSH and ICSH secretion before puberty; C, hypogonadism due to primary failure of the testes.

pearance of the testes (as determined by biopsy); (3) the body habitus (i.e., build and constitution), the presence of secondary sexual characteristics, etc.; (4) the amounts of gonadotropic hormones and 17-ketosteroids in the urine; and (5) the number, appearance, and motility of spermatozoa in the ejaculate (as determined by the microscopic examination of a sample of ejaculate.)*

Examples of the two main types of hypogonadism, in which both spermatozoa production and androgen production fail, are depicted in Fig. 16.32. In one type the primary defect is in the secretion of adenohypophyseal gonadotropic hormones. When the defect occurs before puberty and is accompanied by a defect in the secretion of GH, the result may be pituitary dwarfism (see p. 597). When the defect occurs before puberty and involves only the gonadotropic hormones (as it does more commonly), growth is normal, but secondary testicular activity is impaired, and the result is the so-called *eunuchoid* body habitus, in which normal height is associated with undeveloped secondary sexual characteristics. In the other type the primary defect is in the testes—due to either acquired testicular damage (e.g., from mumps or radiation) or chromosomal disorders such as *Klinefelter's syndrome*, in which the seminiferous tubules begin unaccountably to degenerate after puberty and the karyotype (see Fig. 3.18) reveals, for example, two X chromosomes and a Y chromosome in each body cell.† Leydig cells produce little testosterone. Owing to the lack of suppression of adenohypophyseal function by circulating androgens, urinary gonadotropic hormone levels are high. Body habitus is approximately normal (despite gynecomastia), since growth and the development of secondary sexual characteristics may occur normally. Innumerable variations of this type of hypogonadism have been described, many with bizarre chromosomal

abnormalities. Most cases of faulty spermatogenesis belong to it and are due not to chromosomal disorders but to previous inflammatory disease of the testes. The defect is usually discovered only during investigations of infertility. Sometimes infertility is attributable to subtle hormonal imbalances that may be correctible. For all types of adult hypogonadism, testosterone treatment is invariably attempted. Although such therapy may not restore spermatogenesis, it may be highly beneficial.

REFERENCES AND SUGGESTIONS FOR FURTHER READING

Arey, L. B., *Developmental Anatomy*, 7th ed., Saunders, Philadelphia, 1965.

Austin, C. R., *Fertilization*, Prentice-Hall, Englewood Cliffs, N. J., 1966.

Buxton, C. L., and A. L. Southam, *Human Infertility*, Hoeber, New York, 1958.

Carey, H. M., ed., *Modern Trends in Human Reproductive Physiology*, Butterworths, London, 1963.

Cole, H. H., and P. T. Cupps, *Reproduction in Domestic Animals*, Academic Press, New York, 1959.

Diczfalusy, E., "Endocrine Functions of the Human Fetoplacental Unit," *Federation Proc.*, **23**, 791 (1964).

Donovan, B. T., and J. J. ten Bosch, *Physiology of Puberty*, Williams & Wilkins, Baltimore, 1965.

Dorfman, R. I., "Bioassay of Steroid Hormones," *Physiol. Rev.*, **34**, 138 (1954).

———, and R. A. Shipley, *Androgens: Biochemistry, Physiology, and Clinical Significance*, Wiley, New York, 1956.

———, and F. Ungar, *Metabolism of Steroid Hormones*, Academic Press, New York, 1965.

Edwards, R. G., "Mammalian Eggs in the Laboratory," *Sci. Am.*, **215**, 72 (Aug., 1966).

Emmens, C. W., and A. W. Blackstraw, "Artificial Insemination," *Physiol. Rev.*, **36**, 277 (1956).

Engel, L. L., "Mechanisms of Action of Estrogens," *Vitamins Hormones*, **17**, 205 (1959).

Everett, J. W., "Central Neural Control of Reproductive Functions of the Adenohypophysis," *Physiol. Rev.*, **44**, 373 (1964).

Farris, E. J., *Human Ovulation and Fertility*, Lippincott, Philadelphia, 1959.

Federman, D. D., *Abnormal Sexual Development: Genetic and Endocrine Approach to Differential Diagnosis*, Saunders, Philadelphia, 1967.

Folley, S. J., *The Physiology and Biochemistry of Lactation*, Thomas, Springfield, Ill., 1957.

Giese, A. C., "Comparative Physiology: Annual Reproductive Cycles of Marine Invertebrates," *Ann. Rev. Physiol.*, **21**, 547 (1959).

---

* Some cases of male infertility are due solely to the unexplained production of nonmotile or otherwise abnormal spermatozoa.

† Such abnormalities of chromosome number, *aneuploidy*, result from chromosomal *nondisjunction* (i.e., failure of separation) during mitosis or meiosis. If, for example, in the normal meiosis pictured in Fig. 3.23, the pair of short chromosomes in metaphase I failed to disjoin, one interphase daughter cell would have two short chromosomes, and the other would have none, instead of each having one. Normally a spermatogonium undergoes two meiotic divisions to yield four spermatozoa, two with one X chromosome apiece and two with one Y chromosome apiece. Nondisjunction in the first meiotic division would yield two spermatozoa with an X chromosome *and* a Y chromosome apiece and two with neither. If an X ovum were fertilized by an abnormal XY spermatozoon, the zygote would be XXY, one of the several aneuploid karyotypes of Klinefelter's syndrome.

Hagerman, D. D., and C. A. Villee, "Transport Functions of the Placenta," *Physiol. Rev.*, **40**, 313 (1960).

Hartman, C. G., "Physiological Mechanisms of Conception: An Inventory of Unanswered Questions," *Perspectives Biol. Med.*, **4**, 77 (1960).

Huseby, R. A., O. V. Dominguez, and L. T. Samuels, "Function of Normal and Abnormal Testicular Interstitial Cells in the Mouse," *Recent Progr. Hormone Res.*, **17**, 1 (1961).

Jarabak, J., J. A. Adams, H. G. Williams-Ashman, and P. Talalay, "Purification of a 17β-Hydroxysteroid Dehydrogenase of Human Placenta and Studies of Its Transhydrogenase Function," *J. Biol. Chem.*, **237**, 345 (1962).

Klinefelter, H. F., Jr., E. C. Reifenstein, Jr., and F. Albright, "Syndrome Characterized by Gynecomastia, Aspermatogenesis Without A-Leydigism and Increased Excretion of Follicle-Stimulating Hormone," *J. Clin. Endocrinol. Metab.*, **2**, 615 (1942).

Kon, S. K., and A. T. Cowie, eds., *Milk: The Mammary Gland and Its Secretions*, Academic Press, New York, 1961.

Lang, W. R., ed., "The Vagina," *Ann. N. Y. Acad. Sci.*, **83**, 77 (1959).

Linzell, J. L., "Physiology of the Mammary Glands," *Physiol. Rev.*, **39**, 534 (1959).

Lloyd, C. W., ed., *Human Reproduction and Sexual Behavior*, Lea & Febiger, Philadelphia, 1964.

———, and J. Weisz, "Some Aspects of Reproductive Physiology," *Ann. Rev. Physiol.*, **28**, 267 (1966).

MacLeod, J., and C. Tietze, "Control of Reproductive Capacity," *Ann. Rev. Med.*, **15**, 299 (1964).

Mann, T. *Biochemistry of Semen and the Male Reproductive Tract*, Wiley, New York, 1964.

Masters, W. H., and V. E. Johnson, *Human Sexual Response*, Little, Brown, Boston, 1966.

Meites, J., and C. S. Nicoll, "Adenohypophysis: Prolactin," *Ann. Rev. Physiol.*, **28**, 57 (1966).

Merrill, R. C., "Estriol: A Review," *Physiol. Rev.*, **38**, 463 (1958).

Moore, K. L., ed., *The Sex Chromatin*, Saunders, Philadelphia, 1966.

Mueller, G. C., A. M. Herranen, and K. F. Jervell, "Studies on the Mechanism of Action of Estrogens," *Recent Progr. Hormone Res.*, **14**, 95 (1960).

Nalbandor, A. V. and B. Cook, "Reproduction," *Ann. Rev. Physiol.*, **30**, 245 (1968).

Paschkis, K. E., A. E. Rakoff, and A. Cantarow, *Clinical Endocrinology*, 2nd ed., Hoeber, New York, 1958.

Pincus, G., "Control of Conception by Hormonal Steroids," *Science*, **153**, 493 (1966).

———, "Reproduction," *Ann. Rev. Physiol.*, **24**, 57 (1962).

———, "Progestational Agents and the Control of Fertility," *Vitamins Hormones*, **17**, 307 (1959).

Richardson, G., "Ovarian Physiology," *New Engl. J. Med.*, **274**, 1008, 1064, 1121, 1183 (1966).

Short, R. V., "Reproduction," *Ann. Rev. Physiol.*, **29**, 373 (1967).

Sohval, A. R., "Chromosomes and Sex Chromatin in Normal and Anomalous Sexual Development," *Physiol. Rev.*, **43**, 306 (1963).

Talalay, P., "Enzymatic Mechanisms in Steroid Metabolism," *Physiol. Rev.*, **37**, 362 (1957).

Van Wagenen, G., and M. E. Simpson, *Embryology of the Ovary and Testis*, Yale Univ. Press, New Haven, Conn., 1965.

Velardo, J. T., ed., "The Uterus," *Ann. N. Y. Acad. Sci.*, **75**, 385 (1959).

Villee, C. A., D. D. Hagerman, and P. B. Joel, "An Enzymatic Basis for the Physiologic Functions of Estrogens," *Recent Progr. Hormone Res.*, **16**, 49 (1960).

Wolstenholme, G. E. W., and J. Knight, eds., *Gonadotropins: Physicochemical and Immunological Properties*, Little, Brown, Boston, 1965.

Young, W. C., ed., *Sex and Internal Secretions*, Williams & Wilkins, Baltimore, 1961.

Zuckermann, S., ed., *The Ovary*, Academic Press, New York, 1962.

What makes muscle unique is its mechanical function involving fast and extensive changes. The wing muscle of the housefly, for instance, describes three hundred full cycles of contraction per second. Such rapid motion is possible only if the system is fitted together from small units, arranged with a high degree of regularity and held together, at least in certain directions, by weak forces only. This opens the possibility of taking the whole system to pieces without using drastic methods. The dimensions of our muscles are in the range of an inch [but] it is possible to take the muscle to pieces, step by step, down to the dimensions of 10^{-7} inch, a molecular dimension, studying at every step what the system is still capable of doing and coordinating function with the level of organization. More than that: not only can we pull the system to pieces, but having secured its pieces undamaged, we can put them together again and observe how the biological functions gradually reappear.

Albert Szent-Györgyi, in *Chemistry of Muscular Contraction*, 1951

17 Musculoskeletal System

Introduction

The musculoskeletal system includes the skeleton, whose function is body support, and the voluntary muscles, whose function is body movement. In its reference to the skeleton, this statement suggests a degree of inertness, as though bones were but a crutch for holding up soft flesh. From the viewpoint of contemporary physiology, however, the bones are dynamic structures that metabolize, adapt, respond, and, hard as they are, undergo constant change. Indeed, one of their principal functions is to provide a reservoir of essential minerals, such as calcium and phosphorus.* Thus bones have mechanical *and* metabolic functions.

As for the muscles, much is subsumed in the statement that they move and contract. We shall presently learn that, despite brilliant advances, the mechanism of their movement remains one of physiology's outstanding unsolved problems.

---

* A clue to the evolutionary origin of bone (and evidence of its metabolic importance) is found in the relationship between the composition of the skeleton and that of the blood in vertebrates, since changes in both appear to have occurred together. It is best seen in the three classes of fishes living today: the agnaths, or jawless fishes (such as hagfishes and lampreys), which have little or no bone tissue; the chondrichthyans (such as sharks and rays), whose skeletons are completely cartilaginous; and the osteichthyans (such as trouts and eels), most of which have bony skeletons. Blood salt concentration in agnaths is about 4%; in chondrichthyans, about 2%; and in osteichthyans, about 1%.

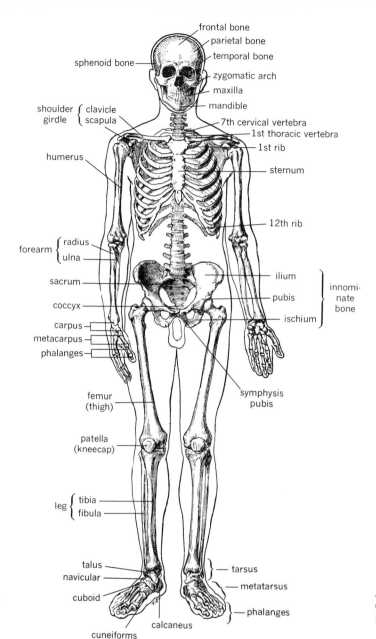

frontal bone
parietal bone
temporal bone
sphenoid bone
zygomatic arch
maxilla
mandible
shoulder girdle { clavicle
scapula
7th cervical vertebra
1st thoracic vertebra
1st rib
humerus
sternum
12th rib
forearm { radius
ulna
ilium
pubis
sacrum
innominate bone
coccyx
ischium
carpus
metacarpus
phalanges
femur (thigh)
symphysis pubis
patella (kneecap)
leg { tibia
fibula
talus
navicular
tarsus
cuboid
metatarsus
phalanges
calcaneus
cuneiforms

17.1 Human skeleton. (From *Gray's Anatomy of the Human Body,* ed. by C. M. Goss. 28th ed., Lea & Febiger, Philadelphia, 1965.)

Architecture

Having evolved—both in ontogeny and in phylogeny—from soft connective tissues, the hard tissues of the skeleton are necessary for welding together and protecting the softer organs, supporting the body, and maintaining its form. Almost all striated muscles attach to the skeleton, whose arrangement of levers and movable joints affords a framework of almost unlimited versatility.

We properly speak of the bones as organs because they are made up of many types of tissue (osseous tissue, cartilage, fibrous tissue, nervous tissue, vascular tissue, marrow, etc.). We reviewed the development and histology of cartilage and bone in Chapter 5. As we noted in our discussion of the parathyroids (see Chapter 15), bone tissue is in constant dynamic chemical equilibrium, with individual bones being continuously remodeled during life.

The skeleton is usually considered to have two major divisions: the *axial skeleton,* which includes the *skull, vertebral column, ribs,* and *sternum;* and the *appendicular skeleton,* which includes *limbs* and the *shoulder* and *pelvic girdles.* A diagram of the entire skeleton appears in Fig. 17.1. An infant has about 350 bones, many of which fuse during growth. The average adult has 206.

AXIAL SKELETON

Skull. The skull contains 29 bones—those of the *cranium* (8 bones), and the *face* (14 bones), the *hyoid* bone (1 bone), and the *ear ossicles* (6 bones).

The bones of the cranium include the *frontal, occipital, sphenoid, ethmoid,* two *parietal,* and two *temporal* bones. Together these make up the domelike vault enclosing the brain. On looking at the skull from the side, we see that the frontal bone is the anterior wall of the cranium and the occipital bone the posterior wall. The parietal bones are the superior and upper lateral walls, and the temporal bones the lower lateral walls. The sphenoid bone occupies a key position. Extending transversely through the center of the skull, it joins anteriorly with the orbital plates of the frontal bone and posteriorly with the central part of the occipital bone. Both the frontal and occipital bones curve far underneath, forming considerable portions of the base of the skull. Wedgelike sections of the temporal bone fit between the occipital and sphenoid bones. The ethmoid bone separates the cranial cavity above from the nasal cavities below.

In the middle of the inferior surface of the occipital bone is a large opening, the *foramen magnum,* through which the spinal cord passes to make connection with the brain stem. On each side of the foramen magnum is a *condyle,* an oval process that is curved anteroposteriorly like a rocker. Fitting into shallow depressions on the upper surface of the *atlas* (the first cervical vertebra), the two condyles form a joint allowing nodding movements of the head.

The facial bones include the *mandible* (the lower jawbone), the two *maxillae* (upper jawbones), the two *zygomatic* bones (cheekbones), the two *lacrimal* bones (anteromedial walls of the orbits), the *vomer* (the bottom of the nasal septum), the two inferior *nasal conchae,* and the *palatine* bone. The U-shaped hyoid bone,

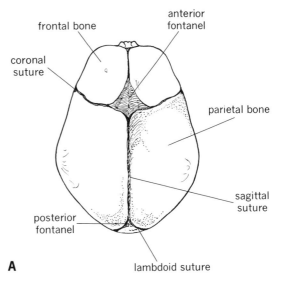

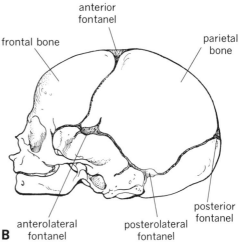

17.2 Skull of a newborn infant, showing sutures and fontanels: A, superior view; B, lateral view.

in the neck (see Fig. 11.2), is the only bone that does not meet other bones to form a joint; only ligaments attach it to extensions of the temporal bones. Some of the muscles of the tongue and mouth insert in the hyoid bone. The ear ossicles were described in Chapter 12.

Compared with the skulls of other animals, the human skull displays a relatively small face and a relatively large cranium. The skull contains five great cavities: the two orbits, the nasal cavity, the oral cavity, and the cranial cavity. Seven pairs of *foramina* in the base of the skull admit the various nerves and blood vessels. Three pairs of intraosseous foramina (*supraorbital*, *intraorbital*, and *mental*) open onto the face.

The joints of the skull, with the exception of the two *temperomandibular joints*, are immovable *sutures*, tightly fused in the adult. As shown in Fig. 17.2, the frontal bone consists of two parts at birth. The spaces between converging bones are called *fontanels*. The newborn infant ordinarily has six, all at angles of the parietal bones. The diamond-shaped *anterior fontanel*, the largest, at the junction of the frontal and parietal bones, usually closes by the eighteenth month. The tri-

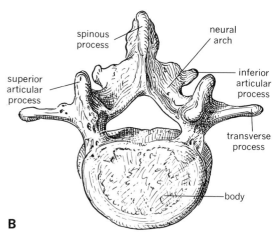

B

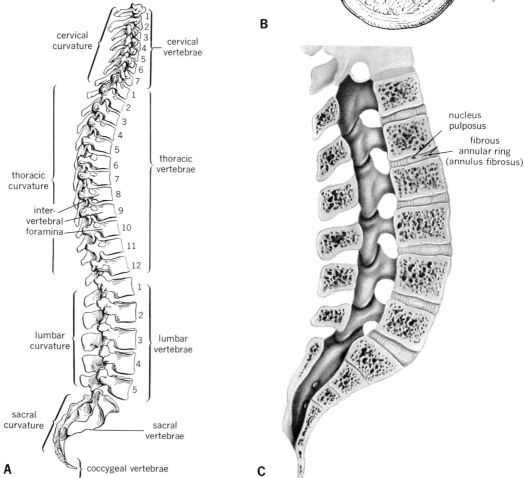

A

C

17.3 Vertebral column (A) and a superior view of a typical lumbar vertebra (B). Sagittal section of lower vertebral column showing intervertebral discs (C). (A from Grant's *Atlas of Anatomy*, 5th ed., Williams & Wilkins, Baltimore, 1965; B from B. J. Anson, *Atlas of Human Anatomy*, 2nd ed., W. B. Saunders Co., Philadelphia, 1963; C from *Therapeutic Notes*, May 1959.)

angular *posterior fontanel*, at the junction of the occipital and parietal bones, usually closes by the second month. The *anterolateral fontanels*, at the junctions of the frontal, parietal, temporal, and sphenoid bones, usually close by the third month. The *posterolateral fontanels*, at the junction of the parietal, occipital, and temporal bones do not close until the second year.

The incomplete ossification (i.e., bone formation) of the skull at birth permits an important adaptive mechanism in parturition. The bones of the skull override each other, so that the head is reduced in diameter and molded as it accommodates itself to the birth canal.

Vertebral column. We were introduced to the vertebral column, or spine, in Chapter 12. It is a strong flexible rod that supports the skull, gives base to the ribs, and protects the spinal cord. As shown in Fig. 17.3, it is composed of 33 (or occasionally 34) bones, the *vertebrae*, which are classified as follows: 7 *cervical*, 12 *thoracic*, 5 *lumbar*, 5 *sacral*, and 4 (or 5) *coccygeal*. In the adult the last two groups are fused into 2 bones, the *sacrum* and the *coccyx*.

Viewed laterally, the vertebral column has four curves, alternately convex and concave. In the embryo it is uniformly concave. The concave curvature persists in the thoracic and sacrococcygeal regions, providing space for the viscera, with the two convex curves arising in the cervical and lumbar regions. The cervical curvature appears when the infant learns to hold his head erect, usually about the third month; the lumbar curvature appears when he has learned to walk, between the twelfth and eighteenth months. The curves increase the weight-bearing strength of the column.

The vertebrae differ in size and shape in the different regions of the column but exhibit a common structural plan. Each has a *body*, which supports weight; a *neural arch*, which protects the spinal cord; a *spinous process* and *right* and *left transverse processes*, three levers on which muscles pull; and four *articular processes*, which restrict movement.

The first two cervical vertebrae, the *atlas* and the *axis*, are specialized to receive the base of the skull and permit pivotal movements of the head. The thoracic vertebrae have special surfaces for attachment to the heads of the ribs. The lumbar vertebrae, the largest, are specially adapted for weight-bearing. The coccyx, the most rudimentary part of the vertebral column, may be regarded as a vestigial tail. In species with a tail a series of caudal bony segments continue into it.

The bodies of adjacent vertebrae are separated from one another by a shock-absorbing *intervertebral disc*. The disc consists of a ringlike mass of white fibrous cartilage surrounding a central semifluid mass, the *nucleus pulposus*. The disc may break down, with extrusion or herniation of the nucleus pulposus. The result is compression of the emerging spinal nerves, one of the most common causes of low back pain.

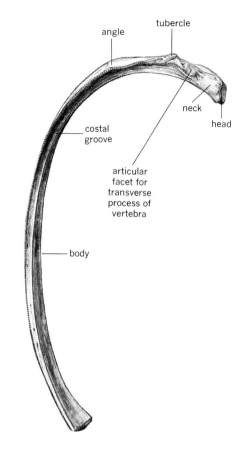

17.4 Rib. (From *Gray's Anatomy of the Human Body*, ed. by C. M. Goss. 28th ed., Lea & Febiger, Philadelphia, 1965.)

Thorax. The thorax is a cage whose walls are formed in back by the *thoracic vertebrae*, at the sides by the *ribs*, and in front by the *costal cartilages* and the *sternum* (see Fig. 17.1).

The 24 ribs, 12 on each side, are long flat bones that are curved and twisted. They are connected to the thoracic vertebrae and continue forward into the costal cartilages. The upper seven pairs are attached directly to the sternum through their costal cartilages and for this reason are called "true" ribs. The remaining five pairs are called "false" ribs. The cartilages of the eighth, ninth, and tenth ribs join to the lower borders of the cartilages of the preceding ribs. The cartilages of the eleventh and twelfth ribs are unattached; hence these ribs are termed "floating" ribs. Each rib slopes downward so that its anterior end is lower than its posterior end.

A central rib (Fig. 17.4) may be considered typical, although certain ribs show variations. The *head* of the

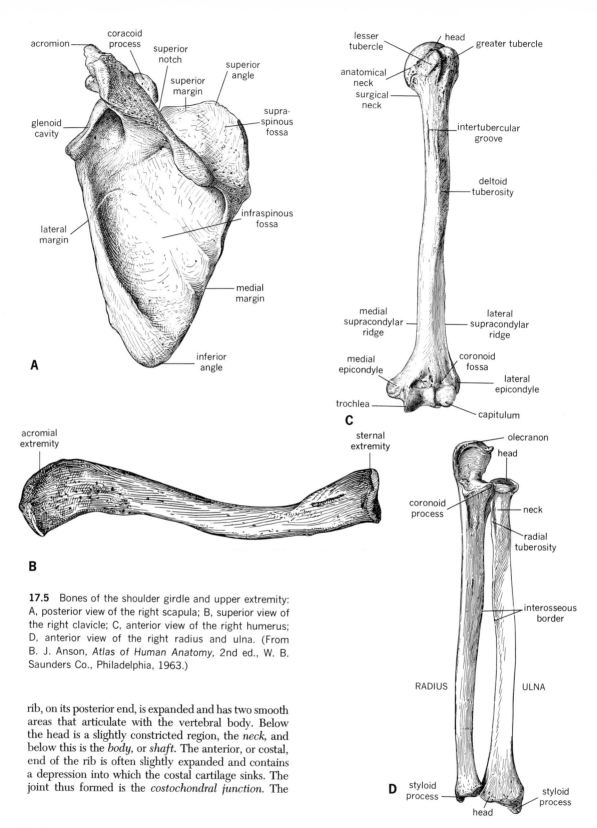

17.5 Bones of the shoulder girdle and upper extremity: A, posterior view of the right scapula; B, superior view of the right clavicle; C, anterior view of the right humerus; D, anterior view of the right radius and ulna. (From B. J. Anson, *Atlas of Human Anatomy*, 2nd ed., W. B. Saunders Co., Philadelphia, 1963.)

rib, on its posterior end, is expanded and has two smooth areas that articulate with the vertebral body. Below the head is a slightly constricted region, the *neck*, and below this is the *body*, or *shaft*. The anterior, or costal, end of the rib is often slightly expanded and contains a depression into which the costal cartilage sinks. The joint thus formed is the *costochondral junction*. The parts are bound together by fusion of the periosteum of the bone with the perichondrium of the cartilage.

The sternum is composed of a *manubrium, body (gladiolus),* and *xiphoid (ensiform) process.*

APPENDICULAR SKELETON

Shoulder girdle and upper extremities. The shoulder girdle consists of two *scapulae (shoulder blades)* and two *clavicles (collarbones).* The scapula is a flat bone of triangular shape. It is attached by muscles to the back of the thorax and constitutes the prominence of the shoulder (Fig. 17.5). Its convex posterior surface has a ridge, the *spine,* and ends laterally in a projection, the *acromion.* The acromion joins the clavicle, an elongated S-shaped bone that supports the shoulder and in turn connects with the sternum, creating a sturdy but flexible girdle.

The *humerus,* the long bone of the upper arm, consists of a shaft and two large extremities. It is attached at one end to the *glenoid cavity* of the scapula and at the other to the two bones of the forearm—the *radius* and the *ulna.* At the proximal end are the *head* and two rounded processes, the *greater* and *lesser tubercles,* enclosing a deep groove, the *intertubercular groove,* for the tendon of the *biceps muscle.* On the posterior side of the shaft is a groove for passage of the radial nerve, and midway down the side of the shaft is a rough area, the *deltoid tuberosity,* into which the *deltoid muscle* inserts. At the distal end a rounded knob, the *capitulum,* connects with the radius; a projection, the *trochlea,* connects with the ulna; a depression, the *olecranon fossa,* accommodates an extension of the ulna when the forearm is extended; and another depression, the *coronoid fossa,* accommodates an extension of the ulna when the forearm is flexed.

The ulna is medial to the radius. Its proximal end connects with the humerus and the radius, continuing beyond the radius in an *olecranon process* (the *elbow*), which contains a *radial notch* fitting the trochlea of the humerus. Its distal end has a small round *head* connecting with the radius and also with the wrist bones by means of a disc of white fibrous cartilage. The radius has a *head* connecting with the capitulum of the humerus and with the radial notch of the ulna; a projection, the *radial tuberosity,* into which the biceps muscle inserts; and a protrusion, the *styloid process,* on the wrist end.

The wrist is composed of 8 small *carpal bones,* and the hand of 5 *metacarpal bones* and 14 *phalanges* (see Fig. 17.1).

Pelvic girdle. On each side of the sacrum is a broad flat *innominate bone* (hipbone). The two innominate bones meet anteriorly to form the pelvic girdle (Fig. 17.6), base for attachment of the bones of the lower extremities and protective frame for the organs of the

urinary and reproductive systems. The large space inside the pelvic girdle is the pelvic cavity.

Each innominate bone is in fact a fusion of three separate bones: the *ilium, ischium,* and *pubis* (see Fig. 17.6B). The ilium is the large uppermost bone, with prominent crests; the pubis is a curved bone anterior and inferior to the ilium; and the ischium is the lowermost bone, whose tuberosity supports the body in a sitting position. At the junction of the three bones is a deep cavity, the *acetabulum,* into which the head of the *femur* (thighbone) fits.

The architecture of the pelvic girdle is particularly interesting because it determines the shape of the birth canal in the female. In the male the pelvis is constructed compactly along lines contributing to motive power and speed. In the female it is wide, light, and capacious. The inlet of the male pelvis is heart-shaped. That of the female pelvis is almost a perfect circle.

It is also fascinating to compare the human pelvis with those of other species. Although the basic pattern of three pelvic bones began with the amphibia, major changes took place as animals acquired the ability to develop intrauterine embryos and to stand erect. To get the skeleton permanently upright required drastic anatomical alterations. If the pelvis and spine of four-legged creatures had merely been rotated 90°, the human female would have an absolutely straight birth canal. With this arrangement abdominal pressure would fall with maximum force upon the pelvic floor, making prolapse of the pelvic viscera even more common than it is now. In fact, the pelvis was rotated only 30 or 40°, so that the birth canal inclines 50 or 60° (see Fig. 17.6C). To get the trunk erect, the lumbar curvature was introduced, and, to get the head erect, the cervical curvature was introduced. As Fig. 17.3 indicates, these two curves depend to a large extent upon the intervertebral discs, whose liability to damage is well known.

Lower extremities. The *femur* is the longest and strongest bone in the body (Fig. 17.7). It is comparable in many ways to the humerus. Its proximal end has a rounded *head* that fits into the acetabulum of the innominate bone. The *medial* and *lateral condyles* are two large eminences at the expanded distal end of the bone whose smooth inferior and posterior surfaces connect with the *tibia.* The *patellar surface* is the smooth shallow concavity on the anterior surface of the condyles that connects with the *patella* (kneecap). Posteriorly, the condyles are separated by a deep, roughened depression, the *intercondyloid fossa.* The somewhat flattened sides of the condyles present bony prominences, the *medial* and *lateral epicondyles.* The tibia and *fibula* are the bones of the lower leg. The tibia, medial to the fibula, has a triangular shaft whose sharp anterior edge is called the *shin.* The proximal end of the tibia has *medial* and *lateral condyles.*

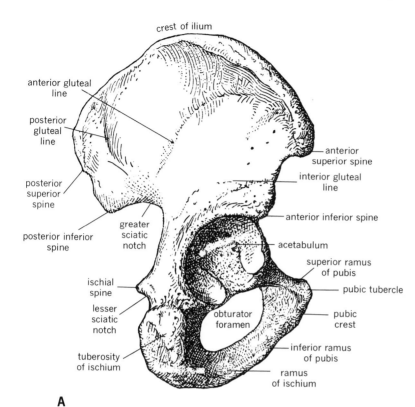

crest of ilium

anterior gluteal
line

posterior
gluteal
line

posterior
superior
spine

posterior inferior
spine

greater
sciatic
notch

ischial
spine

lesser
sciatic
notch

tuberosity
of ischium

anterior
superior spine

interior gluteal
line

anterior inferior spine

acetabulum

superior ramus
of pubis

pubic tubercle

pubic
crest

obturator
foramen

inferior ramus
of pubis

ramus
of ischium

A

17.6 A, Right innominate bone; B, plan of ossification of the hip bone, showing union of the three parts in the acetabulum; C, the regions of the pelvis. (A from T. Jones and W. C. Shepard, *Manual of Surgical Anatomy,* W. B. Saunders Co., Philadelphia, 1945; B from *Gray's Anatomy of the Human Body,* ed. by C. M. Goss. 28th ed., Lea & Febiger, Philadelphia, 1965; C from J. B. DeLee and Greenhill, *Principles and Practice of Obstetrics,* W. B. Saunders Co., Philadelphia, 1943.)

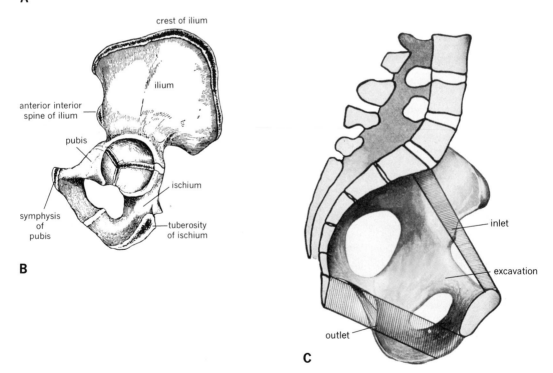

crest of ilium

ilium

anterior interior
spine of ilium

pubis

ischium

symphysis
of
pubis

tuberosity
of ischium

B

inlet

excavation

outlet

C

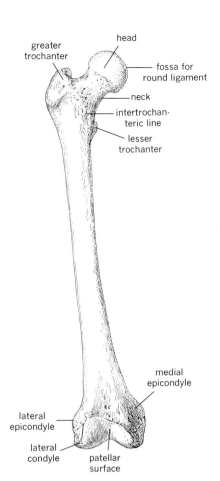

greater trochanter

head

fossa for round ligament

neck

intertrochanteric line

lesser trochanter

medial epicondyle

lateral epicondyle

lateral condyle

patellar surface

A

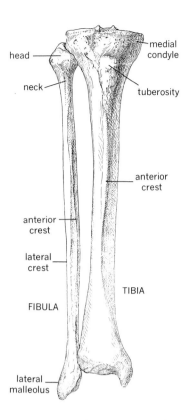

head

medial condyle

neck

tuberosity

anterior crest

anterior crest

lateral crest

TIBIA

FIBULA

lateral malleolus

B

17.7 Long bones of the lower extremity: A, anterior view of the right femur; B, anterior view of the right fibula and tibia. (From T. Jones and W. C. Shepard, *Manual of Surgical Anatomy*, W. B. Saunders Co., Philadelphia, 1945.)

The bony structure of the foot corresponds roughly with that of the hand (see Fig. 17.1). The *tarsal bones* of the heel and back portion of the foot are analogous to the carpal bones of the wrist, though 7 in number rather than 8. The 5 long *metatarsal bones* and the 14 *phalanges* of the toes are analogous to the metacarpals of the hand and the phalanges of the fingers.

The bones of the foot are bound together by ligaments, muscles, and tendons in two *arches*—one longitudinal and the other transverse. The arches are flexible, so that they yield when weight is placed on the foot and spring back into place when the weight is lifted. Defective arches cause *pes planus* (flat foot).

ARTICULATIONS AND BURSAE

Classification of articulations. Classification of *articulations*, or *joints*, is based upon the ab-

sence or presence of joint cavities. Articulations without cavities are called *synarthroses;* they move little or not at all. Those with cavities are called *diarthroses;* they move freely.

In the synarthroses there is a continuous union of bone with fibrous tissue or cartilage. Examples of this group are (1) the sutures of the skull (see Fig. 17.2); (2) the costochondral junctions of the ribs (see Fig. 17.1);* (3) the symphyses, in which bones are separated by a disc of white fibrous cartilage, as the symphysis pubis (see Fig. 17.1) and the intervertebral discs (see Fig. 17.3); and (4) the *syndesmoses,* in which bones are approximated by fibrous or elastic

---

* These are also called *synchondroses.*

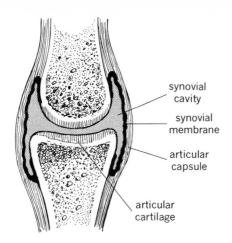

17.8 Typical diarthrosis.

synovial cavity

synovial membrane

articular capsule

articular cartilage

tissue, as the connection between the distal ends of the tibia and fibula (see Fig. 17.1).

In the diarthroses there is a joint cavity between cartilage-covered bone ends (Fig. 17.8). The cavity is encased in a fibrous capsule and lined by a *synovial membrane,* which secretes *synovial fluid.* The capsule and ligaments enclose the articulation and restrain and guide motion. The cartilages provide an elastic area of contact between the opposing bones. A plate of white fibrous cartilage that increases this area and helps to stabilize the moving atriculation is often present. Examples of diarthroses are the main articulations of the limbs. We can further divide them into ball and socket joints (e.g., femoral head in acetabulum), hinge joints (e.g., knee), pivot joints (e.g., skull on atlas and axis), and so on.

Movements of the articulations are described by such terms as *flexion, extension, pronation, supination, adduction,* and *abduction.* These were defined in Chapter 6.

Knee. We cannot deal with the structural details of all the articulations. Therefore, we shall consider one that is a prototype of its class—the knee. As shown in Fig. 17.9, the chief movements of the knee are extension and flexion—characteristic of a hinge joint. A small amount of rotation also occurs when the knee is flexed or semiflexed.

The knee is formed by the articulation of the distal end of the femur with the proximal end of the tibia and the patella (see Fig. 17.1). The fibula is only indirectly involved. A compound diarthrosis, the knee exhibits a gliding type of articulation between the femur

and the patella and a hingelike articulation between the femur and the tibia (see Fig. 17.9).

Though classified anatomically as incongruent, and therefore weak, because its opposing bony surfaces fit together imperfectly, the knee is functionally strong, because its ligaments and other tissues are arranged to withstand the leverage of the femur and tibia without permitting dislocation. The ligaments are both external (the *patellar, tibial collateral, fibular collateral, arcuate popliteal,* and *oblique popliteal ligaments* and the *articular capsule*) and internal (the *anterior cruciate, posterior cruciate,* and *transverse ligaments*) (Fig. 17.10). The collateral ligaments are slack when the knee is flexed and taut when it is extended. The cruciate ligaments cross one another (hence their name), with the posterior one preventing forward displacement of the femur and backward displacement of the tibia and the anterior one preventing backward displacement of the femur and forward displacement of the tibia. In addition, C-shaped white fibrous cartilages, the *medial* and *lateral menisci,* rest upon the tibia, lending depth to its surface and cushioning the knee.

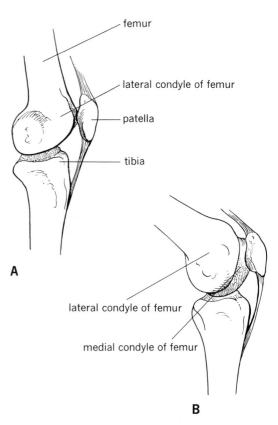

femur

lateral condyle of femur

patella

tibia

A

lateral condyle of femur

medial condyle of femur

B

17.9 Movements of the knee joints: A, extension; B, flexion.

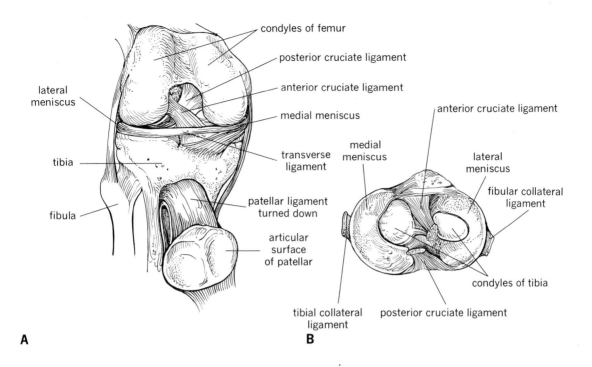

A

B

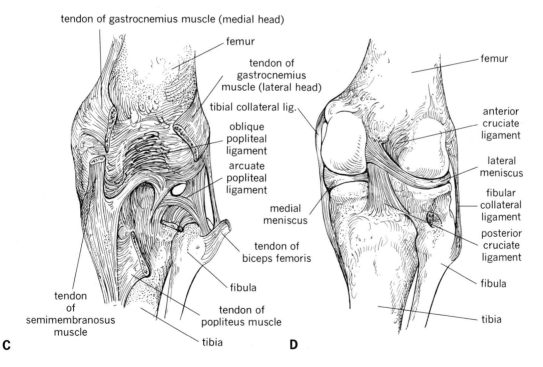

C

D

17.10 Ligaments of the knee joint: A, flexed right knee joint (capsule removed), front view; B, superior view of exposed articular surface of right tibia; C, right knee joint, back view; D, right knee joint (capsule removed), back view.

Many muscles and tendons also cross the knee, act upon it, and support it. The tendon of the *popliteal muscle* crosses behind the articulation and helps to stabilize it posteriorly and laterally. The tendons of the *semimembranosus* and *biceps femoris muscles* and the lateral and medial heads of the *gastrocnemius muscle* provide additional stabilization.*

Classification of bursae. In various places in the body that are subject to friction or pressure, small fluid-containing sacs lined with a synovial-like membrane occur. These are known as *bursae.* They develop in loose connective tissue and, as lubricating devices, are superior to loose connective tissue itself.

The three kinds of bursae are *articular, subcutaneous,* and *subtendinous* bursae. Articular bursae adjoin or relate closely to joint cavities. Several communicate with the joint cavity of the knee, one behind the quadriceps femoris tendon and others behind the gastrocnemius and semimembranosus muscles.

Subcutaneous bursae lie beneath the skin on the convex surfaces of articulations that undergo extreme flexion—where the skin must move freely. Such bursae are prominent at the elbow, hip, knee, and heel. Those at the knee include the large *suprapatellar bursa,* the *infrapatellar bursa* over the proximal end of the patellar ligament, the *deep infrapatellar bursa* between the patellar ligament and the tibia, and the *prepatellar bursa* in front of the patella.

Subtendinous bursae are found wherever tendons rub against resistant structures such as bones, ligaments, or other tendons. A *synovial,* or *tendon, sheath* is a tubular bursa that envelops a tendon.

Synovial membrane and synovial fluid. The synovial membrane lining a joint cavity consists of flattened mesenchymal cells whose structure varies with the character of the underlying connective tissue. Frequently the membrane is thrown into coarse folds that project far into the cavity.

Synovial fluid is a clear, slightly yellow to colorless liquid that does not clot on standing. It contains a mixture of lymphocytes, monocytes, polymorphonuclear leukocytes, and mac-

rophages identical to those of loose connective tissue (see Chapter 5). The average cell count in normal synovial fluid is 60 per cubic millimeter. It also contains the mucopolysaccharide hyaluronic acid (which accounts for its viscosity); proteins in a concentration about one-third that in serum; and electrolytes in concentrations comparable to those in plasma or lymph, except for calcium, whose concentration is higher, presumably owing to its binding by hyaluronic acid.

Hyaluronic acid was isolated from bovine synovial fluid in 1939 by K. Meyer and associates and has recently been proved to be synthesized by the membrane cells. Its repeating unit is an *N*-acetylated glucosamine–glucuronic acid complex.

Aside from its lubricant properties, little is known of the functions of synovial fluid. It provides nutrients for articular cartilage, and motion forcing it against the cartilage appears to be the means by which the nutrients are dispersed to the cartilage cells. Studies of the passage of substances across the blood-fluid barrier indicate fairly free diffusion of water and small molecules. Proteins enter and leave the synovial fluid much as they do the extracellular fluid. Particulate matter injected into a joint cavity is removed primarily through phagocytosis.

Mechanics of the skeleton. The movement of a diarthrosis is produced by contraction of the muscles attached to its bones. The *axis* is an imaginary line about which the movement occurs. Since articular surfaces are irregular, there is usually no single center of movement. The axis shifts slightly during movement just as in a mechanical counterpart in which the opposing surfaces are somewhat irregular.†

The axis represents the fulcrum of a lever made up of bones, their articulations, and associated muscles; the body part—or the part together with the external object it is moving or supporting—constitutes the load or resistance of the system; and the muscles perform the work. Thus the architecture of the skeleton pro-

---

† For example, a ball and socket joint has many axes because it moves in a number of directions. A hinge joint has only a transverse axis, and a pivot joint only a vertical axis. Flexion and extension take place on a transverse axis, adduction and abduction on an anteroposterior axis, and rotation on a vertical axis.

---

* These four structures also bound the diamond-shaped *popliteal space* in back of the knee.

vides levers by which a small force can be made to overcome a large force or the distance through which a load is moved can be greatly increased over the distance through which a force acts.

According to elementary mechanics, there are three general classes of levers, depending upon the relative positions of the load, the fulcrum, and the applied force. In class 1 levers the fulcrum is between the load and the force. Such an arrangement enables the muscles on the back of the neck to raise the head. In class 2 levers, such as the wheelbarrow, the load is between the fulcrum and the force. No levers of this class are found in the body. In class 3 levers the force is between the fulcrum and the load. Levers of this class are the most common in the body and enable muscles to raise, for example, the forearm.

SKELETAL MUSCLES

The skeletal or striated muscles exceed 400 in number and make up about 42% of the total body weight. By contracting, they bring two skeletal regions closer together, producing movement about a joint. Although each muscle has its own connective tissue framework, arterial, venous, and lymphatic vessels, and nerve supply—the whole composing an independent unit—muscles almost always act in groups rather than singly. An individual muscle does not contract in executing a movement. Rather, groups of muscles operate in concert.

Components. Each skeletal muscle has a body and two attachments. The body contains the fleshy muscle tissue, and the attachments usually consist of white fibrous connective tissue. The more stationary attachment of a muscle is its *origin*. The attachment in which the effects of movement are more pronounced is its *insertion*. Generally the origin is proximal, and the insertion is distal.

Muscle may be attached to bone in one of three ways: (1) directly; (2) by a tendon; or (3) by an *aponeurosis*. In a direct attachment the white fibrous connective tissue of the muscle fuses with the fibrous layer of the periosteum covering the bone. A tendon is a band or cord of white fibrous connective tissue (see p. 174).

A tendon may be protected by a synovial An aponeurosis is a heavy sheet of white connective tissue, rather like a flattened t

Mechanics of action. The principal skeletal muscles are displayed in Fig. 17.11. The figure indicates their origins and insertions. The name given to a muscle may be descriptive of its shape (e.g., quadratus, deltoid), general form (e.g., serratus), structure (e.g., semitendinosus), location (e.g., intercostal), number of insertions (e.g., biceps, triceps, quadriceps), mode of action (e.g., adductor magnus), major direction (e.g., rectus abdominis), or contrasting features (e.g., peroneus longus, peroneus brevis).

A number of important principles are illustrated in Fig. 17.11. For example, if the attachments of a muscle are known, its action may be determined by recalling that the insertion moves toward the origin during contraction. In general, muscles can be arranged in opposing or antagonistic groups: flexors and extensors, adductors and abductors, and internal rotators and external rotators. When the muscles for flexing the forearm contract, the muscles for extending it relax, elongating and giving way to the movement.

Figure 17.12 shows some familiar pairs of opposing muscles. The *pectoralis major*, the large triangular muscle covering the upper anterior chest, pulls the arm down toward the chest. It is opposed by the *deltoid*, the short thick muscle located above the articulation of the arm and shoulder, which raises the arm to a horizontal position. Opposition also occurs between the *brachialis* and the *triceps brachii*. Certain muscles of the forearm are classified as pronators and supinators of the hand. These opposing muscles make possible turning of the hand without movement of the elbow or shoulder. The *iliopsoas* and *gluteus maximus* of the hip joint oppose each other, as do the *gastrocnemius*, *tibialis anterior*, and *soleus* of the ankle joint.

The intricate structural organization of muscles and tendons is perhaps nowhere more apparent than in the hand, one of evolution's most remarkable achievements. The primary function of the hand is to grasp, and the range in size of objects that can be grasped tightly is very wide. Moreover, flexion and extension of the fingers

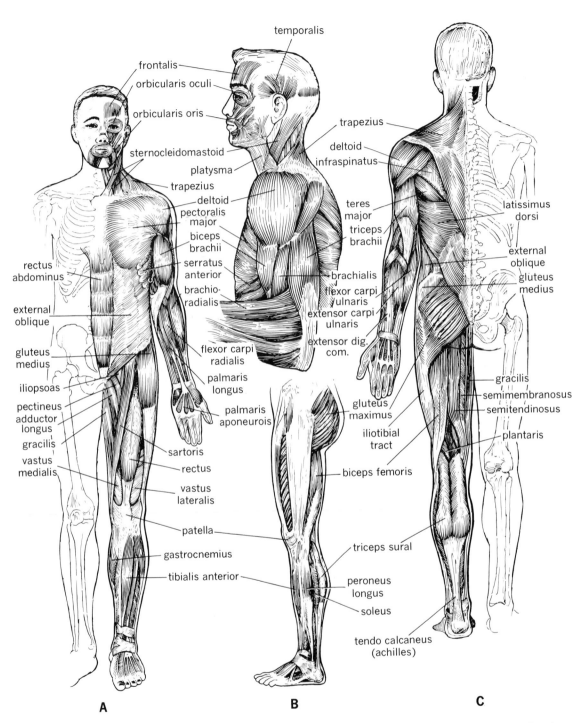

17.11 Structure and location of the superficial skeletal muscles: A, anterior view; B, lateral view; C, posterior view. (From S. Grollman, *The Human Body, Its Structure and Physiology*, 2nd ed., The Macmillan Company, New York, 1969.)

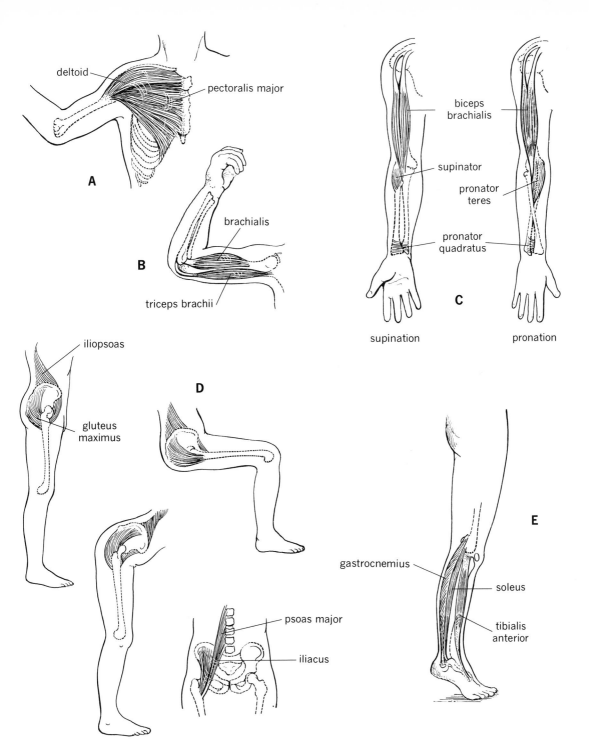

17.12 Opposing actions of antagonistic muscle pairs: A, pectoralis major and deltoid; B, brachialis and triceps brachii; C, pronators and supinators of the hand; D, iliopsoas and gluteus maximus; E, soleus and gastrocnemius opposing the action of tibialis anterior.

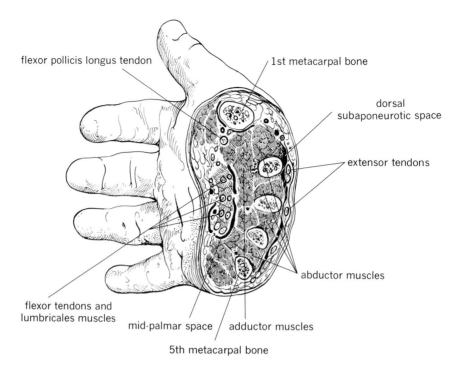

flexor pollicis longus tendon

1st metacarpal bone

dorsal subaponeurotic space

extensor tendons

abductor muscles

flexor tendons and lumbricales muscles

mid-palmar space

adductor muscles

5th metacarpal bone

17.13 Cross section of the hand through the metacarpal bones showing arrangement of muscles and tendons. Note the groupings of tendons and their synovial sheaths. The functions of most of the intrinsic hand muscles are indicated by their names.

are supplemented by adduction and abduction and the special movements of the thumb called opposition (in which the pad of the thumb is placed upon the pad of a finger). Although numerous small muscles originate and insert within the hand itself, many hand and finger movements are controlled by muscles in the forearm whose tendons descend to the hand in orderly array (Fig. 17.13). The tendons travel in two clusters—superficial and deep—and most are encased in their own synovial sheaths.

Physiology of Bone

In earlier chapters we discussed the histology of bone and some of the factors controlling or affecting calcification, including parathyroid hormone, vitamin D, and calcium and phosphate in body fluids (see Chapters 5, 15). We shall now proceed to other aspects of bone physiology.

SKELETAL MATURATION

The skeleton of a young embryo is composed of fibrous membrane and hyaline cartilage. Ossification, or bone formation, begins in these tissues in the eighth week of embryonic life. It may be either *intramembranous* (in fibrous membrane) or *endochondral* (in cartilage). These terms apply solely to bone development and do not imply structural differences in fully formed bones.

Intramembranous ossification. Intramembranous ossification is simpler and more direct than endochondral ossification. The flat bones of the face and cranium and parts of the clavicles form in membrane. In an area in which ossification is about to begin, *fibroblasts*, cells derived from mesenchyme, congregate in long strands running in all directions among many small blood vessels. Soon delicate bundles of *collagen* fibrils produced by the secretory activity of the fibro-

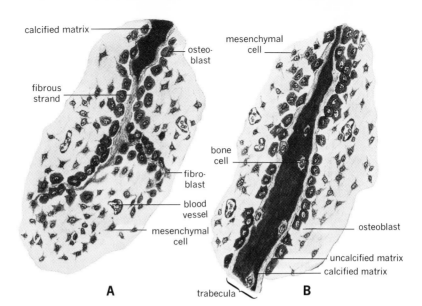

calcified matrix —
osteo-blast —
fibrous strand —
fibro-blast —
blood vessel —
mesenchymal cell —

mesenchymal cell —
bone cell —
osteoblast —
uncalcified matrix —
calcified matrix —
trabecula —

A   B

17.14 Intramembranous ossification: A, early stage, with active osteoblasts ranged along a fibrous strand and laying down calcium; B, later stage showing formation of trabecula. (From B. M. Patten, *Human Embryology*, 3rd ed., Copyright © 1968 by McGraw-Hill, Inc. Used with permission of McGraw-Hill Book Co., New York.)

blasts (see Chapter 5) establish an axis within each strand (Fig. 17.14). The fibrils of the axis become saturated and cemented together with an organic substance known as *osteomucoid*, or *ground substance*. This completes the organic matrix upon which inorganic minerals are deposited.

With the beginning of calcification, or mineralization, the fibroblasts are converted to *osteoblasts*. When a strand is entirely calcified, it is termed a *trabecula*. As successive lamellae of bone are laid down, some osteoblasts are entrapped in their own secretions. Such cells are

called *bone cells*, or *osteocytes;* the spaces that they occupy in bone are called *lacunae* (see Fig. 5.10).

As the trabeculae in an ossification center grow, they fuse into a latticework of *spongy*, or *porous*, or *cancellous*, bone around marrow spaces (Fig. 17.15). Then the mesenchyme enclosing the ossification center differentiates into *periosteum*, and *periosteal ossification* occurs. In this process osteoblasts from the inner layer of the periosteum deposit lamellae of spongy bone, which are transformed into *compact*, or *hard*, bone. Thus a membranous bone ultimately

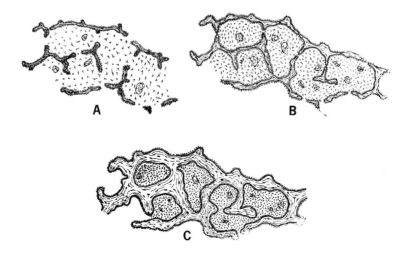

A   B

C

17.15 Stages in the fusion of trabeculae into spongy bone. (From B. M. Patten, *Human Embryology*, 3rd ed., Copyright © 1968 by McGraw-Hill, Inc. Used with permission of McGraw-Hill Book Co., New York.)

consists of two plates of compact bone and a core of spongy bone.

Endochondral ossification. Endochondral ossification is identical with intramembranous ossification except that it involves preliminary cartilage destruction. Most of the bones of the body form in cartilage. They are first represented in the embryo by temporary cartilaginous models.

The endochondral ossification of the tibia is diagramed in Fig. 17.16. The contour of the early cartilaginous model suggests that of the future bone. There is a central shaft, the *diaphysis*, with an *epiphysis* at each end. The cartilage is covered with *perichondrium.* Ossification starts in the diaphysis. The fibroblasts of the perichondrium enlarge, become osteoblasts, and begin to deposit a collar of spongy bone around the middle of the diaphysis. From this stage on perichondrium is referred to as periosteum.

While the periosteum forms bone, changes take place in the cartilage of the diaphysis; its cells swell into vesicles, and the matrix around them calcifies. Portions of the periosteum, fibrous tissue, and blood vessels penetrate the collar of spongy bone and make their way to the interior of the altered cartilage. The calcified matrix surrounding the cartilage cells dissolves, and large spaces open up, which are soon filled with embryonic bone marrow. Cells of the marrow are converted to osteoblasts, which build trabeculae of spongy bone. *Osteoclasts,* large, multinuclear, bone-destroying scavenger cells, probably arising from the fusion of wandering precursor cells,* can be seen in places where new bone is being resorbed.

Meanwhile, the cartilaginous model grows. The periosteum deposits successive lamellae of bone on the outside, gradually extending the collar toward each epiphysis, and cartilage destruction and ossification progress in the same directions on the inside, with the cartilage cells becoming arranged in regular rows. Soon osteoblasts invade the cartilage of the epiphyses and establish secondary ossification centers. One appears in the proximal epiphysis of the tibia soon after birth, and one in its distal epiphysis in the

second year. The result, as in the diaphysis, is spongy bone.

All bone is spongy at first, but later, through destruction and resorption, that at the surface is transformed into compact bone. In the process, the irregular channels in spongy bone are enlarged, and concentric lamellae of bone are laid down on the inside to construct *Haversian systems,* or *osteones* (see Fig. 5.10). Extensive bone destruction leaves a cylindrical space, the *medullary cavity,* in the diaphysis (see Fig. 5.11). Eventually this cavity reaches into the channels near the epiphyses. Growing bone is designed to support body weight and muscular activity, and in every instance maximum strength is gained from minimum bone.

Growth and epiphyseal closure. A disc of *epiphyseal cartilage* lies between the diaphysis and each epiphysis. It persists for some time after birth and provides for longitudinal growth. Since cartilage grows by cell division on the epiphyseal side of a disc and is destroyed and replaced by bone on the diaphyseal side, a disc retains approximately the same thickness. Therefore, the growth zone consists of the disc of epiphyseal cartilage, a region of calcified cartilage, and a region of spongy bone. The bone lengthens as the growth zone moves away from the center.

Ossification is complete when the cartilage cells cease to divide and bone entirely replaces the discs. The epiphyses are then united with the diaphysis, and further longitudinal growth is impossible. Closure of the epiphyses occurs at a different and characteristic age in each bone. In the tibia, for example, it occurs in the distal epiphysis at age 17 or 18 and in the proximal one at age 20. Ossification is not complete in all the bones until about age 25. X-ray examination of the state of the epiphyses permits a reasonably accurate estimate of age.

FORMATION AND RESORPTION OF BONE

We can study the structure of bone at four levels of organization. For the lowest we utilize the techniques of gross anatomy, which distinguish bone as spongy and compact (see Fig. 5.10). For the next we employ the methods of microscopic anatomy or histology, which reveal the

---

* Whether the precursor cells are mesenchymal cells, reticulum cells, macrophages, fibroblasts, or osteoblasts is not known.

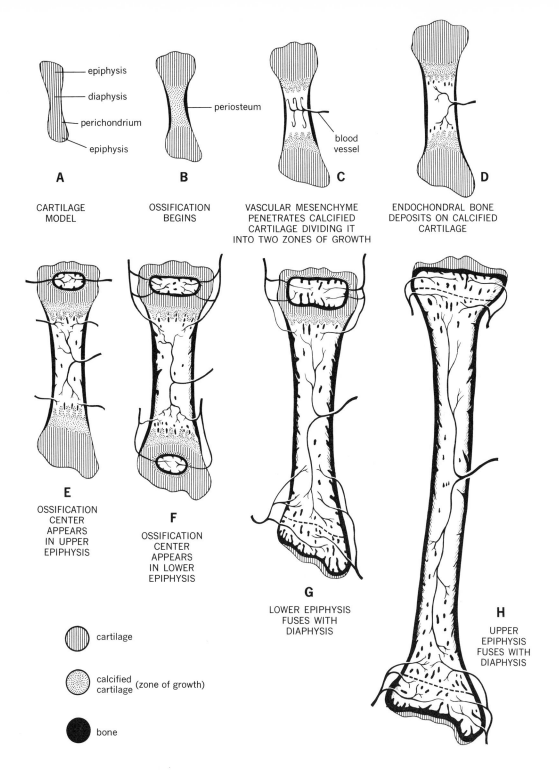

A

epiphysis
diaphysis
perichondrium
epiphysis

CARTILAGE
MODEL

B

periosteum

OSSIFICATION
BEGINS

C

blood
vessel

VASCULAR MESENCHYME
PENETRATES CALCIFIED
CARTILAGE DIVIDING IT
INTO TWO ZONES OF GROWTH

D

ENDOCHONDRAL BONE
DEPOSITS ON CALCIFIED
CARTILAGE

E

OSSIFICATION
CENTER
APPEARS
IN UPPER
EPIPHYSIS

F

OSSIFICATION
CENTER
APPEARS
IN LOWER
EPIPHYSIS

G

LOWER EPIPHYSIS
FUSES WITH
DIAPHYSIS

H

UPPER
EPIPHYSIS
FUSES WITH
DIAPHYSIS

cartilage

calcified
cartilage (zone of growth)

bone

17.16 Ossification and growth of a typical long bone, the tibia.

Haversian systems, trabeculae, and lamellae. For the next we use still higher magnification, which discloses the osteocytes, lacunae, and connective tissue fibers.

These are the levels on which classic anatomy is based. The fourth level—the molecular or ultrastructural level—was unknown to traditional anatomists. From histology we have learned that bone consists of (1) a fibrous framework; (2) an amorphous ground substance; and (3) inorganic crystals—and that these three elements are deposited in the order named. Only recently have we witnessed rapid advances in the study of their ultrastructure.

Organic matrix. The organic matrix of bone, accounting for 30 to 35% of its dry fat-free weight, can be discussed in terms of its two parts: the collagen fibrils, which make up about 95% by weight; and the ground substance, which fills the space around the collagen fibrils and the inorganic crystals. Although this space has been described as amorphous, electron micrographs indicate that it has a highly organized ultrastructure.

We considered the structure of collagen in Chapter 5. We recall that its basic building block, tropocollagen, is secreted in soluble form by fibroblasts and that the macromolecules aggregate in insoluble form as fibrils, with the following characteristics: (1) a distinctive x-ray pattern; (2) an unusual amino acid composition; and (3) an axial periodicity of 700 Å with detailed intraperiod fine structure (see Fig. 5.13).

The ground substance is similar to that of other connective tissues (see Chapter 5). It is coextensive with both the extracellular fluid and the cellular basement membranes—which represent its fluid and condensed portions, respectively. Chemically, ground substance is characterized by its content of mucopolysaccharides containing hexosamines or amino sugars and by its staining reactions. It is the intimate environment of the inorganic crystals, the collagen fibrils, and the cells incorporated within bone, the osteocytes. Its mucopolysaccharides are the products of osteocyte metabolism.

Although the precise function of ground substance has not been established, current investigation reflects a belief that it has an important role in the metabolism and physiology of bone (as well as of other connective tissues). It may contribute to the seeding of inorganic crystals during bone calcification—or, interestingly, to its prevention.

Calcification. Calcification of bone consists in the deposition of hard inorganic crystalline material in or on the organic matrix. We have already dealt with the physiological factors that provide the necessary quantities of calcium and phosphate for bone formation (see Chapter 15). We must now examine the factors that convert soluble inorganic ions into crystals.

The many theories of calcification may be separated into two groups. According to the first, referred to as the "booster theories," a specific enzyme (or enzymes) in the area undergoing calcification splits off inorganic phosphate from an organic substrate.* This local "boosting" raises the concentration of phosphate above the level of spontaneous precipitation and results in crystallization.

According to the other group of theories, first proposed in 1921 and recently revived, the organic matrix initiates crystal formation. Biochemical evidence and the inability of booster theories to explain the localization of crystals at the ultrastructural level lend strong support to this hypothesis.

Opinion as to what substance (or substances) in the organic matrix induces crystallization and by what means has varied. It now appears that the collagen fibrils are involved. As shown in Figs. 5.12 and 5.13, collagen displays the properties of a "crystalline" protein—an ordered aggregation and a high degree of structural regularity. Electron micrographs of compact bone reveal repeating densities along the collagen

---

* The first enzyme implicated in the calcification mechanism was *alkaline phosphatase*. R. Robison demonstrated the enzyme in bone in 1923 and postulated that it liberates phosphate ions from organic combination, thus freeing them to combine with calcium. However, he later abandoned this oversimplified theory. A later theory stated that alkaline phosphatase participates in the production of a "calcifiable" matrix, that is, that its function has to do with calcifiability rather than with calcification per se. The enzyme is present in osteoblasts and osteoclasts. Hence it may be involved in both bone formation and bone destruction. A genetically determined disorder characterized by the absence of alkaline phosphatase from the tissues, *hypophosphatasia*, is invariably associated with defective ossification.

17.17 Nucleation of hydroxyapatite crystals at regular intervals along collagen fibrils of the native type (×70,000) (see Fig. 5.13). (From M. J. Glimcher in *Rev. Mod. Physics, 31,* 359, 1959.)

fibrils that apparently represent the loci of inorganic crystals (Fig. 17.17). At first it was not clear whether the crystals were in the ground substance or the fibrils. However, by high-resolution electron microscopy, M. J. Glimcher established that they are in the fibrils.

These facts led to the conclusion that the aggregation state of collagen, with its recurring regularities, acts as a catalyst for the nucleation of hydroxyapatite crystals.* Thus the mechanism initiating crystallization resides in the stereochemistry of collagen. The positioning of certain reactive groups in the collagen molecule is thought to create highly specific regions that are sites for the nucleation of appropriate crystals from metastable or unstable solutions of the body fluids. Reconstituted native collagen fibrils from normally uncalcified tissues nucleate hydroxyapatite crystals from calcium phosphate solutions. On the other hand, reconstituted collagen fibrils other than the native type (see Fig. 5.13) do not. Therefore, nucleation of hydroxy-

apatite crystals depends not only upon the structural integrity of the collagen molecule; it also depends upon the packing together of groups of molecules in a particular pattern—i.e., a "native" pattern. Such an organization produces specific steric and electrostatic relations among reactive amino acid side chains that promote local nucleation.

One question concerning the mechanism of calcification is why collagenous tissues elsewhere do not calcify under normal conditions but do calcify under various pathological conditions. Glimcher has suggested that the mucopolysaccharides of ground substance inhibit calcification in these areas by selectively binding calcium ions, making them unavailable for crystallization in the collagen fibrils, and that any condition removing or altering these components enhances the likelihood of calcification.

Resorption and remodeling of bone. Much newly formed spongy bone is destroyed and resorbed. In regions of such activity, it is replaced at a rapid rate by bone of the compact type. Spongy bone also undergoes reconstruction; some strands are resorbed, while osteoblasts reinforce others.

---

* It should be remembered that the formation of crystals in a solution is a *phase change*, which can be divided arbitrarily into crystal *nucleation*, the formation of the initial fragments of the new phase, and crystal *growth*, the subsequent enlargement of the fragments into clearly defined crystals.

Osteoclasts are present when bone destruction is taking place and are assumed to be responsible for it. It appears likely that they form in response to various physical or chemical stimuli and secrete a material that liquefies ground substance and liberates bone salts, which then dissolve. Indeed, osteoclasts in the act of resorbing bone have been observed in time-lapse motion pictures of tissue culture preparations.*

Since the organic and inorganic components of bone are resorbed simultaneously, the osteoclasts must release both an acid and a proteolytic enzyme. Perhaps they release a chelating agent instead of an acid, so that solution of the salt crystals may proceed at the neutral pH of the body fluids.

Recent studies have contributed useful information on the mechanisms and functional significance of the continuous remodeling of compact bone (see p. 626). It was once held that remodeling is determined entirely by weight bearing and other functions of the skeleton as a whole. Although this may be true, it is now apparent that remodeling has an important role in keeping a continuous supply of exchangeable bone available for the transfer of calcium and phosphate between bone and blood. As we have noted (see p. 626), the most reactive bone, for both crystal formation and mineral exchange, is the youngest bone that has been laid down.

Investigators have successfully followed the development by resorption of individual tunnels or cavities and their reconstruction into new Haversian systems. An average cavity forms in roughly 3 weeks. It then becomes lined with osteoclasts and filled in with concentric lamellae of bone incorporating lacunae and canaliculi. The building of the structure, including partial calcification of its organic matrix, requires 6 to 12 weeks. Calcification of the matrix to about 70% of the final inorganic content occurs rapidly during and immediately after the deposition of new organic layers. Complete calcification takes many months.

---

* We saw in Chapter 15 that similar tissue culture experiments showed bone resorption to be initiated or hastened by parathyroid extract.

Bone repair and implantation. When a bone is broken, the repair process begins immediately. Cells in the periosteum bordering the fracture proliferate rapidly into the underlying soft tissue and along the bone. Organic matrix soon fills the fracture area, and calcification ensues. The healing tissue in the early stages of calcification is called *callus.* In time, if the broken bone ends have been set in close apposition, the normal bone architecture is restored, and the healed fracture line enters into the existing cycles of bone formation and resorption. The repair may reproduce the original configuration so successfully that the location of the fracture is undetectable.

Attention has lately been focused on the *transplantation,* or *grafting,* of bone as a means of hastening the healing of large fractures or repairing large filling defects of bone. It is still not known whether the new bone and cartilage developing around a graft are formed by cells from the graft or by the adjacent host cells, although most investigators favor the latter possibility.

Physiology of Muscular Contraction

The fundamental problem in muscle physiology is easily stated. Somewhere in the multitude of chemical reactions occurring in an excited fiber, there must be one that proceeds in such a way as to cause the fiber to shorten along one of its axes. The problem, as yet unsolved, is to identify the participants in this reaction and the mechanism of the conversion of chemical energy into mechanical energy. As we consider current views on the subject, we shall find that investigators are at last approaching a solution.

MOLECULAR BASIS OF MUSCULAR CONTRACTION

Ultrastructure of striated muscle. We discussed the histology and ultrastructure of striated muscle in Chapter 5. Fig. 5.16 shows that the basic units of a muscle fiber are myofibrils, parallel elements each 1 to 2 μ in diameter. A myofibril is made up of two kinds of filaments, thick and thin. The thick filaments are 160 Å in diameter by 1.5 μ in length, and the thin ones are 60 Å by 1 μ. Each filament is aligned with

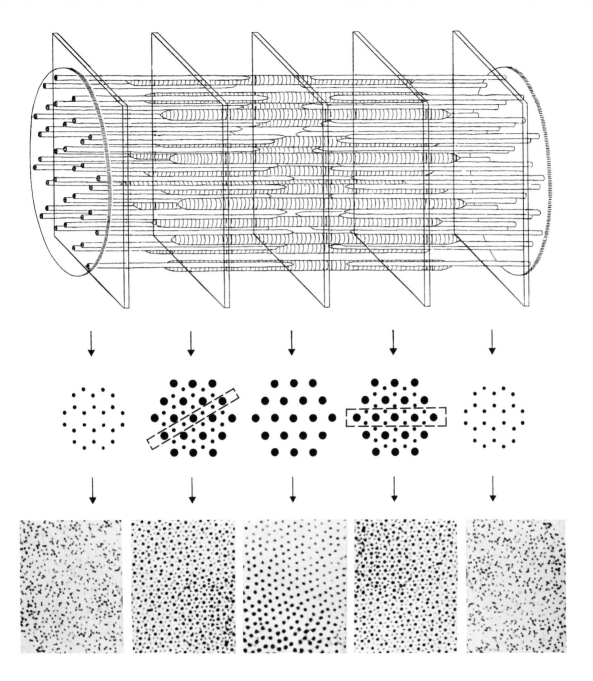

17.18 Transverse section through striated muscle showing hexagonal patterns of thick and thin filaments. Dotted lines show why some longitudinal sections reveal two filaments between each pair of thick filaments and others reveal only one. (From "The Contraction of Muscle" by H. E. Huxley. Copyright © Nov. 1958 by *Scientific American*, Inc. All rights reserved.)

others of the same kind, and the two groups overlap for part of their lengths (see Fig. 5.16). The overlap gives rise to the crossbands or striations of the myofibril. Transverse sections reveal that the filaments lie a few hundred Ångstrom units apart in a remarkably regular hexagonal array (Fig. 17.18). In a section through the dense part of the A band, each thick filament is seen to be surrounded by six thin ones, and each thin filament by three thick ones.

The two kinds of filaments are linked by an intricate system of *cross bridges* (Fig. 17.19). These, the only mechanical connections between filaments, project outward at right angles from each thick filament at regular intervals of 140 to 150 Å over its entire length (except for a central region 1500 to 2000 Å long from which they are absent) and of 60° around its axis. Thus

they form a helix whose repeating unit is six bridges, or about 429 Å, long. This helix joins a thick filament to each of the six surrounding thin filaments once every 429 Å.

The ultrastructure of Z lines, which mark off the sarcomeres, has only recently been elucidated. Electron microscopy of muscle fibers disrupted in a homogenizer discloses thin filaments still attached to a Z line in an orderly manner (Fig. 17.20A) but a Z line that is dense and lacking in fine structure. One must examine intact fibers in the relaxed state (Fig. 17.20B) to discern that the Z line has a zigzag configuration and to delineate its detail. As a thin filament approaches the Z line, it divides into two parts, each uniting with another thin filament in the next sarcomere. The pipe-cleaner model of a Z line in Fig. 17.20C illustrates this arrangement. The important implications of the model are that (1) thin filaments are made up of at least two strands; (2) the strands are continuous across the Z line; and (3) the Z line permits interactions between neighboring thin filaments.

Chemical nature of thick and thin filaments. The biochemical study of muscular contraction began in 1859 with the discovery of W. Kühne that strong potassium chloride solution extracts a viscous gelatinous substance from striated muscle. Kühne named the substance *myosin* and speculated that its gel-like nature was of significance in muscle contraction.

The study was resumed in 1925 by H. H. Weber, who devised an unusual investigative technique based on the fact that myosin is soluble only in salt water. He squirted thin jets of salt solutions of it into distilled water, in which it is insoluble. When the salt diffused away, the myosin formed threads superficially resembling natural muscle fibers. These and other early observations suggested that myosin's molecular structure, at least in part, is like that of a fiber. It was widely believed that myosin was the key to the contractile mechanism and that contraction consisted essentially of a simple folding or pleating of its molecules.

The situation was complicated in 1942 when G. Schramm and Weber resolved myosin into a viscous high–molecular weight component and a low–molecular weight component. F. Straub

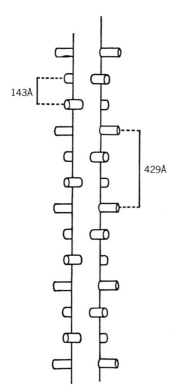

143Å

429Å

17.19 Diagram of arrangement of cross bridges along thick filament. (From H. E. Huxley in *Science, 164,* 1356, 1969.)

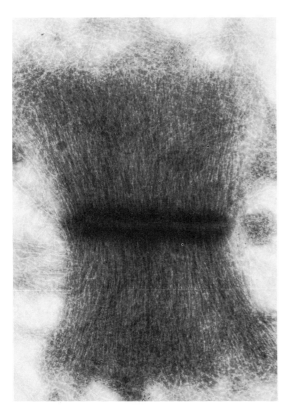

A

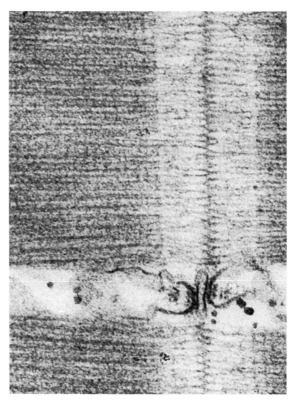

B

17.20 Structure of the Z line: A, electron micrograph of sample of homogenized muscles showing masses of thin filaments in register on both sides of a Z line; B, electron micrograph of a longitudinal section through the Z line of muscle fiber of *Amblystoma* showing zigzag configuration; C, pipe-cleaner model showing how branching of thin filaments accounts for zigzag configuration. (A from H. E. Huxley; B and C from C. Franzini-Armstrong in *Zeitschrift für Zellforschung*, 61, 667, 1964.)

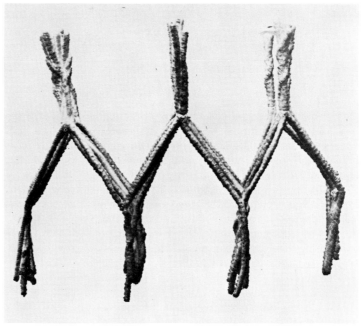

C

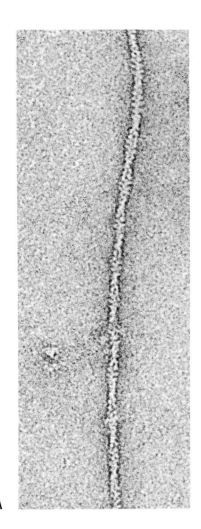

17.21 Structure of actin: A, electron micrograph of F-actin, showing two coils of globular units wound in a double helix; B, model of F-actin, showing spherical G-actin subunits, 55 Å in diameter. (A from H. E. Huxley; B from J. Hanson and J. Lowry, *Biochemistry of Muscle Contraction*, ed. by J. Gergely, Little, Brown and Co., Boston, 1964.)

then demonstrated that the smaller component is a distinctive protein which he designated *actin* and that the larger component is a complex protein made up of actin and another protein which he referred to as *myosin* (or *true myosin,* to distinguish it from Kühne's material). Although it was recognized that the complex, now called *actomyosin,* contains actin and true myosin, it was not clear whether the two combine only during the extraction process, whether actomyosin exists in vivo, and whether a possible in vivo complex of actin and myosin has the same properties as actomyosin in vitro.

When actin and myosin were later characterized, it was found that actin has the unusual property of occurring as a globular protein at low salt concentrations and as a fibrous protein at high salt concentrations.* The former is called *G-actin,* and the latter *F-actin.* G-actin has a molecular weight of 46,000, whereas F-actin ranges in molecular weight from 1,500,000 to more than 3,000,000. Electron microscopy reveals that F-actin is a polymer composed of G-actin subunits, each 55 Å in diameter, arranged in two strands twisted around each other (Fig. 17.21). The number of subunits per turn is between 14 and 15 and the distance between points where the two strands cross is 350 Å. The

---

* The critical salt concentration varies for monovalent and divalent cations.

observation that an isolated F-actin strand has approximately the same diameter as a thin filament in a myofibril implies their identity.

One of several questions surrounding actin concerns its ability to polymerize and depolymerize—i.e., the interconversion of G-actin and F-actin. G-actin polymerizes to F-actin at low salt concentrations if ATP is present. During the reaction ATP is bound to G-actin and dephosphorylated to ADP. Thus the polymerization of n molecules of G-actin (here abbreviated G) to F-actin may be represented as follows:

$$nG + nATP \Longrightarrow n(\text{G-ATP}) \longrightarrow (\text{G-ADP})_n + nP \quad (17.1)$$

Since F-actin includes bound ADP, it may be represented as $(\text{G-ADP})_n$.

The myosin molecule (i.e., true myosin) is 1500 Å long and 20 to 40 Å wide. Treatment with the proteolytic enzyme trypsin yields at least two kinds of subunits, or *meromyosins*. The one sedimenting faster in the ultracentrifuge is called *heavy meromyosin*, or *HMM*, and the other *light meromyosin*, or *LMM*. The molecular

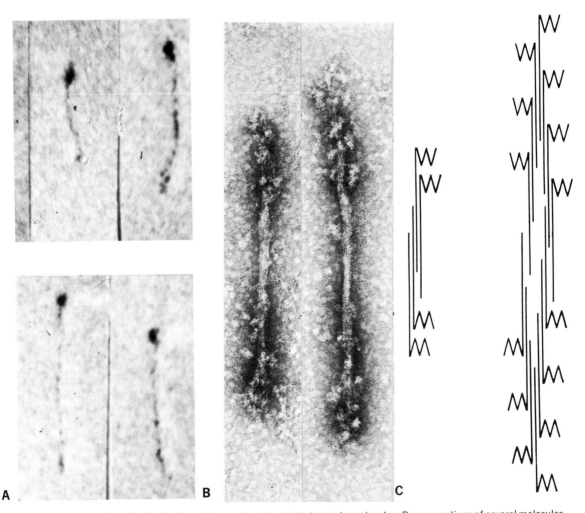

17.22 Structure of myosin: A, electron micrographs of individual myosin molecules; B, aggregations of several molecules of pure myosin showing projecting cross bridges; C, model of possible molecular structure of myosin aggregate, with head of molecule (HMM portion) represented by zigzag line and tail (LMM portion) by straight line. (From "Structural Arrangements and the Contraction Mechanism in Striated Muscle" by H. E. Huxley, in *Proc. Royal Soc. B, 160,* 442, 1964.)

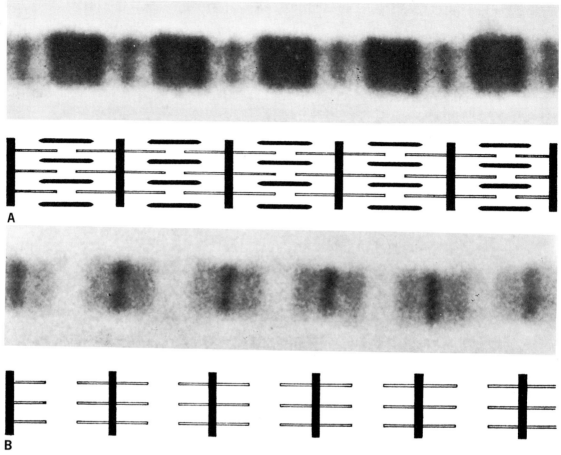

17.23 Evidence that thick filaments consist of myosin: A, electron microscopic appearance and model of normal myofibril; B, myofibril from which myosin has been extracted. (From "The Contraction of Muscle" by H. E. Huxley. Copyright © Nov. 1958 by Scientific American, Inc. All rights reserved.)

weight of intact myosin is between 500,000 and 600,000, and the molecular weight of HMM and LMM are about 350,000 and 150,000, respectively. Hence one myosin molecule must contain one HMM and one LMM plus some other materials. Electron microscopy discloses that an individual myosin molecule is a long fiber with a globular "head" (Fig. 17.22A). Since HMM is a globular fragment and LMM a short fiberlike fragment, it is apparent that HMM and LMM are, respectively, the head and backbone of the intact myosin molecule (Fig. 17.22C).

Purified myosin molecules aggregate into large strands, each having many globular heads at the ends and a bare region in the middle (Fig. 17.22B). In a number of respects, an aggregate resembles a thick filament in a myofibril. For example, both are about 160 Å across; and the globular heads of the aggregate point in one direction for one half its length and in the opposite direction for the other half, as do the cross bridges on opposite sides of the H band in the filament. Indeed, the model in Fig. 17.22C even accounts for the absence of cross bridges in the middle of the filament.

We have evidence, then, that the thick and thin filaments of the myofibrils are mainly myosin and actin, respectively. Additional evidence is seen in Fig. 17.23. If myosin is dissolved out of a myofibril with salt solution, the A band dis-

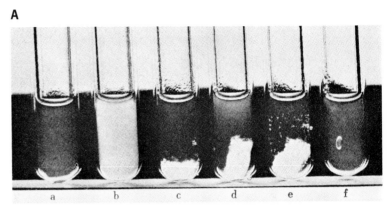

17.24 Effect of ATP upon actomyosin: A, actomyosin fiber (made by mixing actin and myosin) before addition of ATP (above) and after (below); B, effect of ATP on an actomyosin: a, G-actin + myosin; b, G-actin + myosin + ATP (added simultaneously); c, same as b, but ATP added 5 seconds later; d, F-actin + myosin + ATP (added simultaneously); e, same as d, but ATP added 5 seconds later; f, F-actin + myosin. (From A. Szent-Györgyi, *Chemistry of Muscular Contraction,* 2nd ed., Academic Press, New York, 1951.)

appears, and only a "ghost" fibril remains between the Z line and the former H band. If the "ghost" fibril is further treated with potassium iodide solution to extract actin, all visible material between the Z line and the former H band disappears, so that only the Z line is left. Myosin is obviously confined to the A band, whereas actin stretches from the Z line through the I band into the A band, terminating at the H band.

Properties and relations of myosin and actin. We must now take notice of two remarkable properties of myosin. In 1939 the Russian investigators W. A. Engelhardt and M. N. Ljubimova found that purified myosin is a powerful adenosinetriphosphatase, or ATPase, which catalyzes the following reaction:

$$ATP \longrightarrow ADP + P \qquad (17.2)$$

Significantly, the ATPase activity of myosin resides exclusively in its globular head—which, as we have seen, is a cross bridge connecting a thick filament and a thin filament.

A few years later Szent-Györgyi, after separately extracting and purifying myosin and actin

from muscle, discovered that the two compounds combine in solution to yield actomyosin. The complex protein has a viscosity higher than the sum of the viscosities of its two components and a very high molecular weight. He then discovered that artificial fibers prepared from precipitated actomyosin shorten (i.e., shrink) to one-fourth their original length when immersed in a solution of ATP (Fig. 17.24). The shortening is due to *superprecipitation,* a common phenomenon of colloid chemistry in which water is driven out of a gel structure with resulting coalescence of the filaments that originally held it in the solid state.

When the in vitro effect of ATP on actomyosin was first observed, it was assumed to be of great physiological importance, for myosin, actin, and ATP seemed to constitute all the essentials of a contractile system.* Unfortunately many questions were unanswered by

---

* Thrombosthenin, a protein of platelets (see p. 271), resembles actomyosin in the following ways: (1) it shortens in ATP solution in vitro, though it is not established that it does so in vivo; (2) it has ATPase activity; and (3) it can be dissociated into two components that are similar but not identical to actin and myosin.

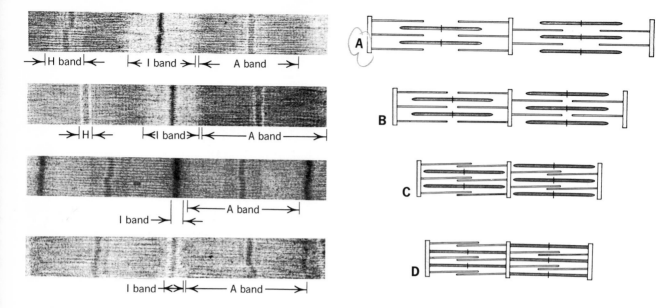

17.25 Change in band pattern during contraction of striated muscle: A and B, H zone closes and I band shortens; C and D, new dense zone develops within A band. Width of A band remains constant. (From "The Mechanism of Muscular Contraction" by H. E. Huxley. Copyright © Dec. 1965 by Scientific American, Inc. All rights reserved.)

theories based on the effect. For example, actomyosin fibers treated with ATP in vitro elongate rather than shorten when required to pull a load. Moreover, they shorten only when their actin and myosin components have been mixed in a random solution (i.e., one without polarity). How then does shortening take place in vivo, where the two proteins are segregated into thick and thin filaments making contact only at the tips of the cross bridges and exhibiting polarity? These and other questions suggest that the superprecipitation of actomyosin in the presence of ATP, though an evocative phenomenon, is not the key to the mechanism of muscular contraction.*

---

* Attempts to explain precisely how ATP induces the superprecipitation of actomyosin have been frustrating. Szent-Györgyi has postulated that more is involved than a simple transfer of bond energy from ATP in many localized reactions (a mechanism difficult to reconcile with the diffuse character of superprecipitation and contraction). According to his novel theory, aggregates of large molecules may act as semiconductors in much the same way as solid-state transistors, which are complexes or "sandwiches" of electron-donor regions and electron-acceptor regions, and the energy released from ATP may be transmitted widely through the aggregates.

The sliding filament theory. The key to muscular contraction was discovered in England in 1953 by H. E. Huxley and Jean Hanson and independently by A. F. Huxley and R. Niedergerke. It had previously been shown by interference and electron microscopy that the width of the A band remains constant when a living isolated muscle fiber is contracted (Fig. 17.25) or stretched beyond its resting length. The I and H bands narrow and then disappear as the Z lines at the ends of the sarcomere come together during contraction.

As soon as the meaning of the bands in striated muscle was clear (see Fig. 5.16), it was obvious that the changing pattern during contraction and stretching would provide insight into the molecular basis of contraction. Since the width of the A band equals the length of the thick filaments, these filaments must be of constant length. Also, although the H band narrows with the I bands, the distance from one H band to another across the Z line remains approximately constant, and so the thin filaments must be of constant length.

The English workers concluded from their observations (and from later electron micrographs) that, when muscle length changes, the two groups of filaments slide past each other (see Fig. 17.25). When a muscle shortens, the thin filaments meet. With further shortening, the ends of the thin filaments overlap. Thus the older view that sarcomere shortening is due to folding or pleating of filaments is incorrect.

With the downfall of the folding theory, investigators were obliged to look for mechanisms or forces capable of inducing the filaments to slide. They found them in the cross bridges on the thick (myosin) filaments. It is now thought that the bridges connect with specific sites on the neighboring thin (actin) filaments. According to one hypothesis (Fig. 17.26), a cross bridge has two centers, A and B, that act as an ATP-splitting center (i.e., an ATPase) and an actin-binding center, respectively; and a thin filament has corresponding loci, A′ and B′. Although myosin alone has powerful ATPase activity, this activity can be sensitively regulated by cations (especially Ca^{++}), but only in the presence of actin. It is postulated, therefore, that the primary event in contraction is the binding of myosin to actin—the formation of a link between B and B′—and that the A′ center of actin then activates the A center of myosin so that energy is released. If such interactions occur simultaneously at many points (Fig. 17.26B), the thin filament moves from right to left (i.e., toward the center of the A band) for a short distance, perhaps 100 Å, and the sarcomere contracts (Fig. 17.26C).

The filament then returns to its original position. Each time a cross bridge goes through such a cycle, a phosphate group is split from a molecule of ATP. To account for the known rates of shortening and energy liberation in muscle, each bridge would have to go through 50 to 100 cycles per second. This figure is compatible with the known rate at which myosin catalyzes the removal of phosphate groups from ATP. We might suppose that a muscle relaxes when the removal of phosphate groups from ATP stops. Various experiments have indicated, in fact, that intact ATP breaks the combination of actin and myosin. The reverse effect—the formation of permanent links between actin and myosin in the absence of ATP—may explain *rigor mortis*, the muscular rigidity that is present for a time after death. With their source of ATP exhausted,

the filaments may bind, or *catch*, like pistons deprived of lubrication.*

No such theory of muscular contraction makes clear, among other things, precisely how a chemical reaction provides the motive force for the molecular movements of contraction. The problem will obviously not be solved independently of the other great biological problems—those pertaining to the structure of proteins, the action of enzymes, and the mode of energy transfer in biological systems.

Metabolic aspects. The dependence of muscular contraction upon ATP implies that muscle cells are equipped to generate it in large quantities. In general, the metabolic pathways responsible for ATP production are those described in Chapter 4—glycolysis and oxidative phosphorylation. In addition, muscle cells have a unique metabolic feature that increases their efficiency in producing and utilizing high-energy phosphate compounds.

In 1930 E. Lundsgaard found that muscles treated with iodoacetate (an inhibitor of glycolysis) can contract repeatedly without forming lactic acid. However, they do form inorganic phosphate. It was then discovered that their contractions are accompanied by the breakdown of an unstable high-energy phosphate compound that had previously gone unnoticed in muscle extracts—creatine phosphate.

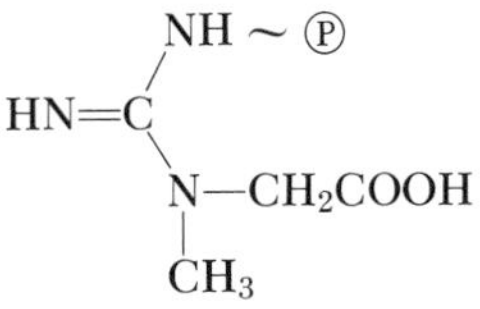

$$
\begin{array}{c}
NH \sim \circledP \\
| \\
HN = C \\
| \\
N — CH_2COOH \\
| \\
CH_3
\end{array}
$$

Later investigation showed that this compound is present only in the muscles of vertebrates. An analogous compound, *arginine phosphate*, is present in the muscles of invertebrates. Such compounds are designated *phosphagens* (i.e., phosphate donors).

---

* The existence of a "catch mechanism" is indicated by many phenomena, among them the ability of the powerful tonic muscles of molluscs (the so-called *catch muscles*) to remain maximally contracted for long periods and the ability of hypnotized humans to remain in a state of muscular rigidity for long periods. These phenomena are tentatively attributed to a temporarily irreversible solidification or phase change (due to increased viscosity) of filament proteins.

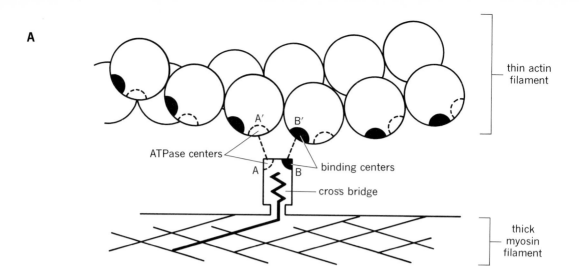

A

thin actin
filament

ATPase centers

binding centers

cross bridge

thick
myosin
filament

A' B'

A B

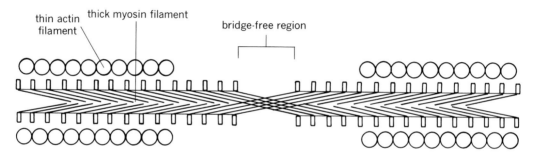

B

thin actin
filament

thick myosin filament

bridge-free region

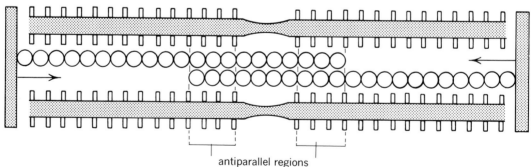

C

antiparallel regions

17.26 Mechanism by which the interaction of myosin and actin causes filaments to slide past one another: A, proposed model of myosin-actin interaction showing two types of interactions between a cross bridge and neighboring actin, a binding interaction (BB') and an ATPase-regulating interaction (AA'); B, the relations of multiple cross bridges to thin filaments in resting myofibril; C, the relative positions of thick and thin filaments during period of maximum contraction, showing overlap of thin filaments.

A phosphagen serves as a ready reserve of the $\sim\text{\textcircled{P}}$ groups. The relatively small amount of free ATP in muscle would soon be consumed if the ADP formed on its cleavage were not rapidly rephosphorylated. The regeneration can be accomplished by the rather cumbersome multienzyme processes of glycolysis and oxidative phosphorylation or by a simple single-enzyme process using the large reservoir of creatine phosphate—four to six times the size of the ATP pool in resting muscle.

$$\text{creatine phosphate} + \text{ADP} \longrightarrow \text{ATP} + \text{creatine} \quad (17.3)$$

The enzyme operative in this conversion is *creatine kinase* (also called *creatine phosphoryltransferase*).*

In summary, muscle metabolism generates ATP by at least three processes, two of which—the creatine phosphate reaction and glycolysis—can occur both aerobically and anaerobically (Fig. 17.27). In glycolysis, muscle glycogen breaks down to pyruvate, liberating a small amount of ATP. In the absence of oxygen, muscle glycogen breaks down only as far as pyruvic acid, which accepts hydrogen from the $NADH_2$ produced earlier to yield lactic acid (see Fig. 4.3). Striated muscle can function normally—i.e., contract and relax—for some time without oxygen. However, fatigue sets in rapidly as lactic acid accumulates. With such an oxygen debt (see p. 575), lactic acid diffuses out into the body fluids and passes to the liver, where it is reoxidized to pyruvic acid. This enters the citric acid cycle for complete dissimilation or is reconverted to glycogen (see Fig. 14.7).

When isolated muscle that has been fatigued anaerobically is exposed to sufficient oxygen, the lactic acid is reoxidized to pyruvic acid, which enters the citric acid cycle and is broken down to carbon dioxide and water, releasing additional ATP. This is the third process by which ATP is generated. It is clear that muscle can perform more work aerobically than anaerobically and that the available oxygen sets an upper limit on the amount of work that can be done in a given time.

* See Fig. 14.12 for the metabolic pathway of creatine synthesis.

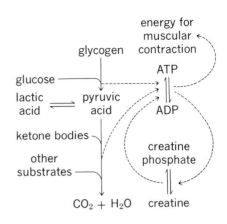

17.27 Chemical changes in muscular activity.

Muscle is able to store small quantities of oxygen by means of the hemoprotein *myoglobin*. This iron-porphyrin complex has a molecular weight one-fourth that of hemoglobin and contains one iron atom per molecule as compared to the four of hemoglobin (see p. 233). Nevertheless, it mimics the action of hemoglobin as an oxygen carrier, forming *oxymyoglobin*. To some extent the oxygen bound by myoglobin postpones the onset of an oxygen debt.† Oxymyoglobin is brilliant red; it accounts for the color of muscle.

Relaxing factors. We have seen that muscle contracts when ATP is present and can be broken down by myosin ATPase. If ATP is absent or if it is present but for some reason cannot be broken down, fiber length remains unchanged. We may now inquire what prevents the ATP in normal unstimulated muscle from exerting its contractile effect.

The studies of E. Bozler in 1952 indicated the existence of a *relaxing factor* in muscle. The same year B. B. Marsh found a substance in muscle extracts that causes previously contracted isolated myofibrils to relax. When the substance is removed, only contraction is possible. Many attempts have been made to identify this factor and to assess its physiological importance. It is now believed to arise in small vesicles or granules

† It is noteworthy that the muscles of aquatic mammals such as seals and whales contain very high concentrations of myoglobin. According to a recent theory, the main function of myoglobin is to accelerate the diffusion of oxygen between blood and mitochondria.

in the *sarcoplasmic reticulum* (i.e., the endoplasmic reticulum of the sarcoplasm) and to inhibit the ATPase activity of myosin by binding the Ca^{++} that normally activates ATPase.

Other relaxing factors have been tentatively reported by other investigators. C. J. Parker, Jr., and J. Gergely recently observed that, during the incubation of relaxing factor with ATP, another substance forms that by itself causes instant relaxation. Although the relevance of their results to living muscle has not been established, it is evident that a relaxing factor, if significant, would operate by suppressing the spontaneous enzymatic cleavage of ATP and the related mechanochemical event. As Gergely has pointed out, such a concept of negative control is consonant with the current view that many enzymes are ordinarily in a state of partial inhibition (see Chapter 4).

MUSCLE DYNAMICS

What is often referred to as the "classic" physiology of striated muscle is concerned with the properties of anatomically intact muscle, including its neural excitation and its physical performance. It nevertheless embodies many fundamental questions relating to events at the molecular level.

Links between excitation and contraction. We learned in Chapter 12 that the impulses delivered by a motor nerve fiber enter a striated muscle fiber via a neuromuscular junction called the *motor endplate.* In the late 1930's it was discovered that an impulse produces a prolonged negative potential at the endplate that is not propagated (i.e., it remains localized in the endplate). This *endplate potential* depolarizes the muscle membrane, whose interior is slightly negative relative to its exterior, by a process involving chemical transmission across the endplate by acetylcholine,* and depolarization waves sweep on into the muscle fiber. As in the case of cardiac muscle (see Chapter 9), depolarization generates a contraction. Even in a frog muscle chilled to 0°C, depolarization causes a contraction within 0.040 seconds.

According to current theory, then, excitation leads to contraction through the widespread propagation of an initially localized zone of depolarization. How depolarization proceeds from the muscle membrane into the fiber was long a puzzle. Since A. V. Hill years ago calculated that the time required for full contraction is too short for a substance to diffuse from the membrane to the central parts of the fiber, a structural component for the transmission of depolarization waves inward has been sought.

Recently electron microscopists have detected the sarcoplasmic reticulum, which is essentially a network of vesicles,† and the transverse *tubu-*

---

† As noted earlier, the sarcoplasmic reticulum is the source of one of the relaxing factors.

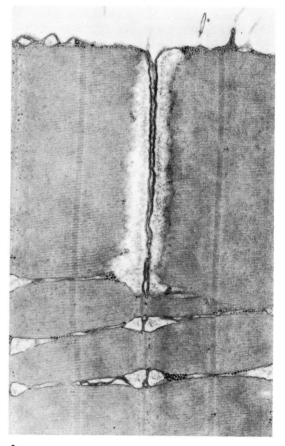

A

---

* The characteristics of the endplate potential are similar to those of electrotonus (see p. 419). With both, local excitation gives rise to propagated disturbance.

lar system or *T system,* a minute system of transverse tubes extending inward from the membrane at every Z line and every A-I junction (Fig. 17.28). The sarcoplasmic reticulum, like the endoplasmic reticulum of other cells, has no connection with the surface membrane. In contrast, the T system consists largely of a series of fingerlike invaginations of the membrane that penetrate the fiber and run across its long axis.

The discovery of the T system ended the search for a structure for the inward transmission of depolarization waves. Potential differences such as those across the surface membrane may be expected to prevail across the membranes of the T system. How a wave is conducted inward, or how a wave in a T tube finally activates a contraction, is not known.

Electromyography. As investigators exploring the electrical properties of living tissue developed electrocardiography, electroencephalography, and electroretinography, it was inevitable that they should also develop *electromyography.* It is based on two primary properties of muscle: its contraction upon electrical stimulation; and its production of current during contraction. Modern electromyography utilizes a needle electrode inserted into muscle (Fig. 17.29). The current is amplified and fed to an oscilloscope, whose wave patterns are photographed.

The purpose of electromyography is to record the potentials of an individual *motor unit* of muscle, consisting of the body of an anterior column cell of the spinal cord, its axon, and all of the muscle fibers supplied by it. Nerve impulses pass from the cell body down the axon and through the motor endplate, in which they initiate a potential that is discharged into each fiber. The oscilloscope registers a synchronous,

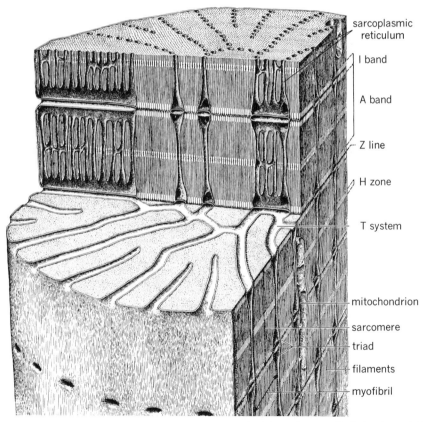

sarcoplasmic reticulum

I band

A band

Z line

H zone

T system

mitochondrion

sarcomere

triad

filaments

myofibril

B

17.28 The T system: A, electron micrograph of longitudinal section through a myofibril showing a lateral view of T tubules originating as invaginations of the sarcolemma; B, diagram showing relations of T system, sarcoplasmic reticulum, and sarcomere. (A from C. Franzini-Armstrong and K. R. Porter, *J. Cell. Biol., 22,* 675, 1964. B from "The Sarcoplasmic Reticulum" by K. R. Porter and C. Franzini-Armstrong. Copyright ⓒ Mar. 1965 by Scientific American, Inc. All rights reserved.)

A

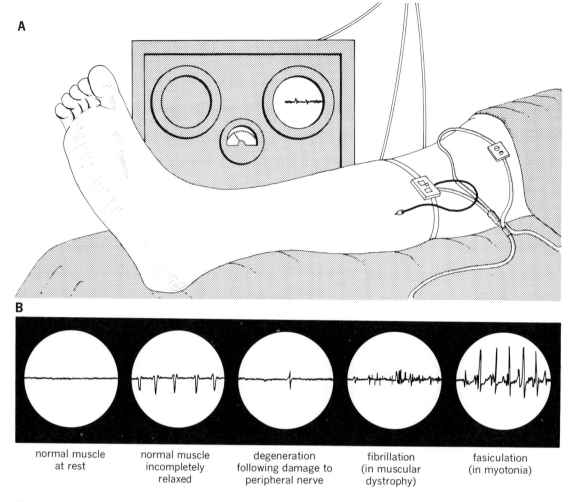

B

| normal muscle at rest | normal muscle incompletely relaxed | degeneration following damage to peripheral nerve | fibrillation (in muscular dystrophy) | fasiculation (in myotonia) |

17.29 Electromyography: A, technique; B, sample records.

usually diphasic wave identified as *simple* motor unit activity. Abnormality of the anterior column cell (as in poliomyelitis) or of nerve impulse transmission (as in nerve lesions) results in an asynchronous, polyphasic wave identified as complex motor unit activity. The oscilloscope shows no sign of electrical activity for resting muscle—only a straight base line.

The main practical value of an electromyograph lies in its ability to demonstrate conclusively muscle denervation and to measure quantitatively functional efficiency in instances of disorders of the central nervous system, peripheral nerves, and neuromuscular junctions. Since muscle contracts involuntarily under electrical stimulation, it also tests muscle excitability.

Fibrillation, localized muscular twitching, follows isolation of an individual muscle fiber from the anterior column cell, spinal motor root, or peripheral nerve.

Far from impairing muscle viability, such denervation causes hyperirritability. *Fasciculation,* coarse and grossly visible muscular twitching resembling tonic spasm, arises from abnormal single impulses affecting intact motor units (in contrast to the spontaneous activity of individual muscle fibers in fibrillation). The condition is due to irritation of a motor neuron. Both fibrillation and fasciculation appear as polyphasic waves on the oscilloscope.

Mechanical aspects of muscular contraction. The basic mechanical changes associated with contraction of intact muscle are the *development of tension* and *fiber shortening.* These have been studied extensively with recording apparatuses such as that shown in Fig. 17.30. When

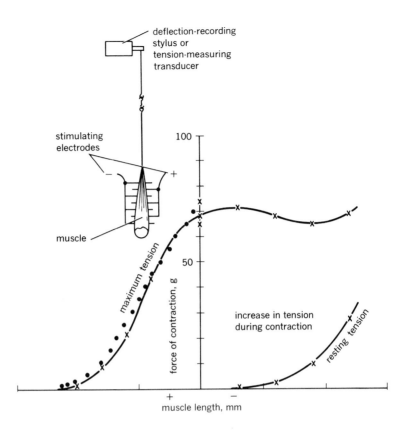

17.30 Relation of muscle length to force of contraction. (From D. R. Wilkie, *Brit. Med. Bulletin, 12,* 177, 1956.)

the muscle lifts a load, its contraction is *isotonic.* Isotonic contraction is usually examined experimentally in the following manner: one end of the muscle is attached to one end of a lever; weights are hung near the fulcrum of the lever; and the response of the muscle to stimulation is recorded by a stylus at the other end of the lever. The load is supported by a stop until the muscle develops sufficient tension to lift it. After that, the tension in the muscle remains constant as the muscle shortens.

When the muscle performs no external work, its contraction is *isometric.* Isometric contraction is examined in the same way as isotonic contraction, except that the contracting muscle is connected to a tension-measuring transducer device that prevents shortening. The muscle can be set at a certain length before stimulation, and its tension at that length can be measured. Tension development is greatest when the muscle is the same length as in the body.

By means of such equipment, a number of significant observations can be made. In a mus-

cle stimulated by a single maximum volley of nerve impulses (or by brief electrical contact), a single sudden twitch occurs (Fig. 17.31A). An isotonic twitch begins and ends more slowly than an isometric twitch, owing to the inertia of the load and its momentum during its return to its resting position, which causes it to overshoot. For this reason physiologists generally describe different muscles in terms of the isometric twitch.

If a second maximum stimulus is applied within the first few milliseconds of the first stimulus, no additional response occurs. If it is applied slightly later but before the first twitch has ended, tension develops further. This *summation* effect may be due to an increase in the number of motor units contracting simultaneously or to an increase in the rate of contraction of each motor unit. If stimuli are applied at progressively shorter intervals, the summation effect is intensified (Fig. 17.31B). At rates between 50 and 100 stimuli per second, a tremulous response, *subtetanus,* occurs; at rates above

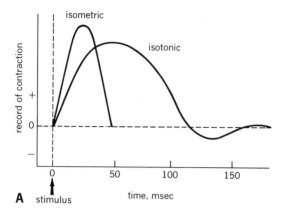

A **stimulus**

time, msec

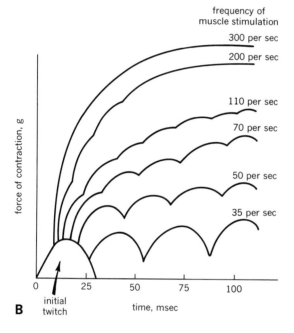

frequency of muscle stimulation

300 per sec

200 per sec

110 per sec

70 per sec

50 per sec

35 per sec

B **initial twitch**

time, msec

17.31 Properties of single and repeated muscle twitches: A, isotonic and isometric; B, summation and tetanus.

relax. Nevertheless, there is no essential difference between the active state of muscle in a twitch and that of muscle in tetanus.

According to the influential physiological school of A. V. Hill, muscle behaves as a two-component system, with a contractile component in series with an elastic component comprising noncontractile connective tissue and tendon. The contractile component shortens to some extent even in isometric contraction, and the so-called series-elastic component lengthens. Thus the development of tension depends upon these two actions, both of which require time. The series-elastic component maintains a constant length in isotonic contraction.

A muscle may contract very rapidly against no load. However, as the load increases, the velocity of contraction decreases: the latent period is prolonged, and the initial speed and maximum degree of shortening are reduced. A plot of the isotonic force against the initial speed of shortening (Fig. 17.32) shows that the speed is zero and contraction is nil when the load equals the maximum isometric force. This curve satisfies the following equation:

$$V = \frac{(p_0 - p)b}{p + a} \qquad (17.4)$$

where V is the initial speed of shortening, p is the actual force acting on the muscle, p_0 is the

100 stimuli per second, a full response, *tetanus,* occurs. In tetanus successive contractions cannot be distinguished from one another, and a maximum level of tension, far exceeding that of the original single twitch, gradually develops. Following a single stimulus, the muscle is active for only a brief period—perhaps 20 to 40 milliseconds—insufficient time for tension to climb to its peak or for a fully contracted muscle to

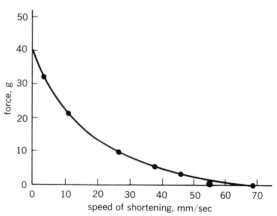

17.32 Relation of force to speed of shortening of muscle. (Adapted from D. R. Wilkie, *Brit. Med. Bulletin,* 12, 177, 1956.)

maximum tension that the muscle can develop, and a and b are constants with the dimensions of force and speed, respectively. Since the initial speed of shortening has a limited value even with no load on the muscle, the muscle does not behave as an undamped elastic body, whose initial speed of shortening would be infinite in the absence of a load. When the speed is zero, the force is maximal, and contraction is isometric. Hill's equation has been found to fit all muscles investigated so far. Such widely diverse structures as the smooth muscles of the snail and the striated muscles of man develop about the same tension per unit of cross-sectional area but differ greatly in initial speed of shortening.

Heat production. One of the principal methods of studying the physiology of muscular contraction has been measurement of the heat liberated during the various phases of contraction. Stimulation of muscle causes its heat production to rise considerably above the resting level. This effect precedes and then accompanies the mechanical events of contraction (Fig. 17.33).*

Hill showed that the heat can be divided into *initial heat* and *heat of recovery*.† Initial heat is liberated during actual contraction. It, too, can be subdivided. Most of it represents the en-

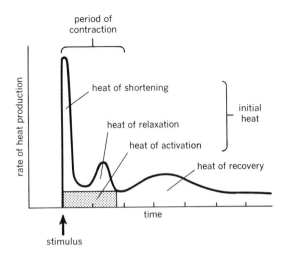

17.33 Heat production during muscular contraction.

---

* Establishing the time relations between heat production and the mechanical events has been difficult because of the relative slowness of techniques for recording heat compared to those for recording movement. However, cooling frog muscle to 0°C slows the mechanical events. Hill also delayed and slowed them by soaking the muscle in hypertonic salt solution. This does no damage to the muscle, which soon recovers its usual rate of contraction when soaked in normal Ringer's solution, but allows a convincing demonstration of the precedence of heat production over the mechanical events.

† Until 1923 heat production was always determined on muscles contracting isometrically rather than on those contracting isotonically, merely because the experimental approach was simpler. It was assumed that a muscle shortened and performed work at the expense of the tension stored in it during isometric contraction. When W. O. Fenn allowed a muscle to do work by lifting a load, he proved that this assumption was in error. In fact, the amount of energy actually released by the muscle (i.e., heat plus external work) varied directly with the load: the larger the load, the more the energy. This phenomenon, known as the *Fenn effect*, illustrates that in some way muscle adjusts its energy output to the work to be performed. This discovery led to the study of the various components of heat.

ergy expended by the muscle in keeping up the active state, the *heat of activation*. In isometric contraction virtually all the heat produced is heat of activation. In isotonic contraction extra heat is produced in proportion to the muscle shortening (the Fenn effect). This *heat of shortening* is presumably related to the energy expended by the cross bridges as the actin filaments slide over the myosin filaments.

If the muscle is permitted to lower its load during relaxation, the potential energy of the load appears as the *heat of relaxation*. If the muscle continues to support the load during relaxation, no heat of relaxation appears.

Heat of recovery is liberated in the resynthesis of ATP by oxidative phosphorylation and glycolysis. When muscle is freely supplied with oxygen, the heat of recovery is about equal to the initial heat. Hill found that the ratio of total energy (initial heat plus heat of recovery) to initial energy (initial heat plus work performed) is 2:1.

The efficiency of an engine is calculated as the percentage of its energy input that is converted to work. Less than 25% of the food energy entering a muscle is converted to work. The remainder becomes heat. Maximum efficiency is realized only when the muscle contracts at moderate speed. If contraction is very slow, a large amount of heat of activation is liberated.

If contraction is very rapid, too much energy is wasted in overcoming friction.

Muscular fatigue. It is well known that prolonged contraction of a muscle leads to fatigue, owing to changes in the muscle that render the contractile and metabolic processes incapable of sustaining further work. There is good evidence that fatigue is not due to impairment of the motor nerves, of the neuromuscular junction, of the spread of the action potential over the muscle fibers, or of the action potential–contraction coupling.

Presumably the changes in the muscle in fatigue are induced by (1) anoxia and (2) the accumulation of waste products. Both of these conditions are offset by blood flow. Hence interruption of blood flow to a muscle greatly hastens the onset of fatigue. There are mechanical problems in maintaining blood flow during prolonged contraction, for the increased intramuscular tension tends to hamper it.

An excessively fatigued muscle is likely to remain contracted, apparently because the small quantities of energy required to promote relaxation after contraction are unavailable. If all energy sources have been exhausted and this energy of relaxation cannot be provided, contraction continues until new ATP has been formed.

REFERENCES AND SUGGESTIONS FOR FURTHER READING

Bennett, H. S., "Structure of Muscle Cells," in J. L. Oncley, ed., *Biophysical Science*, Wiley, New York, 1959.

Bourne, G. H., ed., *The Structure and Function of Muscle*, Academic Press, New York, 1960.

————, ed., *The Biochemistry and Physiology of Bone*, Academic Press, New York, 1956.

Engelhardt, V. A., and M. N. Ljubimova, "Myosin and ATPase," *Nature*, **144**, 668 (1939).

Fourman, P., *Calcium Metabolism and the Bone*, 2nd ed., Oxford Univ. Press, London, 1966.

Franzini-Armstrong, C., and K. R. Porter, "Sarcolemmal Invaginations Constituting the T Systems in Fish Muscle Fibers," *J. Cellular Biol.*, **22**, 675 (1964).

Gergely, J., "Contractile Proteins," *Ann. Rev. Biochem.*, **35**, 691 (1966).

————, ed., *Biochemistry of Muscle Contraction*, Little, Brown, Boston, 1964.

Glimcher, M. J., "Molecular Biology of Mineralized Tissues with Particular Reference to Bone," in J. L. Oncley, ed., *Biophysical Science*, Wiley, New York, 1959.

Hill, A. V., *Trails and Trials in Physiology*, Arnold, London, 1965.

————, "The Heat Production of Muscle and Nerve, 1848–1914," *Ann. Rev. Physiol.*, **21**, 1 (1959).

————, "The Series Elastic Component of Muscle," *Proc. Roy. Soc. (London), Ser. B.*, **137**, 273 (1950).

————, *Muscular Activity*, Williams & Wilkins, Baltimore, 1926.

Huxley, A. F., "Muscle," *Ann. Rev. Physiol.*, **26**, 131 (1964).

————, "Muscle Structure and Theories of Contraction," *Progr. Biophys. Biophys. Chem.*, **7**, 255 (1957).

————, and H. E. Huxley, eds., "A Discussion of the Physical and Chemical Basis of Muscular Contraction," *Proc. Roy. Soc. (London), Ser. B.*, **160**, 433 (1964).

Huxley, H. E., "The Mechanism of Muscular Contraction," *Sci. Am.*, **213**, 18 (Dec., 1965).

————, "Electron Microscopic Studies on the Structure of Natural and Synthetic Protein Filaments from Striated Muscle," *J. Mol. Biol.*, **7**, 281 (1963).

————, "The Contraction of Muscle," *Sci. Am.*, **199**, 66 (Nov., 1958).

Kuffler, S. W., "Relation of Electric Potential Changes to Contracture in Skeletal Muscle," *J. Neurophysiol.*, **9**, 367 (1946).

McLean, F. C., and A. M. Budy, "Connective and Supporting Tissues: Bone," *Ann. Rev. Physiol.*, **21**, 69 (1959).

————, and M. R. Urist, *Bone, An Introduction to the Physiology of Skeletal Tissue*, 2nd ed., Univ. of Chicago Press, Chicago, 1961.

Mommaerts, W. F. H. M., A. J. Brady, and B. C. Abbott, "Major Problems in Muscle Physiology," *Ann. Rev. Physiol.*, **23**, 529 (1961).

Morales, M., J. Botts, J. J. Blum, and T. L. Hill, "Elementary Processes in Muscle Action: An Examination of Current Concepts," *Physiol. Rev.*, **35**, 475 (1955).

Neuman, W. F., and M. W. Newman, *The Chemical Dynamics of the Bone Mineral*, Univ. of Chicago Press, Chicago, 1958.

Paul, W. M., E. E. Daniel, C. M. Kay, and G. Monckton, *Muscle*, Pergamon Press, New York, 1965.

Peachy, L. D., "Muscle" *Ann. Rev. Physiol.*, **30**, 401 (1968).

Perry, S. V., "Relation Between Chemical and Contractile Functions and Structure of Skeletal Muscle Cell," *Physiol. Rev.*, **36**, 1 (1956).

————, "The Structure and Interactions of Myosin," *Progr. Biophys. Mol. Biol.*, **17**, 325 (1967).

Podolsky, R. J., ed., "Symposium on Excitation-Contraction Coupling in Striated Muscle," *Federation Proc.*, **24,** 1112 (1965).

Porter, K. R., and C. Franzini-Armstrong, "The Sarcoplasmic Reticulum," *Sci. Am.*, **212,** 72 (Mar., 1965).

Rees, M. K., and M. Young, "Studies on the Isolation and Molecular Properties of Homogeneous Globular Actin," *J. Biol. Chem.*, **242,** 4449 (1967).

Rodahl, K., J. T. Nicholson, and E. M. Brown, *Bone as a Tissue,* McGraw-Hill, New York, 1960.

Sandow, A., "Excitation-Contraction Coupling in Skeletal Muscle," *Pharmacol. Rev.*, **17,** 265 (1965).

———, "Skeletal Muscle," *Ann. Rev. Physiol.*, **32,** 87 (1970).

Szent-Györgyi, A., "Lost in The Twentieth Century," *Ann. Rev. Biochem.*, **32,** 1 (1963).

———, *Bioenergetics,* Academic Press, New York, 1957.

———, "Structural and Functional Aspects of Myosin," *Advan. Enzymol.*, **16,** 313 (1955).

———, *Chemistry of Muscular Contraction,* 2nd ed., Academic Press, New York, 1951.

Walker, S. M., and G. R. Schrodt, "T System Connexions with the Sarcolemma and Sarcoplasmic Reticulum," *Nature,* **211,** 935 (1966).

Wilkie, D. R., "Muscle," *Ann. Rev. Physiol.*, **28,** 17 (1966).

Young, M., "The Molecular Basis of Muscle Contraction," *Ann. Rev. Biochem.*, **38,** 913 (1969).

Modern biology has vindicated the insight of zoological and botanical systematists that species are important unit products of the evolutionary process. The evolutionary divergence of races is reversible, that of species is irreversible. Races may grow increasingly more distinct in response to the vicissitudes of the environments in the territories which they inhabit. But migration, mixing and miscegenation may undo the divergence; in the melting pot of hybridization races merge into single variable populations. For at least several centuries this has been unquestionably the trend in the human species. Human races are becoming less sharply distinct than they once were.

Theodosius Dobzhansky, in *Species After Darwin*, 1958

18 Integumentary System

Introduction

The *integument* is the outer covering of the body. The *integumentary system* consists of the skin and its accessory structures, hair, nails, and glands. Existing at the interface between the organism and its environment, it is a protective barrier and organ of communication between body and environment. It is also a regulatory agency in the maintenance of homeostasis. In each of these roles, it performs diverse and important functions.

The skin protects the body from the environment in several ways: (1) it is a selective barrier to certain *substances* (e.g., bacteria, chemicals), excluding some and admitting others; (2) it is a

selective barrier to certain forms of *energy* (e.g., heat and ultraviolet, visible, and infrared radiation), excluding some and admitting others; and (3) it is a selective barrier to certain types of *information* (e.g., sensations), excluding some and admitting and transducing others.

The barrier functions of the skin depend to a considerable extent upon the production of a dead horny layer of *keratin*. This aids in guarding the body against mechanical injury, radiation, and microbial invasion. In lower vertebrates with tough hides, such protection is often reinforced with dermal scales or bones. The pigments of the skin, particularly *melanin*, are largely responsible for its action as a radiation barrier, whereas the totality of its coverage (and its blood vessels) and its appendage, hair, account for its capacity to prevent excessive losses

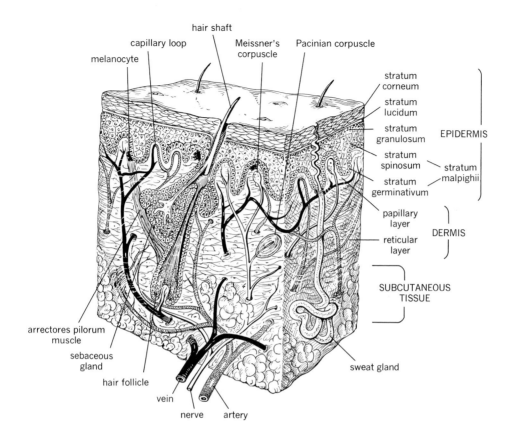

18.1 Anatomy of normal skin.

and gains of thermal energy. Finally, its function as information barrier and transducer is served mainly by its sensory receptors. The skin is probably the area in which sensory and nervous structures originated in ancient metazoans. Though in vertebrates nervous tissues form a discrete organ system remote from the body surface, we have seen that embryologically they still arise in contiguity with skin ectoderm (see p. 431). We have also seen that the skin remains the seat of abundant sensory structures.

The skin regulates itself—grows, differentiates, repairs, and renews itself—and participates in the regulation of other body systems. Thus it aids in the regulation of body water content, body temperature, and body composition. It also helps to regulate calcium metabolism by producing vitamin D upon exposure to sunlight.

Structure and Function

Although the skin may be discussed as a singular entity, it is in fact a highly heterogeneous organ consisting of several functionally discrete "suborgans" and varying greatly in different areas. To appreciate regional variations, we need only compare the skin of the palm, the knuckles, the ear, the cheek, the abdomen, and the scalp. In each of these regions, the skin has characteristic thickness, pliability, surface markings, pigmentation, glandular activity, and hair growth.

As shown in Fig. 18.1, the skin, or *cutis*, is a veneered or stratified tissue with several layers. At the surface is the thin epithelium called *epidermis*. Underneath this is a thicker connective tissue layer called *dermis, corium,* or *true skin.*

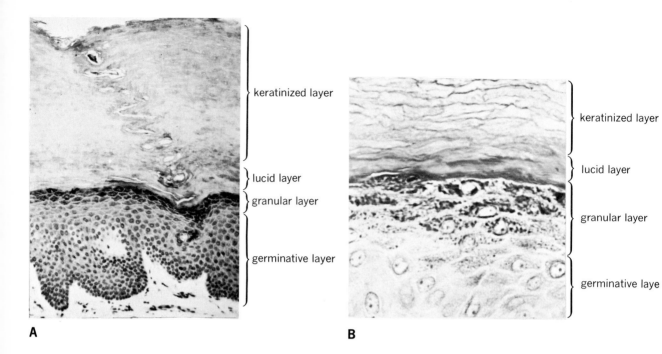

keratinized layer

lucid layer

granular layer

germinative layer

keratinized layer

lucid layer

granular layer

germinative laye

A   B

18.2 Structure of the epidermis: A, from the palm, showing the path of the duct of a sweat gland; B, from the sole. (From W. Montagna, *Structure and Function of the Skin,* 2nd ed., Academic Press, New York, 1962. By permission of the author and the publisher.)

Below the dermis is a bed of loose (areolar) *sub-cutaneous tissue* (also called *tela subcutanea*).

EPIDERMIS AND DERMIS

Epidermis is a cellular layer that derives from embryonic ectoderm. It gives rise to a host of differentiated structures, such as hair (feathers in birds), nails (claws, hoofs, and horns in other animals), and glands. Dermis, on the other hand, derives from embryonic mesoderm. It contains relatively few cells and has a fundamentally fibrous structure.

The capacity of the skin to move and stretch depends upon its thickness, elasticity, and degree of fixation to subcutaneous tissue. The skin of the abdomen has the greatest distensibility but, when stretched too far, becomes damaged. During pregnancy, for example, red streaks known as *striae gravidarum* appear on its surface. These remain as permanent white lines.

In addition to the wrinkles and furrows of wear or age, the skin has congenital flexure lines. These are fixed creases or "skin joints" that indicate firm attach-ments to the underlying tissues. The skin of the palms and soles also has alternating ridges and grooves. These produce prints whose uniqueness and serviceability as a means of identification are well known.*

Structure of the epidermis. It is usually stated that the epidermis is composed of stratified squamous epi-thelium. Actually, however, only the most superficial cells are squamous. Deeper cells are cuboidal or columnar.

It is useful to distinguish four epidermal layers (see Figs. 18.1, 18.2): (1) the deep-lying *germinative layer,* or *stratum malpighii;* (2) the *granular layer,* or *stratum granulosum;* (3) the *lucid layer,* or *stratum lucidum;* and (4) the superficial *keratinized layer,* or *stratum*

---

* The patterns of these ridges and grooves—the combinations of arches, loops, and whorls—are called *dermatoglyphics. Dac-tyloscopy* is the use of the dermatoglyphics of the fingertips for identification. Fingerprints are deposits of water and or-ganic matter left by the fingers upon a foreign surface. The dermatoglyphics of skin areas other than the fingertips are just as satisfactory for identification purposes. That they are under genetic control is evident from their similarity in iden-tical twins. Women's dermatoglyphics have fewer whorls and more arches than men's.

corneum. The term *stratum malpighii* honors the anatomist Marcello Malpighi (1628–1694),* who described it as the "living" part of the skin. It rests upon the dermis and is subdivided into a one-cell thick *basal layer,* or *stratum germinativum* (whose cells are often referred to as *Malpighian cells*), and a layer of variable thickness above it, the *prickle cell layer,* or *stratum spinosum.* Scattered among the cells of the basal layer are the dendritic processes of the melanin-producing *melanocytes,* whose cell bodies lie at the junction of epidermis and dermis.

The cells of the basal layer multiply rapidly and rise progressively to the top. They may be considered the vitally important cells of the epidermis, serving as the "matrix" that provides the successive generations of cells needed for the other layers. Superficial damage to the epithelium is readily repaired. But, when large areas of the basal layer are destroyed, as by a burn, regeneration of the skin fails to occur. We shall say more of wound healing later.

As the cells of the basal layer ascend to the surface, they enlarge and accumulate basophilic cytoplasmic granules, thereby forming the granular layer. The non-staining lucid layer lies above the granular layer and is particularly evident when the epidermis is thick. Finally, the outer keratinized layer, of varying thickness, consists of flattened, scalelike cells. In some regions these are rubbed off or lost piecemeal (e.g., as dandruff).†

The membranes of adjacent epidermal cells make contact by a curious means. Many small cytoplasmic processes, *intercellular bridges,* separated by intercellular spaces, join one cell to another (Fig. 18.3). In fixed preparations of individual cells, these processes stick out like spines; hence the name *prickle* or *spinous cells.* They are best seen in the prickle cell layer.

The nature of the intercellular bridges is not fully understood, although in electron micrographs they seem to be extensions of delicate *tonofibrils* within the cells. At a cell surface tonofibrils are moored to a dense plaquelike structure, a *half-desmosome.* In adjacent epithelial cells two half-desmosomes are opposite each other and together with their interconnecting filaments constitute a *desmosome,* or *macula adherens,* a point of intercellular attachment.

Keratin and keratinization. Broadly speaking, the cells of the epidermis are of two types: the

---

* Other structures named for Malpighi are the *Malpighian corpuscles* of the kidney (see Chapter 10) and the *Malpighian bodies* (i.e., follicles) of lymphoid tissue (see Chapter 8).

† Some snakes seasonally shed the entire layer. The full thickness of the epidermis of fishes and amphibians is living dividing cells. The superficial cells, however, do include some keratin.

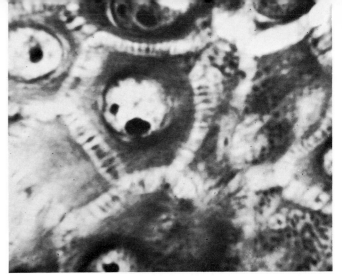

A

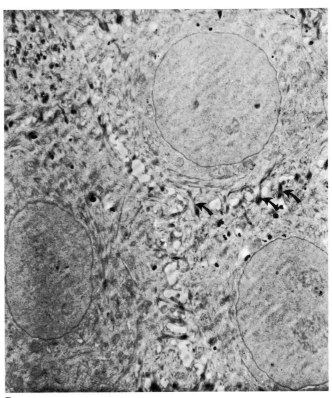

B

18.3 Intercellular bridges and desmosomes: A, from the axilla; B, from the forearm (× 3775). (From W. Montagna, *Structure and Function of the Skin,* 2nd ed., Academic Press, New York, 1962. By permission of the author and the publisher.)

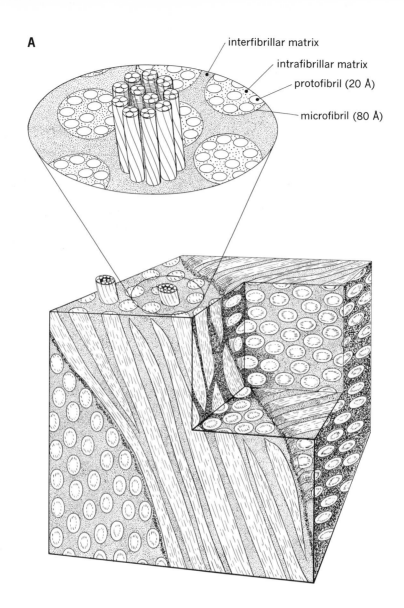

A

interfibrillar matrix

intrafibrillar matrix

protofibril (20 Å)

microfibril (80 Å)

keratinocytes, which form keratin,* and the *melanocytes,* which form melanin. The keratinocytes far exceed the melanocytes in number. Keratinocytes may be regarded as the secretory cells of a huge holocrine gland, the epidermis.

We recall that in holocrine glands secretion-laden cells are discharged intact, dying in the process (see Chapter 5). Thus keratin is a major by-product of epidermal cell metabolism.

Keratin is of great biological and biophysical interest. A typical fibrous protein,† it is exceedingly widespread in nature, occurring in skin,

---

* *Keratos* and *cornu* are, respectively, the Greek and Latin words for horn. *Keratinization* and *cornification* have been used as synonyms, but it has been recommended that the latter term be reserved for the general hardening of surface proteins.

---

† A fibrous protein has an axial ratio (i.e., ratio of the length to the width) greater than 10:1; a globular protein has an axial ratio less than 10:1. Other fibrous proteins are collagen (see p. 178) and F-actin (see p. 704).

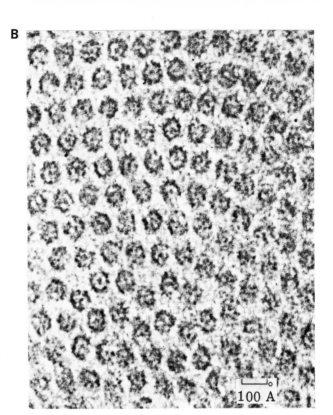

B

100 A

18.4 Ultra structure of α-keratin: A, schematic diagram, showing postulated interrelationship of microfibrils and protofibrils; B, electron micrograph showing fine structure of microfibrils (A, Dr. Robert Scheuplein; B, Dr. G. E. Rogers.)

hair, wool, feathers, nails, etc. As α-keratin it consists of a long peptide chain coiled into an α helix cross-linked by —S—S— groups (see Figs. 2.9, 2.10). Stretching alters the x-ray diffraction pattern of α-keratin, and presumably its helical pattern, and the result is β-keratin. The various forms of α-keratin—in hair, wool, feathers, nails, etc.—represent different packing arrangements of the α helices. Electron micrographs of the packing arrangement and fine structure of α-keratin in wool (a well-studied fiber) have suggested that this α-keratin exists as individual *microfibrils,* each containing 11 *protofibrils* (Fig. 18.4) in the same 9 + 2 configuration as in centrioles (see p. 78) and cilia (see p. 164). However, as shown in Fig. 18.4, high-resolution microscopy has not yet validated this structure. Each protofibril comprises three α helices twisted into a "rope" whose repeating units are 200 Å long.

Keratin can be divided into *high-sulfur* and *low-sulfur* types. High-sulfur keratin, exemplified by one of the two keratins in hair and nails, is unusually rich (15 to 18%) in the sulfur-con-taining amino acid cystine. Low-sulfur keratin, the type in epidermis, is only 2 to 4% cystine. Because of its many —S—S— groups, high-sulfur keratin is extremely insoluble and resistant to attack by the proteolytic enzymes of the alimentary tract. For this reason a substance such as wool cannot serve as a food protein in the human diet. (Interestingly, the alimentary tract of the clothes moth has a high concentration of —SH compounds that can reduce the —S—S— groups of keratin, to yield a substance that is readily broken down by a digestive protease.)

Keratinization of the epidermis apparently begins in the basal layer, but the details are not clear. The basophilic granules of the granular layer are composed of *keratohyalin,* a precursor of epidermal keratin. Vitamin A participates in keratinization of the epidermis. Experiments have revealed that skin cultivated in the presence of an excess of the vitamin becomes mucus-secreting and ciliated instead of keratin-secreting. After removal of the excess, it reverts to keratin formation.

Among the higher vertebrates the keratinized layer of epidermis develops into many special structures. Simplest are *calluses*, thickenings or swellings of surfaces subject to wear (e.g., the soles of the feet). The keratinized layer also produces dermatoglyphics. The horny scales of snakes, the platelike shells of turtles, the beaks of birds, and claws, hoofs, and horns are also derivatives of it.

Dermal–epidermal junction. In section the dermal–epidermal junction appears as an irregular wavy line (see Figs. 18.1, 18.2). Epidermal cones and ridges of different sizes project into the dermis, enclosing vascularized dermal papillae. Recent techniques make it possible to split off the intact epidermis and study the sculpturing of its undersurface. Figure 18.5 illustrates the differences in topography in different regions of the body.

Some investigators have asserted that delicate processes extend from the basal layer into the dermis. However, this claim is disputed. The nature of the submicroscopic membrane between the epidermis and dermis is also in question. Some cement substance must be present, since breakdown or dissolution of the membrane results in the formation of a *vesicle*, or *bulla* (i.e., a blister).

Dermis. The dermis is a sheetlike bed of connective tissue that supports the epidermis and its appendages and provides it with nourishment. It has two layers: a thin, superficial *papillary layer* and a thick, deep *reticular layer*. It is frequently difficult to draw a sharp line of division between them.

The papillary layer is composed of widely separated collagenous, elastic, and reticular fibers, surrounded by abundant viscous ground substance containing hyaluronic acid, chondroitin sulfate, and glycoproteins (see p. 33). The surface of this layer bears the negative imprint of the underside of the epidermis, being molded into intricate grooves and papillae (see Fig. 18.1). Any epidermal structure (such as a hair) that reaches down into the dermis, piercing the reticular layer, is bounded by a sleeve of papillary layer. Scattered through the papillary layer are numerous fibroblasts and a few mast cells, macrophages, and melanocytes.

The reticular layer consists of bundles of collagenous fibers in dense arrangement. They are more or less parallel to the surface, with alternate layers running

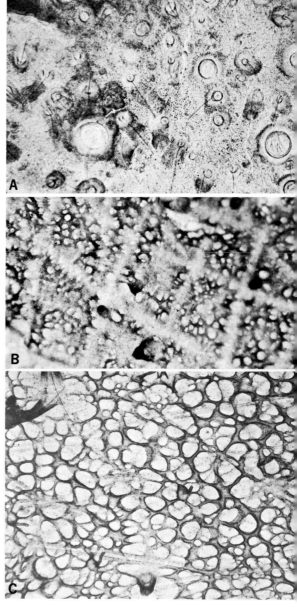

18.5 Undersurface of the epidermis, split from the dermis by trypsin: A, in the cheek; B, in the thigh; C, in the breast. (From W. Montagna, *Structure and Function of the Skin,* 2nd ed., Academic Press, New York, 1962. By permission of the author and the publisher.)

in different directions. The cells in the reticular layer are the same as those in the papillary layer but are more sparsely distributed. In some regions of the body (e.g.,

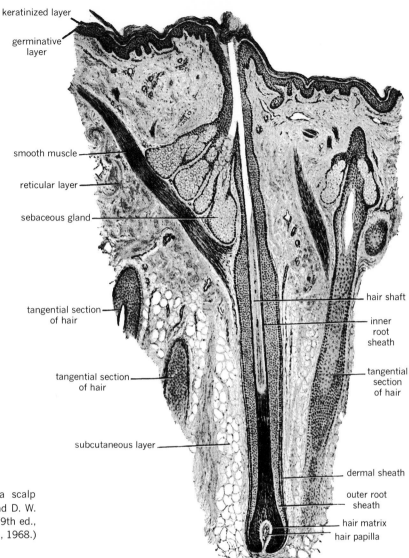

keratinized layer
germinative layer
smooth muscle
reticular layer
sebaceous gland
tangential section of hair
tangential section of hair
subcutaneous layer
hair shaft
inner root sheath
tangential section of hair
dermal sheath
outer root sheath
hair matrix
hair papilla

18.6 Longitudinal section of a scalp hair follicle. (From W. Bloom and D. W. Fawcett, *Textbook of Histology*, 9th ed., W. B. Saunders Co., Philadelphia, 1968.)

nipples, penis, and scrotum), smooth muscle fibers are collected into webs within the reticular layer. The skin of these regions wrinkles when the muscle fibers contract.

The cutaneous nerves and their receptors are in the dermis (see Fig. 12.16), as are rich networks of capillaries that supply the epidermis. The capillary loops are especially prominent in the dermal papillae of friction areas. The capillaries have anastomotic channels similar to those described on p. 325, and significant changes in blood flow can be brought about by vasomotor control, especially in the fingers and toes. By this means the capillaries perform a critical function in temperature and blood pressure regulation. The arterioles and venules serving the capillaries travel in the subcutaneous layer.

PILARY SYSTEM

Though little is known of the evolutionary origin of hair, it is found today only in mammals. Hairs are inert threadlike structures composed of keratinized cells compactly cemented together. As an insulating device, they are mammalian analogues of avian feathers. However, unlike feathers, which have mesodermal components, hairs are purely ectodermal. Human beings can survive and prosper without hair. Thus its biological value to man has diminished to that of an ornamental appendage.

Each hair grows out of a tube of epidermis, the *hair follicle*. This penetrates deep into the dermis (see Fig. 18.1). Since the Latin word for hair is *pilus*, the integumentary subsystem of hairs and hair follicles is called the *pilary system*.* The pilary system together with the sebaceous glands that grow from the follicle is called the *pilosebaceous system*.

Hair follicles. Hair follicles may be thought of as long tubular holocrine glands whose secretions, the hairs, are cylinders of keratinized cells. As shown in Fig. 18.6, the base of a follicle expands into a bulb. This is hollowed out to accommodate the loose connective tissue of the *hair papilla* (Fig. 18.7), through which enter nutrient materials for sustenance of the bulb cells. Adjacent to the follicle and emptying its oily secretion into it is a *sebaceous gland*.

The follicle grows at a slant, and bundles of smooth muscle fibers, the *arrectores pilorum muscles*, extend at an acute angle from the surface of the dermis to a bulge on the side of the follicle below the level of the sebaceous gland. Contraction of the muscles, largely controlled by the sympathetic division of the autonomic nervous system, brings the hair erect—and by its pull on the skin causes goose flesh. Although of little value in man, elevation and depression of the hair can markedly affect the insulating power of the coat in a thickly furred animal.

Lining the follicle are several concentric layers of tissue, collectively known as the *hair sheath* (see Fig. 18.7). This may be subdivided into two multilayered structures, the *inner* and *outer root sheaths*. The *cuticle* of the inner root sheath is a single layer of overlapping flattened, horny cells, directed downward so that they interlock with the outer hair cells, which are directed upward. Overlying the cuticle is *Huxley's layer*, several cells thick, and overlying it is *Henle's layer*, one cell thick. The thick outer root sheath is continuous with the deeper layers of the epidermis. Outside it, a *dermal sheath* of connective tissue is continuous with the papillary layer of the dermis. A glassy *vitreous membrane* separates the outer root sheath from the dermal sheath.

Hair growth and differentiation. It is instructive to examine the hair follicle, for it vividly displays the processes of growth and differentiation. A line drawn across the papilla at its widest point marks off a lower region of undifferentiated cells—the *matrix*—from an upper region of cells differentiating into inner root sheath and hair (see Fig. 18.7A). Actively dividing cells move up in single rows from the matrix, increase in volume, and elongate vertically to become spindle-shaped *cortical cells*. Above the neck of the bulb, they enlarge conspicuously. The cytoplasm of a cortical cell in the inner root sheath soon displays basophilic granules of *trichohyalin*, the counterpart of keratohyalin in epidermal keratin synthesis. At progressively higher levels the granules coalesce into bundles of fine, homogeneous, low-sulfur α-keratin filaments—the tonofibrils that attach to the half-desmosomes. These coarsen as they develop into microfibrils.† The formation of tonofibrils and microfibrils is completed at the level of the boundary between the bottom and middle thirds of the root, the *keratogenous zone*.‡ There high-sulfur keratin appears as amorphous deposits embedding the microfibrils. Thus most cortical cell cytoplasm consists of low-sulfur microfibrils in a high-sulfur ground.

As the follicle cells proceed upward, three types of cells differentiate and fuse together to form the principal components of the hair—the *cuticle, cortex,* and *medulla* (see Figs. 18.7, 18.8). (The hair surface is covered by a thin membrane 100 Å thick, the *epicuticle*.) Hairs from different parts of the body differ in structure. In an ordinary fine scalp hair, the cuticle is a single layer of overlapping, flattened, translucent cells with their free margins directed toward the hair tip. Cuticle cells move up in a single row from the matrix, become fully keratinized in the upper half of the follicle, and then adhere to the underlying cortex. The cortex, comprising the bulk of most hairs, derives from the keratin-filled cortical cells of the keratogenous zone. The medulla, in the center, is discontinuous or absent in very fine hairs. Its cells are large, loosely connected, and not filled with keratin microfibrils. Large intra- and intercellular air spaces in the

---

† It is not known whether trichohyalin is a precursor of the filaments or a cement that glues them together.

‡ At this level the cells of the outer root sheath contain more esterases, β-glucuronidase, glycogen, and phosphorylase than at other levels; the vitreous membrane is very thick; and the surrounding capillary networks are densest. These features suggest that the outer root sheath in this region is metabolically active, and a considerable amount of radioactivity has been recovered from it in a sheep only 6 minutes after the injection of cystine labeled with ^{35}S.

---

* It has even been suggested that mammals be called *Pilifera*.

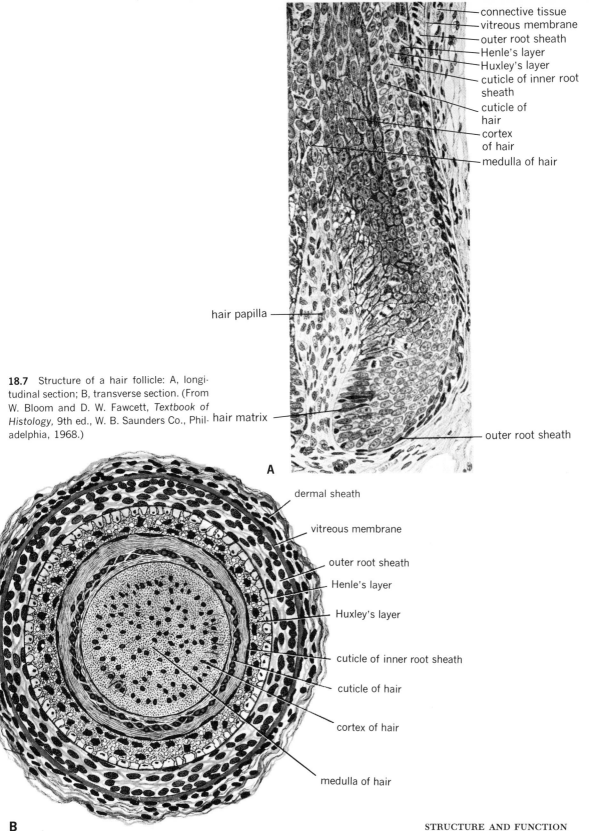

connective tissue
vitreous membrane
outer root sheath
Henle's layer
Huxley's layer
cuticle of inner root sheath
cuticle of hair
cortex of hair
medulla of hair

hair papilla

18.7 Structure of a hair follicle: A, longitudinal section; B, transverse section. (From W. Bloom and D. W. Fawcett, *Textbook of Histology,* 9th ed., W. B. Saunders Co., Philadelphia, 1968.)

hair matrix

outer root sheath

A

dermal sheath

vitreous membrane

outer root sheath

Henle's layer

Huxley's layer

cuticle of inner root sheath

cuticle of hair

cortex of hair

medulla of hair

B

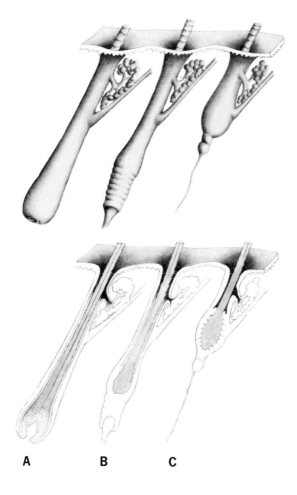

18.8 Hair growth cycle: A, anagen follicle; B, catagen follicle; C, telogen follicle. (From W. Montagna, *Structure and Function of the Skin,* 2nd ed., Academic Press, New York, 1962. By permission of the author and the publisher.)

medulla give sheen and color tones to the hair by modifying the reflections of light. In cross section a hair can be round (straight), rectangular (kinky), or alternately round and oval (wavy).

Almost every hair contains some pigment in its cortex and medulla. Melanin produces brown or black hair. A related pigment produces red hair. Melanin is distributed very precisely in the follicle. Most of the columnar cells lining the dome of the hair papilla are melanocytes that extend between adjacent differentiating cells in the upper bulb.

Hair follicles demonstrate alternating periods of activity and rest. In the human scalp active periods are as long as 4 years, and rest periods are relatively short. The reverse is true in other areas of the body; for example, eyelashes grow actively for only a few months. In man the replacement of hair is continuous and passes unnoticed. In fur-bearing mammals that renew their coats twice yearly, the replacement process is rapid. In a sense, each follicle repeats its life history with each growth cycle.

A rapidly growing hair follicle is termed an *anagen* follicle (Fig. 18.8). As the growth period draws to a close, cell multiplication in the matrix slows down and stops, the elements covering the summit of the papilla become cornified, and the follicle retrogresses. The hair gradually separates from the papilla, moves up to the neck of the follicle, and falls out. The papilla then atrophies. The follicle at this stage is termed a *catagen* follicle.

An inactive, shrunken follicle is termed a *telogen* follicle. It is one-half to one-third the length of an anagen follicle and has no bulb. In time, however, the primordium of a new hair arises in the old sac, and the building of a bulb from indifferent cells commences. The process is similar to that taking place in an embryo, with the papilla presumably inducing differentiation. With the matrix reestablished, cells start to proliferate, and hair growth begins anew.

During its life cycle, hormonal stimulation may cause a hair follicle to undergo drastic changes. The small follicles in the face of a boy and in the axillary and pubic regions of boys and girls, for example, are transformed at puberty into large follicles that produce coarse hairs. Conversely, the large follicles in a balding scalp retrogress to tiny follicles that produce barely visible hairs.*

SEBACEOUS GLANDS

A sebaceous gland pours its secretion into a *pilary canal* and thence into the upper part of

---

* Baldness of the common type is apparently associated with high androgen levels. It is rare in women. Attempts to grow hair on bald heads have long occupied both scientists and laymen, but, although simple irritants sometimes initiate a new growth cycle, a dependable remedy for baldness has not yet been discovered.

the outer root sheath of a hair follicle.* As shown in Figs. 18.1, 18.6, and 18.8, a sebaceous gland consists of a cluster of grapelike acini whose ducts converge toward a central collecting duct. Its gross shape is determined by the abundance of similar glands in the area and by the nature of the dermis in which it grows.

Microscopic anatomy. Sebaceous gland ducts are composed of stratified squamous epithelium continuous with the outer root sheath of the hair follicle and with the epidermis. A sebaceous acinus is lined by small flat cells resembling epidermal cells, which become larger and more misshapen toward the center (Fig. 18.9). The undifferentiated peripheral cells contain many ribosomes, dispersed over the endoplasmic reticulum as in other secretory cells (see Figs. 3.10, 3.11). As the cells differentiate, the ribosomes disappear, leaving agranular reticulum. When it permeates the cytoplasm, numerous lipid granules are scattered through the cytoplasm. The lipids accumulate until they compress the cytoplasm into thin, mitochondria-containing plates. There is evidence that the Golgi complex plays a key role in this process. The cells at the center of an acinus, then, are very large and laden with lipids, their cytoplasm having dwindled to flimsy frameworks surrounding lipid droplets. As in all holocrine glands, the cells die when they have manufactured their product. They then fragment and become part of their semiliquid secretion, called *sebum*.

Secretion of sebum. As sebaceous cells are lost in secretion, they are rapidly replaced, in waves of mitotic activity.† In addition, epithelial buds grow from the duct walls and differentiate into new sebaceous units. These expand, encroach upon nearby ones, and combine with them.

The principal function of the sebaceous glands is to synthesize, store, and secrete lipids. Sebum differs from species to species, with cho-

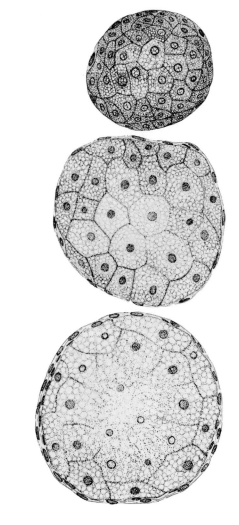

18.9 Maturation of a sebaceous acinus. (From W. Montagna, *Structure and Function of the Skin*, 2nd ed., Academic Press, New York, 1962. By permission of the author and the publisher.)

---

* However, sebaceous glands free from hair follicles are found in the skin of the nipples, labia minora, lips, mouth, and eyelids.

† Any cell of epidermal origin can be transformed into a sebaceous cell. Such a transformation often follows local irritation. Sometimes isolated fragments of hair follicles in the dermis develop into *sebaceous cysts. Comedones* (blackheads) are masses of soft yellowish lipid matter blocking the ducts of sebaceous glands. Usually they are capped at the duct orifices with brownish oxidized debris. Occasionally the plugs are hard and can be expressed as firm grain-shaped bodies. Comedones appear frequently on the face.

lesterol and free fatty acids perhaps the only materials common to nearly all. Other lipids in human sebum are triglycerides, phospholipids, waxes, and cholesterol esters.

It is difficult to imagine what purpose sebum serves. It is not miscible with water, does not form emulsions, is toxic to a number of enzyme systems, and produces skin blemishes. Yet it seems doubtful that it is merely a lubricant.

Sebaceous gland growth and differentiation

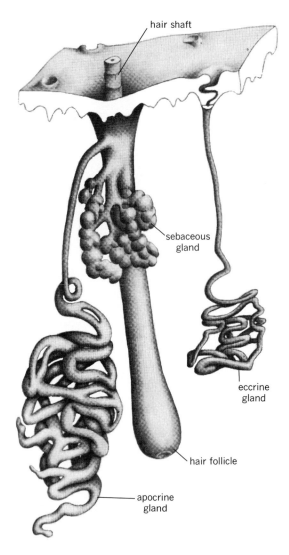

hair shaft

sebaceous
gland

eccrine
gland

hair follicle

apocrine
gland

18.10 Apocrine (left) and eccrine (right) sweat glands. The apocrine sweat gland and the sebaceous gland (middle) empty into a hair follicle. (From W. Montagna, *Structure and Function of the Skin,* 2nd ed., Academic Press, New York, 1962. By permission of the author and the publisher.)

are regulated by hormones. The glands are large in the newborn infant, are relatively small in childhood, and normally become large again at puberty (the glands of eunuchs remain small and underdeveloped). Testosterone and progesterone induce enlargement of the sebaceous glands, whereas removal of the sex glands and

the hypophysis or the adrenals leads to atrophy of the sebaceous glands.

One manifestation of the effect of hormones upon the sebaceous glands is the development of *acne vulgaris* at puberty. Acne is a chronic inflammatory disorder of the sebaceous glands characterized by the presence—mainly on the face, neck, and chest—of many comedones and shallow disfiguring pustules. Apparently heredity and diet predispose an individual to the condition. There is no known cure, but improvement has been attributed to intensive and prolonged antibiotic therapy.

SWEAT GLANDS

Sweat glands are numerous in the skin of man, horse, pig, and cat. Many mammals (e.g., rabbit) have few, or none at all. According to structure and function, sweat glands are separable into *apocrine* and *eccrine* types. In spite of some general similarities, these are profoundly different.

Apocrine sweat glands. Apocrine sweat glands are present in the axillae, external auditory canals, and pubic areas,* and occasionally elsewhere. In the course of evolution, they have been lost, with eccrine sweat glands taking their places.

An apocrine sweat gland is a simple tubular gland with a coiled secretory portion extending to the lower part of the dermis and the subcutaneous tissue (Fig. 18.10). Its duct is close to and parallel to a hair follicle and opens into a pilary canal close to the duct of a sebaceous gland. Rarely the duct of an apocrine sweat gland opens directly onto the surface. The secretory portion is made of simple columnar epithelium, and the duct of two layers of cuboidal cells (Fig. 18.11). The ends of some of the cells in the secretory portion project far into the lumen, but those of others have brush borders. Sandwiched between the secretory epithelium and the thick basement membrane is a layer of spindle-shaped *myoepithelial cells.*

The myoepithelial cells have many of the characteristics of smooth muscle fibers. Their axis is roughly

---

* According to a recent report, those in the pubic area, although structurally typical, are functionally inert, secreting no sweat whatever. This is a novel kind of evolutionary regression. Ordinarily, when function ceases, structure regresses and disappears.

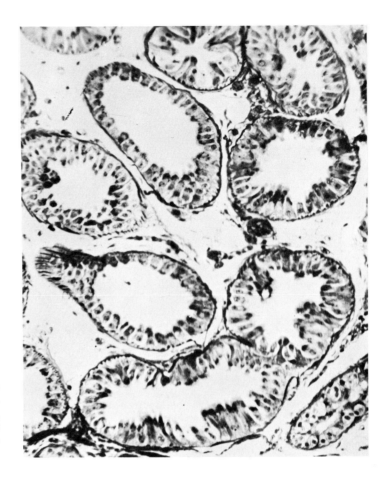

18.11 Structure of an axillary apocrine gland. (From W. Montagna, *Structure and Function of the Skin,* 2nd ed., Academic Press, New York, 1962. By permission of the author and the publisher.)

parallel to the duct, and they are best developed in the gland lined with the tallest epithelium. The myoepithelial cells contract in a peristalsis like muscle cells.

The large spherical nucleus of an apocrine cell is located near the base of the cell (Fig. 18.12). Many large mitochondria surround the nucleus, and many pigmented cytoplasmic granules lie above it. Golgi complexes are prominent.

The production of a *yellow* or *brown pigment* is typical of the apocrine sweat glands. The ceruminous glands of the ears and the axillary glands secrete large amounts. The source, nature, and function of the pigment are not known. The glands contain intraepithelial iron, which appears in some of the pigment granules. Its significance is also unknown.

Although it is generally written that apocrine sweat glands are innervated by sympathetic (adrenergic) nerves, histological evidence of their presence is lacking. It has been shown, however, that epinephrine stimulates apocrine secretion.

Eccrine sweat glands. Eccrine sweat glands are solely responsible for the production of watery sweat. Although many mammals have very few—in dog, cat, and rodent, they occur only in the pads of the paws—man has 2,000,000 to 5,000,000 over his entire body surface, an average of 150 to 350 per square centimeter. The glands are most numerous on the palms and soles, where their openings are regularly spaced in the centers of the epidermal ridges. Other skin areas in decreasing order of gland content are the head, trunk, and extremities.

An eccrine sweat gland is a simple tubular gland also extending from the epidermis to the lower part of the dermis (see Fig. 18.10). It is twisted in the epidermis,

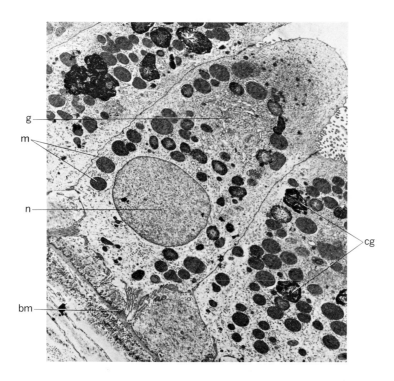

18.12 Electron micrograph of cells of apocrine sweat gland. bm, basement membrane; cg, pigmented cytoplasmic granules; g, Golgi complex; m, mitochondrion; n, nucleus. (From W. Montagna, *Structure and Function of the Skin,* 2nd ed., Academic Press, New York, 1962. By permission of the author and the publisher.)

straight in the upper part of the dermis, and coiled deep in the dermis.

The coiled portion, called the *secretory coil,* is composed of duct and secretory elements in equal parts. The duct is lined with two layers of cuboidal cells of about the same size (Fig. 18.13), and the secretory portion with a single layer of various-sized cells. Large cells are crowded toward the base of the duct, and small ones toward the surface. All cells, however, eventually reach the surface, where they rest upon a layer of myoepithelial cells resembling those of the apocrine sweat glands.

The nucleus of a large eccrine cell is located at the base of the cell, and that of a small eccrine cell near its luminal end. The smaller cells are full of basophilic granules and accordingly have been named *dark cells.* The larger cells contain sparse acidophilic granules and have been named *clear cells.*

A striking feature of an eccrine sweat gland is a system of intercellular canaliculi lined with abundant microvilli. The canaliculi empty into the lumen of the secretory coil. A thick basement membrane surrounds the secretory portion of the gland.

Physiology of sweating. As shown in Table 18.1, the watery secretion of the eccrine sweat glands under ordinary conditions is a weak solution of sodium chloride, with small quantities of

other substances. In fact, its composition is highly variable, depending in part on the rate of sweating. Thus sodium and chloride concentrations in sweat may range from 10 to almost

TABLE 18.1 AVERAGE COMPOSITIONS OF SWEAT AND PLASMA*

Substance	Conc. in sweat, mg/100 ml	Conc. in plasma, mg/100 ml
Sodium	185	325
Potassium	15	15
Calcium	4	10
Magnesium	1	3
Chloride	310	370
Lactate	35	15
Urea nitrogen	20	15
Glucose	2	100
Protein	0	7400

* Adapted from Y. Kuno, *Human Perspiration,* Thomas, Springfield, Ill., 1956.

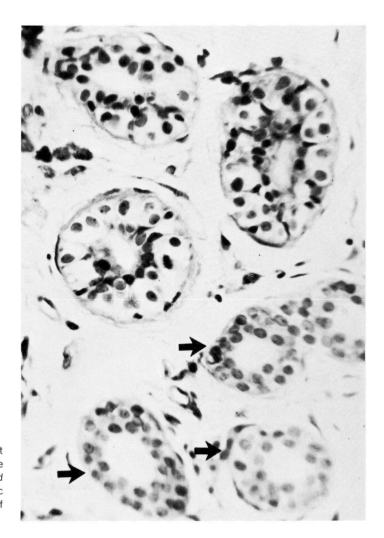

18.13 Structure of an eccrine sweat gland. Arrows denote segments of the ducts. (From W. Montagna, *Structure and Function of the Skin,* 2nd ed., Academic Press, New York, 1962. By permission of the author and the publisher.)

100% of the concentrations in plasma. Sweat has a specific gravity of 1.002 to 1.003 and a pH as reported by different observers of 4.2 to 7.5. Compared to plasma, it (1) has a low concentration of sodium, calcium, magnesium, and chloride; (2) contains almost no glucose and protein; and (3) has a high concentration of lactate (a major anion peculiar to sweat), virtually equivalent to that of bicarbonate in plasma. The potassium concentration is similar to that in plasma; some of the potassium may derive from metabolic processes involved in sweat secretion (e.g., glycogen breakdown). The concentration of urea (the principal nonionic solute) is slightly higher than in plasma but varies widely.

The factors controlling the composition of sweat command interest because the sweat glands act as accessory kidneys, affecting the body salt and water balance (see Chapter 7). It is generally agreed that a *precursor fluid* is removed from plasma in the secretory portion of the secretory coil and modified into sweat in the duct. Presumably precursor fluid is identical to extracellular fluid except that lactate replaces bicarbonate. The coil is apparently impermeable to plasma proteins and glucose.

According to current theories, as precursor fluid enters the duct, it is exposed to a new type of epithelium, which actively reabsorbs some sodium; chloride and lactate follow passively.

Sodium reabsorption is stimulated by mineralocorticoids, which also promote potassium excretion. However, the maximal reabsorptive capacity of the duct is only 20 to 25% the maximal rate of entry of sodium into the precursor fluid.

The structure of the duct is consonant with the functions attributed to it. Its cellular content of glycogen, mitochondria, RNA, and oxidative enzymes points to the existence of active metabolic systems. Moreover, only its basal cells can divide. Since the luminal cells cannot, their endowment of enzymes and other metabolic apparatus must be concerned with other processes. Until we fully understand the mechanism by which the duct participates in the regulation of sweat composition, however, we cannot draw meaningful parallels between it and a renal tubule.

Eccrine sweat glands are innervated chiefly by parasympathetic (cholinergic) nerves.* Two main factors stimulate their secretion: external heat and emotion. Thermal stimulation of sweating is probably mediated by an increase in blood temperature. This acts upon hypothalamic centers that induce sweating through reflex mechanisms. In addition, local reflexes in the skin may have some effects. Visible sweating usually commences after a short latent period at an air temperature between 80 and 90°F. At higher temperatures the body sweats first on the forehead and neck and then on the ventral and dorsal parts of the trunk, lumbar region, and backs of the hands. The palms and soles respond only weakly to thermal stimulation. Profuse sweating in a hot environment may result in large water loss, since evaporation is the body's major cooling device; however, the sweat glands begin to conserve sodium chloride after repeated exposures to heat. This *acclimatization* is attributable to increased adrenocortical activity. The extra aldosterone secreted in some way reduces the sodium concentration in sweat. Indeed, low sodium levels in sweat due to heat are a useful sign of excessive mineralocorticoid production by the adrenal cortex. It should be noted that prolonged thermal stimulation also leads to an eventual decline in the volume of

sweat secreted. This "fatigue" may be due to metabolic exhaustion of the secretory cells, which show losses of glycogen and granules and general atrophy after profuse sweating.

In the sweating of emotional stress, sweat appears primarily on the palms and soles without a preliminary latent period. Gustatory sweating, which also has a neural basis, results from stimulation by extremely spicy foods and affects the face. The sweating that accompanies muscular exercise apparently depends on a combination of thermal and neural elements.

Since ordinary sweating occurs all over the body, it is probably not governed by restricted spinal reflexes. If there are sweat centers in the spinal cord, they must be inactive normally. The centers that control body temperature are in the hypothalamus, and those that control general sweating are probably there, too. The center that controls sweating of the palms and soles is located anteriorly to the motor area of the cortex.

Apocrine sweating differs from eccrine sweating in several ways. It is induced by emotional stress but not by external heat and by epinephrine and norephinephrine but not by acetylcholine, and the volume of its product is but a small fraction of that of eccrine sweating. When care is taken to minimize eccrine secretion, it can be demonstrated that apocrine sweat is a milky, sticky fluid of varying color—whitish, pale gray, yellow, or reddish—that is sterile and odorless when it reaches the skin surface. However, within hours the bacteria on the skin attack it to produce a characteristic odor. In contrast, eccrine sweat resembles clear salt water. Finally, striking specific morphological changes are associated with apocrine secretion. After stimulation nearly all of the secretory cells elongate, protruding cytoplasmic knobs into the lumen, the nuclei are distorted, and the myoepithelial cell layer shows remarkable thickening.

Study of the myoepithelial cell layer of an apocrine sweat gland reveals that peristalsis occurring in response to stimulation discharges the sweat. An apocrine sweat gland can also be milked manually. After it has been emptied, 1 to 2 days must elapse before it is active again. Thus, whereas eccrine sweating is a continuous process, apocrine sweating involves the delivery of preformed sweat to the skin surface as a result

---

* However, since injected acetylcholine or epinephrine causes sweating, they undoubtedly contain sympathetic (adrenergic) nerves as well, despite the lack of histological evidence.

of peristalsis in myoepithelial elements in response to (1) mechanical stimulation, (2) stretch reflexes following overfilling, or (3) sympathetic (adrenergic) stimulation.

NAILS

Gross and microscopic anatomy. The nails are specialized epidermal structures homologous with the claws and hoofs of lower animals (Fig. 18.14). They are dense, translucent plates of cornified cells on the distal phalanges of the fingers and toes. The *nail plate* proper has three layers: (1) an *external*, or *dorsal*, layer that constitutes the body of the nail; (2) an *intermediate* layer of varying thickness; and (3) a *ventral* layer that is exposed beneath the distal nail edge as it grows beyond the end of the digit. As shown in Fig. 18.15, the

18.14 Comparison of nails, claws, and hoofs: A, claw of a carnivore, showing nail bed confined to proximal end; B, hoof of a horse, a shortened broadened claw ensheathing toe; C, nail of an ape, an arboreal, grasping species; D, human nail. (From A. S. Romer, *The Vertebrate Body*, 2nd ed., W. B. Saunders Co., Philadelphia, 1955.)

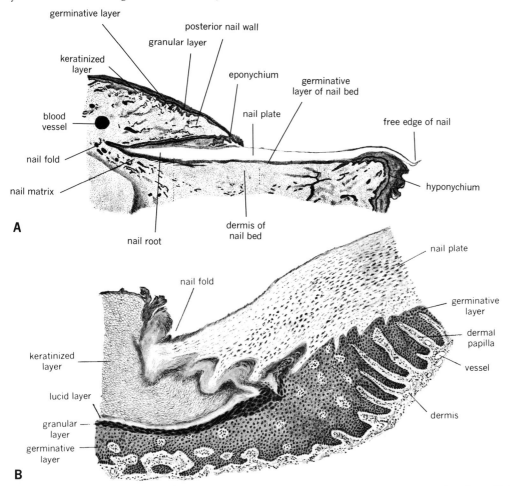

18.15 Structure of the fingernail; A, longitudinal section; B, transverse section of lateral edge. (From W. Bloom and D. W. Fawcett, *Textbook of Histology*, 9th ed., W. B. Saunders Co., Philadelphia, 1968.)

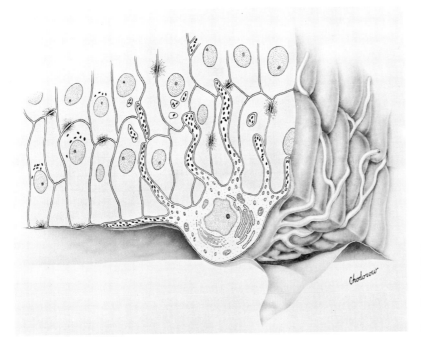

18.16 Location of melanocytes with respect to epidermis and dermis. Note how cytoplasmic process extends into basal layer of epidermis. (Dr. Thomas B. Fitzpatrick, "Hypomelanosis," *Southern Med. Jour.* 57: 995, 1964.)

nails are markedly convex from side to side and slightly convex from proximal edge to distal edge.

The nail lies in the *nail bed*. The skin covering the bed is continuous with the epidermis of the palmar or plantar (i.e., sole) side of the digit. Except for the free distal edge, the edges of the nail plate are tucked into depressions of skin called *nail folds*. The skin over the proximal edge of the nail is the *nail wall*, and the free skin resting directly upon the nail plate is the *eponychium*. Near the *nail root*, under the eponychium, is a crescent-shaped whitish area called the *lunula*. In many individuals it can be seen only in the thumbs unless the eponychia are pushed back. A pad of keratinized cells, the *hyponychium*, grows under the free nail edge.

The surface of the nail plate is indented by delicate longitudinal lines that reflect deep corrugations on the underside. The corrugations fit into grooves in the epidermis of the nail bed. The proximal edge of the nail is a thin wedge in a thick epidermal bed, the *nail matrix*. The matrix extends above and below the proximal edge of the nail, becoming the dorsal matrix which is continuous with the eponychium, and the thick ventral matrix.

The formation of new nail substance takes place in the matrix, which, as a germinative organ, corresponds to the basal layer of the epidermis and the bulb of a hair follicle. Numerous mitotic figures are seen in cells in its deepest layer. The new cells move up in streams to the root, along curving paths.

Growth. In the nail matrix undifferentiated cells at the periphery are gradually transformed to fully keratinized ones. In the process they acquire stainable fibrils as do the hair cells. However, keratinization of the nail yields a product with a distinctive organization of microfibrils and cell nuclei and characteristic chemical properties. For example, the nail plate is composed of several different keratins whose molecular structures are unlike those of keratins elsewhere in the integumentary system and are rich in —S—S— groups. Nevertheless, there are similarities between nail and hair in the formation and arrangement of keratin—the ventral layer of the nail and the inner root sheath can be compared in these respects.

Although it appears that the nail bed is static and that the nail plate glides over it, the two structures are in fact so firmly bound that, if the plate is pulled out, the bed goes with it. The nail bed must therefore grow forward like the inner root sheath of a hair follicle. Usually the epithelium of the nail bed does not become keratinized, but, in certain disorders or after injury, keratohyalin accumulates to produce a horny layer under the nail plate and consequently an abnormally thick and opaque nail.

Pigmentation

We noted on pp. 723 and 730 that the skin and hair papillae contain melanocytes. In the skin they compose a horizontal network at the plane of the dermal–epidermal junction (Fig. 18.16). They are closely integrated with the epidermal cells, establishing contact by means of their numerous "dendrites."

Melanocytes are so named because they secrete melanin, an amorphous brown pigment that is largely responsible for the coloration of the skin, hair, and eyes. Melanin is actually a class of high–molecular weight pigments widely distributed in nature. We shall see in discussing the effects of sunlight upon human skin that without melanin man would be required to live constantly protected from the sun. Hence melanin promotes survival even though, strictly speaking, it is not essential for survival.

DEVELOPMENT AND TRANSFER OF MELANOSOMES

Melanocyte system. After long controversy it was finally shown in 1948 by R. E. Billingham and associates that melanin is formed only in the cytoplasm of melanocytes. By means of transplantation experiments, these workers demonstrated conclusively that melanocytes arise only from melanocytes, by mitotic division, and not from the other epidermal cells, the keratinocytes (see p. 724). It was concluded that, although keratinocytes contain melanin, melanocytes are specific secretory cells and that they somehow transfer melanin to the keratinocytes.* Their fundamental difference from keratinocytes was determined by electron microscopy (Fig. 18.17). A melanocyte contains no tonofibrils. It has the general architecture of a secretory cell, with well-developed endoplasmic reticulum, and its main functions are the elaboration and secretion of melanin. The skin of an adult has approximately 2,000,000,000 melanocytes, which collectively weigh about 1 g.

In 1947 it was established that melanocytes derive embryologically from the neural folds (see Chapter 12). During the formation of the neural tube, cells migrate from the folds toward the surface, and some are recognizable as melanocyte precursors, or *melanoblasts*, by the tenth week of embryonic life (Fig. 18.18). They pene-

---

* The transfer of melanin has been directly observed by electron microscopy. Sagittal sections of a hair follicle at different levels reveal that a portion of the cytoplasmic process of a hair melanocyte is phagocytized by a cortical cell, the cell wall of the process disappears, and the melanin particles are dispersed throughout the cytoplasm of the cortical cell.

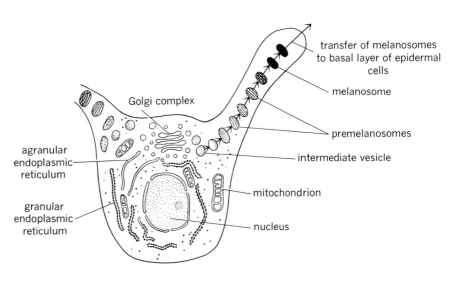

18.17 Melanocyte. (T. B. Fitzpatrick, W. C. Quevedo, M. Seiji, and G. Szabo, "Biology of the Melanin Pigmentary System," in *Dermatology in General Medicine*, T. B. Fitzpatrick, ed. McGraw-Hill, New York, 1970.)

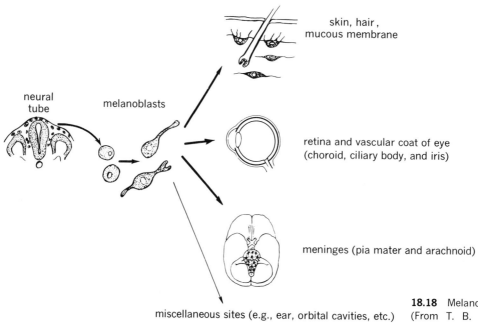

skin, hair, mucous membrane

neural tube

melanoblasts

retina and vascular coat of eye (choroid, ciliary body, and iris)

meninges (pia mater and arachnoid)

miscellaneous sites (e.g., ear, orbital cavities, etc.)

18.18 Melanocyte system. (From T. B. Fitzpatrick, M. Seiji, and A. D. McGugan, "Melanin pigmentation," *New Eng. J. Med., 265,* 328, 1961.)

trate the future epidermis in the eleventh and twelfth weeks and are spread evenly throughout it before the end of the fourth month.

The melanocytes of the skin and mucous membranes are only a part of the body's widely scattered system of pigment cells. Others are present in the vascular tunic (choroid, ciliary body, and iris) and retinal pigment epithelium of the eye and in the thin meninges (pia mater and arachnoid).* All are identical embryologically, morphologically, and functionally. The melanocyte system is in many ways analogous to another system of cells deriving from neural precursors, the chromaffin system (see p. 600).

Melanosomes. The final products of the melanocyte are melanin-containing bodies called *melanosomes.* Four stages are recognized in the formation of melanosomes (Fig. 18.19): (1) an initial stage in which polypeptides of the en-

zyme *tyrosinase* are synthesized on ribosomes and phospholipid-containing proteins are fashioned into membranes by the Golgi complex; (2) a first intermediate stage in which the membranes develop into a small (0.5 μ), spherical, membrane-limited body, the *intermediate vesicle,* within which tyrosinase molecules aggregate into small fibers; (3) a second intermediate stage in which the vesicle acquires the characteristic ultrastructure of a *premelanosome;* and (4) a final stage, the formation of a mature *melanosome,* in which the synthesis and accumulation of melanin begin and, with the disappearance of tyrosinase, suddenly end.

As melanosomes increase in size, they darken in color and become electron dense. The last change may be due to their zinc or copper content. It has been argued that melanosomes are modified mitochondria, but their structure, biochemical properties, and sedimentation behavior appear to contradict this view.

Melanin pigmentation of the skin involves both the production of melanosomes by melanocytes and their distribution to the basal layer cells of the epidermis. The basal layer cells are

---

* Pigment found in neurons in certain areas of the central nervous system (e.g., the *substantia nigra*) is probably not melanin, though it has not been identified. It is not present in melanocytes and is unchanged in the melanocyte disorder albinism.

stage	name and size of organelle	morphology	events
1	(ribosomes, Golgi complex)		biosynthesis of tyrosinase on ribosomes; formation of membrane in Golgi complex.
2	intermediate vesicle, 0.5 μ		development of vesicle from membrane; aggregation of tyrosinase molecules within vesicle into fibers.
3	premelanosome, 0.7 × 0.3 μ		cross-linking of tyrosinase fibers; arrangement of organelle into characteristic structural form.
4	melanosome, 0.7 × 0.3 μ		beginning of melanin formation; accumulation of melanin; completion of structural development of organelle; loss of tyrosinase activity; cessation of melanin formation.

18.19 Stages in the development of melanosomes. (Adapted from M. Seiji, T. B. Fitzpatrick, R. T. Simpson, and M. S. C. Birbeck, *Nature, 197,* 1082, 1963.)

not entirely passive in these operations; indications are that they somehow regulate the rate of melanosome production. A melanocyte together with the basal layer cells to which it transfers melanosomes is termed an *epidermal melanin unit.*

In the transfer of melanosomes, melanocyte dendrites actually penetrate the basal cells and are nipped off in them, the overall process resembling phagocytosis. When it was realized that melanocytes are in fact exocrine glands, the category *cytocrine* was added to the known gland types to distinguish their unique mode of secretion (see Table 5.1).

SYNTHESIS AND FUNCTIONS OF MELANIN

Biosynthetic pathway. Melanin synthesis is dependent upon tyrosinase. Tyrosinase is one of a large group of copper-containing enzymes that catalyze the oxidation of both monohydric and dihydric phenols to orthoquinones. In the reactions catalyzed by tyrosinase, molecular oxygen acts directly as the hydrogen acceptor. Substances such as methylene blue cannot replace oxygen in this capacity. Hence the enzyme is an oxidase, not a dehydrogenase (see Chapter 2).

In melanosomes tyrosinase catalyzes the hydroxylation of the amino acid tyrosine,* the major precursor of melanin, to *dihydroxyphenylalanine.* This compound is usually referred to as *dopa,* a jargon abbreviation of the alternative

---

* We see, then, that tyrosine has a crucial role in metabolism, serving as a precursor of (1) thyroxine and triiodothyronine, thyroid hormones (see Chapter 15); (2) epinephrine and norepinephrine, adrenal medullary hormones (see Chapter 15) and adrenergic transmitters (see Chapter 12); (3) all tissue proteins (see Chapter 2); and (4) melanin.

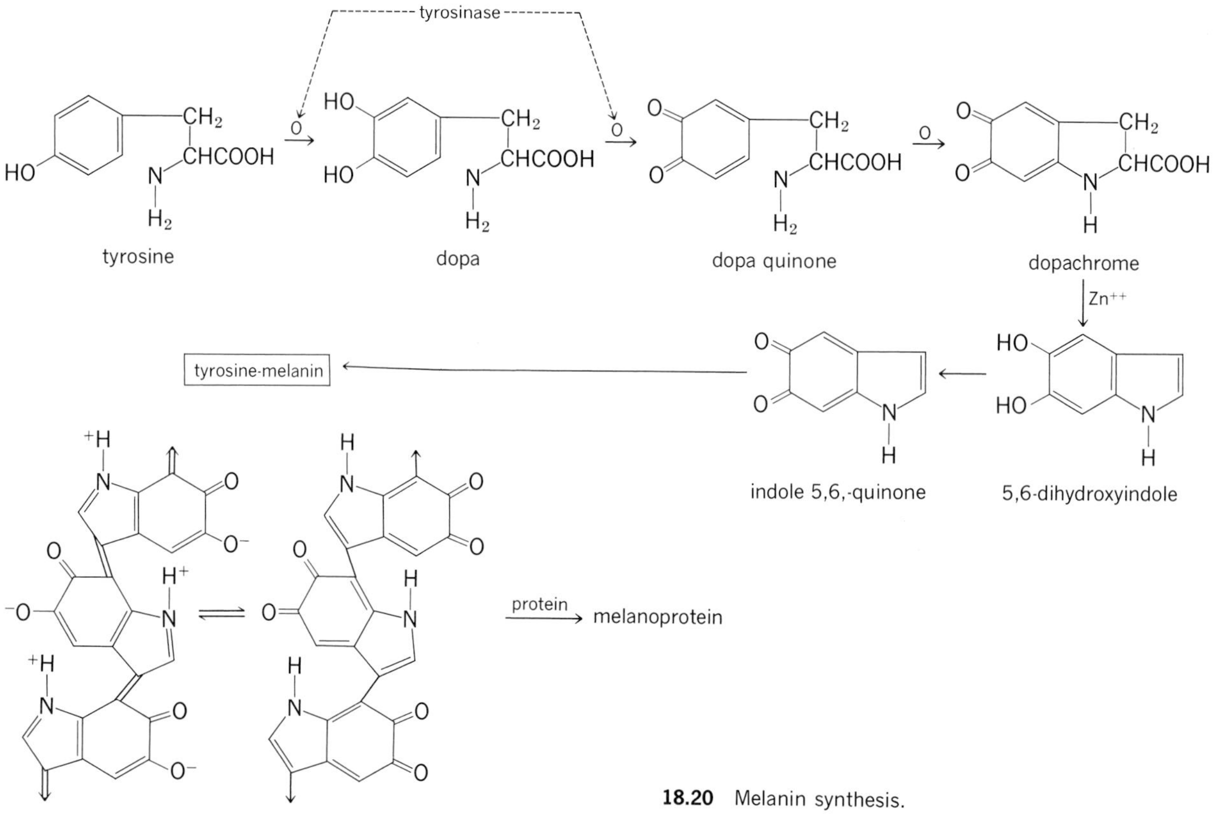

18.20 Melanin synthesis.

name *dioxyphenylalanine*.* Tyrosinase then catalyzes the oxidation of dopa to dopa quinone. Subsequent steps in the synthesis of melanin are outlined in Fig. 18.20. A second ring system develops with the formation of an indole nucleus. Melanin is a polymer of the indole derivatives, though its precise structure is unknown. It should be emphasized that tyrosinase is required only in the hydroxylation of tyrosine and the oxidation of dopa. It is of interest that zinc catalyzes the conversion of dopachrome to the intermediate 5,6-dihydroxyindole. Zinc is present in high concentrations in melanosomes.

The term *tyrosine melanin* is used for human melanin; melanins in lower forms arise from phenolic compounds other than tyrosine.† One reason for our ignorance of the structure of natural tyrosine melanin is its extreme insolubility. We can say only that it is a high–molecular weight "biochrome" formed by the enzymatic oxidation of tyrosine. It is always bound to protein through sulfhydryl or amino groups.

Factors affecting intensity of pigmentation. Skin and hair color depend normally, not upon the numbers of melanocytes present, but upon their activity.‡ By activity, we mean their rate of production and transfer of melanosomes. It can be inferred from the pathway of melanin synthesis in Fig. 18.20 that the rate of melanosome production depends upon the presence in a melanocyte of (1) the substrate tyrosine, (2) molecular

---

* In 1917 B. Bloch demonstrated melanocytes in human skin by exposing a skin section to a solution containing dopa. The enzymes in the melanocytes promoted the formation of melanin, which blackened the cells. This *dopa oxidase* reaction (or a later modification) is a classic test for melanocytes. We now know that dopa oxidase and tyrosinase are the same enzyme.

† Examples are 3,4-dihydroxybenzoic acid, 3,4-dihydroxyphenylacetic acid, and 3,4-dihydroxyphenyllactic acid.

‡ White skin is as well provided with melanocytes as black skin. Racial differences in pigmentation are therefore functional and not anatomical, although obviously genetically determined.

oxygen, (3) the enzyme tyrosinase, and (4) an organized premelanosome. Variations in skin pigmentation are attributable to variations in one or more of these factors.

The tyrosine needed for the synthesis of melanin must be free. Tyrosine within a peptide or protein molecule cannot act as a tyrosinase substrate. The peptidases in pigmented tissues may function in freeing tyrosine bound in peptides.

Tyrosinase activity depends largely upon the availability of substances that activate tyrosinase, particularly those that reduce its copper from the cupric form to the cuprous form essential for the initial production of dopa. Once formed, dopa itself carries out the reduction. The concentration of naturally occurring inhibitors of tyrosinase, such as free phenylalanine, also influences tyrosinase activity.

Pigmentation is further affected by certain physical and chemical changes involving the melanosomes. One of these is darkening, which rapidly follows exposure to ultraviolet light and is believed to result from photo-oxidation, since reduced melanin is light brown and oxidized melanin is dark brown. A second change, aggregation or dispersion of the melanosomes, has been observed only in the specialized melanocytes of vertebrates lower than mammals.

Hormonal control of pigmentation. We were introduced to the hypophyseal melanocyte-stimulating hormones (MSH's) in Chapter 15. We recall that at least two types of MSH, α-MSH and β-MSH, have been isolated, that their structures have been determined, and that α-MSH has been synthesized in the laboratory. ACTH also stimulates melanocytes, presumably because a section of its amino acid chain is identical to a section in each MSH.

It was early observed that the purified MSH's are active in darkening frog skin (Fig. 18.21). Indeed, the color changes induced in vitro in living pieces of frog skin constituted a test of hormone purity. Despite the striking effects on frog skin, it was not certain at first whether MSH's darken human skin. Proof that they do came from experiments showing increased pigmentation in Negroes within 24 hours of the administration of synthetic α-MSH (Fig. 18.22). Whether ACTH darkens human skin has not been established.

Regarding the mechanism of action of the MSH's, it can be said only that, in amphibians and other cold-blooded vertebrates, α-MSH causes the dispersion of aggregated melanosomes throughout the melanocyte cytoplasm

18.21 Demonstration of the effect of hormones on the color of a frog skin. The frog at the bottom has received an injection of α-MSH. The frog at the top is lighter than normal after immersion in water containing melatonin. (Dr. Aaron B. Lerner.)

and that this dispersion alone accounts for darkening of the skin (Fig. 18.23). However, no such movement of melanosomes has been seen in mammalian melanocytes. It is possible that mammals do not have effector cells, or *melanophores* (the name applied by biologists to the amphibian melanocytes containing movable melanosomes).*

In light of the preceding discussion, we can assume that administered MSH or ACTH increases pigmentation by increasing melanosome

---

* Cells containing pigment that can disperse or concentrate are called *chromatophores*. They account for the rapid changes of skin color observed in some species, especially cold-blooded vertebrates, crustaceans, and cephalopods. Melanophores are but one variety of chromatophores.

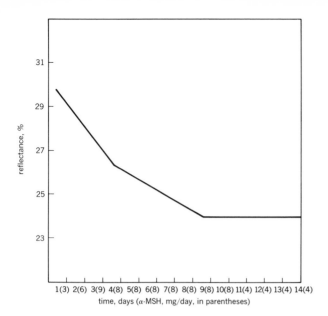

18.22 Darkening of human skin by α-MSH.

reflectance, %

time, days (α-MSH, mg/day, in parentheses)

The MSH's and ACTH are not necessary for normal pigmentation, however. For example, neither congenital panhypopituitary dwarfism nor hypophysectomy is associated with defective pigmentation in Negroes. Nevertheless, hormonal effects may explain certain pigmentation abnormalities. Darkening of the skin accompanies Addison's disease (primary adrenocortical insufficiency) and follows bilateral adrenalectomy. The fact that pigmentation does not increase with adrenocortical insufficiency secondary to primary hypophyseal insufficiency led to the realization that hyperpigmentation occurs only when the hypophysis is competent. We conclude, therefore, that circulating corticosteroids normally inhibit the release of MSH's (Fig. 18.24), that primary adrenocortical insufficiency results in compensatory overactivity of the hypophysis, with increased production of MSH's and ACTH and increased pigmentation, and that in hypopituitarism pigmentation is normal or possibly decreased, owing to a lack of MSH's and ACTH. These conclusions have been confirmed by demonstrations of increased blood ACTH and MSH levels in individuals with Addison's disease and normal blood ACTH and MSH levels in the same individuals after cortisol treatment. The well-known hyperpigmentation of pregnancy is due to a normal increase in hypophyseal activity that takes place despite increased adrenocortical secretion.*

production, by accelerating tyrosinase synthesis, by enhancing tyrosinase activity (perhaps by stimulating biochemical systems in the melanocyte that reduce cupric tyrosinase to cuprous tyrosinase), by promoting phagocytosis of melanosomes by basal epidermal cells, or by darkening formed melanosomes.

---

* Darkening hormones other than the MSH's may be important in pregnancy. Androgens and estrogens darken human skin but not frog skin. Progesterone slightly darkens both.

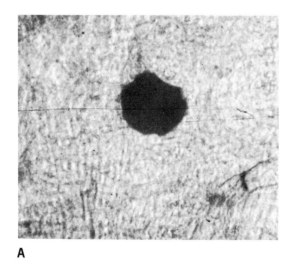
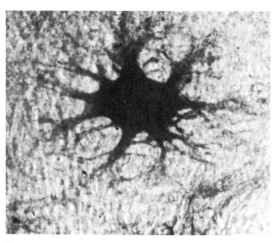

A   **B**

18.23 MSH action in a frog melanocyte: A, melanosome, clumped around the nucleus; B, melanosome, dispersed after the addition of MSH. (Dr. Aaron B. Lerner.)

The MSH's and ACTH are darkening hormones. Investigators have recently discovered a powerful lightening hormone, *melatonin* (*N*-acetyl-5-methoxytryptamine),

$$CH_3O\!-\!-\!CH_2CH_2NHCOCH_3$$

present in the pineal body and peripheral nerves. It has unusual potency in lightening frog skin, being 100,000 times as active as epinephrine and norepinephrine, other lightening agents. It is particularly interesting that melatonin, the most potent known lightener, and α-MSH, the most potent known darkener, are both *N*-acetyl compounds (see p. 600). The significance of melatonin in human physiology is not yet known.

Functions of melanin. The only known function of melanin in man is as an epidermal light filter between the sun and the dermis. As cells move outward from the germinative layer to form the keratinized layer, they carry their melanin with them. The keratinized layer is regu-

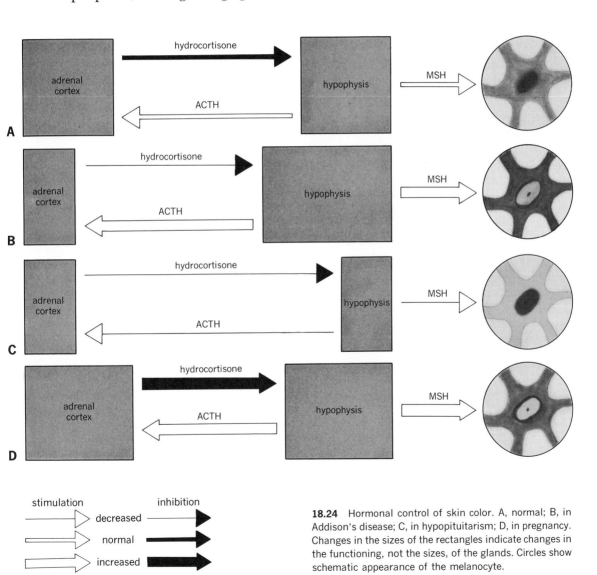

18.24 Hormonal control of skin color. A, normal; B, in Addison's disease; C, in hypopituitarism; D, in pregnancy. Changes in the sizes of the rectangles indicate changes in the functioning, not the sizes, of the glands. Circles show schematic appearance of the melanocyte.

larly flecked with it in Asiatics and Negroes. It is not apparent in Caucasians except after exposure to ionizing and ultraviolet radiation.

We shall briefly consider the photobiology of the skin. Although light is a life-sustaining form of energy transferable through space, it is also capable of inflicting great injury. It can kill bacteria, yeasts, and molds and can cause blindness and major skin diseases in man.

At the earth's surface the spectrum of sunlight extends from a wavelength of about 3000 Å in the ultraviolet region to one of about 23,000 Å in the infrared region. Absorption of light energy by the epidermis is almost complete at 3000 Å, but rapidly diminishes at longer wavelengths. At 4000 Å (the blue end of the visible spectrum), for example, between 50 and 90% of the light energy applied to the surface of the skin is transmitted through the epidermis and into the dermis.

The protective function of melanin may be related to its properties as a stable free radical, a one-dimensional conductor that can store electrons. It has been shown that ultraviolet radiation increases the free radical content of pigmented hair. Melanin may capture electrons generated by the action of ultraviolet light on tissues and guard the skin against their harmful metabolic effects.

The extremely short wavelengths below 3200 Å (less than 0.01% of sunlight) are responsible for *sunburn*. The energy absorbed during a 20-minute exposure to sunburn rays at midday in temperate zones is sufficient to produce *erythema* (redness) in normal, untanned Caucasian skin. The 2- to 6-hour latent period between exposure and appearance of the erythema is thought to represent the time required for the elaboration of vasodilating substances in the epidermis and their passage into the dermis. A second reaction to sunburn is thickening of the keratinized layer.

Sunlight increases the skin's efficiency as a light filter by promoting the formation, migration, and darkening of melanin. Wavelengths between 3000 and 4200 Å cause the darkening of preformed melanin, which commences within minutes of exposure. The formation of new melanin begins a few hours after exposure and continues for several weeks. Meanwhile, preformed melanin and newly formed melanin move upward into the epidermis. Without additional exposure to the sun, the skin maintains a high melanin content for many months.

Light has a number of interesting effects on the skin in addition to those described. For example, it induces the conversion of 7-dehydrocholesterol to vitamin D (see p. 625) and the production of skin cancers and other disorders in susceptible individuals.

Abnormalities of pigmentation. Pigmentary disorders can be divided into three main categories: (1) hypopigmentation, in which the skin is white or lighter than normal; (2) brown or black hyperpigmentation; and (3) blue or gray hyperpigmentation. In general, hypopigmentation is due to a decrease in melanocyte number (e.g., after a severe burn) or activity. Brown hyperpigmentation is due to an increase in melanocyte activity but not number. Blue or gray hyperpigmentation is due to melanin in the dermis—

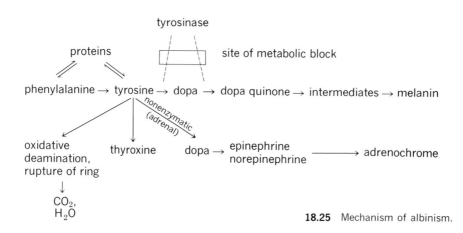

18.25 Mechanism of albinism.

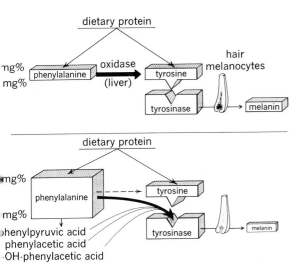

18.26 Comparison of pigmentation mechanisms in a normal individual (A) and in an individual with phenylketonuria (B). (From T. B. Fitzpatrick, P. Brunet, and A. Kukita, *The Biology of Hair Growth,* ed. by Montagna and Ellis, Academic Press, Inc., New York, 1958.)

either ingested by dermal cells or from melanocytes that have migrated into the dermis.*

Albinism is a genetically determined hypopigmentation characterized by white skin and hair and pink, translucent irises. The condition is common throughout the animal kingdom, and its incidence in man has been reported to be from 1:5,000 to 1:25,000. The scarcity or absence of melanin in the melanocyte system of an albino results from a metabolic defect that prevents melanocytes from synthesizing sufficient tyrosinase (Fig. 18.25) (hence they contain unmelanized premelanosomes). Albinism has two disturbing consequences: decreased visual acuity, with poor central vision, and intolerance of sunlight,† with cancers liable to develop on exposed surfaces, especially in Negro albinos living in the tropics. Otherwise, albinos live normal lives.

Phenylketonuria is a genetically determined hypopigmentation due to the inhibition of tyrosinase activity by high concentrations of phenylalanine in the body

fluids (see p. 743). The primary defect in this disorder is a lack of the oxidase that converts phenylalanine to tyrosine. As a result, phenylalanine from dietary protein accumulates (Fig. 18.26), leading to mental deficiency and the urinary excretion of appreciable amounts of diverse phenylalanine derivatives. That the inhibition of tyrosinase activity by phenylalanine is competitive has been proved by in vitro and in vivo experiments. Normal pigmentation is restored after feeding of large quantities of tyrosine.

Vitiligo is a patchy hypopigmentation that is not genetic in origin (Fig. 18.27). Some skin regions are distinguished by abnormal melanocytes that cannot synthesize premelanosomes. The cause is unknown. We have already mentioned the hyperpigmentation due to endocrine malfunction. The phenomenon of *freckling* is another type of hyperpigmentation. Freckles, or *ephelides*, are brown, circumscribed spots scattered irregularly on skin areas exposed to sunlight. They are not present at birth and, curiously, do not usually appear until after infancy, despite repeated exposure. Freckling ordinarily begins in early childhood and may be marked in the sixth to eighth year. The tendency toward it diminishes in adulthood.

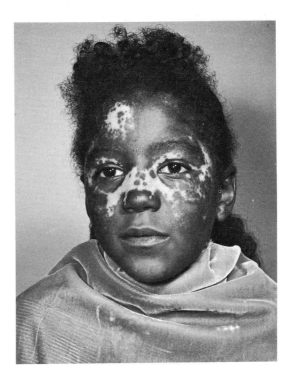

18.27 Appearance of skin in vitiligo. (Dr. Thomas B. Fitzpatrick.)

---

* Blue color is produced by brown particles in the dermis because light passing through the turbid medium of the epidermis is "scattered," the long wavelengths (red) being transmitted and the short wavelengths (blue) being reflected.

† Tolerance can be built up through graded exposure to sunlight, which causes thickening of the keratinized layer.

The pigmentation of a freckle results from a localized increase in melanocyte activity. There is no increase in melanocyte number. Indeed, some evidence points to fewer melanocytes in a freckle than in the surrounding paler areas. The observations of larger, more active melanocytes and juxtaposed smaller, less active ones suggest that freckle melanocytes are type-specific and that their activity is controlled by genetic influences within them.*

Still another type of hyperpigmentation is a *malignant melanoma,* a cancerous proliferation of melanocytes. The lesions are usually blue because of their invasive nature. Sometimes they become so widespread and produce tyrosine oxidation products in such profusion that the skin of the entire body darkens.

Wound Healing

The healing of a superficial skin wound involves more than the mere bridging of a gap by epithelium. In fact, it is a complex process requiring the coordinated functioning of many tissue elements and regulatory agencies. Epithelium grows down into the connective tissue, and the two classes of tissue proliferate and interact, ultimately producing new normal skin. The healing of a deeper wound involves even more complex responses by several classes of tissue, the synthesis of new tissue being accompanied by a remarkable redistribution of old tissue.

PERIODS

For purposes of discussion, it is useful to divide the healing process into four periods: (1) the *latent* period; (2) the period of *contraction;* (3) the period of *epidermal regeneration;* and (4) the period of *cicatrization.* We shall see that restoration of form and strength depends upon two events: *tissue migration* and *tissue synthesis.* We shall see also that wound healing is one of the most specific of biological reactions and that it is not yet understood.

Latent period. The nonspecific local and systemic reactions that follow severe injury are

---

* Large melanocytes similar to those in freckles appear in radiated skin. Conceivably they arise by mutation.

mediated to a large extent by the glucocorticoids of the adrenal cortex and thus are generally catabolic (see Chapter 15). As a result, there is a widespread loss of important cellular constituents. However, healing normally begins despite a negative nitrogen balance. This means that anabolic reactions must occur locally.

The first local reaction to injury is acute *inflammation.* It constitutes both a defensive response to regional tissue damage and the initial stage of repair. The major components of inflammation are as follows: (1) an increase in capillary permeability; (2) the delivery of plasma and leukocytes to the site of the injury, or *exudation;* and (3) the removal of any inciting agent that may be present (e.g., foreign body or bacterium) and the devitalized tissue surrounding it.

The aspect of inflammation that initiates repair has not been identified. It is probably not exudation, since normal tissue is regenerated even when the blood contains almost no leukocytes to infiltrate the injured region. In any event fibroblasts and capillaries develop along the fibrinous network that is early erected in the tissue defect.

The latent period lasts from 1 to 7 days. During this time a wound is completely lacking in tensile strength. Hence the edges of a large wound must be kept together by sutures.

Period of contraction. As the cellular exudation of acute inflammation subsides, fibroblasts rapidly become predominant in a wound. Their source is disputed, but they spread throughout the loose connective tissue at the base of the wound as leukocytes fill its center.

The fibroblasts appear to function in three stages. Proliferation is most noticeable in the first. Its duration depends upon the size of the wound. In the second stage collagen fibrils form. In the final stage, concurrent with the period of cicatrization, a scar forms. During all three stages the fibroblasts evidently secrete mucopolysaccharides.

Second numerically to the fibroblasts in a healing wound are endothelial cells. After injury, hemorrhage and thrombosis occur in the vascular bed directly affected, which becomes a zone of blood sludging and capillary growth.

Closure of an open wound requires a powerful

mobilization of the surrounding tissue as well as regeneration of the epidermal and connective tissue in the defect. The result is wound contraction, which promotes closure. To some extent it is due to the absorption of fluid and shrinkage of the fibrin clot.*

Period of epidermal regeneration. Soon after injury cells at the wound edges begin a swift and extensive regenerative process. A tenfold increase in the number of mitotic cells in these areas and an acceleration of mitosis in them have been demonstrated.

Within a short time tongues of epidermis extend over the denuded surface and join to cover it. This migration starts within a few hours of the infliction of a small epidermal wound and is complete in about 48 hours. Nearly all the dividing cells are found within 1 mm of the edges of a purely epidermal wound, though they are present in the new epidermis over a larger wound as well. When the wound edges are more than 10 to 15 mm apart, full epidermal regeneration is not possible, and skin grafts may be necessary.

In a sutured wound the dermal edges are usually not exactly approximated. Therefore, the epidermis does not grow directly across the defect but down along the cut surface of the dermis until it reaches a level where the tissue is continuous. There the advancing epidermal sheets from opposite edges of the wound meet, to close it. The junction becomes greatly thickened and tends to fill in the defect.

Migrating epidermis undermines the crust, or scab,† over a wound and traverses a bed of healthy, although inflamed, tissue. There is evidence that regenerating epidermis secretes a proteolytic enzyme, specifically a fibrinolysin, that facilitates its movement beneath clot and tissue debris.

Epidermal regeneration is regularly accomplished before connective tissue regeneration. However, the latter process accounts principally for the restoration of tensile strength to injured tissue. Collagen concentration and tensile strength normally reach a maximum value in 12 to 14 days. The density and orientation of collagen fibrils appear to be related in part to mechanical forces.

Period of cicatrization. The formation of a dense, relatively nonvascular scar takes a long time—often many weeks or months. By this stage the fibroblasts have matured to fibrocytes and hence lost their capacity to divide. The scar, or *cicatrix*, fills the defect but lacks specialized epidermal structures such as hair and sweat glands. Gradually the scar loses its many capillaries and accordingly changes color from red to white.

The final phase of healing is *contracture* of the scar (as distinct from wound contraction; see above). Contracture, or shrinkage, of the scar reduces the size of the defect. Sometimes contracture is excessive—particularly in wounds of soft tissue in which there is little mechanical resistance—and serious deformities arise.

Collagen production continues long after the healing process appears complete, and much of the collagen is either not organized into fibers or organized extremely slowly. Occasionally so much collagen is produced that a raised hypertrophic scar, or *keloid*, results. The tendency to keloid formation is apparently genetically determined. Keloids are unsightly lesions that may recur even following removal.

Metabolic effects upon healing. Normal healing depends upon the processes of cellular metabolism; these in turn require adequate cell nutrition. Cells may be nutritionally deprived because of dietary deficiencies, failure in the delivery of nutrients from other parts of the body, or their own inability to use nutritive materials. Thus disorders such as vitamin C deficiency (scurvy) and protein starvation are frequently associated with defective wound healing.

Scurvy is marked by incomplete fibroblast differentiation and impaired collagen and mucopolysaccharide synthesis. Protein starvation is characterized by a retardation of all phases of wound repair; however, healing ultimately takes place—even in animals dying of protein deficiency. Two other factors tending to delay healing are excessive glucocorticoid concentrations and excessive local irradiation. We shall discuss the healing of skin wounds further in connection with tissue grafting in Chapter 19.

---

* Wound contraction should be distinguished from *contracture*, which takes place long after wound contraction has ceased.

† A scab is nothing more than dry fibrinous exudate.

REFERENCES AND SUGGESTIONS
FOR FURTHER READING

Billingham, R. E., and W. K. Silvers, "The Melanocytes of Mammals," *Quart. Rev. Biol.*, **35**, 1 (1960).

Blum, H. F., *Carcinogenesis by Ultraviolet Light*, Princeton Univ. Press, Princeton, N. J., 1959.

———, "Sunburn," in A. Hollaender, ed., *Radiation Biology*, Vol. 2, McGraw-Hill, New York, 1955.

Carruthers, C., and V. Suntzeff, "Biochemistry and Physiology of Epidermis," *Physiol. Rev.*, **33**, 229 (1953).

Chase, H. B., "Growth of Hair," *Physiol. Rev.*, **34**, 113 (1954).

Crewther, W. G., R. D. B. Fraser, F. G. Lennox, and H. Lindley, "The Chemistry of Keratins," *Advan. Protein Chem.*, **20**, 191 (1965).

Della Porta, G., ed., *Structure and Control of the Melanocyte, Proc. Intern. Pigment Cell Conf., 6th*, Sofia, Bulgaria, 1965, Springer-Verlag, Heidelberg, 1966.

Fingerman, M., "Chromatophores," *Physiol. Rev.*, **45**, 296 (1965).

Fitzpatrick, T. B., M. A. Pathak, I. A. Magnus, and W. L. Curwen, "Abnormal Reactions of Man to Light," *Ann. Rev. Med.*, **14**, 195 (1963).

———, W. C. Quevedo, Jr., A. L. Levene, V. J. McGovern, Y. Mishima, and A. G. Oettle, "Terminology of Vertebrate Melanin-Containing Cells: 1965," *Science*, **152**, 88 (1966).

Gordon, M., ed., *Pigment Cell Biol., Proc. Conf. Biol. Normal Atypical Pigment Cell Growth, 4th, Houston, 1957*, Academic Press, New York, 1959.

Hamilton, J. B., ed., "The Growth, Replacement, and Types of Hair," *Ann. N. Y. Acad. Sci.*, **53**, 461 (1951).

Hurley, H. J., and W. B. Shelly, *The Human Apocrine Sweat Gland in Health and Disease*, Thomas, Springfield, Ill., 1960.

Kuno, Y., *Human Perspiration*, Thomas, Springfield, Ill., 1956.

Lerner, A. B., "Hormones and Skin Color," *Sci. Am.*, **205**, 98 (July, 1961).

———, and J. D. Case, "Melatonin," *Federation Proc.*, **19**, 590 (1960).

Montagna, W., "The Skin," *Sci. Am.*, **212**, 56 (Feb., 1965).

———, *The Structure and Function of Skin*, 2nd ed., Academic Press, New York, 1962.

———, and R. E. Billingham, *Advances in the Biology of Skin, Vol. V: Wound Healing*, Pergamon Press, New York, 1964.

———, and R. A. Ellis, eds., *The Biology of Hair Growth*, Academic Press, New York, 1958.

———, and W. C. Lobitz, Jr., eds., *The Epidermis*, Academic Press, New York, 1964.

Potter, B., "The Physiology of the Skin," *Ann. Rev. Physiol.*, **28**, 159 (1966).

Riley, V., and J. G. Fortner, ed., "The Pigment Cell: Molecular, Biological, and Clinical Aspects," *Ann. N. Y. Acad. Sci.*, **100**, 497 (1963).

Robinson, S., and A. H. Robinson, "Chemical Composition of Sweat," *Physiol. Rev.*, **34**, 202 (1954).

Rook, A. J., ed., *Progress in the Biological Sciences in Relation to Dermatology*, Cambridge Univ. Press, Cambridge, 1960.

———, and G. S. Walton, eds., *Comparative Physiology and Pathology of Skin*, Blackwell, Oxford, 1965.

Rothman, S., ed., *The Human Integument: Normal and Abnormal*, American Association for the Advancement of Science, Washington, D. C., 1959.

———, *Physiology and Biochemistry of the Skin*, Univ. of Chicago Press, Chicago, 1954.

Shelley, W. B., and R. P. Arthur, "The Physiology of the Skin," *Ann. Rev. Physiol.*, **20**, 179 (1958).

Shilling, J. A., "Wound Healing," *Physiol. Rev.*, **48**, 374 (1968).

Tregear, R. T., *Physical Functions of Skin*, Academic Press, New York, 1966.

Williamson, M. B., ed., *The Healing of Wounds*, McGraw-Hill, New York, 1957.

Although the inborn differences in human beings are combinational in origin and inner structure, yet the combinants are so numerous, and so generous are the ways in which they may be combined, that every human being is genetically unique: the texture of human diversity is almost infinitely close woven. But what is the "meaning" of this diversity? What intelligible function does it fulfill? . . . The gist of the answer, as it relates to lower organisms, is this. Inborn diversity makes for diversity in evolution. Every living species must provide not only for the present but also for what may happen to it in the future; only those lineages survive to the present day which, in the past, were versatile enough to come to terms with their environment. All organisms must have a genetical system, as they must have immunological and nervous systems, which can cope efficiently with what has not yet been experienced—with what, if they were sentient, we should call the unforeseen. Only inborn diversity, and a genetical system which keeps that diversity permanently in being, can make this possible.

Peter B. Medawar, in *The Uniqueness of the Individual*, 1956

19 *Mechanisms of Immunity*

Introduction

The animal body invaded by foreign materials or organisms defends itself in a variety of ways. Perhaps the most remarkable of these is the development of immunity, in which the body learns from experience of past insults to deal effectively and specifically with present and future ones. The immune response, as we shall see, is one of the body's most important mechanisms for maintaining homeostasis.

Immunology is of obvious medical significance, and until recently most of the scientific attention given to it had some medical aspect. However, it has ranged far beyond the problems of infection and disease. Indeed, the concepts of modern immunology are very near the center of modern biological thought.

HISTORICAL NOTES

Although it was recognized in ancient times that individuals who recover from such diseases as chicken pox, measles, and mumps are immune to future attacks—that is, they are not reinfected by exposure to microorganisms that cause disease in the susceptible—the notion of deliberately induced immunity did not arise until Edward Jenner's discovery of smallpox vaccine in the late eighteenth century. Despite the confusions and controversies surrounding his work, it soon led to the virtual elimination of smallpox from Europe.

By the mid-nineteenth century the principle

was widely accepted that an induced mild form of an infection might protect against a more virulent natural form. With the emergence of the new science of medical bacteriology from the researches of Pasteur, Koch, and their contemporaries, this principle was soon applied to many infectious diseases. Pasteur's early studies of chicken cholera, a bacterial disease of fowls, is a classical example. It was known that the cholera organism grows readily in a simple broth culture and that injection of small doses of the pure culture is fatal to unprotected chickens. Pasteur found that chickens inoculated with cultures that had been left untouched for several months displayed only mild symptoms or none at all. Moreover, when they recovered and were later inoculated with virulent cultures, they survived. Pasteur soon established the regularity of these results and spent much time in working out similar methods for the "attenuation" of the organisms responsible for anthrax, rabies, and other diseases.

During the 1880's two seemingly opposing ideas were put forth. One group of workers, having determined that blood serum frequently kills certain bacteria, concentrated on the bactericidal power of serum. They learned that it is enhanced after experimental infection with the bacteria. Sometimes serum actively destroys the organisms, and sometimes it agglutinates them without actually destroying them. These properties were ascribed to a class of substances, at first wholly hypothetical, called *antibodies*. Further experimentation demonstrated the astonishing specificity of circulating antibodies. For example, those appearing after pneumococcal infection attack only pneumococci. These striking results led many invesigators to consider antibodies the basis of immunity to infection. Because antibodies are soluble agents in the blood—as are all body "humors" according to the ancients—this came to be known as the *humoral* theory of immunity.

A group of workers led by Metchnikoff was impressed with the power of polymorphonuclear leukocytes, monocytes, and other body cells to engulf and destroy bacteria. These workers became early proponents of a *cellular* theory of immunity. We have learned that the numerous phagocytic cells in the spleen, liver, lymph nodes, bone marrow, blood, and other tissues—the cells known collectively as the reticuloendothelial system, or RES (see p. 223) —are capable of seizing and devouring bacteria and other foreign matter. The early students of phagocytosis concluded that these cells are the basis of immunity to infection. As is often the case, proponents of both the humoral and cellular theories of immunity were right. It is now known that body defenses against foreign materials involve at least two types of mechanisms: (1) attacks by soluble antibodies that circulate in the serum; and (2) attacks by certain populations of cells in reactions that are immunological in character, but that do not require the presence of antibodies.

Another impetus to immunological research was the discovery in the late nineteenth century that some bacteria, notably those of diphtheria and tetanus, cause symptoms and death by liberating soluble and diffusible *toxins*. In 1890 it was shown that injection of small nonlethal amounts of a toxin into the body elicits a neutralizing antibody, or *antitoxin*. Evidently, then, antibody production is elicited by bacterial products, as well as by whole bacteria. The studies of diphtheria toxin and antitoxin by Ehrlich, a chemist particularly interested in the quantitative aspects of the toxin–antitoxin reaction, provided the first unified theories on immunity. Because he could titrate one against the other, Ehrlich concluded that antitoxin combines chemically with toxin and thereby neutralizes it. We shall refer later to this theory.

The results of countless experiments since 1890 have made it clear that many substances besides bacteria and bacterial toxins elicit antibodies. Examples are serum or red blood cells from another animal species and purified proteins such as egg albumin. These substances are called *antigens* because they stimulate the body to generate antibodies.

Despite the preoccupation of investigators with the means by which specific immune responses confer protection against infectious disease, it was early observed that immune responses can also produce unpleasant and sometimes dangerous hypersensitivity to the provoking antigen. Protection against infection, it was realized, is not a necessary consequence of the immune response since bacterial infection may induce the formation of many antibodies with no apparent protective value. The explanation for the seeming paradox that was advanced at the beginning of this century by C. F. von Pirquet marked the beginning of modern immunology, for it was put forward in biological rather than clinical terms. In both cases, von Pirquet pointed out, an individual exposed to an antigen develops a changed pattern of reactivity—in one case immunity, in the other hypersensitivity. The common feature is a process of recognition and specific response to the foreignness of a wide range of substances.

Types of immunity. As noted in the foregoing discussion, immunity is based upon both cellular and humoral mechanisms. The phagocytic behavior of the RES is a type of cellular immunity that comprises the most primitive means for dealing with foreign matter. It removes a wide variety of materials (e.g., products of tissue breakdown, bacteria, and other foreign agents). Although RES activity may be enhanced by soluble agents (opsonins) in the serum (see p. 255), it lacks the high specificity of the humoral mechanism.

A much more specific type of cellular immunity is that seen in the phenomenon called *delayed hypersensitivity,* which is illustrated by the *tuberculin reaction.* A small quantity of tuberculin (a mixture of substances extracted from tubercle bacilli) injected into the skin of a man or guinea pig previously infected by these bacilli produces within 24 hours a local reaction consisting of redness and swelling (Fig. 19.1A). Evidence of the cellular nature of the phenomenon is the fact that tuberculin sensitivity can be transferred to a nontubercular animal by inoculation of cells from the spleen or lymph nodes of a sensitive animal. Serum is ineffective in transferring tuberculin sensitivity. A small amount of antibody to tuberculin may be found in the serum; however, it is not essential for the reaction. We conclude therefore that prior exposure to tuberculin bacilli somehow modifies body cells so that in the future they will be hypersensitive to the foreign material.*

Hypersensitivity is delayed in the tuberculin reaction but is immediate in *anaphylaxis,* a form of hypersensitivity associated with antibody production. Anaphylaxis was discovered in 1902 by Portier and Richet. In attempting to immunize dogs against the toxin of "Portuguese man-of-war" tentacles, they observed that a small, normally harmless dose produced immediate collapse and death if given several weeks after an earlier dose. Instead of protection (i.e., prophylaxis), the second dose produced "antiprotection" (i.e., anaphylaxis). Although anaphylaxis in man is not always fatal, it is always dangerous. In contrast to delayed hypersensitivity, immediate hypersensitivity can be transferred to another animal by transferring serum (Fig. 19.1B). The essential point is that, in hypersensitivity, the first contact with the foreign substance is apparently harmless, whereas later exposure produces harmful results. In immediate hypersensitivity, the difficulties are consequences of an antigen–antibody reaction. In delayed hypersensitivity, harmful results occur because of the changed immunological status of the lymphoid cells, a change unaccompanied by demonstrable antibody production.

The humoral type of immunity, which does involve antibodies and antigens, represents a more advanced evolutionary adaptation to a hostile environment. It is often termed "classic" or "orthodox" immunity, because antibodies were among the earliest discoveries in immunology. Quite possibly, more work has been done on the antibodies produced after injection of horses or rabbits with diphtheria toxin than on any other system.

It was formerly taught that humoral immunity is either *acquired* or *natural*—acquired immunity resulting directly from exposure to an antigen, and natural immunity occurring without exposure to an antigen.† The existence of natural immunity has always been difficult to establish, since the possibility of exposure to antigen can seldom be ruled out conclusively. It increasingly appears that all antibodies arise in response to antigens but that some antigens are less easily detected than others. It has been shown, for example, that animals born and kept in a germ-free—and therefore, antigen-free—environ-

---

* It was von Pirquet who established the usefulness of the tuberculin reaction in the early diagnosis of tuberculosis.

† We have earlier encountered the so-called natural blood group agglutinins (see p. 275).

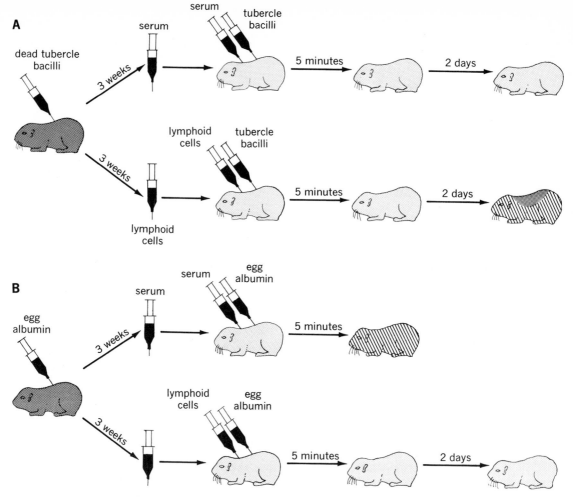

19.1 Delayed and immediate hypersensitivity, with transfer experiments showing cellular and humoral immunity. A. Delayed hypersensitivity to tubercle bacilli is transferrable by lymphoid cell but not by serum. Cells from sensitized guinea pig are transferred to non-sensitive animal which develops local delayed lesion when challenged by tubercle bacilli. B. Immediate hypersensitivity to egg albumin is transferrable with serum but not by lymphoid cells. Serum from a sensitized guinea pig is transferred to non-sensitive animal which develops severe anaphylaxis when given egg albumin.

ment have very low levels of "natural" antibodies.

Acquired humoral immunity may be of two types: *active* and *passive*. In active immunity antibodies are produced in the body's own tissues following exposure to an antigen. In contrast, passive immunity is produced in a body by serum transfer from an immune body. Passive immunity may arise naturally, for example, by transmission to the fetus of maternal antibodies (through the placenta or in the colostrum). Or it may be conferred artificially, for example, by

injection of antibody-containing serum from another human or animal body. The tissues of the passively immune individual takes no part in antibody production. This type of passive immunity is only temporary and fades away in a few weeks. Artificial passive immunity is medically useful because it is immediate, whereas active immunity must built up. The use of diphtheria antitoxin is a familiar example of the application of passive immunity, but passive immunity can be produced by antibacterial as well as antitoxin serum.

Antibodies and Antigens

INTRODUCTION

Antigens, as we have noted, are the substances that stimulate antibody formation. Antibodies are those plasma globulins known collectively as immunoglobulins (see p. 261). An antibody may belong to any one of the five main classes of immunoglobulins—IgG, IgA, IgM, IgD, and IgE (see Table 8.7). As we shall see, each of these classes includes certain distinctive antibody types. Serum containing an antibody to a given antigen is called *immune serum,* or *antiserum.* A great deal of antibody activity resides in a small quantity of immunoglobulin. Therefore, an increase in the amount of a single antibody in an immune serum need not be accompanied by a measurable increase in its total immunoglobulin content. As we shall see, such a serum may sometimes be diluted 10^7 times and still retain detectable antibody activity.

Antigens are most commonly foreign proteins, though DNA and other substances may be antigenic (i.e., when introduced into the body, they excite the formation of neutralizing antibodies). Most proteins and many polysaccharides and lipopolysaccharides are antigenic. It should be noted that antigens may be part of a virus, a bacterium, or a foreign tissue cell, or a soluble derivative of some such structure. Like antibodies, antigens are large molecules.

METHODS OF DEMONSTRATING ANTIBODIES

Introduction. In all scientific inquiries, our concepts and conclusions rest ultimately upon the operations we perform. Knowledge of genes, for example, depends on the results of breeding operations and of enzymes on the results of catalytic operations. This principle is of particular relevance in immunology. Although it had been known or suspected on the basis of clinical observations that one attack of a given disease confers immunity to later attacks, it was not possible to state that antibodies exist until a method could be devised for demonstrating them directly. As techniques for demonstrating them became refined, our concepts of the nature of immune mechanisms became correspondingly more precise.

An antibody is recognizable as such only through the visible consequences of an antibody–antigen reaction. In the absence of such evidence, we cannot infer that an antibody is present. It might be supposed that antibodies could be demonstrated only by their power to destroy or inactivate infectious agents. This is true of many antibodies. For example, if a rabbit is several times inoculated with killed dysentery bacilli, its serum in a few weeks causes suspensions of the organisms to agglutinate visibly.

Antibodies which agglutinate cells (e.g., bacteria, red cells) are *agglutinins* (see Fig. 8.36). Because their capacity to agglutinate antigen-containing cells is so easily demonstrated, agglutinins were among the earliest antibodies detected. By ingenuity, immunologists soon devised means of making visible the effects of the many antibodies that are not agglutinins and that therefore attack their antigens in other ways. The antibodies called *lysins* cause the organisms against which they act to dissolve. The antibodies called *precipitins* act against soluble antigens. When mixed with their antigen in appropriate concentrations, the antigen–antibody complex forms a visible precipitate (Fig. 19.2). Each of these phenomena has proved useful in the detection of antibodies.

It was once believed that each of the above phenomena represented a fundamentally different type of antibody and the distinguishing terms agglutinin, lysin, and precipitin were widely used. It is now apparent that such names add little to our understanding. The points to be emphasized are these: (1) antibody combines with antigen; (2) the mode of combination depends upon the nature of the antigen; (3) with the exceptions to be noted below, antigen–antibody reactions produce visible results which can be exploited as a means of antibody detection; and (4) specificity is a cardinal attribute of antibody–antigen reactions.*

---

* In practice, the demonstration of antibody specificity depends to a large extent upon the purity of the original antigen. The antibody-forming mechanism has such high sensitivity that it is extremely difficult to prepare antigenic materials pure enough to elicit the production of a single antibody that is directed solely against a single antigen. Moreover, as we shall learn later, there can be more than one antigenic determinant (i.e., antibody-eliciting chemical grouping) on an antigen molecule. This means that different portions of a single antigen molecule may each elicit a different antibody specific for that portion of the antigen molecule.

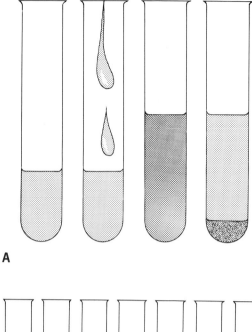

A

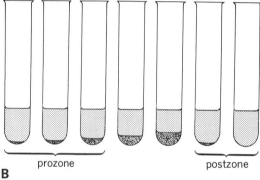

prozone   postzone

B

19.2 The precipitin type of antigen-antibody reaction.
A. Demonstration of a reaction. The tube at left contains
rabbit antibody to bovine serum albumin (BSA). When BSA
is added (second from left), a cloudy suspension is produced (third from left), which settles out in 10 minutes
(right). B. Effect of varying proportions of antigen and
antibody is illustrated. All these test tubes contain the
same amount of antibody to BSA. From left to right,
however, the amount of BSA itself is increased. No precipitate forms when there is too much BSA.

Classic precipitation techniques. Because of
its simplicity and because it is the only method
that measures antibody in absolute units of mass,
the precipitin reaction has been extensively
studied. Let us consider some of its features.

Figure 19.2A illustrates the precipitate
formed when a rabbit antibody to purified
bovine serum albumin (BSA) is mixed with BSA.
It is important to recognize that the mere combination of antibody with antigen does not ensure the formation of a precipitate. An insoluble
precipitate forms only when the two reactants
are mixed in the proper proportions. As shown
in Figure 19.2B, the size of the precipitate increases with an increasing ratio of antigen to
antibody, but an excess of BSA inhibits precipitate formation. When a great excess of antigen is
present, a precipitate does *not* form. We shall
later recall this significant fact.

Determination of the antigen to antibody
ratio that yields the maximum amount of precipitate has provided important clues about how
antigens and antibodies react together. Unlike
the usual situation in which chemical reactants
always combine in definite proportions; the
proportion of antibody to antigen in a precipitate varies with the amount of the two reactants
present. It has been shown experimentally that
antibody is bivalent and antigen multivalent—
that is, each antibody has two combining sites
and each antigen has many. Some antigens (e.g.,
tobacco mosaic virus) have 900 combining sites.
Direct evidence for these conclusions was obtained by electron microscopy of antigen–
antibody aggregates.*

On the basis of these facts, Michael Heidelberger suggested a plausible explanation for the
precipitation of antibody–antigen combinations.
When antigen and antibody are present in correct proportions, it was postulated, antibodies
attach to more than one antigen at a time and
thus build three-dimensional lattices which become too large to remain in solution (Fig.
19.3A). When antigen is present in excess—i.e.,
in the *prozone* (Fig. 19.2B)—small aggregates
are formed because antibodies are utilized in
binding antigens rather than in forming large
lattices (Fig. 19.3B). These small aggregates
remain in solution. When antibody is present in
excess—i.e., in the *postzone* (Fig. 19.2B)—

---

* Convincing evidence of the bivalency of antibody is also
found in the subunit structure of the immunoglobins (see
p. 262). The two antigen-binding sites of the immunoglobulin
molecule are diagrammed in Fig. 8.28B. The properties of
these sites will be discussed later.

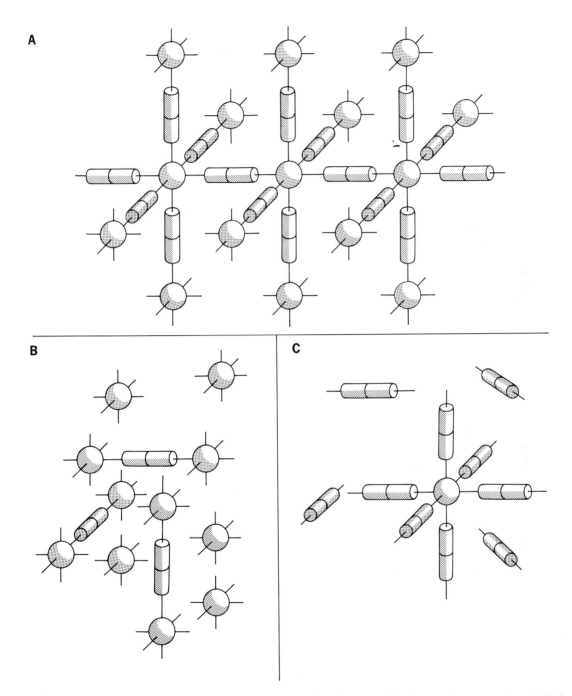

19.3 Lattice hypothesis of antibody-antigen precipitate formation. Antigen molecules are shown as spheres with 6 binding sites; antibodies are cylinders with 2 binding sites. A, when they are present in a ratio of one antigen to three antibodies, an insoluble lattice forms, causing a visible precipitate; B, antigen present in excess (prozone); C, antibody present in excess (postzone). In B and C, aggregates are small and soluble.

another type of water-soluble aggregate develops and lattice formation is again blocked (Fig. 19.3C).*

Classic precipitation techniques have been useful in two areas: (1) in the quantitative assay of antibody; and (2) in the qualitative identification of antigen. The strength, or *titer*, of antibody in a solution has traditionally been assayed by determining the degree to which the solution can be diluted before it just fails to cause visible precipitation. If the dilution is 1:10,000, for example, it is conventional to say that the antibody titer is 10,000. A solution with an antibody titer of 10,000 thus contains more antibody than one with a titer of 100. Serial dilution methods are incapable of great accuracy because the observer is required to choose which dilution is the final active one. Since each dilution is double the preceding one, duplicate determinations often disagree by 50 to 100%. Moreover, dilution methods permit relative comparisons of different antisera active with the same antigen, but do not permit absolute or quantitative measurements of antibody content.

By means of an ingenious application of the precipitin reaction, Heidelberger devised the first quantitative assay for antibody. In this procedure, one measures the nitrogen content of washed antigen–antibody precipitates and calculates results from a graph showing the total nitrogen content of precipitates containing known quantities of antigen and antibody.

The usefulness of the precipitin reaction in the qualitative identification of an unknown antigen may be illustrated by an example. Suppose we wished to determine which of four different antigens—A, B, C, and D—is present in an unknown solution. We would first immunize individual rabbits with pure samples of each antigen and thereby obtain antisera con-

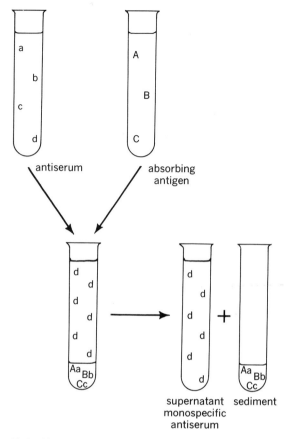

19.4 Absorption technique for preparation of specific antiserum containing only one antibody. Explanation is given in the text.

taining specific antibodies against A, B, C, and D—designated a, b, c, and d, respectively. If the rabbit antiserum containing antibody d precipitated the unknown antigen, we would conclude it is D.

In practice, the antiserum containing antibody d, for example, usually reacts weakly with A, B, and C. Therefore, the rabbit immunized with antigen D also formed small amounts of antibodies a, b, and c. This could be the result of contamination of antigen D with A, B, and C (see footnote on p. 755). It could also occur because a single antigen molecule contains multiple antigenic determinants. If in this case, A, B, C, and D had one or more antigenic determinants in common, a, b, c, and d would *cross-react* to some extent. To obtain an antibody of

---

* S. J. Singer has produced direct evidence in support of this formulation. From ultracentrifugal analyses of the various aggregates in antigen–antibody mixtures, he found that in solutions containing a large excess of antigen (i.e., with each antibody bound to the maximum number of antigen molecules with which it could combine), the dominant peak in the ultracentrifuge pattern corresponded to an aggregate of two antigens with one antibody. In solutions containing progressively smaller excesses of antigen, the aggregates became progressively larger. Thus, it was possible to "see" the growth of antigen–antibody lattices on their way to precipitation.

absolute specificity—i.e., to prepare an antibody such as d so that it is free of a, b, and c—one employs the absorption technique (Fig. 19.4). Antiserum containing antibody d and some a, b, and c is mixed with antigens A, B, and C. The antigens react with a, b, and c, forming the insoluble antigen–antibody complexes Aa, Bb, and Cc, which are eliminated by centrifugation. The remaining supernatant, containing only d, can now be used to identify antigen D in unknown samples.

Antibodies other than precipitins are useful in the identification of antigens and they too, may be purified by absorption methods. However, the precipitin reaction has proved the most versatile. The following sections describe several new techniques based upon the precipitin reaction which have become the indispensable tools of modern immunology.

Diffusion techniques. Diffusion techniques are based on the discovery by J. Oudin in 1946 that when a solution of antigen is layered over a semisolid gel (e.g., agar) containing antibody, antigen diffuses into the gel and a sharp band of precipitate forms at the point where they interact, the thickness of the band being related to the concentrations and diffusion coefficients* of antigen and antibody.

This *single-diffusion* technique was soon superseded by the *double-diffusion* technique developed by O. Ouchterlony. In this procedure, a dish containing solid agar is prepared; two wells are cut with a cork borer (Fig. 19.5A). If antigen is placed in one well and antibody in the other, a sharp milky precipitate line soon forms between the two wells. This is the antigen–antibody aggregate; it appears where antigen and antibody diffusing through the agar encounter each other and react.

The rate at which antigen and antibody diffuse through gels is proportional to their concentrations and diffusion coefficients. Two precipitation lines form when the wells contain two antigens and antibodies. The two lines are at different locations because the reactants producing them have different diffusion coefficients. When two precipitation lines have formed, the techniques can be used to identify the antigens or antibodies. A third well in the agar is filled with a solution containing one of the two antigens or one of the two antibodies (Fig. 19.5B). Precipitation lines formed by reactions between the same antigen–antibody pair join one another to form a continuous chevron-shaped line. Lines that intersect or that do not touch indicate nonidentical reactions. By this method, it is easily established whether two antigens are identical, different, or partially related.

Many modifications of these methods have been devised. Because the rate of band migration depends upon antigen and antibody concentrations, the technique can sometimes be employed for the quantitative assay of antigens and antibodies. It has also proved useful in establishing the complexity of antigens and in following the purification of certain substances. The most useful application is the technique of immunoelectrophoresis. In this procedure, antigens are first separated by electrophoresis (see p. 61) in a semisolid gel (Fig. 19.6). Antiserum containing antibodies to one or more of the antigens is then placed in a well running the length of the electrophoretogram. Antibodies and antigens diffusing through the gel encounter each other and each antigenic component reacts with its specific antibody to form a visible arc of precipitation. Fig. 19.6D illustrates the complex pattern obtained when human serum is separated by electrophoresis and then reacted with the serum of a horse that had been immunized with human serum. Immunoelectrophoresis clearly combines the resolving power of electrophoresis (see Fig. 8.26) with the specificity of the antigen–antibody precipitation reaction. The result is an effective method for the identification of antigens and antibodies.

Fluorescent antibody techniques. A. H. Coons developed a useful technique for the detection of antigenic materials in tissues, by using antibodies that have been combined with a fluorescent dye such as fluorescein isocyanate.† Fluorescent antibody is employed as a specific histochemical stain (Fig. 19.7). Sections of tissues treated with fluorescent antibody are then examined under a microscope which permits the use of ultraviolet light to excite fluorescence.

Specific antigens in the tissue react with and locally bind fluorescent antibody and when illuminated by ultraviolet light display a bright fluorescence whose color depends on which dye was conjugated to the testing antibody. Different antigens can be identified and

---

* The diffusion coefficient of a molecule or particle diffusing through a medium of given viscosity at a given temperature is a function of the molecule's mass and shape. Larger molecules diffuse more slowly, the diffusion constant being inversely proportional to the cube root of the mass.

† The preparation of a fluorescein–antibody complex involves four steps (1) production of an antiserum of high antibody content; (2) conjugation of the antibody with fluorescein isocyanate; (3) purification of the conjugated antibody; and (4) absorption of preparation with ground tissue to rid it of nonspecific antibody material.

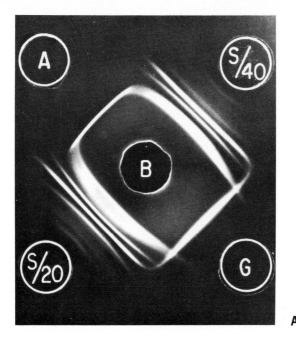

19.5 Principles of gel diffusion. A. Technique for identifying specific reaction lines in a complex pattern. S/20 and S/40 are wells in the agar plate containing dilutions of human serum. B is a well containing rabbit serum containing antibodies to human serum. A is a well containing serum albumin and G is gamma globulin. Note that milky precipitation lines form, one for each antigen-antibody present. Although multiple lines form when whole human serum reacts with antiserum, the locations of albumin and gamma globulin in the mixture are readily identified by their continuity with the lines formed by the pure proteins. B. Cross reactions of serum albumins from a human (HU/50), chimpanzee (CH/50), an ateles monkey (AT/60), and a galago (GA/80), and bovine serum albumin (BSA) with rabbit antiserum to human serum albumin (anti-HSA). Continuous chevron-shaped precipitation lines indicate identity (or great similarity). (Courtesy of Dr. C. A. Williams, The Rockefeller University.)

A

primate serum dilutions

| HU/50 | CH/50 | AT/60 | GA/80 | BSA 0.3 mg/ml |

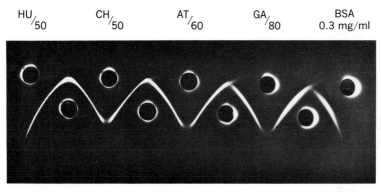

B

rabbit anti HSA

localized within the same cells by employing a different dye for each fluorescent antibody.

The treatment of tissues with fluorescent antibody specific for the antigen under study is called the *direct staining* technique (Fig. 19.7A). The *indirect staining*, or "sandwich," technique (Fig. 19.7B) is based on the fact that a testing antibody can itself serve as an antigen against which a specific antibody (i.e., an anti-γ-globulin) can be prepared. In this technique, tissue is first exposed to an excess of antibody. Then anti-γ-globulin conjugated with fluorescent dye is used to detect the γ-globulin antibody earlier attached to the antigen under study. Because antigen molecules have many sites capable of binding antibody, the indirect method

intensifies staining by causing more fluorescent anti-γ-globulin molecules to react per tissue antigen molecule. Thus the method is more sensitive.

The fluorescent antibody technique conveniently demonstrates the presence of various antigens in cells and their intracellular locations (Fig. 19.7C). We shall later learn that this technique has contributed important information on the location of antibody synthesis.

Complement fixation techniques. Some antibodies combine with their antigens without producing a visible reaction. Many of these can be detected nevertheless because of an interesting

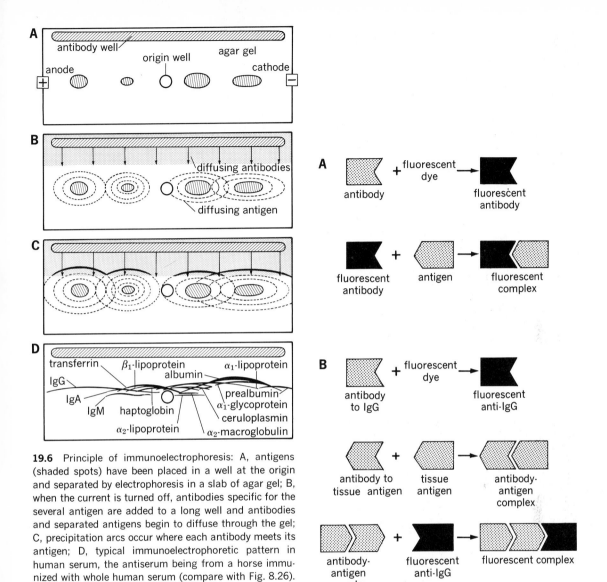

19.6 Principle of immunoelectrophoresis: A, antigens (shaded spots) have been placed in a well at the origin and separated by electrophoresis in a slab of agar gel; B, when the current is turned off, antibodies specific for the several antigen are added to a long well and antibodies and separated antigens begin to diffuse through the gel; C, precipitation arcs occur where each antibody meets its antigen; D, typical immunoelectrophoretic pattern in human serum, the antiserum being from a horse immunized with whole human serum (compare with Fig. 8.26). (From F. W. Putnam, in H. Neurath, ed., *The Proteins*, Vol. III. 2nd ed., Academic Press, New York, 1965).

19.7 Fluorescent antibody technique: A, direct staining method; B, indirect staining method; C, microscopic appearance of plasma cells containing immunoglobulin after indirect staining and exposure to ultraviolet light. Example shows active secondary response to diphtheria toxoid. The cells are from a section of a lymph node of an immunized rabbit on the fourth day after a second dose. The immature plasma cells are revealed by their content of diphtheria antitoxin. (Courtesy of Dr. Albert W. Coons.)

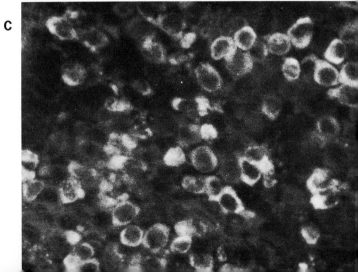

phenomenon discovered in 1901. Serum contains a group of proteins known collectively as *complement,* or the *complement system,* which combines avidly with many antibody–antigen complexes. Thus, the occurrence of an antibody–antigen reaction may be demonstrated even though unaccompanied by visible consequences, if it can be shown that *complement fixation*—i.e., elimination of complement by binding or inactivation—has occurred.

The technique used to demonstrate the fixation of complement by an antibody–antigen reaction is based on the fact that the actions of certain cytolytic (i.e., cell-destroying) antibodies require the presence of the complement system. Complement fixation is itself without visible consequences; it is detected by the addition of a cytolytic antibody–antigen system capable of producing a visible effect only in the presence of complement. A mixture of sheep red cells and an anti-sheep red cell lysin constitutes such a system. If lysin and sheep red cells are added to a mixture of an antigen A, an antibody a, and complement (after enough time has elapsed to allow the reaction to go to completion) and lysis of the red cells occurs, it may be concluded that complement was available; therefore, no reaction took place between A and a. If red cell lysis does not occur, it is concluded that complement was fixed to an antigen–antibody complex and unavailable and therefore that A and a reacted. The complement fixation technique is used in many antibody assays, including the Wasserman test for syphilis.

The nature of the complement system, its function in the body, and its mechanism of action have only recently been elucidated. It is now known to involve at least nine separate globulins (some of which are enzymes), several inhibitors, and two divalent cations—Mg^{++} and Ca^{++}. The components of the system—which are known as C1, C2, C3, through C9*—

interact sequentially to produce immunological lysis of cells, destruction of bacteria, and promotion of certain aspects of inflammation (see p. 748), including an increase in vascular permeability, production of a chemotactic substance (see p. 255), and enhancement of phagocytosis (see pp. 161 and 255).

The sequence in which the components of complement act in red cell lysis is summarized in Fig. 19.8. Certain antigenic sites on the red cell, symbolized by E (for erythrocyte), unite with antibody, denoted by A. A red cell that has bound antibody (EA) is called a sensitized cell; it can "fix," or bind complement. The individual sensitized sites on EA react with the components of the complement system in a precise sequence. When an EA site in complement-containing serum containing Ca^{++} reacts with the "first component of complement," it is converted from an inactive precursor state (C1 or C1q,r,s) to an active enzyme (EAC$\overline{1}$).† Next, the "fourth component of complement" (C4), which unfortunately was numbered before its position in the reaction sequence was determined, interacts with cell-bound C$\overline{1}$ (EAC$\overline{1}$). C4 is activated by cleavage into C4a and C4. Some of the latter is bound to form EAC$\overline{1,4b}$. The rest becomes inactive. The "second component of complement" (C2) is also activated by cleavage and the larger fragment C2a attaches to C4b in the presence of Mg^{++} to form EAC$\overline{1,4b,2a}$, sometimes designated EAC$\overline{1,4,2}$ for simplicity. In the course of the reaction an inactive fragment, (C2a)i is split off. The intermediate EAC$\overline{1,4b,2a}$ has two pathways open to it. It may "decay" back to EAC$\overline{1,4b}$ with release of a hemolytically inactive fragment, C2a^d. EAC$\overline{1,4b}$ can then bind a new molecule of C2 and reenter the cycle. Or, it may interact with the "third component of complement" (C3) to form EAC$\overline{1,4b,2a,3b}$. This intermediate, formed by the interaction of the first four components of the complement system, leads to the elaboration of C3a, also called *anaphylotoxin,* which causes the increased vascular permeability of inflammation.

Subsequent interaction with the next three

---

* In the nomenclature of the complement system, when a component such as C1 becomes enzymatically active it is designated C$\overline{1}$. When components such as C1, C4, C2, and C3 react together, the resulting complex is written, C1,4,2,3. Small letters indicate subcomponents of components, Thus, C1 has three subcomponents, C1q, C1r, and C1s. When a component such as C3 is cleaved proteolytically into two fragments, these are designated C3a and C3b. Inactivated components are designated with the letter i.

---

† The activity of C1, cell-bound or free, against certain synthetic ester substrates (e.g., *p*-toluene sulfonyl L-arginine methyl ester, or TAMe) indicates that C1 is an esterase. Normal serum contains a specific α-globulin inhibitor of C1.

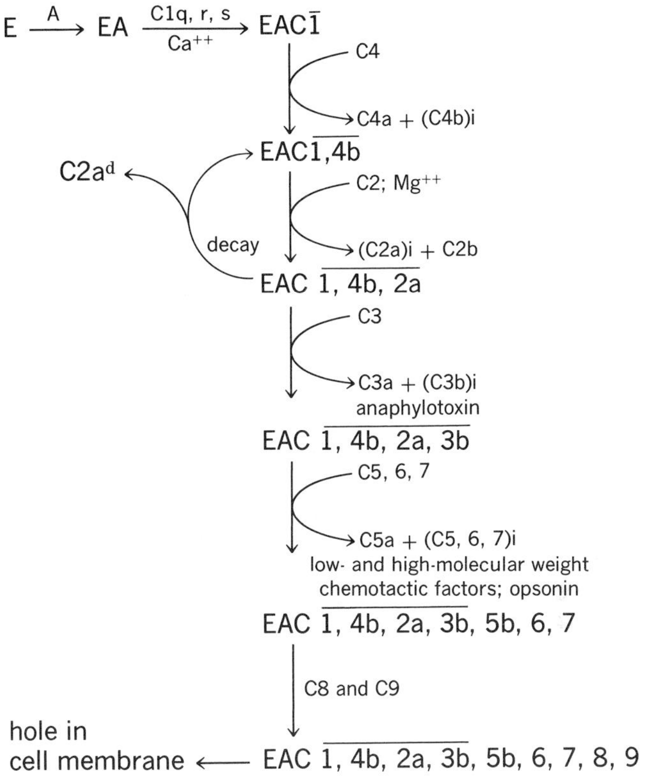

$$E \xrightarrow{A} EA \xrightarrow[\text{Ca}^{++}]{\text{C1q, r, s}} EAC\overline{1}$$

EAC$\overline{1}$ — C4

→ C4a + (C4b)i

→ EAC$\overline{1,4b}$

C2a^d ← ⟵ decay

EAC$\overline{1,4b}$ — C2; Mg^{++}

→ (C2a)i + C2b

EAC $\overline{1, 4b, 2a}$ — C3

→ C3a + (C3b)i
anaphylotoxin

EAC $\overline{1, 4b, 2a, 3b}$ — C5, 6, 7

→ C5a + (C5, 6, 7)i
low- and high-molecular weight
chemotactic factors; opsonin

EAC $\overline{1, 4b, 2a, 3b, 5b, 6, 7}$

C8 and C9

hole in
cell membrane ⟵ EAC $\overline{1, 4b, 2a, 3b, 5b, 6, 7, 8, 9}$

19.8 Current status of the complement sequence, showing the mechanism of membrane damage by antibody and complement. E represents an antigenic site on the membrane surface to which antibody A is directed.

components produces EAC$\overline{1,4b,2a,3b,5b,6,7}$, which leads to the elaboration of *low–molecular weight* and *high–molecular weight chemotactic factors* and *opsonin* that, respectively, attract leukocytes to foreign particles and enhance phagocytosis. Thus, the several components of inflammation occur before the complement reaction has proceeded to completion. When the last two reaction components, C8 and C9, finally interact, the sequence required for a cytolytic reaction is completed, and defects (i.e., holes) appear in the cell membrane (Fig. 19.9).*

Thus, cell lysis, and, probably, bacterial destruction require all nine globulins.

19.9 Membrane defect produced by specific antibody and complement on portion of surface of red blood cells. Defects, visualized by negative staining, appear as dark discs, about 80 Å diameter, surrounded by a clear ring. (× 100,000) (From R. R. Dourmashkin.)

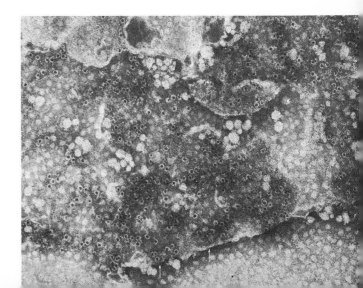

---

* The complement system and the blood clotting system (Fig. 8.32) have many striking similarities. Each is a complex series of serum proteins whose activities are latent. Each requires Ca^{++}. Each proceeds through an elaborate cascade or waterfall sequence in which a component present in latent form is activated, the activated form acting in turn upon the next component to effect its activation. In the clotting system, the ultimate substrate is fibrinogen, and the entire sequence operates to produce thrombin, which converts fibrinogen to fibrin. In the complement system, the ultimate substrate is the cell membrane, and the entire sequence operates to produce a substance or complex that disrupts the membrane.

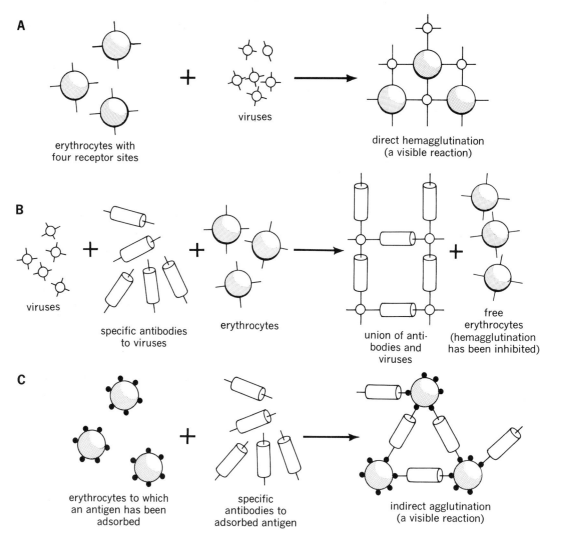

A

erythrocytes with
four receptor sites

viruses

direct hemagglutination
(a visible reaction)

B

viruses

specific antibodies
to viruses

erythrocytes

union of anti-
bodies and
viruses

free
erythrocytes
(hemagglutination
has been inhibited)

C

erythrocytes to which
an antigen has been
adsorbed

specific
antibodies to
adsorbed antigen

indirect agglutination
(a visible reaction)

19.10 Hemagglutination techniques: A, direct hemagglutination of erythrocyte by viruses; B, demonstration of an antibody to virus by its ability to inhibit the direct hemagglutination reaction; C, detection of an antibody by indirect hemagglutination.

Hemagglutination techniques. Despite the extreme sensitivity of classic and modern immunological detection methods, certain antibodies still evade them. A variety of hemagglutination techniques have been developed to deal with these antibodies.

These procedures are based on the fact that certain bacteria and viruses attach themselves to receptors on the surface of erythrocytes. In direct hemagglutination procedures, the attached organisms cause the cells to agglutinate visibly (Fig. 19.10A). The method can be used to detect any antibody which inactivates the viruses or bacteria and thus inhibits hemagglutination

(Fig. 19.10B). In indirect hemagglutination procedures, the erythrocyte receptors nonspecifically bind a soluble antigen without agglutinating. Only when the coated cells are exposed to an antibody to the antigen do they agglutinate (Fig. 19.10C). Here the erythrocyte is merely a carrier which in agglutinating serves as an indicator system. Interestingly, only a few substances— most of them polysaccharides—coat erythrocytes. Proteins do not ordinarily attach to the erythrocyte surface unless this surface is modified. In one of the most sensitive in vitro techniques of immunology, erythrocytes are treated with tannic acid, which alters the surface

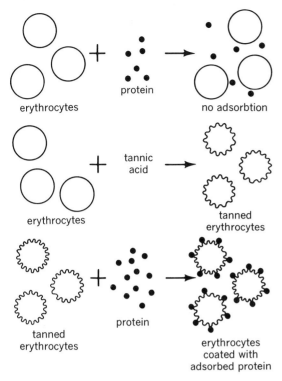

19.11 Tannic acid technique for adsorbing proteins onto erythrocytes. (After S. V. Boyden.)

causing it to adsorb proteins (Fig. 19.11). "Tanned cells" bearing adsorbed proteins are readily agglutinated by specific antibodies.*

**PHYSIOLOGY
OF ANTIBODY PRODUCTION**

Primary and secondary responses. Antibody does not appear in the serum immediately following administration of an antigen, nor does the animal become immune immediately. Instead, there is a latent period, during which the antibody-synthesizing apparatus is preparing itself to elaborate the antibodies that appear later. Although the time required for the appear-

ance of antibodies may vary with the amount of antigen administered, there is ordinarily an abrupt appearance of antibody in the blood within a week after a single initial injection of an antigen (Fig. 19.12). This is the *primary response.* If no further injections of antigen are given, antibody concentration begins to fall off—rapidly at first, more gradually later—although weeks or years may elapse before the specific antibody completely disappears from the serum.

If a second injection of antigen is given before the antibody level begins to drop, some of the antibody present at the moment of injection may be neutralized by the injected antigen; however, new antibody is rapidly produced and the antibody titer reaches a higher level than before. This phenomenon, the *secondary response,* may be elicited repeatedly; the rapidity with which the antibody level is restored increases each time until a limit is reached. In no case is the antibody level maintained at very high levels without periodic injections of antigen.†

As shown in Figure 19.12, the primary response to an antigen injection is slow and of low

---

† This accounts for the practice of administering "booster" immunizations.

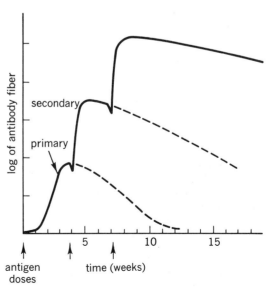

19.12 Curves showing primary, and secondary, antibody responses. (After F. M. Burnet.)

---

* Adaptations of this method have been devised in which the erythrocytes are replaced as antigen-carriers by inert particles (e.g., of latex or bentonite clay) of uniform size. Though the particles are more stable and convenient to use than erythrocytes, the procedures which employ them are less sensitive than those employing erythrocyte agglutination.

titer. Secondary responses are rapid and rise almost logarithmically to a higher titer. These results point strongly to the conclusion that first contact with an antigen sets a "tooling up" process in motion, while secondary contact evokes a rapid and massive production of antibody. Clearly, this implies that antibody-synthesizing cells possess what has been termed *immunological memory*. As we shall see, the nature of this memory is the central question of theoretical immunology.

Site of synthesis. Identifying the body cells which synthesize antibody proved to be an unusually difficult problem. Only recently have we learned the sites and circumstances in which antibody is formed.

Early workers believed that the cells of the RES are the antibody-formers, capturing antigens by phagocytosis. In part, this belief was based on the discovery that phagocytic ingestion of a massive quantity of inert particles blocks further phagocytosis and also depresses antibody formation. Then, Astrid Fagreus produced evidence suggesting that plasma cells are the primary producers of antibody. She observed that within a few days after the injection of an antigen into an experimental animal, *plasmablasts* appear in the spleen and proliferate rapidly, producing many young plasma cells.

Later workers made the following observations: (1) plasma cells are present in all animals that make antibodies; (2) the plasma cell makes its first appearance in infancy at the time that antibody synthesis begins; (3) fluorescent antibody techniques show the bulk of intracellular antibody to be in the plasma cells (see Fig. 19.7C); (4) malignant proliferations of plasma cells such as in multiple myeloma are associated with massive elevations of the serum immunoglobulin level (see Fig. 8.29); and (5) the plasma cell is admirably equipped for the synthesis of large quantities of protein, its ample cytoplasm containing large amounts of granular endoplasmic reticulum—a sign of active protein synthesis (see pp. 74 and 109). Each of these observations strongly suggested that the plasma cell is the site of antibody synthesis. Recently G. J. V. Nossal and, working independently, M. Cohn and E. S. Lennox have demonstrated antibody synthesis by a single plasma cell.

In summary, it is currently believed that antigen introduced into the body does indeed go to the macrophages of the RES. Although these cells do not synthesize antibody or give rise to antibody-forming cells, there are indications that antigen going to the RES plays a role in the induction of antibody production. For example, molecules that are rapidly sequestered into the RES are more antigenic, molecule for molecule, than those that are slowly sequestered. Treatments that enhance antigen capture by the RES (e.g., heat-aggregation, adjuvants) also enhance antibody production. The role of the macrophages is to *process* antigen, i.e., to transfer antigen to lymphoid cells in moderate quantities. A lymphoid cell not previously exposed to antigen, a so-called *virgin lymphocyte,* is killed by very low doses of antigen that contact it directly without the intervention of macrophages.

Interestingly, induction of a primary response, long thought to occur only in vivo, can now be demonstrated in vitro—but only under circumstances which leave the anatomical relationships between RES and lymphoid elements of lymph nodes or spleen intact. The lymphoid cells are antigen-responsive—i.e., they can react to antigenic stimulation in a specific way. The response consists of a series of mitotic divisions and maturation steps (Fig. 19.13). The mature cells emerging from this division sequence are the antibody-forming plasma cells. The 4 or 5 days required for the maturation of certain antigen-responsive lymphoid cells into antibody-synthesizing plasma cells accounts for the 4 or 5 day latent period between antigen administration and primary antibody appearance (see Fig. 19.12).

As we shall presently see, it is not known whether or not individual reactive cells can respond to many or all antigens. It does appear that each antigen stimulates only a few lymphoid cells. Thus, even under conditions of maximal stimulation by a single antigen, the number of responding lymphoid cells is small, perhaps 10^{-5} of all the lymphoid cells in the body. Lymphoid cells capable of responding to an antigen are not fixed, but circulate in the lymph and blood. They are mainly the medium and large lymphocytes, which in contrast with the majority of the small lymphocytes have ultrastructural features of cells capable of protein

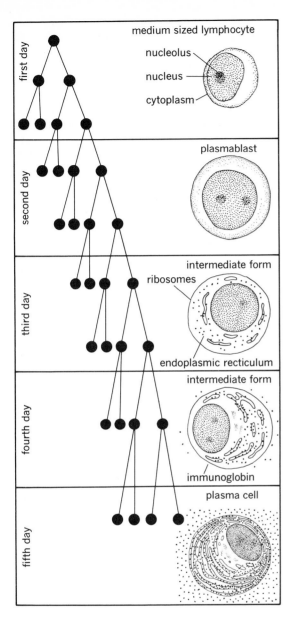

first day
medium sized lymphocyte
nucleolus
nucleus
cytoplasm

second day
plasmablast

third day
intermediate form
ribosomes
endoplasmic recticulum

fourth day
intermediate form
immunoglobin

fifth day
plasma cell

19.13 Clonal development of plasma cells. Right, stages of maturity; left, generations to which they belong. After contact with the antigen the plasmablast takes some 10 hours to divide. Not until the fifth day after contact do the mature plasma cells produce a great deal of antigen.

synthesis (Fig. 19.14). Such lymphoid cells are said to be *immunologically competent.* As we have learned, the ancestors of these cells are produced by Peyer's patches (see p. 225). The circulating small lymphocytes, whose ultrastructure suggests a relatively inactive cell (Fig. 19.14B), perhaps a storehouse of information, depend on the thymus for differentiation and development.

The complex structure of plasma cell cytoplasm contrasts strikingly with the relatively simpler structure of the lymphocyte's scantier cytoplasm (Fig. 19.13). All transitional stages between medium lymphocytes and plasma cells can be observed (Fig. 19.13A), and whether a given cell should be called a lymphocyte or an immature plasma cell is often a subject for debate. We shall later learn that this transformation is given central importance in the several theories of antibody synthesis.

The polyribosomes on the rich endoplasmic reticulum of the plasma cell are the loci of immunoglobulin synthesis, L chains arising on different polyribosomes of different sizes. There is every indication that the basic mechanism of chain synthesis and assembly is similar to that of other proteins.

Interesting evidence in support of the view that plasma cells synthesize antibodies is found in the congenital *antibody deficiency syndromes.* These disorders are associated with an hereditary inability to synthesize immunoglobulins and pronounced susceptibility to bacterial infection. One type is a sex-linked anomaly seen only in males and usually observed in early childhood. There is an absence of demonstrable IgG, IgA, IgM, and IgD. Injection of antigen does not lead to the appearance of circulating antibody. Lymphocytes are present, but significantly plasma cells are absent in the bone marrow and lymph nodes after antigenic stimulation. The defect thus appears to be in the conversion of lymphocytes to plasma cells.*

Fate of antigen. Much work has been done in an attempt to determine the fate of antigen.

---

* Another antibody-deficiency syndrome is due to failure of the thymus to develop during embryonic life. In this fatal disorder, the body contains few lymphocytes and no lymphoid tissues.

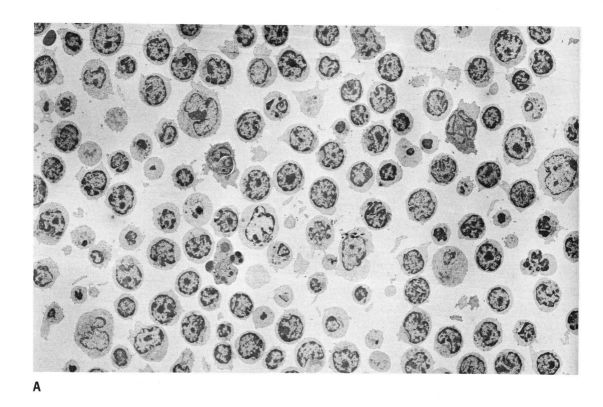

A

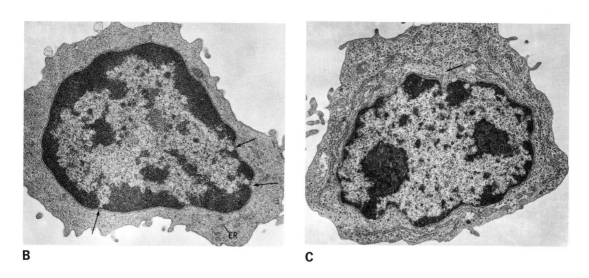

B   C

19.14 Ultrastructure of lymphocyte. A. Low power electron micrograph of rat lymphocytes. The heterogeneity of the cells is apparent. B. Higher magnification of "inactive" small lymphocyte. Ribosomes are scattered singly throughout the cytoplasm, and there is only one profile of rough endoplasmic reticulum (ER). Nucleoli are absent. C. Higher magnification of medium-sized lymphocyte from subject with rheumatoid arthritis. Ribosomes are present in cytoplasm. (From D. Zucker-Franklin, *Seminars in Hematology, 6,* 4, 1969.)

Following the injection of a radioisotope-labeled antigen, an investigator has little difficulty in determining the fate of the label, but such data do not necessarily disclose the fate of the unaltered antigen, since it is difficult to prove that the antigen has not been altered.

Labeled antigens first circulate in the plasma, then leave it for the cells of the RES in the liver, spleen, lymph nodes, and other tissues. As noted earlier, it is probable that antigen is then transferred from the cytoplasm of a reticuloendothelial macrophage to a neighboring antigen-responsive lymphoid cell. It is agreed that most of the antigenic material is rapidly broken down during the ensuing antibody response. What is uncertain, however, is whether or not a small number of antigen molecules (or active fragments thereof) persist in the body for long periods, whether they persist in the immunized body over the entire period of antibody formation, or whether antibody production can later take place in the absence of antigens. This has proved to be an exceedingly difficult experimental problem.

It is evident that small amounts of radioactivity are present in body tissues several months after injection of radioactive antigens—an amount equivalent to several hundred antigen molecules per cell—but it has not been determined whether the retained materials are active or inactive as antigens. Moreover, most of the materials persist in sites not directly involved in antibody formation (e.g., the Kupffer cells of the liver). Even if it is shown that they are still capable of inducing antibody formation, it will nevertheless be necessary to determine whether or not they are essential for continued antibody production.

Fate of antibody. The bulk of the antibacterial and antiviral antibodies is contained in the IgG fraction. However, antibodies against certain microorganisms and toxins are always in the IgM fraction. Interestingly, certain IgM antibodies function best at low temperatures (e.g., at 4°C). Since these *cold agglutinins* probably have no useful functions at body temperature, they are believed to be vestigial remnants of an earlier evolutionary period when organisms were cold-blooded. Certain antibacterial and antiviral antibodies are in the IgA fraction, and

recent evidence suggests that antibodies related to the allergic state are associated with the fifth class of immunoglobulins, IgE. We will discuss them later. A specific immunological function for IgD has not yet been demonstrated.

Immunoglobulin molecules remain intact until they are eliminated or broken down. The locus of the catabolic process is not known. Substantial quantities of immunoglobulin (and other plasma proteins) are secreted into the small intestine. Molecules are eliminated at a constant rate and are randomly chosen for elimination. Thus, the probability that a given newly synthesized molecule will be eliminated is the same as that for an older molecule.

Each of the major classes of immunoglobulins is broken down at a characteristic rate. The approximate half-life of each class is as follows: IgG, 25 days; IgA, 6 days; IgM, 5 days; IgD, 2.8 days; and IgE, 2.3 days. About 7% of the IgG molecules, 25% of the IgA molecules, 18% of the IgM molecules, 37% of the IgD molecules, and 89% of the IgE molecules are catabolized and replaced daily. The serum level of IgM is notably low compared to that of IgG (see Table 8.7); therefore, macroglobulins turn over relatively rapidly. Because of their large size, IgM cannot pass through capillaries and remain almost exclusively within the intravascular compartment.

There is a direct relationship between the rate of breakdown of IgG and its serum concentration. The catabolic rate of IgG is decreased in hypogammaglobulinemia and increased when the serum IgG level is raised by administration of IgG. In contrast, the catabolic rates of IgA and IgM are independent of their serum concentrations. Studies in subjects losing large amounts of IgG from the body, as in proteinuria (see p. 363), have revealed little compensatory increase in the synthesis of IgG. There is no evidence that the rate of synthesis of IgG is affected by its serum concentration.

The mechanism regulating immunoglobulin elimination or destruction is probably triggered by H chains; thus, passively infused H chains increase immunoglobulin catabolism in experimental animals, whereas similarly administered L chains do not. It has been speculated that *protector sites* specific for IgG exist in the body. When intact IgG molecules are bound to these sites, they are protected from catabolism. An increased concentration of IgG would saturate the sites. The remaining unbound molecules would increase the fraction of total body IgG available for catabolism, and thus the fractional catabolic rate would be higher. The reverse would apply when serum levels of IgG are low.

IMMUNOCHEMICAL CONSIDERATIONS

Introduction. A revolution in immunology occurred in the 1930's and 1940's when immunologists such as Landsteiner, Heidelberger, Marrack, Morgan, and Kabat determined to rebuild immunology on chemically rigorous foundations. For the first time, antigen–antibody reactions were studied quantitatively and their kinetics carefully examined in vitro. New classes of antigen were identified and chemically defined. From these beginnings emerged the new discipline of *immunochemistry.*

With the recent elucidation of the structure of immunoglobulins (see Chapter 8), immunochemists have greatly advanced our understanding of the molecular basis of antigen–antibody reactions. Antibodies and antigens have been found to have active sites of chemical activity which are constructed to permit close union with one another. The active *combining site* of the antibody is a small region of the immunoglobulin polypeptide chain that reacts specifically with the antigen. Similarly, the *antigenic determinant* of antigen is a special site (containing as few as three of the several hundred amino acids comprising the antigen molecule) that combines with the antibody. The combination resembles the union of enzyme and substrate. In both combinations, reacting components are chemically unaltered.

Structure of the antibody combining site. After the H and L chains of the immunoglobulins had been separated (see p. 262), efforts were made to assign the antibody combining site to one chain or the other. Isolated L chains were always inactive in the presence of antigen; isolated H chains had varying amounts of activity. For a time it was uncertain whether activity resided only in the H chain or whether the preparations were contaminated with undissociated antibodies. It was also uncertain which of two possible mechanisms accounted for the structural basis of antibody specificity: (1) synthesis of immunoglobulin polypeptide chains with a unique amino acid sequence (i.e., primary structure), which is specific for a particular antigen; or (2) synthesis of immunoglobulin polypeptide chains of a single universal amino acid sequence, followed by the folding of the chains into a unique antigen-specific secondary or tertiary structure that becomes stabilized by hydrogen and disulfide bonds.

It is now clear that H chains of each antibody contain a distinctive amino acid sequence in the N-terminal half (i.e., in the Fd fragment). Presumably, the sequence is specific for the antibody's antigen. The nearby portion of the L chain also has a distinctive amino acid sequence and participates in the antigen–antibody combination. The participating portions of the two chains comprise the Fab fragment. Of the 446 amino acids of each H chain, 115 at the N-terminal end comprise the *variable sequence*, which is thought to provide the uniqueness necessary for combining specificity. The rest of the chain comprises the *constant sequence*. Of the 214 amino acids in each L chain, 108—about half of them—at the N-terminal end comprise the variable sequence; the rest comprise the constant sequence. Together, the two variable sequences make possible many different combinations in the two constituent chains of the Fab fragment. In 1969, G. M. Edelman and co-workers determined for the first time the entire amino acid sequences of both chains of a single IgG molecule.

Structure of antigenic determinants. An antigenic determinant is a portion of an antigen molecule that elicits the production of an antibody and that combines with it in the antigen–antibody reaction. Because of the large number and diversity of antigenic substances, the determination of the structures of antigenic determinants has been a difficult problem.

A significant advance occurred with the discovery, isolation, and characterization of various antigenic pneumococcal polysaccharides by O. T. Avery and W. F. Goebel in the early 1920's. Their work provided immunologically reactive substances of greater structural simplicity than that of protein antigens and offered what appeared to be attractive prospects for the rapid elucidation of the structures of the specific chemical groupings that react with antibody. Some 40 years later it must be reported, however, that the carbohydrate sequences of only a few of these polysaccharides have been determined so far—and even with this knowledge we do not yet know the size and structure of their antigenic determinants. However, much valuable information has been accumulated which has given us partial understanding of the antigenic determinants of many polysaccharides.

Elvin Kabat's studies of the relatively simple polysaccharide dextran have given a clear picture of at least one determinant group. This molecule consists of a linear chain of linked glucose molecules and may be fragmented to give an assortment of smaller chains (oligosaccharides) of 2, 3, 4, 5, or 6 glucose residues. It was found that reaction between dextran and an anti-dextran antibody is completely inhibited if oligosaccharides consisting of 5 or 6 sugar residues (but not of 2, 3, or 4 sugar residues) are added to the antibody before the antigen. This indicates that the combining group of the antibody fits a chain the size of an oligosaccharide containing 5 or 6 sugar residues. Thus, the antigenic determinant of dextran may be considered to be a segment of the polyglucose structure of this length. It is of interest that the antigenic determinants of two of the best studied antigens, blood group substances A and B, are similarly formed by short chains containing 3 to 4 sugar residues (see p. 278).

The recently acquired knowledge concerning the amino acid sequences of many proteins has permitted an assessment of the primary structures of the antigenic determinants of protein antigens. When studies similar to those with dextran were performed with silk fibroin—a linear protein—one antigenic determinant was found to consist of 6 to 7 amino acids. A sequence of only 3 amino acids constitutes the antigenic determinant of insulin.

The increasing availability of proteins of known amino acid sequence should lead to a rapid advance in our knowledge of their antigenic determinant sequences. The observation that fragments of proteins may react with antibodies has greatly facilitated these studies.

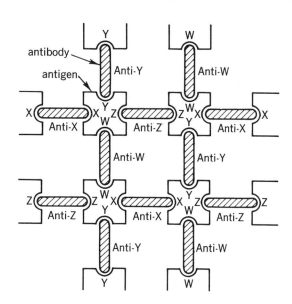

19.15 Antibodies (shown hatched) form in great diversity against a pure antigen. Further complexities arise because each set of antibodies consists of a variety of types of molecule, each with a different affinity for a corresponding antigenic determinant.

Although the determination of a protein's full amino acid sequence is long and difficult, work can be expedited when it is necessary to determine the sequence of only the small fragments capable of reacting with antibody.* C. Lapresle used such a procedure to study human serum albumin and obtained a surprising result. Digestion of albumin with proteolytic enzymes yielded three fragments, each of which reacts with a different portion of the antibody in rabbit antihuman serum albumin. This demonstrated the existence of *multiple* antigenic determinants in the same protein molecule. Similar studies with diphtheria toxin, egg albumin, fibrinogen, and γ-globulin revealed that each also has multiple antigenic determinants (see footnote on p. 755). As a result an antigen-antibody complex may contain a single antigen whose multiple determinants have each reacted with a different antibody (Fig. 19.15).

---

* This approach has obvious similarities to the "fingerprinting" technique by which the altered amino acid sequence of an abnormal protein is quickly localized to a small peptide (see Fig. 8.16).

Procedure: The compounds in the left column (meta-aminobenzoic acid and meta-aminosulfanilic acid) were coupled by diazotization of the amino group to the proteins of horse serum. Antibodies to these modified proteins were obtained by injecting them into rabbits. The resulting rabbit antisera were then mixed with samples of chicken sera which had been coupled with the haptens below. Horse serum was used to produce antibodies and chicken serum in the antibody-antigen test to eliminate any reactions due to the non-hapten portion of the antigen. 0 = no ppt. 4 + = heavy precipitate.

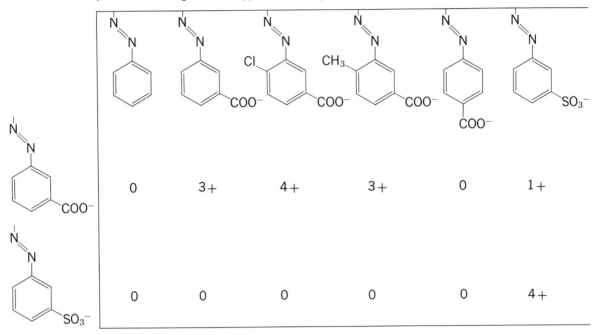

0	3+	4+	3+	0	1+
0	0	0	0	0	4+

19.16 Effect of structural changes in the hapten on antigen-antibody interaction. Note that no interaction occurs unless an acidic group is present on the hapten of the test antigen. The acidic group must be meta to the amino group. Antibodies readily distinguish between $-COO^-$ and $-SO_3^-$. Replacement of a benzene H by Cl or CH_3 has little effect.

Haptens. In one of the classic discoveries of early immunochemistry, Karl Landsteiner used another approach to study the relationship between chemical structure and immunologic specificity. The large size and complexity of known antigen molecules made it difficult to alter their chemical structure in a systematic fashion—as would be necessary in any study of the relationship between structure and specificity. Landsteiner showed, however, that new antigens may be prepared by combining small chemical groups (such as substituted aromatic rings) with proteins. These give rise to antibodies whose specificity depends on the chemical nature of the small conjugated groups, which were called *haptens.*

For example, if w, x, y, and z are a series of chemically related haptens, Aw, Ax, Ay, and Az designate antigens produced from protein A by

conjugation with these haptens, and aw, ax, ay, and az the elicited antibodies. By observing the interactions of the different antibodies with the different antigens, one can determine many of the chemical factors underlying immunological specificity. If, for example, a small molecule xw, containing the hapten w, inhibits the interaction of Aw and aw, it may be concluded that the antibody aw contains sites capable of reacting with the w portion of Aw and that the hapten of xw competes for these sites. An example of such a system is shown in Figure 19.16. Such studies suggest that the specificity of the antibody–antigen reaction depends upon surface complementarity presumably involving an apposition of matching conformations and positive and negative charges.

Investigators have recently given support to such studies by conjugating haptens to synthetic polypeptides of defined amino acid sequence. For example, E. Haber and coworkers have synthesized a peptide of molecular weight 10,000 containing only two amino acids, alanine and lysine, in a ratio of five alanine residues per lysine residue. Since the hapten dinitrophenol conjugates only with lysine residues, and since lysine residues occur at regular intervals of 30 Å, conjugation yielded a polypeptide bearing a hapten at regular intervals of 30 Å. This conjugated polypeptide proved to be highly antigenic in rabbits. Thus the way has been opened for a more refined analysis of the immunological significance of both the structure of haptans and their spacing along the amino acid chain.

Immunoglobulin allotypes. Immunoglobulins are polypeptides and, although they are antibodies, they are also antigens when injected into an individual of another species.* Thus, they contain antigenic determinants as well as antibody-binding sites. They also contain antigenic determinants which differ among individuals of the same species. These are called *allotypes.* Blood group antigens are examples of allotypes (see Chapter 8).

Allotypes are usually detected by immunizing individuals of the same species, but sometimes they can be detected by immunizing other species. Animals and humans injected with IgG from another individual of the same species produce antibodies that react with donor IgG—*if* the injected (i.e., immunized) individual's IgG is of a different allotype than that of the injected IgG. Thus, the ability of one individual to produce antibodies against another's IgG depends on whether or not they have inherited the same or different antigenic variants of immunoglobulin. In man two immunoglobulin allotypic systems are known—the *Gm system* and the *Inv system*—of which nine allotypes have been identified.† Their inheritance is complex. The Gm antigenic determinant is on the H chain of IgG in the *C*-terminal half of the amino acid sequence (in the Fc fragment); it is not present in the immunoglobulins of other classes. The Inv determinant is confined to L chains of the κ type, and hence appears in all immunoglobulin classes. As is true of blood group allotypes, the recognition of human immunoglobulin allotypes has provided anthropologists with a new, useful tool (see p. 279).

---

* As we have seen, the indirect staining fluorescent antibody technique is based on the ability of immunoglobulins to act as antigens which can elicit potent anti-immunoglobulins (see Fig. 19.7B).

† Those most studied are Gm(a), Gm(b), Gm(f), Gm(bʷ), Gm(x), Inv(a), and Inv(b).

Theories of Immunity

SELF-RECOGNITION

Introduction. We have now encountered two of the major questions that must be answered by any theory of immunity. How does the plasma cell "learn" to synthesize antibody globulin molecules which bear a specific combining site complementary to the antigenic determinant of an entirely foreign substance? How do the antibody-producing cells distinguish non-antigenic body components from antigenic foreign matter, "self" from "non-self"?

It is evident that the immunological defenses raised against foreign materials cannot be directed against the body's own components, and it is axiomatic that any theory of immunity must begin by interpreting the capacity of the organism to differentiate foreign chemical configurations from those belonging to itself.

"Self" and "non-self." What constitutes foreign material? A partial answer may be drawn from experience with *skin grafting* and other forms of *tissue transplantation.* It is well known that skin transplanted from the body of one person to that of another (i.e., a *homograft*) will survive only briefly, no matter how competently the procedure is performed. The same is true for the transplantation of other tissues and organs.‡ Ordinarily, the only type of graft that will succeed permanently is an *autograft*—i.e., one in which the donor is also the recipient (e.g., as in the transplantation of skin from one part of the body to another); the only important exception to this generalization is that grafts interchanged between *identical* twins are permanently accepted.§ This fact has occasionally been exploited to save the lives of single members of identical twin pairs who are afflicted

---

‡ Corneal homografts are exempt from this rule.

§ It should be understood that human twins are of several types. *Fraternal* or *nonidentical twins* result from the simultaneous development of two ova. These have separate placentas (except for the rare phenomenon to be described below) and they are as genetically dissimilar as separately born brothers or sisters. *Identical twins* develop from a single ovum.

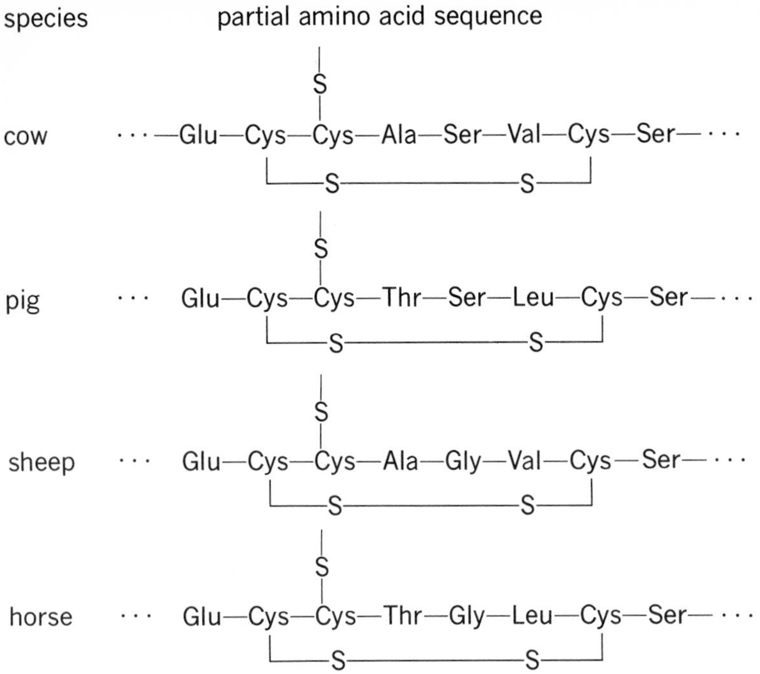

species   partial amino acid sequence

```
                            S
                            |
cow      ···—Glu—Cys—Cys—Ala—Ser—Val—Cys—Ser—···
                      └—S————————S—┘

                            S
                            |
pig      ···  Glu—Cys—Cys—Thr—Ser—Leu—Cys—Ser—···
                      └—S————————S—┘

                            S
                            |
sheep    ···  Glu—Cys—Cys—Ala—Gly—Val—Cys—Ser—···
                      └—S————————S—┘

                            S
                            |
horse    ···  Glu—Cys—Cys—Thr—Gly—Leu—Cys—Ser—···
                      └—S————————S—┘
```

19.17 Differences in amino acid sequence of insulins from different specie. Total amino acid sequence of insulins is shown in Fig. 2.8A.

with severe burns or advanced kidney disease, since skin or a whole kidney can be reliably transplanted from the normal identical twin (see p. 366). Tissues interchanged between ordinary brothers and sisters, between dissimilar twins, or between parents and offspring do not survive unless special measures are taken to suppress the recipient's immune system.

Skin homografts heal satisfactorily at first. However, they soon become inflamed and infiltrated by mononuclear cells, their blood vessels become congested, and the grafts slough away. The entire rejection process takes about two weeks. This failure of homografts is not peculiar to man: it is a general biological phenomenon typical of all higher vertebrates. Goldfish reject homografts no less vigorously than mice, men, chickens, or rabbits. Clearly, the body is rejecting tissue that it recognizes as foreign to itself. The singular fact that animals *within* a species cannot accept grafts from one another implies that individuals carry not only the genes which determine species, but in addition complex sets of allotypic determinants which are responsible for what Peter Medawar has called "the uniqueness of the individual." The possible significance of this immense genetic diversity is suggested in the words of Medawar on page 751. We shall

refer later to the antigens responsible for homograft rejection, antigens whose structure is determined by the individual's genetic makeup.

The fact that most proteins are antigenic to organisms other than the one responsible for their synthesis has led to attempts by immunochemists to compare the antigenic determinants of analogous proteins from different species. For example, insulin produced by one species is mildly antigenic to other species, but not antigenic for the animal that produces it. Most human beings with diabetes receive bovine insulin for years without forming antibodies. However, some do form antibodies to bovine insulin, but when this happens, it can be shown that there are no antibodies to porcine insulin. The small differences between bovine, porcine, and other insulins are shown in Figure 19.17. Of the 51 amino acids in the insulin molecule, 48 have the same arrangement in the insulins of different species; only one segment of three units varies in sequence. An insulin is antigenic for another species when this sequence differs from the corresponding sequence of the animal's own insulin.

Although insulin is a relatively feeble antigen, the fact of its antigenicity again raises the fundamental questions of immunology: how does the body "recognize" the small difference between bovine insulin and its own and then form antibody against the former?

Immunological tolerance. An important clue to the answers to these questions has come from experiments in which an organism is "tricked" into accepting as parts of itself substances or cells that, genetically speaking, have no right to be there. The most impressive examples of this phenomenon come from the rare situations in which genetically dissimilar (i.e., fraternal or nonidentical) human twins share a common placental circulation *in utero*. In this unusual situation each twin continuously receives a large variety of cells from the other, including blood cell precursors capable of colonizing in the bone marrow where they multiply and produce red blood cells. Twins of this kind become and remain *chimeras,* each containing two blood groups: their own and that which was genetically appropriate to their twin (see Chapter 8).* Several pairs of such twins have been recognized in adult life.

Such fraternal twins differ in a second important way from ordinary dissimilar twins who have developed from separate placentas. The latter do not accept skin transplants from each other; but fraternal twins with double blood groups do accept skin grafts from each other exactly as if they were genetically identical twins. To F. M. Burnet, this observation suggested that a foreign antigen introduced early in embryonic life, far from exciting antibody formation, is henceforth regarded as "self" by the body's antibody-producing system. In this view, the ability to recognize self is held to develop sometime during embryonic life. Any configuration present during this period would be, in the future, regarded as self by the antibody-producing systems and therefore nonantigenic.

The implications of this concept are of great importance, and in considering them we must

---

* *Chimerism,* the state in which a single body contains elements of diverse genetic origin, is named for *Chimera,* a mythological fire-breathing monster that was part goat, part lion, and part serpent. Chimerism was first recognized in studies seeking the reason for the frequent appearance of sterile female calves ("freemartins") born as twins of normal males.

recount certain significant results of skin grafting experiments. The fact that the body rejects homografts and accepts autografts raises two parallel sets of questions. First, what is there about a homograft that looks foreign to the body, how is this "foreign-ness" recognized, and how does the body kill the graft? Second, how does the body "know" it is dealing with an autograft that is not to be rejected? To the best of our knowledge, the body's reaction to a homograft involves in large part the type of immune response not associated with the production of classical antibodies. Therefore, let us defer for a moment discussion of the immunological basis of homograft destruction. Rather, let us here consider what these data have to tell us about the problem of self-recognition and the mechanisms of immunological tolerance, a general immunological phenomenon which applies to both protein and carbohydrate antigens, and to humoral and cellular types of immunity.

Skin grafting experiments provided an opportunity to test Burnet's idea about self-recognition. Many strains of mice are now available which, through inbreeding (i.e., repeated brother-sister matings) have become so genetically similar that each individual will accept grafts of skin or other tissue from any other member of its strain. The following illuminating experiments were carried out with two strains of mice that may conveniently be designated A and B.

In an experiment first performed by Peter Medawar (Fig. 19.18A), cells from the spleen, lymph node, or kidney of an embryo of mouse B were inoculated intravenously into a fetal or newborn mouse of strain A. The mouse developed normally. When a piece of B skin was later grafted to the inoculated mouse A (when sufficiently grown), the graft "took" and persisted in healthy condition. When the A mouse was white and the B mouse black, the A mouse displayed an extraordinary anomaly: a patch of healthy black hair.

In the second experiment, performed by R. E. Billingham and L. Brent, another mouseling of strain A was inoculated with lymphoid cells of an adult B mouse, not an embryo. Depending upon the number of cells and the particular pair of mouse strains, the mouseling either died within two or three weeks or developed slowly

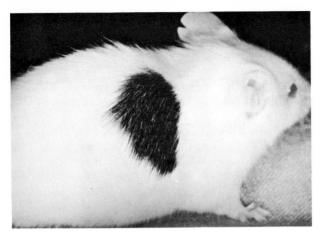

A

B

19.18 Immunological tolerance and the graft-versus-host reaction: A, a white strain-A mouse with tolerance to strain-CBA mouse tissues which has accepted a skin homograft from a black mouse; B, a "runted" strain-A mouse which has received an injection of lymphoid cells from a strain-C57 mouse. (Dr. Paul S. Russell.)

into an undersized, scruffy-looking individual suffering from what has been called *runt disease* (Fig. 19.18B). A similar injection of plasma from an adult *B* mouse had no effect.

The results of the first experiment suggested that host *A* had become *tolerant* of the *B* cells implanted in its tissues before or just after birth. As a result, host *A* subsequently tolerated a graft of *B* skin, presumably because *B* cells were present before the process of "self"-recognition was complete and these were later regarded as "self" by the *A* antibody-producing system. It should be borne in mind that the implanted cells have as much immunological competence as the host cells; if an equilibrium is to be reached, the implanted cells must become tolerant of their foreign host as well as vice versa. As shown above, embryonic *B* cells do become tolerant. But in the second experiment, the adult *B* cells set up their own immune reaction against their host, producing a *graft-versus-host reaction* which manifests itself as runt disease or death. The host is too young to attack the grafted cells (of which it becomes tolerant) but the grafted adult lymphocytes attack the host.

The nature of the graft-versus-host reaction may be illustrated by comparing the effects of interchanging hybrids and their parent pure-line strains as donors and recipients of lymphoid cells. When lymphoid cells, for example, from adult F_1 hybrid offspring (designated *AB*) of two pure-line mouse strains (*AA* and *BB*) are injected into newborn mice of one of the parent strains (say, *AA*), the *AA* mice become tolerant of the foreign antigenic component (*B*), remain healthy, and subsequently accept skin homografts from the F_1 hybrids. If, on the other hand, newborn (or adult) F_1 hybrids are injected with lymphoid cells from one of the parent strains (say, *AA*), the donor cells, lacking some of the host antigens, react against them and produce graft-versus-host reactions. Such a reaction can be induced experimentally in situations where the host cannot destroy a graft of foreign lymphoid cells;* its severity depends upon the extent of antigenic disparity involved. It is clear evidence of the presence of immunologically competent cells in the donor cell inoculum.

In sum, it appears that at some point in embryonic or neonatal life, the body's antibody-producing cells take inventory of the cells and substances then present and classify them as "self." This includes any foreign materials that were introduced accidentally or experimentally. From then on, all other materials are "non-self,"

---

* Such situations include: (1) immunological immaturity of the host, and (2) depression of the host reaction by factors such as radiation, toxic agents, and neonatal thymectomy (see p. 226).

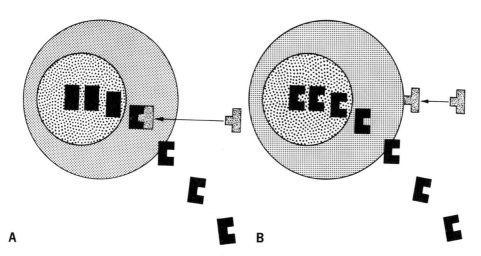

19.19 Diagram showing two theories of immunity: A, classical "instructive" hypothesis in which antigen enters a plasma cell and forms a template from which a complementary antibody is produced; B, the "clonal selection" theory in which the mere contact of a given antigen and a given plasma cell signals DNA in the cell nucleus to start directing production of the corresponding antibody.

and only these will elicit immune responses. "Self" forever receives the dispensation of immunological tolerance. Clearly, its purpose is to prevent the body from reacting against its own constituents—though, as we shall see, it sometimes breaks down. *Autoimmune* reactions do sometimes occur under circumstances in which the dispensation provided by tolerance cannot apply.

How then do antibody-producing cells distinguish "self" from "non-self"?

MECHANISM OF IMMUNITY

Instructive and selective theories. In recent years, it has been customary to classify theories of immunity as *instructive* and *selective*. Instructive theories—supported in the past in various forms by Paul Ehrlich, Karl Landsteiner, and Linus Pauling—hold that the antigen impresses its own pattern, a new and foreign pattern, upon the body's antibody-synthesizing apparatus (Fig. 19.19A). The combining site on the antibody arises, in this view, by being synthesized in contact with an antigenic determinant. The latter is thus visualized as an instructor which tells the protein-synthesizing system how to proceed. According to one instructive theory, the antigen itself is taken into the cell where it

comes into action after the amino acids of the immunoglobulin molecule have been assembled in the processes of protein synthesis and are being folded into globular form. At the folding stage, the globulin is brought into contact with the antigen and, with the antigen acting as a template, is molded into the required complementary pattern. Thus, the antigenic determinant itself supplies the information needed for antibody specificity. Until very recently instructive theories were widely accepted.

Selective theories hold that the function of an antigen is to stimulate the synthesis of an antibody whose specificity preexists in the form of genetic information (Fig. 19.19B). These theories hold that antibody molecules are synthesized in the same way as other proteins—i.e., according to genetic instructions contained in the nuclear DNA—and that information from the outside does not enter the system. Instead, for each one of the thousands of possible foreign antigens, the body already contains a cell or group of cells genetically capable of synthesizing a complementary antibody. Each of these cells or groups of cells already "knows" how to make a specific antibody even if the complementary antigen never enters the body. The function of the antigen is simply to select and stimulate the proliferation of the appropriate group of cells, and thus increase production of the required antibody.*

The clonal selection theory. Selective theories of immunity imply that the mature vertebrate organism possesses an elaborate store of information, a "dictionary," which it consults in deciding whether or not a "word" (i.e., the chemical configuration of a substance) is foreign. This conception, first stated by Niels K. Jerne and later by Joshua Lederberg and others, suggests that the "dictionary" lists only foreign words (all words of its own language having been purged) and, surprisingly, that it lists foreign words without ever having seen them or heard them!

Such a dictionary can be visualized in several ways, but basically it must contain a large number of patterns (words) from among which a specific antibody-combining site can be found that will be complementary to any possible antigenic determinant. As Burnet has pointed out, the number of antigenic determinants is not impossibly large. As we saw in the case of insulin, antigenic determinants are small chemical configurations, perhaps involving no more than three or four amino acids. The number of different three- and four-amino acid combinations from the 20 amino acids found in proteins is respectively 8000 and 160,000—an extremely small number compared to the number of cells in a mouse. David W. Talmage has estimated that only about 10,000 different patterns of reactivity would be needed for the body to respond to all the antigens to which it is exposed. In any event, the selective theory implies that the number of possible variable sequences (i.e., complementary antigen-binding sites) must greatly exceed the number of different antigens to which the body is exposed, since the body

receives no information from the antigen for assembly of the antibody molecule.

How might the body create its foreign-word dictionary—that is, its list of immunological patterns which correspond to antigenic determinants not represented among the body's own constituents? Any theory has three prerequisites: (1) production of each antibody occurs in *clones* of cells, the progeny developing by maturational and divisional events from a single antigen-responsive lymphocyte without exchanging genetic material with other cells (see Fig. 19.13); (2) each cell (and its progeny) makes antibody molecules of only one specificity; and (3) the large number of variable amino acid sequences must be reconciled with present understanding of gene-directed protein synthesis.

One theory attempts to explain the many variable sequences by postulating the presence of a very large number of separate structural genes for variable sequences in germ cells that arose in evolution. Another theory proposes a limited number of "pre-evolved" genes that undergo somatic recombination to form new combinations. A third proposal is that in early embryonic life ancestors of antigen-responsive cells are highly mutable and for a period of time mutate spontaneously in a random fashion, so that genetic patterns for many variable sequences are thereby created (Fig. 19.20A). Each mutated cell, through division, becomes the ancestor of a small clone of identical cells, each of which carries a pattern for one, or at most a few, specific antibodies. Interestingly, the postulated mutations may occur only in the "variable sequence genes," not in those of the constant sequence.

If the postulated mutation process occurred randomly, antibody patterns against "self" antigens would then arise. It is necessary to postulate, therefore, that cells containing "self" patterns are destroyed by contact with their antigens. It is, in fact, well known that high doses of an antigen in an adult do inhibit antibody formation. This is called *immunological paralysis*. It is proposed by the hypermutation form of the clonal selection theory that during an early stage of embryonic life, the "forbidden" clones that match "self"-antigens are eliminated as they arise. Foreign antigens normally cannot reach an embryo; but when they do (as in fra-

---

* To a large extent, this hypothesis proposes that antibody formation resembles induced enzyme synthesis (see p. 153). As we have seen, bacteria produce new enzymes when presented with unusual substances in their medium. Current thought holds that induced enzyme synthesis can occur only if the necessary information is present within a structural gene (see Fig. 4.23). The change of environment merely causes the expression of what was formerly a latent capacity. Despite the attractiveness of the analogy between antibody formation and enzyme induction, the coding arrangements in the structural genes that account for the variable amino acid sequences of H and L chains remain a major genetic puzzle.

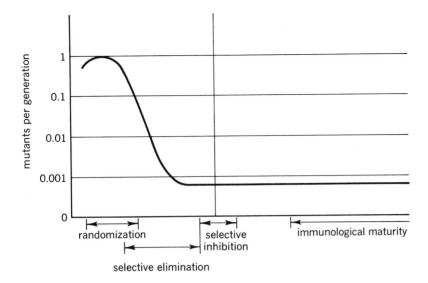

A

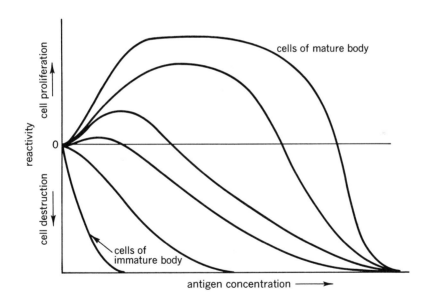

B

19.20 Assumptions concerning mutation rate and immunological reactivity at various stages of maturity according to hypermutation theory. A. Mutation rate per cell-generation of genes carrying antibody patterns is high in early embryonic life (vertical line down chart indicates birth). Then mutation rate slows and most cells carrying "self" antibody patterns are eliminated; later others selectively inhibited. Immunological maturity occurs some time after birth. Mutations continue to appear throughout life, but at a much lower rate than in an individual's early embryonic life. B. Maturity changes reaction of immunological cells to increasing concentration of antigen. The most immature cells are represented by bottom curve; the most mature cells, by top curve. Zero line indicates no reaction to antigen. Above it, cells proliferate, and the thickened curves indicate development of plasma cells and production of antibody. Below the zero line immune cells are first inhibited and then, as the curves indicate, further concentration of antigen can drive them to dormancy and can even destroy the very immature cells. (Adapted from F. M. Burnet.)

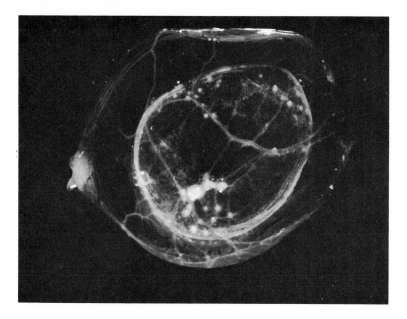

19.21 Chorioallantoic membrane (CAM) assay based on the Simonsen phenomenon. Discrete modules, or pocks, form on CAM of chick embryo due to attack by lymphocytes from a foreign adult chicken. Lymphocytes are dropped onto CAM of 11 day embryo, which is killed 4 days later. Technique isolates the 1 in 20,000 lymphocytes which apparently carries antibody pattern matching membrane antigen. (Courtesy of Dr. Morten Simonsen.)

ternal twins sharing one placenta), they are accepted as "self." In this way the clonal selective theory explains immunological tolerance. If no foreign antigens reach the embryo, it retains all of its randomly created "non-self" patterns.

Later in embryonic life (Fig. 19.20A), the mutation rate in antigen-responsive cells must decrease drastically to the mutation rate found everywhere in the body throughout life.* Since mutation occurs randomly, forbidden clones would continue to arise occasionally, but they would normally be eliminated by their antigens, while the body is still immature. In the mature body, antigen-responsive lymphoid cells containing the pattern for a given foreign antigen are stimulated by it to proliferate into a clone (Fig. 19.13). The fully developed clone includes the plasma cells that synthesize the immunoglobulin molecules that combine with and neutralize the antigens. Fig. 19.20B summarizes the concept that the antigen-responsive cells of the immature body are suppressed or destroyed by their antigens, whereas those of the mature body are stimulated by their antigens to proliferate. The theory is called the clonal selection theory because the action of the antigen is simply to select for proliferation into a clone those cells that contain the genetic information for the synthesis of a complementary antibody.

Many immunologists are skeptical of the view that 10,000 or more different patterns can arise in the body during embryonic life without reference to the foreign antigen determinants with which they are "designed" to react. Others are sympathetic to the basic idea of a clonal selection theory, but are doubtful of the necessity of limiting a given cell or clone to one, two or at most three patterns, preferring a more substantial number, perhaps 10 to 20 related patterns per clone. Some even press the idea to its logical conclusion and assume that every cell that is a potential antibody producer carries its own complete foreign-word dictionary and can therefore recognize any antigenic determinant and through its descendants produce antibody against it.

So far no one has offered an experimental means of differentiating between an instructive theory and the theory that every immunological cell carries all possible antibody patterns. However, several types of experimental evidence support certain elements of the clonal selection theory. These include the direct demonstration in studies of single plasma cells that most cells can produce only one type of antibody (though an occasional cell produces two).

---

* It has been estimated that in an adult about a million body cells undergo mutation each day.

A second line of evidence derives from two phenomena which make it possible to select a small number of cells with a particular antibody pattern from a large population of immunologically competent cells. In one, the chorioallantoic membrane of the chick embryo is inoculated with lymphoid cells from a mature chicken. The inoculation produces one focus, or "pock," for every 20,000 seeded cells (Fig. 19.21). One focus is a clone arising from one of the inoculated cells. The ratio of 1 to 20,000 presumably reflects the proportion of the lymphoid cells with the preformed patterns that correspond to antigens in the embryonic chick membrane that are not present in the mature chicken that provided the cells.

In another procedure, a suspension of cells from the spleen or lymph nodes of an animal immunized against, say, sheep red cells is mixed in a Petri dish with melted agar containing sheep red cells. The agar is allowed to set into a gel. Antibody diffusing from an antibody-producing cell reacts specifically with red cells in the agar in its immediate vicinity, so that when comple-

19.22 Microscopic appearance of an area of an agar plate containing sheep red cells mixed with a small number of spleen cells from a rat immunized with sheep red cells. After brief incubation, the plate was heated with complement. The plaque shows local lysis of red cells by antibody diffusing from the single plasma cell in its center (Dr. Frank W. Fitch.)

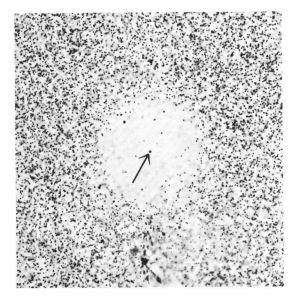

ment is added to the plate only these red cells lyse (Fig. 19.22). The small plaques of the hemolysis identify individual antibody-producing cells present in the agar. The results obtained by this technique, originally devised by Jerne, indicate that one cell per 2 million human spleen cells produces antibody to sheep red cells despite absence of previous exposure to the antigen. According to the clonal selection theory, this would be an estimate of the number of cells capable of forming one specific antibody without previous contact with the relevant antigen.

Although these results are compatible with the clonal selection theory, they do not prove its validity. The theory would be decisively disproved if it were possible to show with antibody-producing cells cultivated in vitro that from a very small initial population of cells any desired antibody could be elicited by stimulation with any antigen. Recently investigators succeeded in growing lymphoid cells in tissue culture so that individual cells gave rise to discrete clones.* Randomly chosen clones were then injected into animals in which the immunological system had been destroyed by radiation. Large lymphoid nodules were subsequently found in the spleens of recipients in numbers proportional to the number of injected lymphoid cells. Thus each nodule was itself a clone. When the cells of different randomly chosen splenic nodules were exposed to multiple antigens they all responded with the same uniform set of immune responses. According to the clonal selection theory, if each cell of a clone has a genetic pattern identical to the clone's parent cell, virtually none of the randomly chosen clones would respond to randomly chosen antigens. These experiments suggest that every lymphoid cell is pluripotential and that selection if it occurs must have a subcellular basis. It remains to be determined, however, whether *every* cell within such a clone is pluripotential or whether a random population has resulted from rapid mutation.

Immunological memory. The characteristic response of the cells of lymphoid tissue to primary antigenic stimulation is proliferation. The proliferation that follows a secondary antigenic stimulus is even more pronounced than that following a primary stimulus. The difference is an expression of the immunological "memory" that is implanted by the primary stimulus.

Which are the "memory cells"—the cells that

---

* Growing lymphoid cells in vitro was for many years extremely difficult. Success was achieved in these experiments because the cells had been treated with *phytohemagglutinin*, a material derived from beans that is a powerful stimulant to lymphocyte proliferation.

are the repositories of immunological memory, and how do they harbor it? The answers are not fully known, but it is possible to make some surmises. Since immunological memory is implanted by the primary stimulus, we may postulate that the antigen-responsive lymphoid cells are progenitors of the memory cells. The event that initiates a primary response is the interaction of antigen with certain cells normally present in the body that possess potential, though yet unrealized, immunological activity, or, in Medawar's term, immunological competence. As a result of the action of antigen, the antigen-responsive cells are transformed into plasmablasts which then evolve into plasma cells, mature antibody-forming cells with a life span of only a few days. When antigen is removed and the plasma cells disappear, the memory cells remain in the resting lymphoid tissues of the immunized subject, always prepared to respond rapidly and intensively to a secondary stimulus by antigen.

Several cell types have been cast in the role of memory cell on the basis of various experimental results, including large lymphocytes, small lymphocytes, plasmablasts, and reticulum cells. Although the question is still open, it appears that whichever cell is the immediate precursor of the plasmablast is likely to be the memory cell. A recent theory has suggested that when plasma cells discharge antibody their cytoplasm "spills" out leaving a form that resembles a naked nucleus. This structure then reforms its cell membrane to produce a memory cell, the initiator of the secondary antibody response. Morphologically, such a cell resembles a small lymphocyte.

This concept is consonant with the discovery, made in labeling experiments, that some small lymphocytes live as long as 10 years (see p. 254). The existence of a relatively dormant cell with an unusually long life span strongly suggests that it is the memory cell.

Role of the thymus. Although the spleen and lymph nodes are major sites of antibody production, they are not the only sites of importance in this respect. After antigenic stimulation, antibody formation also occurs in the bone marrow, small intestine, lung—in lymphoid tissue wherever it occurs in the body, with the notable exception of the thymus.

We encountered the enigmatic thymus as an early source of immunologically competent lymphoid cells in Chapter 8 and as a possible endocrine gland in Chapter 15. Although a role for this lymphoid organ in immunological processes was postulated as early as 1900, the idea has been long discredited for two reasons: (1) neither antibody nor an increased number of plasma cells could be demonstrated in the intact thymus during periods of heightened antibody formation; and (2) thymectomy in the adult animal causes little or no significant depression of the immune response.* As we have seen, however, strong evidence for the old hypothesis was obtained in recent studies of the effects of neonatal thymectomy in mice (see p. 226).

When thymectomy is performed on the first day of life, most of the mice develop normally for three or four weeks. A number of abnormalities then appear. First, the mouse has permanent impairment of the ability to reject skin grafts from donors of foreign strains and, in some instances, of other species. In the majority of mice the grafts remain intact, with luxuriant growth of hair, whereas in normal mice, graft rejection occurs in ten days or less. Most animals thymectomized at birth also have an impaired capacity to produce antibodies. Other effects include development of a wasting disease similar to the runt disease of mice injected neonatally with immunologically competent lymphoid cells (Fig. 19.18B), an inability of the cells from lymphoid organs to induce a graft-versus-host reaction in appropriate recipients, and a deficiency of small lymphocytes throughout the body.

The deficiency of lymphocytes appears to be due to loss of the seeding function of the thymus, whereby it populates lymphoid tissues with the precursors of immunologically competent cells, and to loss of a thymic hormone which regulates

---

* When the thymus is removed from a mouse more than a few weeks old, the only significant effect is a minor reduction in the number of lymphocytes in the blood and in the size of the lymph nodes and spleen. The sole functional effect appears to be a beneficial one: a reduction in the incidence of leukemia in certain strains of mice that are genetically predisposed to this disease.

lymphocyte production and maturation. The existence of a hormone is supported by experiments showing that the immunologic capacity of thymectomized mice can be restored by thymic grafts placed inside chambers permeable to small molecules but not to cells. Clearly, the functional activity of the thymus is at its peak in the first days of life (and perhaps also in the last few days of in utero existence).

Although much remains to be learned about the thymus and the lymphocytes it produces, these discoveries appear to give some support to selective theories of immunity. As we have seen, these theories hold that differentiation and non-specific multiplication of immunologically competent cells take place only in the absence of antigen. Contact of antigen with cells at a certain stage of their differentiation inhibit these processes, whereas contact of antigen with a mature cell having a particular immunological potential specifically induces that cell to grow into a clone. The thymus, an active lymphopoietic tissue—most active of all the lymphoid tissues—that is apparently shielded from the entry of extrinsic antigens, probably provides the most highly specialized environment in which differentiation and nonspecific multiplication of cells with immunological potential can take place. As we shall see, the small lymphocytes arising in and dependent on the thymus have no direct role in antibody production. Rather, the thymus liberates the cells whose descendants are primarily involved in delayed hypersensitivity of the tuberculin type and in the surveillance of the body's cellular integrity.

Autoimmunity. The concept that the normal human or animal body refuses to form antibodies against its own components seems born of necessity. As early as 1900, Ehrlich reached this conclusion on both logical and experimental grounds. In his experiment, he injected goats with the red cells of other goats and then tested the resulting antisera for antibody activity against goat erythrocytes from various sources. An antiserum ordinarily destroyed red cells originating in any other goat, but it never reacted with the animal's own erythrocytes. In today's somewhat confusing terminology, we would say that the immunized animals formed *isoantibodies* (i.e., antibodies against antigens arising from members of the same species) but they failed to produce *autoantibodies* (i.e., antibodies against their own components).* These results prompted Ehrlich to coin the famous term *"horror autotoxicus,"* which denotes the organism's "fear" of developing self-destroying immune processes.

We concluded in discussing immunological tolerance that "self" components arising early in embryonic life and persisting throughout adult life are incapable of stimulating an antibody response. Although this is generally true, exceptional circumstances can arise in which autoantibody may form in response to autoantigen. Instances of this phenomenon have been divided into two categories: (1) autobodies against *inaccessible* antigens; and (2) autobodies against *accessible* antigens.

An antigen is considered inaccessible if it is secluded within the body in a locus anatomically separated from the circulation, or if it remains localized within cell cytoplasm and is thus unable to reach antibody-forming cells in its native form. In 1900, Metalnikoff produced autoantibodies to spermatazoa in guinea pigs by injecting them with their own spermatozoa.† A number of tissue components in the body are well segregated from the antibody-forming cells. These include the proteins of the lens of the eye; certain components of the brain, adrenal, testis, and thyroid; and other components that are restricted to specific organs. When a rabbit is immunized with an extract of one of these organs from another rabbit, or from its own organ, antibodies are formed which can be

---

* Confusion arises from the fact that the prefix "iso" is used differently in the literature of immunology and tissue transplantation. As used by the immunologist, "iso" denotes the same species. Transplantation workers give the prefix "homo" the same meaning (e.g., homograft). These workers reserve "iso" for different individual members of a highly inbred strain. Immunologists have not been troubled to coin a term for this situation, but the transplantationist's "iso" comes closer to the immunologist's "auto" than to his "iso." ("Hetero" refers to a foreign species.) A new terminology has been introduced recently in which "iso" is used to denote members of an inbred species. If the degree of inbreeding were sufficient, "iso" would be in effect synonymous with "auto." In this terminology "allo" refers to members of an outbred or genetically diverse species, and "xeno" refers to a foreign species.

† We now know that during maturation spermatozoa acquire an antigen that is absent from immature germinal cells.

shown to be autoantibodies since they react with extracts of the corresponding organs of the antibody-producing rabbit itself. In other words, the organ-specific antigens, which elicited antibody formation, were also present in the antibody-producer's body, and "isostimulation" was followed by autoantibody formation. When this experiment is repeated with materials that are not inaccessible (red cells, immunoglobulins, etc.), isoantibodies are formed, not autoantibodies—that is, the resulting antisera react with the corresponding antigens originating from other rabbits, not with those obtained from the antibody-producing animal itself.*

Accessible antigens may be defined as those which may easily reach antibody-forming cells and which in turn may easily be reached by antibodies. In this category, one observes (1) the formation of autoantibodies against body components which react with the antigen only at low temperatures (so-called "cold antibodies") and are thus inaccessible themselves under ordinary physiological conditions; (2) the formation of antibodies against body components that have become slightly altered, though it might be questioned whether antibodies against altered autologous antigens should be termed "autoantibodies" in the strict sense of the word; and (3) autoantibodies appearing as a result of frank abnormalities of the immunological apparatus.†

Those diseases produced by immune responses to autoantigens, the so-called *autoimmune diseases*, can originate in three ways: (1) an agent, such as a drug, attaches itself chemically to a cell component and elicits an immune response directed against it and the cell; (2) an immune reaction occurs against a normally inaccessible antigen; and (3) autoimmunity is directed against widespread normal body components.

A well-studied example of the first mechanism is a hemorrhagic disorder due to a decrease in circulating platelets produced by *sedormid*, a sedative drug (see p. 257). This results from the conversion of a hapten (the drug itself) into a complete antigen by loose union with the blood platelets. Antibody to the drug–platelet complex develops in a small proportion of those given the drug. When antibody reaches an adequate level, it destroys the drug-bearing platelets, and hemorrhagic symptoms appear.

An example of the second mechanism, *Hashimoto's thyroiditis*, is a chronic enlargement of the thyroid gland with invasion of thyroid tissue by lymphocytes and plasma cells, and ultimate failure of thyroid function. Antibodies are present which react with thyroid components, particularly thyroglobulin, a normally inaccessible antigen found only with the thyroid follicles (see Fig. 15.17). When, as a result of infection or other pathological change in the gland, thyroglobulin or other thyroid antigens leak into the circulation, they appear as foreign configurations and give rise to antibody and to an immunologically modified population of lymphoid cells. If autoimmunity is responsible for further thyroid destruction, this disorder well illustrates the "vicious cycle" of autoimmune disease—since attack upon thyroid tissue liberates more and more autoantigen.‡

The third group of autoimmune diseases—those in which the antigenic factors are, or appear to be, nonspecialized body components—includes many conditions, e.g., immunohemolytic anemia, lupus erythematosis, and possibly rheumatoid arthritis. The reason for the appearance of an immune response to apparently normal tissue components is not known, but it may involve alteration of their antigenicity by viral and bacterial infection or by other chronic destructive processes.§

---

* The presence of a small amount of autoantibody to an accessible antigen that is always present in large amounts (e.g., albumin) may be difficult to demonstrate by a precipitin reaction. As we observed earlier, a great excess of antigen inhibits an antigen–antibody reaction (see Fig. 19.2B).

† We earlier encountered two examples of the third category, the *LE cell* of *lupus erythematosus* (see p. 255). Sera from individuals suffering from this disease contain antibodies against nuclear DNA, though it is questionable whether attack by this antibody upon body cells accounts for all of the manifestations of the disease. Another is *acquired immunohemolytic anemia* (see Table 8.4) wherein antibodies to the body's own erythrocytes result in abnormal erythrocyte destruction and anemia. It should be emphasized that the presence of an autoantibody in the course of a disease is no proof that the disease is due to an abnormality of the immune mechanism. Rather, the autoantibody may be the consequence of an underlying disease process causing tissue damage. Since the major function of antibody is to clear the blood stream of foreign materials, it appears likely that many of the so-called autoantibodies are serving just this purpose in the event that acute or chronic tissue destruction has released materials into the blood that are not ordinarily present.

---

‡ A second example of the second mechanism is *sympathetic phaco-anaphylactica*, a condition in which injury to one eye liberates lens protein. The result is immunologic damage to the other eye.

§ One interesting variant in this category of autoimmune diseases is provided by rheumatic heart disease and glomerular nephritis. These disorders occur after certain types of streptococcal infection. It has recently been shown that the antigenic components of these streptococci have astonishing

Irrespective of the possible role of autoantibodies in the production of pathological lesions, a successful theory of immunity must account for their occasional production. It now seems likely that in most of these conditions, one is dealing with anomalies of immunologically competent cells rather than with abnormal antigens. None of the antigens against which the antibodies and immunologically active cells are directed in autoimmunity are in any sense abnormal. What is abnormal is (1) the existence of these reactive cells and antibodies; and (2) the breakdown of the homeostatic processes that should have prevented their emergence and pathogenic activity.

The explanation of autoimmunity was particularly obscure when instructive theories of immunity were the only ones available. The clonal selection theory does appear to offer a consistent explanation of these occurrences. It postulates that a "forbidden clone," through mutation or some other mechanism, enjoys an abnormal protection from destruction or inhibition by its corresponding antigen. In autoimmune diseases associated with autoantigens of limited accessibility, failure to eliminate their complementary clone during the embryonic selection period may be postulated. Once the antibody-forming cells start to attack, they break down the cells and tissues containing the antigen, releasing more antigen. The antigen stimulates proliferation of the forbidden clone, which steps up the attack, and the vicious circle of autoimmune disease sets in.

Other theories. Ehrlich's early theory of antibody production—the famous *side-chain theory* —was selective in character. Ehrlich assumed that all foreign antigens—especially the bacterial toxins with which he was primarily concerned—damaged body cells by combining with preexisting chemical configurations (side-chains) that normally were concerned with metabolic processes in the cell. Circulating antibody represented an overproduction of this combining material, whose preexistence implies that it is genetically determined. Clearly, the old side-chain theory suggests the existence of cellular patterns complementary to all possible antigenic determinants.

However, an instructive theory was advanced when Karl Landsteiner discovered antibodies that reacted specifically with synthetic haptens which may never have existed before in nature. He saw no alternative to the conclusion that the foreign antigen impresses a complementary pattern upon an unspecific globulin molecule as it is being synthesized in the cell. Ehrlich's theory would require far too many preexistent receptors, and the theory was therefore discarded by Landsteiner.

In 1940, Linus Pauling gave substance to the Landsteiner hypothesis and proposed the first of the classical instructive theories, one which was entirely consonant with current concepts of protein structure. This conception is diagramed in Fig. 19.19A. In its modern form, this *direct template* theory holds that immunoglobulin is synthesized in the form of polypeptide chains having no immunological reactivity. It acquires its immunological capacity during the processes of secondary and tertiary folding by which the long polypeptide filament is converted into a globular protein. The newly synthesized polypeptide chains, all with identical genetically determined amino-acid sequences, are viewed as flexible entities capable of being folded into many possible configurations of equal probability. However, physical contact with the antigenic determinant pushes them into sterically complementary shapes, which are then made permanent by the formation of hydrogen bonds and disulfide bonds between neighboring segments (see Fig. 2.10). Having given instructions for the molding of the antibody, the antigenic template is released, perhaps to repeat its function elsewhere. As long as the antigen remains in the cell, in this view, it can continue to serve as template for the final folding which confers antibody specificity upon globulin. As we have noted above, it would be experimentally difficult to prove that antigen is *not* present in an antibody-producing cell.

Later research raised three objections to this hypothesis: (1) it is now realized that the pri-

immunologic similarities to normal components in the heart and kidneys. Thus antibody formed against the bacterial components during an infection may also react with the normal tissue antigens to produce further damage. In this situation, it may be said that the body has been deceived into making autoantibody.

mary amino acid sequence of a polypeptide chain specifically predetermines its secondary and tertiary structure and the positioning of its stabilizing hydrogen and disulfide bonds; (2) amino acid sequences *are* different in the combining sites of different antibodies (see p. 770), therefore the preexisting genes of the cells producing them are different; and (3) in vitro studies of pure antibodies treated with agents that unfold the molecule by rupturing hydrogen and disulfide bonds have shown that removal of the agents is followed by refolding of the protein and restoration of antibody activity. Finally, as F. M. Burnet has pointed out, a special disadvantage of instructive theories is their failure to explain why "self" components are non-antigenic.

The virtues of a selective theory were summarized above. We should note, however, that the clonal selection theory advanced by Burnet is but one such theory. Others include the *natural selection* theory of Jerne, the *elective* theory of Lederberg, the *antigen-capture* theory of Monod, and the *coated-cell* theory of Boyden. It is not possible here to present these ideas in detail. Suffice it to say that selective theories appear to be gaining adherents, though it is not yet clear whether one must invoke clones of cells with limited preexisting specificities. The nonclonal theories agree with the concept of preexistent patterns but see no reason for denying the possibility that each lymphoid cell contains 10,000 specific receptors, each of which is attuned to one of the possible antigenic patterns. It is also possible that during embryonic life 5000 or so receptors in *each cell* are eliminated or inhibited by contact with "self" patterns, a view which provides at a subcellular level an explanation of tolerance paralleling that given at the cellular level by the clonal selection theory. Unfortunately, it has not yet been possible to design an experiment capable of distinguishing between one cell that can make any of thousands of different antibodies because it has all the preexisting patterns and merely awaits a proper selective stimulus (as visualized in the nonclonal selective theories) and another cell that can make many antibodies because it uses the antigen as a globulin-shaping mold or template (as visualized in the instructive theories).

Other Types of Immunological Responses

CELLULAR IMMUNITY

Introduction. We have been concerned thus far with humoral mechanisms of immunity. We turn now to the second major category of mechanisms, cellular immunity. We shall find that these processes are less well understood than those associated with the production of antibody.

As we have seen, cellular immunity—as occurs, for example, in the tuberculin reaction (p. 753)—has the following characteristics: (1) the reaction to antigen is delayed, its time course being measured in days; (2) demonstrable circulating antibody is frequently absent; and (3) this type of immunity can be passively transferred to another animal with cells alone and not with serum.

Important questions are raised by these phenomena. How do cells become "sensitized"? How can sensitized cells effect immune responses without the intervention of antibody? How do sensitized cells cause the tissue reaction by which the existence of delayed hypersensitivity is recognized? Let us consider these questions in the light of our earlier discussions.

Cellular transfer of delayed hypersensitivity. Cellular transfer of hypersensitivity in guinea pigs by a method similar to that illustrated in Fig. 19.1A was first demonstrated in 1942 by Karl Landsteiner and Merrill Chase. This phenomenon has since been demonstrated in other species, including man, and with many antigens. It has also been shown that transfer can be effected with lymphoid cells from several sources (e.g., blood, thoracic duct lymph, lymph nodes, etc.).

When cell-transfer experiments were performed with labeled lymphoid cells from *donors* that had received tritiated thymidine (^{3}H-TdR), surprising results were obtained. Although labeled cells were found in the reactive region of antigen injection, they did not accumulate there preferentially. Many appeared in the blood of the recipient, and the number in the local lesion was a constant proportion of the number in the blood. Conversely, if the *recipient* animal was given ^{3}H-TdR for several days before receiving a transfer of unlabeled cells from a sensitized donor, a majority of the cells were labeled in the delayed hypersensitivity reaction subsequently induced by antigen. These results show that (1) cells transferred from the sensitized donor are only a small fraction of the cells infiltrating the local lesion; and (2) infiltrating cells originate within the unsensitized recipient. Later work showed that after contact with the foreign antigen, lymphocytes are carried by the lymphatics to the regional nodes where they multiply and enter the blood stream. Cellular immunity is mediated by these blood-borne sensitized cells.

These conclusions raise questions about the means by which sensitized cells cause local tissue damage at the antigen site. Transferred sensitized cells have been shown to release substances at the site of antigen introduction that amplify the reaction by causing nonsensitized cells to respond to antigens as though they were sensitized—a *recruitment* phenomenon.

Transfer factor. Support for the conclusion that soluble factors produced by sensitized lymphoid cells can recruit nonsensitized cells to respond to a particular antigen came from experiments of H. S. Lawrence and A. M. Pappenheimer, Jr., showing that fragments and extracts of sensitized cells are as effective as whole cells in the transfer of delayed sensitivity. The active transfer factor is both soluble and dialyzable. When active whole cells are mixed in vitro with the specific antigen, they are *desensitized*—that is, they no longer transfer hypersensitivity to normal recipients. Warming sensitized cells to 37°C releases transfer factor.

Attempts to characterize transfer factor have thus far shown that its activity is undiminished following removal of DNA, protein, and part of the RNA. Probably it is a small peptide of molecular weight 10,000. Its effect on human recipients is an actively sensitizing one, for transferred sensitivity appears promptly (body-wide hypersensitivity usually being demonstrable within hours of injection) and remains for months and years. In this respect, it is unlike the transitory passive transfer achieved with living cells.

If active "sensitized" lymphoid cells carry a factor which transfers hypersensitivity to a nonsensitive recipient, an explanation of its mechanism of action must account for the following: (1) it resembles an antibody in that it causes the recipient to undertake immune responses formerly characteristic of the donor, can be removed by mixture with an antigen, and has high antigen specificity; (2) a delay occurs before the recipient becomes hypersensitive; (3) the amounts of material required for effective transfer are extremely minute; and (4) transferred hypersensitivity remains for a long time. These considerations led H. S. Lawrence and others to propose that the cellular transfer factor is an entity produced by an earlier interaction between donor cells and antigen and that this product enters the cells of the recipient body where it is extensively replicated and widely distributed.

Cellular immunity, then, may be regarded as a mechanism for dealing with antigens that do not readily reach lymphoid tissues because they are immobilized at the periphery of the body. It stands in contrast to the immune responses that depend on the transport of antigen to lymphoid tissues and that result in antibody synthesis.

Role of macrophages. The existence of the transfer factor provides one explanation for the relative scarcity of sensitized donor cells at the reaction site in animals with transferred delayed hypersensitivity. Another explanation is that the specific reaction with antigen of even a few cells attracts to the area a much larger number of other effector cells, not specifically sensitized themselves, but with the capability of being triggered into nonspecific activity by the proximity of specific cell–antigen interactions. Macrophages (see pp. 175 and 223) are so numerically prominent in the lesions of delayed hypersensitivity that they have been suspected to be such effector cells.

Several investigators have shown that the migration of macrophages from a piece of lymphoid tissue cultured in vitro is inhibited by adding tuberculin at low concentration if the tissue has been taken from a tuberculin-sensitive animal. In an illuminating analysis of this phenomenon, John David took cells from guinea pigs sensitized to tuberculin and packed them into a capillary tube, which he then fixed horizontally on the bottom of a chamber containing culture fluid. After a day or two of incubation, the cells had migrated fan-wise from the end of the tube, and the extent of migration could be measured by determining the area of the fan of cells.

David showed that addition of suitable antigen to the culture, again in very small amounts, inhibited migration specifically, and in direct proportion to the delayed skin reactivity of the cell donor. The method provides what has hitherto been lacking—a technique of measuring delayed hypersensitivity reproducibly in vitro. The cells whose migration is specifically inhibited are macrophages; and the important observation was made that when cells from a sensitized animal are mixed with cells from a normal animal, the inhibitory effect of antigen on migration affects *all* macrophages, including blood monocytes. It appears therefore that macrophages are "innocent bystanders" in delayed hypersensitivity, being implicated secondarily in immune reactions initiated by lymphoid cells.

There is now evidence that a soluble *macrophage inhibitory factor* (MIF) produced by sensitized lymphoid cells after contact with antigen is responsible for arrest of macrophage migration. MIF is a large molecule (molecular weight, about 60,000) that is not RNA and that is inactivated by proteolytic enzymes. Hence, it is probably a protein.

In summary, the delayed hypersensitivity reaction includes the following steps. After the introduction of an antigen, sensitized circulating lymphocytes passing by are attracted to the antigen depot and come in contact with antigen. Antigen combines with specific antibodies or receptors in the surface of the lymphocyte which then releases MIF. MIF diffuses into the blood and nearby tissues and stops the migration of circulating macrophages (i.e., monocytes), causing them to stick to the endothelial linings of blood vessels. Macrophages thus prevented from leaving the site accumulate and release

lysosomal enzymes that contribute to local tissue injury. They also labor at antigen processing (see p. 766) and thereby enhance the sensitization of lymphocytes. Sensitized lymphocytes activated by antigen undergo transformation and mitosis that leads to the formation of many more sensitized cells. Recruitment of unsensitized lymphocytes proceeds actively as a result of transfer factor production. Hence there is an exponential increase in cell numbers at the site of antigen deposition with ultimate healing and elimination of the antigen. The cohort of sensitized lymphocytes, now greatly expanded and widely distributed, is ready for new engagements with the specific antigen.

Bursa of Fabricius. In chickens, an organ other that the thymus is a source of immunologically competent cells, the *bursa of Fabricius.* The bursa is a blind, saclike structure arising from epithelial tissue and connected by a stalk to the wall of the fowl's cloaca (Fig. 19.23). Though present only in birds, the bursa is of great theoretical importance. It became possible for the first time from studies of the bursa to dissociate the two types of immunity.

During early development of the bursa, epithelial tissue thickens and fills with lymphoid

19.23 Bursa of Fabricius.

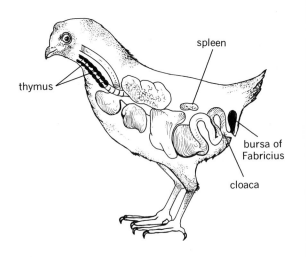

cells—in much the same way that the thymus develops. Since the chicken has a thymus as well, some interesting experiments became possible. Destruction of bursa *and* thymus (by treatment of the developing egg with testosterone) leads to an immunologically incompetent chick that resembles a mouse with runt disease. Removal of the bursa alone (by surgery) produces chicks which at maturity are deficient producers of immunoglobulins but are nonetheless able to manifest delayed hypersensitivity. Removal of the thymus alone results in loss of the ability to display delayed hypersensitivity and to reject homografts, but has no effect on immunoglobulin production. Thus, each type of removal results in loss of a cell system associated with one type of immune response.*

The situation in mammals is not yet as clear-cut. Despite suggestions that neonatal thymectomy more profoundly depresses lymphocytes and delayed hypersensitivity than plasma cell and antibody formation, no precise analogue of the chicken bursa has yet been found. It has been postulated, however, that the antibody-synthesizing component of the immune response may originate in the tonsils, adenoids, appendix, and Peyer's patches—all lymphoid organs that, like the chicken bursa, are associated with the digestive tract.

Is homograft rejection due to cellular immunity? Homograft rejection is an immunological response undertaken by the host. However, it was not unequivocally determined for a long time whether the destruction of the graft is mediated by circulating antibody of the classical type or by the activities of "sensitized" lymphoid cells. The evidence now available preponderantly favors the latter mechanisms. Study of this question has been intense in recent years, because beyond the important theoretical questions at issue lies the promise that one day it will be possible to control or abolish antagonism to another's tissues and thus to graft freely from individual to individual. Before this can be

hoped for, however, it will be necessary to determine the precise nature of the antigens that elicit this immune response of exquisite specificity and the mechanisms that mediate it.

It is reasonably certain that homograft destruction depends upon an actively acquired immune response to foreign antigens present in the grafted tissue. Once elicited, this state of sensitization is present in every part of the recipient body that is served by blood vessels. It is also long lasting. Later homografts from the original donor transplanted to a host that has already been exposed to, and hence sensitized by, tissue cells from that donor, undergo an accelerated rejection—usually referred to as a *"second-set" reaction.* Indeed, this reaction provides an important biological test for homograft sensitization.

Combined genetic and transplantation studies conducted mainly upon genetically identical inbred mice have revealed that the isoantigens responsible for homograft rejection—usually known as *histocompatibility transplantation antigens*—are multiple and determined by dominant Mendelian genes. Not all of them are of equal strength. The most important factor determining the tempo and intensity of the homograft reaction is the degree of genetic disparity between donor and recipient. If this differs only with respect to a single weak factor, the graft may live for many weeks. Evidently, the number of genes determining the histocompatibility antigens in man is large enough to provide for an immense number of possible combinations. Hence, there can be little optimism about the future possibility of being able to "match" individuals with respect to their histocompatibility antigens. Nevertheless, convenient in vitro assays (employing mixed leukocyte cultures) are now available for the major histocompatibility antigen system—the one known as HLA—and it is evident that donor selection on the basis of the degree of HLA compatibility increases the chances of success.

Histocompatibility antigens are *individual-specific*, not tissue- or organ-specific. That is, they are shared in common by *all* of the living cells of a single individual. It is impossible to distinguish between an individual's thyroid, spleen, skin, or testis on the basis of their histocompatibility antigens: each will sensitize a

---

* As we have seen, neonatal thymectomy in the mouse results in a deficiency of small lymphocytes in the blood and lymphoid organs. Neonatal bursectomy in the chicken has no effect on lymphocyte populations, but it does lead to a failure of development of lymphoid follicles and plasma cells.

homologous host in respect of any of the others.*

That homograft sensitivity is fundamentally similar to tuberculin-type delayed hypersensitivity is suggested by the following evidence: (1) both are transferrable with lymphoid cells; (2) a variety of serum antibodies *are* demonstrable in the blood of animals that have rejected homografts, but claims that they are the agents which destroy normal solid tissue homografts are unconvincing (they will not transfer homograft sensitivity and if "cellular" homografts are enclosed in porous chambers made with membranes of different pore diameters and inserted in the body cavities of specifically sensitized animals, they are destroyed only if the holes are large enough to admit host cells); (3) the impressive demonstration by Medawar that guinea pigs, previously sensitized by skin homografts, respond to subsequent intradermal injection of cells or extracts from homologous donor tissues with a delayed-type inflammatory response virtually indistinguishable from the classical reaction which follows intradermal injection of tuberculin in a sensitized individual; and (4) the histologic appearance of skin undergoing rejection resembles that of other delayed-type responses occurring in the skin.

An urgent unsolved problem of transplantation immunology is the immunochemical definition of the antigens. Substances capable of sensitizing mice to skin homografts can be extracted from living tissue cells, and preliminary studies suggest that their antigenic determinant groups may be lipoprotein complexes normally contained in membranous structures (e.g., cell, nuclear, or endoplasmic membranes).

In summary, the weight of evidence indicates that transplantation immunity is provoked when histocompatibility antigens are emitted from a homograft. Vascularization of the graft facilitates its infiltration by sensitized or activated lymphoid cells and macrophages from the recipient. By interaction with the donor cells, such cells bring about the characteristic destructive changes. Surgeons hoping to eliminate this barrier to the free transplantation of tissues from body to body must somehow devise a treatment which will eliminate the intended host's ability to react against the histocompatibility antigens of the intended donor.

Two methods have been found for inducing certain experimental animals to accept homografts. However, neither offers hope of human application. One involves the production of immunological tolerance by injection of newborns with histocompatibility antigens. In the other, adult mice are given x-irradiation in doses high enough to destroy both lymphoid tissue and bone marrow and thus to cause death. The lives of such individuals may be saved, however, by injections of living marrow cells from an unrelated donor. Denuded marrow spaces and lymphoid tissues are histologically reconstructed by the injected cells and these *radiation chimeras* subsequently accept skin or other tissue homografts with the same antigenic constitution as their "borrowed" marrow.† The tolerance of marrow and other tissues depends upon the complete destruction of the host's immunological response machinery. Attempts to produce such results in humans with advanced leukemia have been unsuccessful.

The methods used in humans seek to depress the host's immunological responses. For example, attempts have been made in adults to achieve the same favorable ratio of antigen to lymphoid mass that obtains in newborn babies by diminution of lymphoid tissue. Three methods of doing this have appeared promising. One involves continuous drainage of the thoracic duct and removal of a large proportion of the body's small lymphocytes. The second is the administration of drugs which inhibit nucleic acid synthesis (e.g., 6-mercaptopurine). Such agents, called *immunosuppressive drugs*, unfortunately depress immunity to infectious organisms as well. The third is the use of *antilymphocyte serum* (ALS), an antiserum raised in one species against the lymphocytes of another. In humans, ALS abolishes cell-mediated immune responses, while sparing humoral antibody re-

---

* On the other hand, erythrocytes behave as if they contained no histocompatibility antigens. Since erythrocytes contain no nucleus, this may be evidence that histocompatibility antigens are of nuclear origin. Yet the important and familiar isoagglutinogens they do possess are also distributed widely in tissue cells and are closely related in a genetic sense.

---

† Radiation chimeras rehabilitated with homologous marrow may die after a few weeks from graft-versus-host activity of cells in the transfused marrow. Such reactions can be circumvented by employing cells from very young donors.

sponses. Injection of antigen into experimental animals following such treatments has led to production of long-lasting tolerance and has opened the way to exciting possibilities for transplantation in humans.

Medawar has suggested that the remarkable constancy and wide distribution of the homograft reaction in higher vertebrates must mean that histocompatibility antigens in the individual body play an important role that is entirely unconnected with their role in frustrating the ambitions of surgeons. What this role may be, however, remains unknown. Tissue homotransplantation is an unnatural or artificial procedure that exposes individuals of one genetic constitution to specialized gene-determined products of another. In the intact body, the substances we recognize as histocompatibility antigens are constantly being released by many different types of cell. That they are not antigenic in this situation follows from their presence throughout the body, including its lymphoid tissues. Only when these substances get into the wrong biological context through homografting do they reveal their existence—by behaving as antigens.*

In a sense, mammals throughout evolution have constantly been exposed to "homografts" during pregnancy. Every fetus is a homograft and possesses, even at an early stage, antigens inherited from its father that may be lacking in the mother. The biological success of mammals is strong evidence that special dispensations must apply. A cushion of immunologically inert cells (the trophoblast layer) as well as the vascular quarantine of the fetus is probably the basis of its exemption from immunological rejection. Its histocompatibility antigens may never cross the placental barrier into the mother, and even if they did, there is no vascular pathway for immunologically activated lymphoid cells of maternal origin to enter the fetus and harm it.

ANAPHYLAXIS AND ALLERGY

We stated earlier (p. 753) that anaphylaxis is an immediate form of sensitivity, one which occurs with devastating effects upon the host and which, unlike delayed hypersensitivity, *is* associated with antibody formation. Many immune reactions are associated with antibody production, but few have near-lethal consequences upon later exposure to the antigen. We require an explanation, therefore, for the severity of the events following antigen challenge in anaphylaxis.

Cellular mechanisms in anaphylaxis. That the anaphylactic state is associated with antibody formation can be shown by passive transfer of the sensitive state with small quantities of antibody-containing serum (as in Fig. 19.1B). It now appears that the anaphylactic reaction results from a sequence of events beginning with the attachment of antibody to cells in various parts of the body. Only certain kinds of antibodies become fixed to cell surfaces, thereby sensitizing them, following prolonged immersion of a tissue in an antibody solution. Each species studied to date, including man, produces a specialized immunoglobulin, *anaphylactic antibody*,† capable of sensitizing its own tissues—locally, systemically, or in vitro. Anaphylactic antibody "fits" receptors on the host's own target cells; it does not fit cells of an unrelated species. Therefore it is a *homocytotropic* antibody. Anaphylactic antibody is an IgE—the only antibody thus far associated with this class of immunoglobulins.

When antigen unites with cell-bound antibody of certain cells that have been sensitized with anaphylactic antibody—especially the tissue mast cells (see p. 175)—the reaction triggers the immediate release of pharmacologically active substances, which by their effects on smooth muscle and blood vessels account for the local and general manifestations of anaphylactic shock—generalized urticaria (hives), conjunctival congestion, shock, and shortness of breath. There is nothing novel about the release of pharmacologically active substances by damaged cells. It is well known that an injury is soon followed by local redness, heat and swelling, resulting from dilatation of small blood vessels (see p. 748). All of this is probably due to liberation

---

* The suggestion has been made that homograft sensitivity may be involved in the control of aberrant tissue cells that have arisen by mutation or viral effects (i.e., tumor cells).

† Human anaphylactic antibody was formerly termed *reagin* and *skin-sensitizing antibody*.

of active substances by damaged cells. The problems in anaphylaxis are to explain how the antigen–antibody reaction causes cell damage and to identify the damaged cells.

Two types of evidence point to the possibility that mast cells and their relatives, basophils (see p. 251), are the major cell types damaged. One line of evidence is the discovery that the pharmacologically active substances released in anaphylaxis include histamine, 5-hydroxytryptamine (serotonin), acetylcholine, and heparin.* Each is a prominent component of mast cells and basophils, several occurring within the characteristic cytoplasmic granules. Second, basophils lose these granules in the course of the anaphylactic reaction, suggesting that they and mast cells are target cells which bind anaphylactic antibody. Other cells also undergo morphological changes in anaphylaxis, and it is possible that these, too, become sensitized by binding antibody.

The biochemical events taking place during these reactions are unknown. In attempting to study this problem, investigators have demonstrated the release of histamine from chopped, sensitized guinea pig lung following treatment by specific antigen. It has been suggested that activation of an esterase may be a necessary condition for the release of these two substances. This esterase can be inhibited by diisopropyl fluorophosphate (DFP). Conceivably, DFP-sensitive esterase exists in tissue in a DFP-resistant precursor state until activated by the antigen–antibody interaction. These studies are consistent with the view that anaphylaxis is a complicated reaction involving several steps.

Most studies of anaphylactic antibodies stress their role in producing adverse reactions in the host. There is also evidence that they have a beneficial function. For example, local increases in vascular permeability—similar to those brought about by anaphylotoxin, a product of complement activation (see Fig. 19.8)—may serve as part of a local protective mechanism against bacterial infection.

A striking feature of the immediate hypersensitivity of anaphylaxis is the fact that sensitized animals may undergo *desensitization*. This is accomplished by injecting antigen in doses too small to cause severe reactions. Upon recovery from the minor reaction, the animal shows no reaction at all to subsequent injections of antigen for the next few hours or days. Subsequent doses of antigen are made progressively larger.

Desensitization is believed to occur as a result of one of three mechanisms. Repeated injections of antigen: (1) may use up the anaphylactic antibody; (2) may use up the pharmacologically active substances; and (3) may produce a different kind of antibody which blocks the site sensitized with anaphylactic antibody, preventing its cell-damaging reaction with antigen. Desensitization is not usually feasible in hypersensitivity of the delayed type because of the complexity of the antigens.

The allergic state. The term *allergy*, coined by von Pirquet in a clinical setting, was recently defined by an international commission of "allergologists" as "an acquired, qualitatively altered capacity of living tissue to react, which is induced by a specific allergen." An *allergen* is defined as "any substance capable of producing a state or manifestation of allergy."

In fact, the term allergen is the clinician's synonym for antigen. Therefore, the definition of allergy is similar to that of immunity. In ordinary parlance, the term implies immunity with secondary adverse effects upon the host. The state of allergy is a clinical disorder comprised in part of a mild form of anaphylaxis, in which antigen or allergen reacts with cell-bound homocytotropic antibody with release of histamine and related substances, and in part of delayed hypersensitivity, in which antigen activates cellular immunity. The manifestations of allergy —the bronchospasm of asthma, the exudation of hay fever, the skin wheals of urticaria (hives)— are due to varying combinations of the two mechanisms. They are distinguished from ordinary immune reactions by the fact that they occur at specific sites. What takes place depends

---

* An unidentified substance released from the lungs in anaphylaxis is called *slow-reacting substance* because it produces a slow sustained contraction of the smooth muscle of guinea pig ileum in the classic system used to assay histamine. In contrast histamine causes rapid contraction of smooth muscle. Its cellular source is unknown. It has been implicated as the chemical mediator causing asthma in humans.

on the nature and location of the antigen and sensitive cell, or the amount of antigen introduced into the body and its portal of entry. For example, the pollens which cause hay fever enter through the upper respiratory tract; a variety of food and drug allergens enter through the gastrointestinal tract. In each of these, the clinical manifestations appear directly related to the actions of liberated histamine, heparin, and serotonin.

Much has been made of the fact that some individuals become allergic to one substance and not to another and indeed that most people never become allergic at all. It has been suggested that the tendency to develop allergies is genetically determined, although the exact level at which a genetic abnormality might operate is not known. Since the capacity to make anaphylactic antibody is present in normal individuals and may be stimulated by suitable immunization, the defect may lie in the mechanisms which handle air-borne antigens (i.e., pollens, spores, dust, etc.). Thus, allergic individuals become immunized via air-borne contact and hence sensitized, while normal individuals do not.

Until a deeper understanding is achieved, the medical treatment of the allergic state will consist mainly of desensitization. The *antihistamine drugs* are useful mainly in cutaneous anaphylactic reactions. Although effective against administered histamine, they are of limited value in most allergic reactions.

Another important variety of hypersensitivity reaction mediated by circulating antibody is exemplified by *serum sickness.* In the days before prophylactic immunization against the toxins of diphtheria and tetanus organisms, infections in humans were treated by liberal infusions of an antiserum obtained from horses. Since horse globulin molecules are foreign proteins, they evoked an immune response in the treated individual. When anti-horse-globulin antibodies began to appear, they did so in the presence of large amounts of antigen remaining from the administered serum. Antigen–antibody aggregates formed in large quantities and these produced wide tissue damage and symptoms such as joint aches, skin rashes, and kidney damage. Foreign serum is not the only antigen capable of producing this disease; indeed, any antigen given in sufficient amounts can produce an antibody response sufficient to cause a picture similar to serum sickness.*

Today antitoxins are infrequently used by physicians, but a pattern resembling serum sickness is often seen in conjunction with drug therapy. Here the drug has complexed to a serum protein and has thereby become an antigen. The increasing use of drugs will make this an increasingly important problem in clinical medicine.

REFERENCES AND SUGGESTIONS FOR FURTHER READING

Bloch, K. J., "The Anaphylactic Antibodies of Mammals Including Man," *Prog. Allergy,* **10,** 84 (1967).

Boyd, W. C., *Introduction to Immunological Specificity,* Wiley, New York, 1962.

Burnet, F. M., *The Clonal Selection Theory of Acquired Immunity,* Vanderbilt Univ. Press, Nashville, Tenn., 1959.

————, *The Integrity of the Body,* Harvard University Press, 1962.

————, "The Mechanism of Immunity," *Sci. Am.,* **204,** 58 (Jan. 1961).

————, "Theories of Immunity," *Perspectives Biol. Med.,* **3,** 447 (1960).

Chase, M. W., "The Cellular Transfer of Cutaneous Hypersensitivity to Tuberculin," *Proc. Soc. Exptl. Biol. Med.,* **59,** 134 (1945).

Crowle, A. J., "Interpretation of Immunodiffusion Tests," *Ann. Rev. Microbiol.,* **14,** 161 (1960).

David, J. R., "Macrophage Migration," *Fed. Proc.,* **27,** 61 (1968).

Dubos, R. J., and J. G. Hirsch, eds., *Bacterial and Mycotic Diseases of Man,* 4th ed., Lippincott, Philadelphia, 1965.

Edelman, G. M., and W. E. Gall, "The Antibody Problem," *Ann. Rev. Biochem.,* **38,** 415 (1969).

Fleischman, J. B., "Immunoglobulins," *Ann. Rev. Biochem.,* **35,** 835 (1967).

---

* Maurice Arthus discovered a comparable phenomenon in rabbits. He noted that continuous immunization led to a hemorrhagic reaction at the site of antigen injection at a time when the animals had synthesized large amounts of precipitating antibody. This reaction, the so-called *Arthus reaction,* also results from the formation of large amounts of soluble antigen–antibody complexes in the presence of excess antigen. These complexes lodge in the walls of small blood vessels. The complement sequence is activated, giving rise to chemotactic factors which attract leukocytes. Leukocytes ingest antigen–antibody–complement complex and incidentally release lysosomal enzymes which cause severe local damage in the blood vessel wall.

Fudenberg, H. H., "The Immune Globulins," *Ann. Rev. Microbiol.,* **19,** 301 (1965).

Gatti, R. A., O. Stutman, and R. A. Good, "The Lymphoid System," *Ann. Rev. Physiol.,* **32,** 529 (1970).

Gitlin, D., "Current Aspects of the Structure, Function, and Genetics of the Immunoglobulins," *Ann. Rev. Med.,* **17,** 1 (1966).

Gowans, J. L., and D. D. McGregor, "The Immunological Activities of Lymphocytes," *Prog. Allergy,* **9,** 1 (1965).

Haurowitz, F., "Antibody Formation," *Physiol. Rev.,* **45,** 1 (1965).

Hirsch, J. G., "Phagocytosis," *Ann. Rev. Microbiol.,* **19,** 339 (1965).

Humphrey, J. H., and R. C. White, *Immunology for Students of Medicine,* Blackwell, Oxford, 1963.

Jerne, N. K., "Immunological Speculations," *Ann. Rev. Microbiol.,* **14,** 341 (1960).

Lennox, E. S., and M. Cohn, "Immunoglobulins," *Ann. Rev. Biochem.,* **36,** 365 (1967).

Levine, B. B., "Immunochemical Mechanisms of Drug Allergy," *Ann. Rev. Med.,* **17,** 23 (1966).

Metzger, H., "The Antigen Receptor Problem," *Ann. Rev. Biochem.,* **39,** 889 (1970).

Meuwissen, H. J., O. Stutman, and R. A. Good, "Functions of the Lymphocytes," *Seminars in Hematology,* **6,** 28 (1969).

Mongar, J. L., and H. O. Schild, "Cellular Mechanisms in Anaphylaxis," *Physiol. Rev.,* **42,** 226 (1962).

Müller-Eberhard, H. J., "Complement," *Ann. Rev. Biochem.,* **38,** 389 (1969).

Najjar, V. A., "Some Aspects of Antigen–Antibody Reactions and Theoretical Considerations of the Immunologic Response," *Physiol. Rev.,* **43,** 243 (1963).

Nossal, G. J. V., "Mechanisms of Antibody Production," *Ann. Rev. Med.,* **18,** 81 (1967).

Pauling, L., "A Theory of the Structure and Process of Formation of Antibodies," *J. Am. Chem. Soc.,* **62,** 2643 (1940).

Putnam, F. W., "Immunoglobulin Structure: Variability and Homology," *Science,* **163,** 633 (1969).

Rosenberg, L. T., ""Complement,"" *Ann. Rev. Microbiol.,* **19,** 285 (1965).

Russell, P., and A. Monaco, *The Biology of Tissue Transplantation,* Little, Brown, Boston, 1965.

Talmage, D. W., "Immunological Specificity," *Science,* **129,** 1643 (1959).

Waksman, B. H., "Auto-immunization and the Lesions of Auto-immunity," *Medicine,* **41,** 93 (1962).

———, "Cell Lysis and Related Phenomena in Hypersensitive Reactions, Including Immunohematologic Diseases," *Prog. Allergy,* **5,** 349 (1958).

Williams, C. A., Jr., "Immunoelectrophoresis," *Sci. Am.,* **202,** 130 (March, 1960).

Wolstenholme, G. E. W., and C. M. O'Connor, eds., *Cellular Aspects of Immunity,* Little, Brown, Boston, 1961.

———, and J. Knight, eds., *Ciba Found. Symposium, Complement,* Little, Brown, Boston, 1965.

Zucker-Franklin, D., "The Ultrastructure of Lymphocytes," *Seminars in Hematology,* **6,** 4 (1969).

Index

Carbon monoxide (CO): in lung
capacity measurements, 392
poisoning by, 399–400
Carbonic anhydrase, 52
role in HCl secretion, 525
Carboxyl functional group, 19
hydrogen bonds in, 21
Carboxypeptidase, 511, 523, 542–43
Carcinoid cells, 611
Cardiac arrest, 310
Cardiac catheterization, 313–14
Cardiac cycle (heartbeat), 289, 303–06,
314–17
disorders of, 308–10
Cardiac glands, 518
Cardiac muscle, 183, 287, 289, 303–04
Cardiac output, 319–20
Cardiac pacemaker, 289
electronic, 310n.
Cardiac plexus, 466
Carlson, A. J., 522n., 528, 531
Carnivores, salt in diet of, 209n.
Carotenoid proteins, in photochemistry
of vision, 484–85
Carotid bodies, role in control of
breathing, 402
Carotid sinus, 298
in control of heart rate, 320
Carpal bones, 685
Carpi, Jacopo, 5
Carrier molecule, in active transport,
159
Cartilage, 172–73, 175–76
formation of bone from, 696
of larynx, 373–74
Castle, William B., 219n., 231, 242–43
Castration, effects of, 673–74
Casts (urine sediment), 364
Catabolism, 117. See also Metabolism
Catalase, 325
Catalysis, 47
Catalytic agents, in body, 22–23
Cataract, 473
Catecholamines, 609
parallels with indolamines, 610–11
Catelectrotonus, 420
Cathepsin, 522, 524
Cations, 21
Caudal (inferior) aspect of body, 196
Cavendish, Henry, 199
Cecum, 512, 535
Celiac (solar) plexus, 466
Cells: biosynthesis of constituents of,
140–49
of connective tissue, 175
differentiation of, 84–86, 156
division of, 81–84, 243–44
early theories of, 66–67
epithelial, structure of, 170
irritability of, 162–63
lysis of, in cytolytic reaction, 762–63
membranes of, 79–80, 158–63

metabolism of, 117–40, 149–57
movement of, 163–65
as organisms, 84–86
sex chromatin in, 97
structure of, 71–80
walls of (in plants), 80
Cellular immunity, 752–53, 786–91
Cellulose, 25
Cement (of tooth), 517
Central nervous system. See Nervous
system
Centrifugation, isolation of cell elements by, 74
Centrioles, 72, 78–79
compared with cilia, 165
Centriolar reproduction, 82
Centromeres, 81
Centrosomes, 78
Cephalic (psychic) phase, of digestive
secretion, 525–26
Cerebellum, 439, 443–45
evolution of, 411–13
Cerebral cortex, 445, 450
localized functional areas of, 447–51
role in consciousness and emotion,
461–63
Cerebral ganglia (of invertebrates), as
learning structures, 460
Cerebrosides, 28
Cerebrospinal fluid, 432, 454–56,
(table) 456
Cerebrum, 439, 445–51
cortex of. See Cerebral cortex
embryological development of, 447
evolution of, 413
hemispheres of, 445–47
Ceruloplasmin, 259
Cervical (lymph) nodes, 221
Cervical plexus, 434
Cervical vertebrae, 683
Cervix, 638
Charles' law, 384
Chase, Merrill, 786
Chemical bonds, 21–22
Chemical reactions, 44–47
Chemical senses, 503–07
Chemoreceptors, (table) 425
in control of respiration, 402
in olfaction, 505
in taste sense, 503
Chemotaxis, 255
in cytolytic reaction, 762–63
Chewing, 520
Chimeras, 279, 775, 790
Chloride (Cl$^-$) ions: in body water,
207, 209
concentration of, in pancreas, 542
in gastric juice, 524–25
molecular dimensions of, 259
and regulation of blood pH, 398–99
Chlorophyll, as hemoprotein, 232
Chlorosis, 213

Cholecalciferol (vitamin D$_3$), 127, 625
Cholesterol, 30
in bile, 545
in control of adrenocortical activity,
607
and corticosteroids, 604
role in atherosclerosis, 324
synthesis of, in body, 581–83
Cholinergic synapses, 422, 468
Chorion, 655–56
Chorionic gonadotropin, 645, 657–58
Choroid, 469–70
Choroid plexuses, 452–54
Christmas factor (Factor IX), role in
blood clotting, 269, 273
Chromaffin cells, 600, 611, 740
Chromaffin tumors, 610
Chromatic aberration: in eye, 482
in microscope, 68
Chromatids, 81–82
Chromatin, 72
Chromatography, 57–60
Chromatophores, 743n.
Chromium: dietary requirement of, 572
radioactive (^{51}Cr), in red cell
labeling, 246–47
Chromophils, 594
Chromophobes, 594
Chromophores, 33
in photochemistry of eye, 484
Chromoproteins, 33
Chromosomes, 72
abnormalities of, and infertility, 677
in body and germ cells, 90
mapping of, 99–100
morphology of, 83–84
mutations of, 112
replication of, 104–05
in theory of heredity, 89–90, 93–96
Chylomicrons, in lipid transport, 549,
579–80
Chyme, 521
Chymotrypsin, 511, 523, 542–43
Cicatrization, 749
Cilia (eyelashes), 476
Cilia (flagella), 78–79, 164–65
in epithelial cells, 170
Ciliary body, 470
Circulating granulocyte pool (CGP),
252–54
Circulation: arterial, 297–98, 322–24
in capillaries, 324–26
coronary, 289–91
embryology of, 292–93
Harvey's discovery of, 310–12
hepatic, 541–42
portal, 285, 540–41
pulmonary, 284–85, 296, 386, 390–92
systemic, 284, 296
venous, 298–300, 326–27
Circulatory system, 192, 283–327
action of thyroid hormones on, 618

Daily caloric requirements, 562
Dale, Sir Henry, 421
Dalton's law of partial pressures, 384–85
Dam, Heinrich, 570
Dangerous universal donors, 280
Dark adaptation, of eye, 486–87
Dark-field illumination, 71
Darkening of skin, associated with diseases, 744. *See also* Pigmentation
Darrow-Yanet diagrams, 210
Darwin, Charles, (quoted) 1, 101
Davenport, Horace W., (quoted) 510
David, John, 788
Davis, Bernard D., 117n.
Davy, Sir Humphrey, (quoted) 389–90
De Vries, Hugo, 89
Decibel, 496
Decidua basalis, 654–56
Deep veins, 299
Defecation, 550–51
Defense mechanisms (of body), role of blood (plasma) in, 216, 254–55, 259. *See also* Immunity
Deglutition (swallowing), 520
Dehydration, in metabolic reactions, 119, 138
Delayed hypersensitivity, 753
Delta waves, of brain, 459
Denatured proteins, 32
Dendrites, 414, 426–27
Dentine, 517
Deoxyhemoglobin, 234, 237–38
Deoxyribonuclease, 511
Deoxyribonucleic acid. *See* DNA
Deoxyribonucleotides, 35, 144–46
Deoxyribose, 35
  synthesis of, 144–46
Depth perception, 481
Depolarization, and muscle contraction, 712
Depressant drugs, 463
Dermal-epidermal junction, 726
Dermatoglyphics, 722n.
Dermis, 721, 726–27
Descending colon, 535
Desensitization, to antigens, 792
11-desoxycorticosterone (DOC), 604
  therapeutic effects of, 607
Detergents, emulsification by, 545
Development, action of thyroid hormones on, 618
Dextran, 771
Dextrins, 25, 521
Diabetes insipidus, 355, 592n.
Diabetes mellitus, 355n.
  effect of hypophysectomy on, 597–99
  hyperglycemia due to, 345–46
  steroid, 607
Dialysis, 53
Diamond, L. K., 280

Diapedesis, 249
"Diaphorase," 235–36
Diaphragm, 196, 380–82
Diaphysis, of bone, 696
Diarthroses, 687
Diastole, 314
Diastolic pressure, *311,* 317
Dicumarol, 274–75
Dielectric constant, of water, 200–01
Diencephalon, embryonic, 439, 445
Diet: daily requirements in, 567–68
  reducing, 566–67
Differential count, of white cells, 250
Differential sedimentation, 60–61
Differentiation, of cells, 84–86
  intracellular feedback control in, 156
Diffusion: in lung gases, 384, 386, 390
  in renal tubules, 350
  techniques using, in antibody studies, 759
Diffusion coefficient, of molecule, 759n.
Digestion, 510–11
  chemistry of, 511–12
  gastric, 521–31
  preliminary phases of, 520–21
  in small intestine, 542–46
Digestive enzymes, (table) 511
Digestive system, 192, 510–54
Digitalis: diuretic effect of, 361–62
  use in congestive heart failure, 321
Dihydroxyacetone, 23
Diisopropylfluorophosphate, radioactive (DF^{32}P), in granulocyte life span measurements, 252
Dilution principle, of body water measurement, 203
Dinitrophenol, effect of on oxidative phosphorylation, 133
Dipeptidase, 511, *523*
Diploid number, 90
Disaccharides, 24–25
Disc (gel) electrophoresis, 61, *62*
Distal (anatomical term), 196
Disulfide bonds, 32
  in immunoglobulins, 262
Diuresis, water, 354
Diuretics, 361–63, (table) 362
Diving (deep), physiology of, 404–07
DNA (Deoxyribonucleic acid), 35–38, (table) 35
  in chromatin, 72
  control of protein synthesis by, 106–09
  genetic significance of, 103–04
  replication of, 104–06
DNA polymerase, 105
Dobzhansky, Theodosius, (quoted) 720
Dominant alleles, 88
Dopa (dihydroxyphenylalanine), in melanin synthesis, 741–42
Dopa oxidase reaction, 742n.
Dorfman, R., 604

Dorsal (posterior) aspect of body, 196
Dropsy, 321
Drosophila melanogaster, in genetic studies, 95–98
Drowning, 406
Drug therapy, immunological consequences of, 793
Du Bois, D., 560
Du Bois, E. F., 560
Du Vigneaud, Vincent, 592–93
Duffy blood grouping system, 278
Duodenum, 512, 532
Dura mater, 432, 439
Durnin, J. V. G. A., 562
Dwarfism, pituitary, 597
Dyes, use of in microscopy, 71
Dyspnea, 392, 402–03
Dysproteinemia, 264

Ear, anatomy of, 491–95. *See also* Hearing
Ear ossicles, 681
Eccrine sweat glands, 171, 172, 733–34, 736
Ectoderm, 219
Edelman, G. M., 770
Edema, 210–11, 321, 361–62
Edsall, John T., (quoted) 11
Effectors: in enzyme inhibition, 51
  in neural integration, 409
Efferent (anatomical term), 196
Ehrlich, Paul, 48, 214, 456–57, 752, 777, 783, 785
Ehrlich reaction, 551
Einthoven, Willem, 307
Ejaculation, 672
Electric field, action of on polar molecules, 200–01
Electrical activity: of brain, 458–59
  of heart, 306–10
  of nervous system, 416–19, 427
Electrocardiogram (ECG), 306–08
Electroencephalogram (EEG), 458–59, 461–62
Electrolytes, 21
  in body water, 207
  in plasma, (table) 258
Electroolfactogram, 505
Electromotive force (emf) series, 42–43
Electromyogram, 713–14
Electron microscope, 71
Electrophoresis, 61
  of abnormal hemoglobins, 239
  of multiple myeloma serum, 264–65
  of plasma, 260–61, 761
Electroretinogram, 489
Electrotonus, 419–20, 712n.
Eleostearic acid, 27
Elephantiasis, 221n.
Elution, 59

Globulins: metabolic role of, 581–82
 in plasma, 259–61
 in treatment of Rh incompatibility
  (γ-globulin), 281
Glomerular filtration, 339–41
Glomeruli (of kidneys), 331, 337
Glomeruli (olfactory receptors), 507
Glossopharyngeal (IX cranial) nerve,
 451, 452, 505
Glottis, 373
Glucagon, 579
Glucocorticoids, 355, 601–03
Gluconeogenesis, 575
Glucose, 23
 intestinal absorption of, 547, 548
 metabolism of, 120–29, 574–76
 molecular dimensions of, 259
 reabsorption of, by kidneys, 344–46
 required by heart, 304–05
Glucose G-phosphate, 123–28
Glucose-6-phosphate dehydrogenase:
  feedback activation by, 152
 in glycogen synthesis, 139–40
 in reduction of methemoglobin, 236
 regulatory significance of, 150
 in ribose synthesis, 140–41
Glutamic acid, 29
Glutamine, 29
Glyceraldehyde, stereoisomerism in,
 18n., 23
Glycerol, 25
Glycerophosphatides. See Phospholipids
Glycine, 29
 radioactive, in red cell labeling, 247
Glycogen, 25
 synthesis of, 138–40
 in vaginal epithelium, 640
Glycolysis, 120–21, 123–29
 in red cells, 248
 in white cells, 254
Glycoproteins, 33
Glycosuria, 345–46
Gm system, in immunoglobulins, 773
Goblet cells (mucous glands), 171–72,
 533
Goebel, W. F., 770
Goiter, 621–22
Goldblatt, Harry, 324
Golgi, Camille, 76
Golgi complex (body), 72, 76–77
Gonadotropic hormones, 594, 596,
 642–43
 influence on testes, 673
 secretion of, after menopause, 650
Gonads, as endocrine glands, 589.
  See also Ovaries; Testes
Goose flesh, 728
Gout, 585
Gowans, J. L., 227
Graaf, Regnier de, 643
Graafian follicle, 643
Graft-versus-host reaction, 776

Grafting: of bone, 700
 of skin, 773–76
 See also Transplantation
Gram-equivalent weight, 39n.
Granulocytes, 249–54
 defense functions of, 254–55
Graves, Robert, 622
Graves' disease, 622–23
Gray matter, 432–33
Gregory, R. A., 526
Grosser, O., 655
Growth, effect of thyroid hormones
 on, 618
Growth hormone (GH), 594, 596–99
Guanine, 35, 36
Gums (gingivae), 517
Gut. See Alimentary tract
Gynecomastia, 662n.

H chains, in immunoglobulins, 262,
 769–70
H zone, in ultrastructure of myofibrils,
 181–82
Haber, E., 773
Hageman, D. D., 648
Hageman factor (HF) (Factor XII),
 role in blood clotting, 269,
 270, 273
Hair, growth and differentiation of,
 728–30
Hair cells, of organ of Corti, 495,
 499n.
Haldane, J. S., 396
Half-life, 62
Haller, Albrecht von, 6
Hallucinogens, effect of on emotions,
 463
Hand: bones of, 685
 muscles of, 691–94
Hanson, Jean, 708
Haploid number, 90
Haptens, 772–73
Hard palate, 515
Harden, Arthur, 123
Hargitay, G., 351
Harris, John, 242
Hartline, H. K., 489
Hartsoeker, Niklaas, 85
Harvey, William, 6, (quoted) 283,
 310–12, (quoted) 311, (quoted),
 317
Hashimoto's thyroiditis, 784
Hassall's corpuscles, 226
Haversian systems (osteones) 177, 696
Head, 196
Healing, of wounds, 748–49
Hearing, physiology of, 495–500.
  See also Ear
Heart, 287–94
 all-or-none law of the, 304
 electrical activity of, 306–10
 Harvey's study of, 6

malformations of, 294
 as pump, 310–22
Heart block, 309
Heart failure, congestive, 320–22
Heart-lung machine, 322n.
Heartbeat. See Cardiac cycle
Heat: conversion of foodstuffs to,
 556–57
 in energetics of chemical reactions,
 45
 of fusion, 202
 production of, in muscle contrac-
 tion, 717–18
 of vaporization, 202
Heat capacity, of water, 201–02
Heavy water, in studies of body water,
 203
Hechter, O., 604
Heidelberger, Michael, 756, 758, 770
Heidenhain, Rudolf, 522
Helices, in protein structure, 32
Helmholtz, Hermann L., 487, 497
Hemagglutination techniques, of anti-
 body study, 764–65
Hematin, in assays for hemoglobin,
 234
Hematocrit, determination of, 218–19
Hematology, 213–15
Hematopoiesis: in adult, 220–21
 in embryo, 219–20
 in spleen, 225
Hematuria, 364
Heme, 232
 synthesis of, 244–45
Heme-heme interactions, 234
Hemocytometer, 219
Hemodialysis, 366
Hemodynamics, 312–14
 of glomerular filtration, 340–41
Hemoglobin(s), 232–34, (table) 235
 abnormal, 238–42
 breakdown of, in aged red cells, 248
 in embryo, 220
 role in oxygen transport, 228, 394
 spectrum of, 237
 structure of, 32–33, 35
 synthesis of, 244–45
Hemolysis, 228–29
Hemolytic anemias, 247
Hemolytic disease of the newborn,
 280–81
Hemophilia, 264, 266–68, 273
 sex-linked heredity in, 98
Hemoproteins, 232
Hemorrhage, effect of on red cell
 production, 232
Hemorrhagic diseases, in study of
 blood clotting, 266–67
 See also Bleeding disorders
Hemorrhoids, 536
Hemostasis, 265–67, 271
Hench, Philip, 604
Henderson, Lawrence, 200

Leukocytosis, 255–56
Leukokinetics, 252–54
Leukopenia, 255–56
Leukopoiesis, 220
Levers, of skeleton, 690–91
Levine, Philip, 280
Lewis, G. N., 21, 43n.
Lewis, Sir Thomas, (quoted) 320–21
Lewis, Warren, 161
Lewis blood grouping system, 278
Leydig cells, 668, 673
Li, C. H., 596
Ligaments: of connective tissue, 174
    of knee, 688
Light: adaptation of eye to, 486–87
    effects of on skin, 746
    wavelength of visible, 54
Lightening hormone, 745
Limbic-midbrain system, 462–63
Limbs, bones of, 685–87
Limulus, visual excitation in, 459
Linderstrøm-Lang, Karl, 32
Lingual tonsils, 515–16
Linkage, of genes, 96–97
Linoleic acid, 27
Linolenic acid, 27
Lipases (lipolytic enzymes), 52, 511,
    522, 524, 542–43
Lipemia, 324
Lipids (fats): in body, 23, 25–30
    emulsification of, by bile salts, 545
    energy produced by, 557
    intestinal absorption of, 548–49
    metabolism of, *138*
    of red cell stroma, 230
    required in diet, 567–68
    respiratory quotient (R. Q.) of, 558
    transport of, 579–81
Lipmann, Fritz, 119
Lipoic acid, 127
β-lipoprotein deficiency, congenital,
    549n.
Lipoproteins, 33
    of plasma, 580–81, (table) 581
Liver, 512, 539–42, (table) 544
    embryonic, 514
    excretory function of, 551–54
    metabolic interrelationships in, *577*
    role in circulation, 285
    role in embryonic blood formation,
        220
    role in glucose metabolism, 574–76
    storage of vitamins in, 569
Ljubimova, M. N., 707
Lobotomy, prefrontal, 449
Lochia, 662
Lock-and-key theory, of enzyme ac-
    tion, 48
Loewi, Otto, 421
Loewy, A., 558n.
Long, C. N. H., 602–03
Long bones, structure of, 177
Loop of Henle, 333, 351–54

Lower extremity, 196
Ludwig, Carl, 338
Luteinizing hormone (LH), 594
    in ovarian cycle, 642–44
    See also Interstitial cell-stimulating
        hormone (ICSH)
Luteotropic hormone (LTH), 594
    in ovarian cycle, 642–44
    role in lactation, 662–64
Lumbar plexus, 435
Lumbar vertebrae, 683
Lundsgaard, E., 709
Lungs, 371, 375–78, 386–87
    capillaries in, 295
    embryonic, 514
Lupus erythematosus (LE), 255, 784
Lutheran blood grouping system, 278
Lymphatic system, 221–27, 302
Lymphedema, 221n.
Lymphocytes, 249–51, *Plate I*
    as precursors of plasma cells, 766–68
    of connective tissue, 175
    production of, 225–27
    ultrastructure of, *768*
Lymphoid cells, responses to antigen
        by, 766–68
Lymphoreticular cells, 249
Lyon, Mary, 96
Lysine, 29
Lysins, 755
Lysokinase, 272
Lysosomes, 72, 78
Lysozyme, 477, 522

M line, in ultrastructure of myofibrils,
    182
MacCallum, W., 624
Macfarlane, R. G., 270
Macrocytes, 244
Macroglobulinemia, 265
Macroglobulins, 262
Macromolecules, in body, 33
Macronutrients, 571
Macrophage inhibitory factor (MIF),
    788
Macrophages, 175, 223
    role in antigen processing, 766
    role in cellular immunity, 787–88
Macula (equilibrium receptor), 500
Magnesium: in body, 14
    dietary requirement of, 572
    ions of (Mg^{++}), in body water, 207,
        209
Magoun, H. W., 461
Male: pelvic organs of, *367*
    reproductive system of, 665–77
Male climacteric, 674–75
Malignant melanoma, 748
Malleus, 491, 496
Malpighi, Marcello, 723
Malpighian bodies (of lymphoid
        tissue), 223n.

Malpighian cells (of skin), 723
Malpighian corpuscles (of kidney).
        See Glomerulus
Maltase, 511, 543
Maltose, 24
Malthus, Thomas, 658n.
Mammary glands. *See* Breasts
Mammillary bodies, *441*, 445, 462
Mandible, 681
Manganese, dietary requirement of,
    572
Mann, F. C., 574
Mannose, intestinal absorption of,
    548
Mapping, of chromosomes, 99–100
Marginal granulocyte pool (MGP),
    252–54
Marks, W. B., 457
Marrack, J., 770
Marsh, B. B., 711
Marshall, E. K., Jr., 343, 351
Mass action, law of, 40
Mass spectrometry, 62–63
Massa, Lorenzo, 5
Mast cells, 175, 251
    role in anaphylaxis, 791–92
Mastication (chewing), 520
Matthaei, J. H., 109–10
Maxillae, 681
Maxwell, James Clerk, 487
Mazia, Daniel, 82–83, (quoted) 167
McCance, R. A., 14, 573
Mechanoreceptors, 427–29
Medawar, Peter B., (quoted) 751, 774,
        775, 790, 791
Medial (anatomical term), 196
Median (midsagittal) plane, 196
Mediastinum, 197, 379
Medulla oblongata, 439, 441
Megakaryoblasts, 220
Megakaryocytes, 256–57
Megaloblasts, 244
Meienhofer, J., 578
Meiosis, 90–93, 652n.
Melanin, 720, 730, 739, 741–48
Melanoblasts, 739–40
Melanocyte-stimulating hormones
        (MSH), 594, 599–600, 743–45
Melanocytes, 172, 723, 730, 739–40
Melanophores, 743
Melanosomes, 739–41
Melatonin, 589, 630, 743, 745
Membranes: of cell, 71–73
    semipermeable, 207n.
    of tissue, lipids in, 28
    transport across, 158–61
Memory trace, 460
Menadione, in blood clotting reactions,
    274–75
Mendel, Gregor, 86–89
Mendel, L. B., 568
Mendel's law of independent assort-
        ment, 93–95

distinguished from nerve fibers, 415n.
structure of, 183–85
vasomotor, 322–23
Nervous system, 192, 409–507
action of thyroid hormones on, 618
as irritability machine, 162–63
relation to endocrine system, 586–87
Neural integration, biological significance of, 409–10
Neuroeffector junctions, 429–30
Neurohypophysis, 590–92
hormones secreted by, 588, 592–94
role in urine concentration, 354–55
Neuromuscular junctions, 427, 430
Neurons, 183–85, 413–15
association, 411
circuits involving, 415–30
internuncial, 434
motor (efferent), 437
sensory (afferent), 410
Neurosecretion, 422
Neurospora crassa, in genetic studies, 103
Neutrophils, 249–51
Newburgh, L. H., 566
Newton, Isaac, 4, 487
Niacin (nicotinic acid), as vitamin, 570
Nicotinamide, 126
Nicotinamide adenine dinucleotide. See NAD
Nicotinamide adenine dinucleotide phosphate. See NADP
Nicotinic acid (niacin), as vitamin, 570
Niedergerke, R., 708
Night blindness, 487
Nipples (of breasts), 662
Nirenberg, M. W., 109–10
Nissl bodies, 185
Nitrogen: in amino acid synthesis, 146–47
in body, 15
compounds of, metabolism of, 581–85
excretion of, by kidneys, 346–47
fixation of, 15
in measurement of pulmonary ventilation, 392
Nitrogen narcosis, 406
Nitrous acid (HNO_2): as mutagen, 112–13
in measurement of pulmonary circulation, 392
Nodes of Ranvier, 185, 427–28
Nondisjunction, chromosomal, results of, 677n.
Nonelectrolytes, in body water, 207
Nongranulocytes (white cells), 249
Nonnuclear genes, 113–14
Nonpolar (chemical term), 27
Nonprotein nitrogen, 582
Nonprotein respiratory quotient, 558
Nonsense mutations, 112

Noradrenalin. See Norepinephrine
Norepinephrine: and nerve impulse transmission, 422, 468
synthesis of, 608–11
Norethindrone, 647, 649
Norethynodrel, 647, 649
Normal solution, 39n.
Normoblasts, 230
Nose, 372–73
Nossal, G. J. V., 766
Nostrils (nares), 373
Nucleases (nucleolytic enzymes), 52, 542
Nucleic acids, 23, 33–38, 72n. See also DNA; RNA
Nucleolus, 72
Nucleoproteins, 33–35
Nucleosides, 35
Nucleotidases, 52
Nucleotides, 33–35
Nucleus (of cell), 71, 72
Nutrients, transported by plasma, 257, (table) 258
Nutrition, principles of, 562–73
Nystagmus, 501–03

Obesity, 565, 567
Occipital bone, 681
Oculomotor (III cranial) nerve, 451, 478
Old brain. See Brain stem
Old cerebellum, 443
Olds, J., 462
Oleic acid, 27
Olfactory (I cranial) nerve, 451, 507
Olfactory receptors, 505
Oligosaccharides, 25
Omenta, 534–35
One gene, one enzyme hypothesis, 102–03
Oögenesis, 643
Operator genes, 113
Operator-constitutive mutations, 155–56
Operon, 155
Opposition, of thumb, 694
Opsins, in photochemistry of eye, 484
Opsonins, 255, 753
in cytolytic reactions, 763
Optic disc, 472
Optic (II cranial) nerve, 451
Optic tracts, 490
Optic vesicles, embryonic, 439, 473–75
Optical rotatory activity, of stereoisomers, 18
Optics, physical principles of, 478–80
Oral cavity, 515–16
Oral pharynx, 518
Orbital cavities, 198
Organ of Corti, 495
Organelles, 72–73
Organic compounds, 15–22, (table) 20
in body, 22–39

Organic matrix, of bone, 695, 698
Organs (of body), arrangement into systems, 191
Orgasm, 651
Ornithine, in urea cycle, 147
Orotic acid, in biosynthesis of pyrimidine nucleotides, 142–43
Orthogenesis, 85
Osborne, T. B., 568
Oscillation, in negative feedback systems, 191
Osmol, 207n.
Osmolal concentration, 348n.
Osmolar concentration, 348n.
Osmosis: of body water, 207
in cell membrane, 159
Osmotic fragility, of red cells, 228–29
Osmotic pressure: on cell membranes, 207n.
and abnormality of water distribution, 210
Osseous tissue. See Bone
Ossification, 694–96
Osteoblasts, 695
Osteoclasts, 696
resorption of bone by, 700
role in maintaining calcium level, 626
Osteocytes, 177
Osteones, 696
Otoconia, 500
Ouchterlony, O., 759
Oudin, J., 759
Ova: development of, 641–43
fertilization of, 652–54
Oval window, 491
Ovarian (sexual) cycle, 641–51
Ovaries, 634–38
as endocrine glands, 589
production of haploid cells in, 91
secretion of testosterone by, 673
Overutilization anoxia, 399, 400
Ovulation, 643, 650
Owen, Richard, 623
Oxaloacetate: in hyperketonemia, 576
in transamination reactions, 137–38
Oxidation, 42–44, 138
of fatty acids, 135–37
of hemoglobin, 233
in metabolic reactions, 119
in respiration, 370
Oxidation-reduction reactions, 41–44
Oxidative deamination, 137
Oxidative phosphorylation, 120, 130–34
effect of thyroid hormones on, 618–19
Oxidative shunt. See Pentose phosphate shunt
Oxidoreductases, 52–53
Oxygen: in body, 14
in carbohydrates, 23–25
consumption of (O_2), and metabolic rate, 559

Rouleaux, 217
Round window, 494
Rubner, M., 560–61
Runt disease, 776

Saccule, 493, 500
Sacral vertebrae, 683
Sacrum, 379n., 683
Sagittal plane, 196
St. Martin, Alexis, 521–22
Saliva, 520–21
  water loss via, 206
Salivary amylase, 511, 521
Salivary glands, 512, 516–17
Salt (taste sensation), 503
Salt(s), 21
  in body fluids, 207–10
  excretion of, by kidneys, 347–49
  intestinal absorption of, 549–50
  regulation of, by kidneys, 328–29
  restriction of, and hypertension, 324
  and water balance, role of sweat in,
    735–36
Salt glands, in birds, 348
Salting-out effect, 260
Sanger, Frederick, 30
Sarcolemma, 181
Sarcoplasmic reticulum, 182
  role in muscle contraction, 711–13
Saturated fatty acids, 26
Saturation, of arterial blood, 394
Scabs, 749
Scapulae (shoulder blades), 685
Scars, 749
Schaeffer, Asa, 161
Schleiden, Matthias, 67
Schneider, Howard A., (quoted) 556
Schramm, G., 702
Schultze, Max, 67
Schwann, Theodor, 67, 414
Schwann cells, 185, 414–15
Schwyzer, R., 596
Sclera, 469
Scrotum, 665
Scuba diving, physiology of, 404–07
Scurvy, 571
  and defective wound healing, 749
Sea water, compared with extracellular
    fluid, 209
Sebaceous glands, 172, 730–32
Sebum, 731
Secondary structure, of protein, 32
Secretin, 544, 586, 589
Secretors, 278n.
Secretory granules, 77
Sedatives, effect of on emotions, 463
Sedimentation constants, 60–61
  of plasma proteins, 261
Sedimentation rate, 217
Sedormid, as cause of autoimmune dis-
    ease, 784

Segmentation, rhythmic, of small in-
    testine, 546
Selective theories, of immunity, 777–81
Selenium, dietary requirement of, 572
Self-recognition, and immunity, 773–77
Sella turcica, 372, 590
Semicircular canals, 493, 500–01
Semilunar valves, 289
Semen, 651, 670–72
Seminal vesicles, 665–67
Seminiferous tubules, 665, 668
Semipermeable membranes, 207n.
Senses, of man, 426n.
Sensory area, of cerebral cortex, 449
Sensory (afferent) neurons, 184, 410
  in autonomic nervous system, 464
Serine, 29
Serotonin, 610–11
  in anaphylaxis and allergy, 792–93
  in blood clotting reactions, 256, 266
Serous membrane, 196
Sertoli cells, 668, 670, 673
Serum, 216
  calcium content of, 625
Serum sickness, 793
Sex: chromosomal determination of,
    95–96
  embryonic differentiation of, 637–38
Sex chromatin, 96
Sex hormones: adrenal, 601
  female, 645–49
  male, 673–77
Sex-linked heredity, 98–99
Sexual intercourse (sex act), 651–52,
    672
Sexual reproduction, 86, 95, 634.
  See also Reproduction
Sherrington, Sir Charles, 5n., (quoted)
    409, (quoted) 439
Shin, 685
Shoulder girdle, 685
Sickle cell anemia, 111, 238–39, 242
Side-chain theory, of antibody produc-
    tion, 785
Sideroblasts, 245
Siderophilin, 259n.
Sieber, P., 596
Siebold, Karl von, 67
Sigmoid colon, 535
Simpson, George Gaylord, (quoted)
    189, 190
Singer, S. J., 758n.
Sinus rhythm, of heartbeat, 304
  disorders of, 309–10
Sinuses: of brain, 439
  paranasal, 373
  of Valsava, 289
Sinusoids, of liver, 540
Skeletal muscle. See Striated muscle
Skeleton, 681–91
  maturation of, 694–96
Skin, 721–27
  grafting of, 773–76

pigmentation of, 599–600, 739–48
receptors in, 427–29
water loss through, 206
Skull, 681–83
Slater, E. C., 133
Sleep, 461–62
Sliding filament theory, of muscle
    contraction, 708–09
Small intestine, 512, 531–34
  absorption in, 547–50
  digestion in, 542–46
  movements of, 546–47
  special glands of, 533
Smallpox vaccine, discovery of, 751
Smegma, 668
Smell, 505–07
"Smell brain," 447
Smith, Homer W., (quoted) 370
Smith, P. E., 599
Smooth muscle, 181
Sneezing, 403
Snoring, 404
Soft palate, 515
Sodium: active transport of, 159–61
  balance of, maintained by kidneys,
    342–49
  in body, 14
  in body water, 207, 209–10
  concentration of, in pancreas, 542
  dietary requirement of, 571
  excretion of, 352–56
  molecular dimensions of, 259
  "pumping" of, by cell membrane,
    79–80
  radioactive, in studies of body
    water, 205
  role in heartbeat, 304–06
  in transmission of nerve impulses,
    416–19
Sodium chloride. See Salt(s)
"Sodium pump," 209
  in nerve impulse transmission,
    416–17
  role in sleep, 462n.
Solomon, A. K., 159n.
Solvent drag, in cell membrane, 159
Sound: localization of, 500
  physical principles of, 495–96
  production of, by larynx, 407–08
Sour (taste sensation), 503
Spallanzani, Lazzaro, 521
Specific dynamic action (SDA), of
    food, 561
Specific gravity: of blood, 217
  of urine, 348–49
Specificity, of antibodies, 755
  of enzymes, 48
Specificity theory, of pain, 429
Spectrophotometry, 54–55
Speech: cerebral cortex region govern-
    ing (Broca's area), 448–49
  role of larynx in, 407–08
Spermatic cords, 665

Vagina, 640
Vagus (X cranial) nerve, 451, 452
  control of digestion by, 525–26,
    528–30, 536–37
  role in taste, 505
Valence, 16
Valine, 29
Valves: of heart, 289
  of veins, 295
Valvular insufficiency, 317
Van den Bergh, A. A. Hijmans, 551
Van der Waals' forces, 22
Van Slyke, D. D., 339
Vaporization, heat of, 202
Variable sequence, in antibody
    structure, 770
Varicose veins, 327
Vasa deferentia, 665–66
Vasa vasorum, 294
Vascular system, 295–302
Vasomotor nerves, 322–23
Vasopressin (ADH), 354–55, 592–94
Vassales, G., 624
Vegetative nervous system. See Auto-
    nomic nervous system
Veins, 283, 296
  embryology of, 300–02
  and lymphatic vessels, 302
  See also Circulation
Venae comitantes (companion veins),
    299
Venography, 296
Ventilation, 386, 388–90
Ventilation-to-perfusion ratio, 392
Ventral (anterior) aspect of body, 196
Ventricles (of heart), 283–85, 287
  embryonic, 292
Ventricular fibrillation, 310
Ventricular system (of brain), 452
Ventriloquy, 408
Venules, 296
Verdoperoxidase, 254
Vermiform appendix, *512*, 535
Vertebrae, 379, 683
Vertebral column, 683
Vertebral ganglia, 464
Vertebrates, evolution of kidney in,
    329–31
Vesalius, Andreas, 4–5, 440
Vestibular apparatus (of ear), 500–03
Vicious cycle, 191n.
  in autoimmune diseases, 784
Villee, C., 648
Villi (of small intestine), 533–34
Virchow, Rudolf, 84
Virilizing syndrome, 607
Viruses, genetics of, 103–04
Viscera, 196
Visceral nervous system. See Auto-
    nomic nervous system
Viscosity, of blood, 216–17
Visible spectrum, 487
Vision, 469–91

Visual cortex, 450
  image representation in, 490–91
Visual purple. See Rhodopsin
Vital capacity, 386
Vitamin A (retinol), 127, 569
  in photochemistry of retina, 484–86
  role in keratinization, 725
Vitamin B_1 (thiamine), 124, 570
  required in metabolism, 103n.
Vitamin B_2 (riboflavin), 124, 570
Vitamin B_6 (pyridoxine), 125, 570
Vitamin B_{12} (cyanocobalamin), 125,
    571
  in cell division, 243–44
  in conversion of ribonucleotides,
    144
  discovery of, 242–43
  effects of starvation for, *146*
Vitamin C (L-ascorbic acid), 127, 571
  in control of adrenocortical activity,
    607
  in wound healing, 749
Vitamin D (calciferol derivatives), 30,
    127, 569
  in calcium and phosphorus metabo-
    lism, 624–25
  relation to parathyroid hormone,
    628
Vitamin E (α-tocopherol), 127, 567–70
Vitamin K_1 (phylloquinone), 127, 570
  role in blood clotting, 274–75
Vitamins, (table) 124–27
  as coenzymes, 54n.
  dietary requirement of, 568–71
  function of, in body, 23
  in plasma, (table) 258
  red cell requirements of, 242–43
Vitiligo, 747
Vitreous humor, 472, 475–76
Vocal folds (cords), 373
Voegtlin, Carl, 624
Vogel, F. S., 113
Voice, production of, 407–08
Volkmann, Richard von, 14
Volume, relation to pressure and tem-
    perature (in gases), 384
Volume elasticity, of fibers, 497n.
Vomiting, 531
Vulva, 641

Wakefulness, 461–62
Wald, George, 15n., 484n., 485, 487,
    488
Waller, A. D., 307
Warburg, Otto, 123, 236
Warm-blooded animals, 201–02
Waste, elimination of. See Excretion
Waste products, in plasma, (table)
    258, 259
Water: in composition of body, 14
  incompressibility of, 404–06
  properties of, 199–202

  See also Body water
Water diuresis, 354
Water intoxication, 206–07
Water vapor, partial pressures of,
    (table) 385
Watson, J. D., 37, 104
Weber, H. H., 702
Weeping, 477
Weight, factors affecting, 565–67
Weismann, August, 90
Weiss, Paul (quoted), 116
Weiss, Samuel, 106
White, Abraham, 630
White blood cells. See Leukocytes
White matter, 431–33
White pulp, of spleen, 223
Widdowson, E. M., 14, 573
Wiener, Alexander, 276, 278
Williams-Ashman, H. G., 648
Willis, Thomas, 456
Willmer, E. N., 634
Wilmot, John, (quoted) 634
Wilson, E. B., 71
Wintersteiner, O., 604
Wirz, H., 351–52
Wisdom teeth, 518
Withering, William, 321n.
Witts, L. J., (quoted) 213
Wöhler, Friedrich, 15
Wolf, S., 522n.
Wolff, H. G., 522n.
Woods, P. S., 104
Work, 44
  of breathing, 393
  effect of, on metabolic rate, 561–62
  performed by heart, 319–20
Wounds, healing of, 748–49
Wright's stain, 249n.
Wrist, bones of, 685

X chromosomes, 96
Xanthine derivatives, as diuretics, 362
Xylose, intestinal absorption of, 548
D-xylulose, 23

Y chromosomes, 96
Yawning, 403–04
Yeast, fermentation of glucose by, 123
Yellow marrow, blood formation in,
    220–21
Young, W. J., 123

Z lines, in ultrastructure of myofibrils,
    181, 702
Zachau, H. G., 108
Zinc: in body, 14
  dietary requirement of, 572
Zwitterion, 41
Zygomatic bones, 681
Zygotes, 90
Zymogens, 157

B
C 2
D 3
E 4
F 5
G 6
H 7
I 8
J 9